FOURTH EDITION

Ignacio Bello
Hillsborough Community College/University of South Florida

The McGraw-Hill Companies

INTRODUCTORY ALGEBRA: A REAL-WORLD APPROACH, FOURTH EDITION

Published by McGraw-Hill, a business unit of The McGraw-Hill Companies, Inc., 1221 Avenue of the Americas, New York, NY 10020.

This book is printed on recycled, acid-free paper containing 10% postconsumer waste.

1 2 3 4 5 6 7 8 9 0 QDB/QDB 1 0 9 8 7 6 5 4 3 2 1

ISBN 978–0–07–338439–9
MHID 0–07–338439–9

ISBN 978–0–07–736345–1 (Annotated Instructor's Edition)
MHID 0–07–736345–0

Vice President, Editor-in-Chief: *Marty Lange*
Vice President, EDP: *Kimberly Meriwether David*
Senior Director of Development: *Kristine Tibbetts*
Editorial Director: *Stewart K. Mattson*
Sponsoring Editor: *Mary Ellen Rahn*
Developmental Editor: *Adam Fischer*
Marketing Manager: *Peter A. Vanaria*
Senior Project Manager: *Vicki Krug*
Buyer II: *Sherry L. Kane*
Senior Media Project Manager: *Sandra M. Schnee*
Senior Designer: *David W. Hash*
Cover Designer: *John Joran*
Cover Image: *Santiago de Cuba. Calle Heredia. © Buena Vista Images/Getty Images Inc.*
Senior Photo Research Coordinator: *Lori Hancock*
Compositor: *MPS Limited, a Macmillan Company*
Typeface: *10/12 Times Roman*
Printer: *Quad/Graphics*

Library of Congress Cataloging-in-Publication Data

Bello, Ignacio.
Introductory algebra : a real-world approach / Ignacio Bello. — 4th ed.
p. cm.
Includes index.
ISBN 978–0–07–338439–9 — ISBN 0–07–338439–9 (hard copy : alk. paper) 1. Algebra—Textbooks.
I. Title.
QA152.3.B466 2012
512.9—dc22

2010032247

www.mhhe.com

About the Author

Bello

Ignacio Bello

attended the University of South Florida (USF), where he earned a B.A. and M.A. in Mathematics. He began teaching at USF in 1967, and in 1971 he became a member of the Faculty at Hillsborough Community College (HCC) and Coordinator of the Math and Sciences Department. Professor Bello instituted the USF/HCC remedial program, a program that started with 17 students taking Intermediate Algebra and grew to more than 800 students with courses covering Developmental English, Reading, and Mathematics. Aside from the present series of books (*Basic College Mathematics, Introductory Algebra,* and *Intermediate Algebra*), Professor Bello is the author of more than 40 textbooks including *Topics in Contemporary Mathematics, College Algebra, Algebra and Trigonometry,* and *Business Mathematics*. Many of these textbooks have been translated into Spanish. With Professor Fran Hopf, Bello started the Algebra Hotline, the only live, college-level television help program in Florida. Professor Bello is featured in three television programs on the award-winning Education Channel. He has helped create and develop the USF Mathematics Department Website (http://mathcenter.usf.edu), which serves as support for the Finite Math, College Algebra, Intermediate Algebra, and Introductory Algebra, and CLAST classes at USF. You can see Professor Bello's presentations and streaming videos at this website, as well as at http://www.ibello.com. Professor Bello is a member of the MAA and AMATYC and has given many presentations regarding the teaching of mathematics at the local, state, and national levels.

McGraw-Hill connect™ |MATHEMATICS

McGraw-Hill Connect Mathematics McGraw-Hill conducted in-depth research to create a new and improved learning experience that meets the needs of today's students and instructors. The result is a reinvented learning experience rich in information, visually engaging, and easily accessible to both instructors and students. McGraw-Hill's Connect is a Web-based assignment and assessment platform that helps students connect to their coursework and prepares them to succeed in and beyond the course.

Connect Mathematics enables math instructors to create and share courses and assignments with colleagues and adjuncts with only a few clicks of the mouse. All exercises, learning objectives, videos, and activities are directly tied to text-specific material.

1 *You and your students want a fully integrated online homework and learning management system all in one place.*

McGraw-Hill and Blackboard Inc. Partnership

- McGraw-Hill has partnered with Blackboard Inc. to offer the deepest integration of digital content and tools with Blackboard's teaching and learning platform.
- **Life simplified.** Now, all McGraw-Hill content (text, tools, & homework) can be accessed directly from within your Blackboard course. All with one sign-on.
- **Deep integration.** McGraw-Hill's content and content engines are seamlessly woven within your Blackboard course.
- **No more manual synching!** Connect assignments within Blackboard automatically (and instantly) feed grades directly to your Blackboard grade center. No more keeping track of two gradebooks!

Do More

2 *Your students want an assignment page that is easy to use and includes lots of extra resources for help.*

Efficient Assignment Navigation

- Students have access to immediate feedback and help while working through assignments.
- Students can view detailed step-by-step solutions for each exercise.

Connect. Learn. Succeed.

3 ***Your students want an interactive eBook rich with integrated functionality.***

Integrated Media-Rich eBook

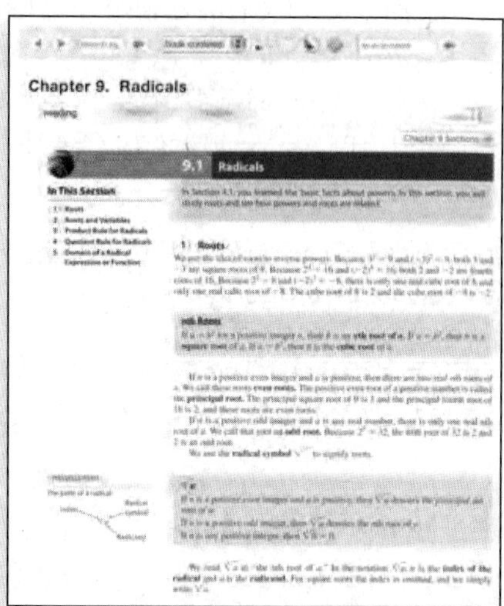

- A Web-optimized eBook is seamlessly integrated within ConnectPlus Mathematics for ease of use.
- Students can access videos, images, and other media in context within each chapter or subject area to enhance their learning experience.
- Students can highlight, take notes, or even access shared instructor highlights/notes to learn the course material.
- The integrated eBook provides students with a cost-saving alternative to traditional textbooks.

4 ***You want a more intuitive and efficient assignment creation process to accommodate your busy schedule.***

Assignment Creation Process

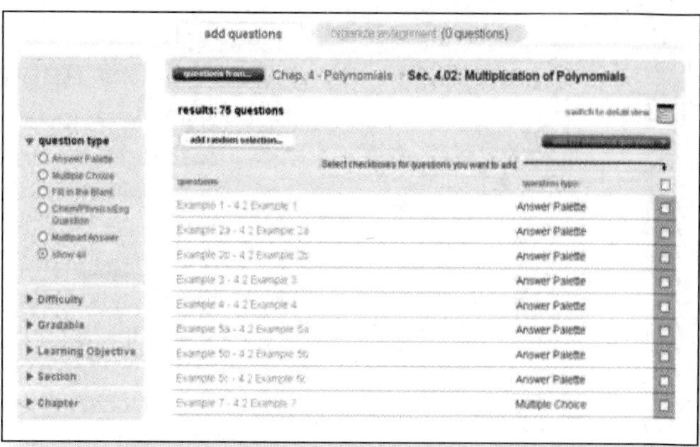

- Instructors can select textbook-specific questions organized by chapter, section, and objective.
- Drag-and-drop functionality makes creating an assignment quick and easy.
- Instructors can preview their assignments for efficient editing.

5 ***You want a gradebook that is easy to use and provides you with flexible reports to see how your students are performing.***

Flexible Instructor Gradebook

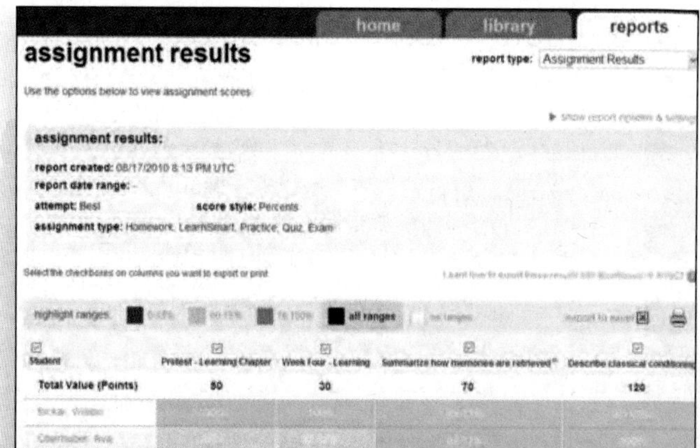

- Based on instructor feedback, Connect Mathematics' straightforward design creates an intuitive, visually pleasing grade management environment.
- View scored work immediately and track individual or group performance with various assignment and grade reports.

www.mcgrawhillconnect.com

Mc Graw Hill — Connect Learn Succeed

Preface

From the Author

The Inspiration for My Teaching

I was born in Havana, Cuba, and I encountered some of the same challenges in mathematics that many of my current students face, all while attempting to overcome a language barrier. In high school, I failed my freshman math course, which at the time was a complex language for me. However, with hard work and perseverance, I scored 100% on the final exam the second time around. While juggling various jobs in high school (roofer, sheetrock installer, and dock worker), I graduated and received a college academic scholarship. I first enrolled in calculus and made a "C." Never one to be discouraged, I became a math major and worked hard to excel in the courses that had previously frustrated me.

While a graduate student at the University of South Florida (USF), I taught at a technical school, Tampa Technical Institute, a decision that contributed to my resolve to teach math and make it come alive for my students the way brilliant instructors such as Jack Britton, Donald Rose, and Frank Cleaver had done for me. My math instructors instilled in me the motivation to work toward success. Through my teaching, I have learned a great deal about the way in which students learn and how the proper guidance through the developmental mathematics curriculum leads to student success. I believe I have developed a strong level of guidance in my textbook series by carefully explaining the language of mathematics and providing my students with the key fundamentals to help them reach success.

A Lively Approach to Build Students' Confidence

Teaching math at the University of South Florida was a great new career for me, but I found that students, professors, including myself, and administrators were disappointed by the rather imposing, mathematically correct but boring book we had to use. So, I took the challenge to write a book on my own, a book that was not only mathematically correct, but **student-oriented** with **interesting applications**—many suggested by the students themselves—and even, dare we say, entertaining! That book's approach and philosophy proved an instant success and was a precursor to my current series.

Students fondly called my class "The Bello Comedy Hour," but they worked hard, and they performed well. When my students ranked among the highest on the common final exam at USF, I knew I had found a way to motivate them through **common-sense language** and humorous, **realistic math applications.** I also wanted to show students they could overcome the same obstacles I had in math and become successful, too. If math has been a subject that some of your students have never felt comfortable with, then they're not alone! This book was written with the **mathanxious** students in mind, so they'll find it contains a jovial tone and explanations that are patient instead of making math seem mysterious, it makes it down-to-earth and easily digestible. For example, after explaining the different methods for simplifying fractions, readers are asked: "Which way should you simplify fractions? The way you understand!" Once students realize that math is within their grasp and not a foreign language, they'll be surprised at how much more confident they feel.

A Real-World Approach: Applications, Student Motivation, and Problem Solving

What is a "real-world approach"? I found that most textbooks put forth "real-world" applications that meant nothing to the real world of my students. How many of my students would really need to calculate the speed of a bullet (unless they are in its way) or cared to know when two trains traveling in different directions would pass by each other (disaster will certainly occur if they are on the same track)? For my students, both traditional and nontraditional, the real world consists of questions such as, "How do I find the best cell phone plan?" and "How will I pay my tuition and fees if they increase by x%?" That is why I introduce mathematical concepts through everyday applications with **real data** and give homework using similar, well-grounded situations (see the Getting Started application that introduces every section's topic and the word problems in every exercise section). Putting math in a real-world context has helped me overcome one of the problems we all face as math educators: **student motivation.** Seeing math in the real world makes students perk up in a math class in a way I have never seen before, and realism has proven to be the best motivator I've ever used. In addition, the real-world approach has enabled me to enhance students' **problem-solving skills** because they are far more likely to tackle a real-world problem that matters to them than one that seems contrived.

Diverse Students and Multiple Learning Styles

We know we live in a pluralistic society, so how do you write one textbook for everyone? The answer is to build a flexible set of teaching tools that instructors and students can adapt to their own situations. Are any of your students members of a **cultural minority?** So am I! Did they learn **English as a second language?** So did I! You'll find my book speaks directly to them in a way that no other book ever has, and fuzzy explanations in other books will be clear and comprehensible in mine.

Do your students all have the same **learning style?** Of course not! That's why I wrote a book that will help students learn mathematics regardless of their personal learning style. **Visual learners** will benefit from the text's clean page layout, careful use of color highlighting, "*Web Its,*" and the video lectures on the text's website. **Auditory learners** will profit from the audio *e-Professor lectures* on the text's website, and both **auditory** and **social learners** will be aided by the *Collaborative Learning* projects. **Applied** and **pragmatic learners** will find a bonanza of features geared to help them: *Pretests* can be found in Connect providing practice problems by every example, and *Mastery Tests* appearing at the end of every section, to name just a few. **Spatial learners** will find the chapter *Summary* is designed especially for them, while **creative learners** will find the *Research Questions* to be a natural fit. Finally, **conceptual learners** will feel at home with features like "*The Human Side of Mathematics*" and the "*Write On*" exercises. Every student who is accustomed to opening a math book and feeling like they've run into a brick wall will find in my books that a number of doors are standing open and inviting them inside.

Listening to Student and Instructor Concerns

McGraw-Hill has given me a wonderful resource for making my textbook more responsive to the immediate concerns of students and faculty. In addition to sending my manuscript out for review by instructors at many different colleges, several times a year McGraw-Hill holds symposia and focus groups with math instructors where the emphasis is *not* on selling products but instead on the **publisher listening** to the needs of faculty and their students. These encounters have provided me with a wealth of ideas on how to improve my chapter organization, make the page layout of my books more readable, and fine-tune exercises in every chapter so that students and faculty will feel comfortable using my book because it incorporates their specific suggestions and anticipates their needs.

R-I-S-E to Success in Math

Why are some students more successful in math than others? Often it is because they know how to manage their time and have a plan for action. Students can use models similar to these tables to make a weekly schedule of their time (classes, study, work, personal, etc.) and a semester calendar indicating major course events like tests, papers, and so on. Have them try to do as many of the suggestions on the **"R-I-S-E"** list as possible. (Larger, printable versions of these tables can be found in MathZone at www.mhhe.com/bello.)

Weekly Time Schedule

Time	S	M	T	W	R	F	S
8:00							
9:00							
10:00							
11:00							
12:00							
1:00							
2:00							
3:00							
4:00							
5:00							
6:00							
7:00							
8:00							
9:00							
10:00							
11:00							

Semester Calendar

Wk	M	T	W	R	F
1					
2					
3					
4					
5					
6					
7					
8					
9					
10					
11					
12					
13					
14					
15					
16					

R—Read and/or view the material before and after each class. This includes the textbook, the videos that come with the book, and any special material given to you by your instructor.

I—Interact and/or practice using the CD that comes with the book or the Web exercises suggested in the sections, or seeking tutoring from your school.

S—Study and/or discuss your homework and class notes with a study partner/group, with your instructor, or on a discussion board if available.

E—Evaluate your progress by checking the odd numbered homework questions with the answer key in the back of the book, using the mastery questions in each section of the book as a selftest, and using the Chapter Reviews and Chapter Practice Tests as practice before taking the actual test.

As the items on this list become part of your regular study habits, you will be ready to **"R-I-S-E"** to success in math.

Are you in need of relevant real-world applications? If so, look no further!

EXAMPLE 8 **How much CO_2 does an acre of trees absorb?**

Suppose you have a lot 80 ft by 540 ft (about an acre) and you plant trees every 8 feet, there will be $\frac{80}{8} = 10$ rows of $\frac{540}{8} \approx 67$ trees. (See diagram: not to scale!)

a. How many trees do you have in your acre lot?

b. If each tree absorbs 50 pounds of CO_2 a year (the amount varies by tree), how many pounds of CO_2 does the acre of trees absorb?

SOLUTION 8

a. You have 10 rows with 67 trees each or $(10)(67) = 670$ trees.

b. Each tree absorbs (-50) lb of CO_2, so the whole acre absorbs $(-50)(10)(67) = -33{,}500$ pounds of CO_2.

Data Source: http://tinyurl.com/yswsdv.

PROBLEM 8

a. If your tree lot has 10 rows of 70 trees, how many trees do you have?

b. How many pounds of CO_2 does the acre of trees absorb?

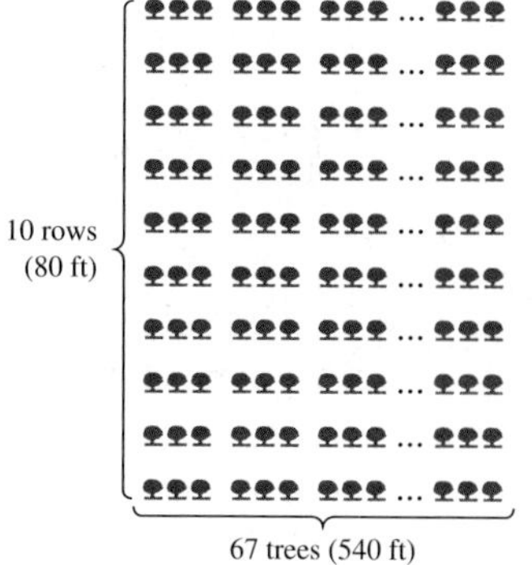

New to this edition of Bello, *Introductory Algebra: A Real-World Approach*

Green Math Applications

We are learning more about the positive and negative impact we can have on the Earth's fragile environment daily. It is everyone's responsibility to help sustain our environment and the best way to get people involved is through awareness and education. The purpose of Green Math Applications is to provide students with the ability to apply mathematics to topics present in all aspects of their lives. Every day people see media reports about the environment, fill their car's tank with gasoline, and make choices about what products they purchase. Green Math Applications teach students how to make and interpret these choices mathematically. Students will understand what it means when they read that this book was printed on paper that is 10% post consumer waste.

"The 'Green Math' applications are a great addition to the text. They answer the question that students always ask "But where will we ever use this stuff?"—Jan Butler, Colorado Community Colleges Online

Preface

Improvements in the Fourth Edition

Based on the valuable feedback of numerous reviewers and users over the years, the following improvements were made to the fourth edition of *Introductory Algebra.*

Green Math Exercises and Examples

- Fifty-eight Green Math Examples and 124 Green Math Exercises were added to the text.

Chapter R

- Added detail clarifying the Procedure of Reducing Fractions to Lowest Terms.
- Added the definition of Prime Numbers following its first introduction.
- Identified the three steps involved in the Rule for Multiplying Fractions.
- Identified the two steps involved in the Rule for Dividing Fractions.

Chapter 2

- Clarified the steps in the examples throughout.
- Created a box identifying the aspects of a linear equation.
- Created a procedure box for finding Least Common Multiple.
- Added Linear Equation exercises with "no solution."

Chapter 3

- Added a notes to Example 6 further explaining how to interpret line graphs.
- Added the objective Solve applications involving inequalities to Section 3.7.

Chapter 4

- Revised the definition of Quotient Rule for Exponents.
- Added a note to further explain negative exponents.
- Added a note to further explain Writing a Number in Scientific Notation ($M \times 10^n$).
- Added objective, Solve applications involving polynomial multiplications to Section 4.7.

Chapter 5

- Modified the procedure box that illustrates and describes FOIL.
- Added objective, Solve applications involving factoring to Section 5.4.
- Added objective, Solve applications involving factoring to Section 5.5.
- Added objective, Solve more applications using quadratics to Section 5.7.

Chapter 6

- Created a box detailing the steps needed to find the values that make a denominator zero.
- Added objective, Solve applications involving fractional equations.

Chapter 7

- Added further explanation for solving a dependent system by elimination.

Chapter 8

- Created information boxes and added color to highlight important material.
- Added objective, solve applications involving the quotient rule to Section 8.2.

Chapter 9

- Created information boxes and added color to highlight important material.

Acknowledgments

Manuscript Review Panels

Teachers and academics from across the country reviewed the various drafts of the manuscript to give feedback on content, design, pedagogy, and organization. This feedback was summarized by the book team and used to guide the direction of the text.

Dr. Mohammed Abella, *Washtenaw Community College*
Harold Arnett, *Highland Community College*
Jean Ashby, *The Community College of Baltimore County*
Michelle Bach, *Kansas City Kansas Community College*
Mark Batell, *Washtenaw Community College*
Randy Burnette, *Tallahassee Community College*
Jan Butler, *Colorado Community Colleges Online*
Edie Carter, *Amarillo College*
Kris Chatas, *Washtenaw Community College*
Amtul Chaudry, *Rio Hondo College*
David DelRossi, *Tallahassee Community College*
Ginny Durham, *Gadsden State Community College*
Kristy Erickson, *Cecil College*
Brandi Faulkner, *Tallahassee Community College*
Angela Gallant, *Inver Hills Community College*
Matthew Gardner, *North Hennepin Community College*
Jane Golden, *Hillsborough Community College*
Lori Grady, *University of Wisconsin–Whitewater*
Jane Gringauz, *Minneapolis Community and Technical College*
Jennie Gurley, *Wallace State Community College–Hanceville*
Lawrence Hahn, *Luzerne County Community College*
Kristen Hathcock, *Barton Community College*
Mary Beth Headlee, *State College of Florida*
Rick Hobbs, *Mission College*
Linda Horner, *Columbia State Community College*
Kevin Hulke, *Chippewa Valley Technical College*
Nancy Johnson, *State College of Florida*
Linda Joyce, *Tulsa Community College*
John Keating, *Massasoit Community College*
Regina Keller, *Suffolk County Community College*
Charyl Link, *Kansas City Kansas Community College*
Debra Loeffler, *The Community College of Baltimore*
Annette Magyar, *Southwestern Michigan College*
Stan Mattoon, *Merced Community College*
Sherry McClain, *College of Central Florida*
Chris McNally, *Tallahassee Community College*
Allan Newhart, *West Virginia University–Parkersburg*
Charles Odion, *Houston Community College*
Laura Perez, *Washtenaw Community College*
Tammy Potter, *Gadsden State Community College*
Linda Prawdzik, *Luzerne County Community College*
Brooke Quinlan, *Hillsborough Community College*
Linda Reist, *Macomb Community College*
Nancy Ressler, *Oakton Community College*
Pat Rhodes, *Treasure Valley Community College*
Neal Rogers, *Santa Ana College*
Lisa Rombes, *Washtenaw Community College*
Nancy Sattler, *Terra Community College*
Joel Sheldon, *Santa Ana College*
Lisa Sheppard, *Lorain County Community College*
James Smith, *Columbia State Community College*
Linda Tremer, *Three Rivers Community College*
Dr. Miguel Uchofen, *Kansas City Kansas Community College*
Sara Van Asten, *North Hennepin Community College*
Alexsis Venter, *Arapahoe Community College*
Josefino Villanueva, *Florida Memorial University*
Ursula Walsh, *Minneapolis Community and Technical College*
Angela Wang, *Santa Ana College*
John Waters, *State College of Florida*
Karen White, *Amarillo College*
Jill Wilsey, *Genesee Community College*
Carol Zavarella, *Hillsborough Community College*
Vivian Zimmerman, *Prairie State College*
Loris Zucca, *Lone Star College–Kingwood*

I would like to thank the following people for their invaluable help:
Randy Welch, with his excellent sense of humor, organization, and very hard work; Dr. Tom Porter, of Photos at Your Place, who improved on some of the pictures I provided; Vicki Krug, one of the most exacting persons at McGraw-Hill, who will always give you the time of day and then solve the problem; Adam Fisher, our developmental editor, great with the numbers, professional, and always ready to help; Josie Rinaldo, who read a lot of the material for content overnight and Pat Steele, our very able copy editor. Finally, thanks to our attack secretary, Beverly DeVine, who still managed to send all materials back to the publisher on time. To everyone, my many thanks.

Features and Supplements

Motivation for a Diverse Student Audience

A number of features exist in every chapter to motivate students' interest in the topic and thereby increase their performance in the course:

The Human Side of Algebra

To personalize the subject of mathematics, the origins of numerical notation, concepts, and methods are introduced through the lives of real people solving ordinary problems.

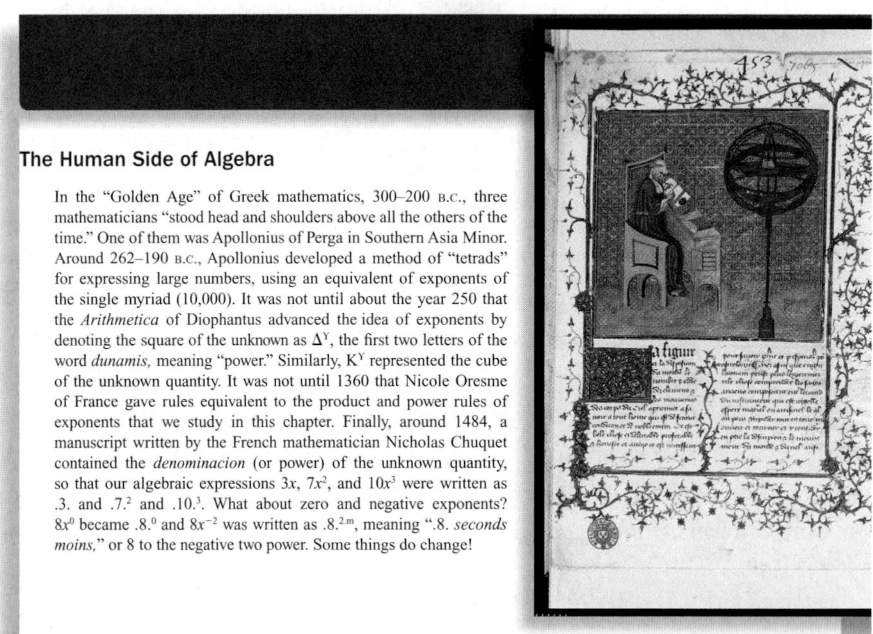

The Human Side of Algebra

In the "Golden Age" of Greek mathematics, 300–200 B.C., three mathematicians "stood head and shoulders above all the others of the time." One of them was Apollonius of Perga in Southern Asia Minor. Around 262–190 B.C., Apollonius developed a method of "tetrads" for expressing large numbers, using an equivalent of exponents of the single myriad (10,000). It was not until about the year 250 that the *Arithmetica* of Diophantus advanced the idea of exponents by denoting the square of the unknown as Δ^Y, the first two letters of the word *dunamis,* meaning "power." Similarly, K^Y represented the cube of the unknown quantity. It was not until 1360 that Nicole Oresme of France gave rules equivalent to the product and power rules of exponents that we study in this chapter. Finally, around 1484, a manuscript written by the French mathematician Nicholas Chuquet contained the *denominacion* (or power) of the unknown quantity, so that our algebraic expressions $3x$, $7x^2$, and $10x^3$ were written as .3. and .7.^{2} and .10.^{3}. What about zero and negative exponents? $8x^0$ became .8.^{0} and $8x^{-2}$ was written as .8.$^{2.m}$, meaning ".8. *seconds moins,*" or 8 to the negative two power. Some things do change!

Getting Started

Each topic is introduced in a setting familiar to students' daily lives, making the subject personally relevant and more easily understood.

Getting Started

Don't Forget the Tip!

Jasmine is a server at CDB restaurant. Aside from her tips, she gets \$2.88/hour. In 1 hour, she earns \$2.88; in 2 hr, she earns \$5.76; in 3 hr, she earns \$8.64, and so on. We can form the set of ordered pairs (1, 2.88), (2, 5.76), (3, 8.64) using the number of hours she works as the first coordinate and the amount she earns as the second coordinate. Note that the ratio of second coordinates to first coordinates is the same number:

$$\frac{2.88}{1} = 2.88, \quad \frac{5.76}{2} = 2.88, \quad \frac{8.64}{3} = 2.88,$$

and so on.

Web It

Appearing in the margin of the section exercises, this URL refers students to the abundance of resources available on the Web that can show them fun, alternative explanations, and demonstrations of important topics.

Exercises 7.4

connect MATHEMATICS — Practice Problems · Self-Tests · Media-rich eBooks · e-Professors · Videos

Web IT go to mhhe.com/bello for more lessons

A Solving Coin and Money Problems
B Solving General Problems

In Problems 1–6, solve the money problems.

1. Mida has \$2.25 in nickels and dimes. She has four times as many dimes as nickels. How many dimes and how many nickels does she have?
2. Dora has \$5.50 in nickels and quarters. She has twice as many quarters as she has nickels. How many of each coin does she have?
3. Mongo has 20 coins consisting of nickels and dimes. If the nickels were dimes and the dimes were nickels, he would have 50¢ more than he now has. How many nickels and how many dimes does he have?
4. Desi has 10 coins consisting of pennies and nickels. Strangely enough, if the nickels were pennies and the pennies were nickels, she would have the same amount of money as she now has. How many pennies and nickels does she have?
5. Don had \$26 in his pocket. If he had only \$1 bills and \$5 bills, and he had a total of 10 bills, how many of each of the bills did he have?
6. A person went to the bank to deposit \$300. The money was in \$10 and \$20 bills, 25 bills in all. How many of each did the person have?

In Problems 7–14, find the solution.

7. The sum of two numbers is 102. Their difference is 16. What are the numbers?
8. The difference between two numbers is 28. Their sum is 82. What are the numbers?

Write On

Writing exercises give students the opportunity to express mathematical concepts and procedures in their own words, thereby internalizing what they have learned.

Write On

71. In the expression "$\frac{1}{2}$ of x," what operation does the word *of* signify?
72. Most people believe that the word *and* always means addition.
    a. In the expression "the sum of x and y," does "and" signify the operation of addition? Explain.
    b. In the expression "the product of 2 and three more than a number," does "and" signify the operation of addition? Explain.
73. Explain the difference between "x divided by y" and "x divided into y."
74. Explain the difference between "a less than b" and "a less b."

Collaborative Learning

Concluding the chapter are exercises for collaborative learning that promote teamwork by students on interesting and enjoyable exploration projects.

Collaborative Learning

How fast can you go?

How fast can you obtain information to solve a problem? Form three groups: library, the Web, and bookstore (where you can look at books, papers, and so on for free). Each group is going to research car prices. Select a car model that has been on the market for at least 5 years. Each of the groups should find:

1. The new car value and the value of a 3-year-old car of the same model
2. The estimated depreciation rate for the car
3. The estimated value of the car in 3 years
4. A graph comparing age and value of the car for the next 5 years
5. An equation of the form $C = P(1 - r)^n$ or $C = rn + b$, where n is the number of years after purchase and r is the depreciation rate

Which group finished first? Share the procedure used to obtain your information so the most efficient research method can be established.

Research Questions

Research questions provide students with additional opportunities to explore interesting areas of math, where they may find the questions can lead to surprising results.

Research Questions

1. In the *Human Side of Algebra* at the beginning of this chapter, we mentioned the Hindu numeration system. The Egyptians and Babylonians also developed numeration systems. Write a report about each of these numeration systems, detailing the symbols used for the digits 1–9, the base used, and the manner in which fractions were written.
2. Write a report on the life and works of Muhammad al-Khwarizmi, with special emphasis on the books he wrote.
3. We have now studied the four fundamental operations. But do you know where the symbols used to indicate these operations originated?
   - **a.** Write a report about Johann Widmann's *Mercantile Arithmetic* (1489), indicating which symbols of operation were found in the book for the first time and the manner in which they were used.
   - **b.** Introduced in 1557, the original equals sign used longer lines to indicate equality. Why were the two lines used to denote equality, what was the name of the person who introduced the symbol, and in what book did the

Abundant Practice and Problem Solving

Bello offers students many opportunities and different skill paths for developing their problem-solving skills.

Pretest

An optional Pretest can be found in MathZone at www.mhhe.com/bello and is especially helpful for students taking the course as a review who may remember some concepts but not others. The answer grid is also found online and gives students the page number, section, and example to study in case they missed a question. The results of the Pretest can be compared with those of the Practice Test at the end of the chapter to evaluate progress and student success.

Practice Test Chapter 1

(Answers on page 108)

Visit www.mhhe.com/bello to view helpful videos that provide step-by-step solutions to several of the following problems.

1. Write in symbols:
   a. The sum of g and h b. g minus h
   c. $6g$ plus $3h$ minus 8
2. Write using juxtaposition:
   a. 3 times g b. $-g$ times h times r
   c. $\frac{1}{5}$ of g d. The product of 9 and g
3. Write in symbols:
   a. The quotient of g and 8
   b. The quotient of 8 and h
4. Write in symbols:
   a. The quotient of $(g + h)$ and r
   b. The sum of g and h, divided by the difference of g and h
5. For $g = 4$ and $h = 3$, evaluate:
   a. $g + h$ b. $g - h$ c. $5g$
6. For $g = 8$ and $h = 4$, evaluate:
   a. $\frac{g}{h}$ b. $2g - 3h$ c. $\frac{2g + h}{h}$
7. Find the additive inverse (opposite) of:
   a. -9 b. $\frac{3}{5}$ c. $0.222\ldots$
8. Find the absolute value.
   a. $\left|-\frac{1}{4}\right|$ b. $|13|$ c. $-|0.92|$

Answers to Practice Test Chapter 1

Answer	If You Missed Question	Review Section	Review Examples	Review Page
1. a. $g + h$ b. $g - h$ c. $6g + 3h - 8$	1	1.1	1	37
2. a. $3g$ b. $-ghr$ c. $\frac{1}{5}g$ d. $9g$	2	1.1	2	37
3. a. $\frac{g}{8}$ b. $\frac{8}{h}$	3	1.1	3a, b	38
4. a. $\frac{g+h}{r}$ b. $\frac{g+h}{g-h}$	4	1.1	3c, d	38
5. a. 7 b. 1 c. 20	5	1.1	4a, b, c	38
6. a. 2 b. 4 c. 5	6	1.1	4d, e	38
7. a. 9 b. $-\frac{3}{5}$ c. $-0.222\ldots$	7	1.2	1, 2	43, 44
8. a. $\frac{1}{4}$ b. 13 c. -0.92	8	1.2	3, 4	45
9. a. 7 b. 0, 7 c. $-8, 0, 7$ d. $-8, \frac{5}{3}, 0, 3.4, 0.333\ldots, -3\frac{1}{2}, 7$ e. $\sqrt{2}, 0.123\ldots$ f. All	9	1.2	5	46–47

Paired Examples/Problems

Examples are placed adjacent to similar problems intended for students to obtain immediate reinforcement of the skill they have just observed. These are especially effective for students who learn by doing and who benefit from frequent practice of important methods. Answers to the problems appear at the bottom of the page.

EXAMPLE 4 Evaluating algebraic expressions

Evaluate the given expressions by substituting 10 for x and 5 for y.

a. $x + y$ b. $x - y$ c. $4y$ d. $\frac{x}{y}$ e. $3x - 2y$

SOLUTION 4

a. Substitute 10 for x and 5 for y in $x + y$.
We obtain: $x + y = 10 + 5 = 15$.
The number 15 is called the **value** of $x + y$.

b. $x - y = 10 - 5 = 5$

c. $4y = 4(5) = 20$

d. $\frac{x}{y} = \frac{10}{5} = 2$

e. $3x - 2y = 3(10) - 2(5) = 30 - 10 = 20$

PROBLEM 4

Evaluate the expressions by substituting 22 for a and 3 for b.

a. $a + b$ b. $2a - b$

c. $5b$ d. $\frac{2a}{b}$

e. $2a - 3b$

RSTUV Method

The easy-to-remember **"RSTUV"** method gives students a reliable and helpful tool in demystifying word problems so that they can more readily translate them into equations they can recognize and solve.

- **R**ead the problem and decide what is being asked.
- **S**elect a letter or □ to represent this unknown.
- **T**ranslate the problem into an equation.
- **U**se the rules you have studied to solve the resulting equation.
- **V**erify the answer.

RSTUV Method for Solving Word Problems

1. Read the problem carefully and decide what is asked for (the unknown).
2. Select a variable to represent this unknown.
3. Think of a plan to help you write an equation.
4. Use algebra to solve the resulting equation.
5. Verify the answer.

Translate This

These boxes appear periodically before word-problem exercises to help students translate phrases into equations, reinforcing the RSTUV method.

TRANSLATE THIS

1. The history of the formulas for calculating ideal body weight W began in **1871** when Dr. P. P. Broca (a French surgeon) created this formula known as Broca's index. The ideal weight W (in pounds) for a woman h inches tall is 100 pounds for the first 5 feet and 5 pounds for each additional inch over 60.
2. The ideal weight W (in pounds) for men h inches tall is 110 pounds for the first 5 feet and 5 pounds for each additional inch over 60.
3. In 1974, Dr. B. J. Devine suggested a formula for the weight W in kilograms (kg) of men h inches tall: 50 plus 2.3 kilograms per inch over 5 feet (60 inches).
4. For women h inches tall, the formula for W is 45.5 plus 2.3 kilograms per inch over 5 feet. By the way, a kilogram (kg) is about 2.2 pounds.
5. In 1983, Dr. J. D. Robinson published a modification of the formula. For men h inches tall, the weight W should be 52 kilograms and 1.9 kilograms for each inch over 60.

In Problems 1–10 **TRANSLATE** the sentence and match the correct translation with one of the equations A–O.

A. $W = 50 + 2.3h - 60$
B. $W = 49 + 1.7(h - 60)$
C. $LBW = B + M + O$
D. $W = 100h + 5(h - 60)$
E. $W = 110 + 5(h - 60)$
F. $LBW = 0.32810C + 0.33929W - 29.5336$
G. $LBW = 0.32810W + 0.33929C - 29.5336$
H. $W = 100 + 5(h - 60)$
I. $LBW = 0.29569W + 0.41813C - 43.2933$
J. $W = 50h + 2.3(h - 60)$
K. $W = 110h + 5(h - 60)$
L. $W = 56.2 + 1.41(h - 60)$
M. $W = 50 + 2.3(h - 60)$
N. $W = 52 + 1.9(h - 60)$
O. $W = 45.5 + 2.3(h - 60)$

6. The Robinson formula W for women h inches tall is 49 kilograms and 1.7 kilograms for each inch over 5 feet.
7. A minor modification of Robinson formula is Miller's formula which defines the weight W for a man h inches tall as 56.2 kilograms added to 1.41 kilograms for each inch over 5 feet
8. There are formulas that suggest your lean body weight **(LBW)** is the sum of the weight of your bones **(B)**, muscles **(M)**, and organs **(O)**. Basically the sum of everything other than fat in your body.
9. For men over the age of 16, C centimeters tall and with weight W kilograms, the lean body weight **(LBW)** is the product of W and 0.32810, plus the product of C and 0.33929, minus 29.5336.
10. For women over the age of 30, C centimeters tall and weighting W kilograms the lean body weight **(LBW)** is the product of 0.29569 and W, plus the product of 0.41813 and C, minus 43.2933

Exercises

A wealth of exercises for each section are organized according to the learning objectives for that section, giving students a reference to study if they need extra help.

Exercises 1.6

connect MATHEMATICS > Practice Problems > Self-Tests > Media-rich eBooks > e-Professors > Videos

mhhe.com/bello for more lessons

⟨A⟩ Identifying the Associative and Commutative Properties In Problems 1–10, name the property illustrated in each statement.

1. $9 + 8 = 8 + 9$
2. $b \cdot a = a \cdot b$
3. $4 \cdot 3 = 3 \cdot 4$
4. $(a + 4) + b = a + (4 + b)$
5. $3 + (x + 6) = (3 + x) + 6$
6. $8 \cdot (2 \cdot x) = (8 \cdot 2) \cdot x$
7. $a \cdot (b \cdot c) = a \cdot (c \cdot b)$
8. $a \cdot (b \cdot c) = (a \cdot b) \cdot c$
9. $a + (b + 3) = (a + b) + 3$
10. $(a + 3) + b = (3 + a) + b$

Guided Tour

Applications

Students will enjoy the exceptionally creative applications in most sections that bring math alive and demonstrate that it can even be performed with a sense of humor.

Applications

82. *Price of a car* The price P of a car is its base price (B) plus destination charges D, that is, $P = B + D$. Tran bought a Nissan in Smyrna, Tennessee, and there was no destination charge.

a. What is D?

b. Fill in the blank in the equation $P = B +$ ______

c. What property tells you that the equation in part **b** is correct?

83. *Area* The area of a rectangle is found by multiplying its length L times its width W.

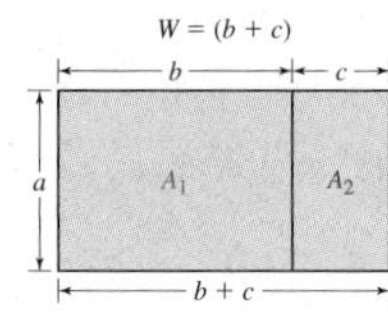

84. *Area* The length of the entire rectangle is a and its width is b.

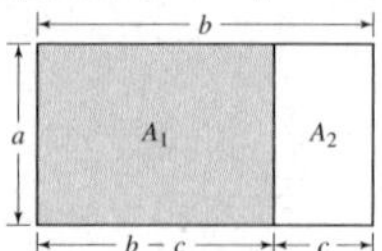

a. What is the area A of the entire rectangle?

b. What is the area of the smaller rectangle A_2?

c. The area of A_1 is the area A of the entire rectangle minus the area of A_2. Write an expression that models this situation.

d. Substitute the results obtained in **a** and **b** to rewrite the equation in part **c**.

85. *Weight* **a.** If you are a woman more than 5 feet (60 inches) tall, your weight W (in pounds) should be $W = 105 + 5(h - 60)$, where h is your height in inches

Using Your Knowledge

Optional, extended applications give students an opportunity to practice what they've learned in a multistep problem requiring reasoning skills in addition to numerical operations.

Using Your Knowledge

Tweedledee and Tweedledum Have you ever read *Alice in Wonderland?* Do you know who the author is? It's Lewis Carroll, of course. Although better known as the author of *Alice in Wonderland,* Lewis Carroll was also an accomplished mathematician and logician. Certain parts of his second book, *Through the Looking Glass,* reflect his interest in mathematics. In this book, one of the characters, Tweedledee, is talking to Tweedledum. Here is the conversation.

Tweedledee: The sum of your weight and twice mine is 361 pounds.

Tweedledum: Contrariwise, the sum of your weight and twice mine is 360 pounds.

41. If Tweedledee weighs x pounds and Tweedledum weighs y pounds, find their weights using the ideas of this section.

Study Aids to Make Math Accessible

Because some students confront math anxiety as soon as they sign up for the course, the Bello system provides many study aids to make their learning easier.

Objectives

The objectives for each section not only identify the specific tasks students should be able to perform, they organize the section itself with letters corresponding to each section heading, making it easy to follow.

3.4 The Slope of a Line: Parallel and Perpendicular Lines

Objectives

A Find the slope of a line given two points.

B Find the slope of a line given the equation of the line.

C Determine whether two lines are parallel, perpendicular, or neither.

D Solve applications involving slope.

To Succeed, Review How To . . .

1. Add, subtract, multiply, and divide signed numbers (pp. 52–56, 60–64).
2. Solve an equation for a specified variable (pp. 137–143).

Getting Started

Facebook and MySpace Visits

Can you tell from the graph the period in which the number of pages per visit declined for MySpace (red graph)? Has Facebook (blue graph) ever had a declining period? You can tell by simply looking at the graph! The pages per visit for MySpace subscribers declined from Jan 07 to Dec 07 (from about 75 pages per visit in Jan 07 to about 35 pages per visit in Dec 07). The **decline** per month was

$$\frac{\text{Difference in pages per visit}}{\text{Number of months}} = \frac{75 - 35}{11} = \frac{40}{11} \approx 4 \text{ (pages per month)}$$

On the other hand, Facebook had a 2-month declining period from March 08 to May 08. Their decline was

$$\frac{\text{Difference in pages per visit}}{\text{Number of months}} = \frac{50 - 40}{2} = 5 \text{ (pages per month)}$$

Reviews

Every section begins with "To succeed, review how to . . . ," which directs students to specific pages to study key topics they need to understand to successfully begin that section.

Concept Checker

This feature has been added to each end-of-section exercises to help students reinforce key terms and concepts.

Concept Checker

Fill in the blank(s) with the correct word(s), phrase, or mathematical statement.

103. If ***m*** and ***n*** are **positive integers,** $x^m \cdot x^n =$ ___________.

104. When **multiplying** numbers with the **same (like) signs,** the **product** is ___________.

105. When **multiplying** numbers with **different (unlike) signs,** the **product** is ___________.

106. When **dividing** numbers with the **same (like) signs,** the quotient is ___________.

107. When **dividing** numbers with **different (unlike) signs,** the quotient is ___________.

108. If ***m*** and ***n*** are **positive integers** with $m > n$, $\left(\frac{x^m}{x^n}\right) =$ ___________.

109. If ***m*** and ***n*** are **positive integers** $(x^m)^n =$ ___________.

an integer	x^{m+n}
0	**negative**
positive	y^m
x^{m-n}	$x^{mk}y^{nk}$
x^{mn}	$\frac{x^m}{y^m}$
$\frac{x^m}{y^n}$	$y^{mk}y^{nk}$

Mastery Test

In Problems 80–87, translate into an algebraic expression.

80. The product of 3 and xy

81. The difference of $2x$ and y

82. The quotient of $3x$ and $2y$

83. The sum of $7x$ and $4y$

84. The difference of b and c divided by the sum of b and c

85. Evaluate the expression $2x + y - z$ for $x = 3$, $y = 4$, and $z = 5$.

86. Evaluate the expression $\frac{p-q}{3}$ for $p = 9$ and $q = 3$.

87. Evaluate the expression $\frac{2x - 3y}{x + y}$ for $x = 10$ and $y = 5$.

Mastery Tests

Brief tests in every section give students a quick checkup to make sure they're ready to go on to the next topic.

Skill Checkers

These brief exercises help students keep their math skills well honed in preparation for the next section.

Skill Checker

In Problems 88–91, add the numbers.

88. $20 + (-20)$

89. $-3.8 + 3.8$

90. $-\frac{2}{7} + \frac{2}{7}$

91. $1\frac{2}{3} + \left(-1\frac{2}{3}\right)$

Calculator Corner

Additive Inverse and Absolute Value

The additive inverse and absolute value of a number are so important that graphing calculators have special keys to handle them. To find the additive inverse, press (−). Don't confuse the additive inverse key with the minus sign key. (Operation signs usually have color keys; the (−) key is gray.)

To find absolute values with a TI-83 Plus calculator, you have to do some math, so press MATH. Next, you have to deal with a special type of number, absolute value, so press ▶ to highlight the NUM menu at the top of the screen. Next press 1, which tells the calculator you want an absolute value; finally, enter the number whose absolute value you want, and close the parentheses. The display window shows how to calculate the additive inverse of −5, the absolute value of 7, and the absolute value of −4.

```
--5
               5
abs (7)
               7
abs (-4)
               4
```

Calculator Corner

When appropriate, optional calculator exercises are included to show students how they can explore concepts through calculators and verify their manual exercises with the aid of technology.

Summary Chapter 1

Section	Item	Meaning	Example
1.1A	Arithmetic expressions	Expressions containing numbers and operation signs	$3 + 4 - 8$, $9 \cdot 4 \div 6$ are arithmetic expressions.
1.1A	Algebraic expressions	Expressions containing numbers, operation signs, and variables	$3x - 4y + 3z$, $2x \div y + 9z$, and $7x - 9y \div 3z$ are algebraic expressions.

Summary

An easy-to-read grid summarizes the essential chapter information by section, providing an item, its meaning, and an example to help students connect concepts with their concrete occurrences.

Review Exercises

Chapter review exercises are coded by section number and give students extra reinforcement and practice to boost their confidence.

Review Exercises Chapter 2

(If you need help with these exercises, look in the section indicated in brackets.)

1. ⟨2.1A⟩ *Determine whether the given number satisfies the equation.*
   a. $5; 7 = 14 - x$
   b. $4; 13 = 17 - x$
   c. $-2; 8 = 6 - x$

2. ⟨2.1B⟩ *Solve the given equation.*
   a. $x - \frac{1}{3} = \frac{1}{3}$
   b. $x - \frac{5}{7} = \frac{2}{7}$
   c. $x - \frac{5}{9} = \frac{1}{9}$

3. ⟨2.1B⟩ *Solve the given equation.*
   a. $-3x + \frac{5}{9} + 4x - \frac{2}{9} = \frac{5}{9}$
   b. $-2x + \frac{4}{7} + 3x - \frac{2}{7} = \frac{6}{7}$
   c. $-4x + \frac{5}{6} + 5x - \frac{1}{6} = \frac{5}{6}$

4. ⟨2.1C⟩ *Solve the given equation.*
   a. $3 = 4(x - 1) + 2 - 3x$
   b. $4 = 5(x - 1) + 9 - 4x$
   c. $5 = 6(x - 1) + 8 - 5x$

Practice Test with Answers

The chapter Practice Test offers students a non-threatening way to review the material and determine whether they are ready to take a test given by their instructor. The answers to the Practice Test give students immediate feedback on their performance, and the answer grid gives them specific guidance on which section, example, and pages to review for any answers they may have missed.

Practice Test Chapter 2

(Answers on page 207)

Visit www.mhhe.com/bello to view helpful videos that provide step-by-step solutions to several of the problems below.

1. Does the number 3 satisfy the equation $6 = 9 - x$?
2. Solve $x - \frac{2}{7} = \frac{3}{7}$.
3. Solve $-2x + \frac{7}{8} + 3x - \frac{5}{8} = \frac{5}{8}$.
4. Solve $2 = 3(x - 1) + 5 - 2x$.
5. Solve $2 + 5(x + 1) = 8 + 5x$.
6. Solve $-3 - 2(x - 1) = -1 - 2x$.

Answers to Practice Test Chapter 2

Answer	If You Missed	Review		
	Question	Section	Examples	Page
1. Yes	1	2.1	1	111
2. $x = \frac{5}{7}$	2	2.1	2	112
3. $x = \frac{3}{8}$	3	2.1	3	113–114
4. $x = 0$	4	2.1	4, 5	115–116
5. No solution	5	2.1	6	117
6. All real numbers	6	2.1	7	117
7. $x = -6$	7	2.2	1, 2, 3	123–126

Cumulative Review

The Cumulative Review covers material from the present chapter and any of the chapters prior to it and can be used for extra homework or for student review to improve their retention of important skills and concepts.

Cumulative Review Chapters 1–2

1. Find the additive inverse (opposite) of -7.
2. Find: $\left|-9\frac{9}{10}\right|$
3. Find: $-\frac{2}{7} + \left(-\frac{2}{9}\right)$
4. Find: $-0.7 - (-8.9)$
5. Find: $(-2.4)(3.6)$
6. Find: $-(2^4)$
7. Find: $-\frac{7}{8} \div \left(-\frac{5}{24}\right)$
8. Evaluate $y \div 5 \cdot x - z$ for $x = 6$, $y = 60$, $z = 3$.
9. Which property is illustrated by the following statement?

   $9 \cdot (8 \cdot 5) = 9 \cdot (5 \cdot 8)$
10. Multiply: $6(5x + 7)$

Supplements for Instructors

Annotated Instructor's Edition

This version of the student text contains **answers** to all odd- and even-numbered exercises in addition to helpful **teaching tips.** The answers are printed on the same page as the exercises themselves so that there is no need to consult a separate appendix or answer key.

Computerized Test Bank (CTB) Online

Available through McGraw-Hill Connect™ Mathematics, this **computerized test bank** utilizes Brownstone Diploma®, an algorithm-based testing software to quickly create customized exams. This user-friendly program enables instructors to search for questions by topic, format, or difficulty level; to edit existing questions or to add new ones; and to scramble questions and answer keys for multiple versions of the same test. Hundreds of text-specific open-ended and multiple-choice questions are included in the question bank. Sample chapter tests and final exams in Microsoft Word® and PDF formats are also provided.

Instructor's Solutions Manual

Available on McGraw-Hill Connect™ Mathematics, the Instructor's Solutions Manual provides comprehensive, **worked-out solutions** to all exercises in the text. The methods used to solve the problems in the manual are the same as those used to solve the examples in the textbook.

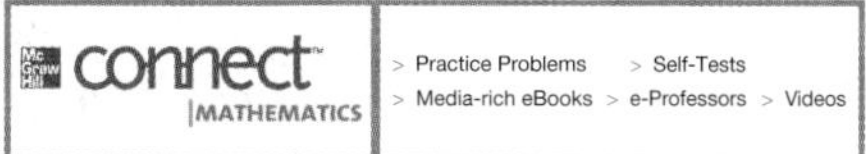

Mcgraw-Hill Connect™ Mathematics is a complete online tutorial and homework management system for mathematics and statistics, designed for greater ease of use than any other system available. Instructors have the flexibility to create and share courses and assignments with colleagues, adjunct faculty, and teaching assistants with only a few clicks of the mouse. All algorithmic exercises, online tutoring, and a variety of video and animations are directly tied to text-specific materials. Completely customizable, Connect Mathematics suits individual instructor and student needs. Exercises can be easily edited, multimedia is assignable, importing additional content is easy, and instructors can even control the level of help available to students while doing their homework. Students have the added benefit of full access to the study tools to individually improve their success without having to be part of a Connect Mathematics course. Connect Mathematics allows for automatic grading and reporting of easy-to assign algorithmically generated homework, quizzes and tests. Grades arc readily accessible through a fully integrated grade book that can be exported in one click to Microsoft Excel, WebCT, or BlackBoard.

Connect Mathematics Offers

- Practice exercises, based on the text's end-of-section material, generated in an unlimited number of variations, for as much practice as needed to master a particular topic.
- Subtitled videos demonstrating text-specific exercises and reinforcing important concepts within a given topic.
- Assessment capabilities, powered through ALEKS, which provide students and instructors with the diagnostics to offer a detailed knowledge base through advanced reporting and remediation tools.

- Faculty with the ability to create and share courses and assignments with colleagues and adjuncts, or to build a course from one of the provided course libraries.
- An Assignment Builder that provides the ability to select algorithmically generated exercises from any McGraw-Hill math textbook, edit content, as well as assign a variety of Connect Mathematics material including an ALEKS Assessment.
- Accessibility from multiple operating systems and Internet browsers.

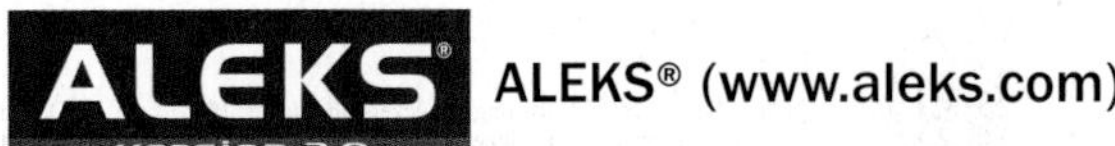

ALEKS® (www.aleks.com)

ALEKS (**A**ssessment and **LE**arning in **K**nowledge **S**paces) is a dynamic online learning system for mathematics education, available over the Web 24/7. ALEKS assesses students, accurately determines their knowledge, and then guides them to the material that they are most ready to learn. With a variety of reports, Textbook Integration Plus, quizzes, and homework assignment capabilities, ALEKS offers flexibility and ease of use for instructors.

- ALEKS uses artificial intelligence to determine exactly what each student knows and is ready to learn. ALEKS remediates student gaps and provides highly efficient learning and improved learning outcomes.
- ALEKS is a comprehensive curriculum that aligns with syllabi or specified textbooks. Used in conjunction with McGraw-Hill texts, students also receive links to text-specific videos, multimedia tutorials, and textbook pages.
- Textbook Integration Plus allows ALEKS to be automatically aligned with syllabi or specified McGraw-Hill textbooks with instructor chosen dates, chapter goals, homework, and quizzes.
- ALEKS with AI-2 gives instructors increased control over the scope and sequence of student learning. Students using ALEKS demonstrate a steadily increasing mastery of the content of the course.
- ALEKS offers a dynamic classroom management system that enables instructors to monitor and direct student progress toward mastery of course objectives.

Supplements for Students

Student's Solutions Manual

This supplement contains complete worked-out solutions to all odd-numbered exercises and all odd- and even-numbered problems in the Review Exercises and Cumulative Reviews in the textbook. The methods used to solve the problems in the manual are the same as those used to solve the examples in the textbook. This tool can be an invaluable aid to students who want to check their work and improve their grades by comparing their own solutions to those found in the manual and finding specific areas where they can do better.

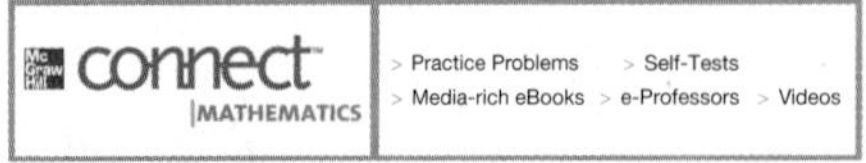

McGraw-Hill Connect Mathematics is a complete online tutorial and homework management system for mathematics and statistics, designed for greater ease of use than any other system available. All algorithmic exercises, online tutoring, and a variety of video and animations are directly tied to text-specific materials.

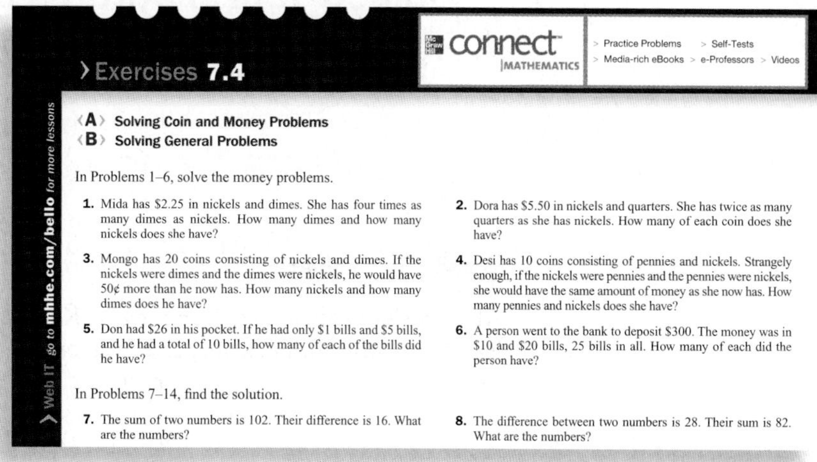

› Exercises **7.4**

connect MATHEMATICS
> Practice Problems > Self-Tests
> Media-rich eBooks > e-Professors > Videos

› Web IT go to mhhe.com/bello for more lessons

‹A› **Solving Coin and Money Problems**
‹B› **Solving General Problems**

In Problems 1–6, solve the money problems.

1. Mida has \$2.25 in nickels and dimes. She has four times as many dimes as nickels. How many dimes and how many nickels does she have?

2. Dora has \$5.50 in nickels and quarters. She has twice as many quarters as she has nickels. How many of each coin does she have?

3. Mongo has 20 coins consisting of nickels and dimes. If the nickels were dimes and the dimes were nickels, he would have 50¢ more than he now has. How many nickels and how many dimes does he have?

4. Desi has 10 coins consisting of pennies and nickels. Strangely enough, if the nickels were pennies and the pennies were nickels, she would have the same amount of money as she now has. How many pennies and nickels does she have?

5. Don had \$26 in his pocket. If he had only \$1 bills and \$5 bills, and he had a total of 10 bills, how many of each of the bills did he have?

6. A person went to the bank to deposit \$300. The money was in \$10 and \$20 bills, 25 bills in all. How many of each did the person have?

In Problems 7–14, find the solution.

7. The sum of two numbers is 102. Their difference is 16. What are the numbers?

8. The difference between two numbers is 28. Their sum is 82. What are the numbers?

ALEKS® (www.aleks.com)

ALEKS (**A**ssessment and **LE**arning in **K**nowledge **S**paces) is a dynamic online learning system for mathematics education, available over the Web 24/7. ALEKS assesses students, accurately determines their knowledge, and then guides them to the material that they are most ready to learn. With a variety of reports, Textbook Integration Plus, quizzes, and homework assignment capabilities, ALEKS offers flexibility and ease of use for instructors.

Bello Video Series

The video series is available online and features the authors introducing topics and working through selected odd-numbered exercises from the text, explaining how to complete them step by step. They are **closed-captioned** for the hearing impaired and are also **subtitled in Spanish.**

Math for the Anxious: Building Basic Skills, by Rosanne Proga

Math for the Anxious: Building Basic Skills is written to provide a practical approach to the problem of math anxiety. By combining strategies for success with a pain-free introduction to basic math content, students will overcome their anxiety and find greater success in their math courses.

Contents

Contents

Chapter three 3

Graphs of Linear Equations, Inequalities, and Applications

Chapter four 4

Exponents and Polynomials

Contents

Chapter five 5

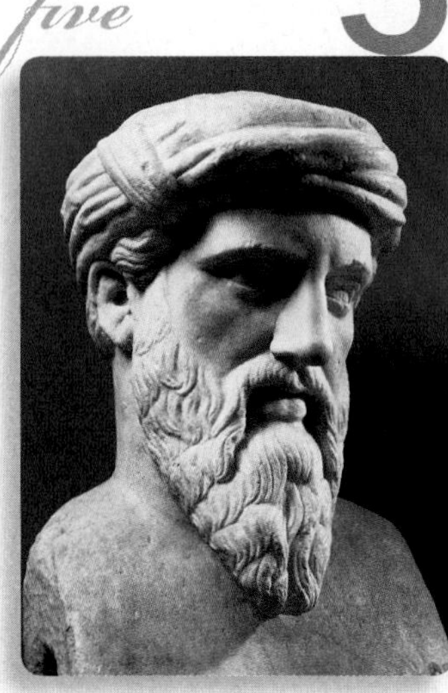

Factoring

Chapter six 6

Rational Expressions

Chapter *seven* 7

Solving Systems of Linear Equations and Inequalities

Chapter *eight* 8

Roots and Radicals

Contents

Chapter *nine* 9

Quadratic Equations

Applications Index

Biology/Health/Life sciences

Business/Finance

Chemistry/Construction

Consumer Applications

Applications Index

Distance/Speed/Time

Education

Food

Geometry

Investment

Politics

Science

Sports

Statistics/Demographics

Transportation

Weather

Bello *Introductory Algebra: A Real-World Approach*

How to Use this Book: A Manual for Success

This brief guide shows you how to use the book effectively: Use It and Succeed!
*It is as easy as **1, 2, 3**.*

BEGINNING OF THE SECTION

1. To succeed: Review the suggested topics at the beginning of each section
2. Objectives: Identify the tasks you should be able to perform (organized by section)
3. Getting Started: Preview the topics being discussed with a familiar application

EXAMPLES AND PAIRED MARGIN PROBLEMS

1. Examples: Explain, expand and help you attain the stated Objectives
2. Green Math Examples: Usually the last Example in each section deals with the environment
3. Paired Margin Problems: Reinforce skills in the Examples. (Answers at the bottom of page)

CHECK FOR MASTERY

1. Concept Checker: To reinforce key terms and get ready for the Mastery Test that follows
2. Mastery Test: Get a quick checkup to make sure you understand the material in the section
3. Skill Checker: Check in advance your understanding of the material in the next section

CONNECT WITH THE EXERCISES

1. Connect: Boost your grade at www.connectmath.com Practice Problems, Self Tests, Videos
2. Exercises (Grouped by Objectives): Practice by doing! Interesting Applications included
3. Green Applications: Marked with a Green bar in the margin; math applied to the Environment

SUMMARY, REVIEW, AND PRACTICE TEST

1. Summary: Easy-to-read grid details the items studied and their meaning and gives an Example
2. Review: Coded by Section number; do them and get extra reinforcement and practice!
3. Practice Test: Answers give Section, Example, and Pages for easy reference to each question

EXTRA, EXTRA

1. Cumulative Review: Covers topics from present and prior chapters. Review all the material!
2. Solutions Manual: Worked out odd numbered solutions, all Reviews and Cumulative Reviews
3. Videos on the Web: Authors working problems from the Practice Test step by step

Section

Chapter

R

Prealgebra Review

R.1 Fractions: Building and Reducing

Objectives

A Write an integer as a fraction.

B Find a fraction equivalent to a given one, but with a specified denominator.

C Reduce fractions to lowest terms.

To Succeed, Review How To . . .

Add, subtract, multiply, and divide natural numbers.

Getting Started

Algebra and Arithmetic

The symbols on the sundial and the Roman clock have something in common: they both use numerals to name the numbers from 1 to 12. Algebra and arithmetic also have something in common: they use the same numbers and the same rules.

In arithmetic you learned about the counting numbers. The numbers used for counting are the **natural numbers:**

$$1, 2, 3, 4, 5, \text{ and so on}$$

These numbers are also used in algebra. We use the **whole numbers**

$$0, 1, 2, 3, 4, \text{ and so on}$$

as well. Later on, you probably learned about the **integers.** The integers include the **positive integers,**

$+1, +2, +3, +4, +5,$ and so on Read "positive one, positive two," and so on.

the **negative integers,**

$-1, -2, -3, -4, -5,$ and so on Read "negative one, negative two," and so on.

and the number **0,** which is neither positive nor negative. Thus, the integers are

$$\ldots, -2, -1, 0, 1, 2, \ldots$$

where the dots (. . .) indicate that the enumeration continues without end. Note that $+1 = 1$, $+2 = 2$, $+3 = 3$, and so on. Thus, the positive integers are the natural numbers.

A > Writing Integers as Fractions

All the numbers discussed can be written as **common fractions** of the form:

$$\text{Fraction bar} \rightarrow \frac{a}{b} \begin{array}{l} \leftarrow \text{Numerator} \\ \leftarrow \text{Denominator} \end{array}$$

When a and b are integers and b is not 0, this ratio is called a **rational number.** For example, $\frac{1}{3}$, $\frac{-5}{2}$, and $\frac{0}{7}$ are rational numbers. In fact, all natural numbers, all whole numbers, and all integers are also rational numbers.

When the numerator a of a fraction is smaller than the denominator b, the fraction $\frac{a}{b}$ is a **proper fraction.** Otherwise, the fraction is **improper.** Improper fractions are often written as **mixed numbers.** Thus, $\frac{9}{5}$ may be written as $1\frac{4}{5}$, and $\frac{13}{4}$ may be written as $3\frac{1}{4}$.

Of course, any integer can be written as a fraction by writing it with a denominator of 1. For example,

$$4 = \frac{4}{1},\quad 8 = \frac{8}{1},\quad 0 = \frac{0}{1},\quad \text{and} \quad -3 = \frac{-3}{1}$$

EXAMPLE 1 Writing integers as fractions

Write the given numbers as fractions with a denominator of 1.

a. 10 **b.** -15

SOLUTION 1

a. $10 = \frac{10}{1}$ **b.** $-15 = \frac{-15}{1}$

PROBLEM 1

Write the given number as a fraction with a denominator of 1.

a. 18 **b.** -24

$\frac{18}{1}$ $\frac{-24}{1}$

The rational numbers we have discussed are part of a larger set of numbers, the set of *real numbers*. The **real numbers** include the *rational numbers* and the *irrational numbers*. The **irrational numbers** are numbers that cannot be written as the ratio of two integers. For example, $\sqrt{2}$, π, $-\sqrt[3]{10}$, and $\frac{\sqrt{3}}{2}$ are irrational numbers. Thus, each real number is either rational or irrational. We shall say more about the irrational numbers in Chapter 1.

B > Equivalent Fractions

In Example 1(a) we wrote 10 as $\frac{10}{1}$. Can you find other ways of writing 10 as a fraction? Here are some:

$$10 = \frac{10}{1} = \frac{10 \times 2}{1 \times 2} = \frac{20}{2}$$

$$10 = \frac{10}{1} = \frac{10 \times 3}{1 \times 3} = \frac{30}{3}$$

and

$$10 = \frac{10}{1} = \frac{10 \times 4}{1 \times 4} = \frac{40}{4}$$

Note that $\frac{2}{2} = 1$, $\frac{3}{3} = 1$, and $\frac{4}{4} = 1$. As you can see, the fraction $\frac{10}{1}$ is **equivalent to** (has the same value as) many other fractions. We can always obtain other fractions equivalent to any given fraction by *multiplying* the numerator and denominator of the original fraction by the *same* nonzero number, a process called **building up** the fraction. This is the same as multiplying the fraction by 1, where 1 is written as $\frac{2}{2}$, $\frac{3}{3}$, $\frac{4}{4}$, and so on. For example,

$$\frac{3}{5} = \frac{3 \times 2}{5 \times 2} = \frac{6}{10}$$

$$\frac{3}{5} = \frac{3 \times 3}{5 \times 3} = \frac{9}{15}$$

and

$$\frac{3}{5} = \frac{3 \times 4}{5 \times 4} = \frac{12}{20}$$

Answers to PROBLEMS

1. a. $\frac{18}{1}$ **b.** $\frac{-24}{1}$

EXAMPLE 2 Finding equivalent fractions

Find a fraction equivalent to $\frac{3}{5}$ with a denominator of 20.

SOLUTION 2 We must solve the problem

$$\frac{3}{5} = \frac{?}{20}$$

Note that the denominator, 5, was multiplied by 4 to get 20. So, we must also multiply the numerator, 3, by 4.

$$\frac{3}{5} = \frac{?}{20}$$ If you multiply the denominator by 4,

Multiply by 4.

Multiply by 4.

$$\frac{3}{5} = \frac{12}{20}$$ you have to multiply the numerator by 4.

Thus, the equivalent fraction is $\frac{12}{20}$.

PROBLEM 2

Find a fraction equivalent to $\frac{4}{7}$ with a denominator of 21.

$\frac{4}{7} = \frac{?}{21}$

$\frac{4}{7} = \frac{4 \times 3}{7 \times 3} = \frac{12}{21}$

$\frac{12}{21}$

Here is a slightly different problem. Can we find a fraction equivalent to $\frac{15}{20}$ with a denominator of 4? We do this in Example 3.

EXAMPLE 3 Finding equivalent fractions

Find a fraction equivalent to $\frac{15}{20}$ with a denominator of 4.

SOLUTION 3 We proceed as before.

$$\frac{15}{20} = \frac{?}{4}$$ 20 was divided by 5 to get 4.

Divide by 5.

Divide by 5.

$$\frac{15}{20} = \frac{3}{4}$$ 15 was divided by 5 to get 3.

PROBLEM 3

Find a fraction equivalent to $\frac{24}{30}$ with a denominator of 5.

$\frac{24}{30} = \frac{}{5}$

$\frac{24}{30} = \frac{24 \div 6}{30 \div 6} = \frac{4}{5}$

We can summarize our work with equivalent fractions in the following procedure.

PROCEDURE

Obtaining Equivalent Fractions

To obtain an **equivalent fraction,** *multiply* or *divide* both numerator and denominator of the fraction by the *same* nonzero number.

C > Reducing Fractions to Lowest Terms

The preceding rule can be used to *reduce* (simplify) fractions to lowest terms. A fraction is *reduced* to lowest terms (simplified) when there is no number (except 1) that will divide the numerator and the denominator exactly. The procedure is as follows.

Answers to PROBLEMS

2. $\frac{12}{21}$ **3.** $\frac{4}{5}$

> **PROCEDURE**
>
> **Reducing Fractions to Lowest Terms**
>
> To **reduce** a fraction to *lowest* terms, divide the numerator and denominator by the *largest* natural number that will divide them exactly.
>
> "Divide them exactly" means that both remainders after division are zero.

To reduce $\frac{12}{30}$ to lowest terms, we divide the numerator and denominator by 6, the *largest* natural number that divides 12 and 30 exactly. [6 is sometimes called the *greatest common divisor* (GCD) of 12 and 30.] Thus,

$$\frac{12}{30} = \frac{12 \div 6}{30 \div 6} = \frac{2}{5}$$

This reduction is sometimes shown like this:

$$\frac{\overset{2}{\cancel{12}}}{\underset{5}{\cancel{30}}} = \frac{2}{5}$$

EXAMPLE 4 Reducing fractions

Reduce to lowest terms:

a. $\frac{15}{20}$ **b.** $\frac{30}{45}$ **c.** $\frac{60}{48}$

SOLUTION 4

a. The *largest* natural number exactly dividing 15 and 20 is 5. Thus,

$$\frac{15}{20} = \frac{15 \div 5}{20 \div 5} = \frac{3}{4}$$

b. The *largest* natural number exactly dividing 30 and 45 is 15. Hence,

$$\frac{30}{45} = \frac{30 \div 15}{45 \div 15} = \frac{2}{3}$$

c. The *largest* natural number dividing 60 and 48 is 12. Therefore,

$$\frac{60}{48} = \frac{60 \div 12}{48 \div 12} = \frac{5}{4}$$

PROBLEM 4

Reduce to lowest terms:

a. $\frac{30}{50}$ **b.** $\frac{45}{60}$ **c.** $\frac{84}{72}$

What if you are unable to see at once that the *largest* natural number dividing the numerator and denominator in, say, $\frac{30}{45}$, is 15? No problem; it just takes a little longer. Suppose you notice that 30 and 45 are both divisible by 5. You then write

$$\frac{30}{45} = \frac{30 \div 5}{45 \div 5} = \frac{6}{9}$$

Now you can see that 6 and 9 are both divisible by 3. Thus,

$$\frac{6}{9} = \frac{6 \div 3}{9 \div 3} = \frac{2}{3}$$

which is the same answer we got in Example 4(b). The whole procedure can be written as

$$\frac{\overset{\overset{2}{\cancel{6}}}{\cancel{30}}}{\underset{\underset{3}{\cancel{9}}}{\cancel{45}}} = \frac{2}{3}$$

We can also reduce $\frac{15}{20}$ using *prime factorization* by writing

$$\frac{15}{20} = \frac{3 \cdot 5}{2 \cdot 2 \cdot 5} = \frac{3}{4}$$

Answers to PROBLEMS

4. a. $\frac{3}{5}$ b. $\frac{3}{4}$ c. $\frac{7}{6}$

(The dot, ·, indicates multiplication.) Note that 15 is written as the *product* of 3 and 5, and 20 is written as the product $2 \cdot 2 \cdot 5$. A **product** is the answer to a multiplication problem. When 15 is written as the product $3 \cdot 5$, the 3 and the 5 are the **factors** of 15. As a matter of fact, 3 and 5 are **prime numbers** (only divisible by itself and 1), so $3 \cdot 5$ is the **prime factorization** of 15; similarly, $2 \cdot 2 \cdot 5$ is the **prime factorization** of 20, because 2 and 5 are primes. In general,

> A natural number is prime if it has *exactly two different factors*, itself and 1.

The first few primes are

2, 3, 5, 7, 11, 13, 17, 19, 23, 29, 31, and so on.

A natural number that is *not* prime is called a **composite number.** Thus, the numbers missing in the previous list, 4, 6, 8, 9, 10, 12, 14, 15, 16, and so on, are **composite.** Note that 1 is considered neither prime (because it does not have *two different* factors) nor composite.

When using prime factorization, we can keep better track of the factors by using a **factor tree.** For example, to reduce $\frac{30}{45}$ using prime factorization, we make two trees to factor 30 and 45 as shown:

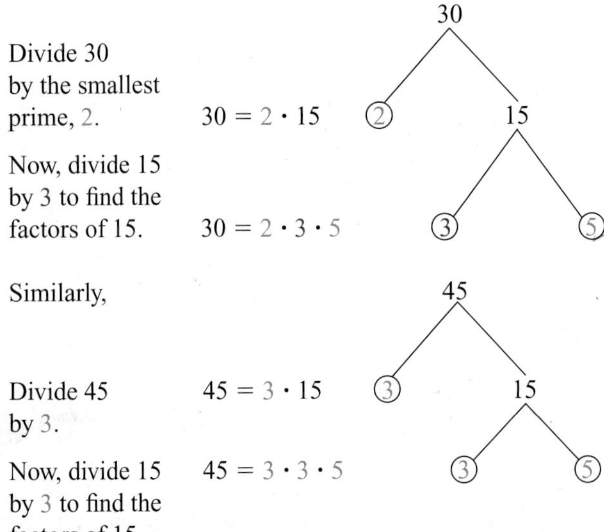
Divide 30 by the smallest prime, 2.
30 = 2 · 15
Now, divide 15 by 3 to find the factors of 15.
30 = 2 · 3 · 5
30
2
15
3
5
Similarly,
Divide 45 by 3.
45 = 3 · 15
Now, divide 15 by 3 to find the factors of 15.
45 = 3 · 3 · 5
45
3
15
3
5

Thus,

$$\frac{30}{45} = \frac{2 \cdot 3 \cdot 5}{3 \cdot 3 \cdot 5} = \frac{2}{3} \quad \text{as before.}$$

Caution: When reducing a fraction to lowest terms, it is easier to look for the *largest common factor* in the numerator and denominator, but if no largest factor is obvious, *any common factor* can be used and the simplification can be done in stages. For example,

$$\frac{200}{250} = \frac{20 \cdot 10}{25 \cdot 10} = \frac{4 \cdot 5 \cdot 10}{5 \cdot 5 \cdot 10} = \frac{4}{5}$$

Which way should you simplify fractions? The way you understand! Make sure you follow the instructor's preferences regarding the procedure used.

> Exercises R.1

⟨ A ⟩ Writing Integers as Fractions In Problems 1–8, write the given number as a fraction with a denominator of 1.

1. 28 **2.** 93 **3.** −42 **4.** −86

5. 0 **6.** 1 **7.** −1 **8.** −17

⟨ B ⟩ Equivalent Fractions In Problems 9–30, find the missing number that makes the fractions equivalent.

9. $\frac{1}{8} = \frac{?}{24}$ **10.** $\frac{7}{1} = \frac{?}{2}$ **11.** $\frac{7}{1} = \frac{?}{6}$ **12.** $\frac{5}{6} = \frac{?}{48}$ **13.** $\frac{5}{3} = \frac{?}{15}$

14. $\frac{9}{8} = \frac{?}{32}$ **15.** $\frac{7}{11} = \frac{?}{33}$ **16.** $\frac{11}{7} = \frac{?}{35}$ **17.** $\frac{1}{8} = \frac{4}{?}$ **18.** $\frac{3}{5} = \frac{27}{?}$

19. $\frac{5}{6} = \frac{5}{?}$ **20.** $\frac{9}{10} = \frac{9}{?}$ **21.** $\frac{8}{7} = \frac{16}{?}$ **22.** $\frac{9}{5} = \frac{36}{?}$ **23.** $\frac{6}{5} = \frac{36}{?}$

24. $\frac{5}{3} = \frac{45}{?}$ **25.** $\frac{21}{56} = \frac{?}{8}$ **26.** $\frac{12}{18} = \frac{?}{3}$ **27.** $\frac{36}{180} = \frac{?}{5}$ **28.** $\frac{8}{24} = \frac{4}{?}$

29. $\frac{18}{12} = \frac{3}{?}$ **30.** $\frac{56}{49} = \frac{8}{?}$

⟨ C ⟩ Reducing Fractions to Lowest Terms In Problems 31–40, reduce the fraction to lowest terms by writing the numerator and denominator as products of primes.

31. $\frac{15}{12}$ **32.** $\frac{30}{28}$ **33.** $\frac{13}{52}$ **34.** $\frac{27}{54}$ **35.** $\frac{56}{24}$

36. $\frac{56}{21}$ **37.** $\frac{22}{33}$ **38.** $\frac{26}{39}$ **39.** $\frac{100}{25}$ **40.** $\frac{21}{3}$

> > > Applications

41. *AOL ad* If you take advantage of the AOL offer and get 1000 hours free for 45 days:

a. How many hours will you get per day? (Assume you use the same number of hours each day and do not simplify your answer.)

b. Write the answer to part **a** in lowest terms.

c. Write the answer to part **a** as a mixed number.

d. If you used AOL 24 hours a day, for how many days would the 1000 free hours last?

42. *Census data* ***Are you "poor"?*** In 1959, the U.S. Census reported that about 40 million of the 180 million people living in the United States were "poor." In the year 2000, about 30 million out of 275 million were "poor" (income less than $8794).

a. What reduced fraction of the people was poor in the year 1959?

b. What reduced fraction of the people was poor in the year 2000?

43. *High school completion rate* In a recent year, about 41 out of 100 persons in the United States had completed 4 years of high school or more. What fraction of the people is that?

44. *High school completion rates* In a recent year, about 84 out of 100 Caucasians, 76 out of 100 African-Americans, and 56 out of 100 Hispanics had completed 4 years of high school or more. What reduced fractions of the Caucasians, African-Americans, and Hispanics had completed 4 years of high school or more?

45. *Pizza consumption* The pizza shown here consists of six pieces. Alejandro ate $\frac{1}{3}$ of the pizza.

a. Write $\frac{1}{3}$ with a denominator of 6.

b. How many pieces did Alejandro eat?

c. If Cindy ate two pieces of pizza, what fraction of the pizza did she eat?

d. Who ate more pizza, Alejandro or Cindy?

Fuel gauge Problems 46–50 refer to the photo of the fuel gauge. What fraction of the tank is full if the needle:

46. Points to the line midway between E and $\frac{1}{2}$ full?

47. Points to the first line to the right of empty (E)?

48. Is midway between empty and $\frac{1}{8}$ of a tank?

49. Is one line past the $\frac{1}{2}$ mark?

50. Is one line before the $\frac{1}{2}$ mark?

Using Your Knowledge

Interpreting Fractions During the 1953–1954 basketball season, the NBA (National Basketball Association) had a problem with the game: it was boring. Danny Biasone, the owner of the Syracuse Nationals, thought that limiting the time a team could have the ball should encourage more shots. Danny figured out that in a fast-paced game, each team should take 60 shots during the 48 minutes the game lasted (4 quarters of 12 minutes each). He then looked at the fraction $\frac{\text{Seconds}}{\text{Shots}}$.

51. a. How many seconds does the game last?

b. How many total shots are to be taken in the game?

c. What is the reduced fraction $\frac{\text{Seconds}}{\text{Shots}}$?

Now you know where the $\frac{\text{Seconds}}{\text{Shots}}$ clock came from!

You have learned how to work with fractions. Now you can use your knowledge to interpret fractions from diagrams. As you see from the diagram (circle graph), "CNN Headline News" devotes the first quarter of every hour $\left(\frac{15}{60} = \frac{1}{4}\right)$ to National and International News.

52. What total fraction of the hour (gold areas) is devoted to: Dollars & Sense?

53. What total fraction of the hour is devoted to: Sports?

54. What total fraction of the hour is devoted to: Local News or People & Places?

55. Which feature uses the most time, and what fraction of an hour does it use?

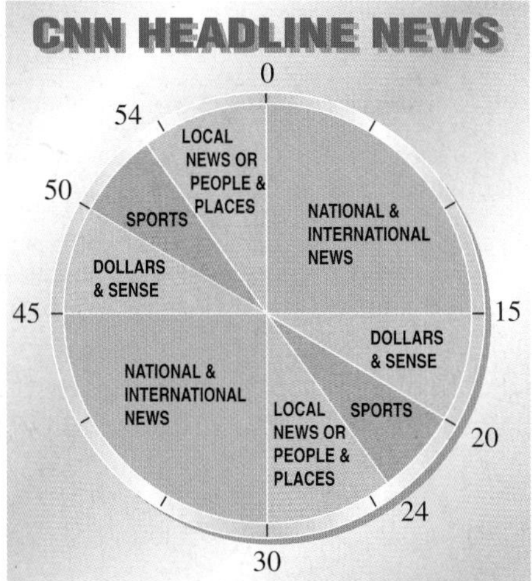

Source: Data from CNN.

As you can see from the illustration, "Bay News 9" devotes $8 + 6 = 14$ minutes of the hour (60 minutes) to News. This represents $\frac{14}{60} = \frac{7}{30}$ of the hour.

56. What fraction of the hour is devoted to Traffic?

57. What fraction of the hour is devoted to Weather?

58. What fraction of the hour is devoted to News & Beyond the Bay?

59. Which features use the most and least time, and what fraction of the hour does each one use?

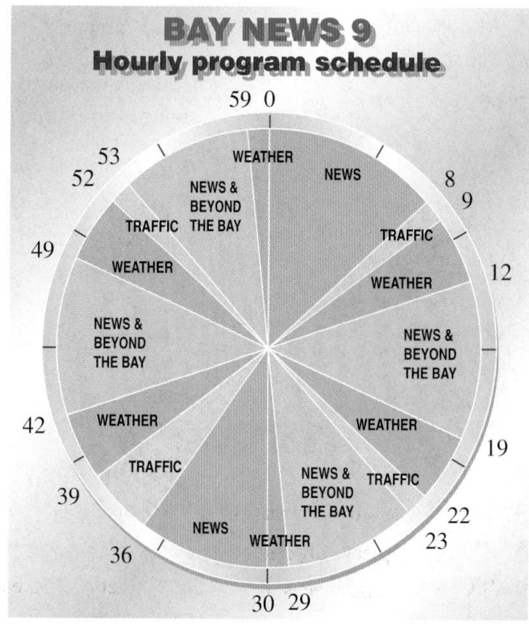

Source: Data from Bay News 9.

Write On

60. Write the procedure you use to reduce a fraction to lowest terms.

61. What are the advantages and disadvantages of writing the numerator and denominator of a fraction as a product of primes before reducing the fraction?

62. Write a procedure to determine whether two fractions are equivalent.

R.2 Operations with Fractions and Mixed Numbers

Objective

A Multiply and divide fractions.

B Add and subtract fractions.

To Succeed, Review How To . . .

1. Reduce fractions (pp. 4–6).
2. Write a number as a product of primes (p. 6).

Getting Started

A Sweet Problem

Each of the cups contains $\frac{1}{4}$ cup of sugar. How much sugar do the cups contain altogether?

To find the answer, we can multiply 3 by $\frac{1}{4}$; that is, we can find $3 \cdot \frac{1}{4}$.

A > Multiplying and Dividing Fractions

How do we perform the multiplication $3 \cdot \frac{1}{4}$ required to determine the total amount of sugar in the three cups? Note that

$$3 \cdot \frac{1}{4} = \frac{3}{1} \cdot \frac{1}{4} = \frac{3 \cdot 1}{1 \cdot 4} = \frac{3}{4}$$

Similarly,

$$\frac{4}{9} \cdot \frac{2}{5} = \frac{4 \cdot 2}{9 \cdot 5} = \frac{8}{45}$$

Here is the general rule for multiplying fractions.

RULE

Multiplying Fractions

To multiply fractions, (1) multiply the numerators, (2) multiply the denominators, and then (3) simplify. In symbols,

$$\frac{a}{b} \cdot \frac{c}{d} = \frac{a \cdot c}{b \cdot d} \qquad b \cdot d \neq 0$$

Note: To avoid repetition, from now on we will assume that the denominators are *not* 0.

Calculator Corner

Multiplying Fractions

To multiply 3 by $\frac{1}{4}$ using a calculator with an [x/y] key, enter [3] [×] [1] [x/y] [4] [ENTER] and you get $\frac{3}{4}$. If your calculator has a [/] or a [÷] key, the strokes will be [3] [×] [1] [÷] [4] [ENTER] and you will get the same answer.

EXAMPLE 1 Multiplying fractions

Multiply: $\frac{9}{5} \cdot \frac{3}{4}$

SOLUTION 1 We use our rule for multiplying fractions:

$$\frac{9}{5} \cdot \frac{3}{4} = \frac{9 \cdot 3}{5 \cdot 4} = \frac{27}{20}$$

PROBLEM 1

Multiply: $\frac{9}{7} \cdot \frac{3}{5}$

When multiplying fractions, we can save time if we divide out common factors *before* we multiply. For example,

$$\frac{2}{5} \cdot \frac{5}{7} = \frac{2 \cdot 5}{5 \cdot 7} = \frac{10}{35} = \frac{2}{7}$$

We can save time by writing

$$\frac{2}{\cancel{5}_1} \cdot \frac{\cancel{5}^1}{7} = \frac{2 \cdot 1}{1 \cdot 7} = \frac{2}{7}$$

This can be done because $\frac{5}{5} = 1$.

CAUTION

Only factors that are common to *both* numerator *and* denominator can be divided out.

Answers to PROBLEMS

1. $\frac{27}{35}$

EXAMPLE 2 Multiplying fractions with common factors

Multiply:

a. $\frac{3}{7} \cdot \frac{7}{8}$ b. $\frac{5}{8} \cdot \frac{4}{15}$

SOLUTION 2

a. $\frac{\overset{1}{\cancel{3}}}{\underset{1}{\cancel{7}}} \cdot \frac{\overset{1}{\cancel{7}}}{8} = \frac{3 \cdot 1}{1 \cdot 8} = \frac{3}{8}$

b. $\frac{\overset{1}{\cancel{5}}}{\underset{2}{\cancel{8}}} \cdot \frac{\overset{1}{\cancel{4}}}{\underset{3}{\cancel{15}}} = \frac{1 \cdot 1}{2 \cdot 3} = \frac{1}{6}$

PROBLEM 2

Multiply:

a. $\frac{4}{9} \cdot \frac{9}{11}$ b. $\frac{5}{14} \cdot \frac{7}{20}$

If we wish to multiply a fraction by a **mixed number,** such as $3\frac{1}{4}$, we must convert the mixed number to a fraction first. The number $3\frac{1}{4}$ (read "3 and $\frac{1}{4}$") means $3 + \frac{1}{4} = \frac{12}{4} + \frac{1}{4} = \frac{13}{4}$. (This addition will be clearer to you after studying the addition of fractions.) For now, we can shorten the procedure by using the following diagram.

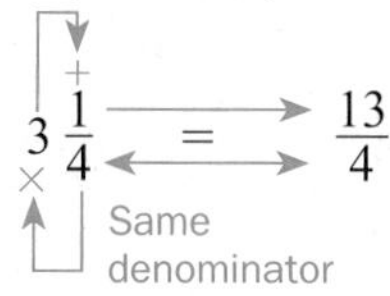

Work clockwise. *First* multiply the denominator 4 by the whole number part 3; add the numerator. This is the new numerator. Use the same denominator.

EXAMPLE 3 Multiplying fractions and mixed numbers

Multiply: $5\frac{1}{3} \cdot \frac{9}{16}$

SOLUTION 3 We first convert the mixed number to a fraction:

$$5\frac{1}{3} = \frac{3 \cdot 5 + 1}{3} = \frac{16}{3}$$

Thus,

$$5\frac{1}{3} \cdot \frac{9}{16} = \frac{\overset{1}{\cancel{16}}}{\underset{1}{\cancel{3}}} \cdot \frac{\overset{3}{\cancel{9}}}{\underset{1}{\cancel{16}}} = \frac{1 \cdot 3}{1 \cdot 1} = 3$$

PROBLEM 3

Multiply: $2\frac{3}{4} \cdot \frac{8}{11}$

If we wish to divide one number by another nonzero number, we can indicate the division by a fraction. Thus, to divide 2 by 5 we can write

$$2 \div 5 = \frac{2}{5} = 2 \cdot \frac{1}{5}$$

Multiply.

The divisor 5 is inverted.

Note that to divide 2 by 5 we multiplied 2 by $\frac{1}{5}$, where the fraction $\frac{1}{5}$ was obtained by *inverting* $\frac{5}{1}$ to obtain $\frac{1}{5}$. (In mathematics, $\frac{5}{1}$ and $\frac{1}{5}$ are called **reciprocals.**) *Note:* Only the 5 (the divisor) is replaced by its reciprocal, $\frac{1}{5}$.

Now let's try the problem $5 \div \frac{5}{7}$. If we do it like the preceding example, we write

$$5 \div \frac{5}{7} = 5 \cdot \frac{7}{5} = \frac{\overset{1}{\cancel{5}}}{1} \cdot \frac{7}{\underset{1}{\cancel{5}}} = 7$$

Multiply. Invert.

In general, to divide $\frac{a}{b}$ by $\frac{c}{d}$, we multiply $\frac{a}{b}$ by the *reciprocal* of $\frac{c}{d}$—that is, $\frac{d}{c}$. Here is the rule.

Answers to PROBLEMS

2. a. $\frac{4}{11}$ **b.** $\frac{1}{8}$ **3.** 2

RULE

Dividing Fractions

To divide fractions, ① multiply the first fraction by the reciprocal of the second fraction and then ② simplify. In symbols,

$$\frac{a}{b} \div \frac{c}{d} = \frac{a}{b} \cdot \frac{d}{c}$$

EXAMPLE 4 Dividing fractions

Divide:

a. $\frac{3}{5} \div \frac{2}{7}$ b. $\frac{4}{9} \div 5$

SOLUTION 4

a. $\frac{3}{5} \div \frac{2}{7} = \frac{3}{5} \cdot \frac{7}{2} = \frac{21}{10}$ b. $\frac{4}{9} \div 5 = \frac{4}{9} \cdot \frac{1}{5} = \frac{4}{45}$

PROBLEM 4

Divide:

a. $\frac{3}{4} \div \frac{5}{7}$ b. $\frac{3}{7} \div 5$

As in the case of multiplication, if the problem involves mixed numbers, we change them to fractions first, like this:

$$2\frac{1}{4} \div \frac{3}{5} = \frac{9}{4} \div \frac{3}{5} = \frac{\overset{3}{\cancel{9}}}{4} \cdot \frac{5}{\underset{1}{\cancel{3}}} = \frac{15}{4}$$

Change. Invert.

EXAMPLE 5 Dividing fractions and mixed numbers

Divide:

a. $3\frac{1}{4} \div \frac{7}{8}$ b. $\frac{11}{12} \div 7\frac{1}{3}$

SOLUTION 5

a. $3\frac{1}{4} \div \frac{7}{8} = \frac{13}{4} \div \frac{7}{8} = \frac{13}{\underset{1}{\cancel{4}}} \cdot \frac{\overset{2}{\cancel{8}}}{7} = \frac{26}{7}$ b. $\frac{11}{12} \div 7\frac{1}{3} = \frac{11}{12} \div \frac{22}{3} = \frac{\overset{1}{\cancel{11}}}{\underset{4}{\cancel{12}}} \cdot \frac{\overset{1}{\cancel{3}}}{\underset{2}{\cancel{22}}} = \frac{1}{8}$

PROBLEM 5

Divide:

a. $2\frac{1}{5} \div \frac{7}{10}$ b. $\frac{11}{15} \div 7\frac{1}{3}$

B > Adding and Subtracting Fractions

Now we are ready to *add* fractions.

The photo shows that 1 quarter plus 2 quarters equals 3 quarters. In symbols,

$$\frac{1}{4} + \frac{2}{4} = \frac{1+2}{4} = \frac{3}{4}$$

In general, to add fractions with the *same* denominator, we add the numerators and keep the denominator.

RULE

Adding Fractions with the Same Denominator

To add fractions with the same denominators, ① add the numerators and ② keep the same denominators. In symbols,

$$\frac{a}{b} + \frac{c}{b} = \frac{a+c}{b}$$

Answers to PROBLEMS

4. a. $\frac{21}{20}$ b. $\frac{3}{35}$

5. a. $\frac{22}{7}$ b. $\frac{1}{10}$

Thus,

$$\frac{1}{5} + \frac{2}{5} = \frac{1+2}{5} = \frac{3}{5}$$

and

$$\frac{3}{8} + \frac{1}{8} = \frac{3+1}{8} = \frac{4}{8} = \frac{1}{2}$$

NOTE

In the last addition we reduced $\frac{4}{8}$ to $\frac{1}{2}$ by dividing the numerator and denominator by 4. When a result involves fractions, we always **reduce** to lowest terms.

Now suppose you wish to add $\frac{5}{12}$ and $\frac{1}{18}$. Since these two fractions do not have the same denominators, the rule does not work. To add them, you must learn how to write $\frac{5}{12}$ and $\frac{1}{18}$ as equivalent fractions with the same denominator. To keep things simple, you should also try to make this denominator as small as possible; that is, you should first find the **least common denominator (LCD)** of the fractions.

The LCD of two fractions is the **least** (*smallest*) **common multiple (LCM)** of their denominators. Thus, to find the LCD of $\frac{5}{12}$ and $\frac{1}{18}$, we find the LCM of 12 and 18. There are several ways of doing this. One way is to select the larger number (18) and find its multiples. The first multiple of 18 is $2 \cdot 18 = 36$, and 12 divides into 36. Thus, 36 is the LCD of 12 and 18.

Unfortunately, this method is not practical in algebra. A more convenient method consists of writing the denominators 12 and 18 in completely *factored* form. We can start by writing 12 as $2 \cdot 6$. In turn, $6 = 2 \cdot 3$; thus,

$$12 = 2 \cdot 2 \cdot 3$$

Similarly,

$$18 = 2 \cdot 9, \quad \text{or} \quad 2 \cdot 3 \cdot 3$$

Now we can see that the smallest number that is a multiple of 12 and 18 must have at least two 2's (there are two 2's in 12) and two 3's (there are two 3's in 18). Thus, the LCD of $\frac{5}{12}$ and $\frac{1}{18}$ is

$$\underbrace{2 \cdot 2}_{\text{Two 2's}} \cdot \underbrace{3 \cdot 3}_{\text{Two 3's}} = 36$$

Fortunately in arithmetic, there is an even shorter way of finding the LCM of 12 and 18. Note that what we want to do is find the *common* factors in 12 and 18, so we can divide 12 and 18 by the smallest divisor common to both numbers (2) and then the next (3), and so on. If we multiply these divisors by the final quotient, the result is the LCM. Here is the shortened version:

```
Divide by 2.   2) 12   18
Divide by 3.   3)  6    9
                   2 —— 3  ——> Multiply 2 · 3 · 2 · 3.
```

The LCM is $2 \cdot 3 \cdot 2 \cdot 3 = 36$.

In general, we use the following procedure to find the LCD of two fractions.

PROCEDURE

Finding the LCD of Two Fractions

1. Write the denominators in a horizontal row and divide each number by a *divisor* common to both numbers.
2. Continue the process until the resulting quotients have no common divisor (except 1).
3. The product of the *divisors* and the *final quotients* is the LCD.

Now that we know that the LCD of $\frac{5}{12}$ and $\frac{1}{18}$ is 36, we can add the two fractions by writing each as an equivalent fraction with a denominator of 36. We do this by multiplying numerators and denominators of $\frac{5}{12}$ by 3 and of $\frac{1}{18}$ by 2. Thus, we get

$$\frac{5}{12} = \frac{5 \cdot 3}{12 \cdot 3} = \frac{15}{36} \longrightarrow \frac{15}{36}$$
$$\frac{1}{18} = \frac{1 \cdot 2}{18 \cdot 2} = \frac{2}{36} \longrightarrow +\frac{2}{36}$$
$$\frac{17}{36}$$

We can also write the results as

$$\frac{5}{12} + \frac{1}{18} = \frac{15}{36} + \frac{2}{36} = \frac{15+2}{36} = \frac{17}{36}$$

EXAMPLE 6 Adding fractions with different denominators

Add: $\frac{1}{20} + 1\frac{1}{18}$

SOLUTION 6 We first find the LCM of 20 and 18.

Divide by 2. $2\,)\,20 \quad 18$

$10 \quad 9 \longrightarrow$ Multiply $2 \cdot 10 \cdot 9$.

Since 10 and 9 have no common divisor, the LCM is

$$2 \cdot 10 \cdot 9 = 180$$

(You could also find the LCM by writing $20 = 2 \cdot 2 \cdot 5$ and $18 = 2 \cdot 3 \cdot 3$; the LCM is $2 \cdot 2 \cdot 3 \cdot 3 \cdot 5 = 180$.)

Now, write the fractions with denominators of 180 and add.

$$\frac{1}{20} = \frac{1 \cdot 9}{20 \cdot 9} = \frac{9}{180} \longrightarrow \frac{9}{180}$$
$$1\frac{1}{18} = \frac{19}{18} = \frac{19 \cdot 10}{18 \cdot 10} = \frac{190}{180} \longrightarrow +\frac{190}{180}$$
$$\frac{199}{180}$$

If you prefer, you can write the procedure like this:

$$\frac{1}{20} + 1\frac{1}{18} = \frac{9}{180} + \frac{190}{180} = \frac{199}{180}, \quad \text{or} \quad 1\frac{19}{180}$$

PROBLEM 6

Add: $\frac{3}{20} + 1\frac{1}{16}$

Can we use the same procedure to add three or more fractions? Almost; but we need to know how to find the LCD for three or more fractions. The procedure is very similar to that used for finding the LCM of two numbers. If we write 15, 21, and 28 in factored form,

$$15 = 3 \cdot 5$$
$$21 = 3 \cdot 7$$
$$28 = 2 \cdot 2 \cdot 7$$

we see that the LCM must contain at least $2 \cdot 2$, 3, 5, and 7. (Note that we select each factor the *greatest* number of times it appears in any factorization.) Thus, the LCM of 15, 21, and 28 is $2 \cdot 2 \cdot 3 \cdot 5 \cdot 7 = 420$.

We can also use the following shortened procedure to find the LCM of 15, 21, and 28.

Answers to PROBLEMS

6. $\frac{97}{80}$ or $1\frac{17}{80}$

PROCEDURE

Finding the LCD of Three or More Fractions

1. Write the denominators in a horizontal row and divide the numbers by a divisor common to two or more of the numbers. If any of the other numbers is not divisible by this divisor, *circle* the number and carry it to the next line.

   Divide by 3. $3\overline{)15 \quad 21 \quad (28)}$

2. Repeat step 1 with the quotients and carrydowns until *no two numbers* have a common divisor (except 1).

   Divide by 7. $7\overline{)(5) \quad 7 \quad 28}$
   $\quad 5 \quad 1 \quad 4$

   Multiply $3 \cdot 7 \cdot 5 \cdot 1 \cdot 4$.

3. The LCD is the *product* of the divisors from the preceding steps and the numbers in the final row.

   The LCD is $3 \cdot 7 \cdot 5 \cdot 1 \cdot 4 = 420$.

Thus, to add

$$\frac{1}{16} + \frac{3}{10} + \frac{1}{28}$$

we first find the LCD of 16, 10, and 28. We can use either method.

Method 1. Find the LCD by writing

$$16 = 2 \cdot 2 \cdot 2 \cdot 2$$
$$10 = 2 \cdot 5$$
$$28 = 2 \cdot 2 \cdot 7$$

and selecting each factor the *greatest* number of times it appears in any factorization. Since 2 appears four times and 5 and 7 appear once, the LCD is

$$\underbrace{2 \cdot 2 \cdot 2 \cdot 2}_{\text{four times}} \cdot \underbrace{5}_{\text{once}} \cdot \underbrace{7}_{\text{once}} = 560$$

Method 2. Use the three-step procedure

Step 1. Divide by 2. $2\overline{)16 \quad 10 \quad 28}$

Step 2. Divide by 2. $2\overline{)\ 8 \quad (5) \quad 14}$
$\quad 4 \quad 5 \quad 7$

Multiply $2 \cdot 2 \cdot 4 \cdot 5 \cdot 7$.

Step 3. The LCD is $2 \cdot 2 \cdot 4 \cdot 5 \cdot 7 = 560$.

We then write $\frac{1}{16}$, $\frac{3}{10}$, and $\frac{1}{28}$ with denominators of 560.

$$\frac{1}{16} = \frac{1 \cdot 35}{16 \cdot 35} = \frac{35}{560} \longrightarrow \quad \frac{35}{560}$$
$$\frac{3}{10} = \frac{3 \cdot 56}{10 \cdot 56} = \frac{168}{560} \longrightarrow \quad +\frac{168}{560}$$
$$\frac{1}{28} = \frac{1 \cdot 20}{28 \cdot 20} = \frac{20}{560} \longrightarrow \quad +\frac{20}{560}$$
$$\overline{\qquad \frac{223}{560}}$$

Or, if you prefer,

$$\frac{1}{16} + \frac{3}{10} + \frac{1}{28} = \frac{35}{560} + \frac{168}{560} + \frac{20}{560} = \frac{223}{560}$$

NOTE

Method 1 for finding the LCD is preferred because it is more easily generalized to algebraic fractions.

Now that you know how to add fractions, *subtraction* is no problem. All the rules we have mentioned still apply! For example, we subtract fractions with the same denominator as shown.

$$\frac{5}{8} - \frac{2}{8} = \frac{5-2}{8} = \frac{3}{8}$$

$$\frac{7}{9} - \frac{1}{9} = \frac{7-1}{9} = \frac{6}{9} = \frac{2}{3}$$

The next example shows how to subtract fractions involving *different* denominators.

EXAMPLE 7 **Subtracting fractions with different denominators**

Subtract: $\frac{7}{12} - \frac{1}{18}$

SOLUTION 7 We first find the LCD of the fractions.

Method 1. Write

$12 = 2 \cdot 2 \cdot 3$

$18 = 2 \cdot 3 \cdot 3$

The LCD is $2 \cdot 2 \cdot 3 \cdot 3 = 36$.

Method 2.

Step 1. $2\overline{)\ 12 \quad 18}$

Step 2. $3\overline{)\ 6 \quad 9}$

$\quad 2 \quad 3$

Step 3. The LCD is $2 \cdot 3 \cdot 2 \cdot 3 = 36$.

We then write each fraction with 36 as the denominator.

$$\frac{7}{12} = \frac{7 \cdot 3}{12 \cdot 3} = \frac{21}{36} \quad \text{and} \quad \frac{1}{18} = \frac{1 \cdot 2}{18 \cdot 2} = \frac{2}{36}$$

Thus,

$$\frac{7}{12} - \frac{1}{18} = \frac{21}{36} - \frac{2}{36} = \frac{21-2}{36} = \frac{19}{36}$$

PROBLEM 7

Subtract: $\frac{7}{18} - \frac{1}{12}$

The rules for adding fractions also apply to subtraction. If mixed numbers are involved, we can use horizontal or vertical subtraction, as illustrated in Example 8.

EXAMPLE 8 **Subtracting mixed numbers**

Subtract: $3\frac{1}{6} - 2\frac{5}{8}$

SOLUTION 8 *Horizontal subtraction*. We convert the mixed numbers to improper fractions, then find the LCM of 6 and 8, which is 24.

$$3\frac{1}{6} = \frac{19}{6} \quad \text{and} \quad 2\frac{5}{8} = \frac{21}{8}$$

Thus,

$$\begin{aligned} 3\frac{1}{6} - 2\frac{5}{8} &= \frac{19}{6} - \frac{21}{8} \\ &= \frac{19 \cdot 4}{6 \cdot 4} - \frac{21 \cdot 3}{8 \cdot 3} \\ &= \frac{76}{24} - \frac{63}{24} \\ &= \frac{13}{24} \end{aligned}$$

Vertical subtraction. Some students prefer to subtract mixed numbers by setting the problem in a column, as shown:

$$\begin{array}{r} 3\frac{1}{6} \\ -2\frac{5}{8} \\ \hline \end{array}$$

PROBLEM 8

Subtract: $4\frac{1}{8} - 2\frac{5}{6}$

Answers to PROBLEMS

7. $\frac{11}{36}$ 8. $\frac{31}{24}$ or $1\frac{7}{24}$

The fractional part is subtracted first, followed by the whole number part. Unfortunately, $\frac{5}{8}$ cannot be subtracted from $\frac{1}{6}$ at this time, so we rename $1 = \frac{6}{6}$ and rewrite the problem as

$$2\frac{7}{6} \qquad 3\frac{1}{6} = 2 + \frac{6}{6} + \frac{1}{6} = 2\frac{7}{6}$$

$$-2\frac{5}{8}$$

Since the LCM is 24, we rewrite $\frac{7}{6}$ and $\frac{5}{8}$ with 24 as the denominator.

$$2\frac{7 \cdot 4}{6 \cdot 4} = 2\frac{28}{24}$$

$$-2\frac{5 \cdot 3}{8 \cdot 3} = 2\frac{15}{24}$$

$$\frac{13}{24}$$

The complete procedure can be written as follows:

$$3\frac{1}{6} \longrightarrow 2\frac{7}{6} \longrightarrow 2\frac{28}{24}$$

$$-2\frac{5}{8} \qquad -2\frac{5}{8} \qquad -2\frac{15}{24}$$

$$\frac{13}{24}$$

Of course, the answer is the same as before.

Now we can do a problem involving both addition and subtraction of fractions.

EXAMPLE 9 Adding and subtracting fractions

Perform the indicated operations: $1\frac{1}{10} + \frac{5}{21} - \frac{3}{28}$

SOLUTION 9 We first write $1\frac{1}{10}$ as $\frac{11}{10}$ and then find the LCM of 10, 21, and 28. (Use either method.)

Method 1.

$$10 = 2 \cdot 5$$
$$21 = 3 \cdot 7$$
$$28 = 2 \cdot 2 \cdot 7$$

The LCD is $2 \cdot 2 \cdot 3 \cdot 5 \cdot 7 = 420$.

Method 2.

Step 1. 2) 10 (21) 28

Step 2. 7) (5) 21 14
5 3 2

Step 3. The LCD is
$2 \cdot 7 \cdot 5 \cdot 3 \cdot 2 = 420$.

Now

$$1\frac{1}{10} = \frac{11}{10} = \frac{11 \cdot 42}{10 \cdot 42} = \frac{462}{420} \longrightarrow \frac{462}{420}$$

$$\frac{5}{21} = \frac{5 \cdot 20}{21 \cdot 20} = \frac{100}{420} \longrightarrow +\frac{100}{420}$$

$$\frac{3}{28} = \frac{3 \cdot 15}{28 \cdot 15} = \frac{45}{420} \longrightarrow -\frac{45}{420}$$

$$\frac{517}{420}$$

Horizontally, we have

$$1\frac{1}{10} + \frac{5}{21} - \frac{3}{28} = \frac{462}{420} + \frac{100}{420} - \frac{45}{420}$$
$$= \frac{462 + 100 - 45}{420}$$
$$= \frac{517}{420} \quad \text{or} \quad 1\frac{97}{420}$$

PROBLEM 9

Perform the indicated operations:

$1\frac{1}{8} + \frac{7}{21} - \frac{3}{28}$

Answers to PROBLEMS

9. $\frac{227}{168}$ or $1\frac{59}{168}$

EXAMPLE 10 Application: Operations with fractions

How many total cups of water and oat bran should be mixed to prepare 3 servings? (Refer to the instructions in the illustration.)

SOLUTION 10 We need $3\frac{3}{4}$ cups of water and $1\frac{1}{2}$ cups of oat bran,

a total of $3\frac{3}{4} + 1\frac{1}{2}$ cups.

Write $3\frac{3}{4}$ and $1\frac{1}{2}$ as improper fractions. $\frac{15}{4} + \frac{3}{2}$

The LCD is 4. Rewrite as $\frac{15}{4} + \frac{6}{4}$

Add the fractions. $\frac{21}{4}$

Rewrite as a mixed number. $5\frac{1}{4}$

Note that this time it is more appropriate to write the answer as a *mixed number*. In general, it is acceptable to write the result of an operation involving fractions as an *improper fraction*.

PROBLEM 10

How many total cups of water and oat bran should be mixed to prepare 1 serving?

Hot Cereal

MICROWAVE AND STOVE TOP AMOUNTS

SERVINGS	1	2	3
WATER	1-1/4 cups	2-1/2 cups	3-3/4 cups
OAT BRAN	1/2 cup	1 cup	1-1/2 cups
SALT (optional)	dash	dash	1/8 tsp

Exercises R.2

Web IT go to mhhe.com/bello for more lessons

A **Multiplying and Dividing Fractions** In Problems 1–20, perform the indicated operations and reduce your answers to lowest terms.

1. $\frac{2}{3} \cdot \frac{7}{3}$ **2.** $\frac{3}{4} \cdot \frac{7}{8}$ **3.** $\frac{6}{5} \cdot \frac{7}{6}$ **4.** $\frac{2}{5} \cdot \frac{5}{3}$

5. $\frac{7}{3} \cdot \frac{6}{7}$ **6.** $\frac{5}{6} \cdot \frac{3}{5}$ **7.** $7 \cdot \frac{8}{7}$ **8.** $10 \cdot 1\frac{1}{5}$

9. $2\frac{3}{5} \cdot 2\frac{1}{7}$ **10.** $2\frac{1}{3} \cdot 4\frac{1}{2}$ **11.** $7 \div \frac{3}{5}$ **12.** $5 \div \frac{2}{3}$

13. $\frac{3}{5} \div \frac{9}{10}$ **14.** $\frac{2}{3} \div \frac{6}{7}$ **15.** $\frac{9}{10} \div \frac{3}{5}$ **16.** $\frac{3}{4} \div \frac{3}{4}$

17. $1\frac{1}{5} \div \frac{3}{8}$ **18.** $3\frac{3}{4} \div 3$ **19.** $2\frac{1}{2} \div 6\frac{1}{4}$ **20.** $3\frac{1}{8} \div 1\frac{1}{3}$

B **Adding and Subtracting Fractions** In Problems 21–50, perform the indicated operations and reduce your answers to lowest terms.

21. $\frac{1}{5} + \frac{2}{5}$ **22.** $\frac{1}{3} + \frac{1}{3}$ **23.** $\frac{3}{8} + \frac{5}{8}$ **24.** $\frac{2}{9} + \frac{4}{9}$

25. $\frac{7}{8} + \frac{3}{4}$ **26.** $\frac{1}{2} + \frac{1}{6}$ **27.** $\frac{5}{6} + \frac{3}{10}$ **28.** $3 + \frac{2}{5}$

29. $2\frac{1}{3} + 1\frac{1}{2}$ **30.** $1\frac{3}{4} + 2\frac{1}{6}$ **31.** $\frac{1}{5} + 2 + \frac{9}{10}$ **32.** $\frac{1}{6} + \frac{1}{8} + \frac{1}{3}$

33. $3\frac{1}{2} + 1\frac{1}{7} + 2\frac{1}{4}$ **34.** $1\frac{1}{3} + 2\frac{1}{4} + 1\frac{1}{5}$ **35.** $\frac{5}{8} - \frac{2}{8}$ **36.** $\frac{3}{7} - \frac{1}{7}$

37. $\frac{1}{3} - \frac{1}{6}$ **38.** $\frac{5}{12} - \frac{1}{4}$ **39.** $\frac{7}{10} - \frac{3}{20}$ **40.** $\frac{5}{20} - \frac{7}{40}$

41. $\frac{8}{15} - \frac{2}{25}$ **42.** $\frac{7}{8} - \frac{5}{12}$ **43.** $2\frac{1}{5} - 1\frac{3}{4}$ **44.** $4\frac{1}{2} - 2\frac{1}{3}$

45. $3 - 1\frac{3}{4}$ **46.** $2 - 1\frac{1}{3}$ **47.** $\frac{5}{6} + \frac{1}{9} - \frac{1}{3}$ **48.** $\frac{3}{4} + \frac{1}{12} - \frac{1}{6}$

49. $1\frac{1}{3} + 2\frac{1}{3} - 1\frac{1}{5}$ **50.** $3\frac{1}{2} + 1\frac{1}{7} - 2\frac{1}{4}$

Answers to PROBLEMS

10. $1\frac{3}{4}$

Applications

51. *Weights on Earth and the moon* The weight of an object on the moon is $\frac{1}{6}$ of its weight on Earth. How much did the Lunar Rover, weighing 450 pounds on Earth, weigh on the moon?

52. *Do you want to meet a millionaire?* Your best bet is to go to Idaho. In this state, $\frac{1}{38}$ of the population happens to be millionaires! If there are about 912,000 persons in Idaho, about how many millionaires are there in the state?

53. *Union membership* The Actors Equity has 28,000 members, but only $\frac{1}{5}$ of the membership is working at any given moment. How many working actors is that?

54. *Battery voltage* The voltage of a standard C battery is $1\frac{1}{2}$ volts (V). What is the total voltage of 6 such batteries?

55. *Excavation* An earthmover removed $66\frac{1}{2}$ cubic yards of sand in 4 hours. How many cubic yards did it remove per hour?

56. *Bond prices* A bond is selling for $\$3\frac{1}{2}$. How many can be bought with \$98?

57. *Household incomes* In a recent survey, it was found that $\frac{9}{50}$ of all American households make more than \$25,000 annually and that $\frac{3}{25}$ make between \$20,000 and \$25,000. What total fraction of American households make more than \$20,000?

58. *Immigration* Between 1971 and 1977, $\frac{1}{5}$ of the immigrants coming to America were from Europe and $\frac{9}{20}$ came from this hemisphere. What total fraction of the immigrants were in these two groups?

59. *Employment* U.S. workers work an average of $46\frac{3}{5}$ hours per week, while Canadians work $38\frac{9}{10}$. How many more hours per week do U.S. workers work?

60. *Bone composition* Human bones are $\frac{1}{4}$ water, $\frac{3}{10}$ living tissue, and the rest minerals. Thus, the fraction of the bone that is minerals is $1 - \frac{1}{4} - \frac{3}{10}$. Find this fraction.

Using Your Knowledge

Hot Dogs and Buns In this section, we learned that the LCM of two numbers is the *smallest* multiple of the two numbers. Can you ever apply this idea to anything besides adding and subtracting fractions? You bet! Hot dogs and buns. Have you noticed that hot dogs come in packages of 10 but buns come in packages of 8 (or 12)?

61. What is the smallest number of packages of hot dogs (10 to a package) and buns (8 to a package) you must buy so that you have as many hot dogs as you have buns? (*Hint:* Think of multiples.)

62. If buns are sold in packages of 12, what is the smallest number of packages of hot dogs (10 to a package) and buns you must buy so that you have as many hot dogs as you have buns?

Write On

63. Write the procedure you use to find the LCM of two numbers.

64. Write the procedure you use to convert a mixed number to an improper fraction.

65. Write the procedure you use to:

a. Multiply two fractions.

b. Divide two fractions.

66. Write the procedure you use to:

a. Add two fractions with the same denominator.

b. Add two fractions with different denominators.

c. Subtract two fractions with the same denominator.

d. Subtract two fractions with different denominators.

Skill Checker

The *Skill Checker* exercises practice skills previously studied. You will need those skills to succeed in the next section! **From now on, the Skill Checker Exercises will appear at the end of every exercise set.**

67. Divide 128 by 16.

68. Divide 2100 by 35.

69. Reduce: $\frac{20}{100}$

70. Reduce: $\frac{75}{100}$

R.3 Decimals and Percents

Objectives

A Write a decimal in expanded form.

B Write a decimal as a fraction.

C Add and subtract decimals.

D Multiply and divide decimals.

E Write a fraction as a decimal.

F Convert decimals to percents and percents to decimals.

G Convert fractions to percents and percents to fractions.

H Round numbers to a specified number of decimal places.

To Succeed, Review How To . . .

1. Divide one whole number by another one.
2. Reduce fractions (pp. 4–6).

Getting Started

Decimals at the Gas Pump

The price of Unleaded Plus gas is \$2.939 (read "two point nine three nine"). This number is called a **decimal.** The word *decimal* means that we count by *tens*—that is, we use *base ten.* The dot in 2.939 is called the *decimal point.* The decimal 2.939 consists of two parts: the whole-number part 2 (the number to the *left* of the decimal point) and the decimal part, 0.939.

A › Writing Decimals in Expanded Form

The number 11.664 can be written in a diagram, as shown. With the help of this diagram, we can write the number in **expanded form,** like this:

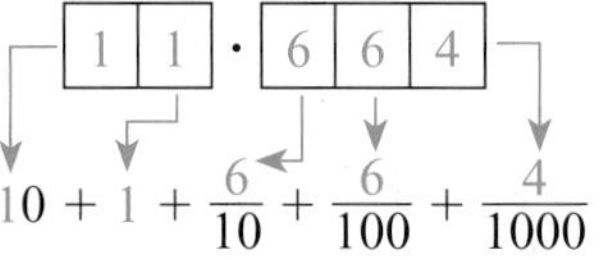

$$10 + 1 + \frac{6}{10} + \frac{6}{100} + \frac{4}{1000}$$

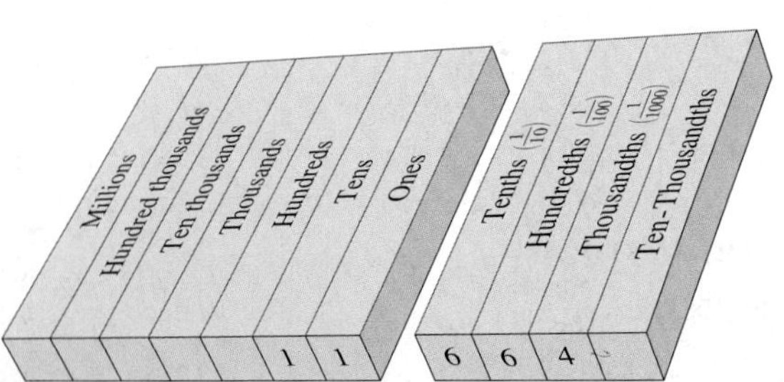

The number 11.664 is read by reading the number to the left of the decimal (*eleven*), using the word *and* for the decimal point, and then reading the number to the right of the decimal point followed by the place value of the *last* digit: *eleven and six hundred sixty-four thousandths.*

EXAMPLE 1 Writing decimals in expanded form

Write in expanded form: 35.216

SOLUTION 1

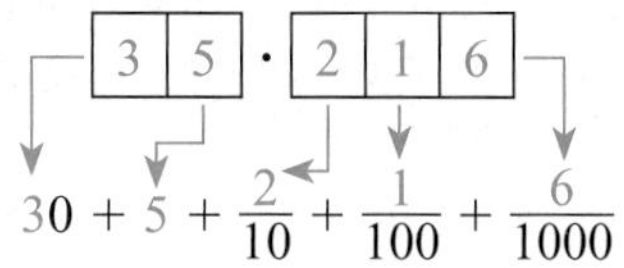

$$30 + 5 + \frac{2}{10} + \frac{1}{100} + \frac{6}{1000}$$

PROBLEM 1

Write in expanded form: 46.325

B > Writing Decimals as Fractions

Decimals can be converted to fractions. For example,

$$0.3 = \frac{3}{10}$$

$$0.18 = \frac{18}{100} = \frac{9}{50}$$

and

$$0.150 = \frac{150}{1000} = \frac{3}{20}$$

Here is the procedure for changing a decimal to a fraction.

PROCEDURE

Changing Decimals to Fractions

1. Write the nonzero digits of the number, omitting the decimal point, as the *numerator* of the fraction.
2. The *denominator* is a 1 *followed by as many zeros as there are decimal digits in the decimal.*
3. Reduce, if possible.

EXAMPLE 2 Changing decimals to fractions

Write as a reduced fraction:

a. 0.035 **b.** 0.0275

SOLUTION 2

a. $0.\underbrace{035}_{\text{3 digits}} = \frac{35}{\underbrace{1000}_{\text{3 zeros}}} = \frac{7}{200}$

b. $0.\underbrace{0275}_{\text{4 digits}} = \frac{275}{\underbrace{10{,}000}_{\text{4 zeros}}} = \frac{11}{400}$

PROBLEM 2

Write as a reduced fraction:

a. 0.045 **b.** 0.0375

If the decimal has a whole-number part, we write all the digits of the number, omitting the decimal point, as the numerator of the fraction. For example, to write 4.23 as a fraction, we write

$$4.\underbrace{23}_{\text{2 digits}} = \frac{423}{\underbrace{100}_{\text{2 zeros}}}$$

EXAMPLE 3 Changing decimals with whole-number parts

Write as a reduced fraction:

a. 3.11 **b.** 5.154

SOLUTION 3

a. $3.11 = \frac{311}{100}$

b. $5.154 = \frac{5154}{1000} = \frac{2577}{500}$

PROBLEM 3

Write as a reduced fraction:

a. 3.13 **b.** 5.258

Answers to PROBLEMS

1. $40 + 6 + \frac{3}{10} + \frac{2}{100} + \frac{5}{1000}$ **2. a.** $\frac{9}{200}$ **b.** $\frac{3}{80}$ **3. a.** $\frac{313}{100}$ **b.** $\frac{2629}{500}$

C > Adding and Subtracting Decimals

Adding decimals is similar to adding whole numbers, as long as we first line up (align) the decimal points by writing them in the *same* column and then make sure that digits of the same place value are in the same column. We then add the columns from right to left, as usual. For example, to add 4.13 + 5.24, we write

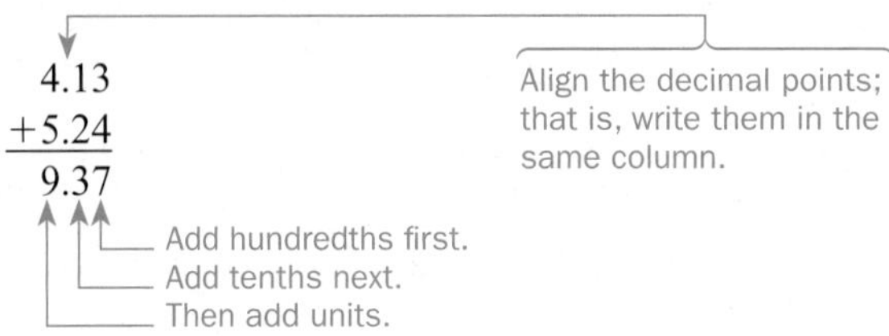

4.13
+5.24
9.37
Align the decimal points; that is, write them in the same column.
Add hundredths first.
Add tenths next.
Then add units.

The result is 9.37.

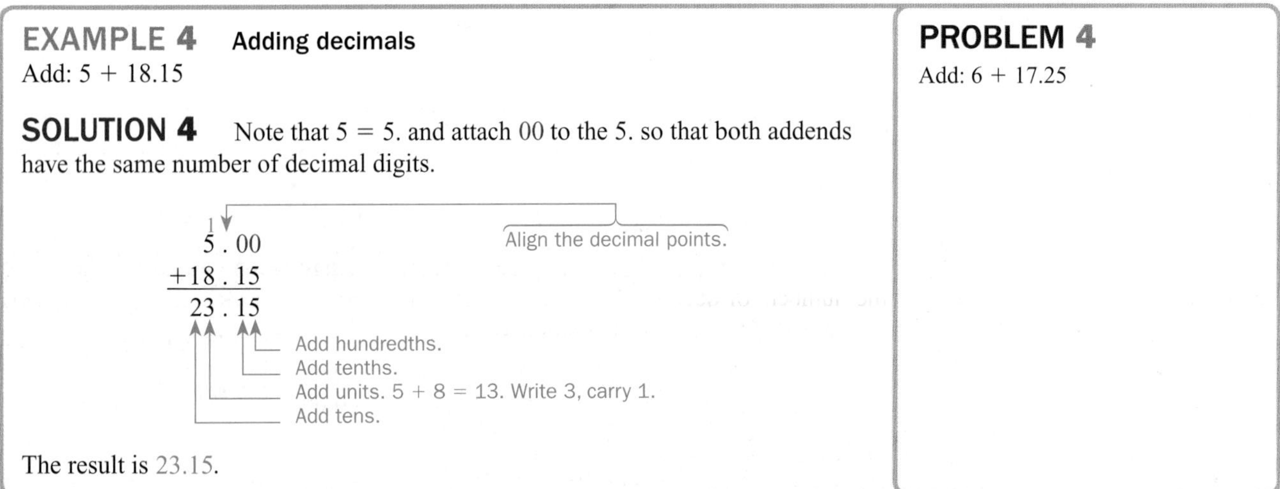

EXAMPLE 4 **Adding decimals**

Add: 5 + 18.15

SOLUTION 4 Note that 5 = 5. and attach 00 to the 5. so that both addends have the same number of decimal digits.

1
5 . 00
+18 . 15
23 . 15
Align the decimal points.
Add hundredths.
Add tenths.
Add units. 5 + 8 = 13. Write 3, carry 1.
Add tens.

The result is 23.15.

PROBLEM 4

Add: 6 + 17.25

Subtraction of decimals is also like subtraction of whole numbers as long as you remember to align the decimal points and attach any needed zeros so that both numbers have the same number of decimal digits. For example, if you earned $231.47 and you made a $52 payment, you have $231.47 − $52. To find how much you have left, we write

```
   1 12 11
$ 2̸ 3̸ 1 . 4 7    Align the decimal points.
−    5 2 . 0 0    Attach two zeros ($52 is the same as 52 dollars and no cents.)
  1 7 9 . 4 7
                  The decimal point is in the same column.
```

Thus, you have $179.47 left. You can check this answer by adding 52 and 179.47 to obtain 231.47.

Calculator Corner

If you have a calculator, you do not have to worry about placing the decimal point or inserting place-holder zeros when adding or subtracting decimal numbers. For example, to add 4.13 + 5.24 simply enter 4 . 1 3 + 5 . 2 4 and press ENTER. The answer 9.37 will appear on the display screen. Similar keystrokes can be used to subtract decimal numbers.

Answers to PROBLEMS

4. 23.25

EXAMPLE 5 **Subtracting decimals**

Subtract: 383.43 − 17.5

SOLUTION 5

```
   7 12 14
 3 8 3 . 4 3     Align the decimal points.
−   1 7 . 5 0    Attach one zero.
 3 6 5 . 9 3     Subtract.
```

The decimal point is in the same column.

PROBLEM 5

Subtract: 273.32 − 14.5

EXAMPLE 6 **Subtracting decimals**

Subtract: 347.8 − 182.231

SOLUTION 6

```
              9
 2 14     7 10 10   Align the decimal points.
 3 4 7 . 8 0 0      Attach two zeros.
−1 8 2 . 2 3 1
 1 6 5 . 5 6 9      Subtract.
```

The decimal point is in the same column.

PROBLEM 6

Subtract: 458.6 − 193.341

D › Multiplying and Dividing Decimals

When multiplying decimals, the number of decimal digits in the product is the *sum* of the numbers of decimal digits in the factors. For example, since 0.3 has *one* decimal digit and 0.0007 has *four,* the product

$$0.3 \times 0.0007 = \frac{3}{10} \times \frac{7}{10{,}000} = \frac{21}{100{,}000} = 0.\underbrace{00021}_{1+4=5 \text{ digits}}$$

has 1 + 4 = 5 decimal digits.

Here is the procedure used to multiply decimals.

PROCEDURE

Multiplying Decimals

1. *Multiply* the two decimal numbers *as if they were whole numbers.*
2. The *number of decimal digits* in the product is the *sum* of the numbers of decimal digits in the factors.

EXAMPLE 7 **Multiplying Decimals**

Multiply:

a. 5.102 × 21.03

b. 5.213 × 0.0012

SOLUTION 7

a.
```
     5.102     3 decimal digits
   × 21.03     2 decimal digits
     15306
     51020
    10204
 107.29506     3 + 2 = 5 decimal digits
```

b.
```
     5.213     3 decimal digits
  × 0.0012     4 decimal digits
     10426
     5213
 0.0062556     3 + 4 = 7 decimal digits
```

Attach two zeros to obtain 7 decimal digits.

PROBLEM 7

Multiply:

a. 6.203 × 31.03

b. 3.123 × 0.0015

Answers to PROBLEMS

5. 258.82 **6.** 265.259

7. a. 192.47909 **b.** 0.0046845

Now suppose you want to divide 52 by 6.5. You can write

$$\frac{52}{6.5}$$

52 ← This is called the dividend.
6.5 ← This is called the divisor.

If we multiply the numerator and denominator of this fraction by 10, we obtain

$$\frac{52}{6.5} = \frac{52 \times 10}{6.5 \times 10} = \frac{520}{65} = 8$$

Thus $\frac{52}{6.5} = 8$, as can be easily checked, since $52 = 6.5 \times 8$. This problem can be shortened by using the following steps.

Step 1. Write the problem in the usual long division form. $6.5\overline{)52}$ (Divisor, Dividend)

Step 2. Move the decimal point in the divisor, 6.5, to the right until a whole number is obtained. (This is the same as multiplying 6.5 by 10.) $65.\overline{)52}$ Multiply by 10.

Step 3. Move the decimal point in the dividend the *same* number of places as in step 2. (This is the same as multiplying the dividend 52 by 10.) Attach zeros if necessary. $65.\overline{)520}$ Multiply by 10.

Step 4. Place the decimal point in the answer directly above the new decimal point in the dividend. $65\overline{)520.}$

Step 5. Divide exactly as you would divide whole numbers. The result is 8, as before.

$$\begin{array}{r} 8. \\ 65\overline{)520.} \\ \underline{520} \\ 0 \end{array}$$

Here is another example. Divide

$$\frac{1.28}{1.6}$$

Step 1. Write the problem in the usual long division form. $1.6\overline{)1.28}$

Step 2. Move the decimal point in the divisor to the right until a whole number is obtained. $16.\overline{)1.28}$ Decimal moved

Step 3. Move the decimal point in the dividend the *same* number of places as in step 2. $16.\overline{)12.8}$ Decimal moved

Step 4. Place the decimal point in the answer directly above the new decimal point in the dividend. $16\overline{)12.8}$

Step 5. Divide exactly as you would divide whole numbers.

$$\begin{array}{r} 0.8 \\ 16\overline{)12.8} \\ \underline{12.8} \\ 0 \end{array}$$

Thus,

$$\frac{1.28}{1.6} = 0.8$$

Calculator Corner

Dividing Decimals

To divide decimals using a calculator, recall that the bar in $\frac{1.28}{1.6}$ means to divide. Accordingly, we enter 1.28  1.6 ENTER. Do not worry about the decimal point; the answer comes out as 0.8 as before.

EXAMPLE 8 Dividing decimals

Divide: $\frac{2.1}{0.035}$

SOLUTION 8 We move the decimal point in the divisor (and also in the dividend) three places to the right. To do this, we need to attach two zeros to 2.1.

$$035.\overline{)2100.}$$

Multiply by 1000.

PROBLEM 8

Divide: $\frac{1.8}{0.045}$

Answers to PROBLEMS

8. 40

We next place the decimal point in the answer directly above the one in the dividend and proceed in the usual manner.

$$\begin{array}{r} 60.\\ 35\overline{)2100.}\\ \underline{210}\\ 00 \end{array}$$

The answer is 60; that is,

$$\frac{2.1}{0.035} = 60$$

CHECK $0.035 \times 60 = 2.100$

E > Writing Fractions as Decimals

Since a fraction indicates a division, we can write a fraction as a decimal by dividing. For example, $\frac{3}{4}$ means $3 \div 4$, so

$$\begin{array}{r} 0.75\\ 4\overline{)3.00}\\ \underline{2\,8}\\ 20\\ \underline{20}\\ 0 \end{array}$$

Note that 3 = 3.00.

Here the division **terminates** (has a 0 remainder). Thus, $\frac{3}{4} = 0.75$, and 0.75 is called a **terminating decimal.**

Since $\frac{2}{3}$ means $2 \div 3$,

$$\begin{array}{r} 0.666\ldots\\ 3\overline{)2.000}\\ \underline{1\,8}\\ 20\\ \underline{18}\\ 20\\ \underline{18}\\ 2 \end{array}$$

The ellipsis (. . .) means that the 6 repeats.

The division continues because the remainder 2 repeats.

Hence, $\frac{2}{3} = 0.666\ldots$. Such decimals (called **repeating decimals**) can be written by placing a bar over the repeating digits; that is, $\frac{2}{3} = 0.\overline{6}$.

EXAMPLE 9 Changing fractions to decimals

Write as a decimal:

a. $\frac{5}{8}$ **b.** $\frac{2}{11}$

SOLUTION 9

a. $\frac{5}{8}$ means $5 \div 8$. Thus,

$$\begin{array}{r} 0.625\\ 8\overline{)5.000}\\ \underline{4\,8}\\ 20\\ \underline{16}\\ 40\\ \underline{40}\\ 0 \end{array}$$

The remainder is 0. The answer is a *terminating decimal*.

$$\frac{5}{8} = 0.625$$

PROBLEM 9

Write as a decimal:

a. $\frac{5}{6}$ **b.** $\frac{1}{4}$

(continued)

Answers to PROBLEMS

9. a. $0.8\overline{3}$ **b.** 0.25

b. $\frac{2}{11}$ means $2 \div 11$. Thus,

$$\begin{array}{r} 0.1818\ldots \\ 11\overline{)2.0000} \\ \underline{1\ 1} \\ 90 \\ \underline{88} \\ 20 \\ \underline{11} \\ 90 \\ \underline{88} \\ 20 \end{array}$$

The remainders 9 and 2 alternately repeat.
The answer 0.1818 . . . is a *repeating decimal.*

$\frac{2}{11} = 0.\overline{18}$ The bar is written over the digits that repeat.

Calculator Corner

Changing Fractions to Decimals

To change a fraction to a decimal using a calculator, just perform the indicated operations. Thus, to convert $\frac{2}{3}$ to a decimal, enter 2 ÷ 3 ENTER and you get the answer 0.666666666 or 0.666666667. Note that you get eight 6's and one 7. How do you know the 6 repeats indefinitely? At this point, you do not. We shall discuss this in the Collaborative Learning section.

F › Converting Decimals and Percents

An important application of decimals is the study of **percents.** The word *percent* means "per one hundred" and is written using the symbol %. Thus,

$$29\% = \frac{29}{100}$$

or "twenty-nine hundredths," that is, 0.29.

Similarly,

$$43\% = \frac{43}{100} = 0.43, \quad 87\% = \frac{87}{100} = 0.87, \quad \text{and} \quad 4.5\% = \frac{4.5}{100} = \frac{45}{1000} = 0.045$$

Note that in each case, the number is divided by 100, which is equivalent to moving the decimal point two places to the *left.*

Here is the general procedure.

PROCEDURE

Converting Percents to Decimals

Move the decimal point in the number *two* places to the *left* and omit the % symbol.

EXAMPLE 10 Converting percents to decimals

Write as a decimal:

a. 98% **b.** 34.7% **c.** 7.2%

SOLUTION 10

a. 98% = 98. = 0.98 Recall that a decimal point follows every whole number.

b. 34.7% = 34.7 = 0.347

c. 7.2% = 07.2 = 0.072 Note that a 0 was inserted so we could move the decimal point two places to the left.

PROBLEM 10

Write as a decimal:

a. 86% **b.** 48.9% **c.** 8.3%

Answers to PROBLEMS

10. a. 0.86 **b.** 0.489 **c.** 0.083

As you can see from Example 10, 0.98 = 98% and 0.347 = 34.7%. Thus, we can reverse the previous procedure to convert decimals to percents. Here is the way we do it.

PROCEDURE

Converting Decimals to Percents

Move the decimal point in the number *two* places to the *right* and attach the % symbol.

EXAMPLE 11 Converting decimals to percents

Write as a percent:

a. 0.53 **b.** 3.19 **c.** 64.7

SOLUTION 11

a. 0.53 = 0.53% = 53%

b. 3.19 = 3.19% = 319%

c. 64.7 = 64.70% = 6470% Note that a 0 was inserted so we could move the decimal point two places to the right.

PROBLEM 11

Write as a percent:

a. 0.49 **b.** 4.17 **c.** 89.2

G › Converting Fractions and Percents

Since % means *per hundred*,

$$5\% = \frac{5}{100}$$

$$7\% = \frac{7}{100}$$

$$23\% = \frac{23}{100}$$

$$4.7\% = \frac{4.7}{100}$$

$$134\% = \frac{134}{100}$$

Thus, we can convert a number written as a percent to a fraction by using the following procedure.

PROCEDURE

Converting Percents to Fractions

Write the *number* over 100, omit the % sign, and *reduce* the fraction, if possible.

EXAMPLE 12 Converting percents to fractions

Write as a fraction:

a. 49% **b.** 75%

SOLUTION 12

a. $49\% = \frac{49}{100}$ **b.** $75\% = \frac{75}{100} = \frac{3}{4}$

PROBLEM 12

Write as a fraction:

a. 37% **b.** 25%

Answers to PROBLEMS

11. a. 49% **b.** 417% **c.** 8920%

12. a. $\frac{37}{100}$ **b.** $\frac{1}{4}$

How do we write a fraction as a percent? If the fraction has a denominator that is a factor of 100, it's easy. To write $\frac{1}{5}$ as a percent, we first multiply numerator and denominator by a number that will make the denominator 100. Thus,

$$\frac{1}{5} = \frac{1 \times 20}{5 \times 20} = \frac{20}{100} = 20\%$$

Similarly,

$$\frac{3}{4} = \frac{3 \times 25}{4 \times 25} = \frac{75}{100} = 75\%$$

Note that in both cases the denominator of the fraction was a factor of 100.

EXAMPLE 13 **Converting fractions to percents**

Write as a percent: $\frac{4}{5}$

SOLUTION 13 We multiply by $\frac{20}{20}$ to get 100 in the denominator.

$$\frac{4}{5} = \frac{4 \times 20}{5 \times 20} = \frac{80}{100} = 80\%$$

PROBLEM 13

Write as a percent: $\frac{3}{5}$

In Example 13 the denominator of the fraction, $\frac{4}{5}$ (the 5), was a *factor* of 100. Suppose we wish to change $\frac{1}{6}$ to a percent. The problem here is that 6 is *not* a factor of 100. But don't panic! We can write $\frac{1}{6}$ as a percent by dividing the numerator by the denominator. Thus, we divide 1 by 6, continuing the division until we have two decimal digits.

$$\begin{array}{r} 0.16 \\ 6\overline{)1.00} \\ \underline{6} \\ 40 \\ \underline{36} \\ 4 \end{array} \quad \text{Remainder}$$

The answer is 0.16 with remainder 4; that is,

$$\frac{1}{6} = 0.16\frac{4}{6} = 0.16\frac{2}{3} = \frac{16\frac{2}{3}}{100} = 16\frac{2}{3}\%$$

Similarly, we can write $\frac{2}{3}$ as a percent by dividing 2 by 3, obtaining

$$\begin{array}{r} 0.66 \\ 3\overline{)2.00} \\ \underline{1\,8} \\ 20 \\ \underline{18} \\ 2 \end{array} \quad \text{Remainder}$$

Thus,

$$\frac{2}{3} = 0.66\frac{2}{3} = \frac{66\frac{2}{3}}{100} = 66\frac{2}{3}\%$$

Here is the procedure we use.

PROCEDURE

Converting Fractions to Percents

Divide the *numerator* by the *denominator* (carry the division to two decimal places), convert the resulting decimal to a percent by *moving the decimal point two places to the right,* and attach the % symbol. (Don't forget to include the remainder, if there is one.)

Answers to PROBLEMS

13. 60%

EXAMPLE 14 Converting fractions to percents

Write as a percent: $\frac{5}{8}$

SOLUTION 14 Dividing 5 by 8, we have

$$\begin{array}{r} 0.62 \\ 8\overline{)5.00} \\ \underline{4\,8} \\ 20 \\ \underline{16} \\ 4 \end{array}$$

Note that the remainder is $\frac{4}{8} = \frac{1}{2}$.

Thus,

$$\frac{5}{8} = 0.62\tfrac{1}{2} = 0.62\tfrac{1}{2}\% = 62\tfrac{1}{2}\%$$

PROBLEM 14

Write as a percent: $\frac{3}{8}$

H > Rounding Numbers

We can shorten some of the decimals we have considered in this section if we use **rounding.** For example, the numbers 107.29506 and 0.062556 can be rounded to three decimal places—that is, to the nearest *thousandth*. Here is the rule we use.

RULE

To Round Numbers

Step 1. **Underline** the digit in the place to which you are rounding.

Step 2. If the first digit to the right of the underlined digit is *5 or more,* add 1 to the underlined digit. Otherwise, do not change the underlined digit.

Step 3. **Drop** all the numbers to the right of the underlined digit (attach 0's to fill in the place values if necessary).

EXAMPLE 15 Rounding decimals

Round to three decimal places:

a. 107.29506 **b.** 0.062556

SOLUTION 15

a. Step 1. Underline the 5. $107.29\underline{5}06$

Step 2. The digit to the right of 5 is 0, so we do not change the 5. $107.29\underline{5}06$

Step 3. Drop all numbers to the right of 5. $107.29\underline{5}$

Thus, when 107.29506 is rounded to three decimal places, that is, to the nearest thousandth, the answer is 107.295.

b. Step 1. Underline the 2. $0.06\underline{2}556$

Step 2. The digit to the right of 2 is 5, so we add 1 to the underlined digit. $0.06\underline{\overset{3}{\not{2}}}556$

Step 3. Drop all numbers to the right of 3. $0.06\underline{3}$

Thus, when 0.062556 is rounded to three decimal places, that is, to the nearest thousandth, the answer is 0.063.

PROBLEM 15

Round to two decimal places:

a. $0.\overline{6}$ **b.** $0.\overline{18}$

Answers to PROBLEMS

14. $37\frac{1}{2}\%$ **15. a.** 0.67 **b.** 0.18

Exercises R.3

Web IT go to mhhe.com/bello for more lessons

⟨A⟩ Writing Decimals in Expanded Form In Problems 1–10, write the given decimal in expanded form.

1. 4.7 **2.** 3.9 **3.** 5.62 **4.** 9.28

5. 16.123 **6.** 18.845 **7.** 49.012

8. 93.038 **9.** 57.104 **10.** 85.305

⟨B⟩ Writing Decimals as Fractions In Problems 11–20, write the given decimal as a reduced fraction.

11. 0.9 **12.** 0.7 **13.** 0.06 **14.** 0.08

15. 0.12 **16.** 0.18 **17.** 0.054 **18.** 0.062

19. 2.13 **20.** 3.41

⟨C⟩ Adding and Subtracting Decimals In Problems 21–40, add or subtract as indicated.

21. \$648.01 + \$341.06 **22.** \$237.49 + \$458.72 **23.** 72.03 + 847.124 **24.** 13.12 + 108.138

25. 104 + 78.103 **26.** 184 + 69.572 **27.** 0.35 + 3.6 + 0.127 **28.** 5.2 + 0.358 + 21.005

29. 27.2 − 0.35 **30.** 4.6 − 0.09 **31.** \$19 − \$16.62 **32.** \$99 − \$0.61

33. 9.43 − 6.406 **34.** 9.08 − 3.465 **35.** 8.2 − 1.356 **36.** 6.3 − 4.901

37. 6.09 + 3.0046 **38.** 2.01 + 1.3045 **39.** 4.07 + 8.0035 **40.** 3.09 + 5.4895

⟨D⟩ Multiplying and Dividing Decimals In Problems 41–60, multiply or divide as indicated.

41. 9.2×0.613 **42.** 0.514×7.4 **43.** 8.7×11 **44.** 78.1×108

45. 7.03×0.0035 **46.** 8.23×0.025 **47.** 3.0012×4.3 **48.** 6.1×2.013

49. 0.0031×0.82 **50.** 0.51×0.0045 **51.** $15\overline{)9}$ **52.** $48\overline{)6}$

53. $5\overline{)32}$ **54.** $8\overline{)36}$ **55.** $8.5 \div 0.005$ **56.** $4.8 \div 0.003$

57. $4 \div 0.05$ **58.** $18 \div 0.006$ **59.** $2.76 \div 60$ **60.** $31.8 \div 30$

⟨E⟩ Writing Fractions as Decimals In Problems 61–76, write the given fraction as a decimal.

61. $\frac{1}{5}$ **62.** $\frac{1}{2}$ **63.** $\frac{7}{8}$ **64.** $\frac{1}{8}$

65. $\frac{3}{16}$ **66.** $\frac{5}{16}$ **67.** $\frac{2}{9}$ **68.** $\frac{4}{9}$

69. $\frac{6}{11}$ **70.** $\frac{7}{11}$ **71.** $\frac{3}{11}$ **72.** $\frac{5}{11}$

73. $\frac{1}{6}$ **74.** $\frac{5}{6}$ **75.** $\frac{10}{9}$ **76.** $\frac{11}{9}$

⟨F⟩ Converting Decimals and Percents In Problems 77–86, write the given percent as a decimal.

77. 33% **78.** 52% **79.** 5% **80.** 9%

81. 300% **82.** 500% **83.** 11.8% **84.** 89.1%

85. 0.5% **86.** 0.7%

In Problems 87–96, write the given decimal as a percent.

87. 0.05 **88.** 0.07 **89.** 0.39 **90.** 0.74

91. 0.416 **92.** 0.829 **93.** 0.003 **94.** 0.008

95. 1.00 **96.** 2.1

⟨G⟩ Converting Fractions and Percents In Problems 97–106, write the given percent as a reduced fraction.

97. 30% **98.** 40% **99.** 6% **100.** 2%

101. 7% **102.** 19% **103.** $4\frac{1}{2}\%$ **104.** $2\frac{1}{4}\%$

105. $1\frac{1}{3}\%$ **106.** $5\frac{2}{3}\%$

In Problems 107–116, write the given fraction as a percent.

107. $\frac{3}{5}$ **108.** $\frac{4}{25}$ **109.** $\frac{1}{2}$ **110.** $\frac{3}{50}$

111. $\frac{5}{6}$ **112.** $\frac{1}{3}$ **113.** $\frac{4}{8}$ **114.** $\frac{7}{8}$

115. $\frac{4}{3}$ **116.** $\frac{7}{6}$

⟨H⟩ Rounding Numbers In Problems 117–126, round each number to the specified number of decimal places.

117. 27.6263; 1 **118.** 99.6828; 1 **119.** 26.746706; 4 **120.** 54.037755; 4

121. 35.24986; 3 **122.** 69.24851; 3 **123.** 52.378; 2 **124.** 6.724; 2

125. 74.846008; 5 **126.** 39.948712; 5

⟩⟩⟩ Applications

Source: www.clickz.com

Coupons Do you clip coupons or do you get them from the Web? Here is a quote from the article Coupons Converge Online: "*The survey finds 95.5 percent of respondents clip or print coupons, while 62.2 percent clip as often as once a week. Just over half (51.3 percent) the respondents redeem most of the coupons they save.*"

127. Write 95.5 percent as a decimal.

128. Write 95.5 percent as a reduced fraction.

129. Write one-half as a fraction, as a decimal, and as a percent.

130. Write 51.3% as a reduced fraction and as a decimal. Is 51.3 percent just over one half? Explain.

Using Your Knowledge

The Pursuit of Happiness *Psychology Today* conducted a survey about happiness. Here are some conclusions from that report.

131. Seven out of ten people said they had been happy over the last six months. What percent of the people is that?

132. Of the people surveyed, 70% expected to be happier in the future than now. What fraction of the people is that?

133. The survey also showed that 0.40 of the people felt lonely.

a. What percent of the people is that?

b. What fraction of the people is that?

134. Only 4% of the men were ready to cry. Write this percent as a decimal.

135. Of the people surveyed, 49% were single. Write this percent as:

a. A fraction.

b. A decimal.

Do you wonder how they came up with some of these percents? They used their knowledge. You do the same and fill in the spaces in the accompanying table, which refers to the marital status of the 52,000 people surveyed. For example, the first line shows that 25,480 persons out of 52,000 were single. This is

$$\frac{25{,}480}{52{,}000} = 49\%$$

Fill in the percents in the last column.

	Marital Status	Number	Percent
	Single	25,480	49%
136.	Married (first time)	15,600	___
137.	Remarried	2600	___
138.	Divorced, separated	5720	___
139.	Widowed	520	___
140.	Cohabiting	2080	___

Skill Checker

141. Find $4(10) - 2(5)$

142. Find $25{,}000 - 4750 - 2(3050)$

Collaborative Learning

Form several groups of four or five students each. The first group will convert the fractions $\frac{1}{2}, \frac{1}{3}, \frac{1}{4}, \frac{1}{5}$, and $\frac{1}{6}$ to decimals. The second group will convert the fractions $\frac{1}{7}, \frac{1}{8}, \frac{1}{9}, \frac{1}{10}$, and $\frac{1}{11}$ to decimals. The next group will convert $\frac{1}{12}, \frac{1}{13}, \frac{1}{14}, \frac{1}{15}$, and $\frac{1}{16}$ to decimals. Ask each group to show the fractions that have terminating decimal representations. You should have $\frac{1}{2}, \frac{1}{4}$, and so on. Now, look at the fractions whose decimal representations are not terminating. You should have $\frac{1}{3}, \frac{1}{6}$, and so on. Can you tell which fractions will have terminating decimal representations? (*Hint:* Look at the denominators of the terminating fractions!)

Practice Test Chapter R

(Answers on page 34)

Visit www.mhhe.com/bello to view helpful videos that provide step-by-step solutions to several of the problems below.

1. Write -18 as a fraction with a denominator of 1.

2. Find a fraction equivalent to $\frac{3}{7}$ with a denominator of 21.

3. Find a fraction equivalent to $\frac{9}{15}$ with a denominator of 5.

4. Reduce $\frac{27}{54}$ to lowest terms.

5. Multiply: $5\frac{1}{4} \cdot \frac{32}{21}$.

6. Divide: $2\frac{1}{4} \div \frac{3}{8}$.

7. Add: $1\frac{1}{10} + \frac{5}{12}$.

8. Subtract: $4\frac{1}{6} - 1\frac{9}{10}$.

9. Write 68.428 in expanded form.

10. Write 0.045 as a reduced fraction.

11. Write 3.12 as a reduced fraction.

12. Add: $847.18 + 29.365$.

13. Subtract: $447.58 - 27.6$.

14. Multiply: 4.315×0.0013.

15. Divide: $\frac{4.2}{0.035}$.

16. **a.** Write $\frac{8}{11}$ as a decimal.

b. Round the answer to two decimal places.

17. Write 84.8% as a decimal.

18. Write 0.69 as a percent.

19. Write 52% as a reduced fraction.

20. Write $\frac{5}{9}$ as a percent.

Answers to Practice Test Chapter R

Answer	If You Missed	Review		
	Question	Section	Examples	Page
1. $\frac{-18}{1}$	1	R.1	1	3
2. $\frac{9}{21}$	2	R.1	2	4
3. $\frac{3}{5}$	3	R.1	3	4
4. $\frac{1}{2}$	4	R.1	4	5
5. 8	5	R.2	1, 2, 3	10–11
6. 6	6	R.2	4, 5	12
7. $\frac{91}{60}$	7	R.2	6	14
8. $\frac{34}{15}$	8	R.2	7, 8, 9	16–17
9. $60 + 8 + \frac{4}{10} + \frac{2}{100} + \frac{8}{1000}$	9	R.3	1	21
10. $\frac{9}{200}$	10	R.3	2	21
11. $\frac{78}{25}$	11	R.3	3	21
12. 876.545	12	R.3	4	22
13. 419.98	13	R.3	5, 6	23
14. 0.0056095	14	R.3	7	23
15. 120	15	R.3	8	24–25
16. **a.** $0.\overline{72}$ **b.** 0.73	16	R.3	9, 15	25–26, 29
17. 0.848	17	R.3	10	26
18. 69%	18	R.3	11	27
19. $\frac{13}{25}$	19	R.3	12	27
20. $55\frac{5}{9}\%$	20	R.3	13, 14	28, 29

Section

Chapter

1

one

Real Numbers and Their Properties

The Human Side of Algebra

The digits 1–9 originated with the Hindus and were passed on to us by the Arabs. Muhammad ibn Musa al-Khwarizmi (ca. A.D. 780–850) wrote two books, one on algebra (*Hisak al-jabr w'almuqabala,* "the science of equations"), from which the name *algebra* was derived, and one dealing with the Hindu numeration system. The oldest dated European manuscript containing the Hindu-Arabic numerals is the *Codex Vigilanus* written in Spain in A.D. 976, which used the nine symbols

1 2 3 4 5 6 7 8 9

Zero, on the other hand, has a history of its own. According to scholars, "At the time of the birth of Christ, the idea of zero as a symbol or number had never occurred to anyone." So, who invented zero? An unknown Hindu who wrote a symbol of his own, a dot he called *sunya,* to indicate a column with no beads on his counting board. The Hindu notation reached Europe thanks to the Arabs, who called it *sifr*. About A.D. 150, the Alexandrian astronomer Ptolemy began using *o* (omicron), the first letter in the Greek word for *nothing,* in the manner of our zero.

1.1 Introduction to Algebra

Objectives

A › Translate words into algebraic expressions.

B › Evaluate algebraic expressions.

To Succeed, Review How To . . .

Add, subtract, multiply, and divide numbers (see Chapter R).

Getting Started

The poster uses the language of algebra to tell you how to be successful. The letters X, Y, and Z are used as *placeholders, unknowns,* or *variables*. The letters t, u, v, w, x, y, and z are frequently used as unknowns, and the letter x is used most often (x, y, and z are used in algebra because they are seldom used in ordinary words).

A › Translating into Algebraic Expressions

In arithmetic, we express ideas using **arithmetic expressions.** How do we express our ideas in algebra? We use **algebraic expressions,** which contain the variables x, y, z, and so on, of course. For example, $3k + 2$, $2x + y$, and $3x - 2y + 7z$ are *algebraic* expressions. See the chart for a comparison.

Arithmetic	Algebra
$9 + 51$	$9 + x$
$4.3 - 2$	$4.3 - y$
7×8.4	$7 \times y$ or $7y$
$\frac{3}{4}$	$\frac{a}{b}$

In algebra, it is better to write $7y$ instead of $7 \times y$ because the multiplication sign $\times$ can be easily confused with the letter x. Besides, look at the confusion that would result if we wrote x multiplied by x as $x \times x$! (You probably know that $x \times x$ is written as x^2.) From now on, we will try to avoid using $\times$ to indicate multiplication.

There are many words that indicate addition or subtraction. We list some of these here for your convenience.

Add (+)	Subtract (−)
Plus, sum, increase, more than	**Minus, difference, decrease, less than**
$a + b$ (read "a plus b") means:	$a - b$ (read "a minus b") means:
1. The sum of a and b	**1.** The difference of a and b
2. a increased by b	**2.** a decreased by b
3. b more than a	**3.** b less than a
4. b added to a	**4.** b subtracted from a
5. The total of a and b	**5.** a take away b

With this in mind, try Example 1.

EXAMPLE 1 **Translations involving addition and subtraction**

Translate into an algebraic expression.

a. The sum of x and y b. x minus y c. $7x$ plus $2a$ minus 3

SOLUTION 1

a. $x + y$ b. $x - y$ c. $7x + 2a - 3$

PROBLEM 1

Translate into an algebraic expression.

a. The sum of p and q

b. q minus p

c. $3q$ plus $5y$ minus 2

How do we write multiplication problems in algebra? We use the raised dot (·) or parentheses ( ). Here are some ways of writing the product of a and b.

A raised dot:	$a \cdot b$
Parentheses:	$a(b)$, $(a)b$, or $(a)(b)$
Juxtaposition (writing a and b next to each other):	ab

In each of these cases, a and b (the variables, or numbers, to be multiplied) are called **factors.** Of course, the last notation (juxtaposition) must not be used when multiplying specific numbers because then "5 times 8" would be written as "58," which looks like fifty-eight. To the right are some words that indicate a multiplication.

Multiply (× or ·)

times, of, product, multiplied by

We will use them in Example 2.

EXAMPLE 2 **Writing products using juxtaposition**

Write the indicated products using juxtaposition.

a. 8 times x b. $-x$ times y times z

c. 4 times x times x d. $\frac{1}{5}$ of x

e. The product of 3 and x

SOLUTION 2

a. 8 times x is written as $8x$. b. $-x$ times y times z is written as $-xyz$.

c. 4 times x times x is written as $4xx$. d. $\frac{1}{5}x$

e. $3x$

PROBLEM 2

Write using juxtaposition.

a. -3 times x

b. a times b times c

c. 5 times a times a

d. $\frac{1}{2}$ of a

e. The product of 5 and a

What about division? In arithmetic we use the division (÷) sign to indicate division. In algebra we usually use fractions to indicate division. Thus, in arithmetic we write $15 \div 3$ (or $3\overline{)15}$) to indicate the quotient of 15 and 3. However, in algebra usually we write

$$\frac{15}{3}$$

Similarly, "the quotient of x and y" is written as

$$\frac{x}{y}$$

[We avoid writing $\frac{x}{y}$ as x/y because more complicated expressions such as $\frac{x}{y+z}$ then need to be written as $x/(y + z)$.]

Here are some words that indicate a division.

Divide (÷ or fraction bar —)

Divided by, quotient

Answers to PROBLEMS

1. a. $p + q$ b. $q - p$ c. $3q + 5y - 2$

2. a. $-3x$ b. abc c. $5aa$ d. $\frac{1}{2}a$ e. $5a$

We will use them in Example 3.

EXAMPLE 3 Translations involving division

Translate into an algebraic expression.

a. The quotient of x and 7

b. The quotient of 7 and x

c. The quotient of $(x + y)$ and z

d. The sum of a and b, divided by the difference of a and b

SOLUTION 3

a. $\frac{x}{7}$ **b.** $\frac{7}{x}$ **c.** $\frac{x+y}{z}$ **d.** $\frac{a+b}{a-b}$

PROBLEM 3

Translate into an algebraic expression.

a. The quotient of a and b

b. The quotient of b and a

c. The quotient of $(x - y)$ and z

d. The difference of x and y, divided by the sum of x and y

B › Evaluating Algebraic Expressions

As we have seen, algebraic expressions contain variables, operation signs, and numbers. If we substitute a value for one or more of the variables, we say that we are **evaluating the expression.** We shall see how this works in Example 4.

EXAMPLE 4 Evaluating algebraic expressions

Evaluate the given expressions by substituting 10 for x and 5 for y.

a. $x + y$ **b.** $x - y$ **c.** $4y$ **d.** $\frac{x}{y}$ **e.** $3x - 2y$

SOLUTION 4

a. Substitute 10 for x and 5 for y in $x + y$.
We obtain: $x + y = 10 + 5 = 15$.
The number 15 is called the **value** of $x + y$.

b. $x - y = 10 - 5 = 5$

c. $4y = 4(5) = 20$

d. $\frac{x}{y} = \frac{10}{5} = 2$

e. $3x - 2y = 3(10) - 2(5)$
$= 30 - 10$
$= 20$

PROBLEM 4

Evaluate the expressions by substituting 22 for a and 3 for b.

a. $a + b$ **b.** $2a - b$

c. $5b$ **d.** $\frac{2a}{b}$

e. $2a - 3b$

The terminology of this section and evaluating expressions are extremely important concepts in everyday life. Examine the federal income tax form excerpt shown in the figure. Can you find the words *subtract* and *multiply?* Suppose we use the variables A, S, and E to represent the adjusted gross income, the standard deduction, and the total *number* of exemptions, respectively. What expression will represent the taxable income, and how can we evaluate it? See Example 5!

Form 1040A (2008) Page **2**

Tax, credits, and payments	22	Enter the amount from line 21 (adjusted gross income).	22	*A*
	23a	Check if: ☐ **You** were born before January 2, 1944, ☐ Blind; ☐ **Spouse** was born before January 2, 1944, ☐ Blind } **Total boxes checked** ▶ 23a ☐		
	b	If you are married filing separately and your spouse itemizes deductions, see page 32 and check here ▶ 23b ☐		
Standard Deduction for— • People who checked any box on line 23a, 23b, or 23c or who can be claimed as a dependent, see page 32.	c	Check if standard deduction includes real estate taxes (see page 32) ▶ 23c ☐		
	24	Enter your **standard deduction** (see left margin).	24	*S*
	25	Subtract line 24 from line 22. If line 24 is more than line 22, enter -0-.	25	
	26	If line 22 is over $119,975, or you provided housing to a Midwestern displaced individual, see page 32. Otherwise, multiply $3,500 by the total number of exemptions claimed on line 6d.	26	*E*
	27	Subtract line 26 from line 25. If line 26 is more than line 25, enter -0-. This is your **taxable income**. ▶	27	

Source: Internal Revenue Service.

Answers to PROBLEMS

3. a. $\frac{a}{b}$ **b.** $\frac{b}{a}$ **c.** $\frac{x-y}{z}$ **d.** $\frac{x-y}{x+y}$ **4. a.** 25 **b.** 41 **c.** 15 **d.** $\frac{44}{3}$ **e.** 35

EXAMPLE 5 **Application: Evaluating expressions**

Look at the part of Form 1040A shown, where A represents the adjusted gross income, S the standard deduction, and E the total *number* of exemptions.

a. Line 25 directs you to subtract line 24 (S) from line 22 (A). What algebraic expression should be written on line 25?

b. Line 26 says to multiply \$3500 by the total number of exemptions (E). What algebraic expression should be written on line 26?

SOLUTION 5

a. Subtract line 24 (S) from line 22 (A) is written as $A - S$.

b. Multiply \$3500 by the total number of exemptions (E) is \$3500E.

PROBLEM 5

a. Line 27 says to subtract line 26 (\$3500E) from $A - S$, which is line 25. What algebraic expression should be entered on line 27?

b. Evaluate the expression in part **a** when $A =$ \$30,000, $S =$ \$5450, and $E = 4$.

Some important and contemporary applications of mathematics concern the environment, ecology and climate change, what we will call "Green Math." These applications will be clearly marked so you can pay special attention to them. Here is one of them.

EXAMPLE 6 **Application: Trees needed to offset car pollution**

If you drive M miles a year and your car gets m miles per gallon, the number of trees you need to offset the carbon dioxide (CO_2) produced by your car in a year is the product of 2 and M divided by 5 times m.

a. Write an expression for the product of 2 and M divided by 5 times m.

b. If you drive 12,000 miles a year and your car gets 30 miles per gallon, how many trees do you need to offset the CO_2 produced? See Problems 57–60 to see where the formula came from!

SOLUTION 6

a. The product of 2 and M divided by 5 times m is $\frac{2M}{5m}$

b. Here $M = 12{,}000$ and $m = 30$ so $\frac{2M}{5m} = \frac{2 \cdot 12{,}000}{5 \cdot 30} = \frac{24{,}000}{150} = 160$.

Thus, you need 160 trees to offset the annual pollution from your car. To see the absorption of a whole acre of trees see http://tinyurl.com/yswsdv.

PROBLEM 6

Some people claim that the formula should be M divided by twice m.

a. Write M divided by twice m.

b. Using this formula, how many trees do you need to offset the CO_2 produced when you drive 12,000 miles a year and your car gets 25 miles per gallon?

› Exercises 1.1

‹ A › Translating into Algebraic Expressions In Problems 1–40, translate into an algebraic expression.

1. The sum of a and c

2. The sum of u and v

3. The sum of $3x$ and y

4. The sum of 8 and x

5. $9x$ plus $17y$

6. $5a$ plus $2b$

7. The difference of $3a$ and $2b$

8. The difference of $6x$ and $3y$

9. $-2x$ less 5

10. $-7y$ less $3x$

11. 7 times a

12. -9 times y

13. $\frac{1}{7}$ of a

14. $\frac{1}{9}$ of y

15. The product of b and d

Answers to PROBLEMS

5. a. $A - S - 3500E$ **b.** \$10,550 **6. a.** $\frac{M}{2m}$ **b.** 240

16. The product of 4 and c

17. xy multiplied by z

18. $-a$ multiplied by bc

19. $-b$ times $(c + d)$

20. $(p + q)$ multiplied by r

21. $(a - b)$ times x

22. $(a + d)$ times $(x - y)$

23. The product of $(x - 3y)$ and $(x + 7y)$

24. The product of $(a - 2b)$ and $(2a - 3b)$

25. $(c - 4d)$ times $(x + y)$

26. x divided by $2y$

27. y divided by $3x$

28. The quotient of $2a$ and b

29. The quotient $2b$ and a

30. The quotient of $2b$ and ac

31. The quotient of a and the sum of x and y

32. The quotient of $(a + b)$ and c

33. The quotient of the difference of a and b, and c

34. The sum of a and b, divided by the difference of x and y

35. The quotient when x is divided into y

36. The quotient when y is divided into x

37. The quotient when the sum of p and q is divided into the difference of p and q

38. The quotient when the difference of $3x$ and y is divided into the sum of x and $3y$

39. The quotient obtained when the sum of x and $2y$ is divided by the difference of x and $2y$

40. The quotient obtained when the difference of x and $3y$ is divided by the sum of x and $3y$

‹B› Evaluating Algebraic Expressions In Problems 41–56, evaluate the expression for the given values.

41. The sum of a and c for $a = 7$ and $c = 9$

42. The sum of u and v for $u = 15$ and $v = 23$

43. $9x$ plus $17y$ for $x = 3$ and $y = 2$

44. $5a$ plus $2b$ for $a = 5$ and $b = 2$

45. The difference of $3a$ and $2b$ for $a = 5$ and $b = 3$

46. The difference of $6x$ and $3y$ for $x = 3$ and $y = 6$

47. $2x$ less 5 for $x = 4$

48. $7y$ less $3x$ for $x = 3$ and $y = 7$

49. 7 times ab for $a = 2$ and $b = 4$

50. 9 times yz for $y = 2$ and $z = 4$

51. The product of b and d for $b = 3$ and $d = 2$

52. The product of 4 and c for $c = 5$

53. xy multiplied by z for $x = 10$, $y = 5$, and $z = 1$

54. a multiplied by bc for $a = 5$, $b = 7$, and $c = 3$

55. The quotient of a and the sum of x and y for $a = 3$, $x = 1$, and $y = 2$

56. The quotient of a plus b divided by c for $a = 10$, $b = 2$, and $c = 3$

Web IT go to mhhe.com/bello for more lessons

››› Applications: Green Math

In Problems 57–60 we will develop the formula used in Example 6.

57. *Gas used in a year* The number of gallons of gas you use in a year depends on how many miles M you drive and how many miles per gallon m your car gets and is given by the quotient of M and m. Write an expression for the quotient of M and m.

58. *CO_2 produced by a car in a year* It is estimated that the amount of CO_2 produced by each gallon G of gas burned by a car in a year is the product of 20 and G. Write the product of 20 and G as an algebraic expression.

59. *CO_2 absorbed by a tree in a year* It is estimated that the amount A of CO_2 absorbed by a tree in a year is A divided by 50. Write an expression for A divided by 50.

60. *Developing a formula by substitution* In Problem 59, the CO_2 absorbed by a tree in a year is given by $\frac{A}{50}$. In Problem 58, the CO_2 produced by a car in a year is $20G$.

a. If a tree absorbs 50 pounds of CO_2 a year, the number of trees needed to absorb A pounds of CO_2 is $\frac{A}{50}$ (Problem 59). Substitute $20G$ for A in the expression $\frac{A}{50}$ (this is the number of trees needed to absorb A pounds of CO_2 in a year).

b. In Problem 57, the number of gallons used in a year is $\frac{M}{m}$. Substitute $\frac{M}{m}$ for G in $\frac{2G}{5}$ (this is the number of trees needed to absorb the CO_2 produced by a car driven M miles a year and getting m miles a gallon). Compare with the formula in Example 6!

Using Your Knowledge

The words we have studied are used in many different fields. Perhaps you have seen some of the following material in your classes! Use the knowledge gained in this section to write it in symbols. The word in italics indicates the field from which the material is taken.

61. *Electricity* The voltage V across any part of a circuit is the product of the current I and the resistance R.

62. *Economics* The total profit TP equals the total revenue TR minus the total cost TC.

63. *Chemistry* The total pressure P in a container filled with gases A, B, and C is equal to the sum of the partial pressures P_A, P_B, and P_C.

64. *Psychology* The intelligence quotient (IQ) for a child is obtained by multiplying his or her mental age M by 100 and dividing the result by his or her chronological age C.

65. *Physics* The distance D traveled by an object moving at a constant rate R is the product of R and the time T.

66. *Astronomy* The square of the period P of a planet's orbit equals the product of a constant C and the cube of the planet's distance R from the sun.

67. *Physics* The energy E of an object equals the product of its mass m and the square of the speed of light c.

68. *Engineering* The depth h of a gear tooth is found by dividing the difference between the major diameter D and the minor diameter d by 2.

69. *Geometry* The square of the hypotenuse c of a right triangle equals the sum of the squares of the sides a and b.

70. *Auto mechanics* The horsepower (hp) of an engine is obtained by multiplying 0.4 by the square of the diameter D of each cylinder and by the number N of cylinders.

Write On

71. In the expression "$\frac{1}{2}$ of x," what operation does the word *of* signify?

72. Most people believe that the word *and* always means addition.

a. In the expression "the sum of x and y," does "and" signify the operation of addition? Explain.

b. In the expression "the product of 2 and three more than a number," does "and" signify the operation of addition? Explain.

73. Explain the difference between "x divided by y" and "x divided into y."

74. Explain the difference between "a less than b" and "a less b."

Concept Checker

This feature is found in every exercise set and is designed to check the student's understanding of the concepts covered in the section.

Fill in the blank(s) with the correct word(s), phrase, or mathematical statement.

75. In symbols, the **sum** of a and b can be written as ______________.

76. In symbols, the **difference** of a and b can be written as ______________.

77. In symbols, the **product** of a and b can be written as ______________.

78. In symbols, the **quotient** of a and b can be written as ______________.

79. If we **substitute a value** for one or more of the variables in an expression, we say that we are ______________ the expression.

solving	ab **or** $a \cdot b$
evaluating	$b - a$
$a - b$	$\frac{a}{b}$ **or** $a \div b$
$a + b$	$\frac{b}{a}$ **or** $b \div a$

Mastery Test

In Problems 80–87, translate into an algebraic expression.

80. The product of 3 and xy

81. The difference of $2x$ and y

82. The quotient of $3x$ and $2y$

83. The sum of $7x$ and $4y$

84. The difference of b and c divided by the sum of b and c

85. Evaluate the expression $2x + y - z$ for $x = 3, y = 4$, and $z = 5$.

86. Evaluate the expression $\frac{p - q}{3}$ for $p = 9$ and $q = 3$.

87. Evaluate the expression $\frac{2x - 3y}{x + y}$ for $x = 10$ and $y = 5$.

Skill Checker

In Problems 88–91, add the numbers.

88. $20 + (-20)$

89. $-3.8 + 3.8$

90. $-\frac{2}{7} + \frac{2}{7}$

91. $1\frac{2}{3} + \left(-1\frac{2}{3}\right)$

1.2 The Real Numbers

Objectives

A Find the additive inverse (opposite) of a number.

B Find the absolute value of a number.

C Classify numbers as natural, whole, integer, rational, or irrational.

D Solve applications using real numbers.

To Succeed, Review How To . . .

Recognize a rational number (see Chapter R).

Getting Started

Temperatures and Integers

To study algebra we need to know about numbers. In this section, we examine *sets* of numbers that are related to each other and learn how to classify them. To visualize these sets more easily and to study some of their properties, we represent (graph) them on a number line. For example, the thermometer shown uses *integers* (not fractions) to measure temperature; the integers are . . . , −3, −2, −1, 0, 1, 2, 3, ... You will find that you use integers every day. For example, when you earn \$20, you *have* 20 dollars; we write this as +20 dollars. When you spend \$5, you *no longer* have it; we write this as −5 dollars. The number 20 is a *positive integer,* and the number −5 (read "negative 5") is a *negative integer.* Here are some other quantities that can be represented by positive and negative integers.

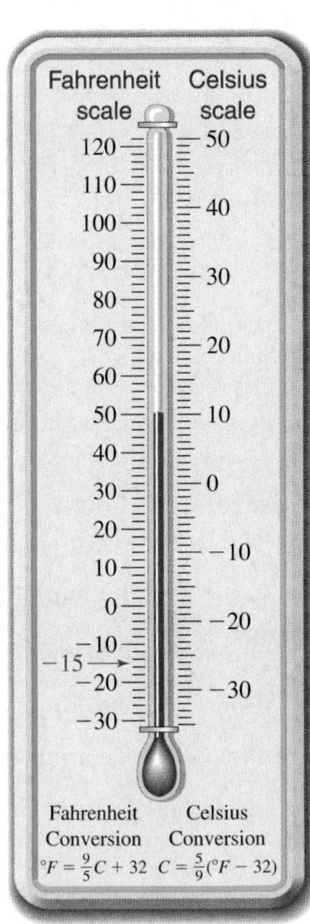

A loss of \$25	−25	A \$25 gain	25
10 ft below sea level	−10	10 ft above sea level	10
15° below zero	−15	15° above zero	15

These quantities are examples of *real numbers.* There are other types of real numbers; we shall learn about them later in this section.

A › Finding Additive Inverses (Opposites)

The temperature 15 degrees below zero (−15°F) is indicated on the Fahrenheit scale of the thermometer in the *Getting Started.* If we take the scale on this thermometer and turn it sideways so that the positive numbers are on the right, the resulting scale is called a

number line (see Figure 1.1). Clearly, on a number line the positive integers are to the right of 0, the negative integers are to the left of 0, and 0 itself is neither positive nor negative.

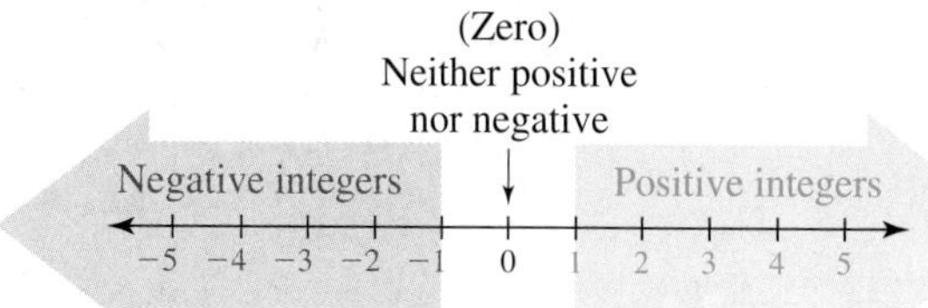

>Figure 1.1

Note that the number line in Figure 1.1 is drawn a little over 5 units long on each side; the arrows at each end indicate that the line could be drawn to any desired length. Moreover, for every *positive* integer, there is a corresponding *negative* integer. Thus, for the positive integer 4, we have the negative integer -4. Since 4 and -4 are the same distance from 0 but in opposite directions, 4 and -4 are called **opposites.** Moreover, since $4 + (-4) = 0$, we call 4 and -4 **additive inverses.** Similarly, the additive inverse (opposite) of -3 is 3, and the additive inverse (opposite) of 2 is -2. Note that $-3 + 3 = 0$ and $2 + (-2) = 0$. In general, we have

> $a + (-a) = (-a) + a = 0$ for any integer a
> This means that the sum of an integer a and its additive inverse $-a$ is always 0.

Figure 1.2 shows the relation between the negative and the positive integers.

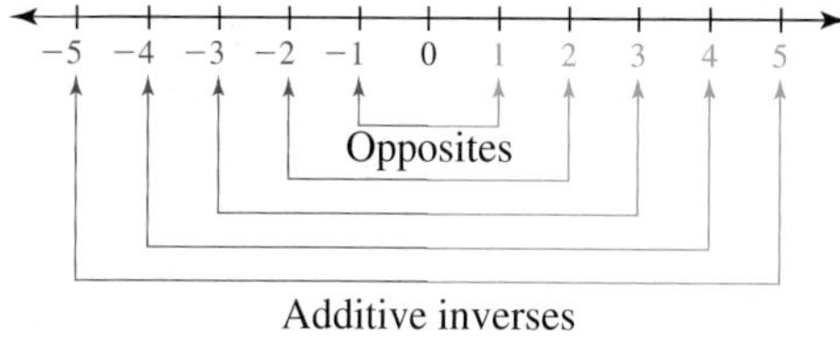

>Figure 1.2

ADDITIVE INVERSE

The **additive inverse** (opposite) of any number a is $-a$.

You can read $-a$ as "the opposite of a" or "the additive inverse of a." Note that a and $-a$ are additive inverses of each other. Thus, 10 and -10 are additive inverses of each other, and -7 and 7 are additive inverses of each other. You can verify this since $10 + (-10) = 0$ and $-7 + 7 = 0$.

EXAMPLE 1 Finding additive inverses of integers

Find the additive inverse (opposite) of:

a. 5 **b.** -4 **c.** 0

SOLUTION 1

a. The additive inverse of 5 is -5 (see Figure 1.3).
b. The additive inverse of -4 is $-(-4) = 4$ (see Figure 1.3).
c. The additive inverse of 0 is 0.

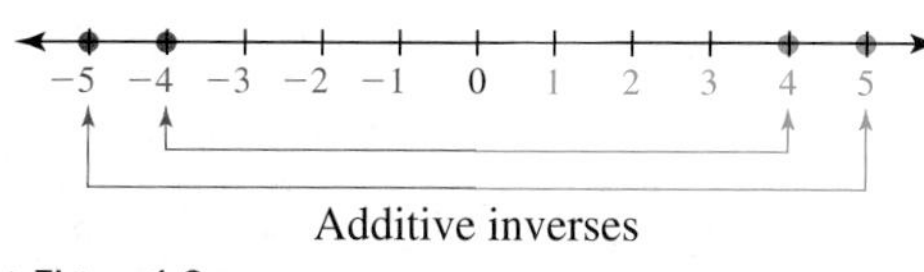

>Figure 1.3

PROBLEM 1

Find the additive inverse of:

a. -8 **b.** 9 **c.** -3

Answers to PROBLEMS

1. a. 8 **b.** -9 **c.** 3

Note that $-(-4) = 4$ and $-(-8) = 8$. In general, we have

$$-(-a) = a \quad \text{for any number } a$$

As with the integers, every rational number—that is, every fraction written as the ratio of two integers—and every decimal has an *additive inverse* (*opposite*). Here are some rational numbers and their additive inverses.

Rational Number	**Additive Inverse (Opposite)**
$\frac{9}{2}$	$-\frac{9}{2}$
$-\frac{3}{4}$	$-\left(-\frac{3}{4}\right) = \frac{3}{4}$
2.9	-2.9
-1.8	$-(-1.8) = 1.8$

EXAMPLE 2 Finding additive inverses of fractions and decimals

Find the additive inverse (opposite) of:

a. $\frac{5}{2}$ **b.** -4.8 **c.** $-3\frac{1}{3}$ **d.** 1.2

SOLUTION 2

a. $-\frac{5}{2}$ **b.** $-(-4.8) = 4.8$

c. $-\left(-3\frac{1}{3}\right) = 3\frac{1}{3}$ **d.** -1.2

These rational numbers and their inverses are graphed on the number line in Figure 1.4. Note that to locate $\frac{5}{2}$, it is easier to first write $\frac{5}{2}$ as the mixed number $2\frac{1}{2}$.

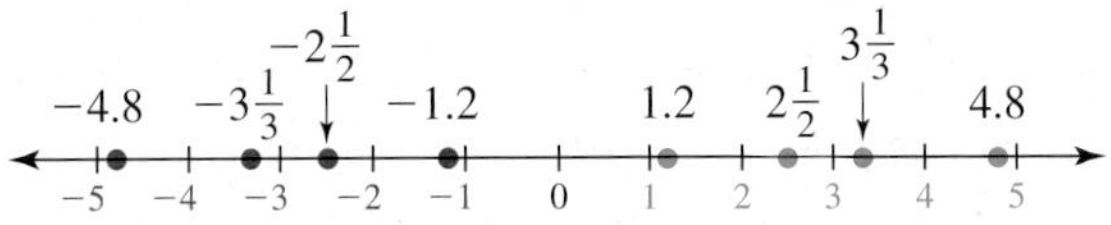

>Figure 1.4

PROBLEM 2

Find the additive inverse of:

a. $\frac{3}{11}$ **b.** -7.4

c. $-9\frac{8}{13}$ **d.** 3.4

B > Finding the Absolute Value of a Number

Let's look at the number line again. What is the distance between 3 and 0? The answer is 3 units. What is the distance between -3 and 0? The answer is *still* 3 units.

The distance between any number n and 0 is called the *absolute value* of the number and is denoted by $|n|$. Thus, $|-3| = 3$ and $|3| = 3$. (See Figure 1.5.)

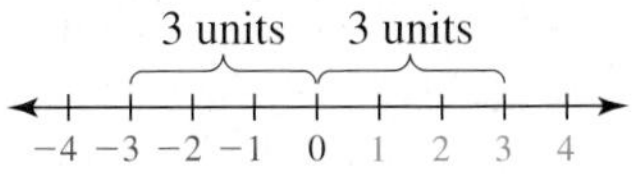

>Figure 1.5

ABSOLUTE VALUE

The **absolute value** of a number n is its distance from 0 and is denoted by $|n|$.

In general,
$$|n| = \begin{cases} n, & \text{if } n \text{ is positive} \\ -n, & \text{if } n \text{ is negative} \\ 0, & \text{if } n = 0 \end{cases}$$

You can think of the absolute value of a number as the number of units it represents disregarding its sign. For example, suppose Pedro and Tamika leave class together.

Answers to PROBLEMS

2. a. $-\frac{3}{11}$ **b.** 7.4

c. $9\frac{8}{13}$ **d.** -3.4

Pedro walks 2 miles east while Tamika walks 2 miles west. Who walked farther? They walked the same distance! They are both 2 miles from the starting point.

NOTE

Since the absolute value of a number can represent a distance and a distance is **never** negative, the absolute value $|a|$ of a nonzero number a is *always* positive. Because of this, $-|a|$ is *always* negative. Thus, $-|6| = -6$ and $-|-8| = -8$.

EXAMPLE 3 Finding absolute values of integers

Find the absolute value:

a. $|-8|$ b. $|7|$ c. $|0|$ d. $-|-3|$

SOLUTION 3

a. $|-8| = 8$ −8 is 8 units from 0.
b. $|7| = 7$ 7 is 7 units from 0.
c. $|0| = 0$ 0 is 0 units from 0.
d. $-|-3| = -3$ −3 is 3 units from 0.

PROBLEM 3

Find the absolute value:

a. $|-5|$ b. $|10|$
c. $|2|$ d. $-|-5|$

Every fraction and decimal also has an absolute value, which is its distance from zero. Thus, $\left|-\frac{1}{2}\right| = \frac{1}{2}$, $|3.8| = 3.8$, and $\left|-1\frac{1}{7}\right| = 1\frac{1}{7}$, as shown in Figure 1.6.

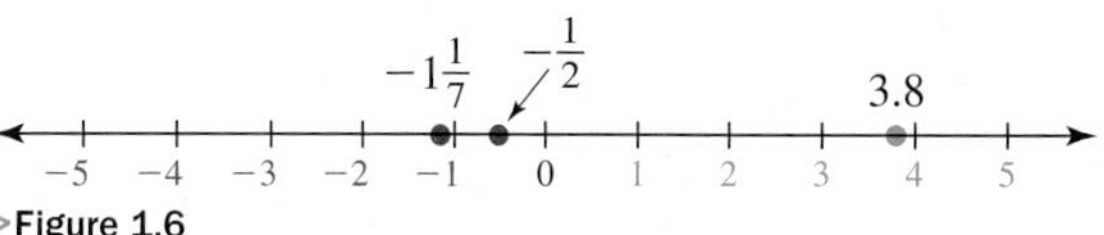

>Figure 1.6

EXAMPLE 4 Finding absolute values of rational numbers

Find the absolute value of:

a. $\left|-\frac{3}{7}\right|$ b. $|2.1|$ c. $\left|-2\frac{1}{2}\right|$

d. $|-4.1|$ e. $-\left|\frac{1}{4}\right|$

SOLUTION 4

a. $\left|-\frac{3}{7}\right| = \frac{3}{7}$ b. $|2.1| = 2.1$ c. $\left|-2\frac{1}{2}\right| = 2\frac{1}{2}$

d. $|-4.1| = 4.1$ e. $-\left|\frac{1}{4}\right| = -\frac{1}{4}$

PROBLEM 4

Find the absolute value of:

a. $\left|-\frac{5}{7}\right|$ b. $|3.4|$ c. $\left|-3\frac{1}{8}\right|$

d. $|-3.8|$ e. $-\left|-\frac{1}{5}\right|$

C > Classifying Numbers

The real-number line is a picture (graph) used to represent the *set* of real numbers. A **set** is a collection of objects called the **members** or **elements** of the set. If the elements can be listed, a pair of braces { } encloses the list with individual elements separated by commas. Here are some sets of numbers contained in the real-number line:

The set of **natural** numbers {1, 2, 3, . . .}
The set of **whole** numbers {0, 1, 2, 3, . . .}
The set of **integers** {. . . , −2, −1, 0, 1, 2, 3, . . .}

The three dots (an ellipsis) at the end or beginning of a list of elements indicates that the list continues indefinitely in the same manner.

As you can see, every natural number is a whole number and every whole number is an integer. In turn, every integer is a *rational number,* a number that can be written

Answers to PROBLEMS

3. a. 5 **b.** 10 **c.** 2 **d.** −5

4. a. $\frac{5}{7}$ **b.** 3.4 **c.** $3\frac{1}{8}$ **d.** 3.8 **e.** $-\frac{1}{5}$

in the form $\frac{a}{b}$, where a and b are integers and b is not zero. Thus, an integer n can always be written as the rational number $\frac{n}{1} = n$. (The word *rational* comes from the word *ratio,* which indicates a quotient.) Since the fraction $\frac{a}{b}$ can be written as a decimal by dividing the numerator a by the denominator b, to obtain either a terminating (as in $\frac{3}{4} = 0.75$) or a repeating (as in $\frac{1}{3} = 0.\overline{3}$) decimal, all terminating or repeating decimals are also rational numbers. The set of rational numbers is described next in words, since it's impossible to make a list containing all the rational numbers.

RATIONAL NUMBERS

The set of rational numbers consists of all numbers that can be written as quotients $\frac{a}{b}$, where a and b are integers and $b \neq 0$.

The set of *irrational numbers* is the set of all real numbers that are not rational.

IRRATIONAL NUMBERS

The set of irrational numbers consists of all real numbers that *cannot* be written as the quotient of two integers.

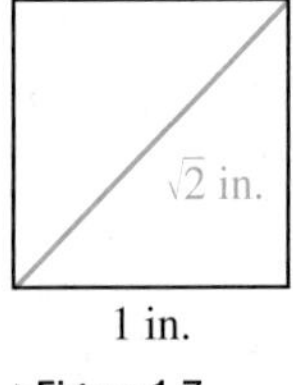

>Figure 1.7

For example, if you draw a square 1 inch on a side, the length of its diagonal is $\sqrt{2}$ (read "the square root of 2"), as shown in Figure 1.7. This number **cannot** be written as a quotient of integers. It is irrational. Since a rational number $\frac{a}{b}$ can be written as a fraction that terminates or repeats, the decimal form of an irrational number never terminates and never repeats. Here are some irrational numbers:

$$\sqrt{2}, \quad -\sqrt{50}, \quad 0.123\ldots, \quad -5.1223334444\ldots, \quad 8.101001000\ldots, \quad \text{and} \quad \pi$$

All the numbers we have mentioned are *real numbers,* and their relationship is shown in Figure 1.8. Of course, you may also represent (graph) these numbers on a number line. Note that a number may belong to more than one category. For example, -15 is an integer, a rational number, and a real number.

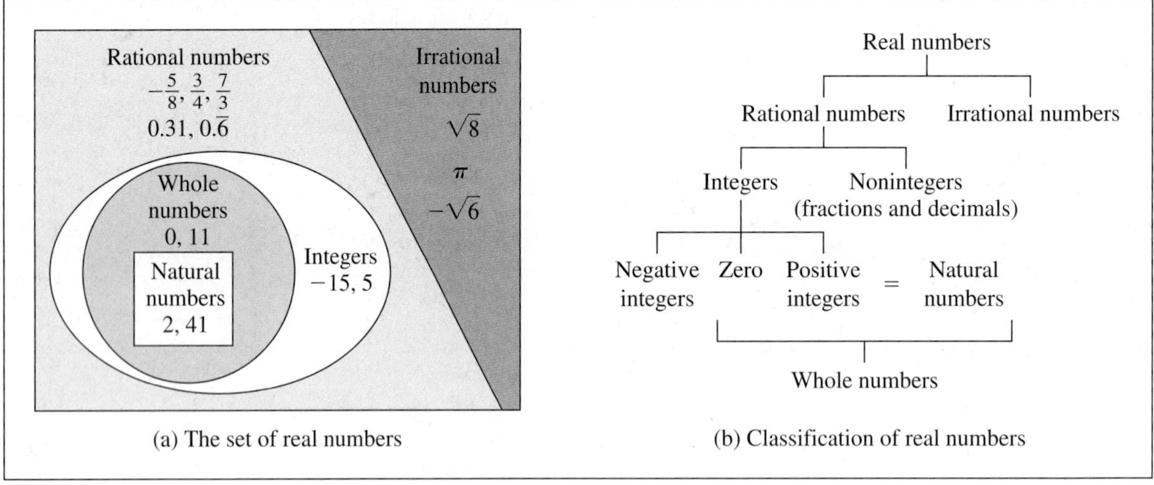

>Figure 1.8

EXAMPLE 5 Classifying numbers

Classify as whole, integer, rational, irrational, or real number:

a. -3 **b.** 0 **c.** $\sqrt{5}$
d. 0.3 **e.** 0.101001000. . . **f.** 0.101001000

PROBLEM 5

Classify as whole, integer, rational, irrational, or real number:

a. -9 **b.** 200
c. $\sqrt{7}$ **d.** 0.9
e. 0.010010001. . . **f.** 0.010010001

Answers to PROBLEMS

5. a. Integer, rational, real **b.** Whole number, integer, rational, real **c.** Irrational, real **d.** Rational, real **e.** Irrational, real **f.** Rational, real

SOLUTION 5

a. -3 is an integer, a rational number, and a real number.
b. 0 is a whole number, an integer, a rational number, and a real number.
c. $\sqrt{5}$ is an irrational number and a real number.
d. 0.3 is a terminating decimal, so it is a rational number and a real number.
e. 0.101001000. . . never terminates and never repeats, so it is an irrational number and a real number.
f. 0.101001000 terminates, so it is a rational number and a real number.

D > Applications Involving Real Numbers

EXAMPLE 6 Using real numbers

Use real numbers to write the quantities in the applications.

a. Average temperatures have *climbed 1.4* degrees Fahrenheit (F) around the world since 1880.
b. The number of glaciers (a huge mass of ice slowly flowing over a land mass) in Glacier National Park in Montana has *decreased by 123* since 1910.

SOLUTION 6

a. Climbed 1.4 degrees Fahrenheit can be written as $+1.4$°F.
b. Decreased by 123 can be written as -123.

Source: http://tinyurl.com/5jlyn.

PROBLEM 6

Use real numbers to write the quantities.

a. Average temperatures have *climbed 0.8* degree Celsius (C) around the world since 1880.
b. Snow cover extent over arctic land areas has *decreased* by about 0.10 over the past 30 years.

EXAMPLE 7 Using + or − to indicate height or depth

The following table lists the altitude (the distance above (+) or below (−) sea level or zero altitude) of various locations around the world.

Location	Altitude (feet)
Mount Everest	+29,029
Mount McKinley	+20,320
Sea level	0
Caspian Sea	−92
Marianas Trench (deepest descent)	−35,813

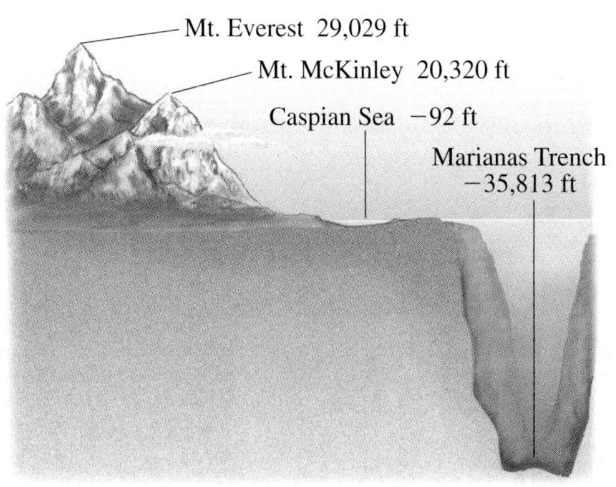

PROBLEM 7

a. Refer to the table. How far above sea level is Mount McKinley?
b. The Dead Sea, Israel-Jordan, is located at altitude -1349 feet. How far below sea level is that?
c. Some scientists claim that the altitude of the Marianas Trench is really $-36,198$ feet. How far below sea level is that?

(continued)

Answers to PROBLEMS

6. a. $+0.8$°C **b.** -0.10 **7. a.** 20,320 feet above sea level **b.** 1349 feet below sea level **c.** 36,198 feet below sea level

Use this information to answer the following questions:

a. How far above sea level is Mount Everest?

b. How far below sea level is the Caspian Sea?

c. How far below sea level was the deepest descent made?

SOLUTION 7

a. 29,029 feet above sea level **b.** 92 feet below sea level

c. 35,813 feet below sea level

Calculator Corner

Additive Inverse and Absolute Value

The additive inverse and absolute value of a number are so important that graphing calculators have special keys to handle them. To find the additive inverse, press (−). Don't confuse the additive inverse key with the minus sign key. (Operation signs usually have color keys; the (−) key is gray.)

To find absolute values with a TI-83 Plus calculator, you have to do some math, so press MATH. Next, you have to deal with a special type of number, absolute value, so press ▶ to highlight the NUM menu at the top of the screen. Next press 1, which tells the calculator you want an absolute value; finally, enter the number whose absolute value you want, and close the parentheses. The display window shows how to calculate the additive inverse of −5, the absolute value of 7, and the absolute value of −4.

```
--5
                5
abs (7)
                7
abs (-4)
                4
```

> Practice Problems > Self-Tests > Media-rich eBooks > e-Professors > Videos

Exercises 1.2

⟨A⟩ Finding Additive Inverses (Opposites) In Problems 1–18, find the additive inverse (opposite) of the given number.

1. 4 **2.** 11 **3.** -49 **4.** -56 **5.** $\frac{7}{3}$

6. $-\frac{8}{9}$ **7.** -6.4 **8.** -2.3 **9.** $3\frac{1}{7}$ **10.** $-4\frac{1}{8}$

11. 0.34 **12.** 0.85 **13.** $-0.\overline{5}$ **14.** $-3.\overline{7}$ **15.** $\sqrt{7}$

16. $-\sqrt{17}$ **17.** π **18.** $\frac{\pi}{2}$

⟨B⟩ Finding the Absolute Value of a Number In Problems 19–38, find the absolute value.

19. $|-2|$ **20.** $|-6|$ **21.** $|48|$ **22.** $|78|$ **23.** $|-(-3)|$

24. $|-(-17)|$ **25.** $\left|-\frac{4}{5}\right|$ **26.** $\left|-\frac{9}{2}\right|$ **27.** $|-3.4|$ **28.** $|-2.1|$

29. $\left|-1\frac{1}{2}\right|$ **30.** $\left|-3\frac{1}{4}\right|$ **31.** $-\left|-\frac{3}{4}\right|$ **32.** $-\left|-\frac{1}{5}\right|$ **33.** $-|-0.\overline{5}|$

34. $-|-3.\overline{7}|$ **35.** $-|-\sqrt{3}|$ **36.** $-|-\sqrt{6}|$ **37.** $-|-\pi|$ **38.** $-\left|-\frac{\pi}{2}\right|$

Web IT go to mhhe.com/bello for more lessons

⟨C⟩ Classifying Numbers In Problems 39–54, classify the given numbers. (See Figure 1.8; some numbers belong in more than one category.)

39. 17 **40.** -8 **41.** $-\frac{4}{5}$ **42.** $-\frac{7}{8}$

43. 0 **44.** 0.37 **45.** 3.76 **46.** $3.\overline{8}$

47. $17.\overline{28}$ **48.** $\sqrt{10}$ **49.** $-\sqrt{3}$ **50.** 0.777. . .

51. $-0.888\ldots$ **52.** 0.202002000. . . **53.** 0.202002000 **54.** $\frac{\pi}{2}$

In Problems 55–62, consider the set $\{-5, \frac{1}{5}, 0, 8, \sqrt{11}, 0.\overline{1}, 2.505005000\ldots, 3.666\ldots\}$. List the numbers in the given set that are members of the specified set.

55. Natural numbers **56.** Whole numbers **57.** Positive integers **58.** Negative integers

59. Nonnegative integers **60.** Irrational **61.** Rational numbers **62.** Real numbers

In Problems 63–74, determine whether the statement is true or false. If false, give an example that shows it is false.

63. The opposite of any positive number is negative.

64. The opposite of any negative number is positive.

65. The absolute value of any real number is positive.

66. The negative of the absolute value of a number is equal to the absolute value of its negative.

67. The absolute value of a number is equal to the absolute value of its opposite.

68. Every integer is a rational number.

69. Every rational number is an integer.

70. Every terminating decimal is rational.

71. Every nonterminating decimal is rational.

72. Every nonterminating and nonrepeating decimal is irrational.

73. A decimal that never repeats and never terminates is a real number.

74. The decimal representation of a real number never terminates and never repeats.

⟨D⟩ Applications Involving Real Numbers In Problems 75–84, use real numbers to write the indicated quantities.

75. *Football* A 20-yard *gain* in a football play.

76. *Football* A 10-yard *loss* in a football play.

77. *Below sea level* The Dead Sea is 1312 feet *below* sea level.

78. *Above sea level* Mount Everest reaches a height of 29,029 feet *above* sea level.

79. *Temperature* On January 22, 1943, the temperature in Spearfish, South Dakota, rose from 4°F *below* zero to 45°F *above* zero in a period of 2 minutes.

80. *Births and deaths* Every hour, there are 460 *births* and 250 *deaths* in the United States.

81. *Internet searches* In a 1-year period the number of U.S. Internet searches *grew* 55%.

82. *Internet searches* In a 1-year period Yahoo searches *declined* by 0.3%.

83. *Internet searches* In a 1-year period MSN searches *went down* 3.1%.

84. *Advertising* In a recent month, the advertising placements in the automotive industry *declined* 20.91%.

Source: www.clickz.com; http://www.clickz.com.

⟩⟩⟩ Applications: Green Math

Household water management Use signed real numbers to write the quantities in the applications.

85. You can **save 1000** gallons of water a month if you run your dishwasher and washing machine only when they are full.

86. A leaky faucet can **waste 140** gallons of water each week.

87. A water-efficient shower head can **save 750** gallons of water each month.

88. Listen for dripping faucets or toilets: they can **waste 300** gallons of water a month or more.

Source: http://www.wateruseitwisely.com/100-ways-to-conserve/index.php.

Using Your Knowledge

Weather According to *USA Today Weather Almanac,* the coldest city in the United States (based on its average annual temperature in degrees Fahrenheit) is International Falls, Minnesota (see Figure 1.9). Note that the lowest ever temperature there was $-46°$F.

89. What was the next lowest ever temperature (February) in International Falls?

90. What was the highest record low?

91. What was the record low in December?

92. What was the lowest average low?

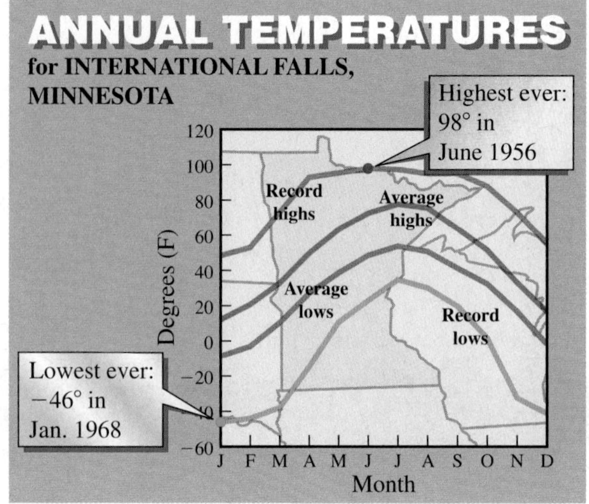

Figure 1.9 Annual temperatures for International Falls, Minnesota (°F)
Source: Data from USA Today Weather Almanac.

Write On

93. What do we mean by the following phrases?

a. the additive inverse of a number

b. the absolute value of a number

94. Explain why every integer is a rational number.

95. The rational numbers have been defined as the set of numbers that can be written in the form $\frac{a}{b}$, where a and b are integers and b is not 0. Define the rational numbers in terms of their decimal representation.

96. Define the set of irrational numbers in terms of their decimal representation.

97. Write a paragraph explaining the set of real numbers as it relates to the natural numbers, the integers, the rational numbers, and the irrational numbers.

Concept Checker

Fill in the blank(s) with the correct word(s), phrase, or mathematical statement.

98. The **additive inverse** (opposite) of any number a is ____________.

99. The **absolute value** of a number n is its distance from ____________.

100. The set of **rational numbers** consists of all numbers that can be written as ____________.

101. The set of ____________ **numbers** consists of all real numbers that ***cannot* be written** as the quotient of two integers.

natural	**−*a***
rational	**−1**
irrational	***n***
0	**numbers**
a	**quotients**

Mastery Test

Find the additive inverse of the number.

102. $0.\overline{7}$ **103.** $-\sqrt{19}$ **104.** $\frac{\pi}{3}$ **105.** $-8\frac{1}{4}$

Find the absolute value.

106. $|\sqrt{23}|$ **107.** $|-0.\overline{4}|$ **108.** $-\left|\frac{3}{5}\right|$ **109.** $-\left|-\frac{1}{2}\right|$

Classify the number as a natural, whole, integer, rational, irrational, or real number. (*Hint:* More than one category may apply.)

110. $\sqrt{21}$ **111.** -2 **112.** 0.010010001. . . **113.** $4\frac{1}{3}$

114. 0.333. . . **115.** 0 **116.** 39

Skill Checker

Add or subtract:

117. $-7.8 + 7.8$ **118.** $8.6 - 3.4$ **119.** $2.3 + 4.1$ **120.** $\frac{5}{8} - \frac{2}{5}$ **121.** $\frac{5}{6} + \frac{7}{4}$

1.3 Adding and Subtracting Real Numbers

Objectives

A Add two real numbers.

B Subtract one real number from another.

C Add and subtract several real numbers.

D Solve application problems involving real numbers.

To Succeed, Review How To . . .

1. Add and subtract fractions (pp. 12–18).
2. Add and subtract decimals (pp. 22–23).

Getting Started

Signed Numbers and Population Changes

Now that we know what real numbers are, we will use the real-number line to visualize the process used to add and subtract them. For example, what is the U.S. population change per hour? To find out, examine the graph and add the births (+456), the deaths (−273), and the new immigrants (+114). The result is

$$456 + (-273) + 114 = 570 + (-273) \quad \text{Add 456 and 114.}$$
$$= +297 \quad \text{Subtract 273 from 570.}$$

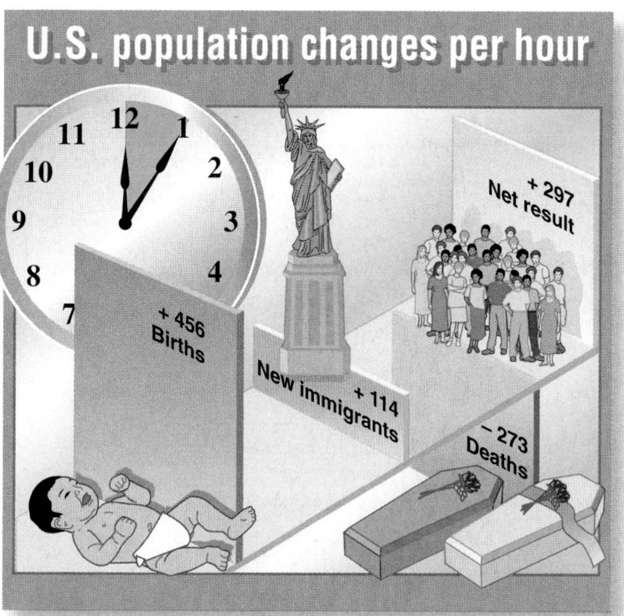

Source: Population Reference Bureau (2000, 2001, and 2002 data).

This answer means that the U.S. population is increasing by 297 persons every hour! Note that

1. The addition of 456 and -273 is written as $456 + (-273)$, instead of the confusing $456 + -273$.
2. To add 570 and -273, we *subtracted* 273 from 570 because subtracting 273 from 570 is the *same* as adding 570 and -273. We will use this idea to define subtraction.

If you want to know what the population change is per day or per minute, you must learn how to multiply and divide real numbers, as we will do in Section 1.4.

A > Adding Real Numbers

The number line we studied in Section 1.2 can help us add real numbers. Here's how we do it.

PROCEDURE

Adding on a Number Line

To add $a + b$ on a number line,

1. Start at zero and move to a (to the *right* if a is *positive,* to the *left* if a is *negative*).
2. **A.** If b is *positive,* move *right* b units.
   B. If b is *negative,* move *left* $|b|$ units.
   C. If b is zero, stay at a.

For example, the sum $2 + 4$, or $(+2) + (+4)$, is found by starting at zero, moving 2 units to the right, followed by 4 more units to the right ending at 6. Thus, $2 + 4 = 6$, as shown in Figure 1.10

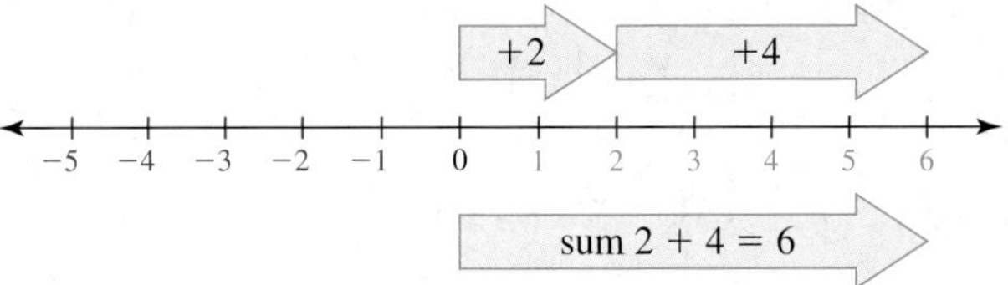

>Figure 1.10

A number along with the sign (+ or −) indicating a direction on the number line is called a **signed number.**

EXAMPLE 1 Adding integers with different signs

Add: $5 + (-3)$

SOLUTION 1 Start at zero. Move 5 units to the right and then 3 units to the left. The result is 2. Thus, $5 + (-3) = 2$, as shown in Figure 1.11.

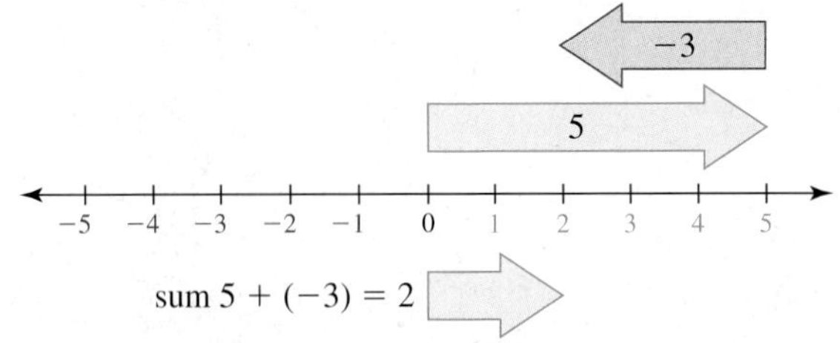

>Figure 1.11

PROBLEM 1

Add: $4 + (-2)$

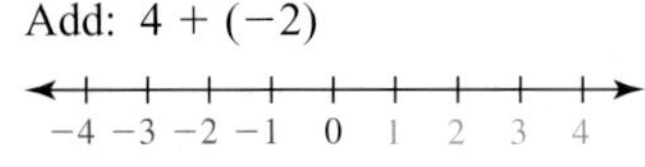

Answers to PROBLEMS

1. 2

Calculator Corner

Adding Integers:

To enter $-3 + (-2)$ using a scientific calculator, enter 3 +/− + 2 +/− ENTER. If your calculator has a set of parentheses, then you can enter parentheses around the -3 and -2.

This same procedure can be used to add two negative numbers. However, we have to be careful when writing such problems. For example, to add -3 and -2, we should write

$$-3 + (-2)$$

Why the parentheses? Because writing

$$-3 + -2$$

is confusing. *Never* write two signs together without parentheses.

EXAMPLE 2 **Adding integers with the same sign**

Add: $-3 + (-2)$

SOLUTION 2 Start at zero. Move 3 units left and then 2 more units left. The result is 5 units left of zero; that is, $-3 + (-2) = -5$, as shown in Figure 1.12.

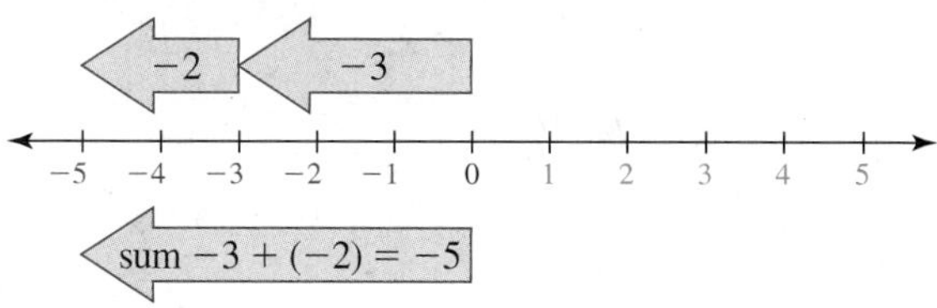

>Figure 1.12

PROBLEM 2

Add: $-2 + (-1)$

−4 −3 −2 −1 0 1 2 3 4

As you can see from Example 2, if we add numbers with the *same* sign (both + or both −), the result is a number with the same sign. Thus,

$$2 + 4 = 6 \quad \text{and} \quad -3 + (-2) = -5$$

If we add numbers with *different* signs, the answer carries the sign of the number with the larger absolute value. Hence, in Example 1,

$5 + (-3) = 2$ The answer is positive because 5 has a larger absolute value than -3. (See Figure 1.11.)

but note that

$-5 + 3 = -2$ The answer is negative because -5 has a larger absolute value than 3.

The following rules summarize this discussion.

RULES

Adding Signed Numbers

1. With the **same (like)** sign: *Add* their absolute values and give the sum the *common* sign.
2. With **different (unlike)** signs: *Subtract* their absolute values and give the difference the sign of the number with the *larger* absolute value.

For example, to add $8 + 5$ or $-8 + (-5)$, we note that the 8 and 5, and -8 and -5 have the **same** signs. So we add their absolute values and give the sum the common sign. Thus,

$$8 + 5 = 13 \quad \text{and} \quad -8 + (-5) = -13$$

Answers to PROBLEMS

2. -3

To add $-8 + 5$, we first notice that the numbers have ***different*** signs. So we subtract their absolute values and give the difference the sign of the number with the larger absolute value. Thus,

$$-8 + 5 = -(8 - 5) = -3$$

Use the sign of the number with the larger absolute value (−).

Subtract the smaller number from the larger one.

Similarly,

$$8 + (-5) = +(8 - 5) = 3$$

Here we have used the sign of the number with the larger absolute value, 8, which is understood to be +.

EXAMPLE 3 **Adding integers with different signs**

Add:

a. $-14 + 6$ **b.** $14 + (-6)$

SOLUTION 3

a. $-14 + 6 = -(14 - 6) = -8$ **b.** $14 + (-6) = +(14 - 6) = 8$

PROBLEM 3

Add:

a. $-10 + 4$ **b.** $10 + (-4)$

The addition of rational numbers uses the same rules for signs, as we illustrate in Examples 4 and 5.

EXAMPLE 4 **Adding decimals**

Add:

a. $-8.6 + 3.4$ **b.** $6.7 + (-9.8)$ **c.** $-2.3 + (-4.1)$

SOLUTION 4

a. $-8.6 + 3.4 = -(8.6 - 3.4) = -5.2$

b. $6.7 + (-9.8) = -(9.8 - 6.7) = -3.1$

c. $-2.3 + (-4.1) = -(2.3 + 4.1) = -6.4$

PROBLEM 4

Add:

a. $-7.8 + 2.5$ **b.** $5.4 + (-7.8)$

c. $-3.4 + (-5.1)$

EXAMPLE 5 **Adding fractions with different signs**

Add:

a. $-\frac{3}{7} + \frac{5}{7}$ **b.** $\frac{2}{5} + \left(-\frac{5}{8}\right)$

SOLUTION 5

a. Note that $\left|\frac{5}{7}\right|$ is larger than $\left|-\frac{3}{7}\right|$; hence

$$-\frac{3}{7} + \frac{5}{7} = +\left(\frac{5}{7} - \frac{3}{7}\right) = \frac{2}{7}$$

b. As usual, we must first find the LCM of 5 and 8, which is 40. We then write

$$\frac{2}{5} = \frac{16}{40} \quad \text{and} \quad -\frac{5}{8} = -\frac{25}{40}$$

Thus,

$$\frac{2}{5} + \left(-\frac{5}{8}\right) = \frac{16}{40} + \left(-\frac{25}{40}\right) = -\left(\frac{25}{40} - \frac{16}{40}\right) = -\frac{9}{40}$$

PROBLEM 5

Add:

a. $-\frac{4}{9} + \frac{5}{9}$ **b.** $\frac{2}{5} + \left(-\frac{3}{4}\right)$

B > Subtracting Real Numbers

We are now ready to subtract signed numbers. Suppose you use *positive* integers to indicate money *earned* and *negative* integers to indicate money *spent* (expenditures). If you earn \$10 and then spend \$12, you owe \$2. Thus,

$$\underbrace{10}_{\text{Earn \$10.}} - \underbrace{12}_{\text{Spend \$12.}} = 10 + (-12) = \underbrace{-2}_{\text{Owe \$2.}}$$

Answers to PROBLEMS

3. a. -6 **b.** 6

4. a. -5.3 **b.** -2.4 **c.** -8.5

5. a. $\frac{1}{9}$ **b.** $-\frac{7}{20}$

Also,

$$-5 - 10 = -5 + (-10) = -15$$ To take away (subtract) earned money is the same as adding an expenditure.

because if you spend \$5 and then spend \$10 more, you now owe \$15. What about $-10 - (-3)$? We claim that

$$-10 - (-3) = -7$$

because if you spend \$10 and then subtract (take away) a \$3 expenditure (represented by -3), you save \$3; that is,

$$-10 - (-3) = -10 + 3 = -7$$ When you take away (subtract) a \$3 expenditure, you save (add) \$3.

In general, we have the following definition.

PROCEDURE

Subtraction

To **subtract** a number b, add its inverse $(-b)$. In symbols,

$$a - b = a + (-b)$$

Thus,

To subtract 8, add its inverse.

$$5 - 8 = 5 + (-8) = -3$$
$$7 - 3 = 7 + (-3) = 4$$
$$-4 - 2 = -4 + (-2) = -6$$
$$-6 - (-4) = -6 + 4 = -2$$

EXAMPLE 6 **Subtracting integers**

Subtract:

a. $17 - 6$
b. $-21 - 4$
c. $-11 - (-5)$
d. $-4 - (-6)$

SOLUTION 6 Use the fact that $a - b = a + (-b)$ to rewrite as an addition.

a. $17 - 6 = 17 + (-6) = 11$
b. $-21 - 4 = -21 + (-4) = -25$
c. $-11 - (-5) = -11 + 5 = -6$
d. $-4 - (-6) = -4 + 6 = 2$

PROBLEM 6

Subtract:

a. $15 - 8$
b. $-23 - 5$
c. $-12 - (-4)$
d. $-5 - (-7)$

EXAMPLE 7 **Subtracting decimals or fractions**

Subtract:

a. $-4.2 - (-3.1)$
b. $-2.5 - (-7.8)$
c. $\frac{2}{9} - \left(-\frac{4}{9}\right)$
d. $-\frac{5}{6} - \frac{7}{4}$

SOLUTION 7

a. $-4.2 - (-3.1) = -4.2 + 3.1 = -1.1$ Note that $-(-3.1) = 3.1$.

b. $-2.5 - (-7.8) = -2.5 + 7.8 = 5.3$ Note that $-(-7.8) = 7.8$.

c. $\frac{2}{9} - \left(-\frac{4}{9}\right) = \frac{2}{9} + \frac{4}{9} = \frac{6}{9} = \frac{2}{3}$ Note that $-\left(-\frac{4}{9}\right) = \frac{4}{9}$.

d. The LCD is 12. Now,

$$-\frac{5}{6} = -\frac{10}{12} \quad \text{and} \quad \frac{7}{4} = \frac{21}{12}$$

Thus,

$$-\frac{5}{6} - \frac{7}{4} = -\frac{5}{6} + \left(-\frac{7}{4}\right) = -\frac{10}{12} + \left(-\frac{21}{12}\right) = -\frac{31}{12}$$

PROBLEM 7

Subtract:

a. $3.2 - (-2.1)$
b. $-3.4 - (-6.9)$
c. $\frac{2}{7} - \left(-\frac{3}{7}\right)$
d. $-\frac{5}{8} - \frac{9}{4}$

Answers to PROBLEMS

6. a. 7 **b.** -28 **c.** -8 **d.** 2

7. a. 5.3 **b.** 3.5 **c.** $\frac{5}{7}$ **d.** $-\frac{23}{8}$

C > Adding and Subtracting Several Real Numbers

Suppose you wish to find $18 - (-10) + 12 - 10 - 17$. Using the fact that $a - b = a + (-b)$, we write

$$\begin{aligned} 18 - (-10) + 12 - 10 - 17 &= 18 + 10 + 12 + (-10) + (-17) \quad [10 + (-10) = 0] \\ &= 18 + 12 + (-17) \\ &= 30 + (-17) \\ &= 13 \end{aligned}$$

EXAMPLE 8 **Adding and subtracting numbers**

Find: $12 - (-13) + 10 - 25 - 13$

SOLUTION 8 First, rewrite as an addition.

$$\begin{aligned} 12 - (-13) + 10 - 25 - 13 &= 12 + 13 + 10 + (-25) + (-13) \quad [13 + (-13) = 0] \\ &= 12 + 10 + (-25) \\ &= 22 + (-25) \\ &= -3 \end{aligned}$$

PROBLEM 8

Find: $14 - (-15) + 10 - 23 - 15$

D > Applications Involving Real Numbers

EXAMPLE 9 **Finding temperature differences**

The greatest temperature variation in a 24-hour period occurred in Browning, Montana, January 23 to 24, 1916. The temperature fell from 44°F to −56°F. How many degrees did the temperature fall?

SOLUTION 9 We have to find the difference between 44 and −56; that is, we have to find $44 - (-56)$:

$$\begin{aligned} 44 - (-56) &= 44 + 56 \\ &= 100 \end{aligned}$$

Thus, the temperature fell 100°F.

PROBLEM 9

The greatest temperature variation in a 12-hour period occurred in Fairfield, Montana. The temperature fell from 63°F at noon to −21°F at midnight. How many degrees did the temperature fall? (When did this happen? December 24, 1924. They certainly did have a cool Christmas!)

Answers to PROBLEMS

8. 1 **9.** 84°F

> Practice Problems > Self-Tests > Media-rich eBooks > e-Professors > Videos

› Exercises 1.3

⟨ A ⟩ **Adding Real Numbers** In Problems 1–40, perform the indicated operations (verify your answer using a number line).

1. $3 + 3$ **2.** $2 + 1$ **3.** $-5 + 1$ **4.** $-4 + 3$

5. $6 + (-5)$ **6.** $5 + (-1)$ **7.** $-2 + (-5)$ **8.** $-3 + (-3)$

9. $3 + (-3)$ **10.** $-4 + 4$ **11.** $-18 + 21$ **12.** $-3 + 5$

13. $19 + (-6)$ **14.** $8 + (-1)$ **15.** $-9 + 11$ **16.** $-8 + 13$

17. $-18 + 9$ **18.** $-17 + 4$ **19.** $-17 + (+5)$ **20.** $-4 + (+8)$

21. $-3.8 + 6.9$ **22.** $-4.5 + 7.8$ **23.** $-7.8 + (3.1)$ **24.** $-6.7 + (2.5)$

25. $3.2 + (-8.6)$ **26.** $4.1 + (-7.9)$ **27.** $-3.4 + (-5.2)$ **28.** $-7.1 + (-2.6)$

29. $-\frac{2}{7} + \frac{5}{7}$ **30.** $-\frac{5}{11} + \frac{7}{11}$ **31.** $-\frac{3}{4} + \frac{1}{4}$ **32.** $-\frac{5}{6} + \frac{1}{6}$

33. $\frac{3}{4} + \left(-\frac{5}{6}\right)$ **34.** $\frac{5}{6} + \left(-\frac{7}{8}\right)$ **35.** $-\frac{1}{6} + \frac{3}{4}$ **36.** $-\frac{1}{8} + \frac{7}{6}$

37. $-\frac{1}{3} + \left(-\frac{2}{7}\right)$ **38.** $-\frac{4}{7} + \left(-\frac{3}{8}\right)$ **39.** $-\frac{5}{6} + \left(-\frac{8}{9}\right)$ **40.** $-\frac{4}{5} + \left(-\frac{7}{8}\right)$

⟨ B ⟩ **Subtracting Real Numbers** In Problems 41–60, perform the indicated operations.

41. $-5 - 11$ **42.** $-4 - 7$ **43.** $-4 - 16$ **44.** $-9 - 11$

45. $7 - 13$ **46.** $8 - 12$ **47.** $9 - (-7)$ **48.** $8 - (-4)$

49. $0 - 4$ **50.** $0 - (-4)$ **51.** $-3.8 - (-1.2)$ **52.** $-6.7 - (-4.3)$

53. $-3.5 - (-8.7)$ **54.** $-6.5 - (-9.9)$ **55.** $4.5 - 8.2$ **56.** $3.7 - 7.9$

57. $\frac{3}{7} - \left(-\frac{1}{7}\right)$ **58.** $\frac{5}{6} - \left(-\frac{1}{6}\right)$ **59.** $-\frac{5}{4} - \frac{7}{6}$ **60.** $-\frac{2}{3} - \frac{3}{4}$

⟨ C ⟩ **Adding and Subtracting Several Real Numbers** In Problems 61–66, perform the indicated operations.

61. $8 - (-10) + 5 - 20 - 10$ **62.** $15 - (-9) + 8 - 2 - 9$ **63.** $-15 + 12 - 8 - (-15) + 5$

64. $-12 + 14 - 7 - (-12) + 3$ **65.** $-10 + 9 - 14 - 3 - (-14)$ **66.** $-7 + 2 - 6 - 8 - (-6)$

⟨ D ⟩ **Applications Involving Real Numbers**

››› Applications: Green Math

67. *Earth temperatures* The temperature in the center core of the Earth reaches +5000°C. In the thermosphere (a region in the upper atmosphere), the temperature is +1500°C. Find the difference in temperature between the center of the Earth and the thermosphere.

68. *Extreme temperatures* The record high temperature in Calgary, Alberta, is +99°F. The record low temperature is −46°F. Find the difference between these extremes.

69. *Temperature variation* The present average global temperature is 56.3°F (56.3 degrees Fahrenheit). It has been predicted that it will *increase* by 1.4°F from its 1800 levels. What will be the new average global temperature?

70. *Temperature variation* The present average global temperature is 13.5°C (13.5 degrees Celsius). It has been predicted that it will *increase* by 0.8°C from its 1800 levels. What will be the new average global temperature?

71. *Temperature variations* Here are the temperature changes (in degrees Celsius) by the hour in a certain city:

1 P.M.	+2
2 P.M.	+1
3 P.M.	−1
4 P.M.	−3

If the temperature was initially 15°C, what was it at 4 P.M.?

Source: http://www.climatechangefacts.info/.

› Web IT go to mhhe.com/bello for more lessons

How many calories do you eat for lunch? How many do you use during exercise? The table shows the number of calories in some fast foods and the number of calories used in different activities.

Food	Calories (+)	Activity	Calories Burned (−)
Hamburger (McD)	280	Bicycling 6 mph	240 cal./hr
Fries (McD)	210	Bicycling 12 mph	410 cal./hr
Hamburger (BK)	310	Jogging 5 mph	740 cal./hr
Fries (BK)	230	Jogging 7 mph	920 cal./hr
Salad (McD)	20	Swimming 25 yd/min	275 cal./hr
Salad (BK)	25	Swimming 50 yd/min	500 cal./hr.

Calories In Problems 72–76 find the number of calories gained or lost in each situation.

72. You have a McDonald's (McD) hamburger and bicycle at 6 mph for 1 hour.

73. You have a Burger King (BK) hamburger with fries and you jog at 7 mph for 1 hour.

74. You have a hamburger, fries, and a salad at McDonald's, then bicycle at 12 mph for an hour and jog at 7 mph for another hour.

75. You have a hamburger, fries, and a salad at BK, and then jog at 7 mph for an hour.

76. You eat fries and a salad at BK then go swimming at 25 yards per minute for an hour. What else can you eat so your caloric intake will be 0?

Using Your Knowledge

A Little History The following chart contains some important historical dates.

Important Historical Dates	
323 B.C.	Alexander the Great died
216 B.C.	Hannibal defeated the Romans
A.D. 476	Fall of the Roman Empire
A.D. 1492	Columbus landed in America
A.D. 1776	The Declaration of Independence signed
A.D. 1939	World War II started
A.D. 2008	Barack Obama elected

We can use negative integers to represent years B.C. For example, the year Alexander the Great died can be written as -323, whereas the fall of the Roman Empire occurred in $+476$ (or simply 476). To find the number of years that elapsed between the fall of the Roman Empire and their defeat by Hannibal, we write

$$476 - (-216) = 476 + 216 = 692$$

Fall of the Roman Empire (A.D. 476) → 476; Hannibal defeats the Romans (216 B.C.) → -216; Years elapsed → 692

Use these ideas to find the number of years elapsed between the following:

77. The fall of the Roman Empire and the death of Alexander the Great

78. Columbus's landing in America and Hannibal's defeat of the Romans

79. Columbus landing in America and the signing of the Declaration of Independence

80. The year Barack Obama was elected and the signing of the Declaration of Independence

81. The start of World War II and the death of Alexander the Great

Write On

82. Explain what the term *signed numbers* means and give examples.

83. State the rule you use to add signed numbers. Explain why the sum is sometimes positive and sometimes negative.

84. State the rule you use to subtract signed numbers. How do you know when the answer is going to be positive? How do you know when the answer is going to be negative?

85. The definition of subtraction is as follows: To subtract a number, add its inverse. Use a number line to explain why this works.

Concept Checker

Fill in the blank(s) with the correct word(s), phrase, or mathematical statement.

86. To **add** signed numbers with the **same** sign, add their absolute value and give the sum the ______________ sign.

87. To add numbers with **different** signs, subtract their absolute value and give the difference the sign of the number with the ______________ absolute value.

88. The **definition of subtraction** states that $a - b = a +$ ______________.

89. To **subtract** a number b from a number a, add the ______________ of b.

reciprocal	**smaller**
inverse	**larger**
(b)	**common**
$(-b)$	**different**

Mastery Test

Add or subtract

90. $7 - 13$

91. $3.8 - 6.9$

92. $\frac{1}{5} - \frac{3}{4}$

93. $-3.5 - 4.2$

94. $-\frac{2}{3} - \frac{4}{5}$

95. $-5 - 15$

96. $-6 - (-4)$

97. $-3.4 - (-4.6)$

98. $-\frac{3}{4} - \left(-\frac{1}{5}\right)$

99. $3.9 + (-4.2)$

100. $\frac{3}{4} + \left(-\frac{1}{6}\right)$

101. $-3.2 + (-2.5)$

102. $7 - (-11) + 13 - 11 - 15$

103. $\frac{3}{4} - \left(-\frac{2}{5}\right) + \frac{4}{5} - \frac{2}{5} + \frac{5}{4}$

104. The temperature in Verkhoyansk, Siberia, ranges from 98°F to −94°F. What is the difference between these temperatures?

Skill Checker

Perform the indicated operation.

105. $\frac{5}{8} \cdot \frac{16}{25}$

106. $5\frac{1}{4} \cdot 3\frac{1}{8}$

107. $6\frac{1}{6} \cdot 5\frac{7}{10}$

108. $6\frac{1}{4} \cdot \frac{20}{21}$

109. $\frac{18}{11} \div 5\frac{1}{2}$

110. $2\frac{1}{4} \div 1\frac{3}{8}$

1.4 Multiplying and Dividing Real Numbers

Objectives

A Multiply two real numbers.

B Evaluate expressions involving exponents.

C Divide one real number by another.

D Solve an application involving multiplying and dividing real numbers.

To Succeed, Review How To . . .

1. Multiply and divide whole numbers, decimals, and fractions (Chapter R, pp. 10–12, 23–24).
2. Find the reciprocal of a number (Chapter R, pp. 11–12).

Getting Started

A Stock Market Loss

In Section 1.3, we learned how to add and subtract real numbers. Now we will learn how to multiply and divide them. Pay particular attention to the notation used when multiplying a number by itself and also to the different applications of the multiplication and division of real numbers. For example, suppose you own 4 shares of stock, and the closing price today is *down* \$3 (written as -3). Your loss then is

$$4 \cdot (-3) \quad \text{or} \quad 4(-3)$$

How do we multiply positive and negative integers? As you may recall, the result of a multiplication is a *product,* and the numbers being multiplied (4 and -3) are called *factors*. Now you can think of multiplication as *repeated addition*. Thus,

$$4 \cdot (-3) = \underbrace{(-3) + (-3) + (-3) + (-3)}_{\text{four } (-3)\text{'s}} = -12$$

Also note that

$$(-3) \cdot 4 = -12$$

So your stock has gone down \$12. As you can see, the product of a *negative* integer and a *positive* integer is negative. What about the product of two negative integers, say $-4 \cdot (-3)$? Look for the pattern in the following table:

The number in this column decreases by 1.

The number in this column increases by 3.

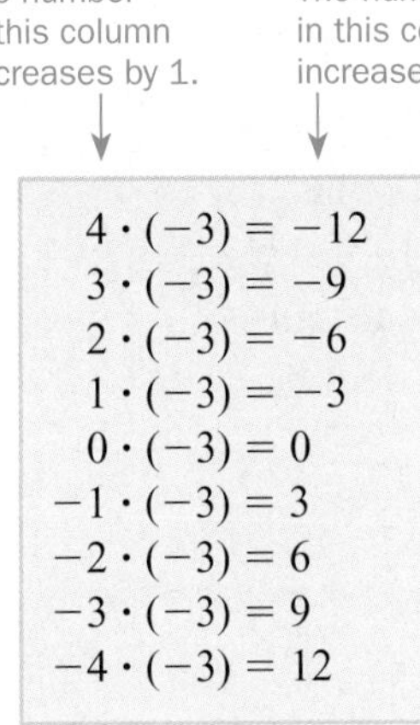

$4 \cdot (-3) = -12$
$3 \cdot (-3) = -9$
$2 \cdot (-3) = -6$
$1 \cdot (-3) = -3$
$0 \cdot (-3) = 0$
$-1 \cdot (-3) = 3$
$-2 \cdot (-3) = 6$
$-3 \cdot (-3) = 9$
$-4 \cdot (-3) = 12$

You can think of $-4 \cdot (-3)$ as subtracting -3 four times; that is,

$$-(-3) - (-3) - (-3) - (-3) = 3 + 3 + 3 + 3 = 12$$

So, when we multiply two integers with *different* (*unlike*) signs, the product is *negative*. If we multiply two integers with the *same* (*like*) signs, the product is *positive*. This idea can be generalized to include the product of any two real numbers, as you will see later.

A > Multiplying Real Numbers

Here are the rules that we used in the *Getting Started*.

RULES

Multiplying Signed Numbers

1. When two numbers with the *same* (*like*) sign are multiplied, the product is *positive* (+).
2. When two numbers with *different* (*unlike*) signs are multiplied, the product is *negative* (−).

Here are some examples.

$$9 \cdot 4 = 36$$
$$-9 \cdot (-4) = 36$$

9 and 4 have the same sign (+); −9 and −4 have the same sign (−); thus, the product is positive.

$$-9 \cdot 6 = -54$$
$$9 \cdot (-6) = -54$$

−9 and 6 have different signs; 9 and −6 have different signs; thus, the product is negative.

EXAMPLE 1 Finding products of integers

Multiply:

a. $7 \cdot 8$ **b.** $-8 \cdot 6$ **c.** $4 \cdot (-8)$ **d.** $-7 \cdot (-9)$

SOLUTION 1

a. $7 \cdot 8 = 56$ (Same sign)

b. $-8 \cdot 6 = -48$ (Different signs; Negative product)

c. $4 \cdot (-8) = -32$ (Different signs; Negative product)

d. $-7 \cdot (-9) = 63$ (Same sign; Positive product)

PROBLEM 1

Multiply:

a. $9 \cdot 6$ **b.** $-7 \cdot 8$

c. $5 \cdot (-4)$ **d.** $-5 \cdot (-6)$

The multiplication of rational numbers also uses the same rules for signs, as we illustrate in Example 2. Remember that $-3.1(4.2)$ means $-3.1 \cdot 4.2$. Parentheses are one of the ways we indicate multiplication.

EXAMPLE 2 Finding products of decimals and fractions

Multiply:

a. $-3.1(4.2)$ **b.** $-1.2(-3.4)$ **c.** $-\frac{3}{4}\left(-\frac{5}{2}\right)$ **d.** $\frac{5}{6}\left(-\frac{4}{7}\right)$

SOLUTION 2

a. -3.1 and 4.2 have different signs. The product is *negative*. Thus,

$$-3.1(4.2) = -13.02$$

b. -1.2 and -3.4 have the same sign. The product is *positive*. Thus,

$$-1.2(-3.4) = 4.08$$

c. $-\frac{3}{4}$ and $-\frac{5}{2}$ have the same sign. The product is *positive*. Thus,

$$-\frac{3}{4}\left(-\frac{5}{2}\right) = \frac{15}{8}$$

d. $\frac{5}{6}$ and $-\frac{4}{7}$ have different signs. The product is *negative*. Thus,

$$\frac{5}{6}\left(-\frac{4}{7}\right) = -\frac{20}{42} = -\frac{10}{21}$$

PROBLEM 2

Multiply:

a. $-4.1 \cdot (3.2)$ **b.** $-1.3(-4.2)$

c. $-\frac{3}{5}\left(-\frac{7}{10}\right)$ **d.** $\frac{5}{9}\left(-\frac{6}{7}\right)$

Answers to PROBLEMS

1. a. 54 **b.** −56 **c.** −20 **d.** 30 **2. a.** −13.12 **b.** 5.46 **c.** $\frac{21}{50}$ **d.** $-\frac{10}{21}$

B > Evaluating Expressions Involving Exponents

Sometimes a number is used several times as a factor. For example, we may wish to find or evaluate the following products:

$$3 \cdot 3 \quad \text{or} \quad 4 \cdot 4 \cdot 4 \quad \text{or} \quad 5 \cdot 5 \cdot 5 \cdot 5$$

In the expression $3 \cdot 3$, the 3 is used as a factor twice. In such cases it's easier to use **exponents** to indicate how many times the number is used as a factor. We then write

3^2 (read "3 squared") instead of $3 \cdot 3$
4^3 (read "4 cubed") instead of $4 \cdot 4 \cdot 4$
5^4 (read "5 to the fourth") instead of $5 \cdot 5 \cdot 5 \cdot 5$

The expression 3^2 uses the exponent 2 to indicate how many times the **base** 3 is used as a factor. Similarly, in the expression 5^4, the 5 is the base and the 4 is the exponent. Now,

$$3^2 = \underbrace{3 \cdot 3} = 9$$ 3 is used as a factor **2** times.

$$4^3 = \underbrace{4 \cdot 4 \cdot 4} = 64$$ 4 is used as a factor **3** times.

and

$$\left(\frac{1}{5}\right)^4 = \underbrace{\frac{1}{5} \cdot \frac{1}{5} \cdot \frac{1}{5} \cdot \frac{1}{5}} = \frac{1}{625}$$ $\frac{1}{5}$ is used as a factor **4** times.

What about $(-2)^2$? Using the definition of exponents, we have

$$(-2)^2 = (-2) \cdot (-2) = 4$$ -2 and -2 have the same sign; thus, their product is positive.

Moreover, -2^2 means $-(2 \cdot 2)$. To emphasize that the multiplication is to be done *first,* we put parentheses around the $2 \cdot 2$. Thus, the placing of the parentheses in the expression $(-2)^2$ is very important. Clearly, since $(-2)^2 = 4$ and $-2^2 = -(2 \cdot 2) = -4$,

$$(-2)^2 \neq -2^2$$ "$\neq$" means "is not equal to."

EXAMPLE 3 Evaluating expressions involving exponents

Evaluate:

a. $(-4)^2$ b. -4^2 c. $\left(-\frac{1}{3}\right)^2$ d. $-\left(\frac{1}{3}\right)^2$

SOLUTION 3

a. $(-4)^2 = (-4)(-4) = 16$ Note that the base is -4.

b. $-4^2 = -(4 \cdot 4) = -16$ Here the base is 4.

c. $\left(-\frac{1}{3}\right)^2 = \left(-\frac{1}{3}\right)\left(-\frac{1}{3}\right) = \frac{1}{9}$ The base is $-\frac{1}{3}$.

d. $-\left(\frac{1}{3}\right)^2 = -\left(\frac{1}{3}\right)\left(\frac{1}{3}\right) = -\frac{1}{9}$ The base is $\frac{1}{3}$.

From parts **a** and **b,** you can see that $(-4)^2 \neq -4^2$.

PROBLEM 3

Evaluate:

a. $(-7)^2$ b. -7^2

c. $\left(-\frac{1}{4}\right)^2$ d. $-\left(\frac{1}{4}\right)^2$

EXAMPLE 4 Evaluating expressions involving exponents

Evaluate:

a. $(-2)^3$ b. -2^3

SOLUTION 4

a. $(-2)^3 = \underbrace{(-2) \cdot (-2)} \cdot (-2)$
$= \quad 4 \quad \cdot (-2)$
$= -8$

b. $-2^3 = -\underbrace{(2 \cdot 2 \cdot 2)}$
$= \quad -(8)$
$= -8$

PROBLEM 4

Evaluate:

a. $(-4)^3$ b. -4^3

Answers to PROBLEMS

3. a. 49 **b.** -49 **c.** $\frac{1}{16}$ **d.** $-\frac{1}{16}$

4. a. -64 **b.** -64

Note that

$$(-4)^2 = 16$$ Negative number raised to an even power; positive result.

but

$$(-2)^3 = -8$$ Negative number raised to an odd power; negative result.

C > Dividing Real Numbers

What about the rules for division? As you recall, a division problem can always be checked by multiplication. Thus, the division

$$3\overline{)18}\ \text{with quotient } 6,\ -18,\ 0 \quad \text{or} \quad \frac{18}{3} = 6$$

is correct because $18 = 3 \cdot 6$. In general, we have the following definition for division.

DIVISION

$$\frac{a}{b} = c \quad \text{means} \quad a = b \cdot c, \quad b \neq 0$$

Note that the operation of division is defined using multiplication. Because of this, the same rules of sign that apply to the multiplication of real numbers also apply to the division of real numbers. Note that the result of the division of two numbers is called the **quotient.**

RULES

Dividing with Signed Numbers

1. When dividing two numbers with the *same* (*like*) sign, give the quotient a *positive* (+) sign.
2. When dividing two numbers with *different* (*unlike*) signs, give the quotient a *negative* (−) sign.

Here are some examples:

$\frac{24}{6} = 4$ 24 and 6 have the same sign; the quotient is positive.

$\frac{-18}{-9} = 2$ −18 and −9 have the same sign; the quotient is positive.

$\frac{-32}{4} = -8$ −32 and 4 have different signs; the quotient is negative.

$\frac{35}{-7} = -5$ 35 and −7 have different signs; the quotient is negative.

EXAMPLE 5 **Finding quotients of integers**

Divide:

a. $48 \div 6$ **b.** $\frac{54}{-9}$ **c.** $\frac{-63}{-7}$ **d.** $-28 \div 4$ **e.** $5 \div 0$

PROBLEM 5

Divide:

a. $56 \div 8$ **b.** $\frac{36}{-4}$ **c.** $\frac{-49}{-7}$

d. $-24 \div 6$ **e.** $-7 \div 0$

(continued)

Answers to PROBLEMS

5. a. 7 **b.** −9 **c.** 7

d. −4 **e.** Not defined

SOLUTION 5

a. $48 \div 6 = 8$ 48 and 6 have the same sign; the quotient is positive.

b. $\frac{54}{-9} = -6$ 54 and −9 have different signs; the quotient is negative.

c. $\frac{-63}{-7} = 9$ −63 and −7 have the same sign; the quotient is positive.

d. $-28 \div 4 = -7$ −28 and 4 have different signs; the quotient is negative.

e. $5 \div 0$ is **not** defined. Note that if we let $5 \div 0$ equal any number, such as a, we have

$$\frac{5}{0} = a \qquad \text{This means } 5 = a \cdot 0 = 0 \text{ or } 5 = 0$$

which, of course, is false. Thus, $\frac{5}{0}$ is *not defined.*

If the division involves real numbers written as fractions, we use the following procedure.

PROCEDURE

Dividing Fractions

To divide $\frac{a}{b}$ by $\frac{c}{d}$, multiply $\frac{a}{b}$ by the reciprocal of $\frac{c}{d}$, that is,

$$\frac{a}{b} \div \frac{c}{d} = \frac{a}{b} \cdot \frac{d}{c} = \frac{ad}{bc} \qquad (b, c, \text{ and } d \neq 0)$$

Of course, the rules of signs still apply!

EXAMPLE 6 Finding quotients of fractions

Divide:

a. $\frac{2}{5} \div \left(-\frac{3}{4}\right)$ **b.** $-\frac{5}{6} \div \left(-\frac{7}{2}\right)$ **c.** $-\frac{3}{7} \div \frac{6}{7}$

SOLUTION 6

a. $\frac{2}{5} \div \left(-\frac{3}{4}\right) = \frac{2}{5} \cdot \left(-\frac{4}{3}\right) = -\frac{8}{15}$

Different signs Negative quotient

b. $-\frac{5}{6} \div \left(-\frac{7}{2}\right) = -\frac{5}{6} \cdot \left(-\frac{2}{7}\right) = \frac{10}{42} = \frac{5}{21}$

Same sign Positive quotient

c. $-\frac{3}{7} \div \frac{6}{7} = -\frac{3}{7} \cdot \frac{7}{6} = -\frac{21}{42} = -\frac{1}{2}$

Different signs Negative quotient

PROBLEM 6

Divide:

a. $\frac{4}{5} \div \left(-\frac{3}{4}\right)$ **b.** $-\frac{5}{6} \div \left(-\frac{7}{4}\right)$

c. $-\frac{4}{7} \div \frac{8}{7}$

Answers to PROBLEMS

6. a. $-\frac{16}{15}$ **b.** $\frac{10}{21}$ **c.** $-\frac{1}{2}$

D > Applications Involving Multiplying and Dividing Real Numbers

When you are driving and push down or let up on the gas or brake pedal, your car changes speed. This change in speed over a period of time is called *acceleration* and is given by

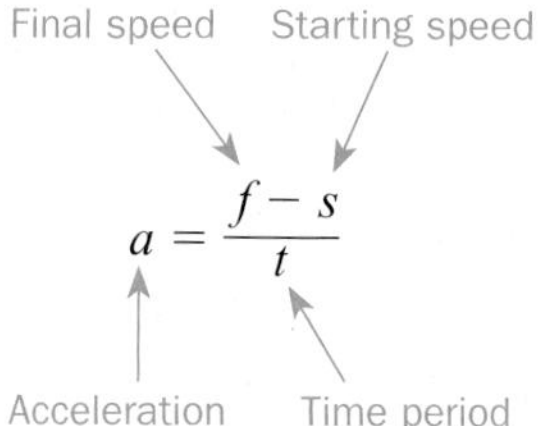

We use this idea in Example 7.

EXAMPLE 7 Acceleration and deceleration

You are driving at 55 miles per hour (mi/hr), and over the next 10 seconds,

a. you *increase* your speed to 65 miles per hour. What is your acceleration?

b. you *decrease* your speed to 40 miles per hour. What is your acceleration?

SOLUTION 7 Your starting speed is $s = 55$ miles per hour and the time period is $t = 10$ seconds (sec).

a. Your final speed is $f = 65$ miles per hour, so

$$a = \frac{(65 - 55)\text{ mi/hr}}{10\text{ sec}} = \frac{10\text{ mi/hr}}{10\text{ sec}} = 1\frac{\text{mi/hr}}{\text{sec}}$$

Thus, your acceleration is 1 mile per hour each second.

b. Here the final speed is 40, so

$$a = \frac{(40 - 55)\text{ mi/hr}}{10\text{ sec}} = \frac{-15\text{ mi/hr}}{10\text{ sec}} = -1\frac{1}{2}\frac{\text{mi/hr}}{\text{sec}}$$

Thus, your acceleration is $-1\frac{1}{2}$ miles per hour every second. When acceleration is *negative,* it is called *deceleration*. Deceleration can be thought of as negative acceleration, so your deceleration is $1\frac{1}{2}$ miles per hour each second.

PROBLEM 7

a. Find your acceleration if you increase your speed from 55 miles per hour to 70 miles an hour over the next 5 seconds.

b. Find your acceleration if you decrease your speed from 50 miles per hour to 40 miles per hour over the next 5 seconds.

What can we do to help the environment? Some of the suggestions include decreasing car CO_2 emissions and planting trees. How many trees do we have to plant? Opinions vary widely, but an acre of trees has been suggested. Find out more in Example 8.

Answers to PROBLEMS

7. a. $3\frac{\text{mi/hr}}{\text{sec}}$ **b.** $-2\frac{\text{mi/hr}}{\text{sec}}$

EXAMPLE 8 How much CO_2 does an acre of trees absorb?

Suppose you have a lot 80 ft by 540 ft (about an acre) and you plant trees every 8 feet, there will be $\frac{80}{8} = 10$ rows of $\frac{540}{8} \approx 67$ trees. (See diagram: not to scale!)

a. How many trees do you have in your acre lot?

b. If each tree absorbs 50 pounds of CO_2 a year (the amount varies by tree), how many pounds of CO_2 does the acre of trees absorb?

SOLUTION 8

a. You have 10 rows with 67 trees each or $(10)(67) = 670$ trees.

b. Each tree absorbs (-50) lb of CO_2, so the whole acre absorbs
$(-50)(10)(67) = -33{,}500$ pounds of CO_2.

Data Source: http://tinyurl.com/yswsdv.

PROBLEM 8

a. If your tree lot has 10 rows of 70 trees, how many trees do you have?

b. How many pounds of CO_2 does the acre of trees absorb?

10 rows (80 ft)

67 trees (540 ft)

> Practice Problems > Self-Tests > Media-rich eBooks > e-Professors > Videos

Exercises 1.4

Web IT go to mhhe.com/bello for more lessons

⟨A⟩ Multiplying Real Numbers In Problems 1–20, perform the indicated operation.

1. $4 \cdot 9$ **2.** $16 \cdot 2$ **3.** $-10 \cdot 4$ **4.** $-6 \cdot 8$

5. $-9 \cdot 9$ **6.** $-2 \cdot 5$ **7.** $-6 \cdot (-3)(-2)$ **8.** $-4 \cdot (-5)(-3)$

9. $-9 \cdot (-2)(-3)$ **10.** $-7 \cdot (-10)(-2)$ **11.** $-2.2(3.3)$ **12.** $-1.4(3.1)$

13. $-1.3(-2.2)$ **14.** $-1.5(-1.1)$ **15.** $\frac{5}{6}\left(-\frac{5}{7}\right)$ **16.** $\frac{3}{8}\left(-\frac{5}{7}\right)$

17. $-\frac{3}{5}\left(-\frac{5}{12}\right)$ **18.** $-\frac{4}{7}\left(-\frac{21}{8}\right)$ **19.** $-\frac{6}{7}\left(\frac{35}{8}\right)$ **20.** $-\frac{7}{5}\left(\frac{15}{28}\right)$

⟨B⟩ Evaluating Expressions Involving Exponents In Problems 21–30, perform the indicated operation.

21. -9^2 **22.** $(-9)^2$ **23.** $(-5)^2$ **24.** -5^2 **25.** -5^3

26. $(-5)^3$ **27.** $(-6)^4$ **28.** -6^4 **29.** $-\left(\frac{1}{2}\right)^5$ **30.** $\left(-\frac{1}{2}\right)^5$

⟨C⟩ Dividing Real Numbers In Problems 31–60, perform the indicated operation.

31. $\frac{14}{2}$ **32.** $10 \div 2$ **33.** $-50 \div 10$ **34.** $\frac{-20}{5}$

35. $\frac{-30}{10}$ **36.** $-40 \div 8$ **37.** $\frac{-0}{3}$ **38.** $-0 \div 8$

39. $-5 \div 0$ **40.** $\frac{-8}{0}$ **41.** $\frac{0}{7}$ **42.** $0 \div (-7)$

Answers to PROBLEMS

8. a. 700 **b.** $-35{,}000$

43. $-15 \div (-3)$

44. $-20 \div (-4)$

45. $\frac{-25}{-5}$

46. $\frac{-16}{-2}$

47. $\frac{18}{-9}$

48. $\frac{35}{-7}$

49. $30 \div (-5)$

50. $80 \div (-10)$

51. $\frac{3}{5} \div \left(-\frac{4}{7}\right)$

52. $\frac{4}{9} \div \left(-\frac{1}{7}\right)$

53. $-\frac{2}{3} \div \left(-\frac{7}{6}\right)$

54. $-\frac{5}{6} \div \left(-\frac{25}{18}\right)$

55. $-\frac{5}{8} \div \frac{7}{8}$

56. $-\frac{4}{5} \div \frac{8}{15}$

57. $\frac{-3.1}{6.2}$

58. $\frac{1.2}{-4.8}$

59. $\frac{-1.6}{-9.6}$

60. $\frac{-9.8}{-1.4}$

D **Applications Involving Multiplying and Dividing Real Numbers** Use the following information in Problems 61–65:

1 all-beef frank	+45 calories
1 slice of bread	+65 calories
Running (1 min)	−15 calories
Swimming (1 min)	−7 calories

61. *Caloric gain or loss* If a person eats 2 beef franks and runs for 5 minutes, what is the caloric gain or loss?

62. *Caloric gain or loss* If a person eats 2 beef franks and runs for 30 minutes, what is the caloric gain or loss?

63. *Caloric gain or loss* If a person eats 2 beef franks with 2 slices of bread and then runs for 15 minutes, what is the caloric gain or loss?

64. *Caloric gain or loss* If a person eats 2 beef franks with 2 slices of bread and then runs for 15 minutes and swims for 30 minutes, what is the caloric gain or loss?

65. *"Burning off" calories* If a person eats 2 beef franks, how many minutes does the person have to run to "burn off" the calories? (*Hint:* You must *spend* the calories contained in the 2 beef franks.)

66. *Automobile acceleration* The highest road-tested acceleration for a standard production car is from 0 to 60 miles per hour in 3.275 seconds for a Ford RS 200 Evolution. What was the acceleration of this car? Give your answer to one decimal place.

67. *Automobile acceleration* The highest road-tested acceleration for a street-legal car is from 0 to 60 miles per hour in 3.89 seconds for a Jankel Tempest. What was the acceleration of this car? Give your answer to one decimal place.

68. *Cost of saffron* Which food do you think is the most expensive? It is saffron, which comes from Spain. It costs \$472.50 to buy 3.5 ounces at Harrods, a store in Great Britain. What is the cost of 1 ounce of saffron?

69. *Long-distance telephone charges* The price of a long-distance call from Tampa to New York is \$3.05 for the first 3 minutes and \$0.70 for each additional minute or fraction thereof. What is the cost of a 5-minute long-distance call from Tampa to New York?

70. *Long-distance telephone charges* The price of a long-distance call from Tampa to New York using a different phone company is \$3 for the first 3 minutes and \$0.75 for each additional minute or fraction thereof. What is the cost of a 5-minute long-distance call from Tampa to New York using this phone company?

Applications: Green Math

Follow the procedure of Example 8b to find the annual absorption of CO_2 for **an acre of trees** (670 trees per acre) of the type specified in each problem.

71. Average trees each absorbing 13 pounds of CO_2 per year

72. Twenty-five-year-old pine trees each absorbing 15 pounds of CO_2 per year

73. Twenty-five-year-old maple trees each absorbing 2.5 pounds of CO_2 per year

74. Trees absorbing 28.6 pounds of CO_2 per year

Using Your Knowledge

Have a Decidedly Lovable Day Have you met anybody *nice* today or did you have an *unpleasant* experience? Perhaps the person you met was *very nice* or your experience *very unpleasant*. Psychologists and linguists have a numerical way to indicate the difference between *nice* and *very nice* or between *unpleasant* and *very unpleasant*. Suppose you assign a positive number (+2, for example) to the adjective *nice,* a negative number (say, −2) to *unpleasant,* and a positive number greater than 1 (say +1.75) to *very*. Then, *very nice* means

Very nice

$$(1.75) \cdot (2) = 3.50$$

Web IT go to mhhe.com/bello for more lessons

and *very unpleasant* means

Very unpleasant

$$(1.75) \cdot (-2) = -3.50$$

Here are some adverbs and adjectives and their average numerical values, as rated by a panel of college students. (Values differ from one panel to another.)

Adverbs		Adjectives	
Slightly	0.54	Wicked	−2.5
Rather	0.84	Disgusting	−2.1
Decidedly	0.16	Average	−0.8
Very	1.25	Good	3.1
Extremely	1.45	Lovable	2.4

Find the value of each.

75. Slightly wicked

76. Decidedly average

77. Extremely disgusting

78. Rather lovable

79. Very good

By the way, if you got all the answers correct, you are 4.495!

Write On

80. Why is $-a^2$ always negative for any nonzero value of a?

81. Why is $(-a)^2$ always positive for any nonzero value of a?

82. Is $(-1)^{100}$ positive or negative? What about $(-1)^{99}$?

83. Explain why $\frac{a}{0}$ is not defined.

Concept Checker

Fill in the blank(s) with the correct word(s), phrase, or mathematical statement.

84. When two numbers with **the *same (like)*** sign are **multiplied,** the **product** is ______________. **positive**

85. When two numbers with ***different (unlike)*** signs are **multiplied,** the **product** is ______________. **negative**

86. When dividing two numbers with the ***same (like)*** sign, give the **quotient** a ______________ sign.

87. When dividing two numbers with ***different (unlike)*** signs, give the **quotient** a ______________ sign.

Mastery Test

Perform the indicated operations.

88. $\frac{-3.6}{1.2}$

89. $\frac{-3.1}{-12.4}$

90. $\frac{6.5}{-1.3}$

91. $-3 \cdot 11$

92. $9 \cdot (-10)$

93. $-5 \cdot (-11)$

94. -9^2

95. $(-8)^2$

96. $\left(-\frac{1}{5}\right)^2$

97. $-2.2(3.2)$

98. $-1.3(-4.1)$

99. $-\frac{3}{7}\left(-\frac{4}{5}\right)$

100. $\frac{6}{7}\left(-\frac{2}{3}\right)$

101. $\frac{1}{5} \div \left(-\frac{4}{7}\right)$

102. $-\frac{6}{7} \div \left(-\frac{3}{5}\right)$

103. $-\frac{4}{5} \div \frac{8}{5}$

104. The driver of a car traveling at 50 miles per hour slams on the brakes and slows down to 10 miles per hour in 5 seconds. What is the acceleration of the car?

Skill Checker

Perform the indicated operations.

105. $\frac{22+4}{8-10}$

106. $\frac{4-8}{6-4}$

107. $\frac{20 \cdot 7}{10}$

108. $\frac{30 \cdot 4}{10}$

1.5 Order of Operations

Objectives

A Evaluate expressions using the correct order of operations.

B Evaluate expressions with more than one grouping symbol.

C Solve an application involving the order of operations.

To Succeed, Review How To . . .

1. Add, subtract, multiply, and divide real numbers (pp. 52–56, 61–64).
2. Evaluate expressions containing exponents (pp. 61–63).

Getting Started

Collecting the Rent

Now that we know how to do the fundamental operations with real numbers, we need to know *in what order* we should do them. Let's suppose all rooms in a motel are taken. (For simplicity, we won't include extra persons in the rooms.) How can we figure out how much money we should collect? To do this, we first multiply the price of each room by the number of rooms available at that price. Next, we add all these figures to get the final answer. The calculations look like this:

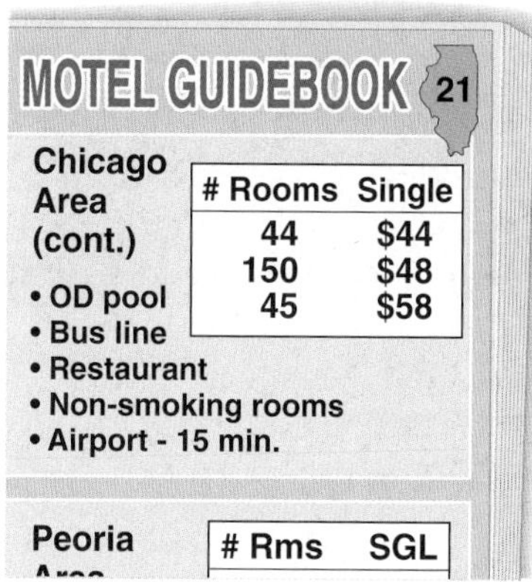

$$44 \cdot \$44 = \$1936$$
$$150 \cdot \$48 = \$7200$$
$$45 \cdot \$58 = \$2610$$
$$\$1936 + \$7200 + \$2610 = \$11{,}746 \quad \text{Total}$$

Note that we *multiplied* before we *added.* This is the correct order of operations. As you will see, changing the order of operations can change the answer!

A > Using the Order of Operations

If we want to find the answer to $3 \cdot 4 + 5$, do we

(1) add 4 and 5 first and then multiply by 3? That is, $3 \cdot 9 = 27$?

or

(2) multiply 3 by 4 first and then add 5? That is, $12 + 5 = 17$?

In (1), the answer is 27. In (2), the answer is 17. What if we write $3 \cdot (4 + 5)$ or $(3 \cdot 4) + 5$? What do the parentheses mean? To obtain an answer we can agree upon, we need the following rules.

RULES

Order of Operations

Operations are always performed in the following order:

1. Do all calculations inside **grouping symbols** such as parentheses ( ) or brackets [ ].
2. Evaluate all exponents.
3. Do multiplications and divisions as they occur from *left to right.*
4. Do additions and subtractions as they occur from *left to right.*

You can remember the order by remembering the underlined letters PEMDAS.

With these conventions

$$\begin{aligned} & 3 \cdot 4 + 5 \\ &= 12 + 5 && \text{First multiply.} \\ &= 17 && \text{Then add.} \end{aligned}$$

Similarly,

$$\begin{aligned} & 3 \cdot (4 + 5) \\ &= 3 \cdot 9 && \text{First add inside parentheses.} \\ &= 27 && \text{Then multiply.} \end{aligned}$$

But

$$\begin{aligned} & (3 \cdot 4) + 5 \\ &= 12 + 5 && \text{First multiply inside parentheses.} \\ &= 17 && \text{Then add.} \end{aligned}$$

Note that multiplications and divisions are done **in order,** from left to right. This means that sometimes you do multiplications first, sometimes you do divisions first, depending on the order in which they occur. Thus,

$$\begin{aligned} & 12 \div 4 \cdot 2 \\ &= 3 \cdot 2 && \text{First divide.} \\ &= 6 && \text{Then multiply.} \end{aligned}$$

But

$$\begin{aligned} & 12 \cdot 4 \div 2 \\ &= 48 \div 2 && \text{First multiply.} \\ &= 24 && \text{Then divide.} \end{aligned}$$

EXAMPLE 1 Evaluating expressions

Find the value of each expression.

a. $8 \cdot 9 - 3$ **b.** $27 + 3 \cdot 5$

SOLUTION 1

a.

$$\begin{aligned} & 8 \cdot 9 - 3 \\ &= 72 - 3 && \text{Do multiplications and divisions in order from left to right } (8 \cdot 9 = 72). \\ &= 69 && \text{Then do additions and subtractions in order from left to right } (72 - 3 = 69). \end{aligned}$$

b.

$$\begin{aligned} & 27 + 3 \cdot 5 \\ &= 27 + 15 && \text{Do multiplications and divisions in order from left to right } (3 \cdot 5 = 15). \\ &= 42 && \text{Then do additions and subtractions in order from left to right } (27 + 15 = 42). \end{aligned}$$

PROBLEM 1

Find the value of each expression.

a. $3 \cdot 8 - 9$ **b.** $22 + 4 \cdot 9$

Answers to PROBLEMS

1. a. 15 **b.** 58

EXAMPLE 2 **Expressions with grouping symbols and exponents**

Find the value of each expression.

a. $63 \div 7 - (2 + 3)$ **b.** $8 \div 2^3 + 3 - 1$

SOLUTION 2

a.

$$
\begin{aligned}
& 63 \div 7 - (2 + 3) \\
&= 63 \div 7 - 5 && \text{First do the operation inside the parentheses.} \\
&= 9 - 5 && \text{Next do the division.} \\
&= 4 && \text{Then do the subtraction.}
\end{aligned}
$$

b.

$$
\begin{aligned}
& 8 \div 2^3 + 3 - 1 \\
&= 8 \div 8 + 3 - 1 && \text{First do the exponentation.} \\
&= 1 + 3 - 1 && \text{Next do the division.} \\
&= 4 - 1 && \text{Then do the addition.} \\
&= 3 && \text{Do the final subtraction.}
\end{aligned}
$$

PROBLEM 2

Find the value of each expression.

a. $56 \div 8 - (3 + 1)$

b. $54 \div 3^3 + 5 - 2$

EXAMPLE 3 **Expression with grouping symbols**

Find the value of: $8 \div 4 \cdot 2 + 3(5 - 2) - 3 \cdot 2$

SOLUTION 3

$$
\begin{aligned}
& 8 \div 4 \cdot 2 + 3(5 - 2) - 3 \cdot 2 \\
&= 8 \div 4 \cdot 2 + 3(3) - 3 \cdot 2 && \text{First do the operations inside the parentheses. Now do the multiplications and divisions in order from left to right:} \\
&= 2 \cdot 2 + 3(3) - 3 \cdot 2 && \text{This means do } 8 \div 4 = 2 \text{ first.} \\
&= 4 + 3(3) - 3 \cdot 2 && \text{Then do } 2 \cdot 2 = 4. \\
&= 4 + 9 - 3 \cdot 2 && \text{Next do } 3(3) = 9. \\
&= 4 + 9 - 6 && \text{And finally, do } 3 \cdot 2 = 6. \\
&= 13 - 6 && \text{We are through with multiplications and divisions. Now do the addition.} \\
&= 7 && \text{The final operation is a subtraction.}
\end{aligned}
$$

PROBLEM 3

Find the value of:

$10 \div 5 \cdot 2 + 4(5 - 3) - 4 \cdot 2$

B › Evaluating Expressions with More than One Grouping Symbol

Suppose a local department store is having a great sale. The advertised items are such good deals that you decide to buy two bedspreads and two mattresses. The price of one bedspread and one mattress is \$14 + \$88. Thus, the price of two of each is $2 \cdot (14 + 88)$. If you then decide to buy a lamp, the total price is

$$[2 \cdot (14 + 88)] + 12$$

Answers to PROBLEMS

2. a. 3 **b.** 5 **3.** 4

We have used two types of grouping symbols in $[2 \cdot (14 + 88)] + 12$, parentheses ( ) and brackets [ ]. There is one more grouping symbol, braces { }. Typically, the order of these grouping symbols is {[( )]}. Here is the rule we use to handle grouping symbols.

RULE

Grouping Symbols

When grouping symbols occur within other grouping symbols, perform the computations in the *innermost* grouping symbols first.

Thus, to find the value of $[2 \cdot (14 + 88)] + 12$, we first add 14 and 88 (the operation inside parentheses, the innermost grouping symbols), then multiply by 2, and finally add 12. Here is the procedure:

$$[2 \cdot (14 + 88)] + 12 \quad \text{Given}$$
$$= [2 \cdot (102)] + 12 \quad \text{Add 14 and 88 inside the parentheses.}$$
$$= 204 + 12 \quad \text{Multiply 2 by 102 inside the brackets.}$$
$$= 216 \quad \text{Do the final addition.}$$

EXAMPLE 4 **Expressions with three grouping symbols**

Find the value of: $20 \div 4 + \{2 \cdot 9 - [3 + (6 - 2)]\}$

SOLUTION 4 The innermost grouping symbols are the parentheses, so we do the operations inside the parentheses, then inside the brackets, and, finally, inside the braces. Here are the details.

$$20 \div 4 + \{2 \cdot 9 - [3 + (6 - 2)]\} \quad \text{Given}$$
$$= 20 \div 4 + \{2 \cdot 9 - [3 + 4]\} \quad \text{Subtract inside the parentheses } (6 - 2 = 4).$$
$$= 20 \div 4 + \{2 \cdot 9 - 7\} \quad \text{Add inside the brackets } (3 + 4 = 7).$$
$$= 20 \div 4 + \{18 - 7\} \quad \text{Multiply inside the braces } (2 \cdot 9 = 18).$$
$$= 20 \div 4 + 11 \quad \text{Subtract inside the braces } (18 - 7 = 11).$$
$$= 5 + 11 \quad \text{Divide } (20 \div 4 = 5).$$
$$= 16 \quad \text{Do the final addition.}$$

PROBLEM 4

Find the value of:

$50 \div 5 + \{2 \cdot 7 - [4 + (5 - 2)]\}$

Answers to PROBLEMS

4. 17

Fraction bars are sometimes used as grouping symbols to indicate an expression representing a single number. To find the value of such expressions, simplify above and below the fraction bars following the order of operations. Thus,

$$\frac{2(3+8)+4}{2(4)-10}$$

$$=\frac{2(11)+4}{2(4)-10}$$ Add inside the parentheses in the numerator ($3+8=11$).

$$=\frac{22+4}{8-10}$$ Multiply in the numerator [$2(11)=22$] and in the denominator [$2(4)=8$].

$$=\frac{26}{-2}$$ Add in the numerator ($22+4=26$), subtract in the denominator ($8-10=-2$).

$$=-13$$ Do the final division. (Remember to use the rules of signs.)

EXAMPLE 5 Using a fraction bar as a grouping symbol

Find the value of:

$$-5^2+\frac{3(4-8)}{2}+10\div 5$$

SOLUTION 5

$$-5^2+\frac{3(4-8)}{2}+10\div 5$$ Given

$$=-5^2+\frac{3(-4)}{2}+10\div 5$$ Subtract inside the parentheses ($4-8=-4$).

$$=-25+\frac{3(-4)}{2}+10\div 5$$ Do the exponentiation ($5^2=25$, so $-5^2=-25$).

$$=-25+\frac{-12}{2}+10\div 5$$ Multiply above the division bar [$3(-4)=-12$].

$$=-25+(-6)+10\div 5$$ Divide ($\frac{-12}{2}=-6$).

$$=-25+(-6)+2$$ Divide ($10\div 5=2$).

$$=-31+2$$ Add [$-25+(-6)=-31$].

$$=-29$$ Do the final addition.

PROBLEM 5

Find the value of:

$$-3^2+\frac{2(3-6)}{3}+20\div 5$$

C > Applications Involving the Order of Operations

EXAMPLE 6 Offsetting your automobile carbon emissions

You can "offset" the carbon emissions from your car by planting trees (or let somebody else do it for you for a fee). How many trees? If you travel 12,000 miles a year, your car gets 30 miles per gallon, produces 20 pounds of CO_2 per mile traveled and each tree absorbs 50 pounds of CO_2 each year, the number of trees you need to plant is

$$\frac{\frac{12{,}000}{30}\cdot 20}{50}$$

Use the order of operations to simplify this expression.

PROBLEM 6

How many trees do you have to plant if you drive 15,000 miles a year, and your car gets 20 miles per gallon?

(continued)

Answers to PROBLEMS

5. −7 **6.** 300

SOLUTION 6

We simplify above the fraction bar using the order of operations and proceeding from left to right, so we do the division $\frac{12{,}000}{30}$ first!

$\dfrac{\frac{12{,}000}{30} \cdot 20}{50}$ Given

$= \dfrac{400 \cdot 20}{50}$ **D:** (Divide $\frac{12{,}000}{30} = 400$.)

$= \dfrac{8000}{50}$ **M:** (Multiply $400 \cdot 20 = 8000$.)

$= 160$ Do the final division $\frac{8000}{50} = 160$.

Here are two carbon calculators with the price for offsetting different carbon emissions: http://tinyurl.com/3x5tvl, http://tinyurl.com/6498zd. Results may be different due to different assumptions.

Calculator Corner

Order of Operations

How does a calculator handle the order of operations? Most calculators perform the order of operations *automatically*. For example, $3 + 4 \cdot 5 = 3 + 20 = 23$; to do this with a calculator, we enter the expression by pressing 3 + 4 × 5 ENTER. The result is shown in Window 1.

To do Example 2(b), you have to know how to enter exponents in your calculator. Many calculators use the ^ key followed by the exponent. Thus, to enter $8 \div 2^3 + 3 - 1$, press 8 ÷ 2 ^ 3 + 3 − 1 ENTER. The result, 3, is shown in Window 1. Note that the calculator follows the order of operations *automatically*.

```
3+4*5
                23
8/2^3+3-1
                 3
```

Window 1

Finally, you must be extremely careful when you evaluate expressions with division bars such as

$$\frac{2(3 + 8) + 4}{2(4) - 10}$$

When fraction bars are used as grouping symbols, they must be entered as sets of parentheses. Thus, you must enter (2 (3 + 8) + 4) ÷ (2 × 4 − 1 0) ENTER to obtain −13 (see Window 2). Note that we didn't use any extra parentheses to enter 2(4).

```
(2(3+8)+4)/(2*4-
10)
               -13
```

Window 2

Exercises 1.5

⟨A⟩ Using the Order of Operations In Problems 1–20, find the value of the given expression.

1. $4 \cdot 5 + 6$

2. $3 \cdot 4 + 6$

3. $7 + 3 \cdot 2$

4. $6 + 9 \cdot 2$

5. $7 \cdot 8 - 3$

6. $6 \cdot 4 - 9$

7. $20 - 3 \cdot 5$

8. $30 - 6 \cdot 5$

9. $48 \div 6 - (3 + 2)$

10. $81 \div 9 - (4 + 5)$

11. $3 \cdot 4 \div 2 + (6 - 2)$

12. $3 \cdot 6 \div 2 + (5 - 2)$

13. $36 \div 3^2 + 4 - 1$

14. $16 \div 2^3 + 3 - 2$

15. $8 \div 2^3 - 3 + 5$

16. $9 \div 3^2 - 8 + 5$

17. $10 \div 5 \cdot 2 + 8 \cdot (6 - 4) - 3 \cdot 4$

18. $15 \div 3 \cdot 3 + 2 \cdot (5 - 2) + 8 \div 4$

19. $4 \cdot 8 \div 2 - 3(4 - 1) + 9 \div 3$

20. $6 \cdot 3 \div 3 - 2(3 - 2) - 8 \div 2$

⟨ B ⟩ Evaluating Expressions with More than One Grouping Symbol In Problems 21–40, find the value of the given expression.

21. $20 \div 5 + \{3 \cdot 4 - [4 + (5 - 3)]\}$

22. $30 \div 6 + \{4 \div 2 \cdot 3 - [3 + (5 - 4)]\}$

23. $(20 - 15) \cdot [20 \div 2 - (2 \cdot 2 + 2)]$

24. $(30 - 10) \cdot [52 \div 4 - (3 \cdot 3 + 3)]$

25. $\{4 \div 2 \cdot 6 - (3 + 2 \cdot 3) + [5(3 + 2) - 1]\}$

26. $-6^2 + \dfrac{4(6 - 8)}{2} + 15 \div 3$

27. $-7^2 + \dfrac{3(8 - 4)}{4} + 10 \div 2 \cdot 3$

28. $-5^2 + \dfrac{6(3 - 7)}{4} + 9 \div 3 \cdot 2$

29. $(-6)^2 \cdot 4 \div 4 - \dfrac{3(7 - 9)}{2} - 4 \cdot 3 \div 2^2$

30. $(-4)^2 \cdot 3 \div 8 - \dfrac{4(6 - 10)}{2} - 3 \cdot 8 \div 2^3$

31. $\left[\dfrac{7 - (-3)}{8 - 6}\right]\left[\dfrac{3 + (-8)}{7 - 2}\right]$

32. $\left[\dfrac{4 - (-3)}{8 - 1}\right]\left[\dfrac{-4 - (-9)}{8 - 3}\right]$

33. $\dfrac{(-3)(-2)(-4)}{3(-2)} - (-3)^2$

34. $\dfrac{(-2)(-4)(-5)}{10(-2)} - (-4)^2$

35. $\dfrac{(-10)(-6)}{(-2)(-3)(-5)} - (-3)^3$

36. $\dfrac{(-8)(-9)}{(-3)(-4)(-2)} - (-2)^3$

37. $\dfrac{(-2)^3(-3)(-9)}{(-2)^2(-3)^2} - 3^2$

38. $\dfrac{(-2)(-3)^2(-10)}{(-2)^2(-3)} - 3^2$

39. $-5^2 - \dfrac{(-2)(-3)(-4)}{(-12)(-2)}$

40. $-6^2 - \dfrac{(-3)(-5)(-8)}{(-3)(-4)}$

⟨ C ⟩ Applications Involving the Order of Operations

41. *Gasoline octane rating* Have you noticed the octane rating of gasoline at the gas pump? This octane rating is given by the equation

$$\frac{R + M}{2}$$

where R is a number measuring the performance of gasoline using the Research Method and M is a number measuring the performance of gasoline using the Motor Method. If a certain gasoline has $R = 92$ and $M = 82$, what is its octane rating?

42. *Gasoline octane rating* If a gasoline has $R = 97$ and $M = 89$, what is its octane rating?

43. *Exercise pulse rate* If A is your age, the minimum pulse rate you should maintain during aerobic activities is $0.72(220 - A)$. What is the minimum pulse rate you should maintain if you are the specified age?

a. 20 years old **b.** 45 years old

44. *Exercise pulse rate* If A is your age, the maximum pulse rate you should maintain during aerobic activities is $0.88(220 - A)$. What is the maximum pulse rate you should maintain if you are the specified age?

a. 20 years old **b.** 45 years old

45. *Weight* Your weight depends on many factors like gender and height. For example, if you are a woman more than 5 ft (60 inches) tall, your weight W (in pounds) should be

$W = 105 + 5(h - 60)$, where h is your height in inches

a. What should the weight of a woman measuring 62 inches be?

b. What should the weight of a woman measuring 65 inches be?

46. *Weight for a man* A man of medium frame measuring h inches should weigh

$W = 140 + 3(h - 62)$ pounds

a. What should the weight of a man measuring 72 inches be?

b. What should the weight of a man measuring 68 inches be?

47. *Cell phone rates* At the present time, the Verizon America Choice 900 plan costs \$59.99 per month and gives you 900 anytime minutes with unlimited night and weekend minutes. After 900 minutes, you pay \$0.40 per minute. The monthly cost C is

$C = \$59.99 + 0.40(m - 900)$, where m is the number of minutes used

a. Find the monthly cost for a talker that used 1000 minutes.

b. Find the monthly cost for a talker that used 945 minutes.

48. *Cell phone rates* The \$39.99 Generic 450 plan provides 450 anytime minutes. After 450 minutes, you pay \$0.45 per minute. The monthly cost C is

$C = \$39.99 + 0.45(m - 450)$, where m is the number of minutes

a. Find the monthly cost for a talker that used 500 minutes.

b. Find the monthly cost for a talker that used 645 minutes.

Web IT go to mhhe.com/bello for more lessons

Applications: Green Math

Planting trees to offset car carbon emissions In Problems 49 and 50 follow the procedure of Example 6 to find the number of trees needed to offset the carbon emissions produced under the given conditions.

	Miles Driven	MPG	CO_2 per Gallon	CO_2 Absorbed by One Tree
49.	10,000	25	20 pounds	50 pounds per year
50.	15,000	30	20 pounds	50 pounds per year

Using Your Knowledge

Children's Dosages What is the corresponding dose (amount) of medication for children when the adult dosage is known? There are several formulas that tell us.

51. *Fried's rule* (for children under 2 years):

(Age in months · Adult dose) ÷ 150 = Child's dose

Suppose a child is 10 months old and the adult dose of aspirin is a 75-milligram tablet. What is the child's dose? [*Hint:* Simplify (10 · 75) ÷ 150.]

52. *Clark's rule* (for children over 2 years):

(Weight of child · Adult dose) ÷ 150 = Child's dose

If a 7-year-old child weighs 75 pounds and the adult dose is 4 tablets a day, what is the child's dose? [*Hint:* Simplify (75 · 4) ÷ 150.]

53. *Young's rule* (for children between 3 and 12):

(Age · Adult dose) ÷ (Age + 12) = Child's dose

Suppose a child is 6 years old and the adult dose of an antibiotic is 4 tablets every 12 hours. What is the child's dose? [*Hint:* Simplify (6 · 4) ÷ (6 + 12).]

Write On

54. State the rules for the order of operations, and explain why they are needed.

55. Explain whether the parentheses are needed when finding

a. 2 + (3 · 4) **b.** 2 · (3 + 4)

56. a. When evaluating an expression, do you *always* have to do multiplications before divisions? Give examples to support your answer.

b. When evaluating an expression, do you *always* have to do additions before subtractions? Give examples to support your answer.

Concept Checker

Fill in the blank(s) with the correct word(s), phrase, or mathematical statement.

57. The **acronym** (abbreviation) to remember the **order of operations** is ________________.

58. In the acronym of Problem 57, the **P** means ________________.

59. In the acronym of Problem 57, the **E** means ________________.

60. In the acronym of Problem 57, the **M** means ________________.

61. In the acronym of Problem 57, the **D** means ________________.

62. In the acronym of Problem 57, the **A** means ________________.

63. In the acronym of Problem 57, the **S** means ________________.

64. When **grouping symbols** occur within **other grouping symbols,** perform the computations in the ________________ grouping symbols **first.**

multiplication	**innermost**
division	**outside**
addition	**exponents**
subtraction	**parentheses**
PEMDAS	

Mastery Test

Find the value of the expressions.

65. $63 \div 9 - (2 + 5)$

66. $-64 \div 8 - (6 - 2)$

67. $16 + 2^3 + 3 - 9$

68. $3 \cdot 4 - 18$

69. $18 + 4 \cdot 5$

70. $12 \div 4 \cdot 2 + 2(5 - 3) - 3 \cdot 4$

71. $15 \div 3 + \{2 \cdot 4 - [6 + (2 - 5)]\}$

72. $-6^2 + \frac{4(6 - 12)}{3} + 8 \div 2$

73. The ideal heart rate while exercising for a person A years old is $[(205 - A) \cdot 7] \div 10$. What is the ideal heart rate for a 35-year-old person?

Skill Checker

Perform the indicated operations.

74. $\frac{1}{9} \cdot 9$

75. $11 \cdot \frac{1}{11}$

76. $17 \cdot \frac{1}{17}$

77. $-2.3 + 2.3$

78. $-1.7 + 1.7$

79. $-2.5 \cdot \frac{1}{-2.5}$

80. $-3.7 \cdot \frac{1}{-3.7}$

1.6 Properties of the Real Numbers

Objectives

A Identify which of the properties (associative or commutative) are used in a statement.

B Use the commutative and associative properties to simplify expressions.

C Identify which of the properties (identity or inverse) are used in a statement.

D Use the properties to simplify expressions.

E Use the distributive property to remove parentheses in an expression.

To Succeed, Review How To . . .

Add, subtract, multiply, and divide real numbers (pp. 52–56, 61–64).

Getting Started

Clarifying Statements

Look at the sign we found hanging outside a lawn mower repair shop. What does it mean? Do the owners want

a small (engine mechanic)?

or

a (small engine) mechanic?

The second meaning is the intended one. Do you see why? The manner in which you *associate* the words makes a difference. Now use parentheses to show how you think the following words should be associated:

Guaranteed used cars

Huge tire sale

If you wrote

Guaranteed (used cars) and Huge (tire sale)

you are well on your way to understanding the associative property. In algebra, if we are multiplying or adding, the way in which we associate (combine) the numbers makes *no* difference in the answer. This is the associative property, and we shall study it and several other properties of real numbers in this section.

A > Identifying the Associative and Commutative Properties

How would you add 17 + 98 + 2? You would probably add 98 and 2, get 100, and then add 17 to obtain 117. Even though the order of operations tells us to add from left to right, we can group the *addends* (the numbers we are adding) any way we want without changing the sum. Thus, (17 + 98) + 2 = 17 + (98 + 2). This fact can be stated as follows.

ASSOCIATIVE PROPERTY OF ADDITION

For any real numbers *a*, *b*, and *c*, changing the *grouping* of two addends does not change the sum. In symbols,

$$a + (b + c) = (a + b) + c$$

The associative property of addition tells us that the *grouping* does not matter in addition.

What about multiplication? Does the grouping matter? For example, is 2 · (3 · 4) the same as (2 · 3) · 4? Since both calculations give 24, the manner in which we group these numbers doesn't matter either. This fact can be stated as follows.

ASSOCIATIVE PROPERTY OF MULTIPLICATION

For any real numbers *a*, *b*, and *c*, changing the *grouping* of two factors does not change the product. In symbols,

$$a \cdot (b \cdot c) = (a \cdot b) \cdot c$$

The associative property of multiplication tells us that the *grouping* does not matter in multiplication.

We have now seen that grouping numbers differently in addition and multiplication yields the same answer. What about order? As it turns out, the order in which we do additions or multiplications doesn't matter either. For example, 2 + 3 = 3 + 2 and 5 · 4 = 4 · 5. In general, we have the following properties.

COMMUTATIVE PROPERTY OF ADDITION

For any numbers *a* and *b*, changing the *order* of two addends does not change the sum. In symbols,

$$a + b = b + a$$

COMMUTATIVE PROPERTY OF MULTIPLICATION

For any numbers *a* and *b*, changing the *order* of two factors does not change the product. In symbols,

$$a \cdot b = b \cdot a$$

The commutative properties of addition and multiplication tell us that *order* doesn't matter in addition or multiplication.

EXAMPLE 1 Identifying properties

Name the property illustrated in each of the following statements:

a. $a \cdot (b \cdot c) = (b \cdot c) \cdot a$

b. $(5 + 9) + 2 = 5 + (9 + 2)$

c. $(5 + 9) + 2 = (9 + 5) + 2$

d. $(6 \cdot 2) \cdot 3 = 6 \cdot (2 \cdot 3)$

PROBLEM 1

Name the property illustrated in each of the following statements:

a. $5 \cdot (4 \cdot 6) = (5 \cdot 4) \cdot 6$

b. $(x \cdot y) \cdot z = z \cdot (x \cdot y)$

c. $(4 + 8) + 7 = 4 + (8 + 7)$

d. $(4 + 3) + 5 = (3 + 4) + 5$

Answers to PROBLEMS

1. a. Associative prop. of mult. **b.** Commutative prop. of mult. **c.** Associative prop. of addition **d.** Commutative prop. of addition

SOLUTION 1

a. We changed the *order* of multiplication. The commutative property of multiplication was used.

b. We changed the *grouping* of the numbers. The associative property of addition was used.

c. We changed the *order* of the 5 and the 9 within the parentheses. The commutative property of addition was used.

d. Here we changed the *grouping* of the numbers. We used the associative property of multiplication.

EXAMPLE 2 Using the properties

Use the correct property to complete the following statements:

a. $-7 + (3 + 2.5) = (-7 + \_\_\_\_) + 2.5$

b. $4 \cdot \frac{1}{8} = \frac{1}{8} \cdot \_\_\_\_$

c. $\_\_\_\_ + \frac{1}{4} = \frac{1}{4} + 1.5$

d. $\left(\frac{1}{7} \cdot 3\right) \cdot 2 = \frac{1}{7} \cdot (3 \cdot \_\_\_\_)$

SOLUTION 2

a. $-7 + (3 + 2.5) = (-7 + \underline{\;3\;}) + 2.5$ Associative property of addition

b. $4 \cdot \frac{1}{8} = \frac{1}{8} \cdot \underline{\;4\;}$ Commutative property of multiplication

c. $\underline{1.5} + \frac{1}{4} = \frac{1}{4} + 1.5$ Commutative property of addition

d. $\left(\frac{1}{7} \cdot 3\right) \cdot 2 = \frac{1}{7} \cdot (3 \cdot \underline{\;2\;})$ Associative property of multiplication

PROBLEM 2

Use the correct property to complete the statement:

a. $5 \cdot \frac{1}{3} = \frac{1}{3} \cdot \_\_\_\_$

b. $\frac{1}{3} \cdot (4 \cdot \_\_\_\_) = \left(\frac{1}{3} \cdot 4\right) \cdot 7$

c. $\_\_\_\_ + \frac{1}{5} = \frac{1}{5} + \frac{1}{3}$

d. $(-8 + \_\_\_\_) + 3.1 = -8 + (4 + 3.1)$

B › Using the Associative and Commutative Properties to Simplify Expressions

The associative and commutative properties are also used to simplify expressions. For example, suppose x is a number: $(7 + x) + 8$ can be simplified by adding the numbers together as follows.

$$(7 + x) + 8 = 8 + (7 + x)$$ Commutative property of addition

$$= (8 + 7) + x$$ Associative property of addition

$$= 15 + x$$ You can also write the answer as $x + 15$ using the commutative property of addition.

Of course, you normally just add the 7 and 8 without going through all these steps, but this example shows *why* and *how* your shortcut can be done.

EXAMPLE 3 Using the properties to simplify expressions

Simplify: $6 + 4x + 8$

SOLUTION 3

$$6 + 4x + 8 = (6 + 4x) + 8$$ Order of operations

$$= 8 + (6 + 4x)$$ Commutative property of addition

$$= (8 + 6) + 4x$$ Associative property of addition

$$= 14 + 4x$$ Add (parentheses aren't needed).

You can also write the answer as $4x + 14$ using the commutative property of addition.

PROBLEM 3

Simplify: $3 + 5x + 8$

Answers to PROBLEMS

2. a. 5 **b.** 7 **c.** $\frac{1}{3}$ **d.** 4 **3.** $11 + 5x$ or $5x + 11$

C › Identifying Identities and Inverses

The properties we've just mentioned are applicable to all real numbers. We now want to discuss two special numbers that have unique properties, the numbers 0 and 1. If we add 0 to a number, the number is unchanged; that is, the number 0 preserves the **identity** of all numbers under addition. Thus, 0 is called the *identity element for addition.* This definition can be stated as follows.

IDENTITY ELEMENT FOR ADDITION

Zero is the **identity element for addition;** that is, for any number *a*,

$$a + 0 = 0 + a = a$$

The number 1 preserves the **identity** of all numbers under multiplication; that is, if we multiply a number by 1, the number remains unchanged. Thus, 1 is called the *identity element for multiplication,* as stated here.

IDENTITY ELEMENT FOR MULTIPLICATION

The number 1 is the **identity element for multiplication;** that is, for any number *a*,

$$a \cdot 1 = 1 \cdot a = a$$

We complete our list of properties by stating two ideas that we will discuss fully later.

ADDITIVE INVERSE (OPPOSITE)

For every number *a*, there exists another number $-a$ called its **additive inverse** (or **opposite**) such that $a + (-a) = 0$.

When dealing with multiplication, the *multiplicative inverse* is called the *reciprocal.*

MULTIPLICATIVE INVERSE (RECIPROCAL)

For every number *a* (except 0), there exists another number $\frac{1}{a}$ called the **multiplicative inverse** (or **reciprocal**) such that $a \cdot \frac{1}{a} = 1$

EXAMPLE 4 Identifying properties

Name the property illustrated in each of the following statements:

a. $5 + (-5) = 0$ **b.** $\frac{8}{3} \cdot \frac{3}{8} = 1$ **c.** $0 + 9 = 9$ **d.** $7 \cdot 1 = 7$

SOLUTION 4

a. $5 + (-5) = 0$ Additive inverse

b. $\frac{8}{3} \cdot \frac{3}{8} = 1$ Multiplicative inverse

c. $0 + 9 = 9$ Identity element for addition

d. $7 \cdot 1 = 7$ Identity element for multiplication

PROBLEM 4

Name the property illustrated:

a. $8 \cdot 1 = 8$ **b.** $\frac{7}{5} \cdot \frac{5}{7} = 1$

c. $3 + 0 = 3$ **d.** $(-4) + 4 = 0$

Answers to PROBLEMS

4. a. Identity element for mult. **b.** Multiplicative inverse **c.** Identity element for add. **d.** Additive inverse

EXAMPLE 5 **Using the properties**

Fill in the blank so that the result is a true statement:

a. ___ $\cdot$ 0.5 = 0.5 b. $\frac{3}{4}$ + ___ = 0 c. ___ + 2.5 = 0

d. ___ $\cdot \frac{3}{5} = 1$ e. 1.8 $\cdot$ ___ = 1.8 f. $\frac{3}{4}$ + ___ = $\frac{3}{4}$

SOLUTION 5

a. $\underline{1} \cdot 0.5 = 0.5$ Identity element for multiplication

b. $\frac{3}{4} + \underline{\left(-\frac{3}{4}\right)} = 0$ Additive inverse property

c. $\underline{-2.5} + 2.5 = 0$ Additive inverse property

d. $\underline{\frac{5}{3}} \cdot \frac{3}{5} = 1$ Multiplicative inverse property

e. $1.8 \cdot \underline{1} = 1.8$ Identity element for multiplication

f. $\frac{3}{4} + \underline{0} = \frac{3}{4}$ Identity element for addition

PROBLEM 5

Fill in the blank so that the result is a true statement:

a. $\frac{7}{3}$ + ___ = $\frac{7}{3}$

b. $\frac{4}{9} \cdot$ ___ = 1

c. $\frac{5}{8}$ + ___ = 0

d. 2.4 $\cdot$ ___ = 2.4

e. 9.1 + ___ = 0

f. 0.2 $\cdot$ ___ = 0.2

D › Using the Identity and Inverse Properties to Simplify Expressions

EXAMPLE 6 **Using the properties of addition**

Use the properties of addition to simplify: $-3x + 5 + 3x$

SOLUTION 6

$$\begin{aligned} -3x + 5 + 3x &= (-3x + 5) + 3x && \text{Order of operations} \\ &= 3x + (-3x + 5) && \text{Commutative property of addition} \\ &= [3x + (-3x)] + 5 && \text{Associative property of addition} \\ &= 0 + 5 && \text{Additive inverse property} \\ &= 5 && \text{Identity element for addition} \end{aligned}$$

PROBLEM 6

Use the addition properties to simplify: $-5x + 7 + 5x$

Keep in mind that after you get enough practice, you won't need to go through all the steps we show here. We have given you all the steps to make sure you understand *why* and *how* we simplify expressions.

E › Using the Distributive Property to Remove Parentheses

The properties we've discussed all contain a single operation. Now suppose you wish to multiply a number, say 7, by the sum of 4 and 5. As it turns out, $7 \cdot (4 + 5)$ can be obtained in two ways:

$7 \cdot (4 + 5)$

$7 \cdot 9$ Add within the parentheses first.

63

Answers to PROBLEMS

5. a. 0 **b.** $\frac{9}{4}$ **c.** $-\frac{5}{8}$ **d.** 1 **e.** -9.1 **f.** 1 **6.** 7

or

$$\underbrace{\underbrace{(7 \cdot 4)}_{28} + \underbrace{(7 \cdot 5)}_{35}}_{63}$$ Multiply and then add.

Thus,

$$7 \cdot (4 + 5) = (7 \cdot 4) + (7 \cdot 5)$$ *First* multiply 4 by 7 and *then* 5 by 7.

The parentheses in $(7 \cdot 4) + (7 \cdot 5)$ can be omitted since, by the order of operations, multiplications must be done before addition. Thus, multiplication distributes over addition as follows:

DISTRIBUTIVE PROPERTY

For any numbers *a*, *b*, and *c*,

$$a(b + c) = ab + ac$$

Note that $a(b + c)$ means $a \cdot (b + c)$, ab means $a \cdot b$, and ac means $a \cdot c$. The distributive property can also be extended to more than two numbers inside the parentheses. Thus, $a(b + c + d) = ab + ac + ad$. Moreover, $(b + c)a = ba + ca$.

EXAMPLE 7 Using the distributive property

Use the distributive property to multiply the following (x, y, and z are real numbers):

a. $8(2 + 4)$ **b.** $3(x + 5)$ **c.** $3(x + y + z)$ **d.** $(x + y)z$

SOLUTION 7

a. $8(2 + 4) = 8 \cdot 2 + 8 \cdot 4$

$= 16 + 32$ Multiply first.

$= 48$ Add next.

b. $3(x + 5) = 3x + 3 \cdot 5$

$= 3x + 15$ This expression cannot be simplified further since we don't know the value of x.

c. $3(x + y + z) = 3x + 3y + 3z$

d. $(x + y)z = xz + yz$

PROBLEM 7

Use the distributive property to multiply:

a. $(a + b)c$ **b.** $5(a + b + c)$

c. $4(a + 7)$ **d.** $3(2 + 9)$

EXAMPLE 8 Using the distributive property

Use the distributive property to multiply:

a. $-2(a + 7)$ **b.** $-3(x - 2)$ **c.** $-(a - 2)$

SOLUTION 8

a. $-2(a + 7) = -2a + (-2)7$

$= -2a - 14$

b. $-3(x - 2) = -3x + (-3)(-2)$

$= -3x + 6$

c. What does $-(a - 2)$ mean? Since

$$-a = -1 \cdot a$$

we have

$$-(a - 2) = -1 \cdot (a - 2)$$
$$= -1 \cdot a + (-1) \cdot (-2)$$
$$= -a + 2$$

PROBLEM 8

Use the distributive property to multiply:

a. $-3(x + 5)$

b. $-5(a - 2)$

c. $-(x - 4)$

Answers to PROBLEMS

7. a. $ac + bc$ **b.** $5a + 5b + 5c$ **c.** $4a + 28$ **d.** 33

8. a. $-3x - 15$ **b.** $-5a + 10$ **c.** $-x + 4$

The distributive property can be used when working with environmental problems such as pollution and population. It is estimated that each of the 310 million people in the United States produces about 4.5 pounds of solid waste (garbage) each and every day. However, the U.S. population of 310 million is increasing by about 2.8 million people every year, so the amount of garbage produced is also increasing. Multiply the amount of garbage generated by each person (4.5 pounds) by the U.S. population $(2.8x + 310)$ million, where x is the number of years after 2010, and we can find the amount of garbage generated each and every day! We will see how much garbage that is in Example 9.

Source: Environmental Protection Agency (EPA), U.S. Census.

EXAMPLE 9 Garbage generated each day in the United States

The amount of garbage generated each day in the United States can be approximated by $4.5(2.8x + 310)$ million pounds, where x is the number of years after 2010.

a. Multiply $4.5(2.8x + 310)$

b. How many pounds of garbage were produced in 2010 $(x = 0)$?

c. How many pounds of garbage would be produced in 2020 $(x = 10)$ according to the model?

SOLUTION 9

a. $4.5(2.8x + 310) = 4.5 \cdot 2.8x + 4.5 \cdot 310$
$= 12.6x + 1395$

This means that $12.6x + 1395$ million pounds of garbage were produced daily!

b. For 2010, $x = 0$ and $12.6x + 1395$ becomes $12.6(0) + 1395 = 1395$.

This means that in 2010 the amount of garbage produced daily was 1395 million pounds!

c. For 2020, $x = 10$ and $12.6x + 1395 = 12.6(10) + 1395 = 126 + 1395$ or 1521 million pounds. This means that in 2020 the amount of garbage produced daily would be 1521 million pounds.

Some people claim that the amount of garbage produced each year is equivalent to burying more than 82,000 football fields 6 feet deep in compacted garbage and that the garbage will fill enough trucks to form a line to the moon.

Here is the source: http://tinyurl.com/yuqlun. You do the math!

PROBLEM 9

A different study estimates that each person in the United States produces about 5.1 pounds of garbage each day.

a. Multiply $5.1(2.8x + 310)$.

b. Use the 5.1 pound estimate to find the amount of garbage produced in 2010 $(x = 0)$.

c. Using the 5.1 pound estimate, how much garbage would be produced in 2020?

Finally, for easy reference, here is a list of all the properties we've studied.

PROPERTIES OF THE REAL NUMBERS

If *a*, *b*, and *c* are real numbers, then the following properties hold.

Addition	Multiplication	Property
$a + b = b + a$	$a \cdot b = b \cdot a$	Commutative property
$a + (b + c) = (a + b) + c$	$a \cdot (b \cdot c) = (a \cdot b) \cdot c$	Associative property
$a + 0 = 0 + a = a$	$a \cdot 1 = 1 \cdot a = a$	Identity property
$a + (-a) = 0$	$a \cdot \frac{1}{a} = 1\ (a \neq 0)$	Inverse property
$a(b + c) = ab + ac$		Distributive property
$(a + b)c = ac + bc$		

Answers to PROBLEMS

9. a. $14.28x + 1581$ **b.** 1581 million pounds **c.** 1723.8 million pounds

CAUTION

The commutative and associative properties apply to addition and multiplication but **not** to subtraction and division. For example,

$$3 - 5 \neq 5 - 3 \quad \text{and} \quad 6 \div 2 \neq 2 \div 6$$

Also,

$$5 - (4 - 2) \neq (5 - 4) - 2 \quad \text{and} \quad 6 \div (3 \div 3) \neq (6 \div 3) \div 3$$

Exercises 1.6

Web IT go to mhhe.com/bello for more lessons

⟨A⟩ Identifying the Associative and Commutative Properties In Problems 1–10, name the property illustrated in each statement.

1. $9 + 8 = 8 + 9$

2. $b \cdot a = a \cdot b$

3. $4 \cdot 3 = 3 \cdot 4$

4. $(a + 4) + b = a + (4 + b)$

5. $3 + (x + 6) = (3 + x) + 6$

6. $8 \cdot (2 \cdot x) = (8 \cdot 2) \cdot x$

7. $a \cdot (b \cdot c) = a \cdot (c \cdot b)$

8. $a \cdot (b \cdot c) = (a \cdot b) \cdot c$

9. $a + (b + 3) = (a + b) + 3$

10. $(a + 3) + b = (3 + a) + b$

⟨B⟩ Using the Associative and Commutative Properties to Simplify Expressions In Problems 11–18, name the property illustrated in each statement, then find the missing number that makes the statement correct.

11. $-3 + (5 + 1.5) = (-3 + \_\_\_) + 1.5$

12. $(-4 + \_\_\_) + \frac{1}{5} = -4 + \left(3 + \frac{1}{5}\right)$

13. $7 \cdot \frac{1}{8} = \frac{1}{8} \cdot \_\_\_$

14. $\frac{9}{4} \cdot \_\_\_ = 3 \cdot \frac{9}{4}$

15. $\_\_\_ + \frac{3}{4} = \frac{3}{4} + 6.5$

16. $7.5 + \_\_\_ = \frac{5}{8} + 7.5$

17. $\left(\frac{3}{7} \cdot 5\right) \cdot \_\_\_ = \frac{3}{7} \cdot (5 \cdot 2)$

18. $\left(\frac{2}{9} \cdot \_\_\_\right) \cdot 4 = \frac{2}{9} \cdot (2 \cdot 4)$

In Problems 19–20, simplify.

19. $5 + 2x + 8$

20. $-10 + 2y + 12$

⟨C⟩ Identifying Identities and Inverses In Problems 21–24, name the property illustrated in each statement.

21. $9 \cdot \frac{1}{9} = 1$

22. $10 \cdot 1 = 10$

23. $8 + 0 = 8$

24. $-6 + 6 = 0$

⟨D⟩ Using the Identity and Inverse Properties to Simplify Expressions In Problems 25–28, simplify.

25. $\frac{1}{5}a - 3 + 3 - \frac{1}{5}a$

26. $\frac{2}{3}b + 4 - 4 - \frac{2}{3}b$

27. $5c + (-5c) + \frac{1}{5} \cdot 5$

28. $-\frac{1}{7} \cdot 7 - 4x + 4x$

⟨E⟩ Using the Distributive Property to Remove Parentheses In Problems 29–38, use the distributive property to fill in the blank.

29. $8(9 + \_\_\_) = 8 \cdot 9 + 8 \cdot 3$

30. $7(\_\_\_ + 10) = 7 \cdot 4 + 7 \cdot 10$

31. $\_\_\_(3 + b) = 15 + 5b$

32. $3(\_\_\_\_ + b) = 3 \cdot 8 + 3b$

33. $\_\_\_\_(b + c) = ab + ac$

34. $3(x + 4) = 3x + 3 \cdot \_\_\_\_$

35. $-4(x + \_\_\_\_) = -4x + 8$

36. $-3(x + 5) = -3x + \_\_\_\_$

37. $-2(5 + c) = -2 \cdot 5 + \_\_\_\_ \cdot c$

38. $-3(\_\_\_\_ + a) = 6 + (-3)a$

In Problems 39–78, use the distributive property to multiply.

39. $6(4 + x)$
40. $3(2 + x)$
41. $8(x + y + z)$
42. $5(x + y + z)$
43. $6(x + 7)$
44. $7(x + 2)$
45. $(a + 5)b$
46. $(a + 2)c$
47. $6(5 + b)$
48. $3(7 + b)$
49. $-4(x + y)$
50. $-3(a + b)$
51. $-9(a + b)$
52. $-6(x + y)$
53. $-3(4x + 2)$
54. $-2(3a + 9)$
55. $-\left(\frac{3a}{2} - \frac{6}{7}\right)$
56. $-\left(\frac{2x}{3} - \frac{1}{5}\right)$
57. $-(2x - 6y)$
58. $-(3a - 6b)$
59. $-(2.1 + 3y)$
60. $-(5.4 + 4b)$
61. $-4(a + 5)$
62. $-6(x + 8)$
63. $-x(6 + y)$
64. $-y(2x + 3)$
65. $-8(x - y)$
66. $-9(a - b)$
67. $-3(2a - 7b)$
68. $-4(3x - 9y)$
69. $0.5(x + y - 2)$
70. $0.8(a + b - 6)$
71. $\frac{6}{5}(a - b + 5)$
72. $\frac{2}{3}(x - y + 4)$
73. $-2(x - y + 4)$
74. $-4(a - b + 8)$
75. $-0.3(x + y - 6)$
76. $-0.2(a + b - 3)$
77. $-\frac{5}{2}(a - 2b + c - 1)$
78. $-\frac{4}{7}(2a - b + 3c - 5)$

Web IT go to mhhe.com/bello for more lessons

Applications: Green Math

Recycled, composted, and combusted garbage

79. Fortunately, not all garbage produced goes to the landfill. According to the EPA about 1.5 pounds of garbage is *recycled* and *composted* by each person in the United States each day. Thus, the total garbage recycled and composted each day is $1.5(2.8x + 310)$ million pounds, where x is the number of years after 2010.

a. Multiply $1.5(2.8x + 310)$.

b. How many pounds of garbage were composted and recycled each day in 2010 ($x = 0$)?

c. How many pounds of garbage would be composted and recycled each day in 2020 ($x = 10$)?

80. According to the EPA, subtracting out what we *recycled* and *composted,* we *combusted* or *discarded* 3 pounds per person per day.

a. Multiply $3(2.8x + 310)$

b. How many pounds of garbage were *combusted* or *discarded* each day in 2010 ($x = 0$)?

c. How many pounds of garbage would be *combusted* or *discarded* each day in 2020 ($x = 10$)?

81. According to the EPA (Problem 80), subtracting out what we *recycled* and *composted,* $1.5(2.8x + 310)$, from the total *garbage generated,* $4.5(2.8x + 310)$ will give the amount of garbage *combusted* or *discarded,* that is,

$$4.5(2.8x + 310) - 1.5(2.8x + 310) = 3(2.8x + 310)$$

Check this by multiplying $4.5(2.8x + 310)$, then subtracting $1.5(2.8x + 310)$ and simplifying. The result should be $8.4x + 930 = 3(2.8x + 310)$. You will learn more about simplifying expressions in the next section!

Applications

82. *Price of a car* The price P of a car is its base price (B) plus destination charges D, that is, $P = B + D$. Tran bought a Nissan in Smyrna, Tennessee, and there was no destination charge.

a. What is D?

b. Fill in the blank in the equation $P = B +$ ______

c. What property tells you that the equation in part **b** is correct?

83. *Area* The area of a rectangle is found by multiplying its length L times its width W.

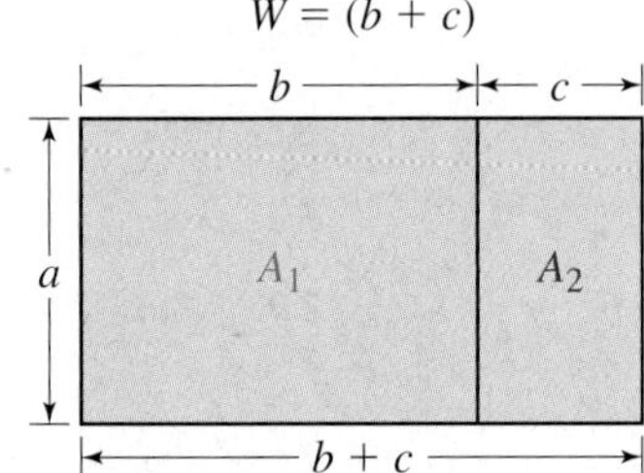

a. The length of the entire rectangle is a and its width is $(b + c)$. What is the area A of the rectangle?

b. The length of rectangle A_1 is a and its width b. What is the area of A_1?

c. The length of rectangle A_2 is a and its width c. What is the area of A_2?

d. The total area A of the rectangle is made up of the areas of rectangles A_1 and A_2, that is, $A = A_1 + A_2$ or $a(b + c) = ab + ac$, since $A = a(b + c)$, $A_1 = ab$ and $A_2 = ac$. Which property does this illustrate?

84. *Area* The length of the entire rectangle is a and its width is b.

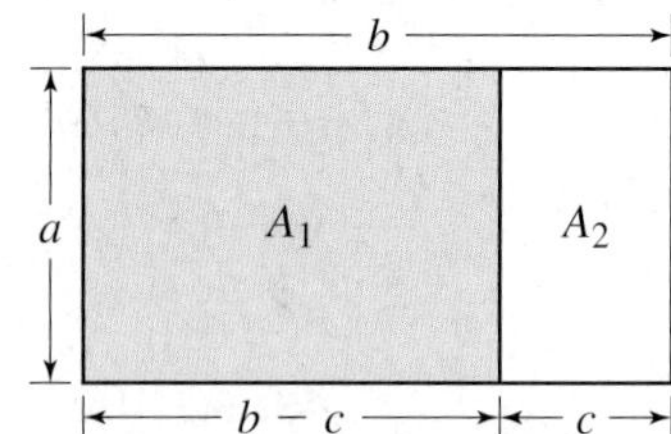

a. What is the area A of the entire rectangle?

b. What is the area of the smaller rectangle A_1?

c. The area of A_1 is the area A of the entire rectangle minus the area of A_2. Write an expression that models this situation.

d. Substitute the results obtained in **a** and **b** to rewrite the equation in part **c**.

85. *Weight* **a.** If you are a woman more than 5 feet (60 inches) tall, your weight W (in pounds) should be

$W = 105 + 5(h - 60)$, where h is your height in inches

Simplify this expression and find the weight of a woman 68 inches tall

b. The formula for the weight W of a man of medium frame measuring h inches is

$$W = 140 + 3(h - 62) \text{ pounds}$$

Simplify this expression and find the weight of a man 72 inches tall.

86. *Parking costs* **a.** The formula for the cost C for parking in the short-term garage at Tampa International airport is

$$C = 2.50(h - 1.5) + 1.25$$

Simplify this formula and find the cost C of parking for 4 hours.

b. The formula for the cost C in the long-term parking garage is

$$C = 1.25(h - 11) + 12.50$$

Simplify this formula and find the cost C of parking for 12 hours.

Using Your Knowledge

Multiplication Made Easy The distributive property can be used to simplify certain multiplications. For example, to multiply 8 by 43, we can write

$$\begin{aligned} 8(43) &= 8(40 + 3) \\ &= 320 + 24 \\ &= 344 \end{aligned}$$

Use this idea to multiply the numbers.

87. 7(38) **88.** 8(23) **89.** 6(46) **90.** 9(52)

Write On

91. Explain why zero has no reciprocal.

92. We have seen that addition is commutative since $a + b = b + a$. Is subtraction commutative? Explain.

93. The associative property of addition states that
$a + (b + c) = (a + b) + c$.
Is $a - (b - c) = (a - b) - c$? Explain.

94. Is $a + (b \cdot c) = (a + b)(a + c)$? Give examples to support your conclusion.

95. Multiplication is distributive over addition since $a \cdot (b + c) = a \cdot b + a \cdot c$. Is multiplication distributive over subtraction? Explain.

Concept Checker

Fill in the blank(s) with the correct word(s), phrase, or mathematical statement.

96. The **associative property of addition** states that for any numbers a, b, and c,

____________________.

97. The **associative property of multiplication** states that for any numbers a, b, and c,

________________.

98. The **commutative property of addition** states that for any numbers a and b

______________.

99. The **commutative property of multiplication** states that for any numbers a and b

______________.

100. The **identity element for addition** states that for any number ***a***, ________________.

101. The **identity element for multiplication** states that for any number ***a***, ________________.

102. The **additive inverse property** states that for any number ***a***, there is an additive inverse ____.

103. The **multiplicative inverse property** states that for any nonzero number ***a***, there is multiplicative inverse ____.

104. The **multiplicative inverse of *a*** is also called the __________ of ***a***.

105. The **distributive property** states that for any numbers a, b, and c, ________________.

$a(b + c) = ab + c$

$a(b + c) = ab + ac$

reciprocal

opposite

$\frac{1}{a}$

$-a$

$\frac{1}{a} \cdot a = a$

$a \cdot 1 = 1 \cdot a = a$

$a \cdot 0 = 0 \cdot a = a$

$a + 0 = 0 + a = a$

$a \cdot b = b \cdot a$

$a + b = b + a$

$a \cdot (b \cdot c) = (a \cdot b) \cdot c$

$a + (b + c) = (a + b) + c$

Mastery Test

Use the distributive property to multiply the expression.

106. $-3(a + 4)$

107. $-4(a - 5)$

108. $-(a - 2)$

109. $4(a + 6)$

110. $2(x + y + z)$

Simplify:

111. $-2x + 7 + 2x$

112. $-4 - 2x + 2x + 4$

113. $7 + 4x + 9$

114. $-3 - 4x + 2$

Fill in the blank so that the result is a true statement.

115. $\frac{2}{5} + \_\_\_ = \frac{2}{5}$

116. $\frac{1}{4} + \_\_\_ = 0$

117. $-3.2 \cdot \_\_\_ = -3.2$

118. $\_\_\_ + 4.2 = 4.2$

119. $3(x + \_\_\_) = 3x + 6$

120. $-2(x + 5) = -2x + \_\_\_$

Name the property illustrated in each of the following statements.

121. $3 \cdot 1 = 3$

122. $\frac{5}{6} \cdot \frac{6}{5} = 1$

123. $2 + 0 = 2$

124. $-3 + 3 = 0$

125. $(6 \cdot x) \cdot 2 = 6 \cdot (x \cdot 2)$

126. $3 \cdot (4 \cdot 5) = 3 \cdot (5 \cdot 4)$

127. $(x + 2) + 5 = x + (2 + 5)$

128. $-3 \cdot 1 = -3$

Name the property illustrated in each statement, then find the missing number that makes the statement correct.

129. $\left(\frac{1}{5} \cdot 4\right) \cdot 2 = \frac{1}{5} \cdot (4 \cdot \_\_\_)$

130. $\frac{1}{3} + 2.4 = \_\_\_ + \frac{1}{3}$

131. $\frac{2}{5} \cdot \_\_\_ = 1 \cdot \frac{2}{5}$

132. $\frac{2}{11} + \_\_\_ = 0$

133. $-2 + (3 + 1.4) = (-2 + \_\_\_) + 1.4$

Skill Checker

Find:

134. $-5 + 3$

135. $-2 + (-5)$

136. $7 + (-9)$

137. $2(x + 1)$

1.7 Simplifying Expressions

Objectives

A Add and subtract like terms.

B Use the distributive property to remove parentheses and then combine like terms.

C Translate words into algebraic expressions and solve applications.

To Succeed, Review How To . . .

1. Add and subtract real numbers (pp. 52–56).
2. Use the distributive property to simplify expressions (pp. 81–83).

Getting Started

Combining Like Terms

If 3 tacos cost \$1.39 and 6 tacos cost \$2.39, then 3 tacos + 6 tacos will cost \$1.39 + \$2.39; that is, 9 tacos will cost \$3.78. Note that

3 tacos + 6 tacos = 9 tacos and

\$1.39 + \$2.39 = \$3.78

A > Adding and Subtracting Like Terms

In algebra, an **expression** is a collection of numbers and letters representing numbers (**variables**) connected by operation signs. The parts to be added or subtracted in these expressions are called **terms.** Thus, xy^2 is an expression with one term; $x + y$ is an expression with two terms, x and y; and $3x^2 - 2y + z$ has three terms, $3x^2$, $-2y$, and z.

The term "3 tacos" uses the number 3 to tell "how many." The number 3 is called the **numerical coefficient** (or simply the **coefficient**) of the term. Similarly, the terms $5x$, y, and $-8xy$ have numerical coefficients of 5, 1, and -8, respectively.

When two or more terms are *exactly alike* (except possibly for their coefficients or the order in which the factors are multiplied), they are called **like terms.** Thus, like terms contain the *same variables* with the *same exponents,* but their coefficients may be different. So 3 tacos and 6 tacos are like terms, $3x$ and $-5x$ are like terms, and $-xy^2$ and $7xy^2$ are like terms. On the other hand, $2x$ and $2x^2$ or $2xy^2$ and $2x^2y$ are *not* like terms. In this section we learn how to simplify expressions using like terms.

In an algebraic expression, like terms can be combined into a single term just as 3 tacos and 6 tacos can be combined into the single term 9 tacos. To combine like terms, make sure that the variable parts of the terms to be combined are identical (only the coefficients may be different), and then add (or subtract) the coefficients and keep the variables. This can be done using the distributive property:

$$\underbrace{3x}_{(x+x+x)} + \underbrace{5x}_{(x+x+x+x+x)} = \underbrace{(3+5)x}_{x+x+x+x+x+x+x+x} = 8x$$

$$\underbrace{2x^2}_{(x^2+x^2)} + \underbrace{4x^2}_{(x^2+x^2+x^2+x^2)} = \underbrace{(2+4)x^2}_{x^2+x^2+x^2+x^2+x^2+x^2} = 6x^2$$

$$\underbrace{3ab}_{(ab+ab+ab)} + \underbrace{2ab}_{(ab+ab)} = \underbrace{(3+2)ab}_{ab+ab+ab+ab+ab} = 5ab$$

But what about another situation, such as combining $-3x$ and $-2x$? We first write

$$\underbrace{-3x}_{(-x)+(-x)+(-x)} + \underbrace{(-2x)}_{(-x)+(-x)}$$

Thus,

$$-3x + (-2x) = [-3 + (-2)]x = -5x$$

Note that we write the addition of $-3x$ and $-2x$ as $-3x + (-2x)$, using parentheses around the $-2x$. We do this to avoid the confusion of writing

$$-3x + -2x$$

Never use two signs of operation together without parentheses.

As you can see, if both quantities to be added are preceded by minus signs, the result is preceded by a minus sign. If they are both preceded by plus signs, the result is preceded by a plus sign. But what about this expression, $-5x + (3x)$? You can visualize this expression as

$$\underbrace{-5x}_{(-x)+(-x)+(-x)+(-x)+(-x)} + \underbrace{3x}_{x+x+x}$$

Since we have two more negative x's than positive x's, the result is $-2x$. Thus,

$$-5x + (3x) = (-5 + 3)x = -2x \quad \text{Use the distributive property.}$$

On the other hand, $5x + (-3x)$ can be visualized as

$$\underbrace{x + x + x + x + x}_{5x} + \underbrace{(-x) + (-x) + (-x)}_{(-3x)}$$

and the answer is $2x$; that is,

$$5x + (-3x) = [5 + (-3)]x = 2x \quad \text{Use the distributive property.}$$

Now here is an easy one: What is $x + x$? Since $1 \cdot x = x$, the coefficient of x is assumed to be 1. Thus,

$$\begin{aligned} x + x &= 1 \cdot x + 1 \cdot x \\ &= (1 + 1)x \\ &= 2x \end{aligned}$$

In all the examples that follow, make sure you use the fact that

$$ac + bc = (a + b)c$$

to combine the numerical coefficients.

EXAMPLE 1 Combining like terms using addition

Combine like terms:

a. $-7x + 2x$ **b.** $-4x + 6x$

c. $-2x + (-5x)$ **d.** $x + (-5x)$

SOLUTION 1

a. $-7x + 2x = (-7 + 2)x = -5x$

b. $-4x + 6x = (-4 + 6)x = 2x$

c. $-2x + (-5x) = [-2 + (-5)]x = -7x$

d. First, recall that $x = 1 \cdot x$. Thus,

$$\begin{aligned} x + (-5x) &= 1x + (-5x) \\ &= [1 + (-5)]x = -4x \end{aligned}$$

PROBLEM 1

Combine like terms:

a. $-9x + 3x$ **b.** $-6x + 8x$

c. $-3x + (-7x)$ **d.** $x + (-6x)$

Subtraction of like terms is defined in terms of addition, as stated here.

SUBTRACTION

To subtract a number b from another number a, add the *additive inverse* (*opposite*) of b to a; that is,

$$a - b = a + (-b)$$

As before, to subtract like terms, we use the fact that $ac + bc = (a + b)c$. Thus,

Additive inverse

$$\begin{aligned} 3x - 5x &= 3x + (-5x) = [3 + (-5)]x = -2x \\ -3x - 5x &= -3x + (-5x) = [(-3) + (-5)]x = -8x \\ 3x - (-5x) &= 3x + (5x) = (3 + 5)x = 8x \\ -3x - (-5x) &= -3x + (5x) = (-3 + 5)x = 2x \end{aligned}$$

Note that

$$-3x - (-5x) = -3x + 5x$$

Answers to PROBLEMS

1. a. $-6x$ **b.** $2x$

c. $-10x$ **d.** $-5x$

In general, we have the following:

PROCEDURE

Subtracting $-b$

$a - (-b) = a + b$ $\quad$ $-(-b)$ is replaced by $+b$, since $-(-b) = +b$.

We can now combine like terms involving subtraction.

EXAMPLE 2 **Combining like terms using subtraction**

Combine like terms:

a. $7ab - 9ab$ b. $8x^2 - 3x^2$

c. $-5ab^2 - (-8ab^2)$ d. $-6a^2b - (-2a^2b)$

SOLUTION 2

a. $7ab - 9ab = 7ab + (-9ab) = [7 + (-9)]ab = -2ab$

b. $8x^2 - 3x^2 = 8x^2 + (-3x^2) = [8 + (-3)]x^2 = 5x^2$

c. $-5ab^2 - (-8ab^2) = -5ab^2 + 8ab^2 = (-5 + 8)ab^2 = 3ab^2$

d. $-6a^2b - (-2a^2b) = -6a^2b + 2a^2b = (-6 + 2)a^2b = -4a^2b$

PROBLEM 2

Combine like terms:

a. $3xy - 8xy$

b. $7x^2 - 5x^2$

c. $-3xy^2 - (-7xy^2)$

d. $-5x^2y - (-4x^2y)$

B › Removing Parentheses

Sometimes it's necessary to remove parentheses before combining like terms. For example, to combine like terms in

$$(3x + 5) + (2x - 2)$$

we have to remove the parentheses *first*. If there is a plus sign (or no sign) in front of the parentheses, we can simply remove the parentheses; that is,

$$+(a + b) = a + b \quad \text{and} \quad +(a - b) = a - b$$

With this in mind, we have

$$\begin{aligned}(3x + 5) + (2x - 2) &= 3x + 5 + 2x - 2 \\ &= 3x + 2x + 5 - 2 && \text{Use the commutative property.} \\ &= (3x + 2x) + (5 - 2) && \text{Use the associative property.} \\ &= 5x + 3 && \text{Simplify.}\end{aligned}$$

Note that we used the properties we studied to group like terms together.

Once you understand the use of these properties, you can then see that the simplification of $(3x + 5) + (2x - 2)$ consists of just adding $3x$ to $2x$ and 5 to -2. The computation can then be shown like this:

Like terms

$$\begin{aligned}(3x + 5) + (2x - 2) &= 3x + 5 + 2x - 2 \\ &= 5x + 3\end{aligned}$$

Like terms

We use this idea in Example 3.

Answers to PROBLEMS

2. **a.** $-5xy$ **b.** $2x^2$ **c.** $4xy^2$ **d.** $-x^2y$

EXAMPLE 3 Removing parentheses preceded by a plus sign

Remove parentheses and combine like terms:

a. $(4x - 5) + (7x - 3)$ **b.** $(3a + 5b) + (4a - 9b)$

SOLUTION 3 We first remove parentheses; then we add like terms.

a. $(4x - 5) + (7x - 3) = 4x - 5 + 7x - 3$

$= 11x - 8$

b. $(3a + 5b) + (4a - 9b) = 3a + 5b + 4a - 9b$

$= 7a - 4b$

PROBLEM 3

Remove parentheses and combine like terms:

a. $(7a - 5) + (8a - 2)$

b. $(4x + 6y) + (3x - 8y)$

To simplify

$$4x + 3(x - 2) - (x + 5)$$

recall that $-(x + 5) = -1 \cdot (x + 5) = -x - 5$. We proceed as follows:

$4x + 3(x - 2) - (x + 5)$	Given
$= 4x + 3x - 6 - (x + 5)$	Use the distributive property.
$= 4x + 3x - 6 - x - 5$	Remove parentheses.
$= (4x + 3x - x) + (-6 - 5)$	Combine like terms.
$= 6x - 11$	Simplify.

EXAMPLE 4 Removing parentheses preceded by a minus sign

Remove parentheses and combine like terms:

$$8x - 2(x - 1) - (x + 3)$$

SOLUTION 4

$8x - 2(x - 1) - (x + 3)$	
$= 8x - 2x + 2 - (x + 3)$	Use the distributive property.
$= 8x - 2x + 2 - x - 3$	Remove parentheses.
$= (8x - 2x - x) + (2 - 3)$	Combine like terms.
$= 5x + (-1)$ or $5x - 1$	Simplify.

PROBLEM 4

Remove parentheses and combine like terms:

$$9a - 3(a - 2) - (a + 4)$$

Sometimes parentheses occur within other parentheses. To avoid confusion, we use different grouping symbols. Thus, we usually don't write $((x + 5) + 3)$. Instead, we write $[(x + 5) + 3]$. To combine like terms in such expressions, remove the *innermost grouping symbols first,* as we illustrate in Example 5.

Answers to PROBLEMS

3. a. $15a - 7$ **b.** $7x - 2y$ **4.** $5a + 2$

EXAMPLE 5 **Removing grouping symbols**

Remove grouping symbols and simplify:

$$[(x^2 - 1) + (2x + 5)] + [(x - 2) - (3x^2 + 3)]$$

SOLUTION 5 We first remove the innermost parentheses and then combine like terms. Thus,

$[(x^2 - 1) + (2x + 5)] + [(x - 2) - (3x^2 + 3)]$

$= [x^2 - 1 + 2x + 5] + [x - 2 - 3x^2 - 3]$ Remove parentheses. Note that $-(3x^2 + 3) = -3x^2 - 3$.

$= [x^2 + 2x + 4] + [-3x^2 + x - 5]$ Combine like terms inside the brackets.

$= x^2 + 2x + 4 - 3x^2 + x - 5$ Remove brackets.

$= -2x^2 + 3x - 1$ Combine like terms.

PROBLEM 5

Remove grouping symbols and simplify:

$$[(a^2 - 2) + (3a + 4)] + [(a - 5) - (4a^2 + 2)]$$

C > Applications: Translating Words into Algebraic Expressions

As we mentioned at the beginning of this section, we express ideas in algebra using expressions. In most applications, problems are first stated in words and have to be translated into algebraic expressions using mathematical symbols. Here's a short mathematics dictionary that will help you translate word problems.

Mathematics Dictionary

Addition (+)	Subtraction (−)
Write: $a + b$ (read "a plus b") Words: Plus, sum, increase, more than, more, added to Examples: a plus b The sum of a and b a increased by b b more than a b added to a	Write: $a - b$ (read "a minus b") Words: Minus, difference, decrease, less than, less, subtracted from Examples: a minus b The difference of a and b a decreased by b b less than a b subtracted from a
Multiplication (× or ·)	**Division (÷ or the bar —)**
Write: $a \cdot b$, ab, $(a)b$, $a(b)$, or $(a)(b)$ (read "a times b" or simply ab) Words: Times, of, product Examples: a times b The product of a and b	Write: $a \div b$ or $\frac{a}{b}$ (read "a divided by b") Words: Divided by, quotient Examples: a divided by b The quotient of a and b

The words and phrases contained in our mathematics dictionary involve the four fundamental operations of arithmetic. We use these words and phrases to translate sentences into *equations*. An **equation** is a sentence stating that two expressions are equal. Here are some words that in mathematics mean "equals":

Equals (=) means is, the same as, yields, gives, is obtained by

We use these ideas in Example 6.

Answers to PROBLEMS

5. $-3a^2 + 4a - 5$

EXAMPLE 6 Translating and finding the number of terms

Write in symbols and indicate the number of terms to the right of the equals sign.

a. The area A of a circle is obtained by multiplying π by the square of the radius r.

b. The perimeter P of a rectangle is obtained by adding twice the length L to twice the width W.

c. The current I across a resistor is given by the quotient of the voltage V and the resistance R.

SOLUTION 6

a. The area A of a circle | is obtained by | multiplying π by the square of the radius r.

$$A = \pi r^2$$

There is one term, πr^2, to the right of the equals sign.

b. The perimeter P of a rectangle | is obtained by | adding twice the length L to twice the width W.

$$P = 2L + 2W$$

There are two terms, $2L$ and $2W$, to the right of the equals sign.

c. The current I across a resistor | is given by | the quotient of the voltage V and the resistance R.

$$I = \frac{V}{R}$$

There is one term to the right of the equals sign.

PROBLEM 6

Write in symbols and indicate the number of terms to the right of the equals sign.

a. The surface area S of a sphere of radius r is obtained by multiplying 4π by the square of the radius r.

b. The perimeter P of a triangle is obtained by adding the lengths a, b, and c of each of the sides.

c. If P dollars are invested for one year at a rate r, the principal P is the quotient of the interest I and the rate r.

GREEN MATH

EXAMPLE 7 Recovered and recycled garbage

The graph shows the total materials (in millions of tons) recovered from garbage (blue) $T = 2.2N + 69$ and the total materials recovered for recycling (red) $A = 1.5N + 52$, where N is the number of years after 2000.

a. Translate and write in symbols: the *difference* of $(2.2N + 69)$ and $(1.5N + 52)$.

b. Simplify this difference (which is the amount left for composting).

c. Estimate the number of tons left for composting in 2010.

SOLUTION 7

a. The *difference* of $(2.2N + 69)$ and $(1.5N + 52)$ is $(2.2N + 69) - (1.5N + 52)$

b.
$$\begin{aligned} & (2.2N + 69) - (1.5N + 52) \\ &= (2.2N + 69) - 1 \cdot (1.5N + 52) \\ &= 2.2N + 69 - 1.5N - 52 && \text{Use the distributive property.} \\ &= 2.2N - 1.5N + 69 - 52 && \text{Rearrange like terms.} \\ &= \mathbf{0.7N + 17} && \text{Simplify.} \end{aligned}$$

c. The year 2010 corresponds to $\mathbf{N = 10}$ (10 years after 2000). When $N = 10$ in $\mathbf{0.7N + 17}$ we have $\mathbf{0.7(10) + 17} = 7 + 17 = 24$. This means that there will be 24 million tons for composting in 2010.

Source: http://tinyurl.com/n3tx9n.

Note: The number of landfills for all these materials has *decreased* from 8000 to 1754 in the last 20 years, so more recycling and composting may be necessary.

PROBLEM 7

The total materials recovered for recycling are $A = 1.5N + 52$ million tons and the amount left for composting is $\mathbf{C = 0.7N + 17}$ where N is the number of years after 2000.

a. Translate and write in symbols: the *sum* of $(1.5N + 52)$ and $\mathbf{(0.7N + 17)}$.

b. Simplify this sum (which is the total materials recovered from garbage).

c. Estimate the amount of total materials recovered from garbage in 2010.

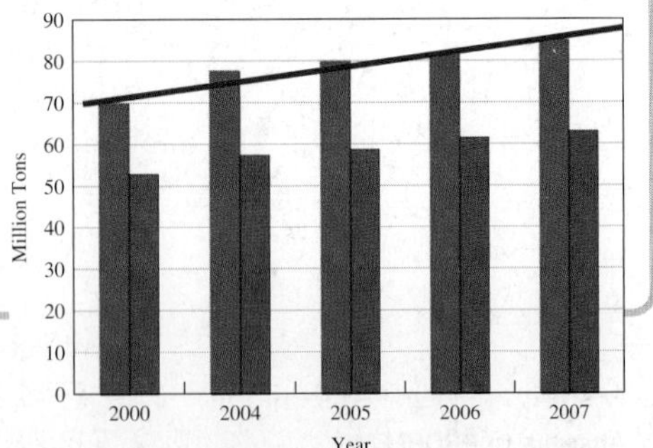

Answers to PROBLEMS

6. a. $S = 4\pi r^2$; one term **b.** $P = a + b + c$; three terms **c.** $P = \frac{I}{r}$; one term

7. a. $(1.5N + 52) + (0.7N + 17)$ **b.** $2.2N + 69$ **c.** 91 million tons

Now that you know how to translate sentences into equations, let's translate symbols into words. For example, in the equation

$$W_T = W_b + W_s + W_e$$ Read "W sub T equals W sub b plus W sub s plus W sub e."

the letters T, b, s, and e are called **subscripts.** They help us represent the total weight W_T of a McDonald's® biscuit, which is composed of the weight of the biscuit W_b, the weight of the sausage W_s, and the weight of the eggs W_e. With this information, how would you translate the equation $W_T = W_b + W_s + W_e$? Here is one way: The total weight W_T of a McDonald's biscuit with sausage and eggs equals the weight W_b of the biscuit plus the weight W_s of the sausage plus the weight W_e of the eggs. (If you look at *The Fast-Food Guide,* you will see that $W_b = 75$ grams, $W_s = 43$ grams, and $W_e = 57$ grams.)

Translation and algebraic expression are used in environmental problems. For example, the total amount of garbage recovered in a recent year G_T consists of the garbage recovered for recycling G_R and the garbage recovered for composting G_C. But these amounts are increasing as the population increases. We will use these ideas in Example 8.

EXAMPLE 8 Recycled and composted garbage

a. Translate into words: The total amount of garbage recovered in a recent year G_T is the amount of garbage recovered for recycling G_R plus the amount of garbage recovered for composting G_C.

b. If $G_R = (2.25N + 58.92)$ million tons and $G_C = (0.55N + 20.48)$ million tons, where N is the number of years after 2005, use the information from part **a.** and find G_T in simplified form.

SOLUTION 8

a. | The total amount of garbage recovered in a recent year G_T | is | The amount recovered for recycling G_R | plus | The amount recovered for composting G_C |
|---|---|---|---|---|
| G_T | = | G_R | + | G_C |

b. Substituting $(2.25N + 58.92)$ for G_R and $(0.55N + 20.48)$ for G_C,

$$G_T = (2.25N + 58.92) + (0.55N + 20.48)$$

$$= 2.25N + 58.92 + 0.55N + 20.48 \quad \text{Remove parentheses.}$$

$$= 2.8N + 79.40 \quad \text{Add like terms.}$$

This means that the amount of garbage recovered N years after 2005 amounts to $2.8N + 79.40$ million tons. For example, in 2005 $(N = 0)$, $2.8(0) + 79.40 = 79.40$ million tons of garbage were recovered. However, 10 years after 2005, in 2015, the amount of garbage recovered would be $2.8(10) + 79.40 = 107.40$ million tons.

Source: http://tinyurl.com/33488mw.

PROBLEM 8

a. Translate into words: The amount of garbage recovered for recycling G_R equals the total amount of garbage recovered G_T minus the amount of garbage recovered for composting G_C.

b. If $G_T = (2.8N + 79.4)$ and $G_C = (0.55N + 20.48)$, where N is the number of years after 2005, write G_R in simplified form.

We gave you a mathematics dictionary. Let's use it on a contemporary topic: extra pounds. Most formulas for your ideal weight W based on your height h contain the phrase: ***for each additional inch over 5 feet (60 inches)***. You translate this as $(h - 60)$. Check it out. If you are 5 foot 3 in. or 63 inches, your additional inches over 60 are $(h - 60) = (63 - 60)$ or 3. Every time you see the phrase ***for each additional inch over 5 feet (60 inches)*** translate it as $(h - 60)$.

Source: http://www.halls.md.

Answers to PROBLEMS

8. a. $G_R = G_T - G_C$

b. $2.25N + 58.92$

TRANSLATE THIS

In Problems 1–10 **TRANSLATE** the sentence and match the correct translation with one of the equations A–O.

1. The history of the formulas for calculating ideal body weight W began in **1871** when Dr. P. P. Broca (a French surgeon) created this formula known as Broca's index. The ideal weight W (in pounds) for a woman h inches tall is 100 pounds for the first 5 feet and 5 pounds for each additional inch over 60.
2. The ideal weight W (in pounds) for men h inches tall is 110 pounds for the first 5 feet and 5 pounds for each additional inch over 60.
3. In 1974, Dr. B. J. Devine suggested a formula for the weight W in kilograms (kg) of men h inches tall: 50 plus 2.3 kilograms per inch over 5 feet (60 inches).
4. For women h inches tall, the formula for W is 45.5 plus 2.3 kilograms per inch over 5 feet. By the way, a kilogram (kg) is about 2.2 pounds.
5. In 1983, Dr. J. D. Robinson published a modification of the formula. For men h inches tall, the weight W should be 52 kilograms and 1.9 kilograms for each inch over 60.

A. $W = 50 + 2.3h - 60$
B. $W = 49 + 1.7(h - 60)$
C. $\text{LBW} = B + M + O$
D. $W = 100h + 5(h - 60)$
E. $W = 110 + 5(h - 60)$
F. $\text{LBW} = 0.32810C + 0.33929W - 29.5336$
G. $\text{LBW} = 0.32810W + 0.33929C - 29.5336$
H. $W = 100 + 5(h - 60)$
I. $\text{LBW} = 0.29569W + 0.41813C - 43.2933$
J. $W = 50h + 2.3(h - 60)$
K. $W = 110h + 5(h - 60)$
L. $W = 56.2 + 1.41(h - 60)$
M. $W = 50 + 2.3(h - 60)$
N. $W = 52 + 1.9(h - 60)$
O. $W = 45.5 + 2.3(h - 60)$

6. The Robinson formula W for women h inches tall is 49 kilograms and 1.7 kilograms for each inch over 5 feet.
7. A minor modification of Robinson formula is Miller's formula which defines the weight W for a man h inches tall as 56.2 kilograms added to 1.41 kilograms for each inch over 5 feet
8. There are formulas that suggest your lean body weight **(LBW)** is the sum of the weight of your bones **(*B*)**, muscles **(*M*)**, and organs **(*O*)**. Basically the sum of everything other than fat in your body.
9. For men over the age of 16, C centimeters tall and with weight W kilograms, the lean body weight **(LBW)** is the product of W and 0.32810, plus the product of C and 0.33929, minus 29.5336.
10. For women over the age of 30, C centimeters tall and weighting W kilograms the lean body weight **(LBW)** is the product of 0.29569 and W, plus the product of 0.41813 and C, minus 43.2933

Try some of the formulas before you go on and see how close to your ideal weight you are!

› Exercises 1.7

› Practice Problems › Self-Tests
› Media-rich eBooks › e-Professors › Videos

› Web IT go to mhhe.com/bello for more lessons

‹A› Adding and Subtracting Like Terms In Problems 1–30, combine like terms (simplify).

1. $19a + (-8a)$
2. $-2b + 5b$
3. $-8c + 3c$
4. $-5d + (-7d)$
5. $4n^2 + 8n^2$
6. $3x^2 + (-9x^2)$
7. $-3ab^2 + (-4ab^2)$
8. $9ab^2 + (-3ab^2)$
9. $-4abc + 7abc$
10. $6xyz + (-9xyz)$
11. $0.7ab + (-0.3ab) + 0.9ab$
12. $-0.5x^2y + 0.8x^2y + 0.3x^2y$
13. $-0.3xy^2 + 0.2x^2y + (-0.6xy^2)$
14. $0.2x^2y + (-0.3xy^2) + 0.4xy^2$
15. $8abc^2 + 3ab^2c + (-8abc^2)$
16. $3xy + 5ab + 2xy + (-3ab)$
17. $-8ab + 9xy + 2ab + (-2xy)$
18. $7 + \frac{1}{2}x + 3 + \frac{1}{2}x$
19. $\frac{1}{5}a + \frac{3}{7}a^2b + \frac{2}{5}a + \frac{1}{7}a^2b$
20. $\frac{4}{9}ab + \frac{4}{5}ab^2 + \left(-\frac{1}{9}ab\right) + \left(-\frac{1}{5}a^2b\right)$
21. $13x - 2x$
22. $8x - (-2x)$
23. $6ab - (-2ab)$
24. $4.2xy - (-3.7xy)$
25. $-4a^2b - (3a^2b)$
26. $-8ab^2 - 4a^2b$
27. $3.1t^2 - 3.1t^2$
28. $-4.2ab - 3.8ab$
29. $0.3x^2 - 0.3x^2$
30. $0 - (-0.8xy^2)$

B Removing Parentheses In Problems 31–55, remove parentheses and combine like terms.

31. $(3xy + 5) + (7xy - 9)$
32. $(8ab - 9) + (7 - 2ab)$
33. $(7R - 2) + (8 - 9R)$
34. $(5xy - 3ab) + (9ab - 8xy)$
35. $(5L - 3W) + (W - 6L)$
36. $(2ab - 2ac) + (ab - 4ac)$
37. $5x - (8x + 1)$
38. $3x - (7x + 2)$
39. $\frac{2x}{9} - \left(\frac{x}{9} - 2\right)$
40. $\frac{5x}{7} - \left(\frac{2x}{7} - 3\right)$
41. $4a - (a + b) + 3(b + a)$
42. $8x - 3(x + y) - (x - y)$
43. $7x - 3(x + y) - (x + y)$
44. $4(b - a) + 3(b + a) - 2(a + b)$
45. $-(x + y - 2) + 3(x - y + 6) - (x + y - 16)$
46. $[(a^2 - 4) + (2a^3 - 5)] + [(4a^3 + a) + (a^2 + 9)]$
47. $(x^2 + 7 - x) + [-2x^3 + (8x^2 - 2x) + 5]$
48. $[(0.4x - 7) + 0.2x^2] + [(0.3x^2 - 2) - 0.8x]$
49. $\left[\left(\frac{1}{4}x^2 + \frac{1}{5}x\right) - \frac{1}{8}\right] + \left[\left(\frac{3}{4}x^2 - \frac{3}{5}x\right) + \frac{5}{8}\right]$
50. $3[3(x + 2) - 10] + [5 + 2(5 + x)]$
51. $2[3(2a - 4) + 5] - [2(a - 1) + 6]$
52. $-2[6(a - b) + 2a] - [3b - 4(a - b)]$
53. $-3[4a - (3 + 2b)] - [6(a - 2b) + 5a]$
54. $-[-(x + y) + 3(x - y)] - [4(x + y) - (3x - 5y)]$
55. $-[-(0.2x + y) + 3(x - y)] - [2(x + 0.3y) - 5]$

C Applications: Translating Words into Algebraic Expressions In Problems 56–70, translate the sentences into equations and indicate the number of terms to the right of the equals sign.

Applications: Green Math

56. *Garbage to landfill*

a. Translate into words: The amount of garbage G_L (in million of tons) that goes to the landfill each year is the total amount of garbage generated (G) minus the total amount of materials recovered G_T minus the amount of garbage burned G_B.

b. If $G = (1.85N + 251)$ million tons, $G_T = 2.8N + 79.40$ million tons, and $G_B = 40$ million tons, where N is the number of years after 2005, find G_L in simplified form.

57. *More garbage to landfill*

a. Referring to Problem 56, how many million tons of garbage went to the landfill in 2005 ($N = 0$)?

b. How many million tons of garbage would go to the landfill in 2015 ($N = 10$)?

Source: http://tinyurl.com/33488mw.

58. *Rocket height* The height h attained by a rocket is the sum of its height a at burnout and

$$\frac{r^2}{20}$$

where r is the speed at burnout.

59. *Temperature* The Fahrenheit temperature F can be obtained by adding 40 to the quotient of n and 4, where n is the number of cricket chirps in 1 min.

60. *Circles* The radius r of a circle is the quotient of the circumference C and 2π.

61. *Interest* The interest received I is the product of the principal P, the rate r, and the time t.

62. *Profit* The total profit P_T equals the total revenue R_T minus the total cost C_T.

63. *Perimeter* The perimeter P is the sum of the lengths of the sides a, b, and c.

64. *Volume* The volume V of a certain gas is the pressure P divided by the temperature T.

65. *Area* The area A of a triangle is the product of the length of the base b and the height h, divided by 2.

66. *Area* The area A of a circle is the product of π and the square of the radius r.

67. *Kinetic energy* The kinetic energy K of a moving object is the product of a constant C, the mass m of the object, and the square of its velocity v.

Web IT go to mhhe.com/bello for more lessons

68. *Averages* The arithmetic mean m of two numbers a and b is found by dividing their sum by 2.

69. *Distance* The distance in feet d that an object falls is the product of 16 and the square of the time in seconds t that the object has been falling.

70. *Horse power* The horse power (hp) that a shaft can safely transmit is the product of a constant C, the speed s of the shaft, and the cube of the shaft's diameter d.

Just before the beginning of these problems we gave some guidelines to find your ideal weight W based on your height h. How can you lose some pounds? Do Problems 71–74 to find out! The secret is the calories you consume and spend each day. *Source:* http://www.annecollins.com.

71. *Calories* For women, the daily calories C needed to *lose* about 1 pound per week can be found by multiplying your weight W (in pounds) by 12 and deducting 500 calories. Translate this sentence and indicate the number of terms to the right of the equal sign.

72. *Calories* For men, the daily calories C needed to *lose* about 1 pound per week can be found by multiplying your weight W (in pounds) by 14 and deducting 500 calories. Translate this sentence and indicate the number of terms to the right of the equal sign.

73. *Calories* How can you "deduct" 500 calories each day? Exercising! Here are some exercises and the number of calories they use per hour.

Type of Exercise	Calories/Hour
Sleeping	55
Eating	85
Housework, moderate	160+
Dancing, ballroom	260
Walking, 3 mph	280
Aerobics	450+

a. How many calories C would you use in h hours by sleeping?

b. How many calories C would you use in h hours by eating?

c. How many calories C would you use in h hours by doing moderate housework?

74. *Calories* Use the table in Problem 73 to find the following:

a. How many calories C would you use in h hours by ballroom dancing?

b. How many calories C would you use in h hours by walking at 3 miles per hour?

c. How many calories C would you use in h hours by doing aerobics?

Using Your Knowledge

Geometric Formulas The ideas in this section can be used to simplify formulas. For example, the **perimeter** (distance around) of the given rectangle is found by following the blue arrows. The perimeter is

$$P = W + L + W + L$$
$$= 2W + 2L$$

Use this idea to find the perimeter P of the given figure; then write a formula for P in symbols and in words.

75. The square of side S

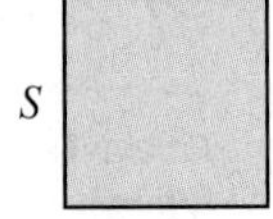

76. The parallelogram of base b and side s

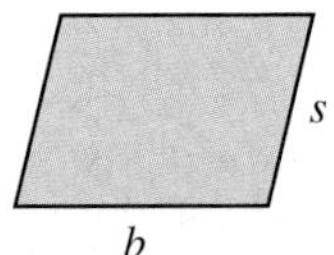

To obtain the actual measurement of certain perimeters, we have to add like terms if the measurements are given in feet and inches. For example, the perimeter of the rectangle shown here is

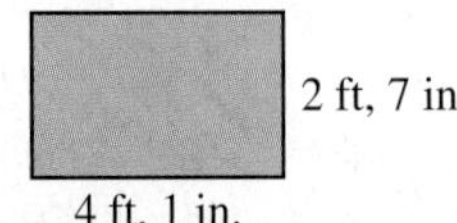

$$P = (2 \text{ ft} + 7 \text{ in.}) + (4 \text{ ft} + 1 \text{ in.}) + (2 \text{ ft} + 7 \text{ in.}) + (4 \text{ ft} + 1 \text{ in.})$$
$$= 12 \text{ ft} + 16 \text{ in.}$$

Since 16 inches = 1 foot + 4 inches,

$$P = 12\text{ ft} + (1\text{ ft} + 4\text{ in.})$$
$$= 13\text{ ft} + 4\text{ in.}$$

Use these ideas to obtain the perimeter of the given rectangles.

77. 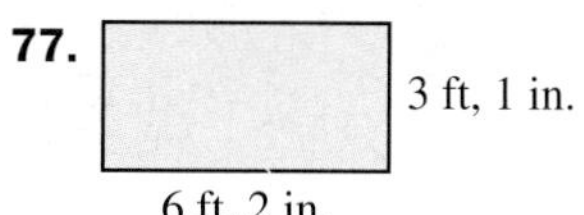

78. 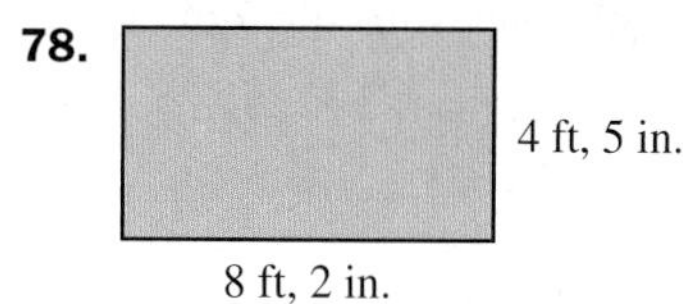

79. The U.S. Postal Service has a regulation stating that "the sum of the length and girth of a package may be no more than 108 inches." What is the sum of the length and girth of the rectangular package below? (*Hint:* The girth of the package is obtained by measuring the length of the red line.)

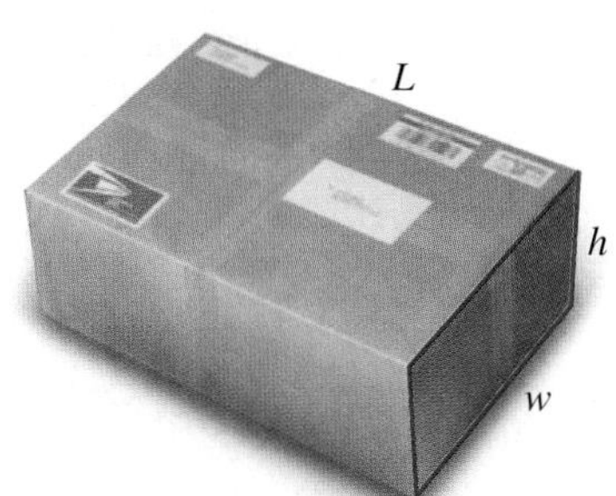

80. Write in simplified form the height of the step block shown.

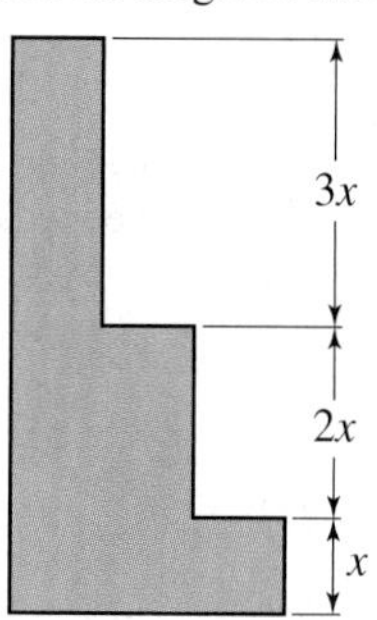

81. Write in simplified form the length of the metal plate shown.

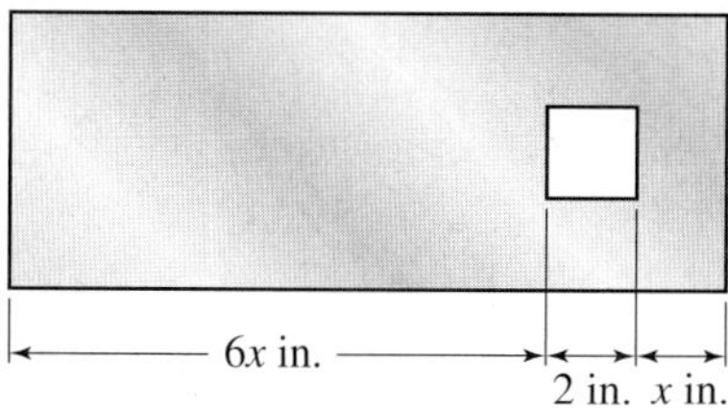

Write On

82. Explain the difference between a factor and a term.

83. Write the procedure you use to combine like terms.

84. Explain how to remove parentheses when no sign or a plus sign precedes an expression within parentheses.

85. Explain how to remove parentheses when a minus sign precedes an expression within parentheses.

Concept Checker

Fill in the blank(s) with the correct word(s), phrase, or mathematical statement.

86. To **subtract** a number b from another number a, add the ______________ of b to a.

87. $a - (-b) =$ ______________.

88. An ______________ is a **sentence** stating that the expressions are **equal.**

89. We use the ______________ sign to indicate that two expressions are equal.

reciprocal	**equation**
$a - b$	**=**
$a + b$	**opposite**

Mastery Test

Write in symbols and indicate the number of terms to the right of the equals sign.

90. When P dollars are invested at r percent, the amount of money A in an account at the end of 1 year is the sum of P and the product of P and r.

91. The area A of a rectangle is obtained by multiplying the length L by the width W.

Remove parentheses and combine like terms.

92. $9x - 3(x - 2) - (2x + 7)$

93. $(8y - 7) + 2(3y - 2)$

94. $(4t + u) + 4(5t - 7u)$

95. $-6ab^3 - (-3ab^3)$

Remove grouping symbols and simplify.

96. $[(2x^2 - 3) + (3x + 1)] + 2[(x - 1) - (x^2 + 2)]$

97. $3[(5 - 2x^2) + (2x - 1)] - 3[(5 - 2x) - (3 + x^2)]$

Skill Checker

Simplify the expression.

98. $\frac{1}{2}(12) - 6$

99. $2(9) - 7$

100. $3(z - 2)$

101. $-2(x + 5)$

Collaborative Learning

Form three groups. The steps to a number trick are shown. Group 1 starts with number 8, Group 2 with -3, and Group 3 with -1.2.

Steps	Group 1	Group 2	Group 3
Pick a number:	8	−3	−1.2
1. ________	10	−1	0.8
2. ________	30	−3	2.4
3. ________	24	−9	−3.6
4. ________	8	−3	−1.2
5. ________	0	0	0

Each group should fill in the blanks in the first column and find out how the trick works. The first group that discovers what the steps are wins! Of course, the other groups can challenge the results. Compare the five steps. Do all groups get the same steps? What are the steps?

The steps given here will give you a different final answer depending on the number you start with.

Number + 2
Ans · 3
Ans − 9
Ans/6

Are you ready for the race? All groups should start with the number 5. What answers do they get? The group that finishes first wins! Then do it again. Start with the number 7 this time. What answer do the groups get now?

Here is the challenge: Write an algebraic expression representing the steps in the table. Again, the group that finishes first wins.

Ultimate challenge: The groups write down the steps that describe the expression $3\left(\frac{x+2}{3} + 4\right) - x + 5$.

Research Questions

1. In the *Human Side of Algebra* at the beginning of this chapter, we mentioned the Hindu numeration system. The Egyptians and Babylonians also developed numeration systems. Write a report about each of these numeration systems, detailing the symbols used for the digits 1–9, the base used, and the manner in which fractions were written.
2. Write a report on the life and works of Muhammad al-Khwarizmi, with special emphasis on the books he wrote.
3. We have now studied the four fundamental operations. But do you know where the symbols used to indicate these operations originated?
   a. Write a report about Johann Widmann's *Mercantile Arithmetic* (1489), indicating which symbols of operation were found in the book for the first time and the manner in which they were used.
   b. Introduced in 1557, the original equals sign used longer lines to indicate equality. Why were the two lines used to denote equality, what was the name of the person who introduced the symbol, and in what book did the notation first appear?
4. In this chapter we discussed mathematical expressions. Two of the signs of operation were used for the first time in the earliest-known treatise on algebra to write an algebraic expression. What is the name of this first treatise on algebra, and what is the name of the Dutch mathematician who used these two symbols in 1514?
5. The *Codex Vigilanus* written in Spain in 976 contains the Hindu-Arabic numerals 1–9. Write a report about the notation used for these numbers and the evolution of this notation.

Summary Chapter 1

Section	Item	Meaning	Example
1.1A	Arithmetic expressions	Expressions containing numbers and operation signs	$3 + 4 - 8$, $9 \cdot 4 \div 6$ are arithmetic expressions.
1.1A	Algebraic expressions	Expressions containing numbers, operation signs, and variables	$3x - 4y + 3z$, $2x \div y + 9z$, and $7x - 9y \div 3z$ are algebraic expressions.
1.1A	Factors	The items to be multiplied	In the expression $4xy$, the factors are 4, x, and y.
1.1B	Evaluate	To substitute a value for one or more of the variables in an expression	Evaluating $3x - 4y + 3z$ when $x = 1$, $y = 2$, and $z = 3$ yields $3 \cdot 1 - 4 \cdot 2 + 3 \cdot 3$ or 4.
1.2A	Additive inverse (opposite)	The additive inverse of any integer a is $-a$.	The additive inverse of 5 is -5, and the additive inverse of -8 is 8.
1.2B	Absolute value of n, denoted by $\|n\|$	The absolute value of a number n is the distance from n to 0.	$\|-7\| = 7$, $\|13\| = 13$, $\|0.2\| = 0.2$, and $\left\|-\frac{1}{4}\right\| = \frac{1}{4}$.

(continued)

Section	Item	Meaning	Example
1.2C	Set of natural numbers	$\{1, 2, 3, 4, 5, \ldots\}$	4, 13, and 497 are natural numbers.
	Set of whole numbers	$\{0, 1, 2, 3, 4, \ldots\}$	0, 92, and 384 are whole numbers.
	Set of integers	$\{\ldots, -2, -1, 0, 1, 2, \ldots\}$	-98, 0, and 459 are integers.
	Rational numbers	Numbers that can be written in the form $\frac{a}{b}$, a and b integers and $b \neq 0$	$\frac{1}{5}, \frac{-3}{4}$, and $\frac{0}{5}$ are rational numbers.
	Irrational numbers	Numbers that cannot be written as the ratio of two integers	$\sqrt{2}, -\sqrt[3]{2}$, and π are irrational numbers.
	Real numbers	The rationals and the irrationals	$3, 0, -9, \sqrt{2}$, and $\sqrt[3]{5}$ are real numbers.
1.3A	Addition of real numbers	If both numbers have the *same* sign, add their absolute values and give the sum the common sign. If the numbers have *different* signs, subtract their absolute values and give the difference the sign of the number with the larger absolute value.	$-3 + (-5) = -8$ $-3 + 5 = +2$ $-3 + 1 = -2$
1.3B	Subtraction of real numbers	$a - b = a + (-b)$	$3 - 5 = 3 + (-5) = -2$ and $4 - (-2) = 4 + 2 = 6$
1.4A, C	Multiplication and division of real numbers	When multiplying or dividing two numbers with the *same* sign, the answer is *positive;* with *different* signs, the answer is *negative*.	$(-3)(-5) = 15$ $\frac{-15}{-3} = 5$ $(-3)(5) = -15$ $\frac{-15}{3} = -5$
1.4B	Exponent	In the expression 3^2, 2 is the exponent.	3^2 means $3 \cdot 3$
	Base	In the expression 3^2, 3 is the base.	
1.5	Order of operations PEMDAS: Parentheses, Exponents, Multiplications, Divisions, Additions, Subtractions	1. Operations inside grouping symbols (like parentheses) 2. Exponents 3. Multiplications and divisions as they occur from *left* to *right* 4. Additions and subtractions as they occur from *left* to *right*	$4^2 \div 2 \cdot 3 - 4 + [2(6 + 1) - 4]$ $= 4^2 \div 2 \cdot 3 - 4 + [2(7) - 4]$ $= 4^2 \div 2 \cdot 3 - 4 + [14 - 4]$ $= 4^2 \div 2 \cdot 3 - 4 + [10]$ $= 16 \div 2 \cdot 3 - 4 + [10]$ $= 8 \cdot 3 - 4 + 10$ $= 24 - 4 + 10$ $= 20 + 10$ $= 30$
1.6A	Associative property of addition	For any numbers a, b, c, $a + (b + c) = (a + b) + c$.	$3 + (4 + 9) = (3 + 4) + 9$
	Associative property of multiplication	For any numbers a, b, c, $a \cdot (b \cdot c) = (a \cdot b) \cdot c$.	$5 \cdot (2 \cdot 8) = (5 \cdot 2) \cdot 8$
	Commutative property of addition	For any numbers a and b, $a + b = b + a$.	$2 + 9 = 9 + 2$
	Commutative property of multiplication	For any numbers a and b, $a \cdot b = b \cdot a$.	$18 \cdot 5 = 5 \cdot 18$

Section	Item	Meaning	Example
1.6 C	Identity element for addition	0 is the identity element for addition.	$3 + 0 = 0 + 3 = 3$
	Identity element for multiplication	1 is the identity element for multiplication.	$1 \cdot 7 = 7 \cdot 1 = 7$
	Additive inverse (opposite)	For any number a, its additive inverse is $-a$.	The additive inverse of 5 is -5 and that of -8 is 8.
	Multiplicative inverse (reciprocal)	The multiplicative inverse of a is $\frac{1}{a}$ if a is not 0 (0 has no reciprocal).	The multiplicative inverse of $\frac{3}{4}$ is $\frac{4}{3}$.
1.6 E	Distributive property	For any numbers a, b, c, $a(b + c) = ab + ac$.	$3(4 + x) = 3 \cdot 4 + 3 \cdot x$
1.7	Expression	A collection of numbers and letters connected by operation signs	xy^3, $x + y$, $x + y - z$, and $xy^3 - y$ are expressions.
	Terms	The parts that are to be added or subtracted in an expression	In the expression $3x^2 - 4x + 5$, the terms are $3x^2$, $-4x$, and 5.
	Numerical coefficient or coefficient	The part of the term indicating "how many"	In the term $3x^2$, 3 is the coefficient. In the term $-4x$, -4 is the coefficient. The coefficient of x is 1.
	Like terms	Terms with the same variables and exponents	$3x$ and $-4x$ are like terms. $-5x^2$ and $8x^2$ are like terms. $-xy^2$ and $3xy^2$ are like terms. *Note:* $-x^2y$ and $-xy^2$ **are not** like terms.
1.7C	Sums and differences	The *sum* of a and b is $a + b$.	The sum of 3 and x is $3 + x$.
	Product	The *difference* of a and b is $a - b$.	The difference of 6 and x is $6 - x$.
	Quotient	The *product* of a and b is $a \cdot b$, $(a)(b)$, $a(b)$, $(a)b$, or ab.	The product of 8 and x is $8x$.
		The *quotient* of a and b is $\frac{a}{b}$.	The quotient of 7 and x is $\frac{7}{x}$.

Review Exercises Chapter 1

(If you need help with these exercises, look in the section indicated in brackets.)

1. ⟨**1.1A**⟩ *Write in symbols:*

a. The sum of a and b

b. a minus b

c. $7a$ plus $2b$ minus 8

2. ⟨**1.1A**⟩ *Write using juxtaposition:*

a. 3 times m

b. $-m$ times n times r

c. $\frac{1}{7}$ of m

d. The product of 8 and m

3. ⟨**1.1A**⟩ *Write in symbols:*

a. The quotient of m and 9

b. The quotient of 9 and n

4. ⟨**1.1A**⟩ *Write in symbols:*

a. The quotient of $(m + n)$ and r

b. The sum of m and n, divided by the difference of m and n

5. ⟨1.1B⟩ *For $m = 9$ and $n = 3$, evaluate:*
a. $m + n$ b. $m - n$ c. $4m$

6. ⟨1.1B⟩ *For $m = 9$ and $n = 3$, evaluate:*
a. $\frac{m}{n}$ b. $2m - 3n$ c. $\frac{2m + n}{n}$

7. ⟨1.2A⟩ *Find the additive inverse (opposite) of:*
a. -5 b. $\frac{2}{3}$ c. 0.37

8. ⟨1.2B⟩ *Find the absolute value of:*
a. $|-8|$ b. $\left|3\frac{1}{2}\right|$ c. $-|0.76|$

9. ⟨1.2C⟩ *Classify each of the numbers as natural, whole, integer, rational, irrational, or real. (Note: More than one category may apply.)*
a. -7 b. 0
c. $0.666\ldots = 0.\overline{6}$ d. $0.606006000\ldots$
e. $\sqrt{41}$ f. $-1\frac{1}{8}$

10. ⟨1.3A⟩ *Add:*
a. $7 + (-5)$ b. $(-0.3) + (-0.5)$
c. $-\frac{3}{4} + \frac{1}{5}$ d. $3.6 + (-5.8)$
e. $\frac{3}{4} + \left(-\frac{1}{2}\right)$

11. ⟨1.3B⟩ *Subtract:*
a. $-16 - 4$ b. $-7.6 - (-5.2)$
c. $\frac{5}{6} - \frac{9}{4}$

12. ⟨1.3C⟩ *Find:*
a. $20 - (-12) + 15 - 12 - 5$
b. $-17 + (-7) + 10 - (-7) - 8$

13. ⟨1.4A⟩ *Multiply:*
a. $-5 \cdot 7$ b. $8(-2.3)$
c. $-6(3.2)$ d. $-\frac{3}{7}\left(-\frac{4}{5}\right)$

14. ⟨1.4B⟩ *Find:*
a. $(-4)^2$ b. -3^2
c. $\left(-\frac{1}{3}\right)^3$ d. $-\left(-\frac{1}{3}\right)^3$

15. ⟨1.4C⟩ *Divide:*
a. $\frac{-40}{10}$ b. $-8 \div (-4)$
c. $\frac{3}{8} \div \left(-\frac{9}{16}\right)$ d. $-\frac{3}{4} \div \left(-\frac{1}{2}\right)$

16. ⟨1.5A⟩ *Find:*
a. $64 \div 8 - (3 - 5)$
b. $27 \div 3^2 + 5 - 8$

17. ⟨1.5B⟩ *Find:*
a. $20 \div 5 + \{2 \cdot 3 - [4 + (7 - 9)]\}$
b. $-6^2 + \frac{2(3 - 6)}{2} + 8 \div (-4)$

18. ⟨1.5C⟩ *The maximum pulse rate you should maintain during aerobic activities is $0.80(220 - A)$, where A is your age. What is the maximum pulse rate you should maintain if you are 30 years old?*

19. ⟨1.6A⟩ *Name the property illustrated in each statement.*
a. $x + (y + z) = x + (z + y)$
b. $6 \cdot (8 \cdot 7) = (6 \cdot 8) \cdot 7$
c. $x + (y + z) = (x + y) + z$

20. ⟨1.6B⟩ *Simplify:*
a. $6 + 4x - 10$
b. $-8 + 7x + 10 - 15x$

21. ⟨ **1.6C, E** ⟩ *Fill in the blank so that the result is a true statement.*

a. $\_\_\_ \cdot \frac{1}{5} = \frac{1}{5}$

b. $\frac{3}{4} + \_\_\_ = 0$

c. $\_\_\_\_ + 3.7 = 0$

d. $\_\_\_ \cdot \frac{3}{2} = 1$

e. $\_\_\_ + \frac{2}{7} = \frac{2}{7}$

f. $3(x + \_\_\_) = 3x + (-15)$

22. ⟨ **1.6E** ⟩ *Use the distributive property to multiply.*

a. $-3(a + 8)$

b. $-4(x - 5)$

c. $-(x - 4)$

23. ⟨ **1.7A** ⟩ *Combine like terms.*

a. $-7x + 2x$

b. $-2x + (-8x)$

c. $x + (-9x)$

d. $-6a^2b - (-9a^2b)$

24. ⟨ **1.7B** ⟩ *Remove parentheses and combine like terms.*

a. $(4a - 7) + (8a + 2)$

b. $9x - 3(x + 2) - (x + 3)$

25. ⟨ **1.7C** ⟩ *Write in symbols and indicate the number of terms to the right of the equals sign.*

a. The number of minutes m you will wait in line at the bank is equal to the number of people p ahead of you divided by the number n of tellers, times 2.75.

b. The normal weight W of an adult (in pounds) can be estimated by subtracting 220 from the product of $\frac{11}{2}$ and h, where h is the person's height (in inches).

Practice Test Chapter 1

(Answers on page 108)

Visit www.mhhe.com/bello to view helpful videos that provide step-by-step solutions to several of the following problems.

1. Write in symbols:
a. The sum of g and h **b.** g minus h
c. $6g$ plus $3h$ minus 8

2. Write using juxtaposition:
a. 3 times g **b.** $-g$ times h times r
c. $\frac{1}{5}$ of g **d.** The product of 9 and g

3. Write in symbols:
a. The quotient of g and 8
b. The quotient of 8 and h

4. Write in symbols:
a. The quotient of $(g + h)$ and r
b. The sum of g and h, divided by the difference of g and h

5. For $g = 4$ and $h = 3$, evaluate:
a. $g + h$ **b.** $g - h$ **c.** $5g$

6. For $g = 8$ and $h = 4$, evaluate:
a. $\frac{g}{h}$ **b.** $2g - 3h$ **c.** $\frac{2g + h}{h}$

7. Find the additive inverse (opposite) of:
a. -9 **b.** $\frac{3}{5}$ **c.** $0.222\ldots$

8. Find the absolute value.
a. $\left|-\frac{1}{4}\right|$ **b.** $|13|$ **c.** $-|0.92|$

9. Consider the set $\{-8, \frac{5}{3}, \sqrt{2}, 0, 3.4, 0.333\ldots, -3\frac{1}{2}, 7, 0.123\ldots\}$. List the numbers in the set that are:
a. Natural numbers **b.** Whole numbers
c. Integers **d.** Rational numbers
e. Irrational numbers **f.** Real numbers

10. Add:
a. $9 + (-7)$ **b.** $(-0.9) + (-0.8)$
c. $-\frac{3}{4} + \frac{2}{5}$ **d.** $1.7 + (-3.8)$
e. $\frac{4}{5} + \left(-\frac{1}{2}\right)$

11. Subtract:
a. $-18 - 6$ **b.** $-9.2 - (-3.2)$
c. $\frac{1}{6} - \frac{5}{4}$

12. Find:
a. $10 - (-15) + 12 - 15 - 6$
b. $-15 + (-8) + 12 - (-8) - 9$

13. Multiply:
a. $-6 \cdot 8$ **b.** $9(-2.3)$
c. $-7(8.2)$ **d.** $-\frac{2}{7}\left(-\frac{3}{5}\right)$

14. Find:
a. $(-7)^2$ **b.** -9^2
c. $\left(-\frac{2}{3}\right)^3$ **d.** $-\left(-\frac{2}{3}\right)^3$

15. Divide:
a. $\frac{-50}{10}$ **b.** $-14 \div (-7)$
c. $\frac{5}{8} \div \left(-\frac{11}{16}\right)$ **d.** $-\frac{3}{4} \div \left(-\frac{7}{9}\right)$

16. Find:
a. $56 \div 7 - (4 - 9)$
b. $36 \div 2^2 + 7$

17. Find the value of:
a. $30 \div 6 + \{3 \cdot 4 - [2 + (8 - 10)]\}$
b. $-7^2 + \frac{4(3 - 9)}{2} + 12 \div (-4)$

18. The handicap H of a bowler with average A is $H = 0.80(200 - A)$. What is the handicap of a bowler whose average is 180?

19. Name the property illustrated in each statement.

a. $x + (y + z) = (x + y) + z$

b. $6 \cdot (8 \cdot 7) = 6 \cdot (7 \cdot 8)$

c. $x + (y + z) = (y + z) + x$

20. Simplify:

a. $7 - 4x - 10$

b. $-9 + 6x + 12 - 13x$

21. Fill in the blank so that the result is a true statement.

a. $9.2 + \_\_\_ = 0$

b. $\_\_\_ \cdot \frac{5}{2} = 1$

c. $\_\_\_ \cdot 0.3 = 0.3$

d. $\frac{1}{4} + \_\_\_ = 0$

e. $\_\_\_ + \frac{3}{7} = \frac{3}{7}$

f. $-3(x + \_\_\_) = -3x + (-15)$

22. Use the distributive property to multiply.

a. $-3(x + 9)$

b. $-5(a - 4)$

c. $-(x - 8)$

23. Combine like terms.

a. $-3x + (-9x)$

b. $-9ab^2 - (-6ab^2)$

24. Remove parentheses and combine like terms.

a. $(5a - 6) - (7a + 2)$

b. $8x - 4(x - 2) - (x + 2)$

25. Write in symbols and indicate the number of terms to the right of the equals sign:

The temperature F (in degrees Fahrenheit) can be found by finding the sum of 37 and one-quarter the number of chirps C a cricket makes in 1 min.

Answers to Practice Test Chapter 1

Answer	If You Missed	Review		
	Question	Section	Examples	Page
1. **a.** $g + h$ **b.** $g - h$ **c.** $6g + 3h - 8$	1	1.1	1	37
2. **a.** $3g$ **b.** $-ghr$ **c.** $\frac{1}{5}g$ **d.** $9g$	2	1.1	2	37
3. **a.** $\frac{g}{8}$ **b.** $\frac{8}{h}$	3	1.1	3a, b	38
4. **a.** $\frac{g+h}{r}$ **b.** $\frac{g+h}{g-h}$	4	1.1	3c, d	38
5. **a.** 7 **b.** 1 **c.** 20	5	1.1	4a, b, c	38
6. **a.** 2 **b.** 4 **c.** 5	6	1.1	4d, e	38
7. **a.** 9 **b.** $-\frac{3}{5}$ **c.** $-0.222\ldots$	7	1.2	1, 2	43, 44
8. **a.** $\frac{1}{4}$ **b.** 13 **c.** -0.92	8	1.2	3, 4	45
9. **a.** 7 **b.** 0, 7 **c.** $-8, 0, 7$ **d.** $-8, \frac{5}{3}, 0, 3.4, 0.333\ldots, -3\frac{1}{2}, 7$ **e.** $\sqrt{2}, 0.123\ldots$ **f.** All	9	1.2	5	46–47
10. **a.** 2 **b.** -1.7 **c.** $-\frac{7}{20}$ **d.** -2.1 **e.** $\frac{3}{10}$	10	1.3	1–5	52–54
11. **a.** -24 **b.** -6.0 **c.** $-\frac{13}{12}$	11	1.3	6, 7	55
12. **a.** 16 **b.** -12	12	1.3	8	56
13. **a.** -48 **b.** -20.7 **c.** -57.4 **d.** $\frac{6}{35}$	13	1.4	1, 2	61
14. **a.** 49 **b.** -81 **c.** $-\frac{8}{27}$ **d.** $\frac{8}{27}$	14	1.4	3, 4	62
15. **a.** -5 **b.** 2 **c.** $-\frac{10}{11}$ **d.** $\frac{27}{28}$	15	1.4	5, 6	63, 64
16. **a.** 13 **b.** 16	16	1.5	1, 2	70, 71
17. **a.** 17 **b.** -64	17	1.5	3–5	71–73
18. 16	18	1.5	6	73–74
19. **a.** Associative property of addition **b.** Commutative property of multiplication **c.** Commutative property of addition	19	1.6	1, 2	78, 79
20. **a.** $-4x - 3$ **b.** $3 - 7x$	20	1.6	3	79
21. **a.** -9.2 **b.** $\frac{2}{5}$ **c.** 1 **d.** $-\frac{1}{4}$ **e.** 0 **f.** 5	21	1.6	4, 5, 7, 8	80–82
22. **a.** $-3x - 27$ **b.** $-5a + 20$ **c.** $-x + 8$	22	1.6	7, 8	82
23. **a.** $-12x$ **b.** $-3ab^2$	23	1.7	1, 2	90, 91
24. **a.** $-2a - 8$ **b.** $3x + 6$	24	1.7	3–5	92, 93
25. $F = 37 + \frac{1}{4}C$; two terms	25	1.7	6, 8	94, 95

Section

Chapter

two

Equations, Problem Solving, and Inequalities

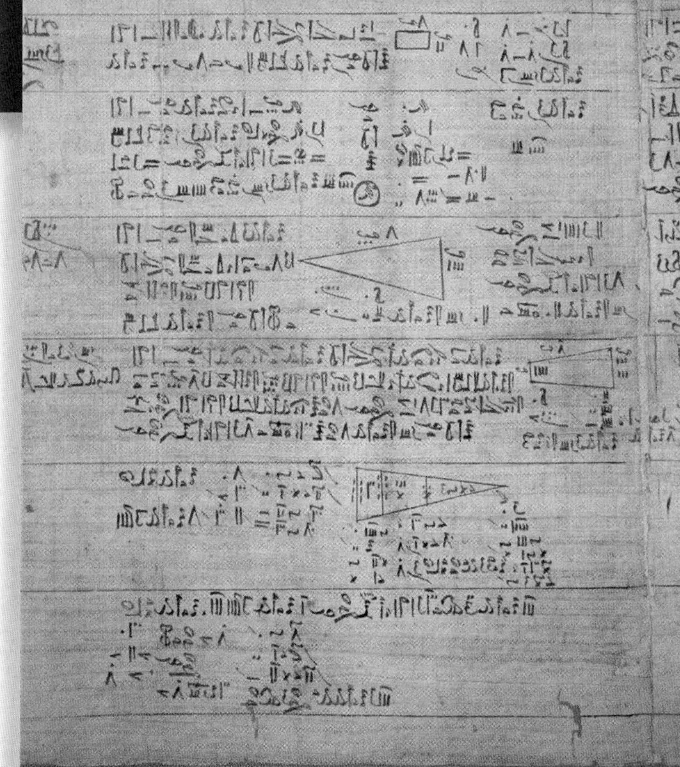

The Human Side of Algebra

Most of the mathematics used in ancient Egypt is preserved in the Rhind papyrus, a document bought in 1858 in Luxor, Egypt, by Henry Rhind. Problem 24 in this document reads:

> A quantity and its $\frac{1}{7}$ added become 19. What is the quantity?

If we let q represent the quantity, we can translate the problem as

$$q + \frac{1}{7}q = 19$$

Unfortunately, the Egyptians were unable to simplify $q + \frac{1}{7}q$ because the sum is $\frac{8}{7}q$, and they didn't have a notation for the fraction $\frac{8}{7}$. How did they attempt to find the answer? They used the method of "false position." That is, they *assumed* the answer was 7, which yields $7 + \frac{1}{7} \cdot 7$ or 8. But the sum as stated in the problem is not 8, it's 19. How can you make the 8 into a 19? By multiplying by $\frac{19}{8}$, that is, by finding $8 \cdot \frac{19}{8} = 19$. But if you multiply the 8 by $\frac{19}{8}$, you should also multiply the assumed answer, 7, by $\frac{19}{8}$ to obtain the true answer, $7 \cdot \frac{19}{8}$. If you solve the equation $\frac{8}{7}q = 19$, you will see that $7 \cdot \frac{19}{8}$ is indeed the correct answer!

2.1 The Addition and Subtraction Properties of Equality

Objectives

A > Determine whether a number satisfies an equation.

B > Use the addition and subtraction properties of equality to solve equations.

C > Use both properties together to solve an equation.

To Succeed, Review How To . . .

1. Add and subtract real numbers (pp. 52–56).
2. Follow the correct order of operations (pp. 69–74).
3. Simplify expressions (pp. 79, 88–95).

Getting Started

A Lot of Garbage!

In this section, we study some ideas that will enable us to solve equations. Do you know what an equation is? Here is an example. How much waste do you generate every day? According to Franklin Associates, LTD, the average American generates about 2.7 pounds of waste daily if we *exclude* paper products! If w and p represent the total amount of waste and paper products (respectively) generated daily by the average American, $w - p = 2.7$. Further research indicates that $p = 1.7$ pounds; thus, $w - 1.7 = 2.7$. The statement

$$w - 1.7 = 2.7$$

is an *equation,* a statement indicating that two expressions are equal. Some equations are *true* $(1 + 1 = 2)$, some are *false* $(2 - 5 = 3)$, and some $(w - 1.7 = 2.7)$ are neither true nor false. The equation $w - 1.7 = 2.7$ is a *conditional* equation that is true for certain values of the *variable* or unknown w. To find the total amount of waste generated daily (w), we have to *solve* $w - 1.7 = 2.7$; that is, we must find the value of the variable that makes the equation a true statement. We learn how to do this next.

A > Verifying Solutions to an Equation

In the equation $w - 1.7 = 2.7$ in the *Getting Started,* the variable w can be replaced by many numbers, but only *one* number will make the resulting statement *true*. This number is called the *solution* of the equation. Can you find the solution of $w - 1.7 = 2.7$? Since w is the total amount of waste, $w = 1.7 + 2.7 = 4.4$, and 4.4 is the solution of the equation.

SOLUTIONS

The solutions of an equation are the replacements of the variable that make the equation a *true* statement. When we find the solution of an equation, we say that we have solved the equation.

How do we know whether a given number solves (or *satisfies*) an equation? We write the number in place of the variable in the given equation and see whether the result is true. For example, to decide whether 4 is a solution of the equation

$$x + 1 = 5$$

we replace x by 4. This gives the true statement:

$$4 + 1 = 5$$

so 4 is a solution of the equation $x + 1 = 5$.

EXAMPLE 1 Verifying a solution

Determine whether the given number is a solution of the equation:

a. 9; $x - 4 = 5$ **b.** 8; $5 = 3 - y$ **c.** 10; $\frac{1}{2}z - 5 = 0$

SOLUTION 1 Remember the rule: if you substitute the number for the variable and the result is *true,* the number is a solution.

a. If x is 9, $x - 4 = 5$ becomes $9 - 4 = 5$, which is a *true* statement. Thus, 9 is a solution of the equation.

b. If y is 8, $5 = 3 - y$ becomes $5 = 3 - 8$, which is a *false* statement. Hence, 8 is *not* a solution of the equation.

c. If z is 10, $\frac{1}{2}z - 5 = 0$ becomes $\frac{1}{2}(10) - 5 = 0$, which is a *true* statement. Thus, 10 is a solution of the equation.

PROBLEM 1

Determine whether the given number is a solution of the equation:

a. 7; $x - 5 = 3$

b. 4; $1 = 5 - y$

c. 6; $\frac{1}{3}z - 2 = 0$

We have learned how to determine whether a number satisfies an equation. Now to find such a number, we must find an *equivalent* equation whose solution is obvious.

EQUIVALENT EQUATIONS

Two equations are equivalent if their solutions are the same.

How do we find these equivalent equations? We use the properties of equality.

B > Using the Addition and Subtraction Properties of Equality

Look at the ad in the illustration. It says that $5 has been cut from the price of a gallon of paint and that the sale price is $6.69. What was the old price p of the paint? Since the old price p was cut by $5, the new price is $p - 5$. Since the new price is $6.69,

$$p - 5 = 6.69$$

To find the old price p, we add back the $5 that was cut. That is,

$$p - 5 + 5 = 6.69 + 5$$ We add 5 to both sides of the equation to obtain an equivalent equation.

$$p = 11.69$$

Thus, the old price was $11.69; this can be verified, since $11.69 - 5 = 6.69$. Note that by adding 5 to both sides of the equation $p - 5 = 6.69$, we produced an equivalent equation,

Answers to PROBLEMS

1. a. No **b.** Yes **c.** Yes

$p = 11.69$, whose solution is obvious. This example illustrates the fact that we can *add* the same number on both sides of an equation and produce an *equivalent* equation—that is, an equation whose solution is identical to the solution of the original one. Here is the property.

THE ADDITION PROPERTY OF EQUALITY

For any number *c*, the equation *a* = *b* is equivalent to

$$a + c = b + c$$

We use this property in Example 2.

EXAMPLE 2 Using the addition property

Solve:

a. $x - 3 = 9$ **b.** $x - \frac{1}{7} = \frac{5}{7}$

SOLUTION 2

a. This problem is similar to our example of paint prices. To solve the equation, we need x by itself on one side of the equation. We can achieve this by adding 3 (the additive inverse of -3) on both sides of the equation.

$$x - 3 = 9$$
$$x - 3 + 3 = 9 + 3 \quad \text{Add 3 to both sides.}$$
$$x = 12$$

Thus, 12 is the solution of $x - 3 = 9$.

CHECK Substituting 12 for x in the original equation, we have $12 - 3 = 9$, a true statement.

b.

$$x - \frac{1}{7} = \frac{5}{7}$$
$$x - \frac{1}{7} + \frac{1}{7} = \frac{5}{7} + \frac{1}{7} \quad \text{Add } \tfrac{1}{7} \text{ to both sides.}$$
$$x = \frac{6}{7} \quad \text{Simplify.}$$

Thus, $\frac{6}{7}$ is the solution of $x - \frac{1}{7} = \frac{5}{7}$.

CHECK $\frac{6}{7} - \frac{1}{7} = \frac{5}{7}$ is a true statement.

PROBLEM 2

Solve:

a. $x - 5 = 7$ **b.** $x - \frac{1}{5} = \frac{3}{5}$

Sometimes it's necessary to simplify an equation before we isolate x on one side. For example, to solve the equation

$$3x + 5 - 2x - 9 = 6x + 5 - 6x$$

we first simplify both sides of the equation by collecting like terms:

$$3x + 5 - 2x - 9 = 6x + 5 - 6x$$

$$(3x - 2x) + (5 - 9) = (6x - 6x) + 5 \quad \text{Group like terms.}$$
$$x + (-4) = 0 + 5 \quad \text{Combine like terms.}$$
$$x - 4 = 5 \quad \text{Rewrite } x + (-4) \text{ as } x - 4.$$
$$x - 4 + 4 = 5 + 4 \quad \text{Now add 4 to both sides.}$$
$$x = 9$$

Thus, 9 is the solution of the equation.

Answers to PROBLEMS

2. a. 12 **b.** $\frac{4}{5}$

CHECK We substitute 9 for x in the original equation. To save time, we use the following diagram where $\stackrel{?}{=}$ means "are they equal?"

$3x + 5 - 2x - 9 \stackrel{?}{=}$	$6x + 5 - 6x$
$3(9) + 5 - 2(9) - 9$	$6(9) + 5 - 6(9)$
$27 + 5 - 18 - 9$	$54 + 5 - 54$
$32 - 18 - 9$	5
$14 - 9$	
5	

Since both sides yield 5, our result is correct.

Now suppose that the price of an article is increased by \$3 and the article currently sells for \$8. What was its old price, p? The equation here is

$$\underbrace{p}_{\text{Old price}} \quad \underbrace{+}_{\text{went up}} \quad \underbrace{3}_{\$3} \quad \underbrace{=}_{\text{and is now}} \quad \underbrace{8}_{\$8.}$$

To solve this equation, we have to bring the price down; that is, we need to subtract 3 on both sides of the equation:

$$p + 3 - 3 = 8 - 3$$
$$p = 5$$

Thus, the old price was \$5. We have *subtracted* 3 on both sides of the equation. Here is the property that allows us to do this.

THE SUBTRACTION PROPERTY OF EQUALITY

For any number c, the equation $a = b$ is equivalent to

$$a - c = b - c$$

This property tells us that we can *subtract* the same number on both sides of an equation to produce an *equivalent* equation. Note that since $a - c = a + (-c)$, you can think of *subtracting* c as *adding* $(-c)$.

EXAMPLE 3 Using the subtraction property

Solve:

a. $2x + 4 - x + 2 = 10$

b. $-3x + \frac{5}{7} + 4x - \frac{3}{7} = \frac{6}{7}$

SOLUTION 3

a. To solve the equation, we need to get x by itself on the left; that is, we want $x = \square$, where $\square$ is a number. We proceed as follows:

$$2x + 4 - x + 2 = 10$$
$$x + 6 = 10 \quad \text{Simplify.}$$
$$x + 6 - 6 = 10 - 6 \quad \text{Subtract 6 from both sides.}$$
$$x = 4$$

Thus, 4 is the solution of the equation.

CHECK

$2x + 4 - x + 2 \stackrel{?}{=}$	10
$2(4) + 4 - 4 + 2$	10
$8 + 2$	
10	

PROBLEM 3

Solve:

a. $5y + 2 - 4y + 3 = 17$

b. $-5z + \frac{3}{8} + 6z - \frac{1}{8} = \frac{5}{8}$

(continued)

Answers to PROBLEMS

3. a. 12 **b.** $\frac{3}{8}$

b. $-3x + \frac{5}{7} + 4x - \frac{3}{7} = \frac{6}{7}$ (combine $-3x$ and $4x$; combine $\frac{5}{7}$ and $-\frac{3}{7}$)

$$x + \frac{2}{7} = \frac{6}{7}$$ Simplify.

$$x + \frac{2}{7} - \frac{2}{7} = \frac{6}{7} - \frac{2}{7}$$ Subtract $\frac{2}{7}$ from both sides.

$$x = \frac{4}{7}$$

Thus, $\frac{4}{7}$ is the solution of the equation.

CHECK

$-3x + \frac{5}{7} + 4x - \frac{3}{7}$	$\stackrel{?}{=} \frac{6}{7}$
$-3\left(\frac{4}{7}\right) + \frac{5}{7} + 4\left(\frac{4}{7}\right) - \frac{3}{7}$	$\frac{6}{7}$
$\underbrace{-\frac{12}{7} + \frac{5}{7}} + \underbrace{\frac{16}{7} - \frac{3}{7}}$	
$-\frac{7}{7} + \frac{13}{7}$	
$\frac{6}{7}$	

C › Using the Addition and Subtraction Properties Together

Can we solve the equation $2x - 7 = x + 2$? Let's try. If we add 7 to both sides, we obtain

$$2x - 7 + 7 = x + 2 + 7$$

$$2x = x + 9$$

But this is not yet a solution. To solve this equation, we must get x by itself on the left—that is, $x = \square$, where $\square$ is a number. How do we do this? We want variables on one side of the equation (and we have them: $2x$) but only specific numbers on the other (here we are in trouble because we have an x on the right). To "get rid of" this x, we subtract x from both sides:

$$2x - x = x - x + 9$$ Remember, $x = 1x$.

$$x = 9$$

Thus, 9 is the solution of the equation.

CHECK

$2x - 7$	$\stackrel{?}{=} x + 2$
$2(9) - 7$	$9 + 2$
11	11

By the way, you do not have to have the variable by itself on the *left* side of the equation. You may have the variables on the *right* side of the equation and your solution may be of the form $\square = x$, where $\square$ is a number.

A recommended procedure is to isolate the variable on the side of the equation that contains the highest coefficient of variables after simplification. Thus, when solving $7x + 3 = 4x + 6$, isolate the variables on the left side of the equation. On the other hand, when solving $4x + 6 = 7x + 3$, isolate the variables on the right side of the equation. Obviously, your solution would be the same in either case. [See Example 4(c) where the variables are on the right.]

Now let's review our procedure for solving equations by adding or subtracting.

PROCEDURE

Solving Equations by Adding or Subtracting

1. Simplify both sides if necessary.
2. Add or subtract the same numbers on both sides of the equation so that one side contains only variables.
3. Add or subtract the same expressions on both sides of the equation so that the other side contains only numbers.

We use these three steps to solve Example 4.

EXAMPLE 4 Solving equations by adding or subtracting

Solve:

a. $3 = 8 + x$

b. $4y - 3 = 3y + 8$

c. $0 = 3(z - 2) + 4 - 2z$

d. $2(x + 1) = 3x + 5$

SOLUTION 4

a. $3 = 8 + x$ Given

1. Both sides of the equation are already simplified. $3 = 8 + x$
2. Subtract 8 on both sides. $3 - 8 = 8 - 8 + x$

$-5 = x$

Step 3 is not necessary here, and the solution is -5. $x = -5$

CHECK

$3 \stackrel{?}{=} 8 + x$	
3	$8 + (-5)$
	3

b. $4y - 3 = 3y + 8$ Given

1. Both sides of the equation are already simplified. $4y - 3 = 3y + 8$
2. Add 3 on both sides. $4y - 3 + 3 = 3y + 8 + 3$

$4y = 3y + 11$

3. Subtract $3y$ on both sides. $4y - 3y = 3y - 3y + 11$

$y = 11$

The solution is 11.

CHECK

$4y - 3 \stackrel{?}{=} 3y + 8$	
$4(11) - 3$	$3(11) + 8$
$44 - 3$	$33 + 8$
41	41

c. $0 = 3(z - 2) + 4 - 2z$ Given

1. Simplify by using the distributive property and combining like terms. $0 = 3z - 6 + 4 - 2z$

$0 = z - 2$

PROBLEM 4

Solve:

a. $5 = 7 + x$

b. $5x - 2 = 4x + 3$

c. $0 = 3(y - 3) + 7 - 2y$

d. $3(z + 1) = 4z + 8$

(continued)

Answers to PROBLEMS

4. a. -2 **b.** 5 **c.** 2 **d.** -5

2. Add 2 on both sides.

$$0 + 2 = z - 2 + 2$$
$$2 = z$$

Step 3 is not necessary, and the solution is 2.

$$z = 2$$

CHECK

$0 \overset{?}{=} 3(z - 2) + 4 - 2z$	
0	$3(2 - 2) + 4 - 2(2)$
	$3(0) + 4 - 4$
	$0 + 4 - 4$
	0

d. $2(x + 1) = 3x + 5$ Given

1. Simplify. $2x + 2 = 3x + 5$

2. Subtract 2 on both sides. $2x + 2 - 2 = 3x + 5 - 2$

$$2x = 3x + 3$$

3. Subtract $3x$ on both sides so all the variables are on the left.

$$2x - 3x = 3x - 3x + 3$$
$$-x = 3$$
$$x = -3$$

Note that if $-x = 3$, then $x = -3$ because the opposite of a number is the number with its sign changed; that is, if the opposite of x is 3, then x itself must be -3. Thus, the solution is -3.

CHECK

$2(x + 1)$	$\overset{?}{=} 3x + 5$
$2(-3 + 1)$	$3(-3) + 5$
$2(-2)$	$-9 + 5$
-4	-4

Keep in mind the following rule; we will use it in Example 5.

PROCEDURE

Solving $-x = a$:

If a is a real number and $-x = a$, then $x = -a$.

EXAMPLE 5 Solving equations by adding or subtracting

Solve: $8x + 7 = 9x + 3$

SOLUTION 5

1. The equation is already simplified. $8x + 7 = 9x + 3$ Given

2. Subtract 7 on both sides. $8x + 7 - 7 = 9x + 3 - 7$

$$8x = 9x - 4$$

3. Subtract $9x$ on both sides.

$$8x - 9x = 9x - 9x - 4$$
$$-x = -4$$
$$x = 4$$

Note that since $-x = -4$, then $x = -(-4) = 4$, so the solution is 4.

CHECK

$8x + 7$	$\overset{?}{=} 9x + 3$
$8(4) + 7$	$9(4) + 3$
$32 + 7$	$36 + 3$
39	39

PROBLEM 5

Solve: $6y + 5 = 7y + 2$

Answers to PROBLEMS

5. 3

The equations in Examples 2–5 each had exactly one solution. For an equation that can be written as $ax + b = c$, there are three possibilities for the solution:

1. The equation has **one** solution. This is a **conditional** equation.
2. The equation has **no** solution. This is a **contradictory** equation.
3. The equation has **infinitely many** solutions. This is an **identity.**

EXAMPLE 6 Solving a contradictory equation

Solve: $3 + 8(x + 1) = 5 + 8x$

SOLUTION 6

$$3 + 8(x + 1) = 5 + 8x \quad \text{Given}$$

1. Simplify by using the distributive property and combining like terms.

$$3 + 8x + 8 = 5 + 8x$$
$$11 + 8x = 5 + 8x$$

2. Subtract 5 on both sides.

$$11 - 5 + 8x = 5 - 5 + 8x$$
$$6 + 8x = 8x$$

3. Subtract $8x$ on both sides.

$$6 + 8x - 8x = 8x - 8x$$
$$6 = 0$$

The statement "6 = 0" is a *false* statement. When this happens, it indicates that the equation has *no* solution—that is, it's a **contradictory** equation and we write "no solution."

PROBLEM 6

Solve: $5 + 3(z - 1) = 4 + 3z$

EXAMPLE 7 Solving an identity

Solve: $7 + 2(x + 1) = 9 + 2x$

SOLUTION 7

$$7 + 2(x + 1) = 9 + 2x \quad \text{Given}$$

1. Simplify by using the distributive property and combining like terms.

$$7 + 2x + 2 = 9 + 2x$$
$$9 + 2x = 9 + 2x$$

You could stop here. Since both sides are *identical,* this equation is an identity. Every real number is a solution. But what happens if you go on? Let's see.

2. Subtract 9 on both sides.

$$9 - 9 + 2x = 9 - 9 + 2x$$
$$2x = 2x$$

3. Subtract $2x$ on both sides.

$$2x - 2x = 2x - 2x$$
$$0 = 0$$

The statement "0 = 0" is a true statement. When this happens, it indicates that *any* real number is a solution. (Try $x = -1$ or $x = 0$ in the original equation.) The equation has **infinitely** many solutions, and we write "all real numbers" for the solution.

PROBLEM 7

Solve: $3 + 4(y + 2) = 11 + 4y$

Answers to PROBLEMS

6. No solution

7. All real numbers

How many miles per gallon (mpg) does your car get? It is going to change! The CAFE (**C**orporate **A**verage **F**uel **E**conomy) standards are federal regulations intended to improve the average fuel economy of cars and light trucks. The combined fuel economy E can be approximated by $E = 0.94N + 27.2$, where N is the number of years after 2010. Based on this formula, can you predict the mileage in 2015? We will do that next.

EXAMPLE 8 CAFE combined mileage in 2015

Use the formula $E = 0.94N + 27.2$, where N is the number of years after 2010, to predict the *combined* fuel economy for passenger cars and light trucks in miles per gallon.

a. 2010 ($N = 0$) **b.** 2015

SOLUTION 8

a. In 2010, $N = \mathbf{0}$ and $E = 0.94(\mathbf{0}) + 27.2 = 27.2$ mpg.

b. The year 2015 is 5 years after 2010, so $N = 5$. When $N = \mathbf{5}$, $E = 0.94(\mathbf{5}) + 27.2 = 4.7 + 27.2 = 31.9$ mpg. Thus, your predicted mpg in 2015 will be 31.9.

There is one problem: critics contend that the CAFE objective of 35.5 mpg for passenger cars by 2015 is not equivalent to the EPA standards. Here is what they say: "A 25–27 mpg EPA rating is equivalent to a 35 mpg CAFE rating" and "today's 27.5 mpg CAFE standard for passenger cars equates to about 21 miles per gallon on an EPA window sticker." The formula in Problem 8 predicts the combined fuel economy for passenger cars using the EPA standards.

Sources: http://tinyurl.com/ossjlx, http://tinyurl.com/233e99x.

PROBLEM 8

The formula $P = 1.1N + 30.4$, where N is the number of years after 2010 is used by the EPA to predict the fuel economy for passenger cars. Use the formula to find the fuel economy for passenger cars in:

a. 2010

b. 2015

Exercises 2.1

Web IT go to mhhe.com/bello for more lessons

A **Verifying Solutions to an Equation** In Problems 1–10, determine whether the given number is a solution of the equation. (Do not solve.)

1. $x = 3; x - 1 = 2$

2. $x = 4; 6 = x - 10$

3. $y = -2; 3y + 6 = 0$

4. $z = -3; -3z + 9 = 0$

5. $n = 2; 12 - 3n = 6$

6. $m = 3\frac{1}{2}; 3\frac{1}{2} + m = 7$

7. $d = 10; \frac{2}{5}d + 1 = 3$

8. $c = 2.3; 3.4 = 2c - 1.4$

9. $a = 2.1; 4.6 = 11.9 - 3a$

10. $x = \frac{1}{10}; 0.2 = \frac{7}{10} - 5x$

B **Using the Addition and Subtraction Properties of Equality** In Problems 11–30, solve and check the given equations.

11. $x - 5 = 9$

12. $y - 3 = 6$

13. $11 = m - 8$

14. $6 = n - 2$

15. $y - \frac{2}{3} = \frac{8}{3}$

16. $R - \frac{4}{3} = \frac{35}{3}$

17. $2k - 6 - k - 10 = 5$

18. $3n + 4 - 2n - 6 = 7$

19. $\frac{1}{4} = 2z - \frac{2}{3} - z$

20. $\frac{7}{2} = 3v - \frac{1}{5} - 2v$

21. $0 = 2x - \frac{3}{2} - x - 2$

22. $0 = 3y - \frac{1}{4} - 2y - \frac{1}{2}$

Answers to PROBLEMS

8. a. 30.4 mpg **b.** 35.9 mpg

23. $\frac{1}{5} = 4c + \frac{1}{5} - 3c$

24. $0 = 6b + \frac{19}{2} - 5b - \frac{1}{2}$

25. $-3x + 3 + 4x = 0$

26. $-5y + 4 + 6y = 0$

27. $\frac{3}{4}y + \frac{1}{4} + \frac{1}{4}y = \frac{1}{4}$

28. $\frac{1}{2}y + \frac{2}{3} + \frac{1}{2}y = \frac{1}{3}$

29. $3.4 = -3c + 0.8 + 2c + 0.1$

30. $1.7 = -3c + 0.3 + 4c + 0.4$

⟨C⟩ Using the Addition and Subtraction Properties Together In Problems 31–55, solve and check the given equations.

31. $6p + 9 = 5p$

32. $7q + 4 = 6q$

33. $3x + 3 + 2x = 4x$

34. $2y + 4 + 6y = 7y$

35. $4(m - 2) + 2 - 3m = 0$

36. $3(n + 4) + 2 = 2n$

37. $5(y - 2) = 4y + 8$

38. $3(z - 1) = 4z + 1$

39. $3a - 1 = 2(a - 4)$

40. $4(b + 1) = 5b - 3$

41. $5(c - 2) = 6c - 2$

42. $-4R + 6 = 5 - 3R + 8$

43. $3x + 5 - 2x + 1 = 6x + 4 - 6x$

44. $6f - 2 - 4f = -2f + 5 + 3f$

45. $-2g + 4 - 5g = 6g + 1 - 14g$

46. $-2x + 3 + 9x = 6x - 1$

47. $6(x + 4) + 4 - 2x = 4x$

48. $6(y - 1) - 2 + 2y = 8y + 4$

49. $10(z - 2) + 10 - 2z = 8(z + 1) - 18$

50. $7(a + 1) - 1 - a = 6(a + 1)$

51. $3b + 6 - 2b = 2(b - 2) + 4$

52. $3b + 2 - b = 3(b - 2) + 5$

53. $2p + \frac{2}{3} - 5p = -4p + 7\frac{1}{3}$

54. $4q + \frac{2}{7} - 6q = -3q + 2\frac{2}{7}$

55. $5r + \frac{3}{8} - 9r = -5r + 1\frac{1}{2}$

››› Applications

56. *Price increases* The price of an item is increased by $7; it now sells for $23. What was the old price of the item?

57. *Average hourly earnings* In a certain year, the average hourly earnings were $9.81, an increase of 40¢ over the previous year. What were the average hourly earnings the previous year?

58. *Consumer Price Index* The Consumer Price Index for housing in a recent year was 169.6, a 5.7-point increase over the previous year. What was the Consumer Price Index for housing the previous year?

59. *Medical costs* The cost of medical care increased 142.2 points in a 6-year period. If the cost of medical care reached the 326.9 mark, what was it 6 years ago?

60. *SAT scores* In the last 10 years, mathematics scores in the Scholastic Aptitude Test (SAT) have declined 16 points, to 476. What was the mathematics score 10 years ago?

››› Applications: Green Math

61. *Waste generation* From 1960 to 2007, the amount of waste generated each year increased by a whopping 166 million tons, ultimately reaching 254.1 million tons! How much waste was generated in 1960? The figure reached 234 million tons in 2000. What was the increase (in million tons) from 1960 to 2000?

62. *Materials recovery* From 1960 to 2007, the amount of materials recovered for recycling increased by 57.7 million tons to 63.3 million tons. What amount of materials was recovered for recycling in 1960? By 2007, 10.4 million more tons were recovered for recycling than in 2000. How many million tons were recovered for recycling in 2000?

Gas Expenditures and Green Car Costs

63. *Motor oil and fuel costs* How much do you spend annually in motor oil and fuel? According to the Bureau of Labor, if you are under 25 the annual expenditures E (in dollars) can be approximated by $E = 111N + 1534$, where N is the number of years after 2005.
   a. Use the formula to find the motor oil and fuel expenditures in 2005.
   b. Use the formula to predict the expenditures in 2015.

64. *Don't forget about the insurance*
   a. Use the formula $I = -67N + 620$, where N is the number of years after 2005 to find the annual insurance cost I in dollars for a person under age 25 in 2005.
   b. According to the formula, what would the cost be 10 years after 2005?
   c. Does the answer to part **b** make sense? Explain.

   Note: You can get a 5%–10% insurance discount if you drive a hybrid car.

› Web IT go to **mhhe.com/bello** *for more lessons*

65. *Green can cost more!* You can buy a natural gas-fueled 2009 Honda Civic GX by paying $6635 dollars **more** than the $18,555 you pay for the gasoline powered 2009 Honda Civic LX-S. How much is the Honda Civic GX?

66. *Green can be less!* If you do buy the natural gas-fueled Honda Civic GX, your cargo space will be 6 cubic feet **less** than for the gasoline powered Honda Civic LX-S, with 12 cubic feet of cargo space. What is the cargo space for the Honda Civic GX?

What's your ideal weight? In general, it depends on your sex and height but there are many opinions and we can approximate all of them!
Source: http://www.halls.md/

67. *Ideal weight* According to Broca's index, the ideal weight W (in pounds) for a woman h inches tall is $W = 100 + 5(h - 60)$.

a. A woman 62 inches tall weighs 120 pounds. Does her weight satisfy the equation?

b. What should her weight be?

68. *Ideal weight* The ideal weight W (in pounds) for a man h inches tall is $W = 110 + 5(h - 60)$.

a. A man 70 inches tall weighs 160 pounds. Does his weight satisfy the equation?

b. Verify that 160 pounds is the correct weight for this man.

69. *Ideal weight* The B. J. Devine formula for the weight W (in kilograms) of a man h inches tall is $W = 50 + 2.3(h - 60)$.

a. A man 70 inches tall weighs 75 kilograms. Does his weight satisfy the equation?

b. How many kilograms overweight is this man?

70. *Ideal weight* For a woman h inches tall, the Devine formula (in kilograms) is $W = 45.5 + 2.3(h - 60)$.

a. A woman 70 inches tall weighs 68.5 kilograms. Does this weight satisfy the equation?

b. Verify that 68.5 kilograms is her correct weight.

71. *Ideal weight* The J. D. Robinson formula suggests a weight $W = 52 + 1.9(h - 60)$ (in kilograms) for a man h inches tall.

a. A 70-inch-tall man weighs 70 kilograms. Does this weight satisfy the equation?

b. How many kilograms underweight is this man?

72. *Ideal weight* The Robinson formula for the weight W of women h inches tall is $W = 49 + 1.7(h - 60)$.

a. A 70-inch-tall woman weighs 65 kilograms. Does that weight satisfy the equation?

b. Is she over or under weight, and by how much?

73. *Ideal weight* Miller's formula defines the weight W of a man h inches tall as $W = 56.2 + 1.41(h - 60)$ kilograms.

a. A 70-inch-tall man weighs 70.3 kilograms. Does his weight satisfy the equation?

b. Verify that 70.3 kilograms is the correct weight for this man.

How many calories can you eat and still lose weight?
Source: http://www.annecollins.com

74. *Caloric intake* The daily caloric intake C needed for a woman to lose 1 pound per week is given by $C = 12W - 500$, where W is her weight in pounds.

a. A woman weighs 120 pounds and consumes 940 calories each day. Does the caloric intake satisfy the equation?

b. Verify that the 940 caloric intake satisfy the equation.

75. *Caloric intake* The daily caloric intake C needed for a man to lose 1 pound per week is given by $C = 14W - 500$, where W is his weight in pounds.

a. A man weighs 150 pounds and eats 1700 calories each day. Does the caloric intake satisfy the equation?

b. How many calories over or under is he?

76. *Caloric intake* The daily caloric intake C for a very active person to lose one pound per week is given by $C = 17W - 500$, where W is the weight of the person.

a. A person weighs 200 pounds and consumes 2900 calories daily. Does the caloric intake satisfy the equation?

b. Verify that the 2900 calories per day satisfy the equation.

Using Your Knowledge

Some Detective Work In this section, we learned how to determine whether a given number *satisfies* an equation. Let's use this idea to do some detective work!

77. Suppose the police department finds a femur bone from a human female. The relationship between the length f of the femur and the height H of a female (in centimeters) is given by

$$H = 1.95f + 72.85$$

If the length of the femur is 40 centimeters and a missing female is known to be 120 centimeters tall, can the bone belong to the missing female?

78. If the length of the femur in Problem 77 is 40 centimeters and a missing female is known to be 150.85 centimeters tall, can the bone belong to the missing female?

79. Have you seen police officers measuring the length of a skid mark after an accident? There's a formula for this. It relates the velocity V_a at the time of an accident and the length L_a of the skid mark at the time of the accident to the velocity and length of a test skid mark. The test skid mark is obtained by driving a car at a predetermined speed V_t, skidding to a stop, and then measuring the length of the skid L_t. The formula is

$$V_a^2 = \frac{L_a V_t^2}{L_t}$$

If $L_t = 36$, $L_a = 144$, $V_t = 30$, and the driver claims that at the time of the accident his velocity V_a was 50 miles per hour, can you believe him?

80. Would you believe the driver in Problem 79 if he says he was going 90 miles per hour?

Write On

81. Explain what is meant by the solution of an equation.

82. Explain what is meant by equivalent equations.

83. Make up an equation that has no solution and one that has infinitely many solutions.

84. If the next-to-last step in solving an equation is $-x = -5$, what is the solution of the equation? Explain.

Concept Checker

Fill in the blank(s) with the correct word(s), phrase, or mathematical statement.

85. According to the **Addition Property of Equality** for any numbers a, b, and c the equation $\boldsymbol{a = b}$ is **equivalent** to the equation ______________.

86. According to the **Subtraction Property of Equality** for any numbers a, b, and c the equation $\boldsymbol{a = b}$ is **equivalent** to the equation ______________.

a − c = b − c **a + c = b + c**

c − a = b − a **b + a = c + a**

Mastery Test

Solve.

87. $5 + 4(x + 1) = 3 + 4x$

88. $2 + 4(x + 1) = 5x + 3$

89. $x - 5 = 4$

90. $x - \frac{1}{5} = \frac{3}{5}$

91. $x - 2.3 = 3.4$

92. $x - \frac{1}{7} = \frac{1}{4}$

93. $2x + 6 - x + 2 = 12$

94. $3x + 5 - 2x + 3 = 7$

95. $-5x + \frac{2}{9} + 6x - \frac{4}{9} = \frac{5}{9}$

96. $5y - 2 = 4y + 1$

97. $0 = 4(z - 3) + 5 - 3z$

98. $2 - (4x + 1) = 1 - 4x$

99. $3(x + 2) + 3 = 2 - (1 - 3x)$

100. $3(x + 1) - 3 = 2x - 5$

Determine whether the given number is a solution of the equation.

101. $-7; \frac{1}{7}z - 1 = 0$

102. $-\frac{3}{5}; \frac{5}{3}x + 1 = 0$

Skill Checker

Perform the indicated operation.

103. $4(-5)$

104. $6(-3)$

105. $-\frac{2}{3}\left(\frac{3}{4}\right)$

106. $-\frac{5}{7}\left(\frac{7}{10}\right)$

Find the reciprocal of each number.

107. $\frac{3}{2}$

108. $\frac{2}{5}$

Find the LCM of each pair of numbers.

109. 6 and 16

110. 9 and 12

111. 10 and 8

112. 30 and 18

2.2 The Multiplication and Division Properties of Equality

Objectives

A Use the multiplication and division properties of equality to solve equations.

B Multiply by reciprocals to solve equations.

C Multiply by LCMs to solve equations.

D Solve applications involving percents.

To Succeed, Review How To . . .

1. Multiply and divide signed numbers (pp. 61, 63–64).
2. Find the reciprocal of a number (pp. 64, 80).
3. Find the LCM of two or more numbers (pp. 13, 15).
4. Write a fraction as a percent, and vice versa (pp. 27–29).

Getting Started

How Good a Deal Is This?

The tire in the ad is on sale at half price. It now costs \$28. What was its old price, p? Since you are paying half price for the tire, the new price is $\frac{1}{2}$ of p—that is, $\frac{1}{2}p$ or $\frac{p}{2}$. Since this price is \$28, we have

$$\frac{p}{2} = 28$$

But what was the old price? Twice as much, of course. Thus, to obtain the old price p, we multiply both sides of the equation by 2, the reciprocal of $\frac{1}{2}$, to obtain

$$2 \cdot \frac{p}{2} = 2 \cdot 28 \quad \text{Note that } \cancel{2} \cdot \frac{p}{\cancel{2}} = 1p \text{ or } p.$$

$$p = 56$$

Hence, the old price was \$56, as can be easily checked, since $\frac{56}{2} = 28$.

This example shows how you can *multiply* both sides of an equation by a nonzero number and obtain an *equivalent* equation—that is, an equation whose solution is the same as the original one. This is the multiplication property, one of the properties we will study in this section.

A › Using the Multiplication and Division Properties of Equality

The ideas discussed in the *Getting Started* can be generalized as the following property.

THE MULTIPLICATION PROPERTY OF EQUALITY

For any nonzero number c, the equation $a = b$ is equivalent to

$$ac = bc$$

This means that we can multiply both sides of an equation by the same nonzero number and obtain an equivalent equation. We use this property in Example 1.

EXAMPLE 1 Using the multiplication property

Solve:

a. $\frac{x}{3} = 2$

b. $\frac{y}{5} = -3$

SOLUTION 1

a.

$\frac{x}{3} = 2$ Given

$3 \cdot \frac{x}{3} = 3 \cdot 2$ Multiply both sides of $\frac{x}{3} = 2$ by 3, the reciprocal of $\frac{1}{3}$.

$\overset{1}{\cancel{3}} \cdot \frac{x}{\cancel{3}} = 6$ Note that $3 \cdot \frac{1}{3} = 1$, since 3 and $\frac{1}{3}$ are reciprocals.

$x = 6$

Thus, the solution is 6.

CHECK

$\frac{x}{3} \overset{?}{=} 2$	
$\frac{6}{3}$	2
2	

b.

$\frac{y}{5} = -3$ Given

$5 \cdot \frac{y}{5} = 5(-3)$ Multiply both sides of $\frac{y}{5} = -3$ by 5, the reciprocal of $\frac{1}{5}$.

$\overset{1}{\cancel{5}} \cdot \frac{y}{\cancel{5}} = -15$ Recall that $5 \cdot (-3) = -15$.

$y = -15$

Thus, the solution is -15.

CHECK

$\frac{y}{5} \overset{?}{=} -3$	
$\frac{-15}{5}$	-3
-3	

PROBLEM 1

Solve:

a. $\frac{x}{5} = 3$ **b.** $\frac{y}{4} = -5$

Suppose the price of an article is doubled, and it now sells for \$50. What was its original price, p? Half as much, right? Here is the equation:

$$2p = 50$$

We solve it by dividing both sides by 2 (to find half as much):

$$\frac{2p}{2} = \frac{50}{2}$$

$$\frac{\overset{1}{\cancel{2}}p}{\cancel{2}} = 25$$

Thus, the original price p is \$25, as you can check:

$$2 \cdot 25 = 50$$

Note that dividing both sides by 2 (the *coefficient* of p) is the same as *multiplying* by $\frac{1}{2}$. Thus, you can also solve $2p = 50$ by multiplying by $\frac{1}{2}$ (the reciprocal of 2) to obtain

$$\frac{1}{2} \cdot 2p = \frac{1}{2} \cdot 50$$

$$p = 25$$

This example suggests that, just as the addition property of equality lets us *add* a number on each side of an equation, the division property lets us *divide* each side of an equation by a (nonzero) number to obtain an *equivalent* equation. We now state this property and use it in the next example.

Answers to PROBLEMS

1. a. 15 **b.** -20

THE DIVISION PROPERTY OF EQUALITY

For any nonzero number c, the equation $a = b$ is equivalent to

$$\frac{a}{c} = \frac{b}{c}$$

This means that we can divide both sides of an equation by the same nonzero number and obtain an equivalent equation. Note that we can also *multiply* both sides of $a = b$ by the *reciprocal* of c—that is, by $\frac{1}{c}$—to obtain

$$\frac{1}{c} \cdot a = \frac{1}{c} \cdot b \quad \text{or} \quad \frac{a}{c} = \frac{b}{c} \qquad \text{Same result!}$$

We shall solve equations by multiplying by reciprocals later in this section.

EXAMPLE 2 Using the division property

Solve:

a. $8x = 24$ **b.** $5x = -20$ **c.** $-3x = 7$

SOLUTION 2

a. We need to get x by itself on the left. That is, we need $x = \square$, where $\square$ is a number.

$$8x = 24 \qquad \text{Given}$$

$$\frac{8x}{8} = \frac{24}{8} \qquad \text{Divide both sides of the equation by 8 (the coefficient of } x\text{).}$$

$$\frac{\overset{1}{\cancel{8}}x}{\cancel{8}} = 3$$

$$x = 3$$

The solution is 3.

CHECK

$8x \overset{?}{=} 24$	
$8 \cdot 3$	24
24	

You can also solve this problem by *multiplying* both sides of $8x = 24$ by $\frac{1}{8}$, the reciprocal of 8.

$$\frac{1}{8} \cdot 8x = \frac{1}{8} \cdot 24$$

$$1x = \frac{1}{\cancel{8}} \cdot \frac{\overset{3}{\cancel{24}}}{1}$$

$$x = 3$$

b. $5x = -20$ Given

$$\frac{\overset{1}{\cancel{5}}x}{\cancel{5}} = \frac{-20}{5} \qquad \text{Divide both sides of } 5x = -20 \text{ by 5 (the coefficient of } x\text{).}$$

$$x = -4$$

The solution is -4.

CHECK

$5x \overset{?}{=} -20$	
$5 \cdot (-4)$	-20
-20	

Of course, you can also solve this problem by *multiplying* both sides by $\frac{1}{5}$, the reciprocal of 5. When solving these types of equations, you always have the option

PROBLEM 2

Solve:

a. $3x = 12$

b. $7x = -21$

c. $-5x = 20$

Answers to PROBLEMS

2. a. 4 **b.** -3 **c.** -4

of dividing both sides of the equation by a specific number or multiplying both sides of the equation by the reciprocal of the number.

$$\frac{1}{5}\cdot 5x = \frac{1}{5}\cdot(-20)$$

$$1x = \frac{1}{\cancel{5}_1}\cdot\frac{\cancel{-20}^{-4}}{1}$$

$$x = -4$$

c. $-3x = 7$ Given

$\frac{\cancel{-3}x}{\cancel{-3}} = \frac{7}{-3}$ Divide both sides of $-3x = 7$ by -3 (the coefficient of x).

$x = -\frac{7}{3}$ Recall that the quotient of two numbers with different signs is negative.

The solution is $-\frac{7}{3}$.

CHECK

$-3x \overset{?}{=} 7$	
$-3\left(-\frac{7}{3}\right)$	7
7	

As you know, this problem can also be solved by multiplying both sides by $-\frac{1}{3}$, the reciprocal of -3.

$$-\frac{1}{3}\cdot(-3x) = -\frac{1}{3}\cdot(7)$$

$$1x = -\frac{1}{3}\cdot\frac{7}{1}$$

$$x = -\frac{7}{3}$$

B > Multiplying by Reciprocals

In Example 2, the coefficients of the variables were integers. In such cases, it's easy to divide each side of the equation by this coefficient. When the coefficient of the variable is a fraction, it's easier to multiply each side of the equation by the reciprocal of the coefficient. Thus, to solve $-3x = 7$, divide each side by -3, but to solve $\frac{3}{4}x = 18$, multiply each side by the reciprocal of the coefficient of x, that is, by $\frac{4}{3}$, as shown next.

EXAMPLE 3 Solving equations by multiplying by reciprocals

Solve:

a. $\frac{3}{4}x = 18$ **b.** $-\frac{2}{5}x = 8$ **c.** $-\frac{3}{8}x = -15$

SOLUTION 3

a. $\frac{3}{4}x = 18$ Given

$\frac{4}{3}\left(\frac{3}{4}x\right) = \frac{4}{3}(18)$ Multiply both sides of $\frac{3}{4}x = 18$ by the reciprocal of $\frac{3}{4}$, that is, by $\frac{4}{3}$.

$$1\cdot x = \frac{4}{\cancel{3}_1}\cdot\frac{\cancel{18}^{6}}{1} = \frac{24}{1}$$

$$x = 24$$

Hence the solution is 24.

PROBLEM 3

Solve:

a. $\frac{3}{5}x = 12$

b. $-\frac{2}{5}x = 6$

c. $-\frac{4}{5}x = -8$

(continued)

Answers to PROBLEMS

3. a. 20 **b.** −15 **c.** 10

CHECK

$$\frac{3}{4}x \stackrel{?}{=} 18$$

$\frac{3}{4}(24)$	18
$\frac{3}{\cancel{4}_1}\cdot\frac{\cancel{24}^6}{1}$	
18	

b. $-\frac{2}{5}x = 8$ Given

$-\frac{5}{2}\left(-\frac{2}{5}x\right) = -\frac{5}{2}(8)$ Multiply both sides of $-\frac{2}{5}x = 8$ by the reciprocal of $-\frac{2}{5}$, that is, by $-\frac{5}{2}$.

$$1 \cdot x = \frac{-5}{\cancel{2}_1}\cdot\frac{\cancel{8}^4}{1} = -\frac{20}{1}$$

$$x = -20$$

The solution is -20.

CHECK

$$-\frac{2}{5}x \stackrel{?}{=} 8$$

$-\frac{2}{5}(-20)$	8
$-\frac{2}{\cancel{5}_1}\left(\frac{\cancel{-20}^{-4}}{1}\right)$	
8	

c. $-\frac{3}{8}x = -15$ Given

$-\frac{8}{3}\left(-\frac{3}{8}x\right) = -\frac{8}{3}(-15)$ Multiply both sides of $-\frac{3}{8}x = -15$ by $-\frac{8}{3}$, the reciprocal of $-\frac{3}{8}$.

$$1 \cdot x = \frac{-8(-15)}{3} = \frac{-8(\cancel{-15}^{-5})}{\cancel{3}_1}$$

$$x = 40$$

The solution is 40.

CHECK

$$-\frac{3}{8}x \stackrel{?}{=} -15$$

$-\frac{3}{8}(40)$	-15
$-\frac{3}{\cancel{8}_1}(\cancel{40}^5)$	
-15	

C › Multiplying by the LCM

Finally, if the equation we are solving contains sums or differences of fractions, we first eliminate these fractions by multiplying each term in the equation by the smallest number that is a multiple of each of the denominators. This number is called the *least common multiple* (or LCM for short) of the denominators. If you forgot about LCMs and fractions, review Sections R.1 and R.2 at the beginning of the book.

Which equation would you rather solve?

$$3x + 2x = 1 \quad \text{or} \quad \frac{x}{2} + \frac{x}{3} = \frac{1}{6}$$

Probably the first! But if you multiply each term in the second equation by the LCM of 2, 3, and 6 (which is 6), you obtain the first equation!

Do you remember how to find the LCM of two numbers? If you don't, here's a quick way to do it. Suppose you wish to solve the equation

$$\frac{x}{6} + \frac{x}{16} = 22$$

To find the LCM of 6 and 16, write the denominators in a horizontal row (see step 1 to the right) and divide each of them by the largest number that will divide *both* of them. In this case, the number is 2. The quotients are 3 and 8, as shown in step 3. Since there are no numbers other than 1 that will divide both 3 and 8, the LCM is the product of 2 and the final quotients 3 and 8, as indicated in step 4.

1. $\underline{|6 \quad 16}$

2. $2\,\underline{|6 \quad 16}$

3. $2\,\underline{|6 \quad 16}$
$\quad 3 \quad 8$

4. $2\,\underline{|6 \quad 16}$
$\quad 3 — 8 \rightarrow 2 \cdot 3 \cdot 8 = 48$ is the LCM.

For an even quicker method, write the multiples of 16 (the larger of the two denominators) until you find a multiple of 16 that is *divisible* by 6. Like this: 16 32 (48)
↑ divisible by 6

The LCM is 48, as before.

Now we can multiply each side of $\frac{x}{6} + \frac{x}{16} = 22$ by the LCM (48):

$$48\left(\frac{x}{6} + \frac{x}{16}\right) = 48 \cdot 22 \qquad \text{To avoid confusion, place parentheses around } \frac{x}{6} + \frac{x}{16}.$$

$$\overset{8}{\cancel{48}} \cdot \frac{x}{\underset{1}{\cancel{6}}} + \overset{3}{\cancel{48}} \cdot \frac{x}{\underset{1}{\cancel{16}}} = 48 \cdot 22 \qquad \text{Use the distributive property. Don't multiply } 48 \cdot 22 \text{ yet, we'll simplify this later.}$$

$$8x + 3x = 48 \cdot 22 \qquad \text{Simplify the left side.}$$

$$11x = 48 \cdot 22 \qquad \text{Combine like terms.}$$

$$\frac{\cancel{11}x}{\cancel{11}} = \frac{48 \cdot \overset{2}{\cancel{22}}}{\underset{1}{\cancel{11}}} \qquad \text{Divide both sides by 11. Do you see why we waited to multiply } 48 \cdot 22\text{?}$$

$$x = 96$$

The solution is 96.

CHECK

$\frac{x}{6} + \frac{x}{16} \overset{?}{=} 22$	
$\frac{96}{6} + \frac{96}{16}$	22
$16 + 6$	
22	

Note that if we wish to clear fractions in

$$\frac{a}{b} + \frac{c}{d} = \frac{e}{f}$$

we can use the following procedure.

PROCEDURE

Clearing Fractions

To clear fractions in an equation, multiply both sides of the equation by the LCM of the denominators, or, equivalently, multiply each term by the LCM.

Thus, if we multiply both sides of $\frac{a}{b} + \frac{c}{d} = \frac{e}{f}$ by the LCM of b, d, and f (which we shall call L), we get

$$L\left(\frac{a}{b} + \frac{c}{d}\right) = L\frac{e}{f}$$ (Where b, d, and $f \neq 0$) Note the added parentheses.

or

$$\frac{La}{b} + \frac{Lc}{d} = \frac{Le}{f}$$ Where b, d, and $f \neq 0$

Thus, to clear the fractions here, we multiply each term by L (using the distributive property).

EXAMPLE 4 Solving equations by multiplying by the LCM

Solve:

a. $\frac{x}{10} + \frac{x}{8} = 9$ **b.** $\frac{x}{3} - \frac{x}{8} = 10$

SOLUTION 4

a. The LCM of 10 and 8 is 40 (since the first four multiples of 10 are 10, 20, 30, and 40, and 8 divides 40). You can also find the LCM by writing

$$2 \,\underline{|10 \quad 8}$$
$$5 — 4 \rightarrow 2 \cdot 5 \cdot 4 = 40$$

Multiplying each term by 40, we have

$$40 \cdot \frac{x}{10} + 40 \cdot \frac{x}{8} = 40 \cdot 9$$
$$4x + 5x = 40 \cdot 9 \quad \text{Simplify.}$$
$$9x = 40 \cdot 9 \quad \text{Combine like terms.}$$
$$x = 40 \quad \text{Divide by 9.}$$

The solution is 40.

CHECK

$\frac{x}{10} + \frac{x}{8}$	$\stackrel{?}{=}$ 9
$\frac{40}{10} + \frac{40}{8}$	9
$4 + 5$	
9	

b. The LCM of 3 and 8 is $3 \cdot 8 = 24$, since the largest number that divides 3 and 8 is 1.

$$1 \,\underline{|3 \quad 8}$$
$$3 — 8 \rightarrow 1 \cdot 3 \cdot 8 = 24$$

Multiplying each term by 24 yields

$$24 \cdot \frac{x}{3} - 24 \cdot \frac{x}{8} = 24 \cdot 10$$
$$8x - 3x = 24 \cdot 10 \quad \text{Simplify.}$$
$$5x = 24 \cdot 10 \quad \text{Combine like terms.}$$
$$\frac{\cancel{5}x}{\cancel{5}} = \frac{24 \cdot \overset{2}{\cancel{10}}}{\underset{1}{\cancel{5}}} \quad \text{Divide by 5.}$$
$$x = 48$$

The solution is 48.

CHECK

$\frac{x}{3} - \frac{x}{8}$	$\stackrel{?}{=}$ 10
$\frac{48}{3} - \frac{48}{8}$	10
$16 - 6$	
10	

PROBLEM 4

Solve:

a. $\frac{x}{10} + \frac{x}{6} = 8$

b. $\frac{x}{4} - \frac{x}{5} = 1$

Answers to PROBLEMS

4. a. 30 **b.** 20

In some cases, the numerators of the fractions involved contain more than one term. However, the procedure for solving the equation is still the same, as we illustrate in Example 5.

EXAMPLE 5 **Solving equations by multiplying by the LCM**

Solve:

a. $\frac{x+1}{3} + \frac{x-1}{10} = 5$ **b.** $\frac{x+1}{3} - \frac{x-1}{8} = 4$

SOLUTION 5

a. The LCM of 3 and 10 is $3 \cdot 10 = 30$, since 3 and 10 don't have any common factors. Multiplying each term by 30, we have

$$\overset{10}{\cancel{30}}\left(\frac{x+1}{\cancel{3}}\right) + \overset{3}{\cancel{30}}\left(\frac{x-1}{\cancel{10}}\right) = 30 \cdot 5 \quad \text{Note the parentheses!}$$

$$10(x+1) + 3(x-1) = 150$$

$$10x + 10 + 3x - 3 = 150 \quad \text{Use the distributive property.}$$

$$13x + 7 = 150 \quad \text{Combine like terms.}$$

$$13x = 143 \quad \text{Subtract 7.}$$

$$x = 11 \quad \text{Divide by 13.}$$

We leave the check for you to verify.

b. Here the LCM is $3 \cdot 8 = 24$. Multiplying each term by 24, we obtain

$$\overset{8}{\cancel{24}}\left(\frac{x+1}{\cancel{3}}\right) - \overset{3}{\cancel{24}}\left(\frac{x-1}{\cancel{8}}\right) = 24 \cdot 4 \quad \text{Note the parentheses!}$$

$$8(x+1) - 3(x-1) = 96$$

$$8x + 8 - 3x + 3 = 96 \quad \text{Use the distributive property.}$$

$$5x + 11 = 96 \quad \text{Combine like terms.}$$

$$5x = 85 \quad \text{Subtract 11.}$$

$$x = 17 \quad \text{Divide by 5.}$$

Be sure you check this answer in the original equation.

PROBLEM 5

Solve:

a. $\frac{x+2}{3} + \frac{x-2}{4} = 6$

b. $\frac{x+2}{5} - \frac{x-2}{3} = 0$

D > Applications Involving Percent Problems

Percent problems are among the most common types of problems, not only in mathematics, but also in many other fields. Basically, there are three types of percent problems.

Type 1 asks you to find a number that is a given percent of a specific number.
Example: 20% (read "20 percent") of 80 is what number?

Type 2 asks you what percent of a number is another given number.
Example: What percent of 20 is 5?

Type 3 asks you to find a number when it's known that a given number equals a percent of the unknown number.
Example: 10 is 40% of what number?

To do these problems, you need only recall how to translate words into equations and how to write percents as fractions (pp. 93–96 and 27–28, respectively).

Now do you remember what 20% means? The symbol % is read as "percent," which means "per hundred." Recall that

$$20\% = \frac{20}{100} = \frac{\overset{1}{\cancel{20}}}{\underset{5}{\cancel{100}}} = \frac{1}{5}$$

Answers to PROBLEMS

5. a. 10 **b.** 8

Similarly,

$$60\% = \frac{60}{100} = \frac{\overset{3}{\cancel{60}}}{\underset{5}{\cancel{100}}} = \frac{3}{5}$$

$$17\% = \frac{17}{100}$$

We are now ready to solve some percent problems.

EXAMPLE 6 Finding a percent of a number (type 1)

a. Twenty percent of 80 is what number?

b. Twenty-five percent of the 260 million tons of garbage generated annually in the United States is recovered for recycling. How many million tons is that?

SOLUTION 6

a. Let's translate this.

20%	of	80	is	what number?
↓	↓	↓	↓	↓
$\frac{20}{100}$	$\cdot$	80	$=$	n

$\frac{1}{5} \cdot 80 = n$ Since $\frac{20}{100} = \frac{1}{5}$

$\frac{80}{5} = n$ Multiply $\frac{1}{5}$ by $\frac{80}{1}$.

$n = 16$ Reduce $\frac{80}{5}$.

Thus, 20% of 80 is 16.

b. We have to find how many tons is 25% of the total 260 million tons generated. The result is the number of million tons r recovered for recycling.

Translating:

25%	of	260	is	what number?
↓	↓	↓	↓	↓
$\frac{25}{100}$	$\cdot$	260	$=$	r

$\frac{1}{4} \cdot \frac{260}{1} = r$ Since $\frac{25}{100} = \frac{1}{4}$.

$\frac{260}{4} = r$ Multiply $\frac{1}{4}$ by $\frac{260}{1}$.

$65 = r$ Reduce $\frac{260}{4} = 65$.

Thus, 25% of 260 is 65, which means that 65 million tons are recovered for recycling.

PROBLEM 6

a. Forty percent of 30 is what number?

b. Fifty-five percent of the 83 million tons of the paper and paperboard are recovered for recycling each year. Find 55% of 83, the amount of paper and paperboard (in millions) recovered for recycling.

GREEN MATH

EXAMPLE 7 Finding a percent (type 2)

a. What percent of 20 is 5?

b. Sixty-five million tons of the 260 million tons of garbage generated annually in the United States is recovered for recycling. What percent of 260 is 65?

PROBLEM 7

a. What percent of 30 is 6?

b. Six million tons of the 16 million tons of steel are recovered for recycling. What percent of 16 is 6?

Answers to PROBLEMS

6. a. 12 **b.** 45.65 **7. a.** 20% **b.** 37.5%

SOLUTION 7

a. Let's translate this.

What percent → x, of → $\cdot$, 20 → 20, is → $=$, 5? → 5

$$x \cdot 20 = 5$$

$$\frac{x \cdot 20}{20} = \frac{5}{20} \qquad \text{Divide by 20.}$$

$$x = \frac{1}{4} \qquad \text{Reduce } \tfrac{5}{20}.$$

But x represents a percent, so we must change $\frac{1}{4}$ to a percent.

$$x = \frac{1}{4} = \frac{25}{100} \text{ or } 25\%$$

Thus, 5 is 25% of 20.

b. We have to find:

What percent → g, of → $\cdot$, the garbage → 260, is → $=$, recovered for recycling? → 65

$$g \cdot 260 = 65$$

$$\frac{g \cdot 260}{260} = \frac{65}{260} \qquad \text{Divide by 260.}$$

$$g = \frac{1}{4} \qquad \text{Reduce } \tfrac{65}{260} = \tfrac{1}{4}.$$

But g represents a percent, so we must change $\frac{1}{4}$ to a percent.

$$g = \frac{1}{4} = \frac{25}{100} = 25\%$$

Thus, 25% of the garbage is recovered for recycling.

GREEN MATH

EXAMPLE 8 Finding a number (type 3)

a. Ten is 40% of what number?

b. Thirty-four percent or 88.4 million tons of materials are recovered each year from the total amount T of garbage generated. What is T, that is, 88.4 is 34% of what number T?

SOLUTION 8

a. First we translate:

10 → 10, is → $=$, 40% → $\frac{40}{100}$, of → $\cdot$, what number? → n

$$10 = \frac{40}{100} \cdot n$$

$$\frac{40}{100} \cdot n = 10 \qquad \text{Rearrange.}$$

$$\frac{2}{5} \cdot n = 10 \qquad \text{Reduce } \tfrac{40}{100}.$$

$$\frac{\cancel{5}}{\cancel{2}} \cdot \frac{\cancel{2}}{\cancel{5}} \cdot n = \frac{5}{\cancel{2}_{1}} \cdot \cancel{10}^{5} \qquad \text{Multiply by } \tfrac{5}{2}.$$

$$n = 25$$

Thus, ten is 40% of 25.

PROBLEM 8

a. Twenty is 40% of what number?

b. Two million tons of textiles are recovered for recycling each year. This represents 16% of the total amount of materials recovered for recycling. Two is 16% of what amount of materials (in millions of tons) recovered for recycling?

Answers to PROBLEMS

8. a. 50 **b.** 12.5

(continued)

b. We have to determine:

88.4 is 34% of what number

$$88.4 = \frac{34}{100} \cdot T \quad \text{(the total generated)?}$$

$$88.4 = \frac{17}{50} \cdot T \qquad \text{Reduce } \frac{34}{100} = \frac{17}{50}.$$

$$\frac{50}{17} \cdot \frac{88.4}{1} = \frac{50}{17} \cdot \frac{17 \cdot T}{50} \qquad \text{Multiply by } \frac{50}{17}.$$

$$260 = T \qquad \frac{50 \cdot 88.4}{17} = \frac{4420}{17} = 260$$

This means that the total amount of garbage generated is 260 million tons.

EXAMPLE 9 Application: Angioplasty versus TPA

The study cited in the margin claims that you can save more lives with angioplasty (a procedure in which a balloon-tipped instrument is inserted in your arteries, the balloon is inflated, and the artery is unclogged!) than with a blood-clot-breaking drug called TPA. Of the 451 patients studied, 226 were randomly assigned for TPA and 225 for angioplasty. After 6 months the results were as follows:

a. 6.2% of the angioplasty patients died. To the nearest whole number, how many patients is that?

b. 7.1% of the drug therapy patients died. To the nearest whole number, how many patients is that?

c. 2.2% of the angioplasty patients had strokes. To the nearest whole number, how many patients is that?

d. 4% of the drug therapy group patients had strokes. To the nearest whole number, how many patients is that?

SOLUTION 9

a. We need to take 6.2% of 225 (the number of angioplasty patients).

$$6.2\% \text{ of } 225 \quad \text{means} \quad 0.062 \cdot 225 = 13.95$$

or 14 when rounded to the nearest whole number.

b. Here, we need 7.1% of 226 = $0.071 \cdot 226 = 16.046$, or 16 when rounded to the nearest whole number.

c. 2.2% of 225 = $0.022 \cdot 225 = 4.95$, or 5 when rounded to the nearest whole number.

d. 4% of 226 = $0.04 \cdot 226 = 9.04$, or 9 when rounded to the nearest whole number.

Now, look at the answers 14, 16, 5, and 9 for parts **a–d,** respectively. Are those answers faithfully depicted in the graphs? What do you think happened? (You will revisit this study in Problems 71–80.)

Saving Lives

Community hospitals without on-site cardiac units can save more lives with angioplasty than with drug treatment, a new study of 451 heart attack victims shows.

Clot-breaking drug
Angioplasty

Number of patients with the following outcome, six weeks after treatment:

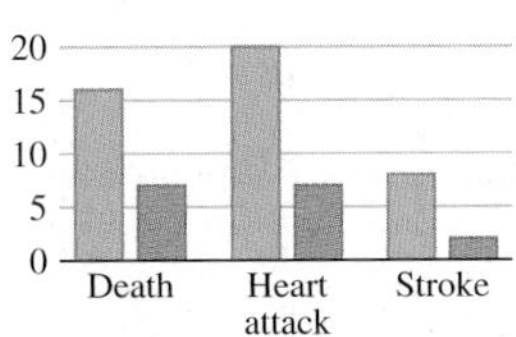

Number of patients with the following outcome, six months after treatment:

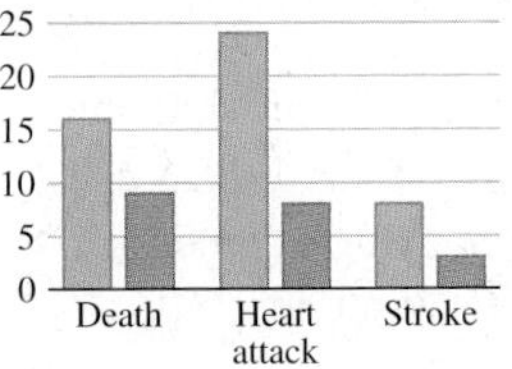

Note: The study was conducted at 11 community hospitals without on-site cardiac surgery units.

PROBLEM 9

The study also said that after six months, 5.3% of the angioplasty patients had a heart attack.

a. To the nearest whole number, how many patients is that?

b. In addition, 10.6% of the drug therapy group had a heart attack. To the nearest whole number, how many patients is that?

Are the answers for **a** and **b** faithfully depicted in the graph?

Answers to PROBLEMS

9. a. 12 **b.** 24

› Exercises 2.2

> Practice Problems > Self-Tests > Media-rich eBooks > e-Professors > Videos

‹A› Using the Multiplication and Division Properties of Equality In Problems 1–26, solve the equations.

1. $\frac{x}{7} = 5$ **2.** $\frac{y}{2} = 9$ **3.** $-4 = \frac{x}{2}$ **4.** $\frac{a}{5} = -6$

5. $\frac{b}{-3} = 5$ **6.** $7 = \frac{c}{-4}$ **7.** $-3 = \frac{f}{-2}$ **8.** $\frac{g}{-4} = -6$

9. $\frac{v}{4} = \frac{1}{3}$ **10.** $\frac{w}{3} = \frac{2}{7}$ **11.** $\frac{x}{5} = \frac{-3}{4}$ **12.** $\frac{-8}{9} = \frac{y}{2}$

13. $3z = 33$ **14.** $4y = 32$ **15.** $-42 = 6x$ **16.** $7b = -49$

17. $-8c = 56$ **18.** $-5d = 45$ **19.** $-5x = -35$ **20.** $-12 = -3x$

21. $-3y = 11$ **22.** $-5z = 17$ **23.** $-2a = 1.2$ **24.** $-3b = 1.5$

25. $3t = 4\frac{1}{2}$ **26.** $4r = 6\frac{2}{3}$

‹B› Multiplying by Reciprocals In Problems 27–40, solve the equations.

27. $\frac{1}{3}x = -0.75$ **28.** $\frac{1}{4}y = 0.25$ **29.** $-6 = \frac{3}{4}C$ **30.** $-2 = \frac{2}{9}F$

31. $\frac{5}{6}a = 10$ **32.** $24 = \frac{2}{7}z$ **33.** $-\frac{4}{5}y = 0.4$ **34.** $0.5x = \frac{-1}{4}$

35. $\frac{-2}{11}p = 0$ **36.** $\frac{-4}{9}q = 0$ **37.** $-18 = \frac{3}{5}t$ **38.** $-8 = \frac{2}{7}R$

39. $\frac{7x}{0.02} = -7$ **40.** $-6 = \frac{6y}{0.03}$

‹C› Multiplying by the LCM In Problems 41–60, solve the equations.

41. $\frac{y}{2} + \frac{y}{3} = 10$ **42.** $\frac{a}{4} + \frac{a}{3} = 14$ **43.** $\frac{x}{7} + \frac{x}{3} = 10$ **44.** $\frac{z}{6} + \frac{z}{4} = 20$

45. $\frac{x}{5} + \frac{x}{10} = 6$ **46.** $\frac{r}{2} + \frac{r}{6} = 8$ **47.** $\frac{t}{6} + \frac{t}{8} = 7$ **48.** $\frac{f}{9} + \frac{f}{12} = 14$

49. $\frac{x}{2} + \frac{x}{5} = \frac{7}{10}$ **50.** $\frac{a}{3} + \frac{a}{7} = \frac{20}{21}$ **51.** $\frac{c}{3} - \frac{c}{5} = 2$ **52.** $\frac{F}{4} - \frac{F}{7} = 3$

53. $\frac{W}{6} - \frac{W}{8} = \frac{5}{12}$ **54.** $\frac{m}{6} - \frac{m}{10} = \frac{4}{3}$ **55.** $\frac{x}{5} - \frac{3}{10} = \frac{1}{2}$ **56.** $\frac{3y}{7} - \frac{1}{14} = \frac{1}{14}$

57. $\frac{x+4}{4} - \frac{x+2}{3} = -\frac{1}{2}$ **58.** $\frac{w-1}{2} + \frac{w}{8} = \frac{7w+1}{16}$ **59.** $\frac{x}{6} + \frac{3}{4} = x - \frac{7}{4}$ **60.** $\frac{x}{6} + \frac{4}{3} = x - \frac{1}{3}$

‹D› Applications Involving Percent Problems Translate into an equation and solve.

61. 30% of 40 is what number? **62.** 17% of 80 is what number?

63. 40% of 70 is what number? **64.** What percent of 40 is 8?

65. What percent of 30 is 15? **66.** What percent of 40 is 4?

67. 30 is 20% of what number? **68.** 10 is 40% of what number?

69. 12 is 60% of what number? **70.** 24 is 50% of what number?

› Web IT go to mhhe.com/bello for more lessons

Medical Study Refer to the study described in Example 9. It was determined that not all 451 patients actually received treatment. As a matter of fact, only 211 patients received the clot-breaking drug (TPA) and 171 received angioplasty. Problems 71–76 refer to the chart shown here. Give your answers to one decimal place.

Clot-breaking drug (211)
Angioplasty (171)

Number of patients with the following outcome, six weeks after treatment:

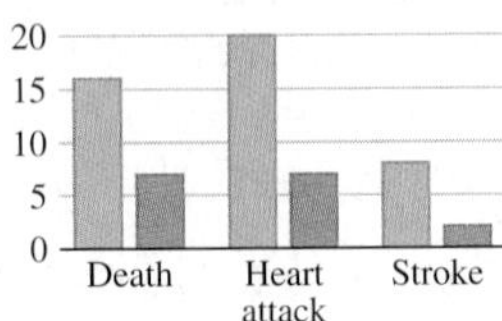

Source: Data from *Journal of the American Medical Association.*

71. As you can see, 16 of the patients receiving the clot-breaking drug died. What percent of the 211 patients is that?

72. What percent of the 171 patients receiving angioplasty died?

73. What percent of the 211 patients receiving the clot-breaking drug had heart attacks?

74. What percent of the 171 patients receiving angioplasty had heart attacks?

75. What percent of the 211 patients receiving the clot-breaking drug had a stroke?

76. What percent of the 171 patients receiving angioplasty had a stroke?

Medical Study Problems 77–80 refer to the chart shown here. Give your answers to one decimal place.

Clot-breaking drug (211)
Angioplasty (171)

Number of patients with the following outcome, six months after treatment:

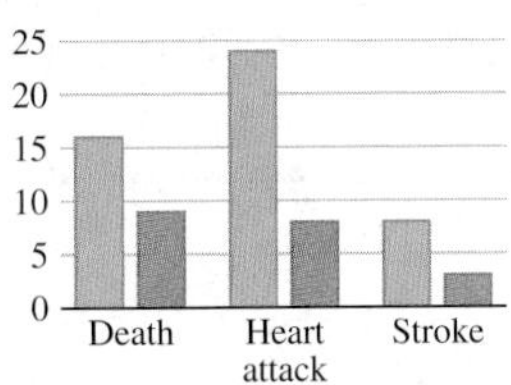

Source: Data from *Journal of the American Medical Association.*

77. As you can see, 9 of the patients receiving angioplasty died. What percent of the 171 patients is that?

78. What percent of the 171 patients receiving angioplasty had a heart attack?

79. What percent of the 211 patients receiving the clot-breaking drug had a stroke?

80. What percent of the 171 patients receiving angioplasty had a stroke?

Type of Exercise	Calories/Hour
Sleeping	55
Eating	85
Housework, moderate	160+
Dancing, ballroom	260
Walking, 3 mph	280
Aerobics	450+

Source: http://www.annecollins.com/.

Burning Calories The table shows the number of calories used in 1 hour of different activities. To lose 1 pound, you need to burn 3500 calories. If you want to do it by *sleeping* h hours at 55 calories per hour (line 1), you have the equation $55h = 3500$, and $h = 63\frac{7}{11}$ hours, so you have to sleep more than 63 hours to lose 1 pound!

In Problems 81–85:

a. Use the chart to write an equation indicating the number h of hours you need to lose 1 pound (3500 calories) by doing the indicated exercise.

b. Solve the equation to find the number of hours needed to lose one pound.

81. Eating

82. Moderate housework

83. Ballroom dancing

84. Walking at 3 mph

85. Aerobics

By the way, if you eat just one McDonald's hamburger (280 calories), you need to walk for exactly 1 hour to burn the calories (see the chart)!

Using Your Knowledge

Applications: Green Math

Plastic Bottles In Examples 6, 7, and 8 we mentioned the huge amounts of materials being recycled but did not mention plastic water bottles. Problems 86–88 deal with plastic bottles.

86. Americans buy an estimated 35 billion single-serving (1 liter or less) plastic water bottles each year (365 days). Almost 80% end up in a landfill or incinerator. What is 80% of 35 billion, the number of bottles ending up in the landfill or incinerator? By the way, that is 888 bottles wasted every second!

Source: http://tinyurl.com/nvlmc2.

87. In a recent year, U.S. bottled water sales were about 9 billion gallons out of the 30 billion gallon market for the U.S. liquid refreshment beverage market. What percent of 30 is 9, the percent of the 30 billion U.S. liquid refreshment markets that is water?

Source: http://en.wikipedia.org/wiki/Bottled_water#Sales.

88. 2.3 billion *pounds* of plastic bottles were recycled in a recent year, a 23% recycling rate. This means that 2.3 billion pounds is 23% of the total number of pounds of plastic bottles produced. What is this number?

Source: http://earth911.com/plastic/plastic-bottle-recycling-facts/.

Write On

89. a. What is the difference between an expression and an equation?

b. What is the difference between simplifying an expression and solving an equation?

90. When solving the equation $-3x = 18$, would it be easier to divide by -3 or to multiply by the reciprocal of -3? Explain your answer and solve the equation.

91. When solving the equation $-\frac{3}{4}x = 15$, would it be easier to divide by $-\frac{3}{4}$ or to multiply by the reciprocal of $-\frac{3}{4}$? Explain your answer and solve the equation.

Concept Checker

Fill in the blank(s) with the correct word(s), phrase, or mathematical statement.

92. According to the **Multiplication Property of Equality,** for any nonzero number c, the equation $a = b$ is equivalent to ________________.

93. According to the **Division Property of Equality,** for any nonzero number c, the equation $a = b$ is equivalent to ________________.

94. The **smallest number** that is a **multiple** of each of the **denominators** in an equation is called the ________________ of the denominators.

95. To **clear fractions** in an equation, **multiply** both sides of the equation by the ________________ of the **denominators.**

LCM	$\frac{a}{c} = \frac{b}{c}$
multiple	$\frac{a}{b} = \frac{c}{a}$
$ab = ac$	**LCD**
$ac = bc$	

Mastery Test

96. 15 is 30% of what number?

97. What percent of 45 is 9?

98. 40% of 60 is what number?

Solve.

99. $\frac{x+2}{4} + \frac{x-1}{5} = 3$

100. $\frac{x+3}{2} - \frac{x-2}{3} = 5$

101. $\frac{y}{6} + \frac{y}{10} = 8$

102. $\frac{y}{5} - \frac{y}{8} = 3$

103. $\frac{4}{5}y = 8$

104. $-\frac{3}{4}y = 6$

105. $-\frac{2}{7}y = -4$

106. $-7y = 16$

107. $\frac{x}{2} = -7$

108. $-\frac{y}{4} = -3$

Skill Checker

Use the distributive property to multiply.

109. $4(x - 6)$

110. $3(6 - x)$

111. $5(8 - y)$

112. $6(8 - 2y)$

113. $9(6 - 3y)$

114. $-3(4x - 2)$

115. $-5(3x - 4)$

Find:

116. $20 \cdot \frac{3}{4}$

117. $24 \cdot \frac{1}{6}$

118. $-5 \cdot \left(-\frac{4}{5}\right)$

119. $-7 \cdot \left(-\frac{3}{7}\right)$

2.3 Linear Equations

Objectives

A Solve linear equations in one variable.

B Solve a literal equation for one of the unknowns.

To Succeed, Review How To . . .

1. Find the LCM of three numbers (p. 15).
2. Add, subtract, multiply, and divide real numbers (pp. 52, 54, 61, 63).
3. Use the distributive property to remove parentheses (pp. 80–81, 92).

Getting Started

Getting the Best Deal

Suppose you want to rent a midsize car. Which is the better deal, paying \$39.99 per day *plus* \$0.20 per mile or paying \$49.99 per day with unlimited mileage? Well, it depends on how many miles you travel! First, let's see how many miles you have to travel before the costs are the same. The total daily cost, C, based on traveling m miles consists of two parts: \$39.99 plus the additional charge at \$0.20 per mile. Thus,

$$\underbrace{C}_{\text{The total daily cost } C} \quad \underbrace{=}_{\text{consists of}} \quad \underbrace{\$39.99}_{\text{fixed cost}} + \underbrace{0.20m}_{\text{mileage.}}$$

If the costs are identical, C must be \$49.99, that is,

$$\underbrace{39.99 + 0.20m}_{\text{The total daily cost } C} \quad \underbrace{=}_{\text{is the same as}} \quad \underbrace{\$49.99}_{\text{the flat rate.}}$$

Solving for m will tell us how many miles we must go for the mileage rate and the flat rate to be the same.

$$39.99 + 0.20m = 49.99 \qquad \text{Given.}$$

$$39.99 - 39.99 + 0.20m = 49.99 - 39.99 \qquad \text{Subtract 39.99.}$$

$$0.20m = 10$$

$$\frac{0.20m}{0.20} = \frac{10}{0.20} \qquad \text{Divide by 0.20.}$$

$$m = 50$$

Thus, if you plan to travel fewer than 50 miles, the mileage rate is better. If you travel 50 miles, the cost is the same for both, and if you travel more than 50 miles the flat rate is better. In this section we shall learn how to solve equations such as $39.99 + 0.20m = 49.99$, a type of equation called a *linear equation in one variable.*

A > Solving Linear Equations in One Variable

Let's concentrate on the idea we used to solve $39.99 + 0.20m = 49.99$. Our main objective is for our solution to be in the form of $m = \square$, where $\square$ is a number. Because of this, we first subtracted 39.99 and then divided by 0.20. This technique works for *linear equations.* Here is the definition.

LINEAR EQUATION

A **linear equation** is an equation that can be written in the form

$$ax + b = c$$

where $a \neq 0$ and a, b, and c are real numbers.

How do we solve linear equations? The same way we solved for m in the *Getting Started.* We use the properties of equality that we studied in Sections 2.1 and 2.2. Remember the steps?

1. The equation is already simplified. $\quad ax + b = c$

2. Subtract b. $\quad ax + b - b = c - b$

$$ax = c - b$$

3. Divide by a. $\quad \frac{ax}{a} = \frac{c - b}{a}$

$$x = \frac{c - b}{a}$$

LINEAR EQUATIONS MAY HAVE

a. *One* solution

b. *No* solution—no value of the variable will make the equation true. ($x + 1 = x + 2$ has no solution.)

c. *Infinitely many* solutions—any value of the variable will make the equation true. [$2x + 2 = 2(x + 1)$ has infinitely many solutions.]

EXAMPLE 1 **Solving linear equations**

Solve:

a. $3x + 7 = 13$ **b.** $-5x - 3 = 1$

SOLUTION 1

a. **1.** The equation is already simplified. $3x + 7 = 13$

2. Subtract 7. $3x + 7 - 7 = 13 - 7$

$3x = 6$

3. Divide by 3 (or multiply by the reciprocal of 3). $\frac{3x}{3} = \frac{6}{3}$

$x = 2$

The solution is 2.

CHECK

$3x + 7 \stackrel{?}{=} 13$	
$3(2) + 7$	13
$6 + 7$	
13	

b. **1.** The equation is already simplified. $-5x - 3 = 1$

2. Add 3. $-5x - 3 + 3 = 1 + 3$

$-5x = 4$

3. Divide by -5 (or multiply by the reciprocal of -5). $\frac{-5x}{-5} = \frac{4}{-5}$

$x = -\frac{4}{5}$

The solution is $-\frac{4}{5}$.

CHECK

$-5x - 3 \stackrel{?}{=} 1$	
$-5\left(-\frac{4}{5}\right) - 3$	1
$4 - 3$	
1	

PROBLEM 1

Solve:

a. $4x + 5 = 17$

b. $-3x - 5 = 2$

Note that the main idea is to place the variables on one side of the equation so you can write the solution in the form $x = \square$ (or $\square = x$), where $\square$ is a number (a constant). Can we solve the equation $5(x + 2) = 3(x + 1) + 9$? This equation is *not* of the form $ax + b = c$, but we can write it in this form if we use the distributive property to remove parentheses, as shown in Example 2.

EXAMPLE 2 **Using the distributive property to solve a linear equation**

Solve: $5(x + 2) = 3(x + 1) + 9$

SOLUTION 2

1. There are no fractions to clear. $5(x + 2) = 3(x + 1) + 9$

2. Use the distributive property. $5x + 10 = 3x + 3 + 9$

$5x + 10 = 3x + 12$ Add 3 and 9.

PROBLEM 2

Solve: $7(x + 1) = 4(x + 2) + 5$

Answers to PROBLEMS

1. a. 3 **b.** $-\frac{7}{3}$ **2.** 2

3. Subtract 10 on both sides. $5x + 10 - 10 = 3x + 12 - 10$

$5x = 3x + 2$

4. Subtract $3x$ on both sides. $5x - 3x = 3x - 3x + 2$

$2x = 2$

5. Divide both sides by 2 (or multiply by the reciprocal of 2). $\frac{2x}{2} = \frac{2}{2}$

$x = 1$

The solution is 1.

6. CHECK

$$\begin{array}{r|l} 5(x+2) & \stackrel{?}{=}\ 3(x+1)+9 \\ \hline 5(1+2) & 3(1+1)+9 \\ 5(3) & 3(2)+9 \\ 15 & 6+9 \\ & 15 \end{array}$$

What if we have fractions in the equation? We can clear the fractions by multiplying by the *l*east *c*ommon *m*ultiple (LCM), as we did in Section 2.2. For example, to solve

$$\frac{3}{4} + \frac{x}{10} - 1$$

we first multiply each term by 20, the LCM of 4 and 10. We can obtain the LCM by noting that 20 is the first multiple of 10 that is divisible by 4, or by writing

$$\begin{array}{r|ll} 2 & 4 & 10 \\ \hline & 2 & 5 \rightarrow 2 \cdot 2 \cdot 5 = 20 \quad \text{is the LCM} \end{array}$$

Now, solve $\frac{3}{4} + \frac{x}{10} = 1$ as follows:

1. Clear the fractions by multiplying by 20, the LCM of 4 and 10. $\overset{5}{\cancel{20}} \cdot \frac{3}{\cancel{4}} + \overset{2}{\cancel{20}} \cdot \frac{x}{\cancel{10}} = 20 \cdot 1$

2. Simplify. $15 + 2x = 20$

3. Subtract 15. $15 - 15 + 2x = 20 - 15$

4. The right side has numbers only. $2x = 5$

5. Divide by 2 (or multiply by the reciprocal of 2). $\frac{2x}{2} = \frac{5}{2}$

The solution is $\frac{5}{2}$. $x = \frac{5}{2}$

6. CHECK

$$\begin{array}{r|l} \frac{3}{4} + \frac{x}{10} & \stackrel{?}{=}\ 1 \\ \hline \frac{3}{4} + \frac{\frac{5}{2}}{10} & 1 \\ \frac{3}{4} + \frac{5}{2} \cdot \frac{1}{10} & \\ \frac{3}{4} + \frac{5}{20} & \\ \frac{3}{4} + \frac{1}{4} & \\ 1 & \end{array}$$

Here is a shortcut for finding the LCM of two or more numbers.

PROCEDURE TO FIND LCMs

1. Check if one number is a multiple of the other. If it is, the larger number is the LCM. For example, the LCM of 7 and 14 is 14 and the LCM of 12 and 24 is 24.
2. If one number is not a multiple of the other, then select the larger number. For an example, using the numbers 4 and 10, select 10. Double it (20), triple it (30), and so on until the smaller number (4) exactly divides the doubled or tripled quantity. Since 4 exactly divides 20, the LCM of 4 and 10 is 20.

For example, if you were looking for the LCM of 15 and 25, select the 25. Double it (50) but 15 does not exactly divide into 50. Triple the 25 (75). Now 15 exactly divides into 75, so the LCM of 15 and 25 is 75.

The procedure we have used to solve the preceding examples can be used to solve any linear equation. As before, what we need to do is isolate the variables on one side of the equation and the numbers on the other so that we can write the solution in the form $x = \square$ or $\square = x$. Here's how we accomplish this, with a step-by-step example shown on the right. (If you have forgotten how to find the LCM for three numbers, see Section R.2.)

PROCEDURE

Solving Linear Equations

$$\frac{x}{4} - \frac{1}{6} = \frac{7}{12}(x - 2) \quad \text{Given}$$

1. Clear any fractions by multiplying each term on both sides of the equation by the LCM of the denominators, 12.

$$12 \cdot \frac{x}{4} - 12 \cdot \frac{1}{6} = 12 \cdot \underbrace{\left[\frac{7}{12}(x - 2)\right]}_{\text{This is one term.}}$$

2. Remove parentheses and collect like terms (simplify) if necessary.

$$3x - 2 = 7(x - 2)$$
$$3x - 2 = 7x - 14$$

3. Add or subtract the same number on both sides of the equation so that the numbers are isolated on one side.

$$3x - 2 + 2 = 7x - 14 + 2$$
$$3x = 7x - 12$$

4. Add or subtract the same term or expression on both sides of the equation so that the variables are isolated on the other side.

$$3x - 7x = 7x - 7x - 12$$
$$-4x = -12$$

5. If the coefficient of the variable is not 1, divide both sides of the equation by the coefficient (or, equivalently, multiply by the reciprocal of the coefficient of the variable).

$$\frac{-4x}{-4} = \frac{-12}{-4}$$
$$x = 3 \quad \text{The solution}$$

6. Be sure to check your answer in the original equation.

6. CHECK

$\frac{x}{4} - \frac{1}{6}$	$\stackrel{?}{=} \frac{7}{12}(x - 2)$
$\frac{3}{4} - \frac{1}{6}$	$\frac{7}{12}(3 - 2)$
$\frac{9}{12} - \frac{2}{12}$	$\frac{7}{12} \cdot 1$
$\frac{7}{12}$	$\frac{7}{12}$

EXAMPLE 3 Solving linear equations

Solve:

a. $\frac{7}{24} = \frac{x}{8} + \frac{1}{6}$ **b.** $\frac{1}{5} - \frac{x}{4} = \frac{7(x+3)}{10}$

SOLUTION 3 We use the six-step procedure.

a. $\frac{7}{24} = \frac{x}{8} + \frac{1}{6}$ Given

1. Clear the fractions; the LCM is 24. $24 \cdot \frac{7}{24} = 24 \cdot \frac{x}{8} + 24 \cdot \frac{1}{6}$
2. Simplify. $7 = 3x + 4$
3. Subtract 4. $7 - 4 = 3x + 4 - 4$
4. The left side has numbers only. $3 = 3x$
5. Divide by 3 (or multiply by the reciprocal of 3). $\frac{3}{3} = \frac{3x}{3}$

$1 = x$

$x = 1$ The solution

6. **CHECK**

$\frac{7}{24} \stackrel{?}{=} \frac{x}{8} + \frac{1}{6}$	
$\frac{7}{24}$	$\frac{1}{8} + \frac{1}{6}$
	$\frac{3}{24} + \frac{4}{24}$
	$\frac{7}{24}$

b. $\frac{1}{5} - \frac{x}{4} = \frac{7(x+3)}{10}$ Given

1. Clear the fractions; the LCM is 20. Remember, you can get the LCM by selecting 10 (the largest of 5, 4, and 10) and doubling it, which gives 20. Since 5 and 4 exactly divide into 20, 20 is the LCM of 5, 4, and 10. $20 \cdot \frac{1}{5} - 20 \cdot \frac{x}{4} = 20 \cdot \frac{7(x+3)}{10}$
2. Simplify and use the distributive law. $4 - 5x = 14(x+3)$

$4 - 5x = 14x + 42$

3. Subtract 4. $4 - 4 - 5x = 14x + 42 - 4$

$-5x = 14x + 38$

4. Subtract 14x. $-5x - 14x = 14x - 14x + 38$

$-19x = 38$

5. Divide by -19 (or multiply by the reciprocal of -19). $\frac{-19x}{-19} = \frac{38}{-19}$

$x = -2$ The solution

6. **CHECK**

$\frac{1}{5} - \frac{x}{4}$ $\stackrel{?}{=}$	$\frac{7(x+3)}{10}$
$\frac{1}{5} - \frac{-2}{4}$	$\frac{7(-2+3)}{10}$
$\frac{1}{5} + \frac{1}{2}$	$\frac{7(1)}{10}$
$\frac{2}{10} + \frac{5}{10}$	$\frac{7}{10}$
$\frac{7}{10}$	

PROBLEM 3

Solve:

a. $\frac{20}{21} = \frac{x}{7} + \frac{x}{3}$

b. $\frac{1}{4} - \frac{x}{5} = \frac{17(x+4)}{20}$

Answers to PROBLEMS

3. a. 2 **b.** -3

B > Solving Literal Equations

The procedures for solving linear equations that we've just described can also be used to solve some *literal equations*. A **literal equation** is an equation that contains known quantities expressed by means of letters.

In business, science, and engineering, literal equations are usually given as formulas such as the area A of a circle of radius r ($A = \pi r^2$), the simple interest earned on a principal P at a given rate r for a given time t ($I = Prt$), and so on. Unfortunately, these formulas are not always in the form we need to solve the problem at hand. However, as it turns out, we can use the same methods we've just learned to solve for a particular variable in such a formula. For example, let's solve for P in the formula $I = Prt$. To keep track of the variable P, we first circle it in color:

$$I = Prt$$

$$\frac{I}{rt} = \frac{Prt}{rt} \quad \text{Divide by } rt.$$

$$\frac{I}{rt} = P \quad \text{Simplify.}$$

$$P = \frac{I}{rt} \quad \text{Rewrite.}$$

Now let's look at another example.

EXAMPLE 4 Solving a literal equation

A trapezoid is a four-sided figure in which only two of the sides are parallel. The area of the trapezoid shown here is

$$A = \frac{h}{2}(b_1 + b_2) \quad \text{Note that } b_1 \text{ and } b_2 \text{ have subscripts 1 and 2.}$$

where h is the altitude and b_1 and b_2 are the bases. Solve for b_2.

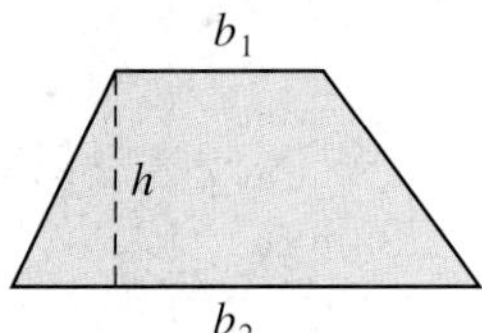

SOLUTION 4 We will circle b_2 to remember that we are solving for it. Now we use our six-step procedure.

$$A = \frac{h}{2}(b_1 + b_2) \quad \text{Given}$$

1. Clear the fraction; the LCM is 2.

$$2 \cdot A = 2 \cdot \frac{h}{2}(b_1 + b_2)$$

$$2A = h(b_1 + b_2)$$

2. Remove parentheses.

$$2A = hb_1 + hb_2$$

3. There are no numbers to isolate.
4. Subtract the same term, hb_1, on both sides.

$$2A - hb_1 = hb_1 - hb_1 + hb_2$$

$$2A - hb_1 = hb_2$$

5. Divide both sides by the coefficient of b_2, h.

$$\frac{2A - hb_1}{h} = \frac{hb_2}{h}$$

$$\frac{2A = hb_1}{h} = b_2$$

$$b_2 = \frac{2A - hb_1}{h} \quad \text{The solution}$$

6. No check is necessary.

PROBLEM 4

The speed S of an ant in centimeters per second is $S = \frac{1}{6}(C - 4)$, where C is the temperature in degrees Celsius. Solve for C.

Answers to PROBLEMS

4. $C = 6S + 4$

EXAMPLE 5 Using a pattern to solve a literal equation

The cost of renting a car is \$40 per day, plus 20¢ per mile m *after* the first 150 miles. (The first 150 miles are free.)

a. What is the formula for the total daily cost C based on the miles m traveled?

b. Solve for m in the equation (m more than 150).

SOLUTION 5

a. Let's try to find a pattern to the general formula.

Miles Traveled	Daily Cost	Per-Mile Cost	Total Cost
50	\$40	0	\$40
100	\$40	0	\$40
150	\$40	0	\$40
200	\$40	0.20(200 − 150) = 10	\$40 + \$10 = \$50
300	\$40	0.20(300 − 150) = 30	\$40 + \$30 = \$70

Yes, there is a pattern. Can you see that the daily cost when traveling more than 150 miles is $C = 40 + 0.20(m - 150)$?

b. Again, we circle the variable we want to isolate.

$C = 40 + 0.20(m - 150)$ Given

$C = 40 + 0.20m - 30$ Use the distributive property.

$C = 10 + 0.20m$ Simplify.

$C - 10 = 10 - 10 + 0.20m$ Subtract 10.

$C - 10 = 0.20m$

$\frac{C - 10}{0.20} = \frac{0.20m}{0.20}$ Divide by 0.20.

$\frac{C - 10}{0.20} = m$

PROBLEM 5

Some people claim that the relationship between shoe size S and foot length L is $S = 3L - 24$, where L is the length of the foot in inches.

a. If a person wears size 12 shoes, what is the length L of their foot?

b. Solve for L.

EXAMPLE 6 Solving for a specified variable

Solve for y: $2x + 3y = 6$

SOLUTION 6

Remember, we want to isolate the y.

$2x + 3y = 6$ Given

$2x - 2x + 3y = 6 - 2x$ Subtract $2x$.

$\frac{3y}{3} = \frac{6 - 2x}{3}$ Divide by 3.

$y = \frac{6 - 2x}{3}$ The solution

PROBLEM 6

Solve for x: $3x + 4y = 7$

Answers to PROBLEMS

5. a. 12 in.

b. $L = \frac{S + 24}{3}$

6. $x = \frac{7 - 4y}{3}$

The red graphs show the thousands of wasted tons of PET (transparent plastic or polyethylene terephthalate) bottles and aluminum cans. If the trend continues, the number of thousands of tons wasted can be approximated by $T = 125N + 1250$, where N is the number of years after 1996. Can we predict how many thousands of tons will be wasted in coming years? We will do exactly that in Example 7.

Source: http://tinyurl.com/n7ewdm.

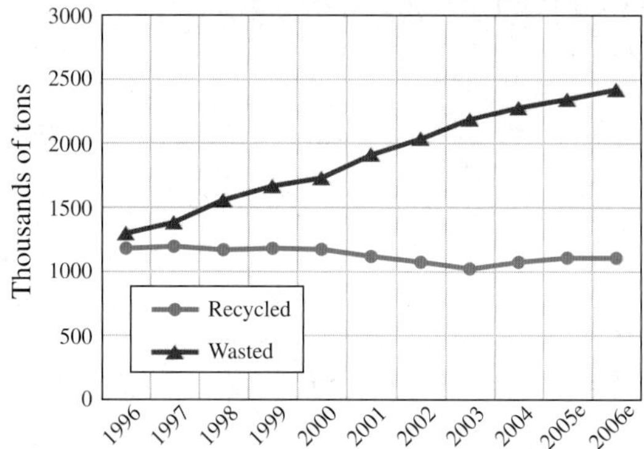

EXAMPLE 7 Wasted PET bottles and aluminum cans

The formula $T = 125N + 1250$, where N is the number of years after 1996, approximates the number of thousands of tons of PET bottles and aluminum cans wasted.

a. How many thousand tons were wasted in 1996? Is the answer close to the approximation in the graph?

b. What is the estimated waste for 2006? Is the answer close to the approximation in the graph?

c. How many years after 1996 will the waste reach 3000 thousand tons?

SOLUTION 7

a. Since N is the number of years after 1996, the year 1996 corresponds to $N = 0$ and $T = 125(\mathbf{0}) + 1250 = 1250$. The answer is indeed close to the approximation in the graph, which shows a number between 1000 and 1500.

b. In 2006, $N = 10$ (2006 – 1996) and $T = 125(\mathbf{10}) + 1250 = 2500$. The answer is slightly more than the one in the graph.

c. We want to find how many years after 1996 the waste will be 3000 thousand tons. This will happen when $T = 125N + 1250 = 3000$

so we now solve for N in $125N + 1250 = 3000$

Subtract 1250 $125N + 1250 - 1250 = 3000 - 1250$

Simplify $125N = 1750$

Divide by 125 $\dfrac{125N}{125} = \dfrac{1750}{125}$

Simplify $N = 14$

This means that **14** years after 1996 (in 1996 + **14** = 2010) 3000 thousand tons of PET bottles and aluminum cans will be wasted.

PROBLEM 7

Use the formula $T = 125N + 1250$ to find

a. The amount of PET bottles and aluminum cans wasted in 1997.

b. The amount of PET bottles and aluminum cans wasted in 2016.

c. How many years after 1996 will the waste reach 4000 thousand tons?

Note: 4000 thousand tons is 4,000,000 or 4 million tons!

Answers to PROBLEMS

7. a. 1375 (thousands of tons) **b.** 3750 (thousands of tons) **c.** 22 years (in 2018)

> Exercises 2.3

> Practice Problems > Self-Tests > Media-rich eBooks > e-Professors > Videos

> Web IT go to mhhe.com/bello for more lessons

⟨A⟩ Solving Linear Equations in One Variable In Problems 1–50, solve the equation.

1. $3x - 12 = 0$
2. $5a + 10 = 0$
3. $2y + 6 = 8$
4. $4b - 5 = 3$
5. $-3z - 4 = -10$
6. $-4r - 2 = 6$
7. $-5y + 1 = -13$
8. $-3x + 1 = -9$
9. $3x + 4 = x + 10$
10. $4x + 4 = x + 7$
11. $5x - 12 = 6x - 8$
12. $5x + 7 = 7x + 19$
13. $4v - 7 = 6v + 9$
14. $8t + 4 = 15t - 10$
15. $6m - 3m + 12 = 0$
16. $10k + 15 - 5k = 25$
17. $10 - 3z = 8 - 6z$
18. $8 - 4y = 10 + 6y$
19. $5(x + 2) = 3(x + 3) + 1$
20. $y - (4 - 2y) = 7(y - 1)$
21. $5(4 - 3a) = 7(3 - 4a)$
22. $\frac{3}{4}y - 4.5 = \frac{1}{4}y + 1.3$
23. $-\frac{7}{8}c + 5.6 = -\frac{5}{8}c - 3.3$
24. $x + \frac{2}{3}x = 10$
25. $-2x + \frac{1}{4} = 2x + \frac{4}{5}$
26. $6x + \frac{1}{7} = 2x - \frac{2}{7}$
27. $\frac{x-1}{2} + \frac{x-2}{2} = 3$
28. $\frac{3x+5}{3} + \frac{x+3}{3} = 12$
29. $\frac{x}{5} - \frac{x}{4} = 1$
30. $\frac{x}{3} - \frac{x}{2} = 1$
31. $\frac{x+1}{4} - \frac{2x-2}{3} = 3$
32. $\frac{z+4}{3} = \frac{z+6}{4}$
33. $\frac{2h-1}{3} = \frac{h-4}{12}$
34. $\frac{5-6y}{7} - \frac{-7-4y}{3} = 2$
35. $\frac{2w+3}{2} - \frac{3w+1}{4} = 1$
36. $\frac{7r+2}{6} + \frac{1}{2} = \frac{r}{4}$
37. $\frac{8x-23}{6} + \frac{1}{3} = \frac{5}{2}x$
38. $\frac{x+1}{2} + \frac{x+2}{3} + \frac{x+4}{4} = -8$
39. $\frac{x-5}{2} - \frac{x-4}{3} = \frac{x-3}{2} - (x - 2)$
40. $\frac{x+1}{2} + \frac{x+2}{3} + \frac{x+3}{4} = 16$
41. $-4x + \frac{1}{2} = 4\left(\frac{1}{8} - x\right)$
42. $-6x + \frac{2}{3} = 4\left(\frac{1}{5} - x\right)$
43. $\frac{1}{2}(8x + 4) - 5 = \frac{1}{4}(4x + 8) + 1$
44. $\frac{1}{3}(3x + 9) + 2 = \frac{1}{9}(9x + 18) + 3$
45. $x + \frac{x}{2} - \frac{3x}{5} = 9$
46. $\frac{5x}{3} - \frac{3x}{4} + \frac{11}{6} = 0$
47. $\frac{4x}{9} - \frac{3}{2} = \frac{5x}{6} - \frac{3x}{2}$
48. $\frac{7x}{2} - \frac{4x}{3} + \frac{2x}{5} = -\frac{11}{6}$
49. a. $\frac{5}{7}(2 - x) = 1 + \frac{9-5x}{7}$
b. $\frac{3x+4}{2} - \frac{1}{8}(19x - 3) = 1 - \frac{7x+18}{12}$
50. a. $\frac{3x-2}{6} = \frac{1}{6}x + \frac{1}{3}(x + 2)$
b. $\frac{11x-2}{3} - \frac{1}{2}(3x - 1) = \frac{17x+7}{6} - \frac{2}{9}(7x - 2)$

⟨B⟩ Solving Literal Equations In Problems 51–60, solve each equation for the indicated variable.

51. $C = 2\pi r$; solve for r.
52. $A = bh$; solve for h.
53. $3x + 2y = 6$; solve for y.
54. $5x - 6y = 30$; solve for y.
55. $A = \pi(r^2 + rs)$; solve for s.
56. $T = 2\pi(r^2 + rh)$; solve for h.
57. $\frac{V_2}{V_1} = \frac{P_1}{P_2}$; solve for V_2.
58. $\frac{a}{b} = \frac{c}{d}$; solve for a.
59. $S = \frac{f}{H - h}$; solve for H.
60. $I = \frac{E}{R + nr}$; solve for R.

Applications

61. *Shoe size and length of foot* The relationship between a person's shoe size S and the length of the person's foot L (in inches) is given by (a) $S = 3L - 22$ for a man and (b) $S = 3L - 21$ for a woman. Solve for L for parts (a) and (b).

62. *Man's weight and height* The relationship between a man's weight W (in pounds) and his height H (in inches) is given by $W = 5H - 190$. Solve for H.

63. *Woman's weight and height* The relationship between a woman's weight W (in pounds) and her height H (in inches) is given by $W = 5H - 200$. Solve for H.

64. *Sleep hours and child's age* The number H of hours a growing child A years old should sleep is $H = 17 - \frac{A}{2}$. Solve for A.

Recall from Problem 61 that the relationship between shoe size S and the length of a person's foot L (in inches) is given by the equations

$$S = 3L - 22 \quad \text{for men}$$

$$S = 3L - 21 \quad \text{for women}$$

65. *Shoe size* If Tyrone wears size 11 shoes, what is the length L of his foot?

66. *Shoe size* If Maria wears size 7 shoes, what is the length L of her foot?

67. *Shoe size* Sam's size 7 tennis shoes fit Sue perfectly! What size women's tennis shoe does Sue wear?

68. *Package delivery charges* The cost of first-class mail is 44 cents for the first ounce and 17 cents for each additional ounce. A delivery company will charge \$5.54 for delivering a package weighing up to 2 lb (32 oz). When would the U.S. Postal Service price, $P = 0.44 + 0.17(x - 1)$, where x is the weight of the package in ounces, be the same as the delivery company's price?

69. *Cost of parking* The parking cost at a garage is modeled by the equation $C = 1 + 0.75(h - 1)$, where h is the number of hours you park and C is the cost in dollars. When is the cost C equal to \$11.50?

70. *Baseball run production* Do you follow major-league baseball? Did you know that the average number of runs per game was highest in 1996? The average number of runs scored per game for the National League can be approximated by the equation $N = 0.165x + 4.68$, where x is the number of years after 1996. For the American League, the approximation is $A = -0.185x + 5.38$. When will $N = A$? That is, when will the National League run production be the same as that of the American League?

Who tends to waste the most time at work, older or younger people? Look at the table and judge for yourself.

Year of Birth	Time Wasted per Day
1950–1959	0.68 hr
1960–1969	1.19 hr
1970–1979	1.61 hr
1980–1985	1.95 hr

Source: www.salary.com http://www.salary.com/.

The number of hours H wasted each day by a person born in the 50s (5), 60s (6), 70s (7), or 80s (8) can be approximated by $H = 0.423x - 1.392$, where x is the decade in which the person was born.

71. *Wasted work time* According to the formula, **a.** how many hours were wasted per day by a person born in the 1950–1959 decade? **b.** What about according to the table?

72. *Wasted work time* According to the formula, **a.** how many hours were wasted per day by a person born in the 1960–1969 decade? **b.** What about according to the table?

73. *Wasted work time* According to the formula, **a.** how many hours were wasted per day by a person born in the 1970–1979 decade? **b.** What about according to the table?

Applications: Green Math

In Example 7, we mentioned the bottles and aluminum cans that were **not** recycled. Let us talk about the materials that **are** recovered from the garbage each day.

74. Of the 4.6 pounds of garbage generated each day by each person in the United States, $0.03x + 1.47$ pounds are recovered *either* for recycling or composting, where x is the number of years after 2005. How many years after 2005 will it be before the amount of total materials recovered for recycling or composting reach 1.77 pounds?

75. Of the total materials recovered for *either* recycling or composting, $0.03x + 1.09$ pounds are recovered **just** for recycling. How many years after 2005 will it be before the amount of materials recovered just for recycling reach 1.69 pounds?

Using Your Knowledge

Car Deals! In the *Getting Started* at the beginning of this section, we discussed a method that would tell us which was the better deal: using the mileage rate or the flat rate to rent a car. In Problems 76–78, we ask similar questions based on the rates given.

76. Suppose you wish to rent a subcompact that costs \$30.00 per day, plus \$0.15 per mile. Write a formula for the cost C based on traveling m miles.

77. How many miles do you have to travel so that the mileage rate, cost C in Problem 76, is the same as the flat rate, which is \$40 per day?

78. If you were planning to travel 300 miles during the week, would you use the mileage rate or the flat rate given in Problems 76 and 77?

79. In a free-market economy, merchants sell goods to make a profit. These profits are obtained by selling goods at a *markup* in price. This markup M can be based on the cost C or on the selling price S. The formula relating the selling price S, the cost C, and the markup M is

$$S = C + M$$

A merchant plans to have a 20% markup on cost.

a. What would the selling price S be?

b. Use the formula obtained in part **a** to find the selling price of an article that cost the merchant \$8.

80. a. If the markup on an article is 25% of the selling price, use the formula $S = C + M$ to find the cost C of the article.

b. Use the formula obtained in part **a** to find the cost of an article that sells for \$80.

Write On

81. The definition of a linear equation in one variable states that a cannot be zero. What happens if $a = 0$?

82. In this section we asked you to solve a formula for a specified variable. Write a paragraph explaining what that means.

83. The simplification of a linear equation led to the following step:

$$3x = 2x$$

If you divide both sides by x, you get $3 = 2$, which indicates that the equation has no solution. What's wrong with this reasoning?

Concept Checker

Fill in the blank(s) with the correct word(s), phrase, or mathematical statement.

84. A **linear equation** is an equation which can be written in the form __________.

85. An **equation** that contains **known quantities expressed by means of letters** is called a ________ equation.

unknown	$x = c$
literal	$ax + b = c$

Mastery Test

86. Solve for b_1 in the equation $A = \frac{h}{2}(b_1 + b_2)$.

87. Solve for m in the equation $50 = 40 + 0.20(m - 100)$.

Solve.

88. $\frac{7}{12} = \frac{x}{4} + \frac{x}{3}$

89. $\frac{1}{3} - \frac{x}{5} = \frac{8(x + 2)}{15}$

90. $10(x + 2) = 6(x + 1) + 18$

91. $-5(x + 2) = -3(x + 1) - 9$

92. $-4x - 5 = 2$

93. $3x + 8 = 11$

Skill Checker

Translate each statement into a mathematical expression.

94. The sum of n and $2n$

95. The quotient of $(a + b)$ and c

96. The quotient of a and $(b - c)$

97. The product of a and the sum of b and c

98. The difference of a and b, times c

99. The difference of a and the product of b and c

2.4 Problem Solving: Integer, General, and Geometry Problems

Objectives

Use the RSTUV method to solve:

A Integer problems

B General word problems

C Geometry word problems

To Succeed, Review How To . . .

1. Translate sentences into equations (pp. 93–96).
2. Solve linear equations (pp. 136–141).

Getting Started

Twin Problem Solving

Now that you've learned how to solve equations, you are ready to apply this knowledge to solve real-world problems. These problems are usually stated in words and consequently are called **word** or **story problems.** This is an area in which many students encounter difficulties, but don't panic; we are about to give you a surefire method of tackling word problems.

To start, let's look at a problem that may be familiar to you. Look at the photo of the twins, Mary and Margaret. At birth, Margaret was 3 ounces heavier than Mary, and together they weighed 35 ounces. Can you find their weights? There you have it, a word problem! In this section we shall study an effective way to solve these problems.

Here's how we solve word problems. It's as easy as 1-2-3-4-5.

PROCEDURE

RSTUV Method for Solving Word Problems

1. **R**ead the problem carefully and decide what is asked for (the unknown).
2. **S**elect a variable to represent this unknown.
3. **T**hink of a plan to help you write an equation.
4. **U**se algebra to solve the resulting equation.
5. **V**erify the answer.

If you really want to learn how to do word problems, this is the method you have to master. Study it carefully, and then use it. *It works!* How do we remember all of these steps? Easy. Look at the first letter in each sentence. We call this the **RSTUV method.**

NOTE

We will present problem solving in a five-step (RSTUV) format. Follow the instructions at each step until you understand them. Now, cover the Example you are studying (a 3 by 5 index card will do) and work the corresponding margin problem. Check the answer and see if you are correct; if not, study the steps in the Example again.

The mathematics dictionary in Section 1.7 also plays an important role in the solution of word problems. The words contained in the dictionary are often the **key** to translating the problem. But there are other details that may help you. Here's a restatement of the RSTUV method that offers hints and tips.

HINTS AND TIPS

Our problem-solving procedure (RSTUV) contains five steps. The numbered steps are listed, with hints and tips following each.

1. **Read the problem.** Mathematics is a language. As such, you have to learn how to read it. You may not understand or even get through reading the problem the first time. That's OK. Read it again and as you do, pay attention to key words or instructions such as *compute, draw, write, construct, make, show, identify, state, simplify, solve,* and *graph.* (Can you think of others?)
2. **Select the unknown.** How can you answer a question if you don't know what the question is? One good way to find the unknown (variable) is to look for the question mark "?" and read the material to its left. Try to determine what is given and what is missing.
3. **Think of a plan. (Translate!)** Problem solving requires many skills and strategies. Some of them are *look for a pattern; examine a related problem; use a formula; make tables, pictures, or diagrams; write an equation; work backward; and make a guess.* In algebra, the plan should lead to writing an equation or an inequality.
4. **Use algebra to solve the resulting equation.** If you are studying a mathematical technique, it's almost certain that you will have to use it in solving the given problem. Look for ways the technique you're studying could be used to solve the problem.
5. **Verify the answer.** Look back and check the results of the original problem. Is the answer reasonable? Can you find it some other way?

Are you ready to solve the problem in the *Getting Started* now? Let's restate it for you:

At birth, Mary and Margaret together weighed 35 ounces. Margaret was 3 ounces heavier than Mary. Can you find their weights?

Basically, this problem is about sums. Let's put our RSTUV method to use.

SUMMING THE WEIGHTS OF MARY AND MARGARET

1. Read the problem. Read the problem slowly—not once, but two or three times. (Reading algebra is *not* like reading a magazine; you may have to read algebra problems several times before you understand them.)

2. Select the unknown. (*Hint:* Let the unknown be the quantity you know nothing about.) Let w (in ounces) represent the weight for Mary which makes Margaret $w + 3$ ounces, since she was 3 ounces heavier.

3. Think of a plan. Translate the sentence into an equation: Together

Margaret	and	Mary	weighed	35 oz.
$(w + 3)$	$+$	w	$=$	35

4. Use algebra to solve the resulting equation.

$(w + 3) + w = 35$	Given
$w + 3 + w = 35$	Remove parentheses.
$2w + 3 = 35$	Combine like terms.
$2w + 3 - 3 = 35 - 3$	Subtract 3.
$2w = 32$	Simplify.
$\frac{2w}{2} = \frac{32}{2}$	Divide by 2.
$w = 16$	Mary's weight

Thus, Mary weighed 16 ounces, and Margaret weighed $16 + 3 = 19$ ounces.

5. Verify the solution. Do they weigh 35 ounces together? Yes, $16 + 19$ is 35. The average birth weight in the United States is 123 ounces, so the twins were quite small!

A > Solving Integer Problems

Sometimes a problem uses words that may be new to you. For example, a popular type of word problem in algebra is the **integer** problem. You remember the integers—they are the positive integers 1, 2, 3, and so on; the negative integers $-1, -2, -3$, and so on; and 0. Now, if you are given any integer, can you find the integer that comes right after it? Of course, you simply add 1 to the given integer and you have the answer. For example, the integer that comes after 4 is $4 + 1 = 5$ and the one that comes after -4 is $-4 + 1 = -3$. In general, if n is any integer, the integer that follows n is $n + 1$. We usually say that n and $n + 1$ are **consecutive integers.** We have illustrated this idea here.

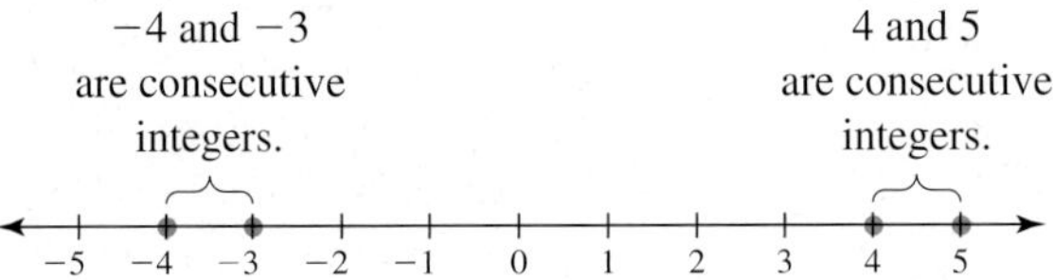
−4 and −3
are consecutive
integers.
4 and 5
are consecutive
integers.
−5 −4 −3 −2 −1 0 1 2 3 4 5

Now suppose you are given an *even* integer (an integer divisible by 2) such as 8 and you are asked to find the next even integer (which is 10). This time, you must add 2 to 8 to obtain the answer. What is the next even integer after 24? $24 + 2 = 26$. If n is even, the next even integer after n is $n + 2$. What about the next *odd* integer after 5? First, recall that the odd integers are $\ldots, -3, -1, 1, 3, \ldots$. The next odd integer after 5 is $5 + 2 = 7$. Similarly, the next odd integer after 21 is $21 + 2 = 23$. In general, if n is odd, the next odd integer after n is $n + 2$. Thus, if you need two consecutive even or odd integers and you call the first one n, the next one will be $n + 2$. We shall use all these ideas as well as the RSTUV method in Example 1.

EXAMPLE 1 A consecutive integer problem

The sum of three consecutive even integers is 126. Find the integers.

SOLUTION 1 We use the RSTUV method.

1. Read the problem. We are asked to find three consecutive *even* integers.

2. Select the unknown. Let n be the first of the integers. Since we want three consecutive even integers, we need to find the next two consecutive even integers. The next even integer after n is $n + 2$, and the one after $n + 2$ is $(n + 2) + 2 = n + 4$. Thus, the three consecutive even integers are n, $n + 2$, and $n + 4$.

PROBLEM 1

The sum of three consecutive odd integers is 129. Find the integers.

Answers to PROBLEMS

1. 41, 43, 45

3. Think of a plan. Translate the sentence into an equation.

The sum of 3 consecutive even integers is 126.

$$n + (n + 2) + (n + 4) = 126$$

4. Use algebra to solve the resulting equation.

$n + (n + 2) + (n + 4) = 126$ Given

$n + n + 2 + n + 4 = 126$ Remove parentheses.

$3n + 6 = 126$ Combine like terms.

$3n + 6 - 6 = 126 - 6$ Subtract 6.

$3n = 120$ Simplify.

$\frac{3n}{3} = \frac{120}{3}$ Divide by 3.

$n = 40$

Thus, the three consecutive even integers are 40, $40 + 2 = 42$, and $40 + 4 = 44$.

5. Verify the solution. Since $40 + 42 + 44 = 126$, our result is correct.

The RSTUV method works for numbers other than consecutive integers. We solve a different type of integer problem in the next example.

EXAMPLE 2 Bottled water sales

The sales of bottled water in the United States are going up, up, up says an article from CRI (Container Recycling Institute). Let us see how.

a. The number of units U sold 3 years ago has *doubled* to 29.8 billion units. How many units were sold 3 years ago?

b. Another way of looking at the change is that even though the annual sales *decreased* by 3.2 billion units in a recent year, the sales S amounted to 7 times the 3.8 billion units sold in 1997. What were the sales S? The CRI conclusion: more PET (plastic) bottles produced, more wasted, and fewer recycled!

SOLUTION 2

a. We use the RSTUV method.

1. Read the problem. We are asked to find the number of units U sold 3 years ago.

2. Select the unknown. U is the unknown.

3. Think of a plan. Translate the sentence into an equation.

The number of units U has doubled to 29.8 billion units

$$2U = 29.8$$

4. Use algebra to solve. $2U = 29.8$

$\frac{2U}{2} = \frac{29.8}{2}$ Divide by 2.

$U = 14.9$

Thus, the number of units sold 3 years ago was 14.9 billion units.

5. Verify the solution. Is double 14.9 the same as 29.8? That is, is $2(14.9) = 29.8$? Yes! The answer is correct.

(continued)

PROBLEM 2

a. The 30% market share of soda aluminum cans represents double the percent P of PET plastic bottled water. What is the percent P of PET plastic bottled water?

b. Four percent less than double the 15% of beer aluminum cans represents double the percent G of glass beer bottles. What is the percent G of glass beer bottles?

Answers to PROBLEMS

2. a. 15% **b.** 13%

b. We use the RSTUV procedure again.

1. R. We are asked for the sales S.

2. S. The unknown is S.

3. T. Translate the sentence into an equation.

4. U. Use algebra to solve.

$$\underbrace{S \text{ decreased by 3.2 billion units}} \quad \underbrace{\text{amounted to}} \quad \underbrace{7 \text{ times } 3.8 \text{ billion}}$$

$S - 3.2$	$= 7 \cdot 3.8$	Translation
$S - 3.2$	$= 26.6$	Multiply 7 by 3.8.
$S - 3.2 + 3.2$	$= 26.6 + 3.2$	Add 3.2.
S	$= 29.8$	Simplify.

Thus, the sales S amounted to 29.8 billion units.
The verification is left for you!

Source: http://container-recycling.org/facts/plastic/.

B > General Word Problems

Many interesting problems can be solved using the RSTUV method. Example 3 is typical.

EXAMPLE 3 Solving a general word problem

Have you eaten at a fast-food restaurant lately? If you eat a cheeseburger and fries, you would consume 1070 calories. As a matter of fact, the fries contain 30 more calories than the cheeseburger. How many calories are there in each food?

SOLUTION 3 Again, we use the RSTUV method.

1. Read the problem. We are asked to find the number of calories in the cheeseburger and in the fries.

2. Select the unknown. Let c represent the number of calories in the cheeseburger. This makes the number of calories in the fries $c + 30$, that is, 30 more calories. (We could instead let f be the number of calories in the fries; then $f - 30$ would be the number of calories in the cheeseburger.)

3. Think of a plan. Translate the problem.

$$\underbrace{\text{The calories in fries}}_{(c+30)} \quad \underbrace{\text{and}}_{+} \quad \underbrace{\text{the calories in cheeseburger}}_{c} \quad \underbrace{\text{total}}_{=} \quad \underbrace{1070.}_{1070}$$

4. Use algebra to solve the equation.

$(c + 30) + c = 1070$	Given
$c + 30 + c = 1070$	Remove parentheses.
$2c + 30 = 1070$	Combine like terms.
$2c + 30 - 30 = 1070 - 30$	Subtract 30.
$2c = 1040$	Simplify.
$\frac{2c}{2} = \frac{1040}{2}$	Divide by 2.
$c = 520$	

Thus, the cheeseburger has 520 calories. Since the fries have 30 calories more than the cheeseburger, the fries have $520 + 30 = 550$ calories.

5. Verify the solution. Does the total number of calories—that is, $520 + 550$—equal 1070? Since $520 + 550 = 1070$, our results are correct.

PROBLEM 3

A McDonald's® cheeseburger and small fries together contain 540 calories. The cheeseburger has 120 more calories than the fries. How many calories are in each food?

Answers to PROBLEMS

3. Cheeseburger, 330; fries, 210

C > Geometry Word Problems

The last type of problem we discuss here deals with an important concept in geometry: angle measure. Angles are measured using a unit called a **degree** (symbolized by °).

Types of Angles

Angle of Measure 1°

One complete revolution around a circle is 360° and $\frac{1}{360}$ of a complete revolution is 1°.

Complementary Angles

Two angles whose sum is 90° are called **complementary angles.** Example: a 30° angle and a 60° angle are complementary since 30° + 60° = 90°.

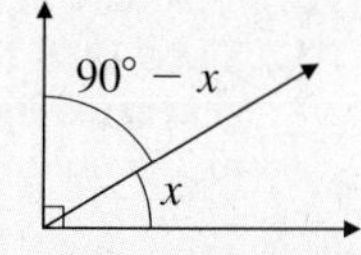

Supplementary Angles

Two angles whose sum is 180° are called **supplementary angles.** Example: a 50° angle and a 130° angle are supplementary since 50° + 130° = 180°.

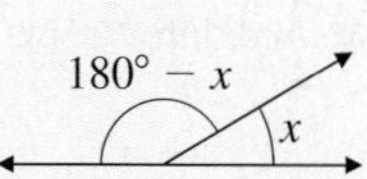

As you can see from the table, if x represents the measure of an angle,

$90 - x$ is the measure of its **complement**

and

$180 - x$ is the measure of its **supplement**

Let's use this information to solve a problem.

EXAMPLE 4 Complementary and supplementary angles

Find the measure of an angle whose supplement is 30° less than 3 times its complement.

SOLUTION 4

As usual, we use the RSTUV method.

1. Read the problem. We are asked to find the measure of an angle. Make sure you understand the ideas of complement and supplement. If you don't understand them, review those concepts.

2. Select the unknown. Let m be the measure of the angle. By definition, we know that

$90 - m$ is its complement

$180 - m$ is its supplement

3. Think of a plan. We translate the problem statement into an equation:

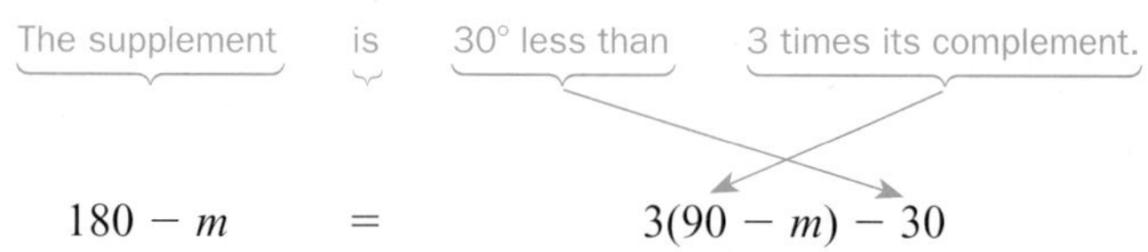

4. Use algebra to solve the equation.

$180 - m = 3(90 - m) - 30$	Given
$180 - m = 270 - 3m - 30$	Remove parentheses.
$180 - m = 240 - 3m$	Combine like terms.
$180 - 180 - m = 240 - 180 - 3m$	Subtract 180.
$-m = 60 - 3m$	Simplify.
$+3m - m = 60 - 3m + 3m$	Add $3m$.
$2m = 60$	Simplify.
$m = 30$	Divide by 2.

PROBLEM 4

Find the measure of an angle whose supplement is 45° less than 4 times its complement.

(continued)

Answers to PROBLEMS

4. 45°

Thus, the measure of the angle is 30°, which makes its complement 90° − 30° = 60° and its supplement 180° − 30° = 150°.

5. Verify the solution. Is the supplement (150°) 30° less than 3 times the complement? That is, is $150 = 3 \cdot 60 - 30$? Yes! Our answer is correct.

Before you do the Problems, practice translations!

TRANSLATE THIS

The third step in the RSTUV procedure is to **TRANSLATE** the information into an equation. In Problems 1–10 **TRANSLATE** the sentence and match the correct translation with one of the equations A–O.

1. Since 1980 **the amount A of garbage generated has increased by 50% of A to 236 million tons** per year.
2. The number n of landfills has **declined** by 6157 to 1767.
3. Since 1990 **the amount of garbage g sent to landfills has decreased by 9 million tons to 131 million tons.**
4. In a recent year, the net per capita discard rate (that's how much garbage you discard) was **3.09 pounds (per person per day), down by 0.05 pounds from the p pounds** discarded the previous year.
5. The total materials recycled **R (in millions of tons) increased by 66.7 million tons to 72.3 million tons** in a 43-year period.

A. $9 - g = 131$
B. $R + 66.7 = 72.3$
C. $33 = 0.14g$
D. $A + 0.50A = 236$
E. $3.09 = p - 0.05$
F. $M + 49.8 = 55.4$
G. $131 = 0.55g$
H. $n - 6157 = 1767$
I. $g - 9 = 131$
J. $35.2\% = t + 23.1\%$
K. $0.93b = 2$
L. $3.09 - p = 0.05$
M. $0.93 = 2b$
N. $35.2\%y = 23.1\%$
O. $6157 - n = 1767$

6. The materials recovered for recycling **M (in millions of tons) increased by 49.8 tons to 55.4 tons** in a 43-year period.
7. In a recent year, about **33 million tons (14%) of the total garbage g** was burned.
8. In a recent year, about **131 million tons (55%) of the total garbage g** generated went to landfills.
9. In a recent year, the largest category in all garbage generated, 35.2% was paper. This **35.2% exceeded the amount of yard trimmings t** (the second largest category) **by 23.1%.**
10. In a recent year, about **93% of the b billion pounds of lead in recycled batteries yielded 2 billion pounds** of lead.

Source: Municipal Solid Waste Generation, Recycling and Disposal in the United States. http://www.epa.gov/.

You will have an opportunity to solve and finish these problems in Problems 41–50.

> Exercises 2.4

> Practice Problems > Self-Tests
> Media-rich eBooks > e-Professors > Videos

⟨A⟩ Solving Integer Problems In Problems 1–20, solve the given problem.

1. The sum of three consecutive even integers is 138. Find the integers.
2. The sum of three consecutive odd integers is 135. Find the integers.
3. The sum of three consecutive even integers is −24. Find the integers.
4. The sum of three consecutive odd integers is −27. Find the integers.
5. The sum of two consecutive integers is −25. Find the integers.
6. The sum of two consecutive integers is −9. Find the integers.
7. Find three consecutive integers (n, $n + 1$, and $n + 2$) such that the last added to twice the first is 23.
8. Maria, Latasha, and Kim live in apartments numbered consecutively (n, $n + 1$, $n + 2$). Kim's apartment number added to twice Maria's apartment number gives 47. What are their apartment numbers?

9. Three lockers are numbered consecutively (n, $n + 1$, $n + 2$) in such a way that the sum of the first and last lockers is the same as twice the middle locker number. What are the locker numbers?

10. Pedro spent 27 more dollars for his math book than for his English book. If his total bill was \$141, what was the price of each of the books?

11. Tyrone bought two used books. One book was \$24 more than the other. If his total purchase was \$64, what was the price of each of the books?

12. The total number of points in a basketball game was 179. The winning team scored 5 more points than the losing team. What was the score?

13. Another basketball game turned out to be a rout in which the winning team scored 55 more points than the losing team. If the total number of points in the game was 133, what was the final score?

14. Sandra and Mida went shopping. Their total bill was \$210, but Sandra spent only two-fifths as much as Mida. How much did each one spend?

15. The total of credit charges on one card is \$4 more than three-eighths the charges on the other. If the total charges are \$147, how much was charged on each card?

16. Marcus, Liang, and Mourad went to dinner. Liang's bill was \$2 more than Marcus's bill, and Mourad's bill was \$2 more than Liang's bill. If the total bill was \$261 before tip, find the amount for each individual bill.

17. The sum of three numbers is 254. The second is 3 times the first, and the third is 5 less than the second. Find the numbers.

18. The sum of three consecutive integers is 17 less than 4 times the smallest of the three integers. Find the integers.

19. The larger of two numbers is 6 times the smaller. Their sum is 147. Find the numbers.

20. Five times a certain fraction yields the same as 3 times 1 more than the fraction. Find the fraction.

B General Word Problems In Problems 21–34, solve the following problems.

21. Do you have Internet access? Polls indicate that 66% of all adults do. About 25% more adults access the Internet from home than from work. If 15% go online from other locations (not work or home), what is the percent of adults accessing the Internet from home? What is the percent of adults accessing the Internet from work?

Source: Harris Interactive poll.

Applications: Green Math

PET beverage bottles recovered The graph will be used in Problems 22–24.

22. Which state recovers the most PET bottles? It is California! The difference between the percent of bottles recovered in California and those recovered in New York is 19%. If New York recovers 11% of its PET bottles, what percent does California recover?

23. The combined percent of PET beverage bottles recovered by New York and Oregon amounts to 16%. If New York recovers 6% more, what percent does Oregon recover?

24. The 39 Non Bottle Bill States recover 4% more bottles than California. If they recover 34% of the bottles, what percent does California recover?

Source: http://www.container-recycling.org/images/graphs/plastic/PETrec-bystate-07.png.

Recovered PET Beverage Bottles by State, 2007

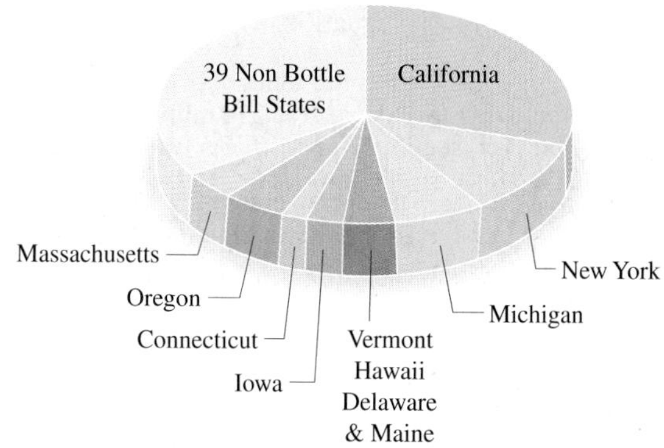

25. The height of the Empire State Building including its antenna is 1472 feet. The building is 1250 feet tall. How tall is the antenna?

26. The cost C for renting a car is given by $C = 0.10m + 10$, where m is the number of miles traveled. If the total cost amounted to \$17.50, how many miles were traveled?

27. A toy rocket goes vertically upward with an initial velocity of 96 feet per second. After t seconds, the velocity of the rocket is given by the formula $v = 96 - 32t$, neglecting air resistance. In how many seconds will the rocket reach its highest point? (*Hint:* At the highest point, $v = 0$.)

28. Refer to Problem 27 and find the number of seconds t that must elapse before the velocity decreases to 16 feet per second.

Web IT go to mhhe.com/bello for more lessons

Web IT go to mhhe.com/bello for more lessons

29. In a recent school election, 980 votes were cast. The winner received 372 votes more than the loser. How many votes did each of the two candidates receive?

30. The combined annual cost of the U.S. and Russian space programs has been estimated at $71 billion. The U.S. program is cheaper; it costs $19 billion less than the Russian program. What is the cost of each program?

31. A direct-dialed telephone call from Tampa to New York is $3.05 for the first 3 minutes and $0.70 for each additional minute or fraction thereof. If Juan's call cost him $7.95, how many minutes was the call?

32. A parking garage charges $1.25 for the first hour and $0.75 for each additional hour or fraction thereof.

a. If Sally paid $7.25 for parking, for how many hours was she charged?

b. If Sally has only $10 with her, what is the maximum number of hours she can park in the garage? (*Hint:* Check your answer!)

33. The cost of a taxicab is $0.95 plus $1.25 per mile.

a. If the fare was $28.45, how long was the ride?

b. If a limo service charges $15 for a 12-mile ride to the airport, which one is the better deal, the cab or the limo?

34. According to the Department of Health and Human Services, the number of heavy alcohol users in the 35-and-older category exceeds the number of heavy alcohol users in the 18–25 category by 844,000. If the total number of persons in these two categories is 7,072,000, how many persons are there in each of these categories?

C **Geometry Word Problems** In Problems 35–40, find the measure of the given angles.

35. An angle whose supplement is 20° more than twice its complement

36. An angle whose supplement is 10° more than three times its complement

37. An angle whose supplement is 3 times the measure of its complement

38. An angle whose supplement is 4 times the measure of its complement

39. An angle whose supplement is 54° less than 4 times its complement

40. An angle whose supplement is 41° less than 3 times its complement

Applications: Green Math

Problems 41–50 are based on the translations you already did in the ***Translate This*** at the beginning of the section. Now, you only have to solve the resulting equations! Use the RSTUV procedure to solve Problems 41–50. Remember, look at the ***Translate This*** for the equation!

41. *Garbage* Since 1980 **the amount A of garbage generated has increased by 50% of A to 236 million tons** per year. Find A to the nearest million ton.

42. *Landfills* The number n of landfills has **declined** by 6157 to 1767. Find n.

43. *Garbage* Since 1990 **the amount of garbage g sent to landfills has decreased by 9 million tons to 131 million tons.** Find g.

44. *Garbage* In a recent year, the net per capita discard rate (that's how much garbage you discard) was **3.09 pounds (per person per day) down by 0.05 pounds from the p pounds** discarded the previous year. Find p.

45. *Recycling* The total materials recycled R **(in millions of tons) increased by 66.7 million tons to 72.3 million tons.** Find R.

46. *Recycling* The materials recovered for recycling M **(in millions of tons) increased by 49.8 tons to 55.4 tons.** Find M.

47. *Garbage* In a recent year, about **33 million tons (14%) of the total garbage g** was burned. Find g to the nearest million ton.

48. *Landfills* In a recent year, about **131 million tons (55%) of the total garbage g** generated went to landfills. Find g to the nearest million ton.

49. *Garbage* In a recent year, the largest category in all garbage generated, 35.2%, was paper. This **35.2% exceeded the amount of yard trimmings t** (the second largest category) **by 23.1%.** Find t.

50. *Recycling* In a recent year, about **93% of the b billion pounds of lead in recycled batteries yielded 2 billion pounds** of lead. Find b to two decimal places.

Using Your Knowledge

Diophantus's Equation If you are planning to become an algebraist (an expert in algebra), you may not enjoy much fame. As a matter of fact, very little is known about one of the best algebraists of all time, the Greek Diophantus. According to a legend, the following problem is in the inscription on his tomb:

One-sixth of his life God granted him youth. After a twelfth more, he grew a beard. After an additional seventh, he married, and 5 years later, he had a son. Alas, the unfortunate son's life span was only one-half that of his father, who consoled his grief in the remaining 4 years of his life.

51. Use your knowledge to find how many years Diophantus lived. (*Hint:* Let x be the number of years Diophantus lived.)

Write On

52. When reading a word problem, what is the first thing you try to determine?

53. How do you verify your answer in a word problem?

54. The "T" in the RSTUV method means that you must "Think of a plan." Some strategies you can use in this plan include *look for a pattern* and *make a picture*. Can you think of three other strategies?

Concept Checker

Fill in the blank(s) with the correct word(s), phrase, or mathematical statement.

55. The procedure used to solve word problems is called the ______________ procedure.

56. n and $n + 1$ are called ______________ integers.

57. Two angles whose sum is 90° are called ______________ angles.

58. Two angles whose sum is 180° are called ______________ angles.

right	**supplementary**
acute	**RSTUV**
complementary	**consecutive**
obtuse	**even**

Mastery Test

59. The sum of three consecutive odd integers is 249. Find the integers.

60. Three less than 4 times a number is the same as the number increased by 9. Find the number.

61. If you eat a single slice of a 16-inch mushroom pizza and a 10-ounce chocolate shake, you have consumed 530 calories. If the shake has 70 more calories than the pizza, how many calories are there in each?

62. Find the measure of an angle whose supplement is 47° less than 3 times its complement.

Skill Checker

Solve.

63. $55T = 100$

64. $88R = 3240$

65. $15T = 120$

66. $81T = 3240$

67. $-75x = -600$

68. $-45x = -900$

69. $-0.02P = -70$

70. $-0.05R = -100$

71. $-0.04x = -40$

72. $-0.03x = 30$

2.5 Problem Solving: Motion, Mixture, and Investment Problems

Objectives

Use the RSTUV method to solve:

A › Motion problems

B › Mixture problems

C › Investment problems

D › Solving Environmental Applications

To Succeed, Review How To . . .

1. Translate sentences into equations (pp. 93–96).
2. Solve linear equations (pp. 136–141).

Getting Started

Birds in Motion

As we have mentioned previously, some of the strategies we can use to help us think of a plan include *use a formula, make a table,* and *draw a diagram.* We will use these strategies as we solve problems in this section. For example, in the cartoon, the bird is trying an impossible task, unless he turns around! If he does, how does Curls know that it will take him less than 2 hours to fly 100 miles? Because there's a formula to figure this out!

By Permission of John L. Hart FLP, and Creators Syndicate, Inc.

If an object moves at a constant rate R for a time T, the distance D traveled by the object is given by

$$D = RT$$

The object could be your car moving at a constant rate R of 55 miles per hour for $T = 2$ hours. In this case, you would have traveled a distance $D = 55 \times 2 = 110$ miles. Here the rate is in miles per *hour* and the time is in *hours.* (Units have to be consistent!) Similarly, if you jog at a constant rate of 5 miles per hour for 2 hours, you would travel $5 \times 2 = 10$ miles. Here the units are in miles per hour, so the time must be in hours. In working with the formula for distance, and especially in more complicated problems, it is often helpful to write the formula in a chart. For example, to figure out how far you jogged, you would write:

	R	$\times$	T	$=$	D
Jogger	5 mi/hr		2 hr		10 mi

As you can see, we used the formula $D = RT$ to solve this problem and organized our information in a chart. We shall continue to use formulas and charts when solving the rest of the problems in this section.

A > Motion Problems

Let's go back to the bird problem, which is a **motion problem.** If the bird turns around and is flying at 5 miles per hour with a tail wind of 50 miles per hour, its rate R would be $50 + 5 = 55$ miles per hour. The wind is helping the bird, so the wind speed must be added to his rate. Clumsy Carp wants to know how long it would take the bird to fly a distance of 100 miles; that is, he wants to find the time T. We can write this information in a table, substituting 55 for R and 100 for D.

	R ×	**T** =	**D**
Bird	55	T	100

Since

$$R \times T = D$$

we have

$$55T = 100 \quad \text{Divide by 55.}$$

$$T = \frac{100}{55} = \frac{20}{11} = 1\frac{9}{11} \text{ hr}$$

Curls was right; it does take less than 2 hours!

Now let's try some other motion problems.

EXAMPLE 1 Finding the speed

The longest regularly scheduled bus route is Greyhound's "Supercruiser" Miami-to-San Francisco route, a distance of 3240 miles. If this distance is covered in about 82 hours, what is the average speed R of the bus rounded to the nearest tenth?

SOLUTION 1

We use our RSTUV method.

1. **Read the problem.** We are asked to find the rate of the bus.
2. **Select the unknown.** Let R represent this rate.
3. **Think of a plan. What type of information do you need? How can you enter it so that $R \times T = D$?** Translate the problem and enter the information in a chart.

	R ×	**T** =	**D**
Supercruiser	R	82	3240

The equation is

$$R \times 82 = 3240 \quad \text{or} \quad 82R = 3240$$

4. **Use algebra to solve the equation.**

$$82R = 3240 \quad \text{Given}$$

$$\frac{82R}{82} = \frac{3240}{82} \quad \text{Divide by 82.}$$

$$R \approx 39.5$$

```
      39.51
82)3240.00
   246
    780
    738
     42 0
     41 0
      1 00
        82
        18
```

By long division (to the nearest tenth) or using a calculator

The bus's average speed is 39.5 miles per hour.

PROBLEM 1

Example 1 is about bus routes in the United States. There is a 6000-mile bus trip from Caracas to Buenos Aires that takes 214 hours, including a 12-hour stop in Santiago and a 24-hour stop in Lima. What is the average speed of the bus rounded to the nearest tenth? *Hint:* Do not count rest time.

(continued)

Answers to PROBLEMS

1. 33.7 mi/hr

5. Verify the solution. Check the arithmetic by substituting 39.5 into the formula $R \times T = D$.

$$39.5 \times 82 = 3239$$

Although the actual distance is 3240, this difference is acceptable because we rounded our answer to the nearest tenth.

Some motion problems depend on the relationship between the distances traveled by the objects involved. When you think of a plan for these types of problems, a diagram can be very useful.

EXAMPLE 2 Finding the time it takes to overtake an object

The Supercruiser bus leaves Miami traveling at an average rate of 40 miles per hour. Three hours later, a car leaves Miami for San Francisco traveling on the same route at 55 miles per hour. How long does it take for the car to overtake the bus?

PROBLEM 2

How long does it take if the car in Example 2 is traveling at 60 miles per hour?

SOLUTION 2

Again, we use the RSTUV method.

1. Read the problem. We are asked to find how many hours it takes the car to overtake the bus.

2. Select the unknown. Let T represent this number of hours.

3. Think of a plan. Translate the given information and enter it in a chart. Note that if the car goes for T hours, the bus goes for $(T + 3)$ hours (since it left 3 hours earlier).

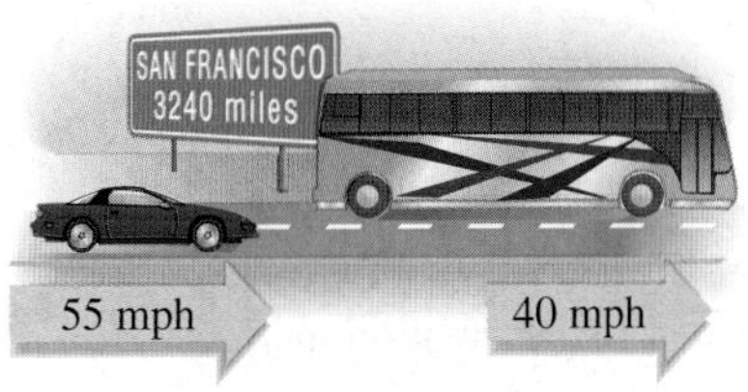

	R	×	T	=	D
Car	55		T		$55T$
Bus	40		$T + 3$		$40(T + 3)$

Since the vehicles are traveling in the same direction, the diagram of this problem looks like this:

Car ($55T$)

Miami → San Francisco

Bus $40(T + 3)$

When the car overtakes the bus, they will have traveled the *same* distance. According to the chart, the car has traveled $55T$ miles and the bus $40(T + 3)$ miles. Thus,

$$\underbrace{55T}_{\text{Distance traveled by car}} \underbrace{=}_{\text{overtakes}} \underbrace{40(T+3)}_{\text{distance traveled by Supercruiser}}$$

4. Use algebra to solve the equation.

$$55T = 40(T + 3) \quad \text{Given}$$

$$55T = 40T + 120 \quad \text{Simplify.}$$

$$55T - 40T = 40T - 40T + 120 \quad \text{Subtract } 40T.$$

Answers to PROBLEMS

2. 6 hr

$$15T = 120 \quad \text{Simplify.}$$

$$\frac{15T}{15} = \frac{120}{15} \quad \text{Divide by 15.}$$

$$T = 8$$

It takes the car 8 hours to overtake the bus.

5. Verify the solution. The car travels for 8 hours at 55 miles per hour; thus, it travels $55 \times 8 = 440$ miles, whereas the bus travels at 40 miles per hour for 11 hours, a total of $40 \times 11 = 440$ miles. Since the car traveled the same distance, it overtook the bus in 8 hours.

In Example 2, the two vehicles moved in the *same* direction. A variation of this type of problem involves motion toward each other, as shown in Example 3.

EXAMPLE 3 Two objects moving toward each other

The Supercruiser leaves Miami for San Francisco 3240 miles away, traveling at an average speed of 40 miles per hour. At the same time, a slightly faster bus leaves San Francisco for Miami traveling at 41 miles per hour. How many hours will it take for the buses to meet?

PROBLEM 3

How long does it take if the faster bus travels at 50 miles per hour?

SOLUTION 3

We use the RSTUV method.

1. Read the problem. We are asked to find how many hours it takes for the buses to meet.

2. Select the unknown. Let T represent the hours each bus travels before they meet.

3. Think of a plan. Translate the information and enter it in a chart.

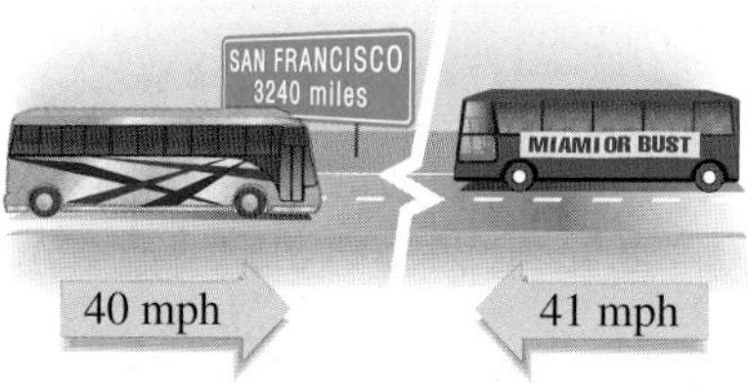

	R	×	T	=	D
Supercruiser	40		T		$40T$
Bus	41		T		$41T$

This time the objects are moving toward each other. The distance the Supercruiser travels is $40T$ miles, whereas the bus travels $41T$ miles, as shown here:

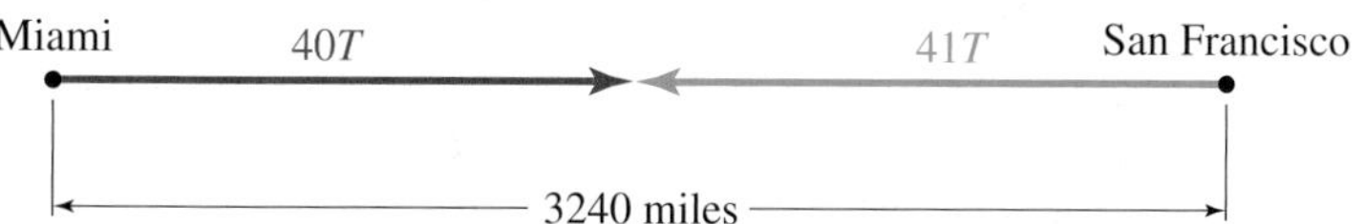

When they meet, the combined distance traveled by *both* buses is 3240 miles. This distance is also $40T + 41T$. Thus, we have

$$\underbrace{\text{Distance traveled by Supercruiser}}_{40T} \quad \underbrace{\text{and}}_{+} \quad \underbrace{\text{distance traveled by other bus}}_{41T} \quad \underbrace{\text{is}}_{=} \quad \underbrace{\text{total distance.}}_{3240}$$

(continued)

Answers to PROBLEMS

3. 36 hr

4. Use algebra to solve the equation.

$$40T + 41T = 3240 \quad \text{Given}$$
$$81T = 3240 \quad \text{Combine like terms.}$$
$$\frac{81T}{81} = \frac{3240}{81} \quad \text{Divide by 81.}$$
$$T = 40$$

Thus, each bus traveled 40 hours before they met.

5. Verify the solution. The Supercruiser travels $40 \times 40 = 1600$ miles in 40 hours, whereas the other bus travels $40 \times 41 = 1640$ miles. Clearly, the total distance traveled is $1600 + 1640 = 3240$ miles.

B > Mixture Problems

Another type of problem that can be solved using a chart is the **mixture problem.** In a mixture problem, two or more things are combined to form a mixture. For example, dental-supply houses mix pure gold and platinum to make white gold for dental fillings. Suppose one of these houses wishes to make 10 troy ounces of white gold to sell for \$1200 per ounce. If pure gold sells for \$1100 per ounce and platinum sells for \$1600 per ounce, how much of each should the supplier mix?

We are looking for the amount of each material needed to make 10 ounces of a mixture selling for \$1200 per ounce. If we let x be the total number of ounces of gold, then $10 - x$ (the balance) must be the number of ounces of platinum. Note that if a quantity T is split into two parts, one part may be x and the other $T - x$. This can be checked by adding $T - x$ and x to obtain $T - x + x = T$. We then enter all the information in a chart. The top line of the chart tells us that if we multiply the price of the item by the ounces used, we will get the total price:

	Price/Ounce	×	Ounces	=	Total Price
Gold	1100		x		$1100x$
Platinum	1600		$10 - x$		$1600(10 - x)$
Mixture	1200		10		12,000

The first line (following the word *gold*) tells us that the price of gold (\$1100) times the amount being used (x ounces) gives us the total price (\$$1100x$). In the second line, the price of platinum (\$1600) times the amount being used ($10 - x$ ounces) gives us the total price of \$$1600(10 - x)$. Finally, the third line tells us that the price of the mixture is \$1200, that we need 10 ounces of it, and that its total price will be \$12,000.

Since the sum of the total prices of gold and platinum in the last column must be equal to the total price of the mixture, it follows that

$$1100x + 1600(10 - x) = 12{,}000$$
$$1100x + 16{,}000 - 1600x = 12{,}000 \quad \text{Simplify.}$$
$$16{,}000 - 500x = 12{,}000$$
$$-500x = 12{,}000 - 16{,}000 \quad \text{Subtract 16,000.}$$
$$-500x = -4000 \quad \text{Simplify.}$$
$$\frac{-500x}{-500} = \frac{-4000}{-500} \quad \text{Divide by } -500.$$
$$x = 8$$

Thus, the supplier must use 8 ounces of gold and $10 - 8 = 2$ ounces of platinum. You can verify that this is correct! (8 ounces of gold at \$1100 per ounce and 2 ounces of platinum at \$1600 per ounce make 10 ounces of the mixture that costs \$12,000.)

EXAMPLE 4 **A mixture problem**

How many ounces of a 50% acetic acid solution should a photographer add to 32 ounces of a 5% acetic acid solution to obtain a 10% acetic acid solution?

SOLUTION 4 We use the RSTUV method.

1. Read the problem. We are asked to find the number of ounces of the 50% solution (a solution consisting of 50% acetic acid) that should be added to make a 10% solution.

2. Select the unknown. Let x stand for the number of ounces of 50% solution to be added.

3. Think of a plan. Remember to write the percents as decimals. To translate the problem, we first use a chart. In this case, the headings should include the percent of acetic acid and the amount to be mixed. The product of these two numbers will then give us the amount of pure acetic acid. Note that the percents have been converted to decimals.

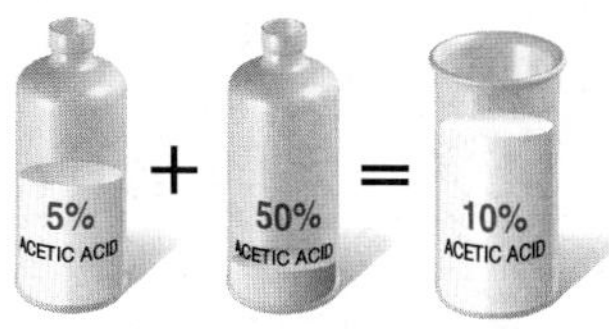

	%	×	Ounces	=	Amount of Pure Acid in Final Mixture
50% solution	0.50		x		$0.50x$
5% solution	0.05		32		1.60
10% solution	0.10		$x + 32$		$0.10(x + 32)$

Since we have x ounces of one solution and 32 ounces of the other, we have $(x + 32)$ ounces of the final mixture.

Since the sum of the amounts of pure acetic acid should be the same as the total amount of pure acetic acid in the final mixture, we have

$$0.50x + 1.60 = 0.10(x + 32) \quad \text{Given}$$

4. Use algebra to solve the equation.

$$10 \cdot 0.50x + 10 \cdot 1.60 = 10 \cdot [0.10(x + 32)] \quad \text{Multiply by 10 to clear the decimals.}$$

$$5x + 16 = 1(x + 32)$$

$$5x + 16 = x + 32$$

$$5x + 16 - 16 = x + 32 - 16 \quad \text{Subtract 16.}$$

$$5x = x + 16 \quad \text{Simplify.}$$

$$5x - x = x - x + 16 \quad \text{Subtract } x.$$

$$4x = 16 \quad \text{Simplify.}$$

$$\frac{4x}{4} = \frac{16}{4} \quad \text{Divide by 4.}$$

$$x = 4$$

Thus, the photographer must add 4 ounces of the 50% solution.

5. Verify the solution. We leave the verification to you.

PROBLEM 4

What if we want to obtain a 30% acetic acid solution?

Answers to PROBLEMS

4. 40 oz

C > Investment Problems

Finally, there's another problem that's similar to the mixture problem—the **investment problem.** Investment problems depend on the fact that the simple interest I you can earn (or pay) on principal P invested at rate r for 1 year is given by the formula

$$I = Pr$$

Now suppose you have a total of \$10,000 invested. Part of the money is invested at 6% and the rest at 8%. If the bank tells you that you have earned \$730 interest for 1 year, how much do you have invested at each rate?

In this case, we need to know how much is invested at each rate. Thus, if we say that we have invested P dollars at 6%, the rest of the money, that is, $10{,}000 - P$, would be invested at 8%. Note that for this to work, the sum of the amount invested at 6%, P dollars, plus the rest, $10{,}000 - P$, must equal \$10,000. Since $P + 10{,}000 - P = 10{,}000$, we have done it correctly. This information is entered in a chart, as shown here:

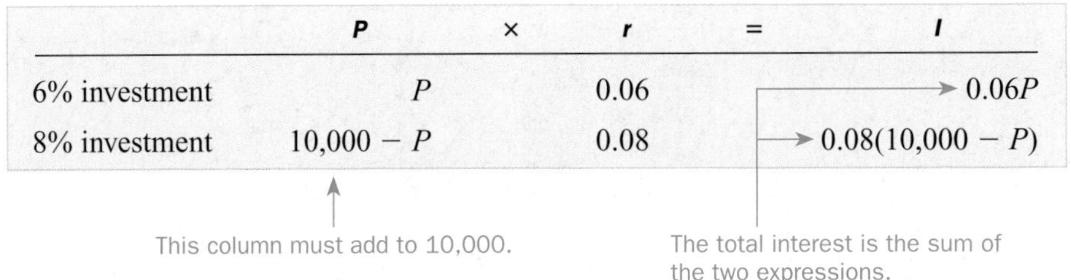

	P	×	r	=	I
6% investment	P		0.06		$0.06P$
8% investment	$10{,}000 - P$		0.08		$0.08(10{,}000 - P)$

This column must add to 10,000.
The total interest is the sum of the two expressions.

From the chart we can see that the total interest is the sum of the expressions in the last column, $0.06P + 0.08(10{,}000 - P)$. Recall that the bank told us that our total interest is \$730. Thus,

$$0.06P + 0.08(10{,}000 - P) = 730$$

You can solve this equation by first multiplying each term by 100 to clear the decimals:

$$\begin{aligned} 100 \cdot 0.06P + 100 \cdot 0.08(10{,}000 - P) &= 100 \cdot 730 \\ 6P + 8(10{,}000 - P) &= 73{,}000 \\ 6P + 80{,}000 - 8P &= 73{,}000 \\ -2P &= -7000 \\ P &= 3500 \end{aligned}$$

You could also solve

$$0.06P + 0.08(10{,}000 - P) = 730$$

as follows:

$$\begin{aligned} 0.06P + 800 - 0.08P &= 730 && \text{Simplify.} \\ 800 - 0.02P &= 730 \\ 800 - 800 - 0.02P &= 730 - 800 && \text{Subtract 800.} \\ -0.02P &= -70 && \text{Simplify.} \\ \frac{-0.02P}{-0.02} &= \frac{-70}{-0.02} && \text{Divide by } -0.02. \\ P &= 3500 \end{aligned}$$

$$\begin{array}{r} 35\,00 \\ 0.02\overline{)70.00} \\ 6 \\ \hline 10 \\ 10 \\ \hline 0 \end{array}$$

Thus, \$3500 is invested at 6% and the rest, 10,000 − 3500, or \$6500, is invested at 8%. You can verify that 6% of \$3500 added to 8% of \$6500 yields \$730.

EXAMPLE 5 **An investment portfolio**

A woman has some stocks that yield 5% annually and some bonds that yield 10%. If her investment totals \$6000 and her annual income from the investments is \$500, how much does she have invested in stocks and how much in bonds?

SOLUTION 5 As usual, we the RSTUV method.

1. Read the problem. We are asked to find how much is invested in stocks and how much in bonds.

2. Select the unknown. Let s be the amount invested in stocks. This makes the amount invested in bonds $(6000 - s)$.

3. Think of a plan. A chart is a good way to visualize this problem. Remember that the headings will be the formula $P \times r = I$ and that the percents must be written as decimals. Now we enter the information:

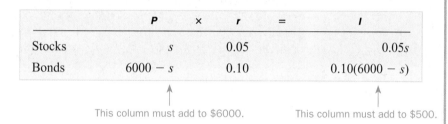

	P	$\times$ r	$=$ I
Stocks	s	0.05	$0.05s$
Bonds	$6000 - s$	0.10	$0.10(6000 - s)$
	↑ This column must add to \$6000.		↑ This column must add to \$500.

The total interest is the sum of the entries in the last column—that is, $0.05s$ and $0.10(6000 - s)$. This amount must be \$500:

$$0.05s + 0.10(6000 - s) = 500$$

4. Use algebra to solve the equation.

$$0.05s + 0.10(6000 - s) = 500 \quad \text{Given}$$

$$0.05s + 600 - 0.10s = 500 \quad \text{Simplify.}$$

$$600 - 0.05s = 500$$

$$600 - 600 - 0.05s = 500 - 600 \quad \text{Subtract 600.}$$

$$-0.05s = -100 \quad \text{Simplify.}$$

$$\frac{-0.05s}{-0.05} = \frac{-100}{-0.05} \quad \text{Divide by } -0.05.$$

$$s = 2000$$

Thus, the woman has \$2000 in stocks and \$4000 in bonds.

5. Verify the solution. To verify the answer, note that 5% of 2000 is 100 and 10% of 4000 is 400, so the total interest is indeed \$500.

PROBLEM 5

What if her income is only \$400?

D › Environmental Applications

On April 20, 2010, an environmental catastrophe occurred in the Gulf of Mexico: the oil rig *Deepwater Horizon* exploded, killing 11 people and triggering a massive oil leak estimated at 5000–60,000 barrels of oil daily (a barrel is 42 gallons). One

Answers to PROBLEMS

5. Stocks, \$4000; bonds, \$2000

fear was that the oil would enter the Gulf Loop Current, which would transport it toward the Florida Keys and southwest Florida. After entering the Loop Current, how many days will it take for the oil to travel from the coasts of Louisiana and Mississippi to the Florida Keys? It depends on the speed of the current, which varies from 50 to 100 miles a day (about 2 mph to 4.2 mph), according to Nan Walker, Director of Louisiana State University's Earth Scan Laboratory.

Source: http://tinyurl.com/2f5ab9k.

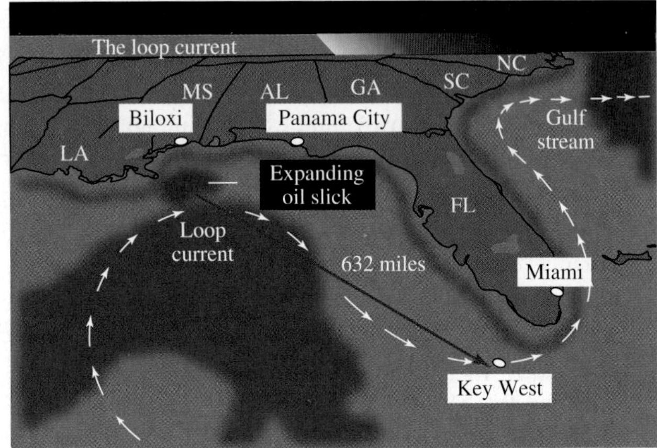

Source: http://tinyurl.com/28h93p8.

We will make our predictions in Example 6.

EXAMPLE 6 **Oil pollution and the Gulf Loop Current**

The distance from the coast of Louisiana to the Florida Keys is about 632 miles. Once the oil enters the Gulf Loop Current, how many hours will it take it to reach Key West, if the current is moving at 4 mph?

SOLUTION 6

Substituting in the formula $D = RT$, with $D = 632$ and $R = 4$ mph, we have

$$632 = 4T$$

$$\frac{632}{4} = T \qquad \text{Divide both sides by 4.}$$

$$158 = T \qquad \text{Simplify.}$$

This means that at this speed, it will take 158 hours for the oil to reach Key West. *The bad news:* 158 hours is about 6.6 days, not much time! *The good news* is that the oil never entered the Gulf Loop Current. See Problems 35 and 36 in the Exercises for further developments.

PROBLEM 6

If the current is moving at 3 mph, how many hours will it take for the oil to travel the 632 miles to Key West?

Before doing the problems, we need to practice more with translation. The problems here correspond to Problems 1–9 in the Exercises. Translate the problem and place the information in a box with the heading: **R** × **T** = **D** Do not solve!

Answers to PROBLEMS

6. 210.7 hours (about 8.8 days)

TRANSLATE THIS

The third step in the RSTUV procedure is to **TRANSLATE** the information into an equation. In Problems 1–10 **TRANSLATE** the sentence and match the correct translation with one of the boxes A–O.

1. The distance from Los Angeles to Sacramento is about 400 miles. A bus covers this distance in about 8 hours.
2. The distance from Boston to New Haven is 120 miles. A car leaves Boston at 7 A.M. and gets to New Haven just in time for lunch, at exactly 12 noon.
3. A 120-VHS Memorex videotape contains 246 meters of tape. When played at standard speed (SP), it will play for 120 minutes.
4. A Laser Writer Select prints one 12-inch page in 6 seconds. What is the rate of output for this printer?
5. The air distance from Miami to Tampa is about 200 miles. A jet flies at an average speed of 400 miles per hour.
6. A freight train leaves the station traveling at 30 miles per hour for T hours.
7. One hour later (see Problem 6), a passenger train leaves the same station traveling at 60 miles per hour in the same direction.
8. An accountant catches a train that travels at 50 miles per hour for T hours.
9. The basketball coach at a local high school left for work on his bicycle traveling at 15 miles per hour for T hours
10. Half an hour later (see Problem 9), his wife noticed that he had forgotten his lunch. She got in her car and took his lunch to him. Luckily, she got to school at exactly the same time as her husband. She made her trip traveling at 60 miles per hour.

	R	×	T	=	D
A.	?		120		246
B.	30		T		$30T$
C.	260		5		?
D.	60		$T-1$		$60(T-1)$
E.	15		T		$15T$
F.	?		5		120
G.	?		8		400
H.	60		$T+1$		$60(T+1)$
I.	?		6		12
J.	60		T		$60T$
K.	60		$T-\frac{1}{2}$		$60(T-\frac{1}{2})$
L.	120		?		246
M.	400		?		200
N.	60		$T+\frac{1}{2}$		$60(T+\frac{1}{2})$
O.	50		T		$50T$

› Exercises 2.5

‹ A › Motion Problems In Problems 1–16, use the RSTUV method to solve the motion problems.

1. The distance from Los Angeles to Sacramento is about 400 miles. A bus covers this distance in about 8 hours. What is the average speed of the bus?
2. The distance from Boston to New Haven is 120 miles. A car leaves Boston at 10 A.M. and gets to New Haven just in time for lunch, at exactly 12 noon. What is the speed of the car?
3. A 120-VHS Memorex videotape contains 246 meters of tape. When played at standard speed (SP), it will play for 120 minutes. What is the rate of play of the tape? Answer to the nearest whole number.
4. A Laser Writer Select prints one 12-inch page in 6 seconds.
   a. What is the rate of output for this printer?
   b. How long would it take to print 60 pages at this rate?

5. The air distance from Miami to Tampa is about 200 miles. If a jet flies at an average speed of 400 miles per hour, how long does it take to go from Tampa to Miami?

6. A freight train leaves the station traveling at 30 miles per hour. One hour later, a passenger train leaves the same station traveling at 60 miles per hour in the same direction. How long does it take for the passenger train to overtake the freight train?

7. A bus leaves the station traveling at 60 kilometers per hour. Two hours later, a student shows up at the station with a briefcase belonging to her absent-minded professor who is riding the bus. If she immediately starts after the bus at 90 kilometers per hour, how long will it be before she reunites the briefcase with her professor?

8. An accountant catches a train that travels at 50 miles per hour, whereas his boss leaves 1 hour later in a car traveling at 60 miles per hour. They had decided to meet at the train station in the next town and, strangely enough, they get there at exactly the same time! If the train and the car traveled in a straight line on parallel paths, how far is it from one town to the other?

9. The basketball coach at a local high school left for work on his bicycle traveling at 15 miles per hour. Half an hour later, his wife noticed that he had forgotten his lunch. She got in her car and took his lunch to him. Luckily, she got to school at exactly the same time as her husband. If she made her trip traveling at 60 miles per hour, how far is it from their house to the school?

10. A jet traveling 480 miles per hour leaves San Antonio for San Francisco, a distance of 1632 miles. An hour later another plane, going at the same speed, leaves San Francisco for San Antonio. How long will it be before the planes pass each other?

11. A car leaves town A going toward B at 50 miles per hour. At the same time, another car leaves B going toward A at 55 miles per hour. How long will it be before the two cars meet if the distance from A to B is 630 miles?

12. A contractor has two jobs that are 275 kilometers apart. Her headquarters, by sheer luck, happen to be on a straight road between the two construction sites. Her first crew left headquarters for one job traveling at 70 kilometers per hour. Two hours later, she left headquarters for the other job, traveling at 65 kilometers per hour. If the contractor and her first crew arrived at their job sites simultaneously, how far did the first crew have to drive?

13. A plane has 7 hours to reach a target and come back to base. It flies out to the target at 480 miles per hour and returns on the same route at 640 miles per hour. How many miles from the base is the target?

14. A man left home driving at 40 miles per hour. When his car broke down, he walked home at a rate of 5 miles per hour; the entire trip (driving and walking) took him $2\frac{1}{4}$ hours. How far from his house did his car break down?

15. The space shuttle *Discovery* made a historical approach to the Russian space station *Mir*, coming to a point 366 feet (4392 inches) away from the station. For the first 180 seconds of the final 366-foot approach, the *Discovery* traveled 20 times as fast as during the last 60 seconds of the approach. How fast was the shuttle approaching during the first 180 seconds and during the last 60 seconds?

16. If, in Problem 15, the commander decided to approach the *Mir* traveling only 13 times faster for the first 180 seconds as for the last 60 seconds, what would be the shuttle's final rate of approach?

⟨B⟩ Mixture Problems In Problems 17–27, use the RSTUV method to solve these mixture problems.

17. How many liters of a 40% glycerin solution must be mixed with 10 liters of an 80% glycerin solution to obtain a 65% solution?

18. How many parts of glacial acetic acid (99.5% acetic acid) must be added to 100 parts of a 10% solution of acetic acid to give a 28% solution? (Round your answer to the nearest whole part.)

19. If the price of copper is $4.00 per pound and the price of zinc is $1.00 per pound, how many pounds of copper and zinc should be mixed to make 80 pounds of brass selling for $2.95 per pound?

20. Oolong tea sells for $19 per pound. How many pounds of Oolong should be mixed with another tea selling at $4 per pound to produce 50 pounds of tea selling for $7 per pound?

21. How many pounds of Blue Jamaican coffee selling at $5 per pound should be mixed with 80 pounds of regular coffee selling at $2 per pound to make a mixture selling for $2.60 per pound? (The merchant cleverly advertises this mixture as "Containing the incomparable Blue Jamaican coffee"!)

22. How many ounces of vermouth containing 10% alcohol should be added to 20 ounces of gin containing 60% alcohol to make a pitcher of martinis that contains 30% alcohol?

23. Do you know how to make manhattans? They are mixed by combining bourbon and sweet vermouth. How many ounces of manhattans containing 40% vermouth should a bartender mix with manhattans containing 20% vermouth so that she can obtain a half gallon (64 ounces) of manhattans containing 30% vermouth?

24. A car radiator contains 30 quarts of 50% antifreeze solution. How many quarts of this solution should be drained and replaced with pure antifreeze so that the new solution is 70% antifreeze?

25. A car radiator contains 30 quarts of 50% antifreeze solution. How many quarts of this solution should be drained and replaced with water so that the new solution is 30% antifreeze?

26. A 12-ounce can of frozen orange juice concentrate is mixed with 3 cans of cold water to obtain a mixture that is 10% juice. What is the percent of pure juice in the concentrate?

27. The instructions on a 12-ounce can of Welch's Orchard fruit juice state: "Mix with 3 cans cold water" and it will "contain 30% juice when properly reconstituted." What is the percent of pure juice in the concentrate? What does this mean to you?

⟨ C ⟩ Investment Problems In Problems 28–36, use the RSTUV method to solve these investment problems.

28. Two sums of money totaling \$15,000 earn, respectively, 5% and 7% annual interest. If the total interest from both investments amounts to \$870, how much is invested at each rate?

29. An investor invested \$20,000, part at 6% and the rest at 8%. Find the amount invested at each rate if the annual income from the two investments is \$1500.

30. A woman invested \$25,000, part at 7.5% and the rest at 6%. If her annual interest from these two investments amounted to \$1620, how much money did she invest at each rate?

31. A man has a savings account that pays 5% annual interest and some certificates of deposit paying 7% annually. His total interest from the two investments is \$1100, and the total amount of money in the two investments is \$18,000. How much money does he have in the savings account?

32. A woman invested \$20,000 at 8%. What additional amount must she invest at 6% so that her annual income is \$2200?

33. A sum of \$10,000 is split, and the two parts are invested at 5% and 6%, respectively. If the interest from the 5% investment exceeds the interest from the 6% investment by \$60, how much is invested at each rate?

34. An investor receives \$600 annually from two investments. He has \$500 more invested at 8% than at 6%. Find the amount invested at each rate.

⟩ ⟩ ⟩ Applications: Green Math

⟨ D ⟩ Environmental Applications: Problems 35 and 36 involve Gulf oil spill calculations.

After the Gulf oil spill of 2010, many people tried to predict the time at which the oil would reach Key West. A map relying on the National Oceanic and Atmospheric Administration computer models predicts the movement of the spill on the water surface more accurately, as shown.

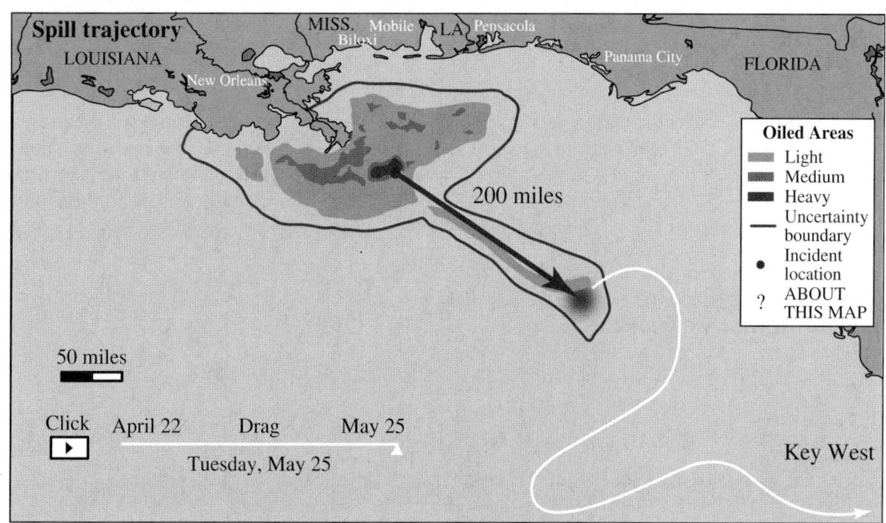

Source: http://www.msnbc.msn.com/id/37133684/ns/gulf_oil_spill/.

35. *Rate of travel* If the oil traveled 200 miles in 10 days (see map), how many miles per day was it traveling?

36. *Predictions*

a. If the oil travels at the rate found in Problem 35, how many days will it take it to travel the 700-mile loop shown in the map to a location south of Key West?

b. If the oil enters the Gulf Loop moving at 3 miles per hour, how many days will it take the oil to travel the 700-mile loop shown?

Web IT go to mhhe.com/bello for more lessons

Using Your Knowledge

Gesselmann's Guessing Some of the mixture problems given in this section can be solved using the *guess-and-correct* procedure developed by Dr. Harrison A. Gesselmann of Cornell University. The procedure depends on taking a guess at the answer and then using a calculator to correct this guess. For example, suppose we have to mix A and B selling for \$400 and \$475 per ounce, respectively, to obtain 10 ounces of a mixture selling for \$415 per ounce. Our first guess is to use *equal amounts* (5 ounces each) of A and B. This gives a mixture with a price per pound equal to the average price of A (\$400) and B (\$475), that is,

$$\frac{400 + 475}{2} = \$437.50 \text{ per ounce}$$

As you can see from the following figure, more of A must be used in order to bring the \$437.50 average down to the desired \$415:

Thus, the correction for the additional amount of A that must be used is

$$\frac{22.50}{37.50} \times 5 \text{ oz}$$

This expression can be obtained by the keystroke sequence

22.50 ÷ 37.50 × 5 ENTER

which gives the correction 3. The correct amount is

First guess	5 oz of A
+Correction	3 oz of A
Total =	8 oz of A

and the remaining 2 ounces is B.

If your instructor permits, use this method to work Problems 19 and 23.

Write On

37. Ask your pharmacist or your chemistry instructor if they mix products of different concentrations to make new mixtures. Write a paragraph on your findings.

38. Most of the problems involved have precisely the information you need to solve them. In real life, however, irrelevant information (called *red herrings*) may be present. Find some problems with red herrings and point them out.

39. The *guess-and-correct* method explained in the *Using Your Knowledge* also works for investment problems. Write the procedure you would use to solve investment problems using this method.

Concept Checker

Fill in the blank(s) with the correct word(s), phrase, or mathematical statement.

40. The formula for the distance D traveled at a rate R in time T is given by ________________.

41. The formula for the annual interest I earned (or paid) on a principal P at a rate r is given by ________________.

$R = DT$	$I = Pr$
$T = DR$	$P = Ir$
$D = RT$	

Mastery Test

42. Billy has two investments totaling \$8000. One investment yields 5% and the other 10%. If the total annual interest is \$650, how much money is invested at each rate?

43. How many gallons of a 10% salt solution should be added to 15 gallons of a 20% salt solution to obtain a 16% solution?

44. Two trains are 300 miles apart, traveling toward each other on adjacent tracks. One is traveling at 40 miles per hour and the other at 35 miles per hour. After how many hours do they meet?

45. A bus leaves Los Angeles traveling at 50 miles per hour. An hour later a car leaves at 60 miles per hour to try to catch the bus. How long does it take the car to overtake the bus?

46. The distance from South Miami to Tampa is 250 miles. This distance can be covered in 5 hours by car. What is the average speed on the trip?

Skill Checker

Find:

47. $2.9(30) + 71$

48. $\frac{5}{9}(95 - 32)$

49. $\frac{1}{2}(20)(10)$

50. $2(38) - 10$

51. $\frac{22 + 1}{3}$

52. $\frac{21 + 1}{3}$

2.6 Formulas and Geometry Applications

Objectives

A Solve a formula for one variable and use the result to solve a problem.

B Solve problems involving geometric formulas.

C Solve geometric problems involving angle measurement.

D Solve an application using formulas.

To Succeed, Review How To . . .

1. Reduce fractions (pp. 4–6).
2. Perform the fundamental operations using decimals (pp. 22–23).
3. Evaluate expressions using the correct order of operations (pp. 62–63, 69–73).

Getting Started

Do You Want Fries with That?

One way to solve word problems is to use a formula. Here's an example in which an incorrect formula has been used. The ad claims that the new Monster Burger has 50% more beef than the Lite Burger because its diameter, 6, is 50% more than 4. But is that the right way to measure them?

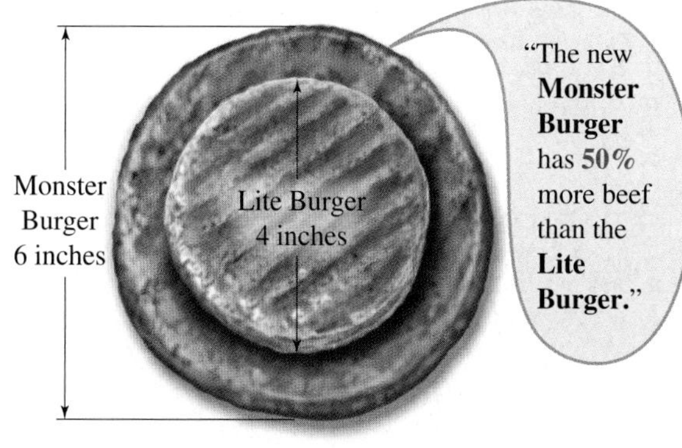
Monster Burger 6 inches
Lite Burger 4 inches
"The new Monster Burger has 50% more beef than the Lite Burger."

We should compare the two hamburgers by comparing the *volume* of the beef in each burger. The volume V of a burger is $V = Ah$, where A is the area and h is the height. The area of the Lite Burger is $L = \pi r^2$, where r is the radius (half the distance across the middle) of the burger. For the Lite Burger, $r = 2$ inches, so its area is

$$L = \pi(2 \text{ in.})^2 = 4\pi \text{ in.}^2$$

For the Monster Burger, $r = 3$ inches, so its area M is given by

$$M = \pi(3 \text{ in.})^2 = 9\pi \text{ in.}^2$$

The volume of the Lite Burger is $V = Ah = 4\pi h$, and for the Monster Burger the volume is $9\pi h$. If the height is the same for both burgers, we can simplify the volume calculation. The difference in volumes is $9\pi h - 4\pi h = 5\pi h$, and the *percent* increase for the Monster Burger is given by

$$\text{Percent increase} = \frac{\text{increase}}{\text{base}} = \frac{5\pi h}{4\pi h} = 1.25 \text{ or } 125\%$$

Thus, based on the discussion, the Monster Burgers actually have 125% more beef than the Lite Burgers. What other assumptions must you make for this to be so? (You'll have an opportunity in the *Write On* to give your opinion.) In this section we shall use formulas (like that for the volume or area of a burger) to solve problems.

Problems in many fields of endeavor can be solved if the proper formula is used. For example, why aren't you electrocuted when you hold the two terminals of your car battery? You may know the answer. It's because the voltage V in the battery (12 volts) is too small. We know this from a formula in physics that tells us that the voltage V is the product of the current I and the resistance R; that is, $V = IR$. To find the current I going through your body when you touch both terminals, solve for I by dividing both sides by R to obtain

$$I = \frac{V}{R}$$

A car battery carries 12 volts and $R = 20{,}000$ ohms, so the current is

$$I = \frac{12}{20{,}000} = \frac{6}{10{,}000} = 0.0006 \text{ amp}$$ Volts divided by ohms yields amperes (amp).

Since it takes about 0.001 ampere to give you a slight shock, your car battery should pose no threat.

A > Using Formulas

EXAMPLE 1 Solving problems in anthropology

Anthropologists know how to estimate the height of a man (in centimeters, cm) by using a bone as a clue. To do this, they use the formula

$$H = 2.89h + 70.64$$

where H is the height of the man and h is the length of his humerus.

a. Estimate the height of a man whose humerus bone is 30 centimeters long.

b. Solve for h.

c. If a man is 163.12 centimeters tall, how long is his humerus?

SOLUTION 1

a. We write 30 in place of h (in parentheses to indicate multiplication):

$$\begin{aligned} H &= 2.89(30) + 70.64 \\ &= 86.70 + 70.64 \\ &= 157.34 \text{ cm} \end{aligned}$$

Thus, a man with a 30-centimeter humerus should be about 157 centimeters tall.

PROBLEM 1

The estimated height of a woman (in inches) is given by $H = 2.8h + 28.1$, where h is the length of her humerus (in inches).

a. Estimate the height of a woman whose humerus bone is 15 inches long.

b. Solve for h.

c. If a woman is 61.7 inches tall, how long is her humerus?

Answers to PROBLEMS

1. a. 70.1 in. **b.** $h = \frac{H - 28.1}{2.8}$ **c.** 12 in.

b. We circle h to track the variable we are solving for.

$$H = 2.89\,h + 70.64 \quad \text{Given}$$

$$H - 70.64 = 2.89\,h \quad \text{Subtract 70.64.}$$

$$\frac{H - 70.64}{2.89} = \frac{2.89\,h}{2.89} \quad \text{Divide by 2.89.}$$

Thus,

$$h = \frac{H - 70.64}{2.89}$$

c. This time, we substitute 163.12 for H in the preceding formula to obtain

$$h = \frac{163.12 - 70.64}{2.89} = 32$$

Thus, the length of the humerus of a 163.12-centimeter-tall man is 32 centimeters.

EXAMPLE 2 Comparing temperatures

The formula for converting degrees Fahrenheit (°F) to degrees Celsius (°C) is

$$C = \frac{5}{9}(F - 32)$$

a. The twentieth-century mean temperature of the earth was 60°F. To the nearest degree, how many degrees Celsius is that?

b. Solve for F.

c. In 1905 the average planet temperature was 14°C. To the nearest degree, how many degrees Fahrenheit is that?

SOLUTION 2

a. In this case, $F = 60$. So

$$C = \frac{5}{9}(F - 32) = \frac{5}{9}(60 - 32)$$

$$= \frac{5}{9}(28) \quad \text{Subtract inside the parentheses first.}$$

$$= \frac{5 \cdot 28}{9}$$

$$= \frac{140}{9}$$

$$\approx 16 \quad \text{To the nearest degree}$$

Thus, the temperature is about 16°C.

b. We circle F to track the variable we are solving for.

$$C = \frac{5}{9}(F - 32) \quad \text{Given}$$

$$9 \cdot C = 9 \cdot \frac{5}{9}(F - 32) \quad \text{Multiply by 9.}$$

$$9C = 5(F - 32) \quad \text{Simplify.}$$

$$9C = 5F - 160 \quad \text{Use the distributive property.}$$

$$9C + 160 = 5F - 160 + 160 \quad \text{Add 160.}$$

$$\frac{9C + 160}{5} = \frac{5F}{5} \quad \text{Divide by 5.}$$

Thus,

$$F = \frac{9C + 160}{5}$$

(continued)

PROBLEM 2

Did you know that the temperature speeds up animals? The hotter the faster.

The formula for the speed of an ant is $S = \frac{1}{6}(C - 4)$ centimeters per second, where C is the temperature in degrees Celsius.

a. If the temperature is 22°C, how fast is the ant moving?

b. Solve for C.

c. What is C when $S = 2$ cm/sec?

Answers to PROBLEMS

2. a. 3 cm/sec **b.** $C = 6S + 4$ **c.** 16°C

c. Substitute 14 for C:

$$F = \frac{9 \cdot 14 + 160}{5}$$
$$= \frac{286}{5}$$
$$\approx 57 \quad \text{To the nearest degree}$$

Thus, the temperature is about 57°F.

EXAMPLE 3 Solving retail problems

The retail selling price R of an item is obtained by adding the original cost C and the markup M on the item.

a. Write a formula for the retail selling price.

b. Find the markup M of an item that originally cost \$50.

SOLUTION 3

a. Write the problem in words and then translate it.

The retail selling price	is obtained	by adding the original cost C and the markup M.
R	$=$	$C + M$

b. Here $C = \$50$ and we must solve for M. Substituting 50 for C in $R = C + M$, we have

$$R = 50 + M$$
$$R - 50 = M \quad \text{Subtract 50.}$$

Thus, $M = R - 50$.

PROBLEM 3

The final cost F of an item is obtained by adding the original cost C and the tax T on the item.

a. Write a formula for the final cost F of the item.

b. Find the tax T on an item that originally cost \$10.

B > Using Geometric Formulas

Many of the formulas we encounter in algebra come from geometry. For example, to find the *area* of a figure, we must find the number of square units contained in the figure. Unit squares look like these:

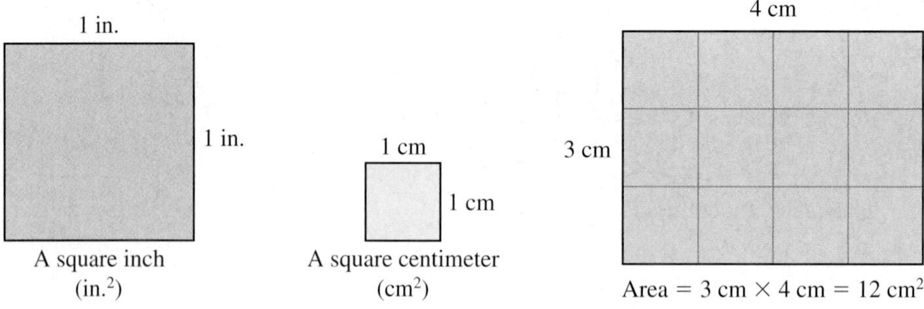

Now, to find the **area** of a figure, say a rectangle, we must find the number of square units it contains. For example, the area of a rectangle 3 centimeters by 4 centimeters is 3 cm × 4 cm = 12 cm^2 (read "12 square centimeters"), as shown in the diagram.

In general, we can find the area A of a rectangle by multiplying its length L by its width W, as given here.

AREA OF A RECTANGLE

The area A of a rectangle of length L and width W is

$$A = LW$$

Answers to PROBLEMS

3. a. $F = C + T$ **b.** $T = F - 10$

What about a rectangle's **perimeter** (distance around)? We can find the perimeter by adding the lengths of the four sides. Since we have two sides of length L and two sides of length W, the perimeter of the rectangle is

$$W + L + W + L = 2L + 2W$$

In general, we have the following formula.

PERIMETER OF A RECTANGLE

The perimeter P of a rectangle of length L and width W is

$$P = 2L + 2W$$

EXAMPLE 4 Finding areas and perimeters

Find:

a. The area of the rectangle shown in the figure.

b. The perimeter of the rectangle shown in the figure.

c. If the perimeter of a rectangle 30 inches long is 110 inches, what is the width of the rectangle?

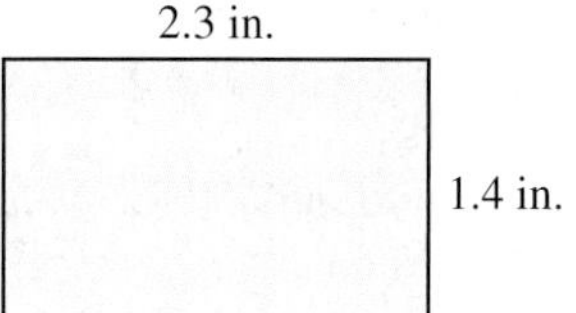

SOLUTION 4

a. The area:

$$\begin{aligned} A &= LW \\ &= (2.3 \text{ in.}) \cdot (1.4 \text{ in.}) \\ &= 3.22 \text{ in.}^2 \end{aligned}$$

b. The perimeter:

$$\begin{aligned} P &= 2L + 2W \\ &= 2(2.3 \text{ in.}) + 2(1.4 \text{ in.}) \\ &= 4.6 \text{ in.} + 2.8 \text{ in.} \\ &= 7.4 \text{ in.} \end{aligned}$$

c. The perimeter:

$$\begin{aligned} P &= 2L + 2W & \\ 110 \text{ in.} &= 2 \cdot (30 \text{ in.}) + 2W & \text{Substitute 30 for } L \text{ and 110 for } P. \\ 110 \text{ in.} &= 60 \text{ in.} + 2W & \text{Simplify.} \\ 110 \text{ in.} - 60 \text{ in.} &= 2W & \text{Subtract 60.} \\ 50 \text{ in.} &= 2W & \\ 25 \text{ in.} &= W & \text{Divide by 2.} \end{aligned}$$

Thus, the width of the rectangle is 25 inches.

Note that the area is given in square units, whereas the perimeter is a length and is given in linear units.

PROBLEM 4

a. What is the area of a rectangle 2.4 by 1.2 inches?

b. What is the perimeter of the rectangle in part **a?**

c. If the perimeter of a rectangle 20 inches long is 100 inches, what is the width of the rectangle?

If we know the area of a rectangle, we can always calculate the area of the shaded triangle:

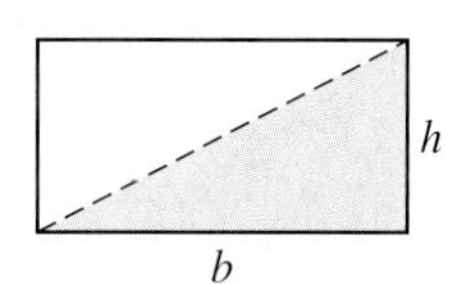

The area of the triangle is $\frac{1}{2}$ the area of the rectangle, which is bh. Thus, we have the following formula.

AREA OF A TRIANGLE

If a triangle has base b and perpendicular height h, its area A is

$$A = \frac{1}{2}bh$$

Answers to PROBLEMS

4. a. 2.88 in.^{2} **b.** 7.2 in. **c.** 30 in.

Note that this time we used b and h instead of L and W. This formula holds true for any type of triangle.

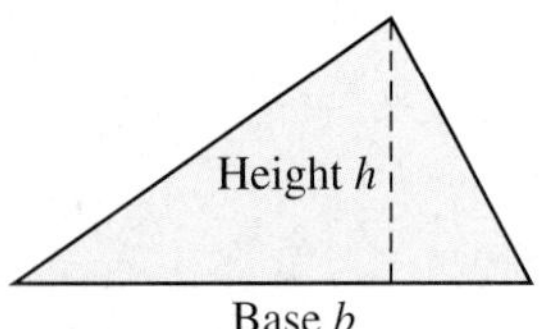

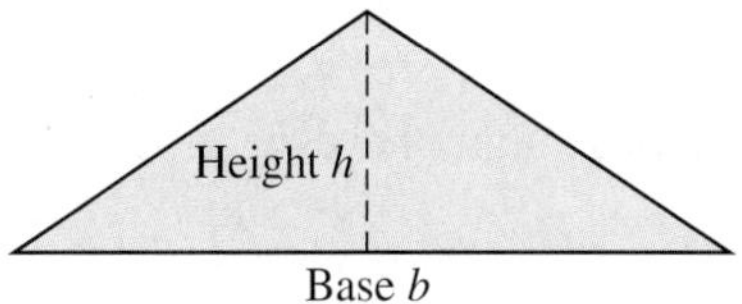

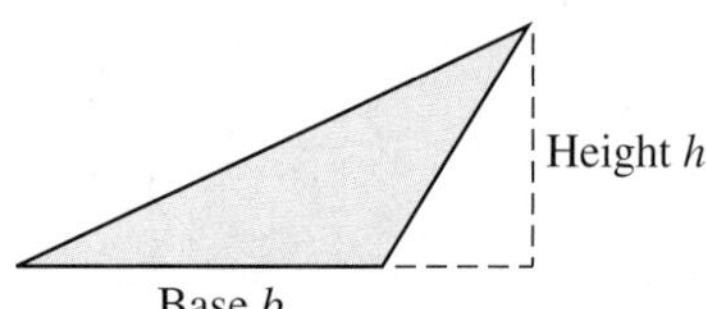

EXAMPLE 5 Finding areas of a triangle

a. Find the area of a triangular piece of cloth 20 centimeters long and 10 centimeters high.

b. The area of the triangular sail on a toy boat is 250 square centimeters. If the base of the sail is 20 centimeters long, how high is the sail?

SOLUTION 5

a. $A = \frac{1}{2}bh$

$= \frac{1}{2}(20 \text{ cm}) \cdot (10 \text{ cm})$

$= 100 \text{ cm}^2$

b. Substituting 250 for A and 20 for b, we obtain

$A = \frac{1}{2}bh$ Becomes

$250 = \frac{1}{2} \cdot 20 \cdot h$

$250 = 10h$ Simplify.

$25 = h$ Divide by 10.

Thus, the height of the sail is 25 centimeters.

PROBLEM 5

a. Find the area of a triangle 30 inches long and 15 inches high.

b. The area of the sail on a boat is 300 square feet. If the base of the sail is 20 ft long, how high is the sail?

The area of a circle can easily be found if we know the **radius** r of the circle, which is the distance from the center of the circle to its edge. Here is the formula.

AREA OF A CIRCLE

The area *A* of a circle of radius *r* is

$$A = \pi \cdot r \cdot r$$
$$= \pi r^2$$

As you can see, the formula for finding the area of a circle involves the number π (read "pie"). The number π is irrational; it cannot be written as a terminating or repeating decimal or a fraction, but it can be *approximated*. In most of our work we shall say that π is about 3.14 or $\frac{22}{7}$; that is, $\pi \approx 3.14$ or $\pi \approx \frac{22}{7}$. The number π is also used in finding the perimeter (distance around) a circle. This perimeter of the circle is called the **circumference** C and is found by using the following formula.

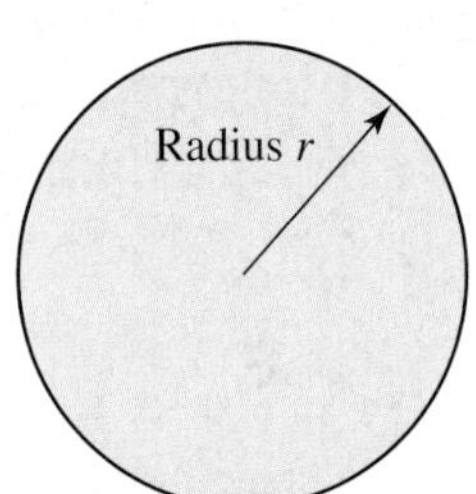

The area A of a circle of radius r is $A = \pi \cdot r \cdot r = \pi r^2$

CIRCUMFERENCE OF A CIRCLE

The circumference *C* of a circle of radius *r* is

$$C = 2\pi r$$

Answers to PROBLEMS

5. a. 225 in.^{2} **b.** 30 ft

Since the radius of a circle is half the diameter, another formula for the circumference of a circle is $C = \pi d$, where d is the diameter of the circle.

EXAMPLE 6 A CD's recorded area and circumference

The circumference C of a CD is 12π centimeters.

a. What is the radius r of the CD?

b. Every CD has a circular region in the center with no grooves, as can be seen in the figure. If the radius of this circular region is 2 centimeters and the rest of the CD has grooves, what is the area of the grooved (recorded) region?

SOLUTION 6

a. The circumference of a circle is

$$C = 2\pi r$$

$$12\pi = 2\pi r \quad \text{Substitute } 12\pi \text{ for } C.$$

$$6 = r \quad \text{Divide by } 2\pi.$$

No sound

Recorded

Thus, the radius of the CD is 6 centimeters.

b. To find the grooved (shaded) area, we find the area of the entire CD and subtract the area of the ungrooved (light blue) region. The area of the entire CD is

$$A = \pi r^2$$

$$A = \pi(6)^2 = 36\pi \quad \text{Substitute } r = 6.$$

The area of the light blue region is

$$A = \pi(2)^2 = 4\pi$$

Thus, the area of the shaded region is $36\pi - 4\pi = 32\pi$ square centimeters.

PROBLEM 6

The circumference C of a CD is 9π inches.

a. What is the radius r of the CD?

b. The radius of the nonrecorded part of this CD is 0.75 inches. What is the area of the recorded region?

C › Solving for Angle Measurements

We've already mentioned complementary angles (two angles whose sum is 90°) and supplementary angles (two angles whose sum is 180°). Now we introduce the idea of *vertical angles.* The figure shows two intersecting lines with angles numbered ①, ❷, ③, and ❹.

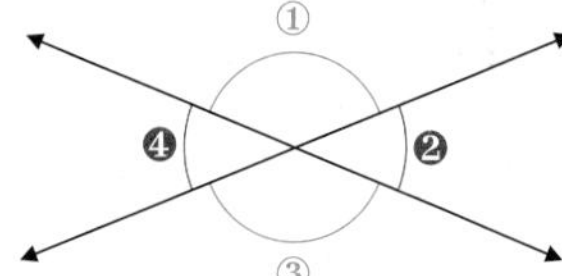

Angles ① and ③ are placed "vertically"; they are called *vertical angles*. Another pair of vertical angles is ❷ and ❹.

VERTICAL ANGLES

Vertical angles have equal measures.

Note that if you add the measures of angles ① and ❷, you get 180°, a **straight angle.** Similarly, the sums of the measures of angles ❷ and ③, ③ and ❹, and ❹ and ① yield straight angles.

Answers to PROBLEMS

6. a. 4.5 in. **b.** 19.6875π in.^{2}

EXAMPLE 7 Finding the measures of angles

Find the measures of the marked angles in each figure:

a.

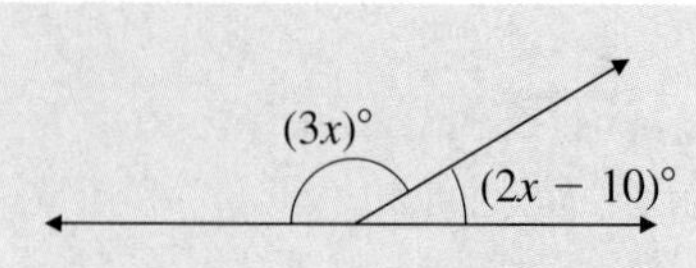

b.

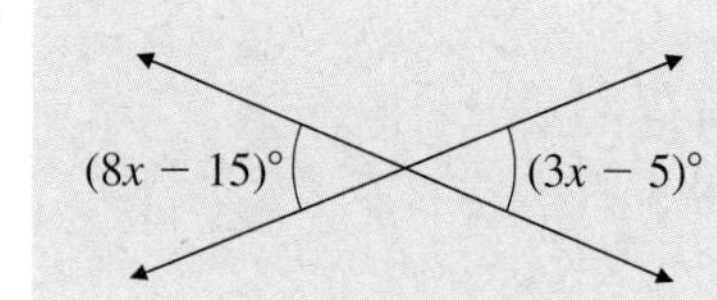

c.

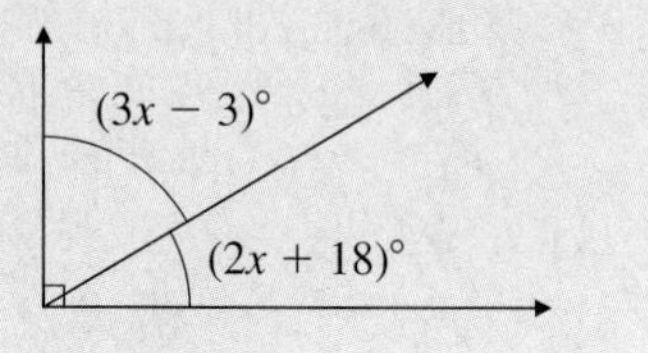

SOLUTION 7

a. The sum of the measures of the two marked angles must be 180°, since the angles are supplementary (they form a straight angle). Thus,

$$(2x - 10) + 3x = 180$$
$$5x - 10 = 180 \quad \text{Simplify.}$$
$$5x = 190 \quad \text{Add 10.}$$
$$x = 38 \quad \text{Divide by 5.}$$

To find the measure of each of the angles, replace x with 38 in $2x - 10$ and in $3x$ to obtain

$$2(38) - 10 = 66 \quad \text{and} \quad 3(38) = 114$$

Thus, the measures are 66° and 114°, respectively. Note that $114 + 66 = 180$, so our result is correct.

b. The two marked angles are vertical angles, so their measures must be equal. Thus,

$$8x - 15 = 3x - 5$$
$$8x = 3x + 10 \quad \text{Add 15.}$$
$$5x = 10 \quad \text{Subtract 3x.}$$
$$x = 2 \quad \text{Divide by 5.}$$

Now, replace x with 2 in $8x - 15$ to obtain $8 \cdot 2 - 15 = 1$. Since the angles are vertical, their measures are both 1°.

c. The sum of the measures of the two marked angles must be 90°, since the angles are complementary. Thus,

$$(3x - 3) + (2x + 18) = 90$$
$$5x + 15 = 90 \quad \text{Simplify.}$$
$$5x = 75 \quad \text{Subtract 15.}$$
$$x = 15 \quad \text{Divide by 5.}$$

Replacing x with 15 in $(3x - 3)$ and $(2x + 18)$, we obtain $3 \cdot 15 - 3 = 42$ and $2 \cdot 15 + 18 = 48$. Thus, the measures of the angles are 42° and 48°, respectively. Note that the sum of the measures of the two angles is 90°, as expected.

PROBLEM 7

a. Find the measures if the angle in the diagram for part **a** is $(8x)°$ instead of $(3x)°$.

b. Find the measures if the angle in the diagram for part **b** is $(4x - 3)°$ instead of $(3x - 5)°$.

c. Find the measures if the angle in the diagram for part **c** is $(x + 17)°$ instead of $(2x + 18)°$.

Answers to PROBLEMS

7. a. 28° and 152° **b.** Both are 9°
c. 54° and 36°

D > Solving Applications by Using Formulas

Many formulas find uses in our daily lives. For example, do you know the *exact* relationship between your shoe size S and the length L of your foot in inches? Here are the formulas used in the United States:

$$L = \frac{22 + S}{3} \quad \text{For men}$$

$$L = \frac{21 + S}{3} \quad \text{For women}$$

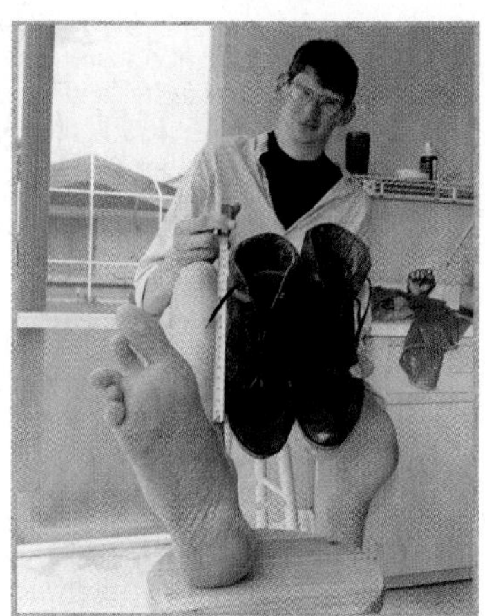

EXAMPLE 8 **Sizing shoes**

Use the shoe sizing formulas to find:

a. The length of the foot corresponding to a size 1 shoe for men.

b. The length of the foot corresponding to a size 1 shoe for women.

c. Solve for S and determine the size shoe needed by Matthew McGrory (the man with the largest feet), whose left foot is 18 inches long.

SOLUTION 8

a. To find the length of the foot corresponding to a size 1 shoe for men, substitute 1 for S in

$$L = \frac{22 + S}{3}$$

$$= \frac{22 + 1}{3}$$

$$= \frac{23}{3} = 7\frac{2}{3} \text{ in.}$$

b. This time we substitute 1 for S in

$$L = \frac{21 + S}{3}$$

$$L = \frac{21 + 1}{3} = \frac{22}{3} = 7\frac{1}{3} \text{ in.}$$

c.

$$L = \frac{22 + S}{3} \quad \text{Given}$$

$$3L = 3 \cdot \frac{22 + S}{3} \quad \text{Multiply by the LCM 3.}$$

$$3L = 22 + S \quad \text{Simplify.}$$

$$3L - 22 = S \quad \text{Subtract 22.}$$

Since the length of McGrory's foot is 18 in., substitute 18 for L to obtain

$$S = 3 \cdot 18 - 22 = 32$$

Thus, McGrory needs a size 32 shoe!

PROBLEM 8

a. Find the length of the foot corresponding to a size 2 shoe for men.

b. Find the length of the foot corresponding to a size 3 shoe for women.

c. McGrory's "smaller" right foot is 17 inches long. What shoe size fits his right foot?

As usual, here are some practice translations.

Answers to PROBLEMS

8. a. 8 in. **b.** 8 in. **c.** 29

TRANSLATE THIS

The third step in the RSTUV procedure is to **TRANSLATE** the information into an equation. In Problems 1–10 **TRANSLATE** the sentence and match the correct translation with one of the equations A–O.

1. The height H of a man (in centimeters) is the sum of 70.64 and the product of 2.89 and h, where h is the length of the man's humerus bone.
2. The formula C for converting degrees Fahrenheit F to degrees Celsius C is $\frac{5}{9}$ times the difference of F and 32.
3. The area A of a square whose side is S units is S squared.
4. The area A of a rectangle of length L and width W is the product of L and W.
5. The perimeter P of a rectangle of length L and width W is the sum of twice the length L and twice the length W.
6. The area A of a triangle with base b and height h is one-half the product of b and h.
7. The area A of a circle of radius r is the product of π and the square of r.
8. The circumference C of a circle of radius r is twice the product of π and r.
9. The length L of a man's foot is obtained by finding the sum of S and 22, and dividing the result by 3.
10. The length L of a woman's foot is the quotient of the sum S plus 21, and 3.

A. $\frac{5}{9}F - 32$

B. $C = 2\pi r$

C. $H = 70.64 + 2.89h$

D. $L = \frac{S + 22}{3}$

E. $C = \frac{5}{9}(F - 32)$

F. $P = 2L + 2W$

G. $L = S + \frac{21}{3}$

H. $C = \frac{5}{9}(32 - F)$

I. $A = S^2$

J. $L = S + \frac{22}{3}$

K. $A = \pi r^2$

L. $L = \frac{S + 21}{3}$

M. $A = \frac{1}{2}bh$

N. $H = (70.64 + 2.89)f$

O. $A = LW$

> Practice Problems > Self-Tests > Media-rich eBooks > e-Professors > Videos

Exercises 2.6

A Using Formulas In Problems 1–10, use the formulas to find the solution.

1. The number of miles D traveled in T hours by an object moving at a rate R (in miles per hour) is given by $D = RT$.
   a. Find D when $R = 30$ and $T = 4$.
   b. Find the distance traveled by a car going 55 miles per hour for 5 hours.
   c. Solve for R in $D = RT$.
   d. If you travel 180 miles in 3 hours, what is R?

2. The rate of travel R of an object moving a distance D in time T is given by

$$R = \frac{D}{T}$$

   a. Find R when $D = 240$ miles and $T = 4$ hours.
   b. Find the rate of travel of a train that traveled 140 miles in 4 hours.
   c. Solve for T in $R = \frac{D}{T}$.
   d. How long would it take the train of part **b** to travel 105 miles?

3. The height H of a man (in inches) is related to his weight W (in pounds) by the formula $W = 5H - 190$.
   a. If a man is 60 inches tall, what should his weight be?
   b. Solve for H.
   c. If a man weighs 200 pounds, how tall should he be?

4. The number of hours H a growing child should sleep is

$$H = 17 - \frac{A}{2}$$

where A is the age of the child in years.
   a. How many hours should a 6-year-old sleep?
   b. Solve for A in $H = 17 - \frac{A}{2}$.
   c. At what age would you expect a child to sleep 11 hours?

> Web IT go to mhhe.com/bello for more lessons

Applications: Green Math

5. The formula for converting degrees Celsius to degrees Fahrenheit is

$$F = \frac{9}{5}C + 32.$$

a. Solve for C in the formula.

b. Climate model projections indicate that the global surface temperature will rise as much as 6°C by the end of the twenty-first century. To the nearest degree, how many degrees Fahrenheit is 6°C?

6. The formula for converting degrees Fahrenheit to degrees Celsius is

$$C = \frac{5}{9}(F - 32).$$

a. Solve for F in the formula.

b. The average temperature of the earth is about 57°F. To the nearest degree, how many degrees Celsius is that?

7. The energy efficiency ratio (EER) for an air conditioner is obtained by dividing the British thermal units (Btu) the air conditioner uses per hour by the watts w.

a. Write a formula that will give the EER of an air conditioner.

b. Find the EER of an air conditioner with a capacity of 9000 British thermal units per hour and a rating of 1000 watts.

c. Solve for the British thermal units in your formula for EER.

d. How many British thermal units does a 2000-watt air conditioner produce if its EER is 10?

8. The capital C of a business is the difference between the assets A and the liabilities L.

a. Write a formula that will give the capital of a business.

b. If a business has \$4800 in assets and \$2300 in liabilities, what is the capital of the business?

c. Solve for L in your formula for C.

d. If a business has \$18,200 in capital and \$30,000 in assets, what are its liabilities?

9. The selling price S of an item is the sum of the cost C and the margin (markup) M.

a. Write a formula for the selling price of a given item.

b. A merchant wishes to have a \$15 margin (markup) on an item costing \$52. What should be the selling price of this item?

c. Solve for M in your formula for S.

d. If the selling price of an item is \$18.75 and its cost is \$10.50, what is the markup?

10. The tip speed S_T of a propeller is equal to π times the diameter d of the propeller times the number N of revolutions per second.

a. Write a formula for S_T.

b. If a propeller has a 2-meter diameter and it is turning at 100 revolutions per second, find S_T. (Use $\pi \approx 3.14$.)

c. Solve for N in your formula for S_T.

d. What is N when $S_T = 275\pi$ and $d = 2$?

B Using Geometric Formulas In Problems 11–15 use the geometric formulas to find the solution.

11. The perimeter P of a rectangle of length L and width W is $P = 2L + 2W$.

a. Find the perimeter of a rectangle 10 centimeters by 20 centimeters.

b. What is the length of a rectangle with a perimeter of 220 centimeters and a width of 20 centimeters?

12. The perimeter P of a rectangle of length L and width W is $P = 2L + 2W$.

a. Find the perimeter of a rectangle 15 centimeters by 30 centimeters.

b. What is the width of a rectangle with a perimeter of 180 centimeters and a length of 60 centimeters?

13. The circumference C of a circle of radius r is $C = 2\pi r$.

a. Find the circumference of a circle with a radius of 10 inches. (Use $\pi \approx 3.14$.)

b. Solve for r in $C = 2\pi r$.

c. What is the radius of a circle whose circumference is 20π inches?

14. If the circumference C of a circle of radius r is $C = 2\pi r$, what is the radius of a tire whose circumference is 26π inches?

15. The area A of a rectangle of length L and width W is $A = LW$.

a. Find the area of a rectangle 4.2 meters by 3.1 meters.

b. Solve for W in the formula $A = LW$.

c. If the area of a rectangle is 60 square meters and its length is 10 meters, what is the width of the rectangle?

Web IT go to mhhe.com/bello for more lessons

Web IT go to mhhe.com/bello for more lessons

‹ C › Solving for Angle Measurements

In Problems 16–29, find the measure of each marked angle.

16.

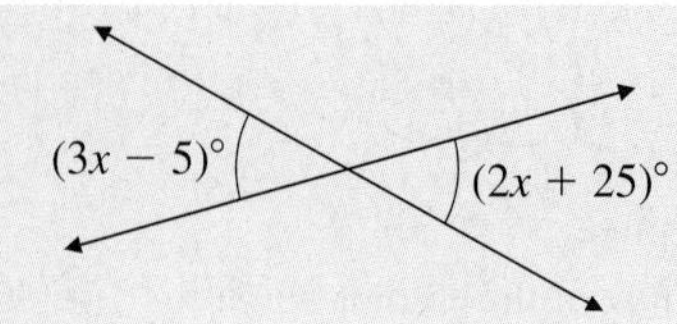

17.

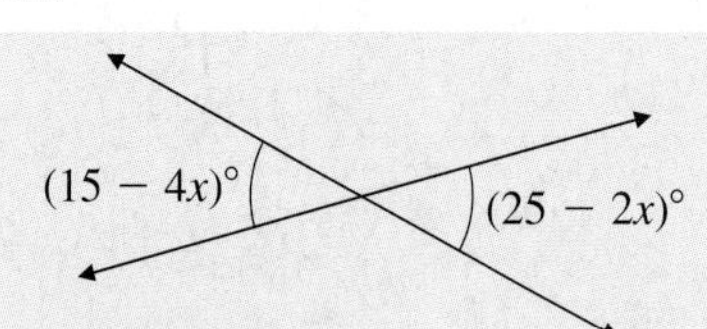

18.

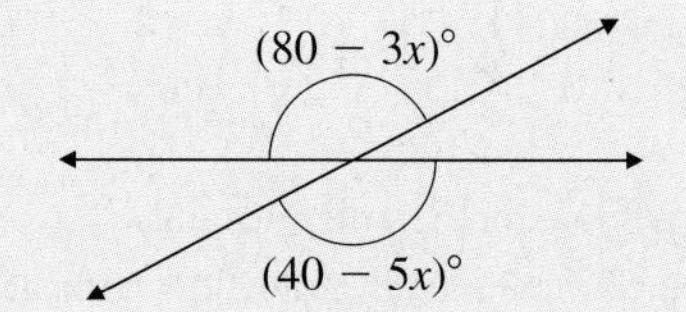

19.

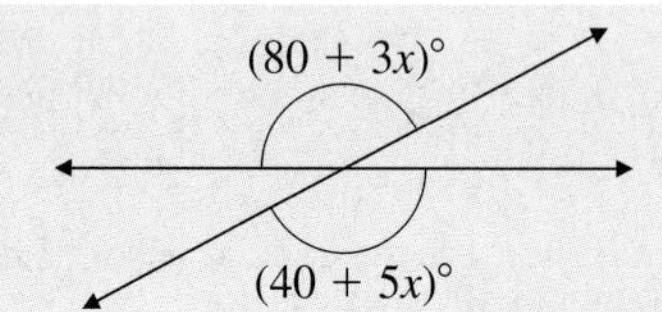

20.

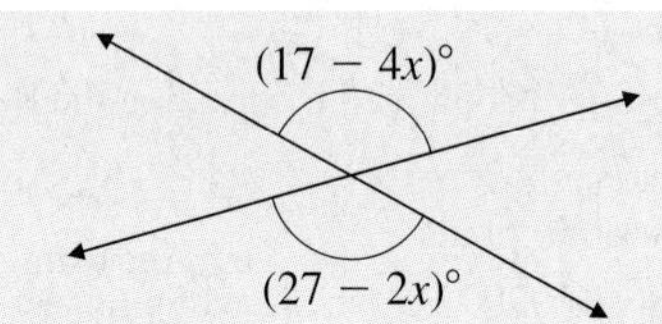

21.

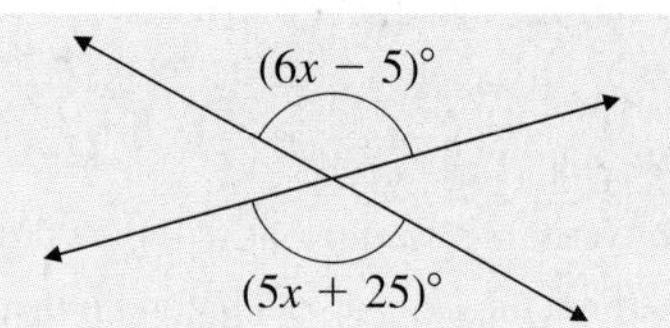

22.

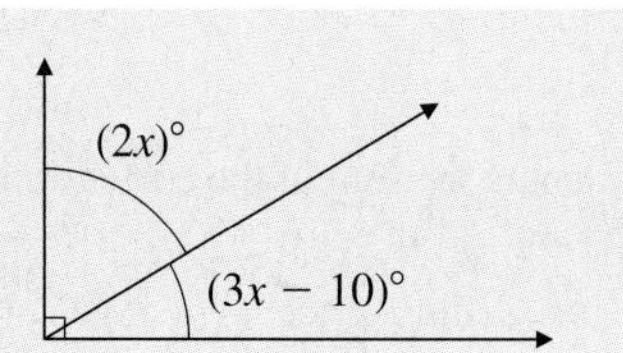

23.

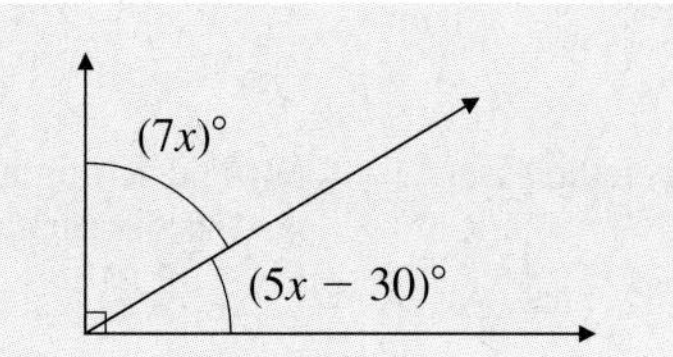

24.

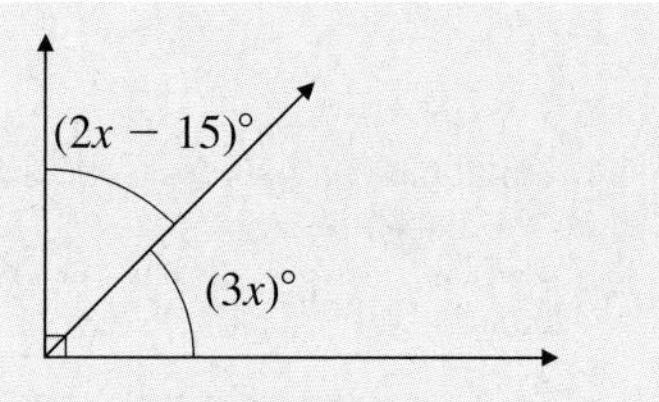

25.

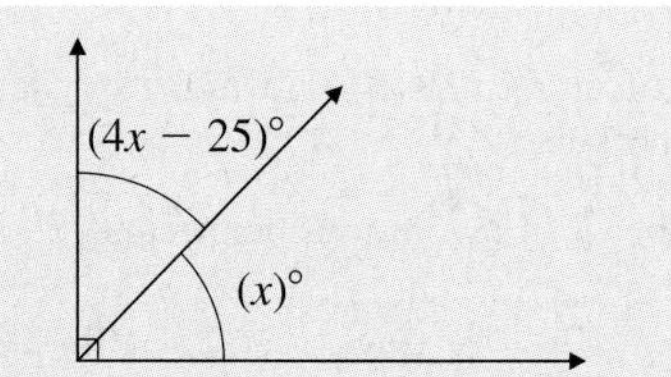

26.

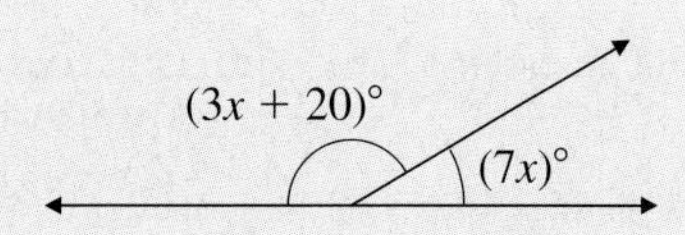

27.

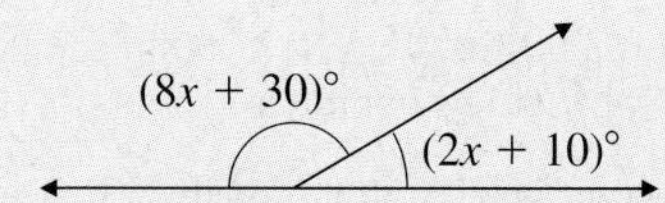

28.

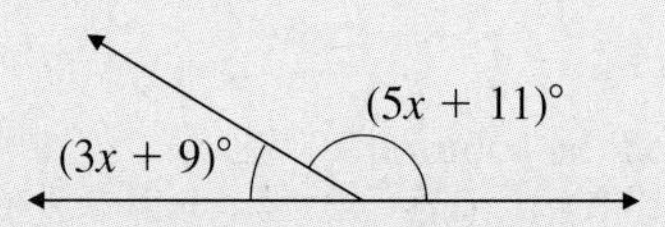

29.

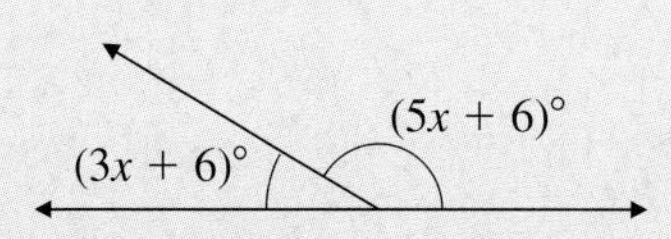

‹ D › Solving Applications by Using Formulas

30. *Supersized omelet* One of the largest rectangular omelets ever cooked was 30 feet long and had an 80-foot perimeter. How wide was it?

31. *Largest pool* If you were to walk around the largest rectangular pool in the world, in Casablanca, Morocco, you would walk more than 1 kilometer. To be exact, you would walk 1110 meters. If the pool is 480 meters long, how wide is it?

32. *Football field dimensions* The playing surface of a football field is 120 yards long. A player jogging around the perimeter of this surface jogs 346 yards. How wide is the playing surface of a football field?

33. *CD diameter* A point on the rim of a CD record travels 14.13 inches each revolution. What is the diameter of this CD? (Use $\pi \approx 3.14$.)

34. *Gigantic pizza* One of the largest pizzas ever made had a 251.2-foot circumference! What was its diameter? (Use $\pi \approx 3.14$.)

35. *Continental suit sizes* Did you know that clothes are sized differently in different countries? If you want to buy a suit in Europe and you wear a size A in America, your continental (European) size C will be $C = A + 10$.

a. Solve for A.

b. If you wear a continental size 50 suit, what would be your American size?

36. *Continental dress sizes* Your continental dress size C is given by $C = A + 30$, where A is your American dress size.

a. Solve for A.

b. If you wear a continental size 42 dress, what is your corresponding American size?

37. *Recreational boats data* According to the National Marine Manufacturers Association, the number N of recreational boats (in millions) has been steadily increasing since 1975 and is given by $N = 9.74 + 0.40t$, where t is the number of years after 1975.

a. What was the number of recreational boats in 1985?

b. Solve for t in $N = 9.74 + 0.40t$.

c. In what year would you expect the number of recreational boats to reach 17.74 million?

38. *NCAA men's basketball teams* The number N of NCAA men's college basketball teams has been increasing since 1980 according to the formula $N = 720 + 5t$, where t is the number of years after 1980.

a. How many teams would you expect in the year 2000?

b. Solve for t.

c. In what year would you expect the number of teams to reach 795?

39. *Vehicle insurance* The amount A spent on vehicle insurance by *persons under 25* can be approximated by the equation $A = 28x + 420$, where x is the number of years after 2000.

a. According to the formula, how much would a person under 25 spend on vehicle insurance in 2010?

b. Solve for x.

c. In how many years would vehicle insurance be \$1000? Answer to the nearest year.

Source: Bureau of Labor Consumer Expenditure Survey.

40. *Vehicle insurance* The amount A spent on vehicle insurance by *persons aged 25–54* can be approximated by the equation $A = 40x + 786$, where x is the number of years after 2000.

a. According to the formula, how much would a person aged 25–54 spend on vehicle insurance in 2010?

b. Solve for x.

c. In how many years would vehicle insurance be \$1000? Answer to the nearest year.

Source: Bureau of Labor Consumer Expenditure Survey.

41. *Height of a man* The height H of a man (in inches) can be estimated by the equation $H = 3.3r + 34$, where r is the length of the radius bone (the bone from the wrist to the elbow).

a. Estimate the height of a man whose radius bone is 10 inches.

b. Solve for r.

c. If a man is 70.3 inches tall, how long is his radius?

Source: Science Safari: The First People.

42. *Height of a woman* The height H of a woman (in inches) can be estimated by $H = 3.3r + 32$, where r is the length of the radius bone (the bone from the wrist to the elbow).

a. Estimate the height of a woman whose radius bone is 10 inches.

b. Solve for r.

c. If a woman is 61.7 inches tall, how long is her radius?

Source: Science Safari: The First People.

43. *Height of a woman* The height H of a woman (in centimeters) can be estimated by $H = 2.9t + 62$, where t is the length of the tibia bone (the bone between the knee and ankle).

a. Estimate the height of a woman whose tibia bone is 30 centimeters.

b. Solve for t.

c. If a woman is 151.9 inches tall, how long is her radius?

Source: Science Safari: The First People.

Using Your Knowledge

Living with Algebra Many practical problems around the house require some knowledge of the formulas we've studied. For example, let's say that you wish to carpet your living room. You need to use the formula for the area A of a rectangle of length L and width W, which is $A = LW$.

44. Carpet sells for \$8 per square yard. If the cost of carpeting a room was \$320 and the room is 20 feet long, how wide is it? (Note that one square yard is nine square feet.)

45. If you wish to plant new grass in your yard, you can buy sod squares of grass that can simply be laid on the ground. Each sod square is approximately 1 square foot. If you have 5400 sod squares and you wish to sod an area that is 60 feet wide, what is the length?

46. If you wish to fence your yard, you need to know its perimeter (the distance around the yard). If the yard is W feet by L feet, the perimeter P is given by $P = 2W + 2L$. If your rectangular yard needs 240 feet of fencing and your yard is 70 feet long, what is the width?

Web IT go to mhhe.com/bello for more lessons

Write On

47. Write an explanation of what is meant by the *perimeter* of a geometric figure.

48. Write an explanation of what is meant by the *area* of a geometric figure.

49. Remember the hamburgers in the *Getting Started?* Write two explanations of how the Monster burger can have 50% more beef than the Lite burger and still look like the one in the picture.

50. To make a fair comparison of the amount of beef in two hamburgers, should you compare the circumferences, areas, or volumes? Explain.

51. The Lite burger has a 4-inch diameter whereas the Monster burger has a 6-inch diameter. How much bigger (in percent) is the circumference of the Monster burger? Can you now explain the claim in the ad? Is the claim correct? Explain.

Concept Checker

Fill in the blank(s) with the correct word(s), phrase, or mathematical statement.

52. The **area** A of a **rectangle** of length L and width W is ___________.

53. The **perimeter** P of a **rectangle** of length L and width W is ___________.

54. The **area** A of a **triangle** with base b and height h is ___________.

55. The **area** A of a **circle** of radius r is ___________.

56. The **circumference** C of a **circle** of radius r is ___________.

57. **Vertical angles** are angles of ___________ **measure.**

LW	πr^2
$2L + 2W$	$2\pi r$
bh	$2\pi r^2$
$\frac{1}{2}bh$	unequal
πr	equal

Mastery Test

Find the measures of the marked angles.

58.

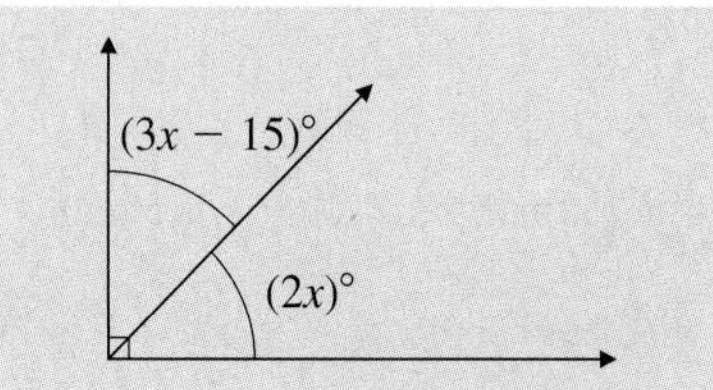

59.

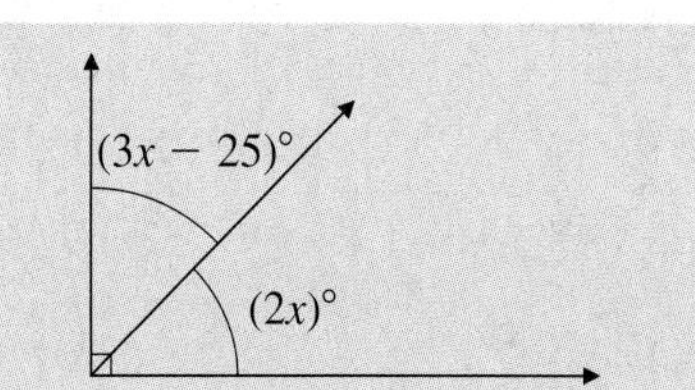

60.

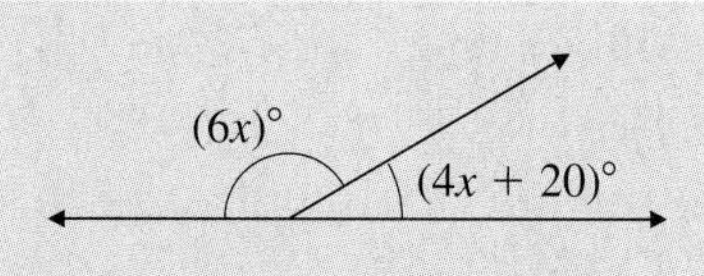

61.

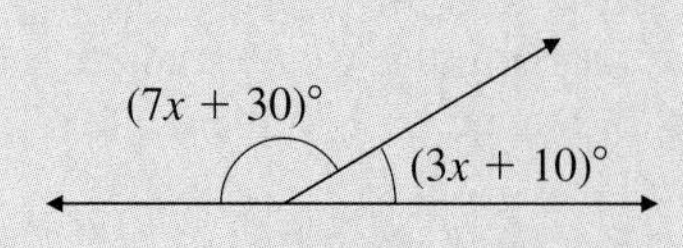

62.

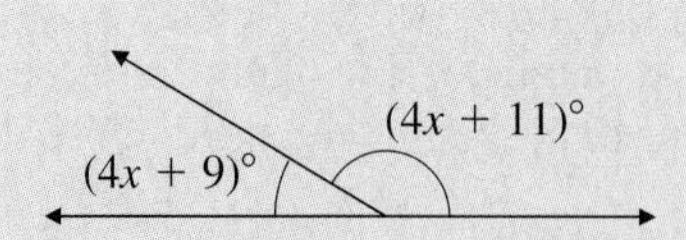

63.

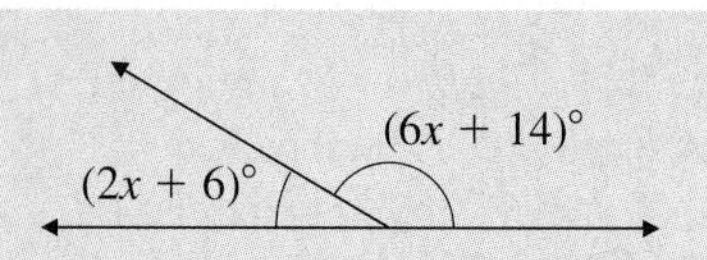

64. a. A circle has a radius of 10 inches. Find its area and its circumference.

b. If the circumference of a circle is 40π inches, what is its radius?

65. The formula for estimating the height H (in centimeters) of a woman using the length h of her humerus as a clue is given by the equation $H = 2.75h + 71.48$.

a. Estimate the height of a woman whose humerus bone is 20 centimeters long.

b. Solve for h.

c. If a woman is 140.23 centimeters tall, how long is her humerus?

66. The total cost T of an item is obtained by adding its cost C and the tax t on the item.

a. Write a formula for the total cost T.

b. Find the tax t on an item that cost \$8 if the total after adding the tax is \$8.48.

67. The area A of a triangle is $A = \frac{1}{2}bh$, where b is the base of the triangle and h is its height.

a. What is the area of a triangle 15 inches long and with a 10-inch base?

b. Solve for b in $A = \frac{1}{2}bh$.

c. The area of a triangle is 18 square inches, and its height is 9 inches. How long is the base of the triangle?

68. According to the Motion Picture Association of America, the number N of motion picture theaters (in thousands) has been growing according to the formula $N = 15 + 0.60t$, where t is the number of years after 1975.

a. How many theaters were there in 1985? (*Hint:* $t = 10$.)

b. Solve for t in $N = 15 + 0.60t$.

c. In what year did the number of theaters total 27,000?

69. The formula for converting degrees Fahrenheit F to degrees Celsius C is

$$C = \frac{5}{9}F - \frac{160}{9}$$

a. Find the Celsius temperature on a day in which the thermometer reads 41°F.

b. Solve for F.

c. What is F when $C = 20$?

Skill Checker

Solve:

70. $3x - 2 = 2(x - 2)$

71. $2x - 1 = x + 3$

72. $3x - 2 = 2(x - 1)$

73. $4(x + 1) = 3x + 7$

74. $\frac{-x}{4} + \frac{x}{6} = \frac{x - 3}{6}$

75. $\frac{x}{3} - \frac{x}{2} = 1$

2.7 Properties of Inequalities

Objectives

A Determine which of two numbers is greater.

B Solve and graph linear inequalities.

C Write, solve, and graph compound inequalities.

D Solve an application involving inequalities.

To Succeed, Review How To . . .

1. Add, subtract, multiply, and divide real numbers (pp. 52, 54, 61, 63).
2. Solve linear equations (pp. 136–141).

Getting Started

Savings on Sandals

After learning to solve linear equations, we need to learn to solve *linear inequalities*. Fortunately, the rules are very similar, but the notation is a little different. For example, the ad says that the price you will pay for these sandals will be cut \$1 to \$3. That is, you will save \$1 to \$3. If x is the amount of money you can save, what can x be? Well, it is at least \$1 and can be as much as \$3; that is, $x = 1$ or $x = 3$ or x is between 1 and 3. In the language of algebra, we write this fact as

$$1 \le x \le 3$$

Read "1 is less than or equal to x and x is less than or equal to 3" or "x is between 1 and 3, inclusive."

The statement $1 \le x \le 3$ is made up of two parts:

$1 \le x$ which means that 1 is *less than or equal to* x (or that x *is greater than or equal to* 1), and

$x \le 3$ which means that x *is less than or equal to* 3 (or that 3 is *greater than or equal to* x).

These statements are examples of *inequalities,* which we will learn how to solve in this section.

In algebra, an **inequality** is a statement with >, <, ≥, or ≤ as its verb. Inequalities can be represented on a number line. Here's how we do it. As you recall, a number line is constructed by drawing a line, selecting a point on this line, and calling it zero (the origin):

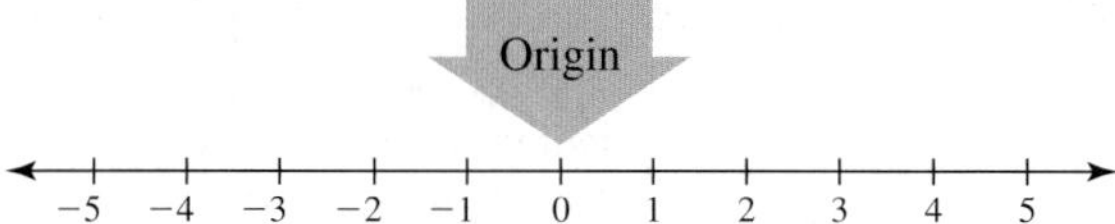

We then locate equally spaced points to the right of the origin on the line and label them with the *positive* integers 1, 2, 3, and so on. The corresponding points to the left of zero are labeled −1, −2, −3, and so on (the *negative* integers). This construction allows the *association of numbers with points on the line.* The number associated with a point is called the **coordinate** of that point. For example, there are points associated with the numbers −2.5, $-1\frac{1}{2}$, $\frac{3}{4}$, and 2.5:

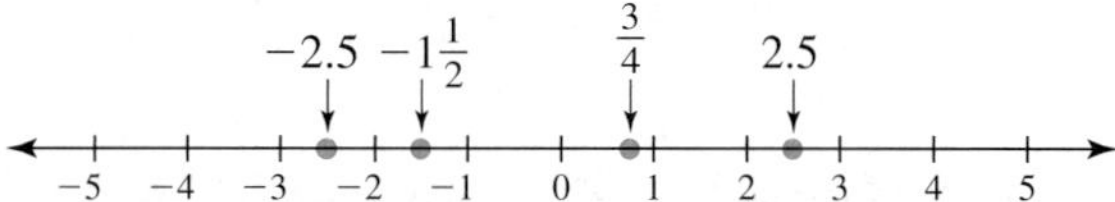

All these numbers are real numbers. As we have mentioned, the real numbers include natural (counting) numbers, whole numbers, integers, fractions, and decimals as well as the irrational numbers (which we discuss in more detail later). Thus, the real numbers can all be represented on the number line.

A > Order of Numbers

As you can see, numbers are placed in order on the number line. *Greater* numbers are always to the *right* of *smaller* ones. (The farther to the *right,* the *greater* the number.) Thus, any number to the *right* of a second number is said to be **greater than** (>) the second number. We also say that the second number is **less than** (<) the first number. For example, since 3 is to the right of 1, we write

$$\underbrace{3}_{3}\ \underbrace{\text{is greater than}}_{>}\ \underbrace{1}_{1} \quad \text{or} \quad \underbrace{1}_{1}\ \underbrace{\text{is less than}}_{<}\ \underbrace{3}_{3}$$

Similarly,

$$-1 > -3 \quad \text{or} \quad -3 < -1$$
$$0 > -2 \quad \text{or} \quad -2 < 0$$
$$3 > -1 \quad \text{or} \quad -1 < 3$$

Note that the inequality signs > and < always point to the smaller number.

EXAMPLE 1 Writing inequalities

Fill in the blank with > or < so that the resulting statement is true.

a. 3 ______ 4 **b.** −4 ______ −3 **c.** −2 ______ −3

PROBLEM 1

Fill in the blank with > or < so that the resulting statement is true.

a. 5 ______ 3 **b.** −1 ______ −4

c. −5 ______ −4

Answers to PROBLEMS

1. a. > **b.** > **c.** <

SOLUTION 1 We first construct a number line containing these numbers. (Of course, we could just think about the number line without actually drawing one.)

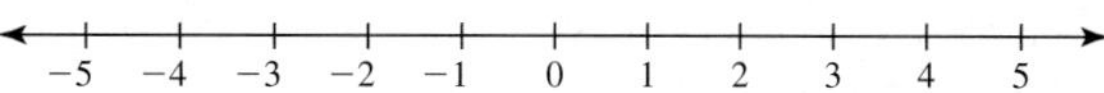

a. Since 3 is to the left of 4, $3 < 4$.
b. Since -4 is to the left of -3, $-4 < -3$.
c. Since -2 is to the right of -3, $-2 > -3$.

B > Solving and Graphing Inequalities

$x < 3$
$2 < 3$
$1\frac{1}{2} < 3$
$1 < 3$
$0 < 3$
$-\frac{1}{2} < 3$

Just as we solved equations, we can also solve inequalities. We do this by extending the addition and multiplication properties of equality (Sections 2.1 and 2.2) to include inequalities. We say that we have *solved* a given inequality when we obtain an inequality equivalent to the one given and in the form $x < \square$ or $x > \square$. (Note that the variable is on the **left** side of the inequality.) For example, consider the inequality

$$x < 3$$

There are many real numbers that will make this inequality a true statement. A few of them are shown in the accompanying table. Thus, 2 is a solution of $x < 3$ because $2 < 3$. Similarly, $-\frac{1}{2}$ is a solution of $x < 3$ because $-\frac{1}{2} < 3$.

As you can see from this table, 2, $1\frac{1}{2}$, 1, 0, and $-\frac{1}{2}$ are *solutions* of the inequality $x < 3$. Of course, we can't list all the real numbers that satisfy the inequality $x < 3$ because there are infinitely many of them, but we can certainly show all the solutions of $x < 3$ *graphically* by using a number line:

$x < 3$

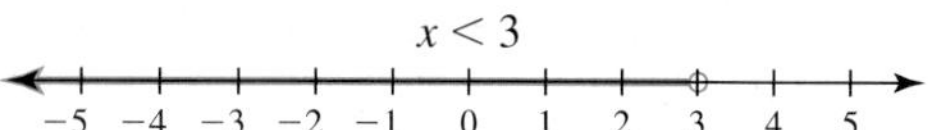

NOTE

$x < 3$ tells you to draw your heavy line to the *left* of 3 because, in this case, the symbol $<$ points left.

This representation is called the **graph** of the solutions of $x < 3$, which are indicated by the heavy line. Note that there is an open circle at $x = 3$ to indicate that 3 is *not* part of the graph of $x < 3$ (since 3 is not less than 3). Also, the colored arrowhead points to the left (just as the $<$ in $x < 3$ points to the left) to indicate that the heavy line continues to the left without end. On the other hand, the graph of $x \geq 2$ should continue to the right without end:

$x \geq 2$

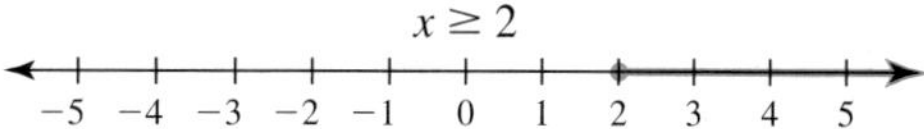

NOTE

$x \geq 2$ tells you to draw your heavy line to the *right* of 2 because, in this case, the symbol $\geq$ points right.

Moreover, since $x = 2$ is included in the graph, a solid dot appears at the point $x = 2$.

EXAMPLE 2 **Graphing inequalities**

Graph the inequality on a number line.

a. $x \geq -1$ **b.** $x < -2$

PROBLEM 2

Graph the inequality on a number line.

a. $x \leq -2$ **b.** $x > -3$

(continued)

Answers to PROBLEMS

2. a. **b.**

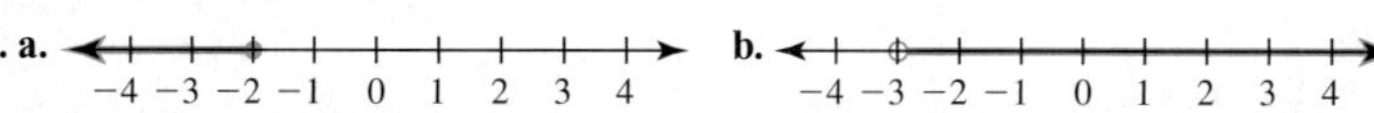

SOLUTION 2

a. The numbers that satisfy the inequality $x \geq -1$ are the numbers that are *greater than or equal to* -1, that is, the number -1 and all the numbers to the right of -1 (remember, $\geq$ points to the right and the dot must be solid). The graph is shown here:

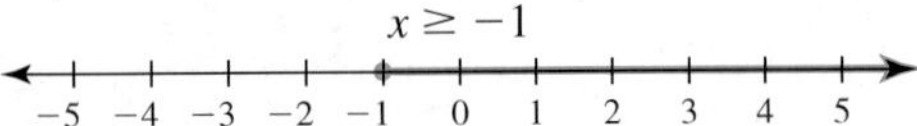

b. The numbers that satisfy the inequality $x < -2$ are the numbers that are *less than* -2, that is, the numbers to the *left of but not including* -2 (note that $<$ points to the left and that the dot is open). The graph of these points is shown here:

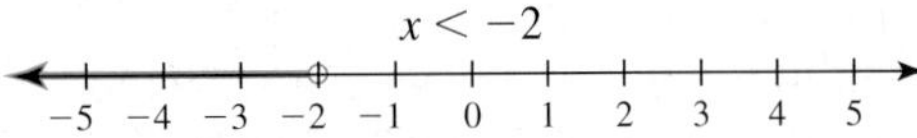

We solve more complicated inequalities just as we solve equations, by finding an equivalent inequality whose solution is obvious. Remember, we have solved a given inequality when we obtain an inequality in the form $x < \square$ or $x > \square$ which is equivalent to the one given. Thus, to solve the inequality $2x - 1 < x + 3$, we try to find an equivalent inequality of the form $x < \square$ or $x > \square$. As before, we need some properties. The first of these are the addition and subtraction properties.

If $3 < 4$, then

$$3 + 5 < 4 + 5 \quad \text{Add 5.}$$
$$8 < 9 \quad \text{True.}$$

Similarly, if $3 > -2$, then

$$3 + 7 > -2 + 7 \quad \text{Add 7.}$$
$$10 > 5 \quad \text{True.}$$

Also, if $3 < 4$, then

$$3 - 1 < 4 - 1 \quad \text{Subtract 1.}$$
$$2 < 3 \quad \text{True.}$$

Similarly, if $3 > -2$, then

$$3 - 5 > -2 - 5 \quad \text{Subtract 5.}$$
$$-2 > -7 \quad \text{True because } -2 \text{ is to the right of } -7.$$

In general, we have the following properties.

ADDITION AND SUBTRACTION PROPERTIES OF INEQUALITIES

You can *add* or *subtract* the same number c on both sides of an inequality and obtain an equivalent inequality. In symbols,

If	$a < b$	**If**	$a > b$
then	$a + c < b + c$	**then**	$a + c > b + c$
or	$a - c < b - c$	**or**	$a - c > b - c$

Note: These properties also hold when the symbols $\leq$ and $\geq$ are used.

NOTE

Since $x - b = x + (-b)$, subtracting b from both sides is the same as adding the inverse of b, $(-b)$, so you can think of subtracting b as adding $(-b)$.

Now let's return to the inequality $2x - 1 < x + 3$. To solve this inequality, we need the variables by themselves (isolated) on one side, so we proceed as follows:

$$2x - 1 < x + 3 \quad \text{Given}$$
$$2x - 1 + 1 < x + 3 + 1 \quad \text{Add 1.}$$
$$2x < x + 4 \quad \text{Simplify.}$$
$$2x - x < x - x + 4 \quad \text{Subtract } x.$$
$$x < 4 \quad \text{Simplify.}$$

Any number less than 4 is a solution. The graph of this inequality is as follows:

$2x - 1 < x + 3$ or, equivalently, $x < 4$

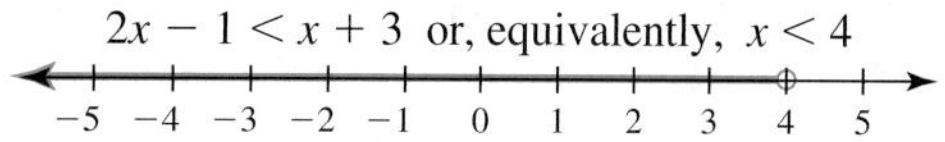

You can check that this solution is correct by selecting any number from the graph (say 0) and replacing x with that number in the original inequality. For $x = 0$, we have $2(0) - 1 < 0 + 3$, or $-1 < 3$, a true statement. Of course, this is only a "partial" check, since we didn't try *all* the numbers in the graph. You can check a little further by selecting a number *not* on the graph to make sure the result is false. For example, when $x = 5$,

$$2x - 1 < x + 3 \quad \text{becomes} \quad 2(5) - 1 < 5 + 3$$
$$10 - 1 < 8$$
$$9 < 8 \quad \text{False.}$$

EXAMPLE 3 Using the addition and subtraction properties to solve and graph inequalities

Solve and graph the inequality on a number line:

a. $3x - 2 < 2(x - 2)$ **b.** $4(x + 1) \geq 3x + 7$

SOLUTION 3

a.

$$3x - 2 < 2(x - 2) \quad \text{Given}$$
$$3x - 2 < 2x - 4 \quad \text{Simplify.}$$
$$3x - 2 + 2 < 2x - 4 + 2 \quad \text{Add 2.}$$
$$3x < 2x - 2 \quad \text{Simplify.}$$
$$3x - 2x < 2x - 2x - 2 \quad \text{Subtract } 2x \text{ (or add } -2x\text{).}$$
$$x < -2 \quad \text{Simplify.}$$

Any number less than -2 is a solution. The graph of this inequality is as follows:

$3x - 2 < 2(x - 2)$ or, equivalently, $x < -2$

−5 −4 −3 −2 −1 0 1 2 3 4 5

b.

$$4(x + 1) \geq 3x + 7 \quad \text{Given}$$
$$4x + 4 \geq 3x + 7 \quad \text{Simplify.}$$
$$4x + 4 - 4 \geq 3x + 7 - 4 \quad \text{Subtract 4.}$$
$$4x \geq 3x + 3 \quad \text{Simplify.}$$
$$4x - 3x \geq 3x - 3x + 3 \quad \text{Subtract } 3x \text{ (or add } -3x\text{).}$$
$$x \geq 3 \quad \text{Simplify.}$$

Any number greater than or equal to 3 is a solution. The graph of this inequality is as follows:

$4(x + 1) \geq 3x + 7$ or, equivalently, $x \geq 3$

−5 −4 −3 −2 −1 0 1 2 3 4 5

PROBLEM 3

Solve and graph:

a. $4x - 3 < 3(x - 2)$

b. $3(x + 2) \geq 2x + 5$

Answers to PROBLEMS

3. a. $x < -3$ **b.** $x \geq -1$

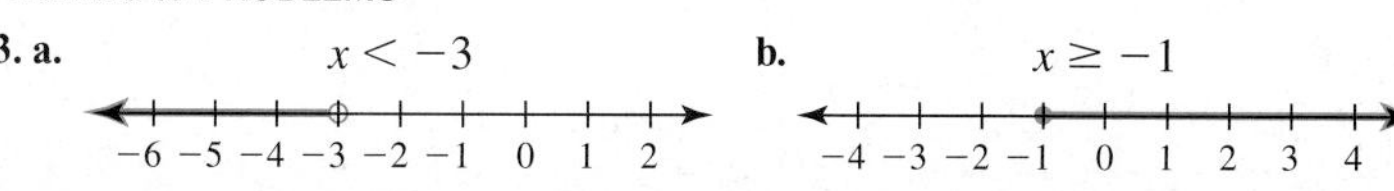

How do we solve an inequality such as $\frac{x}{2} < 3$? If half a number is less than 3, the number must be less than 6. This suggests that you can *multiply* (*or divide*) both sides of an inequality by a *positive* number and obtain an equivalent inequality.

$$\frac{x}{2} < 3 \quad \text{Given}$$

$$2 \cdot \frac{x}{2} < 2 \cdot 3 \quad \text{Multiply by 2.}$$

$$x < 6 \quad \text{Simplify.}$$

And to solve

$$2x < 8$$

$$\frac{2x}{2} < \frac{8}{2} \quad \text{Divide by 2 (or multiply by the reciprocal of 2).}$$

$$x < 4 \quad \text{Simplify.}$$

Any number less than 4 is a solution. Let's try some more examples.

If $3 < 4$, then

$$5 \cdot 3 < 5 \cdot 4$$

$$15 < 20 \quad \text{True}$$

Note that we are multiplying or dividing *both* sides of the inequality by a **positive** number. In such cases, we do not change the inequality symbol.

If $-2 > -10$, then

$$5 \cdot (-2) > 5 \cdot (-10)$$

$$-10 > -50 \quad \text{True}$$

Also, if $6 < 8$, then

$$\frac{6}{2} < \frac{8}{2}$$

$$3 < 4 \quad \text{True}$$

Similarly, if $-6 > -10$, then

$$-\frac{6}{2} > -\frac{10}{2}$$

$$-3 > -5 \quad \text{True}$$

Here are the properties we've just used.

MULTIPLICATION AND DIVISION PROPERTIES OF INEQUALITIES FOR POSITIVE NUMBERS

You can *multiply* or *divide* both sides of an inequality by any *positive* number c and obtain an equivalent inequality. In symbols,

If	$a < b$ (and **c** is positive)	**If**	$a > b$ (and **c** is positive)
then	$ac < bc$	**then**	$ac > bc$
and	$\frac{a}{c} < \frac{b}{c}$	**and**	$\frac{a}{c} > \frac{b}{c}$

Note: These properties also hold when the symbols $\le$ and $\ge$ are used.

NOTE

Since dividing x by a is the same as multiplying x by the reciprocal of a, you can think of dividing by a as multiplying by the reciprocal of a.

EXAMPLE 4 Using the multiplication and division properties with positive numbers

Solve and graph the inequality on a number line:

a. $5x + 3 \le 2x + 9$ **b.** $4(x - 1) > 2x + 6$

PROBLEM 4

Solve and graph:

a. $4x + 3 \le 2x + 5$

b. $5(x - 1) > 3x + 1$

Answers to PROBLEMS

4. a. $x \le 1$ **b.** $x > 3$

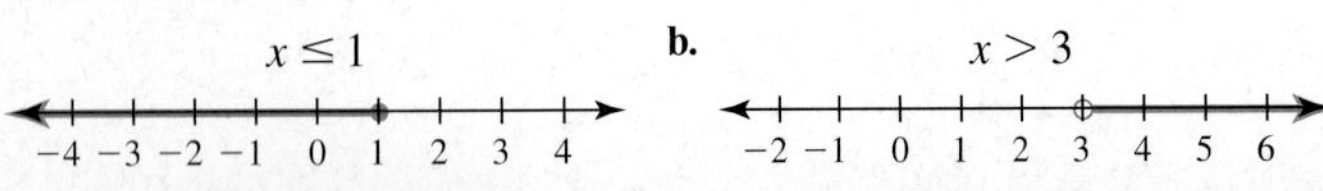

SOLUTION 4

a.

$5x + 3 \le 2x + 9$	Given
$5x + 3 - 3 \le 2x + 9 - 3$	Subtract 3 (or add -3).
$5x \le 2x + 6$	Simplify.
$5x - 2x \le 2x - 2x + 6$	Subtract $2x$ (or add $-2x$).
$3x \le 6$	Simplify.
$\frac{3x}{3} \le \frac{6}{3}$	Divide by 3 (or multiply by the reciprocal of 3).
$x \le 2$	Simplify.

Any number less than or equal to 2 is a solution. The graph is as follows:

$5x + 3 \le 2x + 9$ or, equivalently, $x \le 2$

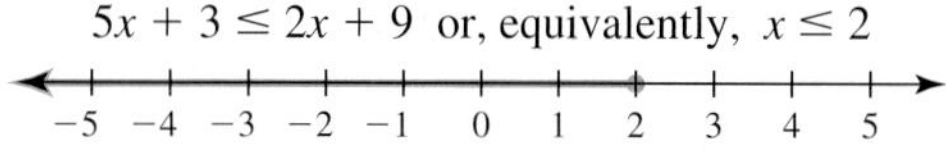

b.

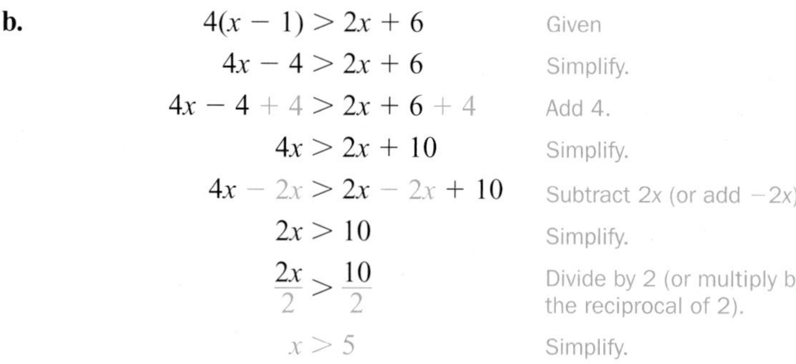

$4(x - 1) > 2x + 6$	Given
$4x - 4 > 2x + 6$	Simplify.
$4x - 4 + 4 > 2x + 6 + 4$	Add 4.
$4x > 2x + 10$	Simplify.
$4x - 2x > 2x - 2x + 10$	Subtract $2x$ (or add $-2x$).
$2x > 10$	Simplify.
$\frac{2x}{2} > \frac{10}{2}$	Divide by 2 (or multiply by the reciprocal of 2).
$x > 5$	Simplify.

Any number greater than 5 is a solution. The graph is as follows:

$4(x - 1) > 2x + 6$ or, equivalently, $x > 5$

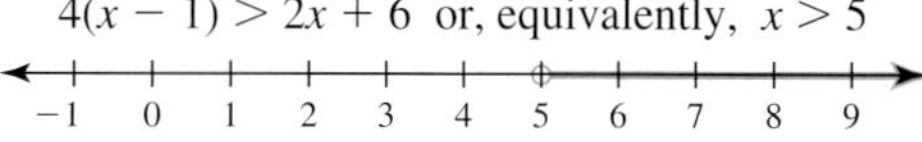

You may have noticed that the multiplication (or division) property allows us to multiply or divide only by a *positive* number. However, to solve the inequality $-2x < 4$, we need to divide by -2 or multiply by $-\frac{1}{2}$, a number that is *not* positive.

Let's first see what happens when we divide both sides of an inequality by a *negative* number. Consider the inequality

$$2 < 4$$

If we divide both sides of this inequality by -2, we get

$$\frac{2}{-2} < \frac{4}{-2} \quad \text{or} \quad -1 < -2$$

which is *not* true. To obtain a true statement, we must *reverse the inequality sign* and write:

$$-1 > -2$$

Similarly, consider

$$-6 > -8$$

$$\frac{-6}{-2} > \frac{-8}{-2}$$

$$3 > 4$$

which again is not true. However, the statement becomes true when we reverse the inequality sign and write $3 < 4$. So if we *divide* both sides of an inequality by a *negative* number, we must reverse the inequality sign to obtain an equivalent inequality. Similarly, if we *multiply* both sides of an inequality by a *negative* number, we must reverse the inequality sign to obtain an equivalent inequality:

$2 < 4$	Given
$-3 \cdot 2 > -3 \cdot 4$	If we multiply by -3, we reverse the inequality sign.
$-6 > -12$	

Note that now we are multiplying or dividing *both* sides of the inequality by a **negative** number. In such cases, we reverse the inequality symbol.

Here are some more examples.

$$3 < 12 \qquad\qquad 8 > 4$$

$$-2 \cdot 3 > -2 \cdot 12 \quad \text{Reverse the sign.} \qquad \frac{8}{-2} < \frac{4}{-2} \quad \text{Reverse the sign.}$$

$$-6 > -24 \qquad\qquad -4 < -2$$

These properties are stated here.

MULTIPLICATION AND DIVISION PROPERTIES OF INEQUALITIES FOR NEGATIVE NUMBERS

You can *multiply* or *divide* both sides of an inequality by a *negative* number c and obtain an equivalent inequality provided you *reverse* the inequality sign. In symbols,

If	$a < b$	(and c is negative)	If	$a > b$	(and c is negative)
then	$ac > bc$	Reverse the sign.	then	$ac < bc$	Reverse the sign.
and	$\frac{a}{c} > \frac{b}{c}$	Reverse the sign.	and	$\frac{a}{c} < \frac{b}{c}$	Reverse the sign.

Note: These properties also hold when the symbols ≤ and ≥ are used.

We use these properties in Example 5.

EXAMPLE 5 Using the multiplication and division properties with negative numbers

Solve:

a. $-3x < 15$ **b.** $\frac{-x}{4} > 2$ **c.** $3(x - 2) \le 5x + 2$

SOLUTION 5

a. To solve this inequality, we need the x by itself on the left; that is, we have to divide both sides by -3. Of course, when we do this, we must reverse the inequality sign.

$$-3x < 15 \quad \text{Given}$$

$$\frac{-3x}{-3} > \frac{15}{-3} \quad \text{Divide by } -3 \text{ and reverse the sign.}$$

$$x > -5 \quad \text{Simplify.}$$

Any number greater than -5 is a solution.

b. Here we multiply both sides by -4 and reverse the inequality sign.

$$\frac{-x}{4} > 2 \quad \text{Given}$$

$$-4\left(\frac{-x}{4}\right) < -4 \cdot 2 \quad \text{Multiply by } -4 \text{ and reverse the inequality sign.}$$

$$x < -8 \quad \text{Simplify.}$$

Any number less than -8 is a solution.

c.

$$3(x - 2) \le 5x + 2 \quad \text{Given}$$

$$3x - 6 \le 5x + 2 \quad \text{Simplify.}$$

$$3x - 6 + 6 \le 5x + 2 + 6 \quad \text{Add 6.}$$

$$3x \le 5x + 8 \quad \text{Simplify.}$$

$$3x - 5x \le 5x - 5x + 8 \quad \text{Subtract } 5x \text{ (or add } -5x\text{).}$$

$$-2x \le 8 \quad \text{Simplify.}$$

$$\frac{-2x}{-2} \ge \frac{8}{-2} \quad \text{Divide by } -2 \text{ (or multiply by the reciprocal of } -2\text{) and reverse the inequality sign.}$$

$$x \ge -4 \quad \text{Simplify.}$$

Thus, any number greater than or equal to -4 is a solution.

PROBLEM 5

Solve:

a. $-4x < 20$

b. $\frac{-x}{3} > 2$

c. $2(x - 1) \le 4x + 1$

Answers to PROBLEMS

5. a. $x > -5$ **b.** $x < -6$ **c.** $x \ge -\frac{3}{2}$

Of course, to solve more complicated inequalities (such as those involving fractions), we simply follow the six-step procedure we use for solving linear equations (p. 140).

EXAMPLE 6 **Using the six-step procedure to solve an inequality**

Solve:

$$\frac{-x}{4} + \frac{x}{6} < \frac{x - 3}{6}$$

SOLUTION 6 We follow the six-step procedure for linear equations.

$$\frac{-x}{4} + \frac{x}{6} < \frac{x - 3}{6} \quad \text{Given}$$

1. Clear the fractions; the LCM is 12. $12 \cdot \left(\frac{-x}{4}\right) + 12 \cdot \frac{x}{6} < 12\left(\frac{x-3}{6}\right)$
2. Remove parentheses (use the distributive property). $-3x + 2x < 2(x - 3)$

   Collect like terms. $-x < 2x - 6$
3. There are no numbers on the left, only the variable $-x$.
4. Subtract $2x$. $-x - 2x < 2x - 2x - 6$

   $-3x < -6$
5. Divide by the coefficient of x, -3, and reverse the inequality sign. $\frac{-3x}{-3} > \frac{-6}{-3}$

   Remember that when you divide both sides of the equation by -3, which is negative, you have to reverse the inequality sign from $<$ to $>$. $x > 2$

Thus, any number greater than 2 is a solution.

6. **CHECK** Try $x = 12$. (It's a good idea to try the LCM. Do you see why?)

$\frac{-x}{4} + \frac{x}{6} \overset{?}{<}$	$\frac{x-3}{6}$
$\frac{-12}{4} + \frac{12}{6}$	$\frac{12-3}{6}$
$-3 + 2$	$\frac{9}{6}$
-1	$\frac{3}{2}$

Since $-1 < \frac{3}{2}$, the inequality is true. Of course, this only partially *verifies* the answer, since we're unable to try *every* solution.

PROBLEM 6

Solve:

$$\frac{-x}{3} + \frac{x}{4} < \frac{x - 8}{4}$$

C > Solving and Graphing Compound Inequalities

What about inequalities like the one at the beginning of this section? Inequalities such as

$$1 \le x \le 3$$

are called **compound inequalities** because they are equivalent to two other inequalities; that is, $1 \le x \le 3$ means $1 \le x$ *and* $x \le 3$. Thus, if we are asked to solve the inequalities

$$2 \le x \quad \text{and} \quad x \le 4$$

we write

$$2 \le x \le 4 \qquad 2 \le x \le 4 \text{ is a compound inequality.}$$

Answers to PROBLEMS

6. $x > 6$

The graph of this inequality consists of all the points between 2 and 4, inclusive, as shown here:

$2 \le x \le 4$

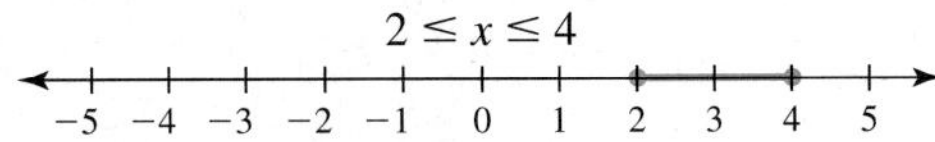

You can also write the solution as [2, 4].

include 2 start end include 4

NOTE

To graph $2 \le x \le 4$, place a solid dot at 2, a solid dot at 4, and draw the line segment *between* 2 and 4. In interval notation, we write this as [2, 4].

To graph $2 < x < 4$ use the same procedure but place an open dot at 2, and at 4

$2 < x < 4$

−3 −2 −1 0 1 2 3 4 5

or (2, 4)

do not include 2 start end do not include 4

The key to solving compound inequalities is to try writing the inequality in the form $a \le x \le b$ (or $a < x < b$), where a and b are real numbers. This form is called the *required* solution. Of course, if we try to write $x < 3$ and $x > 7$ in this form, we get $7 < x < 3$, which is not true.

In general, the notation used to write the solution of compound inequalities and their resulting graphs are as follows:

Symbols	Graph	Interval Notation
$a < x < b$	0 a b	**(a, b)**
$a \le x < b$	a 0 b	**[a, b)**
$a < x \le b$	0 a b	**(a, b]**
$a \le x \le b$	a b 0	**[a, b]**

Note that when an open circle ○ or a parenthesis like (or) is used, the endpoints are **not** included. When a closed circle ● or a bracket like [or] is used, the endpoints **are** included.

EXAMPLE 7 Solving compound inequalities

Solve and graph on a number line:

a. $1 < x$ and $x < 3$

b. $5 \ge -x$ and $x \le -3$

c. $x + 1 \le 5$ and $-2x < 6$

PROBLEM 7

Solve and graph:

a. $2 < x$ and $x < 4$

b. $3 \ge -x$ and $x \le -1$

c. $x + 2 \le 6$ and $-3x < 6$

SOLUTION 7

a. The inequalities $1 < x$ *and* $x < 3$ are written as $1 < x < 3$. Thus, the solution consists of the numbers *between* 1 and 3, as shown here.

Draw an open circle at 1, an open circle at 3, and draw the line segment between 1 and 3.

$1 < x < 3$

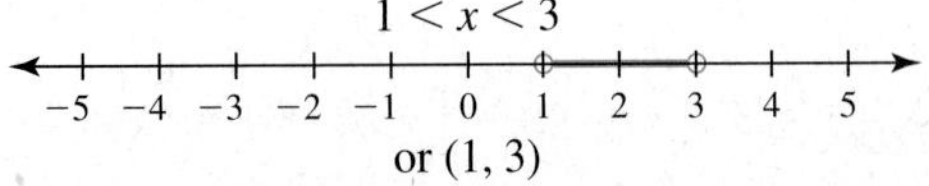

or (1, 3)

Answers to PROBLEMS

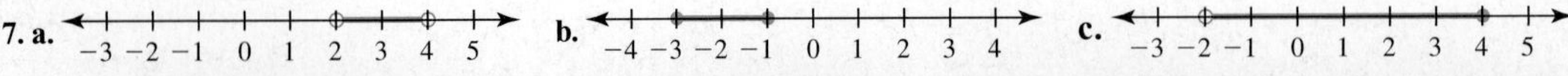

b. Since we wish to write $5 \ge -x$ and $x \le -3$ in the form $a \le x \le b$, we multiply both sides of $5 \ge -x$ by -1 to obtain

$$-1 \cdot 5 \le -1 \cdot (-x)$$

$$-5 \le x$$

We now have

$$-5 \le x \quad \text{and} \quad x \le -3$$

that is,

$$-5 \le x \le -3$$

Thus, the solution consists of all the numbers between -5 and -3 *inclusive.*

Note that -5 and -3 have solid dots.

$$-5 \le x \le -3$$

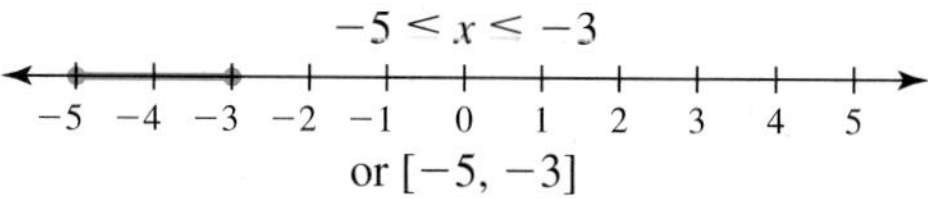

or $[-5, -3]$

c. We solve $x + 1 \le 5$ by subtracting 1 from both sides to obtain

$$x + 1 - 1 \le 5 - 1$$

$$x \le 4$$

We now have

$$x \le 4 \quad \text{and} \quad -2x < 6$$

We then divide both sides of $-2x < 6$ by -2 to obtain $x > -3$. We now have

$$x \le 4 \quad \text{and} \quad x > -3$$

Rearranging these inequalities, we write

$$-3 < x \quad \text{and} \quad x \le 4$$

that is,

$$-3 < x \le 4$$

Here the solution consists of all numbers between -3 and 4 and the number 4 itself:

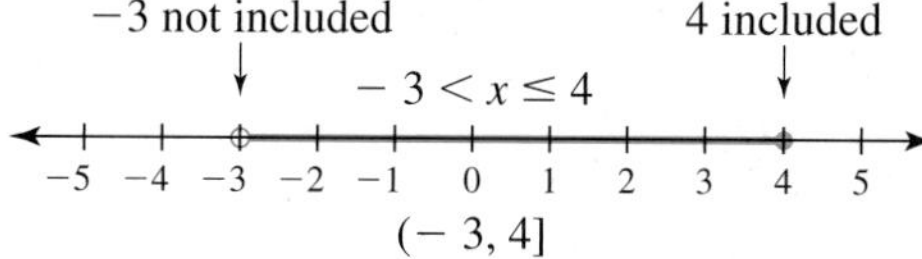

$(-3, 4]$

CAUTION

When graphing **linear inequalities** (inequalities that can be written in the form $ax + b \le c$, where a, b, and c are real numbers and a is not 0), the graph is usually a *ray* pointing in the same direction as the inequality and with the variable on the left-hand side as shown here:

$$x \ge a$$

a

$$x \le a$$

a

On the other hand, when graphing a **compound inequality,** the graph is usually a *line segment:*

$$a \le x \le b$$

a b

D › Solving an Application Involving Inequalities

The **C**orporate **A**verage **F**uel **E**conomy (**CAFE**) regulations in the United States are federal regulations intended to improve the average fuel economy of cars and light trucks (trucks, vans, and sport utility vehicles) sold in the United States. In 2009, the intended goal was to make the mileage more than 36 miles per gallon by 2016. In how many years will you expect the average fuel economy E to be greater than 36? We will answer that question next.

GREEN MATH

EXAMPLE 8 CAFE regulations

The combined car and light truck miles per gallon (mpg) can be approximated by $E = 1.5N + 27$, where N is the number of years after 2010. In what year will E be greater than 36?

SOLUTION 8

We want to find the values of N for which $\mathbf{1.5N + 27 > 36}$.

$$1.5N + 27 > 36 \quad \text{Given}$$

$$1.5N + 27 - 27 > 36 - 27 \quad \text{Subtract 27.}$$

$$1.5N > 9 \quad \text{Simplify.}$$

$$N > \frac{9}{1.5} \quad \text{Divide by 1.5.}$$

$$N > \mathbf{6} \quad \text{Simplify.}$$

This means that when $N = \mathbf{6}$ years after 2010, that is, in 2016, the estimated miles per gallon will be greater than 36, which was exactly the 2009 goal.

The good news: There will be estimated savings of \$400–\$800 a year because of better mileage (assuming the price of gasoline is \$3–\$4 a gallon).

The bad news: New cars will cost about \$1300 more.

PROBLEM 8

In how many years would you expect the combined car and light truck miles per gallon E to be greater than 42 mpg?

Note: There are several options for E. Four estimates based on these options are discussed in Problems 51–54. You can see a comparison of the options at http://tinyurl.com/p9kx2x.

Answers to PROBLEMS

8. In 10 years (in 2020)

› Exercises 2.7

‹ A › Order of Numbers In Problems 1–10, fill in the blank with $>$ or $<$ so that the resulting statement is true.

1. 8 ________ 9

2. -8 ________ -9

3. -4 ________ -9

4. 7 ________ 3

5. $\frac{1}{4}$ ________ $\frac{1}{3}$

6. $\frac{1}{5}$ ________ $\frac{1}{2}$

7. $-\frac{2}{3}$ ________ -1

8. $-\frac{1}{5}$ ________ -1

9. $-3\frac{1}{4}$ ________ -3

10. $-4\frac{1}{5}$ ________ -4

‹ B › Solving and Graphing Inequalities In Problems 11–30, solve and graph the inequalities on a number line.

11. $2x + 6 \le 8$

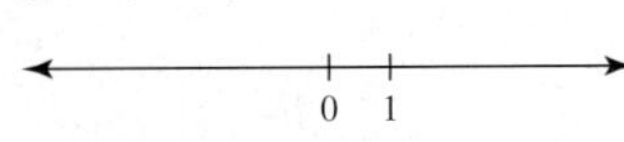

12. $4y - 5 \le 3$

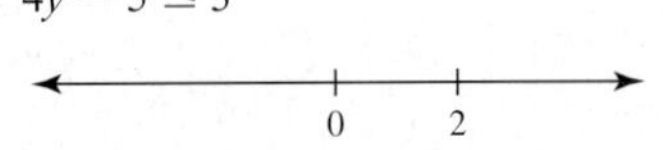

13. $-3y - 4 \ge -10$

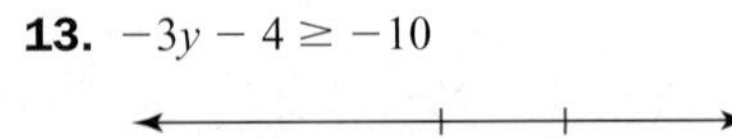

14. $-4z - 2 \geq 6$

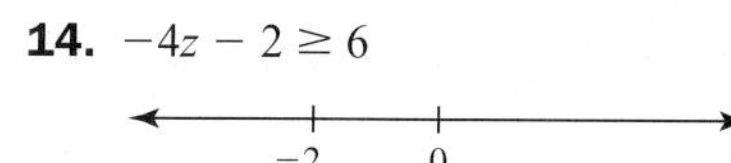

15. $-5x + 1 < -14$

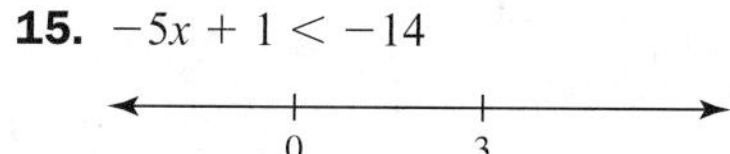

16. $-3x + 1 < -8$

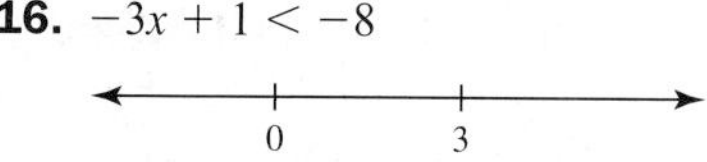

17. $3a + 4 \leq a + 10$

18. $4b + 4 \leq b + 7$

19. $5z - 12 \geq 6z - 8$

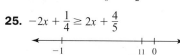

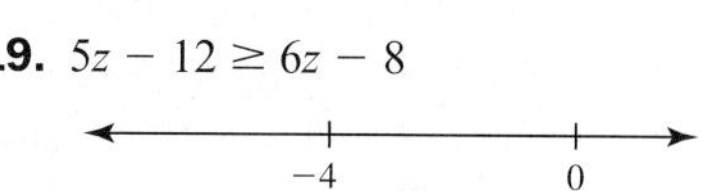

20. $5z + 7 \geq 7z + 19$

21. $10 - 3x \leq 7 - 6x$

22. $8 - 4y \leq -12 + 6y$

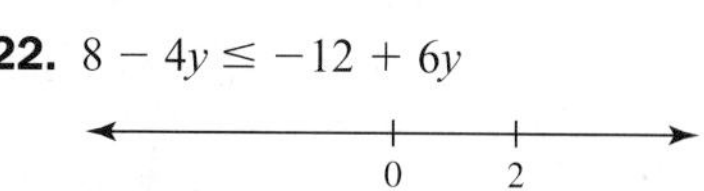

23. $5(x + 2) < 3(x + 3) + 1$

24. $5(4 - 3x) < 7(3 - 4x) + 12$

25. $-2x + \frac{1}{4} \geq 2x + \frac{4}{5}$

26. $6x + \frac{1}{7} \geq 2x - \frac{2}{7}$

27. $\frac{x}{5} - \frac{x}{4} \leq 1$

28. $\frac{x}{3} - \frac{x}{2} \leq 1$

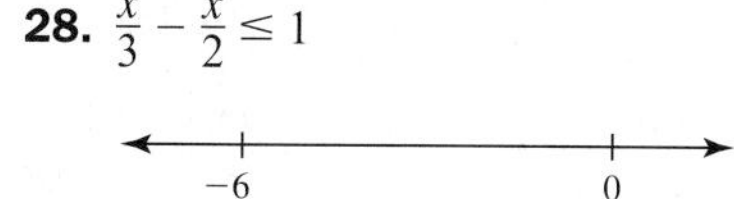

29. $\frac{7x + 2}{6} + \frac{1}{2} \geq \frac{3}{4}x$

30. $\frac{8x - 23}{6} + \frac{1}{3} \geq \frac{5}{2}x$

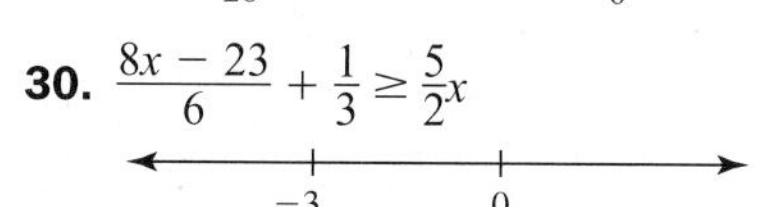

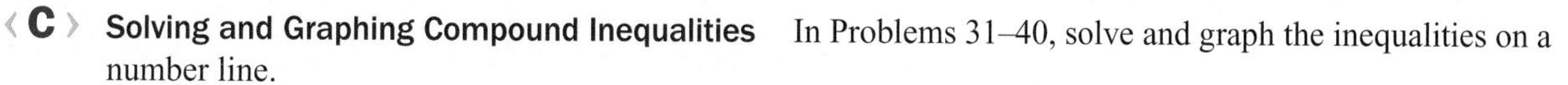

⟨C⟩ Solving and Graphing Compound Inequalities In Problems 31–40, solve and graph the inequalities on a number line.

31. $x < 3$ and $-x < -2$

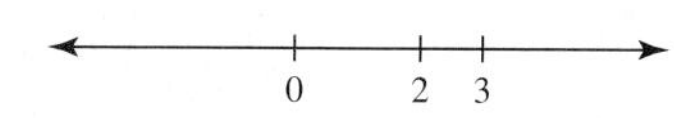

32. $-x < 5$ and $x < 2$

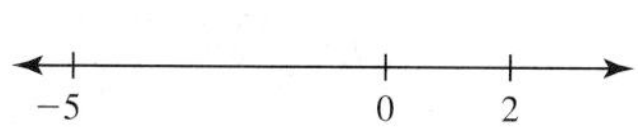

33. $x + 1 < 4$ and $-x < -1$

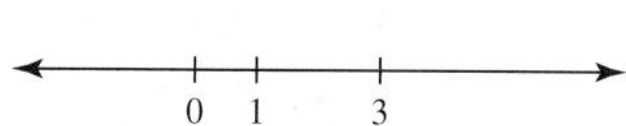

34. $x - 2 < 1$ and $-x < 2$

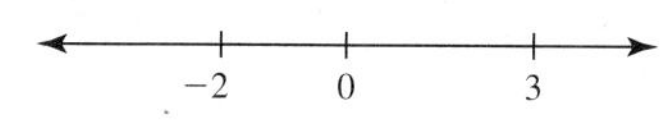

35. $x - 2 < 3$ and $2 > -x$

36. $x - 3 < 1$ and $1 > -x$

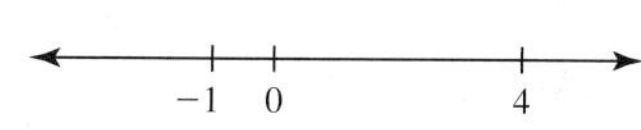

37. $x + 2 < 3$ and $-4 < x + 1$

38. $x + 4 < 5$ and $-1 \leq x + 2$

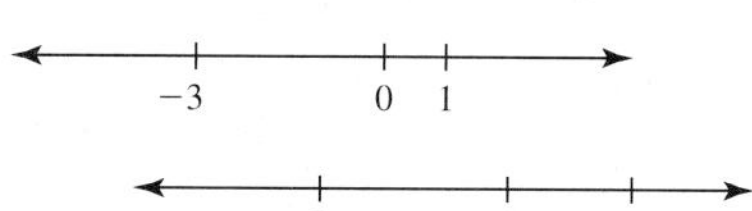

39. $x - 1 \geq 2$ and $x + 7 < 12$

40. $x - 2 > 1$ and $-x \geq -5$

In Problems 41–50, write the given information as an inequality.

41. The temperature t in your refrigerator is between 20°F and 40°F.

42. The height h (in feet) of any mountain is always less than or equal to that of Mount Everest, 29,029 feet.

43. Joe's salary s for this year will be between \$12,000 and \$13,000.

44. Your gas mileage m (in miles per gallon) is between 18 and 22, depending on your driving.

45. The number of possible eclipses e in a year varies from 2 to 7, inclusive.

46. The assets a of the Du Pont family are in excess of \$150 billion.

47. The cost c of ordinary hardware (tools, mowers, and so on) is between \$3.50 and \$4.00 per pound.

48. The range r (in miles) of a rocket is always less than 19,000 miles.

49. The altitude a (in feet) attained by the first liquid-fueled rocket was less than 41 feet.

50. The number of days d a person remained in the weightlessness of space before 1988 did not exceed 370.

Web IT go to mhhe.com/bello for more lessons

D Solving Applications Involving Inequalities

Applications: Green Math

Four estimates for cafe standards At least six options have been proposed for the combined mileage in the CAFE standards; we will discuss **four** of them in Problems 51–54.

51. **First,** a baseline standard in which the federal government did nothing and left the mpg E set at **26 mpg** for the year 2010 and after. Give a formula for E based on this information.

52. **Second,** the Bush proposal "maximizing net societal benefits." Under this proposal E can be approximated by $E = 1.1N + 26$, where N is the number of years after 2010. Using this formula, find N to the nearest year and determine in what year the mileage E will be greater than 36.

53. **Third,** "using a methodology in which net societal benefits equal zero." Here E can be approximated by $E = 1.7N + 27$. Using this formula, find N to the nearest year and determine in what year the mileage E will be greater than 36.

54. The last option is "what can be done if all manufacturers use every fuel economy technology available without regard to cost." Under this assumption, $E = 2.3N + 32$. Use this formula to find N to the nearest year and determine in what year the mileage E will be greater than 36.

Using Your Knowledge

A Question of Inequality Can you solve the problem in the cartoon? Let

J = Joe's height

B = Bill's height

F = Frank's height

S = Sam's height

Translate each statement into an equation or an inequality.

55. Joe is 5 feet (60 inches) tall.

56. Bill is taller than Frank.

57. Frank is 3 inches shorter than Sam.

58. Frank is taller than Joe.

59. Sam is 6 feet 5 inches (77 inches) tall.

60. According to the statement in Problem 56, Bill is taller than Frank, and according to the statement in Problem 58, Frank is taller than Joe. Write these two statements as an inequality of the form $a > b > c$.

61. Based on the answer to Problem 60 and the fact that you can obtain Frank's height by using the results of Problems 57 and 59, what can you really say about Bill's height?

Write On

62. Write the similarities and differences in the procedures used to solve equations and inequalities.

63. As you solve an inequality, when do you have to change the direction of the inequality?

64. A student wrote "$2 < x < -5$" to indicate that x was between 2 and -5. Why is this wrong?

65. Write the steps you would use to solve the inequality $-3x < 15$.

Concept Checker

Fill in the blank(s) with the correct word(s), phrase, or mathematical statement.

66. According to the **Addition Principle of Inequality** if $x < y$ and a is a **real** number, ________________.

67. According to the **Subtraction Principle of Inequality** if $x < y$ and a is a **real** number, ________________.

68. According to the **Multiplication Principle of Inequality** if $x < y$ and a is a **positive** number, ________________.

69. According to the **Multiplication Principle of Inequality** if $x < y$ and a is a **negative** number, ________________.

70. According to the **Division Principle of Inequality** if $x < y$ and a is a **positive** number, ________________.

71. According to the **Division Principle of Inequality** if $x < y$ and a is a **negative** number, ________________.

$x + a < y + a$ $\qquad \frac{-x}{a} < \frac{-y}{a}$

$a - x < a - y$ $\qquad \frac{x}{a} < \frac{y}{a}$

$x - a < y - a$ $\qquad ax < ay$

$ax > ay$ $\qquad \frac{-x}{a} > \frac{-y}{a}$

$\frac{x}{a} > \frac{y}{a}$

Mastery Test

Solve and graph on a number line:

72. $x + 2 \le 6$ and $-3x \le 9$

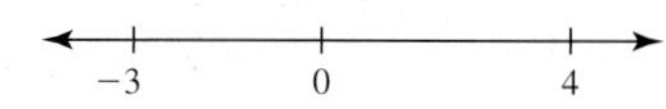

73. 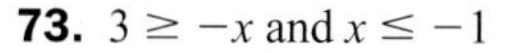

$3 \ge -x$ and $x \le -1$

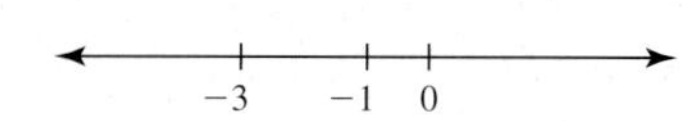

74. $2 < x$ and $x < 4$

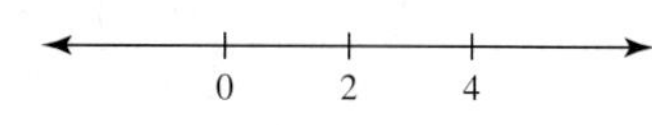

75. $\frac{-x}{3} + \frac{x}{4} < \frac{x - 4}{4}$

0 3

76. $4x + 5 < x + 11$

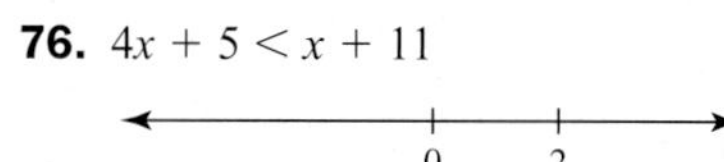

77. $3(x - 1) > x + 3$

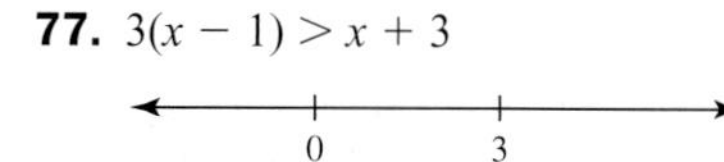

78. $4x - 7 < 3(x - 2)$

0 1

79. $3(x + 1) \ge 2x + 5$

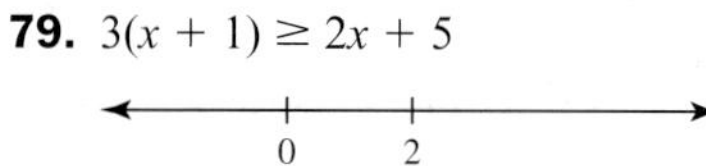

80. $x \ge -2$

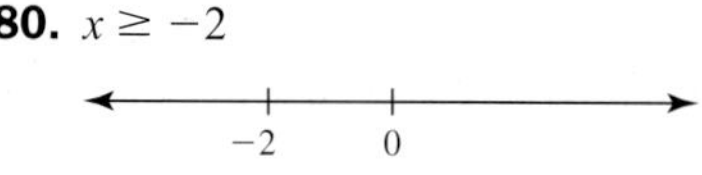

81. $x < 1$

0 1

Fill in the blank with > or < so that the resulting statement is true:

82. 5 ________ 7

83. −2 ________ −1

84. −4 ________ −5

85. $\frac{1}{3}$ ________ −3

86. According to the U.S. Department of Agriculture, the total daily grams of fat F consumed per person is modeled by the equation $F = 181.5 + 0.8t$, where t is the number of years after 2000. After what year would you expect the daily consumption of grams of fat to exceed 189.5?

Skill Checker

In Problems 87–95 graph the number on a number line.

87. 1 −5 −4 −3 −2 −1 0 1 2 3 4 5

88. 4 −5 −4 −3 −2 −1 0 1 2 3 4 5

89. −2 −5 −4 −3 −2 −1 0 1 2 3 4 5

90. −3 −5 −4 −3 −2 −1 0 1 2 3 4 5

91. 0 −5 −4 −3 −2 −1 0 1 2 3 4 5

92. 3 −5 −4 −3 −2 −1 0 1 2 3 4 5

93. 5 −5 −4 −3 −2 −1 0 1 2 3 4 5

94. −1 −5 −4 −3 −2 −1 0 1 2 3 4 5

95. −4 −5 −4 −3 −2 −1 0 1 2 3 4 5

Collaborative Learning

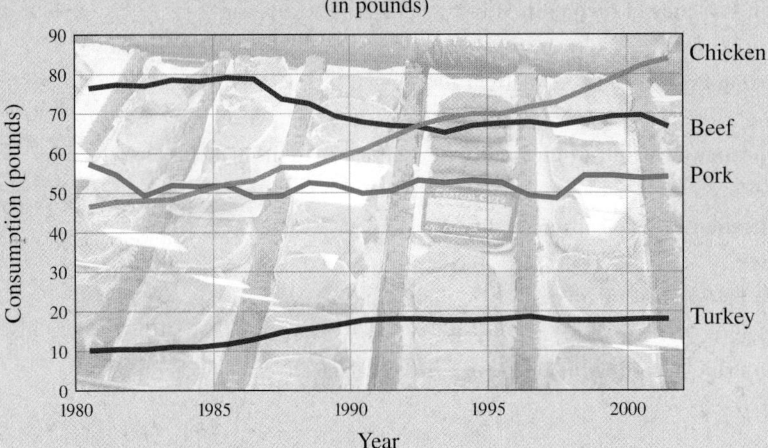

Source: American Meat Institute.

Form four groups: Beefies, porkies, chickens, and turkeys.

1. The annual per capita consumption C of poultry products (in pounds) in the United States t years after 1985 is given by $C = 45 + 2t$. Use the formula to estimate the U.S. consumption of poultry in 2001. How close is the estimate to the one given in the table? (2001 is the last year shown.)
2. From 1990 on, the consumption of beef (B), pork (P), and turkey (T) has been rather steady. Write a consumption equation that approximates the annual per capita consumption of *your product.*
3. The consumption of chicken (C) has been increasing about two pounds per year since 1980 so that $C = 45 + 2t$. In what year is C highest? In what year is it lowest?
4. Write an inequality comparing the annual consumption of *your product* and C.
5. In what year(s) was the annual consumption of *your product* less than C, equal to C, and greater than C?

Research Questions

1. What does the word *papyrus* mean? Explain how the Rhind papyrus got its name.
2. Write a report on the contents and origins of the Rhind papyrus.
3. The Rhind papyrus is one of two documents detailing Egyptian mathematics. What is the name of the other document, and what type of material does it contain?
4. Write a report about the rule of false position and the rule of double false position.
5. Find out who invented the symbols for greater than ($>$) and less than ($<$).
6. What is the meaning of the word *geometry*? Give an account of the origin of the subject.
7. Problem 50 in the Rhind papyrus gives the method for finding the area of a circle. Write a description of the problem and the method used.

Summary Chapter 2

Section	Item	Meaning	Example
2.1	Equation	A statement indicating that two expressions are **equal**	$x - 8 = 9$, $\frac{1}{2} - 2x = \frac{3}{4}$, and $0.2x + 8.9 = \frac{1}{2} + 6x$ are equations.
2.1A	Solutions	The solutions of an equation are the replacements of the variable that make the equation a **true** statement.	4 is a solution of $x + 1 = 5$.
	Equivalent equations	Two equations are **equivalent** if their solutions are the **same.**	$x + 1 = 4$ and $x = 3$ are equivalent.
2.1B	The addition property of equality	$a = b$ is equivalent to $a + c = b + c$.	$x - 1 = 2$ is equivalent to $x - 1 + 1 = 2 + 1$.
	The subtraction property of equality	$a = b$ is equivalent to $a - c = b - c$.	$x + 1 = 2$ is equivalent to $x + 1 - 1 = 2 - 1$.
2.1C	Conditional equation	An equation with **one** solution	$x + 7 = 9$ is a conditional equation whose solution is 2.
	Contradictory equation	An equation with **no** solution	$x + 1 = x + 2$ is a contradictory equation.
	Identity	An equation with **infinitely** many solutions	$2(x + 1) - 5 = 2x - 3$ is an identity. Any real number is a solution.
2.2A	The multiplication property of equality	$a = b$ is equivalent to $ac = bc$ if c is not 0.	$\frac{x}{2} = 3$ is equivalent to $2 \cdot \frac{x}{2} = 2 \cdot 3$.
	The division property of equality	$a = b$ is equivalent to $\frac{a}{c} = \frac{b}{c}$ if c is not 0.	$2x = 6$ is equivalent to $\frac{2x}{2} = \frac{6}{2}$.
2.2B	Reciprocal	The reciprocal of $\frac{a}{b}$ is $\frac{b}{a}$.	The reciprocal of $\frac{5}{2}$ is $\frac{2}{5}$.
2.2C	LCM (least common multiple)	The smallest number that is a multiple of each of the given numbers	The LCM of 3, 8, and 9 is 72.
2.3A	Linear equation	An equation that can be written in the form $ax + b = c$	$5x + 5 = 2x + 6$ is a linear equation (it can be written as $3x + 5 = 6$).
2.3B	Literal equation	An equation that contains letters other than the variable for which we wish to solve	$I = Prt$ and $C = 2\pi r$ are literal equations.
2.4	RSTUV method	To solve word problems, **R**ead, **S**elect a variable, **T**ranslate, **U**se algebra, and **V**erify your answer.	
2.4A	Consecutive integers	If n is an integer, the next consecutive integer is $n + 1$. If n is an even (odd) integer, the next consecutive even (odd) integer is $n + 2$.	4, 5, and 6 are three consecutive integers. 2, 4, and 6 are three consecutive even integers.

Section	Item	Meaning	Example
2.4C	Complementary angles	Two angles whose sum measures 90°	Two angles with measures 50° and 40° are complementary.
	Supplementary angles	Two angles whose sum measures 180°	Two angles with measures 35° and 145° are supplementary.
2.6B	Perimeter	The distance around a geometric figure	The perimeter P of a rectangle with length L and width W is $P = 2L + 2W$.
	Area of a rectangle	The area A of a rectangle of length L and width W is $A = LW$.	The area of a rectangle 8 inches long and 4 inches wide is $A = 8 \text{ in.} \cdot 4 \text{ in.} = 32 \text{ in.}^2$
	Area of a triangle	The area A of a triangle with base b and height h is $A = \frac{1}{2}bh$.	The area of a triangle with base 5 cm and height 10 cm is $A = \frac{1}{2} \cdot 5 \text{ cm} \cdot 10 \text{ cm} = 25 \text{ cm}^2$.
	Area of a circle	The area A of a circle of radius r is $A = \pi r^2$.	The area of a circle whose radius is 5 inches is $A = \pi(5)^2$, that is, $25\pi \text{ in.}^2$
	Circumference of a circle	The circumference of a circle with radius r is $C = 2\pi r$.	The circumference of a circle whose radius is 10 inches is $C = 2\pi(10)$, that is, 20π in.
2.6C	Vertical angles	Angles ① and ②are vertical angles.	① ②
2.7	Inequality	A statement with $>$, $<$, $\geq$, or $\leq$ for its verb	$2x + 1 > 5$ and $3x - 5 \leq 7 - x$ are inequalities.
2.7B	The addition property of inequalities	$a < b$ is equivalent to $a + c < b + c$.	$x - 1 < 2$ is equivalent to $x - 1 + 1 < 2 + 1$.
	The subtraction property of inequalities	$a < b$ is equivalent to $a - c < b - c$.	$x + 1 < 2$ is equivalent to $x + 1 - 1 < 2 - 1$.
	The multiplication property of inequalities	$a < b$ is equivalent to $ac < bc$ if $c > 0$ or $ac > bc$ if $c < 0$	$\frac{x}{2} < 3$ is equivalent to $2 \cdot \frac{x}{2} < 2 \cdot 3$.
	The division property of inequalities	$a < b$ is equivalent to $\frac{a}{c} < \frac{b}{c}$, if $c > 0$ or $\frac{a}{c} > \frac{b}{c}$, if $c < 0$	$-2x < 6$ is equivalent to $\frac{-2x}{-2} > \frac{6}{-2}$.

› Review Exercises Chapter 2

(If you need help with these exercises, look in the section indicated in brackets.)

1. ‹2.1A› *Determine whether the given number satisfies the equation.*
   a. $5; 7 = 14 - x$ b. $4; 13 = 17 - x$
   c. $-2; 8 = 6 - x$

2. ‹2.1B› *Solve the given equation.*
   a. $x - \frac{1}{3} = \frac{1}{3}$ b. $x - \frac{5}{7} = \frac{2}{7}$
   c. $x - \frac{5}{9} = \frac{1}{9}$

3. ‹2.1B› *Solve the given equation.*
   a. $-3x + \frac{5}{9} + 4x - \frac{2}{9} = \frac{5}{9}$
   b. $-2x + \frac{4}{7} + 3x - \frac{2}{7} = \frac{6}{7}$
   c. $-4x + \frac{5}{6} + 5x - \frac{1}{6} = \frac{5}{6}$

4. ‹2.1C› *Solve the given equation.*
   a. $3 = 4(x - 1) + 2 - 3x$
   b. $4 = 5(x - 1) + 9 - 4x$
   c. $5 = 6(x - 1) + 8 - 5x$

5. ‹2.1C› *Solve the given equation.*
   a. $6 + 3(x + 1) = 2 + 3x$
   b. $-2 + 4(x - 1) = -7 - 4x$
   c. $-1 - 2(x + 1) = 3 - 2x$

6. ‹2.1C› *Solve the given equation.*
   a. $5 + 2(x + 1) = 2x + 7$
   b. $-2 + 3(x - 1) = -5 + 3x$
   c. $-3 - 4(x - 1) = 1 - 4x$

7. ‹2.2A› *Solve the given equation.*
   a. $\frac{1}{5}x = -3$ b. $\frac{1}{7}x = -2$
   c. $5x = -10$

8. ‹2.2B› *Solve the given equation.*
   a. $-\frac{3}{4}x = -9$ b. $-\frac{3}{5}x = -9$
   c. $-\frac{2}{3}x = -6$

9. ‹2.2C› *Solve the given equation.*
   a. $\frac{x}{3} + \frac{2x}{4} = 5$ b. $\frac{x}{4} + \frac{3x}{2} = 6$
   c. $\frac{x}{5} + \frac{3x}{10} = 10$

10. ‹2.2C› *Solve the given equation.*
   a. $\frac{x}{3} - \frac{x}{4} = 1$ b. $\frac{x}{2} - \frac{x}{7} = 10$
   c. $\frac{x}{4} - \frac{x}{5} = 2$

11. ‹2.2C› *Solve the given equation.*
   a. $\frac{x - 1}{4} - \frac{x + 1}{6} = 1$
   b. $\frac{x - 1}{6} - \frac{x + 1}{8} = 0$
   c. $\frac{x - 1}{8} - \frac{x + 1}{10} = 0$

12. ‹2.2D› *Solve.*
   a. What percent of 30 is 6?
   b. What percent of 40 is 4?
   c. What percent of 50 is 10?

13. ‹2.2D› *Solve.*
   a. 20 is 40% of what number?
   b. 30 is 90% of what number?
   c. 25 is 75% of what number?

14. ‹2.3A› *Solve.*
   a. $\frac{1}{5} - \frac{x}{4} = \frac{19(x + 4)}{20}$
   b. $\frac{1}{5} - \frac{x}{4} = \frac{6(x + 5)}{5}$
   c. $\frac{1}{5} - \frac{x}{4} = \frac{29(x + 6)}{20}$

15. ⟨**2.3B**⟩ *Solve.*

a. $A = \frac{1}{2}bh$; solve for h.

b. $C = 2\pi r$; solve for r.

c. $V = \frac{bh}{3}$; solve for b.

16. ⟨**2.4A**⟩ *Find the numbers described.*

a. The sum of two numbers is 84 and one of the numbers is 20 more than the other.

b. The sum of two numbers is 47 and one of the numbers is 19 more than the other.

c. The sum of two numbers is 81 and one of the numbers is 23 more than the other.

17. ⟨**2.4B**⟩ *If you eat a fried chicken breast and a 3-ounce piece of apple pie, you have consumed 578 calories.*

a. If the pie has 22 more calories than the chicken breast, how many calories are in each?

b. Repeat part **a** where the number of calories consumed is 620 and the pie has 38 more calories than the chicken breast.

c. Repeat part **a** where the number of calories consumed is 650 and the pie has 42 more calories than the chicken breast.

18. ⟨**2.4C**⟩ *Find the measure of an angle whose supplement is:*

a. 20 degrees less than 3 times its complement.

b. 30 degrees less than 3 times its complement.

c. 40 degrees less than 3 times its complement.

19. ⟨**2.5A**⟩ *Solve.*

a. A car leaves a town traveling at 40 miles per hour. An hour later, another car leaves the same town traveling at 50 miles per hour in the same direction. How long does it take the second car to overtake the first one?

b. Repeat part **a** where the first car travels at 30 miles per hour and the second one at 50 miles per hour.

c. Repeat part **a** where the first car travels at 40 miles per hour and the second one at 60 miles per hour.

20. ⟨**2.5B**⟩ *Solve.*

a. How many pounds of a product selling at \$1.50 per pound should be mixed with 15 pounds of another product selling at \$3 per pound to obtain a mixture selling at \$2.40 per pound?

b. Repeat part **a** where the products sell for \$2, \$3, and \$2.50, respectively.

c. Repeat part **a** where the products sell for \$6, \$2, and \$4.50, respectively.

21. ⟨**2.5C**⟩ *Solve.*

a. A woman invests \$30,000, part at 5% and part at 6%. Her annual interest amounts to \$1600. How much does she have invested at each rate?

b. Repeat part **a** where the rates are 7% and 9%, respectively, and her annual return amounts to \$2300.

c. Repeat part **a** where the rates are 6% and 10%, respectively, and her annual return amounts to \$2000.

22. ⟨**2.6A**⟩ *The cost C of a long-distance call is $C = 3.05m + 3$, where m is the number of minutes the call lasts.*

a. Solve for m and then find the length of a call that cost \$27.40.

b. Repeat part **a** where $C = 3.15m + 3$ and the call cost \$34.50.

c. Repeat part **a** where $C = 3.25m + 2$ and the call cost \$21.50.

23. ⟨**2.6C**⟩ *Find the measures of the marked angles.*

a.

$(3x - 20)°$ $(2x)°$

b.

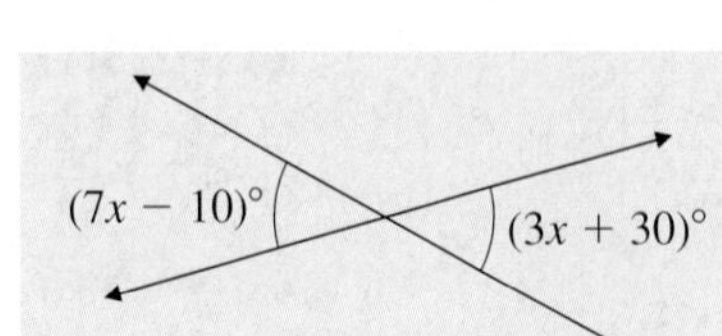

c.

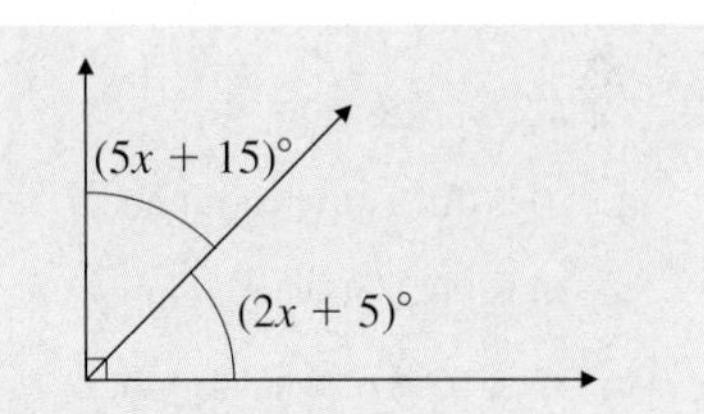

24. **⟨2.7A⟩** *Fill in the blank with the symbol < or > to make the resulting statement true.*

a. -8 _______ -7

b. $\frac{1}{2}$ _______ -3

c. 4 _______ $4\frac{1}{3}$

25. **⟨2.7B⟩** *Solve and graph the given inequality.*

a. $4x - 2 < 2(x + 2)$

0 3

b. $5x - 4 < 2(x + 1)$

0 2

c. $7x - 1 < 3(x + 1)$

0 1

26. **⟨2.7B⟩** *Solve and graph the given inequality.*

a. $6(x - 1) \geq 4x + 2$

0 4

b. $5(x - 1) \geq 2x + 1$

0 2

c. $4(x - 2) \geq 2x + 2$

0 5

27. **⟨2.7B⟩** *Solve and graph the given inequality.*

a. $-\frac{x}{3} + \frac{x}{6} \leq \frac{x - 1}{6}$

0 $\frac{1}{2}$ 1

b. $-\frac{x}{4} + \frac{x}{7} \leq \frac{x - 1}{7}$

0 $\frac{4}{7}$ 1

c. $-\frac{x}{5} + \frac{x}{3} \leq \frac{x - 1}{3}$

0 1 $\frac{5}{3}$ 2

28. **⟨2.7C⟩** *Solve and graph the compound inequality.*

a. $x + 2 \leq 4$ and $-2x < 6$

-3 0 2

b. $x + 3 \leq 5$ and $-3x < 9$

-3 0 2

c. $x + 1 \leq 2$ and $-4x < 8$

-2 0 1

> Practice Test Chapter 2

(Answers on page 207)

Visit www.mhhe.com/bello to view helpful videos that provide step-by-step solutions to several of the problems below.

1. Does the number 3 satisfy the equation $6 = 9 - x$?

2. Solve $x - \frac{2}{7} = \frac{3}{7}$.

3. Solve $-2x + \frac{7}{8} + 3x - \frac{5}{8} = \frac{5}{8}$.

4. Solve $2 = 3(x - 1) + 5 - 2x$.

5. Solve $2 + 5(x + 1) = 8 + 5x$.

6. Solve $-3 - 2(x - 1) = -1 - 2x$.

7. Solve $\frac{2}{3}x = -4$.

8. Solve $-\frac{2}{3}x = -6$.

9. Solve $\frac{x}{4} + \frac{2x}{3} = 11$.

10. Solve $\frac{x}{3} - \frac{x}{5} = 2$.

11. Solve $\frac{x - 2}{5} - \frac{x + 1}{8} = 0$.

12. What percent of 55 is 11?

13. Nine is 36% of what number?

14. Solve $\frac{1}{5} - \frac{x}{3} = \frac{23(x + 5)}{15}$.

15. Solve for h in $S = \frac{1}{3}\pi r^2 h$.

16. The sum of two numbers is 75. If one of the numbers is 15 more than the other, what are the numbers?

17. A man has invested a certain amount of money in stocks and bonds. His annual return from these investments is \$840. If the stocks produce \$230 more in returns than the bonds, how much money does he receive annually from each investment?

18. Find the measure of an angle whose supplement is 50° less than 3 times its complement.

19. A freight train leaves a station traveling at 30 miles per hour. Two hours later, a passenger train leaves the same station traveling in the same direction at 42 miles per hour. How long does it take for the passenger train to catch the freight train?

20. How many pounds of coffee selling for \$1.10 per pound should be mixed with 30 pounds of coffee selling for \$1.70 per pound to obtain a mixture that sells for \$1.50 per pound?

21. An investor bought some municipal bonds yielding 5% annually and some certificates of deposit yielding 7% annually. If her total investment amounts to \$20,000 and her annual return is \$1160, how much money is invested in bonds and how much in certificates of deposit?

22. The cost C of riding a taxi is $C = 1.95 + 0.85m$, where m is the number of miles (or fraction) you travel.

a. Solve for m.

b. How many miles did you travel if the cost of the ride was \$20.65?

23. Find x and the measures of the marked angles.

a.

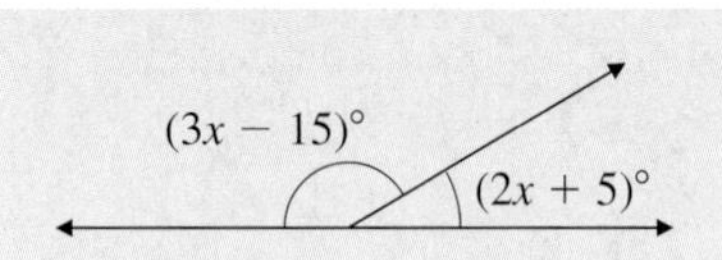

b.

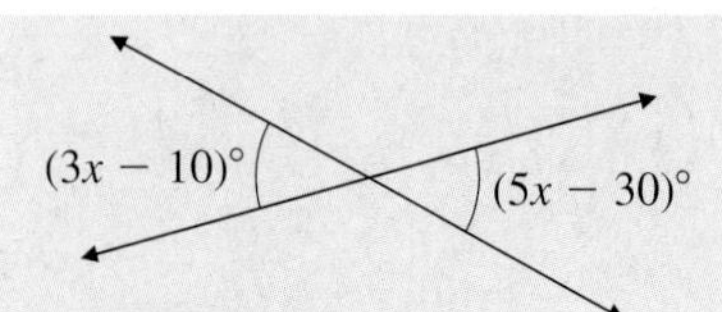

c.

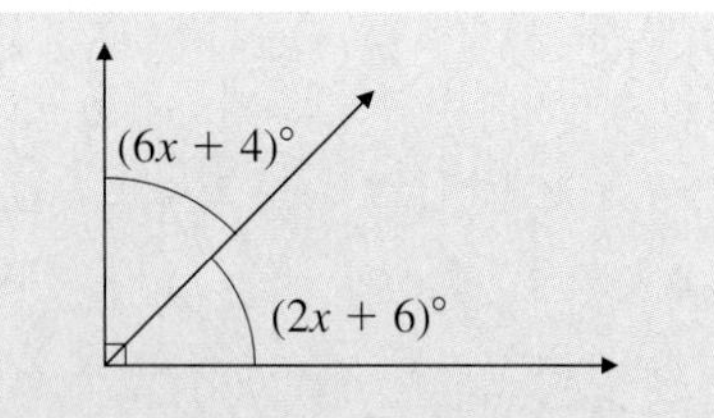

24. Fill in the blank with < or > to make the resulting statement true.

a. -3 ________ -5

b. $-\frac{1}{3}$ ________ 3

25. Solve and graph the inequality.

a. $-\frac{x}{2} + \frac{x}{4} \le \frac{x + 2}{4}$

b. $x + 1 \le 3$ and $-2x < 6$

Answers to Practice Test Chapter 2

Answer	If You Missed	Review		
	Question	Section	Examples	Page
1. Yes	1	2.1	1	111
2. $x = \frac{5}{7}$	2	2.1	2	112
3. $x = \frac{3}{8}$	3	2.1	3	113–114
4. $x = 0$	4	2.1	4, 5	115–116
5. No solution	5	2.1	6	117
6. All real numbers	6	2.1	7	117
7. $x = -6$	7	2.2	1, 2, 3	123–126
8. $x = 9$	8	2.2	1, 2, 3	123–126
9. $x = 12$	9	2.2	4	128
10. $x = 15$	10	2.2	4	128
11. $x = 7$	11	2.2	5	129
12. 20%	12	2.2	7	130–131
13. 25	13	2.2	8	131–132
14. $x = -4$	14	2.3	1, 2, 3	138–141
15. $h = \frac{3S}{\pi r^2}$	15	2.3	4, 5, 6	142–143
16. 30 and 45	16	2.4	1, 2	150–152
17. \$305 from bonds and \$535 from stocks	17	2.4	3	152
18. 20°	18	2.4	4	153–154
19. 5 hours	19	2.5	1, 2, 3	159–162
20. 15 pounds	20	2.5	4	163
21. \$12,000 in bonds, \$8000 in certificates	21	2.5	5	165
22. **a.** $m = \frac{C - 1.95}{0.85}$ **b.** 22 miles	22	2.6	1, 2	172–174
23. **a.** $x = 38$; 99° and 81° **b.** $x = 10$; both are 20° **c.** $x = 10$; 64° and 26°	23	2.6	4, 5, 6, 7	175–178
24. **a.** $>$ **b.** $<$	24	2.7	1	186–187
25. **a.** $x \geq -1$ −5 −4 −3 −2 −1 0 1 2 3 4 5 **b.** $-3 < x \leq 2$ −5 −4 −3 −2 −1 0 1 2 3 4 5	25	2.7	2, 3, 4, 5, 6, 7	187–195

> Cumulative Review Chapters 1–2

1. Find the additive inverse (opposite) of -7.

2. Find: $\left|-9\frac{9}{10}\right|$

3. Find: $-\frac{2}{7} + \left(-\frac{2}{9}\right)$

4. Find: $-0.7 - (-8.9)$

5. Find: $(-2.4)(3.6)$

6. Find: $-(2^4)$

7. Find: $-\frac{7}{8} \div \left(-\frac{5}{24}\right)$

8. Evaluate $y \div 5 \cdot x - z$ for $x = 6$, $y = 60$, $z = 3$.

9. Which property is illustrated by the following statement?

   $9 \cdot (8 \cdot 5) = 9 \cdot (5 \cdot 8)$

10. Multiply: $6(5x + 7)$

11. Combine like terms: $-5cd - (-6cd)$

12. Simplify: $2x - 2(x + 4) - 3(x + 1)$

13. Write in symbols: The quotient of $(a - 4b)$ and c

14. Does the number 4 satisfy the equation $11 = 15 - x$?

15. Solve for x: $5 = 4(x - 3) + 4 - 3x$

16. Solve for x: $-\frac{7}{3}x = -21$

17. Solve for x: $\frac{x}{3} - \frac{x}{5} = 2$

18. Solve for x: $4 - \frac{x}{4} = \frac{2(x + 1)}{9}$

19. Solve for b in the equation $S = 6a^2b$.

20. The sum of two numbers is 155. If one of the numbers is 35 more than the other, what are the numbers?

21. Maria has invested a certain amount of money in stocks and bonds. The annual return from these investments is \$595. If the stocks produce \$105 more in returns than the bonds, how much money does Maria receive annually from each type of investment?

22. Train A leaves a station traveling at 40 mph. Six hours later, train B leaves the same station traveling in the same direction at 50 mph. How long does it take for train B to catch up to train A?

23. Arlene purchased some municipal bonds yielding 12% annually and some certificates of deposit yielding 14% annually. If Arlene's total investment amounts to \$5000 and the annual income is \$660, how much money is invested in bonds and how much is invested in certificates of deposit?

24. Solve and graph: $-\frac{x}{6} + \frac{x}{5} \le \frac{x - 5}{5}$

   0 6

Chapter 3 *three*

Graphs of Linear Equations, Inequalities, and Applications

Section

The Human Side of Algebra

René Descartes, "the reputed founder of modern philosophy," was born March 31, 1596, near Tours, France. His frail health caused his formal education to be delayed until he was 8, when his father enrolled him at the Royal College at La Flèche. It was soon noticed that the boy needed more than normal rest, and he was advised to stay in bed as long as he liked in the morning. Descartes followed this advice and made a lifelong habit of staying in bed late whenever he could.

The idea of analytic geometry came to Descartes while he watched a fly crawl along the ceiling near a corner of his room. Descartes described the path of the fly in terms of its distance from the adjacent walls by developing the *Cartesian coordinate system,* which we study in Section 3.1. Impressed by his knowledge of philosophy and mathematics, Queen Christine of Sweden engaged Descartes as a private tutor. He arrived in Sweden to discover that she expected him to teach her philosophy at 5 o'clock in the morning in the ice-cold library of her palace. Deprived of his beloved morning rest, Descartes caught "inflammation of the lungs," from which he died on February 11, 1650, at age 53.

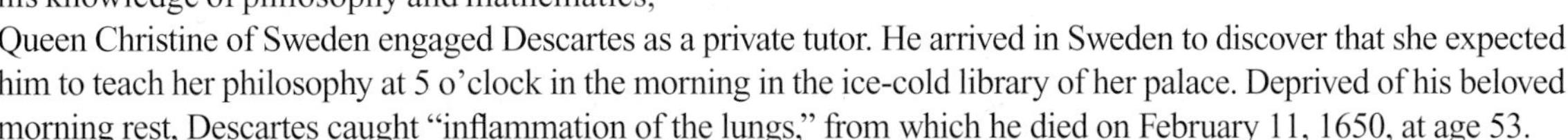

Later on, the first person to use the word *graph* was James Joseph Sylvester, a teacher born in London in 1814, who used the word in an article published in 1878. Sylvester passed the knowledge of graphs to his students, writing articles in *Applied Mechanics* and admonishing his students that they would "do well to graph on squared paper some curves like the following." Unfortunately for his students, Sylvester had a dangerous temper and attacked a student with a sword cane at the University of Virginia. The infraction? The student was reading a newspaper during his class. So now you can learn from this lesson: concentrate on your graphs and do not read newspapers or do texting in the classroom!

3.1 Line Graphs, Bar Graphs, and Applications

Objectives

A Graph (plot) ordered pairs of numbers.

B Determine the coordinates of a point in the plane.

C Read and interpret ordered pairs on a line graph.

D Read and interpret ordered pairs on a bar graph.

E Find the quadrant in which a point lies.

F Given a chart or ordered pairs, create the corresponding line graph.

To Succeed, Review How To . . .

1. Evaluate an expression (pp. 62–63, 69–73).
2. Solve linear equations (pp. 137–141).

Getting Started

Hurricanes and Graphs

The map shows the position of Hurricane Desi. The hurricane is near the intersection of the vertical line indicating 90° longitude and the horizontal line indicating 25° latitude. This point can be identified by assigning to it an **ordered pair** of numbers, called **coordinates,** showing the longitude first and the latitude second. Thus, the hurricane would have the coordinates

(91, 25)

This is the longitude (units right or left). This is the latitude (units up or down).

This method is used to give the position of cities, islands, ships, airplanes, and so on. For example, the coordinates of New Orleans on the map are (90, 30), whereas those of Pensacola are approximately (88, 31). In mathematics we use a system very similar to this one to locate points in a plane. In this section we learn how to graph points in a *Cartesian plane* and then examine the relationship of these points to linear equations.

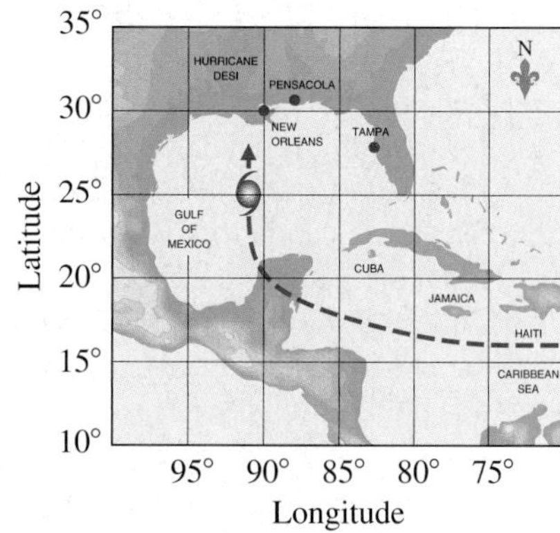

Here is the way we construct a **Cartesian coordinate system** (also called a rectangular coordinate system).

1. Draw a number line (Figure 3.1).
2. Draw another number line perpendicular to the first one and crossing it at 0 (the origin) (see Figure 3.2).

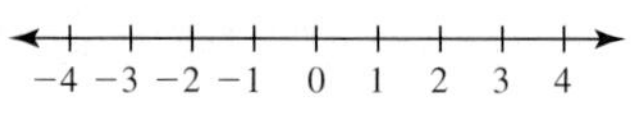

Figure 3.1

On the number line, each point on the graph is a **number.** On a coordinate plane, each point is the **graph** of an ordered pair. The individual numbers in an ordered pair are called **coordinates.** For example, the point P in Figure 3.3 is associated with the ordered pair (2, 3). The first coordinate of P is 2 and the second coordinate is 3. The point $Q(-1, 2)$ has a first coordinate of -1 and a second coordinate of 2. We call the horizontal number line the ***x*-axis** and label it with the letter x; the vertical number line is called the ***y*-axis** and is labeled with the letter y. We can now say that the point $P(2, 3)$ has x-coordinate (**abscissa**) 2, and y-coordinate (**ordinate**) 3.

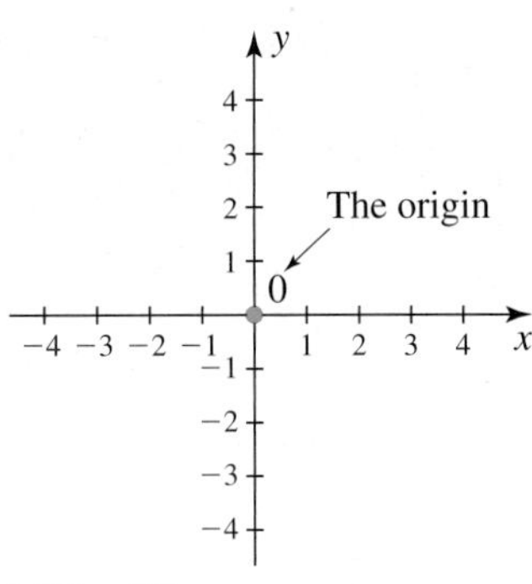

>Figure 3.2

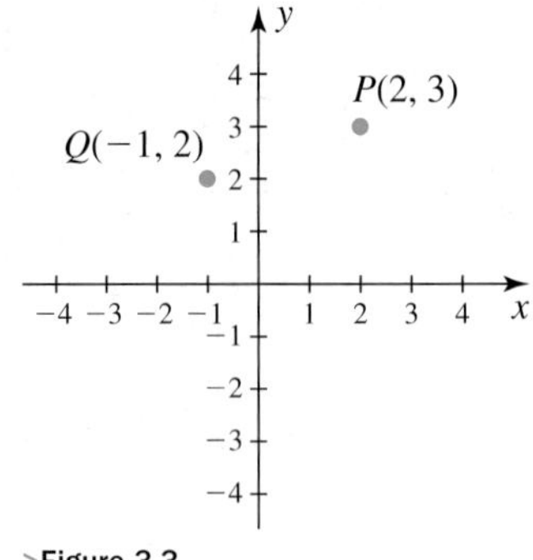

>Figure 3.3

A > Graphing Ordered Pairs

In general, if a point P has coordinates (x, y), we can always locate or graph the point in the coordinate plane. We start at the origin and go x units to the *right* if x is *positive;* we go to the *left* if x is *negative*. We then go y units *up* if y is *positive, down* if y is *negative*.

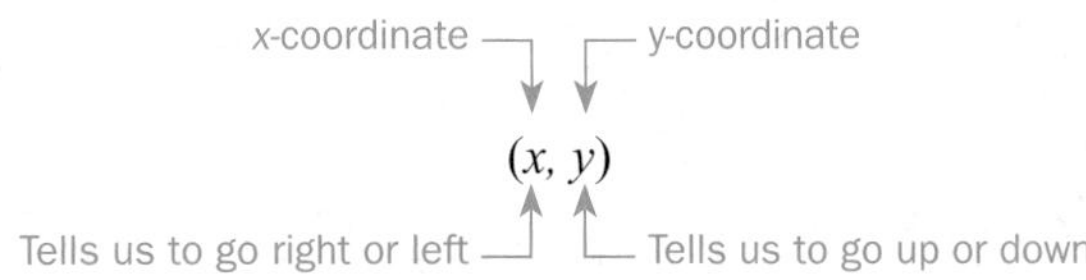

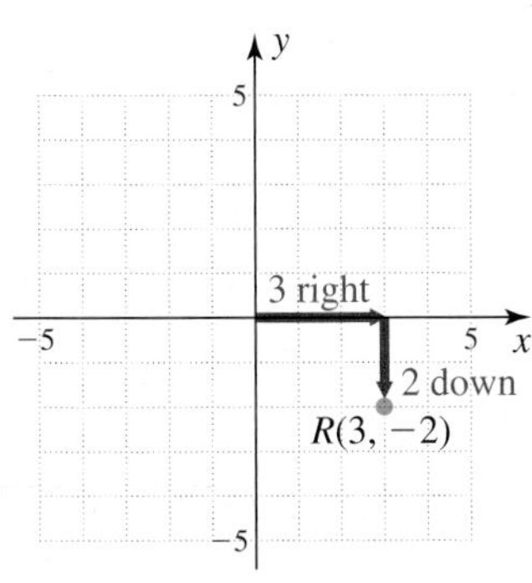

>Figure 3.4

For example, to graph the point $R(3, -2)$, we start at the origin and go 3 units right (since the x-coordinate 3 is positive) and 2 units down (since the y-coordinate -2 is negative). The point is graphed in Figure 3.4.

NOTE

All points on the x-axis have y-coordinate 0 (zero units up or down); all points on the y-axis have x-coordinate 0 (zero units right or left).

EXAMPLE 1 **Graphing points in the coordinate plane**

Graph the points:

a. $A(1, 3)$ **b.** $B(2, -1)$

c. $C(-4, 1)$ **d.** $D(-2, -4)$

SOLUTION 1

a. We start at the origin. To reach point (1, 3), we go 1 unit to the right and 3 units up. The graph of A is shown in Figure 3.5.

b. To graph $(2, -1)$, we start at the origin, go 2 units right and 1 unit down. The graph of B is shown in Figure 3.5.

c. As usual, we start at the origin. The point $(-4, 1)$ means to go 4 units left and 1 unit up, as shown in Figure 3.5.

d. The point $D(-2, -4)$ has both coordinates negative. Thus, from the origin we go 2 units left and 4 units down; see Figure 3.5.

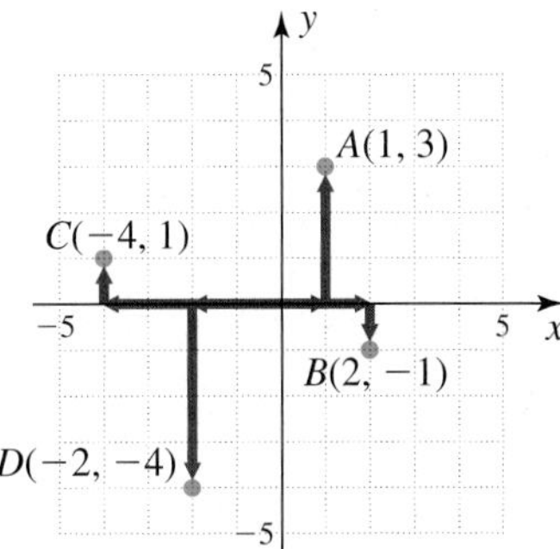

>Figure 3.5

PROBLEM 1

Graph:

a. $A(2, 4)$ **b.** $B(4, -2)$

c. $C(-3, 2)$ **d.** $D(-4, -4)$

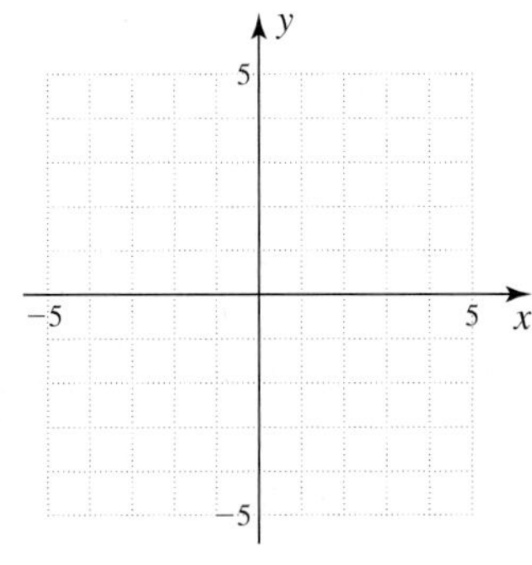

Answer on page 212

B > Finding Coordinates

Since every point P in the plane is associated with an ordered pair (x, y), we should be able to find the coordinates of any point as shown in Example 2.

EXAMPLE 2 Finding coordinates

Determine the coordinates of each of the points in Figure 3.6.

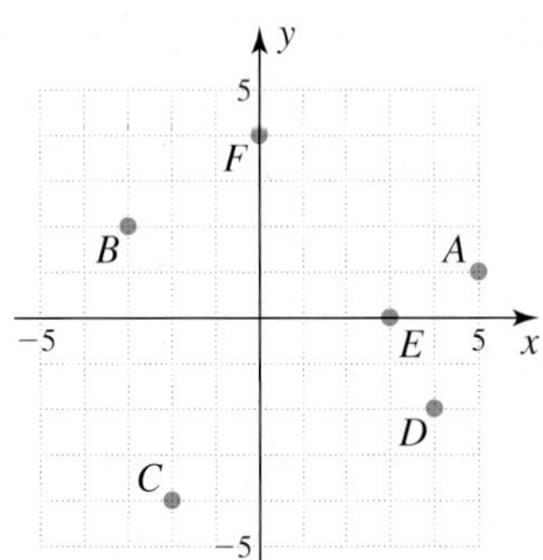

>Figure 3.6

SOLUTION 2 Point A is 5 units to the right of the origin and 1 unit above the horizontal axis. The ordered pair corresponding to A is (5, 1). The coordinates of the other four points can be found in a similar manner. Here is the summary.

Point	Start at the origin, move:	Coordinates
A	5 units *right,* 1 unit *up*	(5, 1)
B	3 units *left,* 2 units *up*	(−3, 2)
C	2 units *left,* 4 units *down*	(−2, −4)
D	4 units *right,* 2 units *down*	(4, −2)
E	3 units *right,* 0 units *up*	(3, 0)
F	0 units *right,* 4 units *up*	(0, 4)

PROBLEM 2

Determine the coordinates of each of the points.

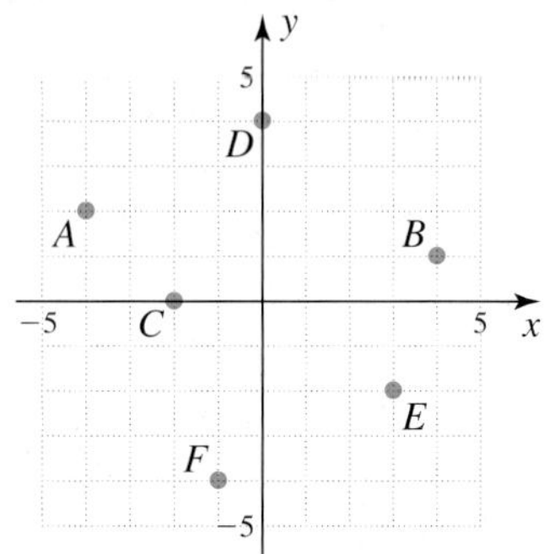

C > Applications: Line Graphs

Now that you know how to find the coordinates of a point, we learn how to read and interpret line graphs.

Answers to PROBLEMS

1.

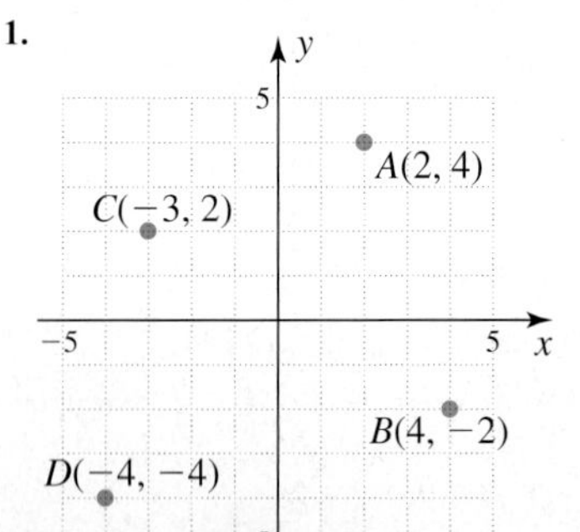

2. $A(-4, 2)$; $B(4, 1)$; $C(-2, 0)$; $D(0, 4)$; $E(3, -2)$; $F(-1, -4)$

EXAMPLE 3 Reading and interpreting line graphs

Figure 3.7 gives the age conversion from human years to dog years. The ordered pair (1, 12) means that 1 human year is equivalent to about 12 dog years.

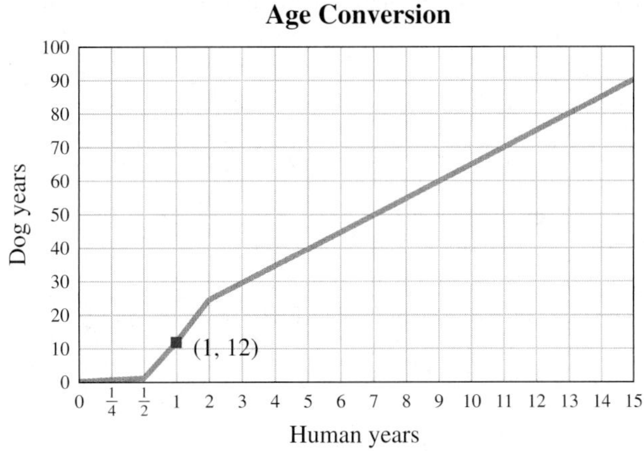

>Figure 3.7

Source: Data from Cindy's K9 Clips.

a. What does the ordered pair (3, 30) represent?

b. If a dog is 9 years old in human years, how old is it in dog years? Write the ordered pair corresponding to this situation.

c. If retirement age is 65, what is the retirement age for dogs in human years? That is, how many human years correspond to 65 dog years? Write the ordered pair corresponding to this situation.

SOLUTION 3

a. The ordered pair (3, 30) means that 3 human years are equivalent to 30 dog years.

b. Start at (0, 0) and move right to 9 on the horizontal axis. (See Figure 3.8.) Now, go up until you reach the graph as shown. The point is 60 units high (the y-coordinate is 60). Thus, 9 years old in human years is equivalent to 60 years old in dog years. The ordered pair corresponding to this situation is (9, 60).

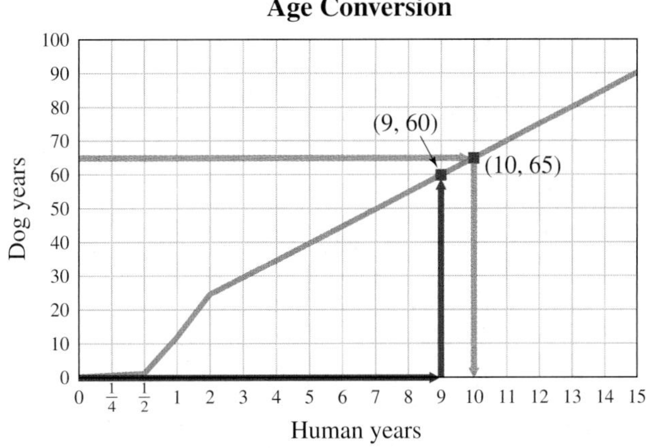

>Figure 3.8

c. We have to find how many human years are represented by 65 dog years. This time, we go to the point at which y is 65 units high, then move right until we reach the graph. At the point for which y is 65 on the graph, x is 10. Thus, the equivalent retirement age for dogs is 10 human years. At that age, they are entitled to Canine Security benefits! The ordered pair corresponding to this situation is (10, 65).

PROBLEM 3

a. What does the ordered pair (5, 40) in Figure 3.7 represent?

b. If a dog is 11 human years old, how old is it in dog years?

c. If the drinking age for humans is 21, what is the equivalent drinking age for dogs in human years? (Answer to the nearest whole number.)

Answers to PROBLEMS

3. a. Five human years are equivalent to 40 dog years. **b.** 70 **c.** 2

D > Applications: Bar Graphs

Another popular use of ordered pairs is **bar graphs,** in which certain *categories* are paired with certain numbers. For example, if you have a $1000 balance at 18% interest and are making the minimum $25 payment each month on your credit card, it will take you forever (actually 5 years) to pay it off. If you decide to pay it off in 12 months, how much would your payment be? The bar graph in Figure 3.9 tells you, provided you know how to read it! First, start at the 0 point and move right horizontally until you get to the **category** labeled 12 months (blue arrow), then go up vertically to the end of the bar (red arrow). According to the vertical scale labeled Monthly payment (the frequency), the arrow is 92 units long, meaning that the monthly payment will be $92 per month. The ordered pair corresponding to this situation is (12, 92).

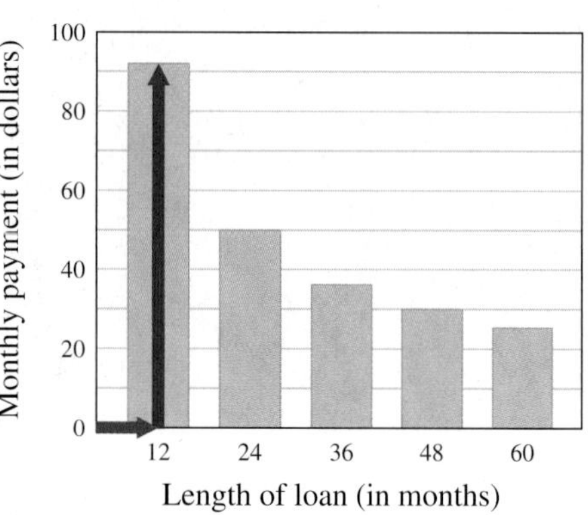

>Figure 3.9

Source: Data from KJE Computer Solutions, LLC.

EXAMPLE 4 Reading and interpreting bar graphs

a. Referring to the graph, what would your payment be if you decide to pay off the $1000 in 24 months?

b. How much would you save if you pay off the $1000 in 24 months?

SOLUTION 4

a. To find the payment corresponding to 24 months, move right on the horizontal axis to the category labeled 24 months and then vertically to the end of the bar in Figure 3.10. According to the vertical scale, the monthly payment will be $50.

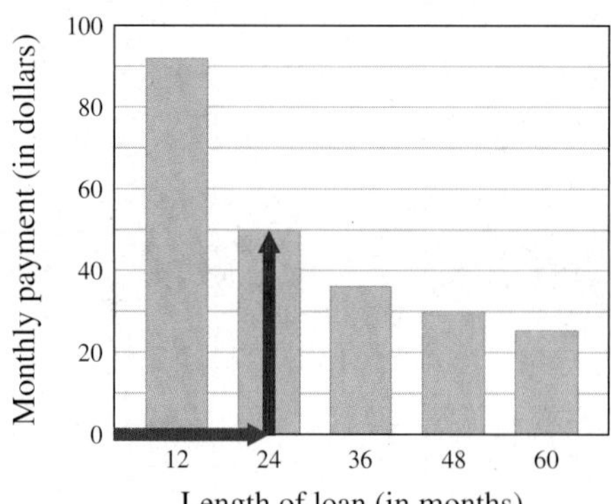

>Figure 3.10

Source: Data from KJE Computer Solutions, LLC.

b. If you pay the minimum $25 payment for 60 months, you would pay $60 \times \$25 = \$1500$. If you pay $50 for 24 months, you would pay $24 \times \$50 = \$1200$ and would save $300 ($1500 − $1200).

PROBLEM 4

a. What would your payment be if you decide to pay off the $1000 in 48 months?

b. How much would you save if you pay off the $1000 in 48 months?

E > Quadrants

Figure 3.11 shows that the x- and y-axes divide the plane into four regions called **quadrants.** These quadrants are numbered in counterclockwise order using Roman numerals and

Answers to PROBLEMS

4. a. $30 per month

b. $60 ($1500 − $1440)

starting in the upper right-hand region. What can we say about the coordinates of the points in each quadrant?

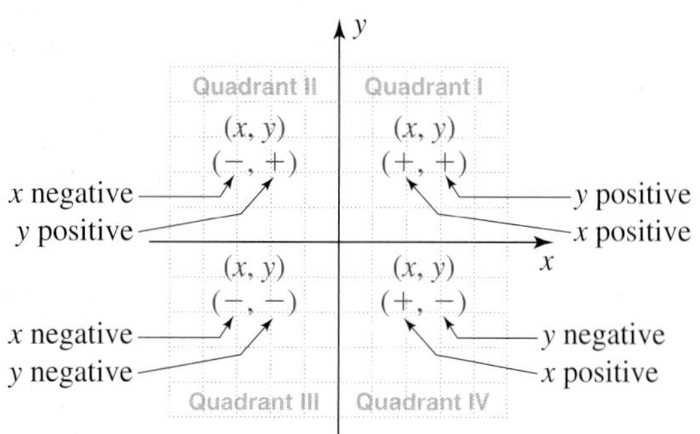

>Figure 3.11

In Quadrant I: Both coordinates positive
In Quadrant II: First coordinate negative, second positive
In Quadrant III: Both coordinates negative
In Quadrant IV: First coordinate positive, second negative

EXAMPLE 5 **Find the quadrant in which each ordered pair lies**

In which quadrant, if any, are the points located?

a. $A(-3, 2)$ **b.** $B(1, -2)$ **c.** $C(3, 4)$
d. $D(-4, -1)$ **e.** $E(0, 3)$ **f.** $F(2, 0)$

SOLUTION 5 The points are shown on the graph.

a. The point $A(-3, 2)$ is in the second quadrant (QII).
b. The point $B(1, -2)$ is in the fourth quadrant (QIV).
c. The point $C(3, 4)$ is in the first quadrant (QI).
d. The point $D(-4, -1)$ is in the third quadrant (QIII).
e. The point $E(0, 3)$ is on the y-axis (*no quadrant*).
f. The point $F(2, 0)$ is on the x-axis (*no quadrant*).

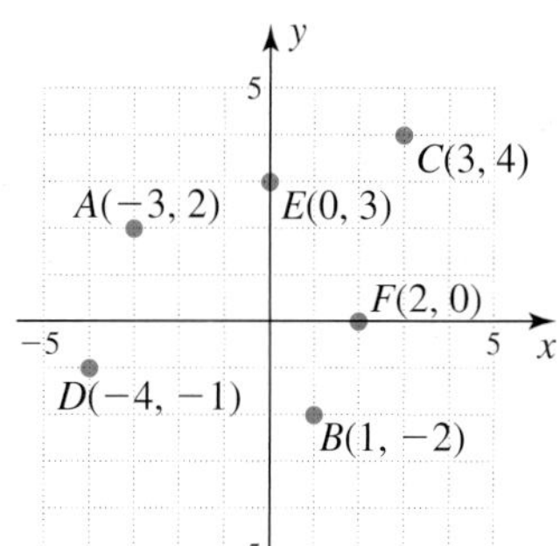

PROBLEM 5

In which quadrant, if any, are the points located?

a. $A(1, 3)$ **b.** $B(-2, -2)$
c. $C(-4, 2)$ **d.** $D(2, -1)$
e. $E(0, 1)$ **f.** $F(3, 0)$

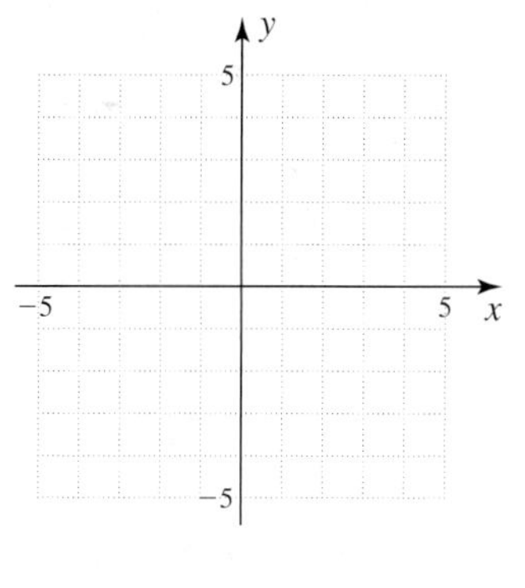

You may be wondering: Why do we need to know about quadrants? There are two reasons:

1. If you are using a graphing calculator, you have to specify the WINDOW (basically, the quadrants) you want to display.
2. If you are graphing ordered pairs, you need to adjust your graph to show the quadrants in which your ordered pairs lie. We will illustrate how this works next.

F > Applications: Creating Line Graphs

Suppose you want to graph the ordered pairs in the table on page 216 (we give equal time to cats!). The ordered pair (c, h) would pair the actual age c of a cat (in "regular" time) to the equivalent human age h. Since both c and h are positive, the graph would have to be in Quadrant I. Values of c would be from 0 to 21 and values of h from 10 to 100, suggesting a scale of 10 units for each equivalent human year. Let us do all this in Example 6.

Answers to PROBLEMS
5. a. QI **b.** QIII
c. QII **d.** QIV
e. y-axis **f.** x-axis

Cat's Actual Age	Equivalent Human Age	Cat's Actual Age	Equivalent Human Age
6 months	10 years	10 years	56 years
8 months	13 years	12 years	64 years
1 year	15 years	14 years	72 years
2 years	24 years	16 years	80 years
4 years	32 years	18 years	88 years
6 years	40 years	20 years	96 years
8 years	48 years	21 years	100 years

It was once thought that 1 year in the life of a cat was equivalent to 7 years of a human life. Recently, a new scale has been accepted: after the first 2 years, the cat's life proceeds more slowly in relation to human life and each feline year is approximately 4 human years. The general consensus is that at about age 7 a cat can be considered "middle-aged," and age 10 and beyond "old."

Source: Data from X Mission Internet.

EXAMPLE 6 Creating and interpreting line graphs

a. Make a coordinate grid to graph the ordered pairs in the table above starting with 1 year and ending at 21.

b. Graph the ordered pairs starting with (1, 15).

c. What does (21, 100) mean?

d. Study the pattern (2, 24), (4, 32), (6, 40). What would be the number h in $(8, h)$?

e. What is the number c in $(c, 56)$?

f. The oldest cat, Spike, lived to an equivalent human age of 140 years. What was Spike's actual age?

SOLUTION 6 The solutions to parts **a** and **b** are shown in the graph. Note that we go from 0 to 22 on the x-axis and from 0 to 100 on the y-axis.

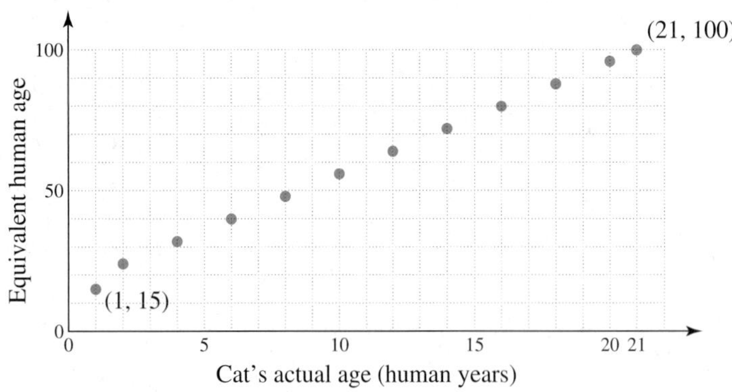

c. (21, 100) means that if a cat's actual age is 21, its equivalent human age is 100 years.

d. The numbers for the second coordinate are 24, 32, and 40 (increasing by 8). The next number would be 48, so the next ordered pair would be (8, 48). You can verify this from the table or the graph. This means that the age of an 8-year-old cat is equivalent to 48 years in a human.

e. From the table or the graph, the ordered pair whose second coordinate is 56 is (10, 56), so $c = 10$.

f. Neither the table nor the graph goes to 140 (the graph stops at 100). But we can look at the last two ordered pairs in the table and follow the pattern, or we can

PROBLEM 6

a. What does the ordered pair (16, 80) in the preceding table mean?

b. What is h in $(12, h)$?

c. What is c in $(c, 72)$?

d. CNN reports that there is a cat living in Texas whose equivalent human age is 148 years. What is the cat's actual age?

Source: CNN.

Answers to PROBLEMS

6. a. (16, 80) means that if a cat's actual age is 16, its equivalent human age is 80.
b. 64 **c.** 14 **d.** 33

extend the graph and find the answer. Here are the last two ordered pairs in the table:

(20, 96)

(21, 100)

.

.

.

(?, 140) ← To go from 100 to 140 we need 10 ($\frac{40}{4} = 10$) increments of 4 for the second coordinate and 10 increments of 1 for the first coordinate. Thus, (31, 140) is the next ordered pair. This means Spike was 31 human years old.

As you can see, *line graphs* can be used to show a certain change over a period of time. If we were to connect the points in the graph of Example 6 starting with (2, 24), we would have a *straight line*. We shall study linear equations and their graphs in Section 3.2 but before we finish, here is one more example. We have already mentioned the CAFE (**C**orporate **A**verage **F**uel **E**conomy) regulations intended to improve the average fuel economy of cars and light trucks sold in the United States (Example 8, Section 2.7). Here are two tables giving the original and a revised proposal. We will graph them in Example 7.

EXAMPLE 7 CAFE standards graph

a. Write the original standards for 2011–2015 as five ordered pairs of the form (year, mpg).

b. Graph the five ordered pairs.

Source: http://tinyurl.com/p9kx2x.

Original		Revised	
Year	**mpg**	**Year**	**mpg**
2011	27	2011	28
2012	29	2012	30
2013	30	2013	32
2014	31	2014	33
2015	32	2015	35

SOLUTION 7

a. For 2011, the goal is 27 mpg (line 1). The ordered pair is (2011, 27).
For 2012, the goal is 29 mpg (line 2). The ordered pair is (2012, 29).
For 2013, the goal is 30 mpg (line 3). The ordered pair is (2013, 30).
The ordered pairs corresponding to 2014 and 2015, respectively, are (2014, 31) and (2015, 32).

PROBLEM 7

a. Write the revised standards for 2011 to 2015 as five ordered pairs of the form (year, mpg).

b. Graph the five ordered pairs in the grid below.

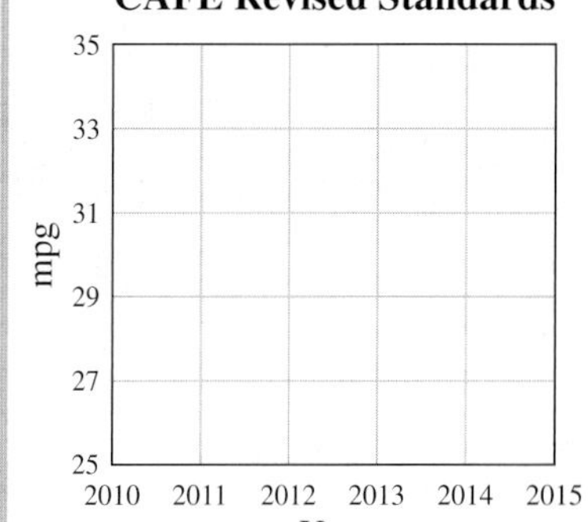

(continued)

Answers to PROBLEMS

7. a. (2011, 28), (2012, 30), (2013, 32), (2014, 33), (2015, 35)

b.

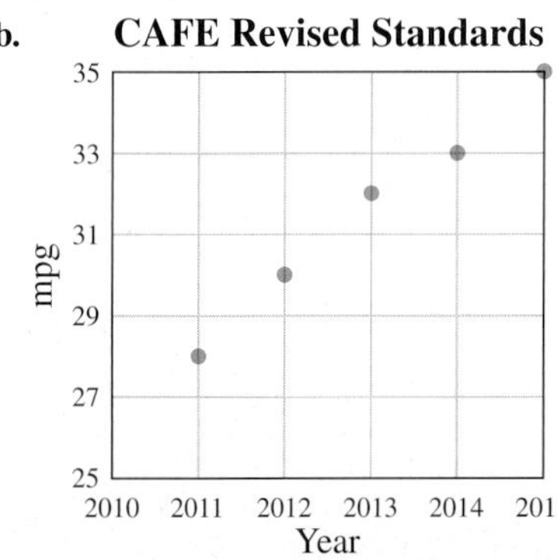

b. We label the horizontal axis starting with the year 2010, then 2011, and so on until we reach 2015. The vertical axis is labeled starting at 25 and ending at 35. (This choice is arbitrary—you can use the interval from 0 to 35 or from 20 to 35).

To graph (2011, 27), start at 2010 and move horizontally to 2011, then go up to 27 and graph the point (2011, 27) indicated by a small circle •. To graph the next point, move horizontally to 2012 and go up to 29. Make a small circle at (2012, 29), the graph of the point. Proceed similarly to graph the other three points, (2013, 30), (2014, 31), and (2015, 32) as shown in the graph.

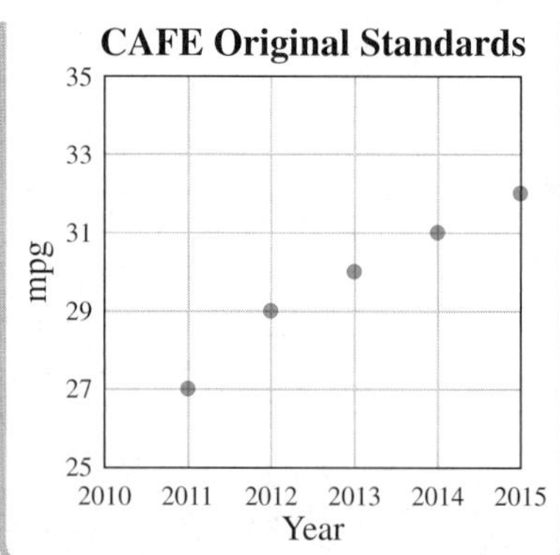

Calculator Corner

Graphing Points

We can do most of the work in this section with our calculators. To do Example 1, first adjust the **viewing screen** or **window** of the calculator. Since the values of x range from -5 to 5, denoted by $[-5, 5]$, and the values of y also range from -5 to 5, we have a $[-5, 5]$ by $[-5, 5]$ viewing rectangle, or window, as shown in Window 1.

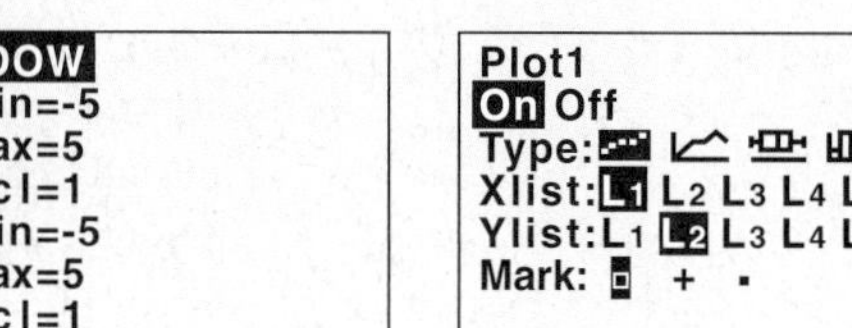

Window 1 Window 2

To graph the points A, B, C, and D of Example 1, set the calculator on the statistical graph mode (2nd Y= 1), turn the plot on (ENTER), select the type of plot, and the list of numbers you are going to use for the x-coordinates (L_1) and the y-coordinates (L_2) as well as the type of mark you want the calculator to make (Window 2). Now, press STAT 1 and enter the x-coordinates of A, B, C, and D under L_1 and the y-coordinates of A, B, C, and D under L_2. Finally, press GRAPH to obtain the points A, B, C, and D in Window 3. Use these ideas to do Problems 1–5 in the exercise set.

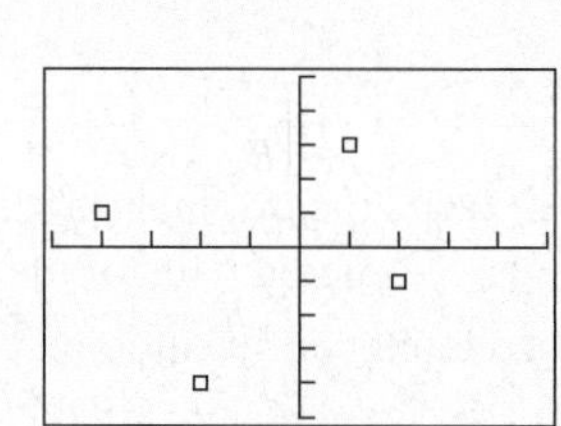

Window 3

› Exercises 3.1

> Practice Problems > Self-Tests > Media-rich eBooks > e-Professors > Videos

‹A› Graphing Ordered Pairs In Problems 1–5, graph the points.

1. a. $A(1, 2)$ **b.** $B(-2, 3)$ **c.** $C(-3, 1)$ **d.** $D(-4, -1)$

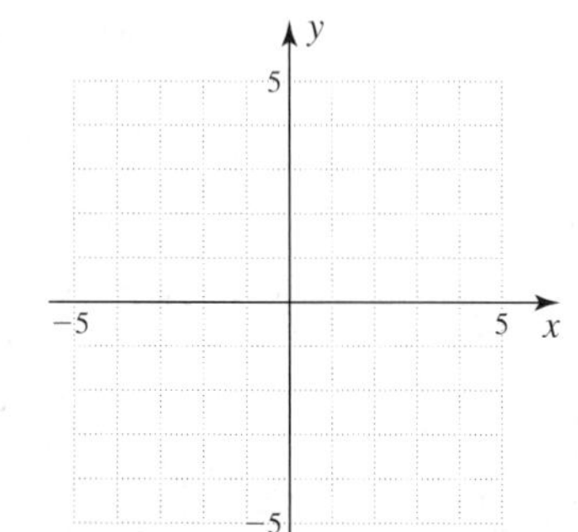

2. a. $A\left(-2\frac{1}{2}, 3\right)$ **b.** $B\left(-1, 3\frac{1}{2}\right)$ **c.** $C\left(-\frac{1}{2}, -4\frac{1}{2}\right)$ **d.** $D\left(\frac{1}{3}, 4\right)$

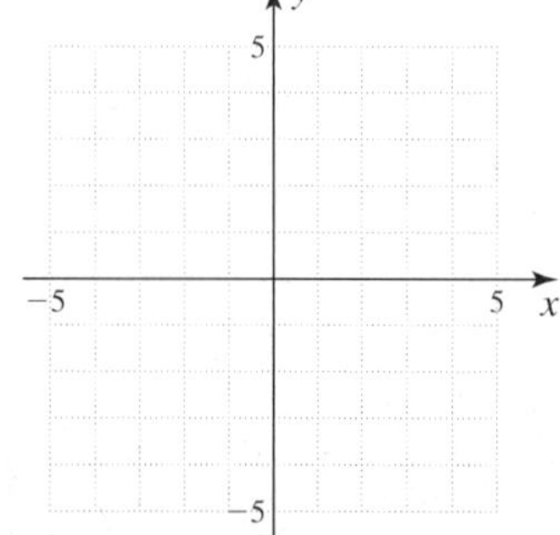

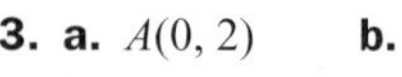

3. a. $A(0, 2)$ **b.** $B(-3, 0)$

c. $C\left(3\frac{1}{2}, 0\right)$ **d.** $D\left(0, -1\frac{1}{4}\right)$

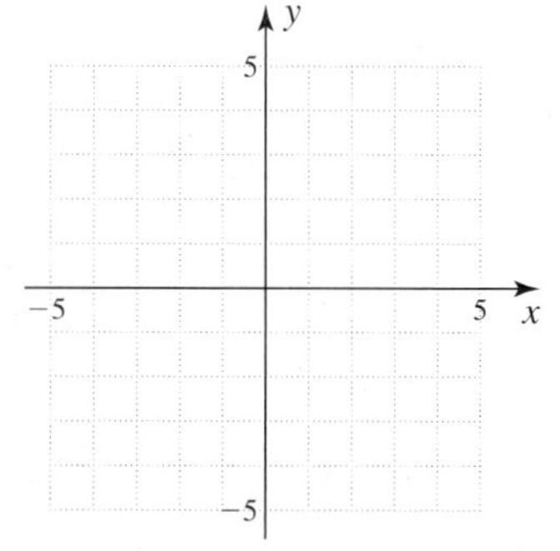

› Web IT go to mhhe.com/bello for more lessons

4. a. $A(20, 20)$ **b.** $B(-10, 20)$
c. $C(35, -15)$ **d.** $D(-25, -45)$

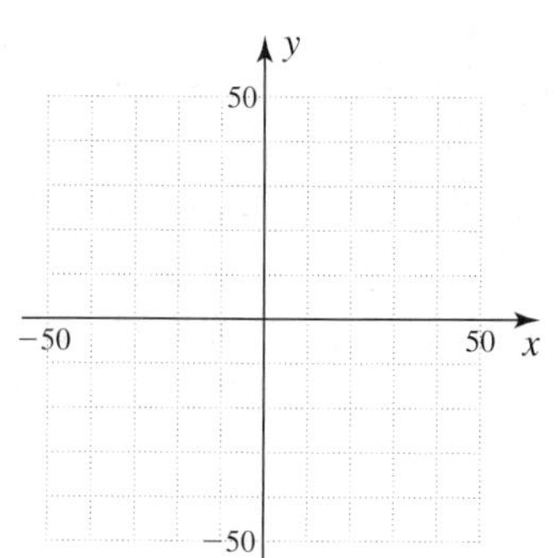

5. a. $A(0, 40)$ **b.** $B(-35, 0)$
c. $C(-40, -15)$ **d.** $D(0, -25)$

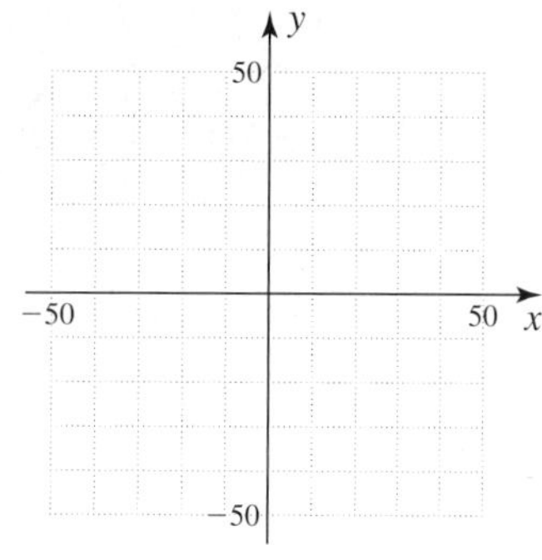

‹B› Finding Coordinates
‹E› Quadrants

In Problems 6–10, give the coordinates of the points and the quadrant in which each point lies.

6.

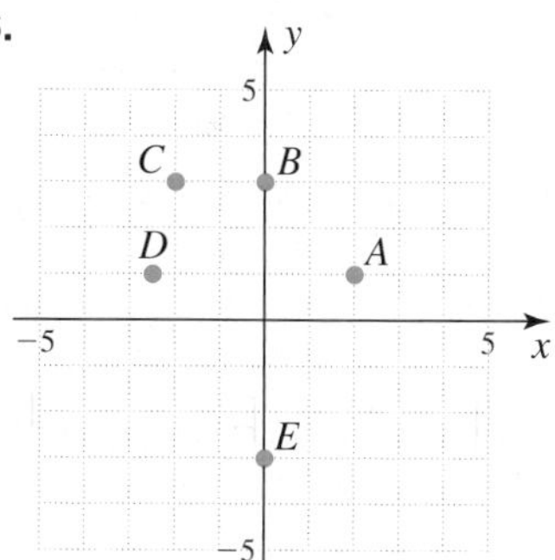

7.

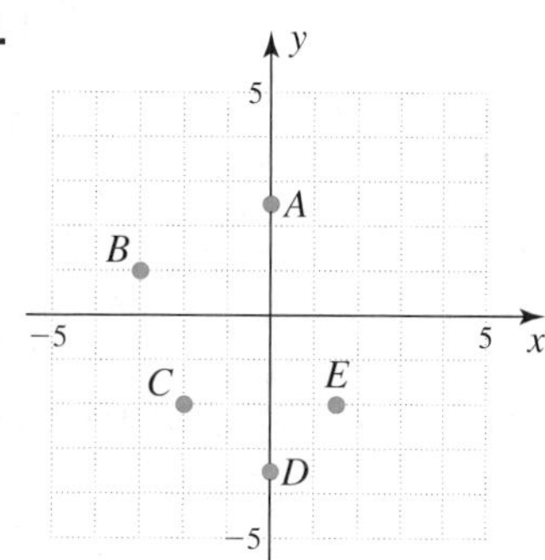

8.

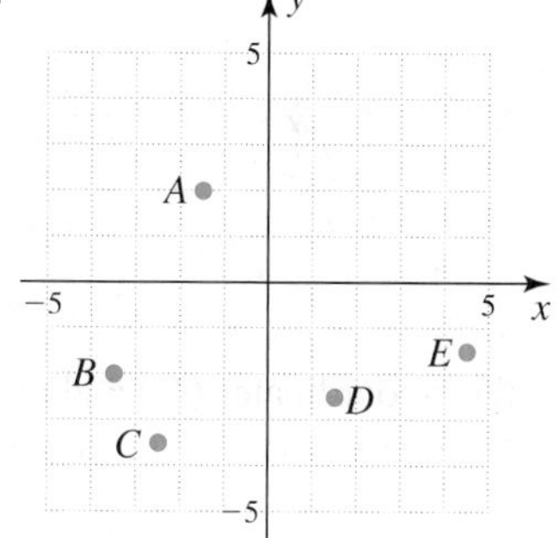

9.

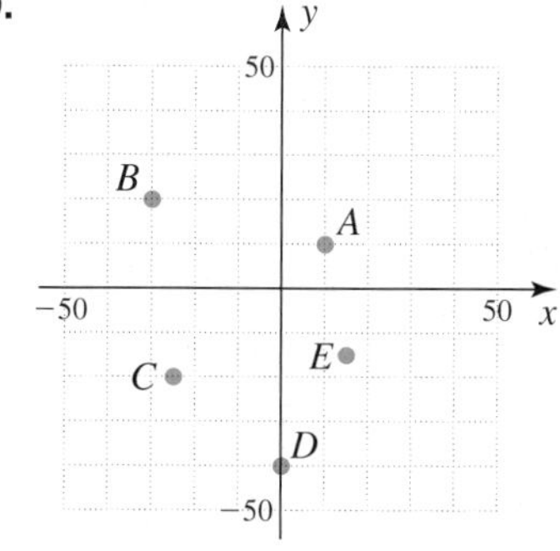

10.

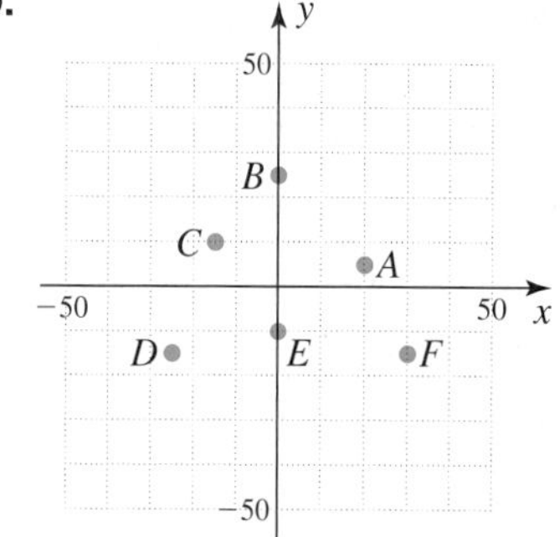

Web IT go to mhhe.com/bello for more lessons

C **Applications: Line Graphs** Use the following graph to work Problems 11–14.

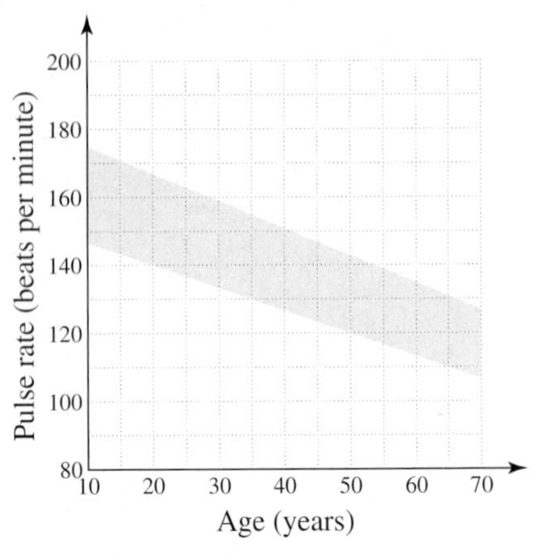

11. What is the lower limit pulse rate for a 20-year-old? Write the answer as an ordered pair.

12. What is the upper limit pulse rate for a 20-year-old? Write the answer as an ordered pair.

13. What is the upper limit pulse rate for a 45-year-old? Write the answer as an ordered pair.

14. What is the lower limit pulse rate for a 50-year-old? Write the answer as an ordered pair.

Applications: Green Math

Use the following graph to work Problems 15–24.

The orange graph shows the pounds of waste generated per capita (per person) each day and the blue graph shows the total waste generated in millions of tons.

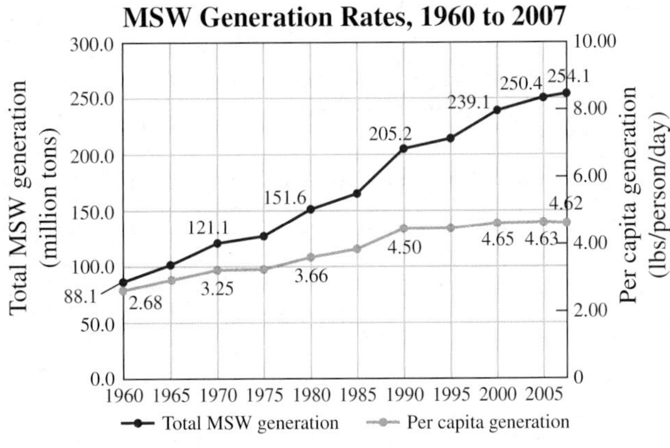

Municipal Solid Waste Generation Rates 1960–2007

Source: www.epa.gov.

15. How many pounds of waste were generated per person each day in 1960?

16. How many pounds of waste were generated per person each day in 2007?

17. How many more pounds were generated per person each day in 2007 than in 1960?

18. How many more pounds were generated per person each day in 2007 than in 1980?

19. Based on the graph, how many pounds of waste would you expect to be generated per person each day in 2008?

20. What was the total waste (in millions of tons) generated in 1960?

21. What was the total waste (in millions of tons) generated in 2007?

22. How many more million tons of waste were generated in 2007 than in 1960?

23. How many more million tons of waste were generated in 2007 than in 2000?

24. Based on your answer to Problem 23, how many million tons of waste would you expect to be generated in 2008?

Use the following graph to work Problems 25–28.

The graph shows the amount owed on a \$1000 debt at an 18% interest rate when the minimum \$25 payment is made or when a new monthly payment of \$92 is made. Thus, at the current monthly payment of \$25, it will take 60 months to pay the \$1000 balance (you got to \$0!). On the other hand, with a new \$92 monthly payment, you pay off the \$1000 balance in 12 months.

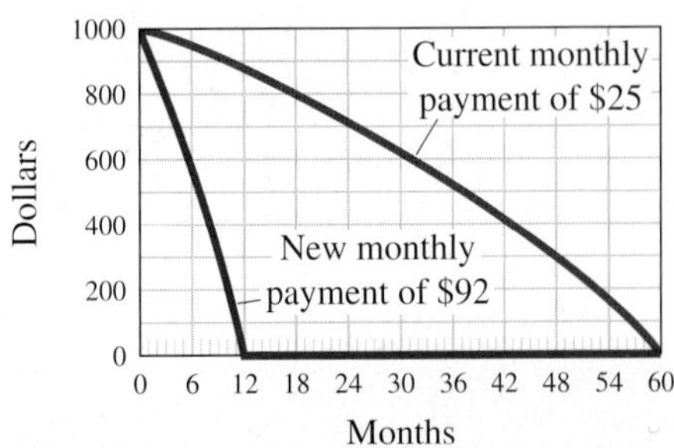

Source: Data from KJE Computer Solutions, LLC.

25. What is your balance after 6 months if you are paying \$25 a month?

26. What is your balance after 6 months if you are paying \$92 per month?

27. What is your balance after 18 months if you are paying \$25 a month?

28. What is your balance after 48 months if you are paying \$25 a month?

Use the following graph to work Problems 29–33.

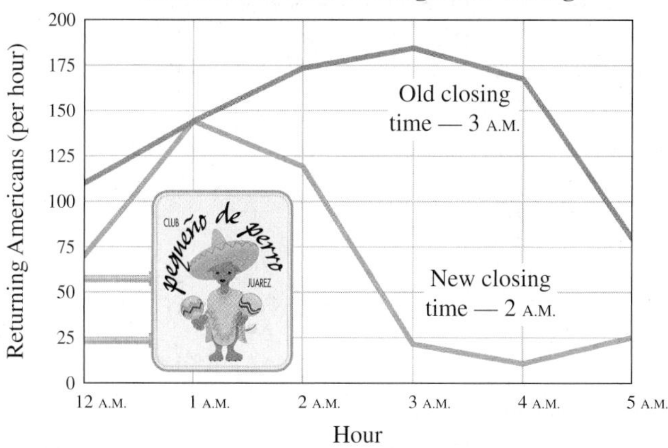

Source: Data from Institute for Public Strategies.

The graph shows the number of young Americans crossing the border at El Paso after the bar closing hour in Juarez, Mexico, changed from 3 A.M. to 2 A.M.

29. About how many people crossed the border at 12 A.M. when the bar closing time was 2 A.M.?

30. About how many people crossed the border at 12 A.M. when the bar closing time was 3 A.M.?

31. At what time was the difference in the number of border crossers greatest? Can you suggest an explanation for this?

32. At what time was the number of border crossers the same?

33. What was the difference in the number of border crossers at 5 A.M.?

‹D› Applications: Bar Graphs

Use the following graph to work Problems 34–38.

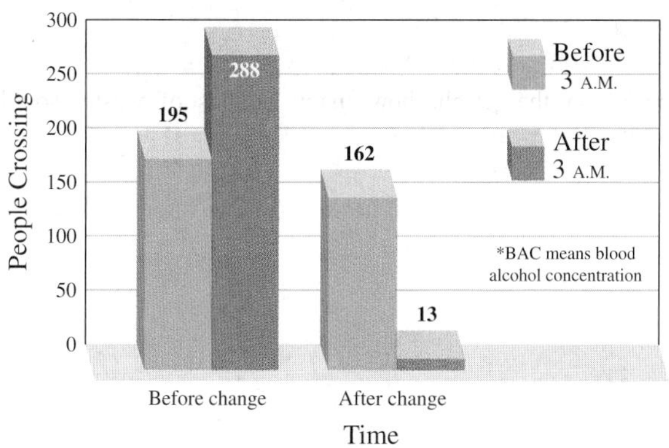

Source: Data from Institute for Public Strategies.

The numbers of crossers before 3 A.M. and after 3 A.M. are shown in the graph.

34. What was the total number of crossers before the change?

35. How many of those were before 3 A.M. and how many after 3 A.M.?

36. How many crossers were there after the change?

37. How many of those were before and how many after 3 A.M.?

38. Why do you think there were so many crossers with high BAC before the change?

››› Applications: Green Math

Use the following graph to work Problems 39–42.

The graph shows the percent of selected materials and the number of tons recycled in each category in a recent year. The total amount of recycled materials was 70 million tons.

39. a. Which material was recycled the most?

b. Which material was recycled the least?

40. What percent of the auto batteries were recycled? What percent of the 70 million tons total were auto batteries?

41. What percent of the tires were recycled? What percent of the 70 million tons total were tires?

42. How many more tons of glass containers than aluminum packaging were recycled?

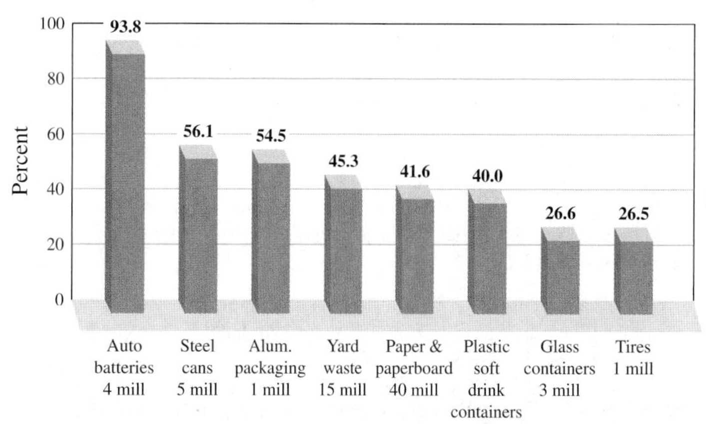

Source: Data from U.S. Environmental Protection Agency.

Web IT go to mhhe.com/bello for more lessons

Use the following graph to work Problems 43–46.

How does the down payment affect the monthly payment?

Purchase price	20,000
Sales tax +	1,050
Fees	40
Total price =	**21,090**
Cash down −	1,500
Net trade in −	1,000
Loan amount =	**18,590**

How Down Payment Affects Monthly Payment

Monthly payment (in dollars)

550
500
450
400
350
300
250

463
438

0 1500 2000 2500

Cash down (in dollars)

Source: Data from KJE Computer Solutions, LLC.

43. What is the monthly payment if the down payment is $0?

44. What is the monthly payment if the down payment is $2000?

45. What is the difference in the monthly payment if you increase the down payment from $1500 to $2500?

46. What is the difference in the annual amount paid if you increase the down payment from $1500 to $2500?

⟨F⟩ Applications: Creating Line Graphs The chart shows the dog's actual age and the equivalent human age for dogs of different weights and will be used in Problems 47–55.

Dog's Age	Weight (pounds)			
	15–30	30–49	50–74	75–100
1	12	12	14	16
2	19	21	23	23
3	25	25	26	29
4	32	32	35	35
5	36	36	37	39
6	40	42	43	45
7	44	45	46	48
8	48	50	52	52
9	52	53	54	56
10	56	56	58	61
11	60	60	62	65
12	62	63	66	70
13	66	67	70	75
14	70	70	75	80
15	72	76	80	85

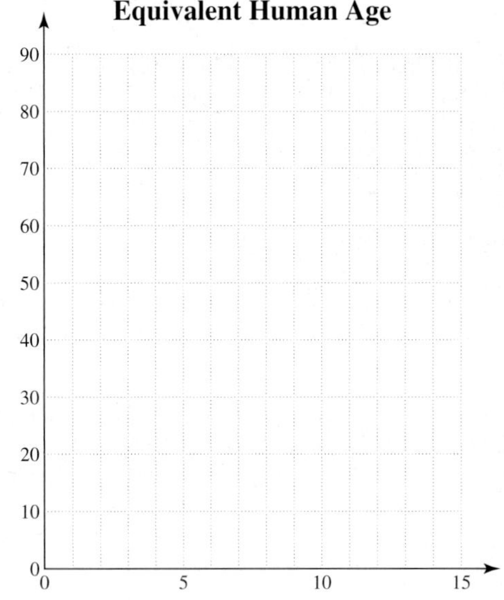

Using different colors, graph the equivalent human age for a dog weighing:

47. Between 15 and 30 pounds

48. Between 30 and 49 pounds

49. Between 50 and 74 pounds

50. Between 75 and 100 pounds

51. If a dog's age is 15, which of the weight scales will make it appear oldest?

52. If a dog's age is 15, which of the weight scales will make it appear youngest?

53. Based on your answers to Problems 51 and 52, the _____ (less, more) a dog weighs, the _____ (younger, older) it appears on the human age scale.

54. If a dog is 6 years old, what is its equivalent human age? *Hint:* There are four answers!

55. From the table, at which dog's age(s) is there no agreement as to what the equivalent human age would be?

56. The red graph shows the balance on a 30-year \$100,000 mortgage at 6.25% paying \$615.72 a month. If you make biweekly payments of \$307.06 you save \$27,027 in interest!

a. What is the approximate balance of the 30-year (red) mortgage after 12 years?

b. What is the approximate balance of the accelerated (blue) mortgage after 12 years?

c. In how many years do you pay off the mortgage under the accelerated plan?

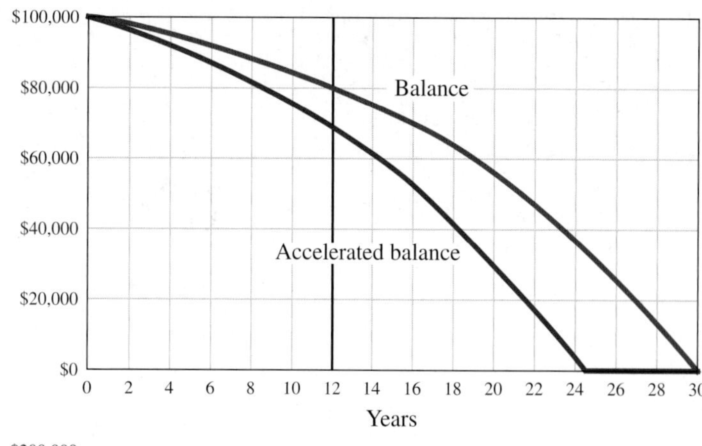

57. The red graph shows the balance on a 30-year \$200,000 mortgage at 7.5% paying \$1398.43 a month. If you make biweekly payments of \$699.22 you save \$80,203 in interest!

a. What is the approximate balance of the 30-year (red) mortgage after 20 years?

b. What is the approximate balance of the accelerated (blue) mortgage after 20 years?

c. In how many years do you pay off the mortgage under the accelerated plan?

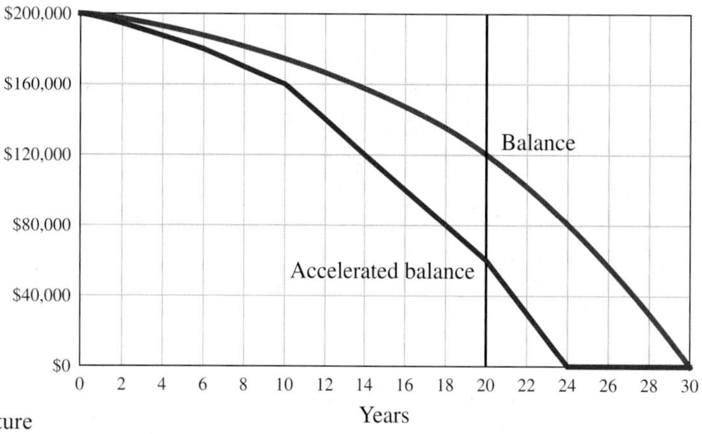

58. The environmental lapse is the rate of decrease of temperature with altitude (elevation): the *higher* you are, the *lower* the temperature. The temperature drops about 4°F (minus 4 degrees Fahrenheit) for each 1000 feet of altitude.

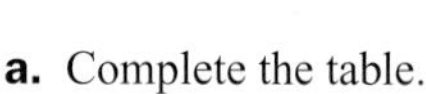
Source: www.answers.com.

Altitude in (1000 ft)	Temperature Change	Ordered Pair
1	−4°F	(1, −4)
2	−8°F	(2, −8)
3		
4		
5		

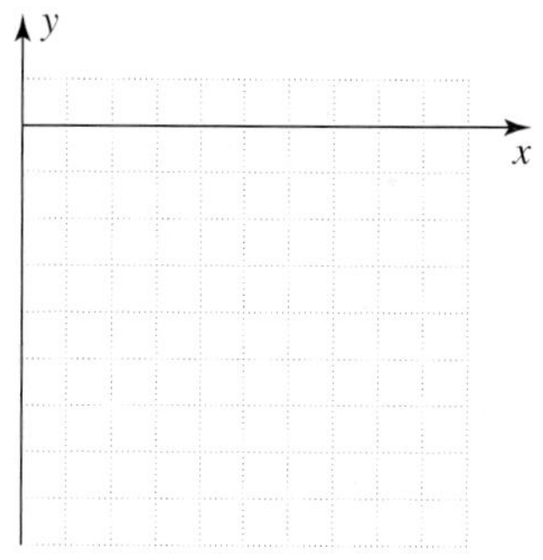

a. Complete the table.

b. Make a graph of the ordered pairs.

59. In the metric system, the environmental lapse is about **7°C** (minus 7 degrees Celsius) for each kilometer of altitude.

Altitude (kilometers)	Temperature Change	Ordered Pair
1	−7°C	(1, −7)
2	−14°C	(2, −14)
3		
4		
5		

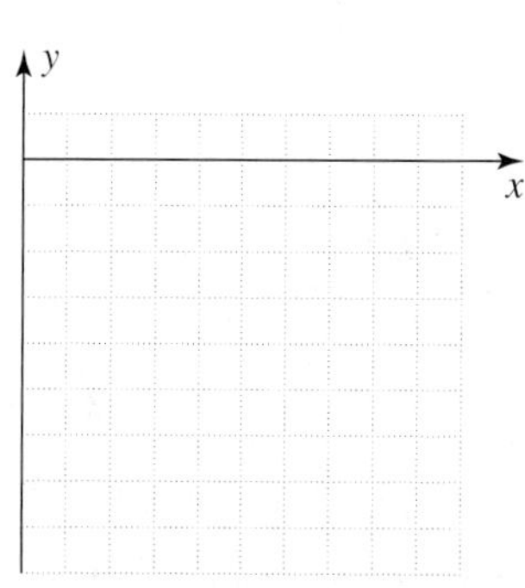

a. Complete the table.

b. Make a graph of the ordered pairs.

60. The graph shows the annual percent of renters among people under age 25.

a. What percent were renters in 2000?

b. What percent were renters in 2004?

c. In what years was the percent of renters decreasing?

d. In what years was the percent of renters unchanged?

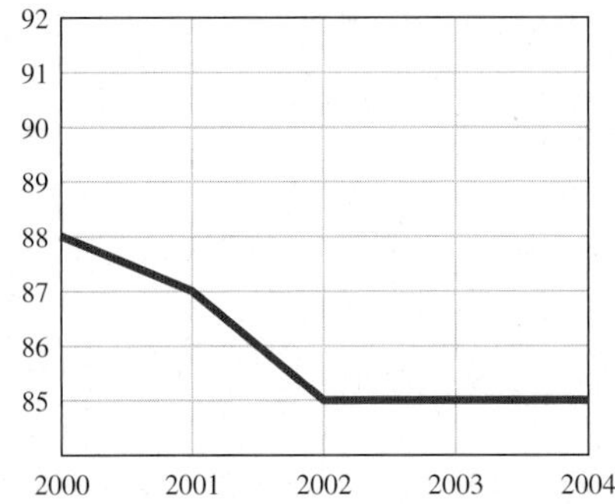

Using Your Knowledge

Graphing the Risks of "Hot" Exercise The ideas we presented in this section are vital for understanding graphs. For example, do you exercise in the summer? To determine the risk of exercising in the heat, you must know how to read the graph on the right.

This can be done by first finding the temperature (the y-axis) and then reading across from it to the right, stopping at the vertical line representing the relative humidity. Thus, on a 90°F day, if the humidity is less than 30%, the weather is in the safe zone.

Use your knowledge to answer the following questions about exercising in the heat.

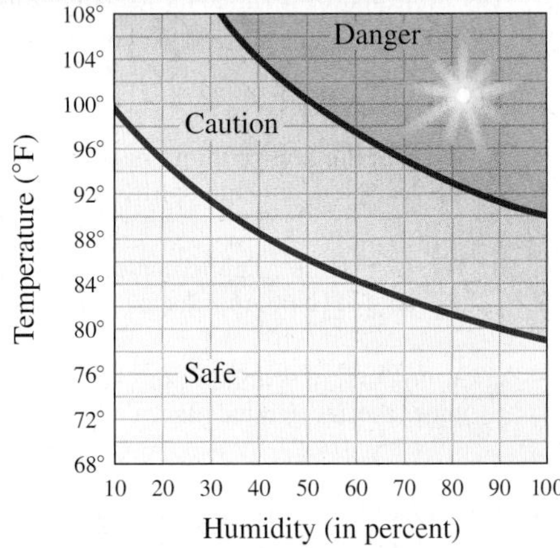

61. If the humidity is 50%, how high can the temperature go and still be in the safe zone for exercising? (Answer to the nearest degree.)

62. If the humidity is 70%, at what temperature will the danger zone start?

63. If the temperature is 100°F, what does the humidity have to be so that it is safe to exercise?

64. Between what temperatures should you use caution when exercising if the humidity is 80%?

65. Suppose the temperature is 86°F and the humidity is 60%. How many degrees can the temperature rise before you get to the danger zone?

Write On

66. Suppose the point (a, b) lies in the first quadrant. What can you say about a and b?

67. Suppose the point (a, b) lies in the second quadrant. What can you say about a and b? Is it possible that $a = b$?

68. In what quadrant(s) can you have points (a, b) such that $a = b$? Explain.

69. Suppose the ordered pairs (a, b) and (c, d) have the same point as their graphs—that is, $(a, b) = (c, d)$. What is the relationship between a, b, c, and d?

70. Now suppose $(a + 5, b - 7)$ and $(3, 5)$ have the same point as their graphs. What are the values of a and b?

Concept Checker

Fill in the blank(s) with the correct word(s), phrase, or mathematical statement.

71. In the **ordered pair** (x, y) the x is called the ________.

72. In the **ordered pair** (x, y) the y is called the ________.

coordinate **ordinance**

abscissa **ordinate**

Mastery Test

Graph the points:

73. $A(4, 2)$

74. $B(3, -2)$

75. $C(-2, 1)$

76. $D(0, -3)$

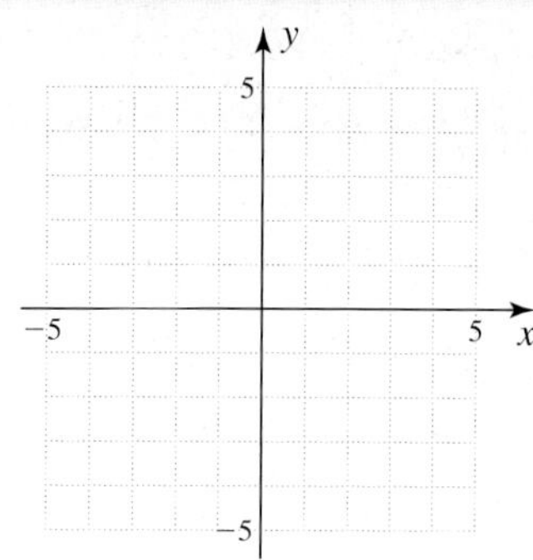

77. The projected cost (in cents per minute) for wireless phone use for the next four years is given in the table. Graph these points.

Year	cost (cents/min)
1	25
2	23
3	22
4	20

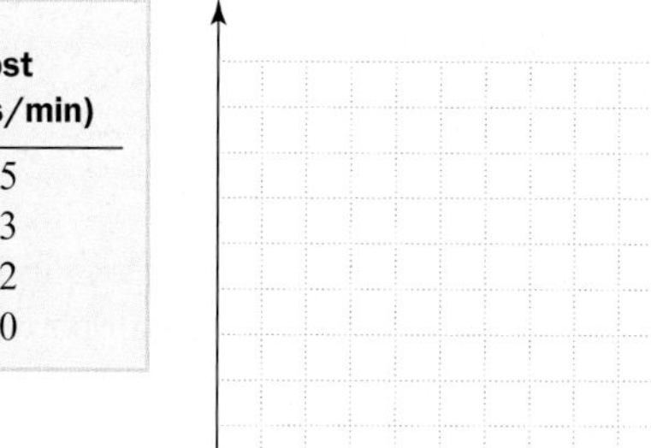

78. Referring to the graph, find an ordered pair giving the equivalent dog age (dog years) for a dog that is actually 7 years old (human years).

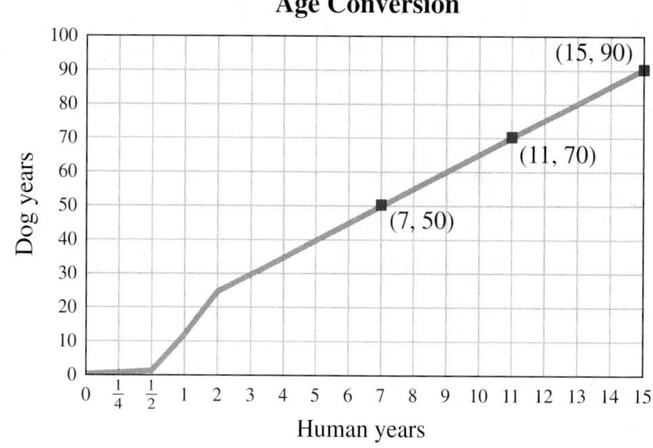

79. Referring to the graph in Problem 78, find the age (human years) of a dog that is 70 years old (dog years).

80. What does the ordered pair (15, 90) in the graph in Problem 78 mean?

81. In which quadrant are the following points?

a. $(-2, 3)$
b. $(3, -2)$
c. $(2, 2)$
d. $(-1, -1)$

82. Give the coordinates of each of the points shown in the graph.

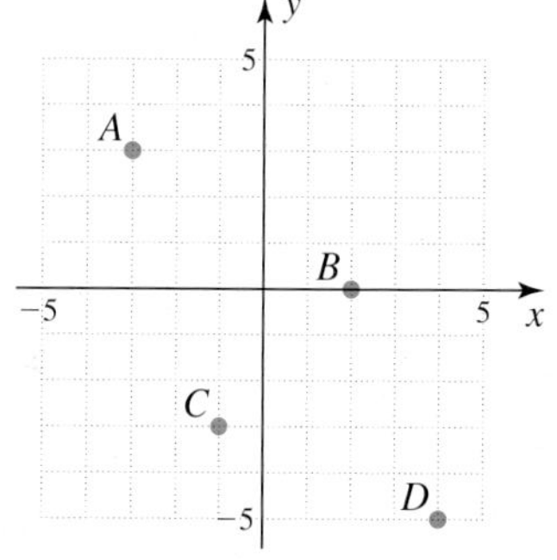

Skill Checker

83. Solve: $y = 3x - 2$ when $y = 7$

84. Solve: $C = 0.10m + 10$ when $m = 30$

85. Solve: $3x + y = 9$ when $x = 0$

86. Solve: $y = 50x + 450$ when $x = 2$

87. Solve: $3x + y = 6$ when $x = 1$

88. Solve: $3x + y = 6$ when $y = 0$

3.2 Graphing Linear Equations in Two Variables

Objectives

A Determine whether a given ordered pair is a solution of an equation.

B Find ordered pairs that are solutions of a given equation.

C Graph linear equations of the form $y = mx + b$ and $Ax + By = C$.

D Solve applications involving linear equations.

To Succeed, Review How To . . .

1. Graph ordered pairs (p. 211).
2. Read and interpret ordered pairs on graphs (pp. 212–215).

Getting Started

Climate Change

The graph shows the 72-hour forecast period from the National Hurricane Center. In the blue line (the actual prediction), the x-coordinate shows the number of ***hours*** (0 to 72), whereas the y-coordinate gives the ***wind speed.***

(x, y)

x is the hours ↲ ↳ y is the wind speed

Thus, at the beginning of the advisory ($x = 0$) the wind speed was 150 mph and 24 hours later ($x = 24$) the wind increased to 165 mph. What ordered pair corresponds to the wind speed $x = 48$ hours later? The answer is (48, 120).

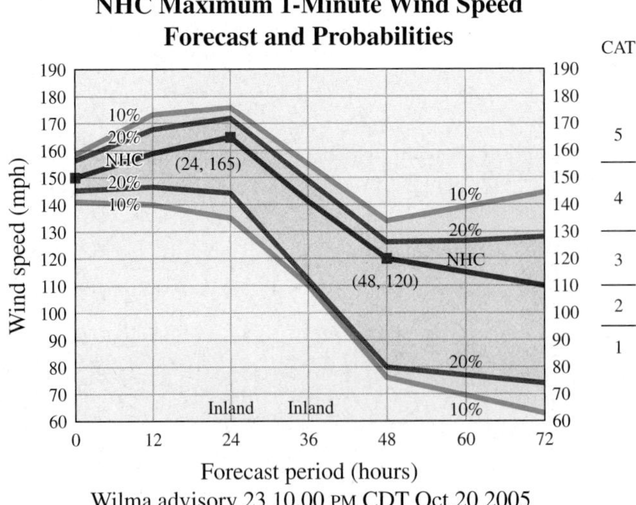

A › Solutions of Equations

How can we construct our own graph showing the wind speed of the hurricane for the first 24 hours? First, we label the x-axis with the hours from 0 to 24 and the y-axis with the wind speed from 0 to 200. We then graph the ordered pairs (x, y) representing the hours x and the wind speed y starting with (0, 150) and (24, 165), as shown in Figure 3.12. Does this line have any relation to algebra? Of course! The line representing the wind speed y can be approximated by the equation

$$y = \frac{5}{8}x + 150$$

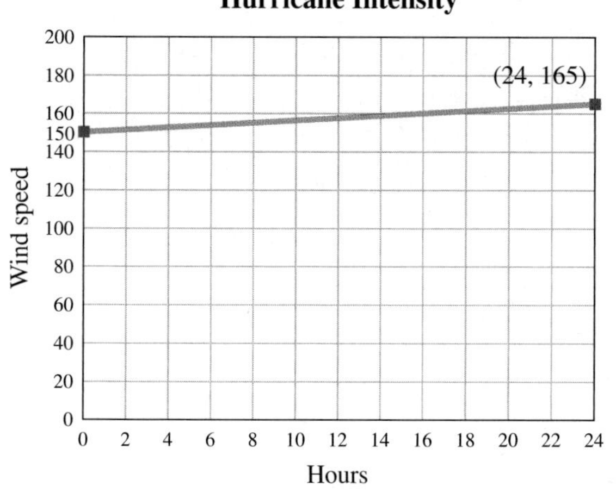

>Figure 3.12

Note that at the beginning of the forecast (when $x = 0$) the wind speed was

$$y = \frac{5}{8} \cdot 0 + 150 = 150 \text{ (mph)}$$

If we write $x = 0$ and $y = 150$ as the ordered pair (0, 150), we say that the ordered pair (0, 150) **satisfies** or is a **solution** of the equation $y = \frac{5}{8}x + 150$. What about $x = 24$ hours after the beginning of the forecast? At that point the wind speed $y = \frac{5}{8} \cdot 24 + 150 = 165$ (mph), hence the ordered pair (24, 165) also satisfies the equation $y = \frac{5}{8}x + 150$ since $165 = \frac{5}{8} \cdot 24 + 150$. Note that the equation $y = \frac{5}{8}x + 150$ has two variables, x and y. Such equations are called **equations in two variables.**

DETERMINING IF AN ORDERED PAIR IS A SOLUTION

To determine whether an ordered pair is a solution of an equation in two variables, we substitute the x-coordinate for x and the y-coordinate for y in the given equation. If the resulting statement is *true,* then the ordered pair is a solution of, or *satisfies,* the equation.

Thus, (2, 5) is a solution of $y = x + 3$, since in the ordered pair (2, 5), $x = 2$, $y = 5$, and

$$y = x + 3$$

becomes

$$5 = 2 + 3$$

which is a true statement.

EXAMPLE 1 Determining whether an ordered pair is a solution

Determine whether the given ordered pairs are solutions of $2x + 3y = 10$.

a. (2, 2) **b.** (−3, 4) **c.** (−4, 6)

SOLUTION 1

a. In the ordered pair (2, 2), $x = 2$ and $y = 2$. Substituting in

$$2x + 3y = 10$$

we get

$$2(2) + 3(2) = 10 \quad \text{or} \quad 4 + 6 = 10$$

which is true. Thus, (2, 2) is a solution of

$$2x + 3y = 10$$

You can summarize your work as shown here:

$2x + 3y$	$\stackrel{?}{=}$ 10
2(2) + 3(2)	10
4 + 6	
10	True

PROBLEM 1

Determine whether the ordered pairs are solutions of $3x + 2y = 10$.

a. (2, 2)

b. (−3, 4)

c. (−4, 11)

(continued)

Answers to PROBLEMS

1. a. Yes; $3(2) + 2(2) = 10$ **b.** No; $3(-3) + 2(4) \neq 10$ **c.** Yes; $3(-4) + 2(11) = 10$

b. In the ordered pair $(-3, 4)$, $x = -3$ and $y = 4$. Substituting in

$$2x + 3y = 10$$

we get

$$2(-3) + 3(4) = 10$$
$$-6 + 12 = 10$$
$$6 = 10$$

which is *not* true. Thus, $(-3, 4)$ is not a solution of the given equation. You can summarize your work as shown here:

$2x + 3y$	$\stackrel{?}{=} 10$
$2(-3) + 3(4)$	10
$-6 + 12$	
6	False

c. In the ordered pair $(-4, 6)$, the x-coordinate is -4 and the y-coordinate is 6. Substituting these numbers for x and y, respectively, we have

$$2x + 3y = 10$$
$$2(-4) + 3(6) = 10$$

or

$$-8 + 18 = 10$$

which is a true statement. Thus, $(-4, 6)$ satisfies, or is a solution of, the given equation. You can summarize your work as shown here.

$2x + 3y$	$\stackrel{?}{=} 10$
$2(-4) + 3(6)$	10
$-8 + 18$	
10	True

B > Finding Missing Coordinates

In some cases, rather than verifying that a certain ordered pair *satisfies* an equation, we actually have to *find* ordered pairs that are solutions of the given equation. For example, we might have the equation

$$y = 2x + 5$$

and would like to find several ordered pairs that satisfy this equation. To do this we substitute any number for x in the equation and then find the corresponding y-value. A good number to choose is $x = 0$. In this case,

$$y = 2 \cdot 0 + 5 = 5$$ Zero is easy to work with because the value of y is easily found when 0 is substituted for x.

Thus, the ordered pair $(0, 5)$ satisfies the equation $y = 2x + 5$. For $x = 1$, $y = 2 \cdot 1 + 5 = 7$; hence, $(1, 7)$ also satisfies the equation.

We can let x be any number in an equation and then find the corresponding y-value. Conversely, we can let y be any number in the given equation and then find the x-value. For example, if we are given the equation

$$y = 3x - 2$$

and we are asked to find the value of x in the ordered pair $(x, 7)$, we simply let y be 7 and obtain

$$7 = 3x - 2$$

We then solve for x by rewriting the equation as

$$\begin{aligned} 3x - 2 &= 7 \\ 3x &= 9 \quad \text{Add 2.} \\ x &= 3 \quad \text{Divide by 3.} \end{aligned}$$

Thus, $x = 3$ and the ordered pair satisfying $y = 3x - 2$ is (3, 7), as can be verified since $7 = 3(3) - 2$.

EXAMPLE 2 Finding the missing coordinate

Complete the given ordered pairs so that they satisfy the equation $y = 4x + 3$.

a. $(x, 11)$ **b.** $(-2, y)$

SOLUTION 2

a. In the ordered pair $(x, 11)$, y is 11. Substituting 11 for y in the given equation, we have

$$\begin{aligned} 11 &= 4x + 3 \\ 4x + 3 &= 11 \quad \text{Rewrite with } 4x + 3 \text{ on the left.} \\ 4x &= 8 \quad \text{Subtract 3.} \\ x &= 2 \quad \text{Divide by 4.} \end{aligned}$$

Thus, $x = 2$ and the ordered pair is (2, 11).

b. Here $x = -2$. Substituting this value in $y = 4x + 3$ yields

$$y = 4(-2) + 3 = -8 + 3 = -5$$

Thus, $y = -5$ and the ordered pair is $(-2, -5)$.

PROBLEM 2

Complete the ordered pairs so that they satisfy the equation $y = 3x + 4$.

a. $(x, 7)$ **b.** $(-2, y)$

EXAMPLE 3 Approximating cholesterol levels

Figure 3.13 shows the decrease in cholesterol with exercise over a 12-week period. If C is the cholesterol level and w is the number of weeks elapsed, the line shown can be approximated by $C = -3w + 215$. (Of course, results vary.)

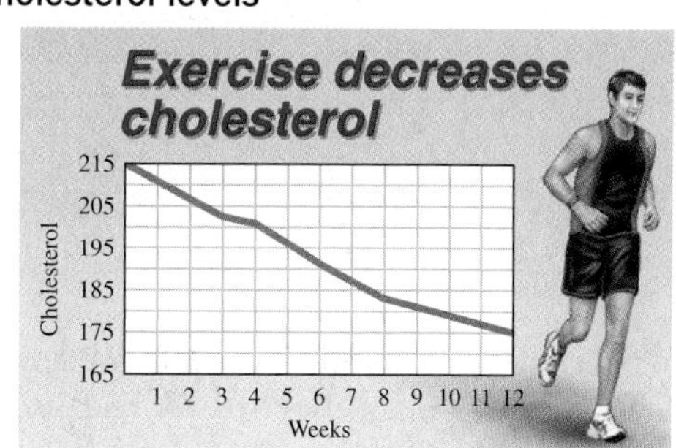

>Figure 3.13

a. According to the graph, what is the cholesterol level at the end of 1 week and at the end of 12 weeks?

b. What is the cholesterol level at the end of 1 week and at the end of 12 weeks using the equation $C = -3w + 215$?

SOLUTION 3

a. The cholesterol level at the end of 1 week corresponds to the point (1, 211) on the graph (the point of intersection of the vertical line above 1 and the graph). Thus, the cholesterol level at the end of 1 week is 211. Similarly, the cholesterol level at the end of 12 weeks corresponds to the point (12, 175). Thus, the cholesterol level at the end of 12 weeks is 175.

b. From the equation $C = -3w + 215$, the cholesterol level at the end of 1 week ($w = 1$) is given by $C = -3(1) + 215 = 212$ (close to the 211 on the graph!) and the cholesterol level at the end of 12 weeks ($w = 12$) is $C = -3(12) + 215 = -36 + 215$, or 179.

PROBLEM 3

a. According to the graph, what is the cholesterol level at the end of 4 weeks?

b. What is the cholesterol level at the end of 4 weeks using the equation in Example 3?

Answers to PROBLEMS

2. a. (1, 7) **b.** (−2, −2)

3. a. About 202 **b.** 203

C > Graphing Linear Equations by Plotting Points

Suppose you want to rent a car that costs \$30 per day plus \$0.20 per mile traveled. If we have the equation for the daily cost C based on the number m of miles traveled, we can graph this equation. The equation is

$$C = \overbrace{0.20m}^{20\text{¢ per mile}} + \overbrace{30}^{\$30\text{ each day}}$$

Now remember that a solution of this equation must be an ordered pair of numbers of the form (m, C). For example, if you travel 10 miles, $m = 10$ and the cost is

$$C = 0.20(10) + 30 = 2 + 30 = \$32$$

Thus, (10, 32) is an ordered pair satisfying the equation; that is, (10, 32) is a *solution* of the equation. If we go 20 miles, $m = 20$ and

$$C = 0.20(20) + 30 = 4 + 30 = \$34$$

Hence, (20, 34) is also a solution.

As you see, we can go on forever finding solutions. It's much better to organize our work, list these two solutions and some others, and then graph the points obtained in the Cartesian coordinate system. In this system the number of miles m will appear on the horizontal axis, and the cost C on the vertical axis. The corresponding points given in the table appear in the accompanying figure.

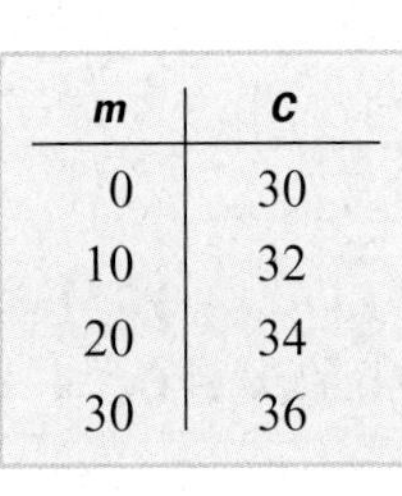

m	C
0	30
10	32
20	34
30	36

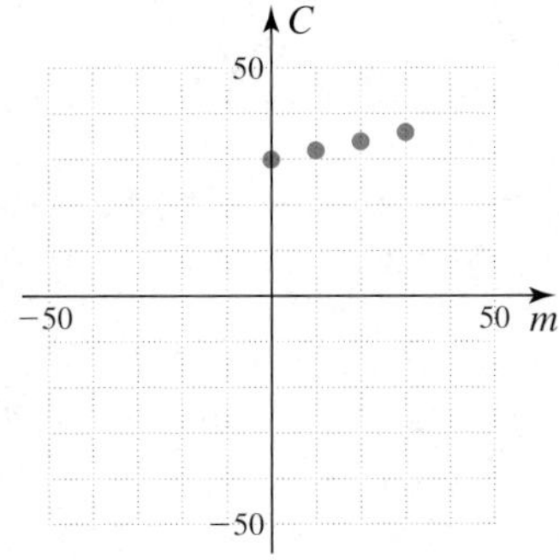

Note that m must be *positive* or *zero*. We have selected values for m that make the cost easy to compute—namely, 0, 10, 20, and 30.

It seems that if we join the points appearing in the graph, we obtain a straight line. Of course, if we knew this for sure, we could have saved time! Why? Because if we know that the graph of an equation is a straight line, we simply find *two* solutions of the equation, graph the two points, and then join them with a straight line. As it turns out, the graph of $C = 0.20m + 30$ *is* a straight line.

The procedure used to graph $C = 0.20m + 30$ can be generalized to graph other lines. Here are the steps.

PROCEDURE

Graphing Lines by Plotting Ordered Pairs

1. Choose a value for one variable, calculate the value of the other variable, and graph the resulting ordered pair.
2. Repeat step 1 to obtain at least two ordered pairs.
3. Graph the ordered pairs and draw a line passing through the points. (You can use a third ordered pair as a check.)

This procedure is called **graphing,** and the following rule lets us know that the graph is indeed a *straight line.*

RULE

Straight-Line Graphs

The graph of a linear equation of the form

$$Ax + By = C \quad \text{or} \quad y = mx + b$$

where A, B, C, m, and b are constants (A and B are not both 0) is a **straight line,** and every straight line has an equation that can be written in one of these forms.

Thus, the graph of $C = 0.20m + 30$ is a straight line and the graph of $3x + y = 6$ is also a straight line, as we show next.

EXAMPLE 4 Graphing lines of the form $Ax + By = C$

Graph: $3x + y = 6$

SOLUTION 4 The equation is of the form $Ax + By = C$, and thus the graph is a straight line. Since two points determine a line, we shall graph two points and join them with a straight line, the graph of the equation. Two easy points to use occur when we let $x = 0$ and find y, and let $y = 0$ and find x. For $x = 0$,

$$3x + y = 6$$

becomes

$$3 \cdot 0 + y = 6 \quad \text{or} \quad y = 6$$

Thus, (0, 6) is on the graph.

When $y = 0$, $3x + y = 6$ becomes

$$3x + 0 = 6$$
$$3x = 6$$
$$x = 2$$

Hence, (2, 0) is also on the graph.

We chose $x = 0$ and $y = 0$ because the calculations are easy. There are infinitely many points that satisfy the equation $3x + y = 6$, but the points (0, 6) and (2, 0)—the *y*- and *x-intercepts*, respectively—are easy to find.

It's a good idea to pick a *third* point as a *check*. For example, if we let $x = 1$, $3x + y = 6$ becomes

$$3 \cdot 1 + y = 6$$
$$3 + y = 6$$
$$y = 3$$

Now we have our third point, (1, 3), as shown in the following table. The points (0, 6), (2, 0), and (1, 3), as well as the completed graph of the line, are shown in Figure 3.14.

x	y	
0	6	← y-intercept
2	0	← x-intercept
1	3	

Note that the point (0, 6) where the line crosses the y-axis is called the ***y*-intercept,** and the point (2, 0) where the line crosses the x-axis is called the ***x*-intercept.**

PROBLEM 4

Graph: $2x + y = 4$

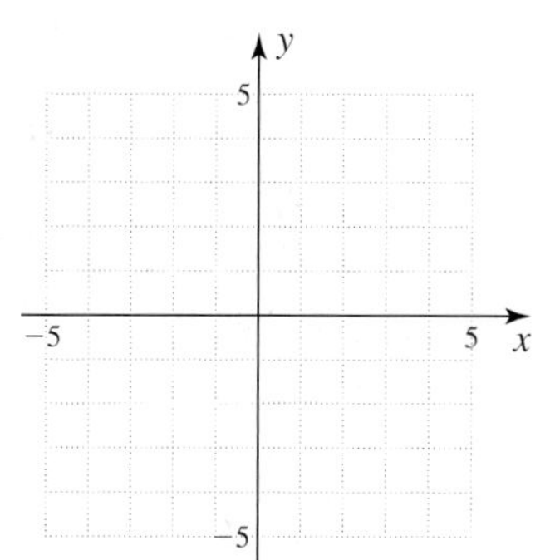

Answers to PROBLEMS

4.

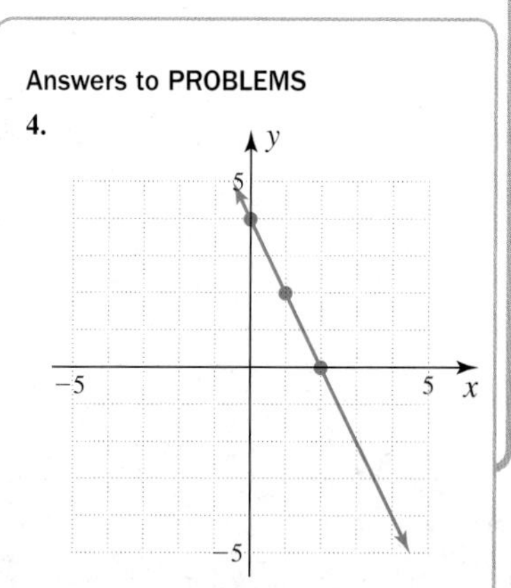

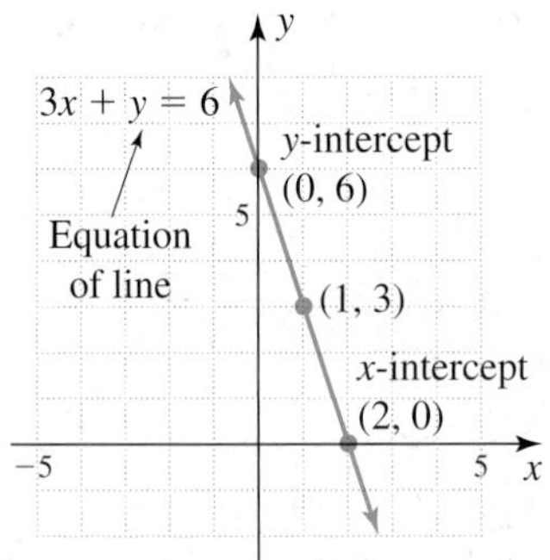

>Figure 3.14

The graph in Figure 3.14 cannot show the entire line, which extends indefinitely in both directions; that is, the graph shown is a part of a line that continues without end in both directions, as indicated by the arrows.

EXAMPLE 5 **Graphing lines of the form $y = mx + b$**

Graph: $y = -2x + 6$

SOLUTION 5 The equation is of the form $y = mx + b$, which is a straight line. Thus, we follow the procedure for graphing lines.

1. Let $x = 0$.

$$y = -2x + 6 \quad \text{becomes}$$
$$y = -2(0) + 6 = 6$$

Thus, the point (0, 6) is on the graph. (See Figure 3.15.)

2. Let $y = 0$.

$$y = -2x + 6 \quad \text{becomes}$$
$$0 = -2x + 6$$
$$2x = 6 \quad \text{Add } 2x.$$
$$x = 3 \quad \text{Divide by 2.}$$

Hence, (3, 0) is also on the line.

3. Graph (0, 6) and (3, 0), and draw a line passing through both points.
To make sure, we will use a third point.
Let $x = 1$.

$$y = -2x + 6 \quad \text{becomes}$$
$$y = -2(1) + 6 = 4$$

Thus, the point (1, 4) is also on the line, as shown in Figure 3.15.

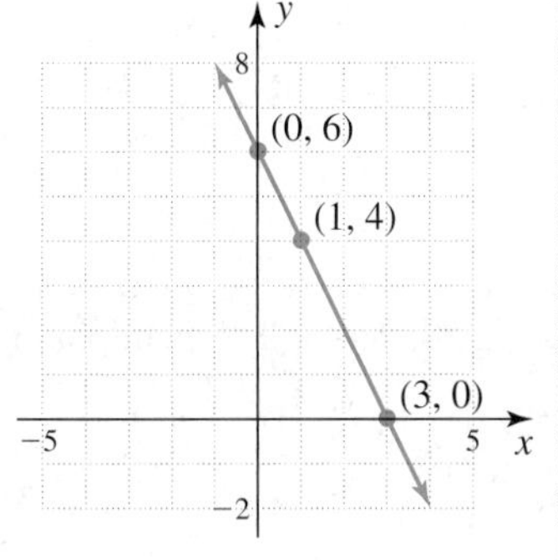

>Figure 3.15

PROBLEM 5

Graph: $y = -3x + 6$

Answers to PROBLEMS

5.

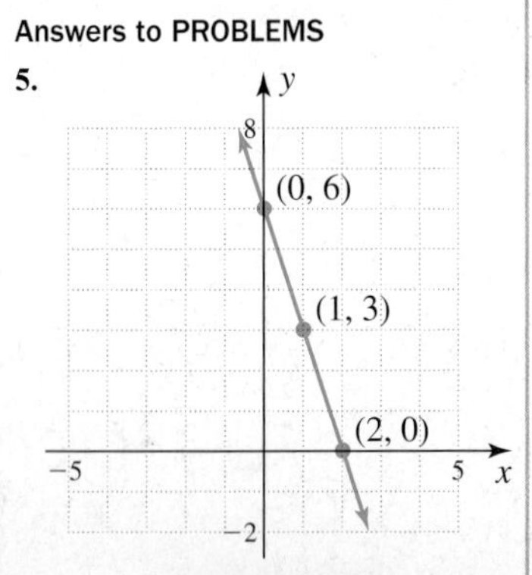

D > Applications Involving Linear Equations

Have you heard the term "wind chill factor"? It is the temperature you actually *feel* because of the wind. Thus, for example, if the temperature is 10 degrees Fahrenheit (10°F) and the wind is blowing at 15 miles per hour, you will *feel* as if the temperature is −7°F. If the wind is blowing at a *constant* 15 miles per hour, the wind chill factor W can be roughly approximated by the equation

$$W = 1.3t - 20, \quad \text{where } t \text{ is the temperature in °F}$$

EXAMPLE 6 Wind chill factor and linear equations

Graph: $W = 1.3t - 20$

SOLUTION 6 First, note that we have to label the axes differently. Instead of y we have W, and instead of x we have t. Next, note that the wind chill factors we will get seem to be negative (try $t = -5$, $t = 0$, and $t = 5$). Thus, we will concentrate on points in the third and fourth quadrants, as shown in the grid. Now let's obtain some points.

For $t = -5$, $W = 1.3(-5) - 20 = -26.5$. Graph $(-5, -26.5)$.
For $t = 0$, $W = 1.3(0) - 20 = -20$. Graph $(0, -20)$.
For $t = 5$, $W = 1.3(5) - 20 = -13.5$. Graph $(5, -13.5)$.

Note that the equation $W = 1.3t - 20$ applies only when the wind is blowing at 15 miles per hour! Moreover, you should recognize that the graph is a line (it is of the form $y = mx + b$ with W instead of y and t instead of x) and that the graph of the line belongs mostly in the third and fourth quadrants.

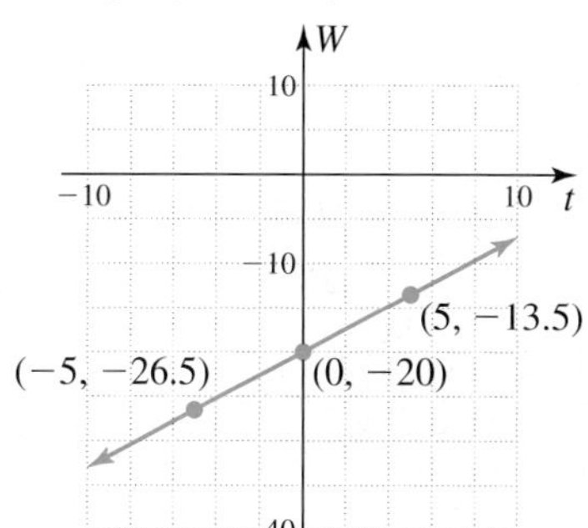

PROBLEM 6

Graph: $W = 1.2t - 20$

Answers to PROBLEMS

6.

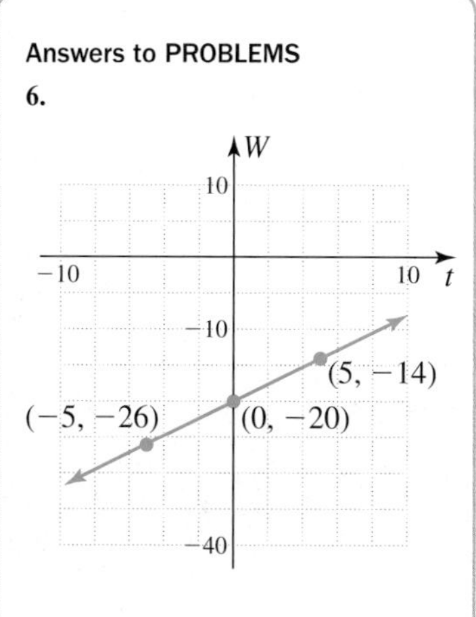

Calculator Corner

Graphing Lines

You can satisfy three of the objectives of this section using a calculator. We will show you how by graphing the equation $2x + 3y = 10$, determining whether the point (2, 2) satisfies the equation, and finding the x-coordinate in the ordered pair $(x, -6)$ using a calculator.

The graph of $2x + 3y = 10$ consists of all points satisfying $2x + 3y = 10$. To graph this equation, first solve for y obtaining

$$y = \frac{10}{3} - \frac{2x}{3}$$

If you want to determine whether (2, 2) is a solution of the equation, turn the statistical plot off (press 2nd Y= 1 and select OFF), then set the window for integers (ZOOM 8 ENTER), and press Y=. Then enter

$$Y_1 = \frac{10}{3} - \frac{2x}{3}$$

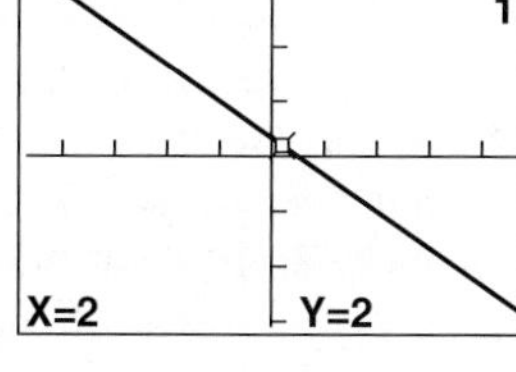

Window 1

Press GRAPH.

Use TRACE to move the cursor around. In Window 1, $x = 2$ and $y = 2$ are shown; thus, (2, 2) satisfies the equation. To find x in $(x, -6)$, use TRACE until you are at the point where $y = -6$; then read the value of x, which is 14. Try it! You can use these ideas to do Problems 1–16.

We do not want to leave you with the idea that all equations you will encounter are linear equations. Later in the book we shall study:

1. Absolute-value equations of the form $y = |x|$
2. Quadratic equations of the form $y = x^2$
3. Cubic equations of the form $y = x^3$

These types of equations can be graphed using the same procedure as that for graphing lines, with one notable exception: in step (3), you draw a smooth curve passing through the points. The results are shown here.

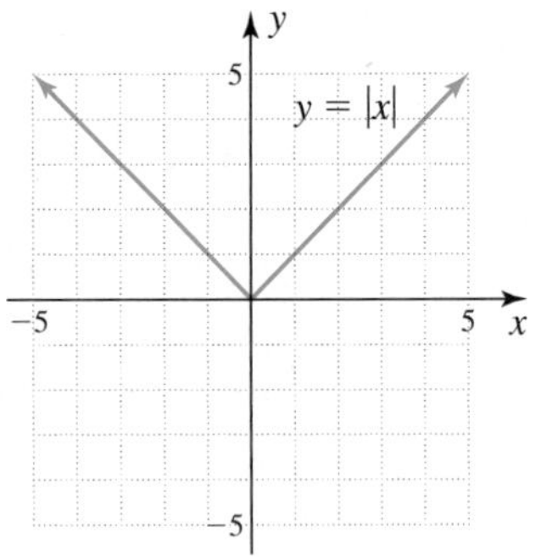

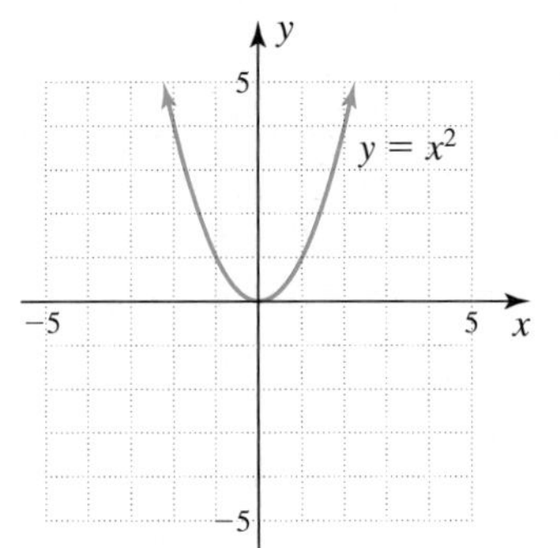

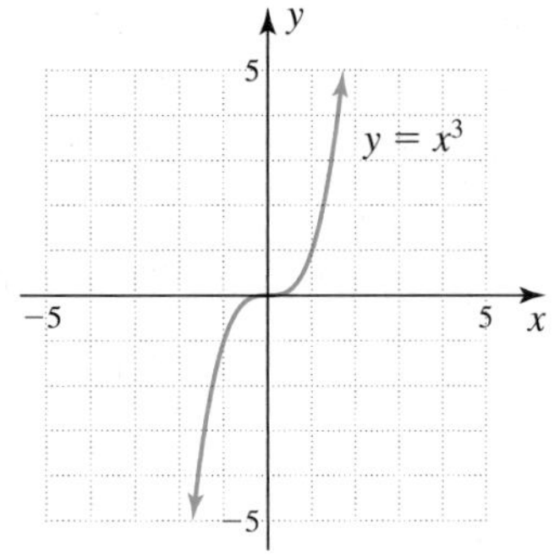

Exercises 3.2

Web IT go to mhhe.com/bello for more lessons

A Solutions of Equations In Problems 1–6, determine whether the ordered pair is a solution of the equation.

1. $(3, 2); x + 2y = 7$

2. $(4, 2); x - 3y = 2$

3. $(5, 3); 2x - 5y = -5$

4. $(-2, 1); -3x = 5y + 1$

5. $(2, 3); -5x = 2y + 4$

6. $(-1, 1); 4y = -2x + 2$

B Finding Missing Coordinates In Problems 7–16, find the missing coordinate.

7. (3, __) is a solution of $2x - y = 6$.

8. (−2, __) is a solution of $-3x + y = 8$.

9. (__, 2) is a solution of $3x + 2y = -2$.

10. (__, −5) is a solution of $x - y = 0$.

11. (0, __) is a solution of $3x - y = 3$.

12. (0, __) is a solution of $x - 2y = 8$.

13. (__, 0) is a solution of $2x - y = 6$.

14. (__, 0) is a solution of $-2x - y = 10$.

15. (−3, __) is a solution of $-2x + y = 8$.

16. (−5, __) is a solution of $-3x - 2y = 9$.

C Graphing Linear Equations by Plotting Points In Problems 17–40, graph the equation.

17. $2x + y = 4$

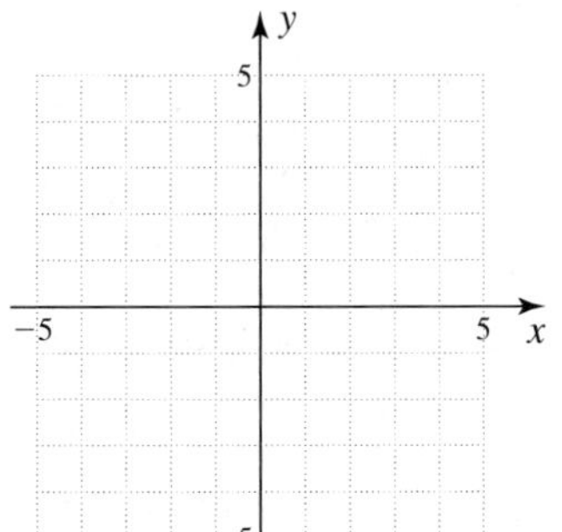

18. $y + 3x = 3$

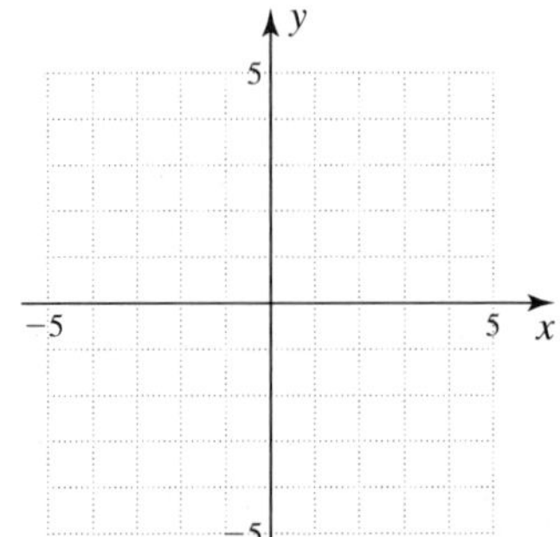

19. $-2x - 5y = -10$

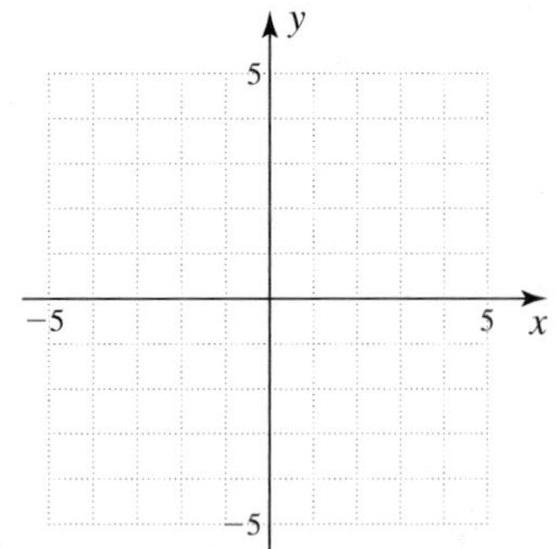

20. $-3x - 2y = -6$

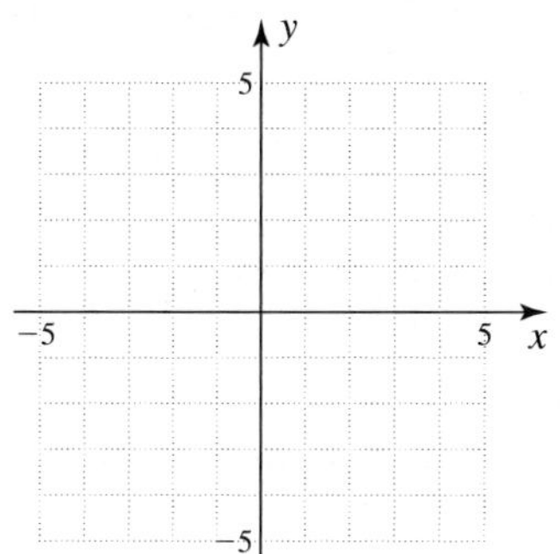

21. $y + 3 = 3x$

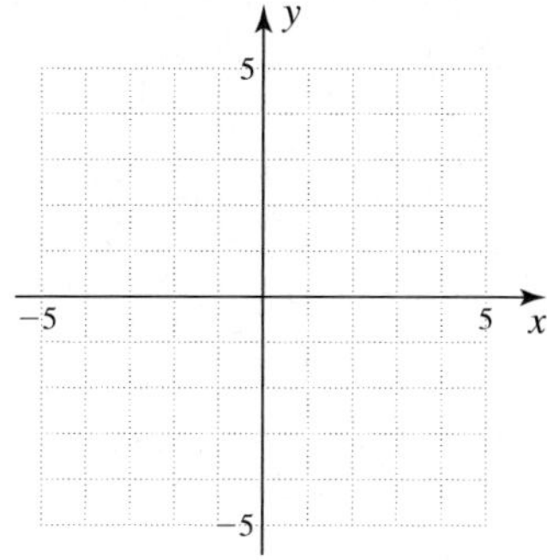

22. $y - 4 = -2x$

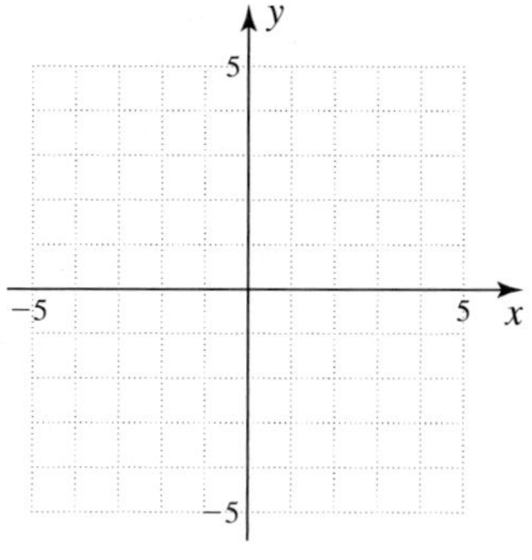

23. $6 = 3x - 6y$

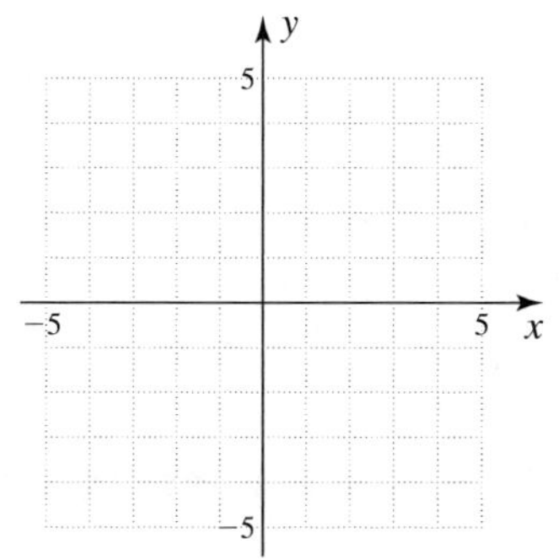

24. $6 = 2x - 3y$

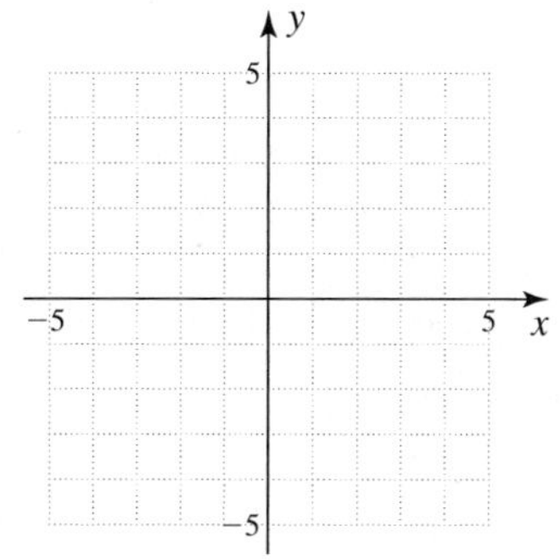

25. $-3y = 4x + 12$

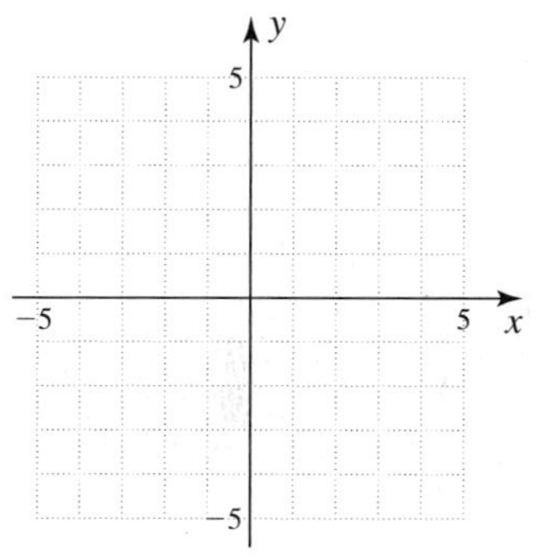

26. $-2x = 5y + 10$

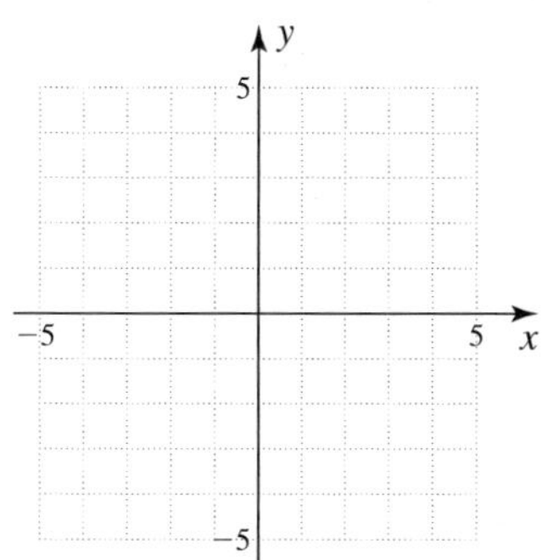

27. $-2y = -x + 4$

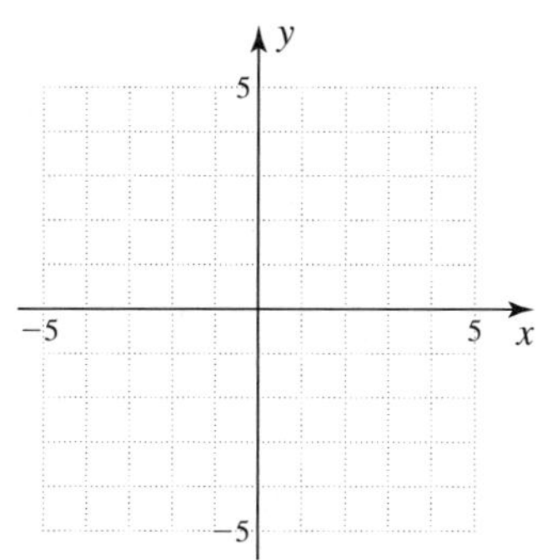

28. $-3y = -x + 6$

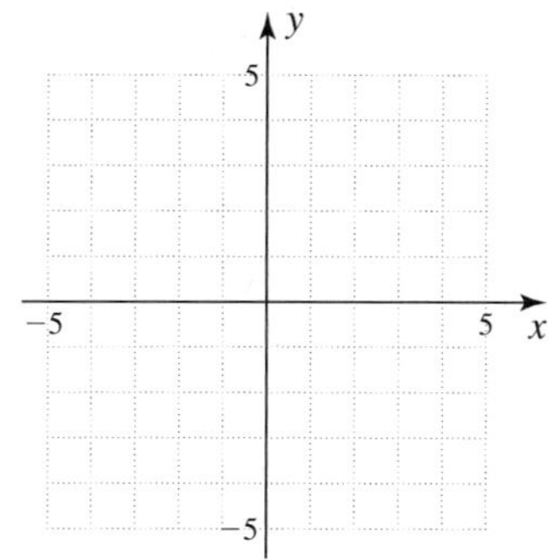

29. $-3y = -6x + 3$

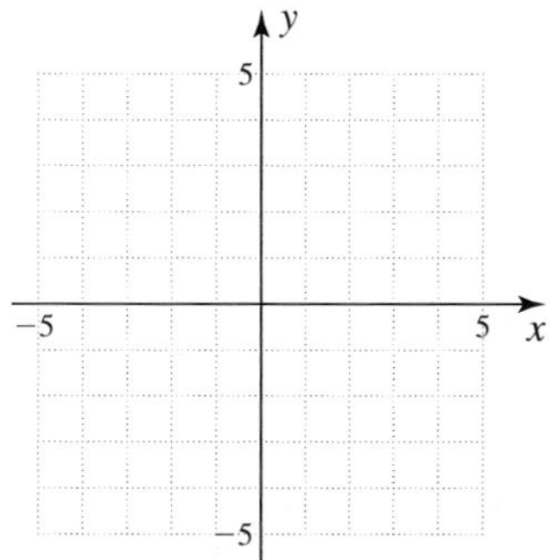

30. $-4y = -2x + 4$

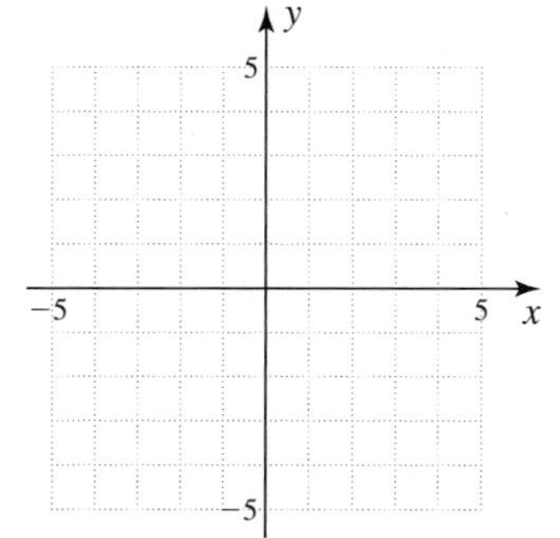

31. $y = 2x + 4$

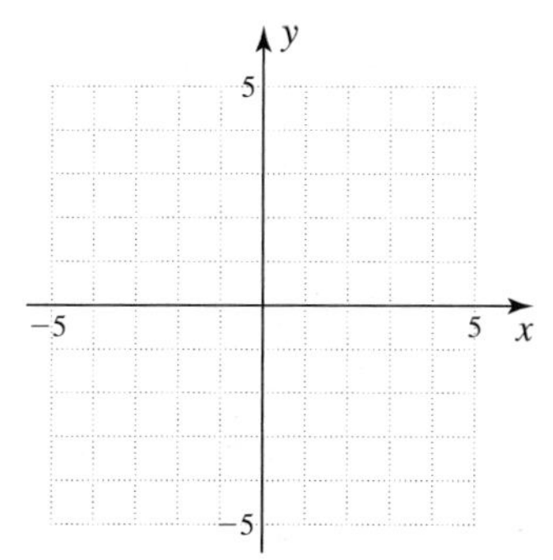

Web IT go to mhhe.com/bello for more lessons

32. $y = 3x + 6$

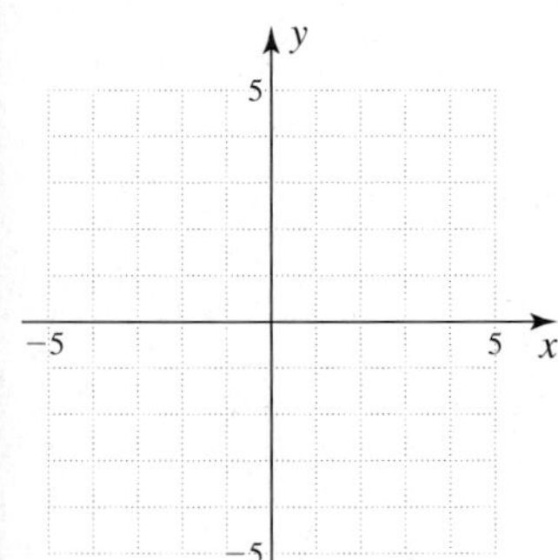

33. $y = -2x + 4$

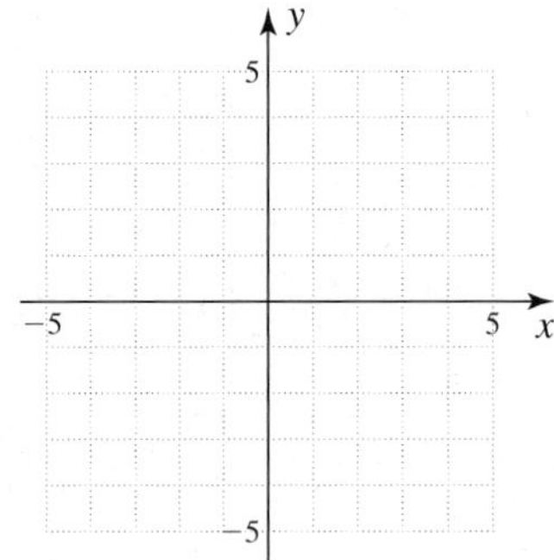

34. $y = -3x + 6$

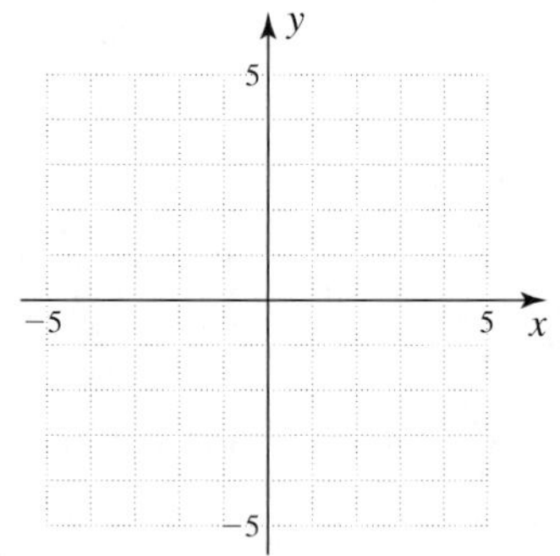

35. $y = -3x - 6$

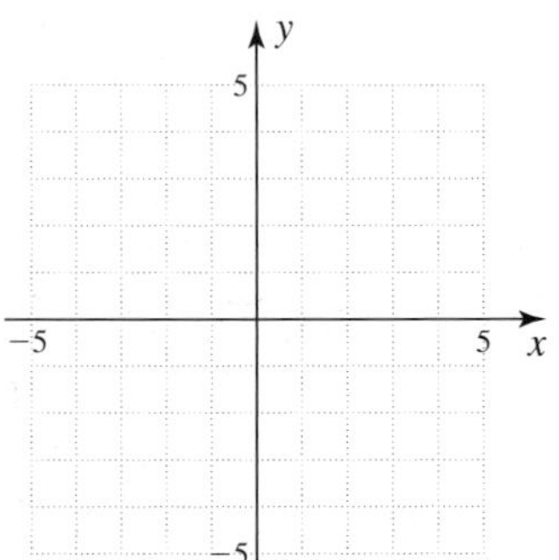

36. $y = -2x - 4$

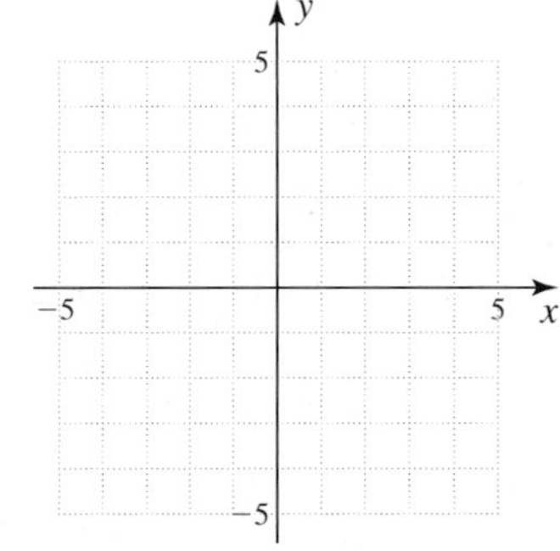

37. $y = \frac{1}{2}x - 2$

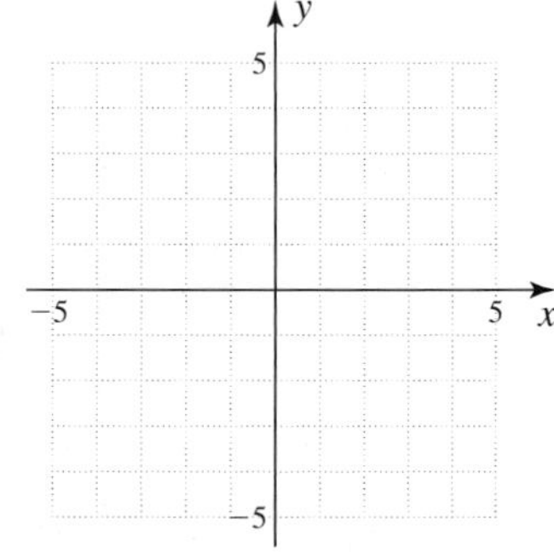

38. $y = \frac{1}{3}x - 1$

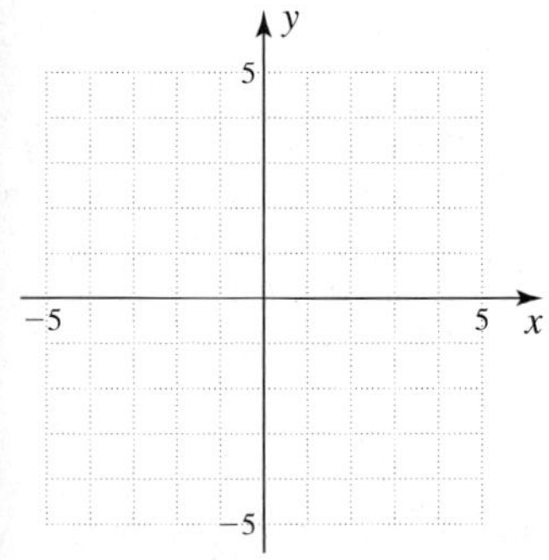

39. $y = -\frac{1}{2}x - 2$

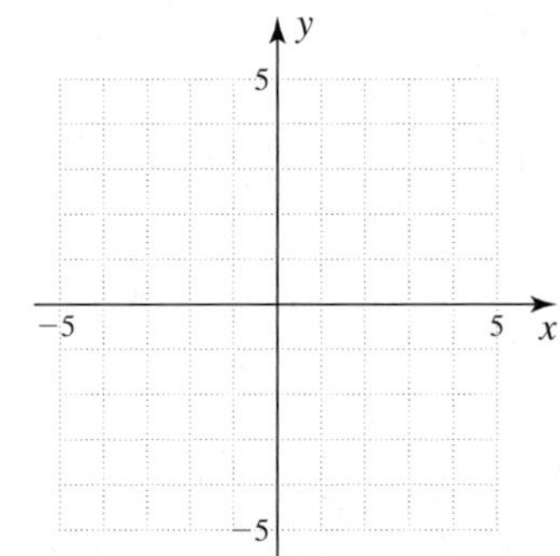

40. $y = -\frac{1}{3}x - 1$

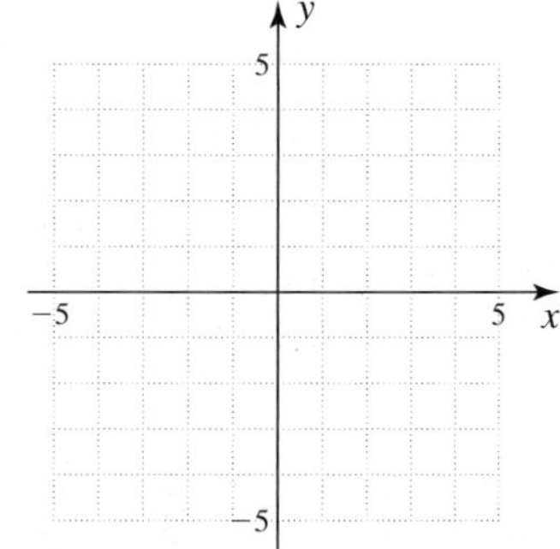

‹ D › Applications Involving Linear Equations

41. *Wind chill factor* When the wind is blowing at a constant 5 miles per hour, the wind chill factor W can be approximated by $W = 1.1t - 9$, where t is the temperature in °F and W is the wind chill factor.

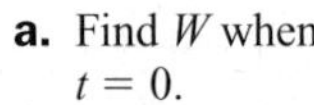

a. Find W when $t = 0$.

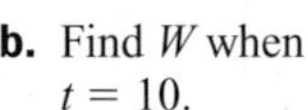

b. Find W when $t = 10$.

c. Graph W.

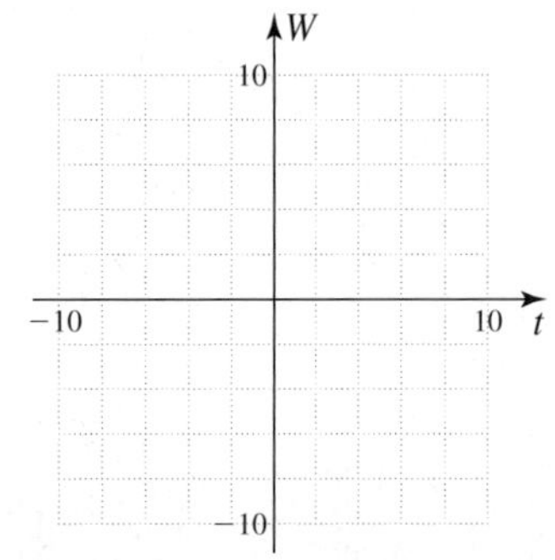

42. *Wind chill factor* When the wind is blowing at a constant 20 miles per hour, the wind chill factor W can be approximated by $W = 1.3t - 21$, where t is the temperature in °F and W is the wind chill factor.

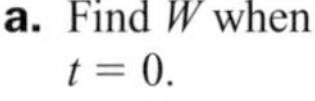

a. Find W when $t = 0$.

b. Find W when $t = 10$.

c. Graph W.

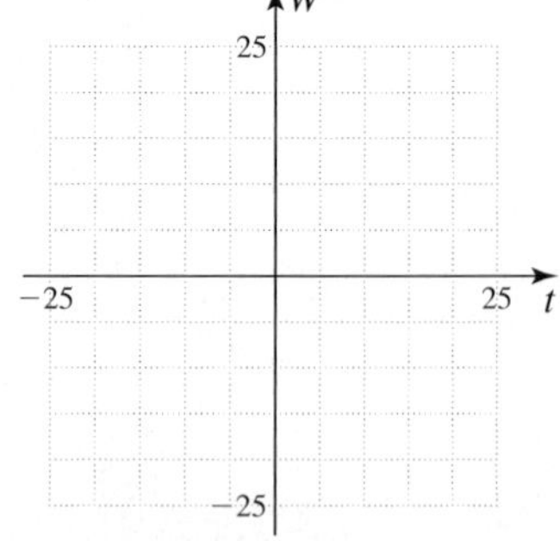

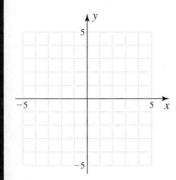

Applications: Green Math

According to the Intergovernmental Panel on Climate Change (IPCC) climate change will be noticeable by various impacts. For example, temperatures and precipitation will change, and sea levels will rise. We will discuss these changes in Problems 43 and 44.

43. If no additional steps are taken to reduce emissions of CO_2 and other problematic gases, then in 2040 the average global air temperature will be 1°C (one degree Celsius) higher than in 2000. In 2100 (100 years after 2000) the temperature will increase to 2.5°C.

Source: http://tinyurl.com/ydbfsbh.

a. Suppose x represents the year and y represents the change in temperature. Write the fact that "in 2040 the temperature will be 1°C higher" using an ordered pair of the form (x, y).

b. Write the fact that "in 2100, the temperature will increase 2.5°C" using an ordered pair of the form (x, y).

c. Graph the points from parts **a** and **b** on the grid shown.

d. The temperature increases can be represented by the equation $y = 0.025x - 50$, where x is the year and y is the temperature increase. Graph this equation on the grid.

Temperature Increases

44. Sea levels will rise by about 18 cm (centimeters) by 2040 and by 48 cm by 2100.

a. Suppose x represents the year and y represents the sea level change. Write the fact that "in 2040 the sea level will rise by 18 cm" using an ordered pair of the form (x, y).

b. Write the fact that "in 2100, the sea level will rise by 48 cm" using an ordered pair of the form (x, y).

c. Graph the points from parts **a** and **b** and the point (2000, 0) on the grid shown.

d. The sea level increases can be represented by the equation $y = 0.5x - 1000$, where x is year and y is the sea level increase. Graph this equation on the grid.

Sea Level Increases

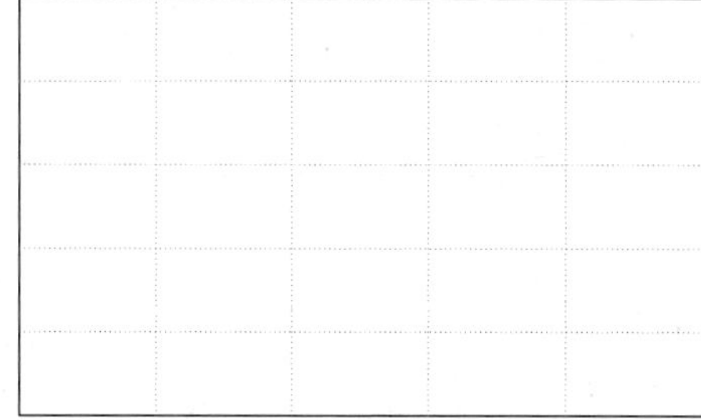

45. *Environmental lapse* Suppose the temperature is 60° Fahrenheit (60°F). As your altitude increases, the temperature y at x (thousand) feet above sea level is given by the equation

$$y = -4x + 60$$

a. What is the temperature at **1** (thousand) feet?

b. What is the temperature at **10** (thousand) feet?

c. What is the temperature at sea level ($x = 0$)?

d. Graph $y = -4x + 60$ using the points you obtained in parts **a**, **b**, and **c**.

e. What happens to the temperature if your altitude is more than **15** (thousand) feet?

Temperature above Sea Level

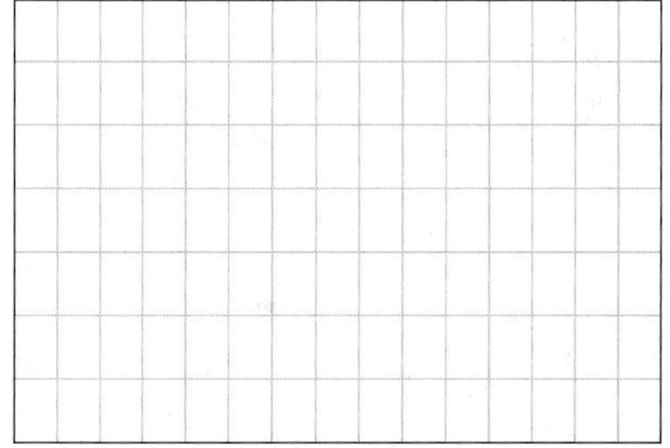

46. *Environmental lapse* Suppose the temperature is 16° Celsius (16°C). As your altitude increases, the temperature y at x kilometers above sea level is given by the equation

$$y = -7x + 14$$

a. What is the temperature when the altitude is **1** kilometer?

b. What is the temperature when the altitude is **2** kilometers?

c. What is the temperature at sea level ($x = 0$)?

d. Graph $y = -7x + 14$ using the points you obtained in parts **a**, **b**, and **c**.

e. What happens to the temperature if your altitude is more than **2** kilometers?

Temperature above Sea Level

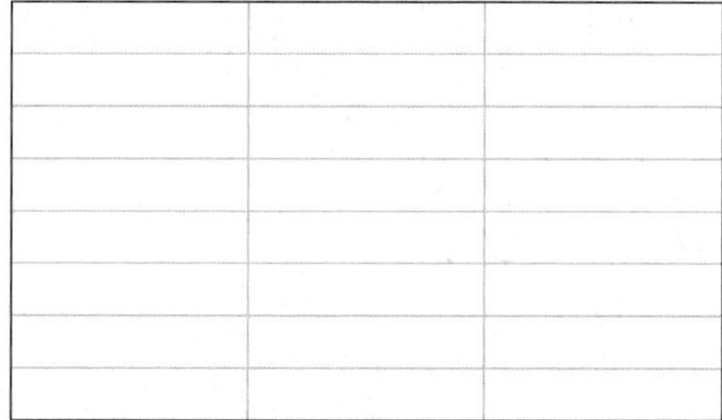

Web IT go to mhhe.com/bello for more lessons

47. *Catering prices* A caterer charges \$40 per person plus a \$200 setup fee. The total cost C for an event can be represented by the equation

$$C = 40x + 200$$

where x is the number of people attending. Assume that fewer than 100 people will be attending.

a. What is the cost when 30 people are attending?

b. What is the cost when 50 people are attending?

c. What is the cost for just setting up? ($x = 0$)

d. Graph $C = 40x + 200$ using the points you obtained in parts **a, b,** and **c.**

Catering Costs

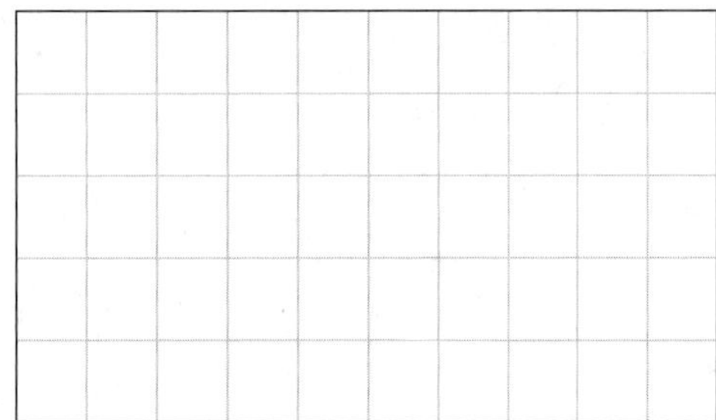

48. *Service staff cost* The service staff for a reception charges \$30 per hour, with a 5-hour minimum (\$150). The cost C can be represented by the equation

$$C = 30h + 150$$

where h is the number of hours beyond 5. Assume your reception is shorter than 8 hours.

a. What is the cost from 0 to 5 hours?

b. What is the cost for 6 hours?

c. What is the cost for 8 hours?

d. Graph $C = 30h + 150$ using the points obtained in parts **a, b,** and **c.**

Service Staff Costs

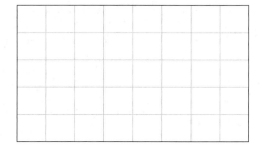

Using Your Knowledge

Anthropology You already know how to determine whether an ordered pair satisfies an equation. Anthropological detectives use this knowledge to estimate the living height of a person using one dried bone as a clue. Suppose a detective finds a 17.9-inch femur bone from a male. To find the height H (in inches) of its owner, use the formula $H = 1.88f + 32.010$ and determine that the owner's height H must have been $H = 1.88(17.9) + 32.010 \approx 66$ inches. Of course, the ordered pair (17.9, 66) satisfies the equation. Now, for the fun part. Suppose you find a 17.9-inch femur bone, but this time you are looking for a missing female 66 inches tall. Can this femur belong to her? No! How do we know? The ordered pair (17.9, 66) *does not* satisfy the equation for female femur bones, which is $H = 1.945f + 28$. Try it! Now, use the formulas to work Problems 49–54.

Living Height (inches)

Male	Female
$H = 1.880f + 32.010$	$H = 1.945f + 28.379$
$H = 2.894h + 27.811$	$H = 2.754h + 28.140$
$H = 3.271r + 33.829$	$H = 3.343r + 31.978$
$H = 2.376t + 30.970$	$H = 2.352t + 29.439$

where f, h, r, and t represent the length of the femur, humerus, radius, and tibia bones, respectively.

49. An 18-inch humerus bone from a male subject has been found. Can the bone belong to a 6′8″ basketball player missing for several weeks?

50. A 16-inch humerus bone from a female subject has been found. Can the bone belong to a missing 5′2″ female student?

51. A radius bone measuring 14 inches is found. Can it belong to Sandy Allen, the world's tallest living woman? (She is 7 feet tall!) How long to the nearest inch should the radius bone be for it to be Sandy Allen's bone?

Source: Guinness World Records.

52. The tallest woman in medical history was Zeng Jinlian. A tibia bone measuring 29 inches is claimed to be hers. If the claim proves to be true, how tall was she, and what ordered pair would satisfy the equation relating the length of the tibia and the height of a woman?

53. A 10-inch radius of a man is found. Can it belong to a man 66.5 inches tall?

54. The longest recorded bone—an amazing 29.9 inches long—is the femur of the German giant Constantine who died in 1902. How tall was he, and what ordered pair would satisfy the equation relating the length of his femur and his height?

Write On

55. Write the procedure you use to show that an ordered pair (a, b) satisfies an equation.

56. Write the procedure you use to determine whether the graph of an equation is a straight line.

57. Write the procedure you use to graph a line.

58. In your own words, write what "wind chill factor" means.

Concept Checker

Fill in the blank(s) with the correct word(s), phrase, or mathematical statement.

59. The **graph** of $Ax + By = C$, where A, B, and C are constants, is the graph of a ______________ line.

60. The **point** at which a line **crosses the** y**-axis** is called the y-__________ and the **point** at which a line **crosses the** x**-axis** is called the x-__________.

straight

intercept

Mastery Test

61. Determine whether the ordered pair is a solution of $3x + 2y = 10$.

a. $(1, 2)$

b. $(2, 1)$

62. Graph: $2x + y = 6$

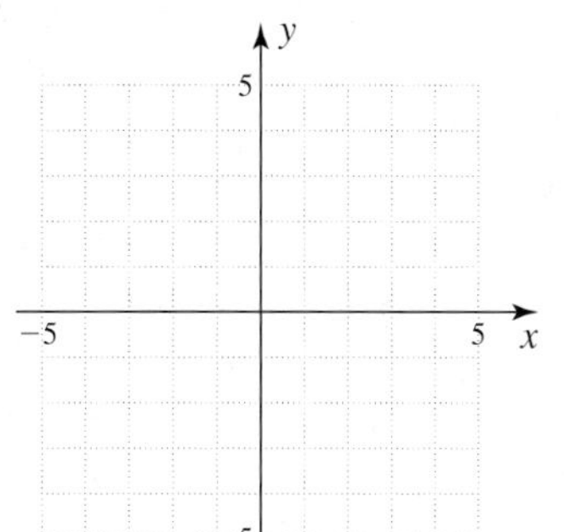

63. Graph: $y = -3x + 6$

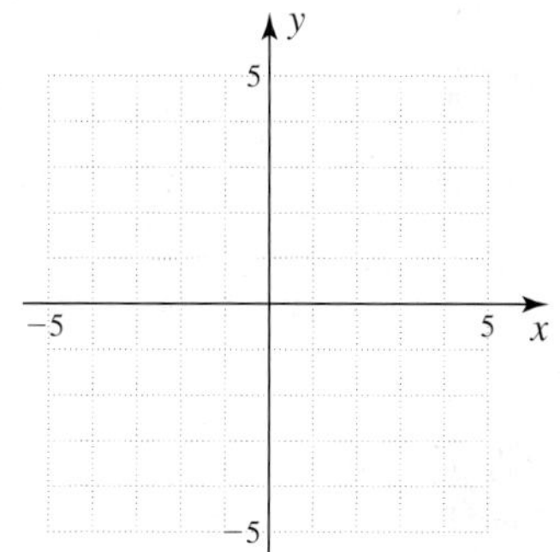

64. Complete the ordered pairs so they satisfy the equation $y = 3x + 2$.

a. $(x, 5)$

b. $(-3, y)$

65. When the wind is blowing at a constant 10 miles per hour, the wind chill factor W can be approximated by $W = 1.3t - 18$, where t is the temperature in °F and W is the wind chill factor.

a. Find W when $t = 0$.

b. Find W when $t = 10$.

c. Graph W.

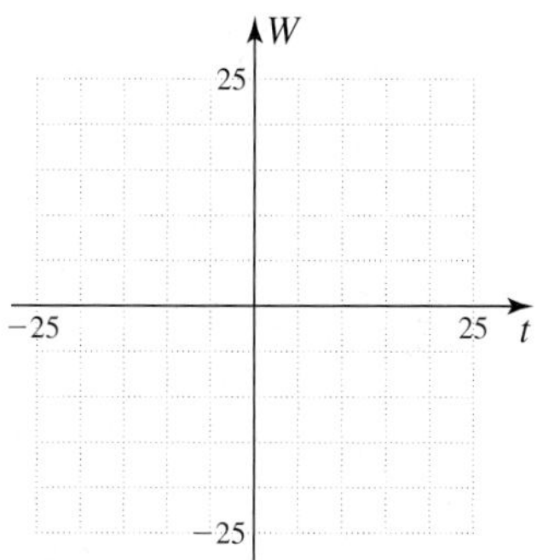

Skill Checker

66. Solve: $2x - 6 = 0$

67. Solve: $2x + 4 = 0$

68. Solve: $3 + 3y = 0$

69. Solve: $0 = 0 - 0.1t$

70. Solve: $0 = 8 - 0.1t$

3.3 Graphing Lines Using Intercepts: Horizontal and Vertical Lines

Objectives

A Graph lines using intercepts.

B Graphs lines passing through the origin.

C Graph horizontal and vertical lines.

D Solve applications involving graphs of lines.

To Succeed, Review How To . . .

1. Solve linear equations (pp. 137–141).
2. Graph ordered pairs of numbers (p. 211).

Getting Started

Education Costs

How much are you paying for your education? The graph shows the tuition and fees in different types of institutions from 1978–1979 to 2008–2009. Which are the least and most expensive? What is the cost of tuition and fees for 08–09? The graph indicates that for 08–09 the cost for private 4-year institutions is \$25,143, for public 4-year institutions \$6585, and for public 2-year institutions \$2402, so private 4-year institutions are the most expensive, and public 2-year institutions are the least expensive. As you can see, the cost for private 4-year institutions is increasing the fastest (the line is steeper), while the cost for public 2-year institutions remains almost steady (horizontal) at about \$2400. It can be approximated by $t = 2400$. From this information, we can predict college costs for the immediate future. In this section, we learn how to graph horizontal and vertical lines and graphs that start from the vertical axis.

Average Published Tuition and Fees

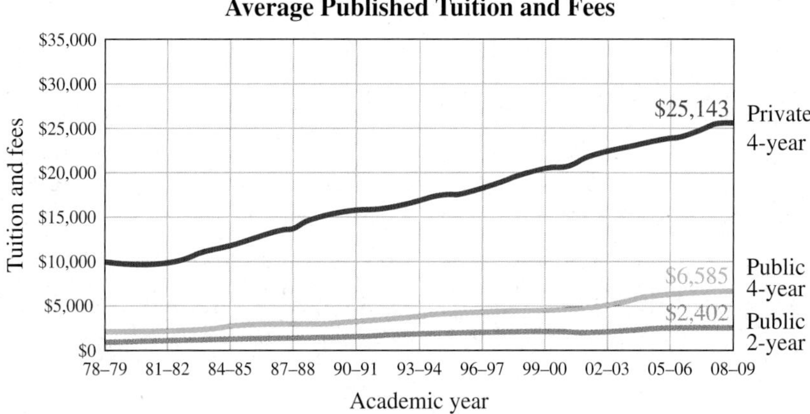

Source: http://professionals.collegeboard.com/profdownload/trends-in-student-aid-2008.pdf.

A > Graphing Lines Using Intercepts

In Section 3.2 we mentioned that the graph of a linear equation of the form $Ax + By = C$ is a *straight line,* and we graphed such lines by finding ordered pairs that satisfied the equation of the line. But there is a quicker and simpler way to graph these equations by using the idea of *intercepts.* Here are the definitions we need.

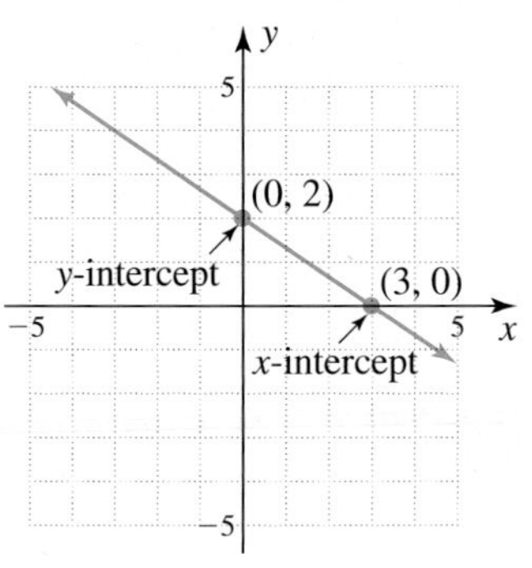

>Figure 3.16

PROCEDURE

1. The ***x*-intercept** $(a, 0)$ of a line is the point at which the line crosses the x-axis. To find the x-intercept, let $y = 0$ and solve for x.
2. The ***y*-intercept** $(0, b)$ of a line is the point at which the line crosses the y-axis. To find the y-intercept, let $x = 0$ and solve for y.

The graph in Figure 3.16 shows a line with y-intercept (0, 2) and x-intercept (3, 0). How do we use these ideas to graph lines? Here is the procedure.

PROCEDURE

To Graph a Line Using the Intercepts

1. Find the x-intercept $(a, 0)$.
2. Find the y-intercept $(0, b)$.
3. Graph the points $(a, 0)$ and $(0, b)$, and connect them with a line.
4. Find a third point to use as a check (make sure the point is on the line!).

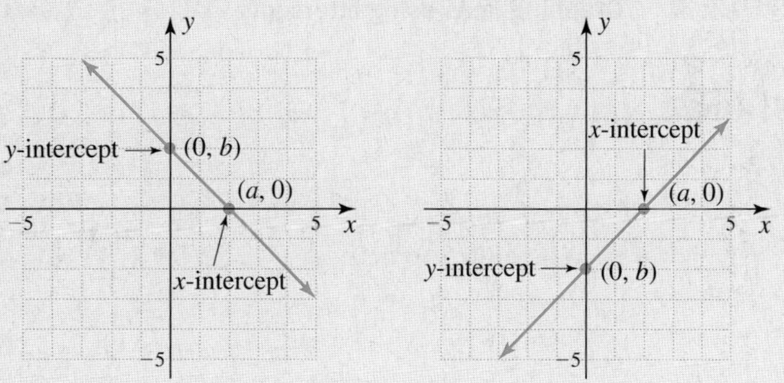

EXAMPLE 1 Graphing lines using intercepts

Graph: $5x + 2y = 10$

SOLUTION 1 First we find the x- and y-intercepts.

1. Let $x = 0$ in $5x + 2y = 10$. Then

$$5(0) + 2y = 10$$
$$2y = 10$$
$$y = 5$$

Hence, (0, 5) is the y-intercept.

2. Let $y = 0$ in $5x + 2y = 10$. Then

$$5x + 2(0) = 10$$
$$5x = 10$$
$$x = 2$$

Hence, (2, 0) is the x-intercept.

3. Now we graph the points (0, 5) and (2, 0) and connect them with a line as shown in Figure 3.17.

PROBLEM 1

Graph: $2x + 5y = 10$

Answers to PROBLEMS

1.

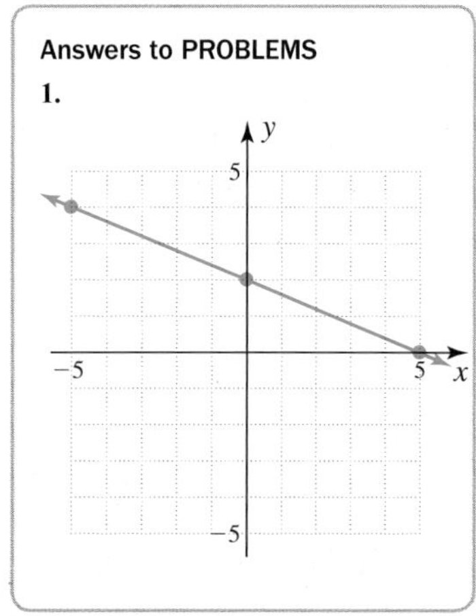

(continued)

4. We need to find a *third* point to use as a *check:* we let $x = 4$ and replace x with 4 in

$$5x + 2y = 10$$
$$5(4) + 2y = 10$$
$$20 + 2y = 10$$
$$2y = -10 \quad \text{Subtract 20.}$$
$$y = -5$$

The three points (0, 5), (2, 0), and (4, −5) as well as the completed graph are shown in Figure 3.17.

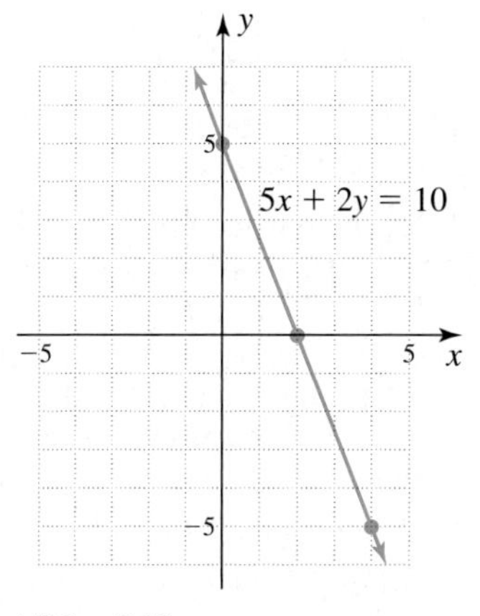

>**Figure 3.17**

The procedure used to graph a line using the intercepts works even when the line is *not* written in the form $Ax + By = C$, as we will show in Example 2.

EXAMPLE 2 Graphing lines using intercepts

Graph: $2x - 3y - 6 = 0$

SOLUTION 2

1. Let $y = 0$. Then

$$2x - 3(0) - 6 = 0$$
$$2x - 6 = 0$$
$$2x = 6 \quad \text{Add 6.}$$
$$x = 3 \quad \text{Divide by 2.}$$

(3, 0) is the x-intercept.

2. Let $x = 0$. Then

$$2(0) - 3y - 6 = 0$$
$$-3y - 6 = 0$$
$$-3y = 6 \quad \text{Add 6.}$$
$$y = -2 \quad \text{Divide by } -3.$$

(0, −2) is the y-intercept.

3. Connect (0, −2) and (3, 0) with a line, the graph of $2x - 3y - 6 = 0$, as shown in Figure 3.18.
4. Use a third point as a check. Examine the line. It seems that for $x = -3$, y is −4. Let us check.

If $x = -3$, we have

$$2(-3) - 3y - 6 = 0$$
$$-6 - 3y - 6 = 0$$
$$-3y - 12 = 0 \quad \text{Simplify.}$$
$$-3y = 12 \quad \text{Add 12.}$$
$$y = -4 \quad \text{Divide by } -3.$$

As we suspected, the resulting point (−3, −4) is on the line. Do you see why we picked $x = -3$? We did so because −3 seemed to be the only x-coordinate that yielded a y-value that was an integer.

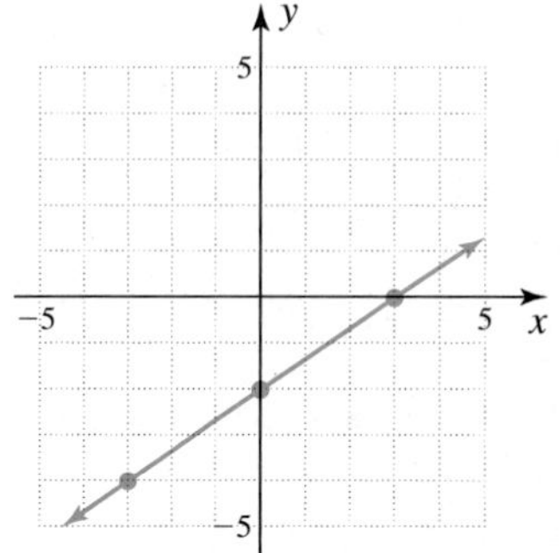

>**Figure 3.18**

PROBLEM 2

Graph: $3x - 2y - 6 = 0$

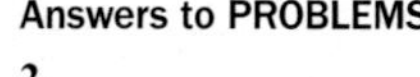

Answers to PROBLEMS

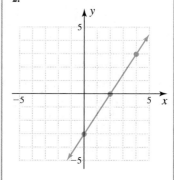

2.

B › Graphing Lines Through the Origin

Sometimes it isn't possible to use the x- and y-intercepts to graph an equation. For example, to graph $x + 5y = 0$, we can start by letting $x = 0$ to obtain $0 + 5y = 0$ or $y = 0$. This means that (0, 0) is part of the graph of the line $x + 5y = 0$. If we now let $y = 0$, we get $x = 0$, which is the same ordered pair. The line $x + 5y = 0$ goes through the origin (0, 0), as seen in Figure 3.19, so we need to find another point. An easy one is $x = 5$. When 5 is substituted for x in $x + 5y = 0$, we obtain

$$5 + 5y = 0 \quad \text{or} \quad y = -1$$

Thus, a second point on the graph is (5, −1). We join the points (0, 0) and (5, −1) to get the graph of the line shown in Figure 3.19. To use a third point as a check, let $x = -5$, which gives $y = 1$ and the point (−5, 1), also shown on the line.

Here is the procedure for recognizing and graphing lines that go through the origin.

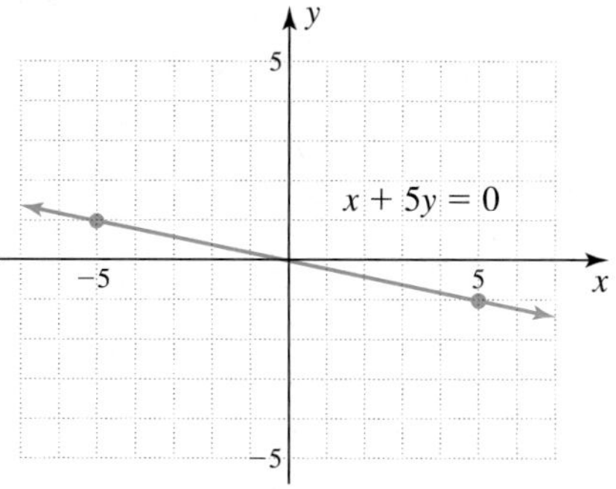

>Figure 3.19

STRAIGHT-LINE GRAPH THROUGH THE ORIGIN

The graph of an equation of the form $Ax + By = 0$, where A and B are constants and not equal to zero, is a straight line that goes through the origin.

PROCEDURE

Graphing Lines Through the Origin

To graph a line through the origin, use the point (0, 0), find another point, and draw the line passing through (0, 0) and this other point. Find a third point and verify that it is on the graph of the line.

EXAMPLE 3 **Graphing lines passing through the origin**

Graph: $2x + y = 0$

SOLUTION 3 The line $2x + y = 0$ is of the form $Ax + By = 0$ so it goes through the origin (0, 0). To find another point on the line, we let $y = 4$ in $2x + y = 0$:

$$2x + 4 = 0$$
$$x = -2$$

Now we join the points (0, 0) and (−2, 4) with a line, the graph of $2x + y = 0$, as shown in Figure 3.20. Note that the check point (2, −4) is on the graph.

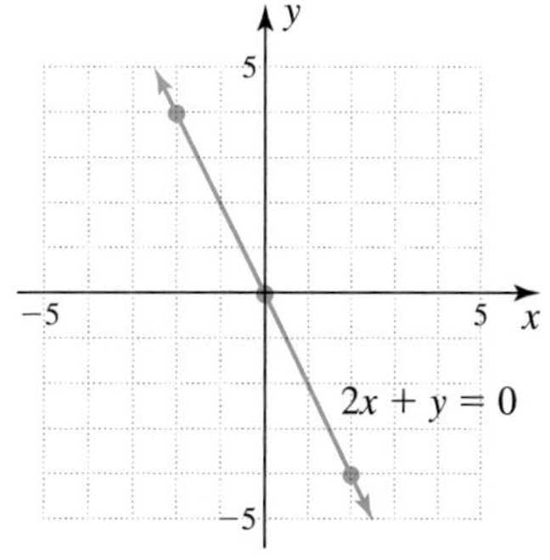

>Figure 3.20

PROBLEM 3

Graph: $3x + y = 0$

Answers to PROBLEMS

3.

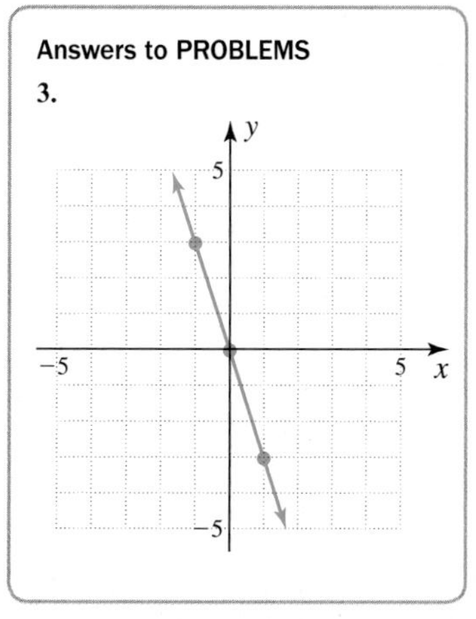

Here is a summary of the techniques used to graph lines that are *not* vertical or horizontal.

To graph $Ax + By = C$ (A, B, and C not 0)	**To graph $Ax + By = 0$ (A and B not 0)**
Find two points on the graph (preferably the x- and y-intercepts) and join them with a line. The result is the graph of $Ax + By = C$.	The graph goes through (0, 0). Find another point and join (0, 0) and the point with a line. The result is the graph of $Ax + By = 0$.
To graph $2x + y = -4$: Let $x = 0$, find $y = -4$, and graph $(0, -4)$. Let $y = 0$, find $x = -2$, and graph $(-2, 0)$.	To graph $x + 4y = 0$, let $x = -4$ in $x + 4y = 0$ $-4 + 4y = 0$ $y = 1$
Join $(0, -4)$ and $(-2, 0)$ to obtain the graph. Use an extra point to check. For example, if $x = -4$, $2(-4) + y = -4$, and $y = 4$. The point $(-4, 4)$ is on the line, so our graph is correct!	Join (0, 0) and $(-4, 1)$ to obtain the graph. Use $x = 4$ as a check. When $x = 4$, $4 + 4y = 0$, and $y = -1$. The point $(4, -1)$ is on the line, so our graph is correct!
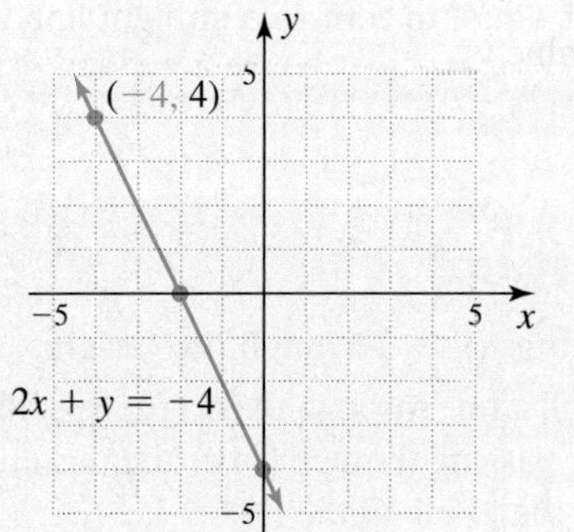	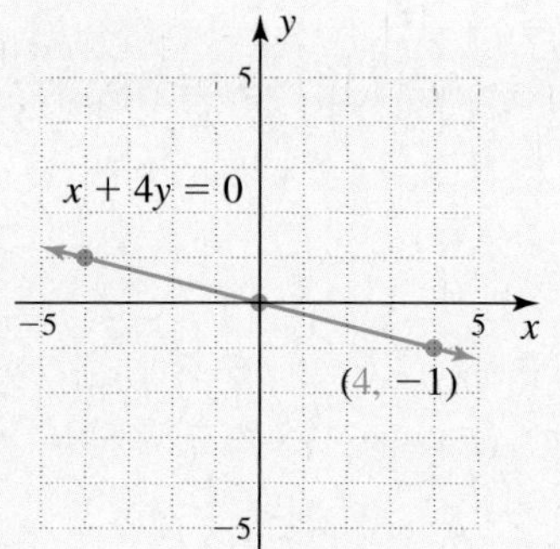

C › Graphing Horizontal and Vertical Lines

Not all linear equations are written in the form $Ax + By = C$. For example, consider the equation $y = 3$. It may seem that this equation is not in the form $Ax + By = C$. However, we can write the equation $y = 3$ as

$$0 \cdot x + y = 3$$

which is an equation written in the desired form. How do we graph the equation $y = 3$? Since it doesn't matter what value we give x, the result is always $y = 3$, as can be seen in the following table:

x	y
0	3
1	3
2	3

In the equation $0 \cdot x + y = 3$, x can be any number; you always get $y = 3$.

These ordered pairs, as well as the graph of $y = 3$, appear in Figure 3.21.

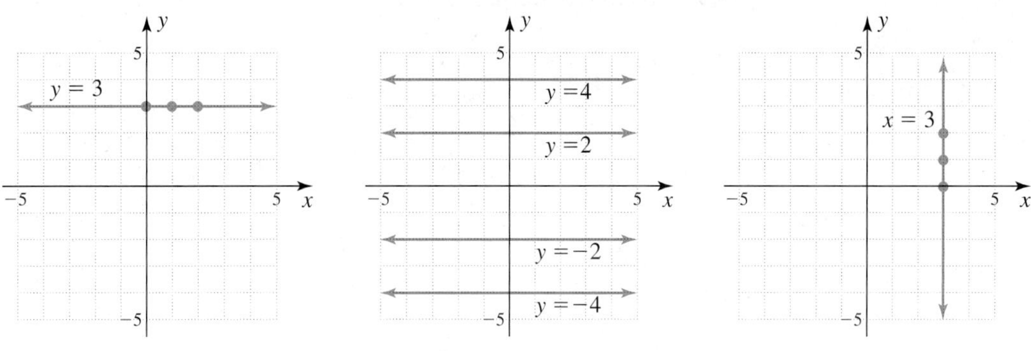

>Figure 3.21 >Figure 3.22 >Figure 3.23

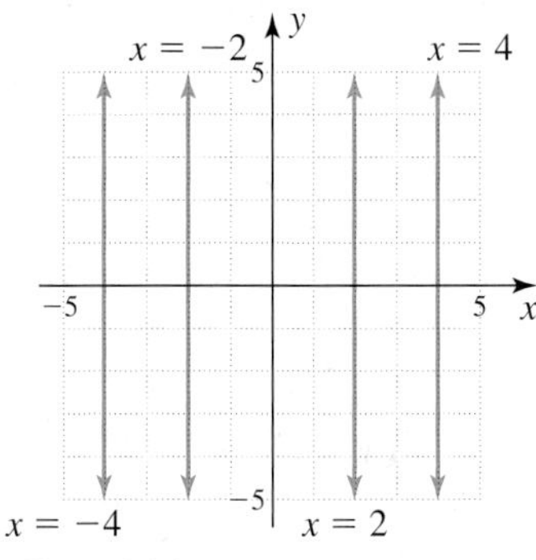

>Figure 3.24

Note that the equation $y = 3$ has for its graph a *horizontal* line crossing the y-axis at $y = 3$. The graphs of some other horizontal lines, all of which have equations of the form $y = k$ (k a constant), appear in Figure 3.22.

If the graph of any equation $y = k$ is a horizontal line, what would the graph of the equation $x = 3$ be? A vertical line, of course! We first note that the equation $x = 3$ can be written as

$$x + 0 \cdot y = 3$$

Thus, the equation is of the form $Ax + By = C$ so that its graph is a straight line. Now for any value of y, the value of x remains 3. Three values of x and the corresponding y-values appear in the following table. These three points, as well as the completed graph, are shown in Figure 3.23. The graphs of other vertical lines $x = -4$, $x = -2$, $x = 2$, and $x = 4$ are given in Figure 3.24.

x	y
3	0
3	1
3	2

Here are the formal definitions.

GRAPH OF $y = k$

The graph of any equation of the form

$y = k$ where k is a constant

is a horizontal line crossing the y-axis at k.

GRAPH OF $x = k$

The graph of any equation of the form

$x = k$ where k is a constant

is a vertical line crossing the x-axis at k.

EXAMPLE 4 Graphing vertical and horizontal lines

Graph: **a.** $2x - 4 = 0$ **b.** $3 + 3y = 0$

SOLUTION 4

a. We first solve for x.

$$\begin{aligned} 2x - 4 &= 0 && \text{Given} \\ 2x &= 4 && \text{Add 4.} \\ x &= 2 && \text{Divide by 2.} \end{aligned}$$

The graph of $x = 2$ is a vertical line crossing the x-axis at 2, as shown in Figure 3.25.

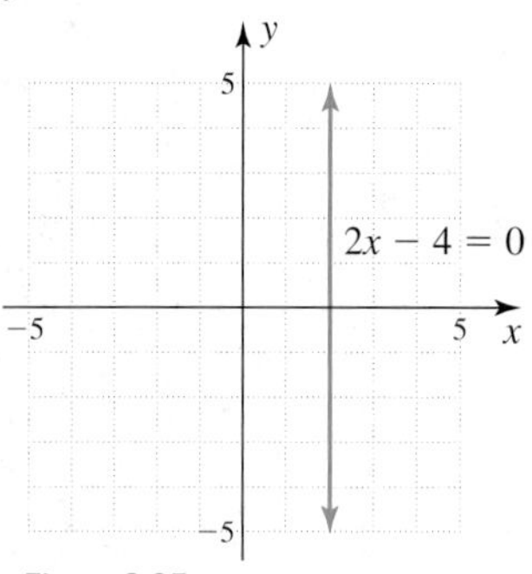

>Figure 3.25

b. In this case, we solve for y.

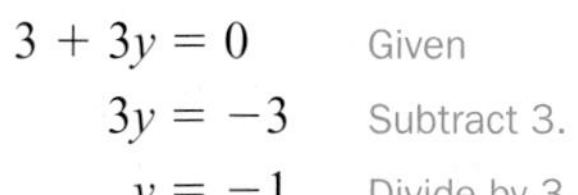

$$\begin{aligned} 3 + 3y &= 0 && \text{Given} \\ 3y &= -3 && \text{Subtract 3.} \\ y &= -1 && \text{Divide by 3.} \end{aligned}$$

The graph of $y = -1$ is a horizontal line crossing the y-axis at -1, as shown in Figure 3.26.

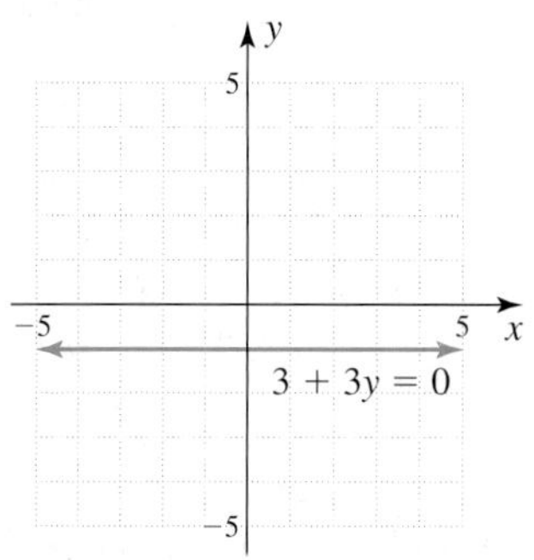

>Figure 3.26

PROBLEM 4

Graph:

a. $3x - 3 = 0$

b. $4 + 2y = 0$

Answers to PROBLEMS

4.

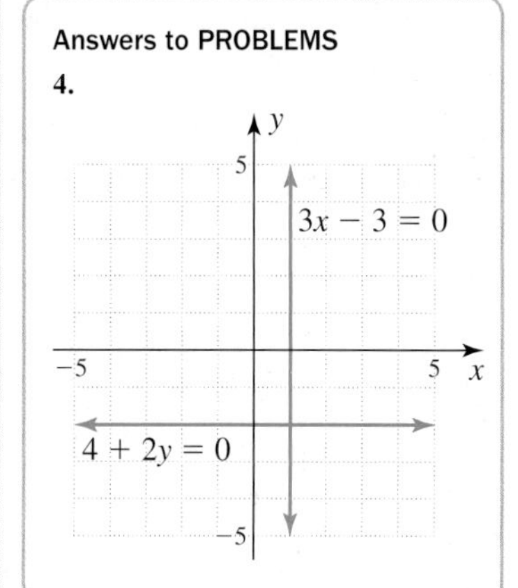

D > Applications Involving Graphs of Lines

As we saw in the *Getting Started,* graphing a linear equation in two variables can be very useful in solving real-world problems. Here is another example.

In Example 7 of Section 3.1, we graphed the points corresponding to the original and revised CAFE (Corporate Average Fuel Economy) regulations. In Example 5 we will graph lines corresponding to these points.

GREEN MATH

EXAMPLE 5 CAFE standards graph

The equation corresponding to the original CAFE standards is $E = 1.2N + 26$, where N is the number of years after 2010. Graph this equation by connecting the E-intercept and the point corresponding to $N = 5$ with a line.

Answers to PROBLEMS

5.

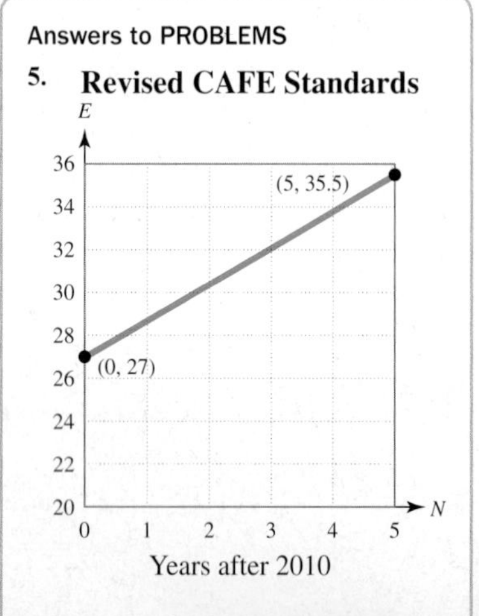

PROBLEM 5

The equation corresponding to the revised CAFE standards is $E = 1.7N + 27$, where N is the number of years after 2010. Graph this equation by connecting the E-intercept and the point corresponding to $N = 5$ with a line.

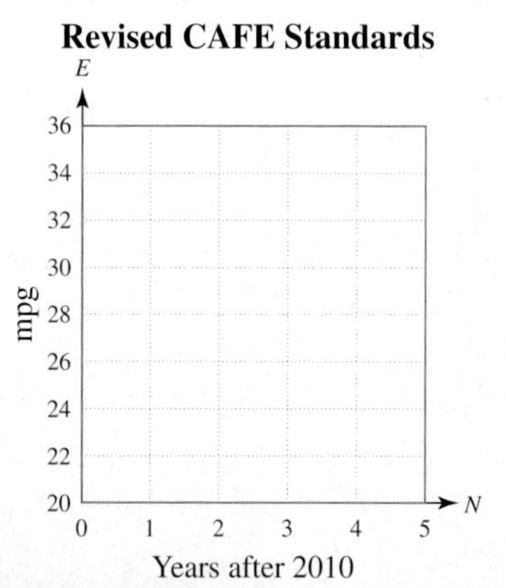

SOLUTION 5 We use the values for the year after 2010 (1, 2, 3, 4, 5) as the N-axis and the miles per gallon (from 25 to 35) as the E-axis, as shown in Figure 3.27. We find the E-intercept by letting $N = \mathbf{0}$, obtaining $E = 1.2(\mathbf{0}) + 26$. Thus, (0, 26) is the E-intercept. Graph the point (0, 26). For $N = \mathbf{5}$, $E = 1.2(\mathbf{5}) + 26 = 32$. Graph the corresponding point (5, 32). Now, join the points (0, 26) and (5, 32) with a line, the graph of the equation $E = 1.2N + 26$.

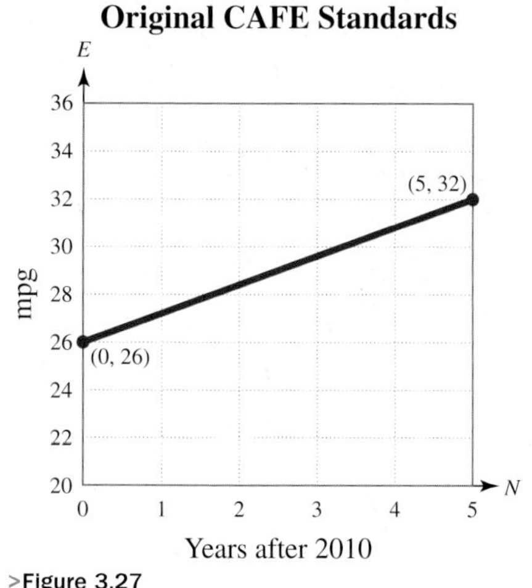

>Figure 3.27

Note: The corresponding value for $N = 1$ in the graph is about 27, which agrees with the value (2011, 27) in Example 7 of Section 3.1. Check the values in the graph for $N = 2$, 3, 4, and 5 and see how close you are to the ordered pairs in the table in Example 7 of Section 3.1.

Calculator Corner

Graphing Lines

Your calculator can easily do the examples in this section, but you have to *know* how to solve for y to enter the given equations. (Even with a calculator, you have to know algebra!) Then to graph $3x + y = 6$ (Example 4 in Section 3.2), we first solve for y to obtain $y = 6 - 3x$; thus, we enter $Y_1 = 6 - 3x$ and press GRAPH to obtain the result shown in Window 1. We used the *default* or *standard* window, which is a $[-10, 10]$ by $[-10, 10]$ rectangle. If you have a calculator, do Examples 2 and 3 of Section 3.3 now. Note that you can graph $3 + 3y = 0$ (Example 4b) by solving for y to obtain $y = -1$, but you *cannot* graph $2x - 4 = 0$ with most calculators. Can you see why?

To graph some other equations, you have to select the appropriate window. Let's see why. Suppose you simply enter $d = 9 - 0.1t$ by using y instead of d and x instead of t. The result is shown in Window 2. To see more of the graph, you have to adjust the window. Algebraically, we know that the t-intercept is (90, 0) and the d-intercept is (0, 9); thus, an appropriate window might be [0, 90] by [0, 10]. Now you only have to select the scales for x and y. Since the x's go from 0 to 90, make Xscl = 10. The y's go from 0 to 10, so we can let Yscl = 1. (See Window 3.) The completed graph appears with a scale that allows us to see most of the graph in Window 4. Press GRAPH to see it! Now do you see how the equation $d = 9 - 0.1t$ can be graphed and why it's graphed in quadrant I?

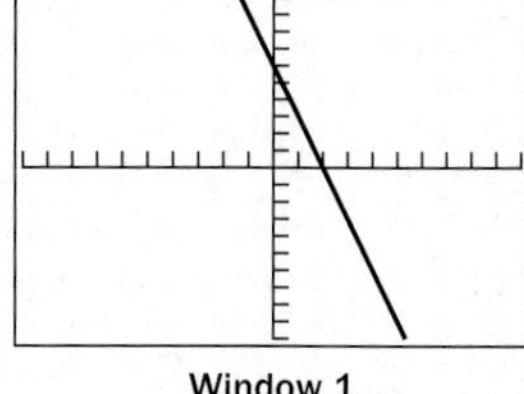
Window 1

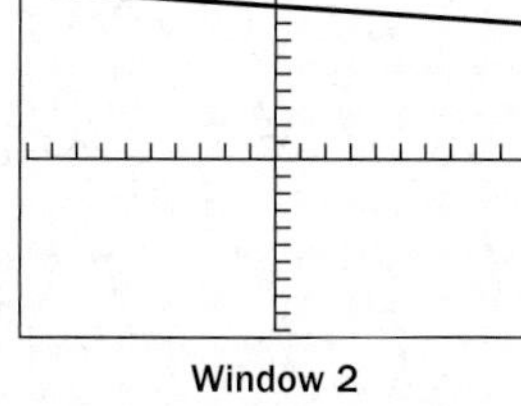
Window 2

WINDOW
Xmin=0
Xmax=90
Xscl=10
Ymin=0
Ymax=10
Yscl=1

Window 3

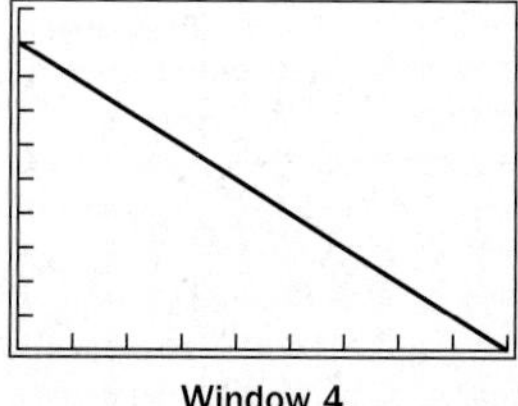
Window 4

You've gotten this far, so you should be rewarded! Suppose you want to find the value of y when given x. Press 2nd TRACE ENTER 30 ENTER to find the value of $y = 9 - 0.1x$ when $X = 30$ $(2000 - 1970 = 30)$. (If your calculator doesn't have this feature, you can use ZOOM and TRACE to find the answer.) Window 5 shows the answer. Can you see what it is? Can you do it without your calculator?

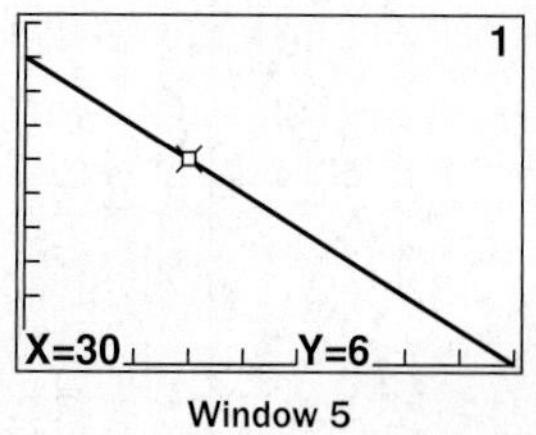

Window 5

Exercises 3.3

Web IT go to mhhe.com/bello for more lessons

⟨A⟩ Graphing Lines Using Intercepts In Problems 1–10, graph the equations.

1. $x + 2y = 4$

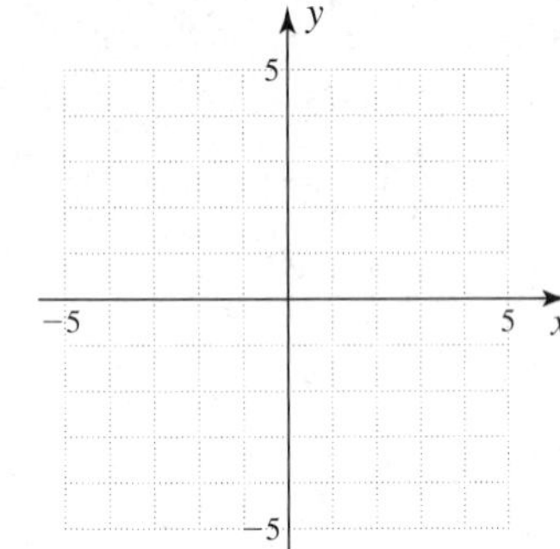

2. $y + 2x = 2$

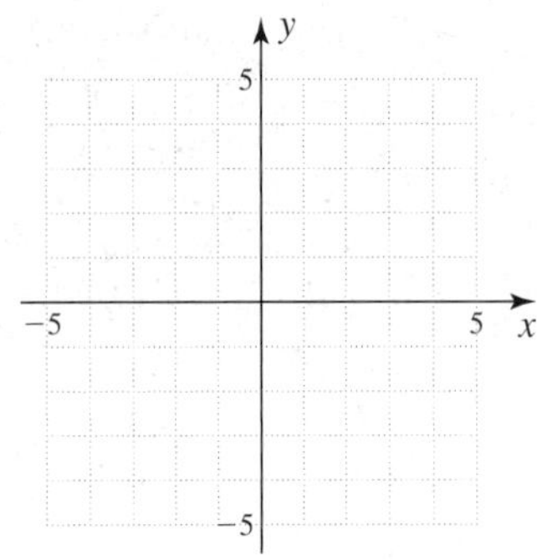

3. $-5x - 2y = -10$

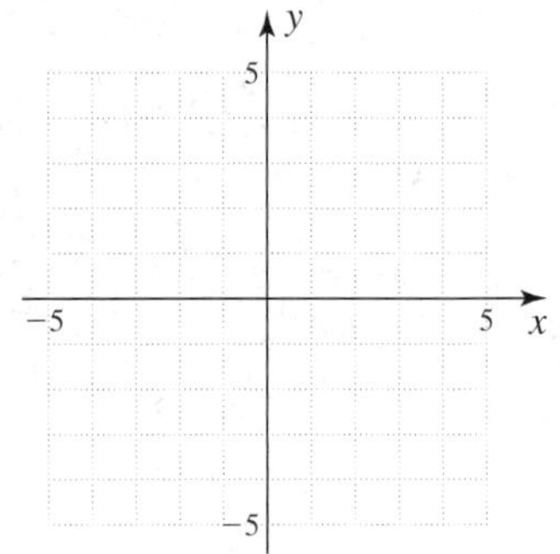

4. $-2x - 3y = -6$

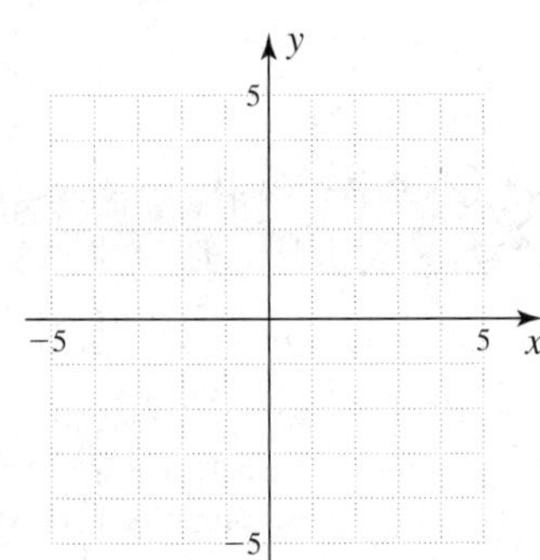

5. $y - 3x - 3 = 0$

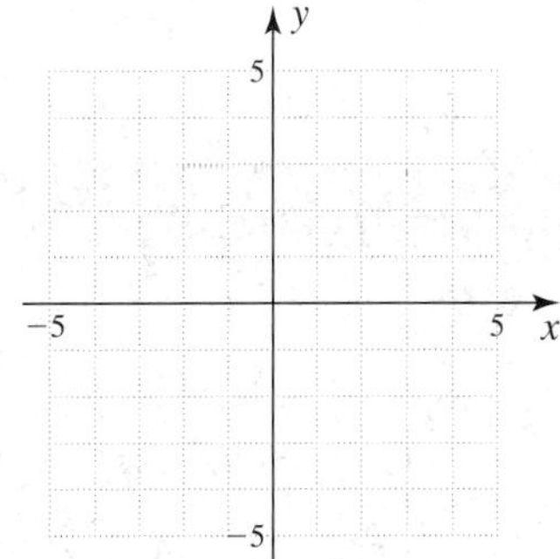

6. $y + 2x - 4 = 0$

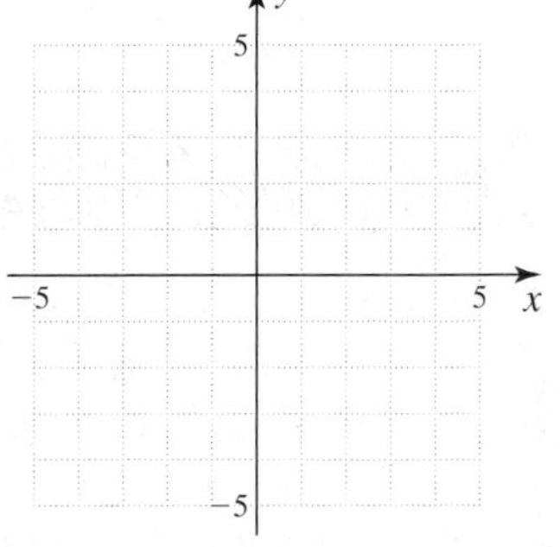

7. $6 = 6x - 3y$

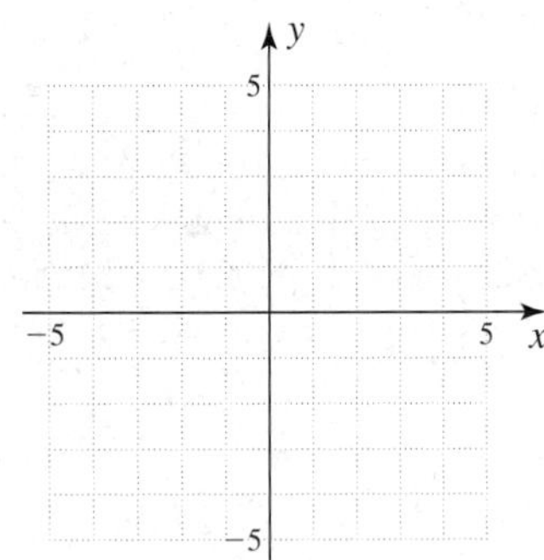

8. $6 = 3y - 2x$

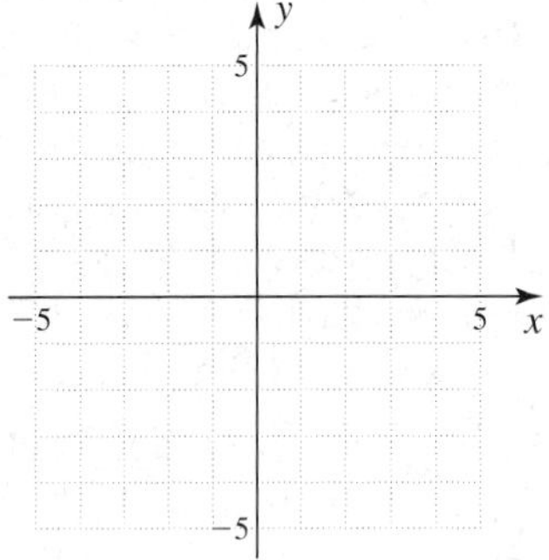

9. $3x + 4y + 12 = 0$

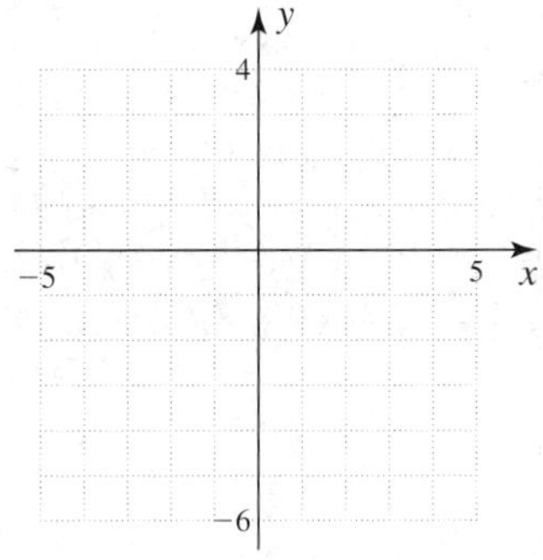

10. $5x + 2y + 10 = 0$

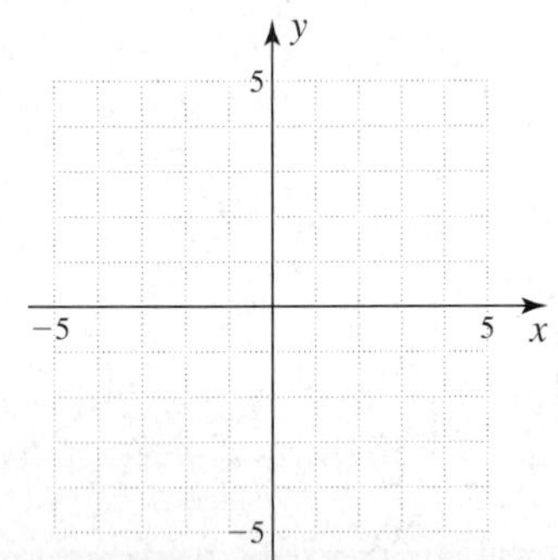

B **Graphing Lines Through the Origin** In Problems 11–20, graph the equations.

11. $3x + y = 0$

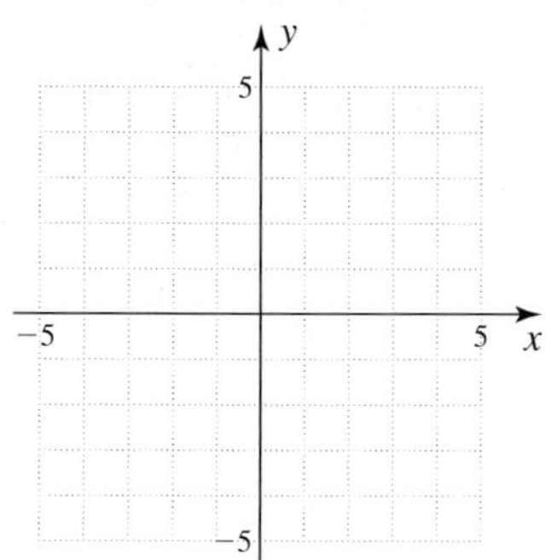

12. $4x + y = 0$

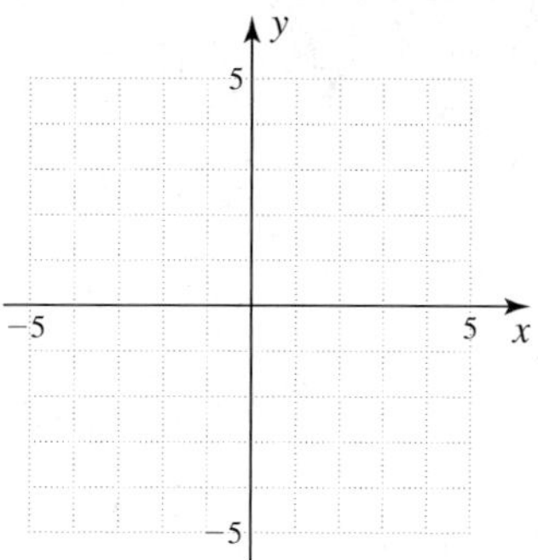

13. $2x + 3y = 0$

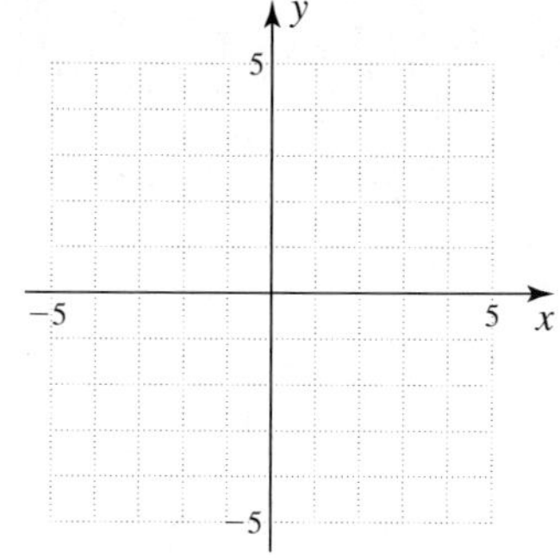

14. $3x + 2y = 0$

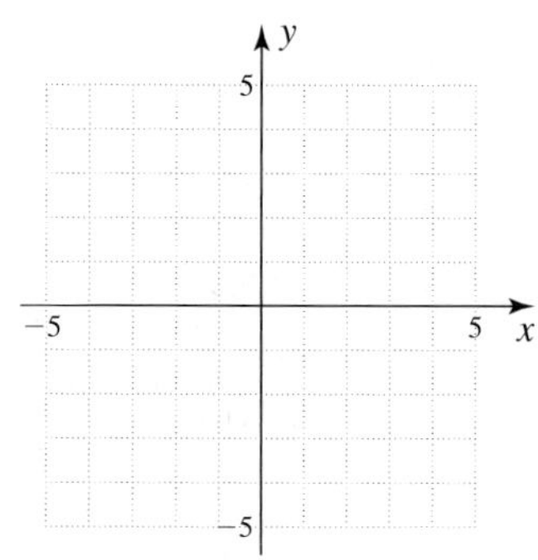

15. $-2x + y = 0$

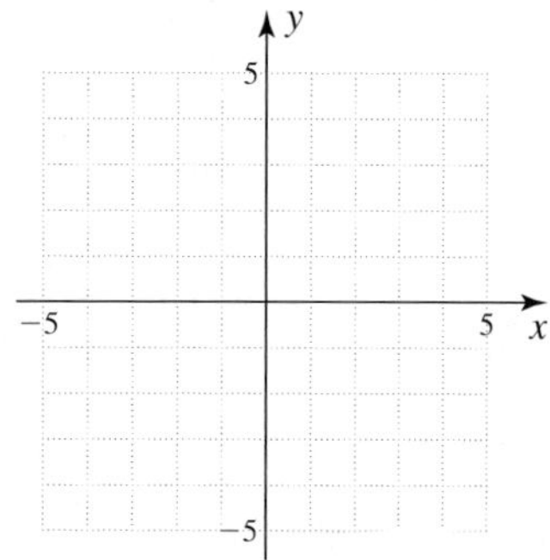

16. $-3x + y = 0$

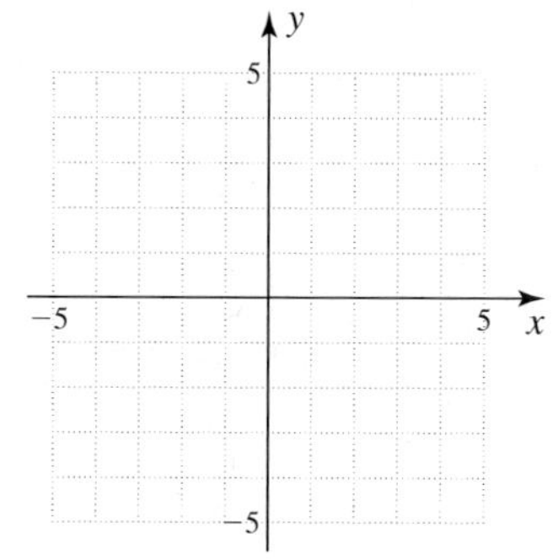

17. $2x - 3y = 0$

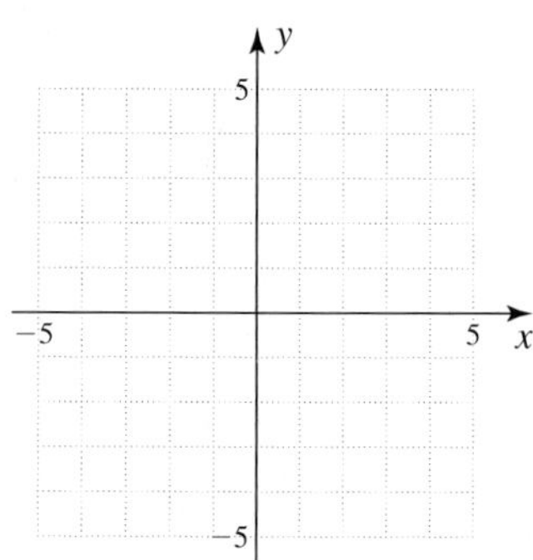

18. $3x - 2y = 0$

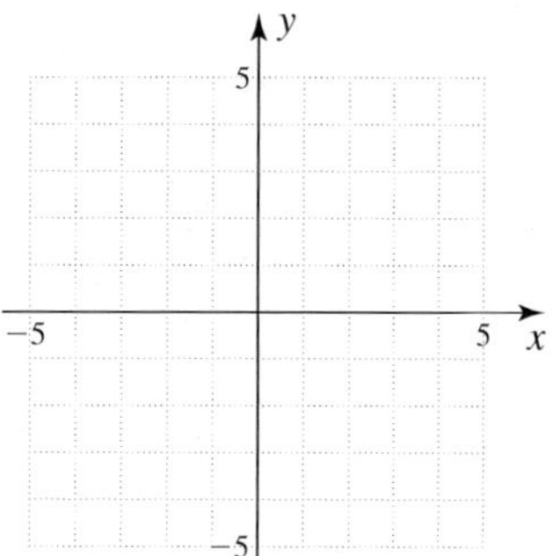

19. $-3x = -2y$

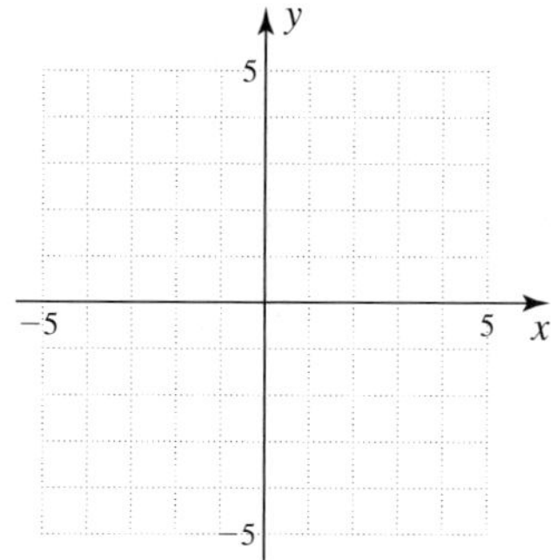

20. $-2x = -3y$

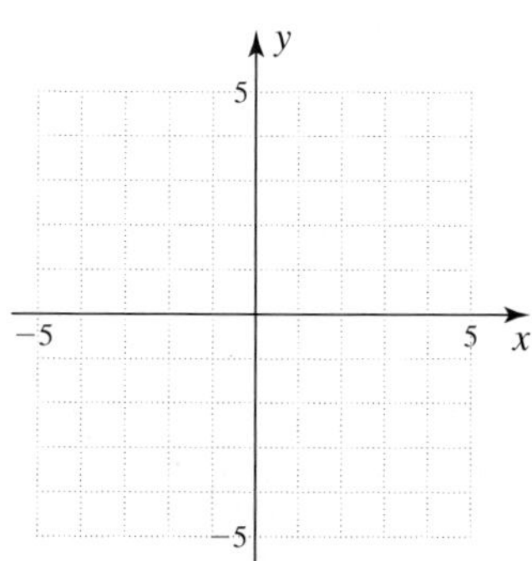

Web IT go to mhhe.com/bello for more lessons

Web IT go to mhhe.com/bello for more lessons

⟨ C ⟩ Graphing Horizontal and Vertical Lines In Problems 21–30, graph the equations.

21. $y = -4$

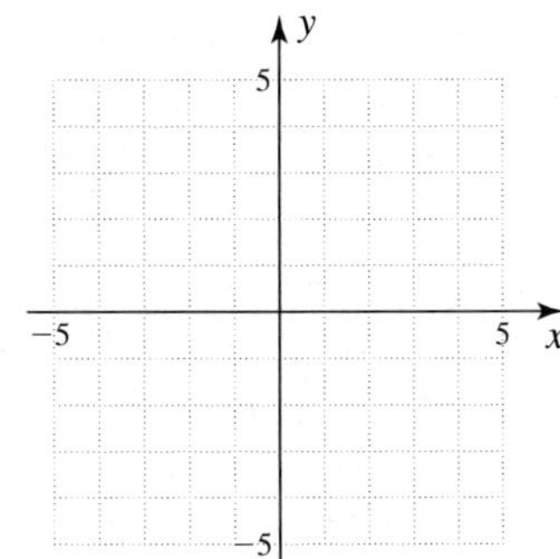

22. $y = -\frac{3}{2}$

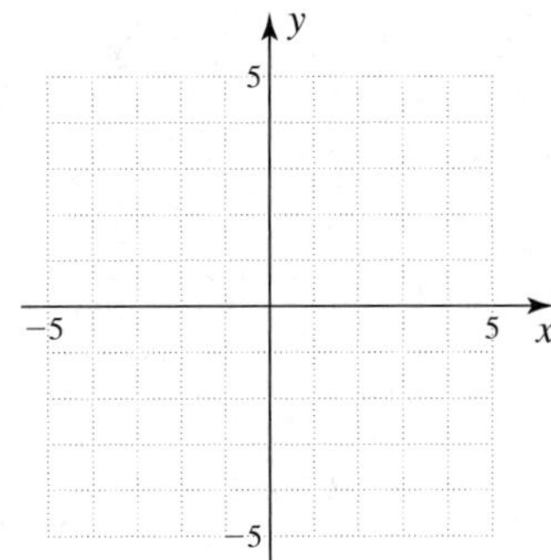

23. $2y + 6 = 0$

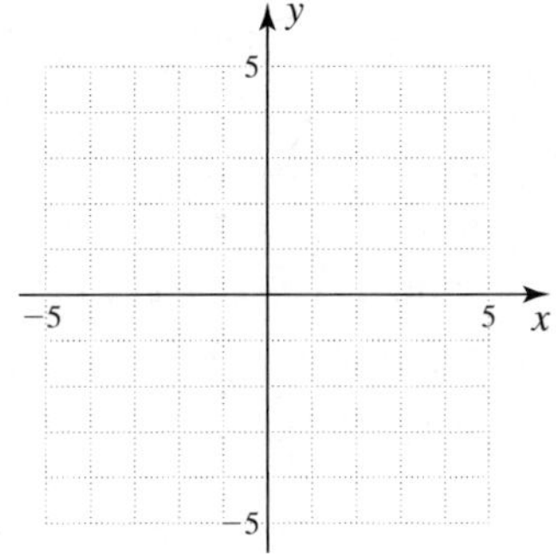

24. $-3y + 9 = 0$

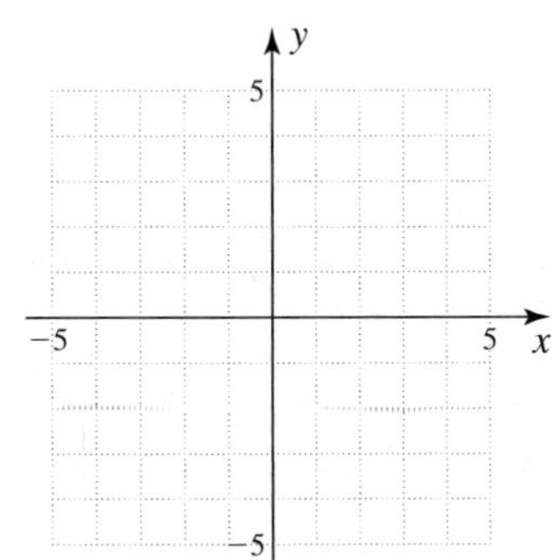

25. $x = -\frac{5}{2}$

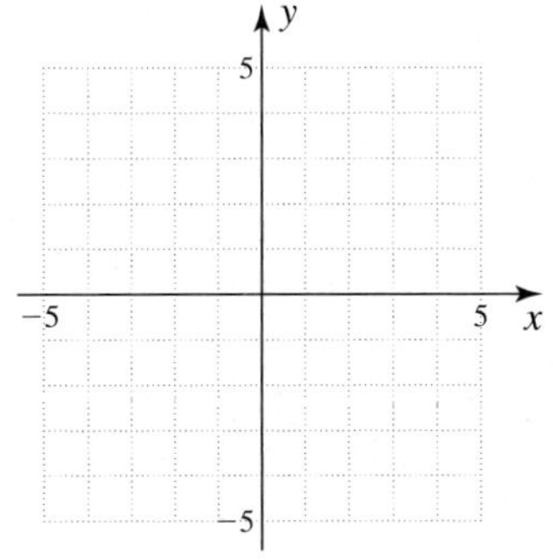

26. $x = \frac{7}{2}$

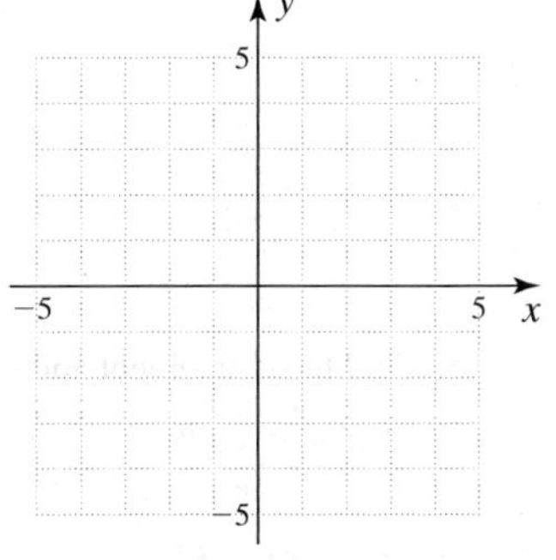

27. $2x + 4 = 0$

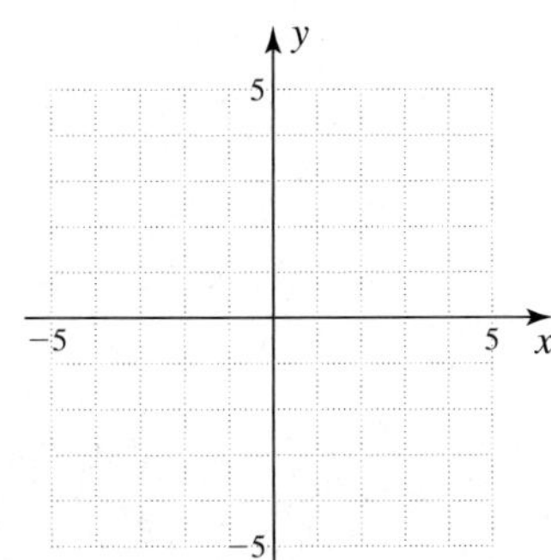

28. $3x - 12 = 0$

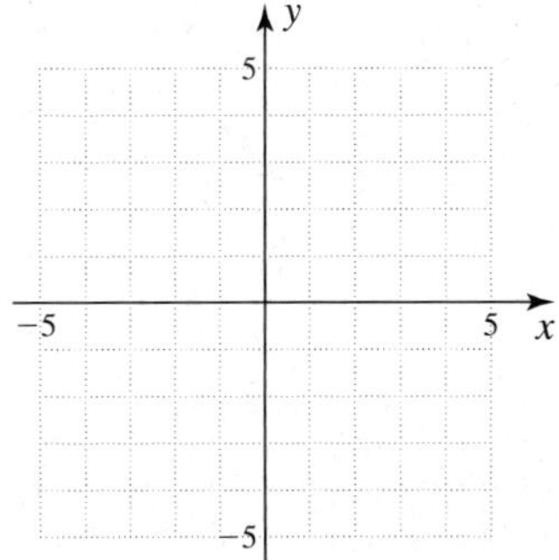

29. $2x - 9 = 0$

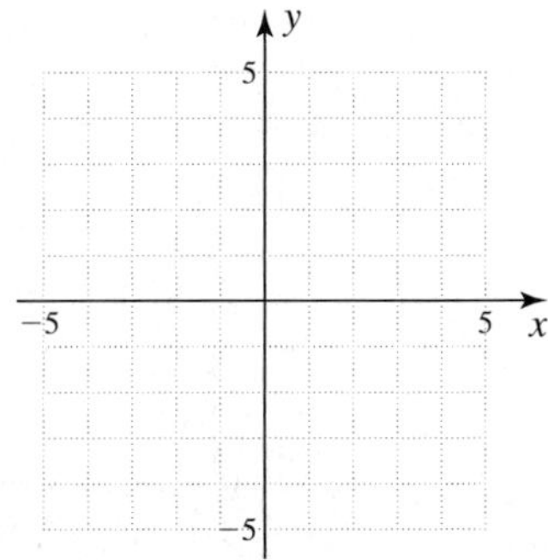

30. $-2x + 7 = 0$

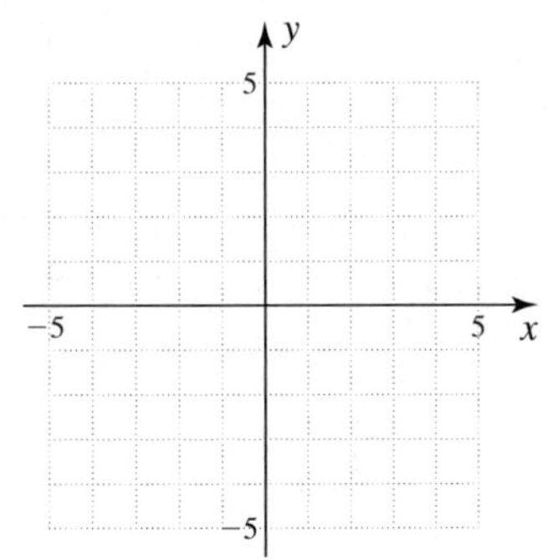

⟨D⟩ Applications Involving Graphs of Lines

⟩⟩⟩ Applications: Green Math

31. The amount A of wasted PET (**polyethylene terephthalate**) beverage bottles and aluminum cans (in thousands of tons) can be approximated by $A = 125N + 1250$, where N is the number of years after 1996.

a. Find the A-intercept and graph it.

b. Graph the point corresponding to $N = 2$.

c. Graph the point corresponding to $N = 6$.

d. Join all the points with a line, the graph of $A = 125N + 1250$.

e. How many thousands of tons were wasted in 2002 (6 years after 1996)?

Source: http://tinyurl.com/n7ewdm.

Wasted PET Bottles and Aluminum Cans

A

N

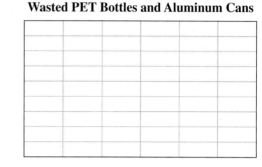

32. The amount of recycled PET beverage bottles and aluminum cans (in thousands of tons) can be approximated by $A = 1100$.

a. Find the A-intercept and graph it.

b. Graph the point corresponding to $N = 6$.

c. What type of line is this?

d. Join all the points with a line, the graph of the horizontal line $A = 1100$.

e. How many thousands of tons were recycled in 2002?

Recycled PET Bottles and Aluminum Cans

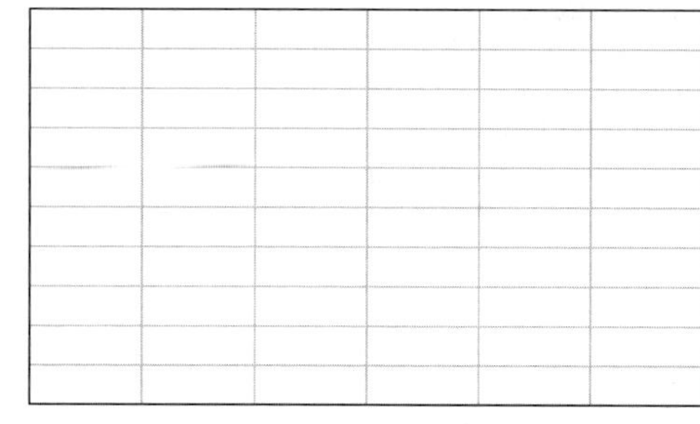

A

N

33. *Daily fat intake* According to the U.S. Department of Agriculture, the total daily fat intake g (in grams) per person can be approximated by $g = 190 + t$, where t is the number of years after 2000.

a. What was the daily fat intake per person in 2000?

b. What was the daily fat intake per person in 2010?

c. What would you project the daily fat intake to be in the year 2020?

d. Use the information from parts **a–c** to graph $g = 190 + t$.

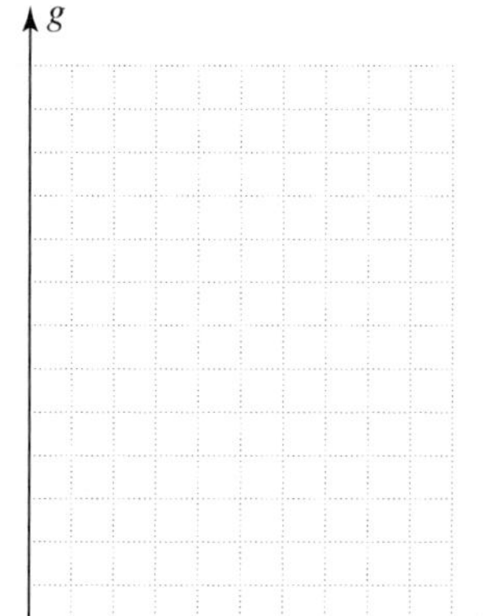

34. *Death rates* According to the U.S. Health and Human Services, the number D of deaths from heart disease per 100,000 population can be approximated by $D = -5t + 290$, where t is the number of years after 1998.

a. How many deaths per 100,000 population were there in 1998?

b. How many deaths per 100,000 population would you expect in 2008?

c. Find the t- and D-intercepts for $D = -5t + 290$ and graph the equation.

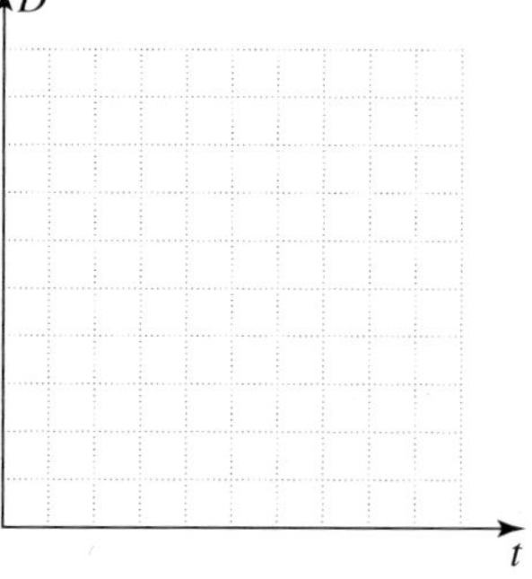

Web IT go to mhhe.com/bello for more lessons

35. *Studying and grades* If you are enrolled for c credit hours, how many hours per week W should you spend outside class studying? For best results, the suggestion is

$$W = 3c$$

a. Suppose you are taking 6 credit hours. How many hours should you spend outside the class studying?

b. Suppose you are a full-time student taking 12 credit hours. How many hours should you spend outside class studying?

c. Graph $y = 3c$ using the points you obtained in parts **a** and **b.**

Source: University of Michigan, Flint.

Study Hours vs. Credit Hours

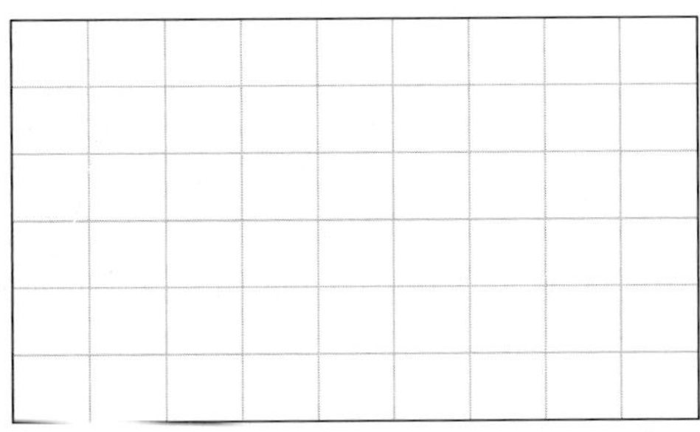

36. *Studying and grades* You may even get away with studying only 2 hours per week W outside class for each credit hour c you are taking.

a. Suppose you are taking 6 credit hours. How many hours should you spend outside the class studying?

b. Suppose you are a full-time student taking 12 credit hours. How many hours should you spend outside class studying?

c. Graph $y = 2c$ using the points you obtained in parts **a** and **b.**

Study Hours vs. Credit Hours

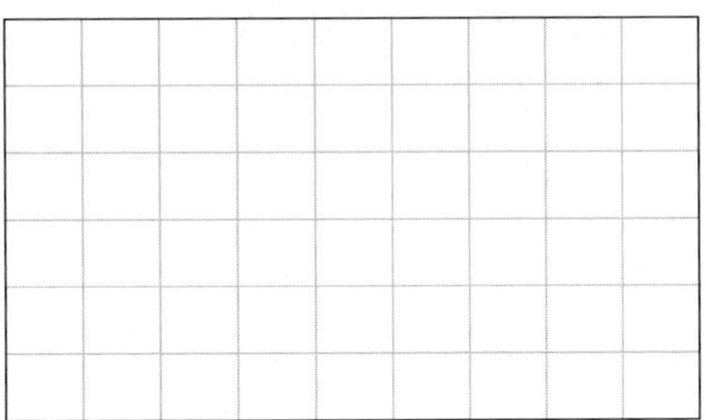

37. *Water pressure* Suppose you take scuba diving. The water pressure P (in pounds per square foot) as you descend to a depth of f feet is given by $P = 62.4f$.

a. What would the pressure P be at sea level ($f = 0$)?

b. What would the pressure P be at a depth of 10 feet?

c. According to PADI (Professional Association of Diving Instructors) the maximum depth for beginners is 40 feet. What would the pressure P be at 40 feet?

d. Graph $P = 62.4f$ using the points you obtained in parts **a–c.**

Pressure at Depth of *f* Feet

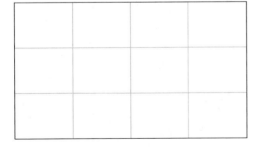

38. *Temperature* When you fly in a balloon, the temperature T decreases 7°F for each 1000 feet f of altitude. Thus, at 3 (thousand) feet, the temperature would decrease by 21°F. If you started at 70°F, now the temperature is 70°F – 21°F, or 49°F.

a. What would the decrease in temperature be when you are 1 (thousand) feet high?

b. What about when you are 5 (thousand) feet high?

c. Graph $T = -7f$ using the points you obtained in parts **a** and **b.**

Decrease in Temperature

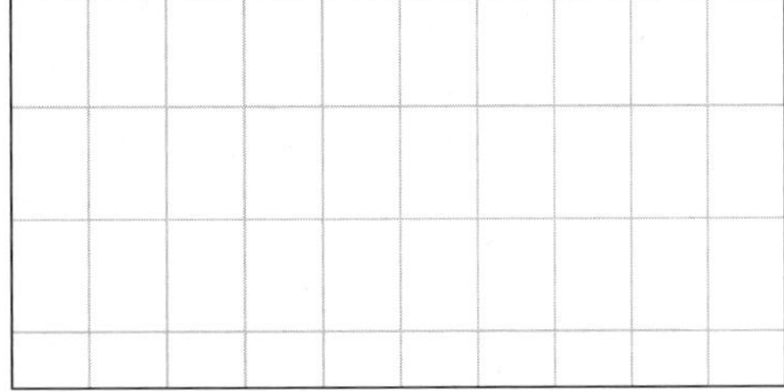

Using Your Knowledge

Cholesterol Level In Example 3 of Section 3.2 (p. 229), we mentioned that the decrease in the cholesterol level C after w weeks elapsed could be approximated by $C = -3w + 215$. (Individual results vary.)

39. According to this equation, what was the cholesterol level initially?

40. What was the cholesterol level at the end of 12 weeks?

41. Use the results of Problems 39 and 40 to graph the equation $C = -3w + 215$.

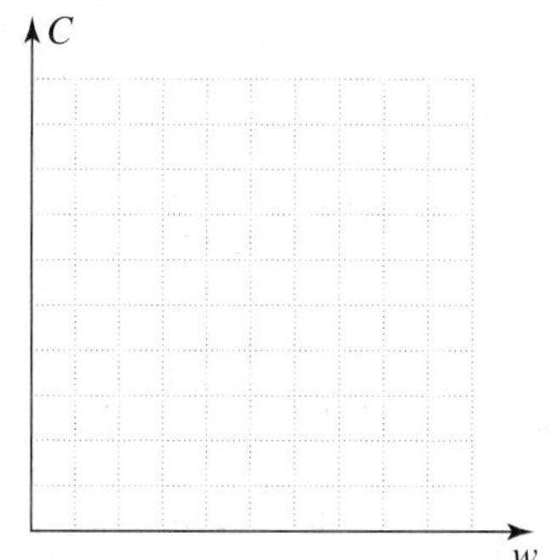

42. According to the graph shown in Example 3 (p. 229), the initial cholesterol level was 215. If we assume that the initial cholesterol level is 230, we can approximate the reduction in cholesterol by $C = -3w + 230$.

a. Find the intercepts and graph the equation on the same coordinate axes as $C = -3w + 215$.

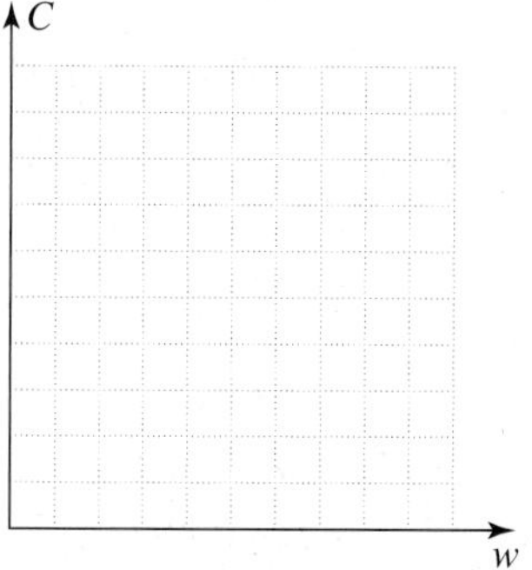

b. How many weeks does it take for a person with an initial cholesterol level of 230 to reduce it to 175?

Write On

43. If in the equation $Ax + By = C$, $A = 0$ and B and C are not zero, what type of graph will result?

44. If in the equation $Ax + By = C$, $B = 0$ and A and C are not zero, what type of graph will result?

45. If in the equation $Ax + By = C$, $C = 0$ and A and B are not zero, what type of graph will result?

46. What are the intercepts of the line $Ax + By = C$, and how would you find them?

47. How many points are needed to graph a straight line?

Concept Checker

Fill in the blank(s) with the correct word(s), phrase, or mathematical statement.

48. The **x-intercept** of a line is the point at which the line **crosses the x-axis** and has **coordinates** _______________.

49. The **y-intercept** of a line is the point at which the line **crosses the y-axis** and has **coordinates** _______________.

50. The **coordinates** of the **origin** are _______________.

51. The **graph of $Ax + By = 0$**, where A and B are constants and not equal to 0, is a **straight line that passes through the** _______________.

52. The **graph of $y = k$**, where k a constant, is a _______________ **line.**

53. The **graph of $x = k$**, where k a constant, is a _______________ **line.**

(0, 0)	**(x, 0)**
origin	**(0, x)**
horizontal	**(y, 0)**
vertical	**(0, y)**
(x, y)	

Mastery Test

Graph:

54. $x + 2y = 4$

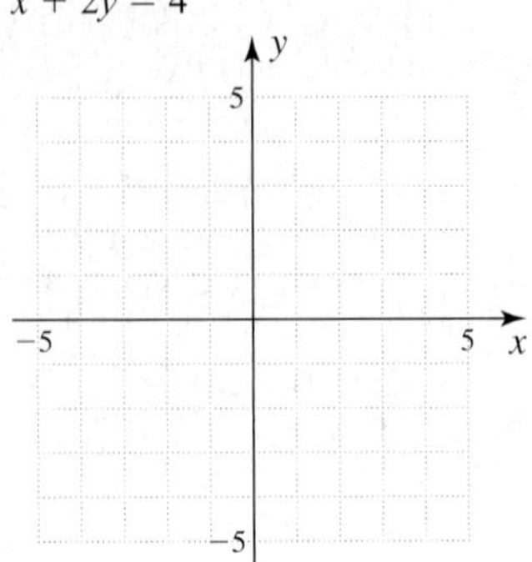

55. $-2x + y = 4$

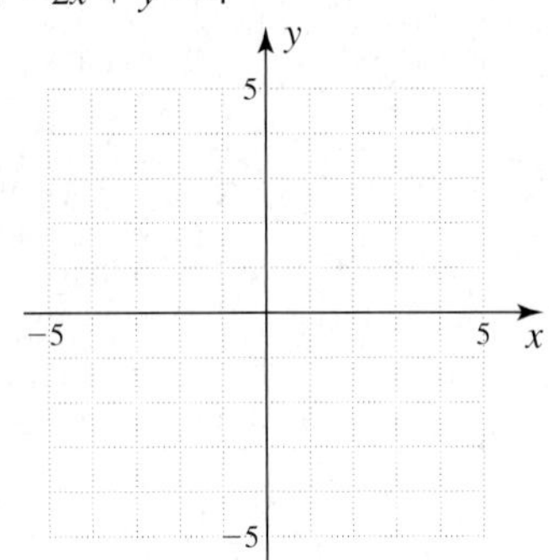

56. $3x - 6 = 0$

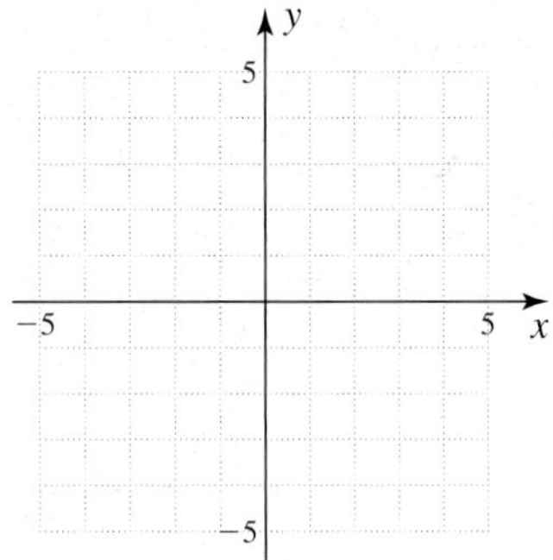

57. $4 + 2y = 0$

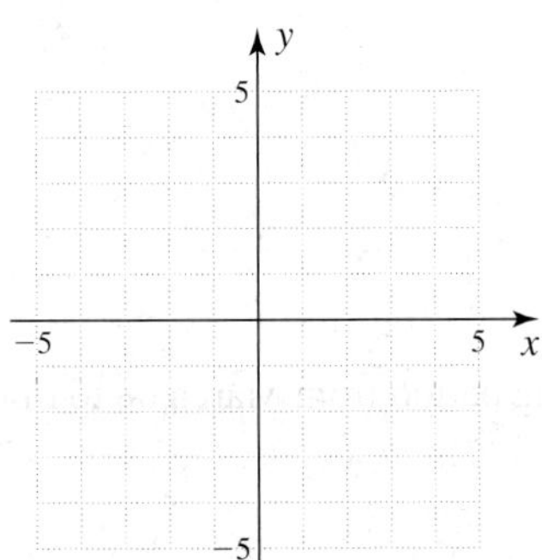

58. $-4x + y = 0$

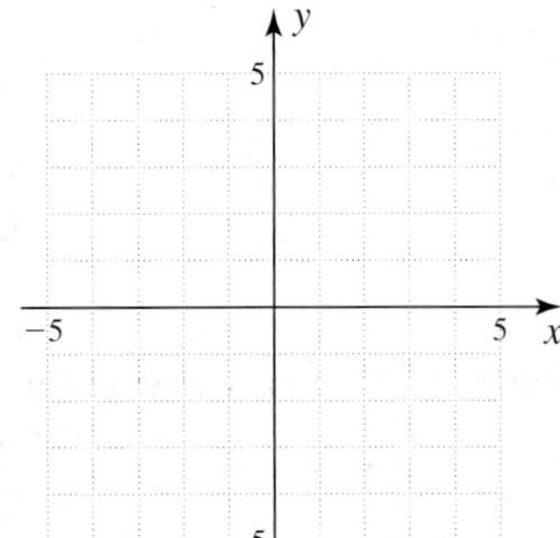

59. $-x + 4y = 0$

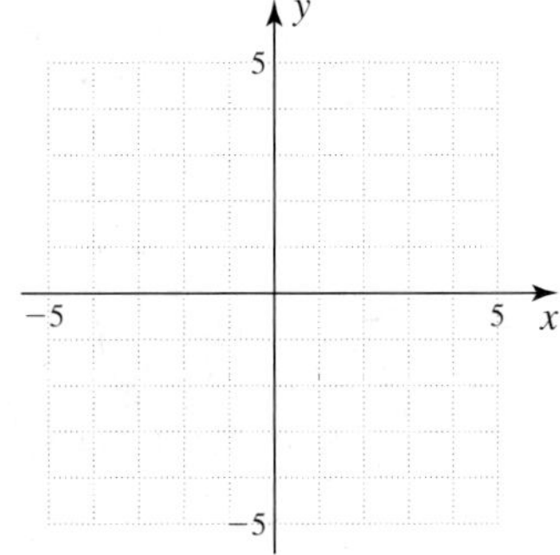

60. The number N (in millions) of recreational boats in the United States can be approximated by $N = 10 + 0.4t$, where t is the number of years after 1975.

a. How many recreational boats were there in 1975?

b. How many would you expect in 1995?

c. Find the t- and N-intercepts of $N = 10 + 0.4t$ and graph the equation.

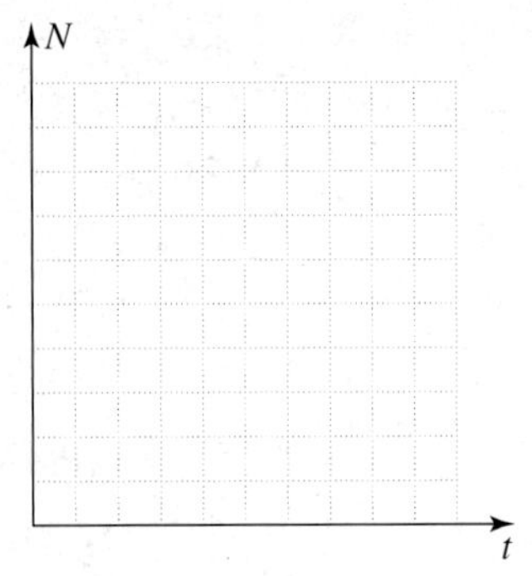

61. The relationship between the Continental dress size C and the American dress size A is $C = A + 30$.

a. Find the intercepts for $C = A + 30$ and graph the equation.

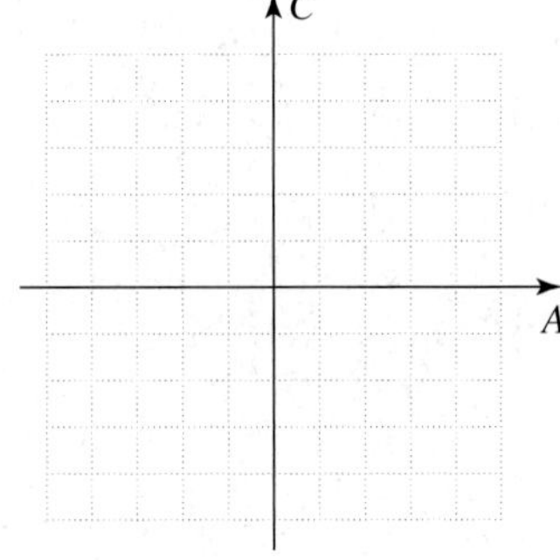

b. In what quadrant should the graph lie?

Skill Checker

Find:

62. $13.5 - 3.5$

63. $3 - (-6)$

64. $5 - (-7)$

65. $-6 - 3$

66. $-5 - 7$

3.4 The Slope of a Line: Parallel and Perpendicular Lines

Objectives

A Find the slope of a line given two points.

B Find the slope of a line given the equation of the line.

C Determine whether two lines are parallel, perpendicular, or neither.

D Solve applications involving slope.

To Succeed, Review How To . . .

1. Add, subtract, multiply, and divide signed numbers (pp. 52–56, 60–64).
2. Solve an equation for a specified variable (pp. 137–143).

Getting Started

Facebook and MySpace Visits

Can you tell from the graph the period in which the number of pages per visit declined for MySpace (red graph)? Has Facebook (blue graph) ever had a declining period? You can tell by simply looking at the graph! The pages per visit for MySpace subscribers declined from Jan 07 to Dec 07 (from about 75 pages per visit in Jan 07 to about 35 pages per visit in Dec 07). The **decline** per month was

$$\frac{\text{Difference in pages per visit}}{\text{Number of months}} = \frac{75 - 35}{11} = \frac{40}{11} \approx 4 \text{ (pages per month)}$$

On the other hand, Facebook had a 2-month declining period from March 08 to May 08. Their decline was

$$\frac{\text{Difference in pages per visit}}{\text{Number of months}} = \frac{50 - 40}{2} = 5 \text{ (pages per month)}$$

As you can see from the graph, the 2-month decline for Facebook was "steeper" but shorter (5 pages per month compared to 4 pages per month). The "steepness" of the declines was calculated by comparing the *vertical* change of the line (usually called the **rise,** but in this case the **fall**) to the *horizontal* change in months (the **run**). Note that the "run" for MySpace was 11 months and that for Facebook was only 2 months, but we can still compare the steepness of the two lines. This measure of steepness is called the **slope** of the line. In this section, we shall learn how to find the **slope** of a line when two points are given or when the equation of the line is given. We will also use the slope to determine when lines are parallel, perpendicular, or neither.

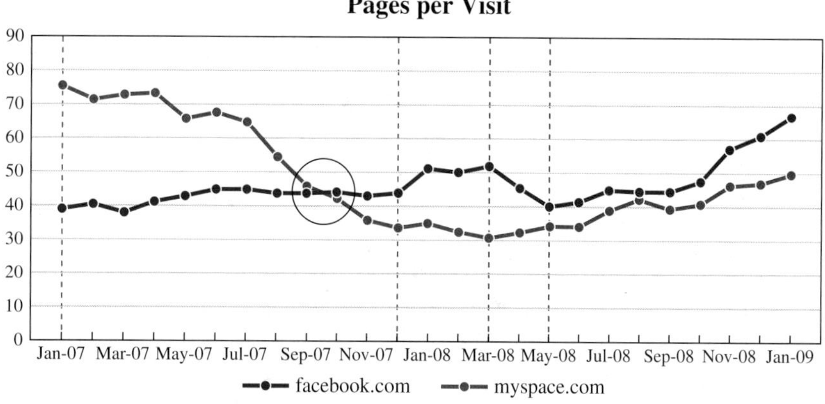

Source: www.Compete.com, graph from http://tinyurl.com/d9tv73.

A > Finding Slopes from Two Points

The steepness of a line can be measured by using the ratio of the **vertical rise** (or **fall**) to the corresponding **horizontal run.** This ratio is called the *slope.* For example, a staircase that rises 3 feet in a horizontal distance of 4 feet is said to have a slope of $\frac{3}{4}$. The definition of slope is as follows.

SLOPE

The **slope** m of the line going through the points (x_1, y_1) and (x_2, y_2), where $x_1 \neq x_2$, is given by

$$m = \frac{y_2 - y_1}{x_2 - x_1} = \frac{\text{rise} \uparrow}{\text{run} \rightarrow}$$

This means that the slope of a line is the change in y (Δy) over the change in x (Δx), that is,

$$m = \frac{\Delta y}{\Delta x} = \frac{y_2 - y_1}{x_2 - x_1}$$

The slope for a vertical line such as $x = 3$ is not defined because all points on this line have the same x-value, 3, and hence,

$$m = \frac{y_2 - y_1}{3 - 3} = \frac{y_2 - y_1}{0}$$

which is undefined because division by zero is undefined (see Example 3). The slope of a horizontal line is zero because all points on such a line have the same y-values. Thus, for the line $y = 7$,

$$m = \frac{7 - 7}{x_2 - x_1} = \frac{0}{x_2 - x_1} = 0$$

See Example 3.

EXAMPLE 1 **Finding the slope given two points: Positive slope**

Find the slope of the line passing through the points $(0, -6)$ and $(3, 3)$ in Figure 3.28.

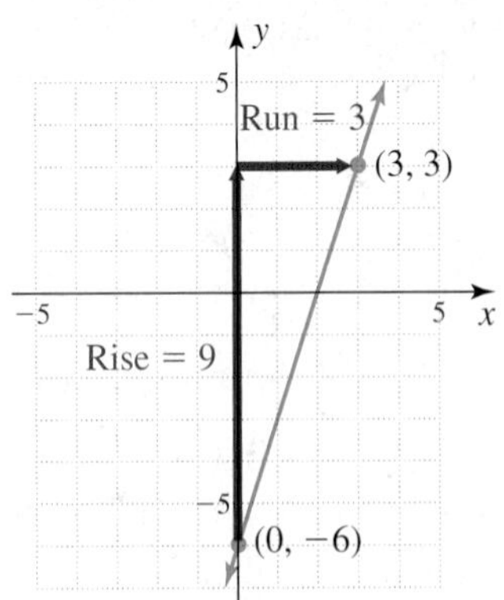

>**Figure 3.28** Line with positive slope: $\frac{\text{Rise}}{\text{Run}} = \frac{9}{3} = 3$

PROBLEM 1

Find the slope of the line going through the points $(0, -6)$ and $(2, 2)$.

Answers to PROBLEMS

1. 4

SOLUTION 1 Suppose we choose $(x_1, y_1) = (0, -6)$ and $(x_2, y_2) = (3, 3)$. Then we use the equation for slope to obtain

$$m = \frac{3 - (-6)}{3 - 0} = \frac{9}{3} = 3$$

If we choose $(x_1, y_1) = (3, 3)$ and $(x_2, y_2) = (0, -6)$, then

$$m = \frac{-6 - 3}{0 - 3} = \frac{-9}{-3} = 3$$

As you can see, it makes no difference which point is labeled (x_1, y_1) and which is labeled (x_2, y_2). Since an interchange of the two points simply changes the sign of both the numerator and the denominator in the slope formula, the result is the same in both cases.

EXAMPLE 2 **Finding the slope given two points: Negative slope**
Find the slope of the line that passes through the points $(3, -4)$ and $(-2, 3)$ in Figure 3.29.

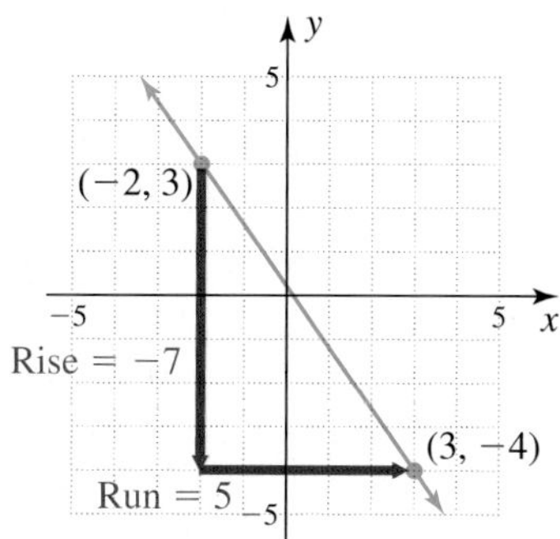

>**Figure 3.29** Line with negative slope: $\frac{\text{Rise}}{\text{Run}} = -\frac{7}{5}$

SOLUTION 2 We take $(x_1, y_1) = (-2, 3)$ so that $(x_2, y_2) = (3, -4)$. Then

$$m = \frac{-4 - 3}{3 - (-2)} = -\frac{7}{5}$$

PROBLEM 2
Find the slope of the line going through the points $(2, -4)$ and $(-3, 2)$.

Examples 1 and 2 are illustrations of the fact that a line that rises from left to right has a *positive slope* and one that falls from left to right has a *negative slope*. What about vertical and horizontal lines? We shall discuss them in Example 3.

EXAMPLE 3 **Finding the slopes of vertical and horizontal lines**
Find the slope of the line passing through the given points.

a. $(-4, -2)$ and $(-4, 3)$ **b.** $(1, 4)$ and $(4, 4)$

SOLUTION 3

a. Substituting $(-4, -2)$ for (x_1, y_1) and $(-4, 3)$ for (x_2, y_2) in the equation for slope, we obtain

$$m = \frac{3 - (-2)}{-4 - (-4)} = \frac{3 + 2}{-4 + 4} = \frac{5}{0}$$

which is undefined. Thus, the slope of a vertical line is *undefined* (see Figure 3.30).

(continued)

PROBLEM 3
Find the slope of the line passing through the given points.

a. $(4, 1)$ and $(-3, 1)$
b. $(-2, 4)$ and $(-2, 1)$

Answers to PROBLEMS
2. $-\frac{6}{5}$ **3. a.** 0 **b.** Undefined

b. This time $(x_1, y_1) = (1, 4)$ and $(x_2, y_2) = (4, 4)$, so

$$m = \frac{4-4}{4-1} = \frac{0}{3} = \mathbf{0}$$

Thus, the slope of a horizontal line is **zero** (see Figure 3.30).

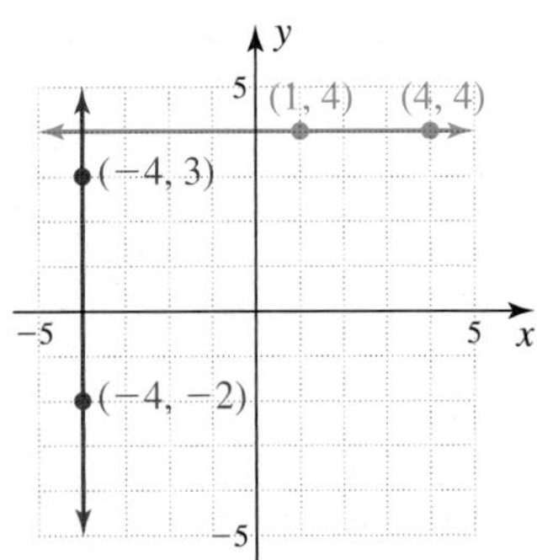

>Figure 3.30

Let's summarize our work with slopes so far.

A line with **positive** slope *rises* from left to right.

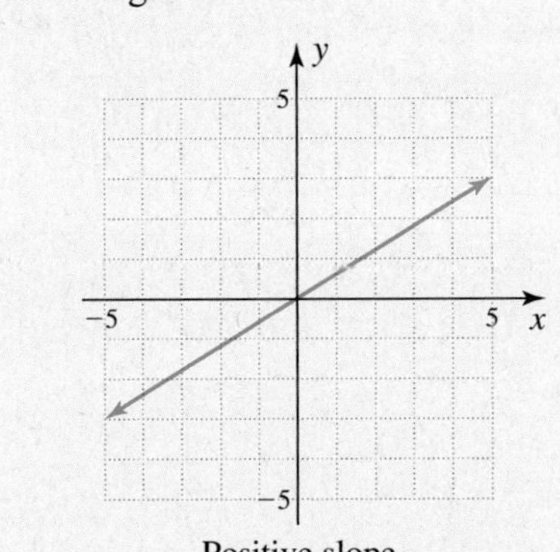

Positive slope

A line with **negative** slope *falls* from left to right.

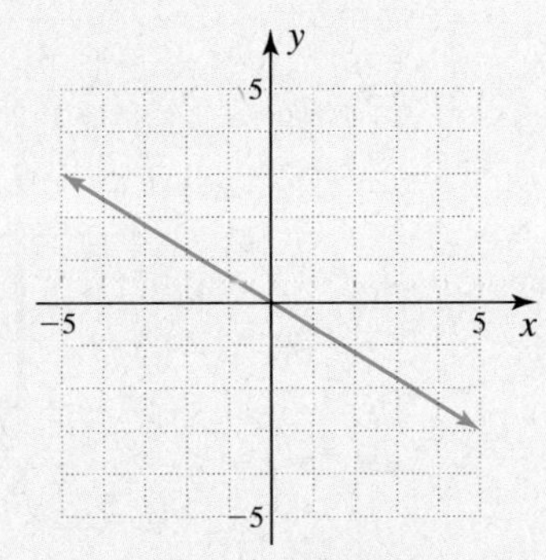

Negative slope

A *horizontal* line with an equation of the form $y = k$ has **zero** slope.

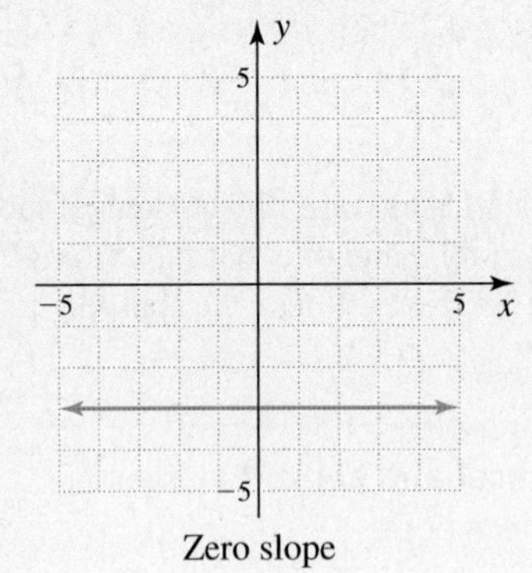

Zero slope

A *vertical* line with an equation of the form $x = k$ has an **undefined** slope.

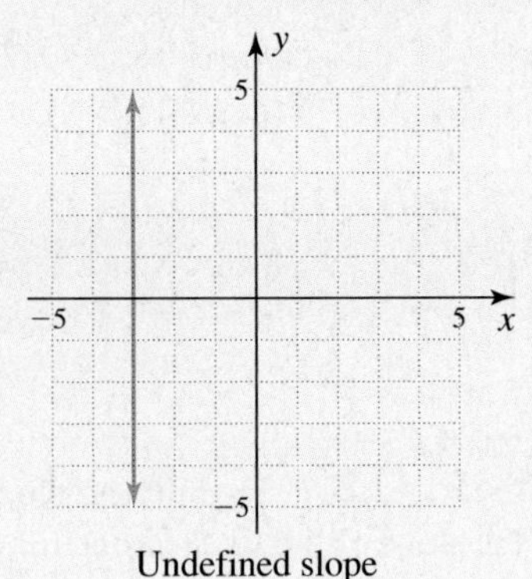

Undefined slope

B > Finding Slopes from Equations

In the *Getting Started,* we found the slope of the line by using a ratio. The slope of a line can also be found from its equation. Thus, if we approximate the number of pages visited by MySpace subscribers from January to December by the equation

$$y = -\frac{40}{11}x + 75,$$

where x is the number of months after month 0 and y is the number of pages per visit, we can find the slope of $y = -\frac{40}{11}x + 75$ by using two values for x and then finding the corresponding y-values.

For $x = 0, y = 75,$ so $(0, 75)$ is on the line.
For $x = 11, y = -\frac{40}{11}(11) + 75 = 35,$ so $(11, 35)$ is on the line.

The slope of $y = -\frac{40}{11}x + 35$ passing through $(0, 75)$ and $(11, 35)$ is

$$m = \frac{75 - 35}{0 - 11} = \frac{40}{-11} = -\frac{40}{11}$$

which is simply the **coefficient** of x in the equation $y = -\frac{40}{11}x + 75$. This idea can be generalized as follows.

SLOPE OF $y = mx + b$

The slope of the line defined by the equation $y = mx + b$ is m.

Thus, if you are given the equation of a line and you want to find its slope, you would use this procedure.

PROCEDURE

Finding a slope

1. Solve the equation for y.
2. The slope is m, the coefficient of x.

EXAMPLE 4 Finding the slope given an equation

Find the slope of the following lines:

a. $2x + 3y = 6$ **b.** $3x - 2y = 4$

SOLUTION 4

a. We follow the two-step procedure.

1. Solve $2x + 3y = 6$ for y.

$$\begin{aligned} 2x + 3y &= 6 && \text{Given} \\ 3y &= -2x + 6 && \text{Subtract } 2x. \\ y &= -\frac{2}{3}x + \frac{6}{3} && \text{Divide each term by 3.} \\ y &= -\frac{2}{3}x + 2 && \text{Simplify.} \end{aligned}$$

2. Since the coefficient of x is $-\frac{2}{3}$, the slope is $-\frac{2}{3}$.

b. We follow the steps.

1. Solve for y.

$$\begin{aligned} 3x - 2y &= 4 && \text{Given} \\ -2y &= -3x + 4 && \text{Subtract } 3x. \\ y &= \frac{-3}{-2}x + \frac{4}{-2} && \text{Divide each term by } -2. \\ y &= \frac{3}{2}x - 2 && \text{Simplify.} \end{aligned}$$

2. The slope is the coefficient of x, so the slope is $\frac{3}{2}$.

PROBLEM 4

Find the slope of the line:

a. $3x + 2y = 6$

b. $2x - 3y = 9$

Answers to PROBLEMS

4. a. $-\frac{3}{2}$ **b.** $\frac{2}{3}$

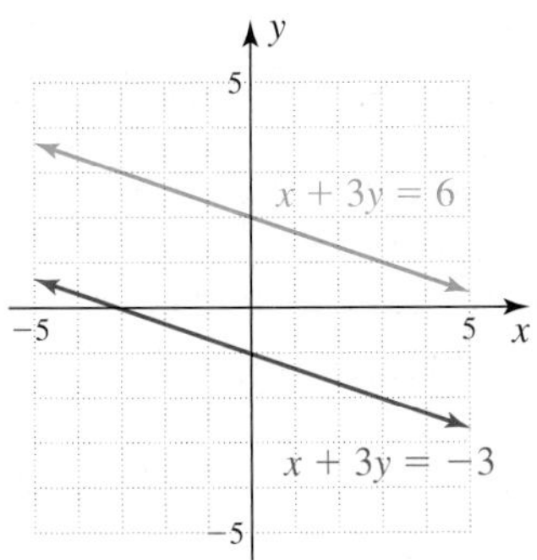

>Figure 3.31

C > Finding Parallel and Perpendicular Lines

Parallel lines are lines in the plane that never intersect. In Figure 3.31, the lines $x + 3y = 6$ and $x + 3y = -3$ appear to be parallel lines. How can we be sure? By solving each equation for y and determining whether both lines have the same slope.

For $x + 3y = 6$: $\quad y = -\frac{1}{3}x + 2 \quad$ The slope is $-\frac{1}{3}$.

For $x + 3y = -3$: $\quad y = -\frac{1}{3}x - 1 \quad$ The slope is $-\frac{1}{3}$.

Since the two lines have the same slope but different y-intercepts, they are parallel lines.

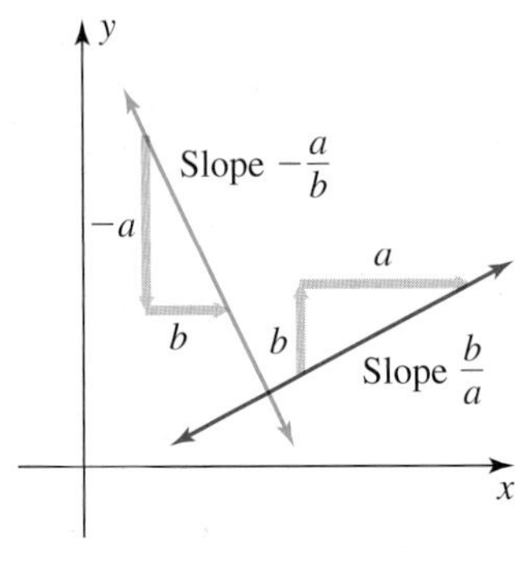

>Figure 3.32

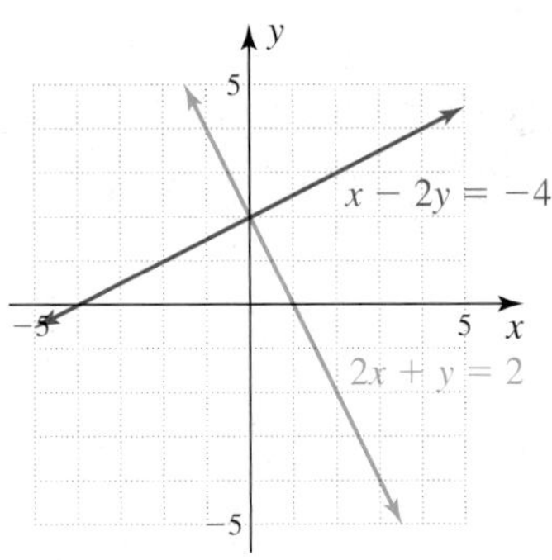

>Figure 3.33

The two lines in Figure 3.32 appear to be **perpendicular;** that is, they meet at a 90° angle. Note that their slopes are *negative reciprocals* and that the product of their slopes is

$$-\frac{a}{b} \cdot \frac{b}{a} = -1$$

The lines $2x + y = 2$ and $x - 2y = -4$ have graphs that also appear to be perpendicular (see Figure 3.33). To show that this is the case, we check their slopes.

For $2x + y = 2$: $\quad y = -2x + 2 \quad$ The slope is -2.

For $x - 2y = -4$: $\quad y = \frac{1}{2}x + 2 \quad$ The slope is $\frac{1}{2}$.

Since the product of the slopes is $-2 \cdot \frac{1}{2} = -1$, the lines are perpendicular.

SLOPES OF PARALLEL AND PERPENDICULAR LINES

Two lines with the **same** slope but different y-intercepts are **parallel.**

Two lines whose slopes have a product of **−1** are **perpendicular.**

EXAMPLE 5 Finding whether two given lines are parallel, perpendicular, or neither

Decide whether the pair of lines are parallel, perpendicular, or neither:

a. $x - 3y = 6$
$2x - 6y = -12$

b. $2x + y = 6$
$x + y = 4$

c. $2x + y = 5$
$x - 2y = 4$

PROBLEM 5

Decide whether the pair of lines are parallel, perpendicular, or neither.

a. $x - 2y = 3$
$2x - 4y = 8$

b. $3x + y = 6$
$x + y = 2$

c. $3x + y = 6$
$x - 3y = 5$

Answers to PROBLEMS

5. a. Parallel **b.** Neither **c.** Perpendicular

SOLUTION 5

a. We find the slope of each line by solving the equations for y.

$x - 3y = 6$	Given	$2x - 6y = -12$	Given
$-3y = -x + 6$	Subtract x.	$-6y = -2x - 12$	Subtract $2x$.
$y = \frac{1}{3}x - 2$	Divide by -3.	$y = \frac{1}{3}x + 2$	Divide by -6.
The slope is $\frac{1}{3}$.		The slope is $\frac{1}{3}$.	

Since both slopes are $\frac{1}{3}$, the slopes are equal and the y-intercepts are different, the lines are parallel, as shown in Figure 3.34.

b. We find the slope of each line by solving the equations for y.

$2x + y = 6$	Given	$x + y = 4$	Given
$y = -2x + 6$	Subtract $2x$.	$y = -x + 4$	Subtract x.
The slope is -2.		The slope is -1.	

Since the slopes are -2 and -1, their product is not -1, so the lines are neither parallel nor perpendicular, as shown in Figure 3.35.

c. We again solve both equations for y to find their slopes.

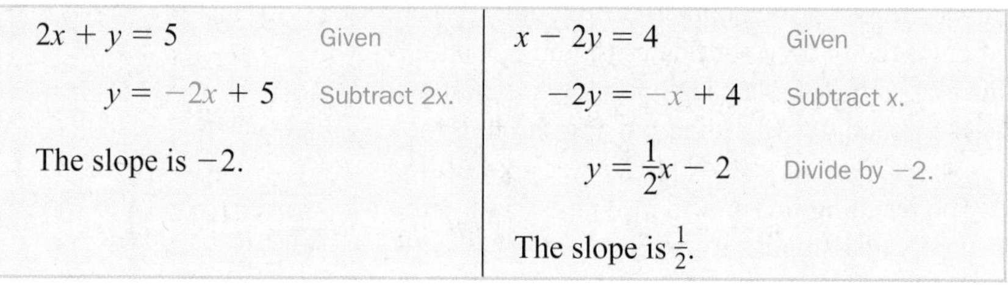

$2x + y = 5$	Given	$x - 2y = 4$	Given
$y = -2x + 5$	Subtract $2x$.	$-2y = -x + 4$	Subtract x.
The slope is -2.		$y = \frac{1}{2}x - 2$	Divide by -2.
		The slope is $\frac{1}{2}$.	

Since the product of the slopes is $-2 \cdot \frac{1}{2} = -1$, the lines are perpendicular, as shown in Figure 3.36.

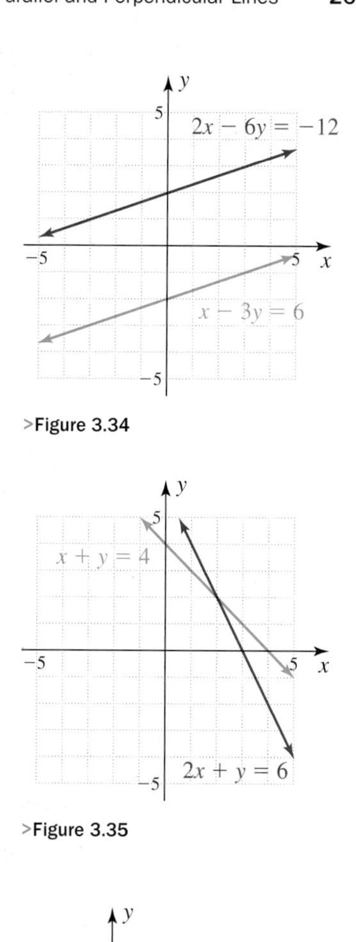

>Figure 3.34

>Figure 3.35

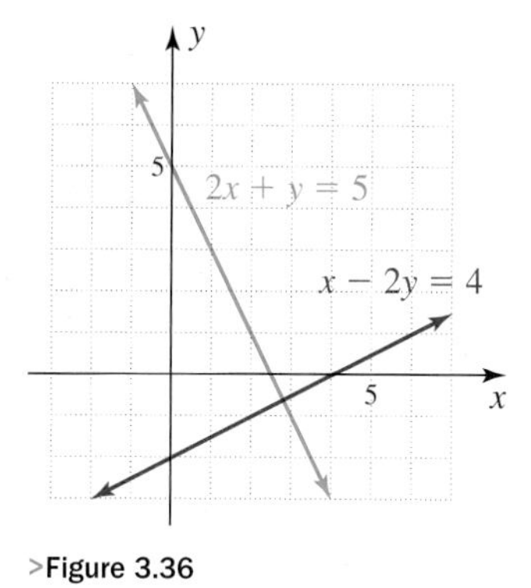

>Figure 3.36

D > Applications Involving Slope

In the *Getting Started,* we saw that the number of pages per visit read by subscribers to Facebook (5 pages per month) *declined* at a faster rate than that of MySpace (4 pages per month). The slope of a line can be used to describe a *rate of change*. For example, the number N of deaths (per 100,000 population) due to heart disease can be approximated by $\mathbf{N = -5t + 300}$, where t is the number of years after 1960. Since the **slope** of the line $\mathbf{N = -5t + 300}$ is $\mathbf{-5}$, this means that the number of deaths per 100,000 population due to heart disease is *decreasing* by 5 every year. What about the amount of garbage recycled each year? We will see if that is increasing or decreasing in Example 6.

EXAMPLE 6 Garbage recycling

The amount R (in millions of tons) of garbage recovered for recycling can be approximated by $R = 2.2N + 69$, where N is the number of years after 2000.

a. What is the slope of $R = 2.2N + 69$?
b. What does the slope represent?

SOLUTION 6

a. The slope of the line $R = 2.2N + 69$ is 2.2.
b. The slope 2.2 means an annual *increase* of 2.2 million tons for the amount of garbage recovered for recycling after the year 2000.

Where do we store all this? See Problem 6 for the real challenge.

PROBLEM 6

In theory, the number of landfills L in the United States can be approximated by $L = -100N + 1964$, where N is the number of years between 2000 and 2002 inclusive.

a. What is the slope of $L = -100N + 1964$?
b. What does the slope represent?

The actual fact is that the number of landfills *after* 2005 has been a constant 1754.

Calculator Corner

Using STAT to find $y = ax + b$

In this section we learned how to find the slope of a line given two points. Your calculator can do better! Let's try Example 1, where we are given the points (0, −6) and (3, 3) and asked to find the slope. Press STAT 1 and enter the two x-coordinates (0 and 3) under L_1 and the two y-coordinates (−6 and 3) under L_2. Your calculator is internally programmed to use a process called *regression* to find the equation of the line passing through these two points. To do so, press STAT, move the cursor right to reach CALC, enter 4 for LinReg (Linear Regression), and then press ENTER. The resulting line shown in Window 1 is written in the slope-intercept form $y = ax + b$. Clearly, the slope is 3 and the y-intercept is −6. See what happens when you use your calculator to do Example 2.

What happens when you do Example 3a? You get the warning shown in Window 2, indicating that the slope is *undefined*. Here you have to *know* some algebra to conclude that the resulting line is a *vertical* line. (Your calculator can't do that for you!)

```
LinReg
 y=ax+b
 a=3
 b=-6
 r=1
```

Window 1

```
ERR:DOMAIN
1:Quit
2:Goto
```

Window 2

Answers to PROBLEMS

6. a. −100
b. That the number of landfills has been *decreasing* by 100 each year between 2000 and 2002 inclusive.

Exercises 3.4

> Practice Problems > Self-Tests > Media-rich eBooks > e-Professors > Videos

⟨A⟩ Finding Slopes from Two Points In Problems 1–14, find the slope of the line that passes through the two given points.

1. (1, 2) and (3, 4)
2. (1, −2) and (−3, −4)
3. (0, 5) and (5, 0)
4. (3, −6) and (5, −6)
5. (−1, −3) and (7, −4)
6. (−2, −5) and (−1, −6)
7. (0, 0) and (12, 3)
8. (−1, −1) and (−10, −10)
9. (3, 5) and (−2, 5)
10. (4, −3) and (2, −3)
11. (4, 7) and (−5, 7)
12. (−3, −5) and (−2, −5)
13. $\left(-\frac{1}{2}, -\frac{1}{3}\right)$ and $\left(-\frac{1}{2}, \frac{1}{3}\right)$
14. $\left(-\frac{1}{5}, 2\right)$ and $\left(-\frac{1}{5}, 1\right)$

⟨B⟩ Finding Slopes from Equations In Problems 15–26, find the slope of the given line.

15. $y = 3x + 7$
16. $y = -4x + 6$
17. $-3y = 2x - 4$
18. $4y = 6x + 3$
19. $x + 3y = 6$
20. $-x + 2y = 3$
21. $-2x + 5y = 5$
22. $3x - y = 6$
23. $y = 6$
24. $x = 7$
25. $2x - 4 = 0$
26. $2y - 3 = 0$

⟨C⟩ Finding Parallel and Perpendicular Lines In Problems 27–37, determine whether the given lines are parallel, perpendicular, or neither.

27. $y = 2x + 5$ and $4x - 2y = 7$
28. $y = 4 - 5x$ and $15x + 3y = 3$
29. $2x + 5y = 8$ and $5x - 2y = -9$
30. $3x + 4y = 4$ and $2x - 6y = 7$
31. $x + 7y = 7$ and $2x + 14y = 21$
32. $y - 5x = 12$ and $y - 3x = 8$
33. $2x + y = 7$ and $-2x - y = 9$
34. $2x - 4 = 0$ and $x - 1 = 0$
35. $2y - 4 = 0$ and $3y - 6 = 0$
36. $3y = 6$ and $2x = 6$
37. $3x = 7$ and $2y = 7$

⟨D⟩ Applications Involving Slope

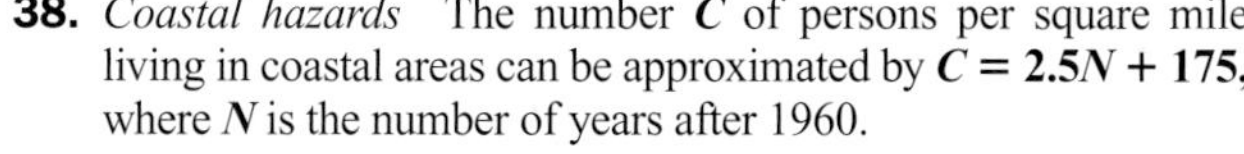

Applications: Green Math

The Intergovernmental Panel on Climate Change (IPCC) forecasts a sea level rise of between 19 and 59 centimeters by 2100. Who will be affected? Nearly half of the U.S. population lives in coastal areas susceptible to coastal hazards! The graph shows that the number of persons per square mile is increasing. How much? We will see next.

38. *Coastal hazards* The number C of persons per square mile living in coastal areas can be approximated by $C = 2.5N + 175$, where N is the number of years after 1960.

a. What is the slope of $C = 2.5N + 175$?

b. What does the slope represent?

c. How many persons per square mile does the equation $C = 2.5N + 175$ predict for the year 2010 (50 years after 1960)?

d. Is the result you get in part **c** close to the one shown in the graph?

Source: http://tinyurl.com/ye94pcv.

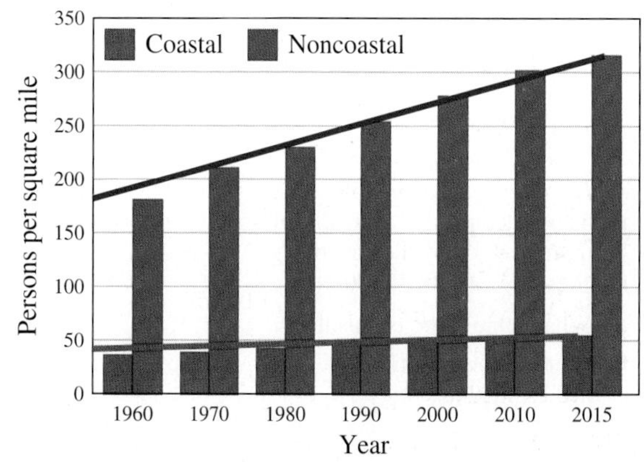

Population Density, 1960–2015

39. *Noncoastal population* The number **NC** of persons per square mile living in noncoastal areas can be approximated by

NC = 0.36N + 30, where N is the number of years after 1960.

a. What is the slope of **NC = 0.36N + 30?**

b. What does the slope represent?

c. How many persons per square mile does the equation **NC = 0.36N + 30** predict for the year 2010 (50 years after 1960)?

d. Is the result you get in part **c** close to the one shown in the graph?

40. *Daily seafood consumption* According to the U.S. Department of Agriculture, the daily consumption T of tuna can be approximated by the equation

$$T = 3.87 - 0.13t \text{ (pounds)}$$

and the daily consumption F of fish and shellfish can be approximated by the equation

$$F = -0.29t + 15.46$$

a. Is the consumption of tuna increasing or decreasing?

b. Is the consumption of fish and shellfish increasing or decreasing?

c. Which consumption (tuna or fish and shellfish) is decreasing faster?

41. *Life expectancy of women* The average life span (life expectancy) y of an American woman is given by the equation $y = 0.15t + 80$, where t is the number of years after 2000.

a. What is the slope of this line?

b. Is the life span of American women increasing or decreasing?

c. What does the slope represent?

Source: U.S. National Center for Health Statistics, *Statistical Abstract of the United States.*

42. *Life expectancy of men* The average life span y of an American man is given by the equation $y = 0.15t + 74$, where t is the number of years after 2000.

a. What is the slope of this line?

b. Is the life span of American men increasing or decreasing?

c. What does the slope represent?

Source: U.S. National Center for Health Statistics, *Statistical Abstract of the United States.*

43. *Velocity of a thrown ball* The speed v of a ball thrown up with an initial velocity of 128 feet per second is given by the equation $v = 128 - 32t$, where v is the velocity (in feet per second) and t is the number of seconds after the ball is thrown.

a. What is the slope of this line?

b. Is the velocity of the ball increasing or decreasing?

c. What does the slope represent?

44. *Velocity of a thrown ball* The speed v of a ball thrown up with an initial velocity of 15 meters per second is given by the equation $v = 15 - 5t$, where v is the velocity (in meters per second) and t is the number of seconds after the ball is thrown.

a. What is the slope of this line?

b. Is the velocity of the ball increasing or decreasing?

c. What does the slope represent?

45. *Daily fat consumption* The number of fat grams f consumed daily by the average American can be approximated by the equation $f = 165 + 0.4t$, where t is the number of years after 2000.

a. What is the slope of this line?

b. Is the consumption of fat increasing or decreasing?

c. What does the slope represent?

Source: U.S. Dept. of Agriculture, *Statistical Abstract of the United States.*

46. *Milk products consumption* The number of gallons of milk products g consumed annually by the average American can be approximated by the equation $g = 24 - 0.2t$, where t is the number of years after 2000.

a. What is the slope of this line?

b. Is the consumption of milk products increasing or decreasing?

c. What does the slope represent?

Source: U.S. Dept. of Agriculture, *Statistical Abstract of the United States.*

Web IT go to mhhe.com/bello for more lessons

Footwear For Problems 47–49, use the information below. The graph shows the annual expenditure on footwear in five consecutive years by persons under 25 years of age.

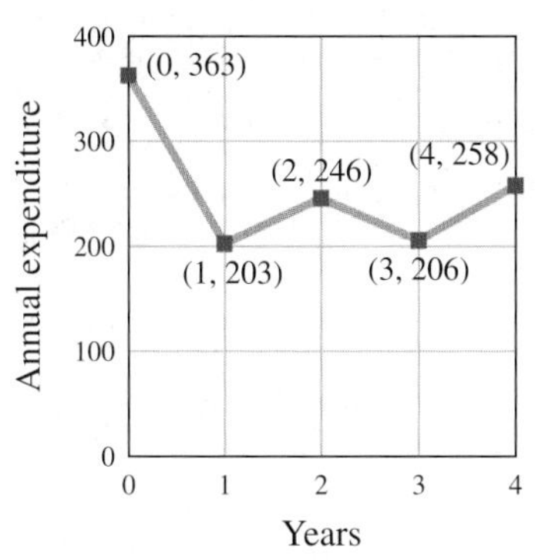

2000 (0)	363
2001 (1)	203
2002 (2)	246
2003 (3)	206
2004 (4)	258

Source: U.S. Department of Labor.

47. a. In what year did the expenditures on footwear decrease the most?

b. What was the decrease from year 0 to year 1?

c. Find the slope m of the line in part **b.**

d. What does the slope m represent?

48. a. In what years did the expenditures on footwear decrease?

b. What was the decrease during year 2?

c. Find the slope m of the line for year 2

d. What does the slope m represent?

49. a. In what year did the expenditures on footwear increase the most?

b. Find the slope m of the line from year 3 to year 4.

c. Find the slope m of the line from year 1 to year 2.

d. Which slope is larger, the slope for year 3 to 4 or the slope for year 1 to 2?

Using Your Knowledge

Up, Down, or Away! The slope of a line can be positive, negative, zero, or undefined. Use your knowledge to sketch the following lines.

50. A line that has a negative slope

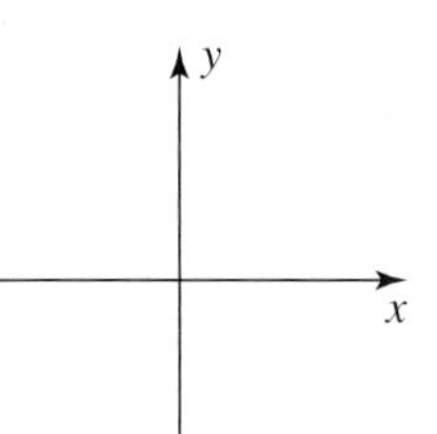

51. A line that has a positive slope

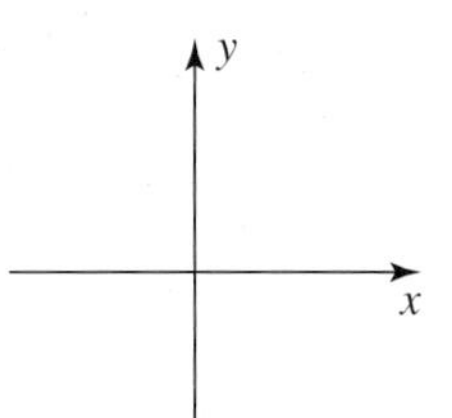

52. A line with zero slope

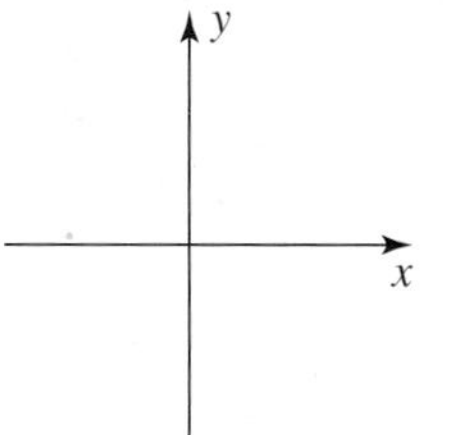

53. A line with an undefined slope

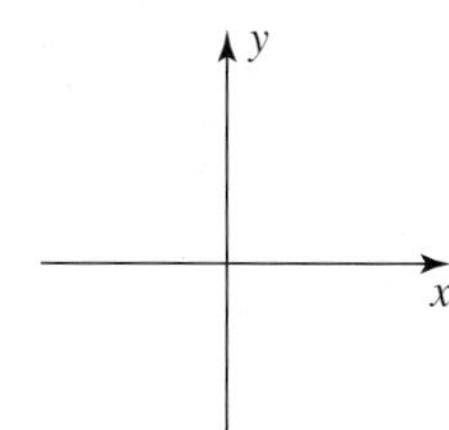

Write On

54. Write in your own words what is meant by the *slope* of a line.

55. Explain why the slope of a horizontal line is zero.

56. Explain why the slope of a vertical line is undefined.

Concept Checker

Fill in the blank(s) with the correct word(s), phrase, or mathematical statement.

57. The **slope** of the line through (x_1, y_1) and (x_2, y_2) is ________.

58. The **slope** of the line $y = mx + b$ is ________.

59. Two **lines** with the **same slope** and **different** y-intercepts are ________ lines.

60. Two lines with **slopes whose product is −1** are ____________ lines.

$\frac{x_2 - x_1}{y_2 - y_1}$	**perpendicular**
b	m
parallel	$\frac{y_2 - y_1}{x_2 - x_1}$

Mastery Test

Find the slope of the line passing through the given points:

61. $(2, -3)$ and $(4, -5)$

62. $(-1, 2)$ and $(4, -2)$

63. $(-4, 2)$ and $(-4, 5)$

64. $(-3, 4)$ and $(-5, 4)$

65. Find the slope of the line $3x + 2y = 6$.

66. Find the slope of the line $-3x + 4y = 12$.

Determine whether the lines are parallel, perpendicular, or neither:

67. $-x + 3y = -6$ and $2x - 6y = -7$

68. $2x + 3y = 5$ and $3x - 2y = 5$

69. $3x - 2y = 6$ and $-2x - 3y = 6$

70. The U.S. population P (in millions) can be approximated by $P = 2.2t + 180$, where t is the number of years after 1960.

a. What is the slope of this line?

b. How fast is the U.S. population growing each year? (State your answer in millions.)

Skill Checker

Simplify:

71. $2[x - (-4)]$

72. $3[x - (-6)]$

73. $-2[x - (-1)]$

3.5 Graphing Lines Using Points and Slopes

Objectives

Find and graph an equation of a line given:

A Its slope and a point on the line.

B Its slope and y-intercept.

C Two points on the line.

To Succeed, Review How To . . .

1. Add, subtract, multiply, and divide signed numbers (pp. 52–56, 60–64).
2. Solve a linear equation for a specified variable (pp. 137–143).

Getting Started

The Formula for Cholesterol Reduction

As we discussed in Example 3, Section 3.2 (p. 229), the cholesterol level C can be approximated by $C = -3w + 215$, where w is the number of weeks elapsed. How did we get this equation? You can see by looking at the graph that the cholesterol level decreases *about* 3 points each week, so the slope of the line is -3. You can make this approximation more exact by using the points $(0, 215)$ and $(12, 175)$ to find the slope

$$m = \frac{215 - 175}{0 - 12} = -\frac{40}{12} \approx -3.3$$

Cholesterol Level Reduction

Since the y-intercept is at 215, you can reason that the cholesterol level starts at 215 and decreases about 3 points each week. Thus, the cholesterol level C based on the number of weeks elapsed is

$$C = 215 - 3w$$

or, equivalently,

$$C = -3w + 215$$

If you want a more exact approximation, you can also write

$$C = -3.3w + 215$$

Thus, if you are given the slope m and y-intercept b of a line, the equation of the line will be $y = mx + b$.

In this section, we shall learn how to find and graph a linear equation when we are given the slope and a point on the line, the slope and the y-intercept, or two points on the line.

A > Using the Point-Slope Form of a Line

We can use the slope of a line to obtain the equation of the line, provided we are given one point on the line. Thus, suppose a line has slope m and passes through the point (x_1, y_1). If we let (x, y) be a second point on the line, the slope of the line shown in Figure 3.37 is given by

$$\frac{y - y_1}{x - x_1} = m$$

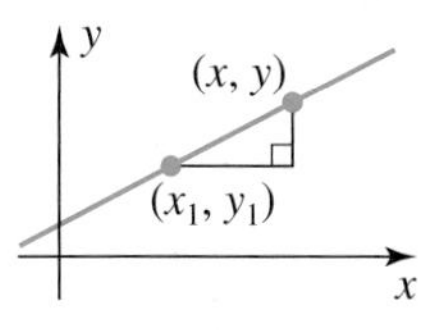

>Figure 3.37

Multiplying both sides by $(x - x_1)$, we get the *point-slope form* of the line.

POINT-SLOPE FORM

The **point-slope form** of the equation of the line going through (x_1, y_1) and having slope m is

$$y - y_1 = m(x - x_1)$$

EXAMPLE 1 Finding an equation for a line given a point and the slope

a. Find an equation of the line that passes through the point $(2, -3)$ and has slope $m = -4$.

b. Graph the line.

SOLUTION 1

a. Using the point-slope form, we get

$$y - (-3) = -4(x - 2)$$
$$y + 3 = -4x + 8$$
$$y = -4x + 5$$

b. To graph this line, we start at the point $(2, -3)$. Since the slope of the line is -4 and, by definition, the slope is

$$\frac{\text{Rise}}{\text{Run}} = -\frac{4}{1}$$

we go 4 units *down* (the rise) and 1 unit *right* (the run), ending at the point $(3, -7)$. We then join the points $(2, -3)$ and $(3, -7)$ with a line, which is the graph of $y = -4x + 5$ (see Figure 3.38). As a final check, does the point $(3, -7)$ satisfy the equation $y = -4x + 5$? When $x = 3$, $y = -4x + 5$ becomes

$$y = -4(3) + 5 = -7 \quad \text{True!}$$

Thus, the point $(3, -7)$ is on the line $y = -4x + 5$. Note that the equation $y + 3 = -4x + 8$ can be written in the standard form $Ax + By = C$.

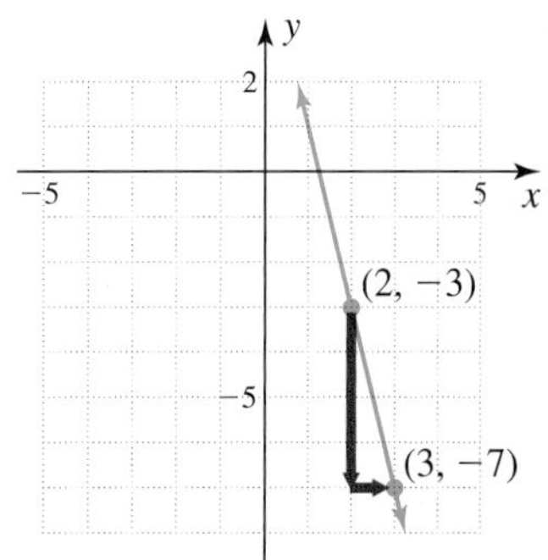

>Figure 3.38

$$y + 3 = -4x + 8 \quad \text{Given}$$
$$4x + y + 3 = 8 \quad \text{Add } 4x.$$
$$4x + y = 5 \quad \text{Subtract 3.}$$

PROBLEM 1

a. Find an equation of the line that goes through the point $(2, -4)$ and has slope $m = -3$.

b. Graph the line.

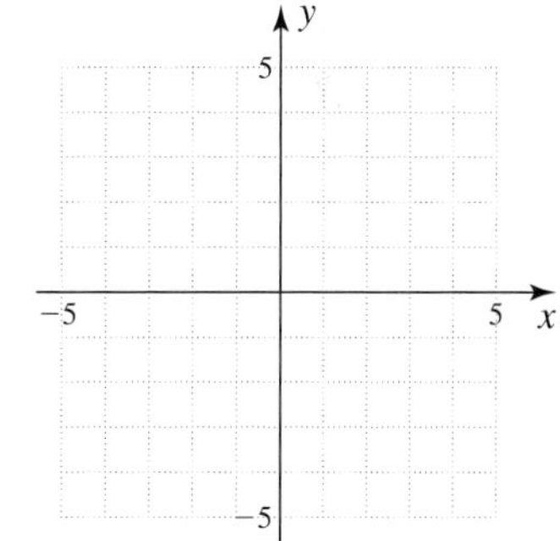

Answer on page 268

B > The Slope-Intercept Form of a Line

An important special case of the point-slope form is that in which the given point is the point where the line intersects the y-axis. Let this point be denoted by $(0, b)$. Then b is called the *y-intercept* of the line. Using the point-slope form, we obtain

$$y - b = m(x - 0)$$

or

$$y - b = mx$$

By adding b to both sides, we get the *slope-intercept form* of the equation of the line.

SLOPE-INTERCEPT FORM

The **slope-intercept form** of the equation of the line having slope m and y-intercept b is

$$y = mx + b$$

EXAMPLE 2 Finding an equation of a line given the slope and the y-intercept

a. Find an equation of the line having slope 5 and y-intercept -4.

b. Graph the line.

SOLUTION 2

a. In this case $m = 5$ and $b = -4$. Substituting in the slope-intercept form, we obtain

$$y = 5x + (-4) \quad \text{or} \quad y = 5x - 4$$

b. To graph this line, we start at the y-intercept $(0, -4)$. Since the slope of the line is 5 and by definition the slope is

$$\frac{\text{Rise}}{\text{Run}} = \frac{5}{1}$$

go 5 units *up* (the rise) and 1 unit *right* (the run), ending at $(1, 1)$. We join the points $(0, -4)$ and $(1, 1)$ with a line, which is the graph of $y = 5x - 4$ (see Figure 3.39). Now check that the point $(1, 1)$ is on the line $y = 5x - 4$.

When $x = 1$, $y = 5x - 4$ becomes

$$y = 5(1) - 4 = 1 \quad \text{True!}$$

Thus, the point $(1, 1)$ is on the line $y = 5x - 4$.

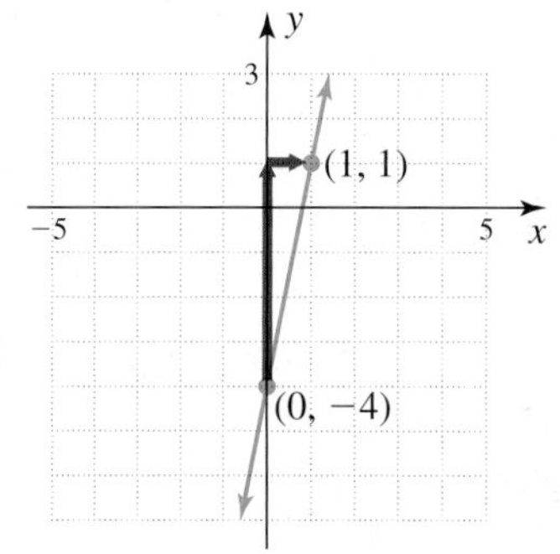

>Figure 3.39

PROBLEM 2

a. Find an equation of the line with slope 3 and y-intercept -6.

b. Graph the line.

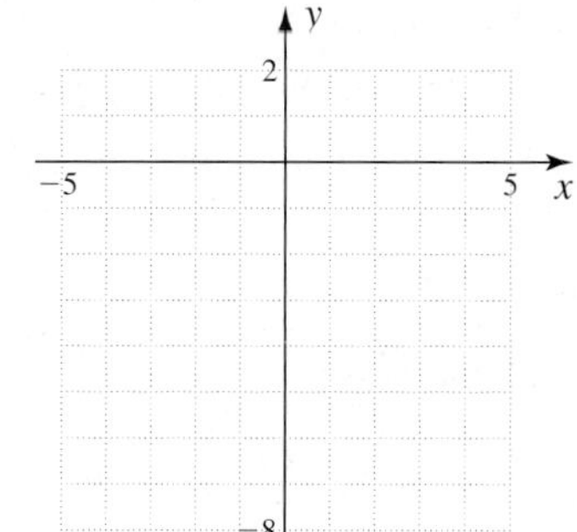

Answers to PROBLEMS

1. a. $y = -3x + 2$ **b.**

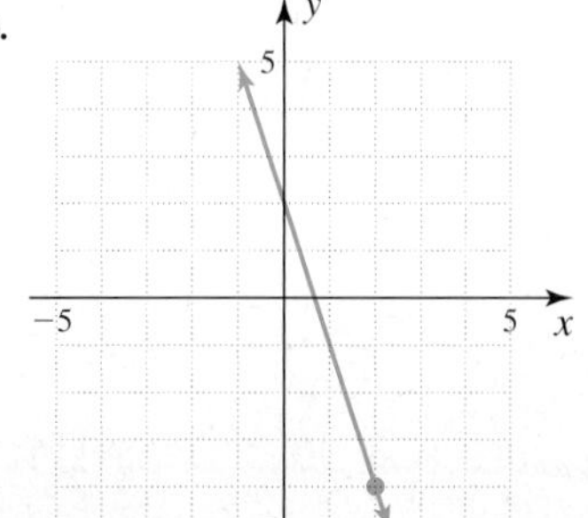

2. a. $y = 3x - 6$ **b.**

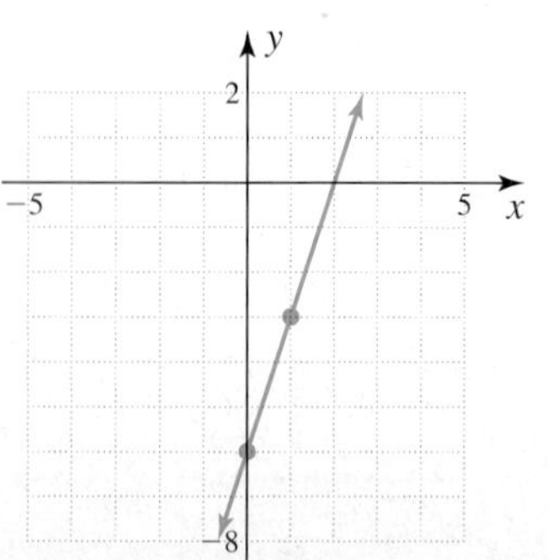

C > The Two-Point Form of a Line

If a line passes through two given points, you can find and graph the equation of the line by using the point-slope form as shown in Example 3, where we also find the equation of a line representing the increase in the number of persons living in coastal areas.

GREEN MATH

EXAMPLE 3 Finding an equation of a line given two points

a. Find an equation of the line passing through the points $(-2, 3)$ and $(1, -3)$.

b. Graph the line.

c. Find an equation of the line passing through the points $(0, 175)$ and $(50, 300)$, the line representing the increase in the number of persons living in coastal areas discussed in Problem 38, Section 3.4.

d. Graph the line.

SOLUTION 3

a. We first find the slope of the line.

$$m = \frac{3 - (-3)}{-2 - 1} = \frac{6}{-3} = -2$$

Now we can use either the point $(-2, 3)$ or $(1, -3)$ and the point-slope form to find the equation of the line. Using the point $(-2, 3) = (x_1, y_1)$ and $m = -2$,

$$y - y_1 = m(x - x_1)$$

becomes

$$\begin{aligned} y - 3 &= -2[x - (-2)] \\ y - 3 &= -2(x + 2) \\ y - 3 &= -2x - 4 \\ y &= -2x - 1 \end{aligned}$$

or, in standard form, $2x + y = -1$.

b. To graph this line, we simply plot the given points $(-2, 3)$ and $(1, -3)$ and join them with a line, as shown in Figure 3.40. To check our results, we make sure that both points satisfy the equation $2x + y = -1$. For $(-2, 3)$, let $x = -2$ and $y = 3$ in $2x + y = -1$ to obtain

$$2(-2) + 3 = -4 + 3 = -1$$

Thus, $(-2, 3)$ satisfies $2x + y = -1$, and our result is correct.

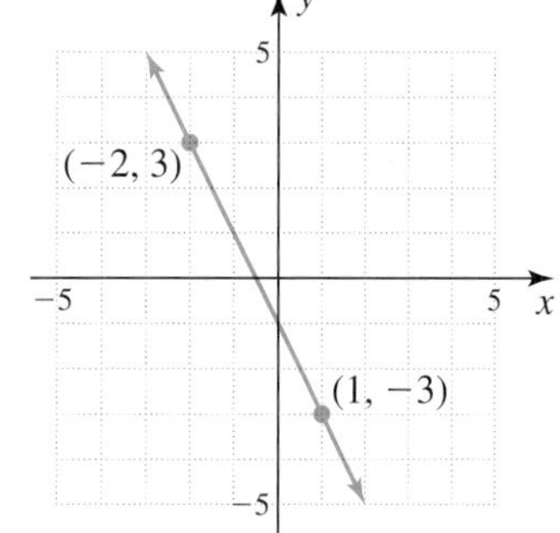

>Figure 3.40

c. The slope of the line is $m = \frac{300 - 175}{50 - 0} = \frac{125}{50} = 2.5$.
Using the point $(0, 175) = (x_1, y_1)$ and $m = 2.5$,

$$y - y_1 = m(x - x_1)$$

becomes $y - 175 = 2.5(x - 0)$

$$\begin{aligned} y - 175 &= 2.5x \\ y &= 2.5x + 175 \end{aligned}$$

Thus, the equation of the line is $y = 2.5x + 175$.

(continued)

PROBLEM 3

a. Find an equation of the line going through $(5, -6)$ and $(-3, 4)$.

b. Graph the line.

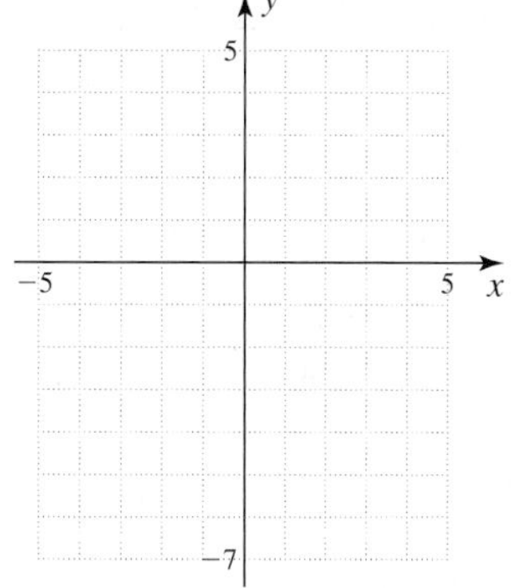

c. Find an equation of the line passing through $(0, 30)$ and $(60, 51)$, the line representing the increase in the number of persons living in noncoastal areas from Problem 39, Section 3.4.

d. Graph the line.

Answer on page 270

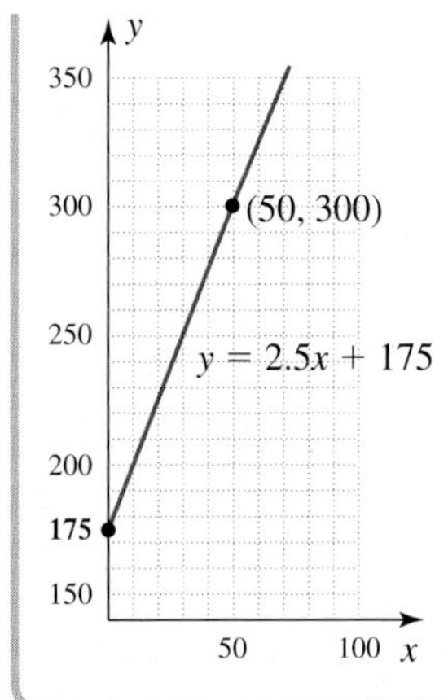

d. To graph this line plot the points (0, 175) and (50, 300) and join them with a line. To check our results make sure that points (0, 175) and (50, 300) satisfy the equation. For (50, 300), let $x = 50$ and $y = 300$ in $y = 2.5x + 175$ obtaining $300 = 2.5(50) + 175$, which is a true statement. Thus, our graph is correct. You can compare this graph with the one in Problem 38, Section 3.4.

At this point, many students ask, "How do we know which form to use?" The answer depends on what information is given in the problem. The following table helps you make this decision. It's a good idea to examine this table closely before you attempt the problems in Exercises 3.5.

Finding the Equation of a Line

Given	Use
A point (x_1, y_1) and the slope m	Point-slope form: $y - y_1 = m(x - x_1)$
The slope m and the y-intercept b	Slope-intercept form: $y = mx + b$
Two points (x_1, y_1) and (x_2, y_2), $x_1 \neq x_2$	Two-point form: $y - y_1 = m(x - x_1)$, where $m = \dfrac{y_2 - y_1}{x_2 - x_1}$

Note that the resulting equation can always be written in the standard form: $Ax + By = C$.

Answers to PROBLEMS

3. a.  $y = -\frac{5}{4}x + \frac{1}{4}$ **b.**

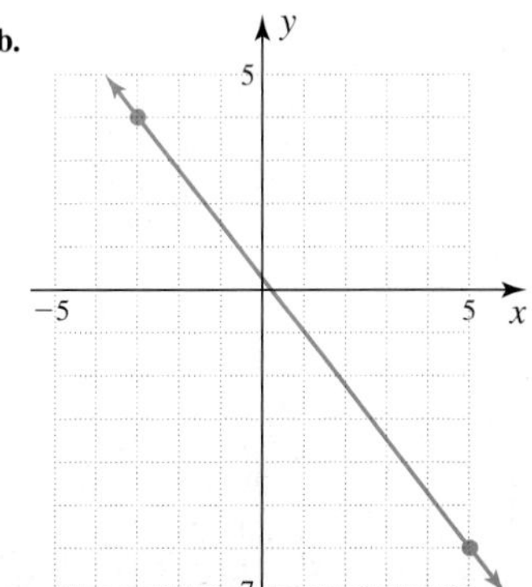

c. $y = \frac{21}{60}x + 30$ **d.**

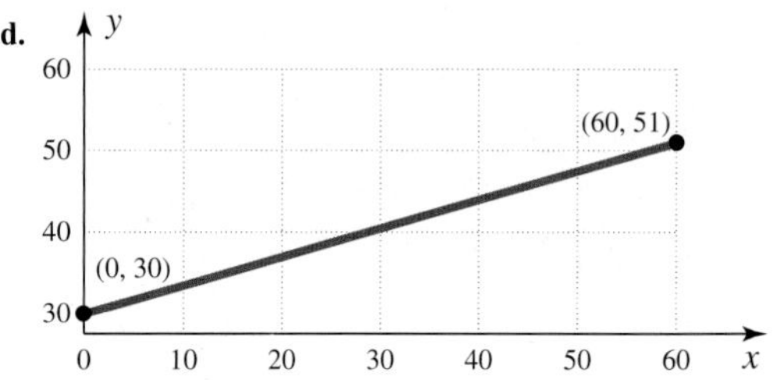

> Exercises 3.5

> Practice Problems > Self-Tests
> Media-rich eBooks > e-Professors > Videos

< A > Using the Point-Slope Form of a Line In Problems 1–6, find the equation of the line that has the given properties (m is the slope), then graph the line.

1. Goes through (1, 2); $m = \frac{1}{2}$

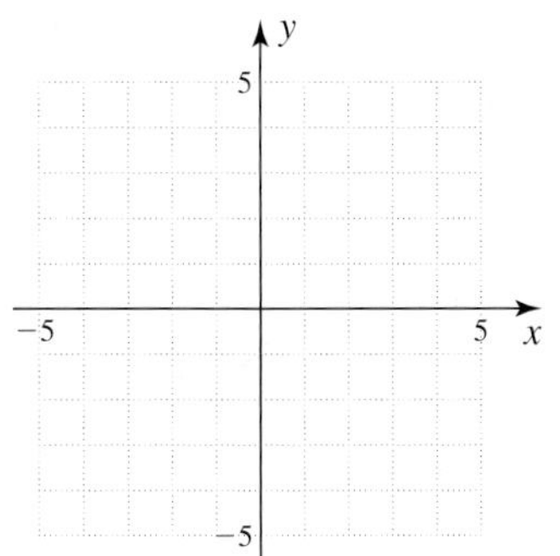

2. Goes through $(-1, -2)$; $m = -2$

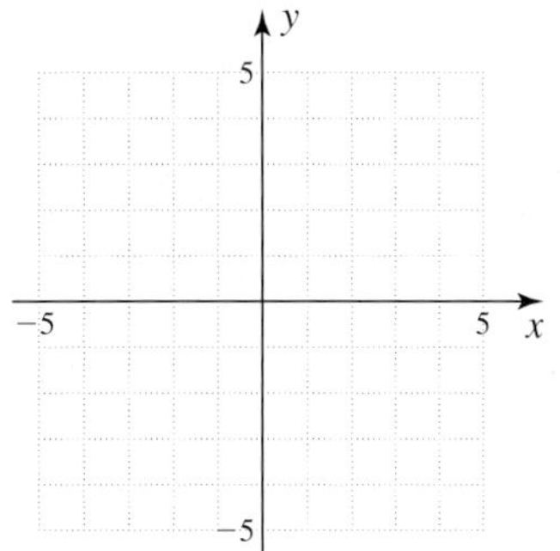

3. Goes through (2, 4); $m = -1$

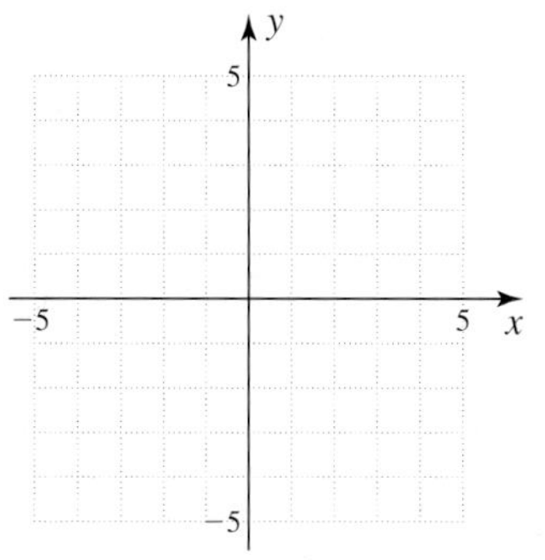

4. Goes through $(-3, 1)$; $m = \frac{3}{2}$

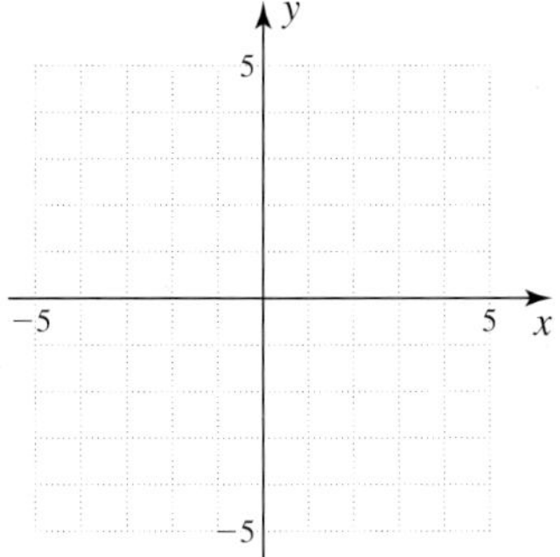

5. Goes through (4, 5); $m = 0$

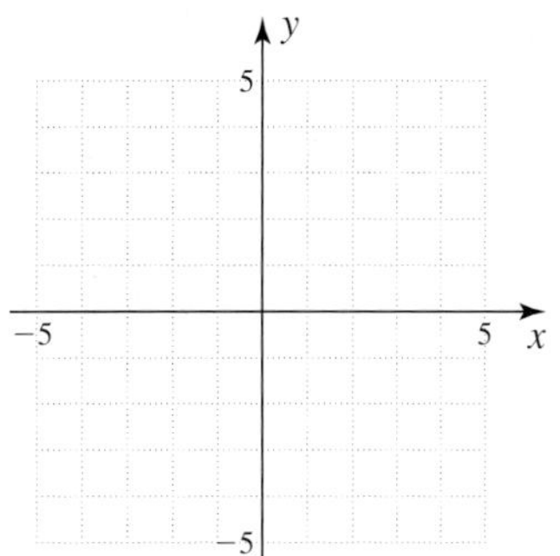

6. Goes through (3, 2); slope is not defined (does not exist)

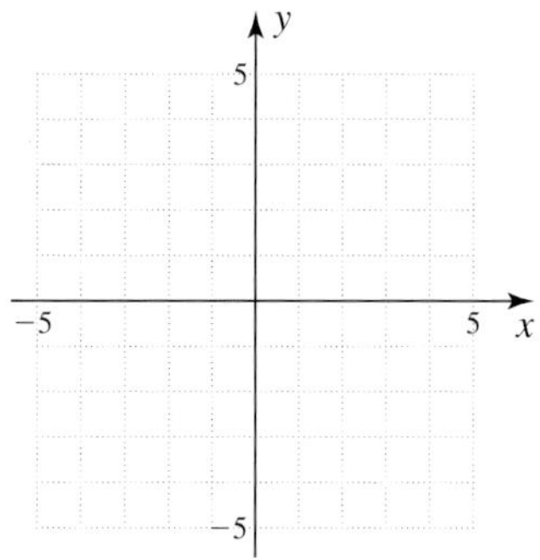

< B > The Slope-Intercept Form of a Line In Problems 7–16, find an equation of the line with the given slope and y-intercept.

7. Slope, 2; y-intercept, -3

8. Slope, 3; y-intercept, -5

9. Slope, -4; y-intercept, 6

10. Slope, -6; y-intercept, -7

11. Slope, $\frac{3}{4}$; y-intercept, $\frac{7}{8}$

12. Slope $\frac{7}{8}$; y-intercept, $\frac{3}{8}$

13. Slope, 2.5; y-intercept, -4.7

14. Slope, 2.8; y-intercept, -3.2

15. Slope, -3.5; y-intercept, 5.9

16. Slope, -2.5; y-intercept, 6.4

> Web IT go to mhhe.com/bello for more lessons

In Problems 17–20, find an equation of the line having slope m and y-intercept b, and then graph the line.

17. $m = \frac{1}{4}, b = 3$

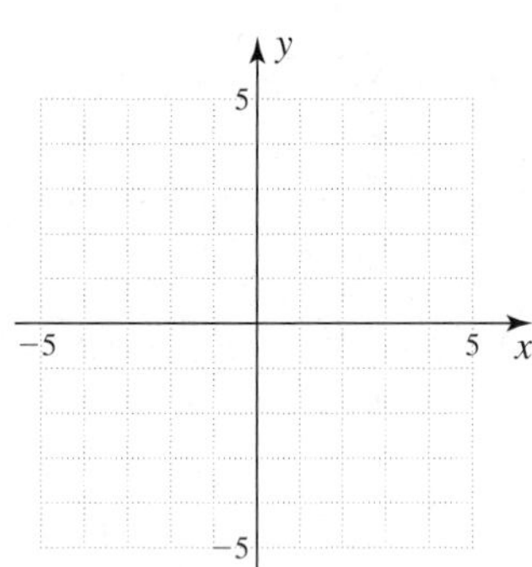

18. $m = -\frac{2}{5}, b = 1$

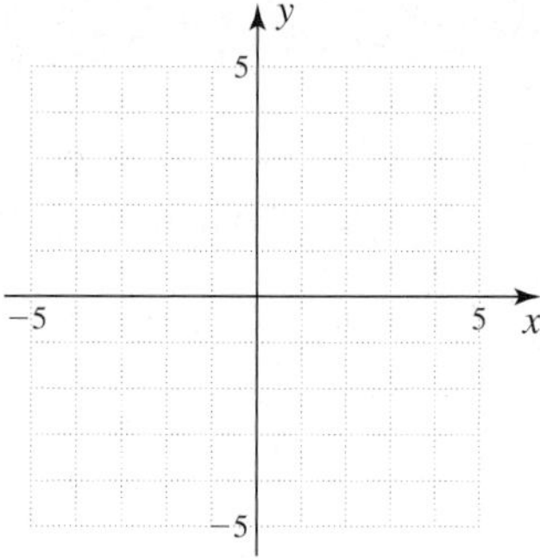

19. $m = -\frac{3}{4}, b = -2$

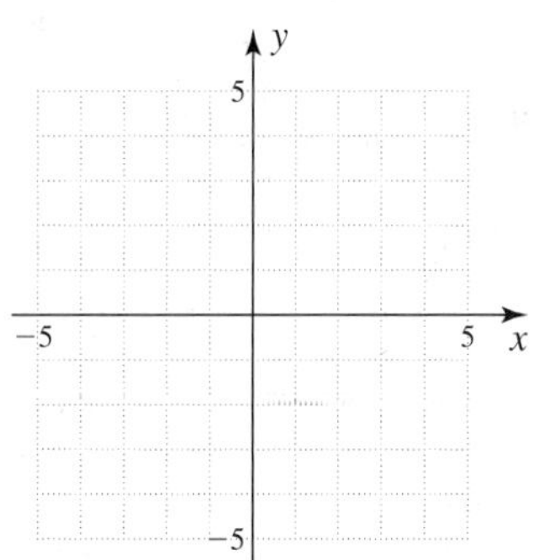

20. $m = -\frac{1}{3}, b = -1$

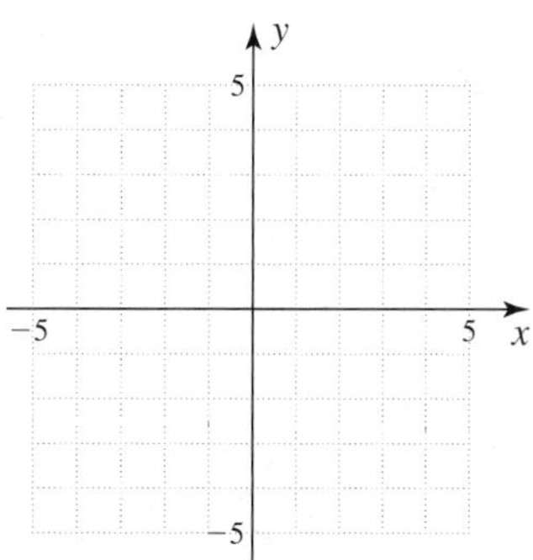

‹ C › The Two-Point Form of a Line In Problems 21–30, find an equation of the line passing through the given points, write the equation in standard form, and then graph the equation.

21. (2, 3) and (0, 1)

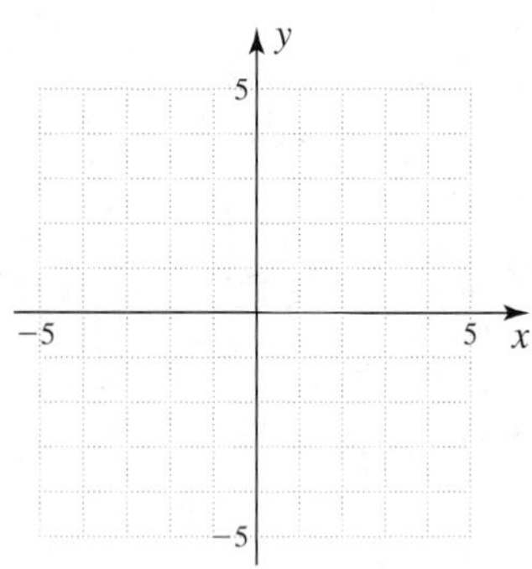

22. (−2, −3) and (1, −6)

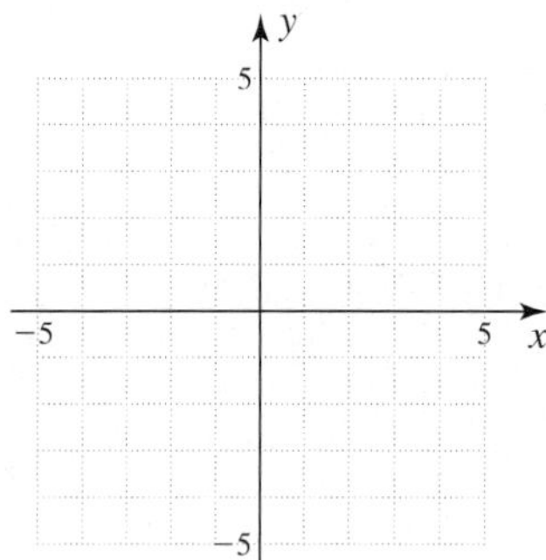

23. (2, 2) and (1, −1)

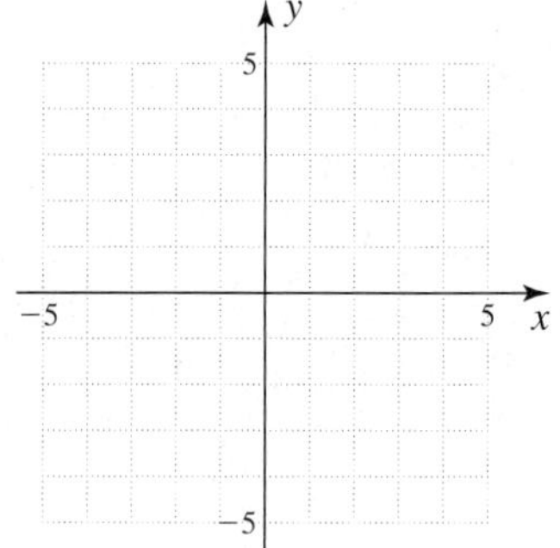

24. (−3, 4) and (−2, 0)

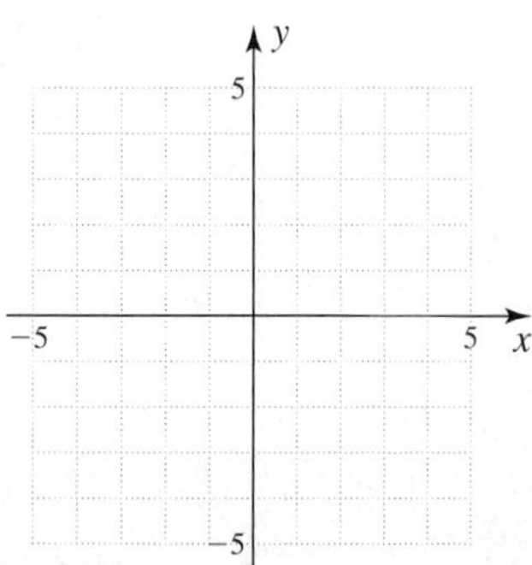

25. (3, 0) and (0, 4)

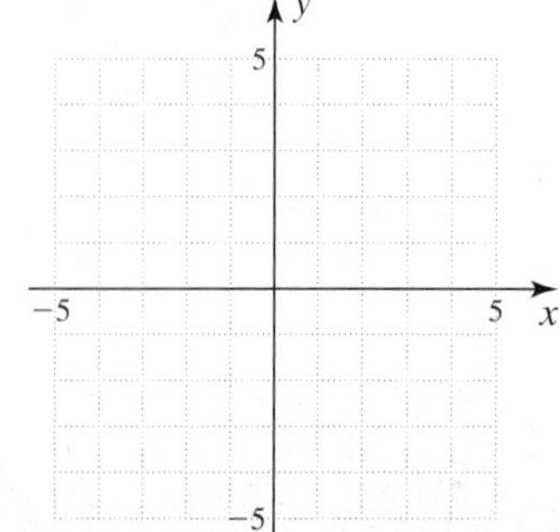

26. (0, −3) and (4, 0)

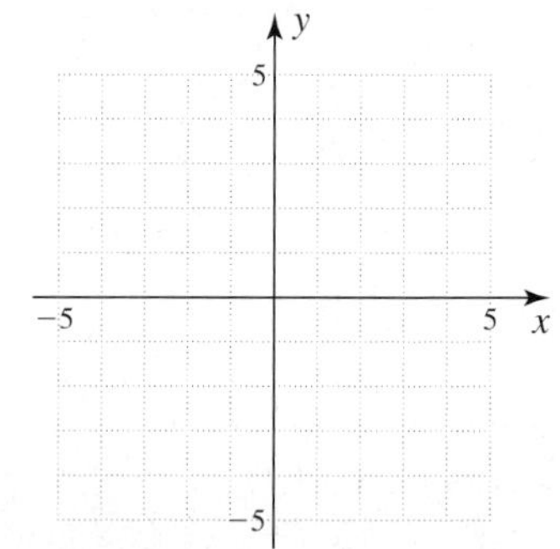

27. (3, 0) and (3, 2)

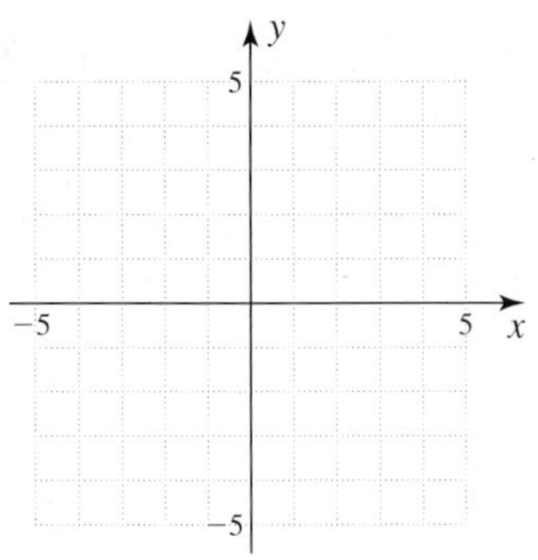

28. (−4, 2) and (−4, 0)

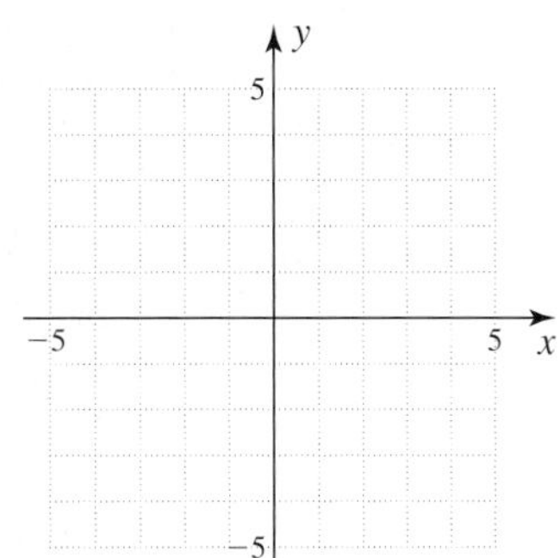

29. (−2, −3) and (1, −3)

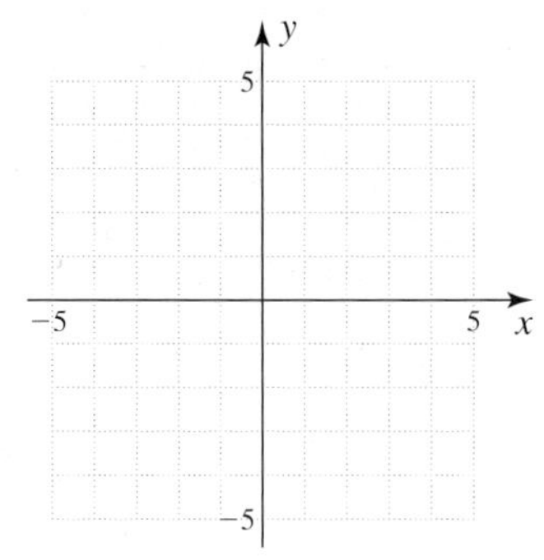

30. (−3, 0) and (−3, 4)

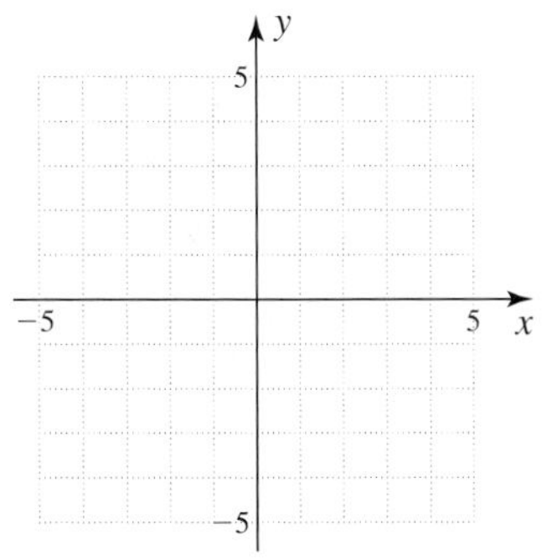

In Problems 31–36, find an equation of the line that satisfies the given conditions.

31. Passes through (0, 3) and (4, 7)

32. Passes through (−1, −2) and (3, 6)

33. Slope 2 and passes through (1, 2)

34. Slope −3 and passes through (−1, −2)

35. Slope −2 and y-intercept −3

36. Slope 4 and y-intercept −1

Web IT go to mhhe.com/bello for more lessons

Using Your Knowledge

Applications: Green Math

The Business of Slopes and Intercepts The slope m and the y-intercept of an equation play important roles in economics, business, and your personal finances! Let's see how.

Do you use regular (incandescent) lightbulbs or fluorescent ones? Incandescent bulbs are cheaper ($0.25) but use more electricity ($0.08 each day for a 100-watt bulb used 8 hours daily). Thus, the total cost y (in dollars) of using an incandescent bulb for x days is

$$\underbrace{y}_{\text{Total cost}} = \underbrace{0.08x}_{\text{Cost per day}} + \underbrace{0.25}_{\text{Cost of bulb}}$$

In general the total cost y of using a bulb costing m dollars each day for x days is

$$y = mx + b$$

where b is the cost of the bulb.

As we have learned, m is the slope of the line and b the y-intercept.

37. *Cost of operating fluorescent bulbs*

a. Find the total cost y for x days of using a MaxLite spiral 25-watt fluorescent bulb costing $0.02 to operate each day for 8 hours if the bulb costs $2.50.

b. If the total cost (in dollars) of using a GE 26-watt spiral bulb for x days is given by $y = 0.03x + 7$, what is the cost of operating the bulb each day and the cost of the bulb?

38. *Cost of operating fluorescent bulbs* Find the total cost y for x days of using a 25-watt fluorescent bulb costing $0.025 to operate each day if the bulb costs $6.

39. *Comparing operating bulb prices*

a. Find the cost for 30 days of using an incandescent bulb costing $y = 0.08x + 0.25$ dollars for x days.

b. Find the cost for 30 days of using a fluorescent bulb costing $y = 0.03x + 1.50$ dollars for x days.

c. In how many days will the cost of using the incandescent of part **a** and the fluorescent of part **b** be the same?

Write On

40. If you are given the equation of a nonvertical line, describe the procedure you would use to find the slope of the line.

41. If you are given a vertical line whose x-intercept is the origin, what is the name of the line?

42. If you are given a horizontal line whose y-intercept is the origin, what is the name of the line?

43. How would you write the equation of a vertical line in the standard form $Ax + By = C$?

44. How would you write the equation of a horizontal line in the standard form $Ax + By = C$?

45. Write an explanation of why a vertical line cannot be written in the form $y = mx + b$.

Concept Checker

Fill in the blank(s) with the correct word(s), phrase, or mathematical statement.

46. The **point-slope form** of the equation of the line passing through (x_1, y_1) and having slope m is ______________.

47. The **slope-intercept** form of the equation of the line having slope m and y-**intercept** b is _________.

48. The **two-point** form of the equation of the line passing through the points (x_1, y_1) and (x_2, y_2) is ________________.

49. The **standard form** of a straight line equation in two variables is ___________.

$x - x_1 = m(y - y_1)$

$x - x_1 = \frac{y_2 - y_1}{x_2 - x_1}(y - y_1)$

$y = mx + b$

$y = ax + m$

$x + y = C$

$y - y_1 = m(x - x_1)$

$Ax + By = C$

$y - y_1 = \frac{y_2 - y_1}{x_2 - x_1}(x - x_1)$

Mastery Test

50. Find an equation of the line with slope 3 and y-intercept -6, and graph the line.

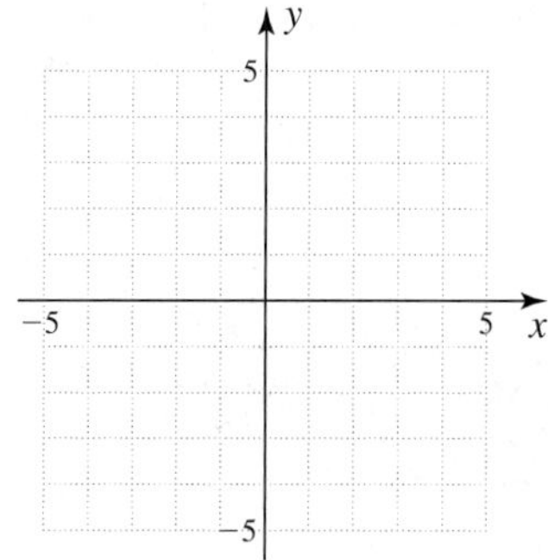

51. Find an equation of the line with slope $-\frac{3}{4}$ and y-intercept 2, and graph the line.

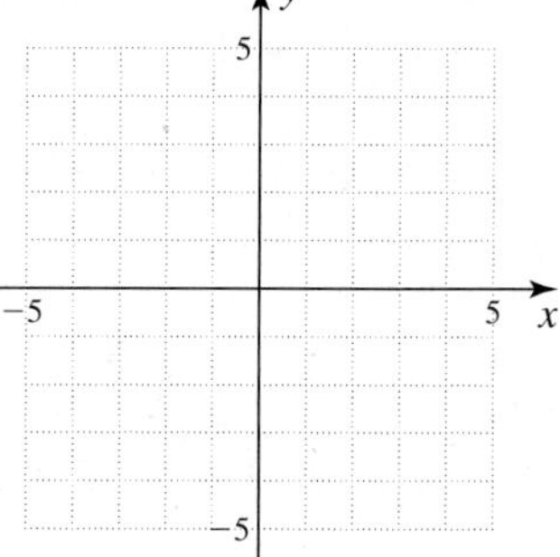

52. Find an equation of the line passing through the point $(-2, 3)$ and with slope -3, and graph the line.

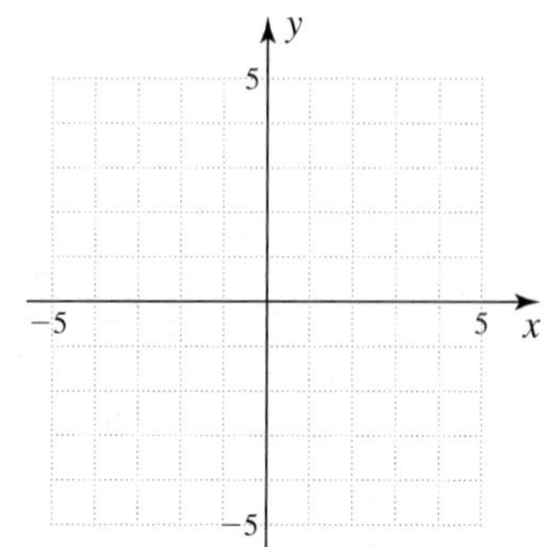

53. Find an equation of the line passing through the point $(3, -1)$ and with slope $\frac{2}{3}$, and graph the line.

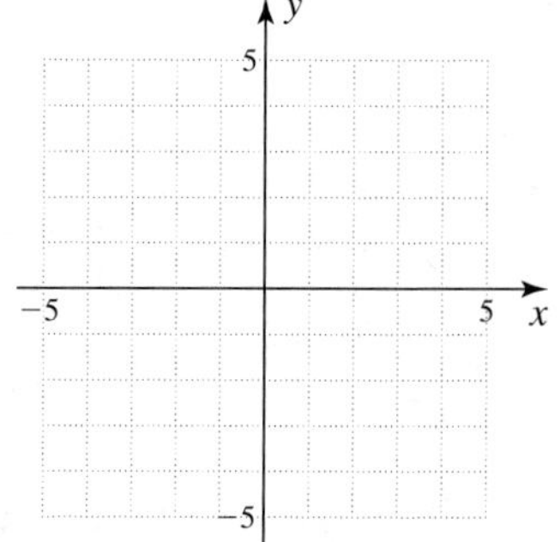

54. Find an equation of the line passing through the points $(-3, 4)$ and $(-2, -6)$, write it in standard form, and graph the line.

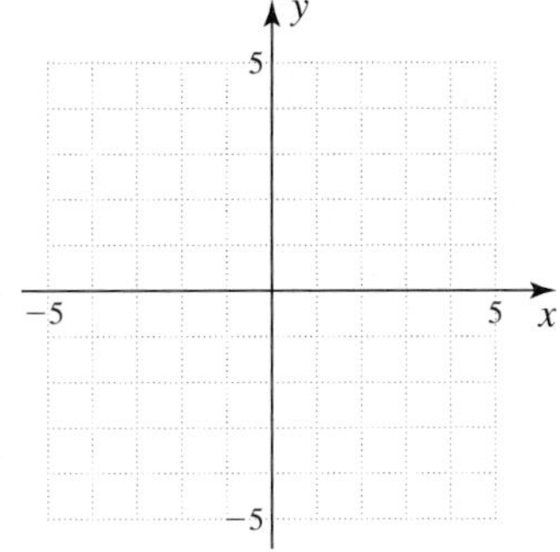

55. Find an equation of the line passing through the points $(-1, -6)$ and $(-3, 4)$, write it in standard form, and graph the line.

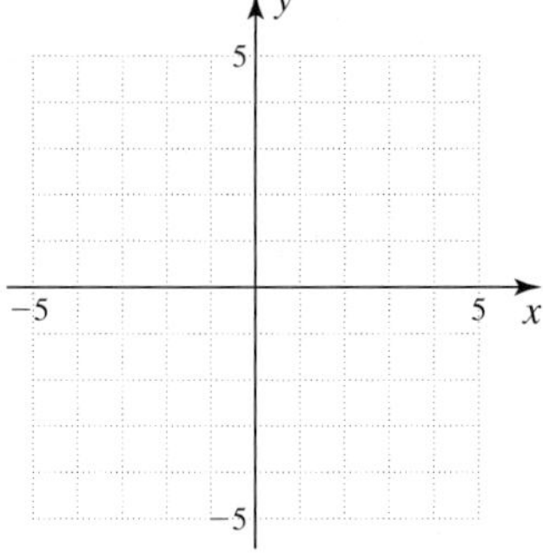

Skill Checker

56. Find the point-slope form of the equation of a line with slope $m = 2.5$ and passing through the point $(0, 1.75)$.

57. Find the slope intercept form of the equation of a line with y-intercept 50 and slope 0.35.

58. Find the slope m of the line passing through $(6, 12)$ and $(8, 20)$.

59. Write in standard form the equation of the line passing through $(6, 12)$ and $(8, 20)$.

3.6 Applications of Equations of Lines

Objectives

A Solve applications involving the point-slope formula.

B Solve applications involving the slope-intercept formula.

C Solve applications involving the two-point formula.

To Succeed, Review How To . . .

1. Find the equation of a line given a point and the slope (p. 267).
2. Find the equation of a line given a point and the y-intercept (p. 268).
3. Find the equation of a line given two points (pp. 269–270).

Getting Started

Taxi! Taxi!

How much is a taxi ride in your city? In Boston, it is \$1.50 for the first $\frac{1}{4}$ mile and \$2 for each additional mile. As a matter of fact, it costs \$21 for a 10-mile ride. Can we find a simple formula based on the number of miles m you ride? If C represents the total cost of the ride and we know that a 10-mile ride costs \$21, we are given one point—namely, (10, 21). We also know that the slope is 2, since the costs are \$2 per additional mile. Using the point-slope form, we have

$$C - 21 = 2(m - 10)$$

or

$$C = 2m - 20 + 21$$

That is,

$$C = 2m + 1$$

Note that this is consistent with the fact that a 10-mile ride costs \$21, since

$$C = 2(10) + 1 = 21$$

In the preceding section, we learned how to find and graph the equation of a line when given a point and the slope, a point and the y-intercept, or two points. In this section we are going to use the appropriate formulas to solve application problems.

A > Applications Involving the Point-Slope Formula

EXAMPLE 1 **Taxicab fare calculations**

Taxi fares in Key West are \$2.25 for the first $\frac{1}{5}$ mile and then \$2.50 for each additional mile. If a 10-mile ride costs \$26.75, find an equation for the total cost C of an m-mile ride. What will be the cost for a 30-mile ride?

SOLUTION 1 Since we know that a 10-mile ride costs \$26.75, and each additional mile costs \$2.50, we are given the point (10, 26.75) and the slope 2.5.

Using the point-slope form, we have

$$C - 26.75 = 2.5(m - 10)$$

or

$$C = 2.5m - 25 + 26.75$$

That is,

$$C = 2.5m + 1.75$$

For a 30-mile ride, $m = 30$ and $C = 2.5(30) + 1.75 = \$76.75$.

PROBLEM 1

A taxicab company charges \$2.50 for the first $\frac{1}{4}$ mile and \$2.00 for each additional mile. If a 10-mile ride costs \$22, find an equation for the total cost C of an m-mile ride. What is the cost of a 25-mile ride?

B > Applications Involving the Slope-Intercept Formula

Do you have a cell phone? Which plan do you have? Most plans have a set number of free minutes for a set fee, after which you pay for additional minutes used. For example, at the present time, Verizon Wireless® has a plan that allows 900 free minutes for \$50 with unlimited weekend minutes. After that, you pay \$0.35 for each additional minute. We will consider such a plan in Example 2.

EXAMPLE 2 **Cell phone plan charges**

Maria subscribed to a cell phone plan with 900 free minutes, a \$50 monthly fee, and \$0.35 for each additional minute. Find an equation for the total cost C of her plan when she uses m minutes after the first 900. What is her cost when she uses 1200 minutes?

SOLUTION 2 Let m be the number of minutes Maria uses after the first 900. If she does not go over the limit, she will pay \$50. The y-intercept will then be 50. Since she pays \$0.35 for each additional minute, the slope is 0.35. Thus, the total cost C when m additional minutes are used is

$$C = 0.35m + 50$$

If she uses 1200 minutes, she has to pay for 300 of them (1200 − 900 free = 300 paid), and the cost will then be

$$\begin{aligned} C &= 0.35(300) + 50 \\ &= 105 + 50 \\ &= \$155 \end{aligned}$$

Thus, Maria will pay \$155 for her monthly payment.

PROBLEM 2

Find the cost of the plan if the monthly fee is \$40 and each minute after the first 900 costs \$0.50. What is the cost when she uses 1000 minutes?

Answers to PROBLEMS

1. $C = 2m + 2$; \$52

2. $C = 0.50m + 40$; \$90

C > Applications Involving the Two-Point Formula

What is the electricity cost of running your computer? The monthly energy cost for a "green" computer (red graph) is claimed to be **\$0.73 per month** (6 hours per day, 7 days a week at a cost of \$0.13/kWh) and the costs of running a "typical" computer (blue graph) is claimed to be *6 times* as much. The graph shows only the ordered pairs (2, 9) and (9, 40), which means the "typical" computer cost \$9 for 2 months and \$40 for 9 months. Now that we know how to make our own graph, we can verify the claims in Example 3.

Source: http://tinyurl.com/ycxacev.

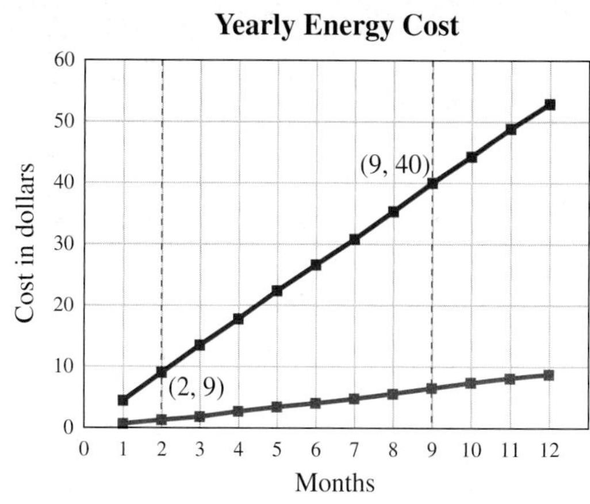

GREEN MATH

EXAMPLE 3 Green versus typical computer annual energy cost

a. The green computer costs \$0.73 each month. Find an equation for the monthly cost y, where x is the number of months.

b. The blue graph passes through the points (2, 9) and (9, 40). Find the equation of the line passing through these two points.

c. Graph the equations obtained in parts **a** and **b.**

SOLUTION 3

a. Since the monthly change is \$0.73 per each month x, the monthly cost y for the green computer is $y = 0.73x$.

b. We use the two-point form $y - y_1 = m(x - x_1)$, $m = \frac{y_2 - y_1}{x_2 - x_1}$

where $(x_1, y_1) = (2, 9)$ and $(x_2, y_2) = (9, 40)$, obtaining

$$m = \frac{40 - 9}{9 - 2} = \frac{31}{7} \quad \text{and} \quad y - 9 = \frac{31}{7}(x - 2)$$

To clear fractions, multiply both sides by 7. $7(y - 9) = 7 \cdot \frac{31}{7}(x - 2)$

Simplify. $7y - 63 = 31x - 62$

Add 63 to both sides. $7y = 31x + 1$

Subtract $31x$. $7y - 31x = 1$

Thus, the equation of the line passing through (2, 9) and (9, 40) is $7y - 31x = 1$, or in standard form $-31x + 7y = 1$.

c. To graph $y = 0.73x$, let $x = 0$ yielding $y = 0$. Then let $x = 9$, which means that $y = 0.73 \cdot 9 = 6.57$. Graph the points (0, 0) and (9, 6.57) and join them with a red line, the graph of $y = 0.73x$. (See margin)

To graph $7y - 31x = 1$, let $x = 0$ yielding $y = \frac{1}{7}$. Then let $x = 9$ which means that $7y - 31 \cdot 9 = 1$, $7y - 279 = 1$, or $7y = 280$.

Thus, $y \approx 40$. Graph the points $(0, \frac{1}{7})$ and (9, 40) and join them with a blue line, the graph of $7y - 31x = 1$. Do the two graphs look like the ones given at the beginning of the discussion? You be the judge!

PROBLEM 3

To verify the claim that the "typical" computer costs 6 times as much as the "green."

a. Find the cost of operating the green computer for 9 months. *Hint:* The calculations are in the example.

b. Find the cost of operating the "typical" computer for 9 months.

c. From the answers to **a** and **b,** is the cost of the typical computer about 6 times as much as the "green" computer?

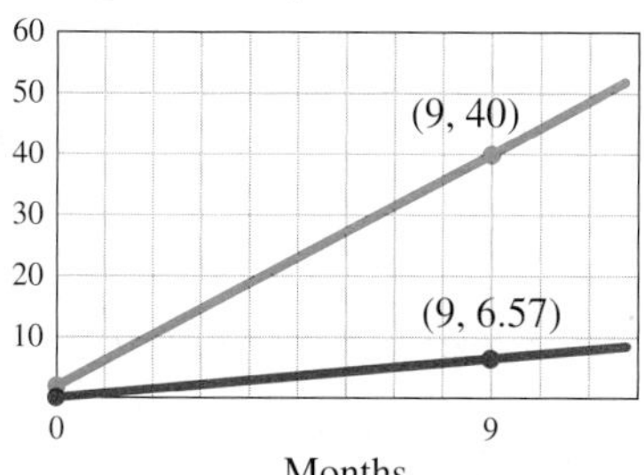

Answers to PROBLEMS

3. a. \$6.57 **b.** \$40

c. 6 times the cost of operating the green computer is \$39.42, about the same as the \$40 cost for the typical computer.

> Exercises 3.6

⟨ A ⟩ Applications Involving the Point-Slope Formula

In Problems 1–10, solve the application.

1. *San Francisco taxi fares* Taxi fares in San Francisco are \$2 for the first mile and \$1.70 for each additional mile. If a 10-mile ride costs \$17.30, find an equation for the total cost C of an m-mile ride. What would the price be for a 30-mile ride?

2. *San Francisco taxi fares* A different taxicab company in San Francisco charges \$3 for the first mile and \$1.50 for each additional mile. If a 20-mile ride costs \$31.50, find an equation for the total cost C of an m-mile ride. What would be the price of a 10-mile ride? Which company is cheaper, this one or the company in Problem 1?

3. *San Francisco taxi fares* Pedro took a cab in San Francisco and paid the fare quoted in Problem 1. Tyrone paid the fare quoted in Problem 2. Amazingly, they paid the same amount! How far did they ride?

4. *New York taxi fares* In New York, a 20-mile cab ride is \$32 and consists of an initial set charge and \$1.50 per mile. Find an equation for the total cost C of an m-mile ride. How much would you have to pay for a 30-mile ride?

5. *San Francisco taxi fares* The cost C for San Francisco fares (Problem 1) is \$2 for the first mile and \$1.70 for each mile thereafter. We can find C by following these steps:

a. What is the cost of the first mile?

b. If the whole trip is m miles, how many miles do you travel after the first mile?

c. How much do you pay per mile after the first mile?

d. What is the cost of all the miles after the first?

e. The total cost C is the sum of the cost of the first mile and the cost of all the miles after the first. What is that cost? Is your answer the same as that in Problem 1?

6. *New York taxi fares* New York fares are easier to compute. They are simply \$2 for the initial set charge and \$1.50 for each mile. Use the procedure in Problem 5 to find the total cost C for an m-mile trip. Do you get the same answer as you did in Problem 4?

7. *Cell phone rental overseas* Did you know that you can rent cell phones for your overseas travel? If you are in Paris, the cost for a 1-week rental, including 60 minutes of long-distance calls to New York, is \$175.

a. Find a formula for the total cost C of a rental phone that includes m minutes of long-distance calls to New York.

b. What is the weekly charge for the phone?

c. What is the per-minute usage charge?

8. *International long-distance rates* Long-distance calls from the Hilton Hotel in Paris to New York cost \$7.80 per minute.

a. Find a formula for the cost C of m minutes of long-distance calls from Paris to New York.

b. How many minutes can you use so that the charges are identical to those you would pay when renting the phone of Problem 7? Answer to the nearest minute.

9. *Wind chill temperatures* The table shows the relationship between the actual temperature (x) in degrees Fahrenheit and the wind chill temperature (y) when the wind speed is 5 miles per hour.

Wind Chill (Wind Speed 5 mi/hr)

Temperature	(x)	20	25	30
Wind Chill	(y)	13	19	25

a. Find the slope of the line (the rate of change of the wind chill temperature) using the points (20, 13) and (25, 19).

b. Find the slope of the line using the points (25, 19) and (30, 25).

c. Are the two slopes the same?

d. Use the point-slope form to find y using the points (20, 13) and (25, 19).

e. Use the point-slope form to find y using the points (25, 19) and (30, 25). Do you get an equation equivalent to the one in part **d?**

f. Use your formula to find the wind chill when the temperature is 5°F and the wind speed is 5 miles per hour.

10. *Heat index values* The table shows the relationship between the actual temperature (x) and the heat index or apparent temperature (y) when the relative humidity is 50%. *Note:* Temperatures above 105°F can cause severe heat disorders with continued exposure!

Relative Humidity (50%)

Temperature	(x)	100	102	104
Heat Index	(y)	118	124	130

a. Find the slope of the line using the points (100, 118) and (102, 124).

b. Find the slope of the line using the points (102, 124) and (104, 130).

c. Are the two slopes the same?

d. Use the point-slope form of the line to find y using the points (100, 118) and (102, 124).

e. Use the point-slope form of the line to find y using the points (102, 124) and (104, 130). Do you get an equation equivalent to the one in part **d?**

f. Use your formula to find the heat index when the temperature is 106°F and the relative humidity is 50%.

⟨B⟩ Applications Involving the Slope-Intercept Formula

In Problems 11–22, solve the application.

11. *Electrician's charges* According to Microsoft Home Advisor®, if you have an electrical failure caused by a blown fuse or circuit breaker you may have to call an electrician. "Plan on spending at least \$100 for a service call, plus the electrician's hourly rate." Assume that the service call is \$100 and the electrician charges \$37.50 an hour.

a. Find an equation for the total cost C of a call lasting h hours.

b. If your bill amounts to \$212.50, for how many hours were you charged?

12. *Appliance technician's charges* Microsoft Home Advisor suggests that the cost C for fixing a faulty appliance is about \$75 for the service call plus the technician's hourly rate. Assume that this rate is \$40 per hour.

a. Find an equation for the total cost C of a call lasting h hours.

b. If your bill amounts to \$195, for how many hours were you charged?

13. *International phone rental charges* The total cost C for renting a phone for 1 week in London, England, is \$40 per week plus \$2.30 per minute for long-distance charges when calling the United States.

a. Find a formula for the cost C of renting a phone and using m minutes of long-distance charges.

b. If the total cost C amounted to \$201, how many minutes were used?

c. The Dorchester Hotel in London charges \$7.20 per minute for long-distance charges to the United States. Find a formula for the cost C of m minutes of long-distance calls from the hotel to the United States.

d. How many minutes can you use so that the charges are identical to those you would pay when renting the phone of part **a** (round to nearest minute)?

14. *Phone rental charges* The rate for incoming calls for a rental phone is \$1.20 per minute. If the rental charge is \$50 per week, find a formula for the total cost C of renting a phone for m incoming minutes. If you paid \$146 for the rental, for how many incoming minutes were you billed?

15. *Cell phone costs* Verizon Wireless has a plan that costs \$50 per month for 900 anytime minutes and unlimited weekend minutes. The charge for each additional minute is \$0.35.

a. Find a formula for the cost C when m additional minutes are used.

b. If your bill was for \$88.50, how many additional minutes did you use?

16. *Cell phone costs* A Verizon competitor has a similar plan, but it charges \$45 per month and \$0.40 for each additional minute.

a. Find a formula for the cost C when m additional minutes are used.

b. If your bill was for \$61, how many additional minutes did you use?

17. *Cell phone costs* How many additional minutes do you have to use so that the costs for the plans in Problems 15 and 16 are identical? After how many additional minutes is the plan in Problem 15 cheaper?

18. *Estimating a man's height* You can estimate a man's height y (in inches) by multiplying the length x of his femur bone by 1.88 and adding 32 to the result.

a. Write an equation for a man's estimated height y in slope-intercept form.

b. What is the slope?

c. What is the y-intercept?

19. *Estimating a female's height* The height y for a female (in inches) can be estimated by multiplying the length x of her femur bone by 1.95 and adding 29 to the result.

a. Write an equation for a woman's height y in slope-intercept form.

b. What is the slope?

c. What is the y-intercept?

20. *Estimating height from femur length* Refer to Problems 18 and 19.

a. What would be the length x of a femur that would yield the same height y for a male and a female?

b. What would the estimated height of a person having such a femur bone be?

21. *Average hospital stay* Since 1970, the number y of days the average person stays in the hospital has steadily decreased from 8 days at the rate of 0.1 day per year. If x represents the number of years after 1970, write a slope-intercept equation representing the average number y of days a person stays in the hospital. What would the average stay in the year 2000 be? In 2010?

22. *Average hospital stay for females* Since 1970, the number y of days the average female stays in the hospital has steadily decreased from 7.5 days at the rate of 0.1 day per year. If x represents the number of years after 1970, write a slope-intercept equation representing the average number y of days a female stays in the hospital. What would the average stay in the year 2000 be? In 2010?

Web IT go to mhhe.com/bello for more lessons

‹C› Applications Involving the Two-Point Formula In Problems 23–26, solve the application.

23. *Dog years vs. human years* It is a common belief that 1 human year is equal to 7 dog years. That is not very accurate, since dogs reach adulthood within the first couple of years. A more accurate formula indicates that when a dog is 3 years old in human years, it would be 30 years old in dog years. Moreover, a 9-year-old dog is 60 years old in dog years.

a. Form the ordered pairs (h, d), where h is the age of the dog in human years and d is the age of the dog in dog years. What does (3, 30) mean? What does (9, 60) mean?

b. Find the slope of the line using the points (3, 30) and (9, 60).

c. Find an equation for the age d of a dog based on the dog's age h in human years ($h > 1$).

d. If a dog is 4 human years old, how old is it in dog years?

e. If a human could retire at age 65, what is the equivalent retirement age for a dog (in human years)?

f. The drinking age for humans is usually 21 years. What is the equivalent drinking age for dogs?

24. *Cat years vs. human years* When a cat is $h = 2$ years old in human years, it is $c = 24$ years old in cat years, and when a cat is $h = 6$ years old in human years, it is $c = 40$ years old in cat years.

a. Form the ordered pairs (h, c), where h is the age of the cat in human years and c is the age of the cat in cat years. What does (2, 24) mean? What does (6, 40) mean?

b. Find the slope of the line using the points (2, 24) and (6, 40).

c. Find an equation for the age c of a cat based on the cat's age h in human years ($h > 1$).

d. If a cat is 4 human years old, how old is it in cat years?

e. If a human could retire at age 60, what is the equivalent retirement age for a cat in human years?

f. The drinking age for humans is usually 21 years. What is the equivalent drinking age for cats?

25. *Blood alcohol concentration* Your blood alcohol level is dependent on the amount of liquor you consume, your weight, and your gender. Suppose you are a 150-pound male and you consume 3 beers (5% alcohol) over a period of 1 hour. Your blood alcohol concentration (BAC) would be 0.052. If you have 5 beers, then your BAC would be 0.103. (You are legally drunk then! Most states regard a BAC of 0.08 as legally drunk.) Consider the ordered pairs (3, 0.052) and (5, 0.103).

a. Find the slope of the line passing through the two points.

b. If b represents the number of beers you had in 1 hour and c represents your BAC, find an equation for c.

c. Find your BAC when you have had 4 beers in 1 hour. Find your BAC when you have had 6 beers in 1 hour.

d. How many beers do you have to drink in 1 hour to be legally drunk (BAC = 0.08)? Answer to the nearest whole number.

26. *Blood alcohol concentration* Suppose you are a 125-pound female and you consume 3 beers (5% alcohol) over a period of 1 hour. Your blood alcohol concentration (BAC) would be 0.069. If you have 5 beers, then your BAC would be 0.136. (You are really legally drunk then!) Consider the ordered pairs (3, 0.069) and (5, 0.136).

a. Find the slope of the line passing through the two points.

b. If b represents the number of beers you had in 1 hour and c represents your BAC, find an equation for c.

c. Find your BAC when you have had 4 beers in 1 hour. Find your BAC when you have had 6 beers in 1 hour.

d. How many beers do you have to drink in 1 hour to be legally drunk (BAC = 0.08)? Answer to the nearest whole number. You can check this information by going to link 7-2-5 on the Bello Website at mhhe.com/bello.

››› Applications: Green Math

27. The energy cost for operating a computer 24/7 can be approximated by $\mathbf{y = 10.50x}$ per month, where x is the number of months. Assume that the cost of one kilowatt-hour (kWH) is 12 cents. If you use the "sleep" mode, the approximation is $\mathbf{y = 1.35x}$.

a. Graph $y = 10.50x$.

b. Graph $y = 1.35x$.

c. Find the cost of operating the computer in both the "regular" and the "sleep" mode for a year and determine the annual savings.

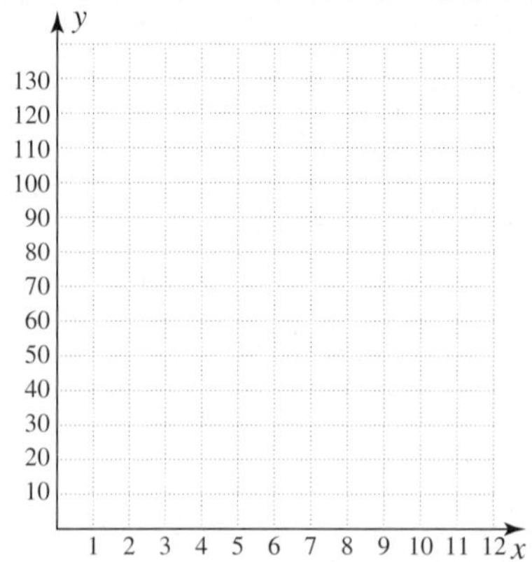

28. If the cost per kilowatt-hour is 10 cents, the cost for operating a computer can be approximated by $y = 8.75x$. In sleep mode, the cost is approximated by $y = 1.15x$, where x is the number of months.

a. Graph $y = 8.75x$.

b. Graph $y = 1.15x$.

c. Find the cost of operating the computer in both the "regular" and the "sleep" mode for a year and determine the annual savings.

Source: http://tinyurl.com/no2rkn.

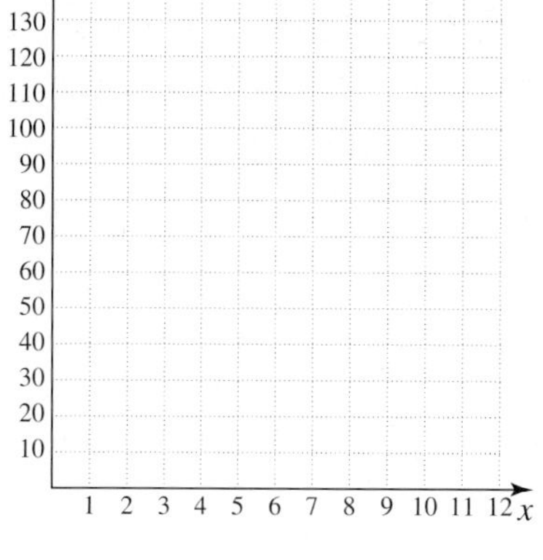

29. *Medical applications* A child's height H (in centimeters) can be estimated using the age A (2–12 years) by using the formula:

Height in centimeters = (age in years) · 6 + 77

a. Write the equation in symbols using H for height and A for age.

b. What is the slope m of $H = 6A + 77$?

c. What is the intercept b of $H = 6A + 77$?

d. Find the estimated height H when $A = 2$. Graph the point.

e. Find the estimated height H when $A = 12$. Graph the point.

f. Use the two points from parts **d** and **e** to graph $H = 6A + 77$.

For convenience, use a y-scale with 50 unit increments.

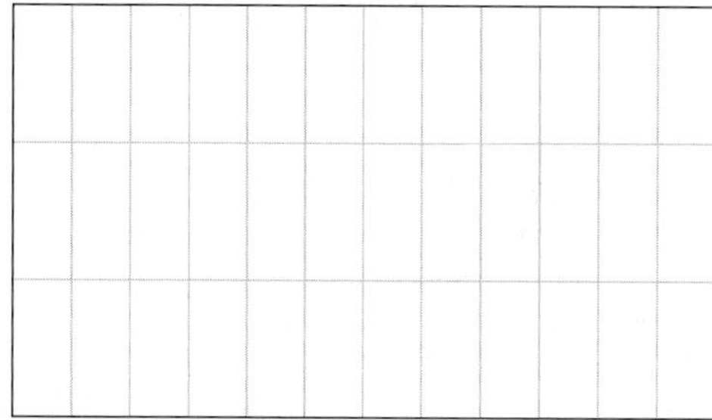

Source: http://www.medal.org.

30. *Medical applications* According to Munoz et al., the most reliable estimate for the height H (in centimeters) of a female is obtained by using the tibia bone and is given by

Height in centimeters = 76.53 + 2.41 (length of tibia in cm)

a. Write the equation in symbols using H for height and t for the length of the tibia in centimeters.

b. What is the slope m of $H = 76.53 + 2.41t$?

c. What is the intercept b of $H = 76.53 + 2.41t$?

Source: http://www.medal.org.

31. *Medical applications* According to Munoz et al., the femur bone gives the most reliable height estimate H (in centimeters) for the height of a male. The equation is

Height in centimeters = 62.92 + 2.39 (length of femur in cm)

a. Write the equation in symbols using H for height and f for the length of the femur in centimeters.

b. What is the slope m of $H = 62.92 + 2.39f$?

c. What is the intercept b of $H = 62.92 + 2.39f$?

Source: http://www.medal.org.

32. *Medical applications* If the gender of the person is unknown, the height H (in centimeters) of the person can be estimated by using the length of the humerus bone and is given by

Height in centimeters = 49.84 + 3.83 (length of humerus in cm)

a. Write the equation in symbols using H for height and h for the length of the humerus in centimeters.

b. What is the slope m of $H = 49.84 + 3.83h$?

c. What is the intercept b of $H = 49.84 + 3.83h$?

Source: http://www.medal.org.

Web IT go to mhhe.com/bello for more lessons

Write On

33. In general, write the steps you use to find the formula for a linear equation in two variables similar to the ones in Problems 11–16.

34. What is the minimum number of points you need to find the formula for a linear equation in two variables like the ones in Problems 23–26?

35. What do the slopes you obtained in Problems 25 and 26 mean?

36. Would the slopes in Problems 25 and 26 be different if the weights of the male and female are different? Explain.

Using Your Knowledge

Internet We have so far neglected the graphs of the applications we have studied, but graphs are especially useful when comparing different situations. For example, the America Online "Light Usage" plan costs $4.95 per month for up to 3 hours and $2.95 for each additional hour. If C is the cost and h is the number of hours:

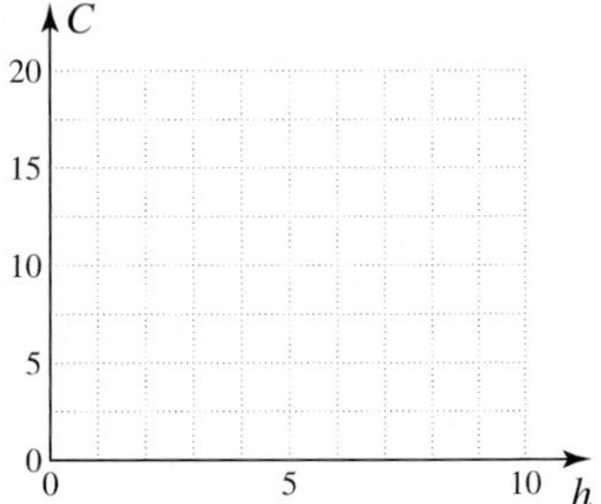

37. Graph C when h is between 0 and 3 hours, inclusive.

38. Graph C when h is more than 3 hours.

39. A different AOL plan costs $14.95 for unlimited hours. Graph the cost for this plan.

40. When is the plan in Problems 37 and 38 cheaper? When is the plan in Problem 39 cheaper?

Mastery Test

41. Irena rented a car and paid $40 a day and $0.15 per mile. Find an equation for the total daily cost C when she travels m miles. What is her cost if she travels 450 miles in a day?

42. In 1990 about 186 million tons of trash were produced in the United States. The amount increases by 3.4 million tons each year after 1990. Use the point-slope formula to find an equation for the total amount of trash y (millions of tons) produced x years after 1990:

a. What would be the slope?

b. What point would you use to find the equation?

c. What is the equation?

43. Somjit subscribes to an Internet service with a flat monthly rate for up to 20 hours of use. For each hour over this limit, there is an additional per-hour fee. The table shows Somjit's first two bills.

Month	Hours of Use	Monthly Fee
March	30	$24
April	33	$26.70

a. What does the point (30, 24) mean?

b. What does the point (33, 26.70) mean?

c. Find an equation for the cost C of x hours ($x > 20$) of use.

Skill Checker

Fill in the blank with $<$ or $>$ so that the result is a true statement:

44. 0 ____ 600

45. 0 ____ −8

46. 0 ____ 6

47. 10 ____ 0

48. −5 ____ 0

49. −1 ____ 0

50. −3 ____ −1

3.7 Graphing Inequalities in Two Variables

Objective

A Graph linear inequalities in two variables.

B Solve applications involving inequalities.

To Succeed, Review How To . . .

1. Use the symbols > and < to compare numbers (pp. 186–187).
2 Graph lines (pp. 215–218, 220–234).
3. Evaluate an expression (pp. 62–63, 69–73).

Getting Started

Renting Cars and Inequalities

Suppose you want to rent a car costing \$30 a day and \$0.20 per mile. The total cost T depends on the number x of days the car is rented and the number y of miles traveled and is given by

$$\underbrace{T}_{\text{Total cost}} = \underbrace{30x}_{\text{Cost per day} \times \text{Number of days}} + \underbrace{0.20y}_{\text{Cost per mile} \times \text{Number of miles}}$$

If you want the cost to be *exactly* \$600 ($T = 600$), we graph the equation $600 = 30x + 0.20y$ by finding the intercepts. For $x = 0$,

$$600 = 30(0) + 0.20y$$

$$\frac{600}{0.20} = y \quad \text{Divide both sides by 0.20.}$$

$$y = 3000$$

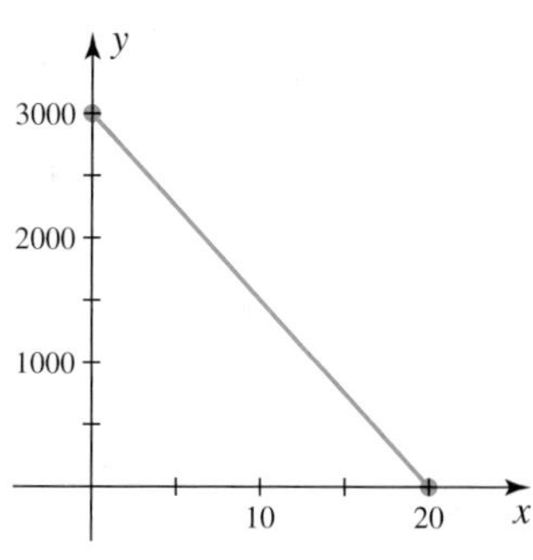

Thus, (0, 3000) is the y-intercept. Now for $y = 0$,

$$600 = 30x + 0.20(0)$$

$$20 = x. \quad \text{Divide both sides by 30.}$$

Thus, (20, 0) is the x-intercept. Join the two intercepts with a line to obtain the graph shown.

If we want to spend *less* than \$600, the total cost T must satisfy (be a solution of) the inequality

$$30x + 0.20y < 600$$

All points satisfying $30x + 0.20y = 600$ are *on* the line. Where are the points so that $30x + 0.20y < 600$? The line $30x + 0.20y = 600$ divides the plane into three regions:

1. The points *below* the line
2. The points *on* the line
3. The points *above* the line

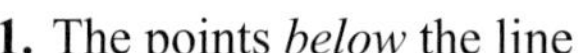

y
3000
2000
1000
10
20
x

The shaded region represents the points for which $30x + 0.20y < 600$.

The line is not part of the graph, so it's shown dashed.

The test point (0, 0) is *below* the line $30x + 0.20y = 600$ and satisfies $30x + 0.20y < 600$. Thus, all the other points *below* the line also satisfy the inequality and are shown shaded in the graph. The line is *not* part of the answer, so it's shown **dashed.** In this section we shall learn how to solve linear inequalities in two variables, and we shall also examine why their solutions are regions of the plane.

A > Graphing Linear Inequalities in Two Variables

The procedure we used in the *Getting Started* can be generalized to graph any linear inequality that can be written in the form $Ax + By < C$. Here are the steps.

PROCEDURE

Graphing a Linear Inequality

1. Determine the line that is the boundary of the region. If the inequality involves $\le$ or $\ge$, draw the line **solid;** if it involves $<$ or $>$, draw a **dashed** line. The points on the *solid* line are part of the solution set.
2. Use any point (a, b) not on the line as a test point. Substitute the values of a and b for x and y in the inequality. If a true statement results, shade the side of the line containing the test point. If a false statement results, shade the other side.

EXAMPLE 1 Graphing linear inequalities where the line is not part of the solution set

Graph: $2x - 4y < -8$

SOLUTION 1

We use the two-step procedure for graphing inequalities.

1. We first graph the boundary line $2x - 4y = -8$.

When $x = 0$, $-4y = -8$, and $y = 2$

When $y = 0$, $2x = -8$, and $x = -4$

x	y
0	2
−4	0

Since the inequality involved is $<$, join the points $(0, 2)$ and $(-4, 0)$ with a dashed line as shown in Figure 3.41.

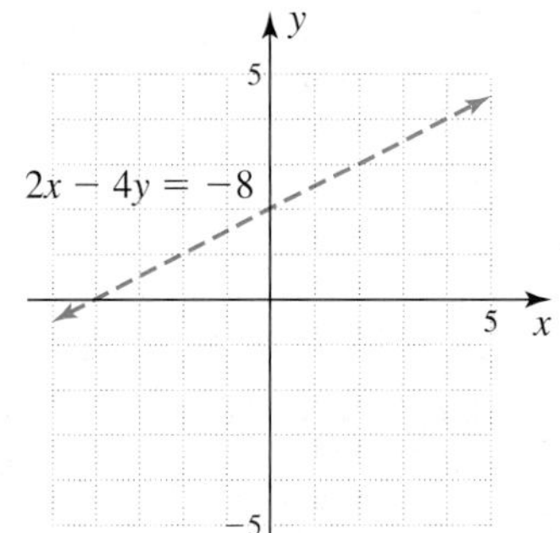

>Figure 3.41

2. Select an easy test point and see whether it satisfies the inequality. If it does, the solution lies on the same side of the line as the test point; otherwise, the solution is on the other side of the line.

An easy point is $(0, 0)$, which is *below* the line. If we substitute $x = 0$ and $y = 0$ in the inequality $2x - 4y < -8$, we obtain

$$2 \cdot 0 - 4 \cdot 0 < -8 \quad \text{or} \quad 0 < -8$$

which is *false*. Thus, the point $(0, 0)$ is not part of the solution. Because of this, the solution consists of the points *above* (on the other side of) the line $2x - 4y = -8$, as shown shaded in Figure 3.42. Note that the line itself is shown **dashed** to indicate that it isn't part of the solution.

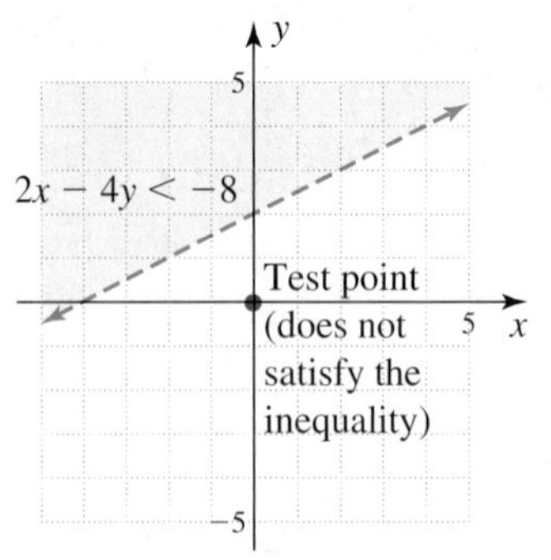

>Figure 3.42

PROBLEM 1

Graph: $3x - 2y < -6$

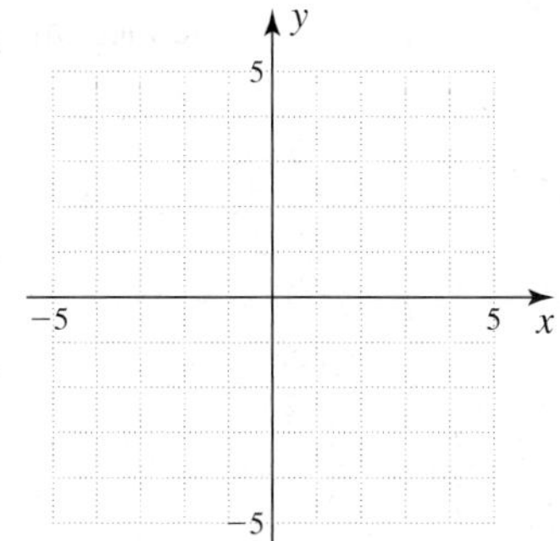

Answers to PROBLEMS

1.

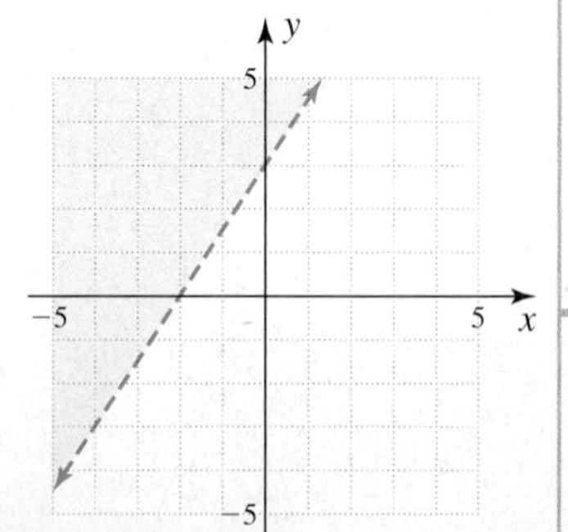

In Example 1, the line was not part of the solution for the given inequality. Next, we give an example in which the line is part of the solution for the inequality.

EXAMPLE 2 **Graphing linear inequalities where the line is part of the solution set**

Graph: $y \le -2x + 6$

SOLUTION 2 As usual, we use the two-step procedure for graphing inequalities.

1. We first graph the line $y = -2x + 6$.

When $x = 0$, $y = 6$
When $y = 0$, $0 = -2x + 6$ or $x = 3$

x	y
0	6
3	0

Since the inequality involved is $\le$, the graph of the line is shown solid in Figure 3.43.

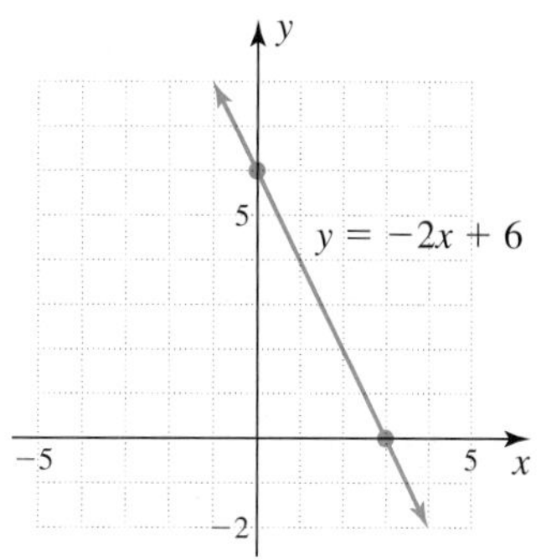

>Figure 3.43

2. Select the point (0, 0) or any other point not on the line as a test point. When $x = 0$ and $y = 0$, the inequality

$$y \le -2x + 6$$

becomes

$$0 \le -2 \cdot 0 + 6 \quad \text{or} \quad 0 \le 6$$

which is *true*. Thus, all the points on the same side of the line as (0, 0)—that is, the points *below* the line—are solutions of $y \le -2x + 6$. These solutions are shown shaded in Figure 3.44.

The line $y = -2x + 6$ is shown solid because it is part of the solution, since $y \le -2x + 6$ allows $y = -2x + 6$. [For example, the point (3, 0) satisfies the inequality $y \le -2x + 6$ because $0 \le -2 \cdot 3 + 6$ yields $0 \le 0$, which is true.]

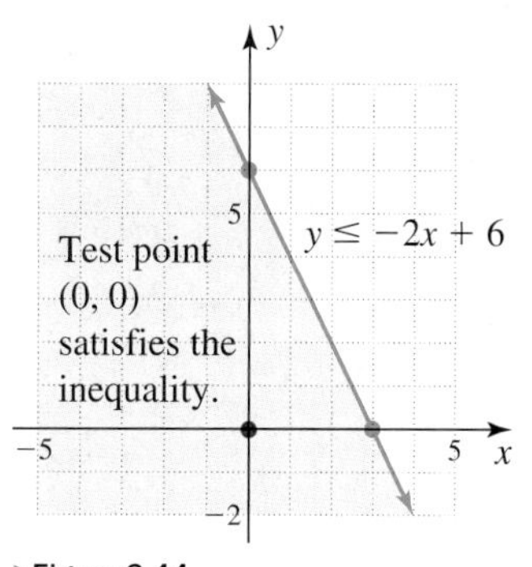

>Figure 3.44

PROBLEM 2

Graph: $y \le -3x + 6$

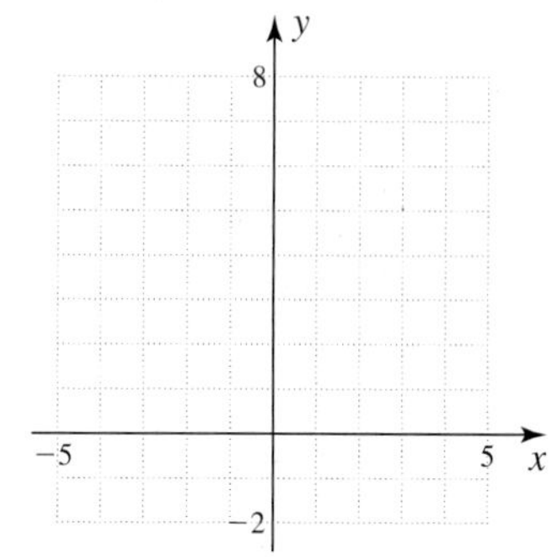

As you recall from Section 3.3, a line with an equation of the form $Ax + By = 0$ passes through the origin, so we cannot use the point (0, 0) as a test point. This is not a problem! Just use any other convenient point, as shown in Example 3.

Answers to PROBLEMS

2.

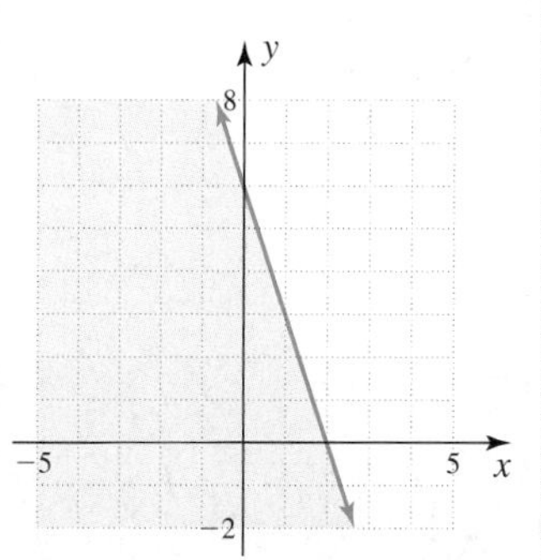

EXAMPLE 3 **Graphing linear inequalities where the boundary line passes through the origin**

Graph: $y + 2x > 0$

SOLUTION 3 We follow the two-step procedure.

1. We first graph the boundary line $y + 2x = 0$:

When $x = 0$, $\quad y = 0$

When $y = -4$, $\quad -4 + 2x = 0$ or $x = 2$

Join the points (0, 0) and (2, −4) with a dashed line as shown in Figure 3.45 since the inequality involved is >.

x	y
0	0
2	−4

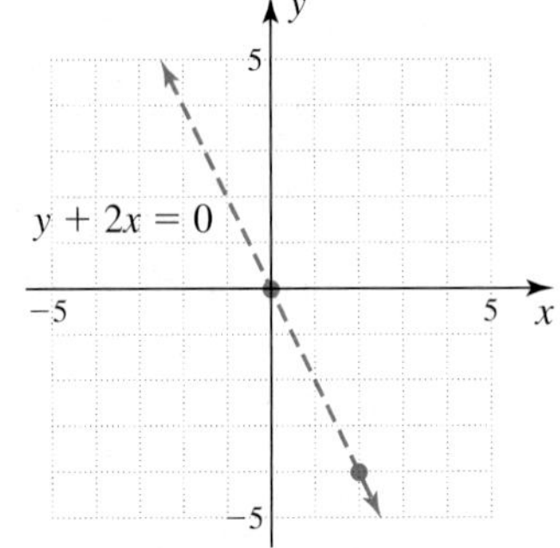

>Figure 3.45

2. Because the line goes through the origin, we cannot use (0, 0) as a test point. A convenient point to use is (1, 1), which is *above* the line. Substituting $x = 1$ and $y = 1$ in $y + 2x > 0$, we obtain

$$1 + 2(1) > 0 \quad \text{or} \quad 3 > 0$$

which is true. Thus, we shade the points *above* the line $y + 2x = 0$, as shown in Figure 3.46.

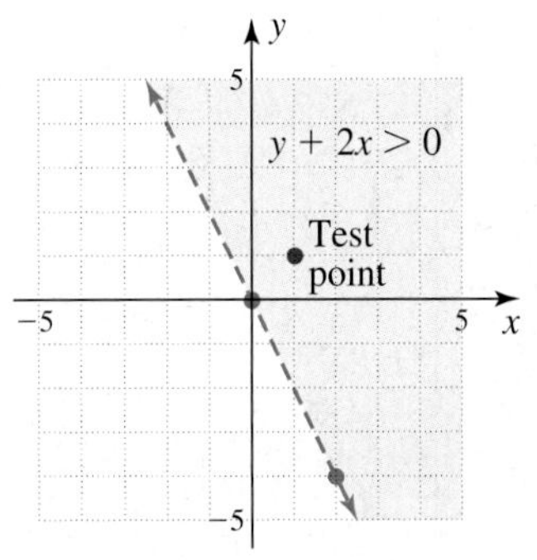

>Figure 3.46

PROBLEM 3

Graph: $y + 3x > 0$

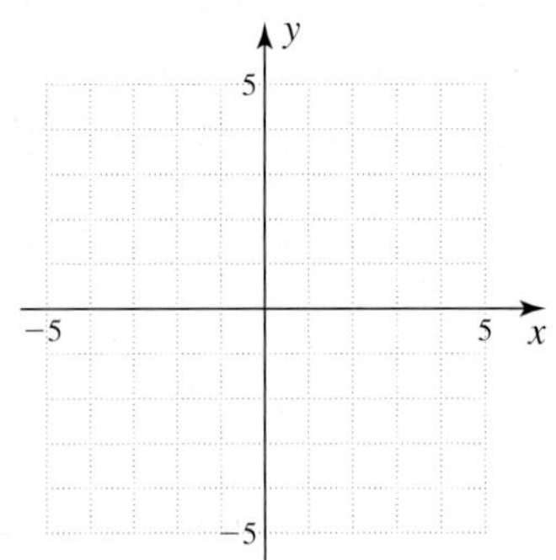

Finally, if the inequalities involve horizontal or vertical lines, we also proceed in the same manner.

Answers to PROBLEMS

3.

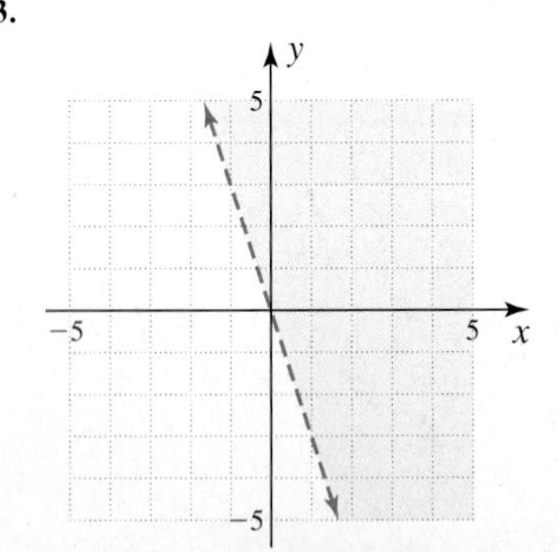

EXAMPLE 4 Graphing linear inequalities involving horizontal or vertical lines

Graph:

a. $x \geq -2$ **b.** $y - 2 < 0$

SOLUTION 4

a. The two-step procedure applies here also.

1. Graph the vertical line $x = -2$ as solid, since the inequality involved is $\geq$.
2. Use (0, 0) as a test point. When $x = 0$, $x \geq -2$ becomes $0 \geq -2$, a true statement. So we shade all the points to the *right* of the line $x = -2$, as shown in Figure 3.47.

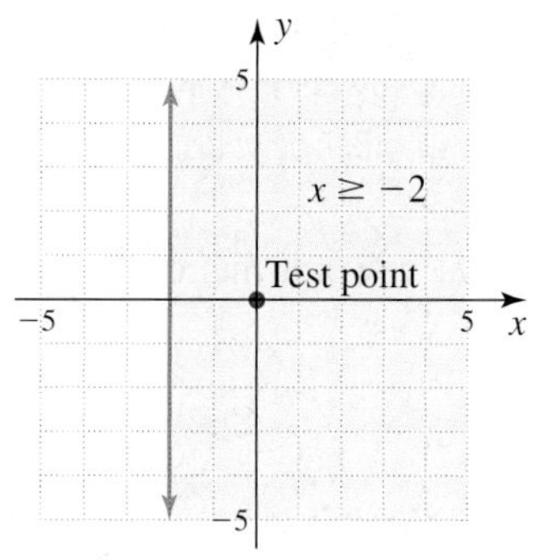

>Figure 3.47

b. We again follow the steps.

1. First, we write the inequality as $y < 2$, and then we graph the horizontal line $y = 2$ as dashed, since the inequality involved is $<$.
2. Use (0, 0) as a test point. When $y = 0$, $y < 2$ becomes $0 < 2$, a true statement. So we shade all the points *below* the line $y = 2$, as shown in Figure 3.48.

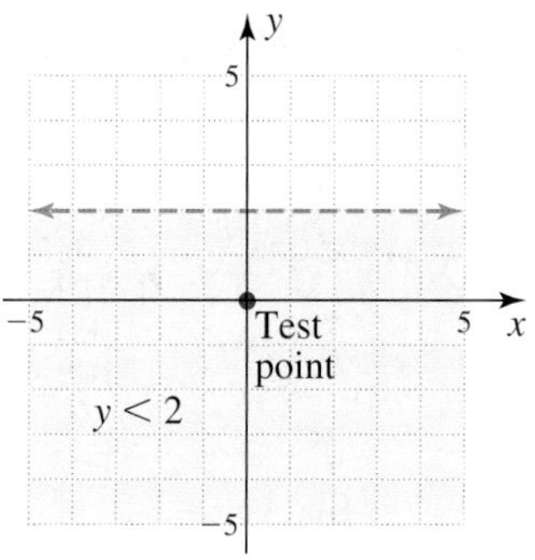

>Figure 3.48

PROBLEM 4

Graph:

a. $x \geq 2$ **b.** $y + 2 > 0$

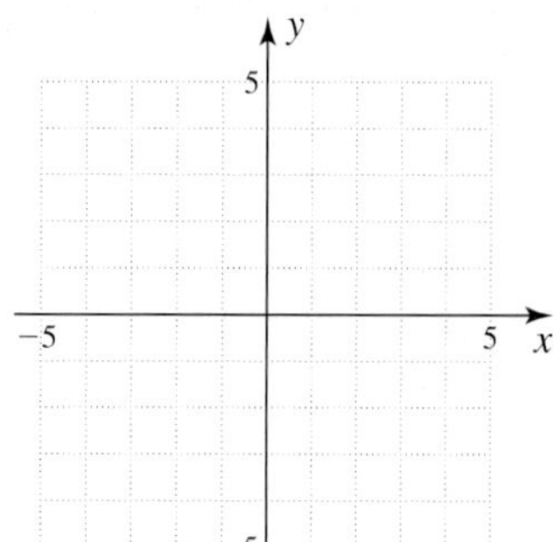

B > Applications Involving Inequalities

The Kyoto Protocol is an international agreement setting targets for industrialized countries and the European community in the reduction of greenhouse gases (GHG) based on their emissions in 1990. One of the main CO_2 polluters is Russia. What are the GHG projections for their future and are they meeting their Kyoto targets? We shall see in Example 5.

Answers to PROBLEMS

4. a.

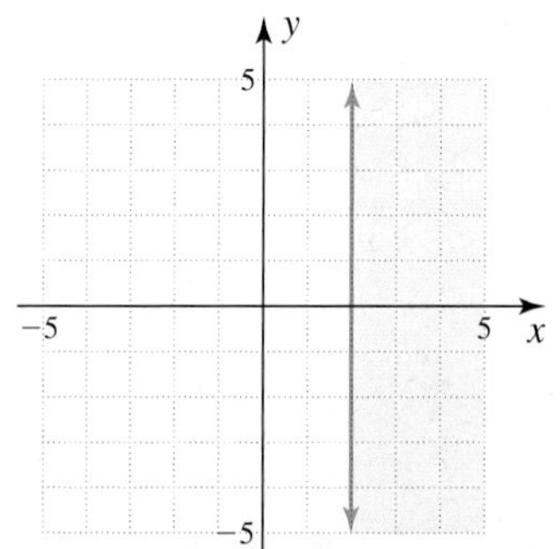

b.

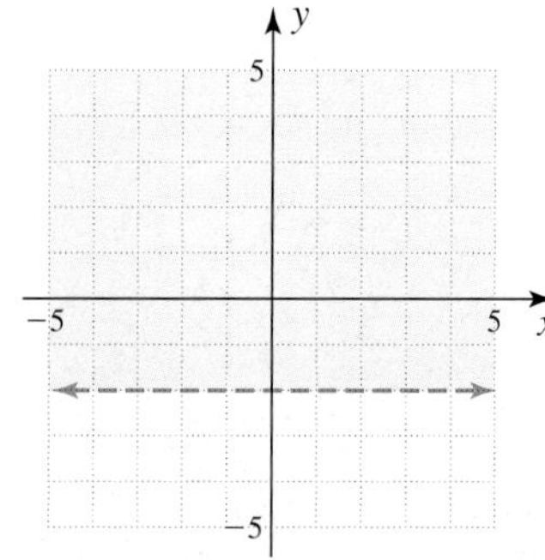

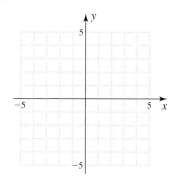

EXAMPLE 5 Kyoto Protocol inequalities

a. The Kyoto target for Russia is approximated by the line $y = 2400$ (megatons). Graph $\mathbf{y = 2400}$ between 2005 and 2020. In the worst-case scenario, emissions can be as high as $y = 60x + 1850$ and as low as $y = 30x + 1750$.

b. Graph $y \le 60x + 1850$, where x is the number of years between 2005 and 2020.

c. Graph $y \ge 30x + 1750$ in the same grid.

d. Use the inequalities in **a** and **b** to define in words the region that describes the worst-case scenario for Russia.

e. Starting with what year (nearest year) will the Russians be above their 2400 megaton target?

SOLUTION 5

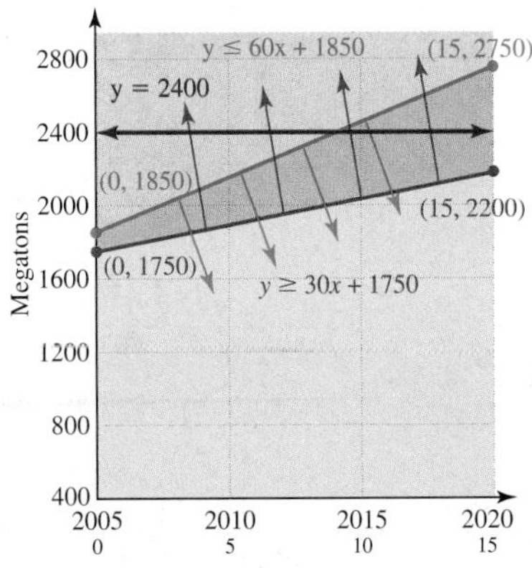

a. Graph the horizontal line $\mathbf{y = 2400}$ (blue).

b. We use the two-step procedure for graphing inequalities:

1. Graph the boundary line of $\mathbf{y \le 60x + 1850.}$ When $x = 0, y = 1850$.
2. When $x = 15, y = 60(15) + 1850 = 2750$.

Join the points (0, 1850) and (15, 2750) with a solid green line since the inequality involved is $\le$. If we use (0, 0) as a test point, $0 \le 1850$ is true, so we shade *below* the line $y = 60x + 1850$.

c. 1. To graph the boundary line of $\mathbf{y \ge 30x + 1750,}$ let $x = 0$ obtaining $y = 1750$.
2. When $x = 15, y = 30(15) + 1750 = 2200$.

Join the points (0, 1750) and (15, 2200) with a solid red line, since the inequality involved is $\ge$. Using (0, 0) as a test point we have $0 \ge 1750$, which is *false,* so we shade *above* and *on* the line $\mathbf{y = 30x + 1750.}$

d. The regions between $\mathbf{y = 30x + 1750}$ and $\mathbf{y = 60x + 1850}$ and **on** both lines (shaded twice in the diagram) describe the situation.

e. The line $\mathbf{y = 60x + 1850}$ is *above* the line $\mathbf{y = 2400,}$ the target line, after about the 9th year (in 2014), so the Russians will be *above* their 2400 megaton emission target after 2014.

If you want to see a more detailed image, go to http://www.newscientist.com/data/images/archive/2418/24185801.jpg.

PROBLEM 5

a. Another possible scenario for the Russian CO_2 emissions can be as low as 1400 megatons. Graph the line $\mathbf{y = 1400}$ between 2005 and 2020.

b. In this scenario, emissions can be as high as $\mathbf{y = 15x + 1700.}$ Graph $\mathbf{y \le 15x + 1700,}$ where x is the number of years after 2005.

c. Use the inequalities in parts **a** and **b** to define in words the region that describes this scenario for Russia.

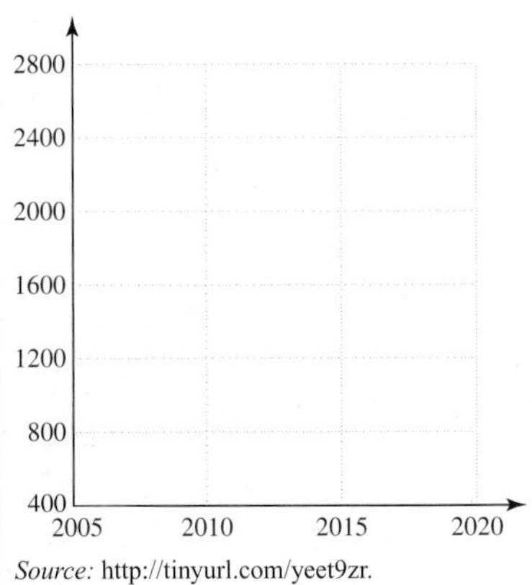

Source: http://tinyurl.com/yeet9zr.

Answers to PROBLEMS

5. a. and **b.**

y ≤ 15x + 1700
(15, 1925)
(0, 1700)
y = 1400
2800, 2400, 2000, 1600, 1200, 800, 400
2005, 2010, 2015, 2020

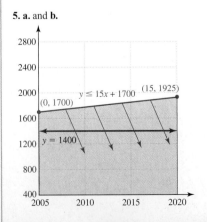

c. The region *above* and *on* the line $y = 1400$ and *below* and *on* the line $y = 15x + 1700$

Calculator Corner

Solving Linear Inequalities

Your calculator can help solve linear inequalities, but you still have to know the algebra. Thus, to do Example 1, you must first solve $2x - 4y = -8$ for y to obtain $y = \frac{1}{2}x + 2$. Now graph this equation using the Zdecimal window, a window in which the x- and y-coordinates are decimals between -4.7 and 4.7. You can do this by pressing ZOOM 4. Now graph $y = \frac{1}{2}x + 2$.

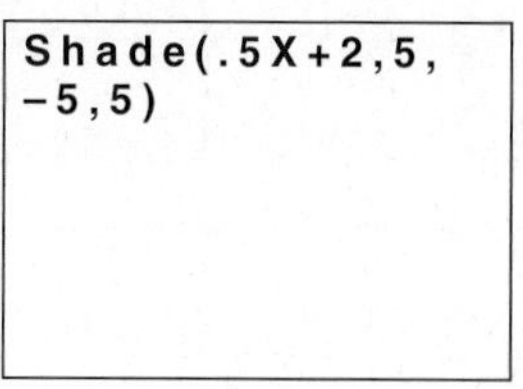

Window 1

To decide whether you need to shade above or below the line, move the cursor *above* the line. In the decimal window, you can see the values for x and y. For $x = 2$ and $y = 5$, $2x - 4y < -8$ becomes

$$2(2) - 4(5) < -8 \quad \text{or} \quad 4 - 20 < -8$$

a true statement. Thus, you need to shade *above* the line. You can do this with the DRAW feature. Press 2nd PRGM 7 and enter the line above which you want to shade—that is, $\frac{1}{2}x + 2$. Now press , and enter the y-range that you want to shade. Let the calculator shade below 5 and above -5 by entering 5 , (–) 5 ,. Finally, enter the value at which you want the shading to end, (say 5), close the parentheses, and press ENTER.

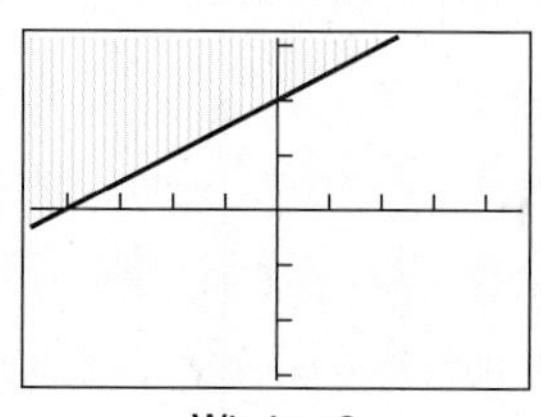
Window 2

The instructions we asked you to enter are shown in Window 1, and the resulting graph is shown in Window 2. What is missing? *You* should know the line itself is not part of the graph, since the inequality $<$ is involved.

Exercises 3.7

> Practice Problems > Self-Tests > Media-rich eBooks > e-Professors > Videos

⟨A⟩ Graphing Linear Inequalities in Two Variables In Problems 1–32, graph the inequalities.

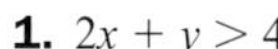

1. $2x + y > 4$

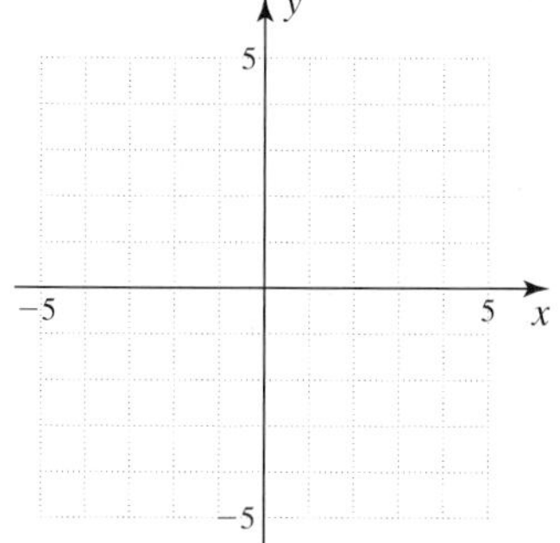

2. $y + 3x > 3$

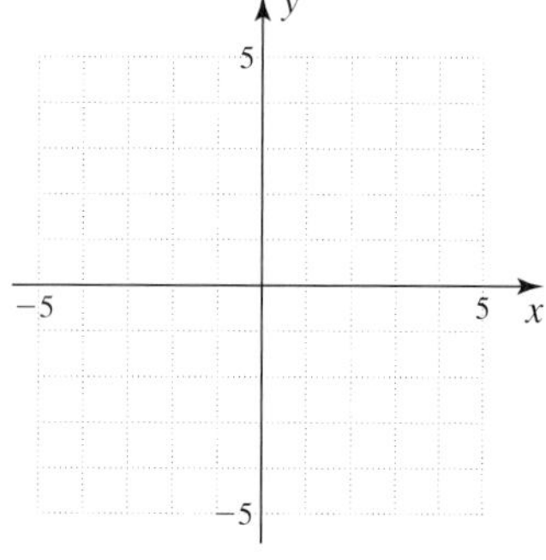

3. $-2x - 5y \le 10$

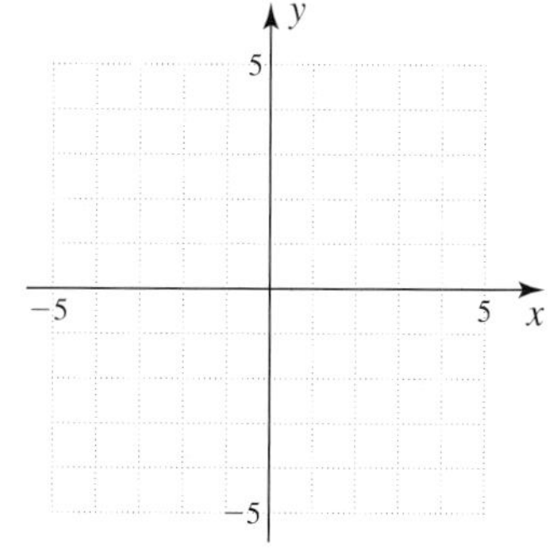

4. $-3x - 2y \le -6$

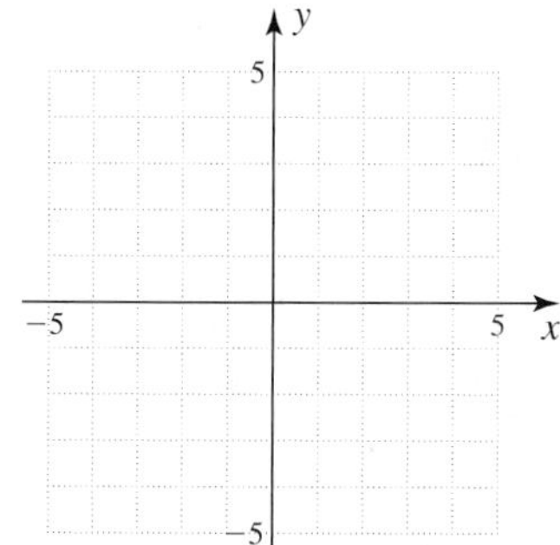

5. $y \ge 3x - 3$

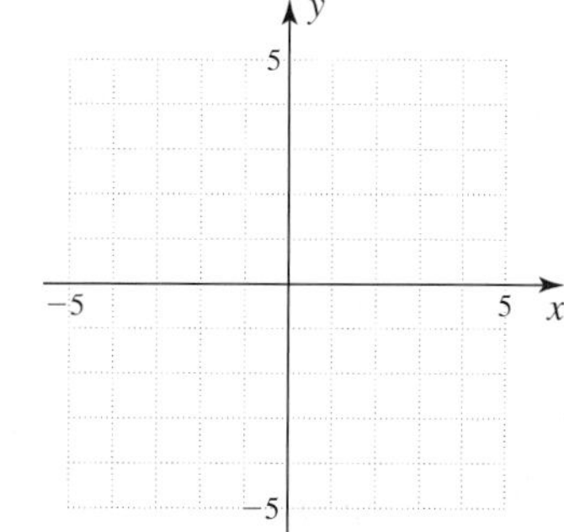

6. $y \ge -2x + 4$

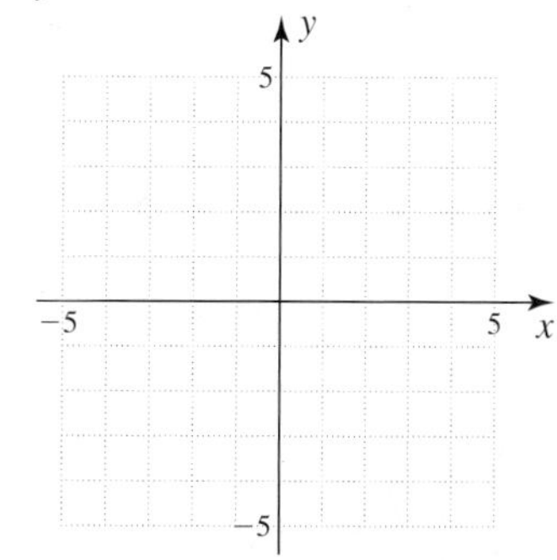

Web IT go to **mhhe.com/bello** for more lessons

Web IT go to **mhhe.com/bello** for more lessons

7. $6 < 3x - 6y$

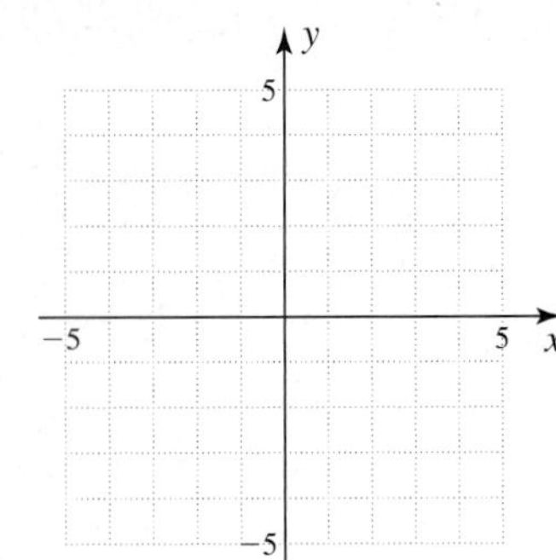

8. $6 < 2x - 3y$

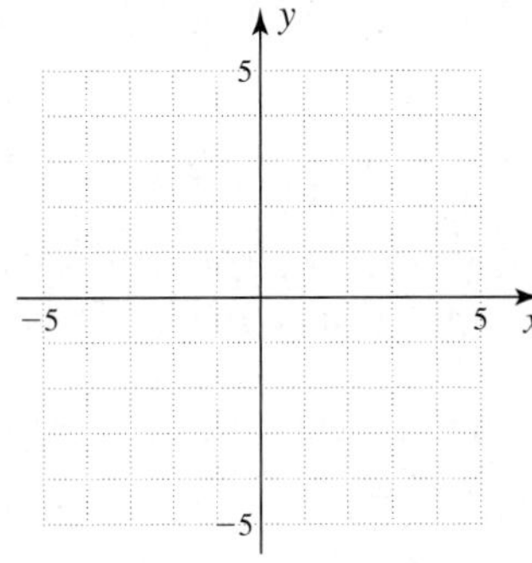

9. $3x + 4y \geq 12$

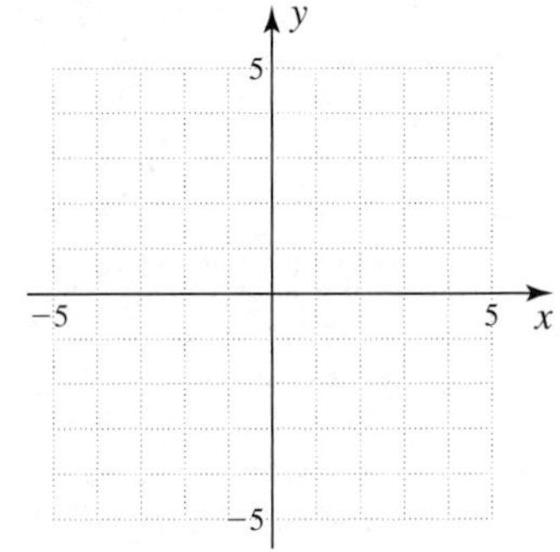

10. $-3y \geq 6x + 6$

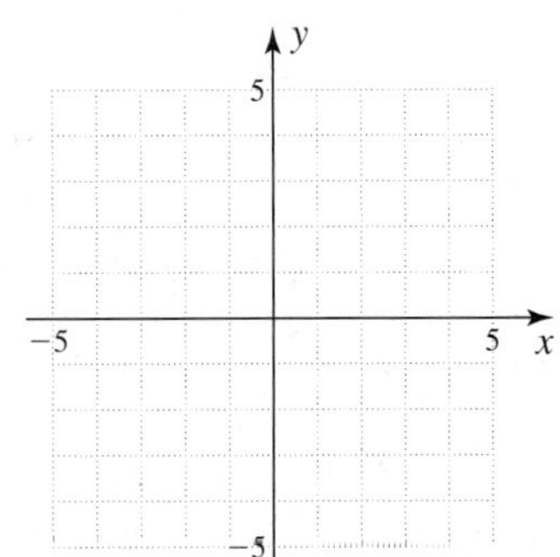

11. $10 < -2x + 5y$

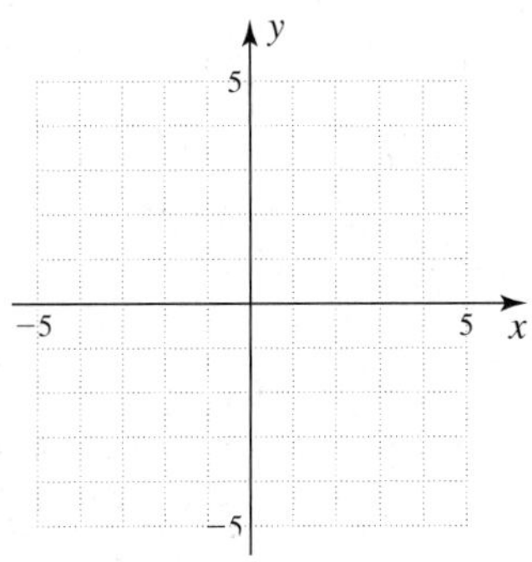

12. $4 < x - y$

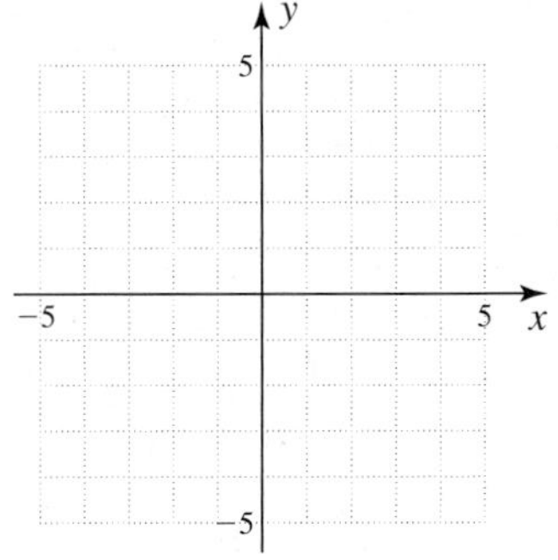

13. $x \geq 2y - 4$

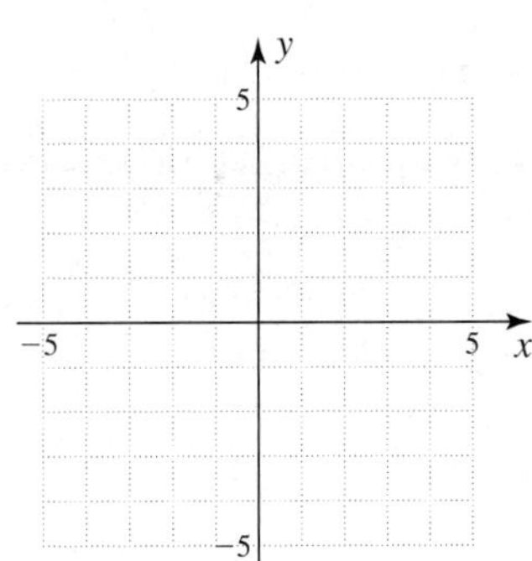

14. $2x \geq 4y + 2$

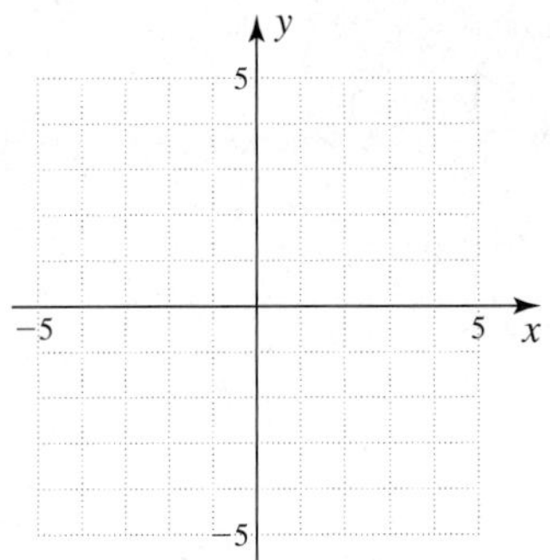

15. $y < -x + 5$

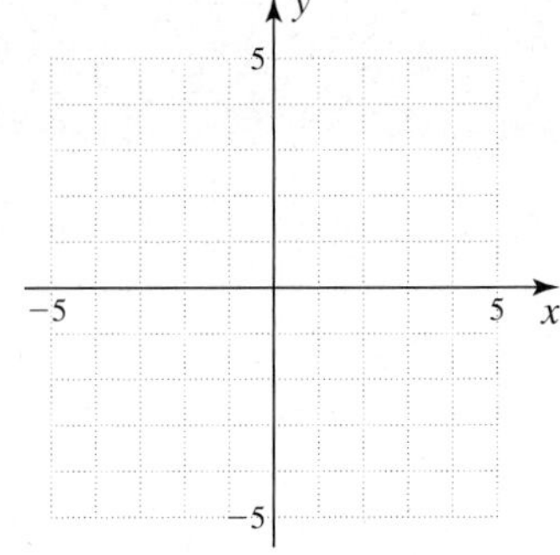

16. $2y < 4x - 8$

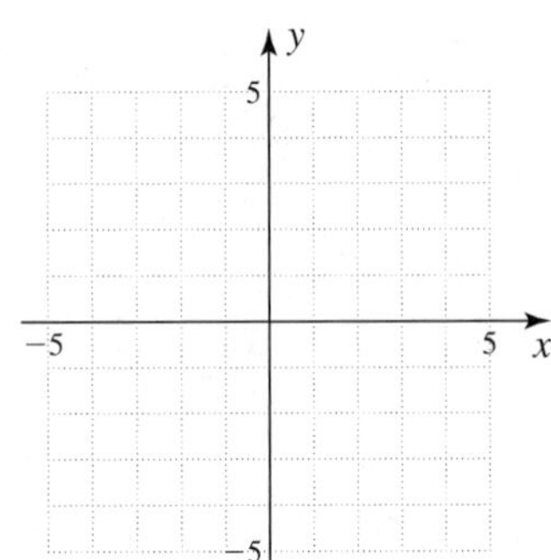

17. $2y < 4x + 5$

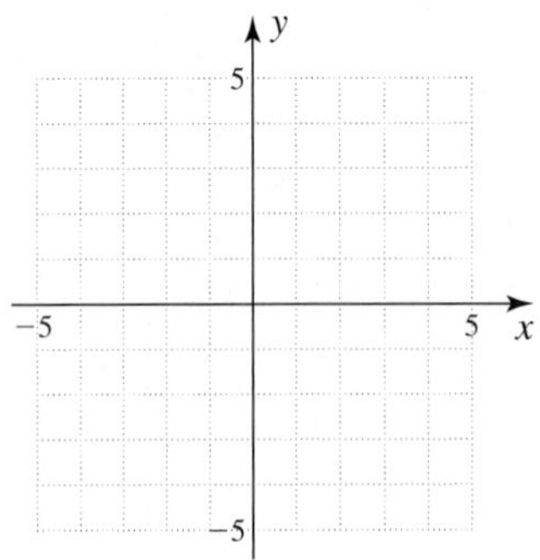

18. $2y \geq 3x + 5$

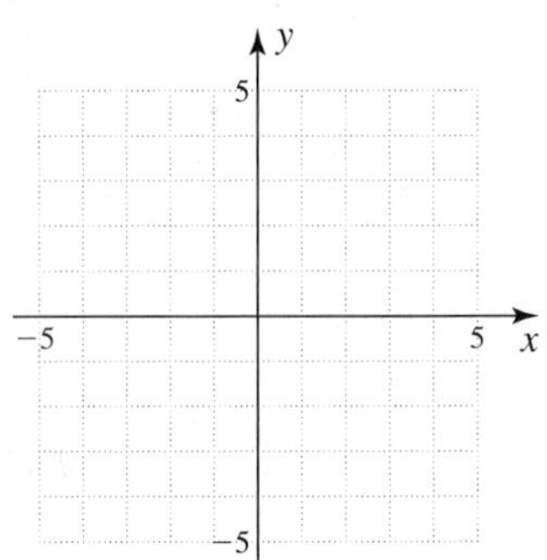

19. $x > 1$

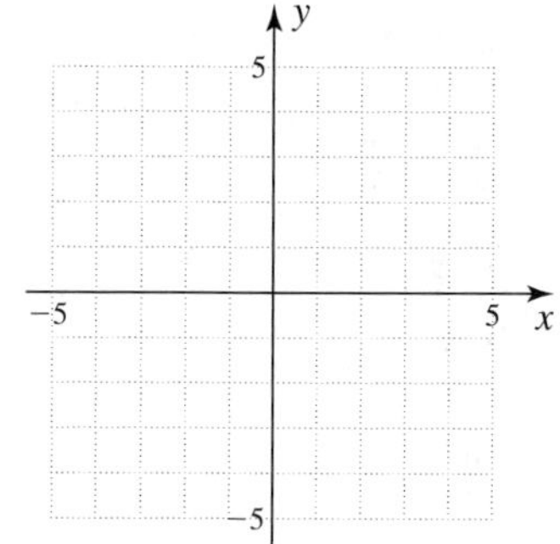

20. $y \le \frac{3}{2}$

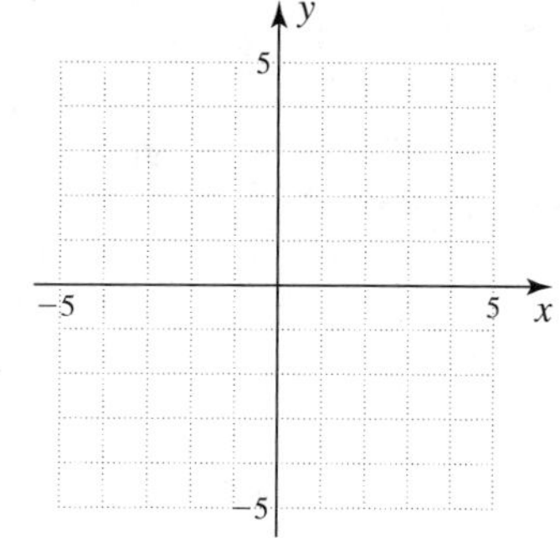

21. $x \le \frac{5}{2}$

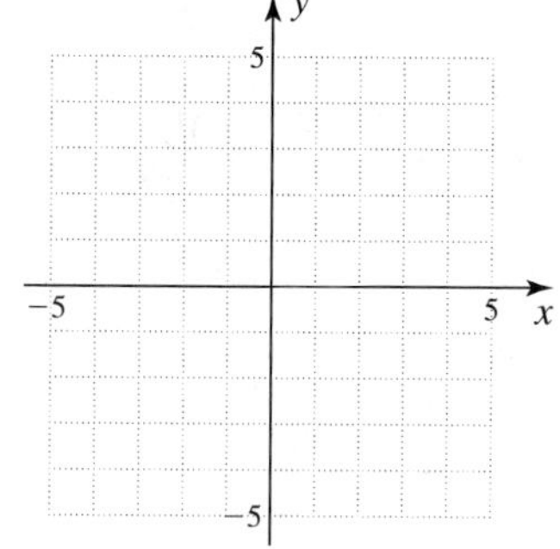

22. $x \ge -\frac{2}{3}$

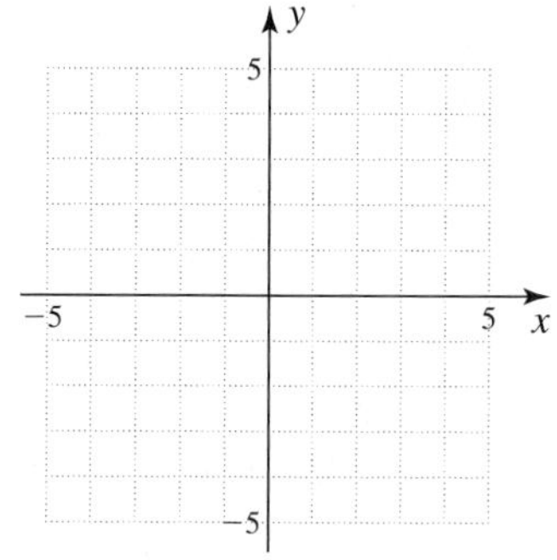

23. $y \le -\frac{3}{2}$

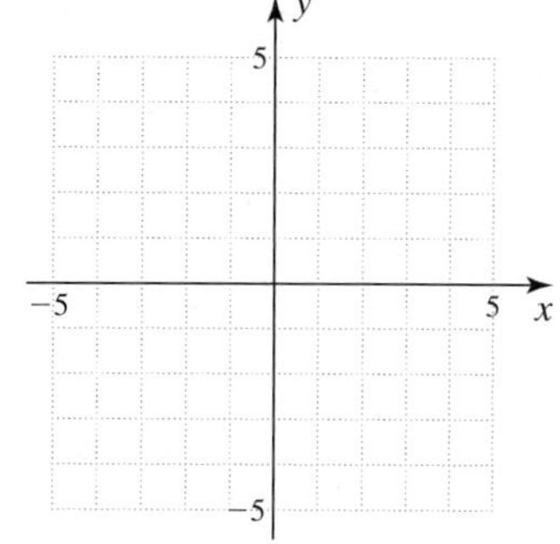

24. $x - \frac{1}{3} \ge 0$

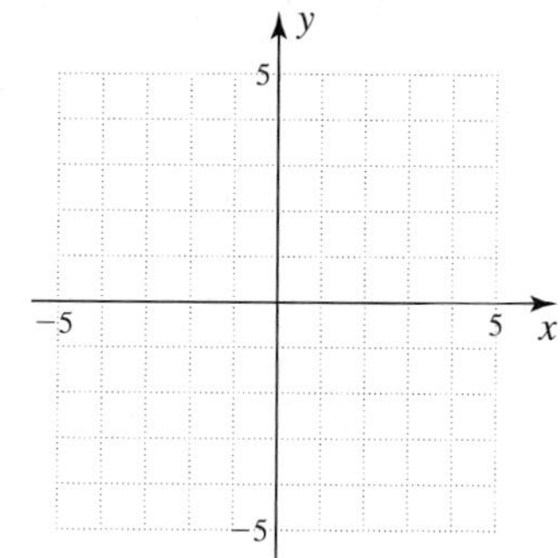

25. $x - \frac{2}{3} > 0$

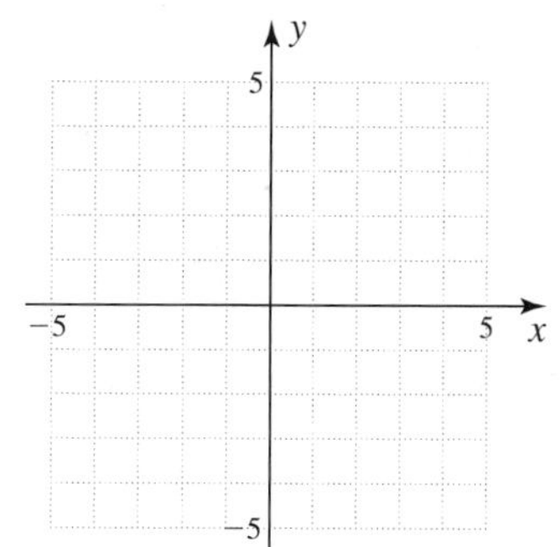

26. $y - \frac{5}{2} > \frac{1}{2}$

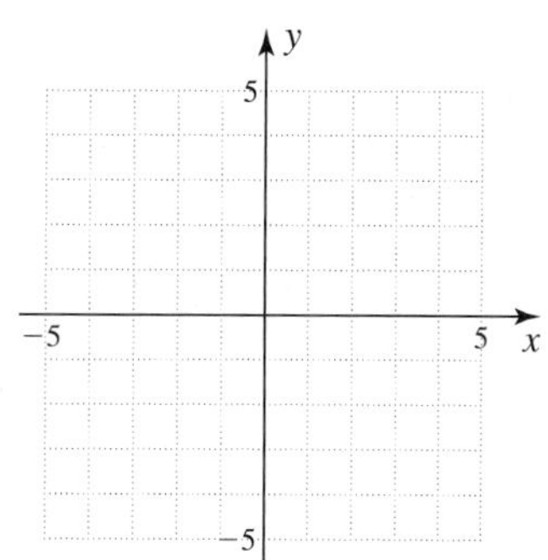

27. $y + \frac{1}{3} \ge \frac{2}{3}$

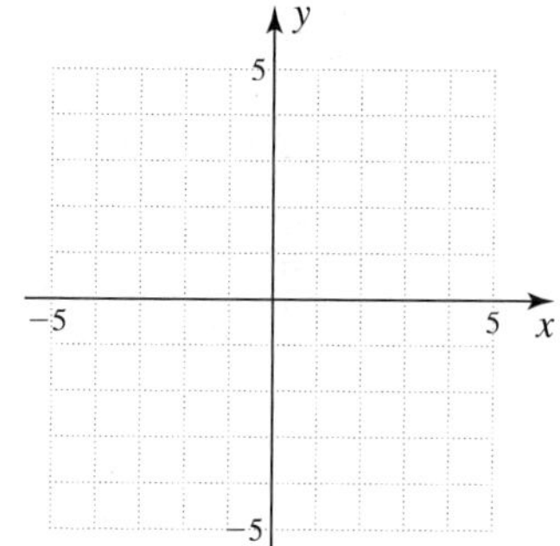

28. $x + \frac{1}{5} < \frac{6}{5}$

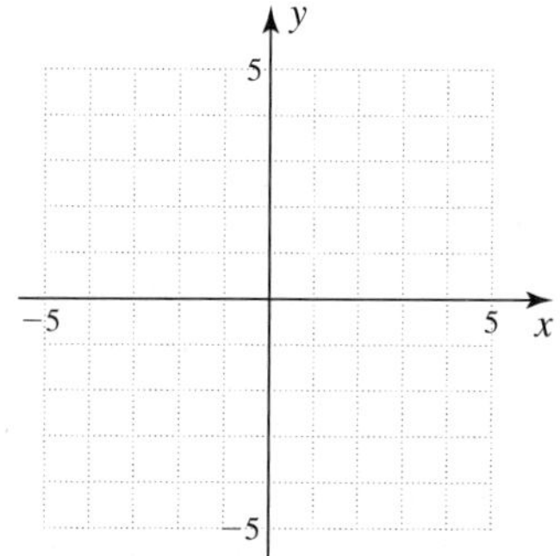

29. $2x + y < 0$

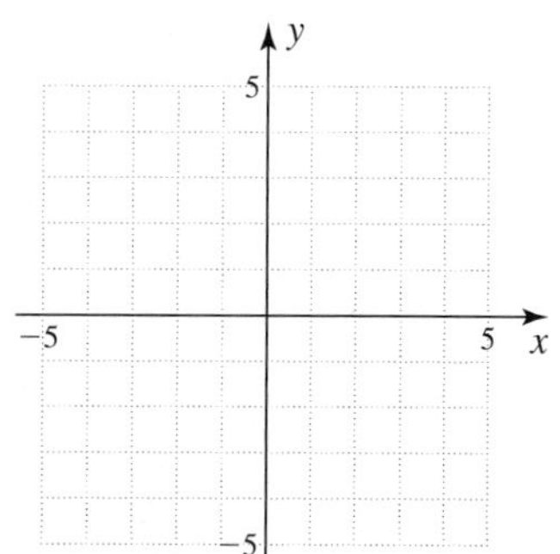

30. $2y + x \ge 0$

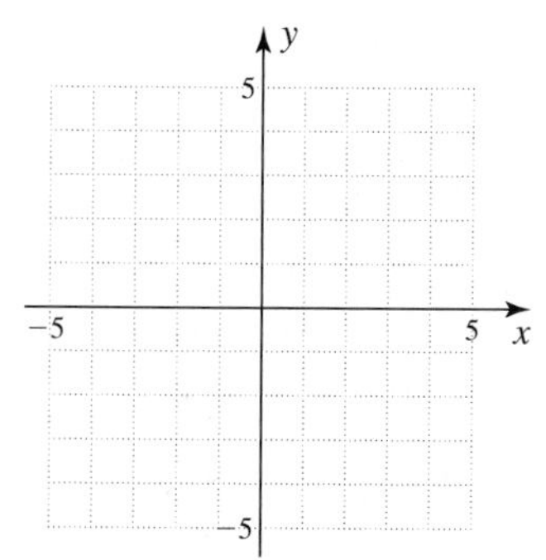

Web IT go to mhhe.com/bello for more lessons

Web IT go to mhhe.com/bello for more lessons

31. $y - 3x > 0$

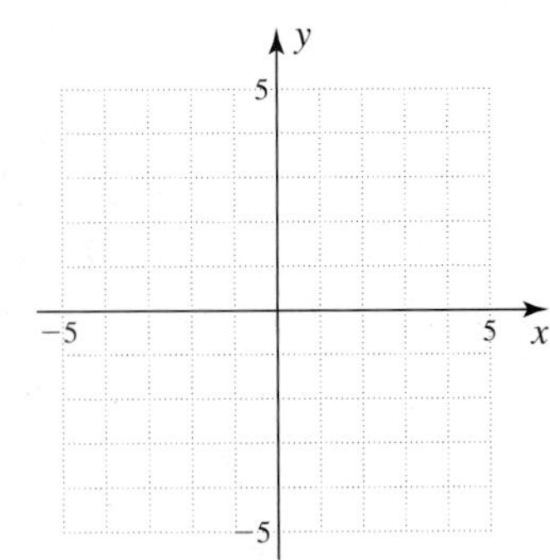

32. $2x - y \le 0$

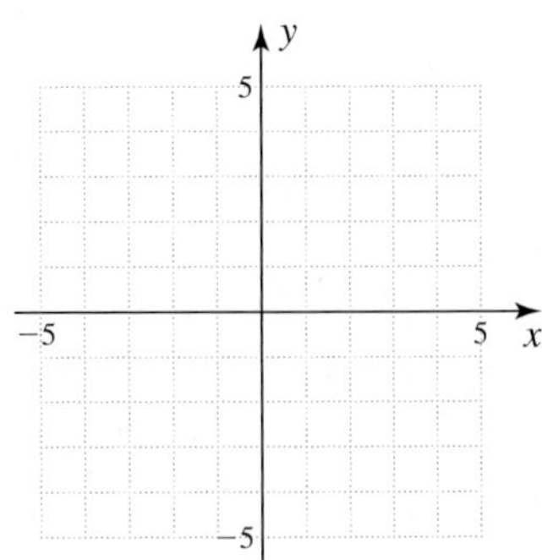

In Problems 33 and 34, the graph of the given system of inequalities defines a geometric figure. Graph the inequalities on the same set of axes and identify the figure.

33. $-x \le -3, \quad -y \ge -4, \quad x \le 4, \quad y \ge 2$

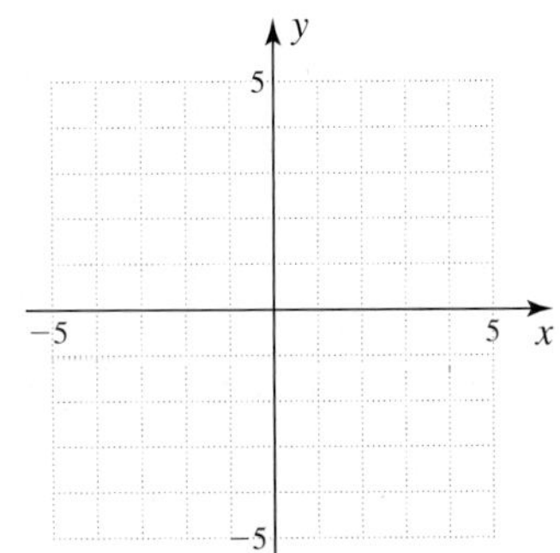

34. $x \ge 2, \quad -x \ge -5, \quad y \le 5, \quad -y \le -2$

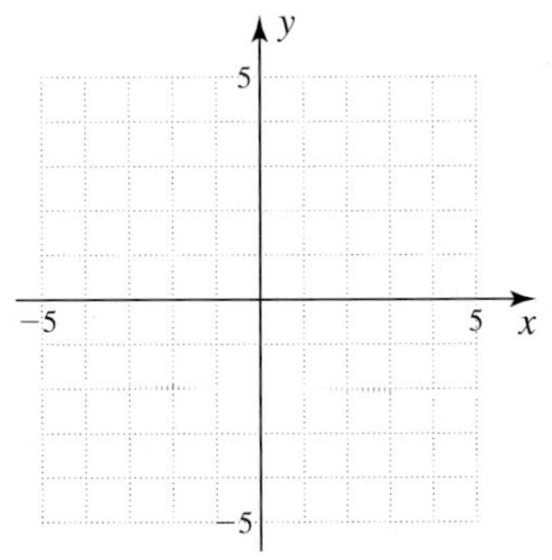

⟨B⟩ Applications Involving Inequalities

Applications: Green Math

35. *Graph carbon dioxide emissions* Let B represent the coordinates of the blue circles • representing the developing countries in the graph (Pakistan, Turkey, South Africa) and P be the coordinates of the red triangles ▲ representing the developed countries in the graph (United States, Japan, Germany). We say that $B < P$ (point B is lower than point P in the graph) from 1990 to just before 2010, that is, in the interval $1990 \le x < 2010$. Use this idea to write an inequality to express the years in which

a. $P > B$

b. $P < B$

c. $P = B$

Source: http://tinyurl.com/ya6nqb3.

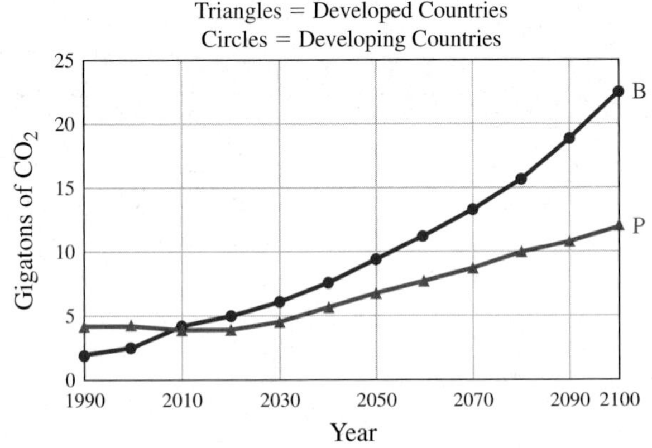

36. *More emissions* Follow the procedure of Problem 35 to write an inequality to express the years in which

a. $B > P$

b. $B < P$

37. *Writing inequalities in words* State in words the result (answer) obtained for the inequality $P > B$ from Problem 35a.

38. *Writing more inequalities in words* State in words the results (answer) obtained for the inequality $B > P$ from Problem 36a.

Using Your Knowledge

Savings on Rentals The ideas we discussed in this section can save you money when you rent a car. Here's how.

Suppose you have the choice of renting a car from company A or from company B. The rates for these companies are as follows:

Company A: \$20 a day plus \$0.20 per mile

Company B: \$15 a day plus \$0.25 per mile

If x is the number of miles traveled in a day and y represents the cost for that day, the equations representing the cost for each company are

Company A: $y = 0.20x + 20$

Company B: $y = 0.25x + 15$

39. Graph the equation representing the cost for company A.

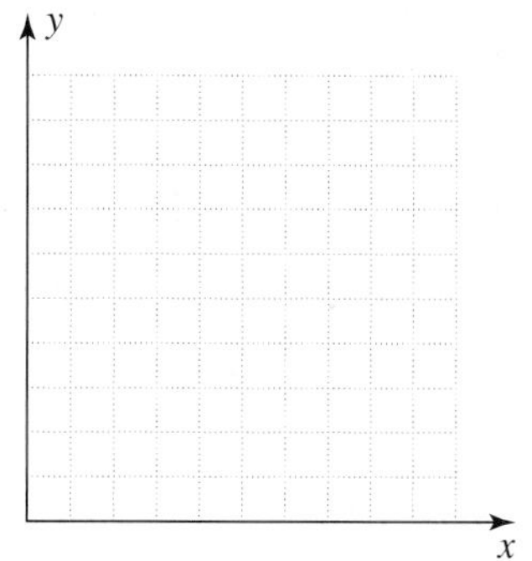

40. On the same coordinate axes you used in Problem 39, graph the equation representing the cost for company B.

41. When is the cost the same for both companies?

42. When is the cost less for company A?

43. When is the cost less for company B?

Write On

44. Describe in your own words the graph of a linear inequality in two variables. How does it differ from the graph of a linear inequality in one variable?

45. Write the procedure you use to solve a linear inequality in two variables. How does the procedure differ from the one you use to solve a linear inequality in one variable?

46. Explain how you decide whether the boundary line is solid or dashed when graphing a linear inequality in two variables.

47. Explain why a point on the boundary line cannot be used as a test point when graphing a linear inequality in two variables.

Concept Checker

Fill in the blank(s) with the correct word(s), phrase, or mathematical statement.

48. A **linear inequality** in two variables x and y involving $<$ can be written in the form ______________.

49. If a **linear inequality** involves $\leq$ or $\geq$ the boundary of the region is drawn as a ______________ line.

50. If a **linear inequality** has a **boundary** that is a **solid line,** the line is in the ______________ **set** of the inequality.

51. If a **linear inequality** involves $<$ or $>$ the **boundary** of the region is drawn as a ______________ line.

solid

dashed

color

$x + y < C$

$Ax + By < C$

solution

Mastery Test

Graph:

52. $x - 4 \le 0$

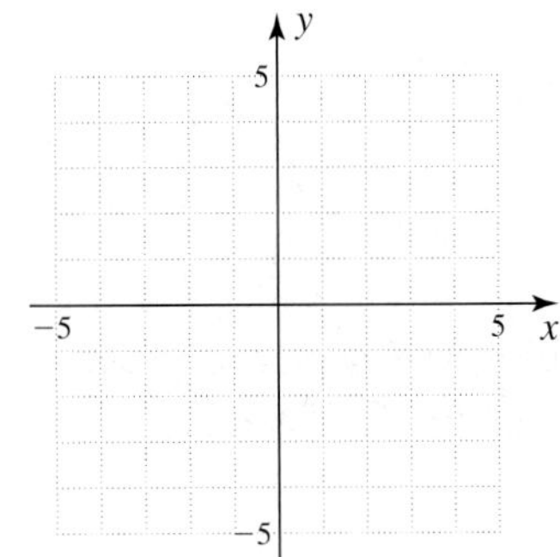

53. $y \ge -4$

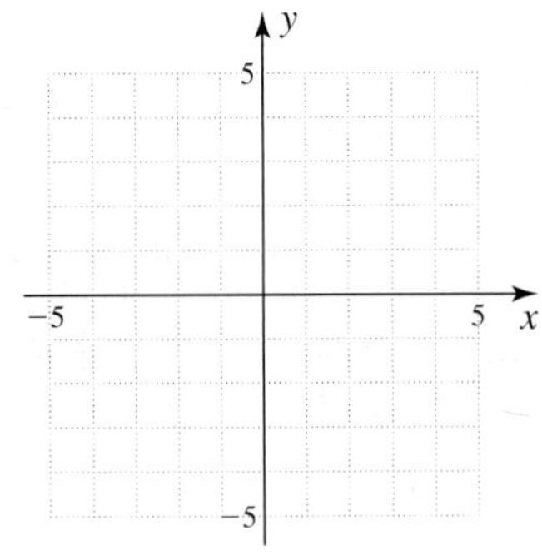

54. $x + 2 < 0$

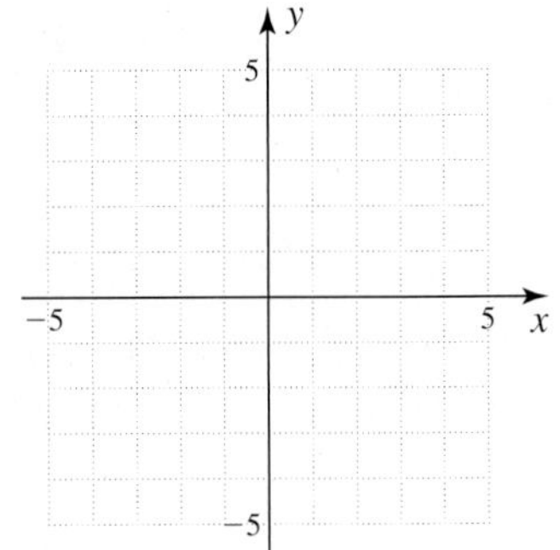

55. $y - 3 > 0$

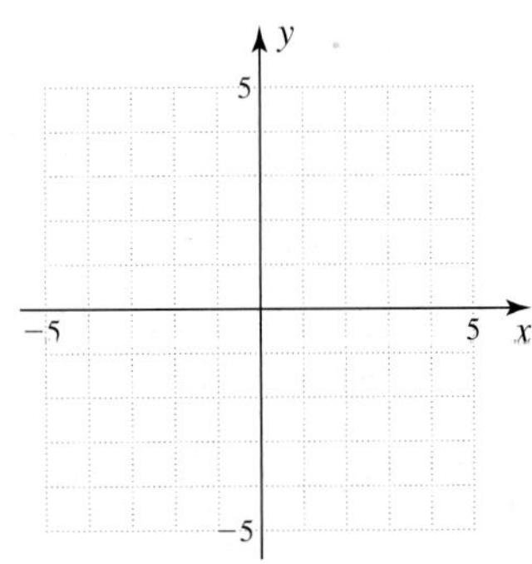

56. $3x - 2y < -6$

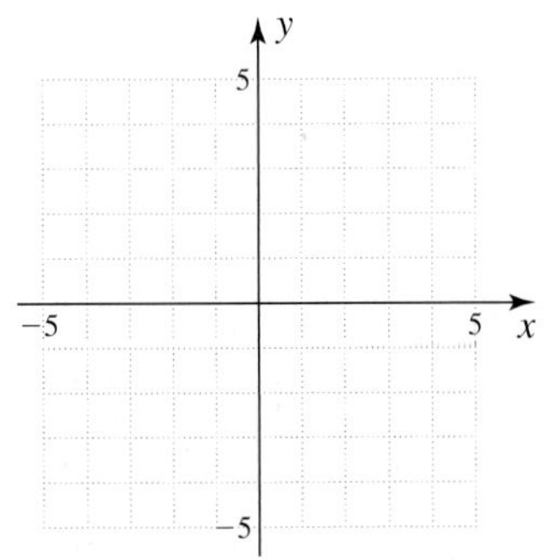

57. $y \le -4x + 8$

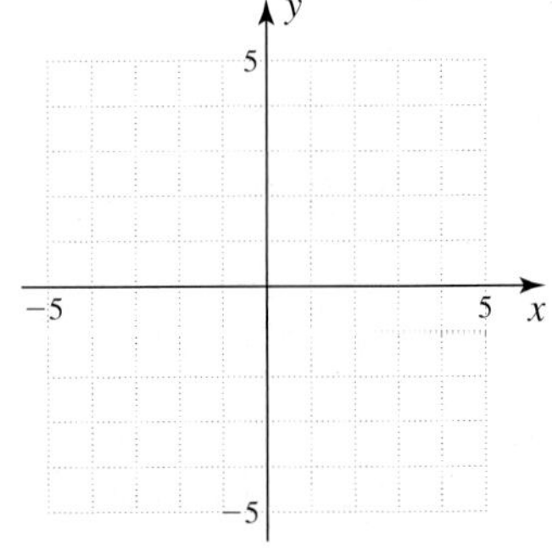

58. $y > 2x$

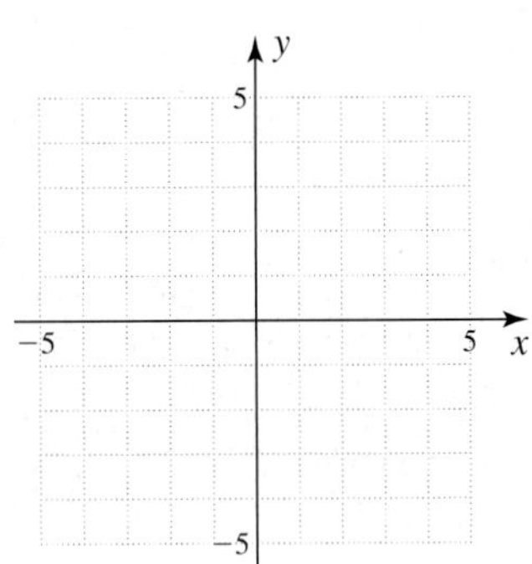

59. $y < 3x$

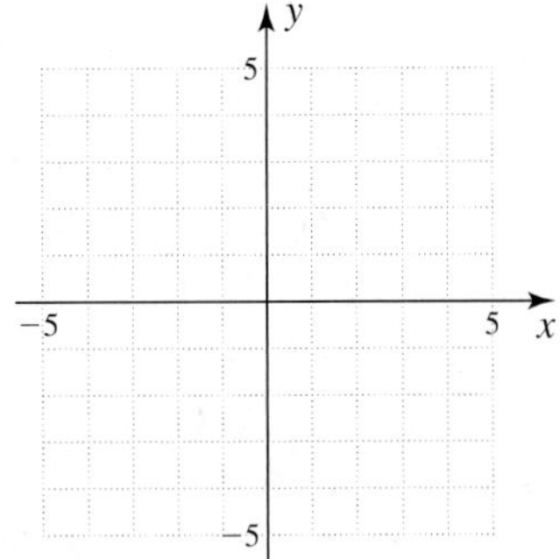

60. $x - 2y > 0$

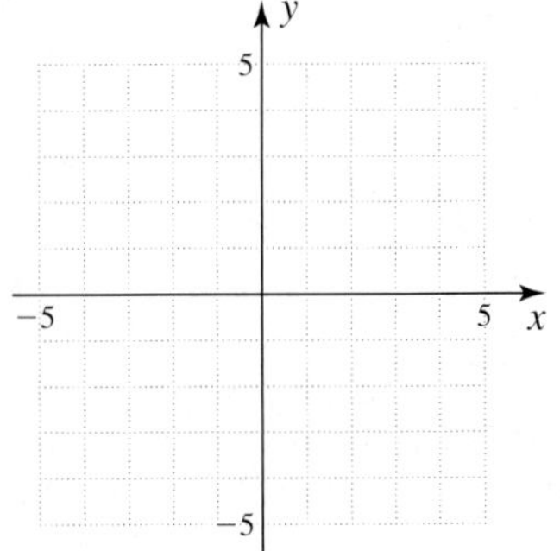

61. $3x + y \le 0$

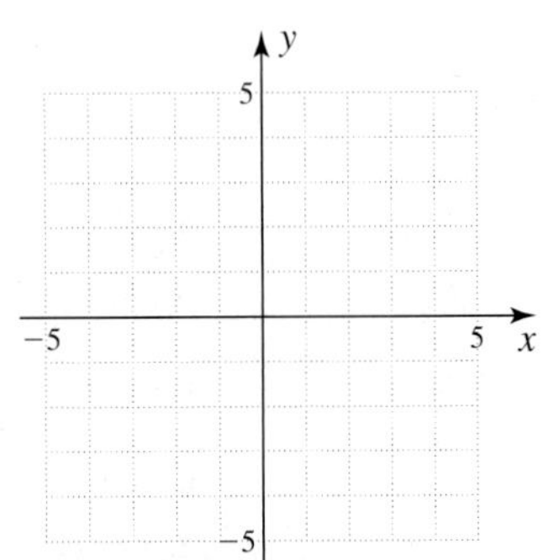

62. $y - 4x > 0$

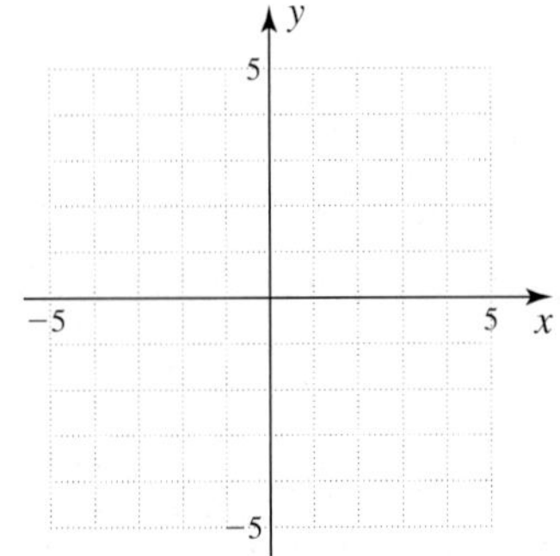

Skill Checker

In Problems 63–65, find the product.

63. $-3 \cdot 8$

64. $(-2)(-7)$

65. $(7)(-3)$

In Problems 66–67, find the quotient.

66. $-\frac{24}{6}$

67. $\frac{24}{-6}$

Collaborative Learning

Hourly Earnings in Selected Industries

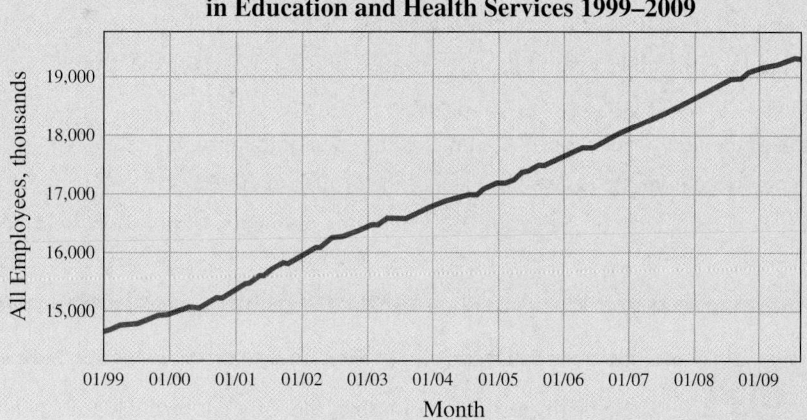

Average Hourly Earnings of Nonsupervisory Workers in Manufacturing Hardware 1999–2009

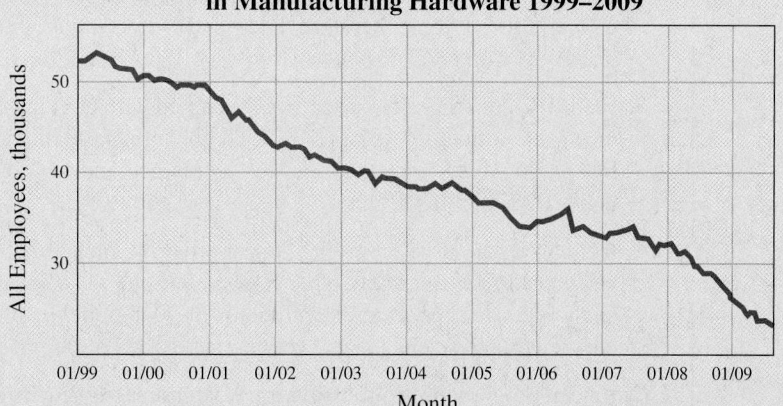

The three charts (third chart on next page) show the average hourly earnings (H) of production workers in Education and Health Services (E), Manufacturing Hardware (M), and Wholesale Office Equipment (W). Form three groups, E, M, and W.

1. What was your group salary (to the nearest dollar) in 1999?
2. What was your group salary (to the nearest dollar) in 2009?

<CONTINUED>

Source: U.S. Department of Labor, Bureau of Labor Statistics. Select the desired industry.

3. Use 1999 = 0. What is $(0, H)$, where H is the hourly earnings to the nearest dollar for each of the groups?
4. What is $(10, H)$, where H is the hourly earnings to the nearest dollar for each of the groups?
5. Use the points obtained in parts 3 and 4 to find the slope of the line for each of the groups.
6. What is the equation of the line for the hourly earnings H for each of the groups?
7. What would be the predicted hourly earnings for each of the groups in 2020?
8. If you were making a career decision based on the equations obtained in Question 6, which career would you choose?

Research Questions

1. Some historians claim that the official birthday of analytic geometry is November 10, 1619. Investigate and write a report on why this is so and on the event that led Descartes to the discovery of analytic geometry.
2. Find out what led Descartes to make his famous pronouncement "*Je pense, donc je suis*" (I think, therefore, I am) and write a report about the contents of one of his works, *La Géométrie*.
3. From 1629 to 1633, Descartes "was occupied with building up a cosmological theory of vortices to explain all natural phenomena." Find out the name of the treatise in which these theories were explained and why it wasn't published until 1654, after his death.
4. Inequalities are used in solving "linear programming" problems in mathematics, economics, and many other fields. Find out what linear programming is and write a short paper about it. Include the techniques involved and the mathematicians and scientists who cooperated in the development of this field.
5. Linear programming problems are sometimes solved using the "simplex" method. Write a few paragraphs describing the simplex method, the people who developed it, and its uses.
6. For many years, scientists have tried to improve on the simplex method. As far back as 1979, the Soviet Academy of Sciences published a paper that did just that. Find the name of the author of the paper as well as the name of the other mathematicians who have supplied and then improved on the proof contained in the paper.

Summary Chapter 3

Section	Item	Meaning	Example
3.1	x-axis y-axis	A horizontal number line on a coordinate plane A vertical number line on a coordinate plane	Q II, Q I, y-axis, x, x-axis, Q III, Q IV, y
	Abscissa	The first coordinate in an ordered pair	The abscissa in the ordered pair (3, 4) is 3.
	Ordinate Quadrant	The second coordinate in an ordered pair One of the four regions into which the axes divide the plane	The ordinate in the ordered pair (3, 4) is 4.
3.2A	Solution of an equation	The ordered pair (a, b) is a solution of an equation if, when the values for a and b are substituted for the variables in the equation, the result is a true statement.	(4, 5) is a solution of the equation $2x + 3y = 23$ because if 4 and 5 are substituted for x and y in $2x + 3y = 23$, the result $2(4) + 3(5) = 23$ is true.
3.2C	Graph of a linear equation Graph of a linear equation	The graph of a linear equation of the form $y = mx + b$ is a straight line. The graph of a linear equation of the form $Ax + By = C$ is a straight line, and every straight line has an equation of this form.	The graph of the equation $y = 3x + 6$ is a straight line. The linear equation $3x + 6y = 12$ has a straight line for its graph.
3.3A	x-intercept y-intercept	The point at which a line crosses the x-axis The point at which a line crosses the y-axis	The x-intercept of the line $3x + 6y = 12$ is (4, 0). The y-intercept of the line $3x + 6y = 12$ is (0, 2).
3.3C	Horizontal line Vertical line	A line whose equation can be written in the form $y = k$ A line whose equation can be written in the form $x = k$	The line $y = 3$ is a horizontal line. The line $x = -5$ is a vertical line.
3.4A	Slope of a line Slope of a line through (x_1, y_1) and (x_2, y_2), $x_1 \neq x_2$	The ratio of the vertical change to the horizontal change of a line $m = \frac{y_2 - y_1}{x_2 - x_1}$	The slope of the line through (3, 5) and (6, 8) is $m = \frac{8 - 5}{6 - 3} = 1$.
3.4B	Slope of the line $y = mx + b$	The slope of the line $y = mx + b$ is m.	The slope of the line $y = \frac{3}{5}x + 7$ is $\frac{3}{5}$.

(continued)

Section	Item	Meaning	Example
3.4C	Parallel lines Perpendicular lines	Two lines are parallel if their slopes are equal and they have different y-intercepts. Two lines are perpendicular if the product of their slopes is -1.	The lines $y = 2x + 5$ and $3y - 6x = 8$ are parallel (both have a slope of 2). The lines defined by $y = 2x + 1$ and $y = -\frac{1}{2}x + 2$ are perpendicular since $2 \cdot (-\frac{1}{2}) = -1$.
3.5A	Point-slope form	The point-slope form of an equation for the line passing through (x_1, x_2) and with slope m is $y - y_1 = m(x - x_1)$.	The point-slope form of an equation for the line passing through $(2, -5)$ and with slope -3 is $y - (-5) = -3(x - 2)$.
3.5B	Slope-intercept form	The slope-intercept form of an equation for the line with slope m and y-intercept b is $y = mx + b$.	The slope-intercept form of an equation for the line with slope -5 and y-intercept 2 is $y = -5x + 2$.
3.5C	Two-point form	The two-point form of an equation for a line going through (x_1, y_1) and (x_2, y_2) is $y - y_1 = m(x - x_1)$, where $m = \frac{y_2 - y_1}{x_2 - x_1}$.	The two-point form of a line going through $(2, 3)$ and $(7, 13)$ is $y - 3 = 2(x - 2)$.
3.7A	Linear inequality	An inequality that can be written in the form $Ax + By < C$ or $Ax + By > C$ (substituting $\le$ for $<$ or $\ge$ for $>$ also yields a linear inequality).	$3x < 2y - 6$ is a linear inequality because it can be written as $3x - 2y < -6$.
3.7A	Graph of a linear inequality	1. Find the boundary. Draw the line solid for $\le$ or $\ge$, dashed for $<$ or $>$. 2. Use (a, b) as a test point and shade the side of the line containing the test point if the resulting inequality is a true statement; otherwise, shade the region that does not contain the test point.	Graph of $3x - 2y < 6$. Using $(0, 0)$ as the test point for graphing $3x - 2y < 6$, we get $3(0) - 2(0) < 6$, true Shade the region containing the test point, which is the region above the dashed line. y, 5, −5, 5, x, −5

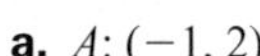

Review Exercises Chapter 3

(If you need help with these exercises, look in the section indicated in brackets.)

1. **〈3.1A〉** *Graph the points.*

a. A: $(-1, 2)$

b. B: $(-2, -1)$

c. C: $(3, -3)$

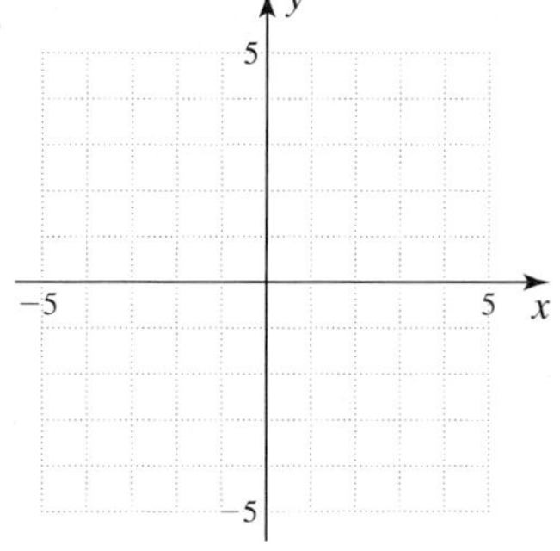

2. **〈3.1B〉** *Find the coordinates of the points shown on the graph.*

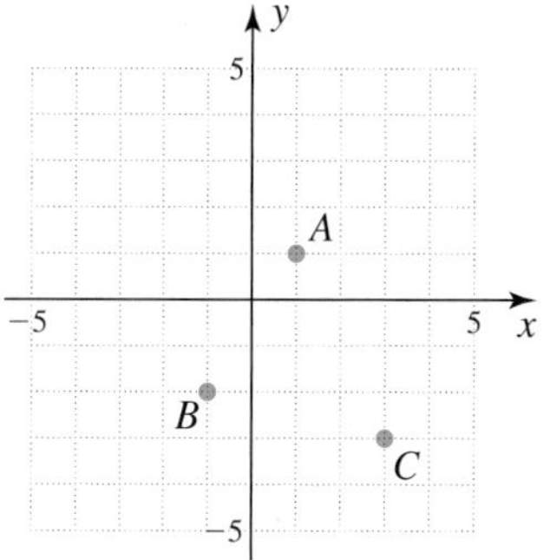

3. **〈3.1C〉** *The graph shows the new wind chill temperatures (top) and the old wind chill temperatures (bottom) for different wind speeds.*

a. On the top graph, what does the ordered pair $(10, -10)$ represent?

b. If the wind speed is 40 miles per hour, what is the approximate new wind chill temperature?

c. If the wind speed is 40 miles per hour, what is the approximate old wind chill temperature?

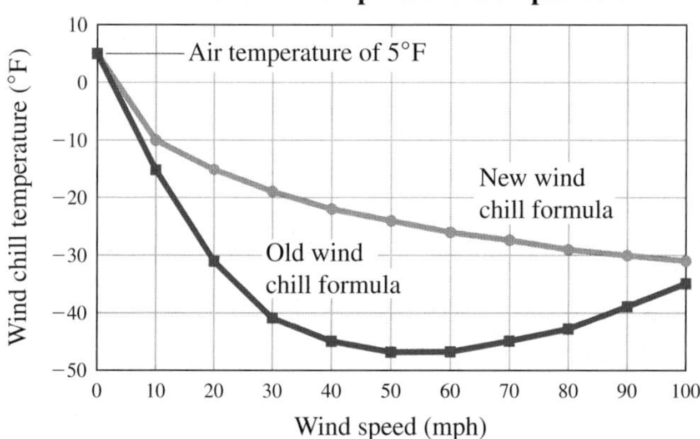

Source: Data from National Oceanic and Atmospheric Administration.

4. **〈3.1D〉** *The bar graph indicates the number of months and monthly payment needed to pay off a \$500 loan at 18% annual interest.*

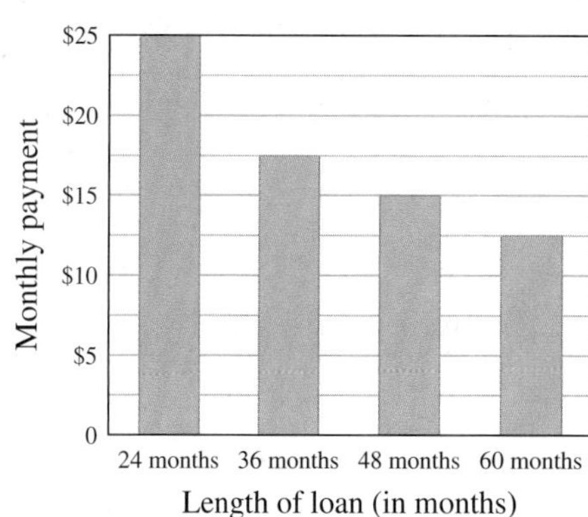

Source: Data from KJE Computer Solutions, LLC.

To the nearest dollar, what is the monthly payment if you want to pay off the loan in:

a. 60 months?

b. 48 months?

c. 24 months?

5. ⟨**3.1E**⟩ *Determine the quadrant in which each of the points is located:*

a. A

b. B

c. C

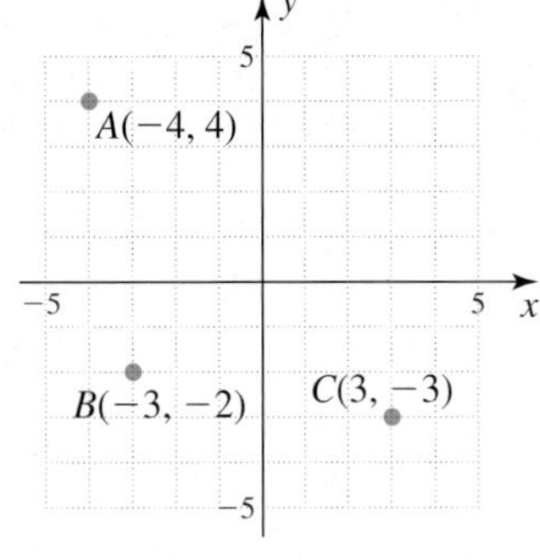

6. ⟨**3.1F**⟩ *The ordered pairs represent the wind chill temperature in degrees Fahrenheit when the wind speed in miles per hour is as indicated.*

Wind Speed	Wind Chill
10	−10
20	−15
30	−18

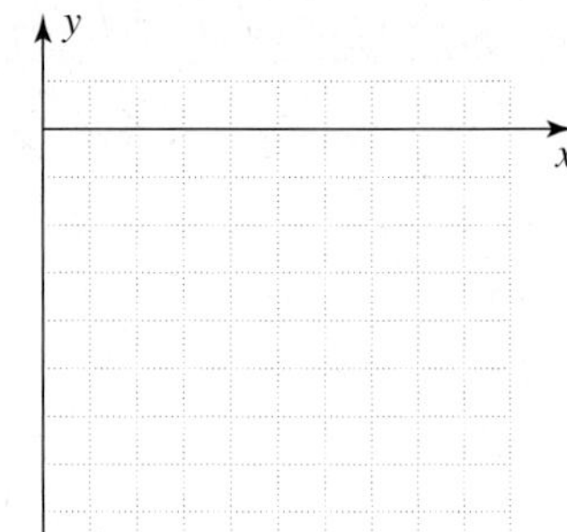

a. Graph the ordered pairs.

b. What does (10, −10) mean?

c. What is the number s in $(s, -15)$?

7. ⟨**3.2A**⟩ *Determine whether the given point is a solution of* $x - 2y = -3$.

a. $(1, -2)$

b. $(2, -1)$

c. $(-1, 1)$

8. ⟨**3.2B**⟩ *Find x in the given ordered pair so that the pair satisfies the equation* $2x - y = 4$.

a. $(x, 2)$

b. $(x, 4)$

c. $(x, 0)$

9. ⟨**3.2C**⟩ *Graph:*

a. $x + y = 4$

b. $x + y = 2$

c. $x + 2y = 2$

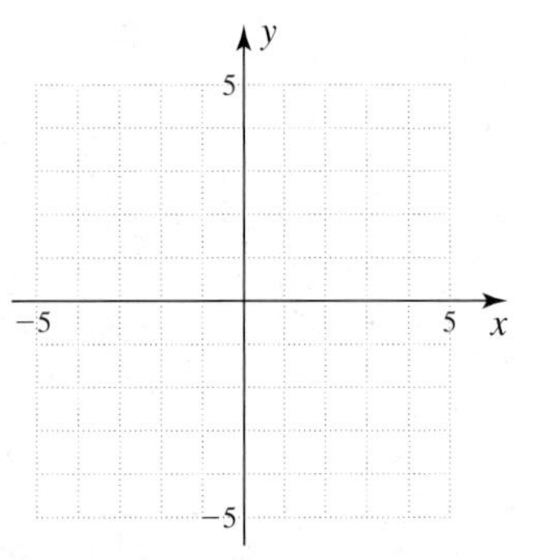

10. ⟨**3.2C**⟩ *Graph:*

a. $y = \frac{3}{2}x + 3$

b. $y = -\frac{3}{2}x + 3$

c. $y = \frac{3}{4}x + 4$

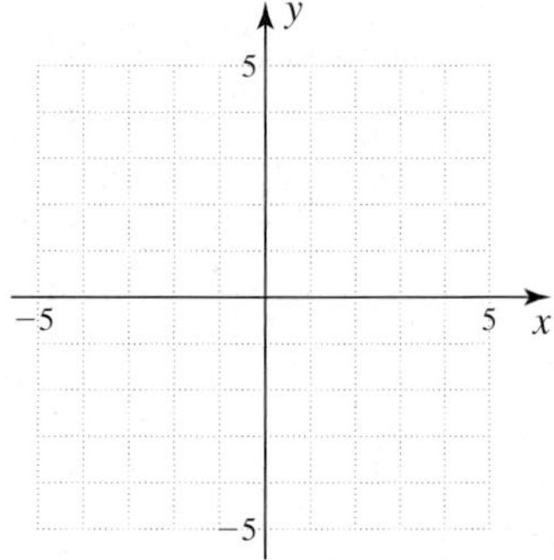

11. ⟨**3.2D**⟩ *The average annual consumption g of milk products (in gallons) per person can be approximated by* $g = 30 - 0.2t$, *where t is the number of years after 1980. Graph:*

a. $g = 30 - 0.2t$

b. $g = 20 - 0.2t$

c. $g = 10 - 0.2t$

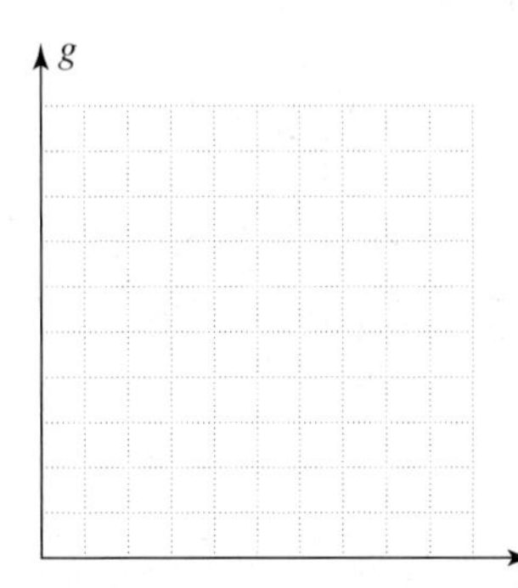

12. ⟨**3.3A**⟩ *Graph:*

a. $2x - 3y - 12 = 0$

b. $3x - 2y - 12 = 0$

c. $2x + 3y + 12 = 0$

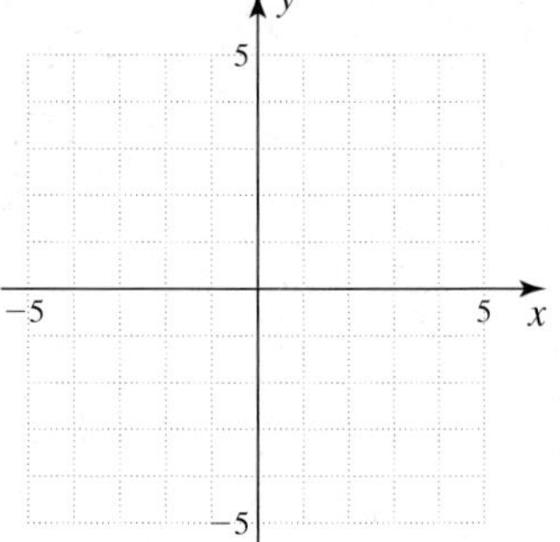

13. ⟨**3.3B**⟩ *Graph:*

a. $3x + y = 0$

b. $-2x + 3y = 0$

c. $-3x + 2y = 0$

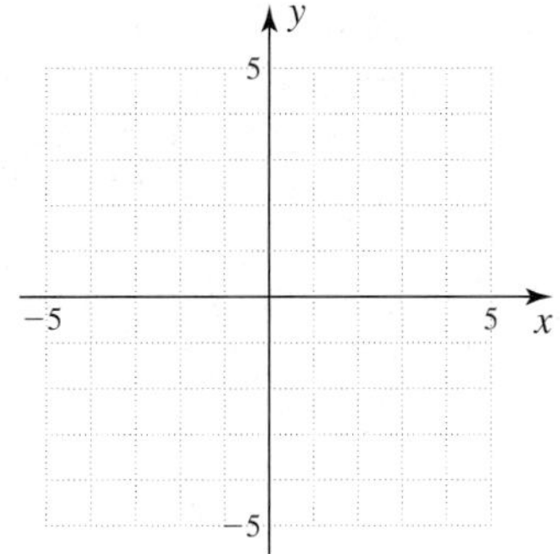

14. ⟨**3.3C**⟩ *Graph:*

a. $2x - 6 = 0$

b. $2x - 2 = 0$

c. $2x - 4 = 0$

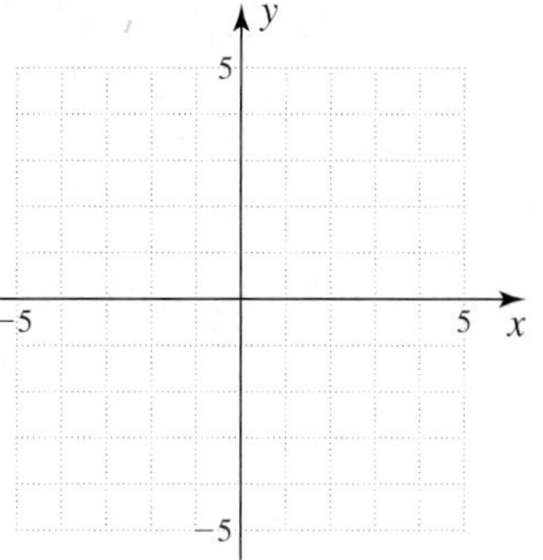

15. ⟨**3.3C**⟩ *Graph:*

a. $y = -1$

b. $y = -3$

c. $y = -4$

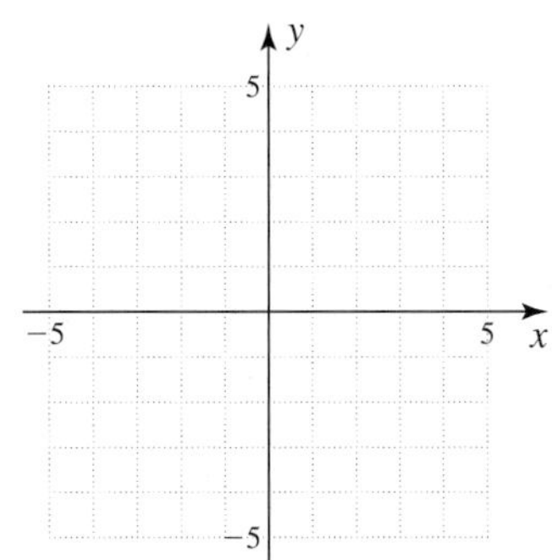

16. ⟨**3.4A**⟩ *Find the slope of the line passing through the given points.*

a. $(-5, 2)$ and $(-5, 4)$

b. $(-3, 5)$ and $(3, 5)$

c. $(-2, -1)$ and $(-2, -5)$

17. ⟨**3.4A**⟩ *Find the slope of the line passing through the given points.*

a. $(1, -4)$ and $(2, -3)$

b. $(5, -2)$ and $(8, 5)$

c. $(3, -4)$ and $(4, -8)$

18. ⟨**3.4B**⟩ *Find the slope of the line.*

a. $3x + 2y = 6$

b. $x + 4y = 4$

c. $-2x + 3y = 6$

19. ⟨**3.4C**⟩ *Decide whether the lines are parallel, perpendicular, or neither.*

a. $2x + 3y = 6$
$6x = 6 - 4y$

b. $3x + 2y = 4$
$-2x + 3y = 4$

c. $2x + 3y = 6$
$-2x - 3y = 6$

20. ⟨**3.4D**⟩ *The number N of theaters t years after 1975 can be approximated by*

$$N = 0.6t + 15 \text{ (thousand)}$$

a. What is the slope of this line?

b. What does the slope represent?

c. How many theaters were added each year?

21. ⟨**3.5A**⟩ *Find an equation of the line going through the point $(3, -5)$ and with the given slope.*

a. $m = -2$

b. $m = -3$

c. $m = -4$

22. ⟨**3.5B**⟩ *Find an equation of the line with the given slope and intercept.*

a. Slope 5, y-intercept -2

b. Slope 4, y-intercept 7

c. Slope 6, y-intercept -4

23. ⟨3.5C⟩ *Find an equation for the line passing through the given points.*

a. $(-1, 2)$ and $(4, 7)$

b. $(-3, 1)$ and $(7, 6)$

c. $(1, 2)$ and $(7, -2)$

24. ⟨3.6A⟩

a. The cost C of a long-distance call that lasts for m minutes is \$3 plus \$0.20 for each minute. If a 10-minute call costs \$5, write an equation for the cost C and find the cost of a 15-minute call.

b. Long-distance rates for m minutes are \$5 plus \$0.20 for each minute. If a 10-minute call costs \$7, write an equation for the cost C and find the cost of a 15-minute call.

c. Long-distance rates for m minutes are \$5 plus \$0.30 for each minute. If a 10-minute call costs \$8, write an equation for the cost C and find the cost of a 15-minute call.

25. ⟨3.6B⟩

a. A cell phone plan costs \$30 per month with 500 free minutes and \$0.40 for each additional minute. Find an equation for the total cost C of the plan when m minutes are used after the first 500. What is the cost when 800 minutes are used?

b. Find an equation when the cost is \$40 per month and \$0.30 for each additional minute. What is the cost when 600 minutes are used?

c. Find an equation when the cost is \$50 per month and \$0.20 for each additional minute. What is the cost when 900 minutes are used?

26. ⟨3.6C⟩

a. The bills for two long-distance calls are \$3 for 5 minutes and \$5 for 10 minutes. Find an equation for the total cost C of the calls when m minutes are used.

b. Repeat part **a** if the charges are \$4 for 5 minutes and \$6 for 10 minutes.

c. Repeat part **a** if the charges are \$5 for 5 minutes and \$7 for 10 minutes.

27. ⟨3.7A⟩ *Graph.*

a. $2x - 4y < -8$

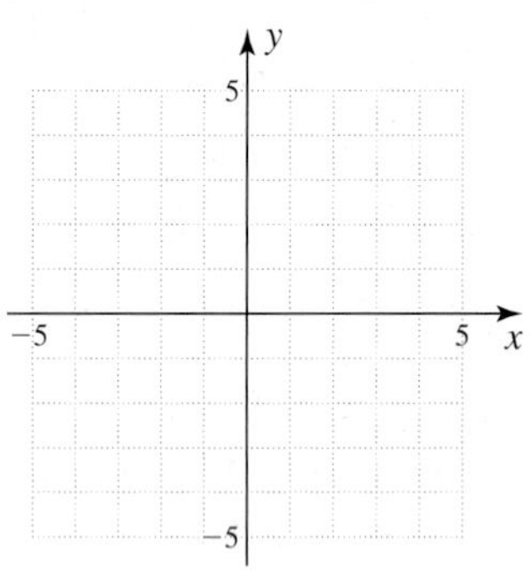

b. $3x - 6y < -12$

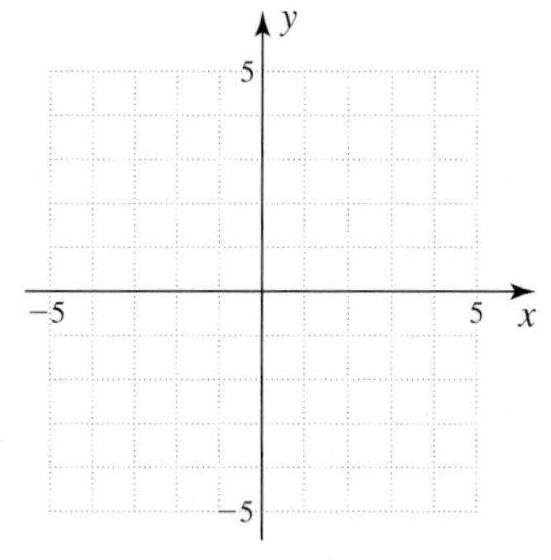

c. $4x - 2y < -8$

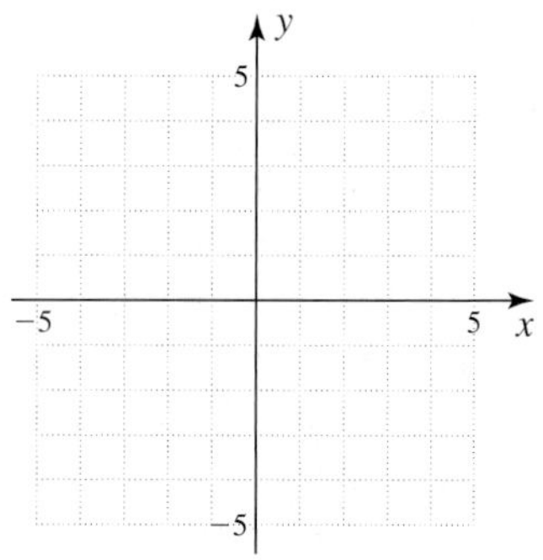

28. ⟨3.7A⟩ *Graph.*

a. $-y \le -2x + 2$

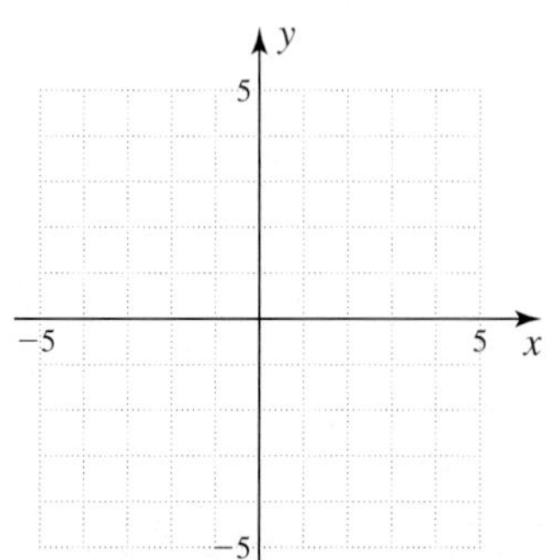

b. $-y \le -2x + 4$

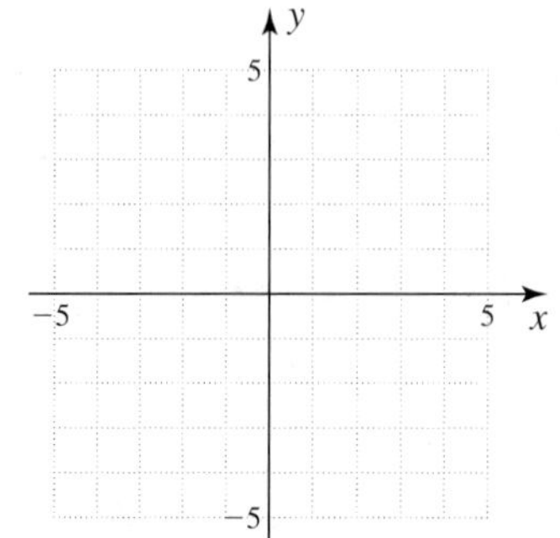

c. $-y \le -x + 3$

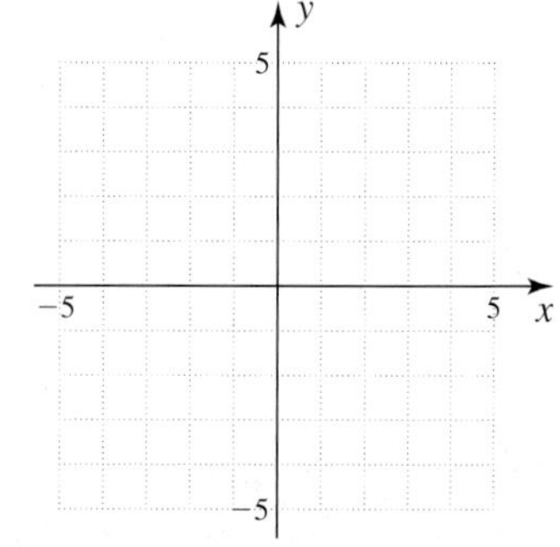

29. ⟨**3.7A**⟩ *Graph.*

a. $2x + y > 0$

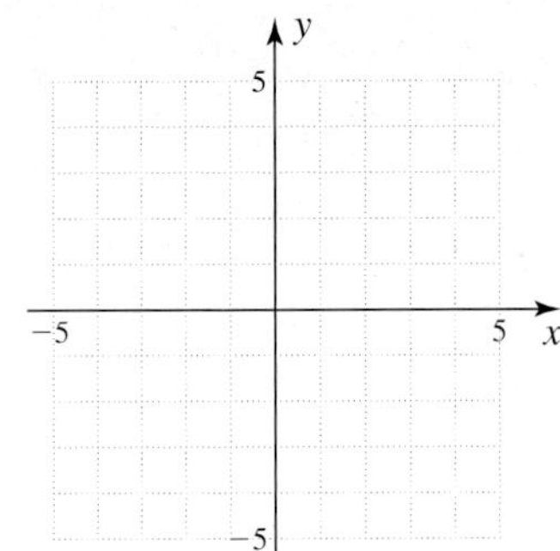

b. $3x + y > 0$

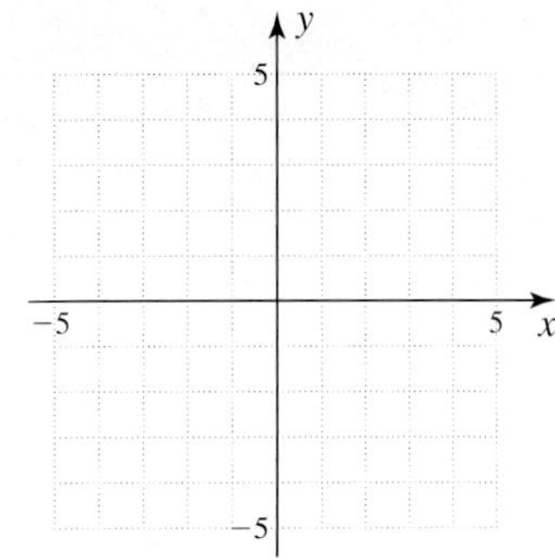

c. $3x - y < 0$

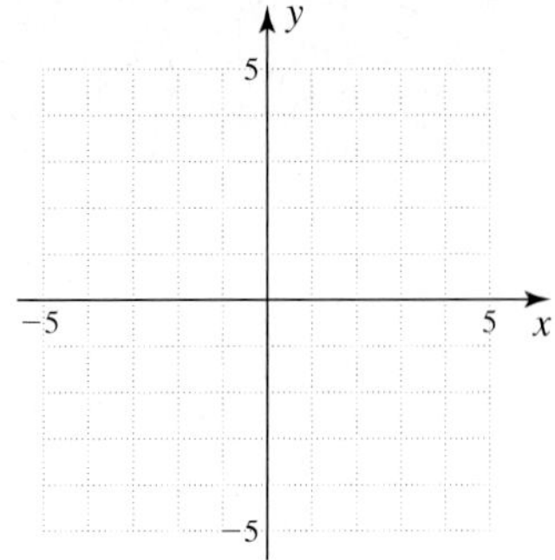

30. ⟨**3.7A**⟩ *Graph.*

a. $x \geq -4$

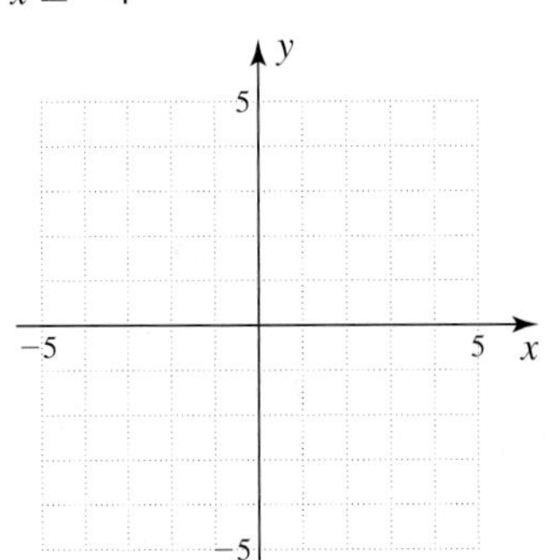

b. $y - 4 < 0$

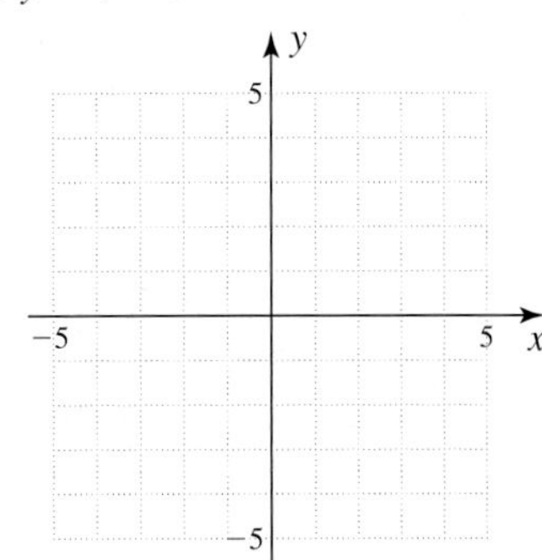

c. $2y - 4 \geq 0$

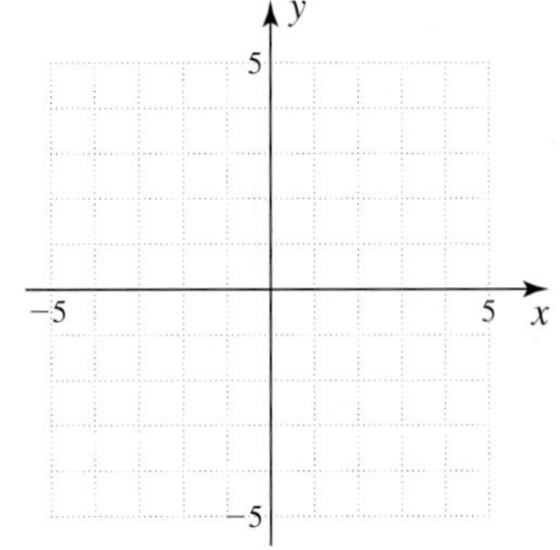

Practice Test Chapter 3

(Answers on pages 308–312)

Visit www.mhhe.com/bello to view helpful videos that provide step-by-step solutions to several of the problems below.

1. Graph the point $(-2, 3)$.

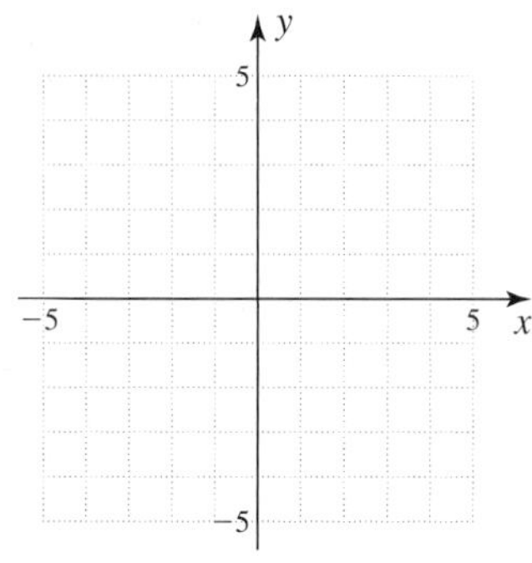

2. Find the coordinates of point A shown on the graph.

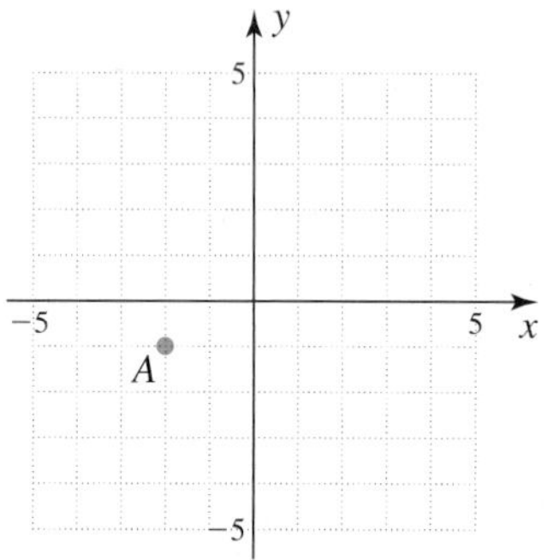

3. The graph shows the new wind chill temperatures (top) and the old wind chill temperatures (bottom) for different wind speeds.

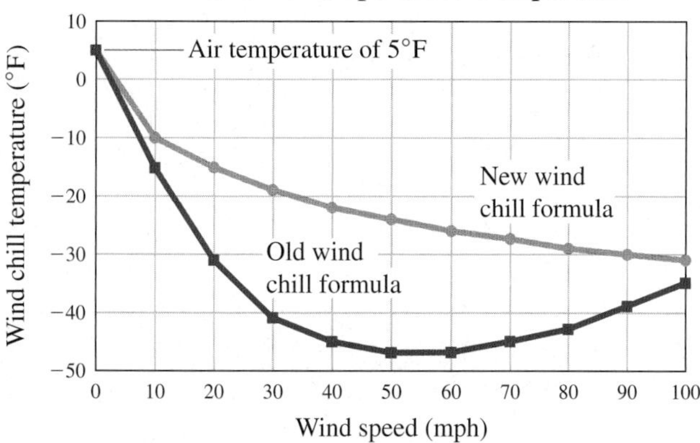

Source: Data from National Oceanic and Atmospheric Administration.

a. On the top graph, what does the ordered pair $(20, -15)$ represent?

b. If the wind speed is 90 miles per hour, what is the approximate new wind chill temperature?

c. If the wind speed is 90 miles per hour, what is the approximate old wind chill temperature?

4. The bar graph indicates the number of months and monthly payment needed to pay off a \$1000 loan at 8% annual interest.

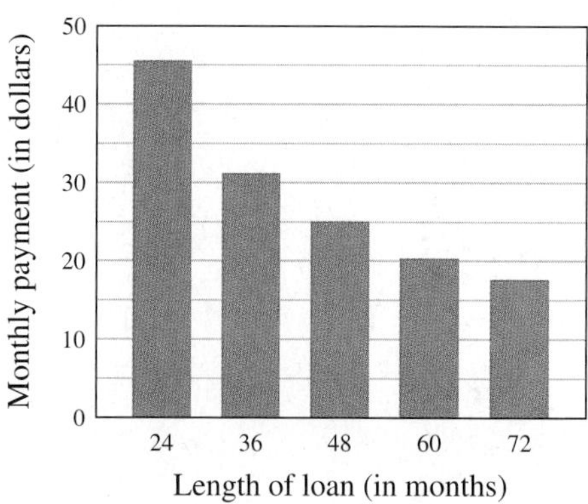

Source: Data from KJE Computer Solutions, LLC.

What is the monthly payment if you want to pay off the loan in the specified number of months?

a. 60 months

b. 48 months

c. 24 months

5. Determine the quadrant in which each of the points is located:

a. A

b. B

c. C

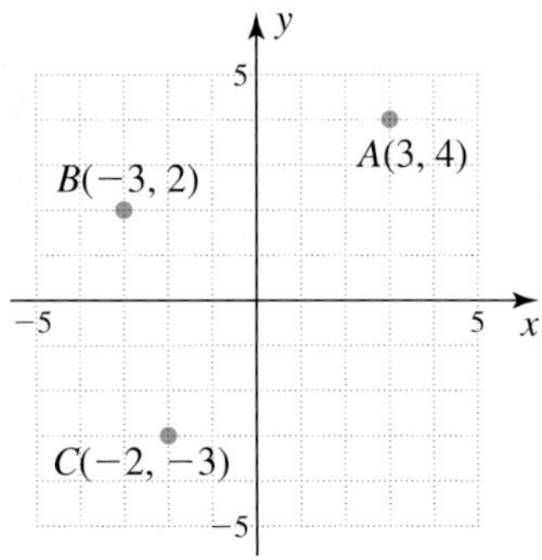

6. The ordered pairs represent the old wind chill temperature in degrees Fahrenheit when the wind speed in miles per hour is as indicated.

Wind Speed	Wind Chill
10	−15
20	−31
30	−41

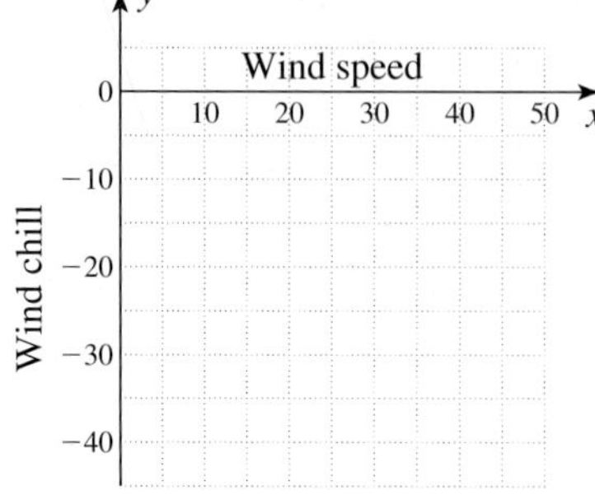

a. Graph the ordered pairs.

b. What does $(10, -15)$ mean?

c. What is the number s in $(s, -15)$?

7. Determine whether the ordered pair $(1, -2)$ is a solution of $2x - y = -2$.

8. Find x in the ordered pair $(x, 2)$ so that the ordered pair satisfies the equation $3x - y = 10$.

9. Graph $x + 2y = 4$.

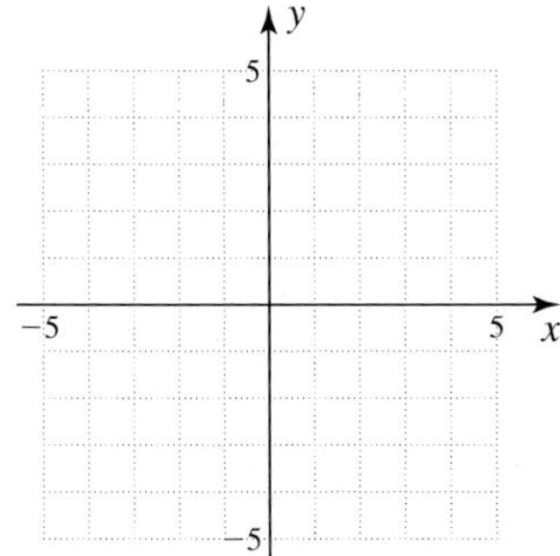

10. Graph $y = -\frac{3}{2}x - 3$.

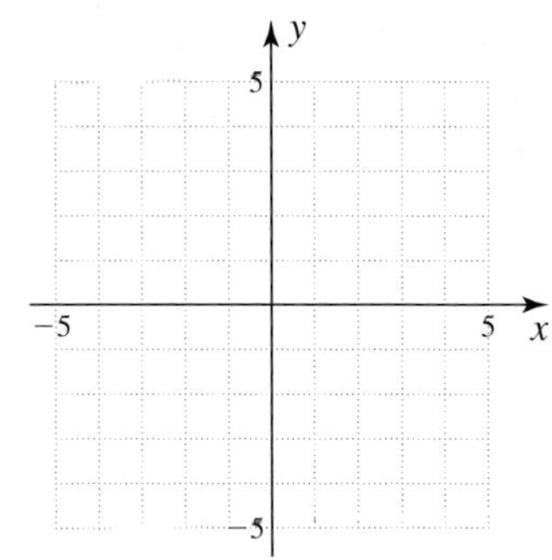

11. The average daily consumption g of protein (in grams) per person can be approximated by $g = 100 + 0.7t$, where t is the number of years after 2000. Graph $g = 100 + 0.7t$.

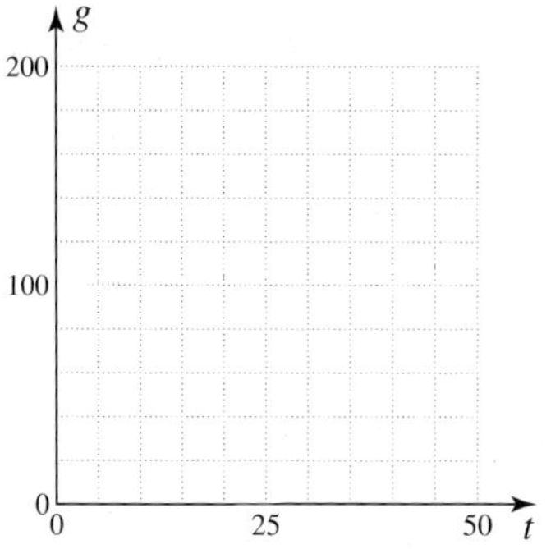

12. Graph $2x - 5y - 10 = 0$.

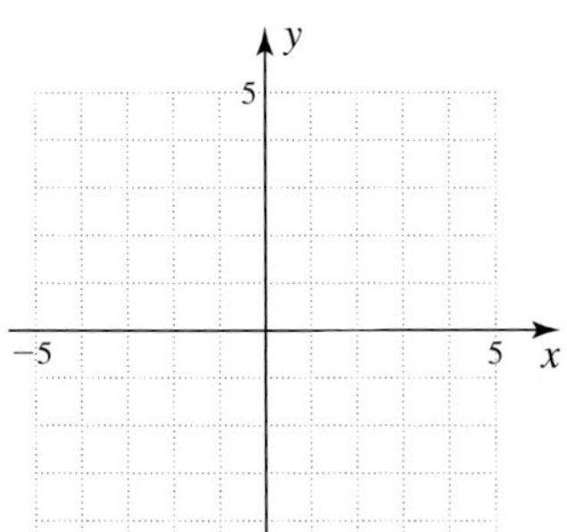

13. Graph $2x - 3y = 0$.

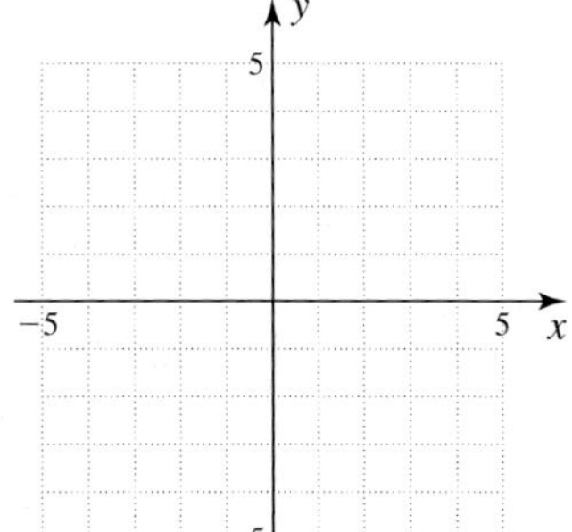

14. Graph $3x - 6 = 0$.

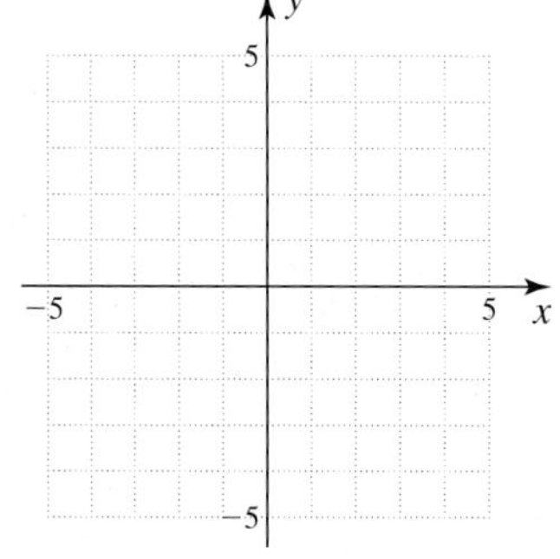

15. Graph $y = -4$.

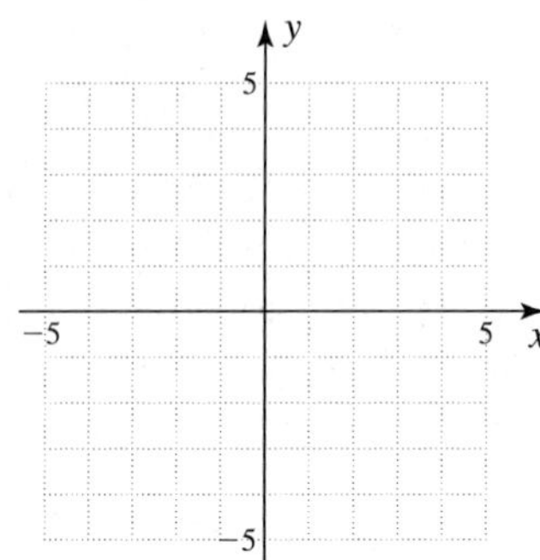

16. Find the slope of the line passing through:

a. $(2, -8)$ and $(-4, -2)$

b. $(6, -8)$ and $(-4, 6)$

17. Find the slope of the line passing through:

a. $(-3, -2)$ and $(-3, 4)$

b. $(2, 4)$ and $(4, 4)$

18. Find the slope of the line

$y - 2x = 6$.

19. Decide whether the lines are parallel, perpendicular, or neither.

a. $3y = x + 5$
$2x - 6y = 6$

b. $3y = -x + 5$
$9x - 3y = 6$

20. The number N of stores in a city t years after 2000 can be approximated by

$$N = 0.8t + 15 \text{ (hundred)}$$

a. What is the slope of this line?

b. What does the slope represent?

c. How many stores were added each year?

21. Find an equation of the line passing through the point $(2, -6)$ with slope -5. Write the answer in point-slope form and then graph the line.

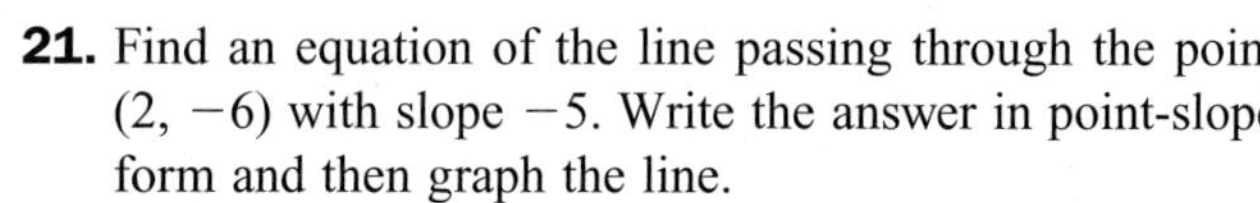

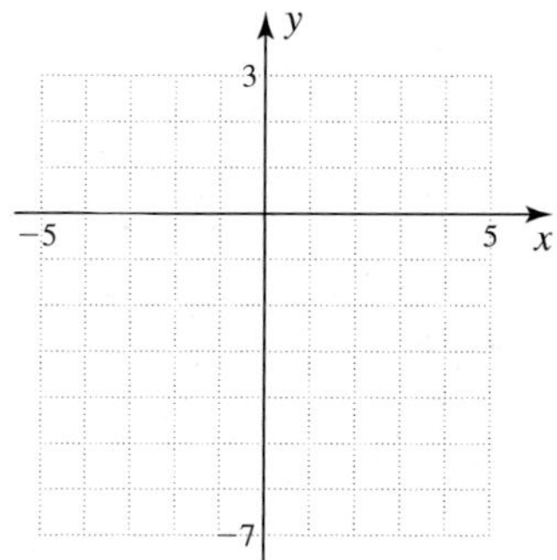

22. Find an equation of the line with slope 5 and y-intercept -4. Write the answer in slope-intercept form and then graph the line.

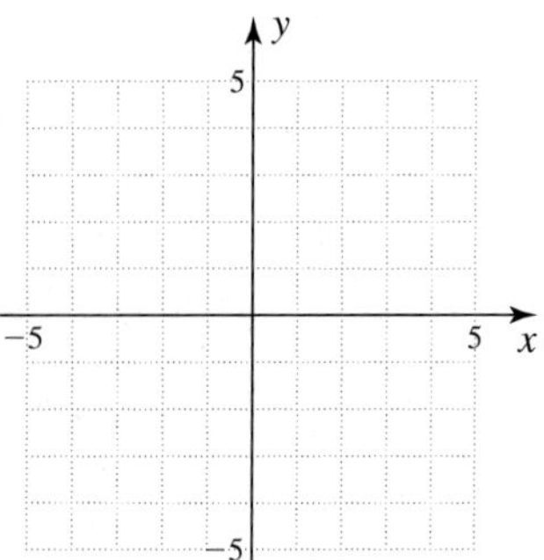

23. Find an equation of the line passing through the points $(2, -8)$ and $(-4, -2)$. Write the answer in standard form.

24. Long-distance rates for m minutes are \$10 plus \$0.20 for each minute. If a 10-minute call costs \$12, write an equation for the total cost C and find the cost of a 15-minute call.

25. A cell phone plan costs \$40 per month with 500 free minutes and \$0.50 for each additional minute. Find an equation for the total cost C of the plan when m minutes are used after the first 500. What is the cost when 800 total minutes are used?

26. The bills for two long-distance calls are \$7 for 5 minutes and \$10 for 10 minutes. Find an equation for the total cost C of calls when m minutes are used.

27. Graph $3x - 2y < -6$.

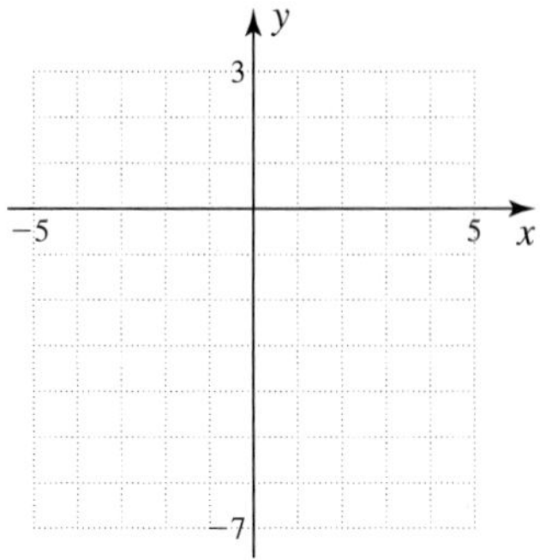

28. Graph $-y \le -3x + 3$.

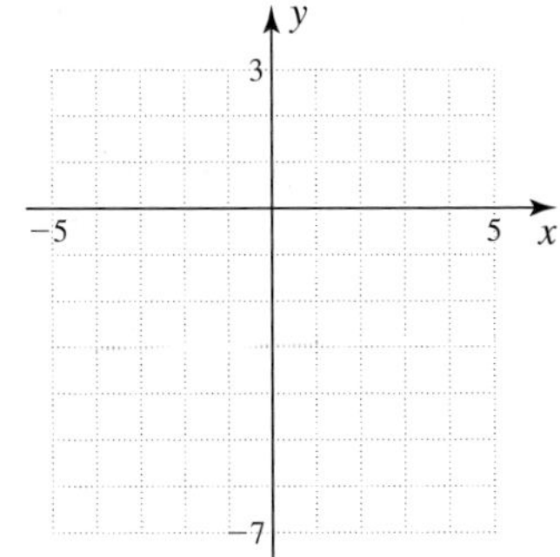

29. Graph $4x - y > 0$.

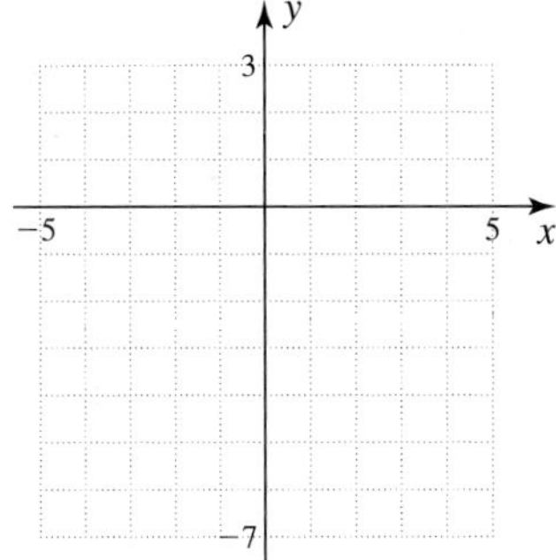

30. Graph $2y - 8 \ge 0$.

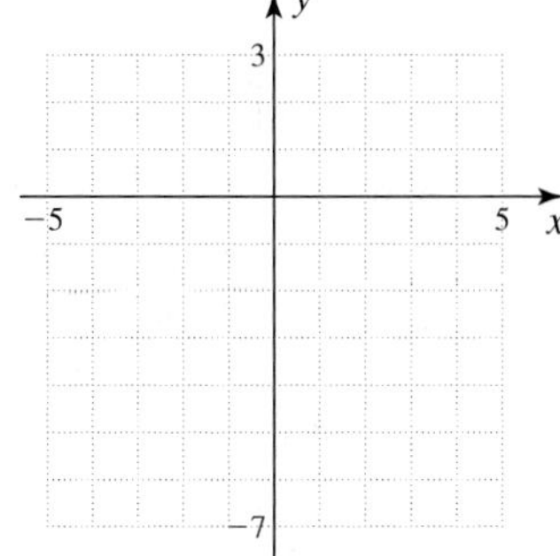

› Answers to Practice Test Chapter 3

Answer	If You Missed Question	Review Section	Review Examples	Review Page
1. 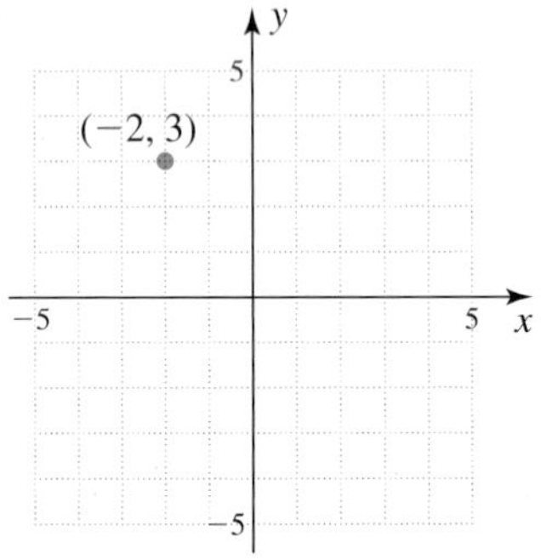	1	3.1	1	211
2. $(-2, -1)$	2	3.1	2	212
3. **a.** When the wind speed is 20 miles per hour, the wind chill temperature is -15 degrees Fahrenheit. **b.** -30 degrees Fahrenheit **c.** -40 degrees Fahrenheit	3	3.1	3	213
4. **a.** \$20 **b.** \$25 **c.** \$45	4	3.1	4	214
5. **a.** Quadrant I **b.** Quadrant II **c.** Quadrant III	5	3.1	5	215
6. **a.** 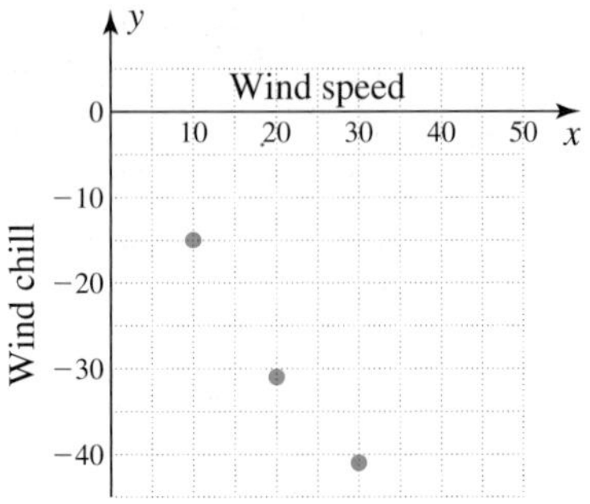 **b.** When the wind speed is 10 mi/hr, the wind chill temperature is -15°F. **c.** 10	6	3.1	6, 7	216–218
7. No	7	3.2	1	227–228

Answer	If You Missed	Review		
	Question	Section	Examples	Page
8. $x = 4$	8	3.2	2	229
9. 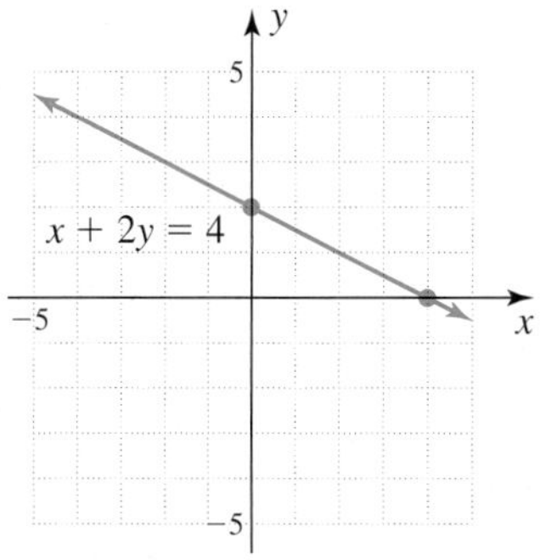	9	3.2	4	231
10. 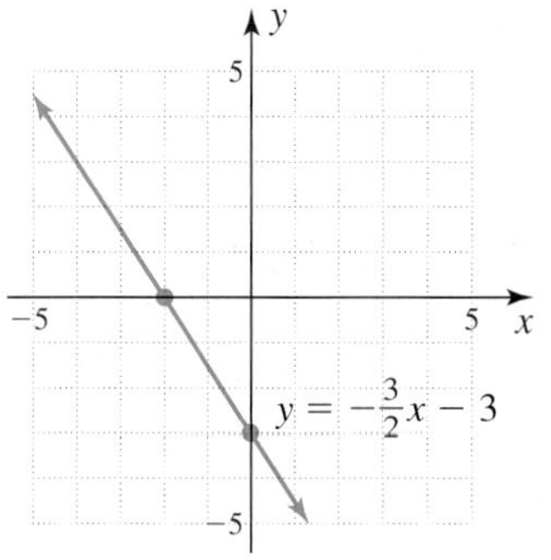	10	3.2	5	232
11. 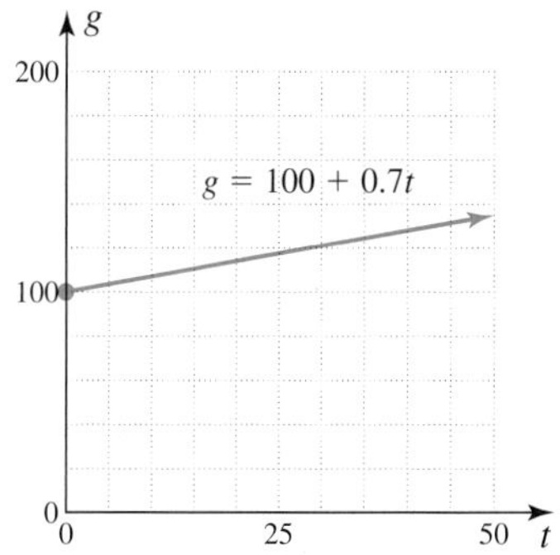	11	3.2	6	233
12. 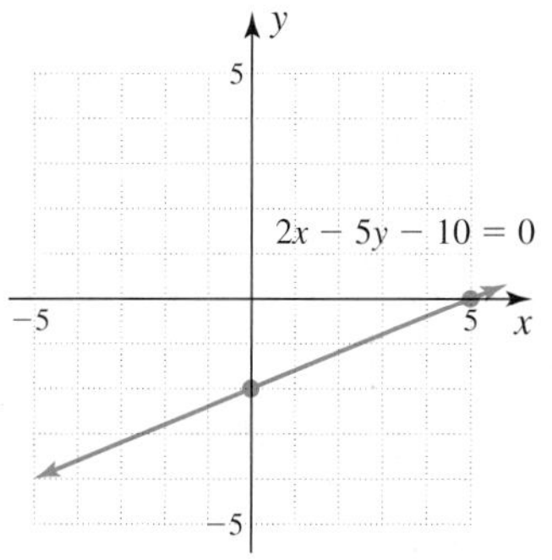	12	3.3	1, 2	241–242

Answer	If You Missed	Review		
	Question	**Section**	**Examples**	**Page**
13. 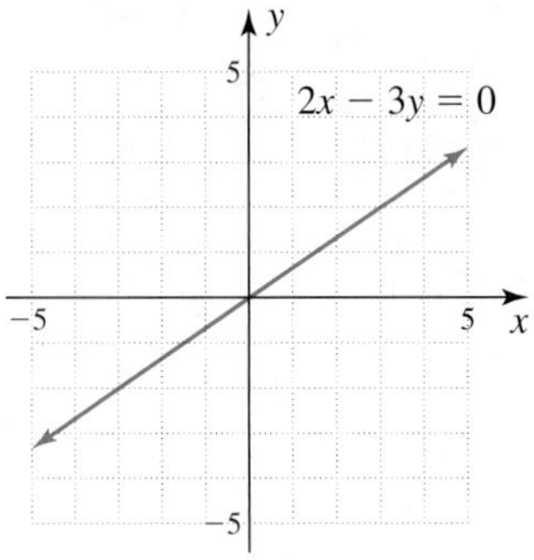 $2x - 3y = 0$	13	3.3	3	243
14. 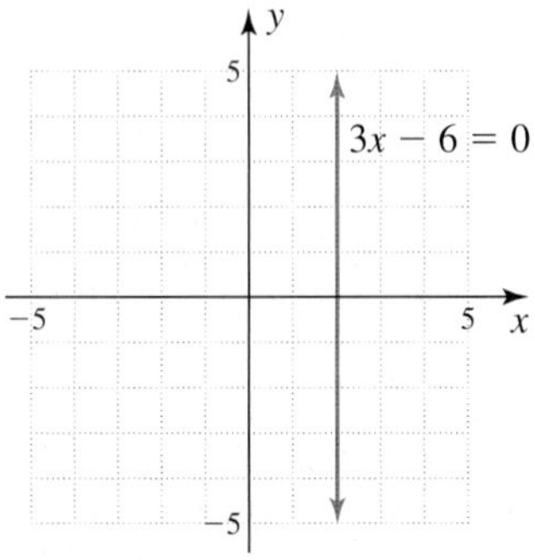 $3x - 6 = 0$	14	3.3	4a	246
15. 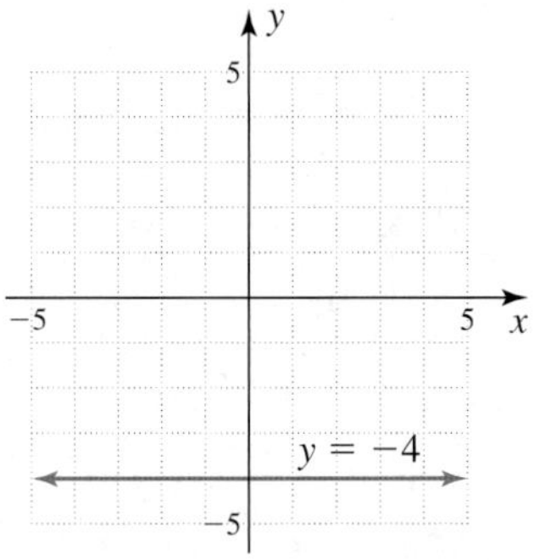 $y = -4$	15	3.3	4b	246
16. a. -1 **b.** $-\frac{7}{5}$	16	3.4	1, 2	256–257
17. a. Undefined **b.** 0	17	3.4	3	257–258
18. 2	18	3.4	4	259
19. a. Parallel **b.** Perpendicular	19	3.4	5	260–261
20. a. 0.8 **b.** Annual increase in the number of stores **c.** 80	20	3.4	6	262

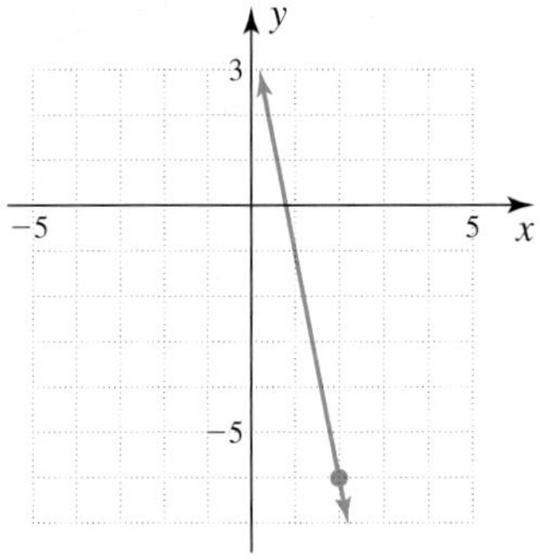

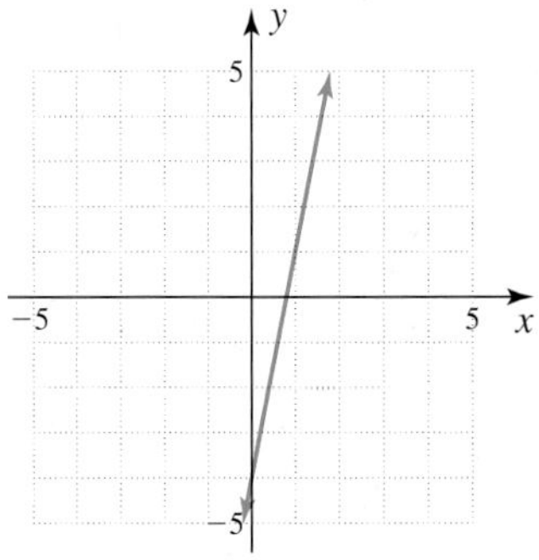

Answer	If You Missed Question	Review Section	Review Examples	Review Page
21. $y + 6 = -5(x - 2)$	21	3.5	1	267
22. $y = 5x - 4$	22	3.5	2	268
23. $x + y = -6$	23	3.5	3	269–270
24. $C = 0.20m + 10$; \$13	24	3.6	1	276
25. $C = 40 + 0.50m$; 190	25	3.6	2	276
26. $C = 0.60m + 4$	26	3.6	3	277
27.	27	3.7	1	284

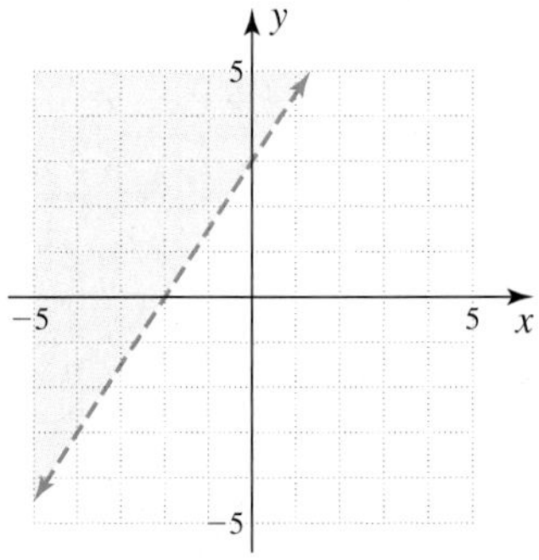

Answer	If You Missed	Review		
	Question	Section	Examples	Page
28. y 5 −5 5 x −5	28	3.7	2	285
29. y 5 −5 5 x −5	29	3.7	3	286
30. y 5 −5 5 x −5	30	3.7	4, 5	287–288

Cumulative Review Chapters 1–3

1. Find the additive inverse (opposite) of -1.

2. Find the absolute value: $\left|-3\frac{1}{7}\right|$

3. Add: $-\frac{1}{6} + \left(-\frac{3}{8}\right)$

4. Subtract: $9.7 - (-3.3)$

5. Multiply: $(-2.6)(7.6)$

6. Multiply: $(-6)^4$

7. Divide: $-\frac{6}{7} \div \left(-\frac{1}{14}\right)$

8. Evaluate $y \div 2 \cdot x - z + 3$ for $x = 3$, $y = 8$, $z = 3$.

9. Which property is illustrated by the following statement?
$$(4 + 3) + 9 = (3 + 4) + 9$$

10. Multiply: $3(6x - 7)$

11. Combine like terms: $-4cd^2 - (-5cd^2)$

12. Simplify: $3x - 3(x + 4) - (x + 2)$

13. Translate into symbols: The quotient of $(m + n)$ and p.

14. Does the number -14 satisfy the equation $1 = 15 - x$?

15. Solve for x: $1 = 5(x - 2) + 5 - 4x$

16. Solve for x: $-\frac{8}{3}x = -24$

17. Solve for x: $\frac{x}{6} - \frac{x}{9} = 3$

18. Solve for x: $4 - \frac{x}{5} = \frac{2(x + 1)}{11}$

19. Solve for d in the equation $S = 7c^2d$.

20. The sum of two numbers is 170. If one of the numbers is 40 more than the other, what are the numbers?

21. Dave has invested a certain amount of money in stocks and bonds. The total annual return from these investments is $625. If the stocks produce $245 more in returns than the bonds, how much money does Dave receive annually from each type of investment?

22. Train A leaves a station traveling at 20 miles per hour. Two hours later, train B leaves the same station traveling in the same direction at 40 miles per hour. How long does it take for train B to catch up to train A?

23. Martin purchased some municipal bonds yielding 10% annually and some certificates of deposit yielding 11% annually. If Martin's total investment amounts to $25,000 and the annual income is $2630, how much money is invested in bonds and how much is invested in certificates of deposit?

24. Graph: $-\frac{x}{7} + \frac{x}{6} \le \frac{x - 6}{6}$

25. Graph the point $C(-3, 4)$.

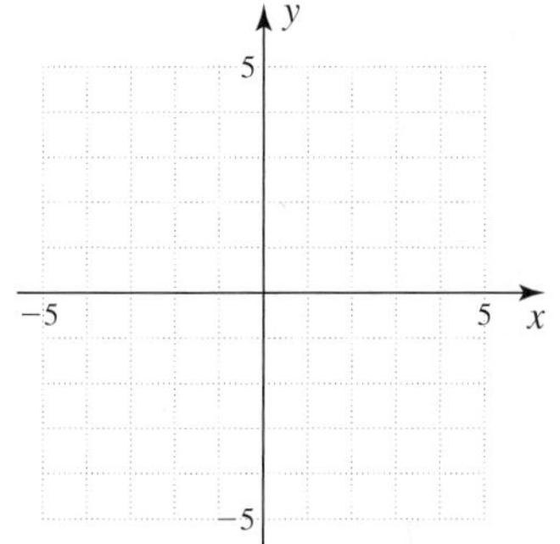

26. What are the coordinates of point A?

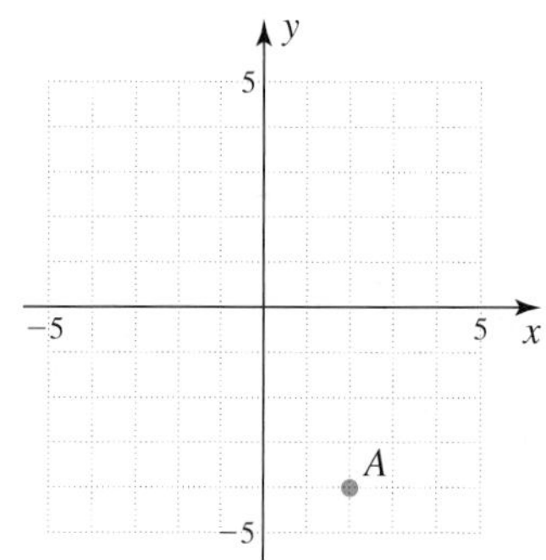

27. Determine whether the ordered pair $(-5, 1)$ is a solution of $4x + y = -21$.

28. Find x in the ordered pair $(x, 2)$ so that the ordered pair satisfies the equation $3x - y = -8$.

29. Graph: $2x + y = 8$

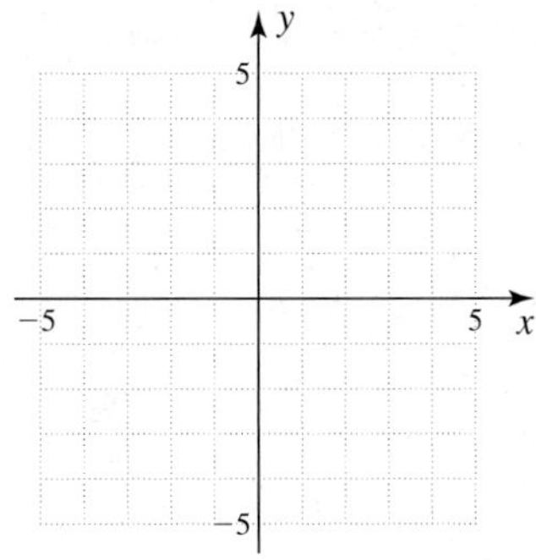

30. Graph: $4x - 8 = 0$

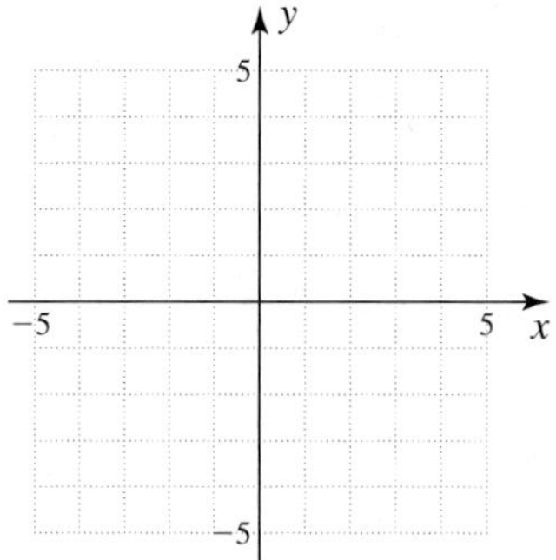

31. Find the slope of the line going through the points $(-5, 3)$ and $(-6, -6)$.

32. What is the slope of the line $4x - 2y = 18$?

33. Find the pair of parallel lines.

(1) $-5y = 4x + 7$

(2) $-15y + 12x = 7$

(3) $12x + 15y = 7$

34. Find the slope of the line passing through the points:

a. $(9, 2)$ and $(9, -5)$

b. $(-3, -7)$ and $(4, -7)$

35. Add: $-\frac{2}{9} + \left(-\frac{1}{8}\right)$

36. Subtract: $4.3 - (-3.9)$

37. Find: $(-2)^4$

38. Divide: $-\frac{1}{5} \div \left(-\frac{1}{10}\right)$

39. Evaluate $y \div 2 \cdot x - z$ for $x = 2$, $y = 8$, $z = 3$.

40. Simplify: $x + 4(x - 2) + (x - 3)$.

41. Write in symbols: The quotient of $(d + 4e)$ and f.

42. Solve for x: $5 = 3(x - 1) + 5 - 2x$

43. Solve for x: $\frac{x}{7} - \frac{x}{9} = 2$

44. The sum of two numbers is 110. If one of the numbers is 40 more than the other, what are the numbers?

45. Susan purchased some municipal bonds yielding 11% annually and some certificates of deposit yielding 14% annually. If Susan's total investment amounts to \$9000 and the annual income is \$1140, how much money is invested in bonds and how much is invested in certificates of deposit?

46. Graph: $-\frac{x}{7} + \frac{x}{4} \geq \frac{x - 4}{4}$

47. Graph the point $C(-2, -3)$.

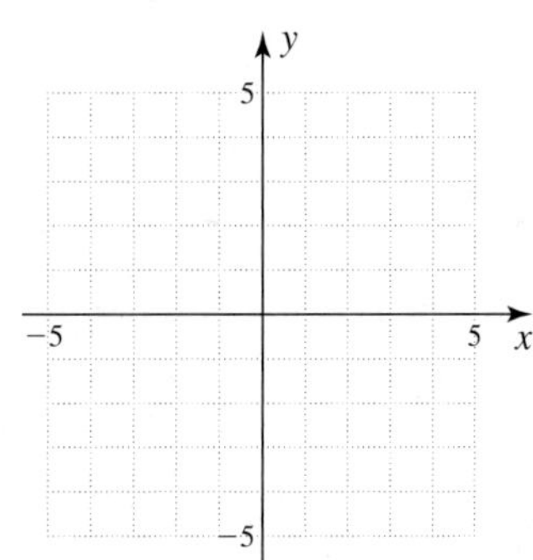

48. Determine whether the ordered pair $(-3, -4)$ is a solution of $5x - y = -19$.

49. Find x in the ordered pair $(x, -1)$ so that the ordered pair satisfies the equation $4x + 2y = -10$.

50. Graph: $3x + y = 3$

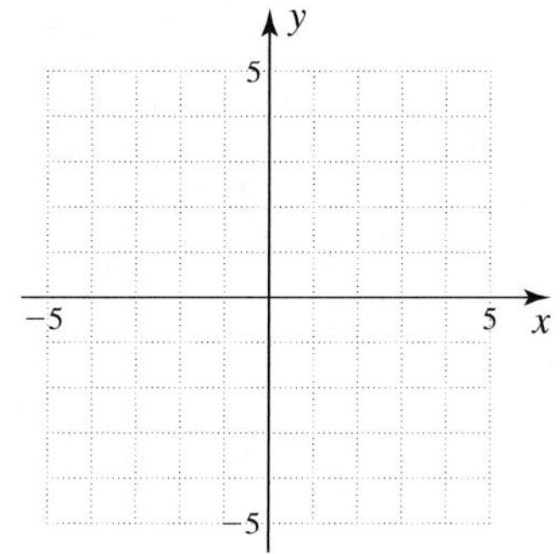

51. Graph: $3x + 9 = 0$

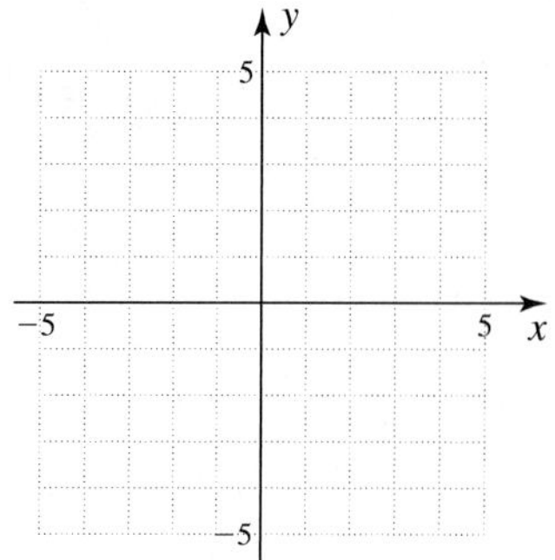

52. Find the slope of the line going through the points $(-4, -6)$ and $(-7, 1)$.

53. What is the slope of the line $12x - 4y = -14$?

54. Find the pair of parallel lines.

(1) $20x - 5y = 2$

(2) $5y + 20x = 2$

(3) $-y = -4x + 2$

55. Find the slope of the line passing through:

a. $(0, 8)$ and $(1, 8)$

b. $(-1, 2)$ and $(-1, 7)$

56. Find an equation of the line that goes through the point $(2, 0)$ and has slope $m = -3$.

57. Find an equation of the line having slope 3 and y-intercept -1.

58. Graph: $6x - y < -6$

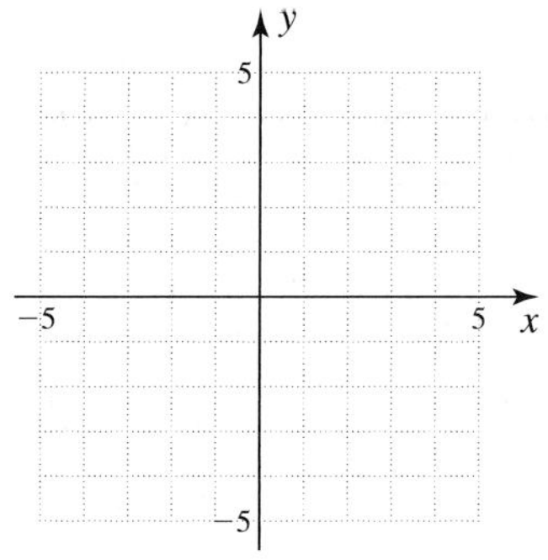

59. Graph: $-y \geq -6x - 6$

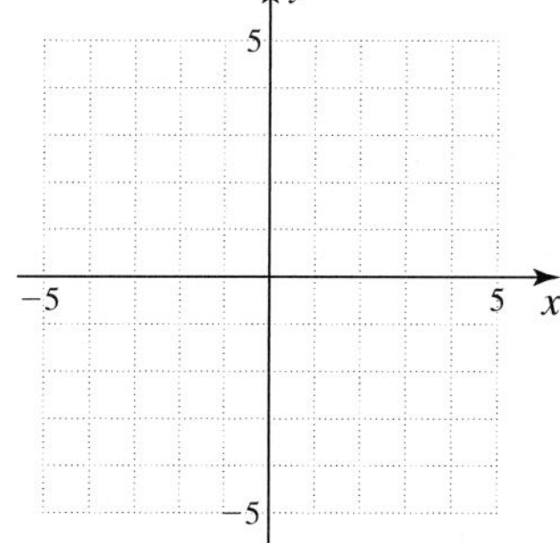

Chapter 4 — four

Exponents and Polynomials

Section

The Human Side of Algebra

In the "Golden Age" of Greek mathematics, 300–200 B.C., three mathematicians "stood head and shoulders above all the others of the time." One of them was Apollonius of Perga in Southern Asia Minor. Around 262–190 B.C., Apollonius developed a method of "tetrads" for expressing large numbers, using an equivalent of exponents of the single myriad (10,000). It was not until about the year 250 that the *Arithmetica* of Diophantus advanced the idea of exponents by denoting the square of the unknown as Δ^Y, the first two letters of the word *dunamis,* meaning "power." Similarly, K^Y represented the cube of the unknown quantity. It was not until 1360 that Nicole Oresme of France gave rules equivalent to the product and power rules of exponents that we study in this chapter. Finally, around 1484, a manuscript written by the French mathematician Nicholas Chuquet contained the *denominacion* (or power) of the unknown quantity, so that our algebraic expressions $3x$, $7x^2$, and $10x^3$ were written as .3. and $.7.^2$ and $.10.^3$. What about zero and negative exponents? $8x^0$ became $.8.^0$ and $8x^{-2}$ was written as $.8.^{2.m}$, meaning ".8. *seconds moins,*" or 8 to the negative two power. Some things do change!

4.1 The Product, Quotient, and Power Rules for Exponents

Objectives

A › Multiply expressions using the product rule for exponents.

B › Divide expressions using the quotient rule for exponents.

C › Use the power rules to simplify expressions.

To Succeed, Review How To . . .

1. Multiply and divide integers (pp. 61, 63–64).
2. Use the commutative and associative properties (p. 79).

Getting Started

Squares, Cubes, and Exponents

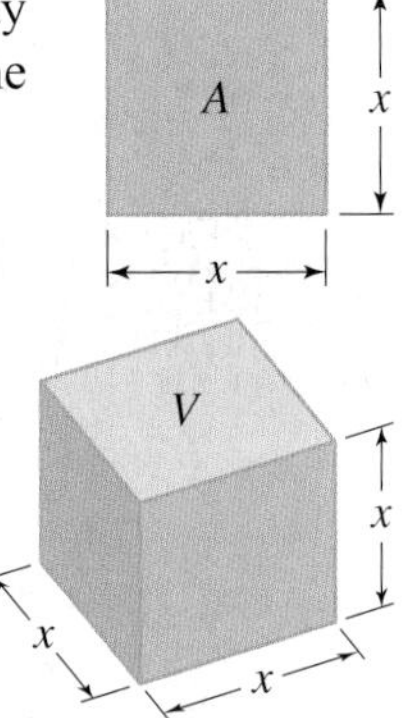

Exponential notation is used to indicate how many times a quantity is to be used as a *factor*. For example, the area A of the *square* in the figure can be written as

$$A = x \cdot x = x^2$$ Read "x squared."

The exponent 2 indicates that the x is used as a factor twice. Similarly, the volume V of the *cube* is

$$V = x \cdot x \cdot x = x^3$$ Read "x cubed."

This time, the exponent 3 indicates that the x is used as a factor three times. Can you think of a way to represent x^4 or x^5 using a geometric figure?

If a variable x (called the **base**) is to be used n times as a factor, we use the following definition:

DEFINITION OF x^n

$$\underbrace{x \cdot x \cdot x \cdot \cdots \cdot x}_{n \text{ factors}} = \underset{\text{Base}}{x}^{n} \leftarrow \text{Exponent}$$

When n is a natural number, some of the powers of x are

$x \cdot x \cdot x \cdot x \cdot x = x^5$ — x to the fifth power

$x \cdot x \cdot x \cdot x = x^4$ — x to the fourth power

$x \cdot x \cdot x = x^3$ — x to the third power (also read "x cubed")

$x \cdot x = x^2$ — x to the second power (also read "x squared")

$x = x^1$ — x to the first power (also read as x)

Note that if the base carries no exponent, the exponent is assumed to be 1. Moreover, for $a \neq 0$, a^0 is defined to be 1.

DEFINITION OF a^1 AND a^0

1. $a = a^1$, $b = b^1$, and $c = c^1$
2. $a^0 = 1 \quad (a \neq 0)$

In this section, we shall learn how to multiply and divide expressions containing exponents by using the product and quotient rules.

A > Multiplying Expressions

We are now ready to multiply expressions involving exponents. For example, to multiply x^2 by x^3, we first write

$$\underbrace{\overbrace{x \cdot x}^{x^2} \cdot \overbrace{x \cdot x \cdot x}^{x^3}}_{x^5}$$

or

Here we've just added the exponents of x^2 and x^3, 2 and 3, to find the exponent of the result, 5. Similarly, $2^2 \cdot 2^3 = 2^{2+3} = 2^5$ and $3^2 \cdot 3^3 = 3^{2+3} = 3^5$. Thus,

$$a^3 \cdot a^4 = a^{3+4} = a^7$$

and

$$b^2 \cdot b^4 = b^{2+4} = b^6$$

From these and similar examples, we have the following product rule for exponents.

PRODUCT RULE FOR EXPONENTS

If m and n are positive integers,

$$x^m \cdot x^n = x^{m+n}$$

This rule means that to multiply expressions with the *same base x,* we keep the base x and *add* the exponents.

NOTE

Note that $x^m \cdot y^n \neq (x \cdot y)^{m+n}$ because the bases x and y are not equal.

NOTE

Before you apply the product rule, make sure that the *bases* are the same.

Of course, some expressions may have numerical coefficients other than 1. For example, the expression $3x^2$ has the numerical coefficient 3. Similarly, the numerical coefficient of $5x^3$ is 5. If we decide to multiply $3x^2$ by $5x^3$, we just multiply numbers by numbers (coefficients) and letters by letters. This procedure is possible because of the *commutative and associative properties of multiplication* we've studied. Using these two properties, we then write

$$\begin{aligned}(3x^2)(5x^3) &= (3 \cdot 5)(x^2 \cdot x^3) && \text{We use parentheses to indicate multiplication.}\\ &= 15x^{2+3} && \text{Add the exponents.}\\ &= 15x^5\end{aligned}$$

and

$$\begin{aligned}&(8x^2y)(4xy^2)(2x^5y^3)\\ &= (8 \cdot 4 \cdot 2) \cdot (x^2 \cdot x^1 \cdot x^5)(y^1 \cdot y^2 \cdot y^3) && \text{Note that } x = x^1 \text{ and } y = y^1.\\ &= 64x^8y^6\end{aligned}$$

NOTE

Be sure you understand the difference between adding and multiplying expressions. Thus,

$$5x^2 + 7x^2 = 12x^2$$

but

$$(5x^2)(7x^2) = 35x^{2+2} = 35x^4$$

EXAMPLE 1 **Multiplying expressions with positive coefficients**

Multiply:

a. $(5x^4)(3x^7)$ **b.** $(3ab^2c^3)(4a^2b)(2bc)$ **c.** $(3 \times 10^8) \times (1.5 \times 10^3)$

SOLUTION 1

a. We use the commutative and associative properties to write the coefficients and the letters together:

$$\begin{aligned}(5x^4)(3x^7) &= (5 \cdot 3)(x^4 \cdot x^7)\\ &= 15x^{4+7}\\ &= 15x^{11}\end{aligned}$$

b. Using the commutative and associative properties, and the fact that $a = a^1$, $b = b^1$, and $c = c^1$, we write

$$\begin{aligned}(3ab^2c^3)(4a^2b)(2bc) &= (3 \cdot 4 \cdot 2)(a^1 \cdot a^2)(b^2 \cdot b^1 \cdot b^1)(c^3 \cdot c^1)\\ &= 24a^{1+2}b^{2+1+1}c^{3+1}\\ &= 24a^3b^4c^4\end{aligned}$$

c. Using the commutative and associative properties of multiplication

$$\begin{aligned}(3 \times 10^8) \times (1.5 \times 10^3) &= (3 \times 1.5) \times (10^8 \times 10^3)\\ &= 4.5 \times 10^{8+3}\\ &= 4.5 \times 10^{11}\\ &= 450{,}000{,}000{,}000 \text{ or } 450 \text{ billion}\end{aligned}$$

This amount represents the amount of garbage produced in the United States every year: (3×10^8) people, each producing (1.5×10^3) pounds each year!

PROBLEM 1

Multiply:

a. $(3x^5)(4x^7)$

b. $(4ab^3c^2)(3ab^3)(2bc)$

c. $(3.3 \times 10^7) \times (1.5 \times 10^3)$

This amount represents the amount of garbage produced in Canada every year.

To multiply expressions involving signed coefficients, we recall the rule of signs for multiplication:

RULES

Signs for Multiplication

1. When multiplying two numbers with the *same* (like) sign, the product is *positive* (+).
2. When multiplying two numbers with *different* (unlike) signs, the product is *negative* (−).

Thus, to multiply $(-3x^5)$ by $(8x^2)$, we first note that the expressions have *different* signs. The product should have a *negative* coefficient; that is,

$$\begin{aligned}(-3x^5)(8x^2) &= (-3 \cdot 8)(x^5 \cdot x^2)\\ &= -24x^{5+2}\\ &= -24x^7\end{aligned}$$

Of course, if the expressions have the *same* sign, the product should have a *positive* coefficient. Thus,

$$(-2x^3)(-7x^5) = (-2)(-7)(x^3 \cdot x^5)$$

Note that the product of −2 and −7 was written as (−2)(−7) and *not* as (−2 · −7) to avoid confusion.

$$= +14x^{3+5}$$

$$= 14x^8$$

Recall that +14 = 14.

Answers to PROBLEMS

1. a. $12x^{12}$ **b.** $24a^2b^7c^3$ **c.** $4.95 \times 10^{10} = 49{,}500{,}000{,}000$ or 49.5 billion pounds

EXAMPLE 2 **Multiplying expressions with negative coefficients**

Multiply:

a. $(7a^2bc)(-3ac^4)$ **b.** $(-2xyz^3)(-4x^3yz^2)$

SOLUTION 2

a. Since the coefficients have *different* signs, the result must have a *negative* coefficient. Hence,

$$\begin{aligned}(7a^2bc)(-3ac^4) &= (7)(-3)(a^2 \cdot a^1)(b^1)(c^1 \cdot c^4)\\ &= -21a^{2+1}b^1c^{1+4}\\ &= -21a^3bc^5\end{aligned}$$

b. The expressions have the *same* sign, so the result must have a *positive* coefficient. That is,

$$\begin{aligned}(-2xyz^3)(-4x^3yz^2) &= [(-2)(-4)](x^1 \cdot x^3)(y^1 \cdot y^1)(z^3 \cdot z^2)\\ &= +8x^{1+3}y^{1+1}z^{3+2}\\ &= 8x^4y^2z^5\end{aligned}$$

PROBLEM 2

Multiply:

a. $(2a^3bc)(-5ac^5)$

b. $(-3x^3yz^4)(-2xyz^4)$

B > Dividing Expressions

We are now ready to discuss the division of one expression by another. As you recall, the same rule of signs that applies to the multiplication of integers applies to the division of integers. We write this rule for easy rcference.

RULES

Signs for Division

1. When dividing two numbers with the *same* (like) sign, the quotient is *positive* (+).
2. When dividing two numbers with *different* (unlike) signs, the quotient is *negative* (−).

Now we know what to do with the numerical coefficients when we divide one expression by another. But what about the exponents? To divide expressions involving exponents, we need another rule. So to divide x^5 by x^3, we first use the definition of exponent and write

$$\frac{x^5}{x^3} = \frac{x \cdot x \cdot x \cdot x \cdot x}{x \cdot x \cdot x} \quad (x \neq 0)$$

Remember, division by zero is not allowed!

Since $(x \cdot x \cdot x)$ is common to the numerator and denominator, we have

$$\frac{x^5}{x^3} = \frac{\cancel{(x \cdot x \cdot x)} \cdot x \cdot x}{\cancel{(x \cdot x \cdot x)}} = x \cdot x = x^2$$

Here the colored x's mean that we divided the numerator and denominator by the common factor $(x \cdot x \cdot x)$. Of course, you can immediately see that the exponent 2 in the answer is simply the difference of the original two exponents, 5 and 3; that is,

$$\frac{x^5}{x^3} = x^{5-3} = x^2 \quad (x \neq 0)$$

Similarly,

$$\frac{x^7}{x^4} = x^{7-4} = x^3 \quad (x \neq 0)$$

and

$$\frac{y^4}{y^1} = y^{4-1} = y^3 \quad (y \neq 0)$$

Answers to PROBLEMS

2. a. $-10a^4bc^6$ **b.** $6x^4y^2z^8$

We can now state the rule for dividing expressions involving exponents.

QUOTIENT RULE FOR EXPONENTS

If m and n are positive integers and m is greater than n then,

$$\frac{x^m}{x^n} = x^{m-n} \quad (x \neq 0)$$

This means that to divide expressions with the same base x, we keep the same base x and subtract the exponents.

NOTE

Before you apply the quotient rule, make sure that the **bases** are the same.

EXAMPLE 3 **Dividing expressions with negative coefficients**

Find the quotient:

$$\frac{24x^2y^6}{-6xy^4}$$

SOLUTION 3 We could write

$$\frac{24x^2y^6}{-6xy^4} = \frac{2 \cdot 2 \cdot \cancel{(2 \cdot 3 \cdot x)} \cdot x \cdot \cancel{(y \cdot y \cdot y \cdot y)} \cdot y \cdot y}{-1 \cdot \cancel{(2 \cdot 3 \cdot x)} \cdot \cancel{(y \cdot y \cdot y \cdot y)}}$$

$$= -2 \cdot 2 \cdot x \cdot y \cdot y$$

$$= -4xy^2$$

but to save time, it's easier to divide 24 by -6, x^2 by x, and y^6 by y^4, like this:

$$\frac{24x^2y^6}{-6xy^4} = \frac{24}{-6} \cdot \frac{x^2}{x} \cdot \frac{y^6}{y^4}$$

$$= -4 \cdot x^{2-1} \cdot y^{6-4}$$

$$= -4xy^2$$

PROBLEM 3

Find the quotient:

$$\frac{25a^3b^7}{-5ab^3}$$

C > Simplifying Expressions Using the Power Rules

Suppose we wish to find $(5^3)^2$. By definition, squaring a quantity means that we multiply the quantity by itself. Thus,

$$(5^3)^2 = 5^3 \cdot 5^3 = 5^{3+3} \quad \text{or} \quad 5^6$$

We could get this answer by multiplying exponents in $(5^3)^2$ to obtain $5^{3 \cdot 2} = 5^6$.

Similarly,

$$(x^2)^3 = x^2 \cdot x^2 \cdot x^2 = x^{2 \cdot 3} = x^6$$

Again, we multiplied exponents in $(x^2)^3$ to get x^6. We use these ideas to state the following power rule for exponents.

POWER RULE FOR EXPONENTS

If m and n are positive integers.

$$(x^m)^n = x^{mn}$$

This means that when raising a *power* to a *power,* we keep the base x and *multiply* the exponents.

Answers to PROBLEMS

3. $-5a^2b^4$

EXAMPLE 4 Raising a power to a power

Simplify:

a. $(2^3)^4$ b. $(x^2)^5$ c. $(y^4)^5$

SOLUTION 4

a. $(2^3)^4 = 2^{3\cdot 4} = 2^{12}$ b. $(x^2)^5 = x^{2\cdot 5} = x^{10}$ c. $(y^4)^5 = y^{4\cdot 5} = y^{20}$

PROBLEM 4

Simplify:

a. $(3^2)^4$ b. $(a^3)^5$ c. $(b^5)^3$

Sometimes we need to raise several factors inside parentheses to a power, such as in $(x^2y^3)^3$. We use the definition of cubing (see the *Getting Started*) and write:

$$\begin{aligned}(x^2y^3)^3 &= x^2y^3 \cdot x^2y^3 \cdot x^2y^3\\ &= (x^2 \cdot x^2 \cdot x^2)(y^3 \cdot y^3 \cdot y^3)\\ &= (x^2)^3(y^3)^3\\ &= x^6y^9\end{aligned}$$

Since we are cubing x^2y^3, we could get the same answer by multiplying each of the exponents in x^2y^3 by 3 to obtain $x^{2\cdot 3}y^{3\cdot 3} = x^6y^9$. Thus, to raise several factors inside parentheses to a power, we raise each factor to the given power, as stated in the following rule.

POWER RULE FOR PRODUCTS

If m, n, and k are positive integers,

$$(x^my^n)^k = (x^m)^k(y^n)^k = x^{mk}y^{nk}$$

This means that to raise a product to a power, we raise each factor in the product to that power.

NOTE

The power rule applies to products only:

$$(x + y)^n \neq x^n + y^n$$

EXAMPLE 5 Using the power rule for products

Simplify:

a. $(3x^2y^2)^3$ b. $(-2x^2y^3)^3$

SOLUTION 5

a. $(3x^2y^2)^3 = 3^3(x^2)^3(y^2)^3$
$= 27x^6y^6$

Note that since 3 is a *factor* in $3x^2y^2$, 3 is also raised to the third power.

b. $(-2x^2y^3)^3 = (-2)^3(x^2)^3(y^3)^3$
$= -8x^6y^9$

PROBLEM 5

Simplify:

a. $(2a^3b^2)^4$ b. $(-3a^3b^2)^3$

Just as we have a product and a quotient rule for exponents, we also have a power rule for products and for quotients. Since the quotient

$$\frac{a}{b} = a \cdot \frac{1}{b},$$

we can use the power rule for products and some of the real-number properties to obtain the following.

Answers to PROBLEMS

4. a. 3^8 **b.** a^{15} **c.** b^{15}

5. a. $16a^{12}b^8$ **b.** $-27a^9b^6$

THE POWER RULE FOR QUOTIENTS

If m is a positive integer,

$$\left(\frac{x}{y}\right)^m = \frac{x^m}{y^m} \quad (y \neq 0)$$

This means that to raise a quotient to a power, we raise the numerator and denominator in the quotient to that power.

EXAMPLE 6 Using the power rule for quotients

Simplify:

a. $\left(\frac{4}{5}\right)^3$ **b.** $\left(\frac{2x^2}{y^3}\right)^4$

SOLUTION 6

a. $\left(\frac{4}{5}\right)^3 = \frac{4^3}{5^3} = \frac{64}{125}$

b. $\frac{(2x^2)^4}{(y^3)^4} = \frac{2^4 \cdot (x^2)^4}{(y^3)^4}$

$= \frac{16x^{2\cdot 4}}{y^{3\cdot 4}}$

$= \frac{16x^8}{y^{12}}$

PROBLEM 6

Simplify:

a. $\left(\frac{2}{3}\right)^3$ **b.** $\left(\frac{3a^2}{b^4}\right)^3$

Here is a summary of the rules we've just studied.

RULES FOR EXPONENTS

If m, n, and k are positive integers, the following rules apply:

Rule		Example
1. Product rule for exponents:	$x^m x^n = x^{m+n}$	$x^5 \cdot x^6 = x^{5+6} = x^{11}$
2. Quotient rule for exponents:	$\frac{x^m}{x^n} = x^{m-n}$ $(m > n, x \neq 0)$	$\frac{p^8}{p^3} = p^{8-3} = p^5$
3. Power rule for products:	$(x^m y^n)^k = x^{mk} y^{nk}$	$(x^4 y^3)^4 = x^{4\cdot 4} y^{3\cdot 4} = x^{16} y^{12}$
4. Power rule for quotients:	$\left(\frac{x}{y}\right)^m = \frac{x^m}{y^m}$ $(y \neq 0)$	$\left(\frac{a^3}{b^4}\right)^6 = \frac{a^{3\cdot 6}}{b^{4\cdot 6}} = \frac{a^{18}}{b^{24}}$

Now let's look at an example where several of these rules are used.

EXAMPLE 7 Using the power rule for products and quotients

Simplify:

a. $(3x^4)^3(-2y^3)^2$ **b.** $\left(\frac{3}{5}\right)^3 \cdot 5^4$

SOLUTION 7

a. $(3x^4)^3(-2y^3)^2 = (3)^3(x^4)^3(-2)^2(y^3)^2$ Use Rule 3.

$= 27x^{4\cdot 3}(4)y^{3\cdot 2}$ Use the Power Rule for Exponents.

$= (27 \cdot 4)x^{12}y^6$

$= 108x^{12}y^6$

PROBLEM 7

Simplify:

a. $(2a^3)^4(-3b^2)^3$ **b.** $\left(\frac{2}{3}\right)^3 \cdot 3^5$

Answers to PROBLEMS

6. a. $\frac{8}{27}$ **b.** $\frac{27a^6}{b^{12}}$ **7. a.** $-432a^{12}b^6$ **b.** 72

b. $\left(\frac{3}{5}\right)^3 \cdot 5^4 = \frac{3^3}{5^3} \cdot 5^4$ Use Rule 4.

$= \frac{3^3}{5^3} \cdot \frac{5^4}{1}$ Since $5^4 = \frac{5^4}{1}$

$= \frac{3^3 \cdot 5^4}{5^3}$ Multiply.

$= 3^3 \cdot 5$ Since $\frac{5^4}{5^3} = 5^{4-3} = 5$

$= 27 \cdot 5$ Since $3^3 = 27$

$= 135$ Multiply.

So far, we have discussed only the theory and rules for exponents. Does anybody need or use exponents? We will show one application in Example 8.

EXAMPLE 8 Application: Volume of an ice cream cone

The volume V of a cone is $V = \frac{1}{3}\pi r^2 h$, where r is the radius of the base (opening) and h is the height of the cone. The volume S of a hemisphere is $S = \frac{2}{3}\pi r^3$. The ice cream cone in the photo is 10 centimeters tall and has a radius of 2 centimeters; the mound of ice cream is 2 centimeters high. To make computation easier, assume that π is about 3; that is, use 3 for π.

a. How much ice cream does it take to fill the cone?

b. What is the volume of the mound of ice cream?

c. What is the total volume of ice cream (cone plus mound) in the photo?

SOLUTION 8

a. The volume V of the cone is $V = \frac{1}{3}\pi r^2 h$, where $r = 2$ and $h = 10$. Thus, $V = \frac{1}{3}\pi(2)^2(10) = \frac{40\pi}{3}$ cubic centimeters (cm^3) or approximately 40 cubic centimeters.

b. The volume of the mound of ice cream is the volume of the hemisphere:

$$S = \frac{2}{3}\pi r^3 = \frac{2}{3}\pi(2)^3 = \frac{16\pi}{3}\text{ cm}^3 \approx 16\text{ cm}^3$$

c. The total amount of ice cream is $\frac{40\pi}{3} + \frac{16\pi}{3}$ cubic centimeters $= \frac{56\pi}{3}$ cubic centimeters or about 56 cubic centimeters.

PROBLEM 8

Answer the same questions as in Example 8 if the cone is 4 inches tall and has a 1-inch radius. Can you see that the hemisphere must have a 1-inch radius?

Calculator Corner

Checking Exponents

Can we check the rules of exponents using a calculator? We can do it if we have only one variable and we agree that **two expressions are equivalent if their graphs are identical.** Thus, to check Example 1(a) we have to check that $(5x^4)(3x^7) = 15x^{11}$. We will consider $(5x^4)(3x^7)$ and $15x^{11}$ separately. As a matter of fact, let $Y_1 = (5x^4)(3x^7)$ and $Y_2 = 15x^{11}$; enter Y_1 in your calculator by pressing Y= and entering $(5x^4)(3x^7)$. (Remember that exponents are entered by pressing ^.) Now, press GRAPH and the graph appears, as shown at the left. Now enter $Y_2 = 15x^{11}$. You get the same graph, indicating that the graphs actually coincide and thus are equivalent. Now, for a numerical quick surprise, press 2nd GRAPH. The screen shows the information on the right. What does it mean? It means that for the values of x in the table (0, 1, 2, 3, etc.), Y_1 and Y_2 have the same values.

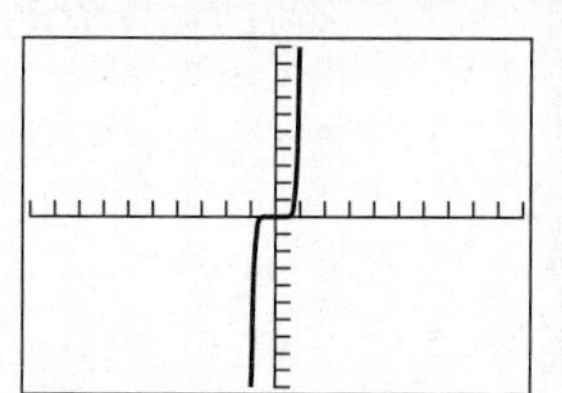

X	Y1	Y2
0	0	0
1	15	15
2	30720	30720
3	2.66E6	2.66E6
4	6.29E7	6.29E7
5	7.32E8	7.32E8
6	5.44E9	5.44E9

X=0

Answers to PROBLEMS

8. a. $\frac{4\pi}{3}$ in.$^3 \approx 4$ in.^{3} **b.** $\frac{2\pi}{3}$ in.$^3 \approx 2$ in.^{3} **c.** 2π in.$^3 \approx 6$ in.^{3}

> Practice Problems > Self-Tests > Media-rich eBooks > e-Professors > Videos

Exercises 4.1

Web IT go to mhhe.com/bello for more lessons

⟨A⟩ Multiplying Expressions In Problems 1–16, find the product.

1. $(4x)(6x^2)$ **2.** $(2a^2)(3a^3)$ **3.** $(5ab^2)(6a^3b)$

4. $(-2xy)(x^2y)$ **5.** $(-xy^2)(-3x^2y)$ **6.** $(x^2y)(-5xy)$

7. $b^3\left(\frac{-b^2c}{5}\right)$ **8.** $-a^3b\left(\frac{ab^4c}{3}\right)$ **9.** $\left(\frac{-5xy^2z}{2}\right)\left(\frac{-3x^2yz^5}{5}\right)$

10. $(-2x^2yz^3)(4xyz)$ **11.** $(-2xyz)(3x^2yz^3)(5x^2yz^4)$ **12.** $(-a^2b)(-0.4b^2c^3)(1.5abc)$

13. $(a^2c^3)(-3b^2c)(-5a^2b)$ **14.** $(ab^2c)(-ac^2)(-0.3bc^3)(-2.5a^3c^2)$ **15.** $(-2abc)(-3a^2b^2c^2)(-4c)(-b^2c)$

16. $(xy)(yz)(xz)(yz)$

⟨B⟩ Dividing Expressions In Problems 17–34, find the quotient.

17. $\frac{x^7}{x^3}$ **18.** $\frac{8a^3}{4a^2}$ **19.** $\frac{-8a^4}{16a^2}$ **20.** $\frac{9y^5}{6y^2}$

21. $\frac{12x^5y^3}{6x^2y}$ **22.** $\frac{18x^6y^2}{9xy}$ **23.** $\frac{-6x^6y^3}{12x^3y}$ **24.** $\frac{8x^8y^4}{4x^5y^2}$

25. $\frac{-14a^8y^6}{-21a^5y^2}$ **26.** $\frac{-2a^5y^8}{-6a^2y^3}$ **27.** $\frac{-27a^2b^8c^3}{-36ab^5c^2}$ **28.** $\frac{-5x^6y^8z^5}{10x^2y^5z^2}$

29. $\frac{3a^3 \cdot a^5}{2a^4}$ **30.** $\frac{y^2 \cdot y^8}{y \cdot y^3}$ **31.** $\frac{(2x^2y^3)(-3x^5y)}{6xy^3}$ **32.** $\frac{(-3x^3y^2z)(4xy^3z)}{6xy^2z}$

33. $\frac{(-x^2y)(x^3y^2)}{x^3y}$ **34.** $\frac{(-8x^2y)(-7x^5y^3)}{-2x^2y^3}$

⟨C⟩ Simplifying Expressions Using the Power Rules In Problems 35–70, simplify.

35. $(2^2)^3$ **36.** $(3^1)^2$ **37.** $(3^2)^1$ **38.** $(2^3)^2$

39. $(x^3)^3$ **40.** $(y^2)^4$ **41.** $(y^3)^2$ **42.** $(x^4)^3$

43. $(-a^2)^3$ **44.** $(-b^3)^5$ **45.** $(2x^3y^2)^3$ **46.** $(3x^2y^3)^2$

47. $(2x^2y^3)^2$ **48.** $(3x^4y^4)^3$ **49.** $(-3x^3y^2)^3$ **50.** $(-2x^5y^4)^4$

51. $(-3x^6y^3)^2$ **52.** $(y^4z^3)^5$ **53.** $(-2x^4y^4)^3$ **54.** $(-3y^5z^3)^4$

55. $\left(\frac{2}{3}\right)^4$ **56.** $\left(\frac{-3}{4}\right)^3$ **57.** $\left(\frac{3x^2}{2y^3}\right)^3$ **58.** $\left(\frac{3a^2}{2b^3}\right)^4$

59. $\left(\frac{-2x^2}{3y^3}\right)^4$ **60.** $\left(\frac{-3x^2}{2y^3}\right)^4$ **61.** $(2x^3)^2(3y^3)$ **62.** $(3a)^3(2b)^2$

63. $(-3a)^2(-4b)^3$ **64.** $(-2x^2)^3(-3y^3)^2$ **65.** $-(4a^2)^2(-3b^3)^2$ **66.** $-(3x^3)^3(-2y^3)^3$

67. $\left(\frac{2}{3}\right)^5 \cdot 3^6$ **68.** $\left(\frac{4}{5}\right)^3 \cdot 5^4$ **69.** $\left(\frac{x}{y}\right)^5 \cdot y^7$ **70.** $\left(\frac{2a^2}{3b}\right)^5 \cdot 3^4b^7$

Applications

Applications: Green Math

Garbage champions In Problems 71–74, find out which nations produce the most garbage annually and how much by **multiplying** the population by the kilograms (kg) per person.

	Country	Population		kg/Person
71.	Norway	(4.6×10^6)	$\times$	(8×10^2)
72.	Ireland	(5.6×10^6)	$\times$	(8×10^2)
73.	Denmark	(5.6×10^6)	$\times$	(6.6×10^2)
74.	Switzerland	(7.6×10^6)	$\times$	(6.5×10^2)

Source: http://tinyurl.com/cfz2e8.

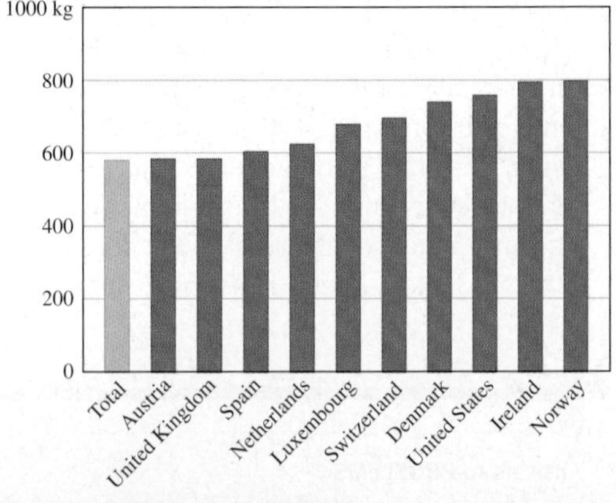

Source: OECD Factbook 2009.

Cuban Sandwiches Problems 75–80 refer to the photo (to the right, top).

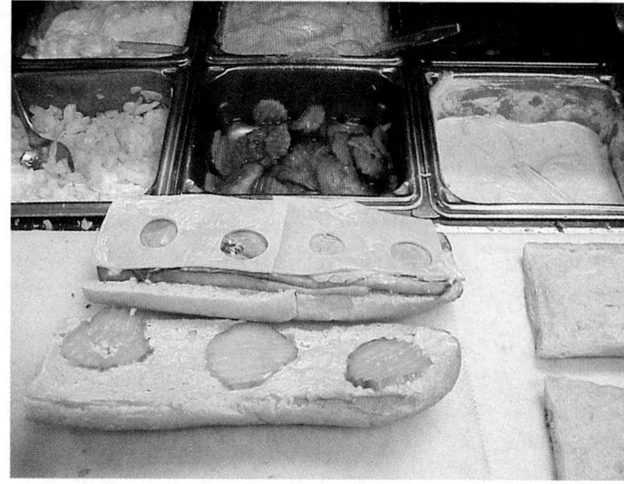

75. A standard pickle container measures x by x by $\frac{2}{3}x$ inches. What is the volume V of the container?

76. Using the formula obtained in Problem 75, what is the volume of the container if $x = 6$ inches?

77. A Cuban sandwich takes about one cubic inch of pickles. How many sandwiches can be made with the contents of the container?

78. A container of pickles holds about 24 ounces. If pickles cost \$0.045 per ounce, what is the cost of the pickles in the container?

79. What is the cost of the pickles in a Cuban sandwich (see Problems 77 and 78)?

80. How many ounces of pickles does a Cuban sandwich take (see Problems 77 and 78)?

Volume Problems 81–85 refer to the photo (to the right, middle).

81. The dimensions of the green bean container are x by $2x$ by $\frac{1}{2}x$. What is the volume V of the container?

82. Using the formula given in Problem 81, what is the volume of the container if $x = 6$?

83. Green beans are served in a small bowl that has the approximate shape of a hemisphere with a radius of 2 inches. If the volume of a hemisphere is $S = \frac{2}{3}\pi r^3$, what is the volume of one serving of green beans? Use 3 for π.

84. Based on your answers to Problems 82 and 83, how many servings of green beans does the whole container hold?

85. If each serving of green beans costs \$0.24 and sells for \$1.99, how much does the restaurant make per container?

Volume Problems 86–90 refer to the photo (to the right, bottom).

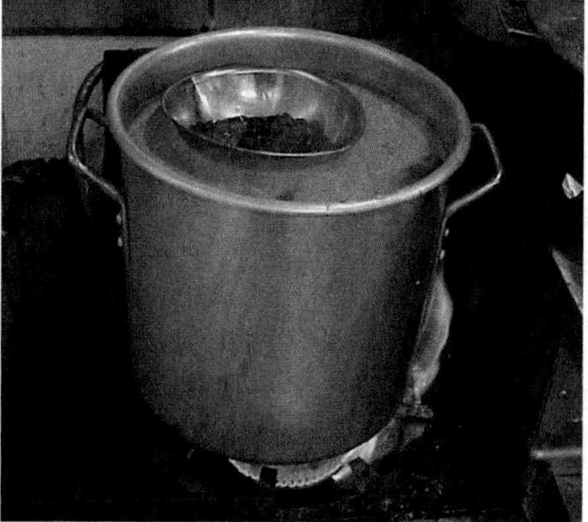

86. The volume V of a cylinder is $V = \pi r^2 h$, where r is the radius and h is the height. If the pot is x inches tall and its radius is $\frac{1}{2}x$, what is the volume of the pot? Use 3 for π.

87. When the pot is used for cooking beans, it is only $\frac{3}{4}$ full. What is the volume of the beans in the pot? (Refer to the formula for the volume of a cylinder given in Problem 86.)

88. If the pot is $\frac{3}{4}$ full and the height of the pot is 10 inches, what is the volume of the beans? (Refer to the formula for the volume of a cylinder given in Problem 86.)

89. A serving of beans is about 1 cup, or 15 in.^{3}, of beans. How many servings of beans are in a pot of beans?

90. How many pots of beans are needed to serve 150 people?

Volume of Solids We can write the formula for the volume of many solids using exponents. In Problems 91–94 the volume of a solid is given in words. Write the formula in symbols.

91. The volume V of a cylinder of radius r and height h is the product of π, the square of r, and h.

Cylinder

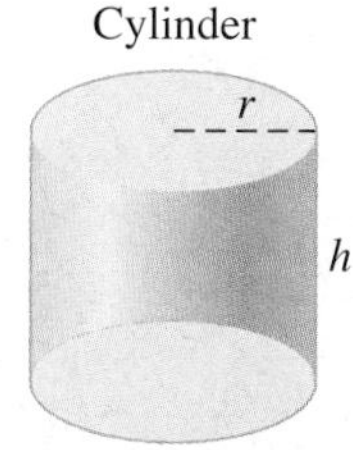

92. The volume V of a sphere of radius r is the product of $\frac{4}{3}\pi$ and the cube of r.

Sphere

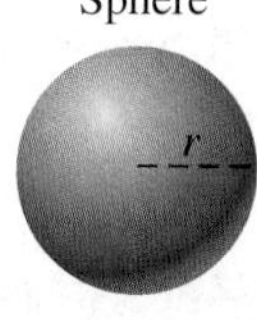

93. The volume V of a cone of height h and radius r is the product of $\frac{1}{3}\pi$, the square of r, and h.

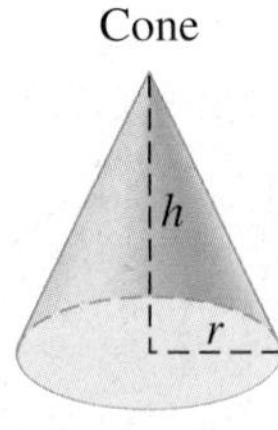

94. The volume V of a cube is the length L of one of the sides cubed.

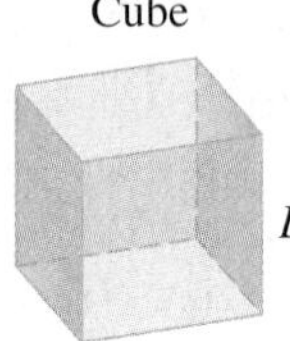

Using Your Knowledge

Follow That Pattern Many interesting patterns involve exponents. Use your knowledge to find the answers to the following problems.

95.
$$1^2 = 1$$
$$(11)^2 = 121$$
$$(111)^2 = 12{,}321$$
$$(1111)^2 = 1{,}234{,}321$$

a. Find $(11{,}111)^2$.

b. Find $(111{,}111)^2$.

96.
$$1^2 = 1$$
$$2^2 = 1 + 2 + 1$$
$$3^2 = 1 + 2 + 3 + 2 + 1$$
$$4^2 = 1 + 2 + 3 + 4 + 3 + 2 + 1$$

a. Use this pattern to write 5^2.

b. Use this pattern to write 6^2.

97.
$$1 + 3 = 2^2$$
$$1 + 3 + 5 = 3^2$$
$$1 + 3 + 5 + 7 = 4^2$$

a. Find $1 + 3 + 5 + 7 + 9$.

b. Find $1 + 3 + 5 + 7 + 9 + 11 + 13$.

98. Can you discover your own pattern? What is the largest number you can construct by using the number 9 three times? (It is not 999!)

Write On

99. Explain why the product rule for exponents does not apply to the expression $x^2 \cdot y^3$.

100. Explain why the product rule for exponents does not apply to the expression $x^2 + y^3$.

101. What is the difference between the product rule and the power rule?

102. Explain why you cannot use the quotient rule stated in the text to conclude that

$$\frac{x^m}{x^m} = 1 = x^0, \quad x \neq 0$$

Concept Checker

Fill in the blank(s) with the correct word(s), phrase, or mathematical statement.

103. If m and n are **positive integers,** $x^m \cdot x^n =$ ______________.

104. When **multiplying** numbers with the **same (like) signs,** the **product** is ______________.

105. When **multiplying** numbers with **different (unlike) signs,** the **product** is ______________.

106. When **dividing** numbers with the **same (like) signs,** the quotient is ______________.

107. When **dividing** numbers with **different (unlike) signs,** the quotient is ______________.

108. If m and n are **positive integers** with $m > n$, $\left(\frac{x^m}{x^n}\right) =$ ______________.

109. If m and n are **positive integers** $(x^m)^n =$ ______________.

110. If m, n, and k are **positive integers** $(x^m y^n)^k =$ ______________.

111. If m is a **positive integer,** $\left(\frac{x}{y}\right)^m =$ ______________.

112. In the quotient $\left(\frac{x}{y}\right)^m$, the *denominator* y cannot be ______________.

an integer	x^{m+n}
0	**negative**
positive	y^m
x^{m-n}	$x^{mk}y^{nk}$
x^{mn}	$\frac{x^m}{y^m}$
$\frac{x^m}{y^n}$	$y^{mk}y^{nk}$
x^m	

Mastery Test

Simplify:

113. $(5a^3)(6a^8)$

114. $(3x^2yz)(2xy^2)(8xz^3)$

115. $(-3x^3yz)(4xz^4)$

116. $(-3ab^2c^4)(-5a^4bc^3)$

117. $\dfrac{15x^4y^7}{-3x^2y}$

118. $\dfrac{-5x^2y^6z^4}{15xy^4z^2}$

119. $\frac{-3a^2b^9c^2}{-12ab^2c}$

120. $(3^2)^2$

121. $(y^5)^4$

122. $(3x^2y^3)^2$

123. $\left(\frac{3y^2}{x^5}\right)^3$

124. $(-2x^2y^3)^2(3x^4y^5)^3$

125. $\left(\frac{1}{2^5}\right) \cdot 2^7$

126. $\left(\frac{x^4}{y^7}\right) \cdot y^9$

Skill Checker

Find:

127. 4^3

128. 5^2

129. $\frac{1}{2^3}$

130. $\frac{1}{3^2}$

131. $\frac{3^4}{2^4}$

4.2 Integer Exponents

Objectives

A Write an expression with negative exponents as an equivalent one with positive exponents.

B Write a fraction involving exponents as a number with a negative power.

C Multiply and divide expressions involving negative exponents.

To Succeed, Review How To . . .

1. Raise numbers to a power (pp. 319–321).
2. Use the rules of exponents (pp. 322–325).

Getting Started

Negative Exponents in Science

In science and technology, negative numbers are used as exponents. For example, the diameter of the DNA molecule pictured on the next page is 10^{-8} meter, and the time it takes for an electron to go from source to screen in a TV tube is 10^{-6} second. So what do the numbers 10^{-8} and 10^{-6} mean? Look at the pattern obtained by *dividing by 10* in each step:

Exponents decrease by 1.

$$10^3 = 1000$$
$$10^2 = 100$$
$$10^1 = 10$$
$$10^0 = 1$$

Divide by 10 at each step.

Hence the following also holds true:

Exponents decrease by 1.

$$10^{-1} = \frac{1}{10} = \frac{1}{10}$$
$$10^{-2} = \frac{1}{100} = \frac{1}{10^2}$$
$$10^{-3} = \frac{1}{1000} = \frac{1}{10^3}$$

Divide by 10 at each step.

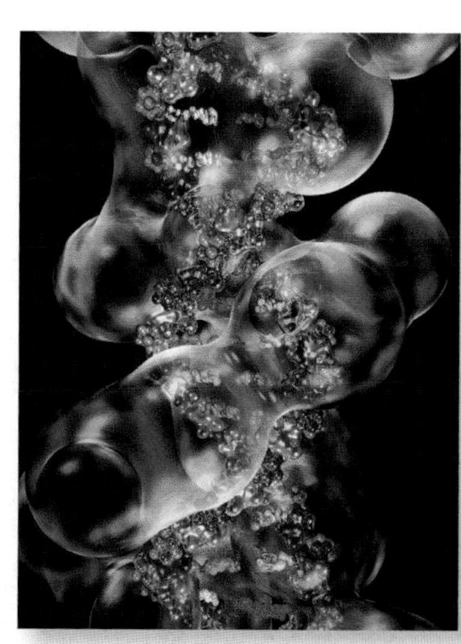

Thus, the diameter of a DNA molecule is

$$10^{-8} = \frac{1}{10^8} = 0.00000001 \text{ meter}$$

and the time elapsed between source and screen in a TV tube is

$$10^{-6} = \frac{1}{10^6} = 0.000001 \text{ second}$$

These are very small numbers and very cumbersome to write, so, for convenience, we use negative exponents to represent them. In this section we shall study expressions involving negative and zero exponents.

A > Negative Exponents

In the pattern used in the *Getting Started,* $10^0 = 1$. In general, we make the following definition.

ZERO EXPONENT

For any nonzero number x,

$$x^0 = 1$$

This means that a nonzero number x raised to the zero power is 1.

Thus, $5^0 = 1$, $8^0 = 1$, $6^0 = 1$, $(3y)^0 = 1$, and $(-2x^2)^0 = 1$.

Let us look again at the pattern in the *Getting Started;* as you can see,

$$10^{-1} = \frac{1}{10^1},\quad 10^{-2} = \frac{1}{10^2},\quad 10^{-3} = \frac{1}{10^3},\quad \text{and} \quad 10^{-n} = \frac{1}{10^n}$$

We also have the following definition.

NEGATIVE EXPONENT

If n is a positive integer, then

$$x^{-n} = \frac{1}{x^n} \qquad (x \neq 0)$$

This definition says that x^{-n} and x^n are **reciprocals**, since

$$x^{-n} \cdot x^n = \frac{1}{x^n} \cdot x^n = 1$$

By definition, then,

$$5^{-2} = \frac{1}{5^2} = \frac{1}{5 \cdot 5} = \frac{1}{25} \quad \text{and} \quad 2^{-3} = \frac{1}{2^3} = \frac{1}{2 \cdot 2 \cdot 2} = \frac{1}{8}$$

When n is positive, we obtain this result:

$$\left(\frac{1}{x}\right)^{-n} = \frac{1}{\left(\frac{1}{x}\right)^n} = \frac{1}{\frac{1^n}{x^n}} = 1 \cdot \frac{x^n}{1^n} = x^n$$

which can be stated as follows.

***n*TH POWER OF A QUOTIENT**

If n is a positive integer, then

$$\left(\frac{1}{x}\right)^{-n} = x^n, (x \neq 0)$$

You can think of a negative exponent as a command to take the **reciprocal** of the base. Thus,

$$\left(\frac{1}{4}\right)^{-3} = 4^3 = 64 \quad \text{and} \quad \left(\frac{1}{5}\right)^{-2} = 5^2 = 25$$

EXAMPLE 1 Rewriting with positive exponents

Use positive exponents to rewrite and simplify:

a. 6^{-2} **b.** 4^{-3} **c.** $\left(\frac{1}{7}\right)^{-3}$ **d.** $\left(\frac{1}{x}\right)^{-5}$

PROBLEM 1

Rewrite and simplify:

a. 2^{-3} **b.** 3^{-3}

c. $\left(\frac{1}{2}\right)^{-3}$ **d.** $\left(\frac{1}{a}\right)^{-4}$

Answers to PROBLEMS

1. a. $\frac{1}{8}$ **b.** $\frac{1}{27}$ **c.** 8 **d.** a^4

SOLUTION 1

a. $6^{-2} = \frac{1}{6^2} = \frac{1}{6 \cdot 6} = \frac{1}{36}$

b. $4^{-3} = \frac{1}{4^3} = \frac{1}{4 \cdot 4 \cdot 4} = \frac{1}{64}$

c. $\left(\frac{1}{7}\right)^{-3} = 7^3 = 343$

d. $\left(\frac{1}{x}\right)^{-5} = x^5$

If we want to rewrite

$$\frac{x^{-2}}{y^{-3}}$$

without negative exponents, we use the definition of negative exponents to rewrite x^{-2} and y^{-3}:

$$\frac{x^{-2}}{y^{-3}} = \frac{\frac{1}{x^2}}{\frac{1}{y^3}} = \frac{1}{x^2} \cdot \frac{y^3}{1} = \frac{y^3}{x^2}$$

so that the expression is in the simpler form,

$$\frac{x^{-2}}{y^{-3}} = \frac{y^3}{x^2}$$

Here is the rule for changing fractions with negative exponents to equivalent ones with positive exponents.

SIMPLIFYING FRACTIONS WITH NEGATIVE EXPONENTS

For any nonzero numbers x and y and any positive integers m and n,

$$\frac{x^{-m}}{y^{-n}} = \frac{y^n}{x^m}$$

This means that to write an equivalent fraction without negative exponents, we interchange numerators and denominators and make the exponents positive.

EXAMPLE 2 Writing equivalent fractions with positive exponents

Write as an equivalent fraction without negative exponents and simplify:

a. $\frac{3^{-2}}{4^{-3}}$ b. $\frac{a^{-7}}{b^{-3}}$ c. $\frac{x^2}{y^{-4}}$

SOLUTION 2

a. $\frac{3^{-2}}{4^{-3}} = \frac{4^3}{3^2} = \frac{64}{9}$

b. $\frac{a^{-7}}{b^{-3}} = \frac{b^3}{a^7}$

c. $\frac{x^2}{y^{-4}} = x^2y^4$

PROBLEM 2

Write as an equivalent expression without negative exponents and simplify:

a. $\frac{2^{-4}}{3^{-3}}$ b. $\frac{x^{-9}}{y^{-4}}$ c. $\frac{a^5}{b^{-8}}$

B > Writing Fractions Using Negative Exponents

As we saw in the *Getting Started,* we can write fractions involving powers in the denominator using negative exponents.

EXAMPLE 3 Writing equivalent fractions with negative exponents

Write using negative exponents:

a. $\frac{1}{5^4}$ b. $\frac{1}{7^5}$ c. $\frac{1}{x^5}$ d. $\frac{3}{x^4}$

SOLUTION 3

We use the definition of negative exponents.

a. $\frac{1}{5^4} = 5^{-4}$

b. $\frac{1}{7^5} = 7^{-5}$

c. $\frac{1}{x^5} = x^{-5}$

d. $\frac{3}{x^4} = 3 \cdot \frac{1}{x^4} = 3x^{-4}$

PROBLEM 3

Write using negative exponents:

a. $\frac{1}{7^6}$ b. $\frac{1}{8^5}$

c. $\frac{1}{a^9}$ d. $\frac{7}{a^6}$

Answers to PROBLEMS

2. a. $\frac{3^3}{2^4} = \frac{27}{16}$ **b.** $\frac{y^4}{x^9}$ **c.** a^5b^8 **3. a.** 7^{-6} **b.** 8^{-5} **c.** a^{-9} **d.** $7a^{-6}$

C > Multiplying and Dividing Expressions with Negative Exponents

In Section 4.1, we multiplied expressions that contained positive exponents. For example,

$$x^5 \cdot x^8 = x^{5+8} = x^{13} \quad \text{and} \quad y^2 \cdot y^3 = y^{2+3} = y^5$$

Can we multiply expressions involving negative exponents using the same idea? Let's see.

$$x^5 \cdot x^{-2} = x^5 \cdot \frac{1}{x^2} = \frac{x^5}{x^2} = x^3$$

Adding the exponents 5 and −2,

$$x^5 \cdot x^{-2} = x^{5+(-2)} = x^3 \qquad \text{Same answer!}$$

Similarly,

$$x^{-3} \cdot x^{-2} = \frac{1}{x^3} \cdot \frac{1}{x^2} = \frac{1}{x^{3+2}} = \frac{1}{x^5} = x^{-5}$$

Adding the exponents −3 and −2,

$$x^{-3} \cdot x^{-2} = x^{-3+(-2)} = x^{-5} \qquad \text{Same answer again!}$$

So, we have the following rule.

PRODUCT RULE FOR EXPONENTS

If m and n are integers,

$$x^m \cdot x^n = x^{m+n}$$

This rule says that when we *multiply* expressions with the same base x, we keep the base x and *add* the exponents.

Note that the rule does *not* apply to $x^7 \cdot y^6$ because the bases x and y are different.

EXAMPLE 4 **Using the product rule**

Multiply and simplify (that is, write the answer without negative exponents):

a. $2^6 \cdot 2^{-4}$ **b.** $4^3 \cdot 4^{-5}$ **c.** $y^{-2} \cdot y^{-3}$ **d.** $a^{-5} \cdot a^5$

SOLUTION 4

a. $2^6 \cdot 2^{-4} = 2^{6+(-4)} = 2^2 = 4$

b. $4^3 \cdot 4^{-5} = 4^{3+(-5)} = 4^{-2} = \frac{1}{4^2} = \frac{1}{16}$

c. $y^{-2} \cdot y^{-3} = y^{-2+(-3)} = y^{-5} = \frac{1}{y^5}$

d. $a^{-5} \cdot a^5 = a^{-5+5} = a^0 = 1$

PROBLEM 4

Multiply and simplify:

a. $3^{-4} \cdot 3^6$ **b.** $2^4 \cdot 2^{-6}$

c. $b^{-3} \cdot b^{-5}$ **d.** $x^{-7} \cdot x^7$

NOTE

We wrote the answers in Example 4 *without* using negative exponents. In algebra, it's customary to write answers *without* negative exponents.

In Section 4.1, we divided expressions with the same base. Thus,

$$\frac{7^5}{7^2} = 7^{5-2} = 7^3 \quad \text{and} \quad \frac{8^3}{8} = 8^{3-1} = 8^2$$

The rule used there can be extended to any exponents that are integers.

Answers to PROBLEMS

4. a. 9 **b.** $\frac{1}{4}$ **c.** $\frac{1}{b^8}$ **d.** 1

QUOTIENT RULE FOR EXPONENTS

If m and n are integers,

$$\frac{x^m}{x^n} = x^{m-n}, \quad x \neq 0$$

This rule says that when we *divide* expressions with the *same* base x, we keep the base x and *subtract* the exponents.

Note that

$$\frac{x^m}{x^n} = x^m \cdot \frac{1}{x^n} = x^m \cdot x^{-n} = x^{m-n}$$

EXAMPLE 5 Using the quotient rule

Simplify:

a. $\frac{6^5}{6^{-2}}$ b. $\frac{x}{x^5}$ c. $\frac{y^{-2}}{y^{-2}}$ d. $\frac{z^{-3}}{z^{-4}}$

SOLUTION 5

a. $\frac{6^5}{6^{-2}} = 6^{5-(-2)} = 6^{5+2} = 6^7$

b. $\frac{x}{x^5} = x^{1-5} = x^{-4} = \frac{1}{x^4}$

c. $\frac{y^{-2}}{y^{-2}} = y^{-2-(-2)} = y^{-2+2} = y^0 = 1$ Recall the definition for the zero exponent.

d. $\frac{z^{-3}}{z^{-4}} = z^{-3-(-4)} = z^{-3+4} = z^1 = z$

PROBLEM 5

Simplify:

a. $\frac{3^5}{3^{-3}}$ b. $\frac{a}{a^{-6}}$

c. $\frac{b^{-5}}{b^{-5}}$ d. $\frac{a^{-4}}{a^6}$

Here is a summary of the definitions and rules of exponents we've just discussed. Please, make sure you read and understand these rules and procedures before you go on.

RULES FOR EXPONENTS

If m, n, and k are integers, the following rules apply:

Rule		Example
Product rule for exponents:	$x^m x^n = x^{m+n}$	$x^{-2} \cdot x^6 = x^{-2+6} = x^4$
Zero exponent:	$x^0 = 1 \quad (x \neq 0)$	$9^0 = 1, y^0 = 1$, and $(3a)^0 = 1$
Negative exponent:	$x^{-n} = \frac{1}{x^n} \quad (x \neq 0)$	$3^{-4} = \frac{1}{3^4} = \frac{1}{81}, y^{-7} = \frac{1}{y^7}$
Quotient rule for exponents:	$\frac{x^m}{x^n} = x^{m-n} \quad (x \neq 0)$	$\frac{p^8}{p^3} = p^{8-3} = p^5$
Power rule for exponents:	$(x^m)^n = x^{mn}$	$(a^3)^9 = a^{3 \cdot 9} = a^{27}$
Power rule for products:	$(x^m y^n)^k = x^{mk} y^{nk}$	$(x^4 y^3)^4 = x^{4 \cdot 4} y^{3 \cdot 4} = x^{16} y^{12}$
Power rule for quotients:	$\left(\frac{x}{y}\right)^m = \frac{x^m}{y^m} \quad (y \neq 0)$	$\left(\frac{a^3}{b^4}\right)^8 = \frac{a^{3 \cdot 8}}{b^{4 \cdot 8}} = \frac{a^{24}}{b^{32}}$
Negative to positive exponent:	$\left(\frac{1}{x}\right)^{-n} = x^n \quad (x \neq 0)$	$\left(\frac{1}{a^2}\right)^{-n} = (a^2)^n = a^{2n}$
Negative to positive exponent:	$\frac{x^{-m}}{y^{-n}} = \frac{y^n}{x^m} \quad (x \neq 0)$	$\frac{x^{-7}}{y^{-9}} = \frac{y^9}{x^7}$

Answers to PROBLEMS

5. a. 3^8 **b.** a^7 **c.** 1 **d.** $\frac{1}{a^{10}}$

Now let's do an example that requires us to use several of these rules.

EXAMPLE 6 **Using the rules to simplify a quotient**

Simplify:

$$\left(\frac{2x^2y^3}{3x^3y^{-2}}\right)^{-4}$$

SOLUTION 6

$$\left(\frac{2x^2y^3}{3x^3y^{-2}}\right)^{-4} = \left(\frac{2x^{2-3}y^{3-(-2)}}{3}\right)^{-4} \quad \text{Use the quotient rule.}$$

$$= \left(\frac{2x^{-1}y^5}{3}\right)^{-4} \quad \text{Simplify.}$$

$$= \frac{2^{-4}x^4y^{-20}}{3^{-4}} \quad \text{Use the power rule.}$$

$$= \frac{3^4x^4y^{-20}}{2^4} \quad \text{Negative to positive}$$

$$= \frac{3^4x^4}{2^4y^{20}} \quad \text{Definition of negative exponent}$$

$$= \frac{81x^4}{16y^{20}} \quad \text{Simplify.}$$

PROBLEM 6

Simplify:

$$\left(\frac{3a^5b^4}{4a^6b^{-3}}\right)^{-3}$$

Let's use exponents in a practical way. What are the costs involved when you are driving a car? Gas, repairs, the car loan payment. What else? Depreciation! The depreciation on a car is the difference between what you paid for the car and what the car is worth now. When you pay P dollars for your car it is then worth 100% of P, but here is what happens to the value as years go by and the car depreciates 10% each year:

Year	Value
0	100% of P
1	90% of $P = (1 - 0.10)P$
2	90% of $(1 - 0.10)P = (1 - 0.10)(1 - 0.10)P$ $= (1 - 0.10)^2P$
3	90% of $(1 - 0.10)^2P = (1 - 0.10)(1 - 0.10)^2P$ $= (1 - 0.10)^3P$

After n years, the value will be

$$(1 - 0.10)^nP$$

But what about x years *ago?* x years ago means $-x$, so the value C of the car x years ago would have been

$$C = (1 - 0.10)^{-x}P$$

There are more examples using exponents in the Using Your Knowledge Exercises.

In a recent year, the popularity of *hybrid* cars has soared (a hybrid car has two motors, an electric one and a gas powered one). They are also the most gasoline-efficient cars, yielding 48 to 60 miles per gallon. On the other hand, they **do** cost more. The highest annual expense for any car is not repairs, insurance, or gas but *depreciation (anywhere from 7% to as high as 45% a year),* but some of the hybrids do not depreciate much. As a matter of fact, in recent years some used hybrids can cost more than new ones!

Answers to PROBLEMS

6. $\frac{64a^3}{27b^{21}}$

EXAMPLE 7 Hybrid car depreciation

Lashonda bought a 3-year-old hybrid for \$17,000. If the car depreciates 13% each year,

a. What will be the value of the car in 2 years?

b. What was the price of the car when it was new?

SOLUTION 7

a. The value in 2 years will be

$$17{,}000(1 - 0.13)^2 = 17{,}000(0.87)^2 = \$12{,}867.30$$

b. The car was new 3 years ago, and it was then worth

$$17{,}000(1 - 0.13)^{-3} = 17{,}000(0.87)^{-3} = \$25{,}816.13$$

PROBLEM 7

A 3-year-old hybrid was bought for \$15,000. If the car depreciates 10% each year,

a. What will be the value of the car in 2 years?

b. What was the price of the car when it was new?

Note: You can find the price of old and new hybrid cars at Kelley Blue Book (www.kbb.com) or at www.edmunds.com.

> Practice Problems > Self-Tests > Media-rich eBooks > e-Professors > Videos

Exercises 4.2

A **Negative Exponents** In Problems 1–16, write using positive exponents and then simplify.

1. 4^{-2} **2.** 2^{-3} **3.** 5^{-3} **4.** 7^{-2}

5. $\left(\frac{1}{8}\right)^{-2}$ **6.** $\left(\frac{1}{6}\right)^{-3}$ **7.** $\left(\frac{1}{x}\right)^{-7}$ **8.** $\left(\frac{1}{y}\right)^{-6}$

9. $\frac{4^{-2}}{3^{-3}}$ **10.** $\frac{2^{-4}}{3^{-2}}$ **11.** $\frac{5^{-2}}{3^{-4}}$ **12.** $\frac{6^{-2}}{5^{-3}}$

13. $\frac{a^{-5}}{b^{-6}}$ **14.** $\frac{p^{-6}}{q^{-5}}$ **15.** $\frac{x^{-9}}{y^{-9}}$ **16.** $\frac{t^{-3}}{s^{-3}}$

B **Writing Fractions Using Negative Exponents** In Problems 17–22, write using negative exponents.

17. $\frac{1}{2^3}$ **18.** $\frac{1}{3^4}$ **19.** $\frac{1}{y^5}$ **20.** $\frac{1}{b^6}$

21. $\frac{1}{q^5}$ **22.** $\frac{1}{t^4}$

C **Multiplying and Dividing Expressions with Negative Exponents** In Problems 23–46, multiply and simplify. (Remember to write your answers without negative exponents.)

23. $3^5 \cdot 3^{-4}$ **24.** $4^{-6} \cdot 4^8$ **25.** $2^{-5} \cdot 2^7$ **26.** $3^8 \cdot 3^{-5}$

27. $4^{-6} \cdot 4^4$ **28.** $5^{-4} \cdot 5^2$ **29.** $6^{-1} \cdot 6^{-2}$ **30.** $3^{-2} \cdot 3^{-1}$

31. $2^{-4} \cdot 2^{-2}$ **32.** $4^{-1} \cdot 4^{-2}$ **33.** $x^6 \cdot x^{-4}$ **34.** $y^7 \cdot y^{-2}$

35. $y^{-3} \cdot y^5$ **36.** $x^{-7} \cdot x^8$ **37.** $a^3 \cdot a^{-8}$ **38.** $b^4 \cdot b^{-7}$

Answers to PROBLEMS

7. a. \$12,150 **b.** \$20,576.13

Web IT go to mhhe.com/bello for more lessons

39. $x^{-5} \cdot x^3$ **40.** $y^{-6} \cdot y^2$ **41.** $x \cdot x^{-3}$ **42.** $y \cdot y^{-5}$

43. $a^{-2} \cdot a^{-3}$ **44.** $b^{-5} \cdot b^{-2}$ **45.** $b^{-3} \cdot b^3$ **46.** $a^6 \cdot a^{-6}$

In Problems 47–60, divide and simplify.

47. $\frac{3^4}{3^{-1}}$ **48.** $\frac{2^2}{2^{-2}}$ **49.** $\frac{4^{-1}}{4^2}$ **50.** $\frac{3^{-2}}{3^3}$

51. $\frac{y}{y^3}$ **52.** $\frac{x}{x^4}$ **53.** $\frac{x}{x^{-2}}$ **54.** $\frac{y}{y^{-3}}$

55. $\frac{x^{-3}}{x^{-1}}$ **56.** $\frac{x^{-4}}{x^{-2}}$ **57.** $\frac{x^{-3}}{x^4}$ **58.** $\frac{y^{-4}}{y^5}$

59. $\frac{x^{-2}}{x^{-5}}$ **60.** $\frac{y^{-3}}{y^{-6}}$

In Problems 61–70, simplify.

61. $\left(\frac{a}{b^3}\right)^2$ **62.** $\left(\frac{a^2}{b}\right)^3$ **63.** $\left(\frac{-3a}{2b^2}\right)^{-3}$ **64.** $\left(\frac{-2a^2}{3b^0}\right)^{-2}$ **65.** $\left(\frac{a^{-4}}{b^2}\right)^{-2}$

66. $\left(\frac{a^{-2}}{b^3}\right)^{-3}$ **67.** $\left(\frac{x^5}{y^{-2}}\right)^{-3}$ **68.** $\left(\frac{x^6}{y^{-3}}\right)^{-2}$ **69.** $\left(\frac{x^{-4}y^3}{x^5y^5}\right)^{-3}$ **70.** $\left(\frac{x^{-2}y^0}{x^7y^2}\right)^{-2}$

Applications

71. *Watch out for fleas* Do you have a dog or a cat? Sometimes they can get fleas! A flea is 2^{-4} inches long. Write 2^{-4} using positive exponents and as a simplified fraction.

72. *Ants in your pantry!* The smallest ant is about $\frac{1}{25}$ of an inch. Write $\frac{1}{25}$ using positive exponents and using negative exponents.

73. *Micron* A **micron** is a unit of length equivalent to 0.001 millimeters.

a. Write 0.001 as a fraction.

b. Write the fraction obtained in part **a** with a denominator that is a power of 10.

c. Write the fraction obtained in part **b** using negative exponents.

d. Write 0.001 using negative exponents.

74. *Nanometer* A **nanometer** is a *billionth* of a meter.

a. Write *one billionth* as a fraction

b. Write the fraction of part **a** with a denominator that is a power of 10.

c. Write the fraction of part **b** using negative exponents.

d. Write *one billionth* using negative exponents.

Using Your Knowledge

Exponential Growth The idea of exponents can be used to measure population growth. Thus, if we assume that the world population is increasing about 2% each year (experts say the rate is between 1% and 3%), we can predict the population of the world next year by *multiplying* the present world population by 1.02 (100% + 2% = 102% = 1.02). If we let the world population be P, we have

$$\text{Population in 1 year} = 1.02P$$
$$\text{Population in 2 years} = 1.02(1.02P) = (1.02)^2P$$
$$\text{Population in 3 years} = 1.02(1.02)^2P = (1.02)^3P$$

75. If the population P in 2000 was 6 billion people, what would it be in 2 years—that is, in 2002? (Round your answer to three decimal places.)

76. What was the population 5 years after 2000? (Round your answer to three decimal places.)

To find the population 1 year from now, we multiply by 1.02. What should we do to find the population 1 year *ago?* We divide by 1.02. Thus, if the population today is P,

$$\text{Population 1 year ago} = \frac{P}{1.02} = P \cdot 1.02^{-1}$$

$$\text{Population 2 years ago} = \frac{P \cdot 1.02^{-1}}{1.02} = P \cdot 1.02^{-2}$$

$$\text{Population 3 years ago} = \frac{P \cdot 1.02^{-2}}{1.02} = P \cdot 1.02^{-3}$$

77. If the population P in 2000 was 6 billion people, what was it in 1990? (Round your answer to three decimal places.)

78. What was the population in 1985? (Round your answer to three decimal places.)

Do you know what inflation is? It is the tendency for prices and wages to rise, making your money worth less in the future than it is today. The formula for the cost C of an item n years from now if the present value (value now) is P dollars and the inflation rate is r% is

$$C = P(1 + r)^n$$

79. In 2008, the cheapest Super Bowl tickets were \$900. If we assume a 3% (0.03) inflation rate, how much would tickets cost:

a. in 2010

b. in 2015

80. In 1967 a Super Bowl ticket was \$12. Assuming a 3% inflation rate:

a. How much should the tickets cost 45 years later? Is the cost more or less than the actual \$1000 price?

b. If a ticket is \$1000 now, how much should it have been 45 years ago?

81. In 1967 the Consumer Price Index (CPI) was 35. Thirty-five years later, it was 180, so ticket prices should be scaled up by a factor of $\frac{180}{35} \approx 5.14$. Using this factor, how much should a \$12 ticket price be 35 years later?

82. The average annual cost for tuition and fees at a 4-year public college is about \$20,000 and rising at a 3.5% rate. How much would you have to pay in tuition and fees 4 years from now?

Applications: Green Math

83. *Car depreciation* Alegria bought a 3-year-old Honda Civic for \$15,000. If the car depreciates 12% each year,

a. What will be the value of the car in 2 years?

b. What was the price of the car when it was new?

84. *Car depreciation* Latrell bought a 3-year-old Ford Fusion for \$10,000. If the car depreciates 15% each year,

a. What will be the value of the car in 2 years?

b. What was the price of the car when it was new?

85. *Car depreciation* Khan bought a 3-year-old Toyota Camry for \$18,000. If the car depreciates 14% each year,

a. What will be the value of the car in 2 years?

b. What was the price of the car when it was new?

86. The number of bacteria in a sample after n hours growing at a 50% rate is modeled by the equation

$$100(1 + 0.50)^n$$

a. What would the number of bacteria be after 6 hours?

b. What was the number of bacteria 2 hours ago?

87. On a Monday morning, you count 40 ants in your room. On Tuesday, you find 60. By Wednesday, they number 90. If the population continues to grow at this rate:

a. What is the percent rate of growth?

b. Write an equation of the form $P(1 + r)^n$ that models the number A of ants after n days.

c. How many ants would you expect on Friday?

d. How many ants would you expect in 2 weeks (14 days)?

e. Name at least one factor that would cause the ant population to slow down its growth.

88. Referring to Problem 87, how many ants (to the nearest whole number) were there on the previous Sunday? On the previous Saturday?

Write On

89. Give three different explanations for why $x^0 = 1$ $(x \neq 0)$.

90. By definition, if n is a positive integer, then

$$x^{-n} = \frac{1}{x^n} \quad (x \neq 0)$$

a. Does this rule hold if n is *any* integer? Explain and give some examples.

b. Why do we have to state $x \neq 0$ in this definition?

91. Does $x^{-2} + y^{-2} = (x + y)^{-2}$? Explain why or why not.

92. Does

$$x^{-1} + y^{-1} = \frac{1}{x+y}?$$

Explain why or why not.

Concept Checker

Fill in the blank(s) with the correct word(s), phrase, or mathematical statement.

93. For any nonzero number x, x^0 = ______________.

94. If n is a **positive integer,** x^{-n} = ______________.

95. If n is a **positive integer,** $\left(\frac{1}{x}\right)^{-n}$ = ______________.

96. For any **nonzero numbers** x and y and any **positive integers** m and n, $\frac{x^{-m}}{y^{-n}}$ = ______________.

97. If m and n are **integers,** $x^m \cdot x^n$ = ______________.

98. If m and n are **integers,** $\frac{x^m}{x^n}$ = ______________.

x^{-n}	x^{m+n}
x^n	n
$\frac{y^n}{x^m}$	x^{n-m}
$\frac{x^n}{y^n}$	x^{m-n}
x^{mn}	$\frac{1}{x^n}$
1	

Mastery Test

Write using positive exponents and simplify:

99. $\frac{7^{-2}}{6^{-2}}$

100. $\frac{p^{-3}}{q^{-7}}$

101. $\left(\frac{1}{9}\right)^{-2}$

102. $\left(\frac{1}{r}\right)^{-4}$

Write using negative exponents:

103. $\frac{1}{7^6}$

104. $\frac{1}{w^5}$

Simplify and write the answer with positive exponents:

105. $5^6 \cdot 5^{-4}$

106. $x^{-3} \cdot x^{-7}$

107. $\frac{z^{-5}}{z^{-7}}$

108. $\frac{r}{r^8}$

109. $\left(\frac{2xy^3}{3x^3y^{-3}}\right)^{-4}$

110. $\left(\frac{3x^{-2}y^{-3}}{2x^3y^{-2}}\right)^{-5}$

111. The price of a used car is \$5000. If the car depreciates (loses its value) by 10% each year:

a. What will the value of the car be in 3 years?

b. What was the value of the car 2 years ago?

Skill Checker

Find:

112. 7.31×10^1

113. 8.39×10^2

114. 7.314×10^3

115. 8.16×10^{-2}

116. 3.15×10^{-3}

4.3 Application of Exponents: Scientific Notation

Objectives

A > Write numbers in scientific notation.

B > Multiply and divide numbers in scientific notation.

C > Solve applications involving scientific notation.

To Succeed, Review How To . . .

1. Use the rules of exponents (pp. 322–325).
2. Multiply and divide real numbers (pp. 61, 63–64).

Getting Started

Sun Facts

How many facts do you know about the sun? Here is some information taken from a NASA source.

Mass: 2.19×10^{27} tons

Temperature: 9.9×10^3 degrees Fahrenheit

Rotation period at poles: 3.6×10 days

All the numbers here are written as products of a number between 1 and 10 and an appropriate power of 10. This is called *scientific notation*. When written in standard notation, these numbers are

2,190,000,000,000,000,000,000,000,000

9900

36

It's easy to see why so many technical fields use scientific notation. In this section we shall learn how to write numbers in scientific notation and how to perform multiplications and divisions using these numbers.

A > Scientific Notation

We define scientific notation as follows:

SCIENTIFIC NOTATION

A number in **scientific notation** is written as

$$M \times 10^n$$

where M is a number between 1 and 10 and n is an integer.

How do we write a number in scientific notation? First, recall that when we *multiply* a number by a power of 10 ($10^1 = 10$, $10^2 = 100$, and so on), we simply move the decimal point as many places to the *right* as indicated by the exponent of 10. Thus,

$$7.31 \times 10^1 = 7.31 = 73.1$$ Exponent 1; move decimal point 1 place right.

$$72.813 \times 10^2 = 72.813 = 7281.3$$ Exponent 2; move decimal point 2 places right.

$$160.7234 \times 10^3 = 160.7234 = 160{,}723.4$$ Exponent 3; move decimal point 3 places right.

On the other hand, if we *divide* a number by a power of 10, we move the decimal point as many places to the *left* as indicated by the exponent of 10. Thus,

$$\frac{7}{10} = 0.7 = 7 \times 10^{-1}$$

$$\frac{8}{100} = 0.08 = 8 \times 10^{-2}$$

and

$$\frac{4.7}{100{,}000} = 0.000047 = 4.7 \times 10^{-5}$$

Remembering the following procedure makes it easy to write a number in scientific notation.

PROCEDURE

Writing a Number in Scientific Notation ($M \times 10^n$)

1. Move the decimal point in the given number so that there is only one nonzero digit to its left. The resulting number is M.
2. Count the number of places you moved the decimal point in step 1. If the decimal point was moved to the *left*, n is *positive;* if it was moved to the *right*, n is *negative*.
3. Write $M \times 10^n$.

For example,

$$5.3 = 5.3 \times 10^0$$ The decimal point in 5.3 must be moved 0 places to get 5.3.

1 $$87 = 8.7 \times 10^1 = 8.7 \times 10$$ The decimal point in 87 must be moved 1 place *left* to get 8.7.

4 $$68{,}000 = 6.8 \times 10^4$$ The decimal point in 68,000 must be moved 4 places *left* to get 6.8.

−1 $$0.49 = 4.9 \times 10^{-1}$$ The decimal point in 0.49 must be moved 1 place *right* to get 4.9.

−2 $$0.072 = 7.2 \times 10^{-2}$$ The decimal point in 0.072 must be moved 2 places *right* to get 7.2.

NOTE

After completing step 1 in the procedure, decide whether you should make the number obtained *larger* (n positive) or *smaller* (n negative). If the number is greater than 1, use a positive exponent; if it is less than 1, use a negative exponent.

EXAMPLE 1 Writing a number in scientific notation

The approximate distance to the sun is 93,000,000 miles, and the wavelength of its ultraviolet light is 0.000035 centimeter. Write 93,000,000 and 0.000035 in scientific notation.

SOLUTION 1

$93{,}000{,}000 = 9.3 \times 10^7$

$0.000035 = 3.5 \times 10^{-5}$

PROBLEM 1

The approximate distance to the moon is 239,000 miles and its mass is 0.012 that of the Earth. Write 239,000 and 0.012 in scientific notation.

EXAMPLE 2 Changing scientific notation to standard notation

A jumbo jet weighs 7.75×10^5 pounds, whereas a house spider weighs 2.2×10^{-4} pound. Write these weights in standard notation.

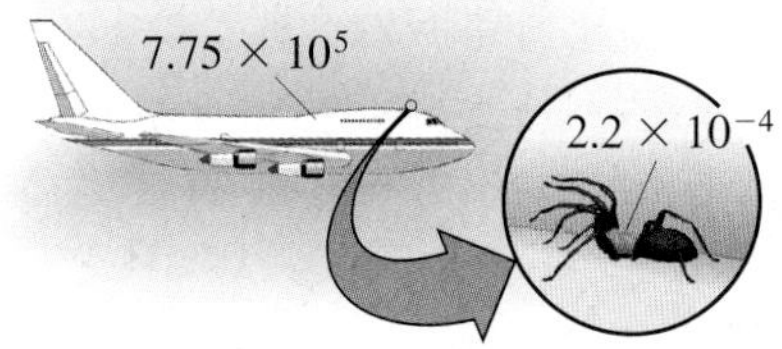

SOLUTION 2

$7.75 \times 10^5 = 775{,}000$ To multiply by 10^5, move the decimal point 5 places right.

$2.2 \times 10^{-4} = 0.00022$ To multiply by 10^{-4}, move the decimal point 4 places left.

PROBLEM 2

In an average year, Carnival Cruise™ line puts more than 10.1×10^6 mints, each weighing 5.25×10^{-2} ounce, on guest's pillows. Write 10.1×10^6 and 5.25×10^{-2} in standard notation.

B > Multiplying and Dividing Using Scientific Notation

Consider the product $300 \cdot 2000 = 600{,}000$. In scientific notation, we would write

$$(3 \times 10^2) \cdot (2 \times 10^3) = 6 \times 10^5$$

To find the answer, we can multiply 3 by 2 to obtain 6 and multiply 10^2 by 10^3, obtaining 10^5. To multiply numbers in scientific notation, we proceed in a similar manner; here's the procedure.

PROCEDURE

Multiplying Using Scientific Notation

1. Multiply the decimal parts first and write the result in scientific notation.
2. Multiply the powers of 10 using the product rule.
3. The answer is the product of the numbers obtained in steps 1 and 2 after simplification.

EXAMPLE 3 Multiplying numbers in scientific notation

Multiply:

a. $(5 \times 10^3) \times (8.1 \times 10^4)$ **b.** $(3.2 \times 10^2) \times (4 \times 10^{-5})$

SOLUTION 3

a. We multiply the decimal parts first, then write the result in scientific notation.

$$5 \times 8.1 = 40.5 = 4.05 \times 10$$

Next we multiply the powers of 10.

$$10^3 \times 10^4 = 10^7$$ Add exponents 3 and 4 to obtain 7.

The answer is $(4.05 \times 10) \times 10^7$, or 4.05×10^8.

PROBLEM 3

Multiply:

a. $(6 \times 10^4) \times (5.2 \times 10^5)$

b. $(3.1 \times 10^3) \times (5 \times 10^{-6})$

(continued)

Answers to PROBLEMS

1. 2.39×10^5; 1.2×10^{-2} **2.** 10,100,000; 0.0525 **3. a.** 3.12×10^{10} **b.** 1.55×10^{-2}

b. Multiply the decimals and write the result in scientific notation.

$$3.2 \times 4 = 12.8 = 1.28 \times 10$$

Multiply the powers of 10.

$$10^2 \times 10^{-5} = 10^{2-5} = 10^{-3}$$

The answer is $(1.28 \times 10) \times 10^{-3}$, or $1.28 \times 10^{1+(-3)} = 1.28 \times 10^{-2}$.

Division is done in a similar manner. For example,

$$\frac{3.2 \times 10^5}{1.6 \times 10^2}$$

is found by dividing 3.2 by 1.6 (yielding 2) and 10^5 by 10^2, which is 10^3. The answer is 2×10^3.

EXAMPLE 4 **Dividing numbers in scientific notation**

Find the quotient: $(1.24 \times 10^{-2}) \div (3.1 \times 10^{-3})$

SOLUTION 4 First divide 1.24 by 3.1 to obtain $0.4 = 4 \times 10^{-1}$. Now divide powers of 10:

$$\begin{aligned} 10^{-2} \div 10^{-3} &= 10^{-2-(-3)} \\ &= 10^{-2+3} \\ &= 10^1 \end{aligned}$$

The answer is $(4 \times 10^{-1}) \times 10^1 = 4 \times 10^0 = 4$.

PROBLEM 4

Find the quotient:

$(1.23 \times 10^{-3}) \div (4.1 \times 10^{-4})$

C > Applications Involving Scientific Notation

Applications involving scientific notation are common because large and small numbers are used in many different fields of study such as astronomy.

EXAMPLE 5 **Energy from the sun**

The total energy received from the sun each minute is 1.02×10^{19} calories. Since the area of the Earth is 5.1×10^{18} square centimeters, the amount of energy received per square centimeter of the Earth's surface every minute (the solar constant) is

$$\frac{1.02 \times 10^{19}}{5.1 \times 10^{18}}$$

Simplify this expression.

SOLUTION 5 Dividing 1.02 by 5.1, we obtain $0.2 = 2 \times 10^{-1}$. Now, $10^{19} \div 10^{18} = 10^{19-18} = 10^1$. Thus, the final answer is

$$(2 \times 10^{-1}) \times 10^1 = 2 \times 10^0 = 2$$

This means that the Earth receives about 2 calories of energy per square centimeter each minute.

PROBLEM 5

The population density for a country is the number of people per square mile. Monaco's population density is

$$\frac{3.3 \times 10^4}{7.5 \times 10^{-1}}$$

Simplify this expression.

Now, let's talk about more "earthly" matters.

Answers to PROBLEMS

4. 3 **5.** 44,000/mi^2

EXAMPLE 6 Printing money

In a recent year, the Treasury Department reported printing the following amounts of money in the specified denominations:

\$3,500,000,000 in \$1 bills	\$2,160,000,000 in \$20 bills
\$1,120,000,000 in \$5 bills	\$250,000,000 in \$50 bills
\$640,000,000 in \$10 bills	\$320,000,000 in \$100 bills

a. Write these numbers in scientific notation.

b. Determine how much money was printed (in billions).

SOLUTION 6

a. $3{,}500{,}000{,}000 = 3.5 \times 10^9$
$1{,}120{,}000{,}000 = 1.12 \times 10^9$
$640{,}000{,}000 = 6.4 \times 10^8$
$2{,}160{,}000{,}000 = 2.16 \times 10^9$
$250{,}000{,}000 = 2.5 \times 10^8$
$320{,}000{,}000 = 3.2 \times 10^8$

b. Since we have to write all the quantities in billions (a billion is 10^9), we have to write all numbers using 9 as the exponent. First, let's consider 6.4×10^8. To write this number with an exponent of 9, we write

$$6.4 \times 10^8 = (0.64 \times 10) \times 10^8 = 0.64 \times 10^9$$

Similarly,

$$2.5 \times 10^8 = (0.25 \times 10) \times 10^8 = 0.25 \times 10^9$$

and

$$3.2 \times 10^8 = (0.32 \times 10) \times 10^8 = 0.32 \times 10^9$$

Writing the other numbers, we get

$$\begin{array}{r} 3.5 \times 10^9 \\ 1.12 \times 10^9 \\ 2.16 \times 10^9 \\ \hline 7.99 \times 10^9 \end{array}$$

Add the entire column.

Thus, 7.99 billion dollars were printed.

PROBLEM 6

How much money was minted in coins? The amounts of money minted in the specified denominations are as follows:

Pennies:	\$1,025,740,000
Nickels:	\$66,183,600
Dimes:	\$233,530,000
Quarters:	\$466,850,000
Half-dollars:	\$15,355,000

a. Write these numbers in scientific notation.

b. Determine how much money was minted (in billions).

Answers to PROBLEMS

6. a. Pennies: 1.02574×10^9
Nickels: 6.61836×10^7
Dimes: 2.3353×10^8
Quarters: 4.6685×10^8
Half-dollars: 1.5355×10^7

b. \$1.8076586 billion

EXAMPLE 7 New planets and scientific notation

In 2006 the International Astronomical Union decided to reclassify Pluto and the newly discovered Eris as "dwarf" planets. Pluto and Eris are now "dwarfs." What makes these planets "dwarf"? Clearly, their size! Here are the sizes of the diameter of the four giant planets—Jupiter, Saturn, Uranus, Neptune—and the two "dwarf" planets.

Jupiter	8.8736×10^4 miles
Neptune	3.0775×10^4 miles
Saturn	7.4978×10^4 miles
Uranus	3.2193×10^4 miles
Pluto	1.422×10^3 miles
Eris	1.490×10^3 miles

a. Which of the two dwarf planets is bigger?

b. Which planet is bigger, Jupiter or Saturn?

PROBLEM 7

a. Ceres is a dwarf planet with a diameter of 5.80×10^2 miles. Is Ceres bigger or smaller than Eris?

b. Which planet is bigger, Neptune or Uranus?

Eris with the sun in the background.

(continued)

Answers to PROBLEMS

7. a. Smaller
b. Uranus is bigger

SOLUTION 7

a. To find the bigger dwarf planet, we compare the diameters of Pluto and Eris: 1.422×10^3 miles and 1.490×10^3 miles.

To do this,

1. Compare the exponents.
2. Compare the decimal parts.

Both have the same exponents (3), but one of the decimal parts (1.490 the decimal part of Eris) is larger, so we can conclude that Eris is larger than Pluto.

b. Again the exponents for Jupiter and Saturn are the same (4) but the decimal part of Jupiter (8.8736) is larger, so Jupiter is larger.

Read more about the "demotion" of Pluto and the new planets at: http://www.gps.caltech.edu/~mbrown/eightplanets/.

Exercises 4.3

> Practice Problems > Self-Tests > Media-rich eBooks > e-Professors > Videos

Web IT go to mhhe.com/bello for more lessons

⟨A⟩ Scientific Notation In Problems 1–10, write the given number in scientific notation.

1. 55,000,000 (working women in the United States)

2. 69,000,000 (working men in the United States)

3. 300,000,000 (U.S. population now)

4. 309,000,000 (estimated U.S. population in the year 2010)

5. 1,900,000,000 (dollars spent on water beds and accessories in 1 year)

6. 0.035 (ounces in a gram)

7. 0.00024 (probability of four-of-a-kind in poker)

8. 0.000005 (the gram-weight of an amoeba)

9. 0.000000002 (the gram-weight of one liver cell)

10. 0.00000009 (wavelength of an X ray in centimeters)

In Problems 11–20, write the given number in standard notation.

11. 1.53×10^2 (pounds of meat consumed per person per year in the United States)

12. 5.96×10^2 (pounds of dairy products consumed per person per year in the United States)

13. 171×10^6 (fresh bagels produced per year in the United States)

14. 2.01×10^6 (estimated number of jobs created in service industries between now and the year 2010)

15. 6.85×10^9 (estimated worth, in dollars, of the five wealthiest women)

16. 1.962×10^{10} (estimated worth, in dollars, of the five wealthiest men)

17. 2.3×10^{-1} (kilowatts per hour used by your TV)

18. 4×10^{-2} (inches in 1 millimeter)

19. 2.5×10^{-4} (thermal conductivity of glass)

20. 4×10^{-11} (energy, in joules, released by splitting one uranium atom)

⟨B⟩ Multiplying and Dividing Using Scientific Notation In Problems 21–30, perform the indicated operations and give your answer in scientific notation.

21. $(3 \times 10^4) \times (5 \times 10^5)$

22. $(5 \times 10^2) \times (3.5 \times 10^3)$

23. $(6 \times 10^{-3}) \times (5.1 \times 10^6)$

24. $(3 \times 10^{-2}) \times (8.2 \times 10^5)$

25. $(4 \times 10^{-2}) \times (3.1 \times 10^{-3})$

26. $(3.1 \times 10^{-3}) \times (4.2 \times 10^{-2})$

27. $\dfrac{4.2 \times 10^5}{2.1 \times 10^2}$

28. $\dfrac{5 \times 10^6}{2 \times 10^3}$

29. $\dfrac{2.2 \times 10^4}{8.8 \times 10^6}$

30. $\dfrac{2.1 \times 10^3}{8.4 \times 10^5}$

⟨ C ⟩ Applications Involving Scientific Notation

⟩⟩⟩ Applications: Green Math

31. *Norwegian garbage* The average Norwegian produces 4.8 pounds of garbage each day. Since there are about 4.6 million Norwegians and 365 days in a year, the annual number of pounds of garbage produced in Norway is $4.8 \times (4.6 \times 10^6) \times (3.65 \times 10^2)$.

a. Write this number in scientific notation.

b. Write this number in standard notation.

32. *Irish garbage* The average Irish person also produces 4.8 pounds of garbage each day. Since there are about 5.6 million Irish people and 365 days in a year, the annual number of pounds of garbage produced in Ireland is $4.8 \times (5.6 \times 10^6) \times (3.65 \times 10^2)$.

a. Write this number in scientific notation.

b. Write this number in standard notation.

33. *Garbage production* America produces 254.1 million tons of garbage each year. Since a ton is 2000 pounds, and there are about 360 days in a year and 310 million Americans, the number of pounds of garbage produced each day of the year for each man, woman, and child in America is

$$\frac{(2.541 \times 10^8) \times (2 \times 10^3)}{(3.1 \times 10^8) \times (3.6 \times 10^2)}$$

Write this number in standard notation to two decimal places.

34. *Velocity of light* The velocity of light can be measured by dividing the distance from the sun to the Earth (1.47×10^{11} meters) by the time it takes for sunlight to reach the Earth (4.9×10^2 seconds). Thus, the velocity of light is

$$\frac{1.47 \times 10^{11}}{4.9 \times 10^2}$$

How many meters per second is that?

35. *Nuclear fission* Nuclear fission is used as an energy source. Do you know how much energy a gram of uranium-235 gives? The answer is

$$\frac{4.7 \times 10^9}{235} \text{ kilocalories}$$

Write this number in scientific notation.

36. *U.S. national debt* The national debt of the United States is about $\$1.19 \times 10^{13}$ (about \$11.9 trillion). If we assume that the U.S. population is 310 million, each citizen actually owes $\frac{1.19 \times 10^{13}}{3.1 \times 10^8}$ dollars. How much money is that? Approximate the answer to the nearest cent. You can see the answer for this very minute at http://www.brillig.com/debt_clock/.

37. *Internet e-mail projections* An Internet travel site sends about 730 million e-mails a year, and that number is projected to grow to 4.38 billion per year. If a year has 365 days, how many e-mails per day are they sending now, and how many are they projecting to send later?

Source: Iconocast.com.

38. *Product information e-mails* Every year, 18.25 billion e-mails requesting product information or service inquiries are sent. If a year has 365 days, how many e-mails a day is that?

Source: Warp 9, Inc.

39. *E-mail proliferation* In 2010, about 310 million people in the United States will be sending 250 billion e-mails every day. How many e-mails will each person be sending per year?

40. *Planets* Do you know how many "giant" planets we have? Here are the distances from the sun (in kilometers) of the four giant planets—Jupiter, Neptune, Saturn, and Uranus:

Jupiter	7.78×10^8 km
Neptune	4.50×10^9 km
Saturn	1.43×10^9 km
Uranus	2.87×10^9 km

a. Which planet is closest to the sun?

b. Which planet is farthest from the sun?

c. Order the four planets from least to greatest distance from the sun.

41. *Terrestrial planets* Do you know the terrestrial planets? Why are they called terrestrial? Here are the distances from the sun (in kilometers) of the four *terrestrial* planets (planets composed primarily of rock and metal)—Earth, Mars, Mercury, and Venus:

Earth	1.49×10^8 km
Mars	2.28×10^8 km
Mercury	5.80×10^7 km
Venus	1.08×10^8 km

a. Which planet is closest to the sun?

b. Which planet is farthest from the sun?

c. Order the four planets from least to greatest distance from the sun.

Web IT go to mhhe.com/bello for more lessons

Web IT go to mhhe.com/bello for more lessons

42. *Pluto demoted!* In August 2006, the International Astronomical Union revised the definition of a planet. Under this new classification Eris, Pluto, and Ceres were classified as "dwarf" planets. The distance from the sun (in kilometers) of each of these planets is shown:

Ceres	4.14×10^8 km
Eris	1.45×10^{10} km
Pluto	5.90×10^9 km

a. Which planet is closest to the sun?

b. Which planet is farthest from the sun?

c. Order the three planets from least to greatest distance from the sun.

43. *Giant planets* The masses of the "giant" planets (in kilograms) are given:

Jupiter	1.90×10^{27} kg
Neptune	1.02×10^{26} kg
Saturn	5.69×10^{26} kg
Uranus	8.68×10^{25} kg

a. Which planet is heaviest?

b. Which planet is lightest?

c. Order the four planets from lightest to heaviest.

44. *Terrestrial planets* The masses of the *terrestrial* planets (in kilograms) are given:

Earth	5.98×10^{24} kg
Mars	6.42×10^{23} kg
Mercury	3.30×10^{23} kg
Venus	4.87×10^{24} kg

a. Which planet is heaviest?

b. Which planet is lightest?

c. Order the four planets from lightest to heaviest.

Using Your Knowledge

Astronomical Quantities As we have seen, scientific notation is especially useful when very large quantities are involved. Here's another example. In astronomy, we find that the speed of light is 299,792,458 meters per second.

45. Write 299,792,458 in scientific notation.

Astronomical distances are so large that they are measured in astronomical units (AU). An astronomical unit is defined as the average separation (distance) of the earth and the sun—that is, 150,000,000 kilometers.

46. Write 150,000,000 in scientific notation.

47. Distances in astronomy are also measured in *parsecs:* 1 parsec $= 2.06 \times 10^5$ AU. Thus, 1 parsec $= (2.06 \times 10^5) \times (1.5 \times 10^8)$ kilometers. Written in scientific notation, how many kilometers is that?

48. Astronomers also measure distances in *light-years,* the distance light travels in 1 year: 1 light-year $= 9.46 \times 10^{12}$ kilometers. The closest star, Proxima Centauri, is 4.22 light-years away. In scientific notation using two decimal places, how many kilometers is that?

49. Since 1 parsec $= 3.09 \times 10^{13}$ kilometers (see Problem 47) and 1 light-year $= 9.46 \times 10^{12}$ kilometers, the number of light-years in a parsec is

$$\frac{3.09 \times 10^{13}}{9.46 \times 10^{12}}$$

Write this number in standard notation rounded to two decimal places.

Write On

50. Explain why the procedure used to write numbers in scientific notation works.

51. What are the advantages and disadvantages of writing numbers in scientific notation?

Concept Checker

Fill in the blank(s) with the correct word(s), phrase, or mathematical statement.

52. A number is written in **scientific notation** if it is written in the form ________________.

53. When a number is written in **scientific notation** in the form $M \times 10^n$, the M is a number between 1 and ________________ and n is a(n) ________________.

54. The **first** step in **multiplying** numbers in **scientific notation** is to **multiply** the ________________ **parts.**

55. The **second** step in multiplying numbers in **scientific notation** is to **multiply** the ________________ **of 10** using the product rule.

0	$M \times 10^n$
10	10^M
whole number	**decimal**
integer	**powers**

Mastery Test

56. The width of the asteroid belt is 1.75×10^8 kilometers. The speed of *Pioneer 10,* a U.S. space vehicle, in passing through this belt was 1.4×10^5 kilometers per hour. Thus, *Pioneer 10* took

$$\frac{1.75 \times 10^8}{1.4 \times 10^5}$$

hours to go through the belt. How many hours is that in scientific notation?

57. The distance to the moon is about 239,000 miles, and its mass is 0.12456 that of the Earth. Write these numbers in scientific notation.

58. The Concorde (a supersonic passenger plane) weighed 4.08×10^5 pounds, and a cricket weighs 3.125×10^{-4} pound. Write these weights in standard notation.

Perform the operations and write the answer in scientific and standard notation.

59. $(2.52 \times 10^{-2}) \div (4.2 \times 10^{-3})$

60. $(4.1 \times 10^2) \times (3 \times 10^{-5})$

61. $(6 \times 10^4) \times (2.2 \times 10^3)$

62. $(3.2 \times 10^{-2}) \div (1.6 \times 10^{-5})$

Skill Checker

Find:

63. $-16(1)^2 + 118$

64. $-16(2)^2 + 118$

65. $-8(3)^2 + 80$

66. $3(2^2) - 5(3) + 8$

67. $-4 \cdot 8 \div 2 + 20$

68. $-5 \cdot 6 \div 2 + 25$

4.4 Polynomials: An Introduction

Objectives

A Classify polynomials.

B Find the degree of a polynomial.

C Write a polynomial in descending order.

D Evaluate polynomials.

To Succeed, Review How To . . .

1. Evaluate expressions (pp. 61–64, 69–73).
2. Add, subtract, and multiply expressions (pp. 78, 89–93, 319–321).

Getting Started

A Diving Polynomial

The diver in the photo jumped from a height of 118 feet. Do you know how many feet above the water the diver will be after t seconds? Scientists have determined a formula for finding the answer:

$$-16t^2 + 118 \quad \text{(feet above the water)}$$

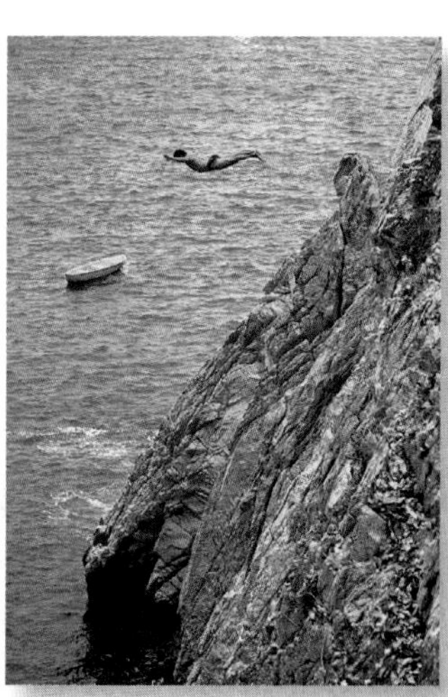

The expression $-16t^2 + 118$ is an example of a *polynomial*. Here are some other polynomials:

$$5x, \quad 9x - 2, \quad -5t^2 + 18t - 4,$$

$$\text{and} \quad y^5 - 2y^2 + \frac{4}{5}y - 6$$

We construct these polynomials by adding or subtracting products of numbers and variables raised to whole-number exponents. Of course, if we use any other operations, the result may not be a polynomial. For example,

$$x^2 - \frac{3}{x} \quad \text{and} \quad x^{-7} + 4x$$

are *not* polynomials (we divided by the variable x in the first one and used negative exponents in the second one). In this section we shall learn how to classify polynomials in one variable, find their degrees, write them in descending order, and evaluate them.

A > Classifying Polynomials

Polynomials can be used to track and predict the amount of waste (in millions of tons) generated annually in the United States. The polynomial approximating this amount is:

$$-0.001t^3 + 0.06t^2 + 2.6t + 88.6$$

where t is the number of years after 1960. But how do we *predict* how much waste will be generated in the year 2010 using this polynomial? We will show you in Example 6! First, let's look at the definition of polynomials.

POLYNOMIAL

A polynomial is an algebraic expression formed by using the operations of addition and subtraction on products of numbers and variables raised to whole-number exponents.

The parts of a polynomial separated by plus signs are called the **terms** of the polynomial. If there are subtraction signs, we can rewrite the polynomial using addition signs, since we know that $a - b = a + (-b)$. Thus,

$5x$ has one term: $5x$

$9x - 2$ has two terms: $9x$ and -2

Recall that $9x - 2 = 9x + (-2)$.

$-5t^2 + 18t - 4$ has three terms: $-5t^2$, $18t$, and -4

$-5t^2 + 18t - 4 = -5t^2 + 18t + (-4)$

Polynomials are classified according to the number of terms they have. Thus,

$5x$ has *one* term; it is called a *monomial*.

$9x - 2$ has *two* terms; it is called a *binomial*.

$-5t^2 + 18t - 4$ has *three* terms; it is called a *trinomial*.

mono means one.
bi means two.
tri means three.

NOTE

1. A polynomial of one term is a **monomial;**
2. A polynomial of two terms, a **binomial;**
3. A polynomial of three terms is a **trinomial;**
4. A polynomial of more than three terms is just a **polynomial.**

EXAMPLE 1 Classifying polynomials

Classify each of the following polynomials as a monomial, binomial, or trinomial.

a. $6x - 1$ **b.** -8 **c.** $-4 + 3y - y^2$ **d.** $5(x + 2) - 3$

SOLUTION 1

a. $6x - 1$ has two terms; it is a binomial.
b. -8 has only one term; it is a monomial.
c. $-4 + 3y - y^2$ has three terms; it is a trinomial.
d. $5(x + 2) - 3$ is a binomial. Note that $5(x + 2)$ is *one* term.

PROBLEM 1

Classify as monomial, binomial, or trinomial.

a. -5
b. $-3 + 4y + 6y^2$
c. $8x - 3$
d. $8(x + 9) - 3(x - 1)$

Answers to PROBLEMS

1. a. Monomial **b.** Trinomial **c.** Binomial **d.** Binomial

B > Finding the Degree of a Polynomial

All the polynomials we have seen contain only one variable and are called *polynomials in one variable*. Polynomials in one variable, such as $x^2 + 3x - 7$, can also be classified according to the *highest* exponent of the variable. The highest exponent of the variable is called the **degree** of the polynomial. To find the degree of a polynomial, you simply examine each term and find the highest exponent of the variable. Thus, the degree of $3x^2 + 5x^4 - 2$ is found by looking at the exponent of the variable in each of the terms.

The exponent in $3x^2$ is 2.

The exponent in $5x^4$ is 4.

The exponent in -2 is 0 because $-2 = -2x^0$. (Recall that $x^0 = 1$.)

Thus, the degree of $3x^2 + 5x^4 - 2$ is 4, the highest exponent of the variable in $3x^2 + 5x^4 - 2$. Similarly, the degree of $4y^3 - 3y^5 + 9y^2$ is 5, since 5 is the highest exponent of the variable present in $4y^3 - 3y^5 + 9y^2$. By convention, a number such as -4 or 7 is called a **polynomial of degree 0,** because if $a \neq 0$, $a = ax^0$. Thus, $-4 = -4x^0$ and $7 = 7x^0$ are polynomials of degree 0. The number 0 itself is called the **zero polynomial** and is *not* assigned a degree. (Note that $0 \cdot x^1 = 0$; $0 \cdot x^2 = 0$, $0 \cdot x^3 = 0$, and so on, so the zero polynomial cannot have a degree.)

EXAMPLE 2 Finding the degree of a polynomial

Find the degree:

a. $-2t^2 + 7t - 2 + 9t^3$ **b.** 8 **c.** $-3x + 7$ **d.** 0

SOLUTION 2

a. The highest exponent of the variable t in the polynomial $-2t^2 + 7t - 2 + 9t^3$ is 3; thus, the degree of the polynomial is 3.

b. The degree of 8 is, by convention, 0.

c. Since $x = x^1$, $-3x + 7$ can be written as $-3x^1 + 7$, making the degree of $-3x^1 + 7$ one.

d. 0 is the zero polynomial; it does not have a degree.

PROBLEM 2

Find the degree:

a. 9 **b.** $-5z^2 + 2z - 8$

c. 0 **d.** $-8y + 1$

C > Writing a Polynomial in Descending Order

The degree of a polynomial is easier to find if we agree to write the polynomial in **descending order;** that is, the term with the *highest* exponent is written *first,* the *second* highest is *next,* and so on. Fortunately, the associative and commutative properties of addition permit us to do this rearranging! Thus, instead of writing $3x^2 - 5x^3 + 4x - 2$, we rearrange the terms and write $-5x^3 + 3x^2 + 4x - 2$ with exponents in the terms arranged in *descending* order. Similarly, to write $-3x^3 + 7 + 5x^4 - 2x$ in descending order, we use the associative and commutative properties and write $5x^4 - 3x^3 - 2x + 7$. Of course, it would not be incorrect to write this polynomial in ascending order (or with no order at all); it is just that we *agree* to write polynomials in descending order for uniformity and convenience.

EXAMPLE 3 Writing polynomials in descending order

Write in descending order:

a. $-9x + x^2 - 17$ **b.** $-5x^3 + 3x - 4x^2 + 8$

PROBLEM 3

Write in descending order:

a. $-4x^2 + 3x^3 - 8 + 2x$

b. $-3y + y^2 - 1$

(continued)

Answers to PROBLEMS

2. a. 0 **b.** 2 **c.** No degree **d.** 1

3. a. $3x^3 - 4x^2 + 2x - 8$ **b.** $y^2 - 3y - 1$

SOLUTION 3

a. $-9x + x^2 - 17 = x^2 - 9x - 17$

b. $-5x^3 + 3x - 4x^2 + 8 = -5x^3 - 4x^2 + 3x + 8$

D > Evaluating Polynomials

Now, let's return to the diver in the *Getting Started.* You may be wondering why his height above the water after t seconds was $-16t^2 + 118$ feet. This expression doesn't even look like a number! But polynomials represent numbers when they are *evaluated.* So, if our diver is $-16t^2 + 118$ feet above the water after t seconds, then after 1 second (that is, when $t = 1$), our diver will be

$$-16(1)^2 + 118 = -16 + 118 = 102 \text{ ft}$$

above the water.

After 2 seconds (that is, when $t = 2$), his height will be

$$-16(2)^2 + 118 = -16 \cdot 4 + 118 = 54 \text{ ft}$$

above the water.

Note that

At $t = 1$, $\quad -16t^2 + 118 = 102$

At $t = 2$, $\quad -16t^2 + 118 = 54$

and so on.

In algebra, polynomials in one variable can be represented by using symbols such as $P(t)$ (read "P of t"), $Q(x)$, and $D(y)$, where the symbol in parentheses indicates the variable being used. Thus, $P(t) = -16t^2 + 118$ is the polynomial representing the height of the diver above the water and $G(t)$ is the polynomial representing the amount of waste generated annually in the United States. With this notation, $P(1)$ represents the value of the polynomial $P(t)$ when 1 is substituted for t in the polynomial; that is,

$$P(1) = -16(1)^2 + 118 = 102$$

and

$$P(2) = -16(2)^2 + 118 = 54$$

and so on.

EXAMPLE 4 **Evaluating polynomials**

When $t = 3$, what is the value of $P(t) = -16t^2 + 118$?

SOLUTION 4 When $t = 3$,

$$P(t) = -16t^2 + 118$$

becomes

$$\begin{aligned} P(3) &= -16(3)^2 + 118 \\ &= -16(9) + 118 \\ &= -144 + 118 \\ &= -26 \end{aligned}$$

PROBLEM 4

Find the value of $P(t) = -16t^2 + 90$ when $t = 2$.

Note that in this case, the answer is *negative,* which means that the diver should be *below* the water's surface. However, since he can't continue to free-fall after hitting the water, we conclude that it took him between 2 and 3 seconds to hit the water.

Answers to PROBLEMS

4. 26

EXAMPLE 5 Evaluating polynomials

Evaluate $Q(x) = 3x^2 - 5x + 8$ when $x = 2$.

SOLUTION 5

When $x = 2$,

$$Q(x) = 3x^2 - 5x + 8$$

becomes

$$\begin{aligned} Q(2) &= 3(2)^2 - 5(2) + 8 \\ &= 3(4) - 5(2) + 8 && \text{Multiply } 2 \cdot 2 = 2^2. \\ &= 12 - 10 + 8 && \text{Multiply } 3 \cdot 4 \text{ and } 5 \cdot 2. \\ &= 2 + 8 && \text{Subtract } 12 - 10. \\ &= 10 && \text{Add } 2 + 8. \end{aligned}$$

Note that to evaluate this polynomial, we followed the order of operations studied in Section 1.5.

PROBLEM 5

Evaluate $R(x) = 5x^2 - 3x + 9$ when $x = 3$.

GREEN MATH

EXAMPLE 6 Generated waste

a. If $G(t) = -0.001t^3 + 0.06t^2 + 2.6t + 88.6$ is the amount of waste (in millions of tons) generated annually in the United States and t is the number of years *after* 1960, how much waste was generated in 1960 ($t = 0$)?

b. How much waste is predicted to be generated in the year 2010?

SOLUTION 6

a. At $t = 0$, $G(t) = -0.001t^3 + 0.06t^2 + 2.6t + 88.6$ becomes

$G(0) = -0.001(0)^{-3} + 0.06(0)^2 + 2.6(0) + 88.6 = 88.6$ (million tons)

Thus, 88.6 million tons were generated in 1960.

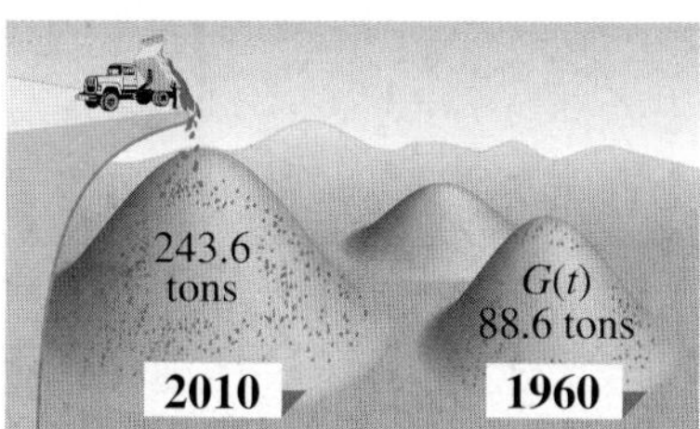

b. The year 2010 is $2010 - 1960 = 50$ years *after* 1960. This means that $t = 50$ and

$$\begin{aligned} G(50) &= -0.001(50)^3 + 0.06(50)^2 + 2.6(50) + 88.6 \\ &= -0.001(125{,}000) + 0.06(2500) + 2.6(50) + 88.6 \\ &= 243.6 \text{ (million tons)} \end{aligned}$$

The prediction is that 243.6 million tons of waste will be generated in the year 2010.

PROBLEM 6

a. How much waste was generated in 1961?

b. How much waste was generated in 2000?

EXAMPLE 7 Blood alcohol level

Do you know how many drinks it takes before you are considered legally drunk? In many states you are drunk if you have a blood alcohol level (BAL) of 0.10 or even lower (0.08). The chart on the next page shows your BAL after consuming 3 ounces

PROBLEM 7

a. Use the graph to find the BAL for a male after 3 hours.

(continued)

Answers to PROBLEMS

5. 45 **6. a.** 91.259 million tons **b.** 224.60 million tons **7. a.** 0.082 **b.** 0.09 **c.** 0.0831 **d.** 0.0892 of alcohol (6 beers with 4% alcohol or 30 ounces of 10% wine or 7.5 ounces of vodka or whiskey) in the time period shown. The polynomial equation $y = -0.0226x + 0.1509$ approximates the BAL for a 150-pound male and $y = -0.0257x + 0.1663$ the BAL of a 150-pound female.

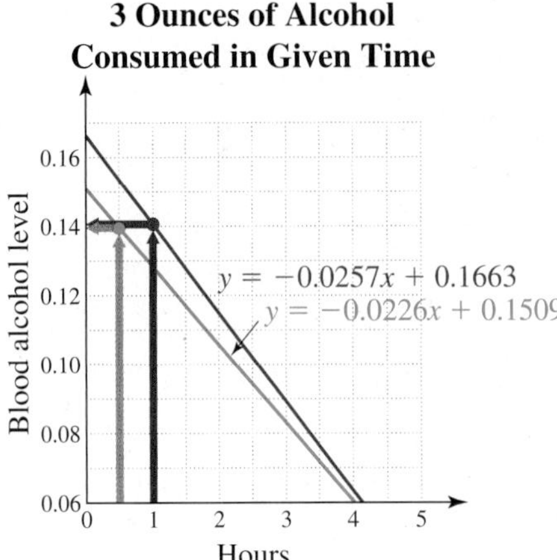

a. Use the graph to find the BAL for a male after 0.5 hour.

b. Use the graph to find the BAL for a female after 1 hour.

c. Evaluate $y = -0.0226x + 0.1509$ for $x = 0.5$. Does your answer coincide with the answer to part **a?**

d. Evaluate $y = -0.0257x + 0.1663$ for $x = 1$. Does your answer coincide with the answer to part **b?**

SOLUTION 7

a. First, locate 0.5 on the x-axis. Move vertically until you reach the blue line and then horizontally (left) to the y-axis. The y-value at that point is approximately 0.14. This means that the BAL of a male 0.5 hour after consuming 3 ounces of alcohol is about 0.14 (legally drunk!).

b. This time, locate 1 on the x-axis, move vertically until you reach the red line, and then move horizontally to the y-axis. The y-value at that point is a little more than 0.14, so we estimate the answer to be 0.141 (legally drunk!).

c. When $x = 0.5$,

$$\begin{aligned} y &= -0.0226x + 0.1509 \\ &= -0.0226(0.5) + 0.1509 = 0.1396 \end{aligned}$$

which is very close to the 0.14 from part **a.**

d. When $x = 1$,

$$\begin{aligned} y &= -0.0257x + 0.1663 \\ &= -0.0257(1) + 0.1663 = 0.1406 \end{aligned}$$

which is also very close to the 0.141 from part **b.**

b. Use the graph to find the BAL for a female after 3 hours.

c. Evaluate $y = -0.0226x + 0.1509$ for $x = 3$.

d. Evaluate $y = -0.0257x + 0.1663$ for $x = 3$.

Calculator Corner

Evaluating Polynomials

If you have a calculator, you can evaluate polynomials in several ways. One way is to make a picture (graph) of the polynomial and use the TRACE and ZOOM keys. Or, better yet, if your calculator has a "value" feature, it will automatically find the value of a polynomial for a given number. Thus, to find the value of $G(t) = -0.001t^3 + 0.06t^2 + 2.6t + 88.6$ when $t = 50$ in Example 6, first graph the polynomial. With a TI-83 Plus, press Y= and enter $-0.001X^3 + 0.06X^2 + 2.6X + 88.6$ for Y_1. (Note that we used X's instead of t's because X's are easier to enter.) If you then press GRAPH, nothing will show in your window! Why? Because a standard window gives values of X only between -10 and 10 and corresponding $Y_1 = G(x)$ values between -10 and 10. Adjust the X- and Y-values to those shown in Window 1 and press GRAPH again. To evaluate $G(X)$ at $X = 50$ with a TI-83 Plus, press 2nd TRACE 1. When the calculator prompts you by showing $X =$, enter 50 and press ENTER. The result is shown in Window 2 as $Y = 243.6$. This means that 50 years after 1960—that is, in the year 2010—243.6 million tons of waste will be generated.

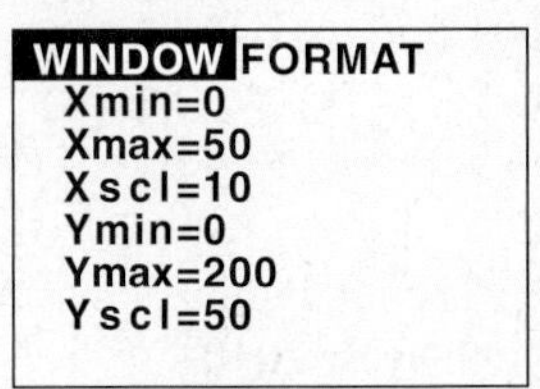

Window 1

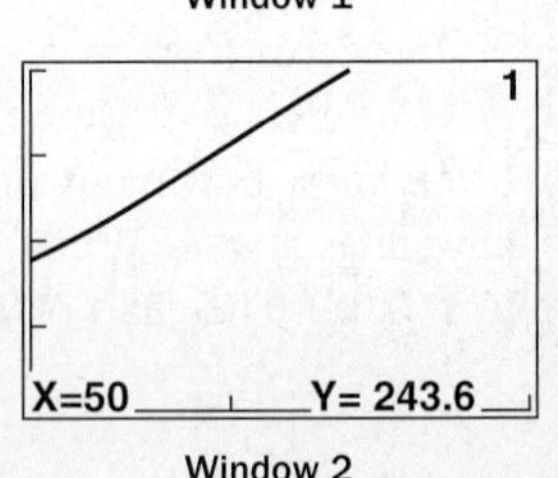

Window 2

You can also evaluate $G(X)$ by first storing the value you wish ($X = 50$) by pressing 50 STO▸ X,T,θ,n ENTER, then entering $-0.001X^3 + 0.06X^2 + 2.6X + 88.6$, and finally pressing ENTER again. The result is shown in Window 3.

The beauty of the first method is that now you can evaluate $G(20)$, $G(50)$, or $G(a)$ for any number a in the chosen interval by simply entering the value of a and pressing ENTER. You don't have to reenter $G(X)$ or adjust the window again!

```
50→X
                 50
-0.001X^3+0.06X^2
+2.6X+88.6
              243.6
```

Window 3

› Exercises 4.4

‹A› Classifying Polynomials
‹B› Finding the Degree of a Polynomial

In Problems 1–10, classify each expression as a monomial, binomial, trinomial, or polynomial and give the degree.

1. $-5x + 7$ **2.** $8 + 9x^3$ **3.** $7x$

4. $-3x^4$ **5.** $-2x + 7x^2 + 9$ **6.** $-x + x^3 - 2x^2$

7. 18 **8.** 0 **9.** $9x^3 - 2x$

10. $-7x + 8x^6 + 3x^5 + 9$

‹B› Finding the Degree of a Polynomial
‹C› Writing a Polynomial in Descending Order

In Problems 11–20, write in descending order and give the degree of each polynomial.

11. $-3x + 8x^3$ **12.** $7 - 2x^3$ **13.** $4x - 7 + 8x^2$

14. $9 - 3x + x^3$ **15.** $5x + x^2$ **16.** $-3x - 7x^3$

17. $3 + x^3 - x^2$ **18.** $-3x^2 + 8 - 2x$ **19.** $4x^5 + 2x^2 - 3x^3$

20. $4 - 3x^3 + 2x^2 + x$

‹D› Evaluating Polynomials

In Problems 21–24, find the value of the polynomial when (a) $x = 2$ and (b) $x = -2$.

21. $3x - 2$ **22.** $x^2 - 3$ **23.** $2x^2 - 1$ **24.** $x^3 - 1$

25. If $P(x) = 3x^2 - x - 1$, find
a. $P(2)$ **b.** $P(-2)$

26. If $Q(x) = 2x^2 + 2x + 1$, find
a. $Q(2)$ **b.** $Q(-2)$

27. If $R(x) = 3x - 1 + x^2$, find
a. $R(2)$ **b.** $R(-2)$

28. If $S(x) = 2x - 3 - x^2$, find
a. $S(2)$ **b.** $S(-2)$

29. If $T(y) = -3 + y + y^2$, find
a. $T(2)$ **b.** $T(-2)$

30. If $U(r) = -r - 4 - r^2$, find
a. $U(2)$ **b.** $U(-2)$

››› Applications

31. *Height of dropped object* If an object drops from an altitude of k feet, its height above ground after t seconds is given by $-16t^2 + k$ feet. If the object is dropped from an altitude of 150 feet, what would be the height of the object after the specified amount of time?

a. t seconds

b. 1 second **c.** 2 seconds

32. *Velocity of dropped object* After t seconds have passed, the velocity of an object dropped from a height of 96 feet is $-32t$ feet per second. What would be the velocity of the object after the specified amount of time?

a. 1 second

b. 2 seconds

33. *Height of dropped object* If an object drops from an altitude of k meters, its height above the ground after t seconds is given by $-4.9t^2 + k$ meters. If the object is dropped from an altitude of 200 meters, what would be the height of the object after the specified amount of time?

a. t seconds

b. 1 second **c.** 2 seconds

34. *Velocity of dropped object* After t seconds have passed, the velocity of an object dropped from a height of 300 meters is $-9.8t$ meters per second. What would be the velocity of the object after the specified amount of time?

a. 1 second

b. 2 seconds

35. *Annual number of robberies* According to FBI data, the annual number of robberies (per 100,000 population) can be approximated by

$$R(t) = 1.76t^2 - 17.24t + 251$$

where t is the number of years after 1980.

a. What was the number of robberies (per 100,000) in 1980 ($t = 0$)?

b. How many robberies per 100,000 would you predict in the year 2000? In 2010?

36. *Annual number of assaults* The number of aggravated assaults (per 100,000) can be approximated by

$$A(t) = -0.2t^3 + 4.7t^2 - 15t + 300$$

where t is the number of years after 2000.

a. What was the number of aggravated assaults (per 100,000) in 2000 ($t = 0$)?

b. How many aggravated assaults per 100,000 would you predict for the year 2020?

››› Applications: Green Math

37. *Saving gas by slowing down* Aggressive driving (speeding, rapid acceleration, and braking) wastes gas! How much? The red graph shows the speed x of a car (mph) and its fuel economy y (mpg) and can be approximated by the polynomial

$$P(x) = -0.01x^2 + x + 7 \text{ (mpg)}$$

a. According to the graph, how many mpg does the car get when driven at 5 mph?

b. Evaluate $P(x) = -0.01x^2 + x + 7$ for $x = 5$. What does the result mean?

c. According to the graph, how many mpg does the car get when driven at 50 mph and at 55 mph?

d. Evaluate $P(x) = -0.01x^2 + x + 7$ for $x = 50$ and $x = 55$. What does the result mean?

e. Are the results you get from reading the graph and from evaluating the polynomial close?

Source: http://www.fueleconomy.gov/FEG/driveHabits.shtml.

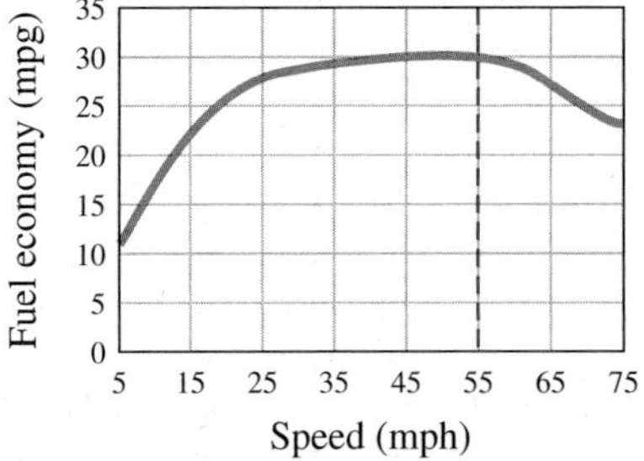

38. a. Based on the graph, at what speed do you get the best mileage?

b. Based on the graph, at what speed do you get the worst mileage?

c. Explain in your own words the relationship between speed and fuel economy.

According to the source cited:
Fuel economy benefit: 7%–23%
Gasoline savings: \$0.18–\$0.59/gallon

39. *Record low temperatures* According to the *USA Today Weather Almanac,* the coldest city in the United States (based on average annual temperature) is International Falls, Minnesota. Record low temperatures (in °F) there can be approximated by $L(m) = -4m^2 + 57m - 175$, where m is the number of the month starting with March ($m = 3$) and ending with December ($m = 12$).

a. Find the record low during July.

b. If $m = 1$ were allowed, what would be the record low in January? Does the answer seem reasonable? Do you see why $m = 1$ is not one of the choices?

40. *Record high temperatures* According to the *USA Today Weather Almanac,* the hottest city in the United States (based on average annual temperature) is Key West, Florida. Record high temperatures there can be approximated by $H(m) = -0.12m^2 + 2.9m + 77$, where m is the number of the month starting with January ($m = 1$) and ending with December ($m = 12$).

a. Find the record high during January. (Answer to the nearest whole number.)

b. In what two months would you expect the highest-ever temperature to have occurred? What is m for each of the two months?

c. Find $H(m)$ for each of the two months of part **b.** Which is higher?

d. The highest temperature ever recorded in Key West was 95°F and occurred in August 1957. How close was your approximation?

41. *Internet use in China* The number of Internet users in China (in millions) is shown in the figure and can be approximated by $N(t) = 0.5t^2 + 4t + 2.1$, where t is the number of years after 1998.

a. Use the graph to find the number of users in 2003.

b. Evaluate $N(t)$ for $t = 5$. Is the result close to the approximation in part **a**?

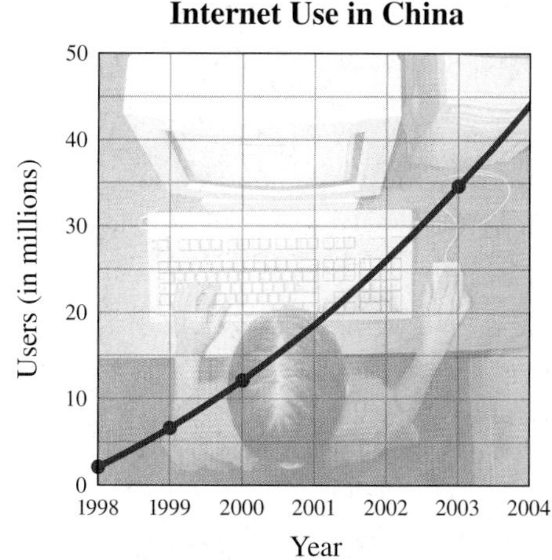

Source: Data from *USA Today,* May 9, 2000.

42. *Stopping distances* The stopping distance needed for a 3000-pound car to come to a complete stop when traveling at the indicated speeds is shown in the figure.

a. Use the graph to estimate the number of feet it takes the car to stop when traveling at 85 miles per hour.

b. Use the quadratic polynomial $D(s) = 0.05s^2 + 2.2s + 0.75$, where s is speed in miles per hour, to approximate the number of feet it takes for the car to stop when traveling at 85 miles per hour.

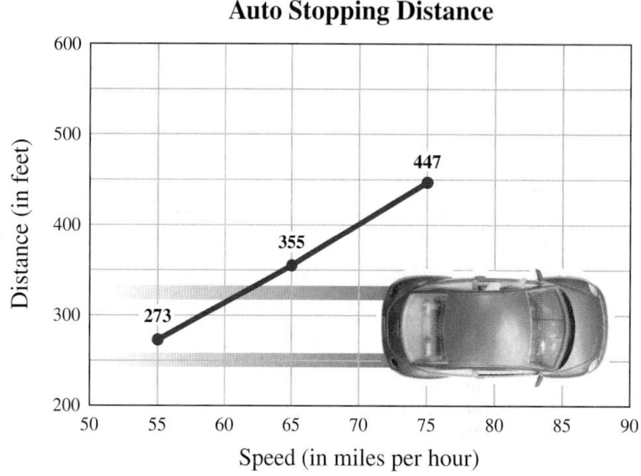

Source: Data from *USA Today*/Foundation for Traffic Safety.

43. *Tuition and fees at 4-year private institutions* The graph shows the tuition and fee charges for 4-year private institutions (blue graph). This cost can be approximated by the binomial $C(t) = 1033t + 16{,}072$, where t is the number of years after 2000.

a. Use the graph to find the cost of tuition and fees in 2005.

b. Evaluate $C(t)$ for $t = 5$. Is the result close to the value in part **a?**

c. Estimate the cost of tuition and fees in 2010 ($t = 10$).

Source: http://www.ed.gov.

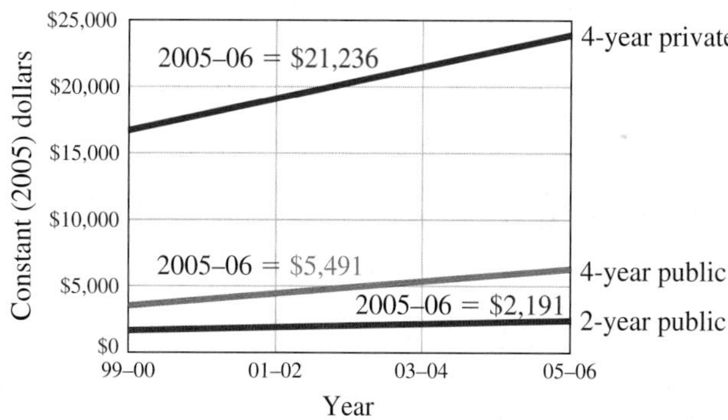

44. *Tuition and fees at 4-year public institutions* The graph shows the tuition and fee charges for 4-year public institutions (red graph). This cost can be approximated by the binomial $C(t) = 397t + 3508$, where t is the number of years after 2000.

a. Use the graph to find the cost of tuition and fees in 2005.

b. Evaluate $C(t)$ for $t = 5$. Is the result close to the value in part **a?**

c. Estimate the cost of tuition and fees in 2010 ($t = 10$).

45. *Tuition and fees at 2-year public institutions* The graph shows the tuition and fee charges for 2-year public institutions (dark red graph). This cost can be approximated by the binomial $C(t) = 110t + 1642$, where t is the number of years after 2000.

a. Use the graph to find the cost of tuition and fees in 2005.

b. Evaluate $C(t)$ for $t = 5$. Is the result close to the value in part **a?**

c. Estimate the cost of tuition and fees in 2010 ($t = 10$).

46. *Tuition, fees, and room and board for 4-year private institutions* The graph shows the tuition, fees, and room and board charges at 4-year private institutions (blue graph). This cost can be approximated by the binomial $C(t) = 1357t + 22{,}240$, where t is the number of years after 2000.

a. Use the graph to find the cost of tuition, fees, and room and board in 2005.

b. Evaluate $C(t)$ for $t = 5$. Is the result close to the value in part **a?**

c. Estimate the cost of tuition, fees, and room and board in 2010 ($t = 10$).

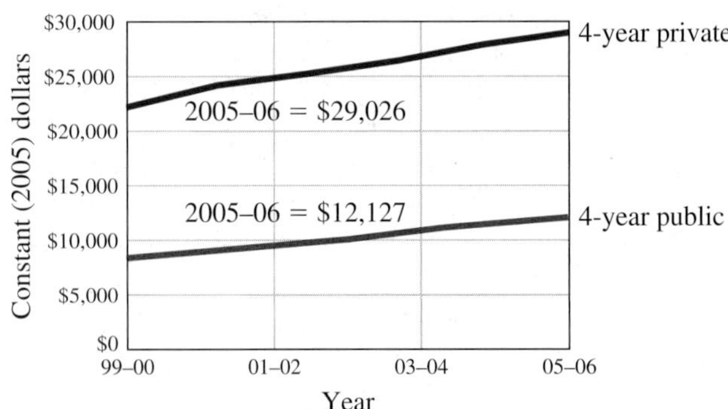

47. *Tuition, fees, and room and board for 4-year public institutions* The graph shows the tuition, fees, and room and board charges at 4-year public institutions (red graph). This cost can be approximated by the binomial $C(t) = 738x + 8439$, where t is the number of years after 2000.

a. Use the graph to find the cost of tuition, fees, and room and board in 2005.

b. Evaluate $C(t)$ for $t = 5$. Is the result close to the value in part **a?**

c. Estimate the cost of tuition, fees, and room and board in 2010 ($t = 10$).

48. *Tuition, fees, and room and board for 4-year public institutions* We can approximate the tuition, fees, and room and board charges at 4-year public institutions (red graph) by using the trinomial $C(t) = 29t^2 + 608t + 8411$, where t is the number of years after 2000.

a. Evaluate $C(t)$ for $t = 5$. Is the result close to the $12,127 value given in the graph?

b. Estimate the cost of tuition, fees, and room and board in 2010 ($t = 10$) and compare with the value obtained in part **c** of Problem 47.

c. You can even approximate the tuition, fees, and room and board charges using $C(t) = -13t^3 + 126t^2 + 431t + 8450$, where t is the number of years after 2000. What will be $C(10)$ and how close is it to the values obtained in **b?**

d. Which is the best approximation, the first-, the second-, or the third-degree polynomial?

Using Your Knowledge

Faster and Faster Polynomials We've already stated that if an object is simply *dropped* from a certain height, its velocity after t seconds is given by $-32t$ feet per second. What will happen if we actually *throw* the object down with an initial velocity, say v_0? Since the velocity $-32t$ is being helped by the velocity v_0, the new final velocity will be given by $-32t + v_0$ (v_0 is *negative* if the object is thrown *downward*).

49. Find the velocity after t seconds have elapsed of a ball thrown downward with an initial velocity of 10 feet per second.

50. What will the velocity of the ball in Problem 49 be after the specified amount of time?

a. 1 second **b.** 2 seconds

51. In the metric system, the velocity after t seconds of an object thrown downward with an initial velocity v_0 is given by the equation

$$-9.8t + v_0 \quad \text{(meters)}$$

What would be the velocity of a ball thrown downward with an initial velocity of 2 meters per second after the specified amount of time?

a. 1 second

b. 2 seconds

52. The height of an object after t seconds have elapsed depends on two factors: the initial velocity v_0 and the height s_0 from which the object is thrown. The polynomial giving this height is given by the equation

$$-16t^2 + v_0t + s_0 \quad \text{(feet)}$$

where v_0 is the initial velocity and s_0 is the height from which the object is thrown. What would be the height of a ball thrown downward from a 300-foot tower with an initial velocity of 10 feet per second after the specified amount of time?

a. 1 second **b.** 2 seconds

Write On

53. Write your own definition of a polynomial.

54. Is $x^2 + \frac{1}{x} + 2$ a polynomial? Why or why not?

55. Is $x^{-2} + x + 3$ a polynomial? Why or why not?

56. Explain how to find the degree of a polynomial in one variable.

57. The degree of x^4 is 4. What is the degree of 7^4? Why?

58. What does "evaluate a polynomial" mean?

Concept Checker

Fill in the blank(s) with the correct word(s), phrase, or mathematical statement.

59. A _______________ is an **algebraic expression** formed by using the operations of **addition and subtraction** on products of numbers and variables raised to whole number exponents.

60. A **polynomial** of **one** term is called a _______________.

61. A **polynomial** of **two** terms is called a _______________.

62. A **polynomial** of **three** terms is called a _______________.

binomial

monomial

polynomial

trinomial

Mastery Test

63. Evaluate $2x^2 - 3x + 10$ when $x = 2$.

64. If $P(x) = 3x^3 - 7x + 9$, find $P(3)$.

65. When $t = 2.5$, what is the value of $-16t^2 + 118$?

Find the degree of each polynomial.

66. $-5y - 3$

67. $4x^2 - 5x^3 + x^8$

68. -9

69. 0

Write each polynomial in descending order.

70. $-2x^4 + 5x - 3x^2 + 9$

71. $-8 + 5x^2 - 3x$

Classify as a monomial, binomial, trinomial, or polynomial.

72. $-4t + t^2 - 8$

73. $-5y$

74. $278 + 6x$

75. $2x^3 - x^2 + x - 1$

76. The amount of waste recovered (in millions of tons) in the United States can be approximated by $R(t) = 0.04t^2 - 0.59t + 7.42$, where t is the number of years after 1960.

a. How many million tons were recovered in 1960?

b. How many million tons would you predict will be recovered in the year 2010?

77. Refer to Example 7.

a. Use the graph to find the BAL for a female after 2 hours.

b. Use the graph to find the BAL for a male after 2 hours.

c. Evaluate $y = -0.0257x + 0.1663$ for $x = 2$. Is the answer close to that of part **a?**

d. Evaluate $y = -0.0226x + 0.1509$ for $x = 2$. Is the answer close to that of part **b?**

Skill Checker

Find:

78. $-5ab + (2ab)$

79. $-3ab + (-4ab)$

80. $-8a^2b + (-5a^2b)$

81. $-3x^2y + 8x^2y - 2x^2y$

82. $-2xy^2 + 7xy^2 - 9xy^2$

83. $5xy^2 - (-3xy^2)$

84. $7x^2y - (-8x^2y)$

4.5 Addition and Subtraction of Polynomials

Objectives

A Add polynomials.

B Subtract polynomials.

C Find areas by adding polynomials.

D Solve applications involving polynomials.

To Succeed, Review How To . . .

1. Add and subtract like terms (pp. 78, 89–93).
2. Remove parentheses in expressions preceded by a minus sign (pp. 91–92).

Getting Started

Wasted Waste

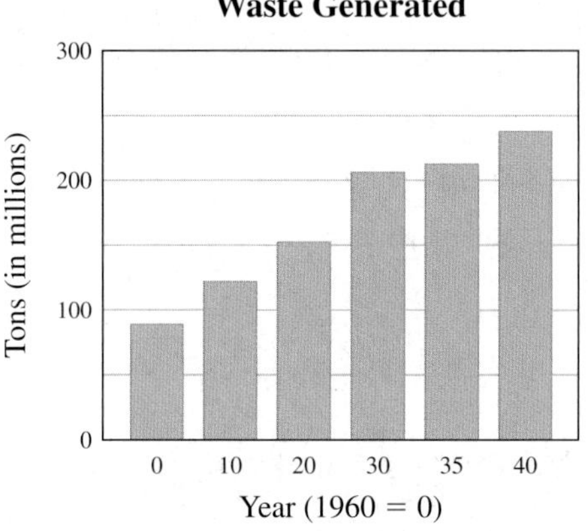

The annual amount of waste (in millions of tons) generated in the United States is approximated by $G(t) = -0.001t^3 + 0.06t^2 + 2.6t + 88.6$, where t is the number of years after 1960. How much of this waste is recovered? That amount can be approximated by $R(t) = 0.06t^2 - 0.59t + 6.4$. From these two approximations, we can estimate that the amount of waste actually "wasted" (not recovered) is $G(t) - R(t)$. To find this difference, we simply subtract like terms. To make the procedure more familiar, we write it in columns:

$$\begin{array}{rrrrr} \text{Generated: } G(t) = & -0.001t^3 & +\ 0.06t^2 & +\ 2.6t & +\ 88.6 \\ \text{Recovered: } \underline{(-)\ R(t)} = & \underline{(-)\qquad} & \underline{0.06t^2} & \underline{-\ 0.59t} & \underline{+\ 6.4} \\ & -0.001t^3 & + \uparrow & +\ 3.19t & +\ 82.2 \end{array}$$

Note that $0.06t^2 - 0.06t^2 = 0$

Thus, the amount of waste generated and *not* recovered is

$$G(t) - R(t) = -0.001t^3 + 3.19t + 82.2.$$

Let's see what this means in millions of tons. Since t is the number of years after 1960, $t = 0$ in 1960, and the amount of waste generated, the amount of waste recycled, and the amount of waste not recovered are as follows:

$$G(0) = -0.001(0)^3 + 0.06(0)^2 + 2.6(0) + 88.6 = 88.6 \quad \text{(million tons)}$$

$$R(0) = 0.06(0)^2 - 0.59(0) + 6.4 = 6.4 \quad \text{(million tons)}$$

$$G(0) - R(0) = 88.6 - 6.4 = 82.2 \quad \text{(million tons)}$$

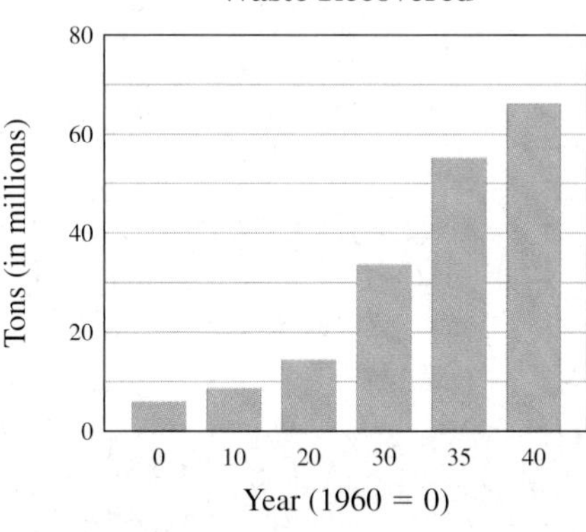

As you can see, there is much more material *not* recovered than material recovered. How can we find out whether the situation is changing? One way is to predict how much waste will be produced and how much recovered, say, in the year 2010. The amount can be approximated by $G(50) - R(50)$. Then we find out if, percentagewise, the situation is getting better. In 1960, the percent of materials recovered was 6.4/88.6, or about 7.2%. What percent would it be in the year 2010? In this section, we will learn how to add and subtract polynomials and use these ideas to solve applications.

A > Adding Polynomials

The addition of monomials and polynomials is a matter of combining like terms (monomials that contain the same variable raised to the same power). For example, suppose we wish to add $3x^2 + 7x - 3$ and $5x^2 - 2x + 9$; that is, we wish to find

$$(3x^2 + 7x - 3) + (5x^2 - 2x + 9)$$

Using the commutative, associative, and distributive properties, we write

$$\begin{aligned}(3x^2 + 7x - 3) + (5x^2 - 2x + 9) &= (3x^2 + 5x^2) + (7x - 2x) + (-3 + 9)\\ &= (3 + 5)x^2 + (7 - 2)x + (-3 + 9)\\ &= 8x^2 + 5x + 6\end{aligned}$$

Similarly, the sum of $4x^3 + \frac{3}{7}x^2 - 2x + 3$ and $6x^3 - \frac{1}{7}x^2 + 9$ is written as

$$\begin{aligned}&\left(4x^3 + \frac{3}{7}x^2 - 2x + 3\right) + \left(6x^3 - \frac{1}{7}x^2 + 9\right)\\ &\qquad = (4x^3 + 6x^3) + \left(\frac{3}{7}x^2 - \frac{1}{7}x^2\right) + (-2x) + (3 + 9)\\ &\qquad = 10x^3 + \frac{2}{7}x^2 - 2x + 12\end{aligned}$$

In both examples, the polynomials have been written in descending order for convenience in combining like terms.

EXAMPLE 1 **Adding polynomials**

Add: $3x + 7x^2 - 7$ and $-4x^2 + 9 - 3x$

SOLUTION 1 We first write both polynomials in descending order and then combine like terms to obtain

$$\begin{aligned}(7x^2 + 3x - 7) + (-4x^2 - 3x + 9) &= (7x^2 - 4x^2) + (3x - 3x) + (-7 + 9)\\ &= 3x^2 + 0 + 2\\ &= 3x^2 + 2\end{aligned}$$

PROBLEM 1

Add: $5x + 8x^2 - 3$ and $-3x^2 + 8 - 5x$

As in arithmetic, the addition of polynomials can be done by writing the polynomials in descending order and then placing like terms in columns. In arithmetic, you add 345 and 678 by writing the numbers in a column:

$$\begin{array}{r} +345\\ \underline{+678}\end{array}$$

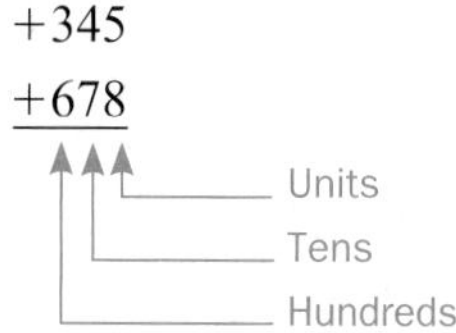

Thus, to add $4x^3 + 3x - 7$ and $7x - 3x^3 + x^2 + 9$, we first write both polynomials in descending order with like terms in the same column, leaving space for any missing terms. We then add the terms in each of the columns:

The x^2 term is missing in $4x^3 + 3x - 7$.

$$\begin{array}{r} 4x^3 \phantom{{}+x^2} + 3x - 7\\ \underline{-3x^3 + x^2 + 7x + 9}\\ x^3 + x^2 + 10x + 2\end{array}$$

Answers to PROBLEMS

1. $5x^2 + 5$

EXAMPLE 2 **Adding polynomials**

Add: $-3x + 7x^2 - 2$ and $-4x^2 - 3 + 5x$

SOLUTION 2 We first write both polynomials in descending order, place like terms in a column, and then add as shown:

$$\begin{array}{r} 7x^2 - 3x - 2 \\ -4x^2 + 5x - 3 \\ \hline 3x^2 + 2x - 5 \end{array}$$

Horizontally, we write:

$$\begin{aligned} &(7x^2 - 3x - 2) + (-4x^2 + 5x - 3) \\ &= \underbrace{(7x^2 - 4x^2)} + \underbrace{(-3x + 5x)} + \underbrace{(-2 - 3)} \\ &= 3x^2 + 2x + (-5) \\ &= 3x^2 + 2x - 5 \end{aligned}$$

PROBLEM 2

Add: $-5y + 8y^2 - 3$ and $-5y^2 - 4 + 6y$

B > Subtracting Polynomials

To subtract polynomials, we first recall that

$$a - (b + c) = a - b - c$$

To remove the parentheses from an expression preceded by a minus sign, we must change the sign of each term *inside* the parentheses. This is the same as multiplying each term inside the parentheses by -1. Thus,

$$\begin{aligned} (3x^2 - 2x + 1) - (4x^2 + 5x + 2) &= 3x^2 - 2x + 1 - 4x^2 - 5x - 2 \\ &= (3x^2 - 4x^2) + (-2x - 5x) + (1 - 2) \\ &= -x^2 + (-7x) + (-1) \\ &= -x^2 - 7x - 1 \end{aligned}$$

Here's how we do it using columns:

$$\begin{array}{r} 3x^2 - 2x + 1 \\ (-)\ 4x^2 + 5x + 2 \\ \hline \end{array} \quad \xrightarrow{\text{is written}} \quad \begin{array}{r} 3x^2 - 2x + 1 \\ (+) -4x^2 - 5x - 2 \\ \hline -x^2 - 7x - 1 \end{array}$$

Note that we changed the sign of *every* term in $4x^2 + 5x + 2$ and wrote $-4x^2 - 5x - 2$.

NOTE

"Subtract b from a" means to find $a - b$.

EXAMPLE 3 **Subtracting polynomials**

Subtract $4x - 3 + 7x^2$ from $5x^2 - 3x$.

SOLUTION 3 We first write the problem in columns, then change the signs and add:

$$\begin{array}{r} 5x^2 - 3x \\ (-)7x^2 + 4x - 3 \\ \hline \end{array} \quad \xrightarrow{\text{is written}} \quad \begin{array}{r} 5x^2 - 3x \\ (+)-7x^2 - 4x + 3 \\ \hline -2x^2 - 7x + 3 \end{array}$$

Thus, the answer is $-2x^2 - 7x + 3$.

PROBLEM 3

Subtract $5y - 4 + 8y^2$ from $6y^2 - 4y$.

Answers to PROBLEMS

2. $3y^2 + y - 7$
3. $-2y^2 - 9y + 4$

To do it horizontally, we write

$(5x^2 - 3x) - (7x^2 + 4x - 3)$

$= 5x^2 - 3x - 7x^2 - 4x + 3$ Change the sign of every term in $7x^2 + 4x - 3$.

$= (5x^2 - 7x^2) + (-3x - 4x) + 3$ Use the commutative and associative properties.

$= -2x^2 - 7x + 3$

Just as in arithmetic, we can add or subtract more than two polynomials. For example, to add the polynomials $-7x + x^2 - 3$, $6x^2 - 8 + 2x$, and $3x - x^2 + 5$, we simply write each of the polynomials in descending order with like terms in the same column and add:

$$\begin{array}{r} x^2 - 7x - 3 \\ 6x^2 + 2x - 8 \\ \underline{-x^2 + 3x + 5} \\ 6x^2 - 2x - 6 \end{array}$$

Or, horizontally, we write

$$\begin{aligned} &(x^2 - 7x - 3) + (6x^2 + 2x - 8) + (-x^2 + 3x + 5) \\ &= (x^2 + 6x^2 - x^2) + (-7x + 2x + 3x) + (-3 - 8 + 5) \\ &= 6x^2 + (-2x) + (-6) \\ &= 6x^2 - 2x - 6 \end{aligned}$$

EXAMPLE 4 Adding polynomials

Add: $x^3 + 2x - 3x^2 - 5$, $-8 + 2x - 5x^2$, and $7x^3 - 4x + 9$

SOLUTION 4 We first write all the polynomials in descending order with like terms in the same column and then add:

$$\begin{array}{rrrr} x^3 & - 3x^2 & + 2x & - 5 \\ & - 5x^2 & + 2x & - 8 \\ \underline{7x^3} & & \underline{- 4x} & \underline{+ 9} \\ 8x^3 & - 8x^2 & & - 4 \end{array}$$

Horizontally, we have

$$\begin{aligned} &(x^3 - 3x^2 + 2x - 5) + (-5x^2 + 2x - 8) + (7x^3 - 4x + 9) \\ &= (x^3 + 7x^3) + (-3x^2 - 5x^2) + (2x + 2x - 4x) + (-5 - 8 + 9) \\ &= 8x^3 + (-8x^2) + 0x + (-4) \\ &= 8x^3 - 8x^2 - 4 \end{aligned}$$

PROBLEM 4

Add: $y^3 + 3y - 4y^2 - 6$, $-9 + 3y - 6y^2$, and $6y^3 - 5y + 8$

C > Finding Areas

Addition of polynomials can be used to find the sum of the areas of several rectangles. To find the total area of the shaded rectangles, add the individual areas.

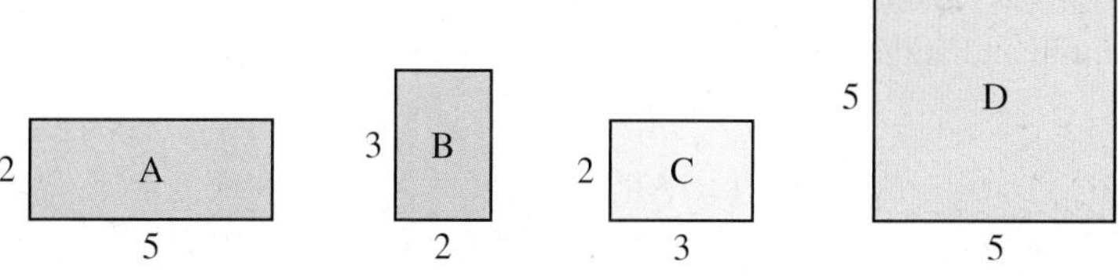

Answers to PROBLEMS

4. $7y^3 - 10y^2 + y - 7$

Since the area of a rectangle is the product of its length and its width, we have

$$\underbrace{5 \cdot 2}_{\text{Area of A}} + \underbrace{2 \cdot 3}_{\text{Area of B}} + \underbrace{3 \cdot 2}_{\text{Area of C}} + \underbrace{5 \cdot 5}_{\text{Area of D}}$$
$$= 10 + 6 + 6 + 25$$

Thus, the total area is

$$10 + 6 + 6 + 25 = 47 \qquad \text{(square units)}$$

This same procedure can be used when some of the lengths are represented by variables, as shown in Example 5.

EXAMPLE 5 **Finding sums of areas**

Find the sum of the areas of the shaded rectangles:

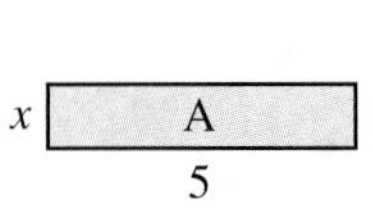

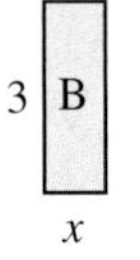

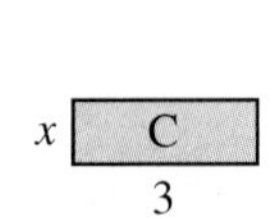

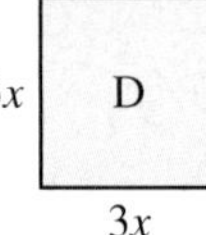

SOLUTION 5 The total area in square units is

$$\underbrace{\underbrace{5x}_{\text{Area of A}} + \underbrace{3x}_{\text{Area of B}} + \underbrace{3x}_{\text{Area of C}}}_{11x} + \underbrace{(3x)^2}_{\text{Area of D}}$$
$$11x + 9x^2$$

or

$$9x^2 + 11x \qquad \text{In descending order}$$

PROBLEM 5

Find the sum of the areas of the shaded rectangles:

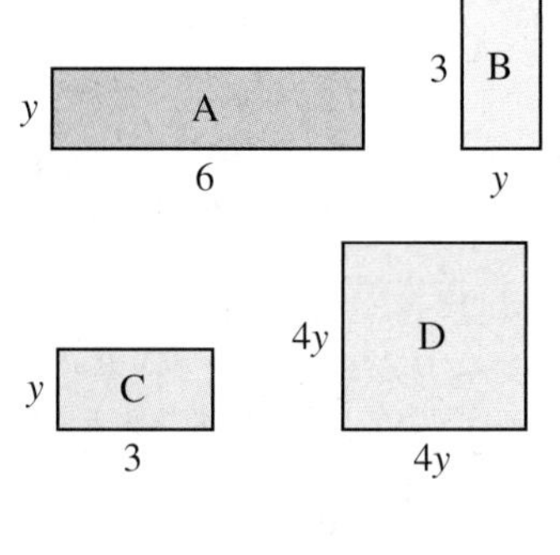

D > Applications Involving Polynomials

In the *Getting Started* section, we subtracted polynomials dealing with garbage production. What about using the addition of polynomials to explore recycling? We do that next.

GREEN MATH

EXAMPLE 6 **Garbage Recycling and Composting**

In a recent year, 254 million tons of municipal solid waste was produced in the United States. How many tons were recovered for recycling? The binomials, $R(t) = 2.25t + 59$ and $C(t) = 0.55t + 20$ represent the amounts of materials recovered for recycling $R(t)$ and composting $C(t)$, where t is the number of years after 2000. If we **add** these two binomials we will know the answer!

a. Add $R(t)$ and $C(t)$.

b. Predict the number of tons of materials recovered for recycling in 2015.

SOLUTION 6

a. $R(t)$ and $C(t)$ are already written in descending order, so we place like terms in a column and add as shown.

$$\begin{aligned} R(t) &= 2.25t + 59 \\ (+)\ C(t) &= \underline{0.55t + 20} \\ &\quad\ 2.80t + 79 \end{aligned}$$

PROBLEM 6

Two of the most common materials recovered for recycling are paper/paperboard and aluminum.

$$P(t) = 0.09t^2 + 0.4t + 38$$

where t is the number of years after 2000, represents the millions of tons of paper/paperboard recovered after 2000 and $A(t) = 0.001t^2 - 0.07t + 0.86$ represents the millions of tons of aluminum recovered after 2000.

a. Express the total amount of paper/paperboard and aluminum recovered after 2000.

Answers to PROBLEMS

5. $16y^2 + 12y$ **6. a.** $0.091t^2 + 0.33t + 38.86$ **b.** 64.285 million tons

Thus, the binomial representing the total amount of solid waste recovered for recycling is $2.80t + 79$ (million tons).

b. To predict the number of tons of materials recovered for recycling in 2015 (15 years after 2000), we let $t = 15$ in $2.80t + 79$ obtaining:

$$2.80(15) + 79 = 121 \text{ million tons}$$

This means that 121 million tons of materials are predicted to be recovered in 2015.

b. Predict the amount of paper/paperboard and aluminum to be recovered in 2015.

Note: Recycling paper conserves resources, saves energy, and creates jobs.

EXAMPLE 7 **Blood alcohol level**

In Example 7 of Section 4.4, we introduced the polynomial equations $y = -0.0226x + 0.1509$ and $y = -0.0257x + 0.1663$ approximating the blood alcohol level (BAL) for a 150-pound male or female, respectively. In these equations, x represents the time since consuming 3 ounces of alcohol.

It is known that the burn-off rate of alcohol is 0.015 per hour (that is, the BAL is reduced by 0.015 per hour if no additional alcohol is consumed). Find a polynomial that would approximate the BAL for a male x hours after consuming the 3 ounces of alcohol.

SOLUTION 7 The initial BAL is $-0.0226x + 0.1509$, but this level is *decreased* by 0.015 each hour. Thus, the actual BAL after x hours is $-0.0226x + 0.1509 - 0.015x$, or $-0.0376x + 0.1509$.

PROBLEM 7

Find a polynomial that would approximate the BAL for a female x hours after consuming 3 ounces of alcohol.

› Exercises 4.5

> Practice Problems > Self-Tests > Media-rich eBooks > e-Professors > Videos

‹ A › Adding Polynomials In Problems 1–30, add as indicated.

1. $(5x^2 + 2x + 5) + (7x^2 + 3x + 1)$

2. $(3x^2 - 5x - 5) + (9x^2 + 2x + 1)$

3. $(-3x + 5x^2 - 1) + (-7 + 2x - 7x^2)$

4. $(3 - 2x^2 + 7x) + (-6 + 2x^2 - 5x)$

5. $(2x + 5x^2 - 2) + (-3 + 5x - 8x^2)$

6. $-3x - 2 + 3x^2$ and $-4 + 5x - 6x^2$

7. $-2 + 5x$ and $-3 - x^2 - 5x$

8. $-4x + 2 - 6x^2$ and $2 + 5x$

9. $x^3 - 2x + 3$ and $-2x^2 + x - 5$

10. $x^4 - 3 + 2x - 3x^3$ and $3x^4 - 2x^2 + 5 - x$

11. $-6x^3 + 2x^4 - x$ and $2x^2 + 5 + 2x - 2x^3$

12. $\frac{1}{2}x^3 + x^2 - \frac{1}{5}x$ and $\frac{3}{5}x + \frac{1}{2}x^3 - 3x^2$

13. $\frac{1}{3} - \frac{2}{5}x^2 + \frac{3}{4}x$ and $\frac{1}{4}x - \frac{1}{5}x^2 + \frac{2}{3}$

14. $0.3x - 0.1 - 0.4x^2$ and $0.1x^2 - 0.1x + 0.6$

15. $0.2x - 0.3 + 0.5x^2$ and $-\frac{1}{10} + \frac{1}{10}x - \frac{1}{10}x^2$

16.
$$\begin{array}{r} -x^2 + 5x + 2 \\ (+)\ 3x^2 - 7x - 2 \\ \hline \end{array}$$

17.
$$\begin{array}{r} -3x^2 + 2x - 4 \\ (+)\quad x^2 - 4x + 7 \\ \hline \end{array}$$

18.
$$\begin{array}{r} -2x^4 \qquad\quad + 2x - 1 \\ (+)\qquad\quad - x^3 - 3x + 5 \\ \hline \end{array}$$

Web IT go to mhhe.com/bello for more lessons

Answers to PROBLEMS

7. $-0.0407x + 0.1663$

Web IT go to **mhhe.com/bello** for more lessons

19. $$\begin{array}{r} 3x^4 \qquad - 3x + 4 \\ (+) \qquad x^3 - 2x - 5 \\ \hline \end{array}$$

20. $$\begin{array}{r} -3x^4 \qquad + 2x^2 - x + 5 \\ (+) \qquad - 2x^3 \qquad + 5x - 7 \\ \hline \end{array}$$

21. $$\begin{array}{r} -5x^4 \qquad - 5x^2 + 3 \\ (+) \qquad 5x^3 + 3x^2 - 5 \\ \hline \end{array}$$

22. $$\begin{array}{r} 3x^3 \qquad + x - 1 \\ x^2 - 2x + 5 \\ (+)\ 5x^3 \qquad - x \\ \hline \end{array}$$

23. $$\begin{array}{r} 5x^3 - x^2 \qquad - 3 \\ 5x + 9 \\ (+) \qquad - 3x^2 \qquad - 7 \\ \hline \end{array}$$

24. $$\begin{array}{r} -\frac{1}{3}x^3 \qquad - \frac{1}{2}x + 5 \\ -\frac{1}{5}x^2 + \frac{1}{2}x - 1 \\ (+)\ \frac{2}{3}x^3 \qquad + x - 2 \\ \hline \end{array}$$

25. $$\begin{array}{r} -\frac{2}{7}x^3 + \frac{1}{6}x^2 \qquad + 2 \\ \frac{1}{7}x^3 \qquad + 5x - 3 \\ (+) \qquad -\frac{5}{6}x^2 \qquad + 1 \\ \hline \end{array}$$

26. $$\begin{array}{r} -\frac{1}{8}x^2 - \frac{1}{3}x + \frac{1}{5} \\ -x^3 + \frac{3}{8}x^2 \qquad - \frac{2}{5} \\ (+)\ -3x^3 \qquad - \frac{2}{3}x + \frac{4}{5} \\ \hline \end{array}$$

27. $$\begin{array}{r} -\frac{1}{7}x^3 \qquad + 2 \\ -\frac{1}{9}x^2 - x - 3 \\ (+)\ -\frac{2}{7}x^3 + \frac{2}{9}x^2 + 2x - 5 \\ \hline \end{array}$$

28. $$\begin{array}{r} -2x^4 + 5x^3 - 2x^2 + 3x - 5 \\ 8x^3 \qquad - 2x + 5 \\ -x^4 \qquad + 3x^2 - x - 2 \\ (+) \qquad 6x^3 \qquad + 2x + 5 \\ \hline \end{array}$$

29. $$\begin{array}{r} - 6x^3 + 2x^2 \qquad + 1 \\ -x^4 + 3x^3 - 5x^2 + 3x \\ - x^3 \qquad - 7x + 2 \\ (+)\ -3x^4 \qquad + 3x - 1 \\ \hline \end{array}$$

30. $$\begin{array}{r} - 3x^4 \qquad + 2x^2 \qquad - 5 \\ x^5 + x^4 - 2x^3 + 7x^2 + 5x \\ 2x^4 \qquad - 2x^2 \qquad + 7 \\ (+)\ 7x^5 \qquad + 2x^3 \qquad - 2x \\ \hline \end{array}$$

⟨ B ⟩ Subtracting Polynomials

In Problems 31–50, subtract as indicated.

31. $(7x^2 + 2) - (3x^2 - 5)$

32. $(8x^2 - x) - (7x^2 + 3x)$

33. $(3x^2 - 2x - 1) - (4x^2 + 2x + 5)$

34. $(-3x + x^2 - 1) - (5x + 1 - 3x^2)$

35. $(-1 + 7x^2 - 2x) - (5x + 3x^2 - 7)$

36. $(7x^3 - x^2 + x - 1) - (2x^2 + 3x + 6)$

37. $(5x^2 - 2x + 5) - (3x^3 - x^2 + 5)$

38. $(3x^2 - x - 7) - (5x^3 + 5 - x^2 + 2x)$

39. $(6x^3 - 2x^2 - 3x + 1) - (-x^3 - x^2 - 5x + 7)$

40. $(x - 3x^2 + x^3 + 9) - (-8 + 7x - x^2 + x^3)$

41. $$\begin{array}{r} 6x^2 - 3x + 5 \\ (-)\ 3x^2 + 4x - 2 \\ \hline \end{array}$$

42. $$\begin{array}{r} 7x^2 + 4x - 5 \\ (-)\ 9x^2 - 2x + 5 \\ \hline \end{array}$$

43. $$\begin{array}{r} 3x^2 - 2x - 1 \\ (-)\ 3x^2 - 2x - 1 \\ \hline \end{array}$$

44. $$\begin{array}{r} 5x^2 \qquad - 1 \\ (-)\ 3x^2 - 2x + 1 \\ \hline \end{array}$$

45. $$\begin{array}{r} 4x^3 \qquad - 2x + 5 \\ (-) \qquad 3x^2 + 5x - 1 \\ \hline \end{array}$$

46. $$\begin{array}{r} - 3x^2 + 5x - 2 \\ (-)\ x^3 - 2x^2 \qquad + 5 \\ \hline \end{array}$$

47. $$\begin{array}{r} 3x^3 \qquad - 2 \\ (-) \qquad 2x^2 - x + 6 \\ \hline \end{array}$$

48. $$\begin{array}{r} x^2 - 2x + 1 \\ (-)\ -3x^3 + x^2 + 5x - 2 \\ \hline \end{array}$$

49. $$\begin{array}{r} -5x^3 \qquad + x - 2 \\ (-) \qquad 5x^2 - 3x + 7 \\ \hline \end{array}$$

50. $$\begin{array}{r} 6x^3 \qquad + 2x - 5 \\ (-) \qquad - 3x^2 - x \\ \hline \end{array}$$

⟨ C ⟩ Finding Areas

In Problems 51–55, find the sum of the areas of the shaded rectangles.

51.

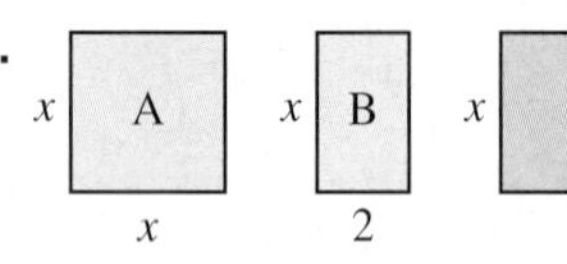

52.

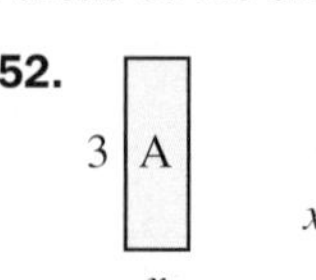

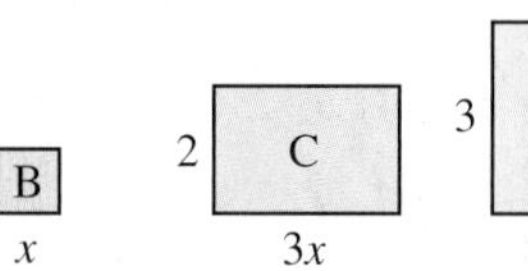

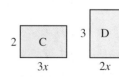

53.

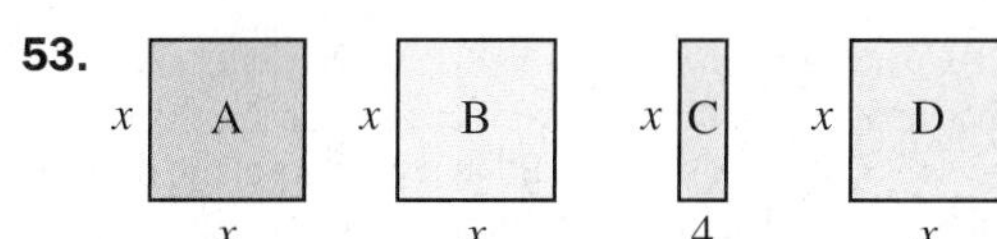

54.

x A 4 x B 4 2x C 2x x D 4

55.

3x A 3x 2x B 5 3x C 3x 3x D 3x

‹D› Applications Involving Polynomials

››› Applications: Green Math

How many miles per gallon (mpg) does your car give? Maybe it will make a difference if you buy a more efficient car or truck. The mileage for new cars and trucks is improving, but you have to be careful because there are *two* estimates for mileage: the **EPA** (Environmental Protection Agency) version and the **revised** for online performance version. What is the difference between them? We will see in Problems 56 and 57.

56. *Light truck mileage predictions* Find the mileage predictions for the year 2010 using:

a. The EPA standard $\mathbf{E(t) = -0.03t^2 + 1.13t + 27}$, where t is the number of years after 2010 ($t = 0$).

b. The revised standard $\mathbf{R(t) = -0.02t^2 + 0.9t + 23}$ for $t = 0$.

c. Which gives you a better mileage, $E(t)$ or $R(t)$?

d. Find the polynomial difference $E(t) - R(t)$.

e. Find the mileage difference for the year 2030 using $E(20) - R(20)$. Use the polynomial of part **d** to find the mileage when $t = 20$.

Note: Some critics of the study complain that the EPA predictions are too high because they are done in the lab rather than under regular road conditions.

57. *New car mileage predictions* Find the predicted car mileage for the year 2010 using:

a. The EPA standard $\mathbf{E(t) = -0.03t^2 + 1.1t + 31}$, where t is the number of years after 2010 ($t = 0$).

b. The revised standard $\mathbf{R(t) = -0.02t^2 + 0.94t + 25}$ for $t = 0$.

c. Which gives a better mileage, $E(t)$ or $R(t)$?

d. Find the polynomial difference $E(t) - R(t)$.

e. Find the mileage difference for the year 2030 using $E(20) - R(20)$. Use the polynomial of part **d** to find the mileage when $t = 20$.

Source: The *Annual Energy Outlook*, Energy Information Administration, Table A7.

58. *College costs* How much are you paying for tuition and fees? In a four-year public institution, the amount $T(t)$ you pay for tuition and fees (in dollars) can be approximated by $T(t) = 45t^2 + 110t + 3356$, where t is the number of years after 2000 ($2000 = 0$).

a. What would you predict tuition and fees to be in 2005?

b. The cost of books t years after 2000 can be approximated by $B(t) = 27.5t + 680$. What would be the cost of books in 2005?

c. What polynomial would represent the cost of tuition and fees and books t years after 2000?

d. What would you predict the cost of tuition and fees and books would be in 2015?

59. *College expenses* The three major college expenses are: tuition and fees, books, and room and board. They can be approximated, respectively, by:

$$T(t) = 45t^2 + 110t + 3356$$
$$B(t) = 27.5t + 680$$
$$R(t) = 32t^2 + 200t + 4730$$

where t is the number of years after 2000 ($2000 = 0$).

a. Write a polynomial representing the total cost of tuition and fees, books, and room and board t years after 2000.

b. What was the cost of tuition and fees, books, and room and board in 2000?

c. What would you predict the cost of tuition and fees, books, and room and board would be in 2015?

60. *Student loans* If you are an undergraduate dependent student you can apply for a Stafford Loan. The amount of these loans can be approximated by $S(t) = 562.5t^2 + 312.5t + 2625$, where t is between 0 and 2 inclusive. If t is between 3 and 5 inclusive, then $S(t) = \$5500$.

a. How much money can you get from a Stafford Loan the first year?

b. What about the second year?

c. What about the fifth year?

61. *College financial aid* Assume that you have to pay tuition and fees, books, and room and board but have your Stafford Loan to decrease expenses. (See Problems 59 and 60.) Write a polynomial that would approximate how much you would have to pay (t between 0 and 2) in your first two years if you start school in 2000.

62. *Dental services* How much do you spend in dental services? The amount can be approximated by $D(t) = -1.8t^2 + 65t + 592$ (in dollars), where t is the number of years after 2000. What about doctors and clinical services? They are more expensive and can be approximated by $C(t) = 4.75t^2 + 35t + 602$ (in dollars).

a. The total annual amount spent on dental and doctors is the sum $D(t) + C(t)$. Find this sum.

b. What was the total amount spent on dental and doctors in 2000?

c. Predict the expenditures on dental and doctor services in 2010.

Source: http://www.census.gov/., Table 121.

63. *Annual expenses for medical services and medicines* The annual amount spent on medical services can be approximated by $M(t) = -t^2 + 12t + 567$ (in dollars), where t is the number of years after 2000. The amount spent on drugs and medical supplies can be approximated by $D(t) = -13t^2 + 61t + 511$ (in dollars).

a. The total amount spent on medical services and medicines is $M(t) + D(t)$. Find this sum.

b. What was the amount spent on medical services and medicines in 2000?

c. Predict the amount spent on medical services and medicines in 2010.

64. *Health expenditures* What are the annual national expenditures for health? They can be approximated by $E(t) = 0.5t^2 + 122t + 1308$ (in billion dollars), where t is the number of years after 2000. Of these, $P(t) = -1.8t^2 + 65t + 592$ are public expenditures and the rest are private.

a. What were the expenditures in 2000?

b. What were the public expenditures in 2000?

c. The private expenditures can be represented by $E(t) - P(t)$. Find this difference.

d. What were the private expenditures in 2000?

e. Predict the private expenditures for 2010.

Source: http://www.census.gov/., Table 118.

65. *Annual wages* According to the Bureau of Labor, the annual wages and salaries (in thousands of dollars) for persons 25–34 years old can be approximated by $W(t) = 0.3t^3 - 2t^2 + 5t + 43$, where t is the number of years after 2000. The federal income tax paid on those wages can be approximated by $T(t) = t^2 - t + 3$, where t is the number of years after 2000.

a. The wages after taxes is the difference of $W(t)$ and $T(t)$. Find this difference.

b. What are the estimated wages after taxes for 2000? For 2010?

Source: Bureau of Labor Consumer Expenditure Survey, http://www.bls.gov/.

66. *Annual wages* According to the Bureau of Labor, the annual wages and salaries (in thousands of dollars) for persons under 25 years old can be approximated by $W(t) = 0.12t^3 - 0.6t^2 + t + 17$, where t is the number of years after 2000. The federal income tax paid on those wages can be approximated by $T(t) = -0.03t^3 + 0.22t^2 - 0.5t + 0.70$, where t is the number of years after 2000.

a. The wages after taxes is the difference of $W(t)$ and $T(t)$. Find this difference.

b. What are the estimated wages after taxes for 2000? For 2010?

Source: Bureau of Labor Consumer Expenditure Survey, http://www.bls.gov/.

Using Your Knowledge

Business Polynomials Polynomials are also used in business and economics. For example, the revenue R may be obtained by subtracting the cost C of the merchandise from its selling price S. In symbols, this is

$$R = S - C$$

Now the cost C of the merchandise is made up of two parts: the *variable cost* per item and the *fixed cost.* For example, if you decide to manufacture Frisbees™, you might spend \$2 per Frisbee in materials, labor, and so forth. In addition, you might have \$100 of fixed expenses. Then the cost for manufacturing x Frisbees is

$$\underbrace{\text{Cost } C \text{ of merchandise}}_{C} \; \underbrace{\text{is}}_{=} \; \underbrace{\text{Frisbee}}_{2x}^{\text{cost per}} \; \underbrace{\text{and}}_{+} \; \underbrace{\text{expenses.}}_{100}^{\text{fixed}}$$

If x Frisbees are then sold for \$3 each, the total selling price S is $3x$, and the revenue R would be

$$\begin{aligned} R &= S - C \\ &= 3x - (2x + 100) \\ &= 3x - 2x - 100 \\ &= x - 100 \end{aligned}$$

Thus, if the selling price S is \$3 per Frisbee, the variable costs are \$2 per Frisbee, and the fixed expenses are \$100, the revenue after selling x Frisbees is given by

$$R = x - 100$$

In Problems 67–69, find the revenue R for the given cost C and selling price S.

67. $C = 3x + 50;\ S = 4x$

68. $C = 6x + 100;\ S = 8x$

69. $C = 7x;\ S = 9x$

70. In Problem 68, how many items were sold if the revenue was zero?

71. If the merchant of Problem 68 suffered a \$40 loss (−\$40 revenue), how many items were sold?

Write On

72. Write the procedure you use to add polynomials.

73. Write the procedure you use to subtract polynomials.

74. Explain the difference between "subtract $x^2 + 3x - 5$ from $7x^2 - 2x + 9$" and "subtract $7x^2 - 2x + 9$ from $x^2 + 3x - 5$." What is the answer in each case?

75. List the advantages and disadvantages of adding (or subtracting) polynomials horizontally or in columns.

Concept Checker

Fill in the blank(s) with the correct word(s), phrase, or mathematical statement.

76. $a - (b + c) =$ ______________.

77. To **subtract** b **from** a means to find ______________.

$a - b + c$ $\quad$ $b - a$

$a - b - c$ $\quad$ $a - b$

Mastery Test

Add:

78. $3x + 3x^2 - 6$ and $-5x^2 + 10 - x$

79. $-5x + 8x^2 - 3$ and $-3x^2 + 4 + 8x$

Subtract:

80. $3 - 4x^2 + 5x$ from $9x^2 - 2x$

81. $9 + x^3 - 3x^2$ from $10 + 7x^2 + 5x^3$

82. Add $2x^3 + 3x - 5x^2 - 2$, $-6 + 5x - 2x^2$, and $6x^3 - 2x + 8$.

83. Find the sum of the areas of the shaded rectangles:

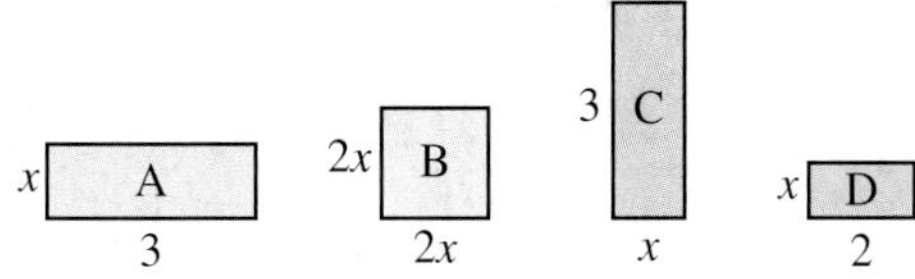

84. The number of robberies (per 100,000 population) can be approximated by $R(t) = 1.85t^2 - 19.14t + 262$, while the number of aggravated assaults is approximated by $A(t) = -0.2t^3 + 4.7t^2 - 15t + 300$, where t is the number of years after 1960.

a. Were there more aggravated assaults or more robberies per 100,000 in 1960?

b. Find the difference between the number of aggravated assaults and the number of robberies per 100,000.

c. What would this difference be in the year 2000? In 2010?

Skill Checker

Simplify:

85. $(-3x^2) \cdot (2x^3)$

86. $(-5x^3) \cdot (2x^4)$

87. $(-2x^4) \cdot (3x^5)$

88. $5(x - 3)$

89. $6(y - 4)$

90. $-3(2y - 3)$

4.6 Multiplication of Polynomials

Objectives

A › Multiply two monomials.

B › Multiply a monomial and a binomial.

C › Multiply two binomials using the FOIL method.

D › Solve an application involving multiplication of polynomials.

To Succeed, Review How To . . .

1. Multiply expressions (pp. 319–321).
2. Use the distributive property to remove parentheses in an expression (pp. 81–83, 91–92).

Getting Started

Deflections on a Bridge

How much does the beam bend (deflect) when a car or truck goes over the bridge? There's a formula that can tell us. For a certain beam of length L, the deflection at a distance x from one end is given by

$$(x - L)(x - 2L)$$

To multiply these two binomials, we must first learn how to do several related types of multiplication.

A › Multiplying Two Monomials

We already multiplied two monomials in Section 4.1. The idea is to use the associative and commutative properties and the rules of exponents, as shown in Example 1.

EXAMPLE 1 Multiplying two monomials

Multiply: $(-3x^2)$ by $(2x^3)$

SOLUTION 1

$(-3x^2)(2x^3) = (-3 \cdot 2)(x^2 \cdot x^3)$ Use the associative and commutative properties.

$= -6x^{2+3}$ Use the rules of exponents.

$= -6x^5$

PROBLEM 1

Multiply: $(-4y^3)$ by $(5y^4)$

B › Multiplying a Monomial and a Binomial

In Sections 1.6 and 1.7, we also multiplied $a(b + c)$, a monomial and a binomial. The procedure was based on the distributive property, as shown next.

EXAMPLE 2 Multiplying a monomial by a binomial

Remove parentheses (simplify):

a. $5(x - 2y)$ **b.** $(x^2 + 2x)3x^4$

PROBLEM 2

Simplify:

a. $4(a - 3b)$ **b.** $(a^2 + 3a)4a^5$

Answers to PROBLEMS

1. $-20y^7$

2. a. $4a - 12b$ **b.** $4a^7 + 12a^6$

SOLUTION 2

a. $5(x - 2y) = 5x - 5 \cdot 2y$

$= 5x - 10y$

b. $(x^2 + 2x)3x^4 = x^2 \cdot 3x^4 + 2x \cdot 3x^4$ Since $(a + b)c = ac + bc$

$= 3 \cdot x^2 \cdot x^4 + 2 \cdot 3 \cdot x \cdot x^4$

$= 3x^6 + 6x^5$

NOTE

You can use the commutative property first and write

$$(x^2 + 2x)3x^4 = 3x^4(x^2 + 2x)$$
$$= 3x^6 + 6x^5 \quad \text{Same answer!}$$

C > Multiplying Two Binomials Using the FOIL Method

Another way to multiply $(x + 2)(x + 3)$ is to use the distributive property $a(b + c) = ab + ac$. Think of $x + 2$ as a, which makes x like b and 3 like c. Here's how it's done.

$$\underbrace{(x + 2)}_{a}\underbrace{(x + 3)}_{(b + c)} = \underbrace{(x + 2)}_{a}\underbrace{x}_{b} + \underbrace{(x + 2)}_{a}\underbrace{3}_{c}$$
$$= x \cdot x + 2 \cdot x + x \cdot 3 + 2 \cdot 3$$
$$= x^2 + 2x + 3x + 6$$
$$= x^2 + 5x + 6$$

Similarly,

$$(x - 3)(x + 5) = (x - 3)x + (x - 3)5$$
$$= x \cdot x + (-3) \cdot x + x \cdot 5 + (-3) \cdot 5$$
$$= x^2 - 3x + 5x - 15$$
$$= x^2 + 2x - 15$$

Can you see a pattern developing? Look at the answers:

$$(x + 2)(x + 3) = x^2 + 5x + 6$$
$$(x - 3)(x + 5) = x^2 + 2x - 15$$

It seems that the *first* term in each answer (x^2) is obtained by multiplying the *first* terms in the factors (x and x). Similarly, the *last* terms (6 and -15) are obtained by multiplying the *last* terms ($2 \cdot 3$ and $-3 \cdot 5$). Here's how it works so far:

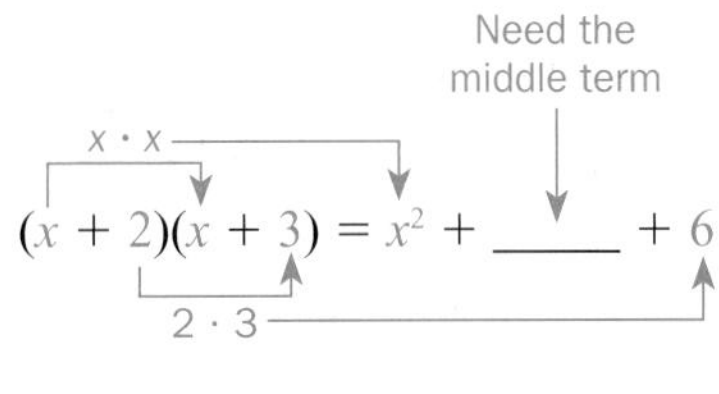

Need the middle term

$x \cdot x$

$(x - 3)(x + 5) = x^2 + \underline{\quad\quad} - 15$

$-3 \cdot 5$

But what about the *middle* terms? In $(x + 2)(x + 3)$, the middle term is obtained by adding $3x$ and $2x$, which is the same as the result we got when we multiplied the *outer* terms (x and 3) and added the product of the inner terms (2 and x). Here's a diagram that shows how the middle term is obtained:

Outer terms: $x \cdot 3$

$$(x + 2)(x + 3) = x^2 + 3x + 2x + 6$$

Inner terms: $2 \cdot x$

Outer terms: $x \cdot 5$

$$(x - 3)(x + 5) = x^2 + 5x - 3x - 15$$

Inner terms: $-3 \cdot x$

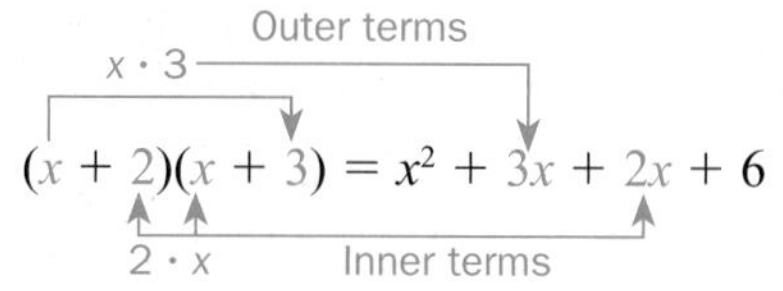

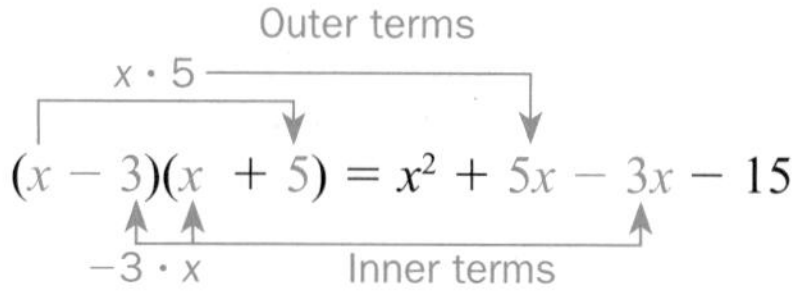

Do you see how it works now? Here is a summary of this method.

PROCEDURE

FOIL Method for Multiplying Binomials

First terms are multiplied first.

Outer terms are multiplied second.

Inner terms are multiplied third.

Last terms are multiplied last.

Of course, we call this method the **FOIL method.** We shall do one more example, step by step, to give you additional practice.

F	$(x + 7)(x - 4) \to x^2$	First: $x \cdot x$
O	$(x + 7)(x - 4) \to x^2 - 4x$	Outer: $-4 \cdot x$
I	$(x + 7)(x - 4) \to x^2 - 4x + 7x$	Inner: $7 \cdot x$
L	$(x + 7)(x - 4) = x^2 - 4x + 7x - 28$	Last: $7 \cdot (-4)$
	$= x^2 + 3x - 28$	

EXAMPLE 3 Using FOIL to multiply two binomials

Use FOIL to multiply:

a. $(x + 5)(x - 2)$ **b.** $(x - 4)(x + 3)$

SOLUTION 3

(First) (Outer) (Inner) (Last)
F O I L

a. $(x + 5)(x - 2) = \underbrace{x \cdot x}_{} - \underbrace{2x + 5x}_{} - \underbrace{5 \cdot 2}_{}$

$= x^2 + 3x - 10$

b. $(x - 4)(x + 3) = \underbrace{x \cdot x}_{} + \underbrace{3x - 4x}_{} - \underbrace{4 \cdot 3}_{}$

$= x^2 - x - 12$

PROBLEM 3

Use FOIL to multiply:

a. $(a + 4)(a - 3)$

b. $(a - 5)(a + 4)$

As in the case of arithmetic, we can use the ideas we've just discussed to do more complicated problems. Thus, we can use the FOIL method to multiply expressions such as $(2x + 5)$ and $(3x - 4)$. We proceed as before; just remember the properties of exponents and the FOIL sequence.

Answers to PROBLEMS

3. a. $a^2 + a - 12$

b. $a^2 - a - 20$

EXAMPLE 4 Using FOIL to multiply two binomials

Use FOIL to multiply:

a. $(2x + 5)(3x - 4)$ **b.** $(3x - 2)(5x - 1)$

SOLUTION 4

(First) F, (Outer) O, (Inner) I, (Last) L

a.
$$(2x + 5)(3x - 4) = \underbrace{(2x)(3x)}_{F} + \underbrace{(2x)(-4)}_{O} + \underbrace{5(3x)}_{I} + \underbrace{(5)(-4)}_{L}$$
$$= 6x^2 - 8x + 15x - 20$$
$$= 6x^2 + 7x - 20$$

b.
$$(3x - 2)(5x - 1) = \underbrace{(3x)(5x)}_{F} + \underbrace{3x(-1)}_{O} - \underbrace{2(5x)}_{I} - \underbrace{2(-1)}_{L}$$
$$= 15x^2 - 3x - 10x + 2$$
$$= 15x^2 - 13x + 2$$

PROBLEM 4

Use FOIL to multiply:

a. $(3a + 5)(2a - 3)$

b. $(2a - 3)(4a - 1)$

Does FOIL work when the binomials to be multiplied contain more than one variable? Fortunately, yes. Again, just remember the sequence and the laws of exponents. For example, to multiply $(3x + 2y)$ by $(2x + 5y)$, we proceed as follows:

$$(2x + 5y)(3x + 2y) = \underbrace{(2x)(3x)}_{F} + \underbrace{(2x)(2y)}_{O} + \underbrace{(5y)(3x)}_{I} + \underbrace{(5y)(2y)}_{L}$$
$$= 6x^2 + 4xy + 15xy + 10y^2$$
$$= 6x^2 + 19xy + 10y^2$$

EXAMPLE 5 Multiplying binomials involving two variables

Use FOIL to multiply:

a. $(5x + 2y)(2x + 3y)$ **b.** $(3x - y)(4x - 3y)$

SOLUTION 5

a.
$$(5x + 2y)(2x + 3y) = \underbrace{(5x)(2x)}_{F} + \underbrace{(5x)(3y)}_{O} + \underbrace{(2y)(2x)}_{I} + \underbrace{(2y)(3y)}_{L}$$
$$= 10x^2 + 15xy + 4xy + 6y^2$$
$$= 10x^2 + 19xy + 6y^2$$

b.
$$(3x - y)(4x - 3y) = \underbrace{(3x)(4x)}_{F} + \underbrace{(3x)(-3y)}_{O} + \underbrace{(-y)(4x)}_{I} + \underbrace{(-y)(-3y)}_{L}$$
$$= 12x^2 - 9xy - 4xy + 3y^2$$
$$= 12x^2 - 13xy + 3y^2$$

PROBLEM 5

Use FOIL to multiply:

a. $(4a + 3b)(3a + 5b)$

b. $(2a - b)(3a - 4b)$

Now one more thing. How do we multiply the expression in the *Getting Started?*

$$(x - L)(x - 2L)$$

We do it in Example 6.

Answers to PROBLEMS

4. a. $6a^2 + a - 15$
b. $8a^2 - 14a + 3$

5. a. $12a^2 + 29ab + 15b^2$
b. $6a^2 - 11ab + 4b^2$

EXAMPLE 6 Multiplying binomials involving two variables

Perform the indicated operation:

$$(x - L)(x - 2L)$$

SOLUTION 6

$$\begin{aligned}(x - L)(x - 2L) &= \underbrace{x \cdot x}_{F} + \underbrace{(x)(-2L)}_{O} + \underbrace{(-L)(x)}_{I} + \underbrace{(-L)(-2L)}_{L} \\ &= x^2 - 2xL - xL + 2L^2 \\ &= x^2 - 3xL + 2L^2\end{aligned}$$

PROBLEM 6

Perform the indicated operation:

$$(y - 2L)(y - 3L)$$

D › Applications Involving Multiplication of Polynomials

Suppose we wish to find out how much is spent annually on hospital care. According to the American Hospital Association, average daily room charges can be approximated by $C(t) = 160 + 14t$ (in dollars), where t is the number of years after 1990. On the other hand, the U.S. National Health Center for Health Statistics indicated that the average stay (in days) in the hospital can be approximated by $D(t) = 7 - 0.2t$, where t is the number of years after 1990. The amount spent annually on hospital care is given by

$$\text{Cost per day} \times \text{Number of days} = C(t) \times D(t)$$

We find this product in Example 7.

EXAMPLE 7 Hospital care costs

Find: $C(t) \times D(t) = (160 + 14t)(7 - 0.2t)$

SOLUTION 7 We use the FOIL method:

$$\begin{aligned}(160 + 14t)(7 - 0.2t) &= \underbrace{160 \cdot 7}_{F} + \underbrace{160 \cdot (-0.2t)}_{O} + \underbrace{14t \cdot 7}_{I} + \underbrace{14t \cdot (-0.2t)}_{L} \\ &= 1120 - 32t + 98t - 2.8t^2 \\ &= 1120 + 66t - 2.8t^2\end{aligned}$$

Thus, the total amount spent annually on hospital care is $1120 + 66t - 2.8t^2$. Can you calculate what this amount was for 2000? What will it be for the year 2010?

PROBLEM 7

Suppose that at a certain hospital, the cost per day is $(200 + 15t)$ and the average stay (in days) is $(10 - 0.1t)$. What is the amount spent annually at this hospital?

Can we predict how much garbage is going to be produced in the year 2020? We can start by predicting the U.S. population: the more people, the more garbage! The U.S. Census estimates that after 2010, the U.S. population will grow by 3 million each year from its base of 310 million in 2010. Thus, the population of the United States after the year 2010 is given by $P(t) = 310 + 3t$, where t is the number of years after 2010. Now, the amount of garbage produced by each person in the United States in the last few years has averaged about 4.6 pounds per day. We do our predictions in Example 8.

Answers to PROBLEMS

6. $y^2 - 5Ly + 6L^2$

7. $-1.5t^2 + 130t + 2000$

EXAMPLE 8 Garbage predictions

If each person in the United States produces 4.6 pounds of garbage each day and there are $(310 + 3t)$ persons, the total amount of garbage produced each day is $4.6(310 + 3t)$ million pounds, where t is the number of years after 2010.

a. Find the product $4.6(310 + 3t)$.

b. Find the amount of garbage produced each day in 2010.

c. How many pounds of garbage will be produced each day in 2020?

SOLUTION 8

a. $4.6(310 + 3t) = 1426 + 13.8t$

b. In 2010, $t = 0$ and $1426 + 13.8(0) = 1426$ million pounds

c. In 2020, $t = 10$ and $1426 + 13.8(10) = 1426 + 138 = 1564$ million pounds

PROBLEM 8

In the last few years, the amount of garbage recovered for recycling each day is about 1.5 pounds per person. If there are $(310 + 3t)$ persons in the United States, the total amount of garbage recovered for recycling is $1.5(310 + 3t)$, where t is the number of years after 2010.

a. Find $1.5(310 + 3t)$.

b. Find the amount of materials recovered for recycling in 2010.

c. How many pounds of materials will be recovered for recycling in 2020?

Some students prefer a **grid method** to multiply polynomials. Thus, to do Example 4(a), $(2x + 5)(3x - 4)$, create a grid separated into four compartments. Place the term $(2x + 5)$ at the top and the term $(3x - 4)$ on the side of the grid. Multiply the rows and columns of the grid as shown. (After you get some practice, you can skip the initial step and write $6x^2$, $15x$, $-8x$, and -20 in the grid.)

	$2x$ +	5
$3x$	$3x \cdot 2x$ $6x^2$	$3x \cdot 5$ $15x$
-4	$-4 \cdot 2x$ $-8x$	$-4 \cdot 5$ -20

Finish by writing the results of each of the grid boxes: $6x^2 + 15x - 8x - 20$

And combining like terms: $6x^2 + 7x - 20$

You can try using this technique in the margin problems or in the exercises!

Calculator Corner

Checking Equivalency

In the Section 4.1 *Calculator Corner,* we agreed that **two expressions are equivalent if their graphs are identical.** Thus, to check Example 1 we have to check that $(-3x^2)(2x^3) = -6x^5$. Let $Y_1 = (-3x^2)(2x^3)$ and $Y_2 = -6x^5$. Press GRAPH and the graph shown here will appear. To confirm the result numerically, press 2nd GRAPH and you get the result in the table.

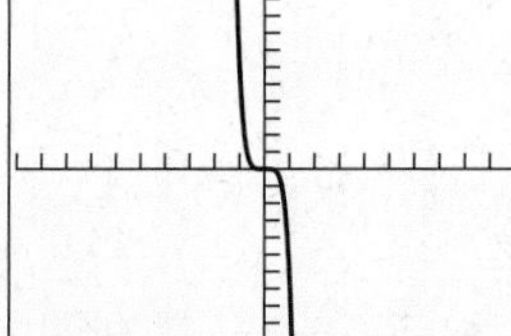

X	Y1	Y2
0	0	0
1	-6	-6
2	-192	-192
3	-1458	-1458
4	-6144	-6144
5	-18750	-18750
6	-46656	-46656

X=0

You can check the rest of the examples except Examples 2(a), 5, and 6. Why?

Answers to PROBLEMS

8. a. $465 + 4.5t$

b. 465 million pounds

c. 510 million pounds

> Exercises 4.6

> Web IT go to mhhe.com/bello for more lessons

⟨ A ⟩ Multiplying Two Monomials In Problems 1–6, find the product.

1. $(5x^3)(9x^2)$
2. $(8x^4)(9x^3)$
3. $(-2x)(5x^2)$
4. $(-3y^2)(4y^3)$
5. $(-2y^2)(-3y)$
6. $(-5z)(-3z)$

⟨ B ⟩ Multiplying a Monomial and a Binomial In Problems 7–20, remove parentheses and simplify.

7. $3(x + y)$
8. $5(2x + y)$
9. $5(2x - y)$
10. $4(3x - 4y)$
11. $-4x(2x - 3)$
12. $-6x(5x - 3)$
13. $(x^2 + 4x)x^3$
14. $(x^2 + 2x)x^2$
15. $(x - x^2)4x$
16. $(x - 3x^2)5x$
17. $(x + y)3x$
18. $(x + 2y)5x^2$
19. $(2x - 3y)(-4y^2)$
20. $(3x^2 - 4y)(-5y^3)$

⟨ C ⟩ Multiplying Two Binomials Using the FOIL Method In Problems 21–56, use the FOIL method to perform the indicated operation.

21. $(x + 1)(x + 2)$
22. $(y + 3)(y + 8)$
23. $(y + 4)(y - 9)$
24. $(y + 6)(y - 5)$
25. $(x - 7)(x + 2)$
26. $(z - 2)(z + 9)$
27. $(x - 3)(x - 9)$
28. $(x - 2)(x - 11)$
29. $(y - 3)(y - 3)$
30. $(y + 4)(y + 4)$
31. $(2x + 1)(3x + 2)$
32. $(4x + 3)(3x + 5)$
33. $(3y + 5)(2y - 3)$
34. $(4y - 1)(3y + 4)$
35. $(5z - 1)(2z + 9)$
36. $(2z - 7)(3z + 1)$
37. $(2x - 4)(3x - 11)$
38. $(5x - 1)(2x - 1)$
39. $(4z + 1)(4z + 1)$
40. $(3z - 2)(3z - 2)$
41. $(3x + y)(2x + 3y)$
42. $(4x + z)(3x + 2z)$
43. $(2x + 3y)(x - y)$
44. $(3x + 2y)(x - 5y)$
45. $(5z - y)(2z + 3y)$
46. $(2z - 5y)(3z + 2y)$
47. $(3x - 2z)(4x - z)$
48. $(2x - 3z)(5x - z)$
49. $(2x - 3y)(2x - 3y)$
50. $(3x + 5y)(3x + 5y)$
51. $(3 + 4x)(2 + 3x)$
52. $(2 + 3x)(3 + 2x)$
53. $(2 - 3x)(3 + x)$
54. $(3 - 2x)(2 + x)$
55. $(2 - 5x)(4 + 2x)$
56. $(3 - 5x)(2 + 3x)$

⟨ D ⟩ Applications Involving Multiplication of Polynomials

57. *Area of a rectangle* The area A of a rectangle is obtained by multiplying its length L by its width W; that is, $A = LW$. Find the area of the rectangle shown in the figure.

$x + 2$ (width), $x + 5$ (length)

58. *Area of a rectangle* Use the formula in Problem 57 to find the area of a rectangle of width $x - 4$ and length $x + 3$.

59. *Height of a thrown object* The height reached by an object t seconds after being thrown upward with a velocity of 96 feet per second is given by $16t(6 - t)$. Use the distributive property to simplify this expression.

60. *Resistance* The resistance R of a resistor varies with the temperature T according to the equation $R = (T + 100)(T + 20)$. Use the distributive property to simplify this expression.

61. *Gas property expression* In chemistry, when V is the volume and P is the pressure of a certain gas, we find the expression $(V_2 - V_1)(CP + PR)$, where C and R are constants. Use the distributive property to simplify this expression.

Area of a rectangle The garage shown is 40 feet by 20 feet. You want to convert it to a bigger garage with two storage areas, S_1 and S_2.

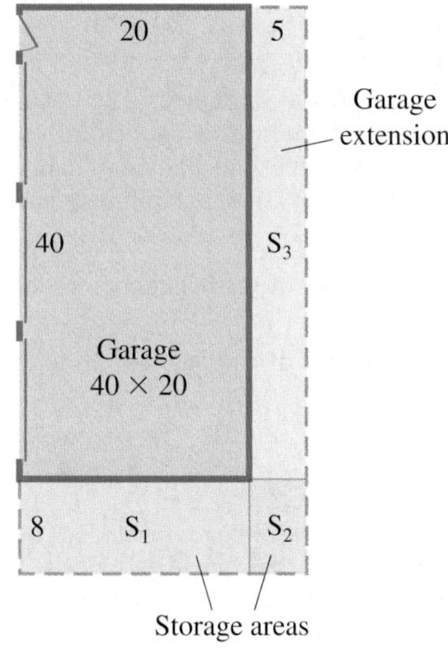

62. What is the area of the existing garage?

63. If you extend the long side by 8 feet and the short side by 5 feet, what is the area of the new garage?

64. Calculate the areas of S_1, S_2, and S_3, and write your answers in the appropriate places in the diagram.

65. Determine the total area of the new garage by adding the area of the original garage to the areas of S_1, S_2, and S_3; that is, add the answers you obtained in Problems 62 and 64.

66. Is the area of the new garage (Problem 63) the same as the answer in Problem 65?

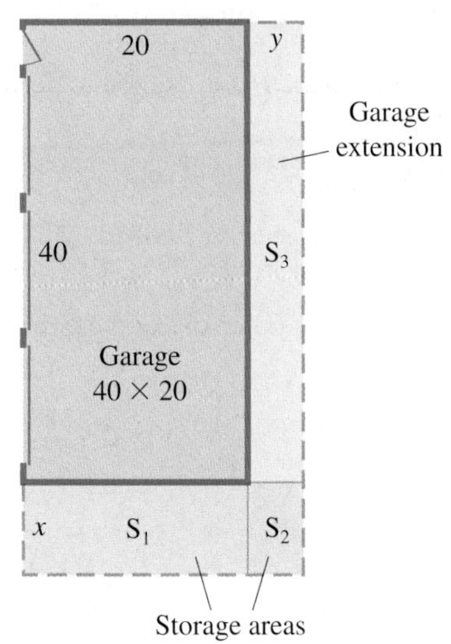

67. If you are not sure how big you want the storage rooms, extend the long side of the garage by x feet and the short side by y feet.

a. Find the area of S_1.

b. Find the area of S_2.

c. Find the area of S_3.

68. The area of the new garage is $(40 + x)(20 + y)$. Simplify this expression.

69. Add the areas of S_1, S_2, S_3, and the area of the original garage. Is the answer the same as the one you obtained in Problem 68?

70. *Blood velocity* The velocity V_r of a blood corpuscle in a vessel depends on the distance r from the center of the vessel and is given by the equation $V_r = V_m\left(1 - \frac{r^2}{R^2}\right)$, where V_m is the maximum velocity and R the radius of the vessel. Multiply this expression.

71. *Crop yield* The yield Y from a grove of Florida orange trees populated with x orange trees per acre is given by the equation $Y = x(1000 - x)$. Multiply this expression.

72. *Dividing lots* A rectangular lot is fenced and divided into two identical fenced lots as shown in the figure. If 1200 feet of fence are used, the area A of the lot is $A = w\left(600 - \frac{3}{2}w\right)$. Multiply this expression.

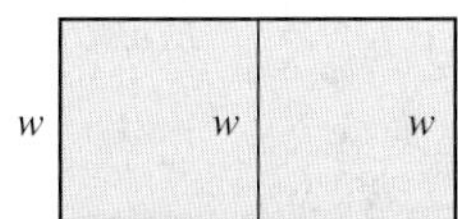

73. *Revenue* If x units of a product are sold at a price p the revenue R is $R = xp$. Suppose that the demand for a product is given by the equation $x = 1000 - 30p$.

a. Write a formula for the revenue R in simplified form.

b. What is the revenue when the price is \$20 per unit?

74. *Revenue* A company manufactures and sells x jogging suits at p dollars every day. If $x = 3000 - 30p$, write a formula for the daily revenue R in simplified form and use it to find the revenue on a day in which the suits were selling for \$40 each.

Using Your Knowledge

Applications: Green Math

75. Of the garbage produced each day by each person in the United States, 0.4 pounds are composted. If the population is $(310 + 3t)$ million, where t is the number of years after 2010, the total amount of garbage composted each day is $0.4(310 + 3t)$ million pounds.

a. Find the product $0.4(310 + 3t)$.

b. Find the amount of garbage composted each day in 2010.

c. How many pounds of garbage will be composted each day in 2020?

Source: http://tinyurl.com/n3tx9n.

76. Of the garbage produced each day by each person in the United States, 2.6 pounds go to the landfill. If the population is $(310 + 3t)$ million, the total amount of garbage going to the landfill each day is $2.6(310 + 3t)$ million pounds.

a. Find the product $2.6(310 + 3t)$.

b. Find the amount of garbage composted each day in 2010.

c. How many pounds of garbage will go to the landfill each day in 2020?

77. The total amount of garbage recovered by each person composting each day in the United States can be approximated by $(0.01t + 0.4)$. If the U.S. population is $(310 + 3t)$ million, where t is the number of years after 2010, the total amount of garbage composted each day is $(0.01t + 0.4)(310 + 3t)$ million pounds.

a. Find the product $(0.01t + 0.4)(310 + 3t)$ and write it in descending order.

b. Find the amount composted each day in 2010.

c. Predict how many pounds of garbage will be composted each day in 2020.

78. The total amount of garbage discarded by each person to the landfill each day in the United States can be approximated by $(-0.02t + 3)$. If the U.S. population is $(310 + 3t)$ million, where t is the number of years after 2010, the total amount of garbage discarded to the landfill each day is $(-0.02t + 3)(310 + 3t)$ million pounds.

a. Find the product $(-0.02t + 3)(310 + 3t)$ and write it in descending order.

b. Find the total amount of garbage discarded to the landfill each day in 2010.

c. Predict how many pounds of garbage will be discarded to the landfill each day in 2020.

Landfill trivia: Only two manmade structures on Earth are large enough to be seen from outer space: the Great Wall of China and the Fresh Kills landfill!

Source: Clean Air Council.

Write On

79. Will the product of two monomials always be a monomial? Explain.

80. If you multiply a monomial and a binomial, will you ever get a trinomial? Explain.

81. Will the product of two binomials (after combining like terms) always be a trinomial? Explain.

82. Multiply:

$$(x + 1)(x - 1) =$$
$$(y + 2)(y - 2) =$$
$$(z + 3)(z - 3) =$$

What is the pattern?

Concept Checker

Fill in the blank(s) with the correct word(s), phrase, or mathematical statement.

83. When using **FOIL** to multiply binomials, the **F** means that the ______________ terms are to be multiplied.

84. When using **FOIL** to multiply binomials, the **O** means that the ______________ terms are to be multiplied.

85. When using **FOIL** to multiply binomials, the **I** means that the ______________ terms are to be multiplied.

86. When using **FOIL** to multiply binomials, the **L** means that the ______________ terms are to be multiplied.

least **outer**

last **first**

inner **final**

odd

Mastery Test

Multiply and simplify:

87. $(-7x^4)(5x^2)$ **88.** $(-8a^3)(-5a^5)$ **89.** $(x + 7)(x - 3)$ **90.** $(x - 2)(x + 8)$

91. $(3x + 4)(3x - 1)$ **92.** $(4x + 3y)(3x + 2y)$ **93.** $(5x - 2y)(2x - 3y)$ **94.** $(x - L)(x - 3L)$

95. $6(x - 3y)$ **96.** $(x^3 + 5x)(-4x^5)$

Skill Checker

Find:

97. $(8x)^2$ **98.** $(4y)^2$ **99.** $(3x)^2$ **100.** $(-A)^2$

101. $(-3A)^2$ **102.** $A(-A)$

4.7 Special Products of Polynomials

Objectives

Expand (simplify) binomials of the form

A $(X + A)^2$

B $(X - A)^2$

C $(X + A)(X - A)$

D Multiply a binomial by a trinomial.

E Multiply any two polynomials.

F Solve applications involving polynomial multiplications.

To Succeed, Review How To . . .

1. Use the FOIL method to multiply polynomials (pp. 369–372).
2. Multiply expressions (pp. 319–321).
3. Use the distributive property to simplify expressions (pp. 81–83, 91–92).

Getting Started

Expanding Your Property

In Section 4.6, we learned how to use the distributive property and the FOIL method to multiply two binomials. In this section we shall develop patterns that will help us in multiplying certain binomial products that occur frequently. For example, do you know how to find the area of this property lot? Since the land is a square, the area is

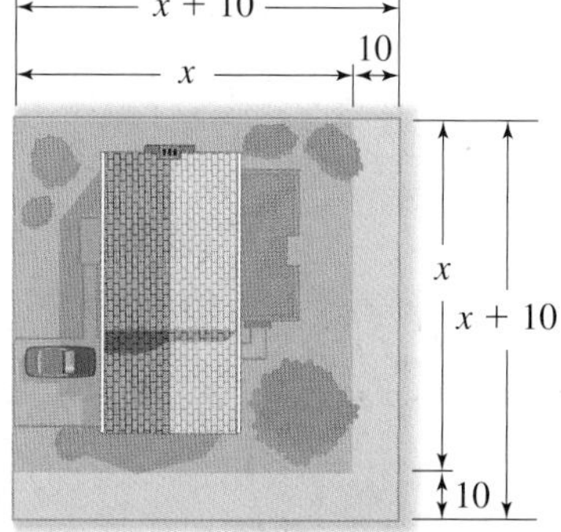

$$(x + 10)(x + 10) = (x + 10)^2$$

The expression $(x + 10)^2$ is the square of the binomial $x + 10$. You can use the FOIL method, of course, but this type of expression is so common in algebra that we have *special products* or formulas that we use to multiply (or expand) them. We are now ready to study several of these special products. You will soon find that we've already studied the first of these special products: the FOIL method is actually special product 1!

As we implied in the *Getting Started,* there's another way to look at the FOIL method: we can format it as a special product. Do you recall the FOIL method?

$$\begin{aligned}(X + A)(X + B) &= \overset{F}{X^2} + \overset{O}{XB} + \overset{I}{AX} + \overset{L}{AB} \\ &= X^2 + (B + A)X + AB \\ &= X^2 + (A + B)X + AB\end{aligned}$$

Thus, we have our first special product.

PRODUCT OF TWO BINOMIALS

Special Product 1

$$(X + A)(X + B) = X^2 + (A + B)X + AB \qquad \textbf{(SP1)}$$

For example, $(x - 3)(x + 5)$ can be multiplied using SP1 with $X = x$, $A = -3$, and $B = 5$. Thus,

$$\begin{aligned}(x - 3)(x + 5) &= x^2 + (-3 + 5)x + (-3)(5) \\ &= x^2 + 2x - 15\end{aligned}$$

Similarly, to expand $(2x - 5)(2x + 1)$, we can let $X = 2x$, $A = -5$, and $B = 1$ in SP1. Hence,

$$\begin{aligned}(2x - 5)(2x + 1) &= (2x)^2 + (-5 + 1)2x + (-5)(1) \\ &= 4x^2 - 8x - 5\end{aligned}$$

Why have we gone to the trouble of reformatting FOIL? Because, as you will see, three very predictable and very handy special products can be derived from SP1.

A > Squaring Sums: $(X + A)^2$

Our second special product (SP2) deals with squaring polynomial sums. Let's see how it is developed by expanding $(X + 10)^2$. First, we start with SP1; we let $A = B$ and note that

$$(X + A)(X + A) = (X + A)^2$$

Now we let $A = 10$, so in our expansion of $(X + 10)^2$ [or $(X + A)^2$], we have

$$(X + A)(X + A) = X \cdot X + (A + A)X + A \cdot A$$

These would be B in SP1.
Note that $(A + A)X = 2AX$.

Thus,

$$(X + A)^2 = X^2 + 2AX + A^2$$

Now we have our second special product, SP2:

THE SQUARE OF A BINOMIAL SUM

Special Product 2

$$(X + A)^2 = X^2 + 2AX + A^2 \qquad \textbf{(SP2)}$$

Note that $(X + A)^2 \neq X^2 + A^2$. (See the *Using Your Knowledge* in Exercises 4.7.)

Here is the pattern used in SP2:

$$\underbrace{(X}_{\text{First term}} + \underbrace{A)^2}_{\text{Second term}} = \underbrace{X^2}_{\text{Square the first term.}} + \underbrace{2AX}_{\text{Multiply the terms and double.}} + \underbrace{A^2}_{\text{Square the last term.}}$$

If you examine this pattern carefully, you can see that SP2 is a result of the FOIL method when a binomial sum is squared. Can we obtain the same product with FOIL?

$$\begin{aligned}(X + A)(X + A) &= \underbrace{X \cdot X}_{F} + \underbrace{AX + AX}_{O\ \ I} + \underbrace{A \cdot A}_{L} \\ &= X^2 + 2AX + A^2\end{aligned}$$

The answer is yes! We are now ready to expand $(X + 10)^2$ using SP2:

Square the first term. | Multiply X by 10; double it. | Square the last term.

$$\begin{aligned}(X + 10)^2 &= X^2 + 2 \cdot 10 \cdot X + 10^2 \\ &= X^2 + 20X + 100\end{aligned}$$

Similarly,

$$\begin{aligned}(x + 7)^2 &= x^2 + 2 \cdot 7 \cdot x + 7^2 \\ &= x^2 + 14x + 49\end{aligned}$$

EXAMPLE 1 Squaring binomial sums

Expand the binomial sum:

a. $(x + 9)^2$ **b.** $(2x + 3)^2$ **c.** $(3x + 4y)^2$

SOLUTION 1

a. $(x + 9)^2 = x^2 + 2 \cdot 9 \cdot x + 9^2 = x^2 + 18x + 81$ Let $A = 9$ in SP2.

b. $(2x + 3)^2 = (2x)^2 + 2 \cdot 3 \cdot 2x + 3^2 = 4x^2 + 12x + 9$ Let $X = 2x$ and $A = 3$ in SP2.

c. $(3x + 4y)^2 = (3x)^2 + 2(4y)(3x) + (4y)^2 = 9x^2 + 24xy + 16y^2$ Let $X = 3x$ and $A = 4y$ in SP2.

PROBLEM 1

Expand the binomial sum:

a. $(y + 6)^2$ **b.** $(2y + 1)^2$

c. $(2x + 3y)^2$

But remember, you do not need to use SP2, you can *always* use FOIL; SP2 simply saves you time and work.

B › Squaring Differences: $(X - A)^2$

Can we also expand $(X - 10)^2$? Of course! But we first have to learn how to expand $(X - A)^2$. To do this, we simply write $-A$ instead of A in SP2 to obtain

$$\begin{aligned}(X - A)^2 &= X^2 + 2(-A)X + (-A)^2 \\ &= X^2 - 2AX + A^2\end{aligned}$$

This is the special product we need, and we call it SP3, the square of a binomial difference.

THE SQUARE OF A BINOMIAL DIFFERENCE

Special Product 3

$$(X - A)^2 = X^2 - 2AX + A^2 \qquad \textbf{(SP3)}$$

Answers to PROBLEMS

1. a. $y^2 + 12y + 36$
   b. $4y^2 + 4y + 1$
   c. $4x^2 + 12xy + 9y^2$

Note that $(X - A)^2 \neq X^2 - A^2$, so the only difference between the square of a sum and the square of a difference is the sign preceding $2AX$. Here is the comparison:

$$(X + A)^2 = X^2 + 2AX + A^2 \quad \textbf{(SP2)}$$
$$(X - A)^2 = X^2 - 2AX + A^2 \quad \textbf{(SP3)}$$

Keeping this in mind,

$$\begin{aligned}(X - 10)^2 &= X^2 - 2 \cdot 10 \cdot X + 10^2 \\ &= X^2 - 20X + 100\end{aligned}$$

Similarly,

$$\begin{aligned}(x - 3)^2 &= x^2 - 2 \cdot 3 \cdot x + 3^2 \\ &= x^2 - 6x + 9\end{aligned}$$

EXAMPLE 2 Squaring binomial differences

Expand the binomial difference:

a. $(x - 5)^2$ **b.** $(3x - 2)^2$ **c.** $(2x - 3y)^2$

SOLUTION 2

a. $(x - 5)^2 = x^2 - 2 \cdot 5 \cdot x + 5^2$ Let $A = 5$ in SP3.
$= x^2 - 10x + 25$

b. $(3x - 2)^2 = (3x)^2 - 2 \cdot 2 \cdot 3x + 2^2$ Let $X = 3x$ and $A = 2$ in SP3.
$= 9x^2 - 12x + 4$

c. $(2x - 3y)^2 = (2x)^2 - 2 \cdot 3y \cdot 2x + (3y)^2$ Let $X = 2x$ and $A = 3y$ in SP3.
$= 4x^2 - 12xy + 9y^2$

PROBLEM 2

Expand the binomial difference:

a. $(y - 4)^2$

b. $(2y - 3)^2$

c. $(3x - 2y)^2$

C › Multiplying Sums and Differences: $(X + A)(X - A)$

We have one more special product, and this one is especially clever! Suppose we multiply the sum of two terms by the difference of the same two terms; that is, suppose we wish to multiply

$$(X + A)(X - A)$$

If we substitute $-A$ for B in SP1, then

$$(X + A)(X + B) = X^2 + (A + B)X + AB$$

becomes

$$\begin{aligned}(X + A)(X - A) &= X^2 + (A - A)X + A(-A) \\ &= X^2 + 0X - A^2 \\ &= X^2 - A^2\end{aligned}$$

This gives us our last very special product, SP4.

THE PRODUCT OF THE SUM AND DIFFERENCE OF TWO TERMS

Special Product 4

$$(X + A)(X - A) = X^2 - A^2 \quad \textbf{(SP4)}$$

Note that $(X + A)(X - A) = (X - A)(X + A)$; so $(X - A)(X + A) = X^2 - A^2$. Thus, to multiply the sum and difference of two terms, we simply square the first term and then subtract from this the square of the last term. Checking this result using the FOIL method, we have

$$\begin{aligned}(X + A)(X - A) &= \overset{F}{X^2} \underbrace{\overset{O}{- AX} + \overset{I}{AX}}_{0} \overset{L}{- A^2} \\ &= X^2 - A^2\end{aligned}$$

Answers to PROBLEMS

2. a. $y^2 - 8y + 16$
b. $4y^2 - 12y + 9$
c. $9x^2 - 12xy + 4y^2$

Since the middle term is *always* zero, we have

$$(x + 3)(x - 3) = x^2 - 3^2 = x^2 - 9$$
$$(x + 6)(x - 6) = x^2 - 6^2 = x^2 - 36$$

Similarly, by the commutative property, we also have

$$(x - 3)(x + 3) = x^2 - 3^2 = x^2 - 9$$
$$(x - 6)(x + 6) = x^2 - 6^2 = x^2 - 36$$

EXAMPLE 3 Finding the product of the sum and difference of two terms

Multiply:

a. $(x + 10)(x - 10)$ **b.** $(2x + y)(2x - y)$
c. $(3x - 5y)(3x + 5y)$

SOLUTION 3

a. $(x + 10)(x - 10) = x^2 - 10^2$ Let $A = 10$ in SP4.
$= x^2 - 100$

b. $(2x + y)(2x - y) = (2x)^2 - y^2$ Let $X = 2x$ and $A = y$ in SP4.
$= 4x^2 - y^2$

c. $(3x - 5y)(3x + 5y) = (3x)^2 - (5y)^2$ Let $X = 3x$ and $A = 5y$ in SP4.
$= 9x^2 - 25y^2$

Note that

$$(3x - 5y)(3x + 5y) = (3x + 5y)(3x - 5y)$$

by the commutative property, so SP4 still applies.

PROBLEM 3

Multiply:

a. $(y + 9)(y - 9)$

b. $(3x + y)(3x - y)$

c. $(3x - 2y)(3x + 2y)$

D > Multiplying a Binomial by a Trinomial

Can we use the FOIL method to multiply any two polynomials? Unfortunately, no. But wait; if algebra is a generalized arithmetic, we should be able to multiply a binomial by a trinomial using the same techniques we employ to multiply, say, 23 by 342.

First let's review how we multiply 23 by 342. Here are the steps.

Step 1	Step 2	Step 3
342	342	342
× 23	× 23	× 23
1026	1026	1026
	6840	6840
		7866
3 × 342 = 1026	20 × 342 = 6840	1026 + 6840 = 7866

Now let's use this same technique to multiply two polynomials, say $(x + 5)$ and $(x^2 + x - 2)$.

Answers to PROBLEMS

3. a. $y^2 - 81$
b. $9x^2 - y^2$
c. $9x^2 - 4y^2$

Step 1 Multiply by 5	Step 2 Multiply by x	Step 3 Add
$\begin{array}{r} x^2 + x - 2 \\ x + 5 \\ \hline 5x^2 + 5x - 10 \end{array}$	$\begin{array}{r} x^2 + x - 2 \\ x + 5 \\ \hline 5x^2 + 5x - 10 \\ x^3 + x^2 - 2x \quad\quad \\ \hline \end{array}$	$\begin{array}{r} x^2 + x - 2 \\ x + 5 \\ \hline 5x^2 + 5x - 10 \\ x^3 + x^2 - 2x \quad\quad \\ \hline x^3 + 6x^2 + 3x - 10 \end{array}$
$5(x^2 + x - 2) = 5x^2 + 5x - 10$	$x(x^2 + x - 2) = x^3 + x^2 - 2x$	$\begin{array}{r} 5x^2 + 5x - 10 \\ x^3 + x^2 - 2x \quad\quad \\ \hline x^3 + 6x^2 + 3x - 10 \end{array}$

Note that in step 3, all *like* terms are placed in the same column so they can be combined.

For obvious reasons, this method is called the **vertical scheme** and can be used when one of the polynomials to be multiplied has *three or more terms*. Of course, we could have obtained the same result by using the distributive property:

$$(a + b)c = ac + bc$$

The procedure would look like this:

$$\begin{aligned} \underbrace{(x + 5)}_{(a+b)} \cdot \underbrace{(x^2 + x - 2)}_{c} &= \underbrace{x}_{a} \cdot \underbrace{(x^2 + x - 2)}_{c} + \underbrace{5}_{b} \cdot \underbrace{(x^2 + x - 2)}_{c} \\ &= x^3 + x^2 - 2x + 5x^2 + 5x - 10 \\ &= x^3 + (x^2 + 5x^2) + (-2x + 5x) - 10 \\ &= x^3 + 6x^2 + 3x - 10 \end{aligned}$$

EXAMPLE 4 Multiplying a binomial by a trinomial

Multiply: $(x - 3)(x^2 - 2x - 4)$

SOLUTION 4

Using the vertical scheme, we proceed as follows:

$$\begin{array}{rl} x^2 - 2x - 4 & \\ x - 3 & \\ \hline -3x^2 + 6x + 12 & \text{Multiply } x^2 - 2x - 4 \text{ by } -3. \\ x^3 - 2x^2 - 4x \quad\quad & \text{Multiply } x^2 - 2x - 4 \text{ by } x. \\ \hline x^3 - 5x^2 + 2x + 12 & \text{Add like terms.} \end{array}$$

Thus, the result is $x^3 - 5x^2 + 2x + 12$. You can also do this problem by using the distributive property $(a - b)c = ac - bc$. The procedure would look like this:

$$\begin{aligned} \underbrace{(x - 3)}_{(a-b)} \cdot \underbrace{(x^2 - 2x - 4)}_{c} &= \underbrace{x}_{a} \cdot \underbrace{(x^2 - 2x - 4)}_{c} - \underbrace{3}_{b} \cdot \underbrace{(x^2 - 2x - 4)}_{c} \\ &= x^3 - 2x^2 - 4x - 3x^2 + 6x + 12 \\ &= x^3 + (-2x^2 - 3x^2) + (-4x + 6x) + 12 \\ &= x^3 - 5x^2 + 2x + 12 \end{aligned}$$

PROBLEM 4

Multiply: $(y - 2)(y^2 - y - 3)$

Note that the same result is obtained in both cases.

Answers to PROBLEMS

4. $y^3 - 3y^2 - y + 6$

E > Multiplying Two Polynomials

Here is the idea we used in Example 4.

PROCEDURE

Multiplying *Any* Two Polynomials (Term-By-Term Multiplication)

To multiply two polynomials, multiply each term of one by every term of the other and add the results.

Now that you've learned all the basic techniques used to multiply polynomials, you should be able to tackle any polynomial multiplication. To do this, you must *first* decide what special product (if any) is involved. Here are the special products we've studied.

SPECIAL PRODUCTS

$(X + A)(X + B) = X^2 + (A + B)X + AB$ **(SP1 or FOIL)**

$(X + A)(X + A) = (X + A)^2 = X^2 + 2AX + A^2$ **(SP2)**

$(X - A)(X - A) = (X - A)^2 = X^2 - 2AX + A^2$ **(SP3)**

$(X + A)(X - A) = X^2 - A^2$ **(SP4)**

Of course, the FOIL method always works for the last three types of equations, but since it's more laborious, learning to recognize special products 2–4 is definitely worth the effort!

EXAMPLE 5 **Using SP1 (FOIL) and the distributive property**

Multiply: $3x(x + 5)(x + 6)$

SOLUTION 5 One way to approach this problem is to save the $3x$ multiplication until last. First, use FOIL (SP1):

$$\begin{aligned} 3x(x + 5)(x + 6) &= 3x(\overset{F}{x^2} + \overset{O}{6x} + \overset{I}{5x} + \overset{L}{30}) \\ &= 3x(x^2 + 11x + 30) \\ &= 3x^3 + 33x^2 + 90x && \text{Use the distributive property.} \end{aligned}$$

Alternatively, you can use the distributive property to multiply $(x + 5)$ by $3x$ and then proceed with FOIL, as shown here:

$$\begin{aligned} 3x(x + 5)(x + 6) &= (3x \cdot x + 3x \cdot 5)(x + 6) && \text{Multiply } 3x(x + 5). \\ &= (3x^2 + 15x)(x + 6) && \text{Simplify.} \\ &= \overset{F}{3x^2 \cdot x} + \overset{O}{3x^2 \cdot 6} + \overset{I}{15x \cdot x} + \overset{L}{15x \cdot 6} && \text{Use FOIL.} \\ &= 3x^3 + 18x^2 + 15x^2 + 90x && \text{Simplify.} \\ &= 3x^3 + 33x^2 + 90x && \text{Collect like terms.} \end{aligned}$$

Of course, both methods produce the same result.

PROBLEM 5

Multiply: $2y(y + 2)(y + 3)$

EXAMPLE 6 **Cubing a binomial**

Expand: $(x + 3)^3$

SOLUTION 6 Recall that $(x + 3)^3 = (x + 3)(x + 3)(x + 3)$. Thus, we can square the sum $(x + 3)$ first and then use the distributive property as shown here.

PROBLEM 6

Expand: $(y + 2)^3$

(continued)

Answers to PROBLEMS

5. $2y^3 + 10y^2 + 12y$ **6.** $y^3 + 6y^2 + 12y + 8$

Square of sum

$$(x + 3)(x + 3)(x + 3) = (x + 3)(x^2 + 6x + 9)$$
$$= x(x^2 + 6x + 9) + 3(x^2 + 6x + 9) \quad \text{Use the distributive property.}$$
$$= x \cdot x^2 + x \cdot 6x + x \cdot 9 + 3 \cdot x^2 + 3 \cdot 6x + 3 \cdot 9$$
$$= x^3 + 9x^2 + 27x + 27$$

EXAMPLE 7 **Squaring a binomial difference involving fractions**

Expand:

$$\left(5t^2 - \frac{1}{2}\right)^2$$

SOLUTION 7 This is the square of a difference. Using SP3, we obtain

$$\left(5t^2 - \frac{1}{2}\right)^2 = (5t^2)^2 - 2 \cdot \frac{1}{2} \cdot (5t^2) + \left(\frac{1}{2}\right)^2$$
$$= 25t^4 - 5t^2 + \frac{1}{4}$$

PROBLEM 7

Expand:

$$\left(3y^2 - \frac{1}{3}\right)^2$$

EXAMPLE 8 **Finding the product of the sum and difference of two terms**

Multiply: $(2x^2 + 5)(2x^2 - 5)$

SOLUTION 8 This time, we use SP4 because we have the product of the sum and difference of two terms.

$$(2x^2 + 5)(2x^2 - 5) = (2x^2)^2 - (5)^2$$
$$= 4x^4 - 25$$

PROBLEM 8

Multiply: $(3y^2 + 2)(3y^2 - 2)$

EXAMPLE 9 **Finding Areas**

In Example 5 of Section 4.5, we used the addition of polynomials to find the total area of several rectangles. As you recall, the area A of a rectangle is the product of its length L and its width W; that is, $A = L \times W$. If the length of the rectangle is $(x + 5)$ and its width is $(x - 3)$, what is its area?

SOLUTION 9 Since the area is $L \times W$, where $L = (x + 5)$ and $W = (x - 3)$, we have

$$A = (x + 5)(x - 3)$$
$$= x^2 - 3x + 5x - 15$$
$$= x^2 + 2x - 15$$

(x − 3) A (x + 5)

PROBLEM 9

The area A of a circle of radius r is $A = \pi r^2$. What is the area of a circle with radius $(x - 2)$?

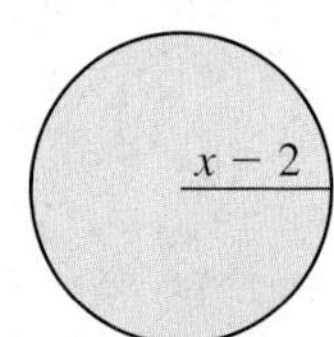

F > Applications Involving Polynomial Multiplication

In recent years, many cities and municipalities have used up their freshwater sources resulting in their pumping water from greater depths and piping water farther to deliver it to residents. But there is an alternative: desalination (removing salt from sea water to make drinking water). We will discuss the costs of desalinated water in Example 10 and yes, the costs do involve polynomials!

Answers to PROBLEMS

7. $9y^4 - 2y^2 + \frac{1}{9}$ **8.** $9y^4 - 4$

9. $\pi(x - 2)^2 = \pi x^2 - 4\pi x + 4\pi$

EXAMPLE 10 Cost of 1000 gallons of desalinated water

The cost of 1000 gallons of desalinated water varies by location but at a specific location can be approximated by **$C(t) = 0.006(t - 40)^2 + 2$ dollars,** where t is the number of years after 1970.

a. Expand $(t - 40)^2$.
b. Simplify $0.006(t - 40)^2 + 2$ and write the result in descending order.
c. Find the cost of 1000 gallons of desalinated water in 2010.
d. What was the cost of 1000 gallons of desalinated water in 1970?
e. Is the price increasing or decreasing?

SOLUTION 10

a. $(t - 40)^2$ is a binomial sum, so we can use SP2 to multiply.

$$(t - 40)^2 = t^2 - 2(40)t + 40^2 = t^2 - 80t + 1600$$

b.
$$\begin{aligned} 0.006(t - 40)^2 + 2 &= 0.006(t^2 - 80t + 1600) + 2 \\ &= 0.006t^2 - 0.48t + 9.6 + 2 \\ &= 0.006t^2 - 0.48t + 11.6 \end{aligned}$$

c. In 2010, $t = 40$ (2010 − 1970).

$$\begin{aligned} \text{For } t = 40,\ C(t) &= 0.006(40 - 40)^2 + 2 \\ &= \$2 \end{aligned}$$

d. In 1970, $t = 0$ and
$$\begin{aligned} C(t) &= 0.006(0 - 40)^2 + 2 \\ &= 0.006(-40)^2 + 2 \\ &= 0.006(1600) + 2 \\ &= \$11.60 \end{aligned}$$

e. The price is decreasing (from \$11.60 in 1970 to \$2 in 2010).

Source: Global Water Intelligence, National Research Council, American Waterworks Association.

PROBLEM 10

In a more expensive desalination plant, the cost of 1000 gallons of water is **$C(t) = 0.006(t - 40)^2 + 4$ dollars,** where t is the number of years after 1970.

a. Simplify $0.006(t - 40)^2 + 4$ and write the result in descending order.
b. Find the cost of 1000 gallons of desalinated water in 2010.
c. What was the cost of 1000 gallons of desalinated water in 1970?
d. Is the price increasing or decreasing?

Water Trivia: The typical American household uses about 127,400 gallons of water per year and their annual water bill is \$413, but the cost of desalination has been decreasing at about 4% per year.

Source: Sandia National Labs.

How do you know which of the special products you should use?

The trick to using these special products effectively, of course, is being able to *recognize* situations where they apply. Here are some tips.

PROCEDURE

Choosing the Appropriate Method for Multiplying Two Polynomials

1. Is the product the square of a binomial? If so, use SP2 or SP3:

$$(X + A)^2 = (X + A)(X + A) = X^2 + 2AX + A^2 \qquad \textbf{(SP2)}$$
$$(X - A)^2 = (X - A)(X - A) = X^2 - 2AX + A^2 \qquad \textbf{(SP3)}$$

Note that both answers have *three* terms.

2. Are the two binomials in the product the sum and difference of the same two terms? If so, use SP4:

$$(X + A)(X - A) = X^2 - A^2 \qquad \textbf{(SP4)}$$

The answer has *two* terms.

(continued)

Answers to PROBLEMS

10. a. $0.006t^2 - 0.48t + 13.6$
b. \$4
c. \$13.60
d. Decreasing

3. Is the binomial product different from those in 1 and 2? If so, use FOIL. The answer will have *three* or *four* terms.
4. Is the product different from all the ones mentioned in 1, 2, and 3? If so, multiply every term of the first polynomial by every term of the second and collect like terms.

Calculator Corner

Checking Results

As we have done previously, we can use a calculator to check the results of the examples in this section. You know the procedure: let the original expression be Y_1 and the answer be Y_2. If the graphs are identical, the answer is correct. Thus, to check Example 5, let $Y_1 = 3x(x + 5)(x + 6)$ and $Y_2 = 3x^3 + 33x^2 + 90x$. Press GRAPH and you get only a partial graph, as shown here on the left. Since we are missing the bottom part of the graph, we need more values of y that are negative, so we have to fix the window we use for the graph. Press WINDOW and let Ymin $= -20$. Still, we do not see the whole graph, so try again with Ymin $= -90$. Press GRAPH and this time we see the complete graph as shown on the right.

You should now check Example 7, but use x instead of t. Can you check Example 1(c)? Why not?

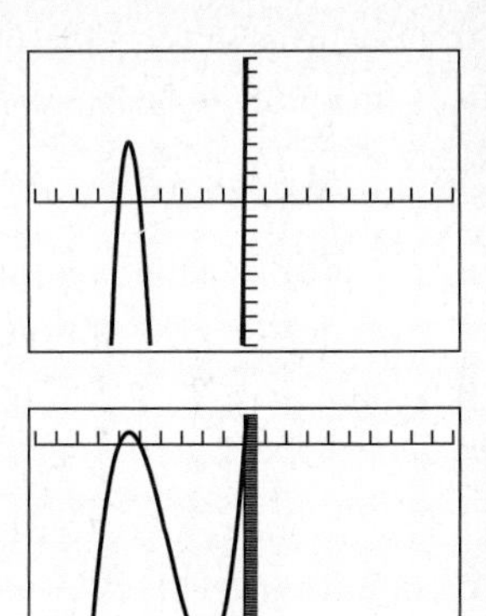

A final word of advice before you do Exercises 4.7: SP1 (the FOIL method) is very important and should be thoroughly understood before you attempt the problems. This is because *all* the special products are *derived* from this formula. As a matter of fact, you can successfully complete most of the problems in Exercises 4.7 if you fully understand the result given in SP1.

> Exercises 4.7

‹ A › Squaring Sums In Problems 1–6, expand the given binomial sum.

1. $(x + 1)^2$
2. $(x + 6)^2$
3. $(2x + 1)^2$
4. $(2x + 5)^2$
5. $(3x + 2y)^2$
6. $(4x + 5y)^2$

‹ B › Squaring Differences In Problems 7–16, expand the given binomial difference.

7. $(x - 1)^2$
8. $(x - 2)^2$
9. $(2x - 1)^2$
10. $(3x - 4)^2$
11. $(3x - y)^2$
12. $(4x - y)^2$
13. $(6x - 5y)^2$
14. $(4x - 3y)^2$
15. $(2x - 7y)^2$
16. $(3x - 5y)^2$

⟨C⟩ Multiplying Sums and Differences In Problems 17–30, find the product.

17. $(x + 2)(x - 2)$

18. $(x + 1)(x - 1)$

19. $(x + 4)(x - 4)$

20. $(3x + y)(3x - y)$

21. $(3x + 2y)(3x - 2y)$

22. $(2x + 5y)(2x - 5y)$

23. $(x - 6)(x + 6)$

24. $(x - 11)(x + 11)$

25. $(x - 12)(x + 12)$

26. $(x - 9)(x + 9)$

27. $(3x - y)(3x + y)$

28. $(5x - 6y)(5x + 6y)$

29. $(2x - 7y)(2x + 7y)$

30. $(5x - 8y)(5x + 8y)$

In Problems 31–40, use the special products to multiply the given expressions.

31. $(x^2 + 2)(x^2 + 5)$

32. $(x^2 - 3)(x^2 + 2)$

33. $(x^2 + y)^2$

34. $(2x^2 + y)^2$

35. $(3x^2 - 2y^2)^2$

36. $(4x^3 - 5y^3)^2$

37. $(x^2 - 2y^2)(x^2 + 2y^2)$

38. $(x^2 - 3y^2)(x^2 + 3y^2)$

39. $(2x + 4y^2)(2x - 4y^2)$

40. $(5x^2 + 2y)(5x^2 - 2y)$

⟨D⟩ Multiplying a Binomial by a Trinomial In Problems 41–52, find the product.

41. $(x + 3)(x^2 + x + 5)$

42. $(x + 2)(x^2 + 5x + 6)$

43. $(x + 4)(x^2 - x + 3)$

44. $(x + 5)(x^2 - x + 2)$

45. $(x + 3)(x^2 - x - 2)$

46. $(x + 4)(x^2 - x - 3)$

47. $(x - 2)(x^2 + 2x + 4)$

48. $(x - 3)(x^2 + x + 1)$

49. $-(x - 1)(x^2 - x + 2)$

50. $-(x - 2)(x^2 - 2x + 1)$

51. $-(x - 4)(x^2 - 4x - 1)$

52. $-(x - 3)(x^2 - 2x - 2)$

⟨E⟩ Multiplying Two Polynomials In Problems 53–80, find the product.

53. $2x(x + 1)(x + 2)$

54. $3x(x + 2)(x + 5)$

55. $3x(x - 1)(x + 2)$

56. $4x(x - 2)(x + 3)$

57. $4x(x - 1)(x - 2)$

58. $2x(x - 3)(x - 1)$

59. $5x(x + 1)(x - 5)$

60. $6x(x + 2)(x - 4)$

61. $(x + 5)^3$

62. $(x + 4)^3$

63. $(2x + 3)^3$

64. $(3x + 2)^3$

65. $(2x + 3y)^3$

66. $(3x + 2y)^3$

67. $(4t^2 + 3)^2$

68. $(5t^2 + 1)^2$

69. $(4t^2 + 3u)^2$

70. $(5t^2 + u)^2$

71. $\left(3t^2 - \frac{1}{3}\right)^2$

72. $\left(4t^2 - \frac{1}{2}\right)^2$

73. $\left(3t^2 - \frac{1}{3}u\right)^2$

74. $\left(4t^2 - \frac{1}{2}u\right)^2$

75. $(3x^2 + 5)(3x^2 - 5)$

76. $(4x^2 + 3)(4x^2 - 3)$

77. $(3x^2 + 5y^2)(3x^2 - 5y^2)$

78. $(4x^2 + 3y^2)(4x^2 - 3y^2)$

79. $(4x^3 - 5y^3)(4x^3 + 5y^3)$

80. $(2x^4 - 3y^3)(2x^4 + 3y^3)$

Web IT go to mhhe.com/bello for more lessons

81. Find the volume V of the cone when the height is $h = x + 4$ and the radius is $r = x + 2$.

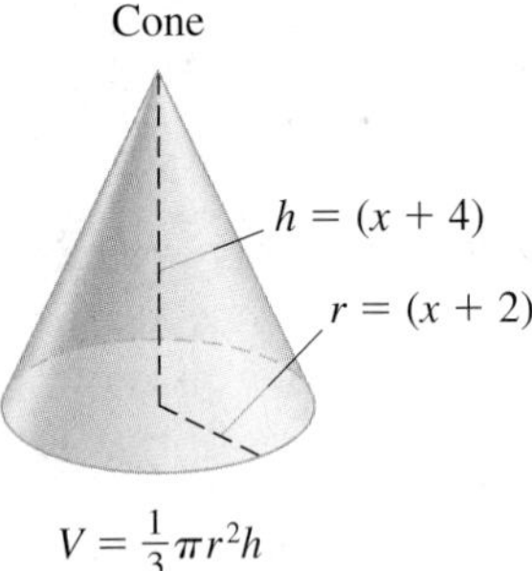

82. Find the volume V of the parallelepiped (box) when the width is $a = x$, the length is $c = x + 5$, and the height is $b = x + 4$.

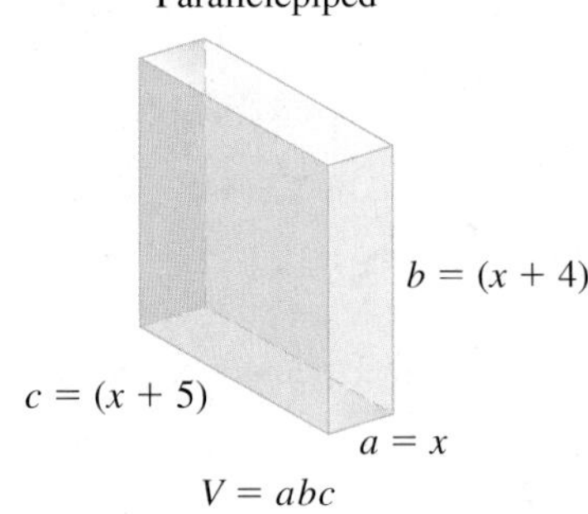

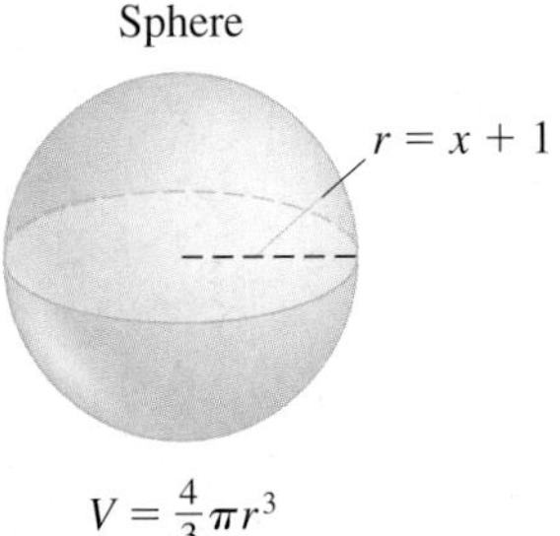

83. Find the volume V of the sphere when the radius is $r = x + 1$.

84. Find the volume V of the hemisphere (half of the sphere) when the radius is $r = x + 2$.

85. Find the volume V of the cylinder when the height is $h = x + 2$ and the radius is $r = x + 1$.

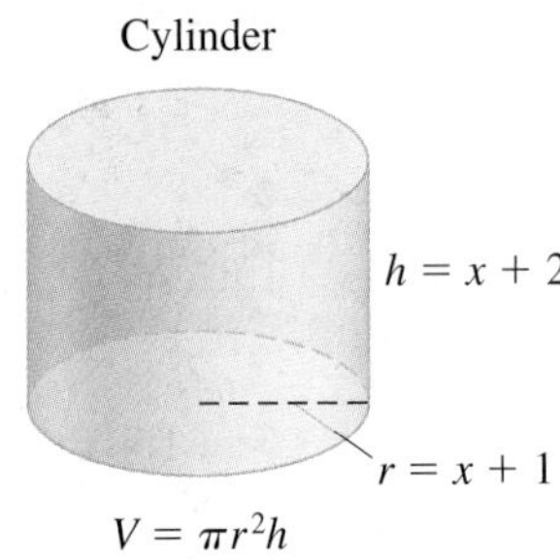

86. Find the volume V of the cube when the length of the side is $a = x + 3$.

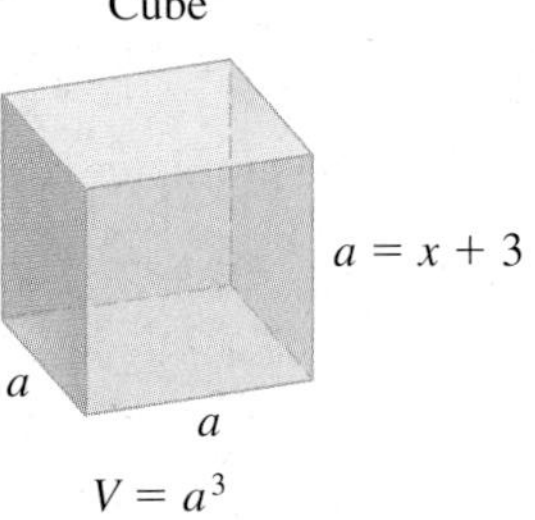

>>> Applications

>>> Applications: Green Math

In Example 10 we mentioned the price of desalinated water. What about the price of 1000 gallons of freshwater? We will explore those costs in Problems 87 and 88.

87. *Freshwater cost* The cost of 1000 gallons of fresh water varies by location but at a specific location cost can be approximated by $C(t) = 0.003(t - 10)^2 + 1$ **dollars,** where t is the number of years after 1970.

a. Expand $(t - 10)^2$.

b. Simplify $0.003(t - 10)^2 + 1$ and write the result in descending order.

c. Find the cost of 1000 gallons of fresh water in 2010.

d. What was the cost of 1000 gallons of fresh water in 1970?

e. Is the price increasing or decreasing?

88. *Freshwater cost* The cost of 1000 gallons of fresh water varies by location but at a specific location cost can be approximated by $C(t) = 0.002(t - 2)^2 + 0.9$ **dollars,** where t is the number of years after 1970.

a. Expand $(t - 2)^2$.

b. Simplify $0.002(t - 2)^2 + 0.9$ and write the result in descending order.

c. Find the cost of 1000 gallons of fresh water in 2010.

d. What was the cost of 1000 gallons of fresh water in 1970?

e. Is the price increasing or decreasing?

Web IT go to mhhe.com/bello for more lessons

89. *Heat transfer* The heat transmission between two objects of temperatures T_2 and T_1 involves the expression

$$(T_1^2 + T_2^2)(T_1^2 - T_2^2)$$

Multiply this expression.

90. *Deflection of a beam* The deflection of a certain beam involves the expression $w(l^2 - x^2)^2$. Expand this expression.

91. *Heat from convection* The heat output from a natural draught convector is given by the equation $K(t_n - t_a)^2$. Expand this expression.

92. *Pressure and volume of gases* When studying the pressure P and the volume V of a gas we deal with the expression $(V_2 - V_1)(CP + PR)$, where C and R are constants. Multiply this expression.

93. *Relation of velocity, acceleration, and distance* The equation relating the velocity, the acceleration a, and the distance s is $(v_i + v_0)(v_i - v_0) = 2as$. Multiply the expression on the left of the equation.

94. *Velocity of fluids* To find the velocity of a fluid through a pipe of radius r and outer diameter D we use the expression $k(D - 2r)(D + 2r)$, where k is a constant. Multiply this expression.

Using Your Knowledge

Binomial Fallacies A common fallacy (mistake) when multiplying binomials is to assume that

$$(x + y)^2 = x^2 + y^2$$

Here are some arguments that should convince you this is *not* true.

95. Let $x = 1, y = 2$.

a. What is $(x + y)^2$?

b. What is $x^2 + y^2$?

c. Is $(x + y)^2 = x^2 + y^2$?

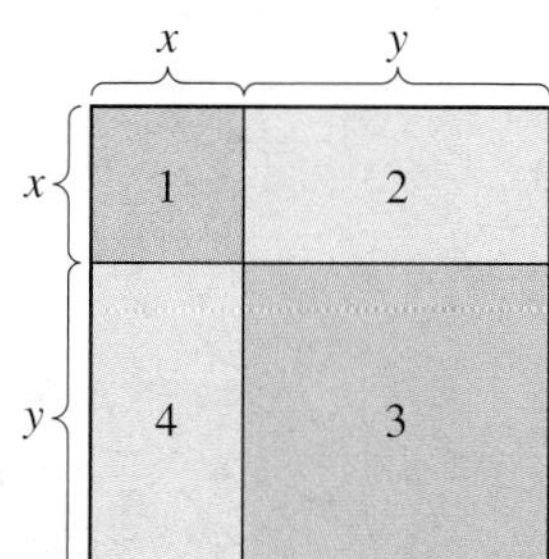

96. Let $x = 2, y = 1$.

a. What is $(x - y)^2$?

b. What is $x^2 - y^2$?

c. Is $(x - y)^2 = x^2 - y^2$?

97. Look at the large square. Its area is $(x + y)^2$. The square is divided into four smaller areas numbered 1, 2, 3, and 4. What is the area of:

a. square 1? **b.** rectangle 2?

c. square 3? **d.** rectangle 4?

98. The total area of the square is $(x + y)^2$. It's also the sum of the four areas numbered 1, 2, 3, and 4. What is the sum of these four areas? (Simplify your answer.)

99. From your answer to Problem 98, what can you say about $x^2 + 2xy + y^2$ and $(x + y)^2$?

100. The sum of the areas of the squares numbered 1 and 3 is $x^2 + y^2$. Is it true that $x^2 + y^2 = (x + y)^2$?

Write On

101. Write the procedure you use to square a binomial sum.

102. Write the procedure you use to square a binomial difference.

103. If you multiply two binomials, you usually get a trinomial. What is the exception? Explain.

104. Describe the procedure you use to multiply two polynomials.

Concept Checker

Fill in the blank(s) with the correct word(s), phrase, or mathematical statement.

105. $(X + A)(X + B) =$ ________________.

106. $(X + A)^2 =$ ________________.

107. $(X - A)^2 =$ ________________.

108. $(X + A)(X - A) =$ ________________.

$X^2 + 2AX - A^2$ $X^2 - A^2$

$X^2 - 2AX - A^2$ $X^2 + 2AX + A^2$

$X^2 - 2AX + A^2$ $X^2 + XB + AX + AB$

$X^2 + A^2$

Mastery Test

Expand and simplify:

109. $(x + 7)^2$

110. $(x - 2y)^2$

111. $(2x + 5y)^2$

112. $(x - 12)^2$

113. $(x^2 + 2y)^2$

114. $(x - 3y)^2$

115. $(3x - 2y)(3x + 2y)$

116. $(3x^2 + 2y)(3x^2 - 2y)$

117. $(3x + 4)(5x - 6)$

118. $(2x + 5)(5x - 2)$

119. $(7x - 2y)(3x - 4y)$

120. $(3y - 4x)(4y - 3x)$

121. $(x + 2)(x^2 + x + 1)$

122. $(x - 3)(x^2 - 3x - 1)$

123. $(x^2 - 2)(x^2 + x + 1)$

124. $(x + 3)^3$

125. $(x + 2y)^3$

126. $5x(x + 2)(x - 4)$

127. $3x(x^2 + 1)(x^2 - 1)$

128. $-4x(x^2 + 1)^2$

Skill Checker

Simplify:

129. $\frac{20x^3}{10x^2}$

130. $\frac{28x^4}{4x^2}$

131. $\frac{-5x^2}{10x^2}$

132. $\frac{10x^2}{5x}$

4.8 Division of Polynomials

Objectives

A Divide a polynomial by a monomial.

B Divide one polynomial by another polynomial.

To Succeed, Review How To . . .

1. Use the distributive property to simplify expressions (pp. 81–83, 91–92).
2. Multiply polynomials (pp. 368–372).
3. Use the quotient rule of exponents (p. 322).

Getting Started

Refrigerator Efficiency

In the preceding sections, we learned to add, subtract, and multiply polynomials. We are now ready for division. For example, do you know how efficient your refrigerator or your heat pump is? Their efficiency E is given by the quotient

$$E = \frac{T_1 - T_2}{T_1}$$

where T_1 and T_2 are the initial and final temperatures between which they operate. Can you do this division?

$$\frac{T_1 - T_2}{T_1}$$

Yes, you can, if you follow the steps:

$$\frac{T_1 - T_2}{T_1} = (T_1 - T_2) \div T_1$$

$$= (T_1 - T_2)\left(\frac{1}{T_1}\right) \quad \text{Division by } T_1 \text{ is the same as multiplication by } \frac{1}{T_1}.$$

$$= T_1\left(\frac{1}{T_1}\right) - T_2\left(\frac{1}{T_1}\right) \quad \text{Use the distributive property.}$$

$$= \frac{T_1}{T_1} - \frac{T_2}{T_1} \quad \text{Multiply.}$$

$$= 1 - \frac{T_2}{T_1} \quad \text{Since } \frac{T_1}{T_1} = 1.$$

In this section, we will learn how to divide a polynomial by a monomial, and then we will generalize this idea, which will enable us to divide one polynomial by another.

A > Dividing a Polynomial by a Monomial

To divide the binomial $4x^3 - 8x^2$ by the monomial $2x$, we proceed as we did in the *Getting Started.*

$$\frac{4x^3 - 8x^2}{2x} = (4x^3 - 8x^2) \div 2x$$

$$= (4x^3 - 8x^2)\left(\frac{1}{2x}\right) \quad \text{Remember that to } \textit{divide} \text{ by } 2x \text{, you multiply by the reciprocal } \frac{1}{2x}.$$

$$= 4x^3\left(\frac{1}{2x}\right) - 8x^2\left(\frac{1}{2x}\right) \quad \text{Use the distributive property.}$$

$$= \frac{4x^3}{2x} - \frac{8x^2}{2x} \quad \text{Multiply.}$$

$$= \frac{\overset{2x^2}{\cancel{4x^3}}}{\cancel{2x}} - \frac{\overset{4x}{\cancel{8x^2}}}{\cancel{2x}} \quad \text{Use the quotient rule.}$$

$$= 2x^2 - 4x \quad \text{Simplify.}$$

We then have

$$\frac{4x^3 - 8x^2}{2x} = \frac{4x^3}{2x} - \frac{8x^2}{2x} = 2x^2 - 4x$$

This example suggests the following rule.

RULE TO DIVIDE A POLYNOMIAL BY A MONOMIAL

To divide a polynomial by a monomial, divide *each* **term** in the polynomial by the *monomial.*

EXAMPLE 1 **Dividing a polynomial by a monomial**

Find the quotient:

a. $\frac{28x^4 - 14x^3}{7x^2}$

b. $\frac{20x^3 - 5x^2 + 10x}{10x^2}$

SOLUTION 1

a. $\frac{28x^4 - 14x^3}{7x^2} = \frac{28x^4}{7x^2} - \frac{14x^3}{7x^2} = 4x^2 - 2x$

b. $\frac{20x^3 - 5x^2 + 10x}{10x^2} = \frac{20x^3}{10x^2} - \frac{5x^2}{10x^2} + \frac{10x}{10x^2} = 2x - \frac{1}{2} + \frac{1}{x}$

PROBLEM 1

Find the quotient:

a. $\frac{27y^4 - 18y^3}{9y^2}$

b. $\frac{8y^3 - 4y^2 + 12y}{4y^2}$

Answers to PROBLEMS

1. a. $3y^2 - 2y$ **b.** $2y - 1 + \frac{3}{y}$

B > Dividing One Polynomial by Another Polynomial

If we wish to divide a polynomial, called the **dividend,** by another polynomial, called the **divisor,** we proceed very much as we did in long division in arithmetic. To show you that this is so, let's go through the division of 337 by 16 and $(x^2 + 3x + 3)$ (dividend) by $(x + 1)$ (divisor) side by side.

1. $\begin{array}{r} 2 \\ 16\overline{)337} \end{array}$	Divide 33 by 16. It goes twice. Write 2 over the 33.	$\begin{array}{r} x \\ x+1\overline{)x^2+3x+3} \end{array}$	Divide x^2 by x $\left(\frac{x^2}{x} = x\right)$ It goes x times. Write x over the $3x$.
2. $\begin{array}{r} 2 \\ 16\overline{)337} \\ -32 \\ \hline 1 \end{array}$	Multiply 2 by 16 and subtract the product 32 from 33 to obtain 1.	$\begin{array}{r} x \\ x+1\overline{)x^2+3x+3} \\ (-)x^2+x \\ \hline 0+2x \end{array}$	Multiply x by $x + 1$ and subtract the product $x^2 + x$ from $x^2 + 3x$ to obtain $0 + 2x$.
3. $\begin{array}{r} 21 \\ 16\overline{)337} \\ -32 \\ \hline 17 \end{array}$	Bring down the 7. Now, divide 17 by 16. It goes once. Write 1 after the 2.	$\begin{array}{r} x+2 \\ x+1\overline{)x^2+3x+3} \\ (-)x^2+x \\ \hline 0+2x+3 \end{array}$	Bring down the 3. Now, divide $2x$ by x $\left(\frac{2x}{x} = 2\right)$ It goes 2 times. Write + 2 after the x.
4. $\begin{array}{r} 21 \\ 16\overline{)337} \\ -32 \\ \hline 17 \\ -16 \\ \hline 1 \end{array}$	Multiply 1 by 16 and subtract the result from 17. The remainder is 1.	$\begin{array}{r} x+2 \\ x+1\overline{)x^2+3x+3} \\ (-)x^2+x \\ \hline 0+2x+3 \\ (-)2x+2 \\ \hline 1 \end{array}$	Multiply 2 by $x + 1$ to obtain $2x + 2$. Subtract this result from $2x + 3$. The remainder is 1.

5. The answer **(quotient)** can be written as 21 R 1 (read "21 remainder 1") or as

$$21 + \tfrac{1}{16}, \text{ which is } 21\tfrac{1}{16}$$

The answer (**quotient**) can be written as $(x + 2)$ R 1 (read "$x + 2$ remainder 1") or as

$$x + 2 + \frac{1}{x + 1}$$

6. You can check this answer by multiplying 21 by 16 (336) and adding the remainder 1 to obtain 337, the dividend.

You can check the answer by multiplying $(x + 2)(x + 1)$, obtaining $x^2 + 3x + 2$, and then adding the remainder 1 to get $x^2 + 3x + 3$, the dividend.

EXAMPLE 2 **Dividing a polynomial by a binomial**

Divide: $x^2 + 2x - 17$ by $x - 3$

SOLUTION 2

x^2 divided by x is x.

$5x$ divided by x is 5.

$$
\begin{array}{r}
x + 5 \\
x - 3\overline{)x^2 + 2x - 17} \\
(-)\,x^2 - 3x \\
\hline
5x - 17 \\
(-)5x - 15 \\
\hline
-2
\end{array}
$$

$\frac{x^2}{x} = x$; $x(x - 3) = x^2 - 3x$

$\frac{5x}{x} = 5$; $5(x - 3) = 5x - 15$

Remainder

Thus, $(x^2 + 2x - 17) \div (x - 3) = (x + 5)$ R -2 or

$$x + 5 + \frac{-2}{x - 3}$$

PROBLEM 2

Divide: $y^2 + y - 7$ by $y - 2$

If there are missing terms in the polynomial being divided, we insert zero coefficients, as shown in Example 3.

EXAMPLE 3 **Dividing a third-degree polynomial by a binomial**

Divide: $2x^3 - 2 - 4x$ by $2 + 2x$

SOLUTION 3 We write the polynomials in *descending* order, inserting $0x^2$ in the dividend, since the x^2 term is missing. We then have

$$
\begin{array}{r}
x^2 - x - 1 \\
2x + 2\overline{)2x^3 + 0x^2 - 4x - 2} \\
(-)\,2x^3 + 2x^2 \qquad\quad \\
\hline
0 - 2x^2 - 4x \quad \\
(-) - 2x^2 - 2x \quad \\
\hline
-2x - 2 \\
(-) - 2x - 2 \\
\hline
0
\end{array}
$$

$\frac{2x^3}{2x} = x^2$; $x^2(2x + 2) = 2x^3 + 2x^2$

$-2x^2$ divided by $2x$ is $-x$. $-x(2x + 2) = -2x^2 - 2x$.

$-2x$ divided by $2x$ is -1. $-1(2x + 2) = -2x - 2$

There is no remainder.

Thus, $(2x^3 - 4x - 2) \div (2x + 2) = x^2 - x - 1$.

PROBLEM 3

Divide: $y^3 + y^2 - 7y - 3$ by $3 + y$

EXAMPLE 4 **Dividing a fourth-degree polynomial by a binomial**

Divide: $x^4 + x^3 - 3x^2 + 1$ by $x^2 - 3$

SOLUTION 4 We write the polynomials in *descending* order, inserting $0x$ for the missing term in the dividend. We then have

$$
\begin{array}{r}
x^2 + x \qquad\qquad\quad \\
x^2 - 3\overline{)x^4 + x^3 - 3x^2 + 0x + 1} \\
(-)\,x^4 \qquad - 3x^2 \qquad\qquad \\
\hline
0 + x^3 \qquad\qquad + 1 \\
(-)x^3 \qquad - 3x \quad \\
\hline
3x + 1
\end{array}
$$

x^4 divided by x^2 is x^2. $x^2(x^2 - 3) = x^4 - 3x^2$

x^3 divided by x^2 is x. $x(x^2 - 3) = x^3 - 3x$

We cannot divide $3x + 1$ by x^2, so we stop. The remainder is $3x + 1$.

(continued)

PROBLEM 4

Divide: $y^4 + y^3 - 2y^2 + y + 1$ by $y^2 - 2$

Answers to PROBLEMS

2. $y + 3 + \frac{-1}{y - 2}$ **3.** $y^2 - 2y - 1$ **4.** $y^2 + y + \frac{3y + 1}{y^2 - 2}$

In general, we stop the division when the degree of the remainder is less than the degree of the divisor or when the remainder is zero. Thus, $(x^4 + x^3 - 3x^2 + 1) \div (x^2 - 3) = (x^2 + x)$ R $(3x + 1)$. You can also write the answer as

$$x^2 + x + \frac{3x + 1}{x^2 - 3}$$

PROCEDURE

Checking the Result

1. Multiply the divisor $x^2 - 3$ by the quotient $x^2 + x$. (Use FOIL.)
2. Add the remainder $3x + 1$.
3. The result must be the dividend

$$x^4 + x^3 - 3x^2 + 1 \quad \text{(and it is!)}$$

$$\underbrace{(x^2 - 3)}_{\text{Divisor}}\underbrace{(x^2 + x)}_{\text{Quotient}} = \begin{array}{r} x^4 + x^3 - 3x^2 - 3x \\ (+) \qquad\qquad\qquad 3x + 1 \\ \hline x^4 + x^3 - 3x^2 \qquad + 1 \end{array}$$

How much is your annual water bill? Can you budget your water expenses for the next 3 years? We will show you how to do that in Example 5, but be warned that water costs fluctuate. Here are some current annual prices: Tampa, Florida, \$156; Pinellas County, Florida, \$456; San Luis Obispo, California, \$300–\$960; and Victorville, California, \$1920. The average national water bill for an American family was \$413 in 2008, possibly \$500 in the next couple of years. If you want better water prices, move to Livonia, New York, where Mayor Cal Lathan reports a \$140 average annual water bill taken from Hemlock Lake and flowing by gravity without pumps. The best part: they are not expecting a raise in rates!

EXAMPLE 5 Budgeting water expenses using polynomials

We can use polynomials to calculate your annual water bill cost for a three-year period. Suppose the cost for the first year is \$500 and there is a 2% increase each year. The expression $\frac{500x^n - 500}{x - 1}$, where x is the sum of 1 and the yearly percent increase, does more: it actually gives the cumulative (total) cost over n years.

a. Use the expression $\frac{500x^n - 500}{x - 1}$ to write the total water cost over a 3-year period.

b. Simplify the expression in part **a** by dividing $500x^3 - 500$ by $x - 1$.

c. When the yearly percent increase is 2%, x is the sum of 1 and 2%; that is, $x = 1 + 2\% = 1.02$. Substitute $x = 1.02$ in $\frac{500x^3 - 500}{x - 1}$ as well as in the simplified version of part **b.** Do you get the same answer? What is the cumulative water cost for three years?

SOLUTION 5

a. Since we want the cost over 3 years, we let $n = 3$ in $\frac{500x^n - 500}{x - 1}$ obtaining $\frac{500x^3 - 500}{x - 1}$.

PROBLEM 5

If the average water bill is \$1000 but the increase is only 1% each year, the expression $\frac{1000x^n - 1000}{x - 1}$, where x is the sum of 1 and the yearly percent increase is the total cost over n years.

a. Use the expression $\frac{1000x^n - 1000}{x - 1}$ to write the total water costs over a 3-year period.

b. Simplify the expression in part **a** by dividing $1000x^3 - 1000$ by $x - 1$.

Answers to PROBLEMS

5. a. $\frac{1000x^3 - 1000}{x - 1}$ **b.** $1000x^2 + 1000x + 1000$ **c.** Yes. \$3030.10

b. We use long division to divide $500x^3 - 500$ by $x - 1$.

The polynomials are in *descending* order, but we insert $0x^2$ and $0x$ in the dividend, since those two terms are missing. We then have

$500x^2 \quad + 500x + 500$	
$x - 1 \overline{)500x^3 + 0x^2 + 0x - 500}$	$\frac{500x^3}{x} = 500x^2$
$(-)\,500x^3 - 500x^2$	$500x^2(x - 1) = 500x^3 - 500x^2$
$0 \quad + 500x^2 + 0x - 500$	$500x^2$ divided by x is $500x$.
$(-)\quad 500x^2 - 500x$	$500x(x - 1) = 500x^2 - 500x$
$0 \quad + 500x - 500$	$500x$ divided by x is 500.
$(-)\quad 500x - 500$	$500(x - 1) = 500x - 500$
0	There is no remainder.

Thus, $\frac{500x^3 - 500}{x - 1} = 500x^2 + 500x + 500$ is the simplified version.

c. For $x = 1.02$, $\frac{500x^3 - 500}{x - 1} = \frac{500(1.02^3 - 1)}{1.02 - 1} = \frac{500(1.061208 - 1)}{0.02}$

$= \$1530.20$

For $x = 1.02$, $500x^2 + 500x + 500 = 500(1.02)^2 + 500(1.02) + 500$

$= 520.20 + 510 + 500$

$= \$1530.20$ (same as before)

Thus, the water cost for the 3 years is \$1530.20.

c. When the yearly percent increase is 1%, x is the sum of 1 and 1%, that is, $x = 1 + 1\% = 1.01$. Substitute $x = 1.01$ in $\frac{1000x^3 - 1000}{x - 1}$ as well as in the simplified version of part **b.** Do you get the same answer? What is the water cost for 3 years?

Calculator Corner

Checking Quotients

Now, how would you check Example 4? The original problem is $Y_1 = (x^4 + x^3 - 3x^2 + 1)/(x^2 - 3)$ and the answer is $Y_2 = x^2 + x + (3x + 1)/(x^2 - 3)$. Be very careful when entering the expressions Y_1 and Y_2. Note the parentheses! Press GRAPH. Did you get the same graph for Y_1 and Y_2? If you are not sure, let us check the numerical values by pressing 2nd GRAPH. Y_1 and Y_2 seem to have the same values, so we are probably correct. Why "probably"? Because we are not able to check that for every value of x (column 1), Y_1 and Y_2 (columns 2 and 3) have the same value. However, the lists give plausible evidence that they do.

X	Y1	Y2
0	-.3333	-.3333
1	0	0
2	13	13
3	13.667	13.667
4	21	21
5	30.727	30.727
6	42.576	42.576
X=0		

› Exercises 4.8

> Practice Problems > Self-Tests > Media-rich eBooks > e-Professors > Videos

‹ A › Dividing a Polynomial by a Monomial In Problems 1–10, divide.

1. $\frac{3x + 9y}{3}$

2. $\frac{6x + 8y}{2}$

3. $\frac{10x - 5y}{5}$

4. $\frac{24x - 12y}{6}$

5. $\frac{8y^3 - 32y^2 + 16y}{-4y^2}$

6. $\frac{9y^3 - 45y^2 + 9}{-3y^2}$

7. $10x^2 + 8x$ by x

8. $12x^2 + 18x$ by x

9. $15x^3 - 10x^2$ by $5x^2$

10. $18x^4 - 24x^2$ by $3x^2$

⟨B⟩ Dividing One Polynomial by Another Polynomial In Problems 11–40, divide.

11. $x^2 + 5x + 6$ by $x + 3$

12. $x^2 + 9x + 20$ by $x + 4$

13. $y^2 + 3y - 11$ by $y + 5$

14. $y^2 + 2y - 16$ by $y + 5$

15. $2x + x^2 - 24$ by $x - 4$

16. $4x + x^2 - 21$ by $x - 3$

17. $-8 + 2x + 3x^2$ by $2 + x$

18. $-6 + x + 2x^2$ by $2 + x$

19. $2y^2 + 9y - 36$ by $7 + y$

20. $3y^2 + 13y - 32$ by $6 + y$

21. $2x^3 - 4x - 2$ by $2x + 2$

22. $3x^3 - 9x - 6$ by $3x + 3$

23. $y^4 - y^2 - 2y - 1$ by $y^2 + y + 1$

24. $y^4 - y^2 - 4y - 4$ by $y^2 + y + 2$

25. $8x^3 - 6x^2 + 5x - 9$ by $2x - 3$

26. $2x^4 - x^3 + 7x - 2$ by $2x + 3$

27. $x^3 - 8$ by $x - 2$

28. $x^3 + 64$ by $x + 4$

29. $8y^3 - 64$ by $2y - 4$

30. $27x^3 - 8$ by $3x - 2$

31. $x^4 - x^2 - 2x + 2$ by $x^2 - x - 1$

32. $y^4 - y^3 - 3y - 9$ by $y^2 - y - 3$

33. $x^5 - x^4 + 6x^2 - 5x + 3$ by $x^2 - 2x + 3$

34. $y^6 - y^5 + 6y^3 - 5y^2 + 3y$ by $y^2 - 2y + 3$

35. $m^4 - 11m^2 + 34$ by $m^2 - 3$

36. $n^3 - n^2 - 6n$ by $n^2 + 3n$

37. $\dfrac{x^3 - y^3}{x - y}$

38. $\dfrac{x^3 + 8}{x + 2}$

39. $\dfrac{x^3 + 8}{x - 2}$

40. $\dfrac{x^5 + 32}{x - 2}$

⟩⟩⟩ Applications

41. *Unit costs* The unit cost $U(x)$ of x key chains is modeled by the equation $U(x) = \frac{2x + 20}{x}$.

a. Divide $2x + 20$ by x.

b. What is the unit cost for one key chain?

c. What is the unit cost for 10 key chains?

42. *Unit costs* The unit cost $U(x)$ of x earrings is modeled by the equation $U(x) = \frac{x^2 + 10}{x}$.

a. Divide $x^2 + 10$ by x.

b. What is the unit cost for one earring?

c. What is the unit cost for 10 earrings?

43. *Weight* A box containing $2x$ books weighs $2x^2 + 4x$ pounds.

a. How much does each book weigh?

b. If the box has 10 books, what is the weight of each book?

⟩⟩⟩ Applications: Green Math

44. *Generalizing Example 5* The expression given in Example 5 can be generalized to any situation in which you need the cumulative amount over a number of **years,** starting with an amount **M** with a **P** percent increase each year. The equivalent expression is

$$\frac{Mx^{\text{years}} - M}{P}, \text{ where } x = 1 + P$$

a. What are the values for M, the years, x, and P if you want to find the cumulative amount of carbon dioxide (CO_2) a car air conditioner produces in a 3-year period, starting at 200 pounds the first year and increasing by 2% each year?

b. Use the information from part **a** to find the total amount of carbon dioxide the air conditioner in a car will produce in a 3-year period.

45. *Carbon sequestration (absorption)*

A newly planted *Acacia angustissima*, 2.5 years old, 15 feet tall, with a trunk 3 inches in diameter sequesters (absorbs) about **20** pounds of CO_2 per year.

a. Use the expression in Problem 44 (or Example 5) to find the amount of CO_2 sequestered in a 3-year period starting with **20** pounds the first year and increasing 1% each year.

b. To the nearest whole number, about how many acacias will you need to absorb the carbon dioxide produced by the air conditioner of Problem 44?

46. *Foam cup generation* According to www.dosomething.org, Americans generate 30 billion foam cups each year.

a. If the rate is increasing by 1%, how many foam cups will be generated in a 3-year period? (Hint: See Problem 44.)

b. If the rate is increasing by 2%, how many foam cups will be generated in a 3-year period?

Styrene facts: 0.025% of the styrene, the plastic used in polystyrene foam cups and other containers, migrates into the beverage it contains. How much? The amount is directly proportional to the fat content of the food and inversely proportional to the size of the container!

Web IT go to mhhe.com/bello for more lessons

Using Your Knowledge

Is It Profitable? In business, the average cost per unit of a product, denoted by $\overline{C(x)}$, is defined by

$$\overline{C(x)} = \frac{C(x)}{x}$$

where $C(x)$ is a polynomial in the variable x, and x is the number of units produced.

47. If $C(x) = 3x^2 + 5x$, find $\overline{C(x)}$.

48. If $C(x) = 30 + 3x^2$, find $\overline{C(x)}$.

The average profit $\overline{P(x)}$ is

$$\overline{P(x)} = \frac{P(x)}{x}$$

where $P(x)$ is a polynomial in the variable x, and x is the number of units sold in a certain period of time.

49. If $P(x) = 50x + x^2 - 7000$ (dollars), find the average profit.

50. If in Problem 49, 100 units are sold in a period of 1 week, what is the average profit?

Write On

51. How can you check that your answer is correct when you divide one polynomial by another? Explain.

52. A problem in a recent test stated: "Find the quotient of $x^2 + 5x + 6$ and $x + 2$." Do you have to divide $x + 2$ by $x^2 + 5x + 6$ or do you have to divide $x^2 + 5x + 6$ by $x + 2$? Explain.

53. When you are dividing one polynomial by another, when do you stop the division process?

Concept Checker

Fill in the blank(s) with the correct word(s), phrase, or mathematical statement.

54. To divide a **polynomial** by a **monomial,** divide each term in the polynomial by the ________________.

55. To **check** the result of a **polynomial division, multiply** the ________________ by the ________________ and add the ________________.

dividend **remainder**
quotient **divisor**
monomial

Mastery Test

Divide:

56. $x^4 + x^3 - 2x^2 + 1$ by $x^2 - 2$

57. $2x^3 + x - 3$ by $x - 1$

58. $x^2 + 4x - 15$ by $x - 2$

59. $\dfrac{24x^4 - 18x^3}{6x^2}$

60. $\dfrac{16x^4 - 4x^2 + 8x}{8x^2}$

61. $\dfrac{-6y^3 + 12y^2 + 3}{-3y^2}$

Skill Checker

Find the LCM of the specified numbers:

62. 23 and 92

63. 20 and 18

64. 30 and 16

65. 40 and 12

66. 10, 18, and 12

67. 20, 30, and 18

68. 40, 15, and 10

Collaborative Learning

How fast can you go?

How fast can you obtain information to solve a problem? Form three groups: library, the Web, and bookstore (where you can look at books, papers, and so on for free). Each group is going to research car prices. Select a car model that has been on the market for at least 5 years. Each of the groups should find:

1. The new car value and the value of a 3-year-old car of the same model
2. The estimated depreciation rate for the car
3. The estimated value of the car in 3 years
4. A graph comparing age and value of the car for the next 5 years
5. An equation of the form $C = P(1 - r)^n$ or $C = rn + b$, where n is the number of years after purchase and r is the depreciation rate

Which group finished first? Share the procedure used to obtain your information so the most efficient research method can be established.

Research Questions

1. In the *Human Side of Algebra* at the beginning of this chapter, we mentioned that in the Golden Age of Greek mathematics, roughly from 300 to 200 B.C., three mathematicians "stood head and shoulders above all the others of the time." Apollonius was one. Who were the other two, and what were their contributions to mathematics?
2. Write a short paragraph about Diophantus and his contributions to mathematics.
3. Write a paper about the *Arithmetica* written by Diophantus. How many books were in this *Arithmetica?* What is its relationship to algebra?
4. It is said that "one of Diophantus' main contributions was the 'syncopation' of algebra." Explain what this means.
5. Write a short paragraph about Nicole Oresme and his mathematical achievements.
6. Expand on the discussion of the works of Nicholas Chuquet described in the *Human Side of Algebra.*
7. Write a short paper about the contributions of François Viète regarding the development of algebraic notation.

Summary Chapter 4

Section	Item	Meaning	Example
4.1A	Product rule for exponents	$x^m \cdot x^n = x^{m+n}$	$x^3 \cdot x^5 = x^{3+5} = x^8$
4.1B	Quotient rule for exponents	$\frac{x^m}{x^n} = x^{m-n}\,(m > n, x \neq 0)$	$\frac{x^8}{x^3} = x^{8-3} = x^5$

Section	Item	Meaning	Example
4.1C	Power rule for exponents	$(x^m)^n = x^{mn}$	$(x^5)^{10} = x^{50}$
	Power rule for products	$(x^m y^n)^k = x^{mk} y^{nk}$	$(x^3y^2)^4 = x^{12}y^8$
	Power rule for quotients	$\left(\frac{x}{y}\right)^m = \frac{x^m}{y^m}$ $(y \neq 0)$	$\left(\frac{x^2}{y^3}\right)^4 = \frac{x^8}{y^{12}}$
4.2A	Zero exponent	For $x \neq 0$, $x^0 = 1$	$3^0 = 1$, $(-8)^0 = 1$, $(3x)^0 = 1$
4.2A	x^{-n}, n a positive integer	$x^{-n} = \frac{1}{x^n}$ $(x \neq 0)$	$2^{-3} = \frac{1}{2^3} = \frac{1}{8}$
4.3A	Scientific notation	A number is in scientific notation when written in the form $M \times 10^n$, where M is a number between 1 and 10 and n is an integer.	3×10^{-3} and 2.7×10^5 are in scientific notation.
4.4A	Polynomial	An algebraic expression formed by using the operations of addition and subtraction on products of numbers and a variable raised to whole-number exponents	$x^2 + 3x - 5$, $2x + 8 - x^3$, and $9x^7 - 3x^3 + 4x^8 - 10$ are polynomials but $\sqrt{x} - 3$, $\frac{x^2 - 2x + 3}{x}$, and $x^{3/2} - x$ are not.
	Terms	The parts of a polynomial separated by plus signs are the **terms.**	The terms of $x^2 - 2x + 3$ are x^2, $-2x$, and 3.
	Monomial	A polynomial with *one* term	$3x$, $7x^2$, and $-3x^{10}$ are monomials.
	Binomial	A polynomial with *two* terms	$3x + x^2$, $7x - 8$, and $x^3 - 8x^7$, are binomials.
	Trinomial	A polynomial with *three* terms	$-8 + 3x + x^2$, $7x - 8 + x^4$, and $x^3 - 8x^7 + 9$ are trinomials.
4.4B	Degree	The degree of a polynomial is the highest exponent of the variable.	The degree of $8 + 3x + x^2$ is 2, and the degree of $7x - 8 + x^4$ is 4.
4.6C	FOIL method (SP1)	To multiply two binomials such as $(x + a)(x + b)$, multiply the **F**irst terms, the **O**uter terms, the **I**nner terms, and the **L**ast terms, then add.	F O I L $(x + 2)(x + 3) = x^2 + 3x + 2x + 6$ $= x^2 + 5x + 6$
4.7A	The square of a binomial sum (SP2)	$(X + A)^2 = X^2 + 2AX + A^2$	$(x + 5)^2 = x^2 + 2 \cdot 5 \cdot x + 5^2$ $= x^2 + 10x + 25$
4.7B	The square of a binomial difference (SP3)	$(X - A)^2 = X^2 - 2AX + A^2$	$(x - 5)^2 = x^2 - 2 \cdot 5 \cdot x + 5^2$ $= x^2 - 10x + 25$
4.7C	The product of the sum and difference of two terms (SP4)	$(X + A)(X - A) = X^2 - A^2$	$(x + 7)(x - 7) = x^2 - 7^2$ $= x^2 - 49$
4.8A	Dividing a polynomial by a monomial	To divide a polynomial by a monomial, divide each term in the polynomial by the monomial.	$\frac{35x^5 - 21x^3}{7x^2} = \frac{35x^5}{7x^2} - \frac{21x^3}{7x^2}$ $= 5x^3 - 3x$

(continued)

Section	Item	Meaning	Example
4.8B	Dividing one polynomial by another	To divide one polynomial by another, use polynomial long division.	Divide $6x^2 + 11x - 35$ by $3x - 5$. $$\begin{array}{r} 2x + 7 \\ 3x - 5\overline{)6x^2 + 11x - 35} \\ (-)6x^2 - 10x \\ \hline 21x - 35 \\ (-)21x - 35 \\ \hline 0 \end{array}$$ Thus, $(6x^2 + 11x - 35) \div (3x - 5)$ is $2x + 7$.

› Review Exercises Chapter 4

(If you need help with these questions, look in the section indicated in brackets.)

1. **‹4.1A›** *Find the product.*

a. **(i)** $(3a^2b)(-5ab^3)$

(ii) $(4a^2b)(-6ab^4)$

(iii) $(5a^2b)(-7ab^3)$

b. **(i)** $(-2xy^2z)(-3x^2yz^4)$

(ii) $(-3x^2yz^2)(-4xy^3z)$

(iii) $(-4xyz)(-5xy^2z^3)$

2. **‹4.1B›** *Find the quotient.*

a. **(i)** $\frac{16x^6y^8}{-8xy^4}$

(ii) $\frac{24x^7y^6}{-4xy^3}$

(iii) $\frac{-18x^8y^7}{9xy^4}$

b. **(i)** $\frac{-8x^9y^7}{-16x^4y}$

(ii) $\frac{-5x^7y^8}{-10x^6y}$

(iii) $\frac{-3x^9y^7}{-9x^8y}$

3. **‹4.1C›** *Simplify.*

a. $(2^2)^3$ **b.** $(2^2)^2$

c. $(3^2)^2$

4. **‹4.1C›** *Simplify.*

a. $(y^3)^2$ **b.** $(x^2)^3$

c. $(a^4)^5$

5. **‹4.1C›** *Simplify.*

a. $(4xy^3)^2$ **b.** $(2x^2y)^3$

c. $(3x^2y^2)^3$

6. **‹4.1C›** *Simplify.*

a. $(-2xy^3)^3$ **b.** $(-3x^2y^3)^2$

c. $(-2x^2y^2)^3$

7. **‹4.1C›** *Simplify.*

a. $\left(\frac{2y^2}{x^4}\right)^2$ **b.** $\left(\frac{3x}{y^3}\right)^3$

c. $\left(\frac{2x^2}{y^4}\right)^4$

8. **‹4.1C›** *Simplify.*

a. $(2x^4)^3(-2y^2)^2$

b. $(3x^2)^2(-2y^3)^3$

c. $(4x^3)^2(-2y^4)^4$

9. **‹4.2A›** *Write using positive exponents and then evaluate.*

a. 2^{-3} **b.** 3^{-4}

c. 5^{-2}

10. **‹4.2A›** *Write using positive exponents.*

a. $\left(\frac{1}{x}\right)^{-4}$ **b.** $\left(\frac{1}{y}\right)^{-3}$

c. $\left(\frac{1}{z}\right)^{-5}$

11. **4.2B** *Write using negative exponents.*

a. $\frac{1}{x^5}$ **b.** $\frac{1}{y^7}$

c. $\frac{1}{z^8}$

12. **4.2C** *Multiply and simplify.*

a. **(i)** $2^8 \cdot 2^{-5}$ **b.** **(i)** $y^{-3} \cdot y^{-5}$

(ii) $2^6 \cdot 2^{-4}$ **(ii)** $y^{-2} \cdot y^{-3}$

(iii) $3^6 \cdot 3^{-3}$ **(iii)** $y^{-4} \cdot y^{-2}$

13. **4.2C** *Divide and simplify.*

a. **(i)** $\frac{x}{x^5}$ **b.** **(i)** $\frac{a^{-2}}{a^{-2}}$ **c.** **(i)** $\frac{x^{-2}}{x^{-3}}$

(ii) $\frac{x}{x^7}$ **(ii)** $\frac{a^{-4}}{a^{-4}}$ **(ii)** $\frac{x^{-5}}{x^{-8}}$

(iii) $\frac{x}{x^9}$ **(iii)** $\frac{a^{-10}}{a^{-10}}$ **(iii)** $\frac{x^{-7}}{x^{-9}}$

14. **4.2C** *Simplify.*

a. $\left(\frac{2x^3y^4}{3x^4y^{-3}}\right)^{-2}$ **b.** $\left(\frac{3x^5y^{-3}}{2x^7y^4}\right)^{-3}$

c. $\left(\frac{3x^{-5}y^{-4}}{2x^{-6}y^{-8}}\right)^{-2}$

15. **4.3A** *Write in scientific notation.*

a. **(i)** 44,000,000

(ii) 4,500,000

(iii) 460,000

b. **(i)** 0.0014

(ii) 0.00015

(iii) 0.000016

16. **4.3B** *Perform the indicated operations and write the answer in scientific notation.*

a. **(i)** $(2 \times 10^2) \times (1.1 \times 10^3)$

(ii) $(3 \times 10^2) \times (3.1 \times 10^4)$

(iii) $(4 \times 10^2) \times (3.1 \times 10^5)$

b. **(i)** $\frac{1.15 \times 10^{-3}}{2.3 \times 10^{-4}}$

(ii) $\frac{1.38 \times 10^{-3}}{2.3 \times 10^{-4}}$

(iii) $\frac{1.61 \times 10^{-3}}{2.3 \times 10^{-4}}$

17. **4.4A** *Classify as a monomial, binomial, or trinomial.*

a. $9x^2 - 9 + 7x$ **b.** $7x^2$

c. $3x - 1$

18. **4.4B** *Find the degree of the given polynomial.*

a. $3x^2 - 7x + 8x^4$ **b.** $-4x + 2x^2 - 3$

c. $8 + 3x - 4x^2$

19. **4.4C** *Write the given polynomial in descending order.*

a. $4x^2 - 8x + 9x^4$

b. $-3x + 4x^2 - 3$

c. $8 + 3x - 4x^2$

20. **4.4D** *Find the value of* $-16t^2 + 300$ *for each value of t.*

a. $t = 1$ **b.** $t = 3$

c. $t = 5$

21. **4.5A** *Add the given polynomials.*

a. $-5x + 7x^2 - 3$ and $-2x^2 - 7 + 4x$

b. $-3x^2 + 8x - 1$ and $3 + 7x - 2x^2$

c. $-4 + 3x^2 - 5x$ and $6x^2 - 2x + 5$

22. **4.5B** *Subtract the first polynomial from the second.*

a. $3x - 4 + 7x^2$ from $6x^2 - 4x$

b. $5x - 3 + 2x^2$ from $9x^2 - 2x$

c. $6 - 2x + 5x^2$ from $2x - 5$

23. **4.6A** *Multiply.*

a. $(-6x^2)(3x^5)$

b. $(-8x^3)(5x^6)$

c. $(-9x^4)(3x^7)$

24. **4.6B** *Remove parentheses (simplify).*

a. $-2x^2(x + 2y)$

b. $-3x^3(2x + 3y)$

c. $-4x^3(5x + 7y)$

25. ⟨**4.6C**⟩ *Multiply.*

a. $(x + 6)(x + 9)$

b. $(x + 2)(x + 3)$

c. $(x + 7)(x + 9)$

26. ⟨**4.6C**⟩ *Multiply.*

a. $(x + 7)(x - 3)$

b. $(x + 6)(x - 2)$

c. $(x + 5)(x - 1)$

27. ⟨**4.6C**⟩ *Multiply.*

a. $(x + 3)(x - 7)$

b. $(x + 2)(x - 6)$

c. $(x + 1)(x - 5)$

28. ⟨**4.6C**⟩ *Multiply.*

a. $(3x - 2y)(2x - 3y)$

b. $(5x - 3y)(4x - 3y)$

c. $(4x - 3y)(2x - 5y)$

29. ⟨**4.7A**⟩ *Expand.*

a. $(2x + 3y)^2$

b. $(3x + 4y)^2$

c. $(4x + 5y)^2$

30. ⟨**4.7B**⟩ *Expand.*

a. $(2x - 3y)^2$

b. $(3x - 2y)^2$

c. $(5x - 2y)^2$

31. ⟨**4.7C**⟩ *Multiply.*

a. $(3x - 5y)(3x + 5y)$

b. $(3x - 2y)(3x + 2y)$

c. $(3x - 4y)(3x + 4y)$

32. ⟨**4.7D**⟩ *Multiply.*

a. $(x + 1)(x^2 + 3x + 2)$

b. $(x + 2)(x^2 + 3x + 2)$

c. $(x + 3)(x^2 + 3x + 2)$

33. ⟨**4.7E**⟩ *Multiply.*

a. $3x(x + 1)(x + 2)$

b. $4x(x + 1)(x + 2)$

c. $5x(x + 1)(x + 2)$

34. ⟨**4.7E**⟩ *Expand.*

a. $(x + 2)^3$

b. $(x + 3)^3$

c. $(x + 4)^3$

35. ⟨**4.7E**⟩ *Expand.*

a. $\left(5x^2 - \frac{1}{2}\right)^2$

b. $\left(7x^2 - \frac{1}{2}\right)^2$

c. $\left(9x^2 - \frac{1}{2}\right)^2$

36. ⟨**4.7E**⟩ *Multiply.*

a. $(3x^2 + 2)(3x^2 - 2)$

b. $(3x^2 + 4)(3x^2 - 4)$

c. $(2x^2 + 5)(2x^2 - 5)$

37. ⟨**4.8A**⟩ *Divide.*

a. $\frac{18x^3 - 9x^2}{9x}$

b. $\frac{20x^3 - 10x^2}{5x}$

c. $\frac{24x^3 - 12x^2}{6x}$

38. ⟨**4.8B**⟩ *Divide.*

a. $x^2 + 4x - 12$ by $x - 2$

b. $x^2 + 4x - 21$ by $x - 3$

c. $x^2 + 4x - 32$ by $x - 4$

39. ⟨**4.8B**⟩ *Divide.*

a. $8x^3 - 16x - 8$ by $2 + 2x$

b. $12x^3 - 24x - 12$ by $2 + 2x$

c. $4x^3 - 8x - 4$ by $2 + 2x$

40. ⟨**4.8B**⟩ *Divide.*

a. $2x^3 - 20x + 8$ by $x - 3$

b. $2x^3 - 21x + 12$ by $x - 3$

c. $3x^3 - 4x + 5$ by $x - 1$

41. ⟨**4.8B**⟩ *Divide.*

a. $x^4 + x^3 - 4x^2 + 1$ by $x^2 - 4$

b. $x^4 + x^3 - 5x^2 + 1$ by $x^2 - 5$

c. $x^4 + x^3 - 6x^2 + 1$ by $x^2 - 6$

Practice Test Chapter 4

(Answers on page 404)

Visit www.mhhe.com/bello to view helpful videos that provide step-by-step solutions to several of the problems below.

1. Simplify:

a. $(2a^3b)(-6ab^3)$ **b.** $(-2x^2yz)(-6xy^3z^4)$

c. $\dfrac{18x^5y^7}{-9xy^3}$

2. Simplify:

a. $(2x^3y^2)^3$

b. $(-3x^2y^3)^2$

3. Simplify:

a. $\left(\dfrac{3}{4}\right)^3$ **b.** $\left(\dfrac{3x^3}{y^4}\right)^3$

4. Simplify and write the answer with positive exponents.

a. $\left(\dfrac{1}{x}\right)^{-7}$ **b.** $\dfrac{x^{-6}}{x^{-6}}$ **c.** $\dfrac{x^{-6}}{x^{-7}}$

5. Simplify.

a. $3x^2x^{-4}$ **b.** $\left(\dfrac{2x^{-3}y^4}{3x^2y^3}\right)^{-2}$

6. Write in scientific notation.

a. 48,000,000 **b.** 0.00000037

7. Perform the indicated operations and write the answers in scientific notation.

a. $(3 \times 10^4) \times (7.1 \times 10^6)$

b. $\dfrac{2.84 \times 10^{-2}}{7.1 \times 10^{-3}}$

8. Classify as a monomial, binomial, or trinomial.

a. $3x - 5$ **b.** $5x^3$ **c.** $8x^2 - 2 + 5x$

9. Write the polynomial $-3x + 7 + 8x^2$ in descending order and find its degree.

10. Find the value of $-16t^2 + 100$ when $t = 2$.

11. Add $-4x + 8x^2 - 3$ and $-5x^2 - 4 + 2x$.

12. Subtract $5x - 2 + 8x^2$ from $3x^2 - 2x$.

13. Remove parentheses (simplify): $-2x^2(x + 3y)$.

14. Multiply $(x + 8)(x - 3)$.

15. Multiply $(x + 4)(x - 6)$.

16. Multiply $(5x - 2y)(4x - 3y)$.

17. Expand $(3x + 5y)^2$.

18. Expand $(2x - 7y)^2$.

19. Multiply $(2x - 5y)(2x + 5y)$.

20. Multiply $(x + 2)(x^2 + 5x + 3)$.

21. Multiply $3x(x + 2)(x + 5)$.

22. Expand $(x + 7)^3$.

23. Expand $\left(3x^2 - \dfrac{1}{2}\right)^2$.

24. Multiply $(3x^2 + 7)(3x^2 - 7)$.

25. Divide $2x^3 - 9x + 5$ by $x - 2$.

Answers to Practice Test Chapter 4

Answer	If You Missed	Review		
	Question	Section	Examples	Page
1. a. $-12a^4b^4$ **b.** $12x^3y^4z^5$ **c.** $-2x^4y^4$	1	4.1	1, 2, 3	320–322
2. a. $8x^9y^6$ **b.** $9x^4y^6$	2	4.1	4, 5	323
3. a. $\frac{27}{64}$ **b.** $\frac{27x^9}{y^{12}}$	3	4.1	6	324
4. a. x^7 **b.** 1 **c.** x	4	4.2	1, 2, 5	330–331, 333
5. a. $\frac{3}{x^2}$ **b.** $\frac{9x^{10}}{4y^2}$	5	4.2	4, 6	332, 334
6. a. 4.8×10^7 **b.** 3.7×10^{-7}	6	4.3	1	341
7. a. 2.13×10^{11} **b.** 4	7	4.3	3, 4	341–342
8. a. Binomial **b.** Monomial **c.** Trinomial	8	4.4	1	348
9. $8x^2 - 3x + 7$; 2	9	4.4	2, 3	349–350
10. 36	10	4.4	4, 5, 6	350–351
11. $3x^2 - 2x - 7$	11	4.5	1, 2	359–360
12. $-5x^2 - 7x + 2$	12	4.5	3	360–361
13. $-2x^3 - 6x^2y$	13	4.6	1, 2	368–369
14. $x^2 + 5x - 24$	14	4.6	3	370
15. $x^2 - 2x - 24$	15	4.6	3	370
16. $20x^2 - 23xy + 6y^2$	16	4.6	5	371
17. $9x^2 + 30xy + 25y^2$	17	4.7	1	379
18. $4x^2 - 28xy + 49y^2$	18	4.7	2	380
19. $4x^2 - 25y^2$	19	4.7	3	381
20. $x^3 + 7x^2 + 13x + 6$	20	4.7	4	382
21. $3x^3 + 21x^2 + 30x$	21	4.7	5	383
22. $x^3 + 21x^2 + 147x + 343$	22	4.7	6	383–384
23. $9x^4 - 3x^2 + \frac{1}{4}$	23	4.7	7	384
24. $9x^4 - 49$	24	4.7	8	384
25. $(2x^2 + 4x - 1)$ R 3	25	4.8	2, 3, 4	393–394

Cumulative Review Chapters 1–4

1. Find the additive inverse (opposite) of -7.

2. Find: $\left|-9\frac{9}{10}\right|$

3. Add: $-\frac{1}{9} + \left(-\frac{1}{6}\right)$

4. Subtract: $7.2 - (-6.4)$

5. Multiply: $(-2.4)(2.6)$

6. Find: $-(2^4)$

7. Divide: $-\frac{3}{4} \div \left(-\frac{7}{8}\right)$

8. Evaluate $y \div 5 \cdot x - z$ for $x = 5$, $y = 50$, $z = 3$.

9. Which property is illustrated by the following statement?
$$6 \cdot (5 \cdot 2) = (6 \cdot 5) \cdot 2$$

10. Multiply: $3(2x - 8)$

11. Combine like terms: $-8xy^3 - (-xy^3)$

12. Simplify: $5x + (x + 4) - 2(x - 3)$

13. Write in symbols: The quotient of $(m + 3n)$ and p

14. Does the number -3 satisfy the equation $12 = 15 - x$?

15. Solve for x: $5 = 5(x - 3) + 5 - 4x$

16. Solve for x: $-\frac{8}{7}x = -56$

17. Solve for x: $\frac{x}{4} - \frac{x}{9} = 5$

18. Solve for x: $5 - \frac{x}{3} = \frac{3(x + 1)}{7}$

19. Solve for b in the equation $S = 6a^2b$.

20. The sum of two numbers is 100. If one of the numbers is 20 more than the other, what are the numbers?

21. Dave has invested a certain amount of money in stocks and bonds. The total annual return from these investments is \$615. If the stocks produce \$245 more in returns than the bonds, how much money does Dave receive annually from each type of investment?

22. Train A leaves a station traveling at 30 mph. Six hours later, train B leaves the same station traveling in the same direction at 40 mph. How long does it take for train B to catch up to train A?

23. Susan purchased some municipal bonds yielding 8% annually and some certificates of deposit yielding 11% annually. If Susan's total investment amounts to \$17,000 and the annual income is \$1630, how much money is invested in bonds and how much is invested in certificates of deposit?

24. Graph: $-\frac{x}{2} + \frac{x}{9} \geq \frac{x - 9}{9}$

0 2

25. Graph the point $C(4, -4)$.

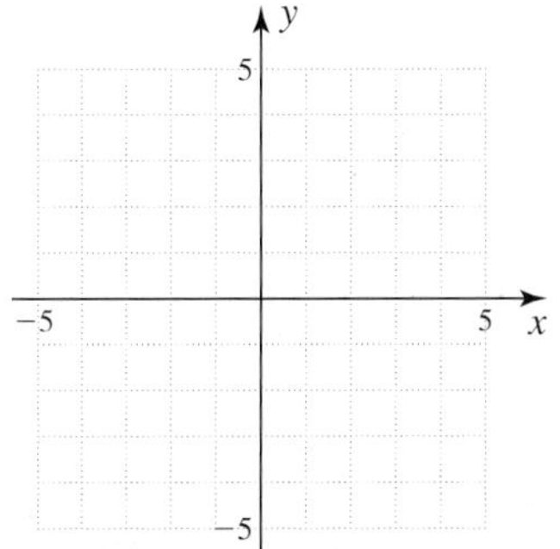

26. What are the coordinates of point A?

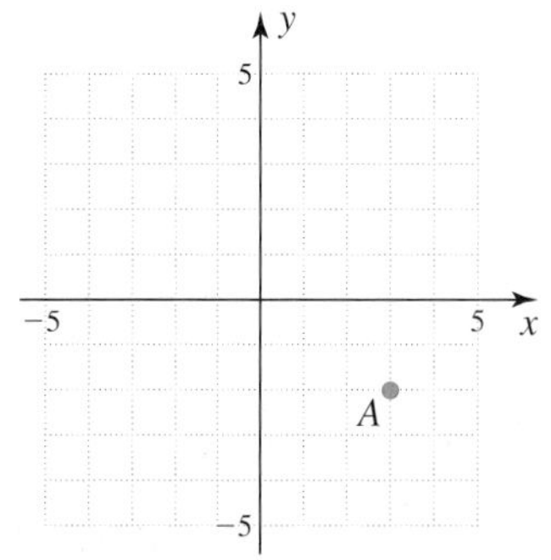

27. Determine whether the ordered pair $(-5, 1)$ is a solution of $5x + 3y = -22$.

28. Find x in the ordered pair $(x, -2)$ so that the ordered pair satisfies the equation $5x - 4y = 13$.

29. Graph: $5x + y = 5$

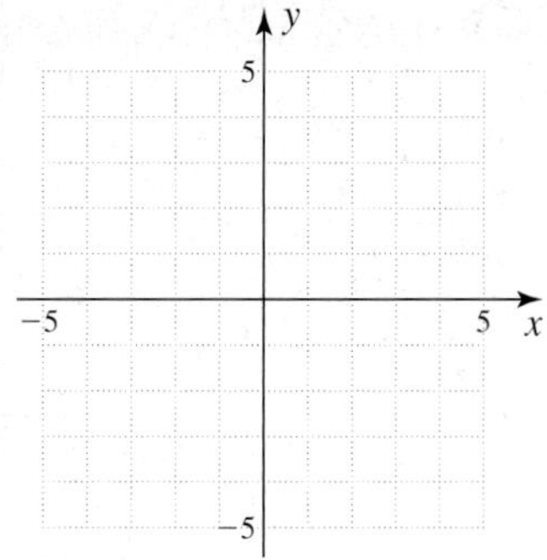

30. Graph: $3y - 15 = 0$

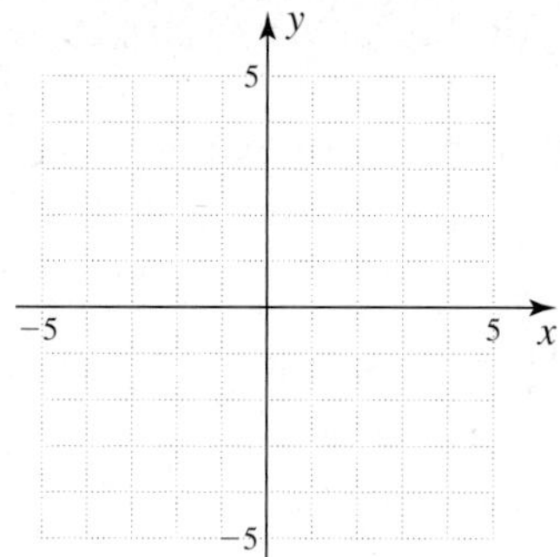

31. Find the slope of the line passing through the points $(-7, -1)$ and $(-2, 6)$.

32. What is the slope of the line $8x - 4y = 14$?

33. Find the pair of parallel lines.

(1) $18y + 24x = 8$

(2) $3y = 4x + 8$

(3) $24x - 18y = 8$

34. Multiply: $(-3x^4y)(-6x^3y^2)$

35. Divide: $\dfrac{25x^2y^5}{-5xy^7}$

36. Divide: $\dfrac{x^{-6}}{x^{-9}}$

37. Multiply and simplify: $x^8 \cdot x^{-3}$

38. Simplify: $(4x^4y^{-4})^3$

39. Write in scientific notation: 8,000,000

40. Divide and express the answer in scientific notation: $(26.04 \times 10^{-5}) \div (6.2 \times 10^3)$

41. Classify as a monomial, binomial, or trinomial. $2x^2$

42. Find the degree of the polynomial: $3x^2 - 3x + 1$

43. Write the polynomial in descending order: $-3x^2 - x - 3x^3 + 4$

44. Find the value of $3x^3 + 2x^2$ when $x = 2$.

45. Add $(x + 6x^3 - 2)$ and $(-2x^3 - 1 + 8x)$.

46. Remove parentheses (simplify): $-3x(3x^2 + 4y)$

47. Find: $(3x + 2y)^2$

48. Find: $(4x - 3y)(4x + 3y)$

49. Find: $\left(5x^2 - \dfrac{1}{2}\right)^2$

50. Find: $(4x^2 + 9)(4x^2 - 9)$

51. Divide $(3x^3 - 20x^2 + 29x - 17)$ by $(x - 5)$.

52. Find an equation of the line passing through $(-2, 6)$ and with slope 4.

53. Find an equation of the line with slope -2 and y-intercept 5.

54. Graph the inequality $2x - 3y < -6$.

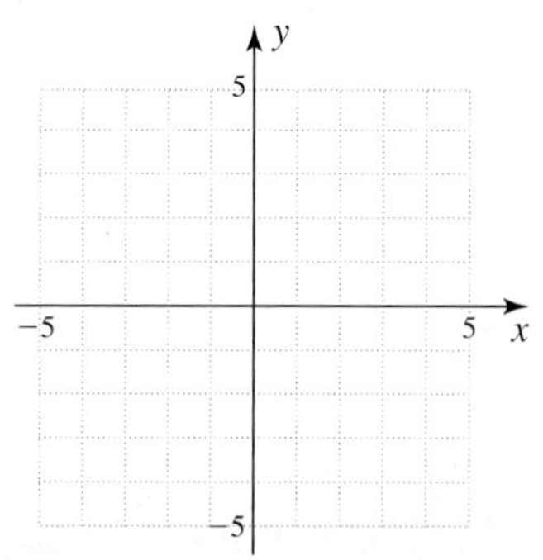

55. Graph the inequality $3x - y \geq 0$.

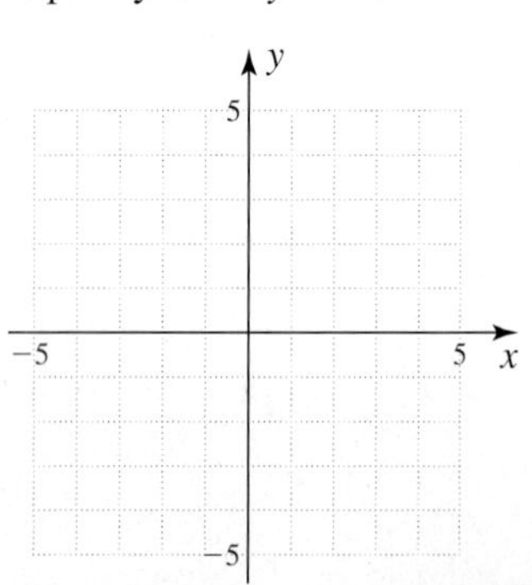

Section

Chapter

Chapter 5 five Factoring

The Human Side of Algebra

One of the most famous mathematicians of antiquity is Pythagoras, to whom has been ascribed the theorem that bears his name (see Section 5.7). It's believed that he was born between 580 and 569 B.C. on the Aegean island of Samos, from which he was later banned by the powerful tyrant Polycrates. When he was about 50, Pythagoras moved to Croton, a colony in southern Italy, where he founded a secret society of 300 young aristocrats called the Pythagoreans. Four subjects were studied: arithmetic, music, geometry, and astronomy. Students attending lectures were divided into two groups: *acoustici* (listeners) and *mathematici*. How did one get to join the *mathematici* in those days? One listened to the master's voice (*acoustici*) from behind a curtain for a period of 3 years! According to Burton's *History of Mathematics,* the Pythagoreans had "strange initiations, rites, and prohibitions." Among them was their refusal "to eat beans, drink wine, pick up anything that had fallen or stir a fire with an iron," but what set them apart from other sects was their philosophy that "knowledge is the greatest purification," and to them, knowledge meant mathematics. You can now gain some of this knowledge by studying Pythagoras' theorem yourself, and evaluating its greatness.

5.1 Common Factors and Grouping

Objectives

A Find the greatest common factor (GCF) of numbers.

B Find the GCF of terms.

C Factor out the GCF.

D Factor a four-term expression by grouping.

To Succeed, Review How To . . .

1. Write a number as a product of primes (p. 6).
2. Write a polynomial in descending order (pp. 349–350).
3. Use the distributive property (pp. 81–83).

Getting Started

Expansion Joints and Factoring

Have you ever noticed as you were going over a bridge that there is always a piece of metal at about the midpoint? This piece of metal is called an *expansion joint,* and it prevents the bridge from cracking when it expands or contracts. We can describe this expansion algebraically. In general, if α is the coefficient of linear expansion, L is the length of the material, and t_2 and t_1 are the high and low temperatures in degrees Celsius, then the linear expansion of a solid is

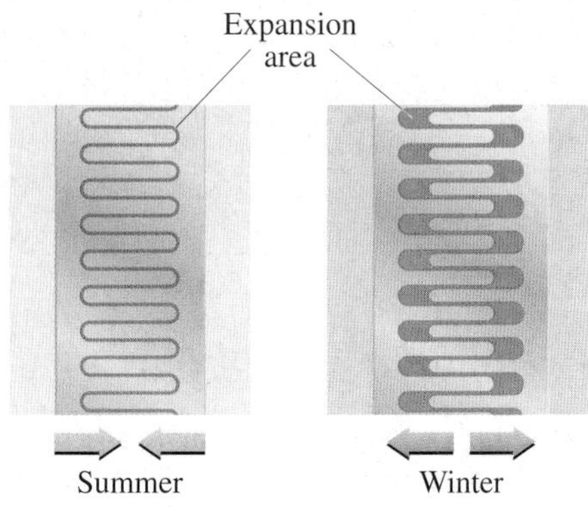

$$e = \alpha L t_2 - \alpha L t_1$$

The expression on the right-hand side of the equation can be written in a more useful way if we *factor* it, that is, write it as a **product** of the expressions $\alpha L \Delta T$. What is ΔT? It is the change in temperature and you will see how to write it when you do Problem 81! In this section, we learn how to factor polynomials by finding a common factor and by grouping.

To *factor* an expression, we write the expression as a *product* of its factors. When two or more numbers are to be multiplied to form a product, the numbers being multiplied are the **factors.** Does this sound complicated? It isn't! Suppose we give you the numbers 3 and 5 and tell you to multiply them. You will probably write

$$3 \times 5 = 15$$

In the reverse process, we give you the product 15 and tell you to **factor** it! (Factoring is the *reverse* of multiplying.) You will then write

$$\underbrace{15}_{\text{Product}} = \underbrace{3 \times 5}_{\text{Factors}}$$

Why? Because you know that multiplying 3 by 5 gives you 15. What about factoring the number 20? Here you can write

$$20 = 4 \times 5 \quad \text{or} \quad 20 = 2 \times 10$$

Note that $20 = 4 \times 5$ and $20 = 2 \times 10$ are not **completely** factored. They contain factors that are *not* prime numbers, numbers divisible only by themselves and 1. The first few prime numbers are 2, 3, 5, 7, 11, and so on. Since neither of these factorizations is *complete,* we note that in the first factorization $4 = (2 \times 2)$, so

$$20 = (2 \times 2) \times 5$$

whereas in the second factorization, $10 = (2 \times 5)$. Thus,

$$20 = 2 \times (2 \times 5)$$

In either case, the complete factorization for 20 is

$$20 = 2 \times 2 \times 5$$

From now on, when we say factor a *number,* we mean factor it *completely.* Moreover, the numerical factors are assumed to be prime numbers unless noted otherwise. Thus, we don't factor 20 as

$$20 = \frac{1}{4} \times 80$$

With these preliminaries out of the way, we can now factor some algebraic expressions. There are different factoring techniques for different situations. We will study most of them in this chapter.

A > Finding the Greatest Common Factor (GCF) of Numbers

When the numbers 15 and 20 are completely factored, we have

$$15 = 3 \cdot 5 \quad \text{and} \quad 20 = 2 \cdot 2 \cdot 5$$

As you can see, 5 is a common factor of 15 and 20. Similarly, 3 is a common factor of 18 and 30, because 3 is a factor of 18 and 3 is a factor of 30. Other common factors of 18 and 30 are 1, 2, and 6. The greatest (largest) of the common factors of 18 and 30 is 6, so 6 is the ***greatest common factor (GCF)*** of 18 and 30. Note that the GCF is the same as the GCD we already studied in Section R.1. In general, we have:

GREATEST COMMON FACTOR OF A LIST OF INTEGERS

The **greatest common factor (GCF)** of a list of integers is the **largest** common factor of the integers in the list.

EXAMPLE 1 Finding the GCF of a List of Numbers

Find the greatest common factor (GCF) of:

a. 45 and 60 **b.** 36, 60, and 108 **c.** 10, 17, and 12

SOLUTION 1

a. To discover the common factors, write each of the numbers as a product of primes using exponents:

$$45 = 3 \cdot 3 \cdot 5 \quad = 3^2 \cdot 5$$
$$60 = 2 \cdot 2 \cdot 3 \cdot 5 = 2^2 \cdot 3 \cdot 5$$

The common factors are 3 and 5; thus, the GCF is $3 \cdot 5 = 15$. Note that to find the GCF, we simply choose the common prime factors with the ***smallest*** exponents and find their product.

(continued)

PROBLEM 1

Find the GCF of:

a. 30 and 45

b. 45, 60, and 108

c. 20, 13, and 18

Answers to PROBLEMS

1. a. 15 **b.** 3 **c.** 1

Here is a shortcut: Write

$$\underline{|45 \quad 60}$$

Divide each number by a prime divisor common to all numbers. The product of all divisors is the GCF.

$$\begin{array}{c|cc} 3 & 45 & 60 \\ \hline 5 & 15 & 20 \\ \hline & 3 & 4 \end{array}$$

GCF $= 3 \cdot 5 = 15$

b. We write each of the integers in factored form using exponents and then select the prime factors with the *smallest* exponents.

Pick the prime with the *smallest* exponent in each column.

$$\begin{aligned} 36 &= 2 \cdot 2 \cdot 3 \cdot 3 &&= 2^2 \cdot 3^2 \\ 60 &= 2 \cdot 2 \cdot 3 \cdot 5 &&= 2^2 \cdot 3 \cdot 5 \\ 108 &= 2 \cdot 2 \cdot 3 \cdot 3 \cdot 3 &&= 2^2 \cdot 3^3 \end{aligned}$$

We select 2^2 from the first column and 3 from the second. Thus, the GCF is $2^2 \cdot 3 = 12$.

Shortcut: Write

$$\underline{|36 \quad 60 \quad 108}$$

Divide by a **prime** divisor common to all numbers:

$$\begin{array}{c|ccc} 2 & 36 & 60 & 108 \\ \hline 2 & 18 & 30 & 54 \\ \hline 3 & 9 & 15 & 27 \\ \hline & 3 & 5 & 9 \end{array}$$

GCF $= 2 \cdot 2 \cdot 3 = 12$ as before.

c. First, write

$$\begin{aligned} 10 &= 2 \cdot 5 &&= 2 \cdot 5 \\ 17 &= 17 &&= 17 \\ 12 &= 2 \cdot 2 \cdot 3 &&= 2^2 \cdot 3 \end{aligned}$$

There are no primes common to all three numbers, so the GCF is 1.

Shortcut: Write

$$\underline{|10 \quad 17 \quad 12}$$

There is no prime divisor common to 10, 17, and 12, so the GCF is 1.

B › Finding the Greatest Common Factor (GCF) of Terms

We can also find the GCF for a list of terms. Simply write the terms in factored form using exponents, and then select the factors with the *smallest* exponents. For example, the GCF of x^5, x^3, and x^6 is x^3 because x^3 is the factor with the *smallest* exponent, 3. Similarly, to find the GCF of $a^3b^4c^2$ and $a^4b^2c^3$ write

$$a^3b^4c^2$$
$$a^4b^2c^3$$

We select the factors with the ***smallest*** exponents in each column: a^3 from the a column, b^2 from the b column, and c^2 from the c column. Thus, the GCF is $a^3 \cdot b^2 \cdot c^2$.

EXAMPLE 2 **Finding the GCF of a List of Terms**

Find the GCF of:

a. $24a^6, 18a^4, -30a^9$ **b.** $a^5b^3, b^6a^4, a^2b^5, a^8$ **c.** $-x^3y, -xy^3$

SOLUTION 2

a. Write:

$$24a^6 = 2^3 \cdot 3a^6$$
$$18a^4 = 2 \cdot 3^2a^4$$
$$-30a^9 = -1 \cdot 2 \cdot 3 \cdot 5a^9$$

The common factors with the smallest exponents are 2, 3, and a^4, so the GCF $= 2 \cdot 3 \cdot a^4 = 6a^4$.

b. Rewrite b^6a^4 as a^4b^6 so we can place all terms in a column, compare exponents, and select the factor with the lowest exponent from each column:

$$a^5b^3$$
$$a^4b^6$$
$$a^2b^5$$
$$a^8$$

There are no factors containing b in a^8, so the GCF is a^2, the factor with the lowest exponent.

c. Write:

$$-x^3y = -1 \cdot 1x^3y$$
$$-xy^3 = -1 \cdot 1xy^3$$

The factors of -1 are -1 and 1, but the ***greatest*** (largest) common factor is 1, so the GCF $= 1 \cdot xy = xy$.

Note that in a list containing negative terms, sometimes a negative common factor is preferred. Thus, in Example 2(c) we may prefer $-xy$ as the common factor. Both answers, xy and $-xy$, are correct.

PROBLEM 2

Find the GCF of:

a. $12x^8$, $9x^5$, and $-30x^8$

b. x^4y^3, y^4x^6, x^3y^8, and y^9

C › Factoring Out the Greatest Common Factor (GCF)

We can use the ideas we have learned about the GCF to write polynomials in factored form. To start, let's compare the multiplications with the factors and see whether we can discover a pattern.

Finding the Product	Finding the Factors
$4(x + y) = 4x + 4y$	$4x + 4y = 4(x + y)$
$5(a - 2b) = 5a - 10b$	$5a - 10b = 5(a - 2b)$
$2x(x + 3) = 2x^2 + 6x$	$2x^2 + 6x = 2x(x + 3)$

What do all these operations have in common? They use the *distributive property*. When multiplying, we have

$$a(b + c) = ab + ac$$

Answers to PROBLEMS

2. a. $3x^5$ **b.** y^3

When factoring, we have

$ab + ac = a(b + c)$ We are factoring a monomial (a) from a binomial ($ab + ac$).

Of course, we are now more interested in the latter operation. It tells us that to factor a binomial, we must find a factor (a in this case) common to all terms. The first step in having a completely factored expression is to select the ***greatest common factor,*** ax^n. Here's how we do it.

GREATEST COMMON FACTOR OF A POLYNOMIAL

The term ax^n is the greatest common factor (GCF) of a polynomial if:

1. **a is the *greatest integer* that divides each of the coefficients of the polynomial, and**
2. **n is the *smallest exponent* of x in all the terms of the polynomial.**

Thus, to factor $6x^3 + 18x^2$, we could write

$$6x^3 + 18x^2 = 3x(2x^2 + 6x)$$

but this is not completely factored because $2x^2 + 6x$ can be factored further. Here the *greatest* integer dividing 6 and 18 is 6, and the smallest exponent of x in all terms is x^2. Thus, the complete factorization is

$6x^3 + 18x^2 = 6x^2(x + 3)$ Where $6x^2$ is the GCF
Note that if you write $6x^3 + 18x^2$ as the sum $2 \cdot 3 \cdot x^3 + 2 \cdot 3 \cdot 3x^2$, you have *not* factored the polynomial; you have only factored the coefficients of the terms. The complete factorization is the product $6x^2(x + 3)$.

You can check this by multiplying $6x^2(x + 3)$. Of course, it might help your accuracy and understanding if you were to write an intermediate step indicating the common factor present in each term. Thus, to factor out the GFC of $6x^3 + 18x^2$, you would write

$$6x^3 + 18x^2 = 6x^2 \cdot x + 6x^2 \cdot 3$$
$$= 6x^2(x + 3)$$

Note that since the 6 is a coefficient, it is *not* written in factored form; that is, we write $6x^2(x + 3)$ and *not* $2 \cdot 3x^2(x + 3)$.

Similarly, to factor $4x - 28$, you would write

$$4x - 28 = 4 \cdot x - 4 \cdot 7$$
$$= 4(x - 7)$$

Remember, if you write $4x - 28$ as $2 \cdot 2 \cdot x - 2 \cdot 2 \cdot 7 \cdot x$, you have factored the terms and *not* the binomial. The factorization of $4x - 28$ is the *product* $4(x - 7)$.

One more thing. When an expression such as $-3x + 12$ is to be factored, we have two possible factorizations:

$$-3(x - 4) \quad \text{or} \quad 3(-x + 4)$$

The first one, $-3(x - 4)$, is the *preferred* one, since in that case the first term of the binomial, $x - 4$, has a **positive** sign.

EXAMPLE 3 **Factoring out a common factor from a binomial**

Factor completely:

a. $8x + 24$ **b.** $-6y + 12$ **c.** $10x^2 - 25x^3$

PROBLEM 3

Factor completely:

a. $6a + 18$ **b.** $-9y + 27$

c. $15a^2 - 45a^3$

Answers to PROBLEMS

3. a. $6(a + 3)$ **b.** $-9(y - 3)$

c. $15a^2(1 - 3a)$ or $-15a^2(3a - 1)$

SOLUTION 3

a. $8x + 24 = 8 \cdot x + 8 \cdot 3$

$= 8(x + 3)$ 8 is the GCF.

b. $-6y + 12 = -6 \cdot y - 6(-2)$

$= -6(y - 2)$ −6 is the GCF.

c. $10x^2 - 25x^3 = 5x^2 \cdot 2 - 5x^2 \cdot 5x$ $5x^2$ is the GCF.

$= 5x^2(2 - 5x)$ or, better yet, $-5x^2(5x - 2)$

Check your results by multiplying the factors obtained, $-5x^2$ and $(5x - 2)$.

We can also factor polynomials with more than two terms, as shown next.

EXAMPLE 4 Factoring out a common factor from a polynomial

Factor completely:

a. $6x^3 + 12x^2 + 18x$

b. $10x^6 - 15x^5 + 20x^4 + 30x^2$

c. $2x^3 + 4x^4 + 8x^5$

SOLUTION 4

a. $6x^3 + 12x^2 + 18x = 6x \cdot x^2 + 6x \cdot 2x + 6x \cdot 3$ 6x is the GCF.

$= 6x(x^2 + 2x + 3)$

b. $10x^6 - 15x^5 + 20x^4 + 30x^7 = 5x^2 \cdot 2x^4 - 5x^2 \cdot 3x^3 + 5x^2 \cdot 4x^2 + 5x^2 \cdot 6$

$= 5x^2(2x^4 - 3x^3 + 4x^2 + 6)$ $5x^2$ is the GCF.

c. $2x^3 + 4x^4 + 8x^5 = 2x^3 \cdot 1 + 2x^3 \cdot 2x + 2x^3 \cdot 4x^2$

$= 2x^3(1 + 2x + 4x^2)$ $2x^3$ is the GCF.

PROBLEM 4

Factor completely:

a. $5x^3 + 15x^2 + 20x$

b. $15x^7 - 20x^6 + 10x^5 + 25x^3$

c. $3a^3 + 9a^4 + 12a^5$

EXAMPLE 5 Factoring out a common factor that is a fraction

Factor completely: $\frac{3}{4}x^2 - \frac{1}{4}x + \frac{5}{4}$

SOLUTION 5

As you can see, we are not working with integers here. This is a very special situation, but this expression can still be factored. Here's how to find the common factor:

$$\frac{3}{4}x^2 - \frac{1}{4}x + \frac{5}{4} = \frac{1}{4} \cdot 3x^2 - \frac{1}{4} \cdot x + \frac{1}{4} \cdot 5$$

$$= \frac{1}{4}(3x^2 - x + 5)$$

PROBLEM 5

Factor completely: $\frac{4}{5}a^2 - \frac{1}{5}a + \frac{2}{5}$

D > Factoring by Grouping

Can we factor $x^3 + 2x^2 + 3x + 6$? It seems that there's no common factor here except 1. However, we can group and factor the first two terms and also the last two terms, and then use the distributive property. Here are the steps we use.

1. Group terms with common factors using the associative property. $x^3 + 2x^2 + 3x + 6 = (x^3 + 2x^2) + (3x + 6)$
2. Factor each resulting binomial. $= x^2(x + 2) + 3(x + 2)$ (Same)
3. Factor out the GCF, $(x + 2)$, using the distributive property. $= (x + 2)(x^2 + 3)$

Answers to PROBLEMS

4. a. $5x(x^2 + 3x + 4)$ **b.** $5x^3(3x^4 - 4x^3 + 2x^2 + 5)$ **c.** $3a^3(1 + 3a + 4a^2)$ **5.** $\frac{1}{5}(4a^2 - a + 2)$

Thus,

$$x^3 + 2x^2 + 3x + 6 = (x + 2)(x^2 + 3)$$

Note that $x^2(x + 2) + 3(x + 2)$ can also be written as $(x^2 + 3)(x + 2)$, since $ac + bc = (a + b)c$. Hence,

$$x^3 + 2x^2 + 3x + 6 = (x^2 + 3)(x + 2)$$

Either factorization is correct. You can check this by multiplying $(x + 2)(x^2 + 3)$ or $(x^2 + 3)(x + 2)$. You will either get $x^3 + 3x + 2x^2 + 6$ or its equivalent $x^3 + 2x^2 + 3x + 6$.

EXAMPLE 6 Factor by grouping

Factor completely:

a. $3x^3 + 6x^2 + 2x + 4$ **b.** $6x^3 - 3x^2 - 4x + 2$

SOLUTION 6

a. We proceed by steps, as before.

1. Group terms with common factors using the associative property. $3x^3 + 6x^2 + 2x + 4 = (3x^3 + 6x^2) + (2x + 4)$

   Same

2. Factor each resulting binomial. $= 3x^2(x + 2) + 2(x + 2)$
3. Factor out the GCF, $(x + 2)$, using the distributive property. $= (x + 2)(3x^2 + 2)$

Note that if you write $3x^3 + 2x + 6x^2 + 4$ in step 1, your answer would be $(3x^2 + 2)(x + 2)$. Since by the commutative property $(3x^2 + 2)(x + 2) = (x + 2)(3x^2 + 2)$, both answers are correct. We will factor polynomials by first writing the terms in *descending* order.

b. Again, we proceed by steps.

1. Group terms with common factors using the associative property. $6x^3 - 3x^2 - 4x + 2 = (6x^3 - 3x^2) - (4x - 2)$

   Note that $-4x + 2 = -(4x - 2)$.

   Same

2. Factor each resulting binomial. $= 3x^2(2x - 1) - 2(2x - 1)$
3. Factor out the GCF, $(2x - 1)$. $= (2x - 1)(3x^2 - 2)$

Thus, $6x^3 - 3x^2 - 4x + 2 = (2x - 1)(3x^2 - 2)$. Note that $2x - 1$ and $3x^2 - 2$ *cannot* be factored any further, so the polynomial is completely factored.

PROBLEM 6

Factor completely:

a. $2a^3 + 6a^2 + 5a + 15$

b. $6a^3 - 2a^2 - 3a + 1$

EXAMPLE 7 Factor by grouping

Factor completely:

a. $2x^3 - 4x^2 - x + 2$ **b.** $6x^4 - 9x^2 + 4x^2 - 6$

SOLUTION 7

a. 1. Group terms with common factors using the associative property. $2x^3 - 4x^2 - x + 2 = (2x^3 - 4x^2) - (x - 2)$

   Same

2. Factor each resulting binomial. $= 2x^2(x - 2) - 1(x - 2)$
3. Factor out the GCF, $(x - 2)$. $= (x - 2)(2x^2 - 1)$

PROBLEM 7

Factor completely:

a. $3a^3 - 9a^2 - a + 3$

b. $6a^4 - 4a^2 + 9a^2 - 6$

Answers to PROBLEMS

6. a. $(a + 3)(2a^2 + 5)$ **b.** $(3a - 1)(2a^2 - 1)$ **7. a.** $(a - 3)(3a^2 - 1)$ **b.** $(3a^2 - 2)(2a^2 + 3)$

b. We proceed as usual.

1. Group terms with common factors using the associative property. $6x^4 - 9x^2 + 4x^2 - 6 = (6x^4 - 9x^2) + (4x^2 - 6)$
2. Factor each resulting binomial. (Same) $= 3x^2(2x^2 - 3) + 2(2x^2 - 3)$
3. Factor out the GCF, $(2x^2 - 3)$. $= (2x^2 - 3)(3x^2 + 2)$

EXAMPLE 8 Factor a polynomial with two variables by grouping

Factor completely: $6x^2 + 2xy - 9xy - 3y^2$

SOLUTION 8

Our steps serve us well in this situation also.

1. Group terms with common factors using the associative property. $6x^2 + 2xy - 9xy - 3y^2 = (6x^2 + 2xy) - (9xy + 3y^2)$

   Note that $-9xy - 3y^2 = -(9xy + 3y^2)$.
2. Factor each resulting binomial. (Same) $= 2x(3x + y) - 3y(3x + y)$
3. Factor out the GCF, $(3x + y)$. $= (3x + y)(2x - 3y)$

PROBLEM 8

Factor completely:

$6a^2 + 3ab - 4ab - 2b^2$

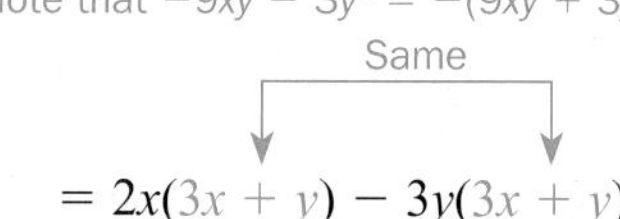

EXAMPLE 9 Polar bear factoring

The size of the polar bear population varies widely but one approximation is $P(t) = 4t^2 + 140t + 5000$, where t is the number of years after 1950. Factor $4t^2 + 140t + 5000$ completely.

SOLUTION 9

$$4t^2 + 140t + 5000 = 4 \cdot t^2 + 4 \cdot 35t + 4 \cdot 1250$$
$$= 4(t^2 + 35t + 1250) \quad \text{4 is the GCF.}$$

Source: http://tinyurl.com/p5qsjg.

PROBLEM 9

Factor $4t^2 + 200t + 5000$ completely.

The approximation in Example 9 is based on the table.

Polar Bear Population Estimates

1950s	5,000
1965–1970	8,000–10,000
1984	25,000
2005	20,000–25,000

Source: New York Times; Covebear.com; International Bear Association; International Wildlife: IUCN, Polar Bear Study Group.

Calculator Corner

Checking Factorization

If you believe a picture is worth a thousand words, you can use your calculator to check factorization problems. The idea is this: If in Example 6(a) you get the same picture (graph) for $3x^3 + 6x^2 + 2x + 4$ and for $(x + 2)(3x^2 + 2)$, these two polynomials must be equal; that is, $3x^3 + 6x^2 + 2x + 4 = (x + 2)(3x^2 + 2)$. Now, use your calculator to show this. Graph the polynomials by pressing Y= and entering $Y_1 = 3x^3 + 6x^2 + 2x + 4$ and $Y_2 = (x + 2)(3x^2 + 2)$. Then press GRAPH. If you get the same graph for Y_1 and Y_2, which means you see only one picture as shown in the window, you probably have the correct factorization.

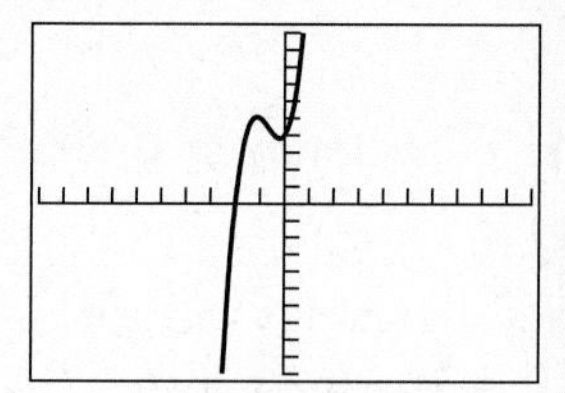

Answers to PROBLEMS

8. $(2a + b)(3a - 2b)$ **9.** $4(t^2 + 50t + 1250)$

› Exercises 5.1

› Web IT go to mhhe.com/bello for more lessons

‹ A › Finding the Greatest Common Factor (GCF) of Numbers

In Problems 1–6, find the GCF.

1. 20, 24 **2.** 30, 70 **3.** 16, 48, 88

4. 52, 26, 130 **5.** 8, 19, 12 **6.** 10, 41, 18

‹ B › Finding the Greatest Common Factor (GCF) of Terms

In Problems 7–20, find the GCF.

7. a^3, a^8 **8.** y^7, y^9 **9.** x^3, x^6, x^{10} **10.** b^6, b^8, b^{10}

11. $5y^6, 10y^7$ **12.** $12y^9, 24y^4$ **13.** $8x^3, 6x^7, 10x^9$ **14.** $9a^3, 6a^7, 12a^5$

15. $9b^2c, 12bc^2, 15b^2c^2$ **16.** $12x^4y^3, 18x^3y^4, 6x^5y^5$ **17.** $9y^2, 6x^3, 3x^3y$ **18.** $15ab^3, 25b^3c^4, 10a^3bc$

19. $18a^4b^3z^4, 27a^5b^3z^4, 81ab^3z$ **20.** $6x^2y^2z^2, 9xy^2z^2, 15x^2y^2z^2$

‹ C › Factoring Out the Greatest Common Factor (GCF)

In Problems 21–56, factor completely.

21. $3x + 15$ **22.** $5x + 45$ **23.** $9y - 18$ **24.** $11y - 33$

25. $-5y + 20$ **26.** $-4y + 28$ **27.** $-3x - 27$ **28.** $-6x - 36$

29. $4x^2 + 32x$ **30.** $5x^3 + 20x$ **31.** $6x - 42x^2$ **32.** $7x - 14x^3$

33. $-5x^2 - 25x^4$ **34.** $-3x^3 - 18x^6$ **35.** $3x^3 + 6x^2 + 9x$

36. $8x^3 + 4x^2 - 16x$ **37.** $9y^3 - 18y^2 + 27y$ **38.** $10y^3 - 5y^2 + 10y$

39. $6x^6 + 12x^5 - 18x^4 + 30x^2$ **40.** $5x^7 - 15x^6 + 10x^3 - 20x^2$ **41.** $8y^8 + 16y^5 - 24y^4 + 8y^3$

42. $12y^9 - 4y^6 + 6y^5 + 8y^4$ **43.** $\frac{4}{7}x^3 + \frac{3}{7}x^2 - \frac{9}{7}x + \frac{3}{7}$ **44.** $\frac{2}{5}x^3 + \frac{3}{5}x^2 - \frac{2}{5}x + \frac{4}{5}$

45. $\frac{7}{8}y^9 + \frac{3}{8}y^6 - \frac{5}{8}y^4 + \frac{5}{8}y^2$ **46.** $\frac{4}{3}y^7 - \frac{1}{3}y^5 + \frac{2}{3}y^4 - \frac{5}{3}y^3$ **47.** $3(x + 4) - y(x + 4)$

48. $5(y - 2) - x(y - 2)$ **49.** $x(y - 2) - (y - 2)$ **50.** $y(x + 3) - (x + 3)$

51. $c(t + s) - (t + s)$ **52.** $p(x - q) - (x - q)$ **53.** $4x^3 + 4x^4 - 12x^5$

54. $5x^6 + 10x^7 - 5x^8$ **55.** $6y^7 - 12y^9 - 6y^{11}$ **56.** $7x^9 - 7x^{13} - 14x^{15}$

‹ D › Factoring by Grouping

In Problems 57–80, factor completely by grouping.

57. $x^3 + 2x^2 + x + 2$ **58.** $x^3 + 3x^2 + x + 3$ **59.** $y^3 - 3y^2 + y - 3$

60. $y^3 - 5y^2 + y - 5$ **61.** $4x^3 + 6x^2 + 2x + 3$ **62.** $6x^3 + 3x^2 + 2x + 1$

63. $6x^3 - 2x^2 + 3x - 1$

64. $6x^3 - 9x^2 + 2x - 3$

65. $4y^3 + 8y^2 + y + 2$

66. $2y^3 - 6y^2 - y + 3$

67. $2a^3 + 3a^2 + 2a + 3$

68. $3a^3 + 2a^2 + 3a + 2$

69. $3x^4 + 12x^2 + x^2 + 4$

70. $2x^4 + 2x^2 + x^2 + 1$

71. $6y^4 + 9y^2 + 2y^2 + 3$

72. $12y^4 + 8y^2 + 3y^2 + 2$

73. $4y^4 + 12y^2 + y^2 + 3$

74. $2y^4 + 2y^2 + y^2 + 1$

75. $3a^4 - 6a^2 - 2a^2 + 4$

76. $4a^4 - 12a^2 - 3a^2 + 9$

77. $6a - 5b + 12ad - 10bd$

78. $3x - 2y + 15xz - 10yz$

79. $x^2 - y - 3x^2z + 3yz$

80. $x^2 + 2y - 2x^2 - 4y$

Applications

81. *Linear expansion*

a. Factor $\alpha Lt_2 - \alpha Lt_1$, where α is the coefficient of linear expansion, L the length of the material, and t_2 and t_1 the high and low temperatures in degrees Celsius.

b. If $e = \alpha L \Delta T$, according to part **a,** what is ΔT?

Applications: Green Math

In Problems 82–84, we will be checking how close the approximations we have used are to the ones given in the table of Example 9.

82. *Approximations using* $4t^2 + 200t + 5000$, where t is the number of years after 1950.

a. What was the bear population in 1950? How does it compare with the table?

b. What was the bear population in 1965? How does it compare with the table?

c. What was the bear population in 2005? How does it compare with the table?

d. This was the approximation from Problem 9. Is it factorable?

83. *Approximations using* $4t^2 + 140t + 5000$, where t is the number of years after 1950.

a. What was the bear population in 1950? How does it compare with the table?

b. What was the bear population in 1965? How does it compare with the table?

c. What was the bear population in 2005? How does it compare with the table?

d. This was the approximation from Example 9. Is it factorable?

84. *Approximations using* $4.1t^2 + 140t + 5000$, where t is the number of years after 1950.

a. What was the bear population in 1950? How does it compare with the table?

b. What was the bear population in 1965? How does it compare with the table?

c. What was the bear population in 2005? How does it compare with the table?

d. This is a new approximation. Is it factorable?

85. *Surface area of cylinder* The surface area A of a right circular cylinder is given by $A = 2\pi rh + 2\pi r^2$, where h is the height of the cylinder and r is its radius. Factor $2\pi rh + 2\pi r^2$.

86. *Area of trapezoid* The area A of a trapezoid is given by $A = \frac{1}{2}b_1h + \frac{1}{2}b_2h$, where h is the altitude of the trapezoid and b_1 and b_2 are the lengths of the bases. Factor $\frac{1}{2}b_1h + \frac{1}{2}b_2h$.

Formulas In Problems 87–92 a formula is given. Factor the expression on the right side of the equation.

	Formula	Used in
87.	$Q_1 = PQ_2 - PQ_1$	Refrigeration
88.	$L = L_0 + L_0at$	Temperature expansion
89.	$a = Vk - PV^2$	Muscle contraction
90.	$d_m = nA - A$	Optics
91.	$f_su = fu + fv_s$	Sound
92.	$T_2W = qT_2 - qT_1$	Energy

Using Your Knowledge

Factoring Formulas Many formulas can be simplified by factoring. Here are a few; factor the expressions given in each problem.

93. The vertical shear at any section of a cantilever beam of uniform cross section is given by

$$-w\ell + wz$$

Factor this expression.

94. The bending moment of any section of a cantilever beam of uniform cross section is given by

$$-P\ell + Px$$

Factor this expression.

95. The surface area of a square pyramid is given by

$$a^2 + 2as$$

Factor this expression.

96. The energy of a moving object is given by

$$800m - mv^2$$

Factor this expression.

97. The height of a rock thrown from the roof of a certain building is given by

$$-16t^2 + 80t + 240$$

Factor this expression. (*Hint:* -16 is a common factor.)

Write On

98. What do we mean by a factored expression, and how can you check whether the result is correct?

99. Explain the procedure you use to factor a monomial from a polynomial. Can you use the definition of GCF to factor $\frac{1}{2}x + \frac{1}{2}$? Explain.

Concept Checker

Fill in the blank(s) with the correct word(s), phrase, or mathematical statement.

100. To **factor** an expression is to write the expression as a _______________ of its factors.

101. **Factoring** is the _______________ of multiplying.

102. The GCF of a **list of integers** is the _______________ **common factor** of the integers in the list.

103. ax^n is the **GCF** of a **polynomial** if n is the _______________ **exponent** of x in all the terms of the polynomial.

largest	**reverse**
smallest	**product**
sum	**inverse**

Mastery Test

Factor:

104. $3x^3 - 6x^2 - x + 2$

105. $6x^4 + 2x^2 - 9x^2 - 3$

106. $2x^3 + 2x^2 + 3x + 3$

107. $6x^3 - 9x^2 - 2x + 3$

108. $\frac{2}{5}x^2 - \frac{3}{5}x - \frac{1}{5}$

109. $3x^6 - 6x^5 + 12x^4 + 27x^2$

110. $7x^3 + 14x^2 - 49x$

111. $12x^2 + 6xy - 10xy - 5y^2$

112. $6x + 48$

113. $-3y + 21$

114. $4x^2 - 32x^3$

115. $5x(x + b) + 6y(x + b)$

116. $3x + 7y - 12x^2 - 28xy$

117. Find the GCF of 14 and 24.

118. Find the GCF of $20x^4$, $35x^5y$, and $40xy^7$.

Skill Checker

Multiply:

119. $(x + 5)(x + 3)$

120. $(x - 5)(x + 2)$

121. $(x - 1)(x - 4)$

122. $(x + 2)(x + 1)$

123. $(x + 1)(2x - 3)$

124. $(x - 2)(x + 4)$

5.2 Factoring $x^2 + bx + c$

Objective

A Factor trinomials of the form $x^2 + bx + c$.

To Succeed, Review How To . . .

1. Expand $(X + A)(X + B)$ (p. 378).
2. Multiply integers (pp. 60–61).
3. Know the definition of a prime number (p. 6).

Getting Started

Supply and Demand

Why do prices go up? One reason is related to the supply of the product. Large supply, low prices; small supply, higher prices. The supply function for a certain item can be stated as

$$p^2 + 3p - 70$$

where p is the price of the product. This trinomial is factorable by using *reverse* multiplication (factoring) as we did in Section 5.1. But why do we need to know how to factor? If we know how to factor $p^2 + 3p - 70$, we can solve the equation $p^2 + 3p - 70 = 0$ by rewriting the left side in factored form as

Supply Affects Price

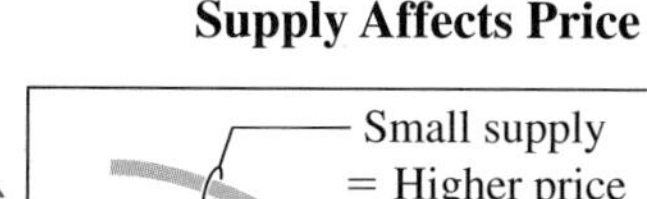

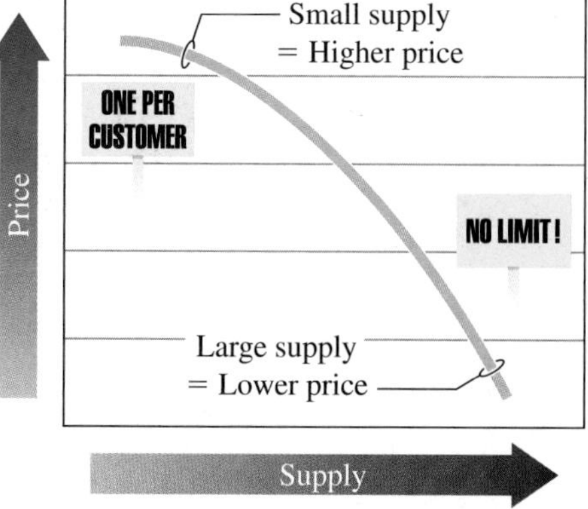

$$(p + 10)(p - 7) = 0$$

Do you see the solutions $p = -10$ and $p = 7$? We will learn how to factor trinomials in this section and then use this knowledge in Sections 5.6 and 5.7 to solve equations and applications.

A > Factoring Trinomials of the Form $x^2 + bx + c$

Since factoring is the reverse of multiplying, we can use the special products that we derived from the FOIL method (Section 4.7), reverse them, and have some equally useful factoring rules. The basis for the factoring rules is special product 1 (SP1), which you also know as the FOIL method. We now rewrite this product as a factoring rule.

FACTORING RULE 1: FACTORING BY REVERSING FOIL

$$X^2 + (A + B)X + AB = (X + A)(X + B) \quad \textbf{(F1)}$$

Thus, to factor $x^2 + bx + c$, we need to find two binomials whose product is $x^2 + bx + c$. Now suppose we wish to factor the polynomial

$$x^2 + 8x + 15$$

To do this, we use F1:

$$\underbrace{X^2}_{x^2} + \underbrace{(A + B)X}_{8x} + \underbrace{AB}_{+\,15}$$

As you can see, 15 is used instead of AB and 8 instead of $A + B$; that is, we have two numbers A and B such that $AB = 15$ and $A + B = 8$. We write the possible factors of $AB = 15$ and their sums in a table.

Factors	Sum
15, 1	16
5, 3	8

The correct numbers are $A = 5$ and $B = 3$. Now

$$\begin{aligned} X^2 + \underbrace{(A + B)}X + \underbrace{AB} &= (X + A)(X + B) \\ x^2 + \underbrace{(5 + 3)}x + \underbrace{5 \cdot 3} &= (x + 5)(x + 3) \\ x^2 + \quad 8x \quad + \quad 15 &= (x + 5)(x + 3) \end{aligned}$$

Remember, to factor $x^2 + 8x + 15$, we need two integers whose product is 15 and whose sum is 8. The integers 5 and 3 will do! So the commutative property enables us to write

$$x^2 + 8x + 15 = (x + 5)(x + 3)$$

as our answer.

What about factoring $x^2 - 8x + 15$? We still need two numbers A and B whose product is 15, but this time their sum must be -8; that is, we need $AB = 15$ and $A + B = -8$. Letting $A = -5$ and $B = -3$ will do it. Check it out:

$$\begin{aligned} (x - 5)(x - 3) &= x^2 - 3x - 5x + (-5)(-3) \\ &= x^2 - 8x + 15 \end{aligned}$$

Do you see how this works? To factor a trinomial of the form $x^2 + bx + c$, we must find two numbers A and B so that $A + B = b$ and $AB = c$. Then,

$$x^2 + bx + c = x^2 + (A + B)x + AB = (x + A)(x + B)$$

For example, to factor

$$x^2 \overset{b}{\boxed{-3}}x \overset{c}{\boxed{-10}}$$

we need two numbers whose product is -10 (c) and whose sum is -3 (b). Here's a table showing the possibilities for the factors and their sum.

Factors	Sum
$-10, 1$	-9
$10, -1$	9
$5, -2$	3
$-5, 2$	-3

← This is the only one in which the sum is -3.

The numbers are -5 and $+2$. Thus,

$$x^2 - 3x - 10 = (x - 5)(x + 2)$$

Note that the answer $(x + 2)(x - 5)$ is also correct by the commutative property. You can check this by multiplying $(x - 5)(x + 2)$ or $(x + 2)(x - 5)$.

Similarly, to factor $x^2 + 5x - 14$, we need two numbers whose product is -14 (so the numbers have different signs) and whose sum is 5 (so the larger one is positive). The numbers are -2 and $+7$ ($-2 \cdot 7 = -14$ and $-2 + 7 = 5$). Thus,

$$x^2 + 5x - 14 = (x - 2)(x + 7)$$

We can also write $x^2 + 5x - 14 = (x + 7)(x - 2)$.

Here are the factorizations we have obtained:

$$x^2 + 8x + 15 = (x + 5)(x + 3)$$
$$x^2 - 8x + 15 = (x - 5)(x - 3)$$
$$x^2 - 3x - 10 = (x - 5)(x + 2)$$
$$x^2 + 5x - 14 = (x - 2)(x + 7)$$

All of them involve factoring a trinomial of the form $x^2 + bx + c$; in each case, we need to find two integers whose product is c and whose sum is b. In general, here is the procedure you need:

PROCEDURE

Factoring $x^2 + bx + c$

Find *two* integers whose product is c and whose sum is b.

1. If b and c are positive, *both* integers must be positive.
2. If c is positive and b is negative, *both* integers must be negative.
3. If c is negative, *one* integer must be positive and *one* negative.

EXAMPLE 1 **Factoring a trinomial with all positive terms**

Factor completely: $6 + 5x + x^2$

SOLUTION 1 Write $6 + 5x + x^2$ in decreasing order as $x^2 + 5x + 6$. To factor $x^2 + 5x + 6$ we need two numbers with product 6 and sum 5. The possibilities are listed in the table. The two factors whose sum is 5 are 3 and 2. Thus,

Factors	Sum
6, 1	7
3, 2	5

$$x^2 + 5x + 6 = (x + 3)(x + 2)$$

CHECK $(x + 3)(x + 2) = x^2 + 2x + 3x + 3 \cdot 2$
$= x^2 + 5x + 6$

PROBLEM 1

Factor completely: $8 + 9x + x^2$

Answers to PROBLEMS

1. $(x + 8)(x + 1)$

EXAMPLE 2 Factoring a trinomial with a negative middle term

Factor completely: $x^2 - 6x + 5$

SOLUTION 2 To factor $x^2 - 6x + 5$ we need two numbers with product 5 and sum -6. In order to obtain the positive product 5, both factors must be negative. There is only one possibility, so the desired numbers are -5 and -1. Thus,

Factors	Sum
$-5, -1$	-6

$$x^2 - 6x + 5 = (x - 5)(x - 1)$$

CHECK $(x - 5)(x - 1) = x^2 - 1x - 5x + (-5) \cdot (-1)$
$= x^2 - 6x + 5$

PROBLEM 2

Factor completely: $x^2 - 8x + 7$

EXAMPLE 3 Factoring a trinomial with two negative terms

Factor completely: $x^2 - 3x - 4$

SOLUTION 3 To factor $x^2 - 3x - 4$ we need two numbers with product -4 and sum -3. To obtain the negative product -4, one number must be positive and one negative. The possibilities are shown in the table. The desired numbers are -4 and 1, so

Factors	Sum
$4, -1$	3
$-4, 1$	-3
$-2, 2$	0

$$x^2 - 3x - 4 = (x - 4)(x + 1)$$

CHECK $(x - 4)(x + 1) = x^2 + 1x - 4x + (-4) \cdot (1)$
$= x^2 - 3x - 4$

PROBLEM 3

Factor completely: $x^2 - 2x - 8$

As you recall from Section R.1, a *prime number* is a number whose only factors are itself and 1. Thus, if we want to factor the number 15 as a product of primes, we write $15 = 3 \cdot 5$. On the other hand, if we want to factor the number 17, we are unable to do it because 17 is prime. If a polynomial cannot be factored using only integers, we say that the polynomial is *prime* and call it a **prime polynomial.** Look at the next example and discover which of the two polynomials is prime.

EXAMPLE 4 Factoring a trinomial with a positive middle term and a negative last term

Factor if possible:

a. $p^2 + 3p - 70$ **b.** $p^2 + 4p - 15$

SOLUTION 4

a. We need two numbers with product -70 and sum 3. Since -70 is negative, the numbers must have different signs. Moreover, since the sum of the two numbers is 3, the *larger* number must be *positive*. Here are the possibilities:

Factors	Sum
$70, -1$	69
$35, -2$	33
$14, -5$	9
$10, -7$	3

← The only ones with sum 3

The numbers we need then are 10 and -7. Thus,

$$p^2 + 3p - 70 = (p + 10)(p - 7)$$

PROBLEM 4

Factor if possible:

a. $y^2 + 4y - 14$

b. $y^2 + 3y - 40$

Answers to PROBLEMS

2. $(x - 7)(x - 1)$ **3.** $(x - 4)(x + 2)$ **4. a.** $y^2 + 4y - 14$ is prime. **b.** $(y + 8)(y - 5)$

b. This time we need two numbers with product -15 and sum 4. Here are the possibilities:

Factors	Sum
15, −1	14
−15, 1	−14
5, −3	2
−5, 3	−2

None of the pairs of factors has a sum of 4. Thus, the trinomial $p^2 + 4p - 15$ cannot be factored using only integers. The polynomial $p^2 + 4p - 15$ is **prime.**

EXAMPLE 5 Factoring a trinomial with two variables

Factor completely: $x^2 + 5ax + 6a^2$

SOLUTION 5 This problem is very similar to Example 1, except here we have two variables, x and a. The procedure, however, is the same. We need two expressions whose product is $6a^2$ and whose sum is $5a$. The possible factors are $6a$ and a, or $3a$ and $2a$. Since $3a + 2a = 5a$, the appropriate factors are $3a$ and $2a$. Thus,

$$x^2 + 5ax + 6a^2 = (x + 3a)(x + 2a)$$

CHECK $(x + 3a)(x + 2a) = x^2 + 2ax + 3ax + 3a \cdot 2a$

$$= x^2 + 5ax + 6a^2$$

PROBLEM 5

Factor completely: $x^2 + 6ax + 8a^2$

Now, do you remember GCFs? Sometimes we have to factor out the GCF in an expression so that we can completely factor it. How? We will show you in Example 6.

EXAMPLE 6 Factoring a trinomial with a GCF

Factor completely:

a. $x^5 - 6x^4 + 9x^3$ **b.** $ax^6 - 3ax^5 - 18ax^4$

SOLUTION 6

a. The GCF of $x^5 - 6x^4 + 9x^3$ is the term with the smallest exponent, that is, x^3. Thus,

$$x^5 - 6x^4 + 9x^3 = x^3(x^2 - 6x + 9)$$

Now, we have to factor $x^2 - 6x + 9$ by finding two numbers whose product is 9 and whose sum is -6. To obtain the positive product 9, both factors must be negative; -3 and -3 will do, so

$$x^2 - 6x + 9 = (x - 3)(x - 3) = (x - 3)^2$$

Here is the complete procedure for you to follow:

$x^5 - 6x^4 + 9x^3 = x^3(x^2 - 6x + 9)$	Factor out the GCF x^3.
$= x^3(x - 3)(x - 3)$	Factor $x^2 - 6x + 9$.
$= x^3(x - 3)^2$	Write $(x - 3)(x - 3)$ as $(x - 3)^2$.

You should check the result!

b. $ax^6 - 3ax^5 - 18ax^4 = ax^4(x^2 - 3x - 18)$	Factor out the GCF, ax^4.
$= ax^4(x + 3)(x - 6)$	Factor $x^2 - 3x - 18$ using 3 and -6, whose product is -18 and whose sum is -3.

We leave the check for you.

PROBLEM 6

Factor completely:

a. $y^5 - 4y^4 + 4y^3$

b. $b^2y^6 - b^2y^5 - 20b^2y^4$

Answers to PROBLEMS

5. $(x + 4a)(x + 2a)$ **6. a.** $y^3(y - 2)^2$ **b.** $b^2y^4(y + 4)(y - 5)$

GREEN MATH

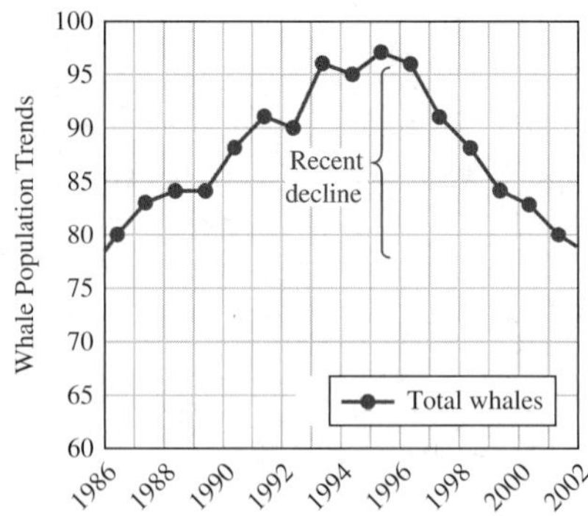

EXAMPLE 7 Killer whales in Puget Sound

From 1986 to 1992, the residents of Puget Sound enjoyed and welcomed the killer whales during their spring and fall residency. After 1996 their joy turned to concern. The number of whales was declining, as shown in the graph and their population can be approximated by $P(t) = -0.3t^2 + 4.8t + 78$, t the number of years after 1986.

Factor $-0.3t^2 + 4.8t + 78$ completely.

SOLUTION 7

$$-0.3t^2 + 4.8t + 78 = -0.3(t^2 - 16t - 260)$$ Factor out the GCF, **−0.3**.

$$= -0.3(t - 26)(t + 10)$$ Factor $t^2 - 16t - 260$ using **−26** and **10**, two numbers whose product is −260 and whose sum −16.

Source: www.nwfsc.noaa.gov/.../kwnewsletter/oct2003.cfm.

PROBLEM 7

Is the polynomial a good model for the graph? Check the values for $t = 0, 1, 2,$ *and* 16 and compare with the values in the graph.

Exercises 5.2

⟨A⟩ Factoring Trinomials of the Form $x^2 + bx + c$

In Problems 1–40, factor completely.

1. $y^2 + 6y + 8$ **2.** $y^2 + 10y + 21$ **3.** $x^2 + 7x + 10$ **4.** $x^2 + 13x + 22$

5. $y^2 + 3y - 10$ **6.** $y^2 + 5y - 24$ **7.** $x^2 + 5x - 14$ **8.** $x^2 + 5x - 36$

9. $x^2 - 6x - 7$ **10.** $x^2 - 7x - 8$ **11.** $y^2 - 5y - 14$ **12.** $y^2 - 4y - 12$

13. $y^2 - 3y + 2$ **14.** $y^2 - 11y + 30$ **15.** $x^2 - 5x + 4$ **16.** $x^2 - 12x + 27$

17. $x^2 + 3x + 4$ **18.** $x^2 - 5x + 6$ **19.** $-7 - 7x + x^2$ **20.** $-5 - 5x + x^2$

Answers to PROBLEMS

7. For $t = 0$, $P(0) = 78$. Graph: 80; for $t = 1$, $P(1) = 82.5$. Graph: 83; for $t = 2$, $P(2) = 86.4$. Graph: 84; for $t = 16$, $P(16) = 78$. Graph: 80

21. $x^2 + 3ax + 2a^2$

22. $x^2 + 12ax + 35a^2$

23. $z^2 + 6bz + 9b^2$

24. $z^2 + 4bz + 16b^2$

25. $r^2 + ar - 12a^2$

26. $r^2 + 3ar - 10a^2$

27. $x^2 + 9ax - 10a^2$

28. $x^2 + 2ax - 8a^2$

29. $-b^2 - 2by + 3y^2$

30. $-b^2 + by + 110y^2$

31. $m^2 - am - 2a^2$

32. $m^2 - 2bm - 15b^2$

33. $2t^3 + 10t^2 + 8t$

34. $3t^3 + 12t^2 + 9t$

35. $a^2x^3 + 3a^2x^2 + 2ax^2$

36. $b^3x^5 + 4b^3x^4 + 2b^3x^3$

37. $b^3x^7 + b^3x^6 - 12b^3x^5$

38. $b^5y^8 + 3b^5y^7 - 10b^5y^6$

39. $2c^5z^6 + 4c^5z^5 - 30c^5z^4$

40. $3z^8 - 12z^7y - 63z^6y$

Applications

In Problems 41–42, the expression

$$-5t^2 + V_0t + h$$

represents the altitude of an object t seconds after being thrown from a height of h meters with an initial velocity of V_0 meters per second.

41. *Altitude of a thrown object*
   a. Find the expression for the altitude of an object thrown upward with an initial velocity V_0 of 5 meters per second from a building 10 meters high.
   b. Factor the expression.
   c. When the product obtained in part **b** is 0, the altitude of the object is 0 and the object is on the ground. What values of t will make the product 0?
   d. Based on the answer to part **c**, how long does it take the object to return to the ground?

42. *Altitude of a thrown object*
   a. Find the expression for the altitude of an object thrown upward with an initial velocity V_0 of 10 meters per second from a building 40 meters high.
   b. Factor the expression.
   c. When the product obtained in part **b** is 0, the altitude of the object is 0 and the object is on the ground. What values of t will make the product 0?
   d. Based on the answer to part **c**, how long does it take the object to return to the ground?

43. *Descent of a rock* The height (in feet) of a rock thrown from the roof of a building after t seconds is given by

$$-16t^2 + 32t + 240$$

   a. Factor this expression.
   b. When the product obtained in part **a** is 0, the altitude of the rock is 0 and the rock is on the ground. What values of t will make the product 0?
   c. Based on the answer to part **b**, how long does it take the rock to return to the ground?

44. *Chlorofluorocarbon production* Do you know what freon is? It is a gas containing chlorofluorocarbons (CFC) used in old air conditioners and linked to the depletion of the ozone layer. Their production (in thousands of tons) can be represented by the expression

$$-0.04t^2 + 2.8t + 120$$

where t is the number of years after 1960.

   a. Factor $-0.04t^2 + 2.8t + 120$ using -0.04 as the GCF.
   b. When will the product obtained in part **a** be 0?
   c. Based on the answer to part **b**, when will the production of CFCs be 0? (The Montreal Protocol called for a 50% decrease by the year 2000.)

45. *Height of a ball* The height H of a ball thrown upward from a height of 32 feet with an initial velocity of 16 feet per second is given by the equation $H = -16t^2 + 16t + 32$.
   a. Factor $-16t^2 + 16t + 32$.
   b. What is the height of the ball after 1 second?
   c. What is the height of the ball after 2 seconds?

46. *Height of toy rocket* The height H of a toy rocket launched from the third floor of a building 48 feet above the street with an initial velocity of 32 feet per second is given by the equation $H = -16t^2 + 32t + 48$.
   a. Factor $-16t^2 + 32t + 48$.
   b. What is the height of the ball after 2 seconds?
   c. What is the height of the ball after 3 seconds?

47. *Number of games* A sponsor wants to organize a tournament consisting of 20 games in which each of the x teams entered plays every other team twice. To find out how many teams you need, we must solve the equation $x^2 - x - 20 = 0$.

a. Factor $x^2 - x - 20$.

b. The number of games played is given by $x(x - 1)$. How many games are played when five teams are entered?

48. *Number of games* If a sponsor wants a tournament consisting of 30 games in which each of the x teams entered plays every other team twice, we have to solve the equation $x^2 - x - 30 = 0$. (See Problem 47.)

a. Factor $x^2 - x - 30$.

b. The number of games played is given by $x(x - 1)$. How many games are played when six teams are entered?

49. *Spring motion* The motion of an object suspended by a helical spring requires the solution of the equation $D^2 + 8D + 12 = 0$. Factor $D^2 + 8D + 12$.

50. *Spring motion* The motion of an object attached to a spring requires the solution of the equation $D^2 + 2kD + k^2 = 0$. Factor $D^2 + 2kD + k^2$.

51. *Beam deflection* The deflection of a beam of length L requires the solution of the equation $4Lx - x^2 - 4L^2 = 0$. Factor $4Lx - x^2 - 4L^2$.

52. *Beam length* To find the length L of a certain beam we have to solve the equation $x^2 - 2xL + L^2 = 0$. Factor $x^2 - 2xL + L^2$.

Applications: Green Math

53. *Alligators on Kiawah Island, South Carolina* The graph shows the estimates of the alligator population along a predetermined route on Kiawah Island (not the actual number of alligators on the entire island) and can be approximated by $A(t) = -16t^2 + 80t + 384$, where t is the number of years after 2005.

a. Factor $-16t^2 + 80t + 384$.

b. Using $A(t)$, find the population estimates for 2005, 2006, and 2007.

c. Using the information in part **b,** is the population increasing or decreasing?

d. Use the polynomial to predict the population in 2010.

Source: http://tinyurl.com/yzza8mb.

54.

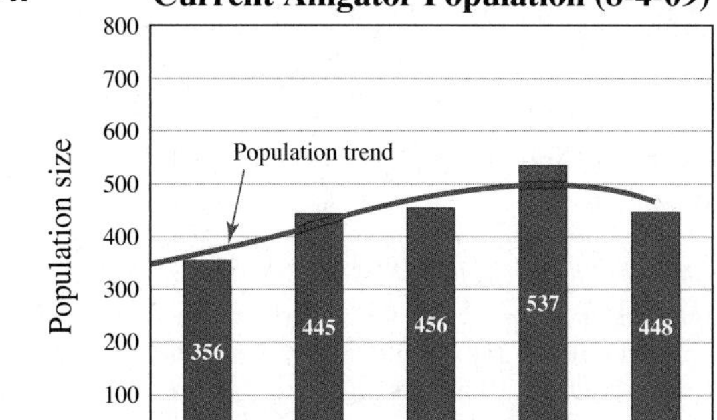

a. Use the graph to find the number of alligators in each of the years 2005–2009.

b. In what years is the population increasing?

c. In what years is the population decreasing?

d. In what year is the prediction from the polynomial $A(t) = -16t^2 + 80t + 384$ and the result from the graph exactly the same?

Write On

55. What should your first step be when factoring any trinomial?

56. Ana says that a polynomial can be factored as $(x - 3)(x - 2)$. Tyrone insists that the answer is really $(x - 2)(x - 3)$. Who is correct?

57. Tyrone says he reworked the problem (see Problem 56) and his new answer is $(2 - x)(3 - x)$. Ana still says she got $(x - 2)(x - 3)$. Who is correct?

58. Explain in your own words what a *prime polynomial* is.

59. When factoring $x^2 + bx + c$ (b and c positive) as $(x + A)(x + B)$, what can you say about A and B?

60. When factoring $x^2 + bx + c$ (b negative, c positive) as $(x + A)(x + B)$, what can you say about A and B?

61. When factoring $x^2 + bx + c$ (c negative) as $(x + A)(x + B)$, what can you say about A and B?

Concept Checker

Fill in the blank(s) with the correct word(s), phrase, or mathematical statement.

62. To **factor** $x^2 + bx + c$ we need two integers whose **product** is _______________ and whose **sum** is _______________.

63. If in problem 62, b and c are **positive**, both integers must be _______________.

64. If in problem 62, c is positive and b is negative, both integers must be _______________.

65. If in problem 62, c is negative, one integer must be _______________ and one _______________.

b

c

positive

negative

Mastery Test

In Problems 66–77, factor completely.

66. $8x^2 - 12xy + 2xy - 3y^2$
67. $3y^3 - 6y^2 - y + 2$
68. $3y^3 + 9y^2 + 2y + 6$
69. $2y^3 - 6y^2 - y + 3$
70. $y^2 + 5y + 6$
71. $x^2 + 7xy + 12y^2$
72. $x^2 + 5x - 14$
73. $x^2 - 7xy + 10y^2$
74. $3y^4 + 6y^5 + 9y^6$
75. $z^2 - 10z - 25$
76. $kx^3 - 2kx^2 - 15kx$
77. $2y^3 + 6y^5 + 10y^7$

Skill Checker

78. Find: **a.** $-8 \cdot 1$ **b.** $-8 + 1$
79. Find: **a.** $-6 \cdot 2$ **b.** $-6 + 2$
80. Find: **a.** $-8 \cdot 2$ **b.** $-8 + 2$
81. Find: **a.** $-4(-1)$ **b.** $-4 + (-1)$
82. Find: **a.** $-4 \cdot 3$ **b.** $-4 + 3$

5.3 Factoring $ax^2 + bx + c, a \neq 1$

Objectives

A Use the *ac* test to determine whether $ax^2 + bx + c$ is factorable.

B Factor $ax^2 + bx + c$ by grouping.

C Factor $ax^2 + bx + c$ using FOIL.

To Succeed, Review How To . . .

1. Expand $(X + A)(X + B)$ (p. 378).
2. Use FOIL to expand polynomials (pp. 369–372).
3. Multiply integers (pp. 60–61).

Getting Started

When There Is Smoke We Need Water

How much water is the fire truck pumping? It depends on many factors, but one of them is the friction loss inside the hose. If the friction loss is 36 pounds per square inch, we have to know how to factor $2g^2 + g - 36$ to find the answer.

This expression is of the form $ax^2 + bx + c$, $a \neq 1$, and we can factor it two ways: by **grouping** or by using **FOIL.** How do we know whether $2g^2 + g - 36$ is even factorable? We will show you how to tell and ask you to factor $2g^2 + g - 36$ in Problem 61.

A > Using the *ac* Test to Determine if a Polynomial Is Factorable

ac TEST FOR $ax^2 + bx + c$

A trinomial of the form $ax^2 + bx + c$ is factorable if there are two integers with product ac and sum b.

Note that a and c are the first and last numbers in $ax^2 + bx + c$ (hence the name *ac* test), and b is the coefficient of x. A diagram may help you visualize this test.

ac TEST

We need two numbers whose product is ac.

$$ax^2 + bx + c$$

The sum of the numbers must be b.

Note: Before you use the *ac* test, factor out the GCF and write the polynomial in descending order.

Thus, to determine whether $2g^2 + g - 36$ is factorable, we need two numbers whose product is $a \cdot c = 2 \cdot (-36) = -72$ and whose sum is $b = 1$. Since $9 \cdot (-8) = -72$ and $9 + (-8) = 1$, $2g^2 + g - 36$ is factorable.

A polynomial that cannot be factored using only factors with integer coefficients is called a **prime polynomial.**

EXAMPLE 1 Using the *ac* test to find whether a polynomial is factorable

Determine whether the given polynomial is factorable:

a. $6x^2 + 7x + 2$ **b.** $2x^2 + 5x + 4$

SOLUTION 1

a. To find out whether $6x^2 + 7x + 2$ is factorable, we use these three steps:

1. Multiply $a = 6$ by $c = 2$ $(6 \cdot 2 = 12)$.
2. Find two integers whose product is $ac = 12$ and whose sum is $b = 7$.

We need two numbers whose product is $6 \cdot 2 = 12$.

$$6x^2 + 7x + 2$$

The sum of the two numbers must be 7.

3. A little searching will produce 4 and 3.

CHECK $\underbrace{4 \times 3}_{\text{Product}} = 12$, $\underbrace{4 + 3}_{\text{Sum}} = 7$

Thus, $6x^2 + 7x + 2$ is factorable.

b. Consider the trinomial $2x^2 + 5x + 4$. Here is the *ac* test for this trinomial.

1. Multiply 2 by 4 $(2 \cdot 4 = 8)$.
2. Find two integers whose product is 8 and whose sum is 5.

PROBLEM 1

Determine whether the polynomial is factorable:

a. $4y^2 + 3y + 2$

b. $3y^2 + 5y + 2$

Answers to PROBLEMS

1. a. Not factorable (prime) **b.** Factorable ($ac = 6$; factors are 3 and 2; sum $b = 3 + 2 = 5$)

3. The factors of 8 are 4 and 2, and 8 and 1. Neither pair adds up to 5 $(4 + 2 = 6, 8 + 1 = 9)$. Thus, the trinomial $2x^2 + 5x + 4$ is *not* factorable using factors containing only integer coefficients; it is a prime polynomial.

B > Factoring $ax^2 + bx + c$ by Grouping

At this point, you should be convinced that the *ac* test really tells you whether a trinomial of the form $ax^2 + bx + c$ is factorable; however, we still don't know *how* to do the actual factorization. But we are in luck; the number *ac* still plays an important part in factoring this trinomial. In fact, the number *ac* is so important that we shall call it the **key number** in the factorization of $ax^2 + bx + c$. To get a little practice, we have found and circled the key numbers of a few trinomials.

	a	*c*	*ac*
$6x^2 + 8x + 5$	6	5	(30)
$2x^2 - 7x - 4$	2	−4	(−8)
$-3x^2 + 2x + 5$	−3	5	(−15)

As shown before, by examining the key number and the coefficient of the middle term, you can determine whether a trinomial is factorable. For example, the key number of the trinomial $6x^2 + 8x + 5$ is 30. But since there are *no* integers with sum 8 whose product is 30, this trinomial is *not* factorable. (The factors of 30 are 6 and 5, 10 and 3, 15 and 2, and 30 and 1. None of these pairs has a sum of 8.)

On the other hand, the key number for $2x^2 - 7x - 4$ is −8, and −8 has two factors (−8 and 1) whose product is −8 and whose sum is the coefficient of the middle term, −7. Thus, $2x^2 - 7x - 4$ is factorable; here are the steps:

1. Find the key number $[2 \cdot (-4) = -8]$.	$2x^2 - 7x - 4$	(−8)
2. Find the factors of the key number and use the appropriate ones to rewrite the middle term.	$2x^2 - 8x + 1x - 4$	−8, 1 $-8(1) = -8$ $-8 + 1 = -7$
3. Group the terms into pairs (as we did in Section 5.1).	$(2x^2 - 8x) + (1x - 4)$	
4. Factor each pair.	$2x(x - 4) + 1(x - 4)$	
5. Note that $(x - 4)$ is the GCF.	$(x - 4)(2x + 1)$	

Thus, $2x^2 - 7x - 4 = (x - 4)(2x + 1)$. You can check to see that this is the correct factorization by multiplying $(x - 4)$ by $(2x + 1)$.

Now a word of warning: You can write the factorization of $ax^2 + bx + c$ in *two* ways. Suppose you wish to factor the trinomial $5x^2 + 7x + 2$. Here's one way:

1. Find the key number $[5 \cdot 2 = 10]$.	$5x^2 + 7x + 2$	
2. Find the factors of the key number; use them to rewrite the middle term.	$5x^2 + 5x + 2x + 2$	5, 2
3. Group the terms into pairs.	$(5x^2 + 5x) + (2x + 2)$	

4. Factor each pair. $5x(x + 1) + 2(x + 1)$
5. Note that $(x + 1)$ is the GCF. $(x + 1)(5x + 2)$

Thus, $5x^2 + 7x + 2 = (x + 1)(5x + 2)$. But there's another way:

1. Find the key number $[5 \cdot 2 = 10]$. $5x^2 + 7x + 2$ (10)
2. Find the factors of the key number and use them to rewrite the middle term. $5x^2 + 2x + 5x + 2$ 2, 5
3. Group the terms into pairs. $(5x^2 + 2x) + (5x + 2)$
4. Factor each pair. $x(5x + 2) + 1(5x + 2)$
5. Note that $(5x + 2)$ is the GCF. $(5x + 2)(x + 1)$

In this case, we found that

$$5x^2 + 7x + 2 = (5x + 2)(x + 1)$$

Is the correct factorization $(x + 1)(5x + 2)$ or $(5x + 2)(x + 1)$? The answer is that *both* factorizations are correct! This is because the multiplication of real numbers is commutative and the variable x, as well as the trinomials involved, also represent real numbers. Thus, the *order* in which the product is written (according to the commutative property of multiplication) *makes no difference in the final answer.*

When factoring a trinomial by grouping, just remember to write first the polynomial in descending order and factor out the GCF (if any).

EXAMPLE 2 Factoring trinomials of the form $ax^2 + bx + c$ by grouping

Factor:

a. $4 - 3x + 6x^2$

b. $4x^2 - 3 - 4x$

SOLUTION 2

a. We first write the polynomial in descending order as $6x^2 - 3x + 4$, then proceed by steps:

1. Find the key number $[6 \cdot 4 = 24]$. $6x^2 - 3x + 4$ 
2. Find the factors of the key number and use them to rewrite the middle term. Unfortunately, it's impossible to find two numbers with product 24 and sum -3. This trinomial is *not* factorable.

b. We first rewrite the polynomial (*in descending order*) as $4x^2 - 4x - 3$, and then proceed by steps.

1. Find the key number $[4 \cdot (-3) = -12]$. $4x^2 - 4x - 3$ 
2. Find the factors of the key number and use them to rewrite the middle term. $4x^2 - 6x + 2x - 3$ −6, 2
3. Group the terms into pairs. $(4x^2 - 6x) + (2x - 3)$
4. Factor each pair. $2x(2x - 3) + 1(2x - 3)$
5. Note that $(2x - 3)$ is the GCF. $(2x - 3)(2x + 1)$

Thus, $4x^2 - 4x - 3 = (2x - 3)(2x + 1)$, as can easily be verified by multiplication.

PROBLEM 2

Factor:

a. $9x^2 - 2 - 3x$

b. $3 - 4x + 2x^2$

Answers to PROBLEMS

2. a. $(3x - 2)(3x + 1)$
b. Not factorable (prime)

Factoring problems in which the third term in step 2 contains a negative number as a coefficient requires that special care be taken with the signs. For example, to factor the trinomial $4x^2 - 5x + 1$, we proceed as follows:

1. Find the key number $[4 \cdot 1 = 4]$. $4x^2 - 5x + 1$ (4)
2. Find the factors of the key number and use them to rewrite the middle term. $4x^2 - 4x - 1x + 1$ $-4, -1$
   Note that the third term has a negative coefficient, -1.
3. Group the terms into pairs. $(4x^2 - 4x) + (-1x + 1)$
4. Factor each pair. $4x(x - 1) - 1(x - 1)$ Recall that $-1(x - 1) = -x + 1$.
5. Note that $(x - 1)$ is the GCF. $(x - 1)(4x - 1)$ If the first pair has $(x - 1)$ as a factor, the second pair will also have $(x - 1)$ as a factor.

Thus, $4x^2 - 5x + 1 = (x - 1)(4x - 1)$.

EXAMPLE 3 Factoring trinomials of the form $ax^2 - bx + c$ by grouping

Factor: $5x^2 - 11x + 2$

SOLUTION 3

1. Find the key number $[5 \cdot 2 = 10]$. $5x^2 - 11x + 2$ (10)
2. Find the factors of the key number and use them to rewrite the middle term. $5x^2 - 10x - 1x + 2$ $-10, -1$
3. Group the terms into pairs. $(5x^2 - 10x) + (-1x + 2)$
4. Factor each pair. $5x(x - 2) - 1(x - 2)$
5. Note that $(x - 2)$ is the GCF. $(x - 2)(5x - 1)$

Thus, the factorization of $5x^2 - 11x + 2$ is $(x - 2)(5x - 1)$.

PROBLEM 3

Factor: $4x^2 - 13x + 3$

So far, we have factored trinomials in one variable only. A procedure similar to the one used for factoring a trinomial of the form $ax^2 + bx + c$ can be used to factor certain trinomials in two variables. We illustrate the procedure in Example 4.

EXAMPLE 4 Factoring trinomials with two variables by grouping

Factor: $6x^2 - xy - 2y^2$

SOLUTION 4

1. Find the key number $[6 \cdot (-2) = -12]$. $6x^2 - xy - 2y^2$ 
2. Find the factors of the key number and use them to rewrite the middle term. $6x^2 - 4xy + 3xy - 2y^2$ $-4, 3$
3. Group the terms into pairs. $(6x^2 - 4xy) + (3xy - 2y^2)$
4. Factor each pair. $2x(3x - 2y) + y(3x - 2y)$
5. Note that $(3x - 2y)$ is the GCF. $(3x - 2y)(2x + y)$

Thus, $6x^2 - xy - 2y^2 = (3x - 2y)(2x + y)$.

PROBLEM 4

Factor: $4x^2 - 4xy - 3y^2$

Answers to PROBLEMS

3. $(x - 3)(4x - 1)$ **4.** $(2x - 3y)(2x + y)$

C > Factoring $ax^2 + bx + c$ by FOIL (Trial and Error)

Sometimes, it's easier to factor a polynomial of the form $ax^2 + bx + c$ by using a FOIL process (*trial and error*). This is especially so when a or c is a prime number such as 2, 3, 5, 7, 11, and so on. Here's how we do this.

PROCEDURE

Factoring $ax^2 + bx + c$ by FOIL (Trial and Error)

1. The product of the numbers in the *first* blanks (F) must be a.

$$ax^2 + bx + c = (\blacksquare x + \_\_)(\blacksquare x + \_\_)$$

2. The coefficients of the *outside* (O) products and the *inside* (I) products must add to b.

$$ax^2 + bx + c = (\blacksquare x + \blacksquare)(\blacksquare x + \blacksquare)$$

3. The product of the numbers in the *last* blanks (L) must be c.

$$ax^2 + bx + c = (\_\_ x + \blacksquare)(\_\_ x + \blacksquare)$$

For example, to factor $2x^2 + 5x + 3$, we write:

$$2x^2 + 5x + 3 = (\_\_ x + \_\_)(\_\_ x + \_\_)$$

3

2

We first look for two numbers whose product is 2. These numbers are 2 and 1, or -2 and -1. We have these possibilities:

$$(2x + \_\_)(x + \_\_) \quad \text{or} \quad (-2x + \_\_)(-x + \_\_)$$

Let's agree that we want the first coefficients inside the parentheses to be *positive*. This eliminates products involving $(-2x + \_\_)$. Now we look for numbers whose product is 3. These numbers are 3 and 1, or -3 and -1, which we substitute into the blanks, to obtain:

$$(2x + 3)(x + 1)$$
$$(2x + 1)(x + 3)$$
$$(2x - 3)(x - 1)$$
$$(2x - 1)(x - 3)$$

Since the final result must be $2x^2 + 5x + 3$, the first expression (shaded) yields the desired factorization:

$$2x^2 + 5x + 3 = (2x + 3)(x + 1)$$

You can save some time if you notice that all coefficients are positive, so the trial numbers must be positive. That leaves 2, 1 and 3, 1 as the only possibilities.

EXAMPLE 5 Factoring by FOIL (trial and error): All terms positive

Factor: $3x^2 + 7x + 2$

SOLUTION 5 Since we want the first coefficients in the factorization to be positive, the only two factors of 3 we consider are 3 and 1. We then look for the numbers whose product will equal 2:

$$3x^2 + 7x + 2 = (3x + \underbrace{___)(x + ___}_{2})$$

These factors are 2 and 1, and the possibilities are

$$(3x + 2)(x + 1) \qquad \text{or} \qquad (3x + 1)(x + 2)$$

$2x$ $\qquad$ x

$3x$ $\qquad$ $6x$

$\overline{5x}$ Add. $\qquad$ $\overline{7x}$ Add.

Since the second product, $(3x + 1)(x + 2)$, yields the correct middle term, $7x$, we have

$$3x^2 + 7x + 2 = (3x + 1)(x + 2)$$

PROBLEM 5

Factor using FOIL (trial and error):

$3x^2 + 5x + 2$

Note that the trial-and-error method is based on FOIL (Section 4.6). Thus, to *multiply* $(2x + 3)(3x + 4)$ using FOIL, we write

$$(2x + 3)(3x + 4) = \overset{\text{F}}{6x^2} + \overset{\text{O}}{8x} + \overset{\text{I}}{9x} + \overset{\text{L}}{12}$$

$$= \underbrace{6x^2}_{\substack{\text{F} \\ 2 \cdot 3}} + \underbrace{17x}_{\substack{\text{O + I} \\ 2 \cdot 4 + 3 \cdot 3}} + \underbrace{12}_{\substack{\text{L} \\ 3 \cdot 4}}$$

Now to *factor* $6x^2 + 17x + 12$, we do the reverse, using trial and error. Since the factors of 6 are 6 and 1, or 3 and 2 (we won't use -6 and -1, or -3 and -2, because then the first coefficients will be negative), the possible combinations are

$$(6x + ___)(x + ___) \qquad (3x + ___)(2x + ___)$$

Since the product of the last two numbers is 12, the possible factors are 12, 1; 6, 2; and 3, 4. The possibilities are

$(6x + 12)(x + 1)^\star$ $\qquad$ $(6x + 1)(x + 12)$

$(6x + 6)(x + 2)^\star$ $\qquad$ $(6x + 2)(x + 6)^\star$

$(6x + 3)(x + 4)^\star$ $\qquad$ $(6x + 4)(x + 3)^\star$

$(3x + 12)(2x + 1)^\star$ $\qquad$ $(3x + 1)(2x + 12)^\star$

$(3x + 6)(2x + 2)^\star$ $\qquad$ $(3x + 2)(2x + 6)^\star$

$(3x + 3)(2x + 4)^\star$ $\qquad$ $(3x + 4)(2x + 3)$

Note that in one binomial each of the starred items has a common factor, but $6x^2 + 17x + 12$ has *no* common factor other than 1. Thus, we can eliminate all starred products.

Thus, $6x^2 + 17x + 12 = (3x + 4)(2x + 3)$.

Note that if there is a common factor, **we must factor it out first.** Thus, to factor $12x^2 + 2x - 2$, we must *first* factor out the common factor 2, as illustrated in Example 6.

EXAMPLE 6 Factoring by FOIL (trial and error): Last term negative

Factor: $12x^2 + 2x - 2$

SOLUTION 6 Since 2 is a common factor, we first factor it out to obtain

$$12x^2 + 2x - 2 = 2 \cdot 6x^2 + 2 \cdot x - 2 \cdot 1$$
$$= 2(6x^2 + x - 1)$$

PROBLEM 6

Factor using FOIL (trial and error):

$18x^2 + 3x - 6$

(continued)

Answers to PROBLEMS

5. $(3x + 2)(x + 1)$

6. $3(2x - 1)(3x + 2)$

Now we factor $6x^2 + x - 1$. The factors of 6 are 6, 1 or 3, 2. Thus,

$$6x^2 + x - 1 = (6x + \_\_\_)(x + \_\_\_)$$

or

$$6x^2 + x - 1 = (3x + \_\_\_)(2x + \_\_\_)$$

The product of the last two terms must be -1. The possible factors are -1, 1. The possibilities are

$$(6x - 1)(x + 1) \qquad (6x + 1)(x - 1)$$
$$(3x - 1)(2x + 1) \qquad (3x + 1)(2x - 1)$$

The only product that yields $6x^2 + x - 1$ is $(3x - 1)(2x + 1)$. Try it! This example shows why this method is sometimes called *trial* and error. Thus,

$$\begin{aligned} 12x^2 + 2x - 2 &= 2(6x^2 + x - 1) \\ &= 2(3x - 1)(2x + 1) \end{aligned}$$

Remember to write the common factor.

EXAMPLE 7 Factoring by FOIL (trial and error): Two variables

Factor: $6x^2 - 11xy - 10y^2$

SOLUTION 7 Since there are no common factors, look for the factors of 6: 6 and 1, or 3 and 2. The possibilities for our factorization are

$$(6x + \_\_\_)(x + \_\_\_) \quad \text{or} \quad (3x + \_\_\_)(2x + \_\_\_)$$

The last term, $-10y^2$, has the following possible factors:

$$10y, -y \qquad -10y, y \qquad 2y, -5y \qquad -2y, 5y$$
$$-y, 10y \qquad y, -10y \qquad -5y, 2y \qquad 5y, -2y$$

Look daunting? Don't despair; just look more closely. Can you see that some trials like $(6x + 10y)(3x - y)$, which correspond to the factors $10y$, $-y$, or $(3x - y)(2x + 10y)$, which correspond to $-y$, $10y$, can't be correct because they contain 2 as a common factor? As you can see, $6x^2 - 11xy - 10y^2$ has no common factors (other than 1), so let's try

$$(3x + 10y)(2x - y)$$

$20xy$
$(+)\ -3xy$
$17xy$ Add.

$17xy$ is not the correct middle term. Next we try

$$(3x - 2y)(2x + 5y)$$

$-4xy$
$(+)\ 15xy$
$11xy$ Add.

Again, $11xy$ is not the correct middle term, but it's close! The result is incorrect but only because of the sign of the middle term. The correct factorization is found by interchanging the signs in the binomials. That is,

$$(3x + 2y)(2x - 5y) = 6x^2 - 11xy - 10y^2$$

$4xy$
$(+)\ -15xy$
$-11xy$ Add.

PROBLEM 7

Factor using FOIL (trial and error):
$6x^2 - 5xy - 6y^2$

Answers to PROBLEMS

7. $(3x + 2y)(2x - 3y)$

How many gallons of gas does your car use in a year? It depends on many factors including your mileage (mpg) and how many miles you drive. The estimate for annual gas consumption varies widely: from 480 to 1100 gallons. Each gallon of gas emits 20 pounds of CO_2, but there is hope: more efficient cars emit less CO_2. In Example 8 we approximate the annual CO_2 emissions under the given conditions.

EXAMPLE 8 Factoring car carbon emissions

Assume your car uses 500 gallons of gas a year and each gallon produces 20 pounds of CO_2. The annual CO_2 produced by the average car can be approximated by $C(t) = -3t^2 - 440t + 10{,}000$, where t is the number of years after 2010 but less than 2015.

a. Factor $-3t^2 - 440t + 10{,}000$.

b. How many pounds of CO_2 will be produced by the average American car in 2013?

SOLUTION 8

a. We have not factored a polynomial with a negative first term, but we can fix that. Use -1 as the GCF and rewrite the polynomial

$-3t^2 - 440t + 10{,}000$ as $-1(3t^2 + \mathbf{440t} - 10{,}000)$. Note the change in sign of every term inside parentheses.

1. Find the key number $[3 \cdot (-10{,}000) = -30{,}000]$.

2. The factors of $-30{,}000$ that add up to **440** are 500 and -60. Rewrite **$440t$** as $500t - 60t$. $\quad -1(3t^2 + 500t - 60t - 10{,}000)$

3. Group the terms into pairs. Note that $-60t - 10{,}000$ is $-(60t + 10{,}000)$. $\quad -1[(3t^2 + 500t) - (60t + 10{,}000)]$

4. Factor each pair. $\quad -1[t(3t + 500) - 20(3t + 500)]$

5. $(3t + 500)$ is the GCF. Factor it out. $\quad -1[(3t + 500)(t - 20)]$

$-(3t + 500)(t - 20)$

The factored form shows $\quad -(\text{Gallons used})(CO_2 \text{ produced})$
the "$-$" means a decrease.

b. In 2013, $t = 3$ (2013 $-$ 2010) and
$$C(3) = -3(3)^2 - 440(3) + 10{,}000$$
$$= -27 - 1320 + 10{,}000$$
$$= 8653 \text{ pounds}$$

Source: http://tinyurl.com/yjg44eu.

PROBLEM 8

Canadians use less gas, about 310 gallons a year, so the annual CO_2 produced by their average car can be approximated by

$$C(t) = -3t^2 - 250t + 6200$$

a. Factor $C(t) = -3t^2 - 250t + 6200$

b. How many pounds of CO_2 will be produced by the average Canadian car in 2013?

How can you help: Keep your car tuned up, check your tire pressure, choose a more fuel-efficient car, and try ride-sharing!

CAUTION

Some students and some instructors prefer the FOIL (trial-and-error) method over the grouping method. You can use either method and your answer will be the same. Which one should you use? The one you understand better or the one your instructor asks you to use!

Answers to PROBLEMS

8. a. $-(3t + 310)(t - 20)$
b. 5423

> Exercises 5.3

⟨ A ⟩ Using the *ac* Test to Determine if a Polynomial Is Factorable
⟨ B ⟩ Factoring $ax^2 + bx + c$ by Grouping
⟨ C ⟩ Factoring $ax^2 + bx + c$ by FOIL (Trial and Error)

In Problems 1–50, determine whether the polynomial is factorable. If the polynomial is factorable, factor it. Use the FOIL method to factor if you wish.

1. $2x^2 + 5x + 3$ **2.** $2x^2 + 7x + 3$ **3.** $6x^2 + 11x + 3$

4. $6x^2 + 17x + 5$ **5.** $6x^2 + 11x + 4$ **6.** $5x^2 + 2x + 1$

7. $2x^2 + 3x - 2$ **8.** $2x^2 + x - 3$ **9.** $3x^2 + 16x - 12$

10. $6x^2 + x - 12$ **11.** $4y^2 - 11y + 6$ **12.** $3y^2 - 17y + 10$

13. $4y^2 - 8y + 6$ **14.** $3y^2 - 11y + 6$ **15.** $6y^2 - 10y - 4$

16. $12y^2 - 10y - 12$ **17.** $12y^2 - y - 6$ **18.** $3y^2 - y - 1$

19. $18y^2 - 21y - 9$ **20.** $36y^2 - 12y - 15$ **21.** $3x^2 + 2 + 7x$

22. $2x^2 + 2 + 5x$ **23.** $5x^2 + 2 + 11x$ **24.** $5x^2 + 3 + 12x$

25. $6x^2 - 5 + 15x$ **26.** $5x^2 - 8 + 6x$ **27.** $3x^2 - 2 - 5x$

28. $5x^2 - 8 - 6x$ **29.** $15x^2 - 2 + x$ **30.** $8x^2 + 15 - 14x$

31. $8x^2 + 20xy + 8y^2$ **32.** $12x^2 + 28xy + 8y^2$ **33.** $6x^2 + 7xy - 3y^2$ **34.** $3x^2 + 13xy - 10y^2$

35. $7x^2 - 10xy + 3y^2$ **36.** $6x^2 - 17xy + 5y^2$ **37.** $15x^2 - xy - 2y^2$ **38.** $5x^2 - 6xy - 8y^2$

39. $15x^2 - 2xy - 2y^2$ **40.** $4x^2 - 13xy - 3y^2$ **41.** $12r^2 + 17r - 5$ **42.** $20s^2 + 7s - 6$

43. $22t^2 - 29t - 6$ **44.** $39u^2 - 23u - 6$ **45.** $18x^2 - 21x + 6$ **46.** $12x^2 - 22x + 6$

47. $6ab^2 + 5ab + a$ **48.** $6bc^2 + 13bc + 6b$ **49.** $6x^5y + 25x^4y^2 + 4x^3y^3$ **50.** $12p^4q^3 + 11p^3q^4 + 2p^2q^5$

In Problems 51–60, first factor out -1. [*Hint:* To factor $-6x^2 + 7x + 2$, the first step will be

$$\begin{aligned} -6x^2 + 7x + 2 &= -1(6x^2 - 7x - 2) \\ &= -(6x^2 - 7x - 2) \end{aligned}$$

then factor inside the parentheses.]

51. $-6x^2 - 7x - 2$ **52.** $-12y^2 - 11y - 2$ **53.** $-9x^2 - 3x + 2$ **54.** $-6y^2 - 5y + 6$

55. $-8m^2 + 10mn + 3n^2$ **56.** $-6s^2 + st + 2t^2$ **57.** $-8x^2 + 9xy - y^2$ **58.** $-6y^2 + 3xy + 2x^2$

59. $-x^3 - 5x^2 - 6x$ **60.** $-y^3 + 3y^2 - 2y$

>>> Applications

61. *Flow rate* To find the flow g (in hundreds of gallons per minute) in 100 feet of $2\frac{1}{2}$-inch rubber-lined hose when the friction loss is 36 pounds per square inch, we need to factor the expression

$$2g^2 + g - 36$$

Factor this expression.

62. *Flow rate* To find the flow g (in hundreds of gallons per minute) in 100 feet of $2\frac{1}{2}$-inch rubber-lined hose when the friction loss is 55 pounds per square inch, we must factor the expression

$$2g^2 + g - 55$$

Factor this expression.

63. *Equivalent resistance* When solving for the equivalent resistance R of two electric circuits, we use the expression

$$2R^2 - 3R + 1$$

Factor this expression.

64. *Rate of ascent* To find the time t at which an object thrown upward at 12 meters per second will be 4 meters above the ground, we must factor the expression

$$5t^2 - 12t + 4$$

Factor this expression.

>>> Applications: Green Math

In Example 8, we discussed the average CO_2 produced by U.S. cars. But there are other nations that also produce CO_2. We discuss some of these in Problems 65 and 66.

Source: http://tinyurl.com/yjeh5t3.

65. *CO_2 emissions in Australia and China*

a. The expression $-3t^2 - 190t + 5000$ corresponds to the amount of CO_2 produced by the average car in Australia. Follow the procedure of Example 8 to factor the expression and discover the number of gallons used per capita in Australia.

b. The expression $-2t^2 + 29t + 220$ corresponds to the amount of CO_2 produced by the average car in China. Follow the procedure of Example 8 to factor the expression and discover the number of gallons used per capita in China.

66. *CO_2 emissions in France and Russia*

a. The expression $-2t^2 - 21t + 1220$ corresponds to the amount of CO_2 produced by the average car in France. Follow the procedure of Example 8 to factor the expression and discover the number of gallons used per capita in France.

b. The expression $-3t^2 - t + 1220$ corresponds to the amount of CO_2 produced by the average car in China. Follow the procedure of Example 8 to factor the expression and discover the number of gallons used per capita in China.

67. *Stopping distance* (See page 355.) If you are traveling at m miles per hour and want to find the speed m that will allow you to stop the car 13 feet away, you must solve the equation $5m^2 + 220m - 1225 = 0$. Factor the left side of the equation.

68. *Stopping distance* If you are traveling at m miles per hour and want to find the speed m that will allow you to stop the car 14 feet away, you must solve the equation $5m^2 + 225m - 1250 = 0$. Factor the left side of the equation.

69. *Height of an object* The height $H(t)$ of an object thrown downward from a height of h meters and initial velocity v_0 is given by the equation $H(t) = -5t^2 + v_0t + h$. If an object is thrown downward at 5 meters per second from a height of 10 meters, then $H(t) = -5t^2 - 5t + 10$. Factor $-5t^2 - 5t + 10$.

70. *Height of an object* The height $H(t)$ of an object thrown downward from a height of h meters and initial velocity v_0 is given by the equation $H(t) = -5t^2 + v_0t + h$. If an object is thrown downward at 4 meters per second from a height of 28 meters, then $H(t) = -5t^2 - 4t + 28$. Factor $-5t^2 - 4t + 28$.

71. *Production cost* The cost $C(x)$ of producing x units of a certain product is given by the equation $C(x) = 3x^2 - 17x + 20$. Factor $3x^2 - 17x + 20$.

72. *Production cost* The cost $C(x)$ of producing x units of a certain product is given by the equation $C(x) = 3x^2 - 12x + 13$. Factor $5x^2 - 28x + 15$.

>>> Using Your Knowledge

Factoring Applications The ideas presented in this section are important in many fields. Use your knowledge to factor the given expressions.

73. To find the deflection of a beam of length L at a distance of 3 feet from its end, we must evaluate the expression

$$2L^2 - 9L + 9$$

Factor this expression.

74. The height after t seconds of an object thrown upward at 12 meters per second is

$$-5t^2 + 12t$$

To determine the time at which the object will be 7 meters above ground, we must solve the equation

$$5t^2 - 12t + 7 = 0$$

Factor $5t^2 - 12t + 7$.

75. In Problem 73, if the distance from the end is x feet, then we must use the expression

$$2L^2 - 3xL + x^2$$

Factor this expression.

Write On

76. Mourad says that the ac key number for $2x^2 + 1 + 3x$ is 2 and hence $2x^2 + 1 + 3x$ is factorable. Bill says ac is 6. Who is correct?

77. When factoring $6x^2 + 11x + 3$ by grouping, student A writes

$$6x^2 + 11x + 3 = 6x^2 + 9x + 2x + 3$$

Student B writes

$$6x^2 + 11x + 3 = 6x^2 + 2x + 9x + 3$$

a. What will student A's answer be?

b. What will student B's answer be?

c. Who is correct, student A or student B? Explain.

78. A student gives $(3x - 1)(x - 2)$ as the answer to a factoring problem. Another student gets $(2 - x)(1 - 3x)$. Which student is correct? Explain.

Concept Checker

Fill in the blank(s) with the correct word(s), phrase, or mathematical statement.

79. A trinomial of the form $ax^2 + bx + c$ is **factorable** if there are two integers with **product** _______________ and **sum** _______________.

80. When factoring $ax^2 + bx + c$ the product ac is called the _______________ number.

main	***b***
key	***c***
ac	***a***
ad	

Mastery Test

Factor:

81. $3x^2 - 4 - 4x$

82. $2x^2 - 11x + 5$

83. $2x^2 - xy - 6y^2$

84. $3x^2 + 5x + 2$

85. $16x^2 + 4x - 2$

86. $3x^3 + 7x^2 + 2x$

87. $3x^4 + 5x^3 - 3x^2$

88. $5x^2 - 2x + 2$

Skill Checker

Expand:

89. $(x + 8)^2$

90. $(x - 7)^2$

91. $(3x - 2)^2$

92. $(2x + 3)^2$

93. $(2x + 3y)^2$

94. $(2x - 3y)^2$

95. $(3x + 5y)(3x - 5y)$

96. $(2x - 5y)(2x + 5y)$

97. $(x^2 + 4)(x^2 - 4)$

98. $(x^2 - 3)(x^2 + 3)$

5.4 Factoring Squares of Binomials

Objectives

A › **Recognize the square of a binomial (a perfect square trinomial).**

B › **Factor a perfect square trinomial.**

C › **Factor the difference of two squares.**

D › **Solve applications involving factoring.**

To Succeed, Review How To . . .

1. Expand the square of a binomial sum or difference (pp. 378–380).
2. Find the product of the sum and difference of two terms (pp. 380–381).

Getting Started

A Moment for a Crane

What is the moment (the product of a quantity and the distance from a perpendicular axis) on the crane? At x feet from its support, the moment involves the expression

$$\frac{w}{2}(x^2 - 20x + 100)$$

where w is the weight of the crane in pounds per foot. The expression $x^2 - 20x + 100$ is the result of *squaring a binomial* and is called a **perfect square trinomial** because $x^2 - 20x + 100 = (x - 10)^2$. Similarly, $x^2 + 12x + 36 = (x + 6)^2$ is the square of a binomial. We can factor these two expressions by using the special products studied in Section 4.7 in reverse. For example, $x^2 - 20x + 100$ has the same form as SP3, whereas $x^2 + 12x + 36$ looks like SP2. In this section we continue to study the *reverse* process of multiplying binomials, that of *factoring* trinomials.

A › Recognizing Squares of Binomials

We start by rewriting the products in SP2 and SP3 so you can use them for factoring.

FACTORING RULES 2 AND 3: PERFECT SQUARE TRINOMIALS

$X^2 + 2AX + A^2 = (X + A)^2$ Note that $X^2 + A^2 \neq (X + A)^2$ **(F2)**

$X^2 - 2AX + A^2 = (X - A)^2$ Note that $X^2 - A^2 \neq (X - A)^2$ **(F3)**

Note that to be the **square of a binomial** (a *perfect square trinomial*), a trinomial must satisfy three conditions:

1. The first and last terms (X^2 and A^2) must be perfect squares.
2. There must be no minus signs before A^2 or X^2.
3. The middle term is twice the product of the expressions being squared in step 1 ($2AX$) or its additive inverse ($-2AX$). Note the X and A are the terms of the binomial being squared to obtain the perfect square trinomial.

EXAMPLE 1 Deciding whether an expression is the square of a binomial

Determine whether the given expression is the square of a binomial:

a. $x^2 + 8x + 16$ **b.** $x^2 + 6x - 9$

c. $x^2 + 4x + 16$ **d.** $4x^2 - 12xy + 9y^2$

SOLUTION 1 In each case, we check the three conditions necessary for having a perfect square trinomial.

a. **1.** x^2 and $16 = 4^2$ are perfect squares.

2. There are no minus signs before x^2 or 16.

3. The middle term is twice the product of the expressions being squared in step 1, x and 4; that is, the middle term is $2 \cdot (x \cdot 4) = 8x$. Thus, $x^2 + 8x + 16$ *is* a perfect square trinomial (the square of a binomial).

b. **1.** x^2 and $9 = 3^2$ are perfect squares.

2. However, there's a minus sign before the 9. Thus, $x^2 + 6x - 9$ is *not* the square of a binomial.

c. **1.** x^2 and $16 = 4^2$ are perfect squares.

2. There are no minus signs before x^2 or 16.

3. The middle term should be $2 \cdot (x \cdot 4) = 8x$, but instead it's $4x$. Thus, $x^2 + 4x + 16$ is *not* a perfect square trinomial.

d. **1.** $4x^2 = (2x)^2$ and $9y^2 = (3y)^2$ are perfect squares.

2. There are no minus signs before $4x^2$ or $9y^2$.

3. The middle term is the additive inverse of twice the product of the expressions being squared in step 1; that is, $-2 \cdot (2x \cdot 3y) = -12xy$. Thus, $4x^2 - 12xy + 9y^2$ *is* a perfect square trinomial.

PROBLEM 1

Determine whether the expression is the square of a binomial:

a. $y^2 + 9y + 9$

b. $y^2 + 4y + 4$

c. $y^2 + 8y - 16$

d. $9x^2 - 12xy + 4y^2$

B > Factoring Perfect Square Trinomials

The formulas given in F2 and F3 can be used to factor any trinomials that are perfect squares. For example, the trinomial $9x^2 + 12x + 4$ can be factored using F2 if we first notice that

1. $9x^2$ and 4 are perfect squares, since $9x^2 = (3x)^2$ and $4 = 2^2$.
2. There are no minus signs before $9x^2$ or 4.
3. $12x = 2 \cdot (2 \cdot 3x)$. (The middle term is twice the product of 2 and $3x$, the expressions being squared in step 1.)

We then write

$$\begin{aligned} 9x^2 + 12x + 4 &= \overset{X^2}{(3x)^2} + \overset{2}{2} \cdot \overset{AX}{(2 \cdot 3x)} + \overset{A^2}{2^2} \\ &= (3x + 2)^2 \end{aligned}$$

We are letting $X = 3x$, $A = 2$ in F2.

Answers to PROBLEMS

1. a. No **b.** Yes **c.** No **d.** Yes

Here are some other examples of this form; study these examples carefully before you continue.

$$X^2 + 2AX + A^2$$

$$9x^2 + 6x + 1 = (3x)^2 + 2 \cdot (1 \cdot 3x) + 1^2 = (3x + 1)^2$$ Letting $X = 3x$, $A = 1$ in F2.

$$16x^2 + 24x + 9 = (4x)^2 + 2 \cdot (3 \cdot 4x) + 3^2 = (4x + 3)^2$$ Here $X = 4x$, $A = 3$.

$$4x^2 + 12xy + 9y^2 = (2x)^2 + 2 \cdot (3y \cdot 2x) + (3y)^2 = (2x + 3y)^2$$ Here $X = 2x$, $A = 3y$.

NOTE

The key for factoring these trinomials is to recognize that the first and last terms are *perfect squares*. (Of course, you have to check the middle term also.)

EXAMPLE 2 Factoring perfect square trinomials

Factor:

a. $x^2 + 16x + 64$ **b.** $25x^2 + 20x + 4$ **c.** $9x^2 + 12xy + 4y^2$

SOLUTION 2

a. We first write the trinomial in the form $X^2 + 2AX + A^2$. Thus,

$$x^2 + 16x + 64 = x^2 + 2 \cdot (8 \cdot x) + 8^2 = (x + 8)^2$$

b. $25x^2 + 20x + 4 = (5x)^2 + 2 \cdot (2 \cdot 5x) + 2^2 = (5x + 2)^2$

c. $9x^2 + 12xy + 4y^2 = (3x)^2 + 2 \cdot (2y \cdot 3x) + (2y)^2 = (3x + 2y)^2$

PROBLEM 2

Factor:

a. $y^2 + 6y + 9$

b. $4x^2 + 28xy + 49y^2$

c. $9y^2 + 12y + 4$

Of course, we use the same technique (but with F3) to factor $x^2 - 16x + 64$ or $25x^2 - 20x + 4$. Do you recall F3?

$$\underbrace{x^2}_{X^2} \underbrace{- 16x}_{-2AX} + \underbrace{64}_{A^2} = x^2 - 2 \cdot (8 \cdot x) + 8^2 = \underbrace{(x - 8)^2}_{(X-A)^2}$$

Similarly,

$$25x^2 - 20x + 4 = (5x)^2 - 2 \cdot (2 \cdot 5x) + 2^2 = (5x - 2)^2$$

EXAMPLE 3 Factoring perfect square trinomials

Factor:

a. $x^2 - 10x + 25$ **b.** $4x^2 - 12x + 9$ **c.** $4x^2 - 20xy + 25y^2$

SOLUTION 3

a. $x^2 - 10x + 25 = x^2 - 2 \cdot (5 \cdot x) + 5^2 = (x - 5)^2$

b. $4x^2 - 12x + 9 = (2x)^2 - 2 \cdot (3 \cdot 2x) + 3^2 = (2x - 3)^2$

c. $4x^2 - 20xy + 25y^2 = (2x)^2 - 2 \cdot (5y \cdot 2x) + (5y)^2 = (2x - 5y)^2$

PROBLEM 3

Factor:

a. $y^2 - 4y + 4$

b. $9y^2 - 12y + 4$

c. $9x^2 - 30xy + 25y^2$

C > Factoring the Difference of Two Squares

Can we factor $x^2 - 9$ as a product of two binomials? Note that $x^2 - 9$ has no middle term. The only special product with no middle term we've studied is the product of the sum and the difference of two terms (SP4). Here is the corresponding factoring rule:

FACTORING RULE 4: THE DIFFERENCE OF TWO SQUARES

$$X^2 - A^2 = (X + A)(X - A) \qquad \textbf{(F4)}$$

Answers to PROBLEMS

2. a. $(y + 3)^2$ **b.** $(2x + 7y)^2$ **c.** $(3y + 2)^2$

3. a. $(y - 2)^2$ **b.** $(3y - 2)^2$ **c.** $(3x - 5y)^2$

We can now factor binomials of the form $x^2 - 16$ and $9x^2 - 25y^2$. To do this, we proceed as follows:

$$\begin{aligned} &X^2 - A^2 = (X + A)(X - A) \\ &x^2 - 16 = (x)^2 - (4)^2 = (x + 4)(x - 4) \end{aligned}$$ Check this by using FOIL.

and

$$\begin{aligned} &X^2 - A^2 = (X + A)(X - A) \\ &9x^2 - 25y^2 = (3x)^2 - (5y)^2 = (3x + 5y)(3x - 5y) \end{aligned}$$

NOTE

$x^2 + A^2$ *cannot* be factored!

EXAMPLE 4 Factoring the difference of two squares

Factor:

a. $x^2 - 4$

b. $25x^2 - 9$

c. $16x^2 - 9y^2$

d. $x^4 - 16$

e. $\frac{1}{4}x^2 - \frac{1}{9}$

f. $7x^3 - 28x$

SOLUTION 4

$X^2 - A^2 = (X + A)(X - A)$

a. $x^2 - 4 = (x)^2 - (2)^2 = (x + 2)(x - 2)$

b. $25x^2 - 9 = (5x)^2 - (3)^2 = (5x + 3)(5x - 3)$

c. $16x^2 - 9y^2 = (4x)^2 - (3y)^2 = (4x + 3y)(4x - 3y)$

d. $x^4 - 16 = (x^2)^2 - (4)^2 = (x^2 + 4)(x^2 - 4)$

But $(x^2 - 4)$ itself is factorable, so

$(x^2 + 4)(x^2 - 4) = (x^2 + 4)(x + 2)(x - 2)$. Thus,

$$x^4 - 16 = \underbrace{(x^2 + 4)}_{\text{Not factorable}}(x + 2)(x - 2)$$

e. We start by writing $\frac{1}{4}x^2 - \frac{1}{9}$ as the difference of two squares. To obtain $\frac{1}{4}x^2$, we must square $\frac{1}{2}x$ and to obtain $\frac{1}{9}$, we must square $\frac{1}{3}$. Thus,

$$\begin{aligned} \frac{1}{4}x^2 - \frac{1}{9} &= \left(\frac{1}{2}x\right)^2 - \left(\frac{1}{3}\right)^2 \\ &= \left(\frac{1}{2}x + \frac{1}{3}\right)\left(\frac{1}{2}x - \frac{1}{3}\right) \end{aligned}$$

f. We start by finding the GCF of $7x^3 - 28x$, which is $7x$. We then write

$$\begin{aligned} 7x^3 - 28x &= 7x(x^2 - 4) \\ &= 7x(x + 2)(x - 2) \end{aligned}$$ Factor $x^2 - 4$ as $(x + 2)(x - 2)$.

PROBLEM 4

Factor:

a. $y^2 - 1$

b. $9y^2 - 25$

c. $9y^2 - 25x^2$

d. $y^4 - 81$

e. $\frac{1}{9}y^2 - \frac{1}{4}$

f. $5y^3 - 45y$

D › Solving Applications Involving Factoring Trinomials

We previously mentioned polar bears and alligators, but do you know how many actual endangered species (a species that is in danger of extinction throughout all or a significant portion of its range) we have? It varies from 1011 to 1556, depending on the source.

Answers to PROBLEMS

4. a. $(y + 1)(y - 1)$
b. $(3y + 5)(3y - 5)$
c. $(3y + 5x)(3y - 5x)$
d. $(y^2 + 9)(y + 3)(y - 3)$
e. $\left(\frac{1}{3}y + \frac{1}{2}\right)\left(\frac{1}{3}y - \frac{1}{2}\right)$
f. $5y(y + 3)(y - 3)$

The graph shows that this is not a new problem (It has been studied since 1980.), but we can use polynomials to try to estimate future trends!

EXAMPLE 5 Perfect square trinomials and endangered species

The graph shows the number of endangered U.S. plant and animal species (blue), which can be approximated by $E(t) = t^2 + 26t + 169$, where t is the number of years after 1980.

a. Factor $t^2 + 26t + 169$.

b. How many endangered species would you predict for 2010?

SOLUTION 5

a. We first write the trinomial in the form $X^2 + 2AX + A^2$. Thus,

$$t^2 + 26t + 169 = t^2 + 2 \cdot (13) \cdot t + 13^2 = (t + 13)^2$$

b. The year 2010 corresponds to $t = 30$ (2010 − 1980).

Letting $t = 30$ in $(t + 13)^2$ yields $(30 + 13)^2 = (43)^2 = 1849$, more than the actual result which ranges from 1011 to 1556 according to our sources.

Sources: http://ecos.fws.gov/tess_public/TESSBoxscore; http://www.earthsendangered.com/continent.asp?gr=&view=&ID=9; http://tinyurl.com/yfjxm8n.

PROBLEM 5

The total number of threatened ***and*** endangered species (the complete blue-green bar) can be approximated by

$$T(t) = t^2 + 28t + 196.$$

a. Factor $t^2 + 28t + 196$.

b. How many threatened and endangered species would you predict for 2010?

The actual results range from 1321 to 1868 depending on the source.

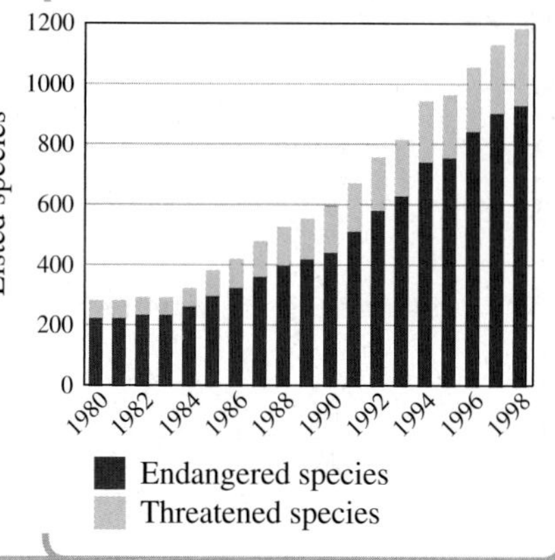

Calculator Corner

More Factoring Checking

Can you use a calculator to check factoring? Yes, but you still have to accept the following: If the graphs of two polynomials are identical, the polynomials are identical. Consider Example 2b, $25x^2 + 20x + 4 = (5x + 2)^2$. Graph $Y_1 = 25x^2 + 20x + 4$ and $Y_2 = (5x + 2)^2$. The graphs are the same! How do you know there are two graphs? Press TRACE and ▼, then ▼. Do you see the little number at the top right of the screen? It tells you which curve is showing. The bottom of the screen shows that for both Y_1 and Y_2 when $x = 0$, $y = 4$. Graph $Y_3 = (5x + 2)^2 + 10$. Press TRACE. Are the values for Y_1, Y_2, and Y_3 the same? They shouldn't be! Do you see why? Use this technique to check other factoring problems.

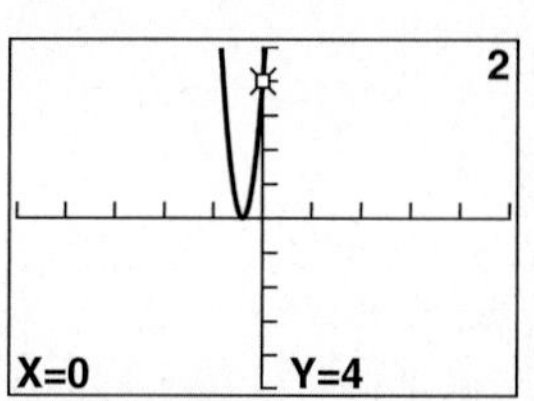

Answers to PROBLEMS

5. a. $(t + 14)^2$

b. 1936 (more than the actual result, which ranges from 1321 to 1868)

Exercises 5.4

⟨A⟩ Recognizing Squares of Binomials In Problems 1–10, determine whether the given expression is a perfect square trinomial (the square of a binomial).

1. $x^2 + 14x + 49$
2. $x^2 + 18x + 81$
3. $25x^2 + 10x - 1$
4. $9x^2 + 12x - 4$
5. $25x^2 + 10x + 1$
6. $9x^2 + 12x + 4$
7. $y^2 - 4y - 4$
8. $y^2 - 20y - 100$
9. $16y^2 - 40yz + 25z^2$
10. $49y^2 - 56yz + 16z^2$

⟨B⟩ Factoring Perfect Square Trinomials In Problems 11–34, factor completely.

11. $x^2 + 2x + 1$
12. $x^2 + 6x + 9$
13. $3x^2 + 30x + 75$
14. $2x^2 + 28x + 98$
15. $9x^2 + 6x + 1$
16. $16x^2 + 8x + 1$
17. $9x^2 + 12x + 4$
18. $25x^2 + 10x + 1$
19. $16x^2 + 40xy + 25y^2$
20. $9x^2 + 30xy + 25y^2$
21. $25x^2 + 20xy + 4y^2$
22. $36x^2 + 60xy + 25y^2$
23. $y^2 - 2y + 1$
24. $y^2 - 6y + 9$
25. $3y^2 - 24y + 48$
26. $2y^2 - 40y + 200$
27. $9x^2 - 6x + 1$
28. $4x^2 - 20x + 25$
29. $16x^2 - 56x + 49$
30. $25x^2 - 30x + 9$
31. $9x^2 - 12xy + 4y^2$
32. $16x^2 - 40xy + 25y^2$
33. $25x^2 - 10xy + y^2$
34. $49x^2 - 56xy + 16y^2$

⟨C⟩ Factoring the Difference of Two Squares In Problems 35–60, factor completely.

35. $x^2 - 49$
36. $x^2 - 121$
37. $9x^2 - 49$
38. $16x^2 - 81$
39. $25x^2 - 81y^2$
40. $81x^2 - 25y^2$
41. $x^4 - 1$
42. $x^4 - 256$
43. $16x^4 - 1$
44. $16x^4 - 81$
45. $\frac{1}{9}x^2 - \frac{1}{16}$
46. $\frac{1}{4}y^2 - \frac{1}{25}$
47. $\frac{1}{4}z^2 - 1$
48. $\frac{1}{9}r^2 - 1$
49. $1 - \frac{1}{4}s^2$
50. $1 - \frac{1}{9}t^2$
51. $\frac{1}{4} - \frac{1}{9}y^2$
52. $\frac{1}{9} - \frac{1}{16}u^2$
53. $\frac{1}{9} + \frac{1}{4}x^2$
54. $\frac{1}{4} + \frac{1}{25}n^2$
55. $3x^3 - 12x$
56. $4y^3 - 16y$
57. $5t^3 - 20t$
58. $7t^3 - 63t$
59. $5t - 20t^3$
60. $2s - 18s^3$

In Problems 61–74, use a variety of factoring methods to factor completely.

61. $49x^2 + 28x + 4$
62. $49y^2 + 42y + 9$
63. $x^2 - 100$
64. $x^2 - 144$
65. $x^2 + 20x + 100$
66. $x^2 + 18x + 81$
67. $9 - 16m^2$
68. $25 - 9n^2$
69. $9x^2 - 30xy + 25y^2$
70. $16x^2 - 40xy + 25y^2$
71. $z^4 - 16$
72. $16y^4 - 81$
73. $3x^3 - 75x$
74. $2y^3 - 72y$

‹D› Solving Applications Involving Factoring Trinomials

75. *Demand function* A business owner finds out that when x units of a product are demanded by consumers, the price per unit is a function of the demand and given by the equation $D(x) = 81 - x^2$. Factor $81 - x^2$.

76. *Demand function* A business estimates that when y units of a product are demanded by consumers, the price per units is a function of the demand and given by the equation $D(y) = 64 - y^2$. Factor $64 - y^2$.

77. *Economic supplies* The relationship between the quantity supplied S and the unit price p is given by the equation $S = C^2 - k^2p^2$, where C and k are constants. Factor $C^2 - k^2p^2$.

78. *Kinetic energy* The change in kinetic energy of a moving object of mass m with initial velocity v_1 and terminal velocity v_2 is given by the expression $\frac{1}{2}mv_1^2 - \frac{1}{2}mv_2^2$. Factor $\frac{1}{2}mv_1^2 - \frac{1}{2}mv_2^2$ completely.

››› Applications: Green Math

79. *Plastic discarded* In a recent year, 31 million pounds of plastics were found in the garbage. These 31 million pounds were expected to grow at a 2% annual rate. What would be the total amount of plastic generated in the next 2 years? If we know that an initial amount I increases at P% each year, after n years the total amount is equal to $\frac{Ix^n - I}{x - 1}$, where $x = 1 + P$.

a. Use $n = 2$, $x = 1 + 2\% = 1.02$, and $I = 31$ to find the total amount of plastics in the garbage during the next two years.

b. Factor $Ix^2 - I$ completely.

c. Use the factored form as the numerator of $\frac{Ix^2 - Ix}{x - 1}$ to simplify this fraction.

d. Use the answer from part **c** to find the total amount of plastics in the garbage during the next 2 years. Is your answer the same as in part **a?**

Landfill facts: Americans buy an estimated 28 billion plastic water bottles every year, but only 23% are recycled. The rest go to the landfill where they take about 700 years to begin to decompose.

Sources: www.wikianswers.com, www.chacha.com

80. *Plastic recycled* The amount of plastic discarded **does not** equal the amount of plastics *recycled.* The amount of plastic recycled is only 2 million pounds but still growing at 2% each year.

a. Use $n = 2$, $x = 1 + 2\% = 1.02$, and $I = 2$ in the formula of Problem 79 to find the total amount of plastics in the garbage during the next 2 years.

b. Factor $2x^2 - 2$ completely.

c. Use the factored form as the numerator of $\frac{Ix^2 - I}{x - 1}$ to simplify this fraction.

d. Use the answer from part **c** to find the total amount of plastics in the garbage during the next 2 years. Is your answer the same as in part **a?**

Recycling facts: What can you do with the recycled bottles? Make more bottles. Just melt the plastic and make new ones (www.wikianswers.com). You can also make a deck or a picnic table and remember: recycling 1 ton of plastic saves the equivalent of 1500 gallons of gasoline (www.agriplasinc.com).

››› Using Your Knowledge

How Does It Function? Many business ideas are made precise by using expressions called **functions.** These expressions are often given in unfactored form. Use your knowledge to factor the given expressions (functions).

81. When x units of an item are demanded by consumers, the price per unit is given by the **demand** function $D(x)$ (read "D of x"):

$$D(x) = x^2 - 14x + 49$$

Factor this expression.

82. When x units are supplied by sellers, the price per unit of an item is given by the **supply** function $S(x)$:

$$S(x) = x^2 + 4x + 4$$

Factor this expression.

83. When x units are produced, the cost function $C(x)$ for a certain item is given by the equation

$$C(x) = x^2 + 12x + 36$$

Factor this expression.

84. When the market price is p dollars, the supply function $S(p)$ for a certain commodity is given by the equation

$$S(p) = p^2 - 6p + 9$$

Factor this expression.

Write On

85. The difference of two squares can be factored. Can you factor $a^2 + b^2$? Can you say that the sum of two squares can *never* be factored? Explain.

86. Can you factor $4a^2 + 16b^2$? Think of the implications for Problem 85.

87. What binomial multiplied by $(x + 2)$ gives a perfect square trinomial?

88. What binomial multiplied by $(2x - 3y)$ gives a perfect square trinomial?

Concept Checker

Fill in the blank(s) with the correct word(s), phrase, or mathematical statement.

89. When written in **factored form,** $X^2 + 2AX + A^2$ = ______________.

90. When written in **factored form,** $X^2 - 2AX + A^2$ = ______________.

91. When written in **factored form,** $X^2 - A^2$ = ______________.

92. The expression $X^2 + A^2$ ______________ be factored.

$(X + A)(X - A)$ **$(X + A)^2$**

can **$(X - A)^2$**

cannot

Mastery Test

Factor, if possible:

93. $x^2 - 1$

94. $9x^2 - 16$

95. $9x^2 - 25y^2$

96. $x^2 - 6x + 9$

97. $9x^2 - 24xy + 16y^2$

98. $9x^2 - 12x + 4$

99. $16x^2 + 24xy + 9y^2$

100. $x^2 + 4x + 4$

101. $9x^2 + 30x + 25$

102. $4x^2 - 20xy + 25y^2$

103. $9x^2 + 4$

104. $\frac{1}{36}x^2 - \frac{1}{49}$

105. $\frac{1}{81} - \frac{1}{4}x^2$

106. $12m^3 - 3mn^2$

107. $18x^3 - 50xy^2$

108. $9x^3 + 25xy^2$

Determine whether the expression is the square of a binomial. If it is, factor it.

109. $x^2 + 6x + 9$

110. $x^2 + 8x + 64$

111. $x^2 + 6x - 9$

112. $x^2 + 8x - 64$

113. $4x^2 - 20xy + 25y^2$

Skill Checker

Multiply:

114. $(A + B)(A - B)$

115. $(R + r)(R - r)$

116. $(P + q)(P - q)$

Factor completely:

117. $6x^2 - 18x - 24$

118. $4x^4 + 12x^3 + 40x^2$

119. $2x^2 - 18$

120. $3x^2 - 27$

5.5 A General Factoring Strategy

Objectives

A Factor the sum or difference of two cubes.

B Factor a polynomial by using the general factoring strategy.

C Solve applications involving factoring.

D Factor expressions whose leading coefficient is -1.

To Succeed, Review How To . . .

Factor a polynomial using the rules for reversing FOIL, perfect square trinomials, and the difference of two squares. (F1–F4) (pp. 420–422; 439–442).

Getting Started

Factoring and Medicine

In an artery (see the cross section in the photo), the speed (in centimeters per second) of the blood is given by

$$CR^2 - Cr^2$$

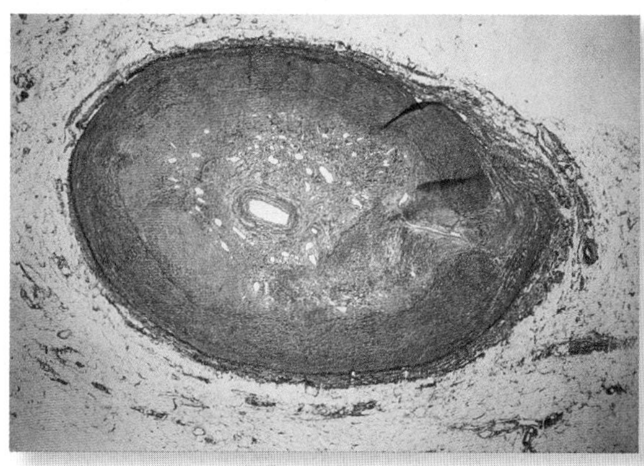

You already know how to factor this expression; using techniques you've already learned, you would proceed as follows:

1. Factor out any common factors (in this case, C). $\qquad CR^2 - Cr^2 = C(R^2 - r^2)$
2. Look at the terms inside the parentheses. You have the difference of two square terms in the expression: $R^2 - r^2$, so you factor it. $\qquad = C(R + r)(R - r)$
3. Make sure the expression is *completely* factored. Note that $C(R + r)(R - r)$ cannot be factored further.

What we've just used here is a *strategy* for factoring polynomials—a logical way to call up any of the techniques you've studied when they fit the expression you are factoring. In this section we shall study one more type of factoring: sums or differences of cubes. We will then examine in more depth the general factoring strategy for polynomials.

A > Factoring Sums or Differences of Cubes

We've already factored the difference of two squares $X^2 - A^2$. Can we factor the difference of two cubes $X^3 - A^3$? Not only can we factor $X^3 - A^3$, we can even factor $X^3 + A^3$! Since factoring is "reverse multiplication," let's start with two multiplication problems: $(X + A)(X^2 - AX + A^2)$ and $(X - A)(X^2 + AX + A^2)$.

$$\begin{array}{lll} & X^2 - AX + A^2 & \\ & \times \quad X + A & \\ & AX^2 - A^2X + A^3 & \leftarrow \text{Multiply } A(X^2 - AX + A^2). \\ X^3 - AX^2 + A^2X & & \leftarrow \text{Multiply } X(X^2 - AX + A^2). \\ \hline X^3 & \qquad\qquad + A^3 & \end{array}$$

$$\begin{array}{lll} & X^2 + AX + A^2 & \\ & \times \quad X - A & \\ & -AX^2 - A^2X - A^3 & \leftarrow \text{Multiply } -A(X^2 + AX + A^2). \\ X^3 + AX^2 + A^2X & & \leftarrow \text{Multiply } X(X^2 + AX + A^2). \\ \hline X^3 & \qquad\qquad - A^3 & \end{array}$$

Thus,

Same Different Same Different

$$X^3 + A^3 = (X + A)(X^2 - AX + A^2) \quad \text{and} \quad X^3 - A^3 = (X - A)(X^2 + AX + A^2).$$

This gives us our final factoring rules.

FACTORING RULES 5 AND 6: THE SUM AND DIFFERENCE OF TWO CUBES

$$X^3 + A^3 = (X + A)(X^2 - AX + A^2) \quad \textbf{(F5)}$$

$$X^3 - A^3 = (X - A)(X^2 + AX + A^2) \quad \textbf{(F6)}$$

NOTE

The trinomials $X^2 - AX + A^2$ and $X^2 + AX + A^2$ cannot be factored further.

EXAMPLE 1 Factoring sums and differences of cubes

Factor completely:

a. $x^3 + 27$ **b.** $8x^3 + y^3$

c. $m^3 - 8n^3$ **d.** $27r^3 - 8s^3$

SOLUTION 1

a. We rewrite $x^3 + 27$ as the sum of two cubes and then use F5:

$$\begin{aligned} x^3 + 27 &= (x)^3 + (3)^3 \\ &= (x + 3)(x^2 - 3x + 3^2) \quad \text{Letting } X = x \text{ and } A = 3 \text{ in F5} \\ &= (x + 3)(x^2 - 3x + 9) \end{aligned}$$

b. This is also the sum of two cubes, so we write:

$$\begin{aligned} 8x^3 + y^3 &= (2x)^3 + (y)^3 \\ &= (2x + y)[(2x)^2 - 2xy + y^2] \quad \text{Letting } X = 2x \text{ and } A = y \text{ in F5} \\ &= (2x + y)(4x^2 - 2xy + y^2) \end{aligned}$$

PROBLEM 1

Factor completely:

a. $y^3 + 8$ **b.** $27y^3 + 8$

c. $y^3 - 27z^3$ **d.** $8a^3 - 27b^3$

Answers to PROBLEMS

1. a. $(y + 2)(y^2 - 2y + 4)$
b. $(3y + 2)(9y^2 - 6y + 4)$
c. $(y - 3z)(y^2 + 3yz + 9z^2)$
d. $(2a - 3b)(4a^2 + 6ab + 9b^2)$

c. Here we have the difference of two cubes, so we use F6. We start by writing the problem as the difference of two cubes:

$$\begin{aligned} m^3 - 8n^3 &= (m)^3 - (2n)^3 \\ &= (m - 2n)[m^2 + m(2n) + (2n)^2] \quad \text{Letting } X = m \text{ and } A = 2n \text{ in F6} \\ &= (m - 2n)(m^2 + 2mn + 4n^2) \end{aligned}$$

d. We write the problem as the difference of two cubes and then use F6.

$$\begin{aligned} 27r^3 - 8s^3 &= (3r)^3 - (2s)^3 \\ &= (3r - 2s)[(3r)^2 + (3r)(2s) + (2s)^2] \quad \text{Letting } X = 3r \text{ and } A = 2s \text{ in F6} \\ &= (3r - 2s)(9r^2 + 6rs + 4s^2) \end{aligned}$$

Note that you can verify all of these results by multiplying the factors in the final answer.

We can also factor $x^3 + 27$ by remembering that the result is a binomial times a trinomial, so we place parentheses accordingly.

1. To get the *binomial factor,* take the cube roots using the same sign.
2. To get the *trinomial factor,* square the terms of the binomial factor and use them as the first and last terms of the trinomial. The middle term of the trinomial is the result of multiplying the two terms of the binomial factor and changing the sign.

Thus,

$$x^3 + 27$$

$$\sqrt[3]{x^3} \quad \sqrt[3]{27}$$

$$(x + 3)(x^2 - 3x + 9)$$

square square

middle term:

(change sign) $-x \cdot 3 = -3x$

B › Using a General Factoring Strategy

We have now studied several factoring techniques. How do you know which one to use? Here is a general factoring strategy that can help you answer this question. Remember that when we say *factor,* we mean *factor completely* using integer coefficients.

PROCEDURE

A General Factoring Strategy

1. Factor out all common factors (the GCF).
2. Look at the number of terms inside the parentheses (or in the original polynomial). If there are

   Four terms: Factor by grouping.

   Three terms: Check whether the expression is a perfect square trinomial. If so, factor it. Otherwise, use the *ac* test to factor.

   Two terms and *squared:* Look for the difference of two squares ($X^2 - A^2$) and factor it. Note that $X^2 + A^2$ is not factorable.

   Two terms and *cubed:* Look for the sum of two cubes ($X^3 + A^3$) or the difference of two cubes ($X^3 - A^3$) and factor it.
3. Make sure the expression is completely factored.

You can check your results by multiplying the factors you obtain.

EXAMPLE 2 Using the general factoring strategy

Factor completely:

a. $6x^2 - 18x - 24$

b. $4x^4 + 12x^3 + 40x^2$

SOLUTION 2

a. We follow the steps in our general factoring strategy.

1. Factor out the common factor: $6x^2 - 18x - 24 = 6(x^2 - 3x - 4)$
2. $x^2 - 3x - 4$ has three terms, and it is factored by finding two numbers whose product is -4 and whose sum is -3. These numbers are 1 and -4. Thus, $x^2 - 3x - 4 = (x + 1)(x - 4)$

   We then have $6x^2 - 18x - 24 = 6(x + 1)(x - 4)$
3. This expression cannot be factored any further.

b.
1. Here the GCF is $4x^2$. Thus, $4x^4 + 12x^3 + 40x^2 = 4x^2(x^2 + 3x + 10)$
2. The trinomial $x^2 + 3x + 10$ is *not* factorable since there are no numbers whose product is 10 with a sum of 3.
3. The complete factorization is simply $4x^4 + 12x^3 + 40x^2 = 4x^2(x^2 + 3x + 10)$

PROBLEM 2

Factor completely:

a. $7a^2 - 14a - 21$

b. $5a^4 + 10a^3 + 25a^2$

EXAMPLE 3 Using the general factoring strategy with four terms

Factor completely: $3x^3 + 9x^2 + x + 3$

SOLUTION 3

1. There are no common factors.
2. Since the expression has four terms, we factor by grouping:

$$\begin{aligned} 3x^3 + 9x^2 + x + 3 &= (3x^3 + 9x^2) + (x + 3) \\ &= 3x^2(x + 3) + 1 \cdot (x + 3) \\ &= (x + 3)(3x^2 + 1) \end{aligned}$$

3. This result cannot be factored any further, so the factorization is complete.

PROBLEM 3

Factor completely:

$3a^3 + 6a^2 + a + 2$

EXAMPLE 4 Using the general factoring strategy with a perfect square trinomial

The heat output from a natural draught convector is $kt_n^2 - 2kt_nt_a + kt_a^2$. ($t_n^2$ is read as "*t* sub *n* squared." The "*n*" is called a **subscript.**) Factor this expression.

SOLUTION 4

As usual, we proceed by steps.

1. The common factor is k. Hence

$$kt_n^2 - 2kt_nt_a + kt_a^2 = k(t_n^2 - 2t_nt_a + t_a^2)$$

2. $t_n^2 - 2t_nt_a + t_a^2$ is a perfect square trinomial, which factors as $(t_n - t_a)^2$. Thus,

$$kt_n^2 - 2kt_nt_a + kt_a^2 = k(t_n - t_a)^2$$

3. This expression cannot be factored further.

PROBLEM 4

Factor completely: $kt_1^2 - 2kt_1t_2 + kt_2^2$

Answers to PROBLEMS

2. a. $7(a + 1)(a - 3)$ **b.** $5a^2(a^2 + 2a + 5)$ **3.** $(a + 2)(3a^2 + 1)$ **4.** $k(t_1 - t_2)^2$

EXAMPLE 5 **Using the general factoring strategy with the difference of two squares**

Factor completely: $D^4 - d^4$

SOLUTION 5

1. There are no common factors.
2. The expression has *two* squared terms separated by a minus sign, so it's the difference of *two* squares. Thus,

$$\begin{aligned} D^4 - d^4 &= (D^2)^2 - (d^2)^2 \\ &= (D^2 + d^2)(D^2 - d^2) \end{aligned}$$

3. The expression $D^2 - d^2$ is also the difference of two squares, which can be factored into $(D + d)(D - d)$. Thus,

$$\begin{aligned} D^4 - d^4 &= (D^2 + d^2)(D^2 - d^2) \\ &= (D^2 + d^2)(D + d)(D - d) \end{aligned}$$

Note that $D^2 + d^2$, which is the *sum* of two squares, *cannot* be factored.

PROBLEM 5

Factor completely: $m^4 - n^4$

EXAMPLE 6 **Using the general factoring strategy with sums and differences of cubes**

Factor completely:

a. $8x^5 - x^2y^3$

b. $8x^5 + x^3y^2$

SOLUTION 6

a. We proceed as usual by steps.

1. The GCF is x^2, so we factor it out. $8x^5 - x^2y^3 = x^2(8x^3 - y^3)$
2. $8x^3 - y^3$ is the difference of two cubes with $X = 2x$ and $A = y$. $= x^2(2x - y)[(2x)^2 + (2x)y + y^2]$
3. Note that the expression cannot be factored further. $= x^2(2x - y)(4x^2 + 2xy + y^2)$

b. The GCF is x^3.

1. Factor the GCF. $8x^5 + x^3y^2 = x^3(8x^2 + y^2)$
2. $8x^2 + y^2$ is the sum of two squares and is not factorable. Thus, $8x^5 + x^3y^2 = x^3(8x^2 + y^2)$
3. Note that the expression cannot be factored further.

PROBLEM 6

Factor completely:

a. $27a^5 - a^2b^3$

b. $64a^5 + a^3b^2$

C > Solving Applications Involving Factoring

GREEN MATH

EXAMPLE 7 **China's CO_2 emissions**

The world's largest polluter of CO_2 gases into the atmosphere is China, with a record 7 billion metric tons in 2008. Assuming a 2% annual increase, after 3 years the amount of pollution produced will be $\frac{7x^3 - 7}{x - 1}$, where $x = 1 + 2\% = 1.02$.

PROBLEM 7

The United States produces almost 6 billion metric tons of gases. Assuming an annual 2% increase, after 4 years the amount of pollution

(continued)

Answers to PROBLEMS

5. $(m^2 + n^2)(m + n)(m - n)$ 6. a. $a^2(3a - b)(9a^2 + 3ab + b^2)$ b. $a^3(64a^2 + b^2)$ 7. a. $6(x^2 + 1)(x + 1)(x - 1)$ b. $6(x^2 + 1)(x + 1)$ c. 24.729648 billion metric tons

a. Factor $7x^3 - 7$ completely.

b. Use the factored form of part **a** as the numerator of $\frac{7x^3 - 7}{x - 1}$ to simplify this fraction.

c. Use the simplified form of the fraction to find the amount of pollution produced after 3 years.

SOLUTION 7

a. We proceed by steps.

1. The GCF is 7, so we factor 7 out. $7x^3 - 7 = 7(x^3 - 1)$
2. $x^3 - 1$ is the difference of two cubes with $X = x$ and $A = 1$. Factor $x^3 - 1$. $= 7(x - 1)(x^2 + x + 1)$

b. $\frac{7x^3 - 7}{x - 1} = \frac{7(x - 1)(x^2 + x + 1)}{x - 1} = 7(x^2 + x + 1)$

c. After 3 years, the amount of pollution is $7(x^2 + x + 1)$.

Letting $x = 1.02$, $7(x^2 + x + 1) = 7[(1.02)^2 + 1.02 + 1)]$

$= 7[1.0404 + 1.02 + 1]$

$= 7[3.0604]$

$= 21.4228$ billion metric tons

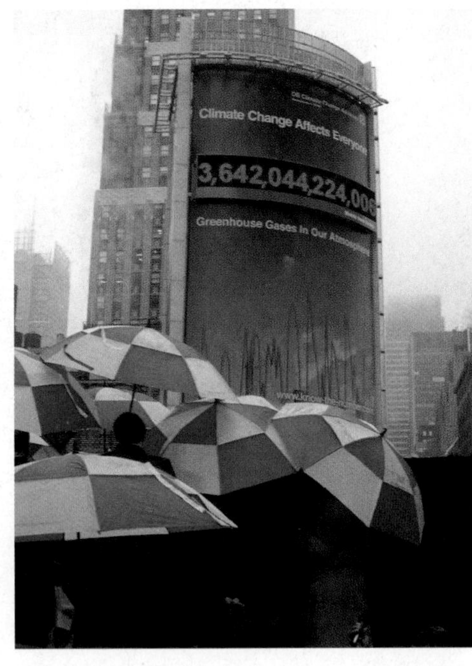

How much is a metric ton? One metric ton of CO_2 is released to the atmosphere for every 100 gallons of gasoline used, so just 20 billion metric tons is equivalent to the release of one **trillion** gallons of gasoline used! The counter near Madison Square Garden tracks the amount of all greenhouse gases (not only CO_2) in the Earth's atmosphere. The result? 3.6 trillion metric tons.

produced will be $\frac{6x^4 - 6}{x - 1}$, where $x = 1 + 2\% = 1.02$.

a. Factor $6x^4 - 6$ completely.

b. Use the factored form of part **a** as the numerator of $\frac{6x^4 - 6}{x - 1}$ to simplify this fraction.

c. Use the simplified form of the fraction to find the amount of pollution produced after 4 years.

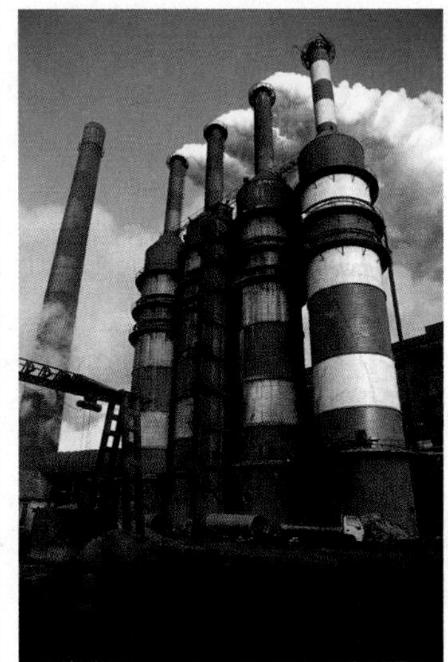

D > Using −1 as a Factor

In the preceding examples, we did not factor expressions in which the leading coefficient is preceded by a minus sign. Can some of these expressions be factored? The answer is yes, but we must first factor −1 from each term. Thus, to factor $-x^2 + 6x - 9$, we first write

$-x^2 + 6x - 9 = -1 \cdot (x^2 - 6x + 9)$

$= -1 \cdot (x - 3)^2$ Note that $x^2 - 6x + 9 = (x - 3)^2$.

$= -(x - 3)^2$ Since $-1 \cdot a = -a$, $-1 \cdot (x - 3)^2 = -(x - 3)^2$.

EXAMPLE 8 Factoring −1 out as the GCF

Factor, if possible:

a. $-x^2 - 8x - 16$ **b.** $-4x^2 + 12xy - 9y^2$
c. $-9x^2 - 12xy + 4y^2$ **d.** $-4x^4 + 25x^2$

SOLUTION 8

a. We first factor out -1 to obtain

$$\begin{aligned} -x^2 - 8x - 16 &= -1 \cdot (x^2 + 8x + 16) \\ &= -1 \cdot (x+4)^2 \quad x^2 + 8x + 16 = (x+4)^2 \\ &= -(x+4)^2 \end{aligned}$$

b.
$$\begin{aligned} -4x^2 + 12xy - 9y^2 &= -1 \cdot (4x^2 - 12xy + 9y^2) \\ &= -1 \cdot (2x - 3y)^2 \quad 4x^2 - 12xy + 9y^2 = (2x-3y)^2 \\ &= -(2x-3y)^2 \end{aligned}$$

c. $-9x^2 - 12xy + 4y^2 = -1 \cdot (9x^2 + 12xy - 4y^2)$

$9x^2 = (3x)^2$ and $4y^2 = (2y)^2$, but $9x^2 + 12xy - 4y^2$ has a minus sign before the last term $4y^2$ and is *not* a perfect square trinomial, so $9x^2 + 12xy - 4y^2$ is not factorable; thus, $-9x^2 - 12xy + 4y^2 = -(9x^2 + 12xy - 4y^2)$.

d.
$$\begin{aligned} -4x^4 + 25x^2 &= -x^2 \cdot (4x^2 - 25) \quad \text{The GCF is } -x^2. \\ &= -x^2 \cdot (2x+5)(2x-5) \quad 4x^2 - 25 = (2x+5)(2x-5) \\ &= -x^2(2x+5)(2x-5) \end{aligned}$$

PROBLEM 8

Factor:

a. $-a^2 - 6a - 9$
b. $-9a^2 + 12ab - 4b^2$
c. $-4a^2 - 12ab + 9b^2$
d. $-4a^4 + 9a^2$

Exercises 5.5

⟨A⟩ Factoring Sums or Differences of Cubes

In Problems 1–10, factor completely.

1. $x^3 + 8$
2. $z^3 + 8$
3. $8m^3 - 27$
4. $8y^3 - 27x^3$
5. $27m^3 - 8n^3$
6. $27x^2 - x^5$
7. $64s^3 - s^6$
8. $t^7 - 8t^4$
9. $27x^4 + 8x^7$
10. $8y^8 + 27y^5$

⟨B⟩ Using a General Factoring Strategy

In Problems 11–46, factor completely.

11. $3x^2 - 3x - 18$
12. $4x^2 - 12x - 16$
13. $5x^2 + 11x + 2$
14. $6x^2 + 19x + 10$
15. $3x^3 + 6x^2 + 21x$
16. $6x^3 + 18x^2 + 12x$
17. $2x^4 - 4x^3 - 10x^2$
18. $3x^4 - 12x^3 - 9x^2$
19. $4x^4 + 12x^3 + 18x^2$
20. $5x^4 + 25x^3 + 30x^2$
21. $3x^3 + 6x^2 + x + 2$
22. $2x^3 + 8x^2 + x + 4$
23. $3x^3 + 3x^2 + 2x + 2$
24. $4x^3 + 8x^2 + 3x + 6$
25. $2x^3 + 2x^2 - x - 1$
26. $3x^3 + 6x^2 - x - 2$

Web IT go to mhhe.com/bello for more lessons

Answers to PROBLEMS

8. a. $-(a+3)^2$ **b.** $-(3a-2b)^2$ **c.** $-(4a^2 + 12ab - 9b^2)$ **d.** $-a^2(2a+3)(2a-3)$

27. $3x^2 + 24x + 48$

28. $2x^2 + 12x + 18$

29. $kx^2 + 4kx + 4k$

30. $kx^2 + 10kx + 25k$

31. $4x^2 - 24x + 36$

32. $5x^2 - 20x + 20$

33. $kx^2 - 12kx + 36k$

34. $kx^2 - 10kx + 25k$

35. $3x^3 + 12x^2 + 12x$

36. $2x^3 + 16x^2 + 32x$

37. $18x^3 + 12x^2 + 2x$

38. $12x^3 + 12x^2 + 3x$

39. $12x^4 - 36x^3 + 27x^2$

40. $18x^4 - 24x^3 + 8x^2$

41. $x^4 - 1$

42. $x^4 - 16$

43. $x^4 - y^4$

44. $x^4 - z^4$

45. $x^4 - 16y^4$

46. $x^4 - 81y^4$

‹ C › Solving Applications Involving Factoring

‹ D › Using −1 as a Factor

In Problems 47–70, factor.

47. $-x^2 - 6x - 9$

48. $-x^2 - 10x - 25$

49. $-x^2 - 4x - 4$

50. $-x^2 - 12x - 36$

51. $-4x^2 - 4xy - y^2$

52. $-9x^2 - 6xy - y^2$

53. $-9x^2 - 12xy - 4y^2$

54. $-4x^2 - 12xy - 9y^2$

55. $-4z^2 + 12zy - 9y^2$

56. $-9x^2 + 12xy - 4y^2$

57. $-18x^3 - 24x^2y - 8xy^2$

58. $-12x^3 - 36x^2y - 27xy^2$

59. $-18x^3 - 60x^2y - 50xy^2$

60. $-12x^3 - 60x^2y - 75xy^2$

61. $-x^3 + x$

62. $-x^3 + 9x$

63. $-x^4 + 4x^2$

64. $-x^4 + 16x^2$

65. $-4x^4 + 9x^2$

66. $-9x^4 + 4x^2$

67. $-2x^4 + 16x$

68. $-24x^4 + 3x$

69. $-16x^5 - 2x^2$

70. $-3x^5 + 24x^2$

››› Applications: Green Math

71. *Alligator population* The number of alligators observed by spotlight survey in Kiawah Island, South Carolina, can be approximated by $-16t^2 + 80t + 384$, where t is the number of years after 2005 and before 2010. Use the general factoring strategy to factor $-16t^2 + 80t + 384$ completely.

72. *Killer whales in Puget Sound* The killer whale population of Puget Sound can be approximated by $-0.3t^2 + 4.8t + 78$, where t is the number of years after 1986 and before 2002. Use the general factoring strategy to factor $-0.3t^2 + 4.8t + 78$, completely.

73. *Hybrid car depreciation* There are many reasons to buy a hybrid car (better gas mileage, fewer emissions), but one that is often overlooked is that they may depreciate less than other cars. If the depreciation rate is r% and the amount you paid for a car is P, after 2 years the value of the car will be $P - 2Pr + Pr^2$.

a. Factor $P - 2Pr + Pr^2$ completely

b. If a Prius hybrid costs \$25,000 now, how much will it be worth in 2 years assuming the depreciation rate r is 15%?

Source: Learn more about depreciation at http://tinyurl.com/ylx4523.

74. *Regular car depreciation* The 2-year depreciation for a \$25,000 Toyota Camry is $25{,}000r^2 - 50{,}000r + 25{,}000$.

a. Factor this expression completely.

b. If the depreciation rate of the Camry is 20%, how much will the car be worth in 2 years?

Using Your Knowledge

Factoring Engineering Problems Many of the ideas presented in this section are used by engineers and technicians. Use your knowledge to factor the given expressions.

75. The bend allowance needed to bend a piece of metal of thickness t through an angle A when the inside radius of the bend is R_1 is given by the expression

$$\frac{2\pi A}{360}R_1 + \frac{2\pi A}{360}Kt$$

where K is a constant. Factor this expression.

76. The change in kinetic energy of a moving object of mass m with initial velocity v_1 and terminal velocity v_2 is given by the expression

$$\frac{1}{2}mv_1^2 - \frac{1}{2}mv_2^2$$

Factor this expression.

77. The parabolic distribution of shear stress on the cross section of a certain beam is given by the expression

$$\frac{3Sd^2}{2bd^3} - \frac{12Sz^2}{2bd^3}$$

Factor this expression.

78. The polar moment of inertia J of a hollow round shaft of inner diameter d_1 and outer diameter d is given by the expression

$$\frac{\pi d^4}{32} - \frac{\pi d_1^4}{32}$$

Factor this expression.

Write On

79. Write the procedure you use to factor the sum of two cubes.

80. Write the procedure you use to factor the difference of two cubes.

81. A student factored $x^4 - 7x^2 - 18$ as $(x^2 + 2)(x^2 - 9)$. The student did not get full credit on the answer. Why?

Concept Checker

Fill in the blank(s) with the correct word(s), phrase, or mathematical expression.

82. In **factored form** $X^3 + A^3 =$ ____________________

83. In **factored form** $X^3 - A^3 =$ ____________________

84. The **first** step in the **general factoring strategy** is to ________ out all **common** factors.

85. The **second** step in the **general factoring strategy** is to ________ at the **number** of terms.

$(X - A)(X^2 + AX + A^2)$ **factor**

$(X - A)(X^2 - AX + A^2)$ **$(X + A)(X^2 + AX + A^2)$**

ignore   **look**

$(X + A)(X^2 - AX + A^2)$

Mastery Test

Factor completely:

86. $8x^2 - 16x - 24$

87. $5x^4 - 10x^3 + 20x^2$

88. $3x^3 + 12x^2 + x + 4$

89. $6x^2 - x - 35$

90. $2x^4 + 7x^3 - 15x^2$

91. $27t^3 - 64$

92. $kt_n^2 + 2kt_nt_a + kt_a^2$

93. $x^4 - 81$

94. $-z^2 - 10z - 25$

95. $-9x^2 - 30xy - 25y^2$

96. $-9x^2 + 30xy - 25y^2$

97. $64y^3 + 27x^3$

98. $-9z^4 + 4z^2$

99. $-x^5 - x^2y^3$

Skill Checker

Use the *ac* test to factor (if possible).

100. $10x^2 + 11x + 6$

101. $10x^2 + 13x - 3$

102. $3x^2 - 5x - 1$

103. $2x^2 - 5x - 3$

104. $2x^2 - 3x - 2$

5.6 Solving Quadratic Equations by Factoring

Objective

A › **Solve quadratic equations by factoring.**

B › **Solve applications involving factoring.**

To Succeed, Review How To . . .

1. Factor an expression of the form $ax^2 + bx + c$ (pp. 420–424).
2. Solve a linear equation (pp. 137–141).

Getting Started

Quadratics and Gravity

BELLOS

Let's suppose the girl throws the ball with an initial velocity of 4 meters per second. If she releases the ball 1 meter above the ground, the *height* of the ball after t seconds is

$$-5t^2 + 4t + 1$$

To find out how long it takes the ball to hit the ground, we set this expression equal to zero. The reason for this is that when the ball *is* on the ground, the height is zero. Thus, we write

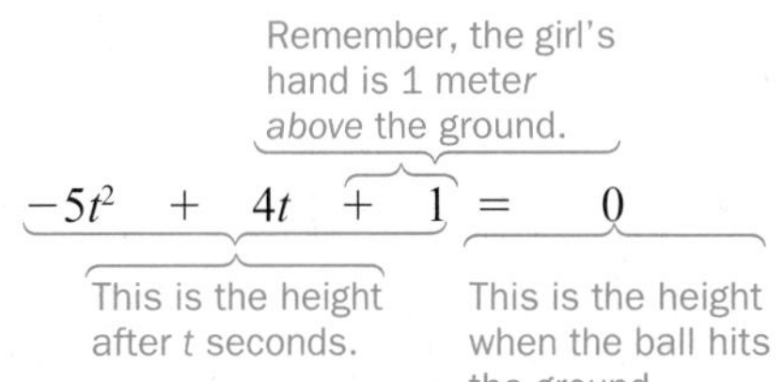
Remember, the girl's hand is 1 meter *above* the ground.
$-5t^2 + 4t + 1 = 0$
This is the height after t seconds.
This is the height when the ball hits the ground.

The equation is a ***quadratic equation,*** an equation in which the greatest exponent of the variable is 2 and which can be written as $\boldsymbol{at^2 + bt + c = 0, a \neq 0}$. We learn how to solve these equations next.

A › Solving Quadratic Equations by Factoring

How can we solve the equation given in the *Getting Started?* For starters, we make the leading coefficient in $-5t^2 + 4t + 1 = 0$ positive by multiplying each side of the equation by -1 to obtain

$$5t^2 - 4t - 1 = 0 \qquad -1(-5t^2 + 4t + 1) = 5t^2 - 4t - 1 \text{ and } -1 \cdot 0 = 0$$

Since $a = 5$ and $c = -1$, the ac number is -5 and we write

$$5t^2 \underbrace{- 5t + 1t}_{-4t} - 1 = 0 \qquad \text{Write } -4t \text{ as } -5t + 1t.$$

$$5t(t - 1) + 1(t - 1) = 0 \qquad \text{The GCF is } (t - 1).$$

$$(t - 1)(5t + 1) = 0$$

(You can also use reverse FOIL or trial and error.) At this point, we note that the product of two expressions $(t - 1)$ and $(5t + 1)$ gives us a result of zero. What does this mean?

We know that if we have two numbers and at least one of them is zero, then their product is zero. For example,

$$-5 \cdot 0 = 0 \qquad \frac{3}{2} \cdot 0 = 0 \qquad x \cdot 0 = 0$$

$$0 \cdot 8 = 0 \qquad 0 \cdot x = 0 \qquad 0 \cdot 0 = 0$$

As you can see, in all these cases at least one of the factors is zero. In general, it can be shown that if the product of the two factors is zero, at least one of the factors *must be* zero. We shall call this idea the **zero product property.**

ZERO PRODUCT PROPERTY

If $A \cdot B = 0$, then $A = 0$ or $B = 0$ (or both A and B are equal to 0).

Now, let's go back to our original equation. We can think of $(t - 1)$ as A and $(5t + 1)$ as B. Then our equation

$$(t - 1)(5t + 1) = 0$$

becomes

$$A \cdot B = 0$$

By the zero product property, if $A \cdot B = 0$, then

$$A = 0 \quad \text{or} \quad B = 0$$

Thus,

$$\begin{aligned} t - 1 &= 0 & \text{or} \qquad 5t + 1 &= 0 \\ t &= 1 & 5t &= -1 \\ t &= 1 & t &= -\frac{1}{5} \end{aligned}$$

We added 1 in the first equation and subtracted 1 in the second.

NOTE

When solving quadratic equations, you usually get two answers, but sometimes in application problems one of the answers must be discarded. It is always a good idea to check that the answers you get apply to the original conditions of the problem.

Thus, the ball reaches the ground after 1 second or after $-\frac{1}{5}$ second. The second answer is negative, which is impossible because we start timing when $t = 0$, so we can see that the ball thrown by the girl at 4 meters per second will reach the ground after $t = 1$ second. You can check this by letting $t = 1$ in the original equation:

$$\begin{aligned} -5t^2 + 4t + 1 &\stackrel{?}{=} 0 \\ -5(1)^2 + 4(1) + 1 &\stackrel{?}{=} 0 \\ -5 + 4 + 1 &\stackrel{?}{=} 0 \\ 0 &= 0 \end{aligned}$$

Similarly, if we want to find how long a ball thrown from level ground at 10 meters per second takes to return to the ground, we need to solve the equation

$$5t^2 - 10t = 0$$

Factoring, we obtain

$$5t(t - 2) = 0$$

By the zero product property,

$$\begin{aligned} 5t &= 0 & \text{or} \qquad t - 2 &= 0 \\ t &= 0 & t &= 2 \end{aligned}$$

If $5t = 0$, then $t = 0$.

Thus, the ball returns to the ground after 2 seconds. (The other possible answer, $t = 0$, indicates that the ball was on the ground when $t = 0$, which is true.)

In the preceding discussion, we solved equations in which the highest exponent of the variable was 2. These equations are called *quadratic equations*. A quadratic equation is an equation in which the greatest exponent of the variable is 2. Moreover, to solve these quadratic equations, one of the sides of the equation must equal zero. These two ideas can be summarized as follows.

QUADRATIC EQUATION IN STANDARD FORM

If *a*, *b*, and *c* are real numbers ($a \neq 0$),

$$ax^2 + bx + c = 0$$

is a quadratic equation in standard form.

To solve a quadratic equation by the method of factoring, the equation must be in *standard form.*

EXAMPLE 1 Solving a quadratic equation by factoring

Solve:

a. $3x^2 + 11x - 4 = 0$ **b.** $6x^2 - x - 2 = 0$

SOLUTION 1

a. The equation is in standard form. To solve this equation, we must first factor the left-hand side. Here's how we do it.

$3x^2 + 11x - 4 = 0$	The key number is -12 ($= 3(-4)$); then determine $12(-1) = -12$ and $12 + (-1) = 11$.
$3x^2 + 12x - 1x - 4 = 0$	Rewrite the middle term, $11x$.
$3x(x + 4) - 1(x + 4) = 0$	Factor each pair.
$(x + 4)(3x - 1) = 0$	Factor out the GCF, $(x + 4)$.
$x + 4 = 0$ or $3x - 1 = 0$	Use the zero product property.
$x = -4$ $\quad 3x = 1$	Solve each equation.
$x = -4$ $\quad x = \frac{1}{3}$	

Thus, the possible solutions are $x = -4$ and $x = \frac{1}{3}$. To verify that -4 is a correct solution, we substitute -4 in the original equation to obtain

$$3(-4)^2 + 11(-4) - 4 = 3(16) - 44 - 4$$
$$= 48 - 44 - 4 = 0$$

We leave it to you to verify that $\frac{1}{3}$ is also a solution.

b. As before, we must factor the left-hand side.

$6x^2 - x - 2 = 0$	The key number is -12.
$6x^2 - 4x + 3x - 2 = 0$	Rewrite the middle term, $-x$.
$2x(3x - 2) + 1(3x - 2) = 0$	Factor each pair.
$(3x - 2)(2x + 1) = 0$	Factor out the GCF, $(3x - 2)$.
$3x - 2 = 0$ or $2x + 1 = 0$	Use the zero product property.
$3x = 2$ $\quad 2x = -1$	Solve each equation.
$x = \frac{2}{3}$ $\quad x = -\frac{1}{2}$	

Thus, the solutions are $\frac{2}{3}$ and $-\frac{1}{2}$. You can check this by substituting these values in the original equation.

PROBLEM 1

Solve:

a. $3x^2 + 8x - 3 = 0$

b. $6x^2 - 7x - 3 = 0$

Answers to PROBLEMS

1. a. $x = -3$ and $x = \frac{1}{3}$ **b.** $x = \frac{3}{2}$ and $x = -\frac{1}{3}$

EXAMPLE 2 Solving a quadratic equation *not* in standard form

Solve: $10x^2 + 13x = 3$

SOLUTION 2 This equation $10x^2 + 13x = 3$ is not in standard form, so we can't solve it as written. However, if we subtract 3 from each side of the equation, we have

$$10x^2 + 13x - 3 = 0$$

which is in standard form. We can now solve by factoring using trial and error or the *ac* test. To use the *ac* test, we write

$10x^2 + \underline{13x} - 3 = 0$	The key number is -30.
$10x^2 + \underline{15x - 2x} - 3 = 0$	Rewrite the middle term, $13x$.
$5x(2x + 3) - 1(2x + 3) = 0$	Factor each pair.
$(2x + 3)(5x - 1) = 0$	Factor out the GCF, $(2x + 3)$.
$2x + 3 = 0$ or $5x - 1 = 0$	Use the zero product property.
$2x = -3$ $5x = 1$	Solve each equation.
$x = -\frac{3}{2}$ $x = \frac{1}{5}$	

Thus, the solutions of the equation are

$$x = -\frac{3}{2} \quad \text{and} \quad x = \frac{1}{5}$$

Check this!

PROBLEM 2

Solve: $6x^2 + 13x = 5$

Sometimes we need to *simplify* the equation before we write it in standard form. For example, to solve the equation

$$(3x + 1)(x - 1) = 3(x + 1) - 2$$

we need to remove parentheses by multiplying the factors involved and write the equation in standard form. Here's how we do it:

$(3x + 1)(x - 1) = 3(x + 1) - 2$	Given
$3x^2 - 2x - 1 = 3x + 3 - 2$	Multiply $(3x + 1)(x - 1)$ and $3(x + 1)$.
$3x^2 - 2x - 1 = 3x + 1$	Simplify on the right.
$3x^2 - 5x - 2 = 0$	Subtract $3x$ and 1 from each side.

Now we factor using the *ac* test. (We could also use trial and error.)

$3x^2 \underline{- 6x + 1x} - 2 = 0$	The key number is -6.
$3x(x - 2) + 1(x - 2) = 0$	Factor each pair.
$(x - 2)(3x + 1) = 0$	Factor out the GCF, $(x - 2)$.
$x - 2 = 0$ or $3x + 1 = 0$	Use the zero product property.
$x = 2$ $3x = -1$	
$x = 2$ or $x = -\frac{1}{3}$	Solve each equation.

Remember to check this by substituting $x = 2$ and then $x = -\frac{1}{3}$ in the original equation.

Answers to PROBLEMS

2. $x = -\frac{5}{2}$ and $x = \frac{1}{3}$

EXAMPLE 3 Solving a quadratic equation by simplifying first

Solve: $(2x + 1)(x - 2) = 2(x - 1) + 3$

SOLUTION 3

$$2x^2 - 3x - 2 = 2x - 2 + 3 \quad \text{Multiply.}$$
$$2x^2 - 3x - 2 = 2x + 1 \quad \text{Simplify.}$$
$$2x^2 - 5x - 2 = 1 \quad \text{Subtract } 2x.$$
$$2x^2 - 5x - 3 = 0 \quad \text{Subtract 1.}$$
$$2x^2 - 6x + 1x - 3 = 0 \quad \text{Rewrite the middle term, } -5x.$$
$$2x(x - 3) + 1(x - 3) = 0 \quad \text{Factor each pair.}$$
$$(x - 3)(2x + 1) = 0 \quad \text{Factor out the GCF, } (x - 3).$$
$$x - 3 = 0 \quad \text{or} \quad 2x + 1 = 0 \quad \text{Use the zero product property.}$$
$$x = 3 \qquad x = -\frac{1}{2} \quad \text{Solve each equation.}$$

Remember to check these answers.

PROBLEM 3

Solve: $(4n + 1)(n - 2) = 4(n + 1) - 3$

Finally, we can always use the general factoring strategy developed in Section 5.5 to solve certain equations. We illustrate this possibility in Example 4.

EXAMPLE 4 Solving a quadratic equation by simplifying first

Solve: $(4x - 1)(x - 1) = 2(x + 2) - 3x - 4$

SOLUTION 4

$$4x^2 - 5x + 1 = 2x + 4 - 3x - 4 \quad \text{Multiply.}$$
$$4x^2 - 5x + 1 = -x \quad \text{Simplify.}$$
$$4x^2 - 4x + 1 = 0 \quad \text{Add } x.$$
$$4x^2 - 2x - 2x + 1 = 0 \quad \text{Rewrite the middle term.}$$
$$2x(2x - 1) - 1(2x - 1) = 0 \quad \text{Factor each pair.}$$
$$(2x - 1)(2x - 1) = 0 \quad \text{Factor out the GCF, } (2x - 1).$$
$$2x - 1 = 0 \quad \text{or} \quad 2x - 1 = 0 \quad \text{Use the zero product property.}$$
$$x = \frac{1}{2} \qquad x = \frac{1}{2} \quad \text{Solve the equations.}$$

Don't forget to check the answer!

PROBLEM 4

Solve:

$(3m + 1)(3m + 2) = 5(m + 1) - 2m - 4$

Note that in Example 4 there is really only *one* solution, $x = \frac{1}{2}$. A lot of work could be avoided if you notice that the expression $4x^2 - 4x + 1 = 0$ is a perfect square trinomial (F3) that can be factored as $(2x - 1)^2$ or, equivalently, $(2x - 1)(2x - 1)$. To avoid extra work, follow the general factoring strategy from Section 5.5.

In Example 5, we will solve several quadratic equations **not** in standard form. To do so, we make the right-hand side of the equation 0 and use the general factoring strategy to factor the left-hand side.

EXAMPLE 5 Solving quadratic equations using the general factoring strategy

Solve:

a. $9z^2 - 16 = 0$ **b.** $y(3y + 7) = -2$ **c.** $m^2 = 3m$

PROBLEM 5

Solve:

a. $16x^2 - 9 = 0$

b. $y(2y + 3) = -1$

c. $n^2 = 4n$

Answers to PROBLEMS

3. $n = -\frac{1}{4}$ and $n = 3$ **4.** $m = -\frac{1}{3}$ **5. a.** $x = \frac{3}{4}$ and $x = -\frac{3}{4}$ **b.** $y = -\frac{1}{2}$ and $y = -1$ **c.** $n = 0$ and $n = 4$

SOLUTION 5

a. The right-hand side is 0, so we use the general factoring strategy to factor $9z^2 - 16$:

$9z^2 - 16 = 0$ Given

$(3z + 4)(3z - 4) = 0$ Factor the difference of squares.

$3z + 4 = 0$ or $3z - 4 = 0$ Use the zero product property.

$z = -\frac{4}{3}$ $z = \frac{4}{3}$ Solve the equations.

Check that the solutions are $-\frac{4}{3}$ and $\frac{4}{3}$ by substituting each number in the original equation.

b. We start by adding 2 to both sides of the equation so that the right-hand side is 0.

$y(3y + 7) + 2 = -2 + 2$ Add 2.

$3y^2 + 7y + 2 = 0$ Simplify.

$(3y + 1)(y + 2) = 0$ Factor.

$3y + 1 = 0$ or $y + 2 = 0$ Use the zero product property.

$y = -\frac{1}{3}$ $y = -2$ Solve.

Check to make sure the solutions are $-\frac{1}{3}$ and -2.

c. Start by subtracting $3m$, then factor the result:

$m^2 = 3m$ Given

$m^2 - 3m = 0$ Subtract $3m$.

$m(m - 3) = 0$ Factor.

$m = 0$ or $m - 3 = 0$ Use the zero product property.

$m = 0$ $m = 3$ Solve.

Check the solutions 0 and 3 in the original equation.

The zero product property can be applied to solve certain **nonquadratic** equations, as long as one side of the equation is 0 and the other side can be written as a product. Example 6 illustrates such a situation.

EXAMPLE 6 **Extending the zero product property to solve other types of equations**

Solve: $(v - 2)(v^2 - v - 12) = 0$

SOLUTION 6 The right-hand side of the equation is 0 as needed. Avoid the temptation of multiplying the expressions on the left-hand side! Remember, what we need is a *product* of factors so we can use the zero product property. With this in mind, factor $v^2 - v - 12$ as shown:

$(v - 2)(v^2 - v - 12) = 0$ Given

$(v - 2)(v - 4)(v + 3) = 0$ Factor $v^2 - v - 12$.

$v - 2 = 0$ or $v - 4 = 0$ or $v + 3 = 0$ Use the zero product property.

$v = 2$ $v = 4$ $v = -3$ Solve.

The solutions are 2, 4, and -3. Check that this is the case by substituting each number in the original equation.

PROBLEM 6

Solve: $(m - 3)(m^2 - m - 2) = 0$

B › Solving Applications Involving Factoring

In Example 5 of Section 5.4 we discussed the expression $t^2 + 26t + 169$ used to *approximate* the number of endangered plants and animals in the United States after 1980. We will now use the expression $t^2 + 66t + 1089$ to *predict* the number of endangered plants and animals in the United States after the year 2000. When will this number reach 1849? Read on.

Answers to PROBLEMS

6. $m = 3$ and $m = 2$ and $m = -1$

EXAMPLE 7 1849 endangered plants and animals

The expression $t^2 + 66t + 1089$ predicts the number of endangered plants and animals **t** years after 2000. When will this number reach 1849? This will happen when **$t^2 + 66t + 1089 = 1849$**. If we solve for t we will know exactly when!

SOLUTION 7

We have to solve the quadratic equation $t^2 + 66t + 1089 = 1849$. We start by subtracting 1849 from both sides so that the right-hand side of the equation is **0.**

$t^2 + 66t + 1089 = 1849$	Given
$t^2 + 66t + 1089 - 1849 = 1849 - 1849$	Subtract 1849.
$t^2 + 66t - 760 = 0$	Simplify.
$(t + 76)(t - 10) = 0$	Factor.
$t + 76 = 0$ or $t - 10 = 0$	Use the zero product property.
$t = -76$ or $t = 10$	Solve.

Thus, the number will reach 1849 in **$t = 10$** years after 2000, that is, in 2010. (Discard the negative solution -76. Do you see why?)

PROBLEM 7

Use our old equation **$t^2 + 26t + 169$** to predict when the number of endangered plants and animals will reach 1849 by solving **$t^2 + 26t + 169 = 1849$**.

Answers to PROBLEMS

7. $t = 30$, that is, in $1980 + 30 =$ 2010 (Same prediction as in Example 7.)

Before you attempt the exercises, we remind you of the steps used to solve quadratic equations by factoring:

PROCEDURE TO SOLVE QUADRATICS BY FACTORING

1. Perform the necessary operations on both sides of the equation so that the right-hand side is 0.
2. Use the general factoring strategy to factor the left side of the equation, if necessary.
3. Use the zero product property and make each factor on the left equal 0.
4. Solve each of the resulting equations.
5. Check the results by substituting the solutions obtained in step 4 in the original equation.

Calculator Corner

Solving Quadratics

Let's look at Example 3, where we have to solve $(2x + 1)(x - 2) = 2(x - 1) + 3$. As we mentioned, *you have to know the algebra* to write the equation in the standard form, $2x^2 - 5x - 3 = 0$. Now graph $Y = 2x^2 - 5x - 3$ (see Window 1). Where are the points where $Y = 0$? They are on the horizontal axis (the x-axis). For any point on the x-axis, its y-value is $y = 0$. The graph appears to have x-values 3 and $-\frac{1}{2}$ at the two points where the graph crosses the x-axis ($y = 0$). You can use your TRACE and ZOOM keys to confirm this.

Zero
X=-.5 Y=0

Window 1

Some calculators have a "zero" feature that tells you when $y = 0$. On a TI-83 Plus, enter 2nd TRACE 2 to activate the zero feature. The calculator asks you to select a left bound—that is, a point on the curve to the left of where the curve crosses the horizontal axis. Pick one near $-\frac{1}{2}$ using your ◀ and ▶ keys to move the cursor. Press ENTER. Select a right bound—that is, a point on the curve to the right of where the curve crosses the horizontal axis, and press ENTER again. The calculator then asks you to guess. Use the calculator's guess by pressing ENTER. The zero is given as $-.5$ (see Window 1). Do the same to find the other zero, $x = 3$.

Use these techniques to solve some of the other examples in this section and some of the problems in Exercises 5.6.

› Exercises 5.6

> Practice Problems > Self-Tests > Media-rich eBooks > e-Professors > Videos

‹ A › Solving Quadratic Equations by Factoring In Problems 1–20, solve the given equation.

1. $2x^2 + 7x + 3 = 0$ **2.** $2x^2 + 5x + 3 = 0$ **3.** $2x^2 + x - 3 = 0$ **4.** $6x^2 + x - 12 = 0$

5. $3y^2 - 11y + 6 = 0$ **6.** $4y^2 - 11y + 6 = 0$ **7.** $3y^2 - 2y - 1 = 0$ **8.** $12y^2 - y - 6 = 0$

9. $6x^2 + 11x = -4$ **10.** $5x^2 + 6x = -1$ **11.** $3x^2 - 5x = 2$ **12.** $12x^2 - x = 6$

13. $5x^2 + 6x = 8$ **14.** $6x^2 + 13x = 5$ **15.** $5x^2 - 13x = -8$ **16.** $3x^2 + 5x = -2$

17. $3y^2 = 17y - 10$ **18.** $3y^2 = 2y + 1$ **19.** $2y^2 = -5y - 2$ **20.** $5y^2 = -6y + 8$

In Problems 21–36, solve (you may use the factoring rules for perfect square trinomials, F2 and F3).

21. $9x^2 + 6x + 1 = 0$ **22.** $x^2 + 14x + 49 = 0$ **23.** $y^2 - 8y = -16$

24. $y^2 - 20y = -100$ **25.** $9x^2 + 12x = -4$ **26.** $25x^2 + 10x = -1$

27. $4y^2 - 20y = -25$ **28.** $16y^2 - 56y = -49$ **29.** $x^2 = -10x - 25$

30. $x^2 = -16x - 64$ **31.** $(2x - 1)(x - 3) = 3x - 5$ **32.** $(3x + 1)(x - 2) = x + 7$

33. $(2x + 3)(x + 4) = 2(x - 1) + 4$ **34.** $(5x - 2)(x + 2) = 3(x + 1) - 7$ **35.** $(2x - 1)(x - 1) = x - 1$

36. $(3x - 2)(3x - 1) = 1 - 3x$

In Problems 37–60, solve.

37. $4x^2 - 1 = 0$ **38.** $9x^2 - 1 = 0$ **39.** $4y^2 - 25 = 0$

40. $25y^2 - 9 = 0$ **41.** $z^2 = 9$ **42.** $z^2 = 25$

43. $25x^2 = 49$ **44.** $9x^2 = 64$ **45.** $m^2 = 5m$

46. $m^2 = 9m$ **47.** $2n^2 = 10n$ **48.** $3n^2 = 12n$

49. $y(y + 11) = -24$ **50.** $y(y + 15) = -56$ **51.** $y(y - 16) = -63$

52. $y(y - 19) = -88$ **53.** $(v - 2)(v^2 + 3v + 2) = 0$ **54.** $(v - 1)(v^2 + 5v + 6) = 0$

55. $(m^2 - 3m + 2)(m - 4) = 0$ **56.** $(m^2 - 4m + 3)(m - 6) = 0$ **57.** $(n^2 - 3n - 4)(n + 2) = 0$

58. $(n^2 - 4n - 5)(n + 8) = 0$ **59.** $(x^2 + 2x - 3)(x - 1) = 0$ **60.** $(x^2 + 3x - 4)(x - 1) = 0$

› Web IT go to mhhe.com/bello for more lessons

Applications

The next four problems are from Section 5.3. In Section 5.3 we asked you to *factor* the equations. Here, we go one step further: *Solve* the equations!

61. *Stopping distance* (See page 355) If you are traveling at m miles per hour and want to find the speed m that will allow you to stop the car 13 feet away, you must solve the equation $5m^2 + 220m - 1225 = 0$.

a. Solve the equation.

b. Since m represents the velocity of the car, can you use both answers? Which is the correct answer?

62. *Stopping distance* If you are traveling at m miles per hour and want to find the speed m that will allow you to stop the car 14 feet away, you must solve the equation $5m^2 + 225m - 1250 = 0$.

a. Solve the equation.

b. Since m represents the velocity of the car, can you use both answers? Which answer is correct?

63. *Height of an object* The height $H(t)$ after t seconds of an object thrown downward from a height of h meters and initial velocity v_0 is modeled by the equation $H(t) = -5t^2 + v_0t + h$. If an object is thrown downward at 5 meters per second from a height of 10 meters, then $H(t) = -5t^2 - 5t + 10$.

a. How long would it take for the object to reach the ground $[H(t) = 0]$?

b. The equation $-5t^2 - 5t + 10 = 0$ has two answers. What are the solutions? Since t represents the number of seconds, can you use both answers? Which answer is correct?

64. *Height of an object* The height $H(t)$ of an object thrown downward from a height of h meters and initial velocity v_0 is modeled by the equation $H(t) = -5t^2 + v_0t + h$. If an object is thrown downward at 4 meters per second from a height of 28 meters, then $H(t) = -5t^2 - 4t + 28$.

a. How long would it take for the object to reach the ground $[H(t) = 0]$?

b. The equation $-5t^2 - 4t + 28 = 0$ has two answers. What are the solutions? Since t represents the number of seconds, can you use both answers? Which answer is correct?

‹ B › Solving Applications Involving Factoring

Using Your Knowledge

Applications: Green Math

"The U.S. government is committed to enacting health insurance reform that provides health care stability and security for all Americans," but you can use your knowledge to explore national health expenditures and make your own predictions by basing results on the historical per capita cost at http://tinyurl.com/mp32zx.

65. *National health expenditures* The annual per capita national health expenditures can be approximated by $H(t) = -3t^2 + 300t + 6300$, where t is the number of years after 2003.

a. In what year will expenditures reach $9000?

b. In what year will expenditures reach $11,100?

66. *Another view* The Kaiser Family Foundation has the approximation shown in the graph. According to Problem 65, the expenditures for 2013 will be $9000.

Source: http://tinyurl.com/cp5xmq.

a. How does this compare to the expenditures for 2013 projected in the graph (blue line)?

b. How do the results using $H(t) = -3t^2 + 300t + 6300$ $(t = 15)$ for the year 2018 and the $13,100 figure predicted in the graph compare?

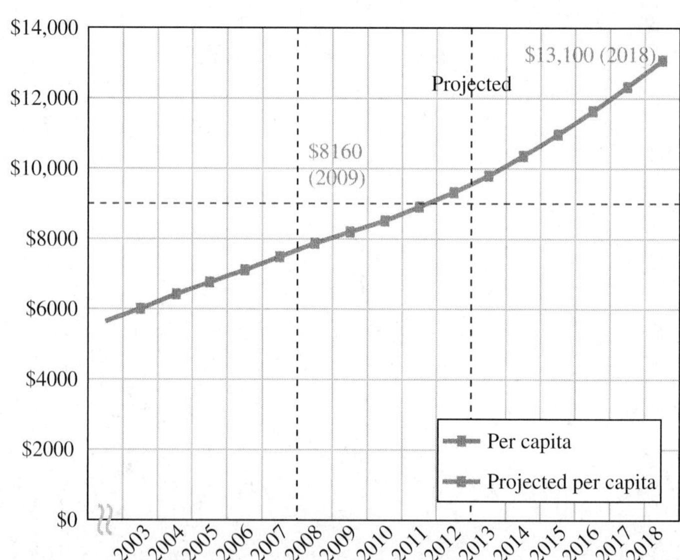

Write On

67. What is a quadratic equation and what procedure do you use to solve these equations?

68. Write the steps you use to solve $x(x - 1) = 0$ and then write the difference between this procedure and the one you would use to solve $3x(x - 1) = 0$.

69. The equation $x^2 + 2x + 1 = 0$ can be solved by factoring. How many solutions does it have? The original equation is equivalent to $(x + 1)^2 = 0$. Why do you think -1 is called a *double root* for the equation?

Concept Checker

Fill in the blank(s) with the correct word(s), phrase, or mathematical statement.

70. An equation of the form $\boldsymbol{ax^2 + bx + c = 0}$, $a \neq 0$, is called a _______________ equation.

71. The zero product property states that if $\boldsymbol{A \cdot B = 0}$ then _______________ or _______________.

linear	$\boldsymbol{B \neq 0}$
$\boldsymbol{A = 0}$	$\boldsymbol{B = 0}$
$\boldsymbol{A \neq 0}$	**quadratic**

Mastery Test

Solve:

72. $10x^2 - 13x = 3$

73. $(3x - 2)(x - 1) = 2(x + 3) + 2$

74. $(9x - 2)(x - 1) = 2(x + 1) - 7x - 1$

75. $5x^2 + 9x - 2 = 0$

76. $3x^2 - 2x - 5 = 0$

77. $x(x - 1) = 0$

78. $2x(x + 3) = 0$

79. $(x - 4)(x^2 + 4x - 5) = 0$

80. $25y^2 - 36 = 0$

81. $m(3m + 5) = -2$

82. $n^2 = 11n$

Skill Checker

Simplify:

83. $H^2 + (3 + H)^2$

84. $H^2 + (6 + H)^2$

Expand:

85. $(H - 9)(H + 3)$

86. $(H - 8)(H + 4)$

5.7 Applications of Quadratics

Objectives

Use the RSTUV procedure to solve:

A › Consecutive integer problems

B › Area and perimeter problems

C › Problems involving the Pythagorean Theorem

D › Motion problems (braking distance)

E › Solve more applications using quadratics

To Succeed, Review How To . . .

1. Solve integer problems (p. 150).
2. Use FOIL to expand polynomials (p. 369–372).
3. Multiply integers (pp. 60–61).

Getting Started

Stop That Car!

The diagram shows the distance needed to stop a car traveling at the indicated speeds. At speed m (in miles per hour), that stopping distance $D(m)$ in feet is given by

$$D(m) = 0.05m^2 + 2.2m + 0.75$$

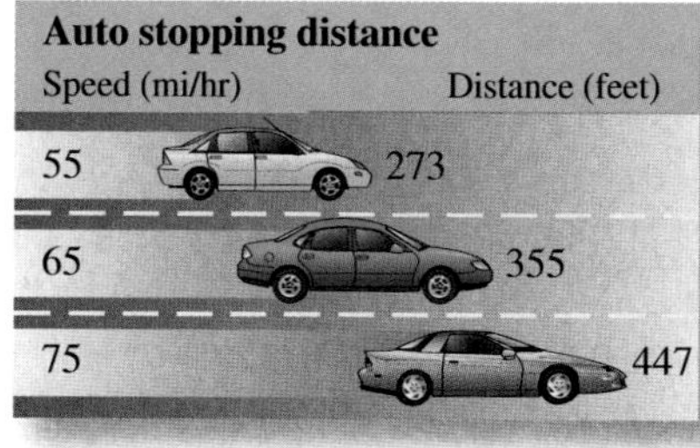

If you are one car length (13 feet) behind a stopped car, how fast can you be traveling and still be able to stop before hitting the car?

In this case, the actual distance is 13 feet, so we have to solve

$$D(m) = 0.05m^2 + 2.2m + 0.75 = 13$$

$$5m^2 + 220m + 75 = 1300 \quad \text{Multiply by 100 (clear the decimals).}$$

$$5m^2 + 220m - 1225 = 0 \quad \text{Subtract 1300 (write in standard form).}$$

$$m^2 + 44m - 245 = 0 \quad \text{Divide by 5.}$$

$$(m - 5)(m + 49) = 0 \quad \text{Factor.}$$

$$m = 5 \quad \text{or} \quad m = -49 \quad \text{Use the zero product property and solve.}$$

So, you can be going 5 miles per hour and stop in 13 feet. If your speed is over 5 miles per hour, you will hit the car!

A › Solving Consecutive Integer Problems

Do you remember the integer problems of Chapter 2? Here's the terminology we need to solve a problem involving integers and quadratic equations.

Terminology	Notation	Examples
Two consecutive integers	$n, n + 1$	$3, 4; -6, -5$
Three consecutive integers	$n, n + 1, n + 2$	$7, 8, 9; -4, -3, -2$
Two consecutive even integers	$n, n + 2$	$8, 10; -6, -4$
Two consecutive odd integers	$n, n + 2$	$13, 15; -21, -19$

As before, we solve this type problem using the RSTUV method.

EXAMPLE 1 **A consecutive integer problem**

The product of two consecutive even integers is 10 more than 7 times the larger of the two integers. Find the integers.

SOLUTION 1

1. **Read the problem.** We are asked to find two consecutive even integers.
2. **Select the unknown.** Let n and $n + 2$ be the integers ($n + 2$ being the larger).
3. **Think of a plan.** We first translate the problem:

$$\underbrace{n(n+2)}_{\text{The product of two consecutive integers}} \underbrace{=}_{\text{is}} \underbrace{10}_{10} \underbrace{+}_{\text{more than}} \underbrace{7(n+2)}_{\text{7 times the larger.}}$$

4. **Use algebra to solve the equation.**

$$n^2 + 2n = 10 + 7n + 14 \quad \text{Use the distributive property.}$$
$$n^2 + 2n = 24 + 7n \quad \text{Simplify.}$$
$$n^2 + 2n - 24 - 7n = 0 \quad \text{Subtract } 24 + 7n.$$
$$n^2 - 5n - 24 = 0 \quad \text{Simplify.}$$
$$(n - 8)(n + 3) = 0 \quad \text{Factor } (-8 \cdot 3 = -24, -8 + 3 = -5).$$
$$n - 8 = 0 \quad \text{or} \quad n + 3 = 0 \quad \text{Use the zero product property.}$$
$$n = 8 \qquad n = -3 \quad \text{Solve each equation.}$$

If $n = 8$ is the first integer, the second is $n + 2 = 8 + 2 = 10$. The solution $n = -3$ is not acceptable because -3 is *not* even. Thus, there is only one pair of integers that satisfies the problem: 8 and 10.

5. **Verify the solution.**

$$\underbrace{8 \cdot 10}_{\text{The product of two consecutive integers}} \underbrace{=}_{\text{is}} \underbrace{10}_{10} \underbrace{+}_{\text{more than}} \underbrace{7 \cdot 10}_{\text{7 times the larger.}} \quad \text{True.}$$

PROBLEM 1

The product of two consecutive odd integers is 10 more than 5 times the smaller of the two integers. Find the integers.

B > Solving Area and Perimeter Problems

Quadratic equations are also used in geometry. As you recall, if L and W are the length and width of a rectangle, then the perimeter P is $P = 2L + 2W$, and the area A is $A = LW$. Let's use these ideas in the next example.

EXAMPLE 2 **Finding the dimensions of a room**

A rectangular room is 4 feet longer than it is wide. The area of the room is numerically equal to its perimeter plus 92. What are the dimensions of the room?

SOLUTION 2

1. **Read the problem.** We are asked to find the dimensions of the room; we also know it is a rectangle.
2. **Select the unknown.** Let W be the width. Since the room is 4 feet longer than it is wide, the length is $W + 4$.
3. **Think of a plan.** Two measurements are involved: the area and the perimeter. Let's start with a picture:

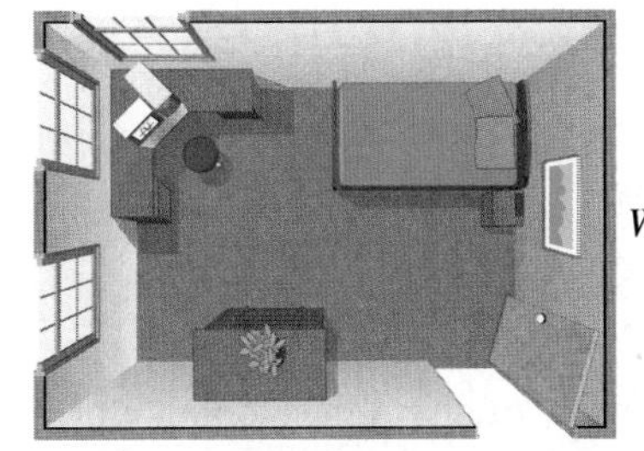

The area of the room is

$$W(W + 4)$$

PROBLEM 2

What are the dimensions of the room in Example 2 if the area exceeds the perimeter by 56?

(continued)

Answers to PROBLEMS

1. 5 and 7 **2.** 8 ft by 12 ft

The perimeter of the room is

$$2(W + 4) + 2W = 4W + 8$$

Look at the wording of the problem. We can restate it as follows:

$$\underbrace{\text{The area of the room}}_{W(W+4)} \quad \underbrace{\text{is equal to}}_{=} \quad \underbrace{\text{its perimeter plus 92.}}_{(4W+8)+92}$$

4. Use algebra to solve the equation.

$W^2 + 4W = 4W + 100$ Simplify both sides.

$W^2 - 100 = 0$ Subtract $4W + 100$.

$(W + 10)(W - 10) = 0$ Factor.

$W + 10 = 0$ or $W - 10 = 0$ Use the zero product property.

$W = -10$ $\quad W = 10$ Solve each equation.

Since a room can't have a negative width, discard the -10, so the width is 10 feet and the length is 4 feet longer or 14 feet. Thus, the dimensions of the room are 10 feet by 14 feet.

5. Verify the solution. The area of the room is $10 \cdot 14 = 140$ and the perimeter is $2 \cdot 10 + 2 \cdot 14 = 20 + 28 = 48$. The area (140) must exceed the perimeter (48) by 92. Does $140 = 48 + 92$? Yes, so our answer is correct.

EXAMPLE 3 Finding the dimensions of a monitor

Denise's monitor is 3 inches wider than it is high and has 130 square inches of viewing area.

a. Find the dimensions of her monitor.

b. Denise wants a Trinitron® monitor, which is 4 inches wider than it is high and has 62 more square inches of viewing area than her current monitor. What are the dimensions of the Trinitron monitor?

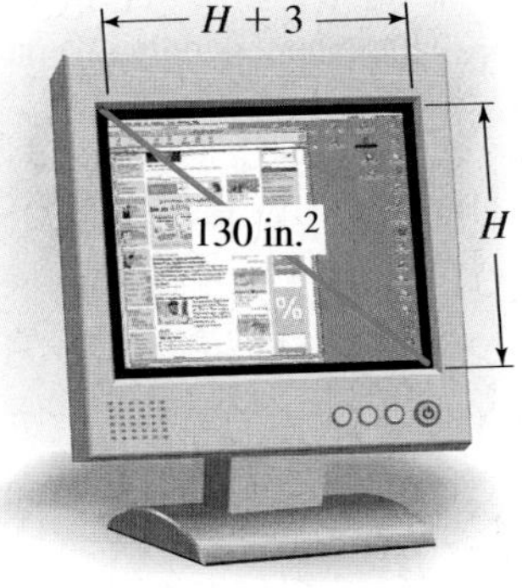

PROBLEM 3

a. What are the dimensions of Denise's current monitor if it is 3 inches wider than it is high and has 108 square inches of viewing area?

b. What are the dimensions of a Cinema monitor, which is 6 inches wider than it is high and has double the viewing area of the monitor in part **a?**

SOLUTION 3

a. **1. Read the problem.** We want the dimensions of the monitor.

2. Select the unknown. Let H be the height of the monitor.

3. Think of a plan. Draw a picture. H is the height. The width is 3 inches more than the height; that is, the width is $H + 3$. The area of the rectangle is the height H times the width, $H + 3$. It is also 130 square inches. Thus,

$$H(H + 3) = 130$$

4. Use algebra to solve the equation.

$H(H + 3) = 130$

$H^2 + 3H = 130$ Simplify.

$H^2 + 3H - 130 = 0$ Subtract 130.

$(H + 13)(H - 10) = 0$ Factor.

$H + 13 = 0$ or $H - 10 = 0$ Use the zero product property.

$H = -13$ or $H = 10$ Solve.

Discard $H = -13$. (Why?) Thus, the dimensions of the monitor are 10 inches by 13 inches.

Answers to PROBLEMS

3. a. 9 in. by 12 in. **b.** 12 in. by 18 in.

b. What are the Trinitron dimensions? Let the height be H, so the width is $H + 4$. The area of the Trinitron screen is $H(H + 4)$, which is 62 square inches more than her old screen; that is,

$$H(H + 4) = 130 + 62 = 192$$

4. Use algebra to solve the equation.

$$H(H + 4) = 192$$

$$H^2 + 4H = 192 \quad \text{Simplify.}$$

$$H^2 + 4H - 192 = 0 \quad \text{Subtract 192.}$$

$$(H - 12)(H + 16) = 0 \quad \text{Use the zero product property.}$$

$$H = 12 \quad \text{or} \quad H = -16 \quad \text{Solve.}$$

Discard $H = -16$. The dimensions of the Trinitron are 12 inches by 16 inches.

5. Verify the solution!

C > Using the Pythagorean Theorem

Suppose you were to buy a 20-inch television set. What does that 20-inch measurement represent? It's the diagonal length of the rectangular screen. The relationship between the length L, the height H, and the diagonal measurement d of the screen can be found by using the Pythagorean Theorem, which is stated here.

PYTHAGOREAN THEOREM

If the longest side of a right triangle (a triangle with a 90° angle) is of length c and the other two sides are of lengths a and b, respectively, then

$$a^2 + b^2 = c^2$$

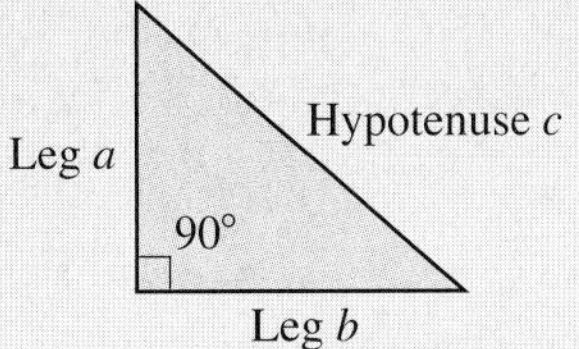

Note that the longest side of the triangle (opposite the 90° angle) is called the **hypotenuse**, and the two shorter sides are called the **legs** of the triangle.

In the case of the television set, $H^2 + L^2 = 20^2$. If the screen is 16 inches long ($L = 16$), can we find its height H? Substituting $L = 16$ into $H^2 + L^2 = 20^2$, we obtain

$$H^2 + 16^2 = 20^2$$

$$H^2 + 256 = 400 \quad \text{Simplify.}$$

$$H^2 - 144 = 0 \quad \text{Subtract 400.}$$

$$(H - 12)(H + 12) = 0 \quad \text{Factor.}$$

$$H - 12 = 0 \quad \text{or} \quad H + 12 = 0 \quad \text{Use the zero product property.}$$

$$H = 12 \qquad H = -12 \quad \text{Solve each equation.}$$

Since H represents the height of the screen, we discard -12 as an answer. (The height cannot be negative.) Thus, the height of the screen is 12 inches.

EXAMPLE 4 Calculating a computer monitor's dimensions

The screen of a rectangular computer monitor is 3 inches wider than it is high, and its diagonal is 6 inches longer than its height. What are the dimensions of the monitor?

SOLUTION 4

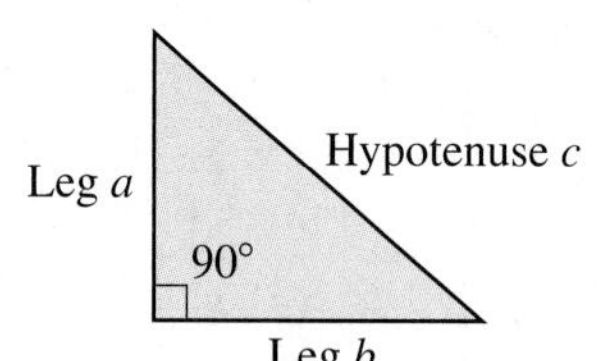

1. Read the problem. We are asked to find the dimensions, which involves finding the width, height, and diagonal of the screen.

2. Select the unknown. Since all measurements are given in terms of the height, let H be the height.

3. Think of a plan. It's a good idea to start with a picture so we can see the relationships among the measurements. Note that the width is $3 + H$ (3 inches wider than the height) and the diagonal is $6 + H$ (6 inches longer than the height). We enter this information in a diagram, as shown here.

4. Use the Pythagorean Theorem to describe the relationship among the measurements and then solve the resulting equation. According to the Pythagorean Theorem:

$$H^2 + (3 + H)^2 = (6 + H)^2$$

$$H^2 + 9 + 6H + H^2 = 36 + 12H + H^2 \quad \text{Expand } (3 + H)^2 \text{ and } (6 + H)^2.$$

$$H^2 + 9 + 6H + H^2 - 36 - 12H - H^2 = 0 \quad \text{Subtract } 36 + 12H + H^2.$$

$$H^2 - 6H - 27 = 0 \quad \text{Simplify.}$$

$$(H - 9)(H + 3) = 0 \quad \text{Factor } (-9 \cdot 3 = -27; -9 + 3 = -6).$$

$$H - 9 = 0 \quad \text{or} \quad H + 3 = 0 \quad \text{Use the zero product property.}$$

$$H = 9 \qquad H = -3 \quad \text{Solve each equation.}$$

Since H is the height, we discard -3, so the height of the monitor is 9 inches, the width is 3 more inches, or 12 inches, and the diagonal is 6 more inches than the height, or $9 + 6 = 15$ inches.

5. Verify the solution! Looking at the diagram and using the Pythagorean Theorem, we see that $9^2 + 12^2$ must be 15^2, that is,

$$9^2 + 12^2 = 15^2$$

Since $81 + 144 = 225$ is a true statement, our dimensions are correct.

PROBLEM 4

A rectangular book is 2 inches higher than it is wide. If the diagonal is 4 inches longer than the width of the book, what are the dimensions of the book?

By the way, television sets claiming to be 25 inches or 27 inches (meaning the length of the screen measured diagonally is 25 or 27 inches) hardly ever measure 25 or 27 inches. This is easy to confirm using the Pythagorean Theorem. For example, a 27-inch Panasonic® has a screen that is 16 inches high and 21 inches long. Can it really be 27 inches diagonally? If this were the case, $16^2 + 21^2$ would equal 27^2. Is this true? Measure a couple of TV or computer screens and see whether the manufacturers' claims are true!

Answers to PROBLEMS

4. 8 in. by 6 in.

D › Solving Motion Problems: Braking Distance

In the *Getting Started,* we gave a formula for the stopping distance for a car traveling at the indicated speeds. That distance involves the **braking distance** b, the distance it takes to stop a car *after* the brakes are applied. This distance is given by

$$b = 0.06v^2$$

where v is the speed of the car when the brakes are applied. It also involves the reaction distance

$$r = 1.5tv$$

where t is the driver's reaction time (in seconds) and v is the speed of the car in miles per hour.

EXAMPLE 5 Finding the speed of a car based on braking distance

A car traveled 96 feet *after* the brakes were applied. How fast was the car going when the brakes were applied?

SOLUTION 5

1. **Read the problem.** We want to find how fast the car was going.
2. **Select the unknown.** Let v represent the velocity.
3. **Think of a plan.** The braking distance b is 96.

The formula for b is $b = 0.06v^2$.

Thus, $0.06v^2 = 96$.

4. **Use algebra to solve the equation.**

$$0.06v^2 = 96$$
$$6v^2 = 9600 \qquad \text{Multiply by 100.}$$
$$6v^2 - 9600 = 0 \qquad \text{Subtract 9600.}$$
$$6(v^2 - 1600) = 0 \qquad \text{Factor out the GCF.}$$
$$6(v + 40)(v - 40) = 0 \qquad \text{Factor.}$$
$$v + 40 = 0 \quad \text{or} \quad v - 40 = 0 \qquad \text{Solve.}$$
$$v = -40 \quad \text{or} \quad v = 40$$

Discard the -40 because the velocity was positive, so the velocity of the car was 40 miles per hour.

5. **Verify the solution.** Substitute $v = 40$ in $b = 0.06v^2$:

$$b = 0.06(40)^2 = 0.06(1600) = 96$$

so the answer is correct.

PROBLEM 5

A car traveled 150 feet *after* the brakes were applied. How fast was the car going when the brakes were applied?

E › Solving More Applications Containing Quadratics

Around the year 2000, Australia was prone to wet weather, but starting in 2003 a long, severe drought, the worst on record, has plagued Australia. Sydney Water has implemented a water conservation program that saves almost 80,000 million liters (ML) of water each year as shown in the graph. How much water can they hope to save in 2009? When can they expect to reach savings of 84,000 ML/year? We will answer those questions in Example 6.

Answers to PROBLEMS

5. 50 mi/hr

GREEN MATH

EXAMPLE 6 Water conservation in Australia

The expression $W(t) = 2t^2 - 13t + 39$, where t is the number of years after 2000, approximates the water savings shown in the graph in thousands of millions of liters (ML).

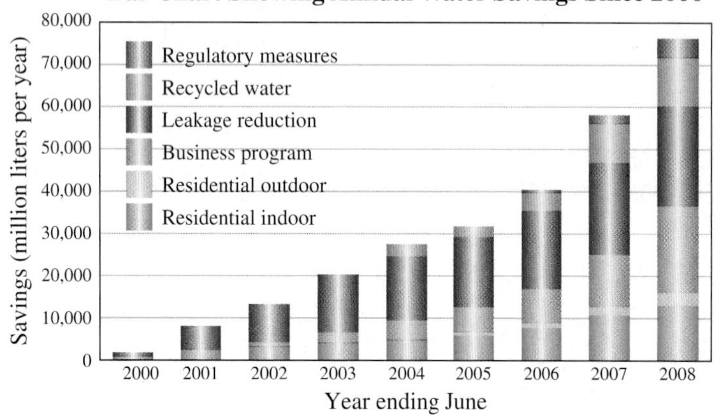

a. What would be the water savings in 10 years (2010)?

b. When will the savings be 84(000) ML/year?

Source: http://tinyurl.com/yjofqye.

PROBLEM 6

a. Use $W(t) = 2t^2 - 13t + 39$ to find the water savings in 2009.

b. Does the answer coincide with the data in part **b** of the example?

Drought in Australia

Does your town, city or school have a water conservation program? You can see the Sydney program at http://tinyurl.com/yl654q5.

SOLUTION 6

a. The water savings in 10 years will be $W(10) = 2 \cdot 10^2 - 13 \cdot 10 + 39$ or 109 thousand million liters (ML) per year.

b. To find when the water savings will be 84(000) ML/year, we have to solve the equation $2t^2 - 13t + 39 = 84$.

$2t^2 - 13t + 39 = 84$	Given
$2t^2 - 13t + 39 - 84 = 84 - 84$	Subtract 84.
$2t^2 - 13t - 45 = 0$	Simplify.
$(2t + 5)(t - 9) = 0$	Factor by trial and error.
$2t + 5 = 0$ or $t - 9 = 0$	Use the zero product property.
$t = -\frac{5}{2}$ or $t = 9$	Solve each equation.

Thus, **9** years after 2000, in 2009, 84,000 ML/year were saved.
We discard the negative answer, since t is a positive number representing the years after 2000.

Answers to PROBLEMS

6. a. 84,000 ML/year

b. Yes, both predict 84,000 ML/year.

Before you attempt to solve the Exercises, practice translating!

TRANSLATE THIS

The third step in the RSTUV procedure is to **TRANSLATE** the information into an equation. In Problems 1–10 **TRANSLATE** the sentence and match the correct translation with one of the equations A–O

1. The product of two consecutive even integers is 20 less than 10 times the larger of the two integers.
2. The product of two consecutive odd integers is 10 more than 5 times the smaller of the two integers.
3. A rectangular room is 5 feet longer than it is wide. What is the area A of the room in terms of the width W?
4. The viewing area of a TV is $A = WH$, where W is the width and H is the height of the rectangular screen. If the height H is 20 inches more than the width, what is the area A?
5. If the width W of a rectangle is 20 inches less than its height H, what is the area of the rectangle?
6. The screen on a rectangular computer monitor is 4 inches wider than it is high. If the diagonal is 8 more inches than its height H, write an equation relating the length of the sides of the screen in terms of H.
7. The braking distance b for a car traveling at v miles per hour is the product of 0.06 and the square of the velocity. If a car travels F feet after the brakes were applied, what is the equation for the braking distance F?
8. Write the area A of a triangle whose base is 2 inches longer than its height h.
9. The hypotenuse of a right triangle is 4 inches longer than the length L of the longer side. If the remaining side is 4 inches shorter than the length L, write an equation relating the sides of the triangle.
10. The sides of a right triangle are consecutive even integers. Write an equation relating the length of the sides.

A. $A = (H - 20)H$

B. $A = W(W + 5)$

C. $A = \frac{1}{2}(h + 2)h$

D. $n(n + 2) = 5n + 10$

E. $(H + 4)^2 = H^2 + (H + 8)^2$

F. $A = (H - 20) + H$

G. $F = 0.06v^2$

H. $(H + 4)^2 + H^2 = (H + 8)^2$

I. $A = (h + 2)h$

J. $L^2 + (L - 4)^2 = (L + 4)^2$

K. $n(n + 2) = 10(n + 2) - 20$

L. $L^2 + (L + 4)^2 = (L - 4)^2$

M. $n^2 + (n + 2)^2 = (n + 4)^2$

N. $n(n + 1) = 10(n + 1) - 20$

O. $A = W(W + 20)$

› Exercises 5.7

> Practice Problems > Self-Tests > Media-rich eBooks > e-Professors > Videos

‹A› Solving Consecutive Integer Problems

In Problems 1–6, use the RSTUV method to solve the integer problems.

1. The product of two consecutive integers is 8 less than 10 times the smaller of the two integers. Find the integers.
2. The product of two consecutive even integers is 4 more than 5 times the smaller of the two integers. Find the integers.
3. The product of two consecutive odd integers is 7 more than their sum. What are the integers?
4. Find three consecutive even integers such that the square of the largest exceeds the sum of the squares of the other two by 12.

››› Applications: Green Math

5. *Water conservation in Australia* How much water does your community save by conserving and recycling? In 2009, the amount of water saved in Sydney, Australia, in 1 year amounted to 84,000 million liters! This represents 4000 more million liters than 4 times the amount saved in 2003. How much water was saved in Sydney in 2003?
6. *More water conservation in Australia* The 84,000 million liters of water saved in Sydney, Australia, in 1 year represents 4000 more million liters than double the amount saved in 2006. How much water was saved in Sydney, Australia, in 2006?

Web IT go to mhhe.com/bello for more lessons

B **Solving Area and Perimeter Problems** Use the information in the table and the RSTUV method to solve Problems 7–14.

Name	Geometric Shapes	Name	Geometric Shapes
Triangle Area $= \frac{1}{2}bh$	h, b	**Rectangle** Area $= LW$	W, L
Trapezoid Area $= \frac{1}{2}h(b_1 + b_2)$	b_1, h, b_2	**Circle** Area $= \pi r^2$	r
Parallelogram Area $= Lh$	h, W, L		

7. The area A of a triangle is 40 square inches. Find its dimensions if the base b is 2 inches more than its height h.

8. The area A of a trapezoid is 105 square centimeters. If the height h is the same length as the smaller side b_1 and 1 inch less than the longer side b_2, find the height h.

9. The area A of a parallelogram is 150 square inches. If the length of the parallelogram is 5 inches more than its height, what are the dimensions?

10. The area of a rectangle is 96 square centimeters. If the width of the rectangle is 4 centimeters less than its length, what are the dimensions of the rectangle?

11. The area of a circle is 49π square units. If the radius r is $x + 3$ units, find x and then find the radius of the circle.

12. A rectangular room is 5 feet longer than it is wide. The area of the room numerically exceeds its perimeter by 100. What are the dimensions of the room?

13. The biggest lasagna ever made had an area of 250 square feet. If it was made in the shape of a rectangle 45 feet longer than wide, what were its dimensions? (It was made in Dublin and weighed 3609 pounds, 10 ounces.)

14. The biggest strawberry shortcake needed 360 square feet of strawberry topping. If this area was numerically 225 more than 3 times its length, what were the dimensions of this rectangular shortcake?

C **Using the Pythagorean Theorem** In Problems 15–20, use the RSTUV procedure to solve the problems.

15. The biggest television ever built is the 289 Sony® Jumbo Tron. (This means the screen measured 289 feet diagonally.) If the length of the screen was 150 feet, how high was the TV.

16. A television screen is 2 inches longer than it is high. If the diagonal length of the screen is 2 inches more than its length, what is the diagonal measurement of this screen?

17. The hypotenuse of a right triangle is 4 inches longer than the shortest side and 2 inches longer than the remaining side. Find the dimensions of the triangle.

18. The hypotenuse of a right triangle is 16 inches longer than the shortest side and 2 inches longer than the remaining side. Find the dimensions of the triangle.

19. One of the sides of a right triangle is 3 inches longer than the shortest side. If the hypotenuse is 3 inches longer than the longer side, what are the dimensions of the triangle?

20. The sides of a right triangle are consecutive even integers. Find their lengths.

D **Solving Motion Problems: Braking Distance** In Problems 21–26, solve the motion problems. Do not round answers resulting in decimals. (*Hint:* For Problems 21–24, use the formula from Example 5.)

21. A car traveled 54 feet after the driver applied the brakes. How fast was the car going when the brakes were applied?

22. A car traveled 216 feet after the brakes were applied. How fast was the car going when the brakes were applied?

23. A car left a 73.5-foot skid mark on the road.

a. How fast was the car going when the brakes were applied and the skid mark was made?

b. If the speed limit was 30 miles per hour, was the car speeding?

24. A police officer measured a skid mark to be 150 feet long. The driver said he was going under the speed limit, which was 45 miles per hour. Was the driver correct? How long would the skid mark have to be if he was indeed going 45 miles per hour when the brakes were applied?

25. Pedro reacts very quickly. In fact, his reaction time is 0.4 second. When driving on a highway, Pedro saw a danger signal ahead and tried to stop. If his car traveled 120 feet before stopping, how fast was he going when he saw the sign? Use

$$d = 1.5tv + 0.06v^2$$

for the stopping distance, where d is in feet, t is in seconds, and v is in miles per hour.

26. Vilmos' reaction time is 0.3 seconds. He tried to avoid hitting a student walking on the road.

a. If his car traveled 33 feet before stopping, how fast was he going?

b. How many feet would he travel at that speed, using the formula given in the *Getting Started?*

‹ E › Solving More Applications Containing Quadratics

Applications

27. *Games in a tournament* The total number of games G played in a tournament in which each of the t teams entered plays every other team twice is modeled by the equation $G = t^2 - t$. Find the number of teams t entered in a 20 game tournament.

28. *Games in a tournament* If in Problem 27 the total number of games ($G = t^2 - t$) played is 30, how many teams were entered in the tournament?

29. *Handshakes* Before a meeting of the Big 10 countries, the attending delegates shook every other delegate's hand only once. If 28 handshakes were exchanged and the number of possible handshakes is modeled by the equation $N = \frac{1}{2}(d^2 - d)$, where d is the number of delegates attending, how many delegates were at the meeting?

30. *Handshakes* A bipartisan committee of c congressmen exchanges 55 handshakes. If the number of possible handshakes is modeled by the equation $N = \frac{1}{2}(c^2 - c)$ and every person shook every other person's hand only once, how many congressmen were in the committee?

31. *Cost of tutoring* The tutorial center spent \$1300 tutoring x students. If the cost of tutoring x students is given by $(x^2 + 10x + 100)$ dollars, how many students were tutored?

32. *Cost of tutoring* The tutorial center will offer individualized tutoring for x students at a price $P = (x^2 + 25x)$ dollars a month. How many students can be tutored when the price P is \$350 per student?

Using Your Knowledge

Motion Problems In Problems 33–36, use the fact that the height $H(t)$ of an object thrown downward from a height of h meters and initial velocity V_0 is modeled by the equation

$$H(t) = -5t^2 - V_0t + h$$

33. An object is thrown downward at 5 meters per second from a height of 10 meters. How long does it take the object to hit the ground?

34. An object is thrown downward from a height of 28 meters with an initial velocity of 4 meters per second. How long does it take the object to hit the ground?

35. An object is thrown downward from a building 15 meters high at 10 meters per second. How long does it take the object to hit the ground?

36. How long does it take a package thrown downward from a plane at 10 meters per second to hit the ground 175 meters below?

Write On

37. In the *Getting Started,* the stopping distance is given by the equation

$$D(m) = 0.05m^2 + 2.2m + 0.75$$

In Problem 25, it is

$$d = 1.5tv + 0.06v^2$$

If you are traveling at 20 miles per hour, which formula can you use to evaluate your stopping distance? Why?

38. What additional information do you need in Problem 37 to use the second formula?

39. Name at least three factors that would influence the stopping distance of a car. Explain.

40. Suppose you are driving at 20 miles per hour. What reaction time will give the same stopping distance for both formulas?

Concept Checker

Fill in the blank(s) with the correct word(s), phrase, or mathematical statement.

41. According to the Pythagorean Theorem, if a, b, and c are the lengths of the sides of a right triangle (with c the hypotenuse), then ______________.

42. The distance it takes to stop a car after the brakes are applied is called the ______________ distance.

$a^2 = b^2 + c^2$ **crashing**

$a^2 + b^2 = c^2$ **braking**

Mastery Test

43. The product of two consecutive odd integers is the difference between 23 and the sum of the two integers. What are the integers?

44. A rectangular room is 2 feet longer than it is wide. Its area numerically exceeds 10 times its longer side by 28. What are the dimensions of the room?

45. The hypotenuse of a right triangle is 8 inches longer than the shortest side and 1 inch longer than the remaining side. What are the dimensions of the triangle?

46. The product of the length of the legs of a right triangle is 325 less than the length of the hypotenuse squared. If the longer side is 5 units longer than the shorter side, what are the dimensions of the triangle?

47. The distance b (in feet) it takes to stop a car *after* the brakes are applied is given by the equation

$$b = 0.06v^2$$

A car traveled 150 feet after the driver applied the brakes. How fast was the car going when the brakes were applied?

Skill Checker

48. Write $\frac{3}{8}$ with a denominator of 16.

49. Write $\frac{5}{6}$ with a denominator of 18.

In Problems 50–54, factor:

50. $6x - 12y$

51. $18x - 36y$

52. $x^2 + 2x - 15$

53. $x^2 + 5x - 6$

54. $x^2 - 9$

Collaborative Learning

Shortcut patterns for solving quadratic equations

In this chapter we have used several methods to solve quadratic equations. Now we are going to develop a shortcut to solve these quadratic equations by letting you discover some applicable patterns.

Form three groups of students. Each of the groups will solve the equations assigned to them and enter the required information.

Recall that to factor $x^2 + bx + c = 0$, we need two numbers whose product is c and whose sum is b.

Group 1	Numbers required to factor	Solutions
$x^2 + 3x + 2 = 0$	2, 1	**−2, −1**
$x^2 - 7x + 12 = 0$		
$x^2 + x - 2 = 0$		
$x^2 - x - 2 = 0$		
Group 2		
$x^2 + 5x + 6 = 0$	2, 3	**−2, −3**
$x^2 - 5x - 6 = 0$		
$x^2 + x - 6 = 0$		
$x^2 - x - 6 = 0$		
Group 3		
$x^2 + 7x + 12 = 0$	3, 4	**−3, −4**
$x^2 - 3x + 2 = 0$		
$x^2 + x - 12 = 0$		
$x^2 - x - 12 = 0$		

Based on the patterns you see, have each of the groups complete the following conjecture (guess):

The factorization of $x^2 + bx + c = 0$ requires two factors F_1 and F_2 whose product is ________ and whose sum is ________.

The solutions of $x^2 + bx + c = 0$ are ________ and ________.

See whether all groups agree.

Is there a pattern that can be used to solve $ax^2 + bx + c = 0$ when $a \neq 1$? First, recall that to factor $ax^2 + bx + c$ we need two factors whose product is ac and whose sum is b. Here are some examples and some work for all groups.

	ac	Factors	Solutions
$4x^2 - 4x + 1 = 0$	4	$-2, -2$	$\frac{1}{2}$
$3x^2 - 5x - 2 = 0$	-6	$-6, 1$	$2, -\frac{1}{3}$
$10x^2 + 13x - 3 = 0$	-30	$15, -2$	$-\frac{3}{2}, \frac{1}{5}$

Based on the patterns you see, have each of the groups complete the following conjecture (guess):

The factorization of $ax^2 + bx + c = 0$ requires two factors F_1 and F_2 whose product is ________ and whose sum is ________.

The solutions of $ax^2 + bx + c = 0$ are ________ and ________.

See whether all groups agree.

Research Questions

1. Proposition 4 in Book II of Euclid's *Elements* states:

   If a straight line be cut at random, the square on the whole is equal to the squares on the segments and twice the rectangles contained by the segments.

   This statement corresponds to one of the rules of factoring we've studied in this chapter. Which rule is it, and how does it relate?

2. Proposition 5 of Book II of the *Elements* also corresponds to one of the factoring formulas we studied. Which one is it, and what does the proposition say?

3. Many scholars ascribe the Pythagorean Theorem to Pythagoras. However, other versions of the theorem exist:

   a. The ancient Chinese proof

   b. Bhaskara's proof

   c. Euclid's proof

   d. Garfield's proof

   e. Pappus's generalization

   Select three of these versions and write a paper giving details, if possible, telling where they appeared, who authored them, and what they said.

4. There are different versions regarding Pythagoras's death. Write a short paper detailing the circumstances of his death, where it occurred, and how.

5. Write a report about Pythagorean triples.

6. In *The Human Side of Algebra,* we mentioned that the Pythagoreans studied arithmetic, music, geometry, and astronomy. Write a report about the Pythagoreans' theory of music.

7. Write a report about the Pythagoreans' theory of astronomy.

Summary Chapter 5

Section	Item	Meaning	Example
5.1A	Greatest common factor (GCF) of numbers	The GCF of a list of integers is the largest common factor of the integers in the list.	The GCF of 45 and 75 is 15.
5.1B	Greatest common factor (GCF) of variable terms	The GCF of a list of terms is the product of each of the variables raised to the lowest exponent to which they occur.	The GCF of x^2y^3, x^4y^2, and x^5y^4 is x^2y^2.
5.1C	Greatest common factor (GCF) of a polynomial	ax^n is the GCF of a polynomial if a divides each of the coefficients and n is the smallest exponent of x in the polynomial.	$3x^2$ is the GCF of $6x^4 - 9x^3 + 3x^2$.
5.2A	Factoring Rule 1: Reversing FOIL (F1)	$X^2 + (A + B)X + AB = (X + A)(X + B)$	$x^2 + 8x + 15 = (x + 5)(x + 3)$
	Factoring a trinomial of the form $x^2 + bx + c$	To factor $x^2 + bx + c$, find two numbers whose product is c and whose sum is b.	To factor $x^2 + 7x + 10$, find two numbers whose product is 10 and whose sum is 7. The numbers are 5 and 2. Thus, $x^2 + 7x + 10 = (x + 5)(x + 2)$.
5.3A	*ac* test	$ax^2 + bx + c$ is factorable if there are two integers with product ac and sum b.	$3x^2 + 8x + 5$ is factorable. (There are two integers whose product is 15 and whose sum is 8: 5 and 3.) $2x^2 + x + 3$ is not factorable. (There are no integers whose product is 6 and whose sum is 1.)
5.4A, B	Factoring squares of binomials (F2 and F3)	$X^2 + 2AX + A^2 = (X + A)^2$ $X^2 - 2AX + A^2 = (X - A)^2$	$x^2 + 10x + 25 = (x + 5)^2$ $x^2 - 10x + 25 = (x - 5)^2$
5.4C	Factoring the difference of two squares (F4)	$X^2 - A^2 = (X + A)(X - A)$	$x^2 - 36 = (x + 6)(x - 6)$
5.5A	Factoring the sum (F5) or difference (F6) of cubes	$X^3 + A^3 = (X + A)(X^2 - AX + A^2)$ $X^3 - A^3 = (X - A)(X^2 + AX + A^2)$	$x^3 + 64 = (x + 4)(x^2 - 4x + 16)$ $x^3 - 64 = (x - 4)(x^2 + 4x + 16)$
5.5B	General factoring strategy	1. Factor out the GCF. 2. Look at the number of terms inside the parentheses or in the original polynomial. *Four terms:* Grouping *Three terms:* Perfect square trinomial or *ac* test *Two terms:* Difference of two squares, sum of two cubes, difference of two cubes 3. Make sure the expression is completely factored.	

Section	Item	Meaning	Example
5.6A	Quadratic equation	An equation in which the greatest exponent of the variable is 2. If a, b, and c are real numbers, $ax^2 + bx + c = 0$ is in standard form.	$x^2 + 3x - 7 = 0$ is a quadratic equation in standard form.
	Zero product property	If $A \cdot B = 0$, then $A = 0$ or $B = 0$.	If $(x + 1)(x + 2) = 0$, then $x + 1 = 0$ or $x + 2 = 0$.
5.7C	Pythagorean Theorem	If the longest side of a right triangle (a triangle with a 90° angle) is of length c and the two other sides are of length a and b, then $a^2 + b^2 = c^2$.	Leg a; Hypotenuse c; 90°; Leg b If the length of leg a is 3 inches and the length of leg b is 4 inches, then the length h of the hypotenuse is $3^2 + 4^2 = h^2$ Thus, $9 + 16 = h^2$ $25 = h^2$ $5 = h$
5.7D	Braking distance $b = 0.06v^2$	If a car is moving v miles per hour, b is the distance (in feet) needed to stop the car *after* the brakes are applied.	The braking distance b for a car moving at 20 miles per hour is $b = 0.06(20^2) = 24$ feet.

› Review Exercises **Chapter 5**

(If you need help with these exercises, look in the section indicated in brackets.)

1. ‹ **5.1A** › *Find the GCF of:*

a. 60 and 90

b. 12 and 18

c. 27, 80, and 17

2. ‹ **5.1B** › *Find the GCF of:*

a. $24x^7, 18x^5, -30x^{10}$

b. $18x^8, 12x^9, -20x^{10}$

c. $x^6y^4, y^7x^5, x^3y^6, x^9$

3. ‹ **5.1C** › *Factor.*

a. $20x^3 - 55x^5$

b. $14x^4 - 35x^6$

c. $16x^7 - 40x^9$

4. ‹ **5.1C** › *Factor.*

a. $\frac{3}{7}x^6 - \frac{5}{7}x^5 + \frac{2}{7}x^4 - \frac{1}{7}x^2$

b. $\frac{4}{9}x^7 - \frac{2}{9}x^6 + \frac{2}{9}x^5 - \frac{1}{9}x^3$

c. $\frac{3}{8}x^9 - \frac{7}{8}x^8 + \frac{3}{8}x^7 - \frac{1}{8}x^5$

5. ‹ **5.1D** › *Factor.*

a. $3x^3 - 21x^2 - x + 7$

b. $3x^3 + 18x^2 + x + 6$

c. $4x^3 - 8x^2y + x - 2y$

6. ‹ **5.2A** › *Factor.*

a. $x^2 + 8x + 7$

b. $x^2 - 8x - 9$

c. $x^2 + 6x + 5$

7. ⟨**5.2A**⟩ *Factor.*

a. $x^2 - 7x + 10$

b. $x^2 - 9x + 14$

c. $x^2 + 2x - 8$

8. ⟨**5.3B**⟩ *Factor.*

a. $6x^2 - 6 + 5x$

b. $6x^2 - 1 + x$

c. $6x^2 - 5 + 13x$

9. ⟨**5.3B**⟩ *Factor.*

a. $6x^2 - 17xy + 5y^2$

b. $6x^2 - 7xy + 2y^2$

c. $6x^2 - 11xy + 4y^2$

10. ⟨**5.4B**⟩ *Factor.*

a. $x^2 + 4x + 4$

b. $x^2 + 10x + 25$

c. $x^2 + 8x + 16$

11. ⟨**5.4B**⟩ *Factor.*

a. $9x^2 + 12xy + 4y^2$

b. $9x^2 + 30xy + 25y^2$

c. $9x^2 + 24xy + 16y^2$

12. ⟨**5.4B**⟩ *Factor.*

a. $x^2 - 4x + 4$

b. $x^2 - 6x + 9$

c. $x^2 - 12x + 36$

13. ⟨**5.4B**⟩ *Factor.*

a. $4x^2 - 12xy + 9y^2$

b. $4x^2 - 20xy + 25y^2$

c. $4x^2 - 28xy + 49y^2$

14. ⟨**5.4C**⟩ *Factor.*

a. $x^2 - 36$

b. $x^2 - 49$

c. $x^2 - 81$

15. ⟨**5.4C**⟩ *Factor.*

a. $16x^2 - 81y^2$

b. $25x^2 - 64y^2$

c. $9x^2 - 100y^2$

16. ⟨**5.5A**⟩ *Factor.*

a. $m^3 + 125$

b. $n^3 + 64$

c. $y^3 + 8$

17. ⟨**5.5A**⟩ *Factor.*

a. $8y^3 - 27x^3$

b. $64y^3 - 125x^3$

c. $8m^3 - 125n^3$

18. ⟨**5.5B**⟩ *Factor.*

a. $3x^3 - 6x^2 + 27x$

b. $3x^3 - 6x^2 + 30x$

c. $4x^3 - 8x^2 + 32x$

19. ⟨**5.5B**⟩ *Factor.*

a. $2x^3 - 2x^2 - 4x$

b. $3x^3 - 6x^2 - 9x$

c. $4x^3 - 12x^2 - 16x$

20. ⟨**5.5B**⟩ *Factor.*

a. $2x^3 + 8x^2 + x + 4$

b. $2x^3 + 10x^2 + x + 5$

c. $2x^3 + 12x^2 + x + 6$

21. ⟨**5.5B**⟩ *Factor.*

a. $9kx^2 + 12kx + 4k$

b. $9kx^2 + 30kx + 25k$

c. $4kx^2 + 20kx + 25k$

22. ⟨**5.5D**⟩ *Factor.*

a. $-3x^4 + 27x^2$

b. $-4x^4 + 64x^2$

c. $-5x^4 + 20x^2$

23. ⟨**5.5D**⟩ *Factor.*

a. $-x^3 - y^3$

b. $-8m^3 - 27n^3$

c. $-64n^3 - m^3$

24. ⟨**5.5D**⟩ *Factor.*

a. $-y^3 + x^3$

b. $-8m^3 + 27n^3$

c. $-64t^3 + 125s^3$

25. ⟨**5.5D**⟩ *Factor.*

a. $-4x^2 - 12xy + 9y^2$

b. $-25x^2 - 30xy + 9y^2$

c. $-16x^2 - 24xy + 9y^2$

26. ⟨**5.6A**⟩ *Solve.*

a. $x^2 - 4x - 5 = 0$

b. $x^2 - 5x - 6 = 0$

c. $x^2 - 6x - 7 = 0$

27. ⟨**5.6A**⟩ *Solve.*

a. $2x^2 + x = 10$

b. $2x^2 + 3x = 5$

c. $2x^2 + x = 3$

28. ⟨**5.6A**⟩ *Solve.*

a. $(3x + 1)(x - 2) = 2(x - 1) - 4$

b. $(2x + 1)(x - 4) = 6(x - 4) - 1$

c. $(2x + 1)(x - 1) = 3(x + 2) - 1$

29. ⟨**5.7A**⟩ *Find the integers if the product of two consecutive even integers is:*

a. 4 more than 5 times the smaller of the integers

b. 4 more than twice the smaller of the integers

c. 10 more than 11 times the smaller of the integers

30. ⟨**5.7C**⟩ *The hypotenuse of a right triangle is 6 inches longer than the longest side of the triangle and 12 inches longer than the remaining side. What are the dimensions of the triangle?*

Practice Test Chapter 5

(Answers on page 483)

Visit www.mhhe.com/bello to view helpful videos that provide step-by-step solutions to several of the problems below.

1. Find the GCF of 40 and 60.

2. Find the GCF of $18x^2y^4$ and $30x^3y^5$.

3. Factor $10x^3 - 35x^5$.

4. Factor $\frac{4}{5}x^6 - \frac{3}{5}x^5 + \frac{2}{5}x^4 - \frac{1}{5}x^2$.

5. Factor $2x^3 + 6x^2y + x + 3y$.

6. Factor $x^2 - 8x + 12$.

7. Factor $6x^2 - 3 + 7x$.

8. Factor $6x^2 - 11xy + 3y^2$.

9. Factor $4x^2 + 12xy + 9y^2$.

10. Factor $x^2 - 14x + 49$.

11. Factor $9x^2 - 12xy + 4y^2$.

12. Factor $x^2 - 100$.

13. Factor $16x^2 - 25y^2$.

14. Factor $125t^3 + 27s^3$.

15. Factor $8y^3 - 125x^3$.

16. Factor $3x^3 - 6x^2 + 24x$.

17. Factor $2x^3 - 8x^2 - 10x$.

18. Factor $2x^3 + 6x^2 + x + 3$.

19. Factor $4kx^2 + 12kx + 9k$.

20. Factor $-9x^4 + 36x^2$.

21. Factor $-9x^2 - 24xy - 16y^2$.

22. Solve $x^2 - 3x - 10 = 0$.

23. Solve $2x^2 - x = 15$.

24. Solve $(2x - 3)(x - 4) = 2(x - 1) - 1$.

25. Solve $y(2y + 7) = -3$.

26. The product of two consecutive odd integers is 13 more than 10 times the larger of the two integers. Find the integers.

27. The product of two consecutive integers is 14 less than 10 times the smaller of the two integers. What are the integers?

28. The area of a rectangle is numerically 44 more than its perimeter. If the length of the rectangle is 8 inches more than its width, what are the dimensions of the rectangle?

29. A rectangular 10-inch television screen (measured diagonally) is 2 inches wider than it is high. What are the dimensions of the screen?

30. A car traveled 24 feet after the brakes were applied. How fast was the car going when the brakes were applied? ($b = 0.06v^2$)

Answers to Practice Test Chapter 5

Answer	If You Missed	Review		
	Question	Section	Examples	Page
1. 20	1	5.1	1	409–410
2. $6x^2y^4$	2	5.1	2	411
3. $5x^3(2 - 7x^2)$	3	5.1	3, 4	412–413
4. $\frac{1}{5}x^2(4x^4 - 3x^3 + 2x^2 - 1)$	4	5.1	5	413
5. $(x + 3y)(2x^2 + 1)$	5	5.1	6–8	414–415
6. $(x - 2)(x - 6)$	6	5.2	1–4	421–423
7. $(3x - 1)(2x + 3)$	7	5.3	2, 3, 5, 6	430–431, 433–434
8. $(3x - y)(2x - 3y)$	8	5.3	4, 7	431, 434
9. $(2x + 3y)^2$	9	5.4	2	441
10. $(x - 7)^2$	10	5.4	3a, b	441
11. $(3x - 2y)^2$	11	5.4	3c	441
12. $(x + 10)(x - 10)$	12	5.4	4a, b	442
13. $(4x + 5y)(4x - 5y)$	13	5.4	4c	442
14. $(5t + 3s)(25t^2 - 15st + 9s^2)$	14	5.5	1a, b	448
15. $(2y - 5x)(4y^2 + 10xy + 25x^2)$	15	5.5	1c, d	449
16. $3x(x^2 - 2x + 8)$	16	5.5	2	450
17. $2x(x + 1)(x - 5)$	17	5.5	2	450
18. $(x + 3)(2x^2 + 1)$	18	5.5	3	450
19. $k(2x + 3)^2$	19	5.5	4	450
20. $-9x^2(x + 2)(x - 2)$	20	5.5	8	453
21. $-(3x + 4y)^2$	21	5.5	8	453
22. $x = 5$ or $x = -2$	22	5.6	1	458
23. $x = 3$ or $x = -\frac{5}{2}$	23	5.6	2	459
24. $x = 5$ or $x = \frac{3}{2}$	24	5.6	3, 4	460
25. $y = -3$ or $y = -\frac{1}{2}$	25	5.6	5	460–461
26. 11 and 13 or −3 and −1	26	5.7	1	467
27. 7 and 8 or 2 and 3	27	5.7	1	467
28. 6 in. by 14 in.	28	5.7	2, 3	467–469
29. 6 in. by 8 in.	29	5.7	4	470
30. 20 mi/hr	30	5.7	5	471

Cumulative Review Chapters 1–5

1. Add: $-\frac{3}{8} + \left(-\frac{1}{6}\right)$

2. Subtract: $6.6 - (-9.8)$

3. Multiply: $(-5.5)(5.7)$

4. Find: $-(5^2)$

5. Divide: $-\frac{5}{6} \div \left(-\frac{5}{18}\right)$

6. Evaluate $y \div 5 \cdot x - z$ for $x = 6, y = 60, z = 3$.

7. Which property is illustrated by the following statement? $7 \cdot (6 \cdot 4) = (7 \cdot 6) \cdot 4$

8. Combine like terms: $-8xy^4 - (-9xy^4)$

9. Simplify: $2x - (x + 3) - 2(x + 4)$

10. Write in symbols: The quotient of $(d + 5e)$ and f

11. Solve for x: $5 = 3(x - 1) + 5 - 2x$

12. Solve for x: $\frac{x}{7} - \frac{x}{9} = 2$

13. Solve for x: $8 - \frac{x}{3} = \frac{6(x + 1)}{7}$

14. The sum of two numbers is 105. If one of the numbers is 25 more than the other, what are the numbers?

15. Train A leaves a station traveling at 50 miles per hour. Two hours later, train B leaves the same station traveling in the same direction at 60 miles per hour. How long does it take for train B to catch up to train A?

16. Susan purchased some municipal bonds yielding 11% annually and some certificates of deposit yielding 12% annually. If Susan's total investment amounts to \$6000 and the annual income is \$680, how much money is invested in bonds and how much is invested in certificates of deposit?

17. Graph: $-\frac{x}{6} + \frac{x}{2} \ge \frac{x - 2}{2}$

18. Graph the point $C(3, -3)$.

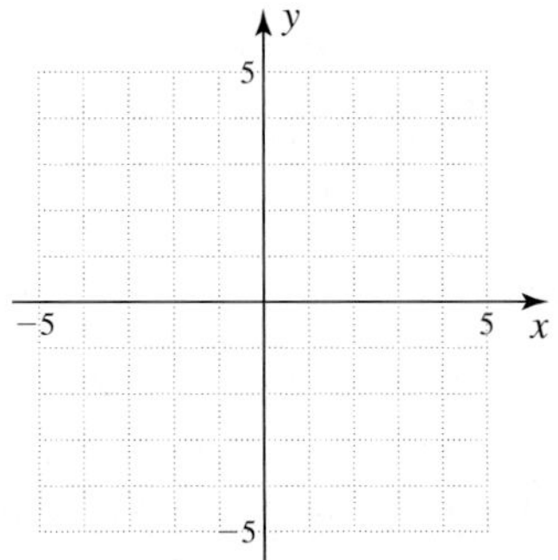

19. Determine whether the ordered pair $(-3, -3)$ is a solution of $5x - y = -18$.

20. Find x in the ordered pair $(x, 3)$ so that the ordered pair satisfies the equation $2x - 3y = -5$.

21. Graph: $x + y = 4$

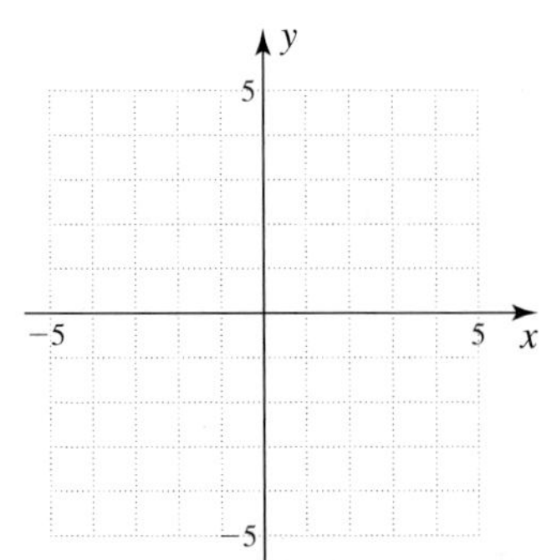

22. Graph: $4y - 20 = 0$

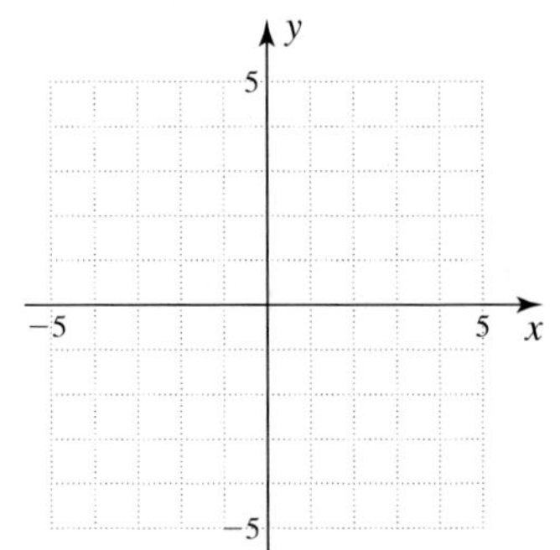

23. Find the slope of the line passing through the points $(7, 7)$ and $(-4, 5)$.

24. What is the slope of the line $6x - 2y = 15$?

25. Find the pair of parallel lines.

(1) $4y = x - 7$

(2) $7x - 28y = -7$

(3) $28y + 7x = -7$

26. Find an equation of the line passsing through $(-5, 3)$ and with slope 6.

27. Find the equation of the line with slope -3 and y intercept 4.

28. Graph $4x - 3y < -12$.

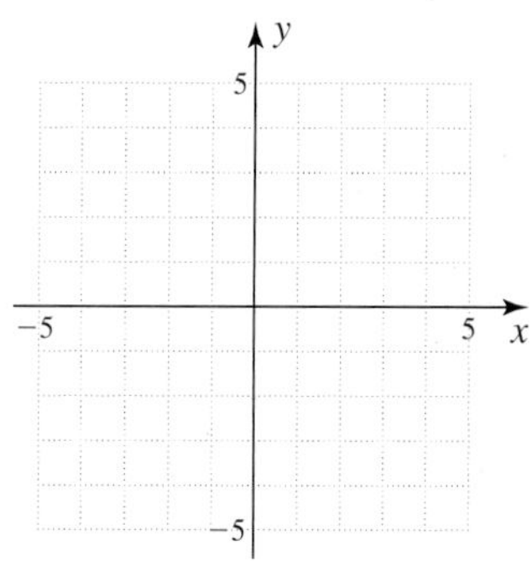

29. Graph $2x - y \geq 0$.

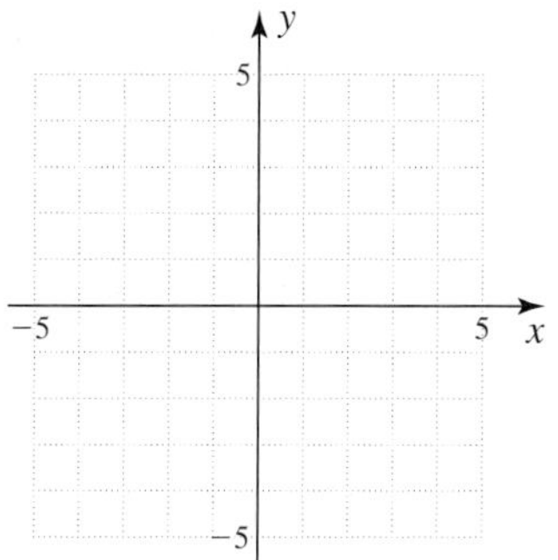

30. Simplify: $\dfrac{20x^6y^3}{-5x^4y^7}$

31. Simplify: $\dfrac{x^{-7}}{x^{-9}}$

32. Multiply and simplify: $x \cdot x^{-5}$

33. Simplify: $(3x^3y^{-3})^{-3}$

34. Write in scientific notation: 0.000048

35. Divide and express the answer in scientific notation: $(5.98 \times 10^{-3}) \div (1.3 \times 10^2)$

36. Find the degree of the polynomial: $6x^2 + x + 2$.

37. Find the value of $x^3 + 2x^2 + 1$ when $x = -2$.

38. Add $(-3x^2 - 2x^3 - 6)$ and $(5x^3 - 7 - 4x^2)$.

39. Simplify: $-4x^4(8x^2 + 4y)$

40. Expand: $(3x + 2y)^2$

41. Multiply: $(5x - 7y)(5x + 7y)$

42. Expand: $\left(2x^2 - \frac{1}{5}\right)^2$

43. Multiply: $(5x^2 + 9)(5x^2 - 9)$

44. Divide: $(2x^3 + x^2 - 5x - 9)$ by $(x - 2)$.

45. Factor completely: $12x^6 - 14x^9$

46. Factor completely: $\frac{4}{5}x^7 - \frac{3}{5}x^6 + \frac{4}{5}x^5 - \frac{2}{5}x^3$

47. Factor completely: $x^2 - 12x + 27$

48. Factor completely: $20x^2 - 23xy + 6y^2$

49. Factor completely: $25x^2 - 49y^2$

50. Factor completely: $-5x^4 + 80x^2$

51. Factor completely: $3x^3 - 6x^2 - 9x$

52. Factor completely: $2x^2 + 5x + 6x + 15$

53. Factor completely: $9kx^2 + 6kx + k$

54. Solve for x: $4x^2 + 17x = 15$

Section

Chapter

six

Rational Expressions

The Human Side of Algebra

The concept of whole number is one of the oldest in mathematics. The concept of rational numbers (so named because they are *ratios* of whole numbers) developed much later because nonliterate tribes had no need for such a concept. Rational numbers evolved over a long period of time, stimulated by the need for certain types of measurement. For example, take a rod of length 1 unit and cut it into two equal pieces. What is the length of each piece? One-half, of course. If the same rod is cut into four equal pieces, then each piece is of length $\frac{1}{4}$. Two of these pieces will have length $\frac{2}{4}$, which tells us that we should have $\frac{2}{4} = \frac{1}{2}$.

It was ideas such as these that led to the development of the arithmetic of the rational numbers.

During the Bronze Age, Egyptian hieroglyphic inscriptions show the reciprocals of integers by using an elongated oval sign. Thus, $\frac{1}{8}$ and $\frac{1}{20}$ were respectively written as

 and 

In this chapter, we generalize the concept of a rational number to that of a *rational expression*—that is, the quotient of two polynomials.

6.1 Building and Reducing Rational Expressions

Objectives

A Determine the values that make a rational expression undefined.

B Build fractions.

C Reduce (simplify) a rational expression to lowest terms.

To Succeed, Review How To . . .

1. Factor polynomials (pp. 420–421, 429–435, 439–442, 448–453).
2. Write a fraction with specified denominator (pp. 3–4).
3. Simplify fractions (pp. 4–6).

Getting Started

Recycling Waste and Rational Expressions

We've already mentioned that algebra is generalized arithmetic. In arithmetic, we study the natural numbers, the whole numbers, the integers, and the rational numbers. In algebra, we've studied expressions and polynomials, and we have also discussed how they follow rules similar to those used with the real numbers. We will find that rational expressions in algebra follow the same rules as rational numbers in arithmetic.

In arithmetic, a **rational number** is a number that can be written in the form $\frac{a}{b}$, where a and b are integers and b is not zero. As usual, a is called the *numerator* and b, the *denominator*. A similar approach is used in algebra.

In algebra, an expression of the form $\frac{A}{B}$, where A and B are *polynomials* and B is not zero, is called an *algebraic fraction,* or a *rational expression.* So if

$$G(t) = -0.027t^3 + 0.17t^2 + 2.3t + 239$$

is a polynomial representing the amount of waste generated in the United States (in millions of tons), $R(t) = 0.056t^2 + 1.8t + 69$ is a polynomial representing the amount of waste recovered (in millions of tons), and t is the number of years after 2000, then

$$\frac{R(t)}{G(t)}$$

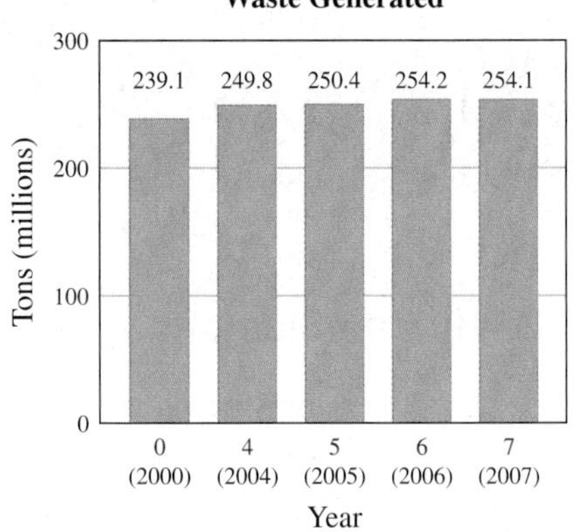

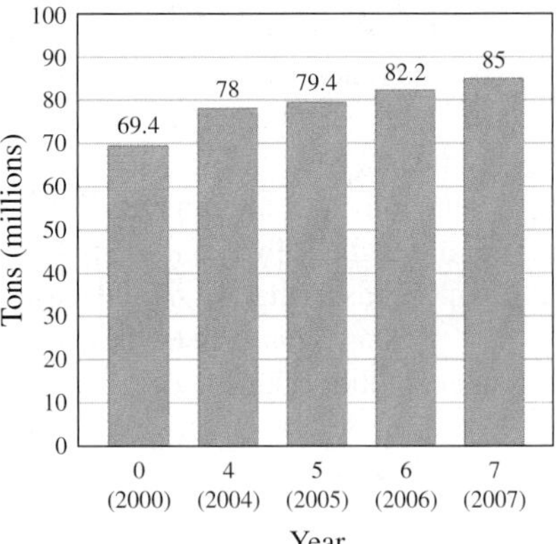

Source: http:/www.epa.gov/waste/nonhaz/municipal/pubs/msw07-rpt.pdf.

is a *rational expression* representing the *fraction* of the waste recovered in those years. Thus, in 2000 (when $t = 0$), the fraction is

$$\frac{R(0)}{G(0)} = \frac{0.056(0)^2 + 1.8(0) + 69}{-0.027(0)^3 + 0.17(0)^2 + 2.3(0) + 239} = \frac{69}{239}$$

which is about 29%. What percent of the waste will be recovered in the year 2010 when $t = 2010 - 2000$? To find the answer you have to calculate $\frac{R(10)}{G(10)}$.

In this section we shall learn how to determine when rational expressions are undefined and how to write them with a given denominator and then simplify them.

A > Undefined Values of Rational Expressions

DEFINITION OF RATIONAL EXPRESSION

An algebraic fraction or a rational expression is an expression of the form $\frac{A}{B}$ where *A* and *B* are *polynomials* and $B \neq 0$.

The expressions

$$\frac{8}{x}, \quad \frac{x^2 + 2x + 3}{x + 5}, \quad \frac{y}{y - 1}, \quad \text{and} \quad \frac{x^2 + 3x + 9}{x^2 - 4x + 4}$$

are rational expressions. Of course, since we don't want the denominators of these expressions to be zero, we must place some restrictions on these denominators. Let's see what these restrictions must be.

For $\frac{8}{x}$, x cannot be 0 ($x \neq 0$) because then we would have $\frac{8}{0}$, which is not defined. For

$$\frac{x^2 + 2x + 3}{x + 5}, \quad x \neq -5$$

If x is -5, $\frac{x^2 + 2x + 3}{x + 5} = \frac{25 - 10 + 3}{0}$, which is undefined.

For

$$\frac{y}{y - 1}, \quad y \neq 1$$

If y is 1, $\frac{y}{y - 1} = \frac{1}{0}$, which is undefined.

and for

$$\frac{x^2 + 3x + 9}{x^2 - 4x + 4} = \frac{x^2 + 3x + 9}{(x - 2)^2}, \quad x \neq 2$$

If x is 2, $\frac{x^2 + 3x + 9}{x^2 - 4x + 4} = \frac{4 + 6 + 9}{4 - 8 + 4} = \frac{19}{0}$—again, undefined.

To avoid stating repeatedly that the denominators of rational expressions must not be zero, we make the following rule.

RULE

Avoiding Zero Denominators

The variables in a rational expression must not be replaced by numbers that make the denominator zero.

To find the value or values that make the denominator zero:

1. Set the denominator equal to zero and

2. Solve the resulting equation for the variable.

For example, we found that

$$\frac{x^2 + 2x + 3}{x + 5}$$

is undefined for $x = -5$ by letting $x + 5 = 0$ and solving for x, to obtain $x = -5$. Similarly, we can find the values that make

$$\frac{1}{x^2 + 3x - 4}$$

undefined by letting

$$x^2 + 3x - 4 = 0$$

$$(x + 4)(x - 1) = 0 \quad \text{Factor.}$$

$$x + 4 = 0 \quad \text{or} \quad x - 1 = 0 \quad \text{Solve each equation.}$$

$$x = -4 \qquad\qquad x = 1$$

Thus,

$$\frac{1}{x^2 + 3x - 4}$$

is undefined when $x = -4$ or $x = 1$. (Check this out!)

EXAMPLE 1 **Finding values that make rational expressions undefined**

Find the values for which the rational expression is undefined:

a. $\dfrac{n}{2n - 3}$ **b.** $\dfrac{x + 1}{x^2 + 5x - 6}$ **c.** $\dfrac{m + 2}{m^2 + 2}$

SOLUTION 1

a. We set the denominator equal to zero (that makes the expression undefined) and solve:

$$2n - 3 = 0$$
$$2n = 3 \quad \text{Add 3.}$$
$$n = \frac{3}{2} \quad \text{Divide by 2.}$$

Thus,

$$\frac{n}{2n - 3}$$

is undefined for $n = \frac{3}{2}$.

b. We proceed as before and set the denominator equal to zero:

$$x^2 + 5x - 6 = 0$$
$$(x + 6)(x - 1) = 0 \quad \text{Factor.}$$
$$x + 6 = 0 \quad \text{or} \quad x - 1 = 0 \quad \text{Solve each equation.}$$
$$x = -6 \qquad\qquad x = 1$$

Thus,

$$\frac{x + 1}{x^2 + 5x - 6}$$

is undefined for $x = -6$ or $x = 1$.

c. Setting the denominator equal to zero, we have

$$m^2 + 2 = 0$$

so that $m^2 = -2$. But the square of any real number is not negative, so there are no real numbers m for which the denominator is zero. (Note that m^2 is greater than or equal to zero, so $m^2 + 2$ is always positive.) There are *no values* for which

$$\frac{m + 2}{m^2 + 2}$$

is undefined.

PROBLEM 1

Find values for which the rational expression is undefined:

a. $\dfrac{m}{2m + 3}$ **b.** $\dfrac{y + 1}{y^2 + 4y - 5}$

c. $\dfrac{n + 3}{n^2 + 3}$

B › Building Fractions

Now that we know that we must avoid zero denominators, recall what we did with fractions in arithmetic. First, we learned how to recognize which ones are equal, and then we used this idea to reduce or build fractions. Let's talk about equality first.

Answers to PROBLEMS

1. a. $m = -\frac{3}{2}$

b. $y = -5$ or $y = 1$

c. No values

What does this picture tell you (if you write the value of the coins using fractions)? It states that

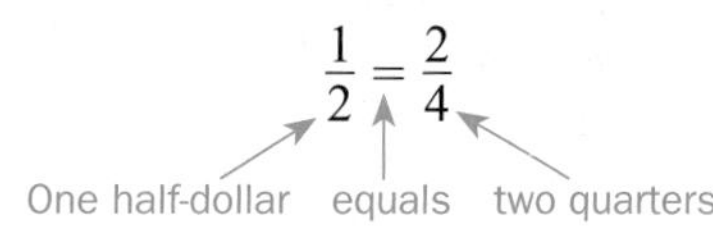

As a matter of fact, the following is also true:

$$\frac{1}{2} = \frac{2}{4} = \frac{3}{6} = \frac{4}{8} = \frac{x}{2x} = \frac{x^2}{2x^2}$$

and so on. Do you see the pattern? Here is another way to write it:

$$\frac{1}{2} = \frac{1 \cdot 2}{2 \cdot 2} = \frac{2}{4} \qquad \text{Note that } \frac{2}{2} = 1.$$

$$\frac{1}{2} = \frac{1 \cdot 3}{2 \cdot 3} = \frac{3}{6} \qquad \text{Here, } \frac{3}{3} = 1.$$

$$\frac{1}{2} = \frac{1 \cdot 4}{2 \cdot 4} = \frac{4}{8}$$

$$\frac{1}{2} = \frac{1 \cdot x}{2 \cdot x} = \frac{x}{2x}$$

$$\frac{1}{2} = \frac{1 \cdot x^2}{2 \cdot x^2} = \frac{x^2}{2x^2}$$

We can always obtain a rational expression that is equivalent to another rational expression $\frac{A}{B}$ by multiplying the numerator and denominator of the given rational expression by the same nonzero number or expression, C. Here is this rule stated in symbols.

RULE

Fundamental Rule of Rational Expressions

$$\frac{A}{B} = \frac{A \cdot C}{B \cdot C} \qquad (B \neq 0, C \neq 0)$$

Note that

$$\frac{A}{B} = \frac{A \cdot C}{B \cdot C} \quad \text{because} \quad \frac{C}{C} = \mathbf{1}$$

and multiplying by **1** does not change the value of the expression.

This idea is very important for adding or subtracting rational expressions (Section 6.3). For example, to add $\frac{1}{2}$ and $\frac{3}{4}$, we write

$$\begin{array}{r} \frac{1}{2} = \frac{2}{4} \\ +\frac{3}{4} = \frac{3}{4} \\ \hline \frac{5}{4} \end{array} \qquad \text{Note that } \frac{1}{2} = \frac{1 \cdot 2}{2 \cdot 2} = \frac{2}{4}.$$

Here we've used the fundamental rule of rational expressions to write the $\frac{1}{2}$ as an equivalent fraction with a denominator of 4, namely, $\frac{2}{4}$. Now, suppose we wish to write $\frac{3}{8}$ with a denominator of 16. How do we do this? First, we write the problem as

$$\frac{3}{8} = \frac{?}{16} \qquad \text{Note that } 16 = 8 \cdot 2.$$

Multiply by 2.

and notice that to get 16 as the denominator, we need to multiply the 8 by 2. Of course, we must do the same to the numerator 3; we obtain

Multiply by 2.

$$\frac{3}{8} = \frac{6}{16}$$

By the fundamental rule of rational expressions, if we multiply the denominator by 2, we must multiply the numerator by 2.

Note that $\frac{3 \cdot 2}{8 \cdot 2} = \frac{6}{16}$.

Similarly, to write

$$\frac{5x}{3y}$$

with a denominator of $6y^3$, we first write the new equivalent expression

$$\frac{?}{6y^3}$$

with the old denominator $3y$ factored out:

$$\frac{5x}{3y} = \frac{?}{6y^3} = \frac{?}{3y(2y^2)}$$

Write $6y^3$ as $3y(2y^2)$.

Multiply by $2y^2$.

Since the multiplier is $2y^2$, we have

Multiply by $2y^2$.

$$\frac{5x}{3y} = \frac{5x(2y^2)}{3y(2y^2)} = \frac{10xy^2}{6y^3}$$

Thus,

$$\frac{5x}{3y} = \frac{10xy^2}{6y^3}$$

We multiplied the denominator by $2y^2$, so we have to multiply the numerator by $2y^2$.

EXAMPLE 2 Rewriting rational expressions with a specified denominator

Write:

a. $\frac{5}{6}$ with a denominator of 18

b. $\frac{2x}{9y^2}$ with a denominator of $18y^3$

c. $\frac{3x}{x-1}$ with a denominator of $x^2 + 2x - 3$

SOLUTION 2

a. $\frac{5}{6} = \frac{?}{18}$

Multiply by 3.

Multiply by 3.

$$\frac{5}{6} = \frac{15}{18}$$

Note that $\frac{5 \cdot 3}{6 \cdot 3} = \frac{15}{18}$.

PROBLEM 2

Write:

a. $\frac{4}{7}$ with a denominator of 14

b. $\frac{2y}{7x^2}$ with a denominator of $14x^3$

c. $\frac{2x}{x+2}$ with a denominator of $x^2 + x - 2$

Answers to PROBLEMS

2. a. $\frac{8}{14}$ **b.** $\frac{4xy}{14x^3}$ **c.** $\frac{2x^2 - 2x}{x^2 + x - 2}$

b. Since $18y^3 = 9y^2(2y)$,

$$\frac{2x}{9y^2} = \frac{?}{9y^2(2y)}$$

Multiply by 2y.

Multiply by 2y.

$$\frac{2x}{9y^2} = \frac{2x(2y)}{9y^2(2y)} = \frac{4xy}{18y^3}$$

c. We first note that $x^2 + 2x - 3 = (x - 1)(x + 3)$. Thus,

$$\frac{3x}{x-1} = \frac{?}{(x-1)(x+3)}$$

Multiply by (x + 3).

Multiply by (x + 3).

$$\frac{3x}{x-1} = \frac{3x(x+3)}{(x-1)(x+3)} = \frac{3x^2+9x}{x^2+2x-3}$$

C > Reducing Rational Expressions

Now that we know how to build up rational expressions, we are now ready to use the reverse process—that is, to **reduce** them. In Example 2(a), we wrote $\frac{5}{6}$ with a denominator of 18; that is, we found out that

$$\frac{5}{6} = \frac{5 \cdot 3}{6 \cdot 3} = \frac{15}{18}$$

Of course, you will probably agree that $\frac{5}{6}$ is written in a "simpler" form than $\frac{15}{18}$. Certainly you will eventually agree—though it is hard to see at first glance—that

$$\frac{5}{3} = \frac{5(x+3)(x^2-4)}{3(x+2)(x^2+x-6)}$$

and that $\frac{5}{3}$ is the "simpler" of the two expressions. In algebra, the process of removing all factors common to the numerator and denominator is called **reducing to lowest terms,** or **simplifying** the expression. How do we reduce the rational expression

$$\frac{5(x+3)(x^2-4)}{3(x+2)(x^2+x-6)}$$

to lowest terms? We proceed by steps.

PROCEDURE

Reducing Rational Expressions to Lowest Terms (Simplifying)

1. Write the numerator and denominator of the expression in factored form.
2. Find the factors that are common to the numerator and denominator.
3. Replace the quotient of the common factors by the number 1, since $\frac{a}{a} = 1$.
4. Rewrite the expression in simplified form.

Let's use this procedure to reduce

$$\frac{5(x+3)(x^2-4)}{3(x+2)(x^2+x-6)}$$

to lowest terms:

1. Write the numerator and denominator in factored form. $\frac{5(x+3)(x+2)(x-2)}{3(x+2)(x+3)(x-2)}$
2. Find the factors that are common to the numerator and denominator (we rearranged them so the common factors are in columns). $\frac{5(x+2)(x+3)(x-2)}{3(x+2)(x+3)(x-2)}$
3. Replace the quotient of the common factors by the number 1. Remember, $\frac{a}{a}=1$. $\frac{5\cancel{(x+2)}^{1}\cancel{(x+3)}^{1}\cancel{(x-2)}^{1}}{3\cancel{(x+2)}\cancel{(x+3)}\cancel{(x-2)}}$
4. Rewrite the expression in simplified form. $\frac{5}{3}$

The whole procedure can be written as

$$\frac{5(x+3)(x^2-4)}{3(x+2)(x^2+x-6)} = \frac{5\cancel{(x+3)}^{1}\cancel{(x+2)}^{1}\cancel{(x-2)}^{1}}{3\cancel{(x+2)}\cancel{(x+3)}\cancel{(x-2)}} = \frac{5}{3}$$

EXAMPLE 3 Reducing rational expressions to lowest terms

Simplify:

a. $\frac{-5x^2y}{15xy^3}$ b. $\frac{-8(x^2-y^2)}{-4(x+y)}$

SOLUTION 3

a. We use our four-step procedure.

1. Write in factored form. $\frac{-5x^2y}{15xy^3} = \frac{(-1)\cdot 5\cdot x\cdot x\cdot y}{3\cdot 5\cdot x\cdot y\cdot y\cdot y}$
2. Find the common factors. $= \frac{5\cdot x\cdot y(-1)\cdot x}{5\cdot x\cdot y\cdot 3\cdot y\cdot y}$
3. Replace the quotient of the common factors by 1. Remember, $\frac{a}{a}=1$. $= \frac{1\cdot(-1)\cdot x}{3\cdot y\cdot y}$
4. Rewrite in simplified form. $= \frac{-x}{3y^2}$

The whole process can be written as

$$\frac{-5x^2y}{15xy^3} = -\frac{\overset{-1x}{\cancel{5x^2y}}}{\underset{3\ \ y^2}{\cancel{15xy^3}}} = \frac{-x}{3y^2}$$

b. We again use our four-step procedure.

1. Write the fractions in factored form. $\frac{-8(x^2-y^2)}{-4(x+y)} = \frac{(-1)\cdot 2\cdot 2\cdot 2(x+y)(x-y)}{(-1)\cdot 2\cdot 2(x+y)}$
2. Find the common factors. $= \frac{(-1)\cdot 2\cdot 2\cdot (x+y)(x-y)\cdot 2}{(-1)\cdot 2\cdot 2\cdot (x+y)}$

PROBLEM 3

Simplify:

a. $\frac{-3x^2y}{6xy^4}$

b. $\frac{-6(m^2-n^2)}{-3(m-n)}$

Answers to PROBLEMS

3. a. $\frac{-x}{2y^3}$ **b.** $2(m+n)$

3. Replace the quotient of the common factors by 1. $= 1 \cdot (x - y) \cdot 2$

4. Rewrite in simplified form. $= 2(x - y)$

The abbreviated form is

$$\frac{-8(x^2 - y^2)}{-4(x + y)} = \frac{\overset{2}{\cancel{-8}}\overset{1}{\cancel{(x + y)}}(x - y)}{\cancel{-4}\cancel{(x + y)}} = 2(x - y)$$

You may have noticed that in Example 3(a) we wrote the answer as

$$\frac{-x}{3y^2}$$

It could be argued that since

$$\frac{-5}{15} = -\frac{1}{3}$$

the answer should be

$$-\frac{x}{3y^2}$$

However, to avoid confusion, we agree to write

$$-\frac{x}{3y^2} \quad \text{as} \quad \frac{-x}{3y^2}$$

with the *negative* sign in the numerator. In general, since a fraction has **three** possible signs (numerator, denominator, and the sign of the fraction itself), we use the following conventions.

FORMS OF A FRACTION

$-\frac{a}{b}$ is written as $\frac{-a}{b}$ $\qquad$ $\frac{-a}{-b}$ is written as $\frac{a}{b}$

$\frac{a}{-b}$ is written as $\frac{-a}{b}$ $\qquad$ $-\frac{-a}{b}$ is written as $\frac{a}{b}$

$-\frac{-a}{-b}$ is written as $\frac{-a}{b}$ $\qquad$ $-\frac{a}{-b}$ is written as $\frac{a}{b}$

STANDARD FORM OF A FRACTION

The forms

$$\frac{a}{b} \quad \text{and} \quad \frac{-a}{b}$$

are called the **standard forms** of a fraction.

The forms $\frac{a}{b}$ and $\frac{-a}{b}$ are the preferred forms to write answers involving fractions, as we shall see in Example 4.

EXAMPLE 4 Reducing rational expressions to lowest terms

Simplify:

a. $\dfrac{6x - 12y}{18x - 36y}$ b. $-\dfrac{y}{y + xy}$ c. $\dfrac{x + 3}{-(x^2 - 9)}$

SOLUTION 4

a. $\dfrac{6x - 12y}{18x - 36y} = \dfrac{6(x - 2y)}{18(x - 2y)}$

$= \dfrac{\overset{1}{\cancel{6}}\overset{1}{\cancel{(x - 2y)}}}{\underset{3}{\cancel{18}}\underset{1}{\cancel{(x - 2y)}}}$

$= \dfrac{1}{3}$

b. $-\dfrac{y}{y + xy} = -\dfrac{y}{y(1 + x)}$

$= -\dfrac{\overset{1}{\cancel{y}}}{\cancel{y}(1 + x)}$

$= \dfrac{-1}{1 + x}$

c. $\dfrac{x + 3}{-(x^2 - 9)} = \dfrac{(x + 3)}{-(x + 3)(x - 3)}$

$= \dfrac{\overset{1}{\cancel{(x + 3)}}}{-\cancel{(x + 3)}(x - 3)}$

$= \dfrac{1}{-(x - 3)}$

$= \dfrac{-1}{x - 3}$ In standard form

PROBLEM 4

Simplify:

a. $\dfrac{3x - 12y}{12x - 48y}$

b. $-\dfrac{x}{x + xy}$

c. $\dfrac{y - 3}{-(y^2 - 9)}$

Is there a way in which

$$\frac{-1}{x - 3}$$

can be simplified further? The answer is yes. See whether you can see the reasons behind each step:

$$\frac{-1}{x - 3} = \frac{-1}{-(3 - x)} \qquad -(3 - x) = -3 + x = x - 3$$

$$= \frac{1}{3 - x}$$

Thus,

$$\frac{-1}{x - 3} = \frac{1}{3 - x}$$

Note that $\frac{-1}{x - 3}$ involves finding the additive inverse of 1 in the numerator and subtracting 3 from x in the denominator, whereas $\frac{1}{3 - x}$ only involves subtracting x from 3 in the denominator. Less work!

And why, you might ask, is this answer simpler than the other? Because

$$\frac{-1}{x - 3}$$

Answers to PROBLEMS

4. a. $\frac{1}{4}$ b. $\frac{-1}{1 + y}$ c. $\frac{-1}{y + 3}$

is a quotient involving a subtraction $(x - 3)$ and an additive inverse (-1), but

$$\frac{1}{3 - x}$$

is a quotient that involves only a subtraction $(3 - x)$. Be aware of these simplifications when writing your answers!

Here is a situation that occurs very frequently in algebra. Can you reduce this expression?

$$\frac{a - b}{b - a}$$

Look at the following steps. Note that the original denominator, $b - a$, can be written as $-(a - b)$ since $-(a - b) = -a + b = b - a$. Thus, $b - a = -(a - b)$.

$$\frac{a - b}{b - a} = \frac{\overset{1}{\cancel{(a - b)}}}{-\cancel{(a - b)}} \quad \text{Write } b - a \text{ as } -(a - b) \text{ in the denominator.}$$

$$= \frac{1}{-1}$$

$$= -1$$

QUOTIENT OF ADDITIVE INVERSES

For any real numbers a and b, where $b - a \neq 0$,

$$\frac{a - b}{b - a} = -1$$

This means that a fraction whose numerator and denominator are additive inverses (such as $a - b$ and $b - a$) equals -1. We use this idea in Example 5.

EXAMPLE 5 Reducing rational expressions to lowest terms

Reduce to lowest terms:

a. $\frac{x^2 - 16}{4 + x}$ b. $-\frac{x^2 - 16}{x + 4}$ c. $\frac{x^2 + 2x - 15}{3 - x}$

SOLUTION 5

a. $\frac{x^2 - 16}{4 + x} = \frac{\overset{1}{\cancel{(x + 4)}}(x - 4)}{\cancel{x + 4}}$ $\quad 4 + x = x + 4$ by the commutative property.

$= x - 4$

b. $-\frac{x^2 - 16}{x + 4} = -\frac{(x + 4)(x - 4)}{(x + 4)}$

$= -\frac{\overset{1}{\cancel{(x + 4)}}(x - 4)}{\cancel{(x + 4)}}$ $\quad$ Note that $\frac{x + 4}{x + 4} = 1$.

$= -(x - 4)$

$= -x + 4$

$= 4 - x$

c. $\frac{x^2 + 2x - 15}{3 - x} = \frac{(x + 5)(x - 3)}{3 - x}$

$= \frac{\overset{-1}{\cancel{(x - 3)}}(x + 5)}{\cancel{(3 - x)}}$ $\quad$ Note that $\frac{x - 3}{3 - x} = -1$.

$= -(x + 5)$

PROBLEM 5

Reduce to lowest terms:

a. $\frac{y^2 - 9}{3 + y}$

b. $\frac{y^2 - 9}{y + 3}$

c. $\frac{y^2 + y - 12}{3 - y}$

(continued)

Answers to PROBLEMS

5. a. $y - 3$ **b.** $y - 3$

c. $-(y + 4)$

We have used the fact that

$$\frac{a-b}{b-a} = -1$$

in the second step to simplify

$$\frac{(x-3)}{(3-x)}$$

Note that the final answer is written as $-(x + 5)$ instead of $-x - 5$, since $-(x + 5)$ is considered to be simpler.

EXAMPLE 6 Global Warming and Gas Emissions

Are you planning to buy a new car? Which model would be most beneficial to the environment? The one that produces the least gas emissions: the Escape hybrid. If the initial 4 tons of gas emissions increases at 1% each year, after 3 years the total amount of gases emitted is equal to $\frac{4x^3 - 4x}{x - 1}$, $x = 1 + P = 1.01$.

a. Reduce $\frac{4x^3 - 4}{x - 1}$ to lowest terms.

b. What is the total amount of gases emitted for the 3-year period?

SOLUTION 6

a.

$$\frac{4x^3 - 4}{x - 1} = \frac{4(x^3 - 1)}{x - 1} \quad \text{Factor out the GCF 4.}$$

$$= \frac{4(x-1)(x^2 + x + 1)}{x - 1} \quad \text{Factor } x^3 - 1.$$

$$= \frac{4\cancel{(x-1)}(x^2 + x + 1)}{\cancel{x-1}} \quad \text{Note that } \frac{(x-1)}{x-1} = 1.$$

$$= 4(x^2 + x + 1)$$

b. We let $x = 1.01$ in $4(x^2 + x + 1)$ obtaining $4[(1.01)^2 + (1.01) + 1]$ or 12.1204 tons for the 3-year period.

PROBLEM 6

For a 2-year period, the emissions for the Escape hybrid will be $\frac{4x^2 - 4}{x - 1}$.

a. Reduce $\frac{4x^2 - 4}{x - 1}$ to lowest terms.

b. What is the total amount of gases emitted for the 2-year period?

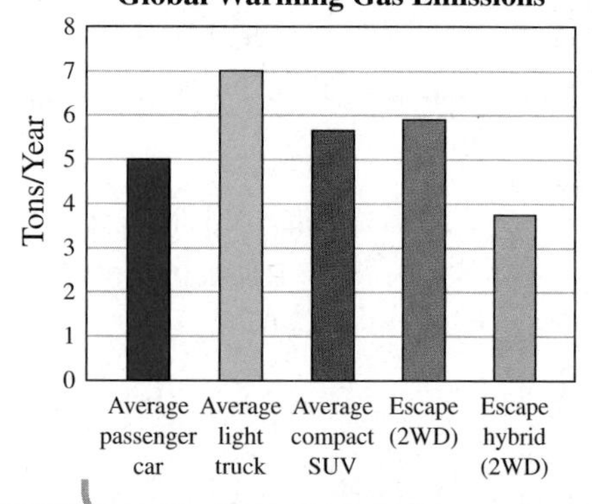

Calculator Corner

Checking Quotients

How can you check that your answers are correct? If you have a calculator, you can do it two ways:

1. *By looking at the graph* If the graph of the original problem and the graph of the answer coincide, the answer is probably correct.

   Thus, to check the results of Example 5(c), enter

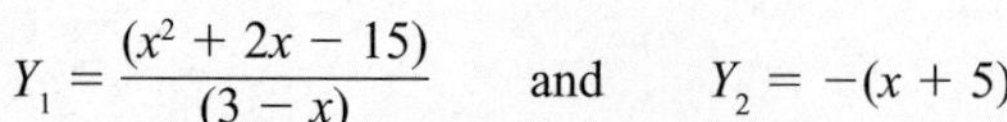

$$Y_1 = \frac{(x^2 + 2x - 15)}{(3 - x)} \quad \text{and} \quad Y_2 = -(x + 5)$$

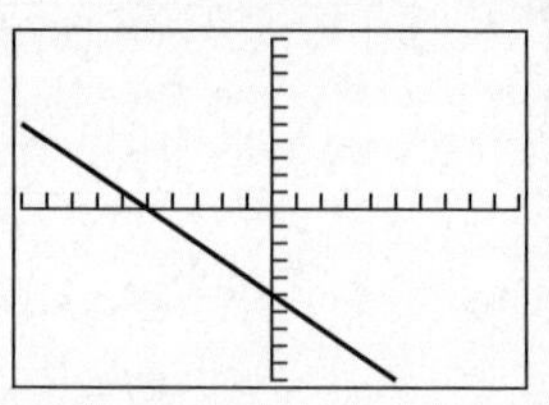

   Note the parentheses in the numerator and denominator of Y_1, and remember that the "−" in $-(x + 5)$ has to be entered with the (−) key.

2. *By looking at the numerical values of Y_1 and Y_2* (Press 2nd GRAPH.) The graph and the numerical values are as shown.

   One more thing: When you are looking at the table of values, what happens when $x = 3$? Why?

X	Y1	Y2
0	-5	-5
1	-6	-6
2	-7	-7
3	ERROR	-8
4	-9	-9
5	-10	-10
6	-11	-11
X=3		

Answers to PROBLEMS

6. a. $4(x + 1)$ **b.** 8.04 tons

Exercises 6.1

A Undefined Values of Rational Expressions In Problems 1–20, find the value(s) for which the rational expression is undefined.

1. $\frac{x}{x-7}$ **2.** $\frac{3y}{2y+9}$ **3.** $\frac{y-5}{2y+8}$

4. $\frac{y+4}{3y-9}$ **5.** $\frac{x+9}{x^2-9}$ **6.** $\frac{y+4}{y^2-16}$

7. $\frac{y+3}{y^2+5}$ **8.** $\frac{y+4}{y^2+1}$ **9.** $\frac{2y+9}{y^2-6y+8}$

10. $\frac{3x+5}{x^2-7x+6}$ **11.** $\frac{x^2-4}{x^3+8}$ **12.** $\frac{y^2-9}{y^3+1}$

13. $\frac{x^2+4}{x^3-27}$ **14.** $\frac{x^2+16}{x^3-8}$ **15.** $\frac{x-1}{x^2+6x+9}$

16. $\frac{x-4}{x^2+8x+16}$ **17.** $\frac{y-3}{y^2-6y-16}$ **18.** $\frac{y-5}{y^2-10y+25}$

19. $\frac{x+4}{x^3+2x^2+x}$ **20.** $\frac{x+5}{x^3+4x^2+4x}$

B Building Fractions In Problems 21–32, write the given fraction as an equivalent one with the indicated denominator.

21. $\frac{3}{7}$ with a denominator of 21 **22.** $\frac{5}{9}$ with a denominator of 36

23. $-\frac{8}{11}$ with a denominator of 22 **24.** $\frac{-5}{17}$ with a denominator of 51

25. $\frac{5x}{6y^2}$ with a denominator of $24y^3$ **26.** $\frac{7y}{5x^3}$ with a denominator of $10x^4$

27. $\frac{-3x}{7y}$ with a denominator of $21y^4$ **28.** $\frac{-4y}{7x^2}$ with a denominator of $28x^3$

29. $\frac{4x}{x+1}$ with a denominator of x^2-x-2 **30.** $\frac{5y}{y-1}$ with a denominator of y^2+2y-3

31. $\frac{-5x}{x+3}$ with a denominator of x^2+x-6 **32.** $\frac{-3y}{y-4}$ with a denominator of y^2-2y-8

C Reducing Rational Expressions In Problems 33–70, reduce to lowest terms (simplify).

33. $\frac{7x^3y}{14xy^4}$ **34.** $\frac{24xy^3}{6x^3y}$ **35.** $\frac{-9xy^5}{3x^2y}$

36. $\frac{-24x^3y^2}{48xy^4}$ **37.** $\frac{-6x^2y}{-12x^3y^4}$ **38.** $\frac{-9xy^4}{-18x^5y}$

39. $\frac{-25x^3y^2}{-5x^2y^4}$ **40.** $\frac{-30x^2y^2}{-6x^3y^5}$ **41.** $\frac{6(x^2-y^2)}{18(x+y)}$

42. $\frac{12(x^2-y^2)}{48(x+y)}$ **43.** $\frac{-9(x^2-y^2)}{3(x+y)}$ **44.** $\frac{-12(x^2-y^2)}{3(x-y)}$

45. $\frac{-6(x+y)}{24(x^2-y^2)}$ **46.** $\frac{-8(x+3)}{40(x^2-9)}$ **47.** $\frac{-5(x-2)}{-10(x^2-4)}$

Web IT go to mhhe.com/bello for more lessons

Web IT go to mhhe.com/bello for more lessons

48. $\frac{-12(x-2)}{-60(x^2-4)}$

49. $\frac{-3(x-y)}{-3(x^2-y^2)}$

50. $\frac{-10(x+y)}{-10(x^2-y^2)}$

51. $\frac{4x-4y}{8x-8y}$

52. $\frac{6x+6y}{2x+2y}$

53. $\frac{4x-8y}{12x-24y}$

54. $\frac{15y-45x}{5y-15x}$

55. $-\frac{6}{6+12y}$

56. $-\frac{4}{8+12x}$

57. $-\frac{x}{x+2xy}$

58. $-\frac{y}{2y+6xy}$

59. $-\frac{6y}{6xy+12y}$

60. $-\frac{4x}{8xy+16x}$

61. $\frac{3x-2y}{2y-3x}$

62. $\frac{5y-2x}{2x-5y}$

63. $\frac{x^2+4x-5}{1-x}$

64. $\frac{x^2-2x-15}{5-x}$

65. $\frac{x^2-6x+8}{4-x}$

66. $\frac{x^2-8x+15}{3-x}$

67. $\frac{2-x}{x^2+4x-12}$

68. $\frac{3-x}{x^2+3x-18}$

69. $-\frac{3-x}{x^2-5x+6}$

70. $-\frac{4-x}{x^2-3x-4}$

Applications

71. *Annual advertising expenditures* How much is spent annually on advertising? The total amount is given by the polynomial $S(t) = -0.3t^2 + 10t + 50$ (millions), where t is the number of years after 1980. The amounts spent on national and local advertisement (in millions) are given by the respective polynomials

$$N(t) = -0.13t^2 + 5t + 30$$

and

$$L(t) = -0.17t^2 + 5t + 20$$

a. What was the total amount spent on advertising in 1980 ($t = 0$)? In 2000? In 2010?

b. What amount was spent on national advertising in 1980? In 2000? In 2010?

c. What amount was spent on local advertising in 1980? In 2000? In 2010?

d. What does the rational expression

$$\frac{N(t)}{S(t)}$$

represent?

72. *Annual advertising expenditures* Use the information given in Problem 71 to answer these questions:

a. What percent of the total amount spent in 1980 was for national advertising? In 2000?

b. What percent of the total amount spent in 1980 was for local advertising? In 2000?

c. What rational expression represents the percent spent for national advertising?

d. What percent of all advertising would be spent on national advertising in the year 2010?

73. *Expenditures for television advertising* The amount spent annually on television advertising can be approximated by the polynomial $T(t) = -0.07t^2 + 2t + 11$ (millions), where t is the number of years after 1980. Use the information in Problem 71 to find what percent of the total amount spent annually on advertising would be

a. Spent on television in the year 2010.

b. Spent on local advertising in the year 2010.

c. What does the rational fraction

$$\frac{T(t)}{S(t)}$$

represent?

74. *Television spot advertising* The estimated fees for spot advertising on television can be approximated by the polynomial $C(t) = 0.04t^3 - t^2 + 6t + 2.5$ (billions), where t is the number of years after 1980. Of this amount, automotive spot advertising fees can be approximated by the polynomial $A(t) = 0.02t^3 - 0.5t^2 + 3t + 0.3$ (billions).

a. What was the total amount spent on TV spot advertising in 1980?

b. What was the amount spent on automotive TV spot advertising in 1980?

c. What was the total amount spent on TV spot advertising in 1990? In 2000? What will it be in 2010?

d. What was the amount spent on automotive TV spot advertising in 1990? In 2000? What will it be in 2010?

e. What does the rational fraction

$$\frac{A(t)}{C(t)}$$

represent?

>>> Applications: Green Math

75. *Trenton, New Jersey, landfill* How do we estimate the cost of reducing environmental pollution? The rational expression $\frac{2.25x}{100 - x}$ gives the cost (in millions) of cleaning x percent (x a whole number) of a contaminated landfill.

a. What is the cost of cleaning 95% of the landfill?

b. What would the cost be to clean 98% of the landfill contamination?

c. For what number is the expression undefined and what does it mean?

Source: Trenton Legal Newsline; http://tinyurl.com/ykpbw38.

76. *Reducing nitrogen oxide emissions* In 2010, TECO energy will renovate one of its coal-fired plants to reduce nitrogen oxide emissions by 85% from levels recorded in 1998, making the plant one of the cleanest coal-fired plants in the nation. The cost (in millions) is given by $\frac{58x}{100 - x}$, where x is the percent of emissions removed.

a. What is the cost for removing 85% of the nitrogen oxide emissions?

b. What happens to the cost as x gets closer to 100?

Source: http://en.wikipedia.org/wiki/TECO_Energy.

77. *H1N1 Vaccine administration* The World Health Organization (WHO) is donating 200 million doses of the H1N1 flu vaccine so that member nations can vaccinate 10% of their population. The cost (in millions) of the vaccinations is given by the rational expression $\frac{3600x}{1 - x}$, where x is the percent of people being vaccinated.

a. Use the expression to find the cost.

b. Find the cost if they decide to double the percent of the population to be vaccinated to 20%.

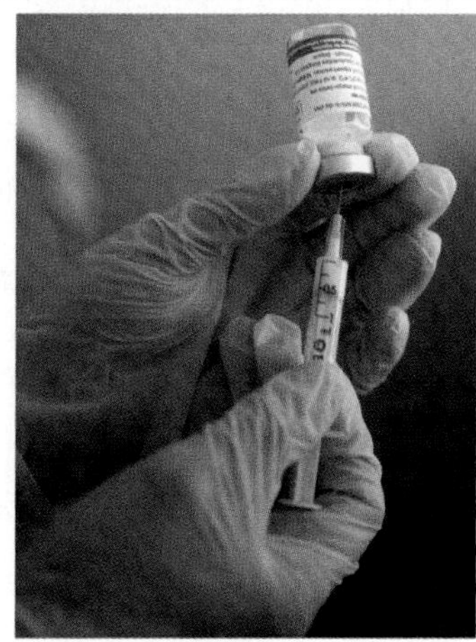

Source: CNN International.

U.S. population As of October 2006 the U.S. population reached 300 million. What happens after that? We can project the U.S. population by using the polynomial $P(t) = 2.8t + 281$ (million), where t is the number of years after 2000. The Census Bureau uses the projections shown in the graph.

Total Population and Older Population: United States, 1950–2050

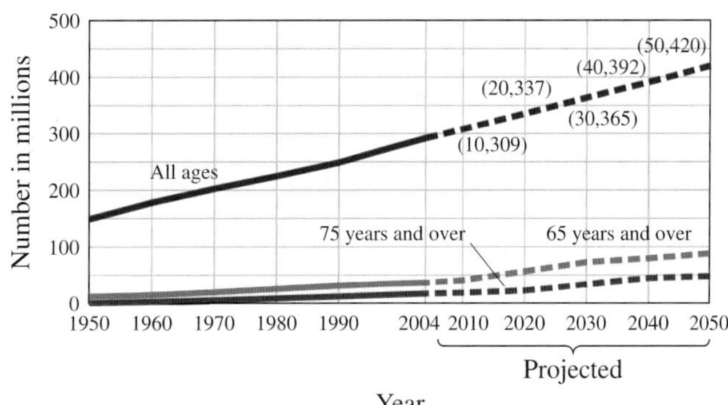

78. Use $P(t) = 2.8t + 281$ (million), where t is the number of years after 2000 to find the projected population in the year:

a. 2010 **b.** 2020

c. 2030 **d.** 2040

e. 2050

f. Are the results close to those shown on the graph?

79. The polynomial $S(t) = 1.1t + 33$ (million), where t is the number of years after 2000, projects the population over age 65 (65+).

a. Write the rational expression $\frac{S(t)}{P(t)}$.

b. Use the rational expression obtained in part **a** to project, to the nearest percent, the percent of the 65+ population in the year 2000 and in the year 2050.

c. Is the percent of the 65+ population increasing or decreasing?

80. The polynomial $T(t) = 0.13t + 20$ (million), where t is the number of years after 2000, projects the population of persons 20–24 (T).

a. Write the rational expression $\frac{T(t)}{P(t)}$.

b. Use the rational expression obtained in part **a** to project, to the nearest percent, the percent of the population 20–24 in the year 2000 and in the year 2050.

c. Is the percent of the population 20–24 years old increasing or decreasing?

Using Your Knowledge

Ratios There is an important relationship between fractions and *ratios.* In general, a **ratio** is a way of comparing two or more numbers. For example, if there are 10 workers in an office, 3 women and 7 men, the ratio of women to men is 3 to 7 or

$$\frac{3}{7} \quad \begin{matrix} \leftarrow \text{Number of women} \\ \leftarrow \text{Number of men} \end{matrix}$$

On the other hand, if there are 6 men and 4 women in the office, the **reduced ratio** of women to men is

$$\frac{4}{6} = \frac{2}{3} \quad \begin{matrix} \leftarrow \text{Number of women} \\ \leftarrow \text{Number of men} \end{matrix}$$

Use your knowledge to solve the following problems.

81. A class is composed of 40 men and 60 women. Find the reduced ratio of men to women.

82. Do you know the teacher-to-student ratio in your school? Suppose your school has 10,000 students and 500 teachers.

a. Find the reduced teacher-to-student ratio.

b. If the school wishes to maintain a $\frac{1}{20}$ ratio and the enrollment increases to 12,000 students, how many teachers are needed?

83. The transmission ratio in your automobile is defined by

$$\text{Transmission ratio} = \frac{\text{Engine speed}}{\text{Drive shaft speed}}$$

a. If the engine is running at 2000 revolutions per minute and the drive shaft speed is 500 revolutions per minute, what is the reduced transmission ratio?

b. If the transmission ratio of a car is 5 to 1, and the drive shaft speed is 500 revolutions per minute, what is the engine speed?

We will examine ratios more closely in Section 6.6.

Write On

84. Write the procedure you use to determine the values for which the rational expression

$$\frac{P(x)}{Q(x)}$$

is undefined.

85. Consider the rational expression

$$\frac{1}{x^2 + a}$$

a. Is this expression always defined when a is positive? Explain. (*Hint:* Let $a = 1, 2$, and so on.)

b. Is this expression always defined when a is negative? Explain. (*Hint:* Let $a = -1, -2$, and so on.)

86. Write the procedure you use to reduce a fraction to lowest terms.

87. If

$$\frac{P(x)}{Q(x)}$$

is equal to -1, what is the relationship between $P(x)$ and $Q(x)$?

Concept Checker

Fill in the blank(s) with the correct word(s), phrase, or mathematical statement.

88. The **variables** in a rational expression **must not** be replaced by numbers that **make the denominator** ______________.

89. The **fundamental rule of rational expressions** states that $\frac{A}{B} =$ ______________.

90. The **standard form** of the fraction $-\frac{a}{b}$ is ______________.

91. The **standard form** of the fraction $\frac{a}{-b}$ is ______________.

92. The **standard form** of the fraction $-\frac{-a}{-b}$ is ______________.

93. The **quotient** of **additive inverses** $\frac{a-b}{b-a} =$ ______________.

1

−1

0

$\frac{A \cdot C}{B \cdot C}$

$\frac{-a}{b}$

Mastery Test

Reduce to lowest terms (simplify):

94. $\frac{x^2 - 9}{3 + x}$

95. $\frac{x^2 - 9}{x + 3}$

96. $\frac{x^2 - 3x - 10}{5 - x}$

97. $\frac{10x - 15y}{4x - 6y}$

98. $-\frac{x}{xy + x}$

99. $\frac{x + 4}{-(x^2 - 16)}$

100. $\frac{-3xy^2}{12x^2y}$

101. $\frac{-6(x^2 - y^2)}{-3(x - y)}$

102. Write $\frac{7}{8}$ with a denominator of 16.

103. Write $\frac{3x}{8y^2}$ with a denominator of $24y^3$.

104. Write $\frac{4x}{x + 2}$ with a denominator of $x^2 - x - 6$.

105. Find the values for which $\frac{x^2 + 1}{x^2 - 4}$ is undefined.

106. Find the values for which $\frac{x + 4}{x^2 - 6x + 8}$ is undefined.

Skill Checker

Multiply:

107. $\frac{3}{2} \cdot \frac{4}{9}$

108. $\frac{3}{5} \cdot \frac{10}{9}$

Factor:

109. $x^2 + 2x - 3$

110. $x^2 + 7x + 12$

111. $x^2 - 7x + 10$

112. $x^2 + 3x - 4$

6.2 Multiplication and Division of Rational Expressions

Objectives

A Multiply two rational expressions.

B Divide one rational expression by another.

To Succeed, Review How To . . .

1. Multiply, divide, and reduce fractions (pp. 4–5, 10–12).
2. Factor trinomials (pp. 420–421, 429–435, 439–441).
3. Factor the difference of two squares (pp. 441–442).

Getting Started

Gearing for Multiplication

How fast can the last (smallest) gear in this compound gear train go? It depends on the speed of the first gear, and the number of teeth in all the gears! The formula that tells us the number of revolutions per minute (rpm) the last gear can turn is

$$\text{rpm} = \frac{T_1}{t_1} \cdot \frac{T_2}{t_2} \cdot R$$

where T_1 and T_2 are the numbers of teeth in the driving gears, t_1 and t_2 are the numbers of teeth in the gears being driven, and R is the number of revolutions per minute the first driving gear is turning. Many useful formulas require that we know how to multiply and divide rational expressions, and we shall learn how to do so in this section.

A > Multiplying Rational Expressions

Can you simplify this expression?

$$\frac{T_1}{t_1} \cdot \frac{T_2}{t_2} \cdot R$$

Of course you can, if you remember how to multiply fractions in arithmetic.* As you recall, in arithmetic the product of two fractions is another fraction whose numerator is the product of the original numerators and whose denominator is the product of the original denominators. Here's how we state this rule in symbols.

RULE

Multiplying Rational Expressions

$$\frac{A}{B} \cdot \frac{C}{D} = \frac{AC}{BD}, \qquad B \neq 0, D \neq 0$$

Thus, the formula in the *Getting Started* can be simplified to

$$\text{rpm} = \frac{T_1T_2R}{t_1t_2}$$

If we assume that $R = 12$ and then count the teeth in the gears, we get $T_1 = 48$, $T_2 = 40$, $t_1 = 24$, and $t_2 = 24$. To find the revolutions per minute, we write

$$\text{rpm} = \frac{48}{24} \cdot \frac{40}{24} \cdot \frac{12}{1}$$

Then we have the following:

1. Reduce each fraction. $\frac{\overset{2}{\cancel{48}}}{\underset{1}{\cancel{24}}} \cdot \frac{\overset{5}{\cancel{40}}}{\underset{3}{\cancel{24}}} \cdot \frac{12}{1} = \frac{2}{1} \cdot \frac{5}{3} \cdot \frac{12}{1}$

2. Multiply the numerators. $\frac{120}{1 \cdot 3 \cdot 1}$

3. Multiply the denominators. $\frac{120}{3}$

4. Reduce the answer. 40

Thus, the speed of the final gear is 40 revolutions per minute. Here is what we have done.

PROCEDURE

Multiplying Rational Expressions

1. Reduce each expression if possible.
2. Multiply the numerators to obtain the new numerator.
3. Multiply the denominators to obtain the new denominator.
4. Reduce the answer if possible.

Note that you could also write

$$\frac{2}{1} \cdot \frac{5}{\underset{1}{\cancel{3}}} \cdot \frac{\overset{4}{\cancel{12}}}{1}$$

and obtain $2 \cdot 5 \cdot 4 = 40$ as before.

* Multiplication and division of arithmetic fractions is covered in Section R.2.

EXAMPLE 1 Multiplying rational expressions

Multiply:

a. $\frac{x}{6} \cdot \frac{7}{y}$

b. $\frac{3x^2}{2} \cdot \frac{4y}{9x}$

SOLUTION 1

a. $\frac{x}{6} \cdot \frac{7}{y} = \frac{7x}{6y}$ ← Multiply numerators. ← Multiply denominators.

b. $\frac{3x^2}{2} \cdot \frac{4y}{9x} = \frac{12x^2y}{18x}$ ← Multiply numerators. ← Multiply denominators.

$= \frac{\overset{2 \cdot x}{\cancel{12x^2}}y}{\underset{3 \cdot 1}{\cancel{18x}}}$ Reduce the answer.

$= \frac{2xy}{3}$

PROBLEM 1

Multiply:

a. $\frac{m}{4} \cdot \frac{5}{n}$

b. $\frac{5y^3}{2} \cdot \frac{4x}{15y}$

The procedure used to find the product in Example 1(b) can be shortened if we do some reduction beforehand. Thus, we can write

$$\frac{\overset{1x}{\cancel{3x^2}}}{\underset{1}{\cancel{2}}} \cdot \frac{\overset{2}{\cancel{4}}y}{\underset{3}{\cancel{9x}}} = \frac{2xy}{3}$$

We use this idea in Example 2.

EXAMPLE 2 Multiplying rational expressions involving signed numbers

Multiply:

a. $\frac{-6x}{7y^2} \cdot \frac{14y}{12x^2}$

b. $8y^2 \cdot \frac{9x}{16y^2}$

SOLUTION 2

a. Since $\frac{14y}{12x^2} = \frac{7y}{6x^2}$, we write $\frac{7y}{6x^2}$ instead of $\frac{14y}{12x^2}$:

$$\frac{-6x}{7y^2} \cdot \frac{14y}{12x^2} = \frac{\overset{-1}{\cancel{-6x}}}{\underset{1y}{\cancel{7y^2}}} \cdot \frac{\overset{1}{\cancel{7y}}}{\underset{1x}{\cancel{6x^2}}}$$

$$= \frac{-1}{xy}$$

b. Since $8y^2 = \frac{8y^2}{1}$, we have

$$8y^2 \cdot \frac{9x}{16y^2} = \frac{\overset{1}{\cancel{8y^2}}}{1} \cdot \frac{9x}{\underset{2}{\cancel{16y^2}}}$$

$$= \frac{9x}{2}$$

PROBLEM 2

Multiply:

a. $\frac{-7m}{5n^2} \cdot \frac{15n}{21m^2}$

b. $5x^2 \cdot \frac{7y}{10x^2}$

Answers to PROBLEMS

1. a. $\frac{5m}{4n}$ b. $\frac{2xy^2}{3}$

2. a. $\frac{-1}{mn}$ b. $\frac{7y}{2}$

Can all problems be done as in these examples? Yes, but note the following.

NOTE

When the numerators and denominators involved are binomials or trinomials, it isn't easy to do the reductions we just did in the examples *unless* the numerators and denominators involved are *factored*.

Thus, to multiply

$$\frac{x^2+2x-3}{x^2+7x+12}\cdot\frac{x+4}{x+5}$$

we *first* factor and then multiply. The result is,

Factor first.

$$\frac{x^2+2x-3}{x^2+7x+12}\cdot\frac{x+4}{x+5}=\frac{(x-1)\overset{1}{\cancel{(x+3)}}}{\underset{1}{\cancel{(x+4)}}\underset{1}{\cancel{(x+3)}}}\cdot\frac{\overset{1}{\cancel{(x+4)}}}{(x+5)}$$

$$=\frac{x-1}{x+5}$$

Thus, when multiplying fractions involving trinomials, we factor the trinomials and reduce the answer if possible. But note the following.

NOTE

Only **factors** can be divided out (canceled), never terms. Thus,

$$\frac{x\overset{1}{\cancel{y}}}{\cancel{y}}=x$$

but

$$\frac{x+y}{y}$$

cannot be reduced (simplified) further.

EXAMPLE 3 Factoring and multiplying rational expressions

Multiply:

a. $(x-3)\cdot\dfrac{x+5}{x^2-9}$

b. $\dfrac{x^2-x-20}{x-1}\cdot\dfrac{1-x}{x+4}$

SOLUTION 3

a. Since $(x-3)=\dfrac{x-3}{1}$ and $x^2-9=(x+3)(x-3)$,

$$(x-3)\cdot\frac{x+5}{x^2-9}=\frac{\cancel{(x-3)}}{1}\cdot\frac{x+5}{(x+3)\cancel{(x-3)}}$$

$$=\frac{x+5}{x+3}$$

b. Since $x^2-x-20=(x+4)(x-5)$ and $\dfrac{1-x}{x-1}=-1$,

$$\frac{x^2-x-20}{x-1}\cdot\frac{1-x}{x+4}=\frac{\overset{1}{\cancel{(x+4)}}(x-5)}{\cancel{(x-1)}}\cdot\frac{\overset{-1}{\cancel{(1-x)}}}{\underset{1}{\cancel{x+4}}}$$

Remember that $\frac{a-b}{b-a}=-1$.

$$=-1(x-5)$$
$$=-x+5$$
$$=5-x$$

PROBLEM 3

Multiply:

a. $(m+2)\cdot\dfrac{m+3}{m^2-4}$

b. $\dfrac{y^2-y-12}{y-2}\cdot\dfrac{2-y}{y+3}$

Answers to PROBLEMS

3. a. $\dfrac{m+3}{m-2}$ **b.** $4-y$

B > Dividing Rational Expressions

What about dividing rational expressions? We are in luck! The division of rational expressions uses the same rule as in arithmetic.

RULE

Dividing Rational Expressions

$$\frac{A}{B} \div \frac{C}{D} = \frac{A}{B} \cdot \frac{D}{C} = \frac{AD}{BC}, \quad B, C, \text{ and } D \neq 0$$

Thus, to divide $\frac{A}{B}$ by $\frac{C}{D}$, we simply **invert** $\frac{C}{D}$ (interchange the numerator and denominator) and multiply. That is, to divide $\frac{A}{B}$ by $\frac{C}{D}$, we multiply $\frac{A}{B}$ by the **reciprocal** (inverse) of $\frac{C}{D}$. For example, to divide

$$\frac{x+4}{x-5} \div \frac{x^2+3x-4}{x^2-7x+10}$$

we use the given rule and write

Invert.

$$\frac{x+4}{x-5} \div \frac{x^2+3x-4}{x^2-7x+10} = \frac{x+4}{x-5} \cdot \frac{x^2-7x+10}{x^2+3x-4}$$

$$= \frac{\overset{1}{\cancel{x+4}}}{\underset{1}{\cancel{x-5}}} \cdot \frac{\overset{1}{\cancel{(x-5)}}(x-2)}{\underset{1}{\cancel{(x+4)}}(x-1)} \qquad \text{First factor numerator and denominator.}$$

$$= \frac{x-2}{x-1} \qquad \text{Reduce.}$$

Here is another example.

EXAMPLE 4 **Dividing rational expressions involving the difference of two squares**

Divide:

a. $\frac{x^2-16}{x+3} \div (x+4)$

b. $\frac{x+5}{x-5} \div \frac{x^2-25}{5-x}$

SOLUTION 4

a. Since $(x+4) = \frac{(x+4)}{1}$,

Invert.

$$\frac{x^2-16}{x+3} \div \frac{(x+4)}{1} = \frac{x^2-16}{x+3} \cdot \frac{1}{(x+4)}$$

$$= \frac{\overset{1}{\cancel{(x+4)}}(x-4)}{x+3} \cdot \frac{1}{\underset{1}{\cancel{(x+4)}}} \qquad \text{Factor } x^2 - 16.$$

$$= \frac{x-4}{x+3} \qquad \text{Reduce.}$$

(continued)

PROBLEM 4

Divide:

a. $\frac{y^2-9}{y+5} \div (y-3)$

b. $\frac{y+4}{y-4} \div \frac{y^2-16}{4-y}$

Answers to PROBLEMS

4. a. $\frac{y+3}{y+5}$ **b.** $\frac{-1}{y-4} = \frac{1}{4-y}$

Note: $\frac{1}{4-y}$ is preferred!

Invert.

b. $\frac{x+5}{x-5} \div \frac{x^2-25}{5-x} = \frac{x+5}{x-5} \cdot \frac{5-x}{x^2-25}$

$= \frac{\cancel{x+5}}{\cancel{x-5}} \cdot \frac{\cancel{5-x}}{\cancel{(x+5)}(x-5)}$ Factor $x^2 - 25$ and note that $\frac{5-x}{x-5} = -1$.

$= \frac{-1}{x-5}$ Reduce.

Of course, this answer *can* be simplified further (to show fewer negative signs), since

$$\frac{-1}{x-5} = \frac{(-1)(-1)}{(-1)(x-5)}$$ Multiply numerator and denominator by (-1).

$$= \frac{1}{-x+5} = \frac{1}{5-x}$$

Here's another example.

EXAMPLE 5 Factoring and dividing rational expressions

Divide:

a. $\frac{x^2+5x+4}{x^2-2x-3} \div \frac{x^2-4}{x^2-6x+8}$ **b.** $\frac{x^2-1}{x^2+x-6} \div \frac{x^2-4x+3}{x^2-4}$

SOLUTION 5

Invert.

a. $\frac{x^2+5x+4}{x^2-2x-3} \div \frac{x^2-4}{x^2-6x+8} = \frac{x^2+5x+4}{x^2-2x-3} \cdot \frac{x^2-6x+8}{x^2-4}$

$= \frac{(x+4)\cancel{(x+1)}}{(x-3)\cancel{(x+1)}} \cdot \frac{(x-4)\cancel{(x-2)}}{(x+2)\cancel{(x-2)}}$ Factor.

$= \frac{(x+4)(x-4)}{(x-3)(x+2)}$

$= \frac{x^2-16}{x^2-x-6}$

Invert.

b. $\frac{x^2-1}{x^2+x-6} \div \frac{x^2-4x+3}{x^2-4} = \frac{x^2-1}{x^2+x-6} \cdot \frac{x^2-4}{x^2-4x+3}$

$= \frac{(x+1)\cancel{(x-1)}}{(x+3)\cancel{(x-2)}} \cdot \frac{(x+2)\cancel{(x-2)}}{\cancel{(x-1)}(x-3)}$ Factor.

$= \frac{(x+1)(x+2)}{(x+3)(x-3)}$

$= \frac{x^2+3x+2}{x^2-9}$

PROBLEM 5

Divide:

a. $\frac{y^2-3y+2}{y^2-4y+3} \div \frac{y^2-49}{y^2-5y-14}$

b. $\frac{y^2-1}{y^2-y-6} \div \frac{y^2-3y+2}{y^2-9}$

Answers to PROBLEMS

5. a. $\frac{y^2-4}{y^2+4y-21}$

b. $\frac{y^2+4y+3}{y^2-4}$

A final word of warning! Be very careful when you reduce fractions.

CAUTION

You may cancel factors, but you must not cancel terms.

$$\frac{\cancel{a}b}{\cancel{a}c} = \frac{b}{c} \quad \text{Yes!}$$

$$\frac{\cancel{a}+b}{\cancel{a}+c} \quad \text{No!}$$

Thus,

$$\frac{\cancel{x^2}+5x+6}{\cancel{x^2}+8x+15} = \frac{5x+6}{8x+15} \quad \text{is wrong!}$$

Note that x^2 is a **term,** *not* a factor. The correct way is to *factor* first and then cancel. Thus,

$$\frac{x^2+5x+6}{x^2+8x+15} = \frac{\cancel{(x+3)}(x+2)}{\cancel{(x+3)}(x+5)} = \frac{x+2}{x+5}$$

Of course,

$$\frac{x+2}{x+5}$$

cannot be reduced further. To write

$$\frac{\cancel{x}+2}{\cancel{x}+5} = \frac{2}{5} \quad \text{is wrong!}$$

Again, x is a term, not a factor. (If you had

$$\frac{2x}{5x} = \frac{2}{5}$$

that would be correct. In the expressions $2x$ and $5x$, x is a *factor* that may be canceled.) Why is

$$\frac{x+2}{x+5} \neq \frac{2}{5}?$$

Try it when x is 4.

$$\frac{x+2}{x+5} = \frac{4+2}{4+5} = \frac{6}{9} = \frac{2}{3}$$

Thus, the answer *cannot* be $\frac{2}{5}$!

EXAMPLE 6 Offsetting your car emissions

How many trees do you need to offset the carbon dioxide emitted by your car? It depends on several factors, but we can find out by multiplying and dividing expressions. We shall do it by steps.

a. If you divide the annual number of miles M you drive by the miles per gallon m your car gets, you get the number of gallons of gas you use annually. Write the quotient of M and m.

b. Since a gallon of gas emits 20 pounds of CO_2, the product of the number of gallons you use and 20 is the annual amount of CO_2 your car produces. Write an expression for the product of 20 and the quotient of M and m.

PROBLEM 6

a. Suppose your car produces 22 pounds of carbon dioxide for each gallon of gas used and an acacia tree absorbs 22 pounds of CO_2 a year. Follow the procedure of Example 6 and find an expression that estimates the number of acacias needed to offset the CO_2 emissions of your car.

(continued)

Answers to PROBLEMS

6. a. $\frac{M}{m}$ **b.** 1000 acacias

c. Different trees absorb different amounts of CO_2 (anywhere from 13 to 50 pounds a year). To find the number of trees we need, divide the answer from part **b** by **50** and simplify. This is the expression representing the number of trees needed to offset your car emissions!

d. If you drive 20,000 miles a year ($M = 20{,}000$), and your car gets 25 miles per gallon ($m = 25$), use the expression of part **c** to estimate the number of trees needed to offset the CO_2 emissions of your car.

SOLUTION 6

a. The quotient of M and m is $\frac{M}{m}$.

b. The product of 20 and the quotient of M and m is $20 \cdot \frac{M}{m} = \frac{20M}{m}$.

c. The answer to part **b,** which is $\frac{20M}{m}$, divided by 50 is

$$\frac{20M}{m} \div \frac{50}{1} = \frac{20M}{m} \cdot \frac{1}{50} = \frac{20M}{50m} = \frac{2M}{5m}$$

d. If $M = 20{,}000$ and $m = 25$, then $\frac{2M}{5m} = \frac{2(20{,}000)}{5(25)} = 320$ trees.

You don't have to worry about planting the trees yourself; there is a carbon dioxide emission calculator that will help you gain an idea of how much carbon dioxide some of your activities generate and how many trees it would take to offset those emissions.

Source: http://tinyurl.com/6498zd.

b. If you drive 20,000 miles a year and your car gets 20 miles per gallon, how many acacias do you need to offset the CO_2 emissions of your car?

To find out how the amount of carbon dioxide different trees absorb in a year is measured, see our source.

Source: http://tinyurl.com/yganbv8.

> Exercises 6.2

> Web IT go to mhhe.com/bello for more lessons

< A > Multiplying Rational Expressions In Problems 1–30, multiply and simplify.

1. $\frac{x}{3} \cdot \frac{8}{y}$

2. $\frac{-x}{4} \cdot \frac{7}{y}$

3. $\frac{-6x^2}{7} \cdot \frac{14y}{9x}$

4. $\frac{-5x^3}{6y^2} \cdot \frac{18y}{-10x}$

5. $7x^2 \cdot \frac{3y}{14x^2}$

6. $11y^2 \cdot \frac{4x}{33y}$

7. $\frac{-4y}{7x^2} \cdot 14x^3$

8. $\frac{-3y^3}{8x^2} \cdot -16y$

9. $(x - 7) \cdot \frac{x + 1}{x^2 - 49}$

10. $3(x + 1) \cdot \frac{x + 2}{x^2 - 1}$

11. $-2(x + 2) \cdot \frac{x - 1}{x^2 - 4}$

12. $-3(x - 1) \cdot \frac{x - 2}{x^2 - 1}$

13. $\frac{3}{x - 5} \cdot \frac{x^2 - 25}{x + 1}$

14. $\frac{1}{x - 2} \cdot \frac{x^2 - 4}{x - 1}$

15. $\frac{x^2 - x - 6}{x - 2} \cdot \frac{2 - x}{x - 3}$

16. $\frac{x^2 + 3x - 4}{3 - x} \cdot \frac{x - 3}{x + 4}$

17. $\frac{x - 1}{3 - x} \cdot \frac{x + 3}{1 - x}$

18. $\frac{2x - 1}{5 - 3x} \cdot \frac{3x - 5}{1 - 2x}$

19. $\frac{3(x - 5)}{14(4 - x)} \cdot \frac{7(x - 4)}{6(5 - x)}$

20. $\frac{7(1 - x)}{10(x - 5)} \cdot \frac{5(5 - x)}{14(x - 1)}$

21. $\frac{6x^3}{x^2 - 16} \cdot \frac{x^2 - 5x + 4}{3x^2}$

22. $\frac{3a^4}{a^2 - 4} \cdot \frac{a^2 - a - 2}{9a^3}$

23. $\frac{y^2 + 2y - 3}{y - 5} \cdot \frac{y^2 - 3y - 10}{y^2 + 5y - 6}$

24. $\frac{f^2 + 2f - 8}{f^2 + 7f + 12} \cdot \frac{f^2 + 2f - 3}{f^2 - 3f + 2}$

25. $\frac{2y^2 + y - 3}{6 - 11y - 10y^2} \cdot \frac{5y^3 - 2y^2}{3y^2 - 5y + 2}$

26. $\frac{3x^2 - x - 2}{2 - x - 6x^2} \cdot \frac{2x^4 - x^3}{3x^2 - 2x - 1}$

27. $\frac{15x^2 - x - 2}{2x^2 + 5x - 18} \cdot \frac{2x^2 + x - 36}{3x^2 - 11x - 4}$

28. $\frac{6x^2 + x - 1}{3x^2 + 5x + 2} \cdot \frac{3x^2 - x - 2}{2x^2 - x - 1}$

29. $\frac{27y^3 + 8}{6y^2 + 19y + 10} \cdot \frac{4y^2 - 25}{9y^2 - 6y + 4}$

30. $\frac{8y^3 + 27}{10y^2 + 19y + 6} \cdot \frac{25y^2 - 4}{4y^2 - 6y + 9}$

⟨B⟩ Dividing Rational Expressions In Problems 31–64, divide and simplify.

31. $\frac{x^2 - 1}{x + 2} \div (x + 1)$

32. $\frac{x^2 - 4}{x - 3} \div (x + 2)$

33. $\frac{x^2 - 25}{x - 3} \div 5(x + 5)$

34. $\frac{x^2 - 16}{8(x - 3)} \div 4(x + 4)$

35. $(x + 3) \div \frac{x^2 - 9}{x + 4}$

36. $4(x - 4) \div \frac{8(x^2 - 16)}{5}$

37. $\frac{-3}{x - 4} \div \frac{6(x + 3)}{5(x^2 - 16)}$

38. $\frac{-6}{x - 2} \div \frac{3(x - 1)}{7(x^2 - 4)}$

39. $\frac{-4(x + 1)}{3(x + 2)} \div \frac{-8(x^2 - 1)}{6(x^2 - 4)}$

40. $\frac{-10(x^2 - 1)}{6(x^2 - 4)} \div \frac{5(x + 1)}{-3(x + 2)}$

41. $\frac{x + 3}{x - 3} \div \frac{x^2 - 1}{3 - x}$

42. $\frac{4 - x}{x + 1} \div \frac{x - 4}{x^2 - 1}$

43. $\frac{x^2 - 4}{7(x^2 - 9)} \div \frac{x + 2}{14(x + 3)}$

44. $\frac{x^2 - 25}{3(x^2 - 1)} \div \frac{5 - x}{6(x + 1)}$

45. $\frac{3(x^2 - 36)}{14(5 - x)} \div \frac{6(6 - x)}{7(x^2 - 25)}$

46. $\frac{6(x^2 - 1)}{35(x^2 - 4)} \div \frac{12(1 - x)}{7(2 - x)}$

47. $\frac{x + 2}{x - 1} \div \frac{x^2 + 5x + 6}{x^2 - 4x + 4}$

48. $\frac{x - 3}{x + 2} \div \frac{x^2 - 4x + 3}{x^2 - x - 6}$

49. $\frac{x - 5}{x + 3} \div \frac{5(x - 5)}{x^2 + 9x + 18}$

50. $\frac{x - 3}{x + 4} \div \frac{2(x - 3)}{x^2 + 2x - 8}$

51. $\frac{x^2 + 2x - 3}{x - 5} \div \frac{x^2 + 6x + 9}{x^2 - 2x - 15}$

52. $\frac{x^2 - 3x + 2}{x^2 - 5x + 6} \div \frac{x^2 - 5x + 4}{x^2 - 7x + 12}$

53. $\frac{x^2 - 1}{x^2 + 3x - 10} \div \frac{x^2 - 3x - 4}{x^2 - 25}$

54. $\frac{x^2 - 4x - 21}{x^2 - 10x + 25} \div \frac{x^2 + 2x - 3}{x^2 - 6x + 5}$

55. $\frac{x^2 + 3x - 4}{x^2 + 7x + 12} \div \frac{x^2 + x - 2}{x^2 + 5x + 6}$

56. $\frac{x^2 + x - 2}{x^2 + 6x - 7} \div \frac{x^2 - 3x - 10}{x^2 + 5x - 14}$

57. $\frac{x^2 - y^2}{x^2 - 2xy} \div \frac{x^2 + xy - 2y^2}{x^2 - 4y^2}$

58. $\frac{x^2 + xy - 2y^2}{x^2 - 4y^2} \div \frac{x^2 - y^2}{x^2 - 2xy}$

59. $\frac{x^2 + 2xy - 3y^2}{y^2 - 7y + 10} \div \frac{x^2 + 5xy - 6y^2}{y^2 - 3y - 10}$

60. $\frac{x^2 + 2xy - 8y^2}{x^2 + 7xy + 12y^2} \div \frac{x^2 - 3xy + 2y^2}{x^2 + 2xy - 3y^2}$

61. $\frac{2x^2 - x - 28}{3x^2 - x - 2} \div \frac{4x^2 + 16x + 7}{3x^2 + 11x + 6}$

62. $\frac{15x^2 - x - 2}{2x^2 + 5x - 18} \div \frac{3x^2 - 11x - 4}{2x^2 + x - 36}$

63. $\frac{(a^3 - 27)(a^2 - 9)}{(a - 3)^2(a + 3)^3} \div \frac{a^2 + 3a + 9}{a^2 + 3a}$

64. $\frac{(y^3 - 8)(y^2 - 4)}{(y + 2)^2(y - 2)^3} \div \frac{y^2 + 2y + 4}{y^2 - 2y}$

Web IT go to mhhe.com/bello for more lessons

››› Applications: Green Math

65. *Extra miles by hybrid* According to Example 6, **320** trees are needed to offset the CO_2 produced by a car driven **20,000** miles a year and getting **25** miles per gallon. But what if you have a hybrid car getting **35** miles per gallon? Use $\frac{2M}{5m} = 320$ to find the following:

a. How many annual miles M can you drive with the **35 mpg** hybrid and still offset the emissions using the **320** trees?

b. If gas is \$3 per gallon, what are your gas savings if you drive 28,000 miles with the hybrid instead of 20,000 with the regular car?

66. *Most efficient hybrid* The 2010 Toyota Prius claims a 50 mpg combined fuel efficiency. According to Example 6, **320** trees are needed to offset the CO_2 produced by a car driven **20,000** miles a year and getting **25** miles per gallon. What about with the Prius? Use $\frac{2M}{5m} = 320$ to find the following:

a. How many annual miles M can you drive with the **50 mpg** hybrid and still offset the emissions using the **320** trees?

b. If gas is \$3 per gallon, compare the cost and the savings of driving the Prius 40,000 miles with the cost of driving a regular car getting **25** mpg for 20,000 miles.

››› Applications

67. *Price and demand* If the price P for x units of a product is given by the expression

$$\frac{5x + 10}{2}$$

and the demand D is given by the expression

$$\frac{400}{x^2 + 2x}$$

find PD, the product of the price and the demand.

68. *Current, voltage, and resistance* In a simple electrical circuit, the current I is the quotient of the voltage E and the resistance R. If the resistance changes with the time t according to

$$R = \frac{t^2 + 9}{t^2 + 6t + 9}$$

and the voltage changes according to the formula

$$E = \frac{4t}{t + 3}$$

find the current I.

69. *Current, voltage, and resistance* If in Problem 68 the resistance is given by the equation

$$R = \frac{t^2 + 5}{t^2 + 4t + 4}$$

and the voltage is given by the equation

$$E = \frac{5t}{t + 2}$$

find the current I.

70. *Current, voltage, and resistance* In a simple electric circuit, the current is modeled by the equation $I = \frac{E}{R}$, where E is the voltage and R is the resistance. If

$$R = \frac{t^2 + 4}{t^2 + 4t + 4}$$

and

$$E = \frac{3t}{t + 2}$$

find I.

››› Using Your Knowledge

Resistance, Molecules, and Reordering

71. In the study of parallel resistors, the expression

$$R \cdot \frac{R_T}{R - R_T}$$

occurs, where R is a known resistance and R_T a required one. Perform the multiplication.

72. The molecular model predicts that the pressure of a gas is given by

$$\frac{2}{3} \cdot \frac{mv^2}{2} \cdot \frac{N}{v}$$

where m, v, and N represent the mass, velocity, and total number of molecules, respectively. Perform the multiplication.

73. Suppose a store orders 3000 items each year. If it orders x units at a time, the number N of reorders is

$$N = \frac{3000}{x}$$

If there is a fixed \$20 reorder fee and a \$3 charge per item, the cost of each order is

$$C = 20 + 3x$$

The yearly reorder cost C_R is then given by

$$C_R = N \cdot C$$

Find C_R.

Write On

74. Write the procedure you use to multiply two rational expressions.

75. Write the procedure you use to divide one rational expression by another.

76. Explain why you cannot "cancel" the x's in

$$\frac{x+5}{x+2}$$

to obtain an answer of $\frac{5}{2}$ but you can cancel the x's in $\frac{5x}{2x}$.

77. Explain what the statement "you can cancel factors but you cannot cancel terms" means and give examples.

Concept Checker

Fill in the blank(s) with the correct word(s), phrase, or mathematical statement.

78. $\frac{A}{B} \cdot \frac{C}{D} =$ _____.

79. $\frac{A}{B} \div \frac{C}{D} =$ _____.

$\frac{AD}{BC}$ $\frac{AC}{BD}$ $\frac{AD}{AC}$ $\frac{BC}{AD}$

Mastery Test

Divide and simplify:

80. $\frac{x^2 - 3x + 2}{x^2 - 4x + 3} \div \frac{x^2 - 49}{x^2 + 5x - 14}$

81. $\frac{x^2 - 1}{x^2 - x - 6} \div \frac{x^2 - 3x + 2}{x^2 - 9}$

82. $\frac{x^2 - 25}{x + 1} \div (x + 5)$

83. $\frac{x + 6}{x - 6} \div \frac{x^2 - 36}{6 - x}$

84. $\frac{x^3 + 8}{x - 2} \div \frac{x^2 - 2x + 4}{x^2 - 4}$

85. $\frac{x^3 - 27}{x + 3} \div \frac{x^2 + 3x + 9}{x^2 - 9}$

Multiply and simplify:

86. $(x - 4) \cdot \frac{x + 8}{x^2 - 16}$

87. $\frac{x^2 + x - 6}{x - 3} \cdot \frac{3 - x}{x + 3}$

88. $\frac{-3x}{4y^2} \cdot \frac{18y}{12x^2}$

89. $\frac{6x}{11y^2} \cdot 22y^2$

90. $\frac{x^3 - 64}{x + 1} \cdot \frac{x^2 - 1}{x^2 + 4x + 16}$

91. $\frac{2x^2 - x - 28}{3x^2 - x - 2} \cdot \frac{3x^2 + 11x + 6}{4x^2 + 16x + 7}$

Skill Checker

Add:

92. $\frac{7}{12} + \frac{1}{12}$

93. $\frac{7}{8} + \frac{2}{5}$

94. $\frac{7}{12} + \frac{1}{18}$

Subtract:

95. $\frac{7}{8} - \frac{2}{5}$

96. $\frac{7}{12} - \frac{1}{18}$

97. $\frac{5}{2} - \frac{1}{6}$

98. $\frac{8}{3} - \frac{3}{4}$

6.3 Addition and Subtraction of Rational Expressions

Objectives

A Add and subtract rational expressions with the same denominator.

B Add and subtract rational expressions with different denominators.

C Solve applications involving rational expressions.

To Succeed, Review How To . . .

1. Find the LCD of two or more fractions (pp. 13–15).
2. Add and subtract fractions (pp. 12–18).

Getting Started

Tennis, Anyone?

The racket hits the ball with such tremendous force that the ball is distorted. Can we find out how much force? The answer is

$$\frac{mv}{t} - \frac{mv_0}{t}$$

where

m = mass of ball
v = velocity of racket
v_0 = "initial" velocity of racket
t = time of contact

Since the expressions involved have the same denominator, subtracting them is easy. As in arithmetic, we simply subtract the numerators and keep the same denominator. Thus,

$$\frac{mv}{t} - \frac{mv_0}{t} = \frac{mv - mv_0}{t}$$ ← Subtract numerators. ← Keep the denominator.

In this section, we shall learn how to add and subtract rational expressions.

A › Adding and Subtracting Rational Expressions with the Same Denominator

As you recall from Section R.2, $\frac{1}{5} + \frac{2}{5} = \frac{3}{5}$, $\frac{1}{7} + \frac{4}{7} = \frac{5}{7}$, and $\frac{1}{11} + \frac{8}{11} = \frac{9}{11}$. The same procedure works for rational expressions. For example,

$$\frac{3}{x} + \frac{5}{x} = \frac{3+5}{x} = \frac{8}{x}$$ ← Add numerators. ← Keep the denominator.

Similarly,

$$\frac{5}{x+1} + \frac{2}{x+1} = \frac{5+2}{x+1} = \frac{7}{x+1}$$ ← Add numerators. ← Keep the denominator.

and

$$\frac{5}{7(x-1)} + \frac{2}{7(x-1)} = \frac{5+2}{7(x-1)} = \frac{\overset{1}{\cancel{7}}}{\underset{1}{\cancel{7}}(x-1)} = \frac{1}{x-1}$$

For subtraction,

$$\frac{8}{x+5} - \frac{2}{x+5} = \frac{8-2}{x+5} = \frac{6}{x+5}$$ ← Subtract numerators. ← Keep the denominator.

and

$$\frac{8}{9(x-3)} - \frac{2}{9(x-3)} = \frac{8-2}{9(x-3)} = \frac{\overset{2}{\cancel{6}}}{\underset{3}{\cancel{9}}(x-3)} = \frac{2}{3(x-3)}$$

EXAMPLE 1 **Adding and subtracting rational expressions: Same denominator**

Add or Subtract:

a. $\frac{8}{3(x-2)} + \frac{1}{3(x-2)}$ b. $\frac{7}{5(x+4)} - \frac{2}{5(x+4)}$

SOLUTION 1

a. $\frac{8}{3(x-2)} + \frac{1}{3(x-2)} = \frac{8+1}{3(x-2)} = \frac{\overset{3}{\cancel{9}}}{\underset{1}{\cancel{3}}(x-2)} = \frac{3}{x-2}$ Remember to reduce the answer.

b. $\frac{7}{5(x+4)} - \frac{2}{5(x+4)} = \frac{7-2}{5(x+4)} = \frac{\overset{1}{\cancel{5}}}{\underset{1}{\cancel{5}}(x+4)} = \frac{1}{x+4}$

PROBLEM 1

Add or Subtract:

a. $\frac{4}{5(y-1)} + \frac{1}{5(y-1)}$

b. $\frac{9}{7(y+2)} - \frac{2}{7(y+2)}$

B › Adding and Subtracting Rational Expressions with Different Denominators

Not all rational expressions have the same denominator. To add or subtract rational expressions with different denominators, we again rely on our experiences in arithmetic. Let's practice adding rational numbers before we try rational expressions.

EXAMPLE 2 **Adding fractions: Different denominators**

Add: $\frac{7}{12} + \frac{5}{18}$

SOLUTION 2 We first must find a common denominator—that is, a *multiple* of 12 and 18. Of course, it's more convenient to use the smallest one available. In general, the **lowest common denominator (LCD)** of two fractions is the **smallest** number that is a multiple of *both* denominators. To find the LCD, we can use successive divisions as in Section R.2.

2	12	18
3	6	9
	2	3 →

The LCD is $2 \times 3 \times 2 \times 3 = 36$. Better yet, we can also factor both numbers and write each factor in a column to obtain

Pick the number with the highest exponent.

$12 = 2 \cdot 2 \cdot 3 = 2^2 \cdot 3^1$

$18 = 2 \cdot 3 \cdot 3 = 2^1 \cdot 3^2$ Note that all the 2's and all the 3's are written in separate columns.

(continued)

PROBLEM 2

Add: $\frac{5}{12} + \frac{7}{18}$

Answers to PROBLEMS

1. a. $\frac{1}{y-1}$ **b.** $\frac{1}{y+2}$ **2.** $\frac{29}{36}$

Note that since we need a number that is a multiple of 12 and 18, we select the factors raised to the *highest* power in each column—that is, 2^2 and 3^2. The product of these factors is the LCD. Thus, the LCD of 12 and 18 is $2^2 \cdot 3^2 = 4 \cdot 9 = 36$, as before. We then write each fraction with a denominator of 36 and add.

$$\frac{7}{12} = \frac{7 \cdot 3}{12 \cdot 3} = \frac{21}{36}$$ We multiply the denominator 12 by 3 (to get 36), so we do the same to the numerator.

$$\frac{5}{18} = \frac{5 \cdot 2}{18 \cdot 2} = \frac{10}{36}$$ Here we multiply the denominator 18 by 2 to get 36, so we do the same to the numerator.

$$\frac{7}{12} + \frac{5}{18} = \frac{21}{36} + \frac{10}{36} = \frac{31}{36}$$ Can you see how this one is done?

The procedure can also be written as

$$\begin{array}{r} \dfrac{7}{12} = \dfrac{21}{36} \\ + \dfrac{5}{18} = \dfrac{10}{36} \\ \hline \dfrac{31}{36} \end{array}$$

NOTE

Make sure you know how to do this type of problem before you go on. The idea in adding and subtracting rational expressions is the same as the idea used with rational numbers. In fact, Problems 1–20 in Exercises 6.3 have two parts: one with rational numbers and one with rational expressions. If you know how to do one, you should be able to do the other one.

Now, let's practice subtraction of fractions.

EXAMPLE 3 Subtracting fractions: Different denominators

Subtract: $\frac{11}{15} - \frac{5}{18}$

SOLUTION 3 We can use successive divisions to find the LCD, writing

$$\begin{array}{r|rr} 3 & 15 & 18 \\ \hline & 5 & 6 \end{array}$$

The LCD is $3 \times 5 \times 6 = 90$. Better yet, we can factor the denominators and write them as follows:

$$15 = 3 \cdot 5 \quad = \quad 3 \cdot 5$$
$$18 = 2 \cdot 3 \cdot 3 = 2 \cdot 3^2$$

Note that all the 2's, all the 3's, and all the 5's are in separate columns.

As before, the LCD is

$$2 \cdot 3^2 \cdot 5 = 2 \cdot 9 \cdot 5 = 90$$

Then we write $\frac{11}{15}$ and $\frac{5}{18}$ as equivalent fractions with a denominator of 90.

$$\frac{11}{15} = \frac{11 \cdot 6}{15 \cdot 6} = \frac{66}{90}$$ Multiply the numerator and denominator by 6.

$$\frac{5}{18} = \frac{5 \cdot 5}{18 \cdot 5} = \frac{25}{90}$$ Multiply the numerator and denominator by 5.

PROBLEM 3

Subtract: $\frac{13}{15} - \frac{7}{12}$

Answers to PROBLEMS

3. $\frac{17}{60}$

and then subtract

$$\frac{11}{15} - \frac{5}{18} = \frac{66}{90} - \frac{25}{90}$$

$$= \frac{66 - 25}{90}$$

$$= \frac{41}{90}$$

Of course, it's possible that the denominators involved have no common factors. In this case, the LCD is the *product* of the denominators. Thus, to add $\frac{3}{5}$ and $\frac{4}{7}$, we use $5 \cdot 7 = 35$ as the LCD and write

$$\frac{3}{5} = \frac{3 \cdot 7}{5 \cdot 7} = \frac{21}{35}$$ Multiply the numerator and denominator by 7.

$$\frac{4}{7} = \frac{4 \cdot 5}{7 \cdot 5} = \frac{20}{35}$$ Multiply the numerator and denominator by 5.

Thus,

$$\frac{3}{5} + \frac{4}{7} = \frac{21}{35} + \frac{20}{35} = \frac{41}{35}$$

Similarly, the expression

$$\frac{4}{x} + \frac{5}{3}$$

has $3x$ as the LCD. We then write $\frac{4}{x}$ and $\frac{5}{3}$ as equivalent fractions with $3x$ as the denominator and add:

$$\frac{4}{x} = \frac{4 \cdot 3}{x \cdot 3} = \frac{12}{3x}$$ Multiply the numerator and denominator by 3.

$$\frac{5}{3} = \frac{5 \cdot x}{3 \cdot x} = \frac{5x}{3x}$$ Multiply the numerator and denominator by x.

Thus,

$$\frac{4}{x} + \frac{5}{3} = \frac{12}{3x} + \frac{5x}{3x}$$

$$= \frac{12 + 5x}{3x}$$

Are you ready for a more complicated problem? First, let's state a generalized procedure for adding and subtracting fractions with different denominators.

PROCEDURE

Adding (or Subtracting) Fractions with Different Denominators

1. Find the LCD.
2. Write all fractions as equivalent ones with the LCD as the denominator.
3. Add (or subtract) numerators and keep denominators.
4. Reduce if possible.

Let's use these steps to add

$$\frac{x + 1}{x^2 + x - 2} + \frac{x + 3}{x^2 - 1}$$

1. We first find the LCD of the denominators. To do this, we factor the denominators.

$$x^2 + x - 2 = (x + 2)(x - 1)$$
$$x^2 - 1 = (x - 1)(x + 1)$$
$$(x + 2)(x - 1)(x + 1) \quad \text{The LCD}$$

2. We then write

$$\frac{x + 1}{x^2 + x - 2} \quad \text{and} \quad \frac{x + 3}{x^2 - 1}$$

as equivalent fractions with $(x + 2)(x - 1)(x + 1)$ as denominator.

$$\frac{x + 1}{x^2 + x - 2} = \frac{x + 1}{(x + 2)(x - 1)} = \frac{(x + 1)(x + 1)}{(x + 2)(x - 1)(x + 1)}$$

$$\frac{x + 3}{x^2 - 1} = \frac{x + 3}{(x + 1)(x - 1)} = \frac{(x + 3)(x + 2)}{(x + 1)(x - 1)(x + 2)}$$

$$= \frac{(x + 3)(x + 2)}{(x + 2)(x - 1)(x + 1)}$$

3. Add the numerators and keep the denominator.

$$\frac{x + 1}{x^2 + x - 2} + \frac{x + 3}{x^2 - 1} = \frac{(x + 1)(x + 1)}{(x + 2)(x - 1)(x + 1)} + \frac{(x + 3)(x + 2)}{(x + 2)(x - 1)(x + 1)}$$

$$= \frac{(x^2 + 2x + 1) + (x^2 + 5x + 6)}{(x + 2)(x - 1)(x + 1)}$$

$$= \frac{2x^2 + 7x + 7}{(x + 2)(x - 1)(x + 1)}$$

Note that the denominator is left as an indicated product.

4. The answer is not reducible, since there are no factors common to the numerator and denominator.

We use this procedure to add and subtract rational expressions in Example 4.

EXAMPLE 4 Adding and subtracting rational expressions: Different denominators

Add or Subtract:

a. $\frac{7}{8} + \frac{2}{x}$

b. $\frac{2}{x - 1} - \frac{1}{x + 2}$

SOLUTION 4

a. Since 8 and x don't have any common factors, the LCD is $8x$. We write $\frac{7}{8}$ and $\frac{2}{x}$ as equivalent fractions with $8x$ as denominator and add.

$$\frac{7}{8} = \frac{7 \cdot x}{8 \cdot x} = \frac{7x}{8x}$$

$$\frac{2}{x} = \frac{2 \cdot 8}{x \cdot 8} = \frac{16}{8x}$$

Thus,

$$\frac{7}{8} + \frac{2}{x} = \frac{7x}{8x} + \frac{16}{8x}$$

$$= \frac{7x + 16}{8x}$$

Note that when the denominators A and B do not have any common factors, the common denominator is AB.

PROBLEM 4

Add or Subtract:

a. $\frac{3}{5} + \frac{2}{y}$

b. $\frac{2}{y - 2} - \frac{1}{y + 1}$

Answers to PROBLEMS

4. a. $\frac{3y + 10}{5y}$ b. $\frac{y + 4}{(y - 2)(y + 1)}$

b. Since $(x - 1)$ and $(x + 2)$ don't have any common factors, the LCD of

$$\frac{2}{x-1} \quad \text{and} \quad \frac{1}{x+2}$$

is $(x - 1)(x + 2)$. We then write

$$\frac{2}{x-1} \quad \text{and} \quad \frac{1}{x+2}$$

as equivalent fractions with $(x - 1)(x + 2)$ as the denominator.

$$\frac{2}{x-1} = \frac{2 \cdot (x+2)}{(x-1)(x+2)}$$ Multiply numerator and denominator by $(x + 2)$.

$$\frac{1}{x+2} = \frac{1 \cdot (x-1)}{(x+2)(x-1)} = \frac{(x-1)}{(x-1)(x+2)}$$ Multiply numerator and denominator by $(x - 1)$.

Hence,

$$\frac{2}{x-1} - \frac{1}{x+2} = \frac{2 \cdot (x+2)}{(x-1)(x+2)} - \frac{(x-1)}{(x-1)(x+2)}$$

$$= \frac{2(x+2) - (x-1)}{(x-1)(x+2)}$$ ← Subtract numerators. ← Keep denominator.

$$= \frac{2x + 4 - x + 1}{(x-1)(x+2)}$$ Remember that $-(x - 1) = -x + 1$.

$$= \frac{x+5}{(x-1)(x+2)}$$ ← Simplify numerator. ← Keep denominator.

EXAMPLE 5 **Subtracting rational expressions: Different denominators**

Subtract:

$$\frac{x-2}{x^2 - x - 6} - \frac{x+3}{x^2 - 9}$$

SOLUTION 5 We use the four-step procedure.

1. To find the LCD, we factor the denominators to obtain

$$x^2 - x - 6 = (x-3)(x+2)$$
$$x^2 - 9 = (x+3)(x-3)$$
$$(x+3)(x-3)(x+2)$$ The LCD

2. We write each fraction as an equivalent one with the LCD as the denominator. Hence,

$$\frac{x-2}{x^2 - x - 6} = \frac{(x-2)(x+3)}{(x-3)(x+2)(x+3)}$$

$$\frac{x+3}{x^2 - 9} = \frac{(x+3)(x+2)}{(x+3)(x-3)(x+2)}$$

(continued)

PROBLEM 5

Subtract:

$$\frac{x-3}{(x+1)(x-2)} - \frac{x+3}{x^2 - 4}$$

Answers to PROBLEMS

5. $\frac{-5x - 9}{(x+1)(x+2)(x-2)}$

3. $\frac{x-2}{x^2-x-6} - \frac{x+3}{x^2-9} = \frac{(x-2)(x+3)}{(x+3)(x-3)(x+2)} - \frac{(x+3)(x+2)}{(x+3)(x-3)(x+2)}$

$$= \frac{(x^2+x-6)-(x^2+5x+6)}{(x+3)(x-3)(x+2)}$$

$$= \frac{x^2+x-6-x^2-5x-6}{(x+3)(x-3)(x+2)}$$

Note that $-(x^2+5x+6) = -x^2-5x-6$.

$$= \frac{-4x-12}{(x+3)(x-3)(x+2)}$$

$$= \frac{-4(x+3)}{(x+3)(x-3)(x+2)}$$

Factor the numerator and keep the denominator.

4. Reduce.

$$\frac{-4\cancel{(x+3)}}{\cancel{(x+3)}(x-3)(x+2)} = \frac{-4}{(x-3)(x+2)}$$

C > Applications Involving Rational Expressions

GREEN MATH

EXAMPLE 6 Total materials recovered for recycling each day

The percent of total materials recovered for recycling each day is the sum $\frac{R(t)}{G(t)} + \frac{C(t)}{G(t)}$, where $R(t)$ is the amount recovered for recycling, $C(t)$ is the amount recovered for composting, and t is the number of years after 2005.

a. Write this sum as a single rational expression.

b. If $R(t) = 0.03t + 0.94$, $C(t) = 0.005t + 0.4$, and $G(t) = -0.005t + 4.66$ (in pounds), and t is the number of years after 2005, write a simplified expression for $\frac{R(t)}{G(t)} + \frac{C(t)}{G(t)}$.

c. Use the answer for part **b** and the year 2005 ($t = 0$) to find the percent of total materials recovered.

SOLUTION 6

a. $\frac{R(t)}{G(t)}$ and $\frac{C(t)}{G(t)}$ have the same denominator so we add numerators obtaining $\frac{R(t) + C(t)}{G(t)}$.

b. Substituting $0.03t + 0.94$ for $R(t)$, $0.005t + 0.4$ for $C(t)$, and $-0.005t + 4.66$ for $G(t)$, $\frac{R(t) + C(t)}{G(t)} = \frac{(0.03t + 0.94) + (0.005t + 0.4)}{-0.005t + 4.66}$

$$= \frac{0.035t + 1.34}{-0.005t + 4.66}$$

c. When $t = 0$, $\frac{0.035t + 1.34}{-0.005t + 4.66} = \frac{0.035(0) + 1.34}{-0.005(0) + 4.66} \approx 0.2876 \approx 29\%$.

The actual percent was 31.7!

PROBLEM 6

This time use

$G(t) = 1.85t + 251$ (in tons),

$R(t) = 2.25t + 59$ (in tons), and

$C(t) = 0.55t + 20.5$ (in tons)

to write:

a. A simplified expression for $\frac{R(t)}{G(t)} + \frac{C(t)}{G(t)}$.

b. Use the answer you get in part **a** and the year 2005 ($t = 0$) to find the percent of materials recovered. Is the result close to the actual result of 31.7%?

Landfill facts: About 54% of this garbage actually goes to the landfill. There is enough space to take it all in now, but some experts say that by 2022 we will run out of landfill space.

Answers to PROBLEMS

6. a. $\frac{2.8t + 79.5}{1.85t + 251}$ b. $0.3167 \approx 31.7\%$ (Same as the actual result!)

And now, a last word before you go on to the exercise set. At this point, you can see that there are great similarities between algebra and arithmetic. In fact, in this very section we use the arithmetic addition of fractions as a model to do the algebraic addition of rational expressions. To show that these similarities are very strong and also to give you more practice, Problems 1–20 in the exercise set consist of two similar problems, an arithmetic one and an algebraic one. Use the practice and experience gained in working one to do the other.

> Exercises 6.3

> Web IT go to mhhe.com/bello for more lessons

⟨A⟩ Adding and Subtracting Rational Expressions with the Same Denominator
⟨B⟩ Adding and Subtracting Rational Expressions with Different Denominators

In Problems 1–40, perform the indicated operations.

1. a. $\frac{2}{7} + \frac{3}{7}$
b. $\frac{3}{x} + \frac{8}{x}$

2. a. $\frac{5}{9} + \frac{2}{9}$
b. $\frac{9}{x-1} + \frac{2}{x-1}$

3. a. $\frac{8}{9} - \frac{2}{9}$
b. $\frac{6}{x} - \frac{2}{x}$

4. a. $\frac{4}{7} - \frac{2}{7}$
b. $\frac{6}{x+4} - \frac{2}{x+4}$

5. a. $\frac{6}{7} + \frac{8}{7}$
b. $\frac{3}{2x} + \frac{7}{2x}$

6. a. $\frac{1}{9} + \frac{2}{9}$
b. $\frac{3}{8(x-2)} + \frac{1}{8(x-2)}$

7. a. $\frac{8}{3} - \frac{2}{3}$
b. $\frac{11}{3(x+1)} - \frac{9}{3(x+1)}$

8. a. $\frac{3}{8} - \frac{1}{8}$
b. $\frac{7}{15(x-1)} - \frac{2}{15(x-1)}$

9. a. $\frac{8}{9} + \frac{4}{9}$
b. $\frac{7x}{4(x+1)} + \frac{3x}{4(x+1)}$

10. a. $\frac{15}{14} + \frac{3}{14}$
b. $\frac{29x}{15(x-3)} + \frac{4x}{15(x-3)}$

11. a. $\frac{3}{4} - \frac{1}{3}$
b. $\frac{7}{x} - \frac{3}{8}$

12. a. $\frac{5}{7} - \frac{2}{5}$
b. $\frac{x}{3} - \frac{7}{x}$

13. a. $\frac{1}{5} + \frac{1}{7}$
b. $\frac{4}{x} + \frac{x}{9}$

14. a. $\frac{1}{3} + \frac{1}{9}$
b. $\frac{5}{x} + \frac{6}{3x}$

15. a. $\frac{2}{5} - \frac{4}{15}$
b. $\frac{4}{7(x-1)} - \frac{3}{14(x-1)}$

16. a. $\frac{9}{2} - \frac{5}{8}$
b. $\frac{8}{9(x+3)} - \frac{5}{36(x+3)}$

17. a. $\frac{4}{7} + \frac{3}{8}$
b. $\frac{3}{x+1} + \frac{5}{x-2}$

18. a. $\frac{2}{9} + \frac{4}{5}$
b. $\frac{2x}{x+2} + \frac{3x}{x-4}$

19. a. $\frac{7}{8} - \frac{1}{3}$
b. $\frac{6}{x-2} - \frac{3}{x+1}$

20. a. $\frac{6}{7} - \frac{2}{3}$
b. $\frac{4x}{x+1} - \frac{4x}{x+2}$

21. $\frac{x+1}{x^2+3x-4} + \frac{x+2}{x^2-16}$

22. $\frac{x-2}{x^2-9} + \frac{x+1}{x^2-x-12}$

23. $\frac{3x}{x^2+3x-10} + \frac{2x}{x^2+x-6}$

24. $\frac{x+3}{x^2-x-2} + \frac{x-1}{x^2+2x+1}$

25. $\frac{1}{x^2-y^2} + \frac{5}{(x+y)^2}$

26. $\frac{3x}{(x+y)^2} + \frac{5x}{x-y}$

27. $\frac{2}{x-5} - \frac{3x}{x^2-25}$

28. $\frac{x+3}{x^2-x-2} - \frac{x-1}{x^2+2x+1}$

29. $\frac{x-1}{x^2+3x+2} - \frac{x+7}{x^2+5x+6}$

30. $\frac{2}{x^2+3xy+2y^2} - \frac{1}{x^2-xy-2y^2}$

31. $\frac{y}{y^2-1} + \frac{y}{y+1}$

32. $\frac{3y}{y^2-4} - \frac{y}{y+2}$

33. $\frac{3y+1}{y^2-16} - \frac{2y-1}{y-4}$

34. $\frac{2y+1}{y^2-4} - \frac{3y-1}{y+2}$

35. $\frac{x+1}{x^2-x-2} + \frac{x-1}{x^2+2x+1}$

36. $\frac{y+3}{y^2+y-6} + \frac{y-2}{y^2+3y-10}$

37. $\frac{a}{a-w} - \frac{w}{a+w} - \frac{a^2+w^2}{a^2-w^2}$

38. $\frac{x}{x-y} + \frac{y}{x+y} - \frac{x^2+y^2}{x^2-y^2}$

39. $\frac{1}{a^3+8} + \frac{a+1}{a^2-2a+4}$

40. $\frac{c}{c^3-1} + \frac{2}{c^2+c+1}$

Web IT go to mhhe.com/bello for more lessons

‹C› Applications Involving Rational Expressions

41. *Odds in favor of an event* If the odds in favor of an event are f to u, the probability p of the event happening is $p = \frac{f}{f+u}$ and the probability of the event **not** happening is $q = \frac{u}{f+u}$.

a. Write $p + q$ as a single rational expression.

b. Simplify the fraction.

42. *Measuring noise* The noise measure M of a system is given by the equation $M = \frac{GF}{G-1} - \frac{G}{G-1}$, where F is the noise figure and G is the associated gain of the device. Perform the subtraction and write M as a single rational expression in reduced form.

43. *Noise analysis* When analyzing the noise in an electronic device, we need to add the expressions $\frac{1}{g_m} + \frac{8}{g_m^2}$, where g_m is the maximum noise conductance. Write $\frac{1}{g_m} + \frac{8}{g_m^2}$ as a single rational expression in reduced form.

››› Applications: Green Math

44. *Material for composting* The pounds per day for different categories of garbage are as follows:

$T(t) = 0.04t + 1$ (total materials for recovery),

$R(t) = 0.03t + 0.94$ (materials for recycling), and

$G(t) = -0.005t + 4.66$ (garbage generated),

where t is the number of years after 2005.

a. The expression $\frac{T(t)}{G(t)} - \frac{R(t)}{G(t)}$ represents the percent of materials recovered for composting. Substitute $T(t) = 0.04t + 1$, $R(t) = 0.03t + 0.94$, and $G(t) = -0.005t + 4.66$ and write the result as an expression in simplified form.

b. What percent of material was recovered for composting in 2005 ($t = 0$)?

c. What percent will be recovered for composting in 2015 ($t = 10$)?

45. *More composting* The tons per year for different categories of garbage are

$T(t) = 2.8t + 79$ (in tons),

$R(t) = 2.25t + 59$ (in tons), and

$G(t) = 1.85t + 251$ (in tons),

where t is the number of years after 2005.

a. The expression $\frac{T(t)}{G(t)} - \frac{R(t)}{G(t)}$ represents the percent of materials recovered for composting. Substitute $T(t) = 2.8t + 79$, $R(t) = 2.25t + 59$, and $G(t) = 1.85t + 251$ and write the result as an expression in simplified form.

b. What percent of material was recovered for composting in 2005 ($t = 0$)?

c. What percent will be recovered for composting in 2015 ($t = 10$)?

46. *Second law of motion* When working with Kepler's second law of motion, we have the expression

$$\frac{GM}{(1-\varepsilon)a} - \frac{GM}{(1+\varepsilon)a}$$

where G is the gravitational constant, M the mass of the object, a the object's semimajor axis, and ε is the specific orbital energy. Perform the subtraction and write the answer as a single fraction in reduced form.

47. *Moment of a beam* The moment M of a cantilever beam of length L, x units from the end is given by the expression

$$-\frac{w_0x^3}{6L} + \frac{w_0Lx}{2} - \frac{w_0L^2}{3}$$

Write this expression as a single rational expression in reduced form.

48. *Deflection of a beam* The deflection d of the beam of Problem 47 involves the expression

$$\frac{-x^4}{24L} + \frac{Lx^2}{4} - \frac{L^2x}{3}$$

Write this expression as a single rational expression in reduced form.

49. *Planetary Motion* In astronomy, planetary motion is given by the expression.

$$\frac{p^2}{2mr^2} - \frac{gmM}{r}$$

Write this expression as a single rational expression in reduced form.

50. *Pendulum* The motion of a pendulum is given by the expression

$$\frac{P_1^2 + P_2^2}{2(h_1 + h_2)} + \frac{P_1^2 - P_2^2}{2(h_1 - h_2)}$$

Write this expression as a single rational expression in reduced form.

Web IT go to mhhe.com/bello for more lessons

Using Your Knowledge

Continuing the Study of Fractions In Chapter 1, we mentioned that some numbers cannot be written as the ratio of two integers. These numbers are called **irrational numbers.** We can approximate irrational numbers by using a type of fraction called a **continued fraction.** Here's how we do it. From a table of square roots, or a calculator, we find that

$$\sqrt{2} \approx 1.4142$$ $\approx$ means "approximately equal."

Can we find some continued fraction to approximate $\sqrt{2}$?

51. Try $1 + \frac{1}{2}$ (write it as a decimal).

52. Try $1 + \frac{1}{2 + \frac{1}{2}}$ (write it as a decimal).

53. Try $1 + \frac{1}{2 + \frac{1}{2 + \frac{1}{2}}}$ (write it as a decimal).

54. Look at the pattern for the approximation of $\sqrt{2}$ given in Problems 51–53. What do you think the next approximation (when written as a continued fraction) will be?

55. How close is the approximation for $\sqrt{2}$ in Problem 53 to the value $\sqrt{2} \approx 1.4142$?

Write On

56. Write the procedure you use to find the LCD of two rational expressions.

57. Write the procedure you use to find the sum of two rational expressions:

a. with the same denominator.

b. with different denominators.

58. Write the procedure you use to find the difference of two rational expressions:

a. with the same denominator.

b. with different denominators.

Concept Checker

Fill in the blank(s) with the correct word(s), phrase, or mathematical statement.

59. The **first step** in **adding** or **subtracting** fractions with **different denominators** is to **find** the ________________ of the fractions.

60. The **second step** in **adding** or **subtracting** fractions with **different denominators** is to **write** all fractions as **equivalent fractions** with the ________________ as the denominator.

GCF

LCD

Mastery Test

61. The fraction of the waste recovered in the United States can be approximated by

$$\frac{R(t)}{G(t)}$$

Of this,

$$\frac{P(t)}{G(t)}$$

is paper and paperboard. If

$$P(t) = 0.02t^2 - 0.25t + 6$$

$$G(t) = 0.04t^2 + 2.34t + 90$$

$$R(t) = 0.04t^2 - 0.59t + 7.42$$

and t represents the number of years after 1960, find the fraction of the waste recovered that is *not* paper and paperboard.

Perform the indicated operations.

62. $\frac{x-3}{x^2-x-2} - \frac{x+3}{x^2-4}$

63. $\frac{5}{x-1} - \frac{3}{x+3}$

64. $\frac{4}{5} + \frac{3}{x}$

65. $\frac{4}{5(x-2)} + \frac{6}{5(x-2)}$

66. $\frac{11}{3(x+2)} - \frac{2}{3(x+2)}$

67. $\frac{x+3}{x^2-x-2} - \frac{x-3}{x^2-4}$

Skill Checker

Perform the indicated operations:

68. $2 + \frac{2}{9}$

69. $1 \div \frac{20}{9}$

70. $1 \div \frac{30}{7}$

71. $12x\left(\frac{2}{x} + \frac{3}{2x}\right)$

72. $12x\left(\frac{4}{3x} - \frac{1}{4x}\right)$

73. $x^2\left(1 - \frac{1}{x^2}\right)$

74. $x^2\left(1 + \frac{1}{x}\right)$

6.4 Complex Fractions

Objective

A › **Simplify a complex fraction using one of two methods.**

To Succeed, Review How To . . .

Add, subtract, multiply, and divide fractions (pp. 10–18).

Getting Started

Planetary Models and Complex Fractions

We've already learned how to do the four fundamental operations using rational expressions. In some instances, we want to find the quotient of two expressions that contain fractions in the numerator or the denominator or in both. For example, the model of the planets in our solar system shown here is similar to the model designed by the seventeenth-century Dutch mathematician and astronomer Christian Huygens. The gears used in the model were especially difficult to design since they had to make each of the planets revolve around the sun at different rates. For example, Saturn goes around the sun in

$$29 + \frac{1}{2 + \frac{2}{9}} \text{ yr}$$

The expression

$$\frac{1}{2 + \frac{2}{9}}$$

is a *complex fraction*. In this section we shall learn how to simplify complex fractions.

A › Simplifying Complex Fractions

Before we simplify *complex fractions,* we need a formal definition.

COMPLEX FRACTION

A **complex fraction** is a fraction that has one or more fractions in its numerator, denominator, or both.

Complex fractions can be simplified in either of two ways:

PROCEDURE

Simplifying Complex Fractions

1. Multiply numerator and denominator by the LCD of the fractions involved, or
2. Perform the operations indicated in the numerator and denominator of the complex fraction, and then divide the simplified numerator by the simplified denominator.

We illustrate these two methods by simplifying the complex fraction

$$\frac{1}{2 + \frac{2}{9}}$$

Method 1. Multiply numerator and denominator by the LCD of the fractions involved (in our case, by 9).

$$\frac{1}{2 + \frac{2}{9}} = \frac{9 \cdot 1}{9\left(2 + \frac{2}{9}\right)}$$ Note that $9\left(2 + \frac{2}{9}\right) = 9 \cdot 2 + 9 \cdot \frac{2}{9} = 18 + 2$.

$$= \frac{9}{18 + 2}$$

$$= \frac{9}{20}$$

Method 2. Perform the operations indicated in the numerator and denominator of the complex fraction and then divide the numerator by the denominator.

$$\frac{1}{2 + \frac{2}{9}} = \frac{1}{\frac{18}{9} + \frac{2}{9}} = \frac{1}{\frac{20}{9}}$$ Add $2 + \frac{2}{9}$.

$$= 1 \div \frac{20}{9}$$ Write $\frac{1}{\frac{20}{9}}$ as $1 \div \frac{20}{9}$.

$$= 1 \cdot \frac{9}{20}$$ Multiply by the reciprocal of $\frac{20}{9}$.

$$= \frac{9}{20}$$

Either procedure also works for more complicated rational expressions. In Example 1 we simplify a complex fraction using both methods. Compare the results and see which method you prefer!

EXAMPLE 1 **Simplifying complex fractions: Both methods**

Simplify:

$$\frac{\frac{1}{a} + \frac{2}{b}}{\frac{3}{a} - \frac{1}{b}}$$

SOLUTION 1

Method 1. The LCD of the fractions involved is ab, so we multiply the numerator and denominator by ab to obtain

$$\frac{ab \cdot \left(\frac{1}{a} + \frac{2}{b}\right)}{ab \cdot \left(\frac{3}{a} - \frac{1}{b}\right)} = \frac{ab \cdot \frac{1}{a} + ab \cdot \frac{2}{b}}{ab \cdot \frac{3}{a} - ab \cdot \frac{1}{b}}$$ Note that $\cancel{a}b \cdot \frac{1}{\cancel{a}} = b$, $a\cancel{b} \cdot \frac{2}{\cancel{b}} = 2a$, $\cancel{a}b \cdot \frac{3}{\cancel{a}} = 3b$, and $a\cancel{b} \cdot \frac{1}{\cancel{b}} = a$.

$$= \frac{b + 2a}{3b - a}$$

PROBLEM 1

Simplify:

$$\frac{\frac{2}{a} - \frac{3}{b}}{\frac{1}{a} + \frac{2}{b}}$$

Answers to PROBLEMS

1. $\frac{2b - 3a}{b + 2a}$

Method 2. Add the fractions in the numerator and subtract the fractions in the denominator. In both cases, the LCD of the fractions is ab. Hence,

$$\frac{\frac{1}{a}+\frac{2}{b}}{\frac{3}{a}-\frac{1}{b}} = \frac{\frac{b}{ab}+\frac{2a}{ab}}{\frac{3b}{ab}-\frac{a}{ab}}$$ Write the fractions with their LCD.

$$= \frac{\frac{b+2a}{ab}}{\frac{3b-a}{ab}}$$ Add in the numerator, subtract in the denominator.

$$= \frac{b+2a}{ab} \cdot \frac{ab}{3b-a}$$ Multiply by the reciprocal of $\frac{3b-a}{ab}$.

$$= \frac{b+2a}{3b-a}$$ Simplify.

EXAMPLE 2 Simplifying complex fractions: Method 1

Simplify:

$$\frac{\frac{2}{x}+\frac{3}{2x}}{\frac{4}{3x}-\frac{1}{4x}}$$

SOLUTION 2 We must first find the LCD of x, $2x$, $3x$, and $4x$. Now

$x =$			x
$2x = 2$			$\cdot\, x$
$3x =$		3	$\cdot\, x$
$4x = 2^2$			$\cdot\, x$

Write the factors in columns.

The LCD is $2^2 \cdot 3 \cdot x = 12x$. Multiplying numerator and denominator by $12x$, we have

$$\frac{\frac{2}{x}+\frac{3}{2x}}{\frac{4}{3x}-\frac{1}{4x}} = \frac{12x \cdot \left(\frac{2}{x}+\frac{3}{2x}\right)}{12x \cdot \left(\frac{4}{3x}-\frac{1}{4x}\right)}$$

$$= \frac{12x \cdot \frac{2}{x} + 12x \cdot \frac{3}{2x}}{12x \cdot \frac{4}{3x} - 12x \cdot \frac{1}{4x}}$$ Use the distributive property.

$$= \frac{12 \cdot 2 + 6 \cdot 3}{4 \cdot 4 - 3 \cdot 1}$$ Simplify.

$$= \frac{24+18}{16-3}$$

$$= \frac{42}{13}$$

PROBLEM 2

Simplify:

$$\frac{\frac{1}{4x}+\frac{2}{3x}}{\frac{3}{2x}-\frac{1}{x}}$$

EXAMPLE 3 Simplifying complex fractions: Method 1

Simplify:

$$\frac{1-\frac{1}{x^2}}{1+\frac{1}{x}}$$

PROBLEM 3

Simplify:

$$\frac{1-\frac{1}{x^2}}{1-\frac{1}{x}}$$

(continued)

Answers to PROBLEMS

2. $\frac{11}{6}$ **3.** $\frac{x+1}{x}$

SOLUTION 3 Here the LCD of the fractions involved is x^2. Thus,

$$\frac{1-\frac{1}{x^2}}{1+\frac{1}{x}} = \frac{x^2\cdot\left(1-\frac{1}{x^2}\right)}{x^2\cdot\left(1+\frac{1}{x}\right)}$$ Multiply numerator and denominator by x^2.

$$= \frac{x^2\cdot 1 - x^2\cdot\frac{1}{x^2}}{x^2\cdot 1 + x^2\cdot\frac{1}{x}}$$ Use the distributive property.

$$= \frac{x^2-1}{x^2+x}$$ Simplify.

$$= \frac{(x+1)(x-1)}{x(x+1)}$$ Factor.

$$= \frac{x-1}{x}$$ Simplify.

How many pounds of garbage do you produce each day? Each American produces about 4.6 pounds of garbage each day. What percent of that is recycled? We shall see in Example 4 but there is hope: the number of pounds produced each day is ***decreasing*** and the amount recovered for recycling is ***increasing***!

Source: http://tinyurl.com/n3tx9n.

EXAMPLE 4 **Garbage recovered for recycling**

The percent of garbage recovered for recycling is given by

$$\frac{\left(30+\frac{1090}{t}\right)}{\left(-5+\frac{4630}{t}\right)}$$

where t is the number of years after 2005.

a. Simplify this complex fraction.

b. Use the simplified form to find what ***percent*** of the garbage will be recovered for recycling in 2005 and 2010.

SOLUTION 4

a. We use method 1 and multiply numerator and denominator by t.

$$\frac{t\left(30+\frac{1090}{t}\right)}{t\left(-5+\frac{4630}{t}\right)} = \frac{30t+1090}{-5t+4630}$$ Use the distributive property.

b. The year 2005 corresponds to $t = 0$. Substituting $t = 0$ in the expression, we get

$$\frac{30t+1090}{-5t+4630} = \frac{30(0)+1090}{-5(0)+4630} = \frac{1090}{4630} \approx 24\%.$$

The year 2010 corresponds to $t = 5$, since t is the number of years after 2005. Substituting $t = 5$ in the expression gives

$$\frac{30t+1090}{-5t+4630} = \frac{30(5)+1090}{-5(5)+4630} = \frac{150+1090}{-25+4630} = \frac{1240}{4605} \approx 27\%.$$

PROBLEM 4

Unfortunately, not all garbage is recycled. The percent going to the landfill is given by $\frac{\left(-10+\frac{2580}{t}\right)}{\left(-5+\frac{4630}{t}\right)}$ where t is the number of years after 2005.

a. Simplify this complex fraction.

b. Use the simplified form to find what ***percent*** of the garbage will be going to the landfill in 2005 and 2010.

Answers to PROBLEMS

4. a. $\frac{-10t+2580}{-5t+4630}$ **b.** 56%, 55%

Exercises 6.4

Web IT go to mhhe.com/bello for more lessons

A Simplifying Complex Fractions

In Problems 1–24, simplify.

1. $\dfrac{\frac{1}{2}}{2+\frac{1}{2}}$ 2. $\dfrac{\frac{1}{4}}{3+\frac{1}{4}}$ 3. $\dfrac{\frac{1}{2}}{2-\frac{1}{2}}$ 4. $\dfrac{\frac{1}{4}}{3-\frac{1}{4}}$ 5. $\dfrac{a-\frac{a}{b}}{1+\frac{a}{b}}$

6. $\dfrac{1-\frac{1}{a}}{1+\frac{1}{a}}$ 7. $\dfrac{\frac{1}{a}+\frac{1}{b}}{\frac{1}{a}-\frac{1}{b}}$ 8. $\dfrac{\frac{2}{a}+\frac{1}{b}}{\frac{2}{a}-\frac{1}{b}}$ 9. $\dfrac{\frac{1}{2a}+\frac{1}{3b}}{\frac{4}{a}-\frac{3}{4b}}$ 10. $\dfrac{\frac{1}{2a}+\frac{1}{4b}}{\frac{2}{a}-\frac{3}{5b}}$

11. $\dfrac{\frac{1}{3}+\frac{3}{4}}{\frac{3}{8}-\frac{1}{6}}$ 12. $\dfrac{\frac{1}{5}+\frac{3}{2}}{\frac{5}{8}-\frac{3}{10}}$ 13. $\dfrac{2+\frac{1}{x}}{4-\frac{1}{x^2}}$ 14. $\dfrac{3+\frac{1}{x}}{9-\frac{1}{x^2}}$ 15. $\dfrac{2+\frac{2}{x}}{1+\frac{1}{x}}$

16. $\dfrac{5+\frac{5}{x^2}}{1+\frac{1}{x^2}}$ 17. $\dfrac{\frac{1}{y}+\frac{1}{x}}{\frac{x}{y}-\frac{y}{x}}$ 18. $\dfrac{\frac{1}{x}+\frac{1}{y}}{\frac{y}{x}-\frac{x}{y}}$ 19. $\dfrac{x-2-\frac{8}{x}}{x-3-\frac{4}{x}}$ 20. $\dfrac{x-2-\frac{15}{x}}{x-3-\frac{10}{x}}$

21. $\dfrac{\frac{1}{x+5}}{\frac{4}{x^2-25}}$ 22. $\dfrac{\frac{1}{x-3}}{\frac{2}{x^2-9}}$ 23. $\dfrac{\frac{1}{x^2-16}}{\frac{2}{x+4}}$ 24. $\dfrac{\frac{3}{x^2-64}}{\frac{4}{x+8}}$

Applications: Green Math

25. *Garbage to be composted* The percent of garbage that is not recycled and is composted is given by $\dfrac{\left(5+\frac{400}{t}\right)}{\left(-5+\frac{4630}{t}\right)}$, where t is the number of years after 2005.
   a. Simplify this complex fraction.
   b. Use the simplified form to find to one decimal place what ***percent*** of the garbage will be composted in 2005 and in 2010.

26. *Total materials recycled* The percent of total materials recycled is given by $\dfrac{\left(40+\frac{1000}{t}\right)}{\left(-5+\frac{4630}{t}\right)}$, where t is the number of years after 2005.
   a. Simplify this complex fraction.
   b. Use the simplified form to find to one decimal place the ***percent*** of total materials recycled in 2005 and in 2010.

Using Your Knowledge

Around the Sun in Complex Fractions In the *Getting Started* of this section, we mentioned that Saturn takes

$$29+\frac{1}{2+\frac{2}{9}}\text{ yr}$$

to orbit the sun. Since we have shown that

$$\frac{1}{2+\frac{2}{9}}=\frac{9}{20}$$

we know that Saturn takes $29+\frac{9}{20}=29\frac{9}{20}$ years to orbit the sun.

In Problems 27–32, use your knowledge to simplify the number of years it takes the following planets to orbit the sun.

27. Mercury, $\dfrac{1}{4+\frac{1}{6}}$ yr

28. Venus, $\dfrac{1}{1+\frac{2}{3}}$ yr

In Problems 29–32, write your answer as a mixed number.

29. Jupiter, $11 + \dfrac{1}{1 + \dfrac{7}{43}}$ yr

30. Mars, $1 + \dfrac{1}{1 + \dfrac{3}{22}}$ yr

31. Uranus, $84 + \dfrac{1}{90 + \dfrac{10}{11}}$

32. Neptune, $164 + \dfrac{1}{1 + \dfrac{21}{79}}$

As you may know, there are three new "dwarf" planets: Ceres, Eris, and the newly reclassified Pluto. How long does it take (in years) these planets to orbit the sun? We will see in Problems 33–36. Write your answer as a mixed number.

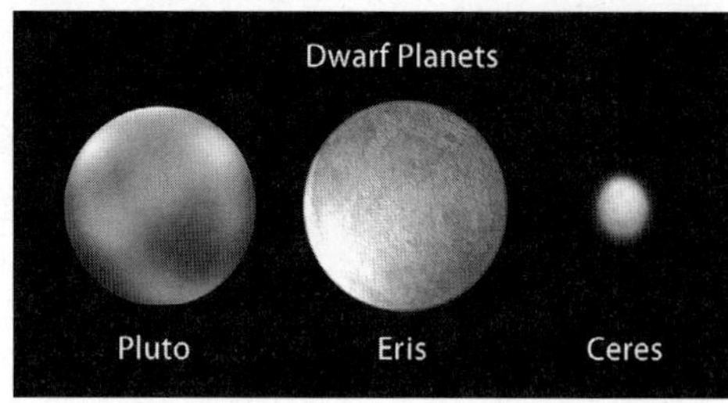

33. Ceres, $4 + \dfrac{1}{1 + \dfrac{2}{3}}$

34. Eris, $556 + \dfrac{1}{1 + \dfrac{3}{7}}$

35. Pluto, $248 + \dfrac{1}{1 + \dfrac{23}{27}}$

36. Look at the results of Problems 27–35.

a. Which planet takes the longest time to orbit the sun?

b. Which planet takes the shortest time to orbit the sun?

Write On

37. What is a complex fraction?

38. List the advantages and disadvantages of Method 1 when simplifying a complex fraction.

39. List the advantages and disadvantages of Method 2 when simplifying a complex fraction.

40. Which method do you prefer to simplify complex fractions? Why?

41. How do you know which method to use when simplifying complex fractions?

Concept Checker

Fill in the blank(s) with the correct word(s), phrase, or mathematical statement.

42. A ______________ **fraction** is a **fraction** that has **one or more fractions** in its **numerator, denominator,** or **both.**

43. The **first** step to **simplify** a **complex fraction** is to **multiply its numerator** and **denominator** by the ______________ of all the fractions involved.

simplified **GCF**
complex **LCD**

Mastery Test

Simplify:

44. $\dfrac{\dfrac{2}{a} - \dfrac{3}{b}}{\dfrac{1}{a} + \dfrac{2}{b}}$

45. $\dfrac{\dfrac{1}{4x} + \dfrac{2}{3x}}{\dfrac{3}{2x} - \dfrac{1}{x}}$

46. $\dfrac{1 - \dfrac{1}{x^2}}{1 - \dfrac{1}{x}}$

47. $\dfrac{w + 2 - \dfrac{18}{w - 5}}{w - 1 - \dfrac{12}{w - 5}}$

48. $\dfrac{\dfrac{8x}{3x + 1} - \dfrac{3x - 1}{x}}{\dfrac{4x}{3x + 1} - \dfrac{2x - 1}{x}}$

49. $\dfrac{\dfrac{3}{m - 4} - \dfrac{16}{m - 3}}{\dfrac{2}{m - 3} - \dfrac{15}{m + 5}}$

Skill Checker

Solve:

50. $17x = 408$

51. $19w = 2356$

52. $18L = 2232$

53. $9x + 24 = x$

54. $10x + 36 = x$

55. $5x = 4x + 3$

56. $6x = 5x + 5$

6.5 Solving Equations Containing Rational Expressions

Objectives

A Solve equations that contain rational expressions.

B Solve a fractional equation for a specified variable.

C Solve applications involving fractional equations.

To Succeed, Review How To . . .

1. Solve linear equations (pp. 137–141).
2. Find the LCD of two or more rational expressions (pp. 515–520).
3. Solve quadratic equations (pp. 456–462).

Getting Started

Salute One of the Largest Flags

This American flag (one of the largest ever) was displayed in J. L. Hudson's store in Detroit. By law, the ratio of length to width of the American flag should be $\frac{19}{10}$. If the length of this flag was 235 feet, what should its width be to conform with the law? To solve this problem, we let W be the width of the flag and set up the equation

$$\frac{19}{10} = \frac{235}{W} \quad \begin{array}{l}\leftarrow \text{Length} \\ \leftarrow \text{Width}\end{array}$$

This equation is an example of a *fractional equation*. A **fractional equation** is an equation that contains one or more rational expressions. To solve this equation, we must clear the denominators involved. We did this in Chapter 2 (Section 2.3) by multiplying each term by the LCD. Since the LCD of $\frac{19}{10}$ and $\frac{235}{W}$ is $10W$, we have

$$10W \cdot \frac{19}{10} = \frac{235}{W} \cdot 10W \quad \text{Multiply by the LCD.}$$

$$19W = 2350 \quad \text{Simplify.}$$

$$W = \frac{2350}{19} \quad \text{Divide by 19.}$$

$$W \approx 124 \quad \text{Approximate the answer.}$$

$$\begin{array}{r} 123.6 \\ 19\overline{)2350.0} \\ \underline{19} \\ 45 \\ \underline{38} \\ 70 \\ \underline{57} \\ 130 \\ \underline{114} \\ 16 \end{array}$$

(By the way, the flag was only 104 feet long and weighed 1500 pounds. It was *not* an official flag.) In this section we shall learn how to solve fractional equations.

A > Solving Fractional Equations

The first step in solving fractional equations is to multiply each side of the equation by the **least common multiple (LCM)** of the denominators present. This is equivalent to multiplying *each term* by the LCD because if you have the equation

$$\frac{a}{b} + \frac{c}{d} = \frac{e}{f}$$

then multiplying each side by L, the LCM of the denominators, gives

$$L \cdot \left(\frac{a}{b} + \frac{c}{d}\right) = L \cdot \frac{e}{f}$$

or, using the distributive property,

$$L \cdot \frac{a}{b} + L \cdot \frac{c}{d} = L \cdot \frac{e}{f}$$

Thus, we have the following result.

LEAST COMMON MULTIPLE

Multiplying each side of the equation

$$\frac{a}{b} + \frac{c}{d} = \frac{e}{f} \text{ (where } b, d, \text{ and } f \neq 0)$$

by L is equivalent to multiplying each *term* by L, that is,

$$L\left(\frac{a}{b} + \frac{c}{d}\right) = L\left(\frac{e}{f}\right) \text{ is equivalent to}$$

$$L \cdot \frac{a}{b} + L \cdot \frac{c}{d} = L \cdot \frac{e}{f}.$$

The procedure for solving fractional equations is similar to the one used to solve linear equations. You may want to review the procedure for solving linear equations (see pp. 137–141) before continuing with this section.

EXAMPLE 1 Solving fractional equations

Solve:

$$\frac{3}{4} + \frac{2}{x} = \frac{1}{12}$$

SOLUTION 1 The LCD of $\frac{3}{4}$, $\frac{2}{x}$, and $\frac{1}{12}$ is $12x$.

1. Clear the fractions; the LCD is $12x$. $\quad 12x \cdot \frac{3}{4} + 12x \cdot \frac{2}{x} = 12x \cdot \frac{1}{12}$
2. Simplify. $\quad 9x + 24 = x$
3. Subtract 24 from each side. $\quad 9x = x - 24$
4. Subtract x from each side. $\quad 8x = -24$
5. Divide each side by 8. $\quad x = -3$

The answer is -3.

6. Here is the check:

$$\frac{3}{4} + \frac{2}{x} \stackrel{?}{=} \frac{1}{12}$$

$\frac{3}{4} + \frac{2}{-3}$	$\frac{1}{12}$
$\frac{3 \cdot 3}{4 \cdot 3} + \frac{2 \cdot 4}{-3 \cdot 4}$	
$\frac{9}{12} - \frac{8}{12}$	
$\frac{1}{12}$	

PROBLEM 1

Solve:

$$\frac{4}{5} + \frac{1}{x} = \frac{3}{10}$$

Answers to PROBLEMS

1. $x = -2$

In some cases, the denominators involved may be more complicated. Nevertheless, the procedure used to solve the equation remains the same. Thus, we can also use the six-step procedure to solve

$$\frac{2x}{x-1} + 3 = \frac{4x}{x-1}$$

as shown in Example 2.

EXAMPLE 2 **Solving fractional equations**

Solve:

$$\frac{2x}{x-1} + 3 = \frac{4x}{x-1}$$

SOLUTION 2 Since $x - 1$ is the only denominator, it must be the LCD. We then proceed by steps.

1. Clear the fractions; the LCD is $(x - 1)$. $\quad (x-1)\cdot\frac{2x}{x-1} + 3(x-1) = (x-1)\cdot\frac{4x}{x-1}$
2. Simplify. $\quad 2x + 3x - 3 = 4x$
   $\quad 5x - 3 = 4x$
3. Add 3. $\quad 5x = 4x + 3$
4. Subtract $4x$. $\quad x = 3$
5. Division is not necessary.
6. You can easily check this by substitution.

Thus, the answer is $x = 3$, as can easily be verified by substituting 3 for x in the original equation.

PROBLEM 2

Solve:

$$\frac{8}{x-1} - 4 = \frac{2x}{x-1}$$

So far, the denominators used in the examples have *not* been factorable. In the cases where they are, it's very important that we factor them *before* we find the LCD. For instance, to solve the equation

$$\frac{x}{x^2-16} + \frac{4}{x-4} = \frac{1}{x+4}$$

we first note that

$$x^2 - 16 = (x+4)(x-4)$$

We then write

$$\frac{x}{x^2-16} + \frac{4}{x-4} = \frac{1}{x+4}$$

The denominator $x^2 - 16$ has been factored as $(x + 4)(x - 4)$.

as

$$\frac{x}{(x+4)(x-4)} + \frac{4}{x-4} = \frac{1}{x+4}$$

The solution to this equation is given in the next example.

EXAMPLE 3 **Solving fractional equations**

Solve:

$$\frac{x}{x^2-16} + \frac{4}{x-4} = \frac{1}{x+4}$$

PROBLEM 3

Solve:

$$\frac{x}{x^2-16} + \frac{4}{x-4} = \frac{1}{x+4}$$

(continued)

Answers to PROBLEMS

2. $x = 2$ **3.** $x = -5$

SOLUTION 3 Since $x^2 - 16 = (x + 4)(x - 4)$, we write the equation with $x^2 - 16$ factored as

$$\frac{x}{(x+4)(x-4)} + \frac{4}{x-4} = \frac{1}{x+4}$$

1. Clear the fractions; the LCD is $(x + 4)(x - 4)$.

$$(x+4)(x-4)\cdot\frac{x}{(x+4)(x-4)} + (x+4)(x-4)\cdot\frac{4}{x-4} = (x+4)(x-4)\cdot\frac{1}{x+4}$$

2. Simplify. $x + 4(x + 4) = x - 4$

$x + 4x + 16 = x - 4$

$5x + 16 = x - 4$

3. Subtract 16. $5x = x - 20$

4. Subtract x. $4x = -20$

5. Divide by 4. $x = -5$

Thus, the solution is $x = -5$.

6. You can easily check this by substituting -5 for x in the original equation.

By now, you've probably noticed that we always recommend checking the possible or proposed solution by *direct substitution* into the original equation. This is done to avoid **extraneous solutions,** that is, possible or proposed solutions that do not satisfy the original equation when substituted for the variable.

EXAMPLE 4 Solving fractional equations: No-solution case

Solve:

$$\frac{x}{x+4} - \frac{2}{5} = \frac{-4}{x+4}$$

SOLUTION 4 Here the denominators are 5 and $(x + 4)$.

1. Clear the fractions; the LCD is $5(x + 4)$.

$$5(x+4)\cdot\frac{x}{x+4} - \frac{2}{5}\cdot 5(x+4) = \frac{-4}{x+4}\cdot 5(x+4)$$

2. Simplify. $5x - 2x - 8 = -20$

$3x - 8 = -20$

3. Add 8. $3x = -12$

4. The variable is already isolated, so step 4 is not necessary.

5. Divide by 3. $x = -4$

Thus, the solution seems to be $x = -4$.

6. But now let's do the check. If we substitute -4 for x in the original equation, we have

$$\frac{-4}{-4+4} - \frac{2}{5} = \frac{-4}{-4+4}$$

or

$$\frac{-4}{0} - \frac{2}{5} = \frac{-4}{0}$$

Division by zero is not defined.

Two of the terms are not defined. Thus, this equation has no solution.

PROBLEM 4

Solve:

$$\frac{x}{x-2} + \frac{3}{4} = \frac{2}{x-2}$$

Answers to PROBLEMS

4. No solution

NOTE

Remember, no matter how careful you are when you get an answer, call it a *possible* or a *proposed* answer. A given number is *not* an answer until you show that, when the number is substituted for the variable in the original equation, the result is a true statement.

Finally, we must point out that the equations resulting when clearing denominators are not *always* linear equations—that is, equations that can be written in the form $ax + b = c$ $(a \neq 0)$. For example, to solve the equation

$$\frac{x^2}{x+2} = \frac{4}{x+2}$$

we first multiply by the LCD $(x + 2)$ to obtain

$$(x+2) \cdot \frac{x^2}{x+2} = (x+2) \cdot \frac{4}{x+2}$$

or

$$x^2 = 4$$

In this equation, the variable x has a 2 as an exponent; thus, it is a **quadratic** equation and can be solved when written in standard form—that is, by writing the equation as

$x^2 - 4 = 0$ Recall that a quadratic equation is an equation that can be written in standard form as $ax^2 + bx + c = 0$ $(a \neq 0)$.

$(x + 2)(x - 2) = 0$ Factor.

$x + 2 = 0$ or $x - 2 = 0$ Use the zero product property.

$x = -2$ or $x = 2$ Solve each equation.

Thus, $x = 2$ is a solution since

$$\frac{2^2}{2+2} = \frac{4}{2+2}$$

However, for $x = -2$,

$$\frac{x^2}{x+2} = \frac{2^2}{-2+2} = \frac{4}{0}$$

and the denominator $x + 2$ becomes 0. Thus, $x = -2$ is *not* a solution; -2 is called an *extraneous* solution. The only solution is $x = 2$.

EXAMPLE 5 **Solving fractional equations: Extraneous solution case**

Solve:

$$1 + \frac{3}{x-2} = \frac{12}{x^2-4}$$

SOLUTION 5 Since $x^2 - 4 = (x + 2)(x - 2)$, the LCD is $(x + 2)(x - 2)$. We then write the equation with the denominator $x^2 - 4$ in factored form and multiply each term by the LCD as before. Here are the steps.

$$(x+2)(x-2) \cdot 1 + (x+2)(x-2) \cdot \frac{3}{x-2} = (x+2)(x-2) \cdot \frac{12}{(x+2)(x-2)}$$ The LCD is $(x + 2)(x - 2)$.

PROBLEM 5

Solve:

$$1 - \frac{4}{x^2-1} = \frac{-2}{x-1}$$

(continued)

Answers to PROBLEMS

5. $x = -3$

$$(x^2 - 4) + 3(x + 2) = 12$$ Simplify.

$$x^2 - 4 + 3x + 6 = 12$$

$$x^2 + 3x + 2 = 12$$

$$x^2 + 3x - 10 = 0$$ Subtract 12 from both sides to write in standard form.

$$(x + 5)(x - 2) = 0$$ Factor.

$$x + 5 = 0 \quad \text{or} \quad x - 2 = 0$$ Use the zero product property.

$$x = -5 \qquad x = 2$$ Solve each equation.

Since $x = 2$ makes the denominator $x - 2$ equal to zero, the only possible solution is $x = -5$. This solution can be verified in the original equation.

B › Solving Fractional Equations for a Specified Variable

In Section 2.6, we solved a formula for a specified variable. We can also solve fractional equations for a specified variable. Here S_n is the sum of an arithmetic sequence:

$$S_n = \frac{n(a_1 + a_n)}{2}$$

(By the way, don't be intimidated by subscripts such as n in S_n, which is read "S sub n"; they are simply used to distinguish one variable from another—for example, a_1 is different from a_n because the subscripts are different.) So, to solve for n in this equation, we proceed as follows:

$$S_n = \frac{n(a_1 + a_n)}{2}$$ Given. Since the only denominator is 2, the LCD is 2.

1. Clear any fractions; the LCD is 2: $2 \cdot S_n = 2 \cdot \frac{n(a_1 + a_n)}{2}$
2. Simplify. $2S_n = n(a_1 + a_n)$
3. Since we want n by itself, divide by $(a_1 + a_n)$: $\frac{2S_n}{a_1 + a_n} = n$

Thus, the solution is $n = \frac{2S_n}{a_1 + a_n}$.

EXAMPLE 6 **Solving for a specified variable**

The sum S_n of a geometric sequence is

$$S_n = \frac{a_1(1 - r^n)}{1 - r}$$

Solve for a_1.

SOLUTION 6 This time, the only denominator is $1 - r$, so it must be the LCD. As before, we proceed by steps.

$$S_n = \frac{a_1(1 - r^n)}{1 - r}$$ Given

1. Clear any fractions; the LCD is $1 - r$. $(1 - r)S_n = (1 - r) \cdot \frac{a_1(1 - r^n)}{1 - r}$
2. Simplify. $(1 - r)S_n = a_1(1 - r^n)$
3. Divide both sides by $1 - r^n$ to isolate a_1. $\frac{(1 - r)S_n}{1 - r^n} = a_1$

Thus, the solution is $a_1 = \frac{(1 - r)S_n}{1 - r^n}$.

PROBLEM 6

Solve for a in

$$S = \frac{a(r^n - 1)}{r - 1}$$

Answers to PROBLEMS

6. $a = \frac{S(r - 1)}{r^n - 1}$

You will have more opportunities to practice solving for a specified variable in the *Using Your Knowledge.*

C > Applications Involving Fractional Equations

The 37th largest producer of air pollution in the United States (11 million pounds of toxic chemicals released annually into the air) has decided to invest \$330 million dollars to renovate one of its power stations and make it one of the cleanest coal-fired power plants in the nation. What percent of the nitrogen oxide emissions can they reduce with the \$330 million dollar investment? We can use fractional equations to find the answer!

Source: http://en.wikipedia.org/wiki/TECO_Energy.

EXAMPLE 7 Applications of fractional equations to remove oxides

The equation $\frac{60x}{1-x} = 330$ gives the cost (in millions) of removing x percent of nitrogen oxide emissions.

Find x to determine what percent of the emissions (to the nearest percent) can be removed with \$330 million.

SOLUTION 7

We proceed by steps.

1. Clear the fractions. $\cancel{(1-x)}\frac{60x}{\cancel{1-x}} = 330(1-x)$ Multiply both sides by $(1 - x)$.
2. Simplify. $60x = 330 - 330x$
3. Add $330x$ to both sides. $390x = 330$
4. Divide by 390. $x = \frac{330}{390} \approx 85\%$

Thus, they can reduce 85% of the emissions with \$330 million.

PROBLEM 7

To the nearest percent, what percent of the nitrogen oxide can be removed if they wanted to invest \$300 million?

> Exercises 6.5

< A > Solving Fractional Equations In Problems 1–50, solve (if possible).

1. $\frac{x}{4} = \frac{3}{2}$
2. $\frac{x}{8} = \frac{-7}{4}$
3. $\frac{3}{x} = \frac{3}{4}$
4. $\frac{6}{x} = \frac{-2}{7}$
5. $-\frac{8}{3} = \frac{16}{x}$
6. $-\frac{5}{6} = \frac{10}{x}$
7. $\frac{4}{3} = \frac{x}{9}$
8. $-\frac{3}{7} = \frac{x}{14}$
9. $\frac{2}{5} + \frac{3}{x} = \frac{23}{20}$
10. $\frac{6}{7} + \frac{2}{x} = \frac{3}{21}$
11. $\frac{3}{x} - \frac{2}{7} = \frac{11}{35}$
12. $\frac{4}{x} - \frac{2}{9} = \frac{22}{63}$
13. $\frac{3}{5} + \frac{7x}{10} = 2$
14. $\frac{2}{7} + \frac{4x}{21} = \frac{2}{3}$
15. $\frac{3x}{4} - \frac{1}{5} = \frac{13}{10}$
16. $\frac{2x}{3} - \frac{1}{4} = -1$
17. $\frac{3}{x+2} = \frac{4}{x-1}$
18. $\frac{2}{x-2} = \frac{5}{x+1}$

Answers to PROBLEMS

7. 83%

19. $\frac{-1}{x+1} = \frac{3}{x+5}$

20. $\frac{2}{x-1} = \frac{-3}{x+9}$

21. $\frac{3x}{x-3} + 2 = \frac{5x}{x-3}$

22. $\frac{2x}{x-2} + 18 = \frac{8x}{x-2}$

23. $\frac{5x}{x+1} - 6 = \frac{3x}{x+1}$

24. $\frac{5x}{x+1} - 2 = \frac{2x}{x+1}$

25. $\frac{x}{x^2-25} + \frac{5}{x-5} = \frac{1}{x+5}$

26. $\frac{x}{x^2-64} + \frac{8}{x-8} = \frac{1}{x+8}$

27. $\frac{x}{x^2-49} + \frac{7}{x-7} = \frac{1}{x+7}$

28. $\frac{x}{x^2-1} + \frac{1}{x-1} = \frac{1}{x+1}$

29. $\frac{x}{x+3} + \frac{3}{4} = \frac{-3}{x+3}$

30. $\frac{1}{5} + \frac{x}{x-2} = \frac{2}{x-2}$

31. $\frac{x}{x-4} - \frac{2}{7} = \frac{4}{x-4}$

32. $\frac{x}{x-8} - \frac{1}{5} = \frac{8}{x-8}$

33. $1 + \frac{2}{x-1} = \frac{4}{x^2-1}$

34. $1 + \frac{2}{x-3} = \frac{5}{x^2-9}$

35. $2 - \frac{6}{x^2-1} = \frac{-3}{x-1}$

36. $2 - \frac{4}{x^2-4} = \frac{-1}{x-2}$

37. $\frac{4}{x-3} - \frac{2}{x-1} = \frac{2}{x+2}$

38. $\frac{5}{x+1} - \frac{1}{x+2} = \frac{13}{x+5}$

39. $\frac{2x}{x^2-1} + \frac{4}{x-1} = \frac{1}{x-1}$

40. $\frac{3x-2}{x^2-4} + \frac{4}{x+2} = \frac{1}{x-2}$

41. $\frac{2z+7}{z-3} + 1 = \frac{z+5}{z-6} + 2$

42. $\frac{4x-2}{x+4} - 3 = \frac{2-5x}{x-2} + 6$

43. $\frac{2y-5}{2} - \frac{1}{y-1} = y + 1$

44. $\frac{4}{x+1} - \frac{3}{x} = \frac{1}{x-2}$

45. $\frac{5}{2v-1} + \frac{2}{v} = \frac{18v}{4v^2-1}$

46. $\frac{3y}{y^2-9} - \frac{4}{y+3} = \frac{6-y}{y^2+3y}$

47. $\frac{z+7}{z-1} - \frac{z+3}{z-2} = \frac{3}{z-6}$

48. $\frac{x+1}{x-3} - \frac{x+5}{x-2} = \frac{-3}{x-1}$

49. $\frac{2}{x^2-4x+3} - \frac{5}{x^2-x-6} = \frac{x-7}{(x-1)(x-3)(x+2)}$

50. $\frac{y-5}{y^2-4} - \frac{1}{y^2+2y-8} = \frac{y^2}{(y^2-4)(y+4)}$

‹ B › Solving Fractional Equations for a Specified Variable

Use the following information for Problems 51–53. Suppose you inflate a balloon for a party. How can you make it smaller? One way is to squeeze it. This will increase the pressure P of the air inside the balloon and decrease its volume V. The law stating the relationship between the pressure P and the volume V is called Boyle's law after Robert Boyle, an Irish physicist and chemist.

The law simply states that for a fixed amount of gas the product of the pressure P and the volume V is a constant, that is, $PV =$ constant provided the temperature is not changed.

If you start heating the balloon, the pressure will change!

This means that if you have two different balloons with two different pressures and volumes, the product of their respective pressures and volumes is the same constant. Thus,

$$P_1V_1 = P_2V_2$$

51. Solve for V_2.

52. Suppose you have a 2-liter (2-L) volume balloon so $V_1 = 2$ at the regular atmospheric pressure $P_1 = 1$ [1 atmosphere (atm) = 14.7 pounds per square inch]. If you go on a submarine with your balloon and the pressure increases to $P_2 = 2$ atmospheres, what is the volume V_2 of the balloon? (*Hint:* Use the results of Problem 51.)

53. Now, suppose you go on the space shuttle with your 2-liter balloon ($V_1 = 2$ L and $P_1 = 1$ atm) and the pressure on the space shuttle is only half as much as on earth ($P_2 = \frac{1}{2}$ atmosphere). What is the volume V_2 of your balloon?

Web IT go to mhhe.com/bello for more lessons

Use the following information for Problem 54. How can you make your balloon smaller without squeezing it, that is, leaving the pressure constant? You can freeze it! The law stating the relationship between the volume of a gas and its temperature is called Charles' law after the French chemist, physicist, and mathematician Jacques Charles. The law simply states that for a fixed pressure, the quotient of the volume V of a gas and its temperature T (in Kelvins) is a constant. Thus,

$$\frac{V_1}{T_1} = \frac{V_2}{T_2}$$

54. a. Solve for V_2.

b. Consider a balloon with a volume $V_1 = 21$ liters at $T_1 = 273$ Kelvins. What will be the new volume V_2 if the temperature increases to $T_2 = 299$ Kelvins?

Applications: Green Math

The 1990 Clean Air Act Amendment sets a cap on total sulfur dioxide emissions. New total controls costs (the costs of measures used to mitigate pollution) saved an estimated $750 million to $1.5 billion by using new technology.

Source: http://www.econlib.org/library/Enc/PollutionControls.html.

55. *Reducing emissions* Use the equation $\frac{250x}{1-x} = 750$ million to find what percent x of the emissions can be reduced with $750 million.

You can actually buy "pollution permits" for certain pollutants! See the source to learn how this works.

56. *More emission reduction* The more you pay, the more pollution you can reduce! Suppose you are willing to pay 1000 million dollars (that's a billion) to reduce pollution. Use the equation $\frac{250x}{1-x} = 1000$ million to find what percent of the emissions can then be reduced.

Using Your Knowledge

Looking for Variables in the Right Places Use your knowledge of fractional equations to solve the given problem for the indicated variable.

57. The area A of a trapezoid is

$$A = \frac{h(b_1 + b_2)}{2}$$

Solve for h.

58. When studying an electric circuit, we have to work with the equation

$$\frac{1}{R} = \frac{1}{R_1} + \frac{1}{R_2}$$

Solve for R.

59. In refrigeration we find the formula

$$\frac{Q_1}{Q_2 - Q_1} = P$$

Solve for Q_1.

60. When studying the expansion of metals, we find the formula

$$\frac{L}{1 + at} = L_0$$

Solve for t.

61. Manufacturers of camera lenses use the formula

$$\frac{1}{f} = \frac{1}{a} + \frac{1}{b}$$

Solve for f.

Write On

62. Explain the difference between adding two rational expressions such as

$$\frac{1}{x} + \frac{1}{2}$$

and solving an equation such as

$$\frac{1}{x} + \frac{1}{2} = 1$$

63. Write the procedure you use to solve a fractional equation.

64. In Section 6.1, we used the fundamental rule of rational expressions to multiply the numerators and denominators of *fractions* so that the resulting fractions have the LCD as their denominator. In this section we multiplied each *term* of an equation by the LCD. Explain the difference.

Concept Checker

Fill in the blank(s) with the correct word(s), phrase, or mathematical statement.

false	letter
true	term

65. **Multiplying** each side of $\frac{a}{b} + \frac{c}{d} = \frac{e}{f}$ (where b, d, and $f \neq 0$) by the **least common multiple** L of the denominators is **equivalent** to multiplying each ____________ in the equation by L.

66. A given number ***n*** **is not** a solution (it is only a ***proposed*** solution) of an equation until you show that **when *n* is substituted for the variable** in the **original** equation, the result is a ____________ statement.

Mastery Test

Solve:

67. $1 - \frac{4}{x^2 - 1} = \frac{-2}{x - 1}$

68. $\frac{x}{x - 2} + \frac{3}{4} = \frac{2}{x - 2}$

69. $\frac{x}{x^2 - 9} + \frac{3}{x - 3} = \frac{1}{x + 3}$

70. $\frac{8}{x - 1} - 4 = \frac{2x}{x - 1}$

71. $\frac{4}{5} + \frac{1}{x} = \frac{3}{10}$

72. $\frac{1}{x^2 + 2x - 3} + \frac{1}{x^2 - 9} = \frac{1}{(x - 1)(x^2 - 9)}$

Solve for the specified variable:

73. $C = \frac{5}{9}(F - 32)$; F

74. $A = P(1 + r)$; r

Skill Checker

Solve:

75. $4(x + 3) = 45$

76. $\frac{d}{3} + \frac{d}{4} = 1$

77. $\frac{h}{4} + \frac{h}{6} = 1$

78. $60(R - 5) = 40(R + 5)$

79. $30(R - 5) = 10(R + 15)$

80. $\frac{n}{90} = \frac{1}{15}$

6.6 Ratio, Proportion, and Applications

Objectives

A › **Solve proportions.**

B › **Solve applications involving ratios and proportions.**

To Succeed, Review How To . . .

1. Find the LCD of two or more fractions (pp. 13–15).
2. Solve linear equations (pp. 137–141).
3. Use the RSTUV method to solve word problems (pp. 148–149).

Getting Started

Coffee, Ratio, and Proportion

The manufacturer of this jar of instant coffee claims that 4 ounces of its instant coffee is equivalent to 1 pound (16 ounces) of regular coffee. In mathematics, we say that the ratio of instant coffee used to regular coffee used is 4 to 16. The ratio 4 to 16 can be written as the fraction $\frac{4}{16}$ or in symbols as 4:16.

In this section we shall learn how to use ratios to solve proportions and to solve different applications involving these proportions.

A › Solving Proportions

DEFINITION OF A RATIO

A ratio is a quotient of two numbers. There are *three* ways in which the ratio of a number *a* to another number *b* can be written:

1. *a* to *b*
2. *a*:*b*
3. $\frac{a}{b}$

Let's look more closely at the coffee jar in the *Getting Started*. The label on the back claims that 4 ounces of instant coffee will make 60 cups of coffee. Thus, the ratio of ounces to cups, when written as a fraction, is

$$\frac{4}{60} = \frac{1}{15} \quad \begin{matrix} \leftarrow \text{Ounces} \\ \leftarrow \text{Cups} \end{matrix}$$

The fraction $\frac{1}{15}$ is called the *reduced ratio* of ounces of coffee to cups of coffee. It tells us that 1 ounce of coffee will make 15 cups of coffee. Now suppose you want to make 90 cups of coffee. How many ounces do you need? The ratio of ounces to cups is $\frac{1}{15}$, and we need to know how many ounces will make 90 cups. Let n be the number of ounces needed. Then

$$\frac{1}{15} = \frac{n}{90} \quad \begin{matrix} \leftarrow \text{Ounces} \\ \leftarrow \text{Cups} \end{matrix}$$

Note that in both fractions the numerator indicates the number of ounces and the denominator indicates the number of cups. The equation $\frac{1}{15} = \frac{n}{90}$ is an equality between two ratios. In mathematics, an equality between ratios is called a **proportion.** Thus, $\frac{1}{15} = \frac{n}{90}$ is a proportion. To solve this proportion, which is simply a fractional equation, we proceed as before. First, since the LCM of 15 and 90 is 90:

1. Multiply by the LCM. $90 \cdot \frac{1}{15} = 90 \cdot \frac{n}{90}$

2. Simplify. $6 = n$

Thus, we need 6 ounces of instant coffee to make 90 cups.

We use the same ideas in Example 1.

EXAMPLE 1 Writing ratios and solving proportions

A car travels 140 miles on 8 gallons of gas.

a. What is the reduced ratio of miles to gallons? (Note that another way to state this ratio is to use miles *per* gallon.)

b. How many gallons will be needed to travel 210 miles?

SOLUTION 1

a. The ratio of miles to gallons is

$$\frac{140}{8} = \frac{35}{2} \quad \begin{matrix} \leftarrow \text{Miles} \\ \leftarrow \text{Gallons} \end{matrix}$$

b. Let g be the gallons needed. The ratio of miles to gallons is $\frac{35}{2}$; it is also $\frac{210}{g}$. Thus,

$$\frac{210}{g} = \frac{35}{2}$$

Multiplying by $2g$, the LCD, we have

$$2g \cdot \frac{210}{g} = 2g \cdot \frac{35}{2}$$

$$420 = 35g$$

$$g = \frac{420}{35} = 12$$

Hence 12 gallons of gas will be needed to travel 210 miles.

PROBLEM 1

A car travels 150 miles on 9 gallons of gas. How many gallons does it need to travel 900 miles?

Proportions are so common that we use a shortcut method to solve them. The method depends on the fact that if two fractions $\frac{a}{b}$ and $\frac{c}{d}$ are equivalent, their **cross products** are also equal. Here is the rule.

RULE

Cross Products

If $\frac{a}{b} = \frac{c}{d}$, then $ad = bc$ $(b \neq 0 \text{ and } d \neq 0)$

Note that if $\frac{a}{b} = \frac{c}{d}$

then $bd \cdot \frac{a}{b} = bd \cdot \frac{c}{d}$

$ad = bc$

Answers to PROBLEMS

1. 54 gal

Thus, since $\frac{1}{2} = \frac{2}{4}$, $1 \cdot 4 = 2 \cdot 2$, and since $\frac{3}{9} = \frac{1}{3}$, $3 \cdot 3 = 9 \cdot 1$. To solve the proportion

$$\frac{210}{g} = \frac{35}{2}$$

of Example 1(b), we use cross products and write:

$$210 \cdot 2 = 35g$$

$$\frac{210 \cdot 2}{35} = g \qquad \text{Dividing by 35 yields } g\text{, as before.}$$

$$12 = g$$

> **NOTE**
>
> This technique avoids having to find the LCD first, but it applies only when you have *one* term on each side of the equation!

B > Applications Involving Ratios and Proportions

Ratios and proportions can be used to solve work problems. For example, suppose a worker can finish a certain job in 3 days, whereas another worker can do it in 4 days. How many days would it take both workers working together to complete the job? Before we solve this problem, let's see how ratios play a part in the problem itself.

Since the first worker can do the job in 3 days, she does $\frac{1}{3}$ of the job in 1 day.
The second worker can do the job in 4 days, so he does $\frac{1}{4}$ of the job in 1 day.
Working together they do the job in d days, and in 1 day, they do $\frac{1}{d}$ of the job.

Here's what happens in 1 day:

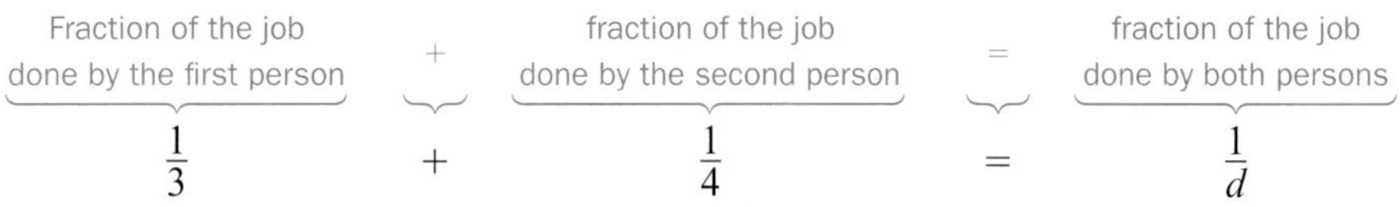

Fraction of the job done by the first person
+
fraction of the job done by the second person
=
fraction of the job done by both persons
1/3 + 1/4 = 1/d

This is a fractional equation that can be solved by multiplying each term by the LCD, $3 \cdot 4 \cdot d = 12d$. Note that we can't "cross multiply" here because we have three terms! Thus, we do it like this:

$$\frac{1}{3} + \frac{1}{4} = \frac{1}{d} \qquad \text{Given}$$

$$12d \cdot \frac{1}{3} + 12d \cdot \frac{1}{4} = 12d \cdot \frac{1}{d} \qquad \text{Multiply each term by the LCD, } 12d.$$

$$4d + 3d = 12 \qquad \text{Simplify.}$$

$$7d = 12 \qquad \text{Combine like terms.}$$

$$d = \frac{12}{7} = 1\frac{5}{7} \qquad \text{Divide by 7.}$$

Thus, if they work together they can complete the job in $1\frac{5}{7}$ days.

EXAMPLE 2 Work problems and ratios

A computer can do a job in 4 hours. Computer sharing is arranged with another computer that can finish the job in 6 hours. How long would it take for both computers operating simultaneously to finish the job?

SOLUTION 2

We use the RSTUV method.

1. Read the problem. We are asked to find the time it will take both computers working together to complete the job.

2. Select the unknown. Let h be the number of hours it takes to complete the job when both computers are operating simultaneously.

3. Think of a plan. Translate the problem:

The first computer does $\frac{1}{4}$ of the job in 1 hour.

The second computer does $\frac{1}{6}$ of the job in 1 hour.

When both work together, they do $\frac{1}{h}$ of the job in 1 hour.

The sum of the fractions of the job done in 1 hour by each computer, $\frac{1}{4} + \frac{1}{6}$, must equal the fraction of the job done each hour when they are operating simultaneously, $\frac{1}{h}$. Thus,

$$\frac{1}{4} + \frac{1}{6} = \frac{1}{h}$$

4. Use algebra to solve the problem. To solve this equation, we proceed as usual.

$$12h \cdot \frac{1}{4} + 12h \cdot \frac{1}{6} = 12h \cdot \frac{1}{h} \quad \text{Multiply by the LCD, } 12h.$$

$$3h + 2h = 12 \quad \text{Simplify.}$$

$$5h = 12$$

$$h = \frac{12}{5} = 2\frac{2}{5} \quad \text{Divide by 5.}$$

$$= 2.4 \text{ hr}$$

Thus, both computers working together take 2.4 hours to do the job.

5. Verify the solution. We leave the verification to you.

PROBLEM 2

A worker can finish a report in 5 hours. Another worker can do it in 8 hours. How many hours will it take to finish the report if both workers work on it?

Other types of problems often require ratios and proportions for their solution—for example, distance, rate, and time problems. We've already mentioned that the formula relating these three variables is

$$D = RT$$

(Distance = D, Rate = R, Time = T)

We use this formula to solve a motion problem in Example 3.

EXAMPLE 3 Boat speed and proportions

Suppose you are cruising down a river one sunny afternoon in your powerboat. Before you know it, you've gone 60 miles. Now, it's time to get back. Perhaps you can make it back in the same time? Wrong! This time you cover only 40 miles in the same time. What happened? You traveled 60 miles downstream in the same time it took to travel 40 miles upstream! Oh yes, the current—it was flowing at 5 miles per hour. What was the speed of your boat in still water?

PROBLEM 3

A freight train travels 120 miles in the same time a passenger train covers 140 miles. If the passenger train is 5 miles per hour faster, what is the speed of the freight train?

Answers to PROBLEMS

2. $\frac{40}{13} = 3\frac{1}{13}$ hr **3.** 30 mi/hr

SOLUTION 3 We use the RSTUV method.

1. **Read the problem.** You are asked to find the speed of your boat in still water.
2. **Select the unknown.** Let R be the speed of the boat in still water.

speed downstream: $R + 5$ Current helps.
speed upstream: $R - 5$ Current hinders.

3. **Think of a plan.** Translate the problem. The time taken downstream and upstream must be the same:

$$T_{up} = T_{down}$$

Since $D = RT$,
$$T = \frac{D}{R}$$

$$\underbrace{\frac{60}{R+5}}_{T_{up}} = \underbrace{\frac{40}{R-5}}_{T_{down}}$$

4. **Use the cross-product rule to solve the problem.**

$60(R - 5) = 40(R + 5)$ Cross multiply.
$60R - 300 = 40R + 200$ Simplify.
$60R = 40R + 500$ Add 300.
$20R = 500$ Subtract $40R$.
$R = 25$ Divide by 20.

Thus, the speed of the boat in still water is 25 miles per hour.

5. **Verify the solution.** Substitute 25 for R in

$$\frac{60}{R+5} = \frac{40}{R-5}$$
$$\frac{60}{25+5} = \frac{40}{25-5}$$
$$2 = 2$$ A true statement

How much air pollution does your car create? A car that gets 20 miles per gallon driven 12,000 miles a year will use 600 gallons of gas and will produce 11,400 (19 × 600) pounds of CO_2. We find out how many trees are needed to absorb this CO_2 next.

GREEN MATH

EXAMPLE 4 **Using trees to absorb CO_2**

A tree absorbs 48 pounds of CO_2 per year. At that rate, how many trees are needed to absorb 11,400 pounds of CO_2?

SOLUTION 4 We use the RSTUV procedure.

1. **Read the problem.** We want to find how many trees are needed to absorb 11,400 pounds of CO_2.
2. **Select the unknown.** Let t be the number of trees needed to absorb the 11,400 pounds of CO_2.
3. **Think of a plan.**

1 tree absorbs 48 pounds. The rate of absorption is $\frac{1}{48}$.

t trees need to absorb 11,400 pounds at a $\frac{t}{11{,}400}$ rate.

To maintain the rate, $\frac{t}{11{,}400} = \frac{1}{48}$.

PROBLEM 4

If your car gets 30 miles per gallon, you need only 400 gallons of gas a year and you will then produce 19 × 400 or 7600 pounds of CO_2. Write the proportion you need to maintain the 1-tree rate of absorption and the number of trees needed to absorb the 7600 pounds of carbon.

Driving tips: You can pollute less by driving fewer miles or having a more efficient car.

(continued)

Answers to PROBLEMS

4. $\frac{t}{7600} = \frac{1}{48}$; $158.\overline{3}$ or 159 trees

4. Use cross products to solve the problem.

Cross multiply. $48t = 11{,}400$

Divide by 48 $t = \frac{11{,}400}{48} = 237.5$ or 238

Thus, you need 238 trees to absorb the CO_2.

5. Verify the answer.

In everyday life many objects are *similar* but *not* the same; that is, they have the same shape but are not necessarily the same size. Thus, a penny and a nickel are similar, and two computer or television screens may be similar. In geometry, figures that have exactly the same shape but not necessarily the same size are called **similar figures.** Look at the two similar triangles:

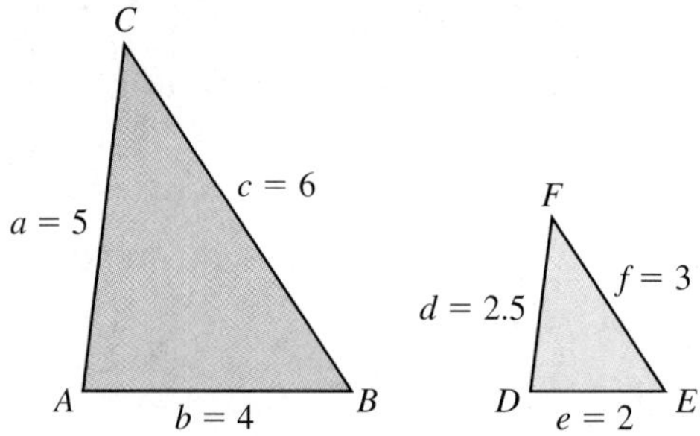

Side *a* corresponds to side *d*.
Side *b* corresponds to side *e*.
Side *c* corresponds to side *f*.

To show that corresponding sides are in proportion, we write the ratios of the corresponding sides:

$$\frac{a}{d} = \frac{5}{2.5} = 2, \quad \frac{b}{e} = \frac{4}{2} = 2, \quad \text{and} \quad \frac{c}{f} = \frac{6}{3} = 2$$

As you can see, corresponding sides are proportional. We use this idea to solve Example 5.

EXAMPLE 5 Similar triangles

Two similar triangles measured in centimeters (cm) are shown. Find *f* for the triangle on the right:

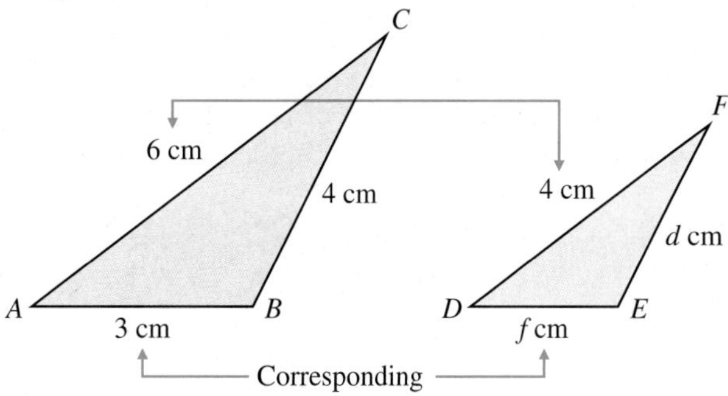

SOLUTION 5 Since the triangles are similar, the corresponding sides must be proportional. Thus,

$$\frac{f}{3} = \frac{4}{6}$$

Since $\frac{4}{6} = \frac{2}{3}$, we have

$$\frac{f}{3} = \frac{2}{3}$$

$$3f = 6 \quad \text{Cross multiply.}$$

Solving this equation for *f*, we get

$$f = \frac{6}{3} = 2 \text{ cm}$$

Thus, *f* is 2 centimeters long.

PROBLEM 5

Find *d* in the diagram.

Answers to PROBLEMS

5. $d = \frac{8}{3} = 2\frac{2}{3}$ cm

› Exercises 6.6

> Practice Problems > Self-Tests > Media-rich eBooks > e-Professors > Videos

‹A› Solving Proportions In Problems 1–13, solve the proportion problems.

1. Do you think your gas station sells a lot of gas? The greatest number of gallons sold through a single pump is claimed by the Downtown Service Station in New Zealand. It sold 7400 Imperial gallons in 24 hours. At that rate, how many Imperial gallons would it sell in 30 hours?

2. Do you like lemons? Bob Blackmore ate three whole lemons in 24 seconds. At that rate, how many lemons could he eat in 1 minute (60 seconds)?

3. Michael Cisneros ate 25 tortillas in 15 minutes. How many could he eat in 1 hour at the same rate?

4. If your lawn is yellowing, it may need "essential minor elements" administered at the rate of 2.5 pounds per 100 square feet of lawn. If your lawn covers 150 square feet, how many pounds of essential minor elements do you need?

5. If you have roses, they may need Ironite® administered at the rate of 4 pounds per 100 square feet. If your rose garden is 15 feet by 10 feet, how many pounds of Ironite do you need?

6. The directions for a certain fertilizer recommend that 35 pounds of fertilizer be applied per 1000 square feet of lawn.

a. If your lawn covers 1400 square feet, how many pounds of fertilizer do you need?

b. If fertilizer comes in 40-pound bags, how many bags do you need?

7. You can fertilize your flowers by using 6-6-6 fertilizer at the rate of 2 pounds per 100 square feet of flower bed. If your flower bed is 12 feet by 15 feet, how many pounds of fertilizer do you need?

8. Petunias can be planted in a mixture consisting of two parts peat moss to five parts of potting soil. If you have a 20-pound bag of potting soil, how many pounds of peat moss do you need to make a mixture to plant your petunias?

9. According to the U.S. Bureau of the Census, there were 400 homeless persons "visible on the street" for every 1200 homeless in shelters in Dallas.

a. If the homeless persons visible on the street grew to 750, how many would you expect to find in shelters?

b. If each shelter houses 50 homeless persons, how many shelters would be needed when there are 600 homeless persons visible on the street?

10. In Minneapolis, there were 30 homeless persons visible on the street for every 1050 homeless in shelters. If the number of homeless visible on the street grew to 75, how many would you expect to find in shelters?

11. What is the ratio of "homeless in shelters" to "homeless visible on the street" in

a. Dallas? (See Problem 9.)

b. Minneapolis? (See Problem 10.)

c. Where is the ratio of "homeless in shelters" to "homeless visible on the street" greater, Dallas or Minneapolis? Why do you think this is so?

12. According to the U.S. Bureau of the Census, about 40,000 persons living in Alabama were born in a foreign country. If Alabama had 4 million people, how many persons born in a foreign country would you expect if the population of Alabama increased to 5 million people?

13. In the year 2005, the U.S. population was about 295 million people, 36 million of which were foreign born. If the population in 2007 increased to 300 million people, how many foreign-born persons would you expect to be living in the United States? Answer to the nearest million.

‹B› Applications Involving Ratios and Proportions

14. Mr. Gerry Harley, of England, shaved 130 men in 60 minutes. If another barber takes 5 hours to shave the 130 men, how long would it take both men working together to shave the 130 men?

15. Mr. J. Moir riveted 11,209 rivets in 9 hours (a world record). If it takes another man 12 hours to do this job, how long would it take both men to rivet the 11,209 rivets?

› Web IT go to mhhe.com/bello for more lessons

16. It takes a printer 3 hours to print a certain document. If a faster printer can print the document in 2 hours, how long would it take both printers working together to print the document?

17. A computer can send a company's e-mail in 5 minutes. A faster computer does it in 3 minutes. How long would it take both computers operating simultaneously to send out the company's e-mail?

18. Two secretaries working together typed the company's annual report in 4 hours. If one of them could type the report by herself in 6 hours, how long would it take the other secretary working alone to type the report?

19. Two fax machines can together send all the office mailings in 2 hours. When one of the machines broke down, it took 3 hours to send all the mailings. If the number of faxes sent was the same in both cases, how many hours would it take the remaining fax machine to send all the mailings?

20. The train *Trés Grande Vitesse* covers the 264 miles from Paris to Lyons in the same time a regular train covers 124 miles. If the *Trés Grande Vitesse* is 70 miles per hour faster, how fast is it?

21. The strongest current on the East Coast of the United States is at St. Johns River in Pablo Creek, Florida, where the current reaches a speed of 6 miles per hour. A motorboat can travel 4 miles downstream on Pablo Creek in the same time it takes to go 16 miles upstream. (No, it is *not* wrong! The St. Johns flows *up*stream.) What is the speed of the boat in still water?

22. The strongest current in the United States occurs at Pt. Kootzhahoo in Chatam, Alaska. If a boat that travels at 10 miles per hour in still water takes the same time to travel 2 miles upstream in this area as it takes to travel 18 miles downstream, what is the speed of the current?

23. One of the fastest point-to-point trains in the world is the *New Tokaido* from Osaka to Okayama. This train covers 450 miles in the same time a regular train covers 180 miles. If the *Tokaido* is 60 miles per hour faster than the regular train, how fast is it?

24. The world's strongest current, reaching 18 miles per hour, is the Saltstraumen in Norway. A motorboat can travel 48 miles downstream in the Saltstraumen in the same time it takes to go 12 miles upstream. What is the speed of the boat in still water?

Applications: Green Math

Have you ever heard of Tesco, a megasupermarket in England? They use carbon footprint in their labels. What is the exact amount of greenhouse gas emissions generated in the production of a single roll of toilet paper? You will find out in a moment.

25. *Sheets in recycled content toilet paper roll* When making one "sheet" of Tesco-recycled content toilet paper, 1.1 grams of CO_2 are produced as a byproduct. If 220 grams of CO_2 result when producing the whole roll of toilet paper, how many sheets are there in a roll?

26. *Sheets in nonrecycled content toilet paper roll* When making one "sheet" of Tesco's nonrecycled content toilet paper roll, 1.8 grams of CO_2 are produced as a byproduct. If 360 grams of CO_2 result when producing the whole roll of toilet paper, how many sheets are there in a roll?

27. *CO_2 byproducts from juice production* 360 grams of CO_2 byproduct result when producing 250 milliliters (8.5 cups) of juice. How many grams of CO_2 will result when producing 4 cups of juice? Answer to the nearest gram.

28. *Carbon footprint of one carton of orange juice* How many sheets of recycled-content toilet paper (1.1 grams per sheet) are needed to match the carbon footprint of one carton of orange juice (360 grams)? Answer to the nearest sheet.

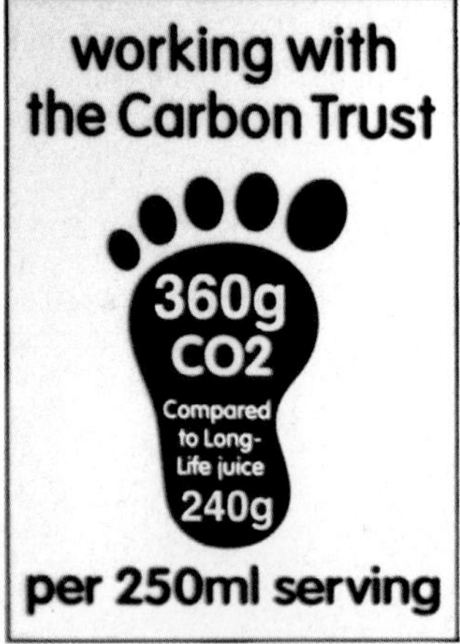

In Problems 29–41, the pairs of triangles are similar. Find the lengths of the indicated unknown sides.

29.

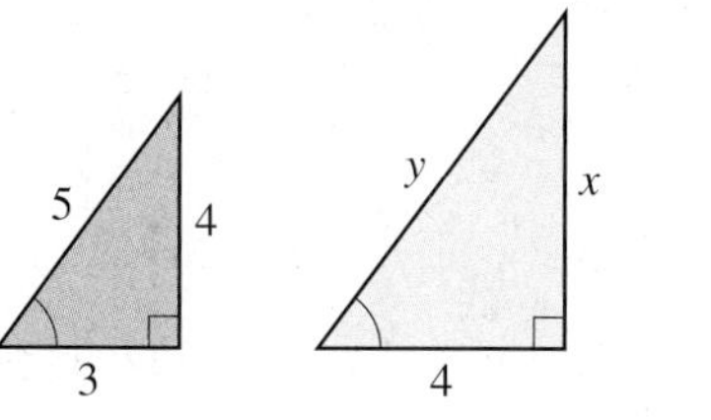

30. Side *DE*

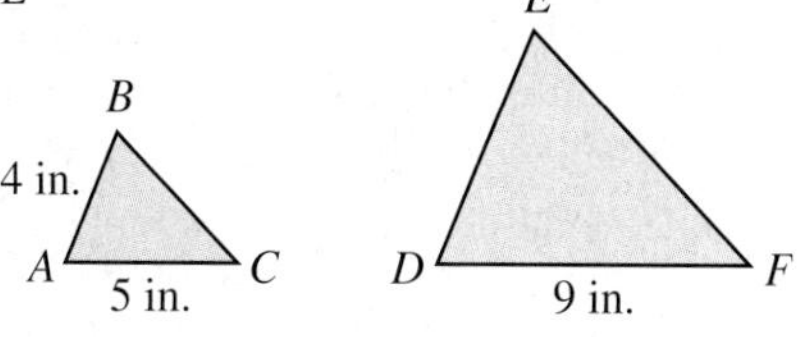

31. Side *DE*

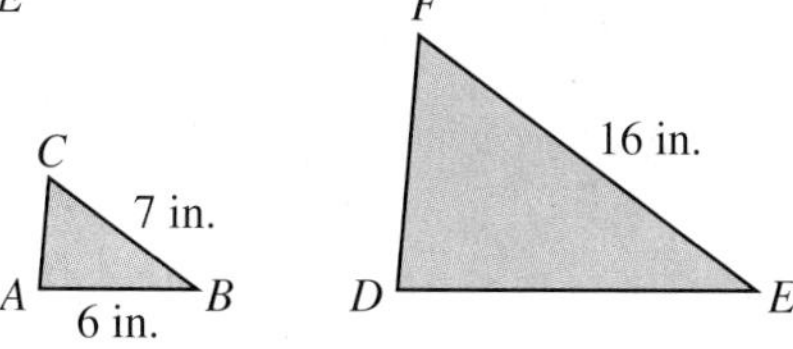

32.

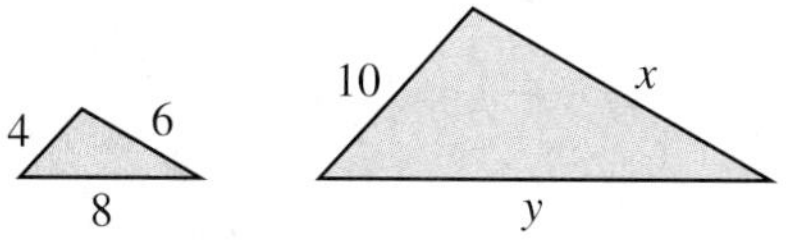

33.

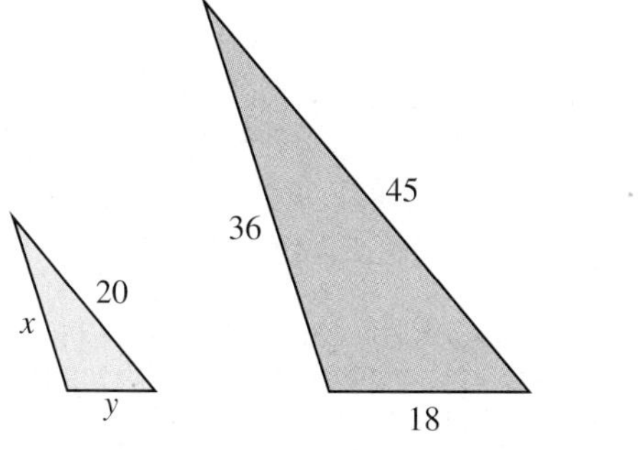

34.

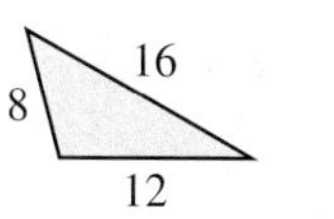

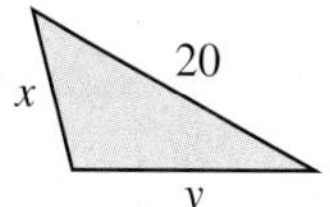

35.

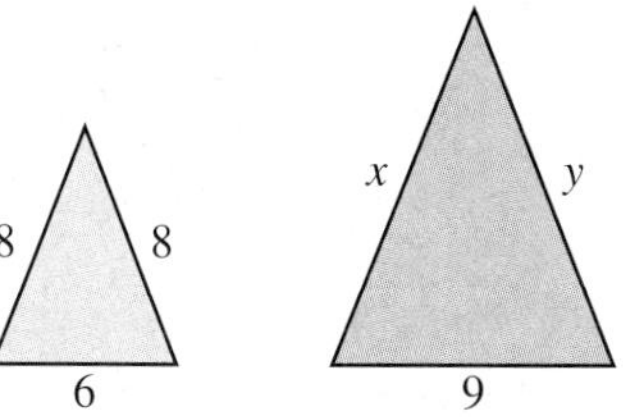

36.

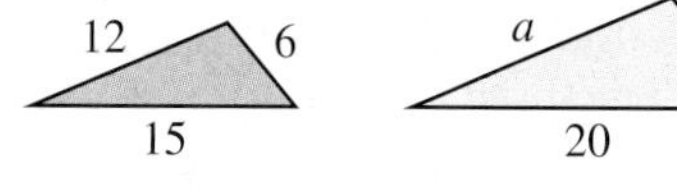

37.

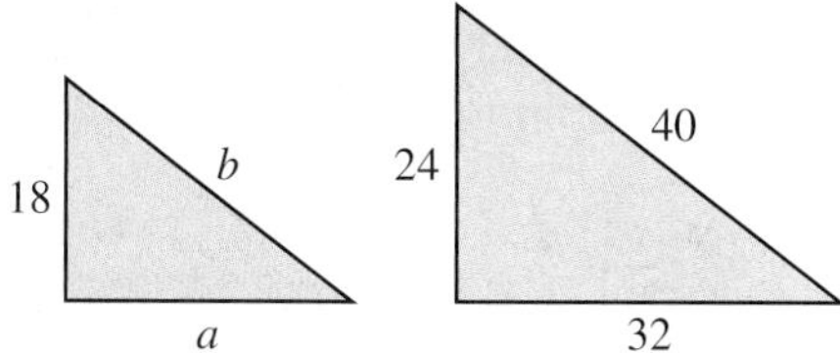

38.

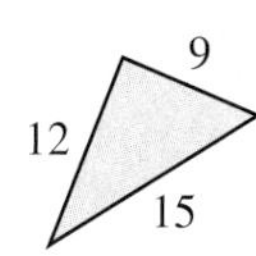

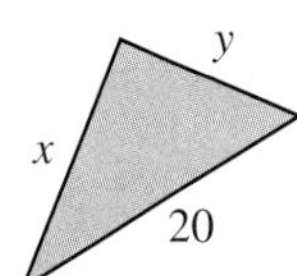

39.

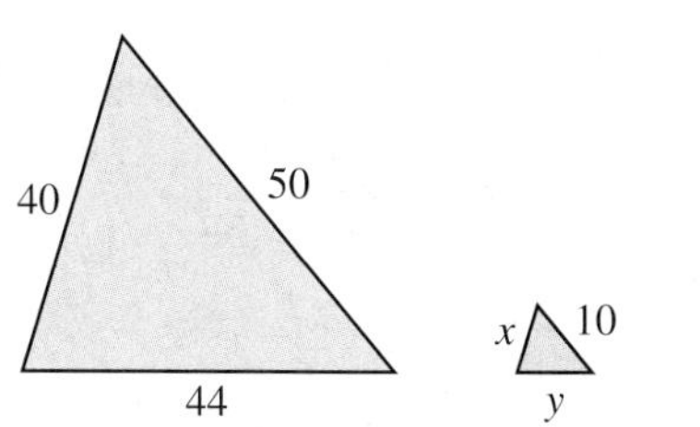

40.

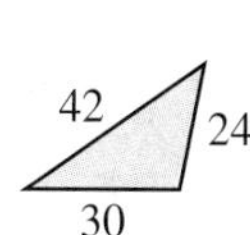

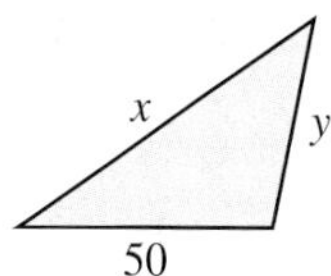

41.

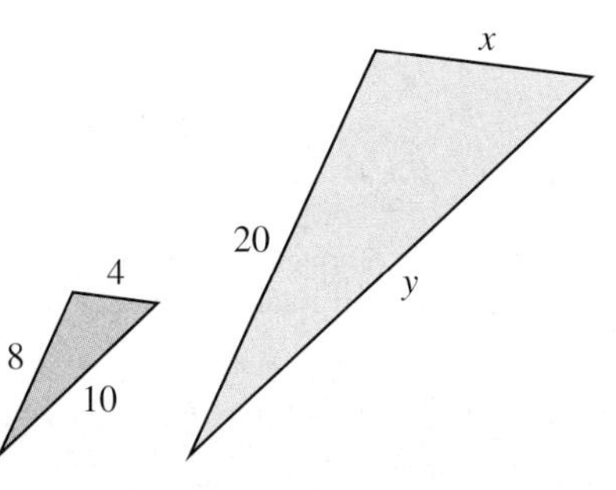

Applications

42. *Film processing* A film processing department can process nine rolls of film in 2 hours. At that rate, how long will it take them to process 20 rolls of film?

43. *Blueprints* If a blueprint uses a scale of $\frac{1}{2}$ inch = 3 feet, find the scaled-down dimensions of a room that is 9 feet × 12 feet.

44. *Travel* Latrice and Bob want to drive from Miramar, Florida, to Pittsburgh, Pennsylania, a distance of approximately 1000 miles. Each day they plan to drive 7 hours and cover 425 miles.

a. At that rate, how many driving hours should the trip take?

b. At that rate, how many days should the trip take?

45. *Cycling* Marquel is an avid bicyclist and averages 16 miles per hour when he rides. He likes to ride 80 miles per day. He plans to take a 510-mile, round trip, between New Orleans and Panama City.

a. At that rate, how many hours of riding will it take Marquel to complete the trip?

b. At that rate, how many days should it take Marquel to complete the trip?

46. *Batting average* Early in the season a softball player has a 0.250 batting average (5 hits in 20 times at bat). How many consecutive hits would she need to bring her average to 0.348?

Using Your Knowledge

A Matter of Proportion Proportions are used in many areas other than mathematics. The following problems are typical.

47. In a certain experiment, a physicist stretched a spring 3 inches by applying a 7-pound force to it. How many pounds of force would be required to make the spring stretch 8 inches?

48. Suppose a photographer wishes to enlarge a 2-inch by 3-inch picture so that the longer side is 10 inches. How wide does she have to make it?

49. In carpentry the pitch of a rafter is the ratio of the rise to the run of the rafter. What is the rise of a rafter having a pitch of $\frac{2}{5}$ if the run is 15 feet?

50. In an automobile the rear axle ratio is the ratio of the number of teeth in the ring gear to the number of teeth in the pinion gear. A car has a 3-to-1 rear axle ratio, and the ring gear has 60 teeth. How many teeth does the pinion gear have?

51. A zoologist took 250 fish from a lake, tagged them, and released them. A few days later 53 fish were taken from the lake, and 5 of them were found to be tagged. Approximately how many fish were originally in the lake?

Write On

52. What is a proportion?

53. Write the procedure you use to solve a proportion.

54. We can solve proportions by using their cross products. Can you solve the equation

$$\frac{x}{2} + \frac{x}{4} = 3$$

by using cross products? Explain.

55. Find some examples of similar figures and then write your own definition of similar figures.

Concept Checker

Fill in the blank(s) with the correct word(s), phrase, or mathematical statement.

56. A _______________ is a **quotient of two numbers.**

57. Using the **cross-product** rule, if $\frac{a}{b} = \frac{c}{d}$ (where $b \neq 0$ and $d \neq 0$), then _______________.

58. If R and T are the **rate** and **time,** respectively, the **distance** D is given by _______________.

59. Figures that have **exactly the same shape** but **not** necessarily the **same size** are called _______________ figures.

$ac = bd$	$\frac{R}{T}$
$ad = bc$	equal
ratio	similar
RT	

Mastery Test

60. A freight train travels 90 miles in the same time a passenger train travels 105 miles. If the passenger train is 5 miles per hour faster, what is the speed of the freight train?

61. A typist can finish a report in 5 hours. Another typist can do it in 8 hours. How many hours will it take to finish the report if both work on it?

62. A car travels 150 miles on 9 gallons of gas. How many gallons does the car need to travel 800 miles?

63. A baseball player has 160 singles in 120 games. If he continues at this rate, how many singles will he hit in 150 games?

64. Find $\frac{d}{2}$ for the triangle in Example 5.

Skill Checker

65. Solve: $600 = 120k$

66. Solve: $300 = 15k$

67. Solve: $60 = \frac{k}{4}$

68. Solve: $23 = \frac{k}{5}$

6.7 Direct and Inverse Variation: Applications

Objectives

A Find and solve equations of direct variation given values of the variables.

B Find and solve equations of inverse variation given values of the variables.

C Solve applications involving variation.

To Succeed, Review How To . . .

Solve linear equations (pp. 137–141).

Getting Started

Don't Forget the Tip!

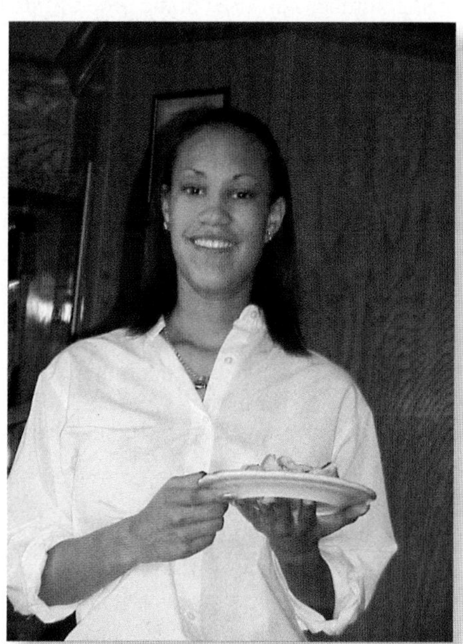

Jasmine is a server at CDB restaurant. Aside from her tips, she gets \$2.88/hour. In 1 hour, she earns \$2.88; in 2 hr, she earns \$5.76; in 3 hr, she earns \$8.64, and so on. We can form the set of ordered pairs (1, 2.88), (2, 5.76), (3, 8.64) using the number of hours she works as the first coordinate and the amount she earns as the second coordinate. Note that the ratio of second coordinates to first coordinates is the same number:

$$\frac{2.88}{1} = 2.88, \quad \frac{5.76}{2} = 2.88, \quad \frac{8.64}{3} = 2.88,$$

and so on.

When the ratio of ordered pairs of numbers is constant, we say that there is a **direct variation.** In this case, the earnings E **vary directly** (or are **directly proportional**) to the number of hours h, that is,

$$\frac{E}{h} = 2.88 \quad \text{(a constant)} \quad \text{or} \quad E = 2.88h$$

Note that in the graph of $E = 2.88h$ or the more familiar $y = 2.88x$, the constant $k = 2.88$ is the **slope** of the line.

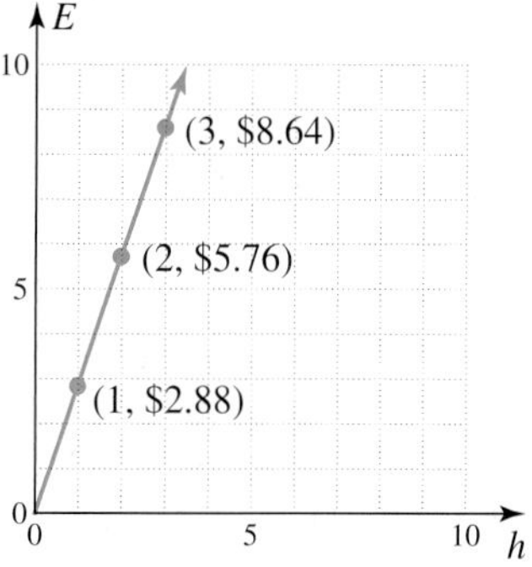

A > Direct Variation

In general, if we have a situation in which a variable **varies directly** or **is directly proportional** to another variable, we make the following definition.

DEFINITION

y* varies directly as *x (or *y* is **proportional** to ***x***) if there is a constant *k* such that $y = kx$. The constant *k* is the constant of variation or proportionality.

In real life, we can find *k* by experimenting. Suppose you are dieting and want to eat some McDonald's® fries, but only 100 calories worth. The fries come in small, medium, or large sizes with 210, 450, or 540 calories, respectively. You could use **estimation** and eat approximately half of a small order $\left(\frac{210}{2} \approx 105 \text{ calories}\right)$, but you can estimate better! The numbers of fries in the small, medium, and large sizes are 45, 97, and 121; thus, the calories per fry are:

$$\frac{210}{45} \approx 4.7 \text{ (small)}, \quad \frac{450}{97} \approx 4.6 \text{ (medium)}, \quad \frac{540}{121} \approx 4.5 \text{ (large)}$$

Source: Author (counted and ate!).

EXAMPLE 1 Using direct variation

The number *C* of calories in an order of McDonald's french fries varies directly as the number *n* of fries in the order. Suppose you buy a large order of fries (121 fries, 540 calories).

a. Write an equation of variation.

b. Find *k*.

c. If you want to eat 100 calories worth, how many fries would you eat?

SOLUTION 1

a. If *C* varies directly as *n*, then $C = kn$.

b. A large order of fries has $C = 540$ calories and $n = 121$ fries. Thus,

$$C = kn \quad \text{becomes} \quad 540 = k(121) \quad \text{or} \quad k = \frac{540}{121} \approx 4.5$$

c. If you want to eat $C = 100$ calories worth,

$$C = kn = 4.5n \quad \text{becomes} \quad 100 = 4.5n$$

Solving for *n* by dividing both sides by 4.5, we get

$$n = \frac{100}{4.5} \approx 22$$

So, if you eat about 22 fries, you will have consumed about 100 calories. Note that this is indeed about half of a small order!

PROBLEM 1

The number *C* of calories in an order of Burger King® fries also varies directly as the number *n* of fries in the order. An order of king-size fries has 590 calories and 101 fries.

a. Write an equation of variation.

b. Find *k*.

c. If you want to eat 100 calories worth, how many fries would you eat?

Now that we have eaten some calories, let's see how we can spend (burn) some calories. Read on.

EXAMPLE 2 Using direct variation

If you weigh about 160 pounds and you jog (5 miles per hour) or ride a bicycle (12 miles per hour), the number *C* of calories used is proportional to the time *t* (in minutes).

a. Find an equation of variation.

b. If jogging for 15 minutes uses 150 calories, find *k*.

c. How many calories would you use if you jog for 20 minutes?

d. To lose a pound, you have to use about 3500 calories. How many minutes do you have to jog to lose 1 pound?

PROBLEM 2

If you weigh 120 pounds and you jog (5 miles per hour) or ride a bicycle (12 miles per hour), the number *C* of calories used is proportional to the time *t* (in minutes).

a. Find an equation of variation.

b. If jogging for 15 minutes uses 105 calories, find *k*.

Answers to PROBLEMS

1. a. $C = kn$ **b.** $k = \frac{590}{101} \approx 5.8$ **c.** $\frac{100}{5.8} \approx 17$ (about 17 fries) **2. a.** $C = kt$ **b.** $C = 7t$ **c.** 140 calories **d.** 500 minutes

SOLUTION 2

a. Since the number C of calories used is proportional to the time t (in minutes), the equation of variation is $C = kt$.

b. If jogging for 15 minutes ($t = 15$) uses $C = 150$ calories, then

$C = kt$ becomes $150 = k(15)$

Dividing both sides by 15 $10 = k$

Note that now $C = kt$ becomes $C = 10t$

c. We want to know how many calories C you would use if you jog for $t = 20$ minutes. Substitute 20 for t in $C = 10t$

obtaining $C = 10(20) = 200$

Thus, you will use 200 calories if you jog for 20 minutes.

d. To lose 1 pound, you need to use $C = 3500$ calories.

$C = 10t$ becomes $3500 = 10t$

Dividing both sides by 10 $350 = t$

Thus, you need 350 minutes of jogging to lose 1 pound! By the way, if you jog for 350 minutes at 5 miles per hour, you will be 29 miles away!

c. How many calories would you use if you jog for 20 minutes?

d. To lose a pound, you have to use about 3500 calories. How many minutes do you have to jog to lose 1 pound?

B > Inverse Variation

Sometimes, as one quantity increases, a related quantity decreases proportionately. For example, the more time we spend practicing a task, the less time it will take us to do the task. In this case, we say that the quantities **vary inversely** as each other.

DEFINITION

y* varies inversely as *x (or is **inversely proportional** to *x*) if there is a constant *k* such that

$$y = \frac{k}{x}$$

EXAMPLE 3 Inverse proportion problem

The rate of speed v at which a car travels is inversely proportional to the time t it takes to travel a given distance.

a. Write the equation of variation.

b. If a car travels at 60 miles per hour for 3 hours, what is k, and what does it represent?

SOLUTION 3

a. The equation is

$$v = \frac{k}{t}$$

b. We know that $v = 60$ when $t = 3$. Thus,

$$60 = \frac{k}{3}$$

$$k = 180$$

In this case, k represents the distance traveled, and the new equation of variation is

$$v = \frac{180}{t}$$

PROBLEM 3

Suppose a car travels at 55 miles per hour for 2 hours.

a. Find k and explain what it represents.

b. Write the new equation of variation.

Answers to PROBLEMS

3. a. $k = 110$ and represents the distance traveled. **b.** $v = \frac{110}{t}$

EXAMPLE 4 Boom boxes and inverse proportion

Have you ever heard one of those loud "boom" boxes or a car sound system that makes your stomach tremble? The loudness L of sound is inversely proportional to the square of the distance d that you are from the source.

a. Write an equation of variation.

b. The loudness of rap music coming from a boom box 5 feet away is 100 decibels (dB). Find k.

c. If you move to 10 feet away from the boom box, how loud is the sound?

SOLUTION 4

a. The equation is

$$L = \frac{k}{d^2}$$

b. We know that $L = 100$ for $d = 5$, so that

$$100 = \frac{k}{5^2} = \frac{k}{25}$$

Multiplying both sides by 25, we find that $k = 2500$, and the new equation of variation is

$$L = \frac{2500}{d^2}$$

c. When $d = 10$,

$$L = \frac{2500}{10^2} = 25 \text{ dB}$$

PROBLEM 4

Suppose the loudness is 80 decibels at 5 feet.

a. Find k.

b. Write the new equation of variation.

c. If you move to 10 feet away, how loud is the sound now?

Answers to PROBLEMS

4. a. $k = 2000$

b. $L = \frac{2000}{d^2}$

c. 20 dB

5. $g = \frac{3}{2}A$

C > Applications Involving Variation

Spring snow-melt is a major source of water supply to areas in temperate zones near mountains that catch and hold winter snow, especially those with a prolonged dry summer.

EXAMPLE 5 Using direct variation: water from snow

Figure 6.1 shows the number of gallons of water, g (in millions), produced by an inch of snow in different cities. Note that the larger the area of the city, the more gallons of water are produced, so g is directly proportional to A, the area of the city (in square miles).

a. Write an equation of variation.

b. If the area of St. Louis is about 62 square miles, what is k?

c. Find the amount of water produced by 1 inch of snow falling in Anchorage, Alaska, with an area of 1700 square miles.

SOLUTION 5

a. Since g is directly proportional to A, $g = kA$.

b. From Figure 6.1 we can see that $g = 100$ (in millions) is the number of gallons of water produced by 1 inch of snow in St. Louis. Since it is given that $A = 62$, $g = kA$ becomes

$$100 = k \cdot 62 \quad \text{or} \quad k = \frac{100}{62} = \frac{50}{31}$$

c. For Anchorage, $A = 1700$, thus, $g = \frac{50}{31} \cdot 1700 \approx 2742$ million gallons of water.

PROBLEM 5

If the area of the city of Boston is about 40 square miles and 60 (million) gallons of water are produced by 1 inch of snow, find an equation of variation for Boston.

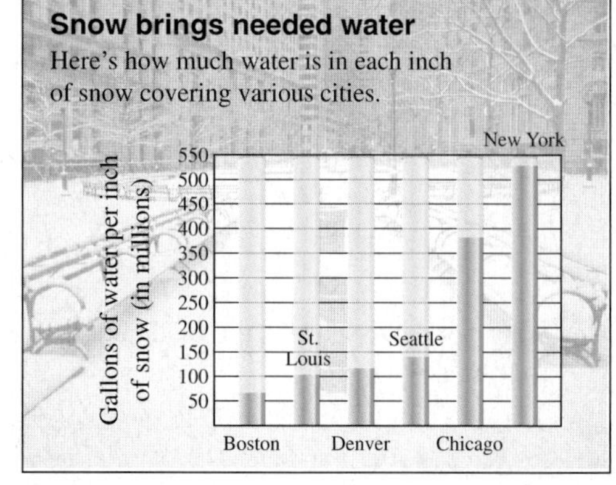

>Figure 6.1

Source: Data from USA Today, 1994.

> Exercises 6.7

> Web IT go to mhhe.com/bello for more lessons

⟨A⟩ Direct Variation In Problems 1–7, solve the direct variation problems.

1. Do you know how much your skin weighs? Your weight W (in pounds) is proportional to the weight S of your skin, which can be obtained by dividing W by 16.

a. Write an equation of variation.

b. What is k?

c. If you weigh 160 pounds, what is the weight of your skin?

2. How much blood do you have? Your weight W (in kilograms) is directly proportional to your blood volume V (in liters) and can be obtained by dividing W by 12.

a. Write an equation of variation.

b. What is k?

c. If you weigh 64 kilograms (about 140 pounds), what is your blood volume?

3. Your weight W (in pounds) is directly proportional to your basal metabolic rate R (number of calories you burn when at rest) and can be obtained by dividing W by 10.

a. Write an equation of variation.

b. What is k?

c. If you weigh 160 pounds, what is the value of your basal metabolic rate?

4. The amount I of annual interest received on a savings account is directly proportional to the amount m of money you have in the account.

a. Write an equation of variation.

b. If \$480 produced \$26.40 in interest, what is k?

c. How much annual interest would you receive if the account had \$750?

5. The number R of revolutions a record makes as it is being played varies directly with the time t that it is on the turntable.

a. Write an equation of variation.

b. A record that lasted $2\frac{1}{2}$ minutes made 112.5 revolutions. What is k?

c. If a record makes 108 revolutions, how long does it take to play the entire record?

6. The distance d an automobile travels after the brakes have been applied varies directly as the square of its speed s.

a. Write an equation of variation.

b. If the stopping distance for a car going 30 miles per hour is 54 feet, what is k?

c. What is the stopping distance for a car going 60 miles per hour ?

7. The weight of a person varies directly as the cube of the person's height h (in inches). The **threshold weight** T (in pounds) for a person is defined as the "crucial weight, above which the mortality risk for the person rises astronomically."

a. Write an equation of variation relating T and h.

b. If $T = 196$ when $h = 70$, find k to five decimal places.

c. To the nearest pound, what is the threshold weight for a person 75 inches tall?

⟨B⟩ Inverse Variation In Problems 8–15, solve the inverse variation problems.

8. To remain popular, the number S of new songs a rock band needs to produce each year is inversely proportional to the number y of years the band has been in the business.

a. Write an equation of variation.

b. If, after 3 years in the business, the band needs 50 new songs, how many songs will it need after 5 years?

9. When a camera lens is focused at infinity, the f-stop on the lens varies inversely with the diameter d of the aperture (opening).

a. Write an equation of variation.

b. If the f-stop on a camera is 8 when the aperture is $\frac{1}{2}$ inch, what is k?

c. Find the f-stop when the aperture is $\frac{1}{4}$ inch.

10. Boyle's law states that if the temperature is held constant, then the pressure P of an enclosed gas varies inversely as the volume V. If the pressure of the gas is 24 pounds per square inch when the volume is 18 cubic inches, what is the pressure if the gas is compressed to 12 cubic inches?

11. For the gas of Problem 10, if the pressure is 24 pounds per square inch when the volume is 18 cubic inches, what is the volume if the pressure is increased to 40 pounds per square inch?

Web IT go to mhhe.com/bello for more lessons

12. The weight W of an object varies inversely as the square of its distance d from the center of the Earth.

a. Write an equation of variation.

b. An astronaut weighs 121 pounds on the surface of the Earth. If the radius of the Earth is 3960 miles, find the value of k for this astronaut. (Do not multiply out your answer.)

c. What will this astronaut weigh when she is 880 miles above the surface of the Earth?

13. One of the manuscript pages of this book had about 600 words and was typed using a 12-point font. Suppose the average number w of words that can be printed on a manuscript page is inversely proportional to the font size s.

a. Write an equation of variation.

b. What is k?

c. How many words could be typed on the page if a 10-point font was used?

14. The price P of oil varies inversely with the supply S (in million barrels per day). In the year 2000, the price of one barrel of oil was \$26.00 and OPEC production was 24 million barrels per day.

a. Write an equation of variation.

b. What is k?

c. If OPEC plans to increase production to 28 million barrels per day, what would the price of one barrel be? Answer to the nearest cent.

15. According to the National Center for Health Statistics, the number b of births (per 1000 women) is inversely proportional to the age a of the woman. The number b of births (per 1000 women) for 27-year-olds is 110.

a. Write an equation of variation.

b. What is k?

c. What would you expect the number b of births (per 1000 women) to be for 33-year-old women?

‹C› Applications Involving Variation

16. *Miles per gallon* The number of miles m you can drive in your car is directly proportional to the amount of fuel g in your gas tank.

a. Write an equation of variation.

b. The greatest distance yet driven without refueling on a single fill in a standard vehicle is 1691.6 miles. If the twin tanks used to do this carried a total of 38.2 gallons of fuel, what is k (round to two decimal places)?

c. How many miles per gallon is this?

17. *Distance and speed of car* The distance d (in miles) traveled by a car is directly proportional to the average speed s (in miles per hour) of the car, even when driving in reverse!

a. Write an equation of variation.

b. The highest average speed attained in any nonstop reverse drive of more than 500 miles is 28.41 miles per hour. If the distance traveled was 501 miles, find k (round to two decimal places).

c. What does k represent in this situation?

18. *Responses to radio call-in contest* Have you called in on a radio contest lately? According to Don Burley, a radio talk-show host in Kansas City, the listener response to a radio call-in contest is directly proportional to the size of the prize.

a. If 40 listeners call when the prize is \$100, write an equation of variation using N for the number of listeners and P for the prize in dollars.

b. How many calls would you expect for a \$5000 prize?

19. *Cricket chirps and temperature* The number C of chirps a cricket makes each minute is directly proportional to 37 less than the temperature F in degrees Fahrenheit.

a. If a cricket chirps 80 times when the temperature is 57°F, what is the equation of variation?

b. How many chirps per minute would the cricket make when the temperature is 90°F?

››› Applications: Green Math

20. *Atmospheric carbon dioxide concentration* The concentration of carbon dioxide (CO_2) in the atmosphere has been increasing due to automobile emissions, electricity generation, and deforestation. In 1965, CO_2 concentration was 319.9 parts per million (ppm), and 23 years later, it had increased to 351.3 ppm. The *increase* I, of carbon dioxide concentration in the atmosphere is directly proportional to the number n of years elapsed since 1965.

a. Write an equation of variation for I.

b. Find k (round to two decimal places).

c. What would you predict the CO_2 concentration to be in the year 2000 (round to two decimal places)?

21. *Water from melting snow* The number of gallons of water g (in millions) produced by an inch of snow covering the city of Denver is directly proportional to the 155 square mile area A of the city.

a. Write an equation of variation.

b. If 100 million gallons of water are produced by every inch of snow melting in Denver, what is k?

c. Find the amount of water produced by 2 inches of snow covering Denver.

22. *Water depth and temperature* At depths of more than 1000 meters (a kilometer), water temperature T (in degrees Celsius) in the Pacific Ocean varies inversely as the water depth d (in meters). If the water temperature at 4000 meters is 1°C, what would it be at 8000 meters?

23. *Blood alcohol concentration* You might think that the BAC is directly proportional to how many beers you drink in an hour. Strangely enough, for both males and females of a specific weight, the BAC is directly proportional to $(N - 1)$, which is one less than the number of beers consumed during the last hour.

a. Write an equation of variation.

b. For a 150-pound man, the average BAC after 3 beers is 0.052. Find k.

c. What is the BAC after 5 beers?

d. How many beers can the man drink before going over the 0.08 limit?

24. *Blood alcohol concentration* For a 130-pound female, the average BAC is also directly proportional to one less than the number of beers consumed during the last hour $(N - 1)$.

a. Write an equation of variation.

b. For a 130-pound female, the average BAC after 3 beers is 0.066. Find k.

c. What is the BAC after 5 beers?

d. How many beers can the woman drink before going over the 0.08 limit?

25. *BAC and weight* If you drank 3 beers in the last hour, your blood alcohol content (BAC) is inversely proportional to your weight W. For a 130-pound male, the BAC after drinking 3 beers is 0.06.

a. Write an equation of variation.

b. What is the BAC of a 260-pound male after drinking 3 beers?

c. In most states, you are legally drunk if your BAC is 0.08 or higher. What is the weight of a male whose BAC is exactly 0.08 after drinking 3 beers in the last hour?

d. What would your BAC be if you weigh more than the male in part **c**?

26. *BAC and weight* If you drank 3 beers in the last hour, your blood alcohol content (BAC) is inversely proportional to your weight W. For a 130-pound female, the BAC after drinking 3 beers is 0.066.

a. Write an equation of variation.

b. What is the BAC of a 260-pound female after drinking those 3 beers?

c. In most states, you are legally drunk if your BAC is 0.08 or higher. What is the weight of a female whose BAC is exactly 0.08 after drinking 3 beers in the last hour?

d. What would your BAC be if you weigh more than the female in part **c**?

Web IT go to mhhe.com/bello for more lessons

Using Your Knowledge

27. Do you know what your Sun Protection Factor (SPF) is? The SPF of a sunscreen indicates the time period you can stay in the sun without burning, based on your complexion. The time T you are in the sun is directly proportional to your SPF (S). For example, suppose you can stay in the sun for 15 minutes without burning. Then, $T = 15S$ and we assume that your S is 1. If you want to stay in the sun for 30 minutes, $T = 30 = 15S$ and now you need a sunscreen with an SPF of 2, written as SPF2. Thus, you can stay out twice as much time in the sun $(2 \times 15 = 30)$ without burning.

a. Write an equation of variation relating the time T and the SPF S.

b. What SPF do you need if you want to stay out for an hour without burning?

Write On

28. Write in your own words what it means for two variables to be directly proportional. Give examples of variables that are directly proportional.

29. Write in your own words what it means for two variables to be inversely proportional. Give examples of variables that are inversely proportional.

30. Explain in your own words the type of relationship (direct or inverse) that should exist between the blood alcohol concentration (BAC) and:

a. The weight of the person

b. The number of beers the person has had in the last hour

c. The gender of the person

31. In Problems 23 and 24, we stated that for a specific weight the BAC is directly proportional to one less than the number of beers consumed in the last hour $(N - 1)$. Why do you think it is not directly proportional to the number N of beers consumed in the last hour? (You can do a Web search to explore this further.)

32. In the definition of direct variation, there is a constant of proportionality, k. In your own words, what is this k and what does it represent?

Concept Checker

Fill in the blank(s) with the correct word(s), phrase, or mathematical statement.

33. y varies ________ as x if there is a constant k such that $y = kx$.

34. y varies ________ as x if there is a constant k such that $y = \frac{k}{x}$.

inversely

directly

Mastery Test

35. The number of calories C in an order of KFC Colonel's Crispy Strips varies directly as the number n of strips in the order. Suppose you buy an order of the Strips (3 strips, 300 calories).

a. Write an equation of variation.

b. Find k.

c. If you want to eat 200 calories, how many strips would you eat?

36. If you weigh about 160 pounds and ride a bicycle at 12 miles per hour, the number C of calories used is proportional to the time t (in minutes) you ride.

a. Find an equation of variation.

b. If bicycling for 15 minutes burns about 105 calories, find k.

c. How many calories would you burn if you bicycled for 20 minutes?

d. To lose a pound, you have to burn about 3500 calories. How many minutes do you have to bicycle to lose 1 pound?

37. The rate of speed v at which a train travels is inversely proportional to the time t it takes the train to travel a given distance.

a. Write the equation of variation.

b. If a train travels at 30 miles per hour for 4 hours, find k.

c. What does the constant k represent in this situation?

38. If you are in the front row of a rock concert (10 feet away) the loudness L of the music is inversely proportional to the square of the distance d you are from the source.

a. Write an equation of variation.

b. If the loudness of the music in the front row (10 feet away) is 110 decibels, find k.

c. If you move 25 feet away from the music, how loud is the sound?

Skill Checker

In Problems 39–42, graph the equation.

39. $x + 2y = 4$

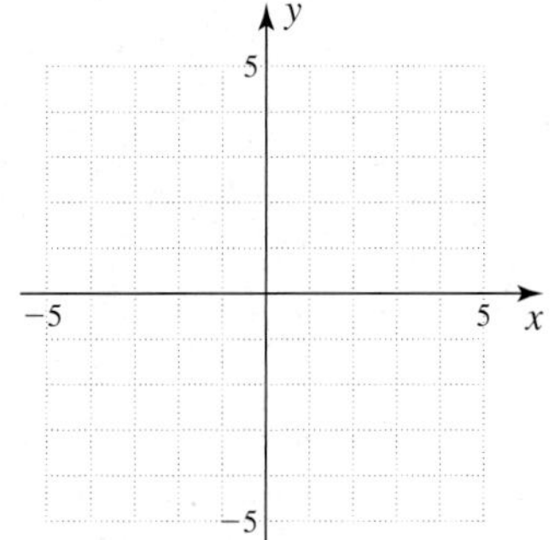

40. $2y - x = 0$

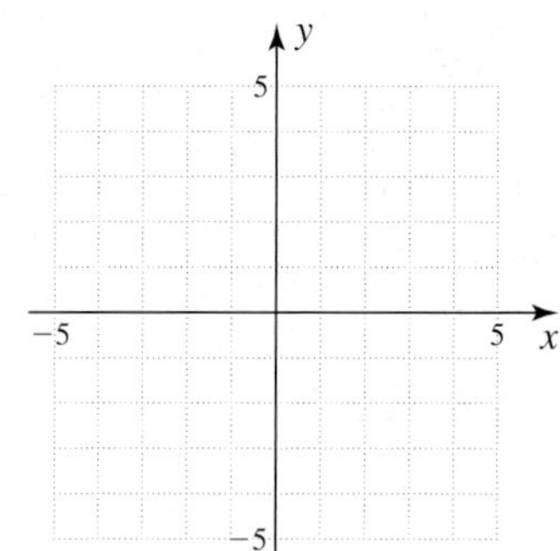

41. $y - 2x = 4$

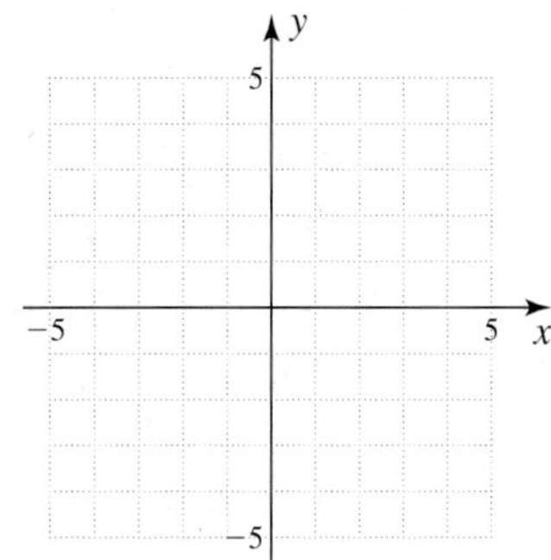

42. $2y - 4x = 4$

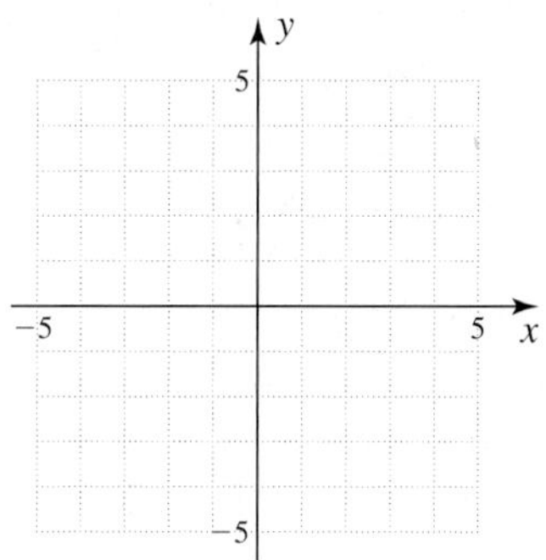

In Problems 43–44, determine whether the lines are parallel.

43. $x + y = 4$
$2x - y = -1$

44. $x + 2y = 4$
$2x + 4y = 6$

Collaborative Learning

BAC Meters

In this section we have stated that the blood alcohol concentration (BAC) is directly proportional to one less than the number of beers consumed during the last hour ($N - 1$). No, there will be no beer drinking, but we want to corroborate this information.

1. Form several groups and find websites that will automatically tell you the BAC based on the number of beers consumed during the last hour. (Try an online search for "BAC meter.") How do the BAC values compare with those given for the 150-pound man and 130-pound woman mentioned in Problems 23 and 24?
2. Do the meters you found corroborate the information in Problems 23 and 24?
3. What are the factors considered by the BAC meter your group is using?
4. Compare the results you get with different online BAC meters. Are the results different? Why do you think that is?

Collaborative Learning

Golden Rectangles

A **golden rectangle** is described as one of the most pleasing shapes to the human eye. Fill in the blanks in the table.

Item	Length	Width	Ratio of length to width	Golden rectangle?
3×5 card				
8.5×11 sheet of paper				
Desktop				
Textbook				
Teacher's desk				
ID card				
Anything else of interest . . .				

Source: Oswego (NY) City School District Regents Exam Prep Center.

1. Which is the ratio of length to width in a golden rectangle?
2. Form several groups. Each group select a topic:

   The golden rectangle in architecture

   The golden section in art

   The golden section in music

   Magical properties of the golden section

   Make a report to the rest of the group about your findings.

Research Questions

In this chapter, we have written rational numbers either as fractions ($\frac{2}{7}$) or decimals (0.285714285714285714 . . .). The ancient Egyptians used a number system based on *unit fractions*—fractions with a 1 in the numerator. This idea let them represent numbers such as $\frac{1}{7}$ easily enough; other numbers such as $\frac{2}{7}$ were represented as sums of unit fractions (e.g., $\frac{2}{7} = \frac{1}{4} + \frac{1}{28}$). Further, the same fraction could not be used twice (so $\frac{2}{7} = \frac{1}{7} + \frac{1}{7}$ is not allowed). We call a formula representing a sum of distinct unit fractions an *Egyptian fraction*. (*Source*: David Eppstein, ICS, University of California, Irvine.) Do a Web search for Egyptian fractions to answer the following questions.

(continued)

1. Write $\frac{3}{7}$ as an Egyptian fraction.
2. There are many algorithms (procedures) that will show how to write a fraction as an Egyptian fraction. Name three of these algorithms.
3. One of the algorithms uses *continued fractions*. Describe what a *continued fraction* is and how they are used to write fractions as Egyptian fractions.
4. The Indian mathematician Aryabhata (d. 550 A.D.) used a continued fraction to solve a linear *indeterminate equation*. What is an *indeterminate equation?*
5. Write a short paragraph about Aryabhata and his contributions to mathematics.
6. Can you write $\frac{5}{6}$ as an Egyptian fraction? Find two sites that will do it for you.
7. Use the sites you found in question 6 to write $\frac{5}{11}$ as an Egyptian fraction. Do you get the same answer at both sites?

Summary Chapter 6

Section	Item	Meaning	Example
6.1	Rational number	A number that can be written as $\frac{a}{b}$, a and b integers and $b \neq 0$	$\frac{3}{4}$, $-\frac{6}{5}$, 0, -8, and $1\frac{1}{3}$ are rational numbers.
	Rational expression	An expression of the form $\frac{A}{B}$, where A and B are polynomials, $B \neq 0$	$\frac{x^2+3}{x-1}$ and $\frac{y^2+y-1}{3y^3+7y+8}$ are rational expressions.
6.1B	Fundamental rule of rational expressions	$\frac{A}{B} = \frac{A \cdot C}{B \cdot C}$ $B \neq 0, C \neq 0$	$\frac{3}{4} = \frac{3 \cdot 5}{4 \cdot 5}$ and $\frac{x}{x^2+2} = \frac{5 \cdot x}{5 \cdot (x^2+2)}$
6.1C	Reducing a rational expression	The process of removing a common factor from the numerator and denominator of a rational expression	$\frac{8}{6} = \frac{2 \cdot 4}{2 \cdot 3} = \frac{4}{3}$ is reduced.
	Standard form of a fraction	$\frac{a}{b}$ and $\frac{-a}{b}$ $(b \neq 0)$ are the standard forms of a fraction.	$-\frac{-a}{-b}$ is written as $\frac{-a}{b}$ and $\frac{a}{-b}$ is written as $\frac{-a}{b}$.
6.2A	Multiplication of rational expressions	$\frac{A}{B} \cdot \frac{C}{D} = \frac{A \cdot C}{B \cdot D}$ $B \neq 0, D \neq 0$	$\frac{3}{4} \cdot \frac{7}{5} = \frac{3 \cdot 7}{4 \cdot 5} = \frac{21}{20}$
6.2B	Division of rational expressions	$\frac{A}{B} \div \frac{C}{D} = \frac{A \cdot D}{B \cdot C}$ $B \neq 0, C \neq 0, D \neq 0$	$\frac{3}{7} \div \frac{5}{4} = \frac{3 \cdot 4}{7 \cdot 5} = \frac{12}{35}$
6.3B	Addition of rational expressions with different denominators	$\frac{A}{B} + \frac{C}{D} = \frac{AD + BC}{BD}$ $B \neq 0, D \neq 0$	$\frac{1}{6} + \frac{1}{5} = \frac{5+6}{6 \cdot 5} = \frac{11}{30}$ $\frac{x}{6} + \frac{x}{5} = \frac{11x}{30}$
	Subtraction of rational expressions with different denominators	$\frac{A}{B} - \frac{C}{D} = \frac{AD - BC}{BD}$ $B \neq 0, D \neq 0$	$\frac{1}{5} - \frac{1}{6} = \frac{6-5}{5 \cdot 6} = \frac{1}{30}$ $\frac{x}{5} - \frac{x}{6} = \frac{x}{30}$

Section	Item	Meaning	Example
6.4	Complex fraction	A fraction that has fractions in the numerator, denominator, or both	$\dfrac{\frac{x}{x+2}}{x^2+x+1}$ is a complex fraction.
	Simplifying complex fractions	You can simplify a complex fraction by: 1. Multiplying numerator and denominator by the LCD of all fractions involved or 2. Performing the indicated operations in the numerator and denominator and then dividing the numerator by the denominator	Simplify $\dfrac{\frac{1}{x}+1}{\frac{1}{x}-1}$. Method 1: $\dfrac{x\left(\frac{1}{x}+1\right)}{x\left(\frac{1}{x}-1\right)}=\dfrac{1+x}{1-x}$ Method 2: $\dfrac{\frac{1}{x}+\frac{x}{x}}{\frac{1}{x}-\frac{x}{x}}=\dfrac{\frac{1+x}{x}}{\frac{1-x}{x}}=\dfrac{1+x}{x}\cdot\dfrac{x}{1-x}=\dfrac{1+x}{1-x}$
6.5	Fractional equation	An equation containing one or more rational expressions	$\frac{x}{2}+\frac{x}{3}=1$ is a fractional equation.
6.6A	Ratio	A quotient of two numbers	3 to 4, 3:4, and $\frac{3}{4}$ are ratios.
	Proportion	An equality between ratios	3:4 as x:6, or $\frac{3}{4}=\frac{x}{6}$
	Cross products	If $\frac{a}{b}=\frac{c}{d}$ (where $b \neq 0$ and $d \neq 0$), then $ad = bc$. ad and bc are the cross products.	If $\frac{3}{4}=\frac{x}{6}$, then $3\cdot 6 = 4\cdot x$. $3\cdot 6$ and $4\cdot x$ are the cross products.
6.6B	Similar figures	Figures that have the same shape but not necessarily the same size	▭ and ▭ are similar figures.
6.7A	Direct variation	y varies directly as x if $y = kx$. We also say y is proportional to x, where k is the constant of proportionality.	The cost C of bagels is proportional to the number n you buy. $C = kn$ and k is the price of one bagel.
6.7B	Inverse variation	y varies inversely as x if $y=\frac{k}{x}$, where k is the constant of proportionality.	The acceleration a of an object is inversely proportional to its mass m: $a=\frac{k}{m}$.

› Review Exercises Chapter 6

(If you need help with these exercises, look in the section indicated in brackets.)

1. ‹ **6.1B** › *Write the given fraction with the indicated denominator.*

   a. $\dfrac{5x}{8y}$ with a denominator of $16y^2$

   b. $\dfrac{3x}{4y^2}$ with a denominator of $16y^3$

   c. $\dfrac{2y}{3x^3}$ with a denominator of $15x^5$

2. ‹ **6.1C** › *Reduce the given fraction to lowest terms.*

   a. $\dfrac{-9(x^2-y^2)}{3(x+y)}$

   b. $\dfrac{-10(x^2-y^2)}{5(x+y)}$

   c. $\dfrac{-16(x^2-y^2)}{-4(x+y)}$

3. ⟨**6.1A, C**⟩ *Determine the values for which the given fraction is undefined, then simplify the expression.*

a. $\frac{-x}{x^2 + x}$

b. $\frac{-x}{x^2 - x}$

c. $\frac{-x}{x - x^2}$

4. ⟨**6.1C**⟩ *Reduce to lowest terms.*

a. $\frac{x^2 - 3x - 18}{6 - x}$

b. $\frac{x^2 - 2x - 8}{4 - x}$

c. $\frac{x^2 - 2x - 15}{5 - x}$

5. ⟨**6.2A**⟩ *Multiply.*

a. $\frac{3y^2}{7} \cdot \frac{14x}{9y}$

b. $\frac{7y^2}{5} \cdot \frac{15x}{14y}$

c. $\frac{6y^3}{7} \cdot \frac{28x}{3y}$

6. ⟨**6.2A**⟩ *Multiply.*

a. $(x - 3) \cdot \frac{x + 2}{x^2 - 9}$

b. $(x - 5) \cdot \frac{x + 1}{x^2 - 25}$

c. $(x - 4) \cdot \frac{x + 5}{x^2 - 16}$

7. ⟨**6.2B**⟩ *Divide.*

a. $\frac{x^2 - 9}{x + 2} \div (x + 3)$

b. $\frac{x^2 - 16}{x + 1} \div (x + 4)$

c. $\frac{x^2 - 25}{x + 4} \div (x + 5)$

8. ⟨**6.2B**⟩ *Divide.*

a. $\frac{x + 5}{x - 5} \div \frac{x^2 - 25}{5 - x}$

b. $\frac{x + 1}{x - 1} \div \frac{x^2 - 1}{1 - x}$

c. $\frac{x + 2}{x - 2} \div \frac{x^2 - 4}{2 - x}$

9. ⟨**6.3A**⟩ *Add.*

a. $\frac{3}{2(x - 1)} + \frac{1}{2(x - 1)}$

b. $\frac{5}{6(x - 2)} + \frac{7}{6(x - 2)}$

c. $\frac{3}{4(x + 1)} + \frac{1}{4(x + 1)}$

10. ⟨**6.3A**⟩ *Subtract.*

a. $\frac{7}{2(x + 1)} - \frac{3}{2(x + 1)}$

b. $\frac{11}{5(x + 2)} - \frac{1}{5(x + 2)}$

c. $\frac{17}{7(x + 3)} - \frac{3}{7(x + 3)}$

11. ⟨**6.3B**⟩ *Add.*

a. $\frac{2}{x + 2} + \frac{1}{x - 2}$

b. $\frac{3}{x + 1} + \frac{1}{x - 1}$

c. $\frac{4}{x + 3} + \frac{1}{x - 3}$

12. ⟨**6.3B**⟩ *Subtract.*

a. $\frac{x - 1}{x^2 + 3x + 2} - \frac{x + 7}{x^2 + 5x + 6}$

b. $\frac{x + 3}{x^2 - x - 2} - \frac{x - 1}{x^2 + 2x + 1}$

c. $\frac{x - 1}{x^2 + 3x + 2} - \frac{x + 1}{x^2 + x - 2}$

13. ⟨**6.4A**⟩ *Simplify.*

a. $\dfrac{\frac{3}{2x} - \frac{1}{x}}{\frac{2}{3x} + \frac{3}{4x}}$

b. $\dfrac{\frac{3}{2x} - \frac{1}{3x}}{\frac{2}{x} + \frac{1}{4x}}$

c. $\dfrac{\frac{3}{2x} - \frac{1}{x}}{\frac{3}{4x} + \frac{4}{3x}}$

14. ⟨**6.5A**⟩ *Solve.*

a. $\frac{2x}{x - 1} + 3 = \frac{4x}{x - 1}$

b. $\frac{6x}{x - 5} + 7 = \frac{8x}{x - 5}$

c. $\frac{5x}{x - 4} + 6 = \frac{7x}{x - 4}$

15. ⟨ **6.5A** ⟩ *Solve.*

a. $\frac{x}{x^2 - 4} + \frac{2}{x - 2} = \frac{x - 3}{x^2 - x - 6}$

b. $\frac{x}{x^2 - 16} + \frac{4}{x - 4} = \frac{x - 5}{x^2 - x - 20}$

c. $\frac{x}{x^2 - 25} + \frac{5}{x - 5} = \frac{x - 6}{x^2 - x - 30}$

16. ⟨ **6.5A** ⟩ *Solve.*

a. $\frac{x}{x + 6} - \frac{1}{7} = \frac{-6}{x + 6}$

b. $\frac{x}{x + 7} - \frac{1}{8} = \frac{-7}{x + 7}$

c. $\frac{x}{x + 8} - \frac{1}{9} = \frac{-8}{x + 8}$

17. ⟨ **6.5A** ⟩ *Solve.*

a. $3 + \frac{5}{x - 4} = \frac{50}{x^2 - 16}$

b. $4 + \frac{6}{x - 5} = \frac{84}{x^2 - 25}$

c. $5 + \frac{7}{x - 6} = \frac{126}{x^2 - 36}$

18. ⟨ **6.5B** ⟩ *Solve for the indicated variable.*

a. $A = \frac{a_1(1 - b)}{1 - b^n}; a_1$

b. $B = \frac{b_1(1 - c^n)}{1 - c}; b_1$

c. $C = \frac{c_1(1 - d)^n}{d + 1}; c_1$

19. ⟨ **6.6A** ⟩ *A car travels 160 miles on 7 gallons of gas.*

a. How many gallons will it need to travel 240 miles?

b. Repeat the problem where the car travels 180 miles on 9 gallons of gas and we wish to go 270 miles.

c. Repeat the problem where the car travels 200 miles on 12 gallons and we wish to go 300 miles.

20. ⟨ **6.6A** ⟩ *Solve using cross products.*

a. $\frac{x + 3}{6} = \frac{7}{2}$

b. $\frac{x + 4}{8} = \frac{9}{5}$

c. $\frac{x + 5}{2} = \frac{6}{5}$

21. ⟨ **6.6B** ⟩ *A person can do a job in 6 hours. Another person can do it in 8 hours.*

a. How long would it take to do the job if both of them work together?

b. Repeat the problem where the first person takes 10 hours and the second person takes 8 hours.

c. Repeat the problem where the first person takes 9 hours and the second person takes 6 hours.

22. ⟨ **6.6B** ⟩ *A boat can travel 10 miles against a current in the same time it takes to travel 30 miles with the current. What is the speed of the boat in still water if the current flows at:*

a. 2 miles per hour?

b. 4 miles per hour?

c. 6 miles per hour?

23. ⟨ **6.6B** ⟩ *A baseball player has 30 home runs in 120 games. At that rate, how many home runs will he have in:*

a. 128 games?

b. 140 games?

c. 160 games?

24. ⟨ **6.6B** ⟩ *A company wants to produce 1500 items in 1 year (12 months). To attain this goal, how many items should be produced by the end of:*

a. September (the 9th month)?

b. October (the 10th month)?

c. November (the 11th month)?

25. ⟨ **6.6B** ⟩ *Find the unknown in the given similar triangle.*

a. Find x.

b. Find b.

c. Find s.

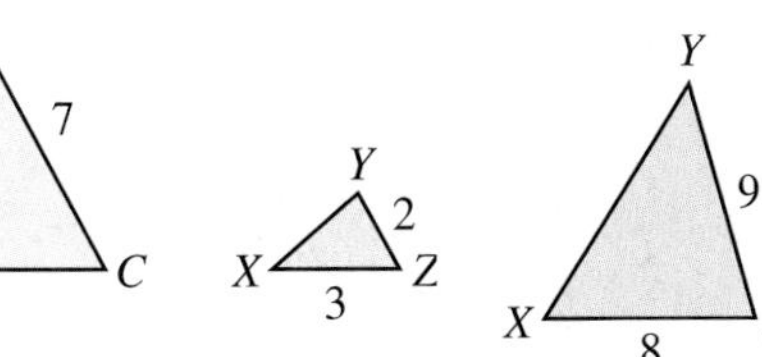

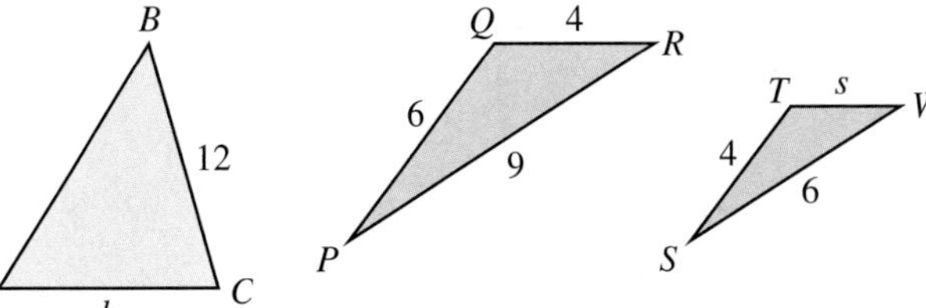

26. ⟨ 6.7A ⟩

a. If you walk for m minutes, the number C of calories used is proportional to the time you walk. If 30 calories are used when you walk for 12 minutes, write an equation of variation and find the number of calories used when you walk for an hour.

b. If you row for m minutes, the number C of calories used is proportional to the time you row. If 140 calories are used when you row for 20 minutes, write an equation of variation and find the number of calories used when you row for an hour.

c. If you play tennis for m minutes, the number C of calories used is proportional to the time you play. If 180 calories are used when you play for 180 minutes, write an equation of variation and find the number of calories used when you play for 45 minutes.

27. ⟨ 6.7B ⟩

a. The amount F of force you exert on a wrench handle to loosen a rusty bolt varies inversely with the length L of the handle. If $k = 30$, write an equation of variation and find the force needed when the handle is 6 inches long.

b. What force is needed when the handle is 10 inches long?

c. What force is needed when the handle is 15 inches long?

Practice Test Chapter 6

(Answers on page 566)

Visit www.mhhe.com/bello to view helpful videos that provide step-by-step solutions to several of the problems below.

1. Write $\frac{3x}{7y}$ with a denominator of $21y^3$.

2. Reduce $\dfrac{-6(x^2 - y^2)}{3(x - y)}$ to lowest terms.

3. Determine the values for which the expression $\dfrac{-x}{x + x^2}$ is undefined then simplify.

4. Reduce to lowest terms $\dfrac{x^2 + 2x - 8}{2 - x}$.

In Problems 5–12, perform the indicated operations and simplify.

5. Multiply $\dfrac{2y^2}{7} \cdot \dfrac{21x}{4y}$.

6. Multiply $(x - 2) \cdot \dfrac{x + 3}{x^2 - 4}$.

7. Divide $\dfrac{x^2 - 4}{x + 5} \div (x - 2)$.

8. Divide $\dfrac{x + 3}{x - 3} \div \dfrac{x^2 - 9}{3 - x}$.

9. Add $\dfrac{5}{2(x - 2)} + \dfrac{1}{2(x - 2)}$.

10. Subtract $\dfrac{7}{3(x + 1)} - \dfrac{1}{3(x + 1)}$.

11. Add $\dfrac{2}{x + 1} + \dfrac{1}{x - 1}$.

12. Subtract $\dfrac{x + 1}{x^2 + x - 2} - \dfrac{x + 2}{x^2 - 1}$.

13. Simplify $\dfrac{\frac{1}{x} - \frac{2}{3x}}{\frac{3}{4x} + \frac{1}{2x}}$.

14. Solve $\dfrac{3x}{x - 2} + 4 = \dfrac{5x}{x - 2}$.

15. Solve $\dfrac{x}{x^2 - 9} + \dfrac{3}{x - 3} = \dfrac{1}{x + 3}$.

16. Solve $\dfrac{x}{x + 5} - \dfrac{1}{6} = \dfrac{-5}{x + 5}$.

17. Solve $2 + \dfrac{4}{x - 3} = \dfrac{24}{x^2 - 9}$.

18. Solve for d_1 in

$$D = \frac{d_1(1 - d^n)}{(1 + d)^n}.$$

19. Solve $\dfrac{x + 5}{7} = \dfrac{11}{6}$.

20. A car travels 150 miles on 9 gallons of gas. How many gallons will it need to travel 400 miles?

21. A woman can paint a house in 5 hours. Another one can do it in 8 hours. How long would it take to paint the house if both women work together?

22. A boat can travel 10 miles against a current in the same time it takes to travel 30 miles with the current. If the speed of the current is 8 miles per hour, what is the speed of the boat in still water?

23. A baseball player has 20 singles in 80 games. At that rate, how many singles will he have in 160 games?

24. A conversion van company wants to finish 150 vans in 1 year (12 months). To attain this goal, how many vans should be finished by the end of April (the fourth month)?

25. Find the unknown in the given similar triangles.

a.

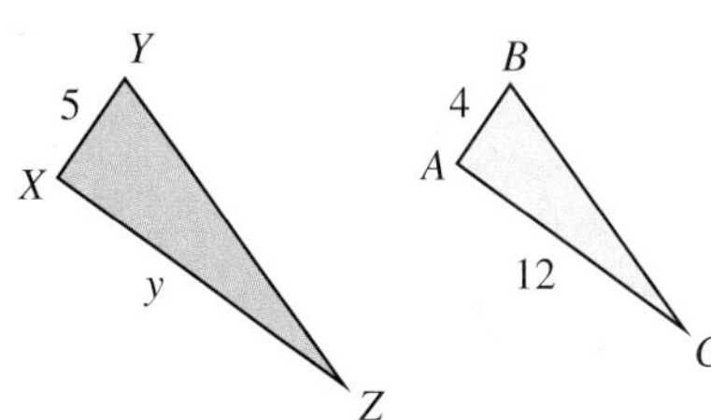

b.

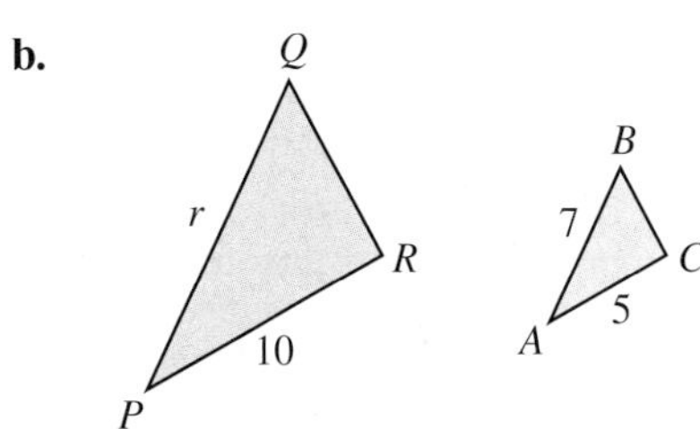

26. Have you raked leaves lately? If you rake leaves for m minutes, the number C of calories used is proportional to the time you rake. If 60 calories are used when you rake for 30 minutes, write an equation of variation and find the number of calories used when you rake for $2\frac{1}{2}$ hours.

27. The maximum weight W that can be supported by a 2-by-4-inch piece of pinewood varies inversely with its length L. If the maximum weight W that can be supported by a 10-foot-long 2-by-4 piece of pine is 500 pounds, find an equation of variation and the maximum weight W that can be supported by a 25-foot length of 2-by-4 pine.

Answers to Practice Test Chapter 6

Answer	If You Missed Question	Review Section	Examples	Page
1. $\frac{9xy^2}{21y^3}$	1	6.1	2	492–493
2. $-2(x + y)$ or $-2x - 2y$	2	6.1	3	494–495
3. $\frac{-1}{1 + x}$; undefined for $x = 0$ and $x = -1$	3	6.1	4, 1	496, 490
4. $-(x + 4)$ or $-x - 4$	4	6.1	5	497–498
5. $\frac{3xy}{2}$	5	6.2	1, 2	505
6. $\frac{x + 3}{x + 2}$	6	6.2	3	506
7. $\frac{x + 2}{x + 5}$	7	6.2	4	507–508
8. $\frac{-1}{x - 3}$ or $\frac{1}{3 - x}$	8	6.2	4, 5	507–508
9. $\frac{3}{x - 2}$	9	6.3	1a	515
10. $\frac{2}{x + 1}$	10	6.3	1b	515
11. $\frac{3x - 1}{x^2 - 1} = \frac{3x - 1}{(x + 1)(x - 1)}$	11	6.3	2, 4a	515–516, 518
12. $\frac{-2x - 3}{(x + 2)(x + 1)(x - 1)}$	12	6.3	3, 4b, 5	516–517, 518–520
13. $\frac{4}{15}$	13	6.4	1, 2	526–527
14. $x = 4$	14	6.5	1, 2	532–533
15. $x = -4$	15	6.5	3	533–534
16. No solution	16	6.5	4	534
17. $x = -5$	17	6.5	5, 7	535–536, 537
18. $d_1 = \frac{D(1 + d)^n}{(1 - d^n)}$	18	6.5	6	536
19. $x = \frac{47}{6}$	19	6.6	1	542
20. 24	20	6.6	1	542
21. $3\frac{1}{13}$ hr	21	6.6	2	544
22. 16 mi/hr	22	6.6	3	544–545
23. 40	23	6.6	4	545–546
24. 50	24	6.6	4	545–546
25. **a.** $y = 15$ **b.** $r = 14$	25	6.6	5	546
26. $C = 2m$; 300	26	6.7	1, 2	552–553
27. $W = \frac{5000}{L}$; 200 lb	27	6.7	3, 4	553–554

Cumulative Review Chapters 1–6

1. Add: $-\frac{1}{6} + \left(-\frac{2}{5}\right)$

2. Subtract: $-5.6 - (-8.3)$

3. Find: $(-5)^2$

4. Divide: $-\frac{7}{8} \div \left(-\frac{7}{16}\right)$

5. Evaluate $y \div 5 \cdot x - z$ for $x = 6$, $y = 60$, $z = 3$.

6. Simplify: $x + 4(x - 3) + (x - 2)$

7. Write in symbols: The quotient of $(a - b)$ and c.

8. Solve for x: $5 = 3(x - 1) + 4 - 2x$

9. Solve for x: $\frac{x}{7} - \frac{x}{9} = 2$

10. The sum of two numbers is 95. If one of the numbers is 35 more than the other, what are the numbers?

11. Susan purchased some municipal bonds yielding 8% annually and some certificates of deposit yielding 10% annually. If Susan's total investment amounts to \$12,000 and the annual income is \$1020, how much money is invested in bonds and how much is invested in certificates of deposit?

12. Graph: $-\frac{x}{2} + \frac{x}{4} \geq \frac{x - 4}{4}$

13. Graph the point $C(-1, -3)$.

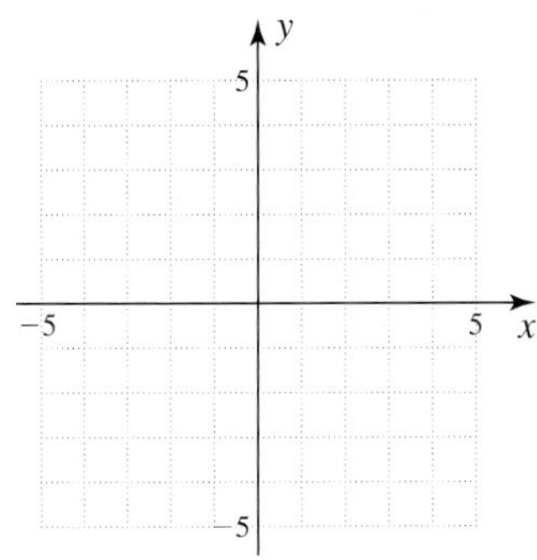

14. Determine whether the ordered pair $(-3, -3)$ is a solution of $5x - y = -12$.

15. Find x in the ordered pair $(x, -2)$ so that the ordered pair satisfies the equation $3x - y = 11$.

16. Graph: $2x + y = 2$

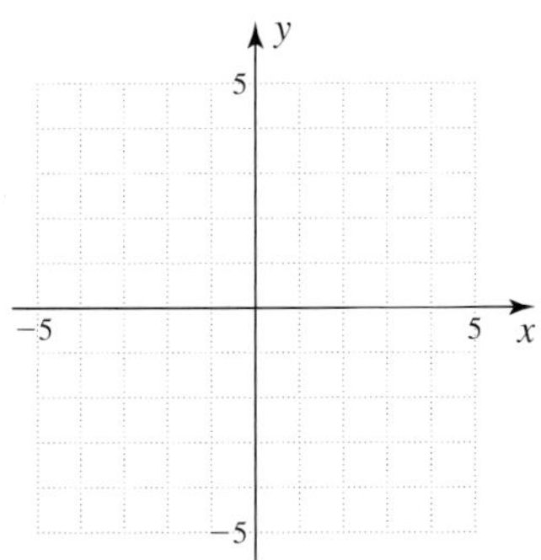

17. Graph: $3y + 6 = 0$

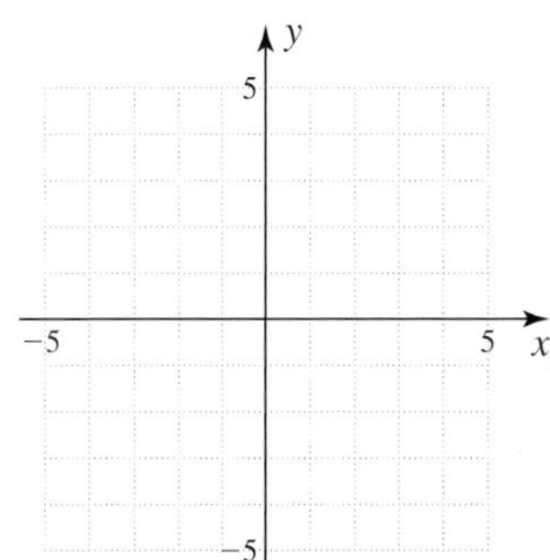

18. Find the slope of the line passing through the points $(-7, -9)$ and $(9, -2)$.

19. What is the slope of the line $6x - 3y = -10$?

20. Find the pair of parallel lines.

(1) $15y + 20x = 8$

(2) $20x - 15y = 8$

(3) $3y = 4x + 8$

21. Find an equation of the line passing through $(-4, 3)$ and with slope 6.

22. Find an equation of the line with slope -2 and y-intercept 5.

23. Graph $3x - 4y < -12$.

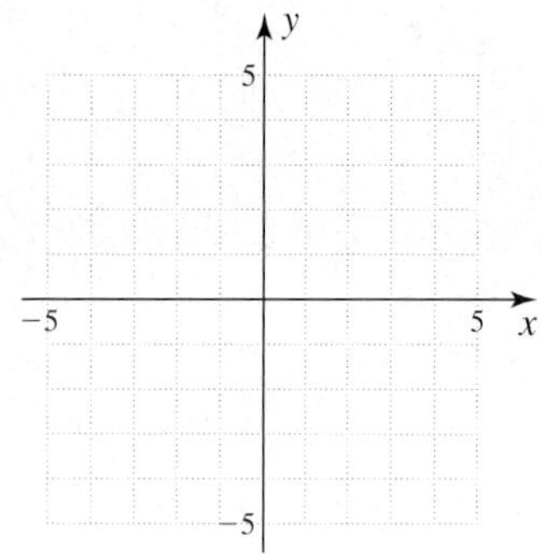

24. Graph $3x - y \geq 0$.

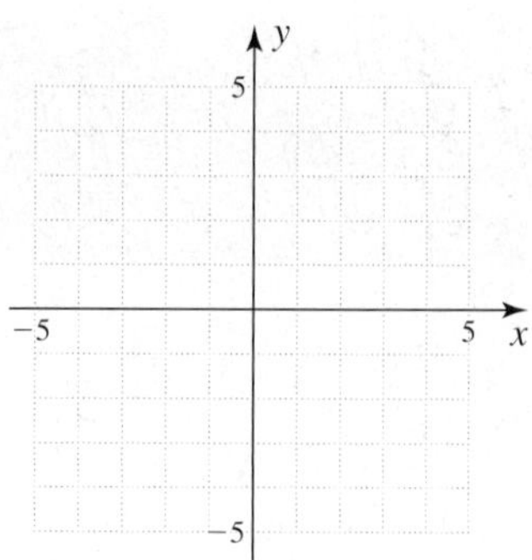

25. Simplify: $\frac{x^{-8}}{x^{-9}}$

26. Multiply and simplify: $x^5 \cdot x^{-7}$

27. Simplify: $(3x^3y^{-2})^{-4}$

28. Write in scientific notation: 0.00036

29. Divide and express the answer in scientific notation: $(14.04 \times 10^{-3}) \div (7.8 \times 10^2)$

30. Find the value of $x^2 + 3x - 2$ when $x = -2$.

31. Add $(2x^4 - 6x^6 - 7)$ and $(-6x^6 + 3 + 7x^4)$.

32. Find: $\left(5x^2 - \frac{1}{5}\right)^2$

33. Multiply: $(9x^2 + 2)(9x^2 - 2)$

34. Divide $(3x^3 + 11x^2 + 18)$ by $(x + 4)$.

35. Factor completely: $6x^2 - 9x^4$

36. Factor completely: $\frac{2}{5}x^7 - \frac{4}{5}x^6 + \frac{4}{5}x^5 - \frac{1}{5}x^3$

37. Factor completely: $x^2 - 15x + 56$

38. Factor completely: $15x^2 - 37xy + 20y^2$

39. Factor completely: $16x^2 - 9y^2$

40. Factor completely: $-4x^4 + 4x^2$

41. Factor completely: $3x^3 - 6x^2 - 9x$

42. Factor completely: $4x^2 + 4x + 3x + 3$

43. Factor completely: $16kx^2 - 8kx + k$

44. Solve for x: $2x^2 + x = 15$

45. Write $\frac{7x}{6y}$ with a denominator of $12y^3$.

46. Reduce to lowest terms: $\frac{-8(x^2 - y^2)}{4(x - y)}$

47. Reduce to lowest terms: $\frac{x^2 + 4x - 5}{1 - x}$

48. Multiply: $(x - 7) \cdot \frac{x + 3}{x^2 - 49}$

49. Divide: $\frac{x + 3}{x - 3} \div \frac{x^2 - 9}{3 - x}$

50. Add: $\frac{3}{2(x + 8)} + \frac{9}{2(x + 8)}$

51. Subtract: $\frac{x + 5}{x^2 + x - 30} - \frac{x + 6}{x^2 - 25}$

52. Simplify: $\dfrac{\frac{2}{x} + \frac{3}{2x}}{\frac{1}{3x} - \frac{1}{4x}}$

53. Solve for x: $\frac{3x}{x - 2} + 3 = \frac{4x}{x - 2}$

54. Solve for x: $\frac{x}{x^2 - 49} + \frac{7}{x - 7} = \frac{1}{x + 7}$

55. Solve for x: $\frac{x}{x + 4} - \frac{1}{5} = \frac{-4}{x + 4}$

56. Solve for x: $1 + \frac{2}{x - 3} = \frac{12}{x^2 - 9}$

57. A car travels 120 miles on 6 gallons of gas. How many gallons will it need to travel 460 miles?

58. Solve for x: $\frac{x + 8}{9} = \frac{2}{11}$

59. Maria can paint a kitchen in 6 hours, and James can paint the same kitchen in 7 hours. How long would it take for both working together to paint the kitchen?

60. An enclosed gas exerts a pressure P on the walls of a container. This pressure is directly proportional to the temperature T of the gas. If the pressure is 5 pounds per square inch when the temperature is 250° F. find k.

61. If the temperature of a gas is held constant, the pressure P varies inversely as the volume V. A pressure of 1960 pounds per square inch is exerted by 7 cubic feet of air in a cylinder fitted with a piston. Find k.

Section

Chapter

7 seven

Solving Systems of Linear Equations and Inequalities

中國人民郵政 8 分
祖冲之(公元429-500)數學家,精確
算出圓周率爲3.14159265
紀33.4—2
(126)1955

The Human Side of Algebra

The first evidence of a systematic method of solving systems of linear equations is provided in the *Nine Chapters of the Mathematical Arts,* the oldest arithmetic textbook in existence. The method for solving a system of three equations with three unknowns occurs in the 18 problems of the eighth chapter, entitled "The Way of Calculating by Arrays." Unfortunately, the original copies of the *Nine Chapters* were destroyed in 213 B.C. However, the Chinese mathematician Liu Hui wrote a commentary on the *Nine Chapters* in A.D. 263, and information concerning the original work comes to us through this commentary.

In modern times, when a large number of equations or inequalities has to be solved, a method called the **simplex method** is used. This method, based on the **simplex algorithm,** was developed in the 1940s by George B. Dantzig. It was first used by the Allies of World War II to solve logistics problems dealing with obtaining, maintaining, and transporting military equipment and personnel.

7.1 Solving Systems of Equations by Graphing

Objectives

A Solve a system of two equations in two variables by graphing.

B Determine whether a system of equations is consistent, inconsistent, or dependent.

C Solve an application involving systems of equations.

To Succeed, Review How To . . .

1. Graph the equation of a line (pp. 230–231, 240–245).
2. Determine whether an ordered pair is a solution of an equation (pp. 226–228).

Getting Started

Supply, Demand, and Intersections

Can you tell from the graph when the energy supply and the demand were about the same? This happened where the line representing the supply and the line representing the demand intersect, or way back in 1980. When the demand and the supply are *equal,* prices reach *equilibrium.* If x is the year and y the number of millions of barrels of oil per day, the point (x, y) at which the demand is the same as the supply—that is, the point at which the graphs *intersect*—is (1980, 85). We can graph a pair of linear equations and find a point of intersection if it exists. This point of intersection is an ordered pair of numbers such as (1980, 85) and is a **solution** of both equations. This means that when you substitute 1980 for x and 85 for y in the original equations, the results are true statements. In this section we learn how to find the solution of a system of two equations in two variables by using the graphical method, which involves graphing the equations and finding their point of intersection, if it exists.

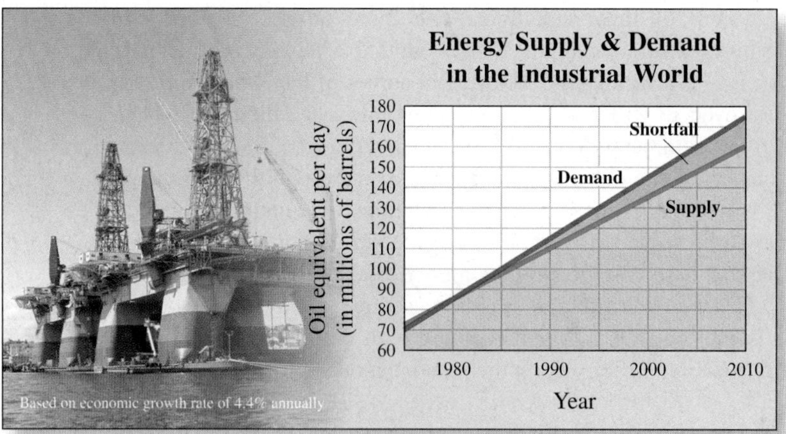

A › Solving a System by Graphing

The solution of a linear equation in one variable, as studied in Chapter 2, was a single number. Thus if we solve *two* linear equations in two variables simultaneously, we expect to get **two** numbers. We write this solution as an ordered pair. For example, the solution of

$$x + 2y = 4$$
$$2y - x = 0$$

is (2, 1). This can be checked by letting $x = 2$ and $y = 1$ in both equations:

$$\begin{aligned} x + 2y &= 4 \\ 2 + 2(1) &= 4 \\ 2 + 2 &= 4 \\ 4 &= 4 \end{aligned} \qquad \begin{aligned} 2y - x &= 0 \\ 2(1) - 2 &= 0 \\ 2 - 2 &= 0 \\ 0 &= 0 \end{aligned}$$

Clearly, a true statement results in both cases. We call a system of two linear equations a **system of simultaneous equations.** To solve one of these systems, we need to find (if possible) all ordered pairs of numbers that satisfy **both** equations. Thus, to *solve* the system

$$\begin{aligned} x + 2y &= 4 \\ 2y - x &= 0 \end{aligned}$$

This system is called a system of *simultaneous* equations because we have to find a solution that satisfies both equations.

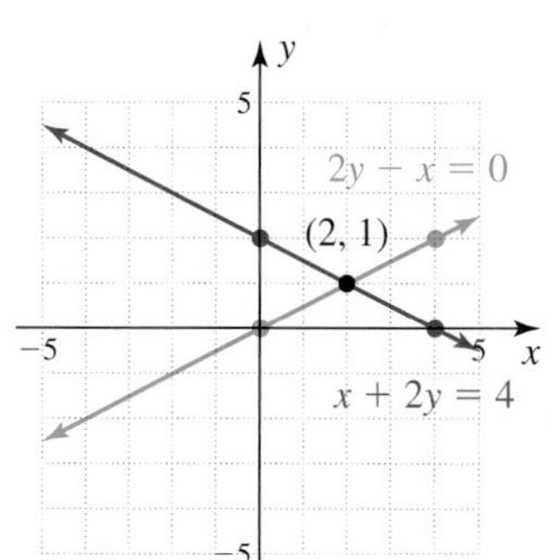

>Figure 7.1

we graph each of the equations in the same coordinate axis in the usual way. To graph $x + 2y = 4$, we find the intercepts using the following table:

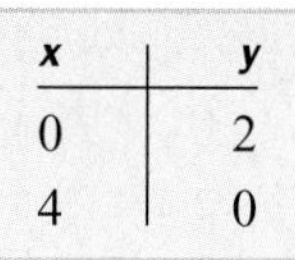

x	y
0	2
4	0

The intercepts (0, 2) and (4, 0), as well as the completed graph, are shown in red in Figure 7.1.

The equation $2y - x = 0$ is graphed similarly using the following table:

x	y
0	0

Note that we need another point in the table. We can pick any x we want and then find y. If we pick $x = 4$, then $2y - 4 = 0$, or $y = 2$, giving us the point (4, 2). The graph of the equation $2y - x = 0$ is shown in blue. The lines intersect at (2, 1), which is the solution of the system of equations. (Recall that we checked that this point satisfies the equations by substituting 2 for x and 1 for y in each equation.)

EXAMPLE 1 Using the graphical method to solve a system

Use the graphical method to find the solution of the system:

$$\begin{aligned} 2x + y &= 4 \\ y - 2x &= 0 \end{aligned}$$

SOLUTION 1 We first graph the equation $2x + y = 4$ using the following table:

x	y
0	4
2	0

The two points and the complete graph are shown in blue in Figure 7.2. We then graph $y - 2x = 0$ using the following table:

x	y
0	0
2	4

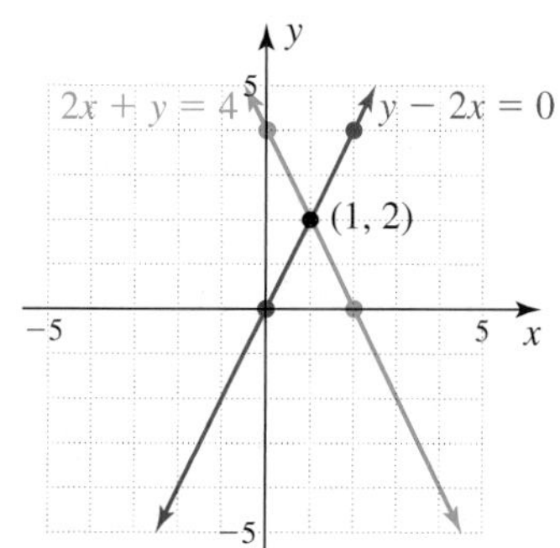

>Figure 7.2

PROBLEM 1

Use the graphical method to solve the system:

$$\begin{aligned} x + 2y &= 4 \\ 2x - 4y &= 0 \end{aligned}$$

(continued)

Answers to PROBLEMS

1.

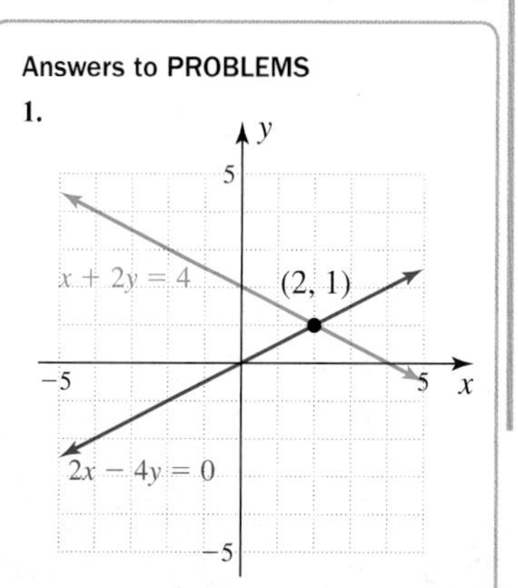

The graph of $y - 2x = 0$ is shown in red. The lines intersect at (1, 2). Note that (1, 2) is part of both lines, so (1, 2) is the **solution** of the system of equations. We can check this by letting $x = 1$ and $y = 2$ in

$$2x + y = 4$$
$$y - 2x = 0$$

thus obtaining the true statements

$$2(1) + 2 = 4$$
$$2 - 2(1) = 0$$

EXAMPLE 2 **Solving an inconsistent system**

Use the graphical method to find the solution of the system:

$$y - 2x = 4$$
$$2y - 4x = 12$$

SOLUTION 2 We first graph the equation $y - 2x = 4$ using the following table:

x	y
0	4
−2	0

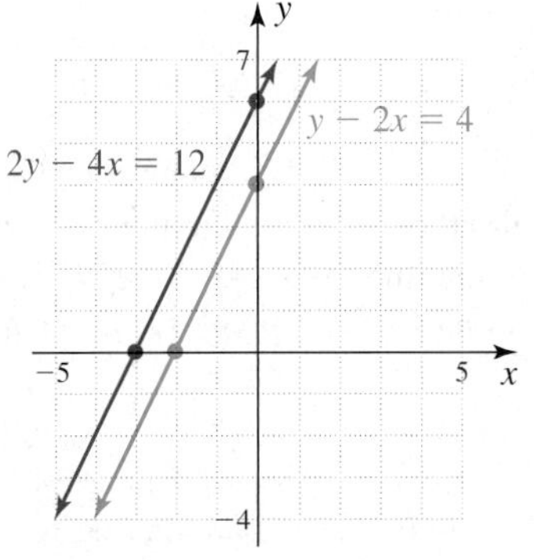

>Figure 7.3

The two points, as well as the completed graph, are shown in blue in Figure 7.3. We then graph $2y - 4x = 12$ using the following table:

x	y
0	6
−3	0

The graph of $2y - 4x = 12$ is shown in red. The two lines appear to be parallel; they do not intersect. If we examine the equations more carefully, we see that by dividing both sides of the second equation by 2, we get $y - 2x = 6$. Thus, one equation says $y - 2x = 4$, and the other says that $y - 2x = 6$. Hence, both equations cannot be true at the same time, and their graphs cannot intersect.

To confirm this, note that $y - 2x = 6$ is equivalent to $y = 2x + 6$ and $y - 2x = 4$ is equivalent to $y = 2x + 4$. Since $y = 2x + 6$ and $y = 2x + 4$ both have slope 2 but different y-intercepts, their graphs are **parallel lines.** Thus, there is *no solution* for this system, since the two lines do not have any points in common; the system is said to be **inconsistent.**

PROBLEM 2

Use the graphical method to solve the system:

$$y - 3x = 3$$
$$2y - 6x = 12$$

Answers to PROBLEMS

2. No solution

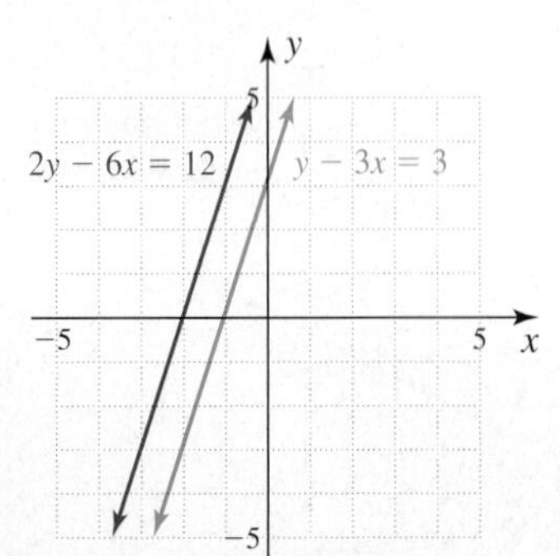

EXAMPLE 3 **Solving a dependent system**

Use the graphical method to solve the system:

$$2x + y = 4$$
$$2y + 4x = 8$$

SOLUTION 3 We use the table

x	y
0	4
2	0

to graph $2x + y = 4$, which is shown in color in Figure 7.4. To graph $2y + 4x = 8$, we first let $x = 0$ to obtain $2y = 8$ or $y = 4$. For $y = 0$, $4x = 8$ or $x = 2$. Thus, the two points in our second table will be

x	y
0	4
2	0

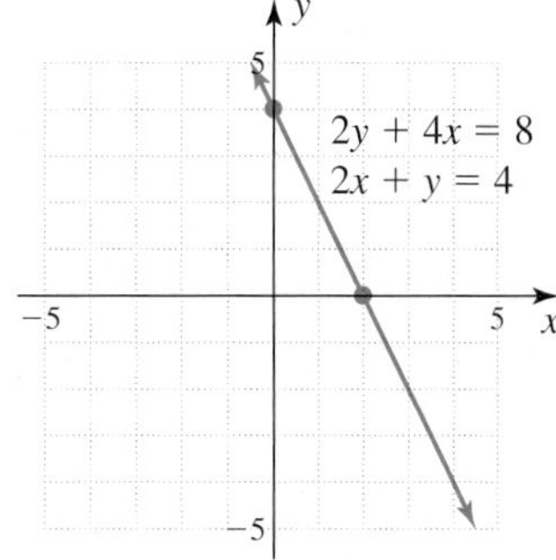

>Figure 7.4

But these points are exactly the same as those obtained in the first table! What does this mean? It means that the graphs of the lines $2x + y = 4$ and $2y + 4x = 8$ **coincide** (are the same). Thus, a solution of one equation is *automatically* a solution for the other. In fact, there are *infinitely many* solutions; every point on the graph is a solution of the system. Such a system is said to be **dependent.** In a dependent system, one of the equations is a **constant multiple** of the other. (If you multiply both sides of the first equation by 2, you get the second equation.)

PROBLEM 3

Use the graphical method to solve the system:

$$x + 2y = 4$$
$$4y + 2x = 8$$

B > Finding Consistent, Inconsistent, and Dependent Systems of Equations

As you can see from the examples we've given, a system of equations can have exactly *one* solution (when the lines *intersect,* as in Figure 7.5), *no* solution (when the lines are *parallel,* as in Figure 7.6), or *infinitely many* solutions (when the graphs of the two lines are *identical,* as in Figure 7.7). These examples illustrate the three possible solutions to a system of simultaneous equations.

Case 1

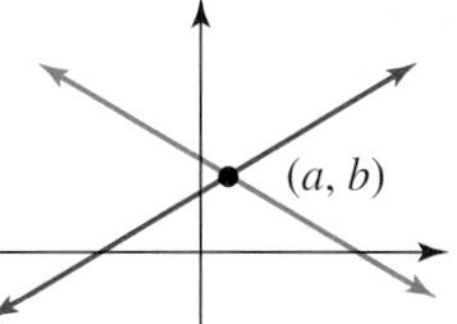

Consistent and independent (one solution: (a, b))

>Figure 7.5

Case 2

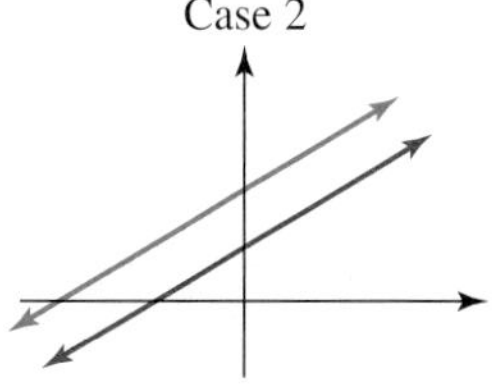

Inconsistent; parallel lines (no solution)

>Figure 7.6

Case 3

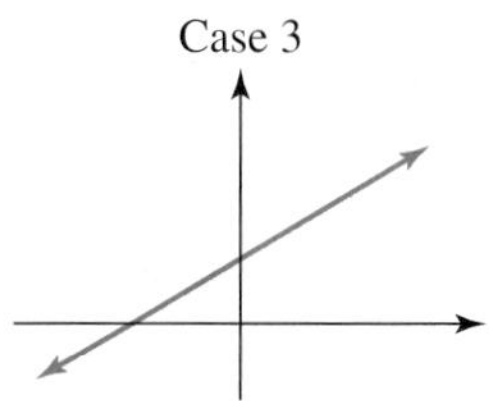

Dependent; lines coincide (infinitely many solutions)

>Figure 7.7

Answers to PROBLEMS

3. Infinitely many solutions

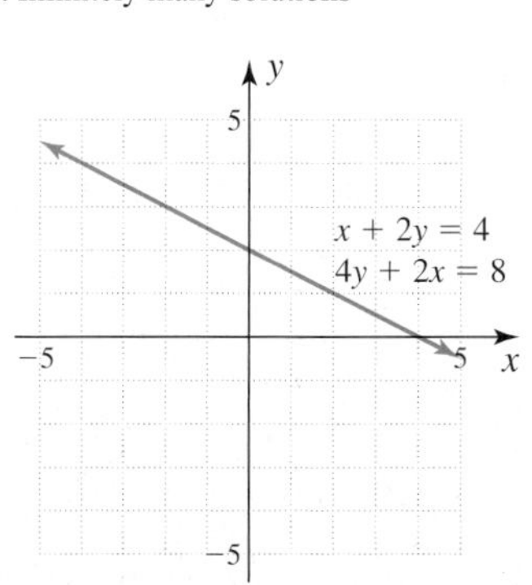

POSSIBLE SOLUTIONS TO A SYSTEM OF SIMULTANEOUS EQUATIONS

1. **Consistent and independent systems:** The graphs of the equations intersect at *one* point, whose coordinates give the solution of the system.
2. **Inconsistent systems:** The graphs of the equations are *parallel* lines; there is *no* solution for the system.
3. **Dependent systems:** The graphs of the equations *coincide* (are the same). There are *infinitely many* solutions for the system.

The following table will help you further.

Type of Lines	Slopes	y-Intercept	Number of Solutions	Type of System
Intersecting	Different	Same or different	One	Consistent
Parallel	Same	Different	None	Inconsistent
Coinciding	Same	Same	Infinite	Dependent

EXAMPLE 4 Classifying a system by graphing

Use the graphical method to solve the given system of equations. Classify each system as consistent (one solution), inconsistent (no solution), or dependent (infinitely many solutions).

a. $x + y = 4$
$2y - x = -1$

b. $x + 2y = 4$
$2x + 4y = 6$

c. $x + 2y = 4$
$4y + 2x = 8$

SOLUTION 4

a. The respective tables for $x + y = 4$ and $2y - x = -1$ are

x	y
0	4
4	0

x	y
0	$\frac{-1}{2}$
1	0

The graphs of these two lines are shown in Figure 7.8. As you can see, the solution is (3, 1). (Check this!) The system is *consistent.*

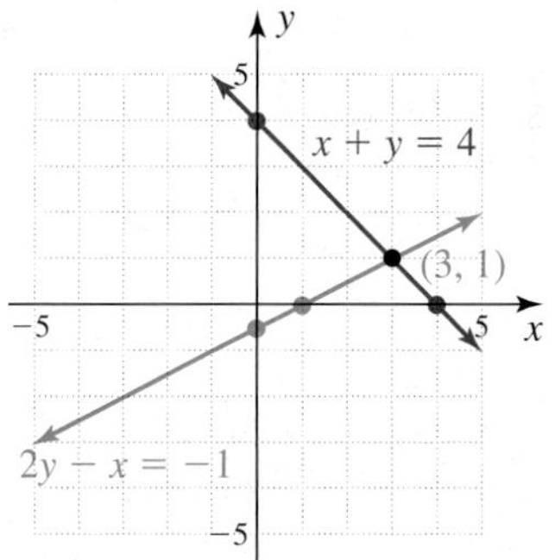

>Figure 7.8

PROBLEM 4

Use the graphical method to solve the given system. Classify each system as consistent (one solution), inconsistent (no solution), or dependent (infinitely many solutions).

a. $x + y = 4$
$2y - x = 2$

b. $2x + y = 4$
$2y + 4x = 6$

c. $2x + y = 4$
$2y + 4x = 8$

Answers to PROBLEMS

4. a. Consistent; solution (2, 2)

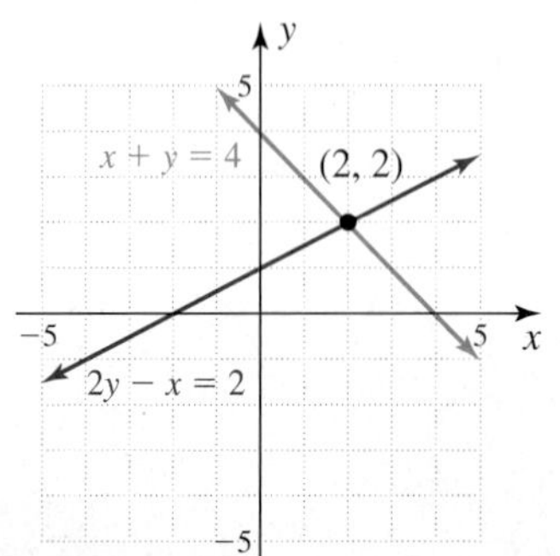

b. Inconsistent; no solution

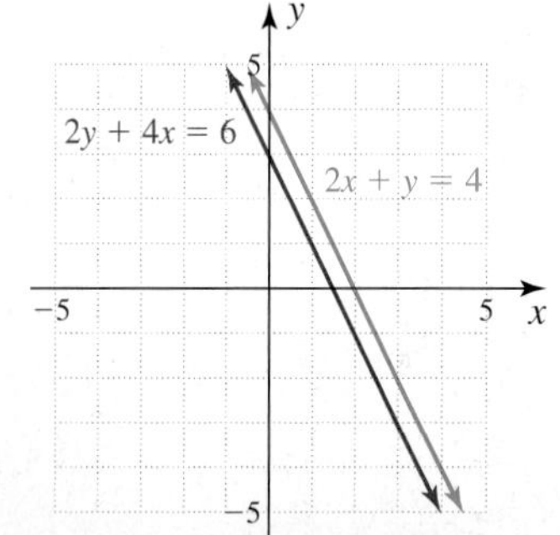

c. Dependent; infinitely many solutions

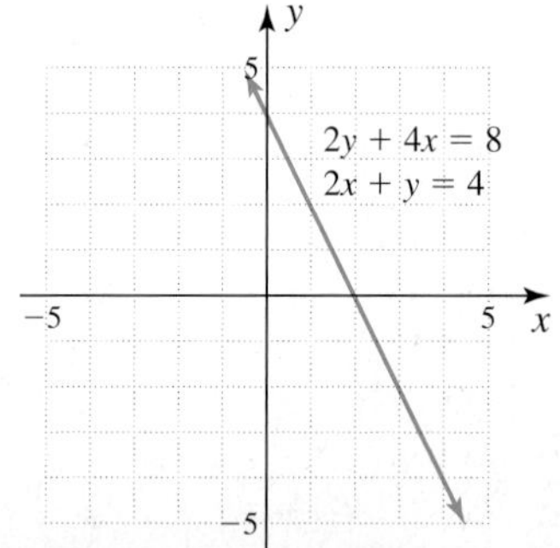

b. The respective tables for $x + 2y = 4$ and $2x + 4y = 6$ are

x	y
0	2
4	0

x	y
0	$\frac{3}{2}$
3	0

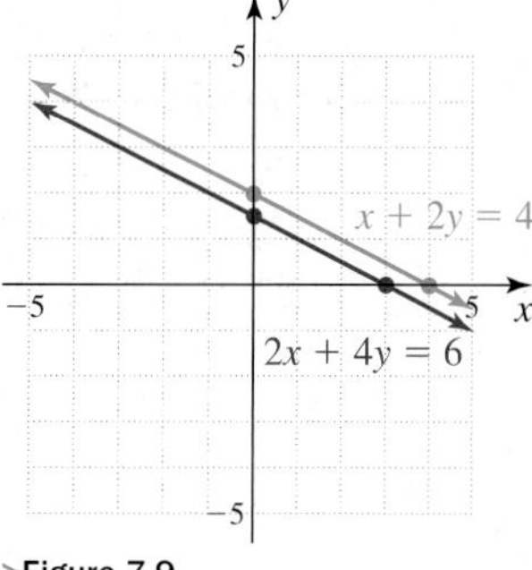

>Figure 7.9

The graphs of the two lines are shown in Figure 7.9. There is no solution because the lines are parallel. To see this, we solve $2x + 4y = 6$ and $x + 2y = 4$ for y to obtain

$$y = -\frac{1}{2}x + \frac{3}{2} \qquad \text{and} \qquad y = -\frac{1}{2}x + 2$$

These equations represent two lines with the same slope and different y-intercepts. Thus, the lines are parallel, there is no solution, and the system is *inconsistent.*

c. The respective tables for $x + 2y = 4$ and $4y + 2x = 8$ are

x	y
0	2
4	0

x	y
0	2
4	0

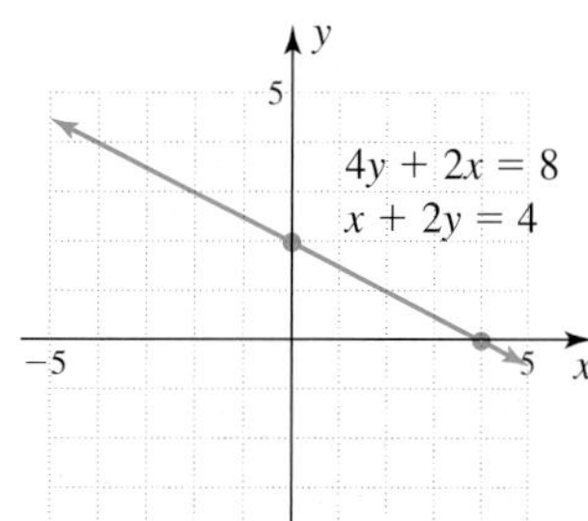

>Figure 7.10

and they are identical. So, there are infinitely many solutions because the lines coincide (see Figure 7.10). The system is *dependent,* and the solutions are all the points on the graph. For example, (0, 2), (4, 0), and (2, 1) are solutions.

Note that in a *dependent* system, one equation is a constant *multiple* of the other. Thus,

$$x + 2y = 4 \qquad \text{and} \qquad 4y + 2x = 8$$

are a dependent system because $4y + 2x = 8$ is a constant *multiple* of $x + 2y = 4$. Note that

$$2(x + 2y) = 2(4)$$

becomes

$$2x + 4y = 8 \qquad \text{or} \qquad 4y + 2x = 8$$

which is the second equation.

A HELPFUL HINT

If you want to know what *type* of solutions you are going to have, solve both equations for y to obtain the system

$$y = m_1x + b_1$$
$$y = m_2x + b_2$$

When $m_1 \neq m_2$, the lines intersect (there is **one** solution).
When $m_1 = m_2$ and $b_1 = b_2$, there is only one line (**infinitely many** solutions).
When $m_1 = m_2$ and $b_1 \neq b_2$, the lines are parallel (there is **no** solution).

C > Applications Involving Systems of Equations

Most of the problems that we've discussed use x- and y-values ranging from -10 to 10. This is not the case when working real-life applications! For example, the prices for two printers and their ink cartridges will be discussed in Example 5 but we will have to use a different type of coordinate system (grid) to compare costs.

Have you bought ink cartridges lately? Kodak claims you can save \$110/year on ink (see their comparison calculator at http://tinyurl.com/yamj4za) but Hewlett-Packard (HP) denies it. (See their counterclaims at http://tinyurl.com/yh8s2so). Let us compare these claims, taking into account the printer used.

SAVE on avg. **\$110/year on INK**

Printer	Price	Ink Cost for Year	Total Cost for x Years
Kodak ESP7	\$150	\$ 70	$150 + 70x$ (dollars)
PhotoSmart	\$100	\$180	$100 + 180x$ (dollars)

Next, we look at the graph, the costs, and the savings.

EXAMPLE 5 Comparing printing costs

a. Graph the annual costs y for the Kodak (K) and the PhotoSmart (P).

K: $y = 150 + 70x$
P: $y = 100 + 180x$

b. Which has a less expensive start-up?

c. What is the 1-year cost for the Kodak and the PhotoSmart?

d. How much are your savings the first year?

SOLUTION 5

a. We first graph the equation $y = 150 + 70x$ using the table.

x	y
0	150
1	220

The x-axis will go from 0 to 1 and the y-axis from 0 to 280. Graph the points (0, 150) and (1, 220) and join them with a green line, the graph of the Kodak, $y = 150 + 70x$.

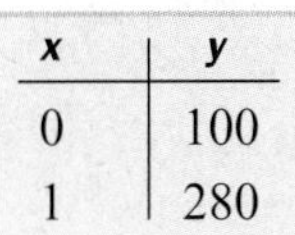

x	y
0	100
1	280

To graph $y = 100 + 180x$ graph the points (0, 100) and (1, 280) and join them with a red line, the graph of the PhotoSmart, $y = 100 + 180x$. See Figure 7.11.

b. When $x = 0$, the Kodak cost is \$150 and the PhotoSmart cost is \$100, so the PhotoSmart has the less expensive start-up.

c. When $x = 1$, the Kodak costs \$220, as indicated by the point (1, 220), and the PhotoSmart costs \$280, so the Kodak is cheaper after 1 year.

d. By using the Kodak you save \$280 − \$220 = \$60 a year.

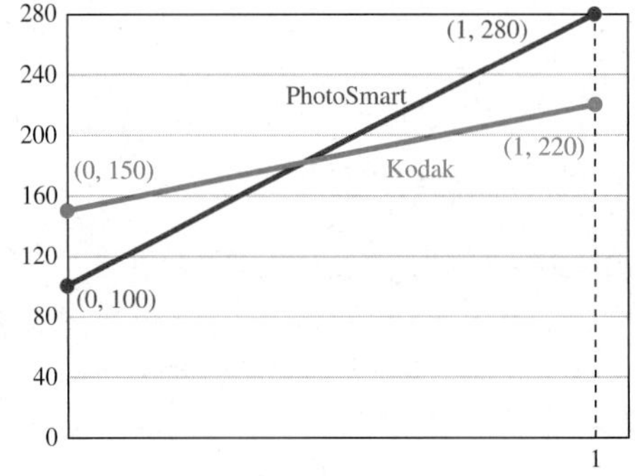

>Figure 7.11

PROBLEM 5

a. Graph the annual cost y for a Kodak and a PhotoSmart bought at different stores.

K: $y = 170 + 100x$
P: $y = 160 + 180x$

b. Which has a less expensive start-up?

c. What is the 1-year cost for the Kodak and the PhotoSmart?

d. How much are your savings the first year?

Some facts about cartridges:

1. More than 13 cartridges are discarded in the United States every second.
2. According to PrintCountry, only 5% of empty printer cartridges are recycled.
3. Forty-six percent of laser jet cartridges and 84% of inkjet cartridges are dumped in a landfill after one use.
4. Thirty-four percent of HP laser jet cartridges and 78% of their inkjet cartridges end up in landfills after one use.

Source: HP study.

Answers to PROBLEMS

5. a.

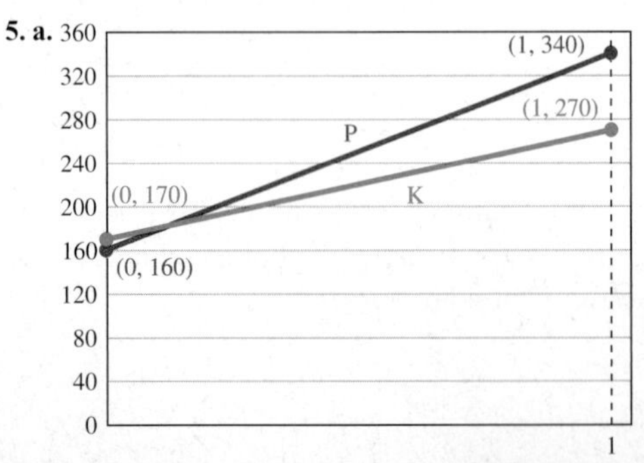

b. The PhotoSmart

c. \$340 for the PhotoSmart and \$270 for the Kodak

d. You save \$70 by selecting the Kodak.

Calculator Corner

Solving Systems of Equations

If you have a calculator, you can solve a linear system of equations very easily. However, you must know how to solve an equation for a specified variable. Thus, in Example 1, you can solve $2x + y = 4$ for y to obtain $y = -2x + 4$, and graph $Y_1 = -2x + 4$. Next, solve $y - 2x = 0$ for y to obtain $y = 2x$ and graph $Y_2 = 2x$. To find the intersection, use a decimal window and the trace and zoom features of your calculator, or better yet, if you have an intersection feature, press 2nd TRACE 5 and follow the prompts. (The graph is shown in Window 1.)

Example 2 is done similarly. Solve $y - 2x = 4$ for y to obtain $y = 2x + 4$. Next, solve $2y - 4x = 12$ for y to get $y = 2x + 6$. You have the equations $y = 2x + 4$ and $y = 2x + 6$, but you don't need a graph to know that there's no solution! Since the lines have the same slope, algebra tells you that the lines are parallel! (Sometimes algebra is better than your calculator.) To verify this, graph $y = 2x + 4$ and $y = 2x + 6$. The results are shown in Window 2.

To solve Example 3, solve $2x + y = 4$ for y to obtain $y = -2x + 4$. Next, solve $2y + 4x = 8$ for y to get $y = -2x + 4$. Again, you don't need a calculator to see that the lines are the same! Their graph appears in Window 3.

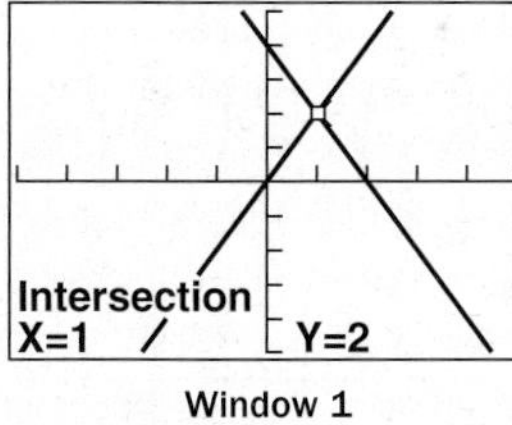

Window 1

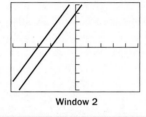

Window 2

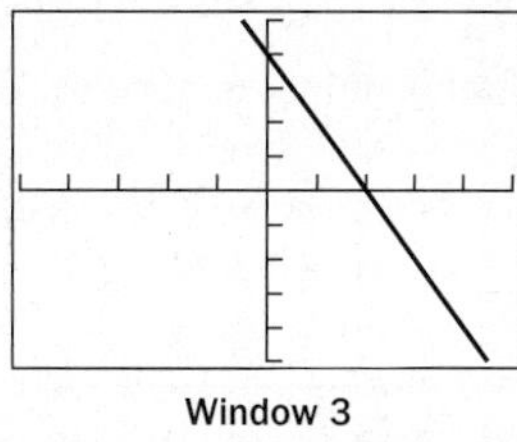

Window 3

Exercises 7.1

Practice Problems · Self-Tests · Media-rich eBooks · e-Professors · Videos

Web IT go to mhhe.com/bello for more lessons

‹A› Solving a System by Graphing
‹B› Finding Consistent, Inconsistent, and Dependent Systems of Equations

In Problems 1–30, solve by graphing. Label each system as consistent (write the solution), inconsistent (no solution), or dependent (infinitely many solutions).

1. $x + y = 4$
$x - y = -2$

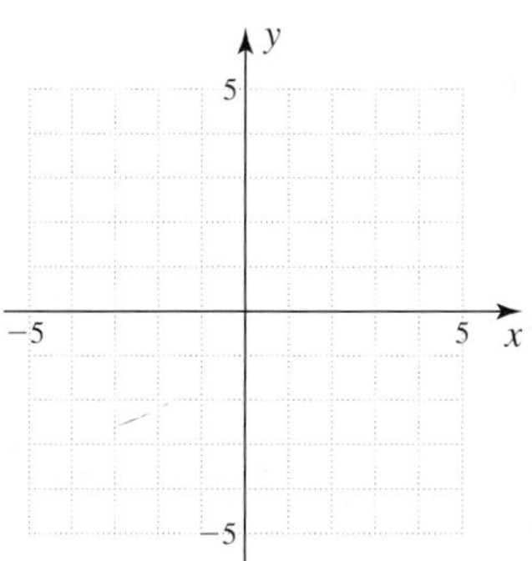

2. $x + y = 3$
$x - y = -5$

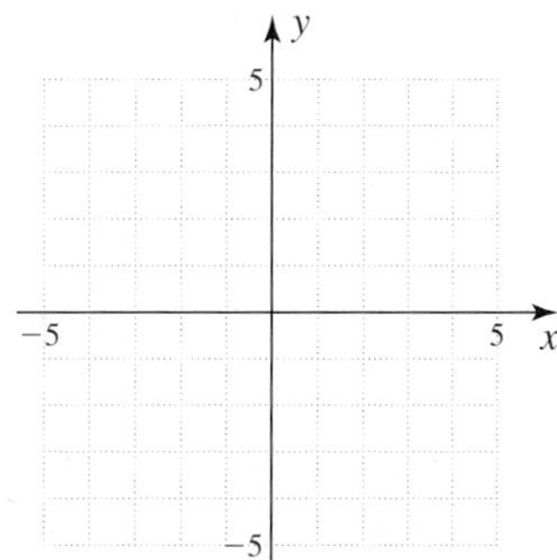

3. $x + 2y = 0$
$x - y = -3$

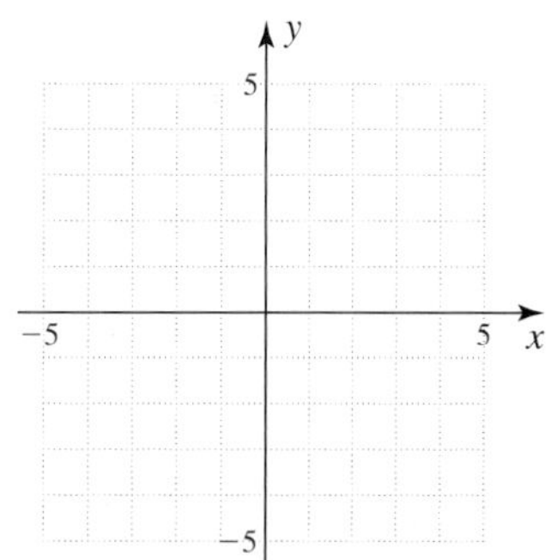

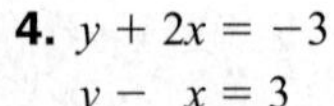

4. $y + 2x = -3$
$y - x = 3$

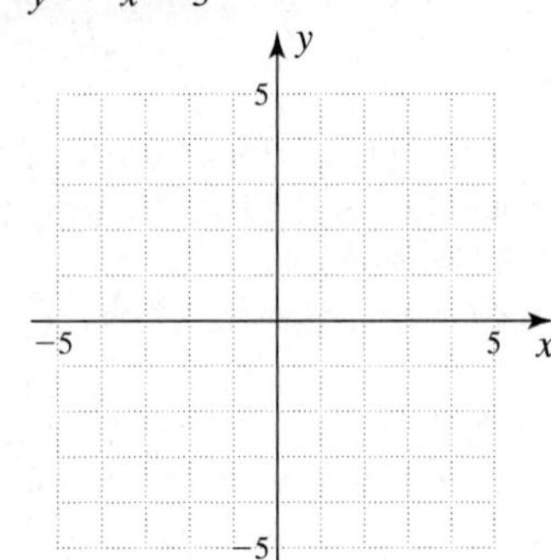

5. $3x - 2y = 6$
$6x - 4y = 12$

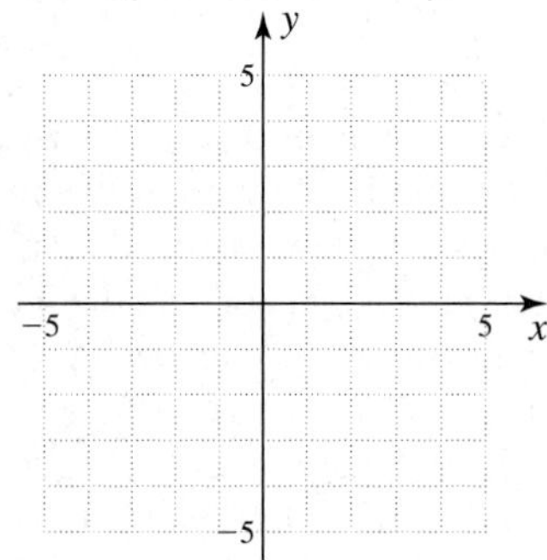

6. $2x + y = -2$
$8x + 4y = 8$

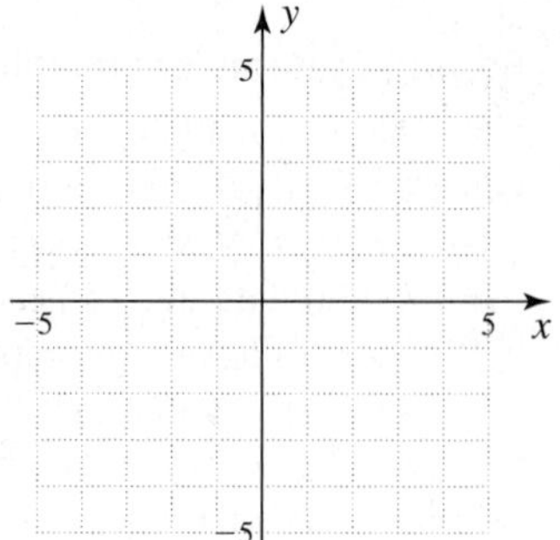

7. $3x - y = -3$
$y - 3x = 3$

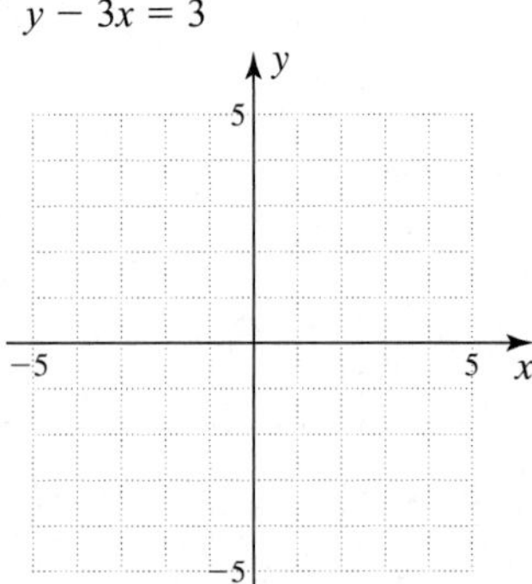

8. $4x - 2y = 8$
$y - 2x = -4$

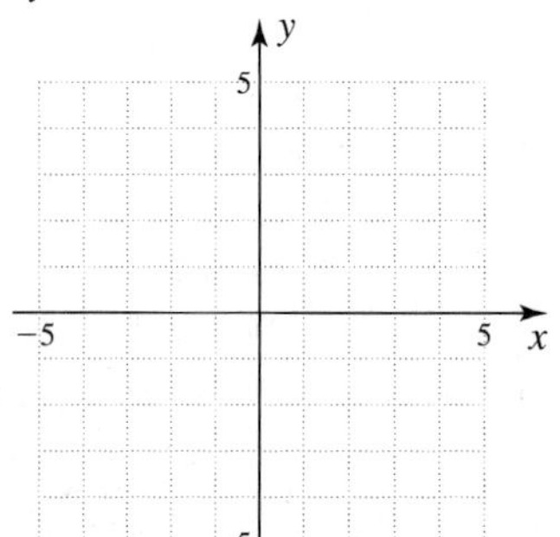

9. $2x - y = -2$
$y = 2x + 4$

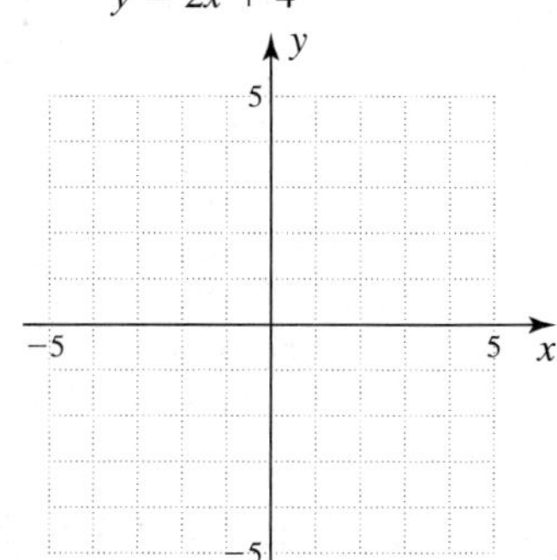

10. $2x + y = -2$
$y = -2x + 4$

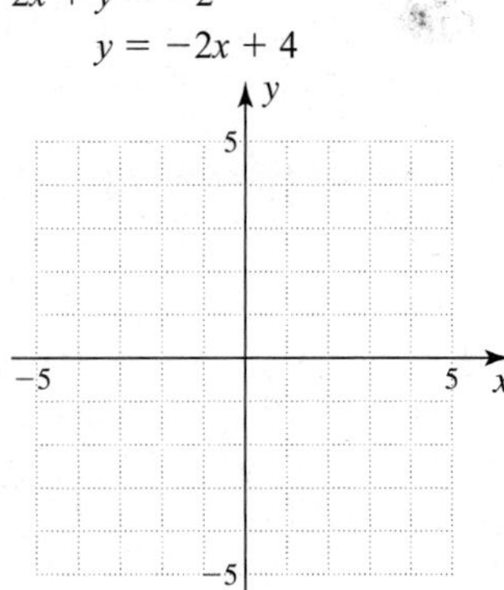

11. $y = -2$
$2y = x - 2$

12. $3y = 6 - x$
$y = 3$

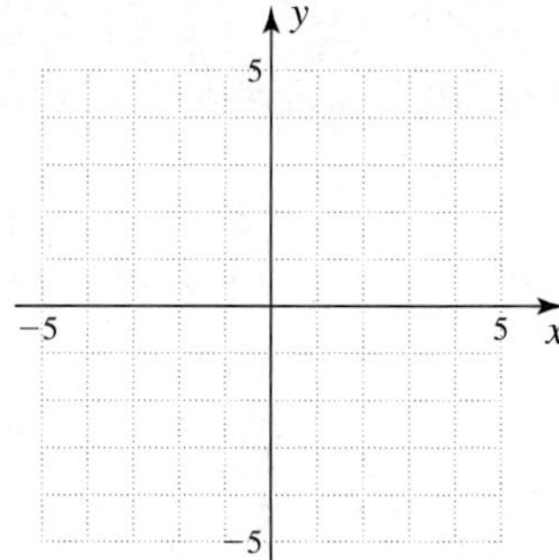

13. $x = 3$
$y = 2x - 4$

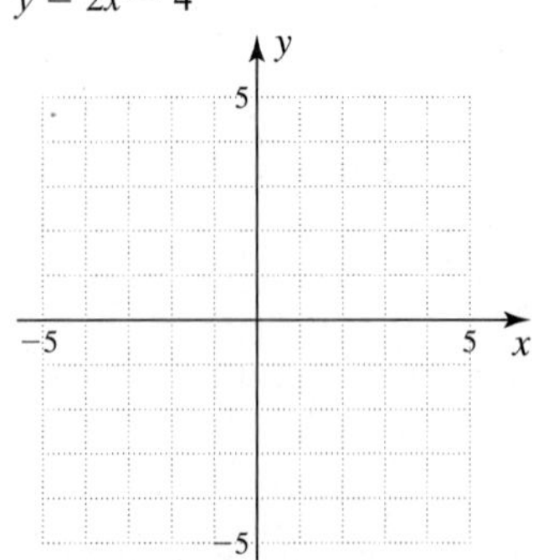

14. $y = -x + 2$
$x = -1$

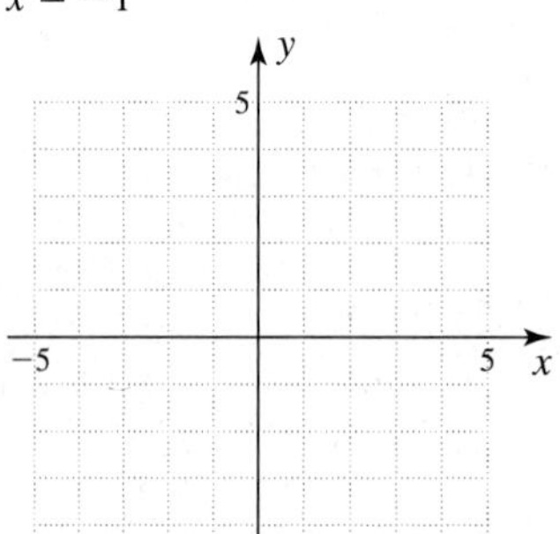

15. $x + y = 3$
$2x - y = 0$

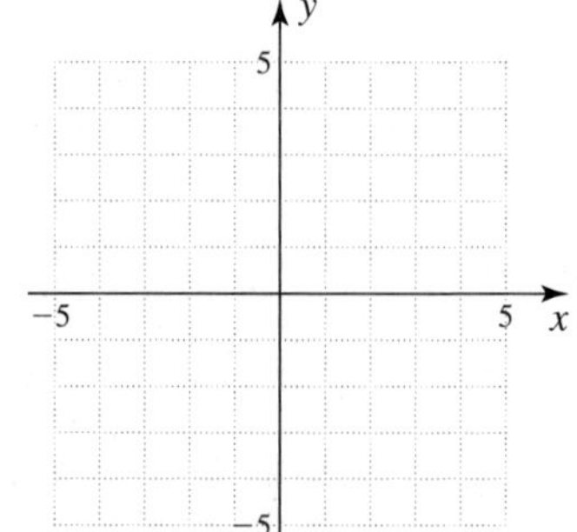

16. $x + y = 5$
$x - 4y = 0$

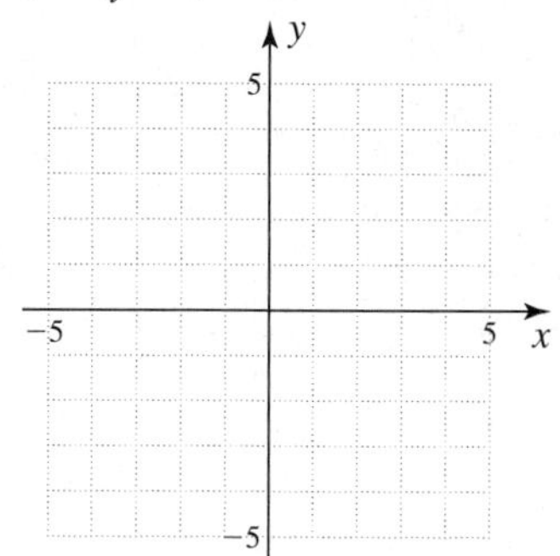

17. $5x + y = 5$
$5x = 15 - 3y$

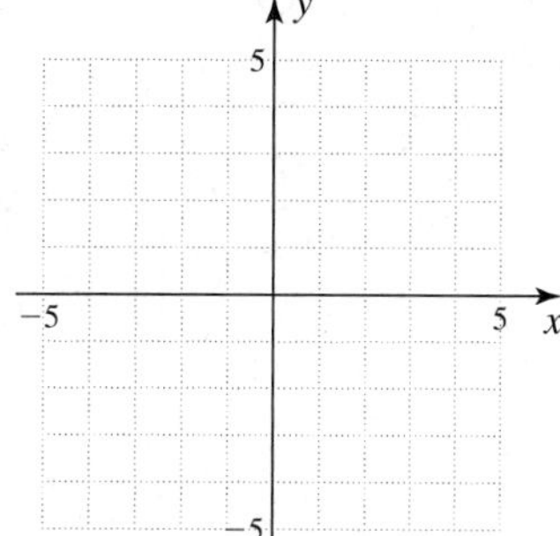

18. $2x - y = -4$
$4x = 4 + 2y$

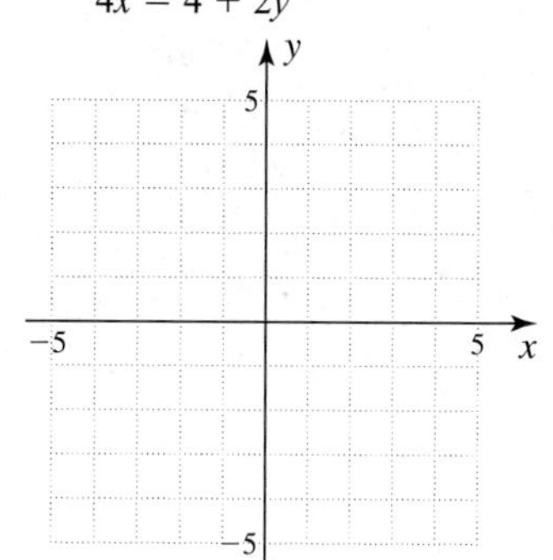

19. $3x + 4y = 12$
$8y = 24 - 6x$

20. $2x - 3y = 6$
$6x = 18 + 9y$

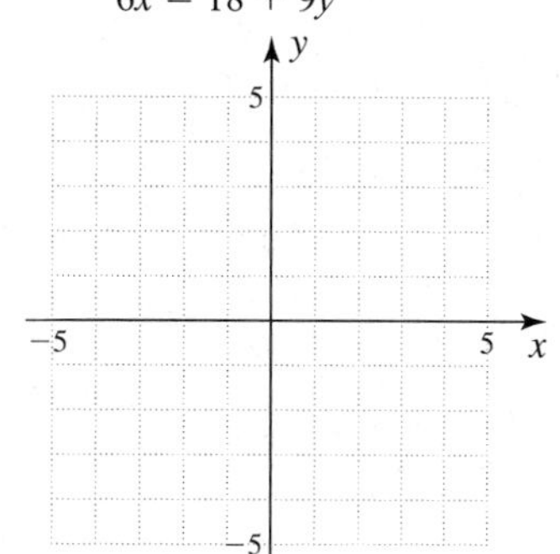

21. $y = x + 3$
$y = -x + 3$

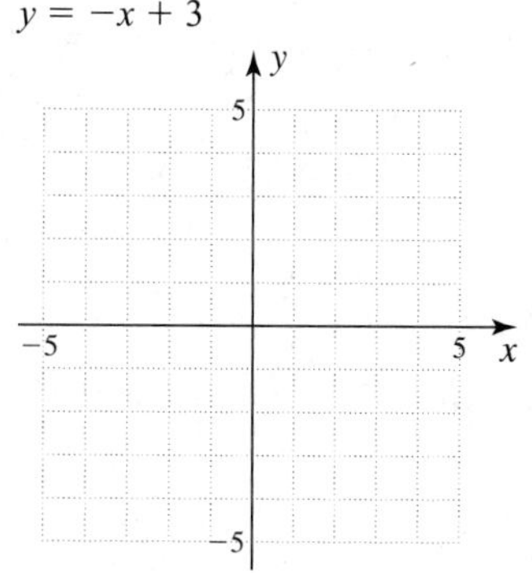

22. $y = 3x + 6$
$y = -2x - 4$

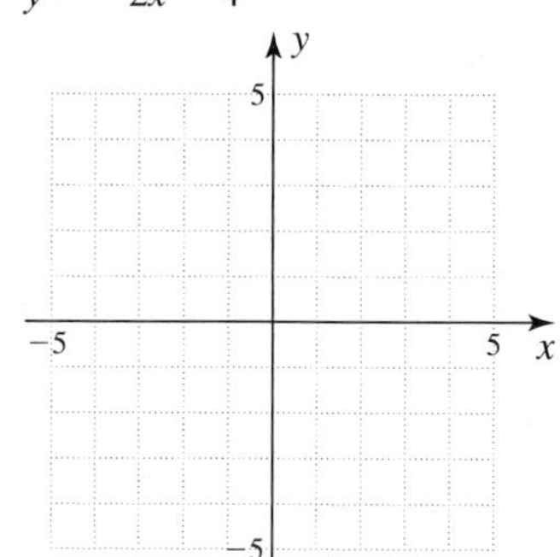

23. $y = 2x - 2$
$y = -3x + 3$

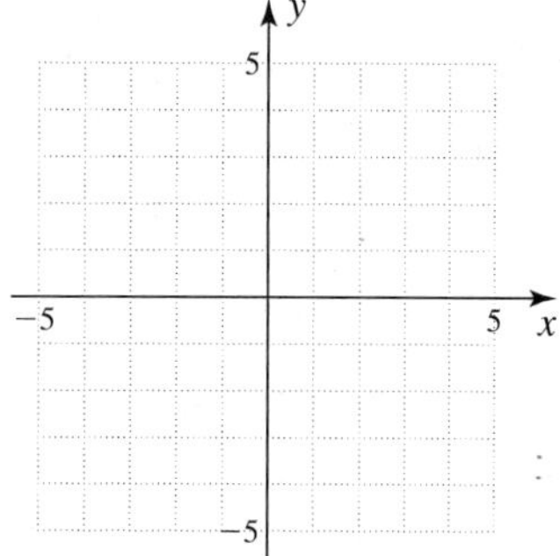

24. $3x = 6$
$y = -2$

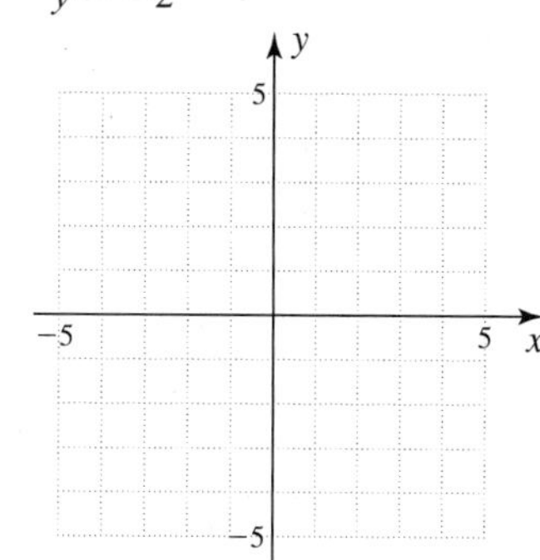

25. $-2x = 4$
$y = -3$

26. $y = 2$
$y = 2x - 4$

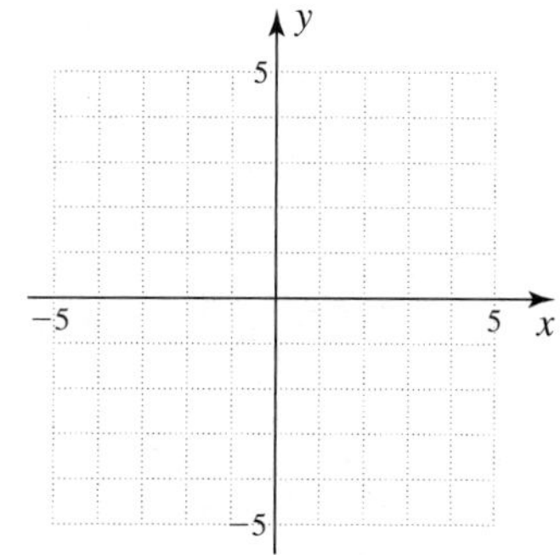

27. $y = -3$
$y = -3x + 6$

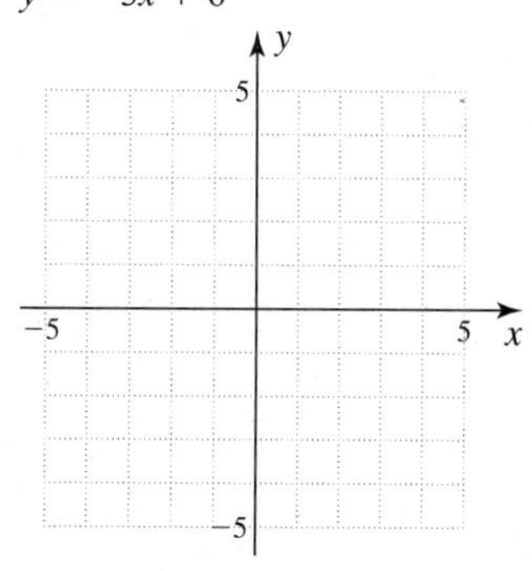

Web IT go to mhhe.com/bello for more lessons

28. $y = -\frac{1}{3}x + 2$
$3y + x = 6$

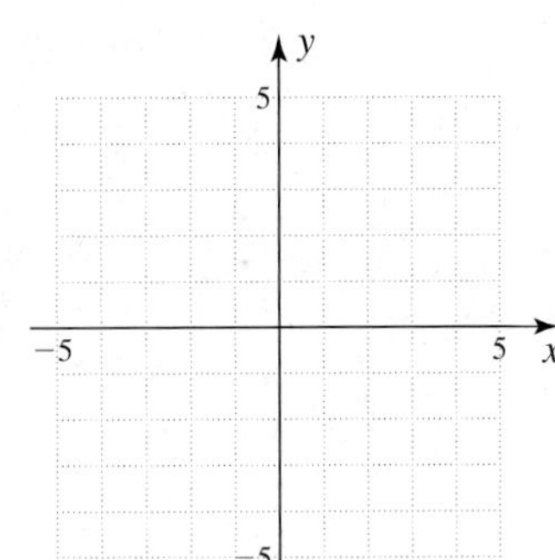

29. $x + 4y = 4$
$y = -\frac{1}{4}x + 2$

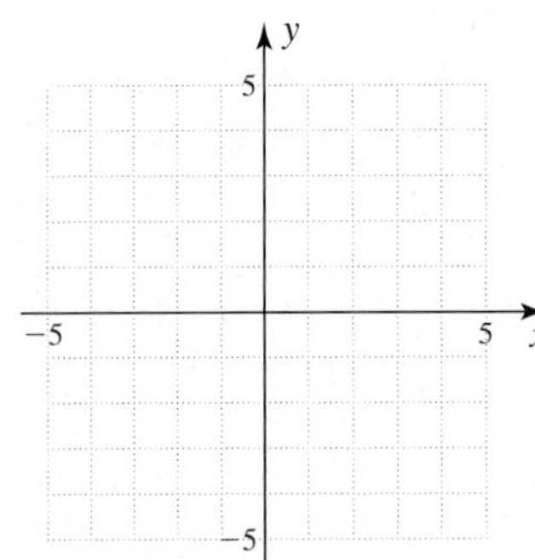

30. $2x - y = 2$
$y = \frac{1}{2}x + 1$

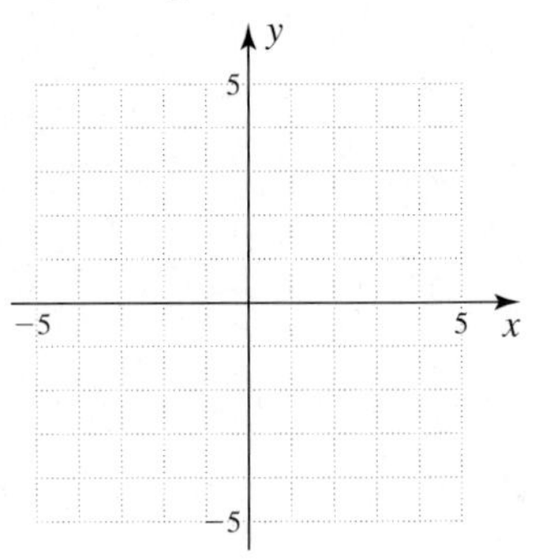

Web IT go to mhhe.com/bello for more lessons

⟨C⟩ Applications Involving Systems of Equations In Problems 31–35, use the following information. You want to watch 10 movies at home each month. You have two options:

OPTION 1 Get cable service. The cost is $20 for the installation fee and $35 per month.

OPTION 2 Buy a DVD player and rent movies. The cost is $200 for a DVD player and $25 a month for movie rental fees.

31. *Cost of cable service*

a. If C is the cost of installing cable service plus the monthly fee for m months, write an equation for C in terms of m.

b. Complete the following table where C is the cost of cable service for m months:

m	C
6	
12	
18	

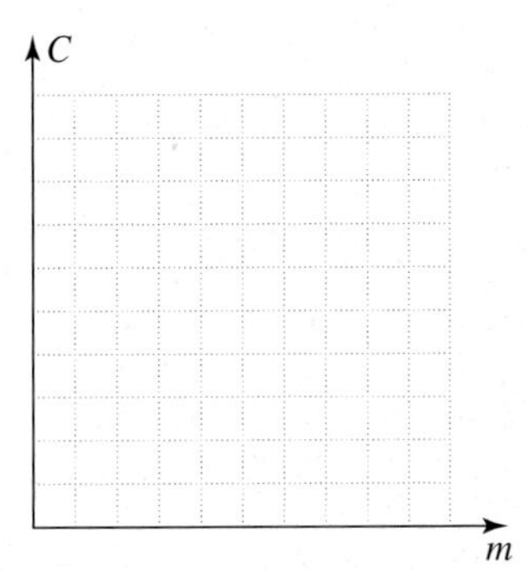

c. Graph the information obtained in parts **a** and **b.** (*Hint:* Let m run from 0 to 30 and C run from 0 to 1000 in increments of 100.)

32. *Cost of renting movies*

a. If C is the total cost of buying the DVD player plus renting the movies for m months, write an equation for C in terms of m.

b. Complete the following table where C is the cost of buying a DVD player and renting movies for m months:

m	C
6	
12	
18	

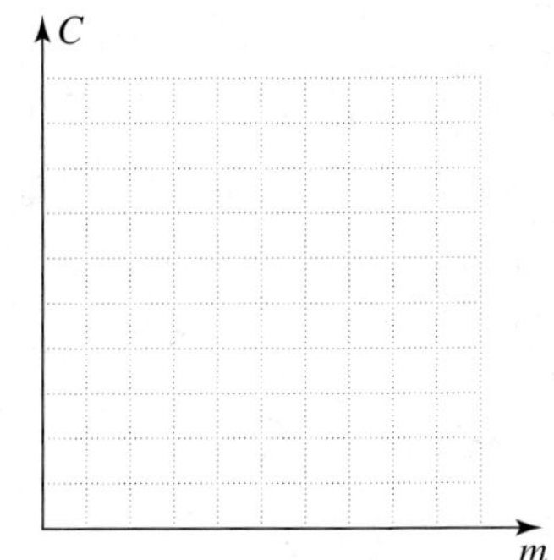

c. Graph the information obtained in parts **a** and **b.**

33. *Graphical comparison* Make a graph of the information obtained in Problems 31 and 32 on the same coordinate axis.

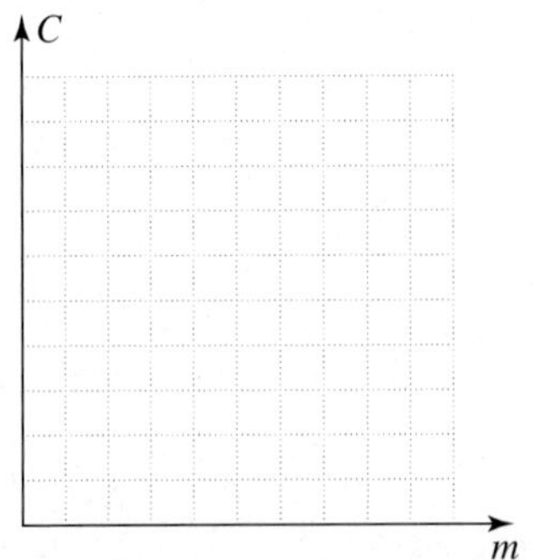

34. *Cable service* Based on the graph for Problem 33, when is the cable service cheaper?

35. *When is renting movies cheaper?* Based on the graph for Problem 33, when is the DVD player and rental option cheaper?

36. *Lower wages, higher tips* At Grady's restaurant, servers earn \$80 a week plus tips, which amount to \$5 per table.

a. Write an equation for the weekly wages W based on serving t tables.

b. Complete the following table where W is the wages and t is the number of tables served:

t	W
5	
10	
15	
20	

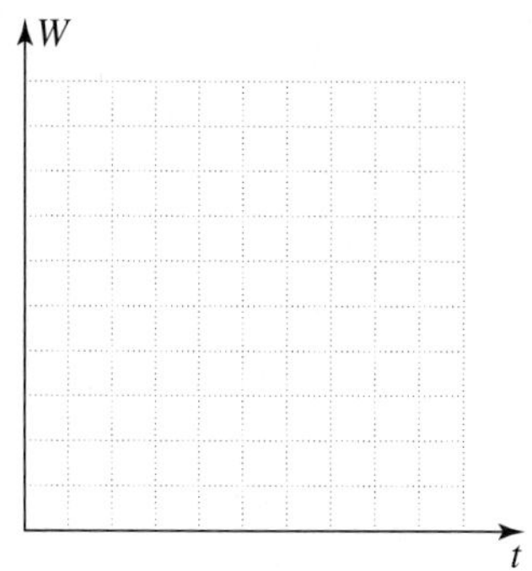

c. Graph the information obtained in parts **a** and **b.**

37. *Higher wages, lower tips* At El Centro restaurant, servers earn \$100 a week, but the average tip per table is only \$3.

a. Write an equation for the weekly wages W based on serving t tables.

b. Complete the following table where W is the wages and t is the number of tables served:

t	W
5	
10	
15	
20	

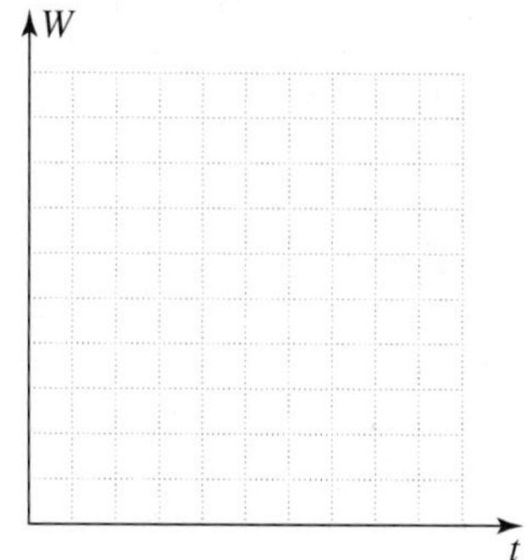

c. Graph the information obtained in parts **a** and **b.**

38. *Graphical comparison* Graph the information from Problems 36 and 37 on the same coordinate axis. Based on the graph, answer the following questions.

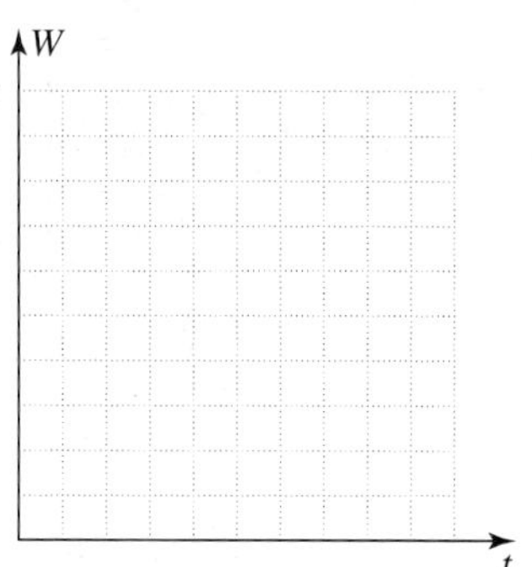

a. When does a server at Grady's make more money than a server at El Centro?

b. When does a server at El Centro make more money than a server at Grady's?

39. *Cell phone plans* How much do you pay a month? At the present time, two companies have cell phone plans that cost \$19.95 per month. However, plan A costs \$0.60 per minute of airtime during peak hours, while plan B costs \$0.45 per minute of airtime during peak hours. Plan A offers a free phone with its plan, while plan B's phone costs \$45. For comparison purposes, since the monthly cost is the same for both plans, the cost C is based on the price of the phone plus the number m of minutes of airtime used.

a. Write an equation for the cost C of plan A.

b. Write an equation for the cost C of plan B.

c. Using the same coordinate axis, make a graph for the costs of plans A and B. (*Hint:* Let m and C run from 0 to 500.)

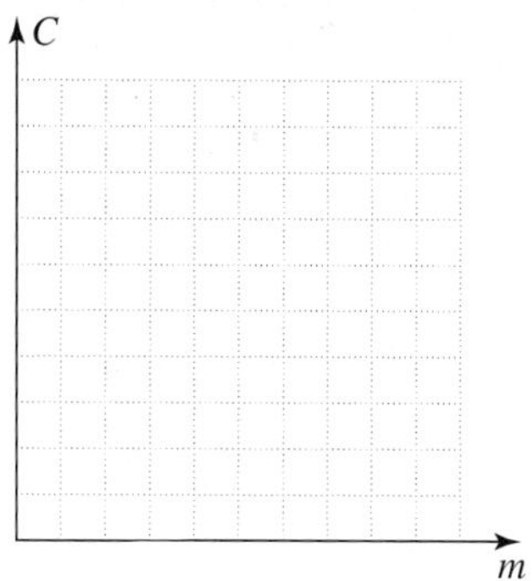

40. *Better buy?* Based on the graphs obtained in Problem 39, which plan would you buy, A or B? Explain.

41. *Catering at school* Here are the actual catering prices for Jefferson City Public Schools.

Jefferson City Public Schools

Basket Lunch	Deli Buffet	Hot Buffet
\$5.00 per person	\$5.50 per person	\$6.00 per person

Suppose x students buy Basket Lunches costing \$5 each and y students buy the Hot Buffet costing \$6 each.

a. Write an equation for the cost C_B of the x Basket Lunches.

b. Write an equation for the cost C_H of the y Hot Buffets.

c. Write an expression that represents the total number of lunches purchased. If this total is 50, write an equation for the total number of lunches purchased.

d. The total bill for the 50 lunches is $5x + 6y$. If this total is \$270, write an equation for the total bill.

e. Graph the equations $x + y = 50$ (the total number of lunches purchased) and $5x + 6y = 270$ (the total cost) on the same coordinate axis and find the number of Basket Lunches and Hot Buffet lunches purchased.

Lunches at Jefferson City

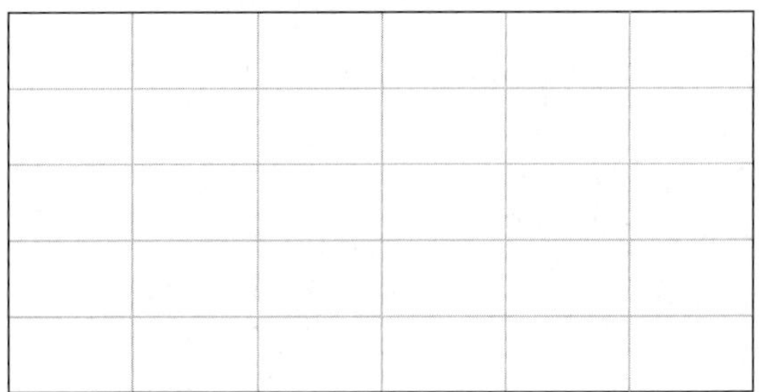

42. *Breakfast at restaurant* Here are some breakfast prices at a restaurant.

Breakfast

Continental I	$\$7.99 \approx \$8.00$
Continental II	$\$8.99 \approx \$9.00$

Suppose x guests buy the Continental I breakfast costing \$8 each and y guests buy the Continental II costing \$9 each.

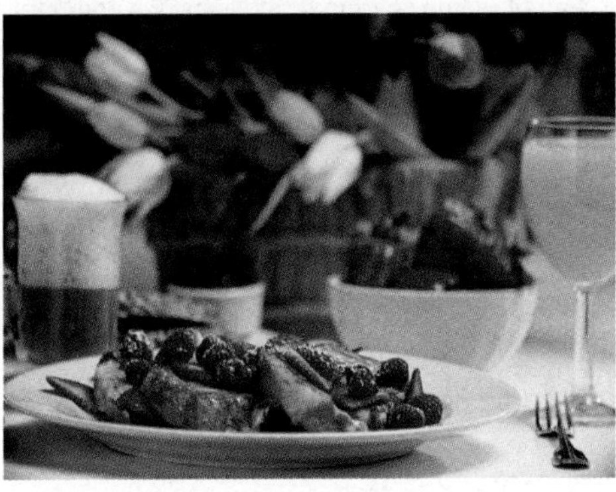

a. Write an equation for the cost C_{I} of the x Continental I breakfasts.

b. Write an equation for the cost C_{II} of the y Continental II breakfasts.

c. Write an expression that represents the total number of breakfasts purchased. If this total is 26, write an equation for the total number of breakfasts purchased.

d. The total bill for the 26 breakfasts is $8x + 9y$. If this total is \$216, write an equation for the total bill.

e. Graph the equations $x + y = 26$ (the total number of breakfasts purchased) and $8x + 9y = 216$ (the total cost) on the same coordinate axis and find the number of Continental I and Continental II breakfasts purchased.

Breakfast at Restaurant

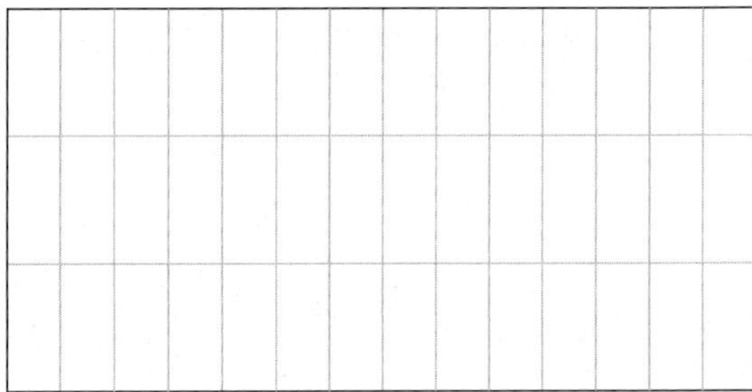

How do we find the y? Remember that $x + y = 26$ and from the graph we can see that $x = 18$. Thus, $x + y = 26$ becomes $18 + y = 26$, or $y = 8$.

43. *Breakfast at College* Here are some breakfast prices at College.

Breakfast

Breakfast croissant	\$6
Cheddar cheese omelet	\$7
Egg and sausage burritos	\$7.25

Suppose x guests buy the breakfast croissant costing \$6 each and y guests buy the cheddar cheese omelet costing \$7 each.

a. Write an equation for the cost C_B of the x breakfast croissants.

b. Write an equation for the cost C_C of the y breakfast omelets.

c. Write an expression that represents the total number of breakfasts purchased. If this total is 20, write an equation for the total number of breakfasts purchased.

d. The total bill for the 20 breakfasts is $6x + 7y$. If this total is \$126, write an equation for the total bill.

e. Graph the equations $x + y = 20$ (the total number of breakfasts purchased) and $6x + 7y = 126$ (the total cost) on the same coordinate axis and find the number of croissant and omelet breakfasts purchased.

Breakfast at College

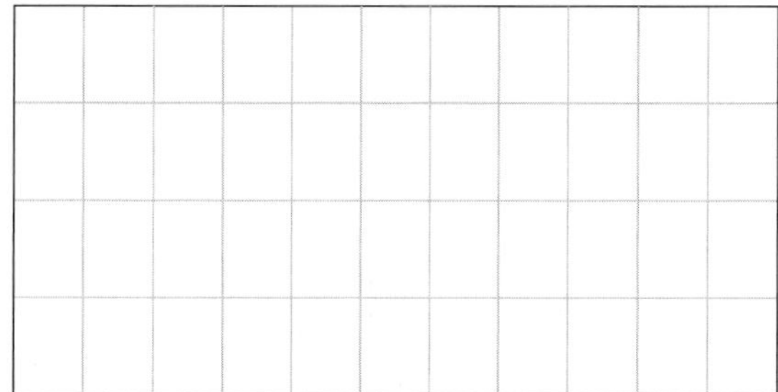

To find y after you know from the graph that $x = 14$, substitute the 14 in the equation $x + y = 20$ obtaining $14 + y = 20$, which means that $y = 6$.

44. *Saturated fats* According to *Restaurant Confidential* by Jacobson and Hurley "all burgers are not created equal. You can't rely on calorie-counting guides that don't make the distinction among burgers served at fast-food establishments, family restaurants, and dinner houses." What's the problem?

The saturated fats! If x is the saturated fat content (in grams) in a McDonald's Quarter Pounder and y is the amount of saturated fat in a family-style restaurant, when you eat two Quarter Pounders ($2x$ saturated fat grams) and only one of the family-style burgers (y saturated fat grams), the amount of saturated fat is **30 grams,** 8 over the recommended daily allowance for a female between the ages of 19 and 50. If you add the saturated fats in the McDonald's (x grams) and the family-style restaurant (y grams) the result $x + y$ is exactly **22 grams,** the recommended amount of saturated fat you should have the whole day, even if you ate nothing else!

a. Write an equation for the amount of saturated fats when you eat one McDonald's and one family-style restaurant burger.

b. Write an equation for the amount of saturated fats P_Q contained in two Quarter Pounders.

c. Write an equation for the amount of saturated fats P_F contained in one family-style restaurant burger.

d. The total grams you eat when consuming two Quarter Pounders and one family-style restaurant burger is $2x + y$, or **30** grams. Write an equation for the total grams of fat in the meal.

e. We have not said how many grams of saturated fats there are in each burger, but we can find out! Graph $x + y = 22$ and $2x + y = 30$ on the same coordinate axis and find x (saturated fat grams in the McDonald's) and y (saturated fat grams in the family-style restaurant).

Saturated Fat in Burgers

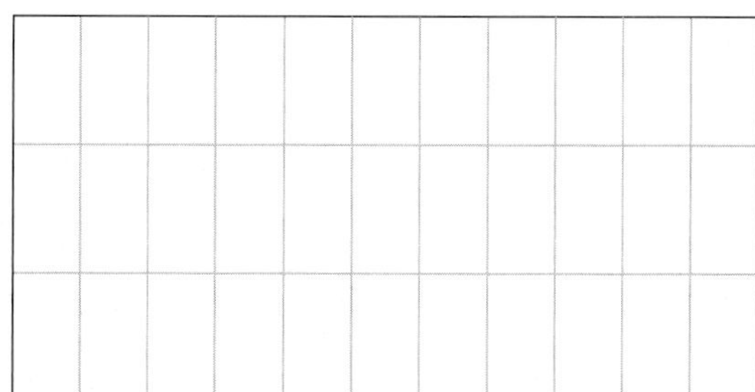

Note that when $x = 8$, $x + y = 22$ becomes $8 + y = 22$, that is, $y = 14$.

Web IT go to mhhe.com/bello for more lessons

Web IT go to mhhe.com/bello for more lessons

45. *Comparing McDonald's and Burger King* If x is the saturated fat content (in grams) in a McDonald's Quarter Pounder and y is the amount of saturated fat in a Burger King Whopper Jr, when you eat two Quarter Pounders ($2x$ saturated fat grams) and only one Whopper Jr. burger (y saturated fat grams), the amount of saturated fat is **24 grams.** If you add the saturated fats in the McDonald's (x grams) and the Whopper Jr. (y grams) the result $x + y$ is only **16 grams,** under the recommended amount of saturated fats you should have the whole day.

a. Write an equation for the amount of saturated fats when you eat one McDonald's and one Whopper Jr. burger.

b. Write an equation for the amount of saturated fats P_Q contained in two Quarter Pounders.

c. Write an equation for the amount of saturated fats P_W contained in one Whopper Jr. burger.

d. The total grams you eat when consuming two Quarter Pounders and one Whopper Jr. burger is $2x + y$, or 24 grams. Write an equation for the total grams of fat in the meal.

e. We have not said how many grams of saturated fats are in each burger. Graph $x + y = 16$ and $2x + y = 24$ on the same coordinate axis and find x (saturated fat grams in the McDonald's) and y (saturated fat grams in the Whopper Jr.) and find out!

McDonald's vs. Burger King

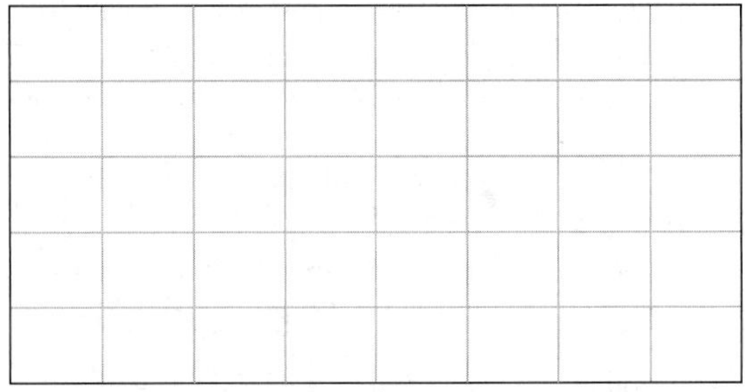

If you find from the graph that $x = 8$ and you have $x + y = 16$, we can substitute 8 for x, obtaining $8 + y = 16$, which means $y = 8$.

46. *Comparing McDonald's and Wendy's* Any burgers with less saturated fat than McDonald's? Let's try Wendy's Jr. Bacon Cheeseburger. If x is the saturated fat content (in grams) in a McDonald's Quarter Pounder and y is the amount of saturated fat in Wendy's Jr. Bacon Cheeseburger, when you eat two Quarter Pounders ($2x$ saturated fat grams) and only one Wendy's burger (y saturated fat grams), the amount of saturated fat is **23 grams.** If you add the saturated fats in the McDonald's (x grams) and the Wendy's burger (y grams), the result $x + y$ is only **15 grams,** under the recommended amount of saturated fats you should have in a day.

a. Write an equation for the amount of saturated fats when you eat one McDonald's and one Wendy's Jr. Bacon Cheeseburger.

b. Write an equation for the amount of saturated fats P_Q contained in two Quarter Pounders.

c. Write an equation for the amount of saturated fats P_B contained in one Bacon Cheeseburger.

d. The total grams you eat when consuming two Quarter Pounders and one Bacon Cheeseburger is $2x + y$, or 23 grams. Write an equation for the total grams of fat in the meal.

e. We have not said how many grams of saturated fats are in each burger. Graph $x + y = 15$ and $2x + y = 23$ on the same coordinate axis and find x (saturated fat grams in the McDonald's) and y (saturated fat grams in the Bacon Cheeseburger) and find out!

McDonald's vs. Wendy's

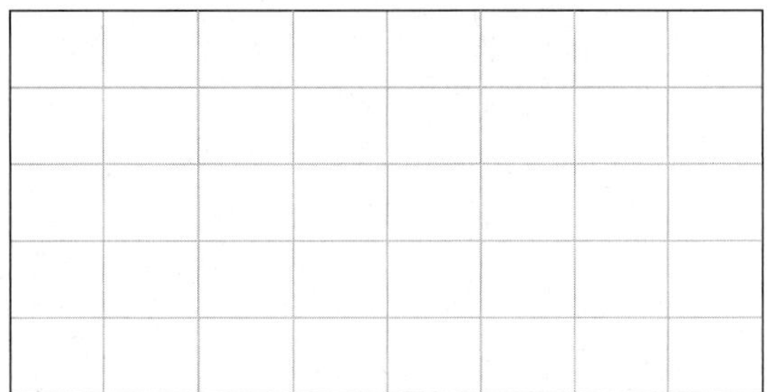

By the way, after you find $x = 8$, how do you find y? Since we know that $x + y = 15$, substitute 8 for x in $x + y = 15$, obtaining $8 + y = 15$ or $y = 7$.

Using Your Knowledge

Applications: Green Math

Comparing Printing Costs The table shows the cost of several printers and their ink cartridges. Note that in some cases (Lexmark) the price of the printer is low but the price of the ink is high.

Printer	Price	Ink Cost per Year	Total Cost per x Years
Canon MP560	\$100	\$ 80	$100 + 80x$ (dollars)
Epson Artisan	\$150	\$ 70	$150 + 70x$ (dollars)
Brother MFC 490	\$130	\$ 70	$130 + 70x$ (dollars)
Lexmark X4650	\$ 60	\$150	$60 + 150x$ (dollars)

47. Graph the total cost $y = 100 + 80x$ for the Canon.

48. Graph the total cost $y = 150 + 70x$ for the Epson.

49. Refer to the graphs for Problems 47 and 48.

a. Which has a less expensive start-up, Canon or Epson?

b. What is the 1-year cost for Canon and for Epson?

c. How much are your savings the first year?

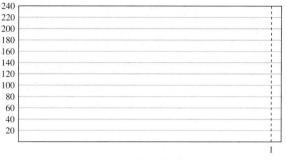

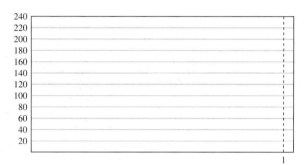

Answers for 47–48

50. Graph the total cost $y = 130 + 70x$ for the Brother.

51. Graph the total cost $y = 60 + 150x$ for the Lexmark.

52. Refer to the graphs for Problems 51 and 52.

a. Which has a less expensive start-up, Brother or Lexmark?

b. What is the 1-year cost for Brother and for Lexmark?

c. How much are your savings the first year?

Answers for 51–52

53. The cost y of using an incandescent bulb costing just one quarter ($0.25) is $\mathbf{y = 0.25 + 0.08x}$, where x is the number of days the bulb is used. The cost y of using a fluorescent bulb costing $1.50 is $\mathbf{y = 1.50 + 0.03x}$.

a. Graph both costs in the same coordinate system.

b. Which has a less expensive start-up?

c. In how many days will the cost be the same?

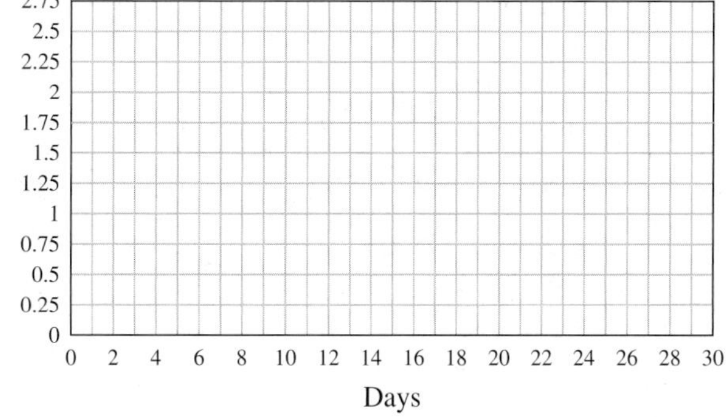

Write On

54. What does the solution of a system of linear equations represent?

55. Define a consistent, an inconsistent, and a dependent system of equations.

56. How can you tell graphically whether a system is consistent, inconsistent, or dependent?

Suppose you have a system of equations and you solve both equations for y to obtain:

$$y = m_1x + b_1$$
$$y = m_2x + b_2$$

What can you say about the graph of the system when

57. $m_1 = m_2$ and $b_1 \neq b_2$? How many solutions do you have? Explain.

58. $m_1 = m_2$ and $b_1 = b_2$? How many solutions do you have? Explain.

59. $m_1 \neq m_2$? How many solutions do you have? Explain.

Concept Checker

Fill in the blank(s) with the correct word(s), phrase, or mathematical statement.

finite	**solution**
intersect	**no**
many	**infinitely**
perpendicular	**parallel**
one	**coincide**

60. The ordered pair **(a, b)** that is the point of **intersection** of a **system of equations** consisting of two lines in the plane is the ___________ of the system.

61. In a **consistent** and **independent** system of equations, the **graph of the equations** intersect at _________ point.

62. In an **inconsistent** system of equations, the **graph of the equations** are ___________ lines.

63. In an **inconsistent** system of equations, there is ________ solution for the system.

64. In a **dependent** system of equations, the graph of the equations ___________.

65. A **dependent** system of equations has ___________ many solutions.

Mastery Test

Use the graphical method to find the solution of the system of equations (if it exists).

66. $x + 2y = 4$
$2x - 4y = 0$

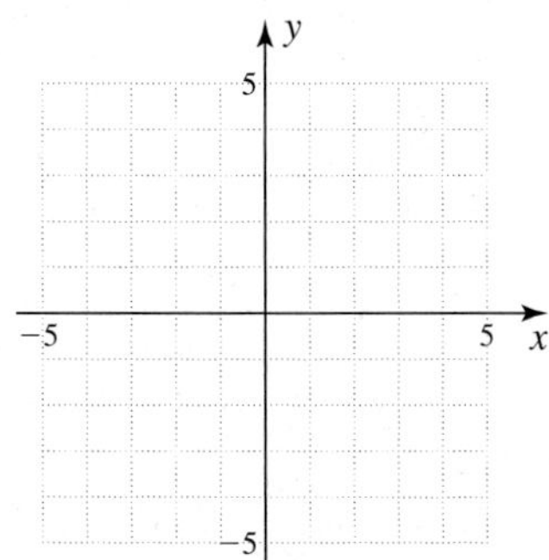

67. 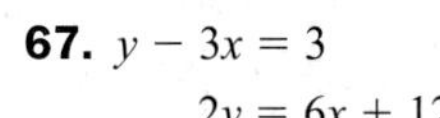
$y - 3x = 3$
$2y = 6x + 12$

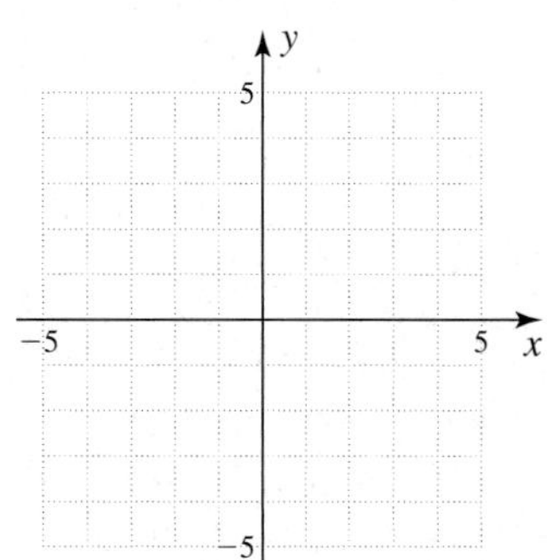

68. 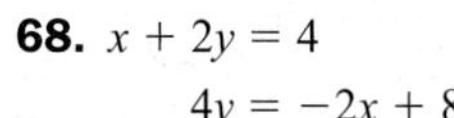
$x + 2y = 4$
$4y = -2x + 8$

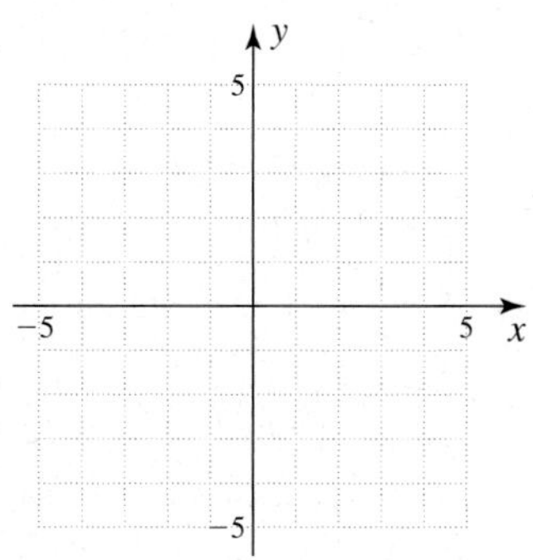

Graph and classify each system as consistent, inconsistent, or dependent; if the system is consistent, find the solution.

69. $x + y = 4$
$2x - y = 2$

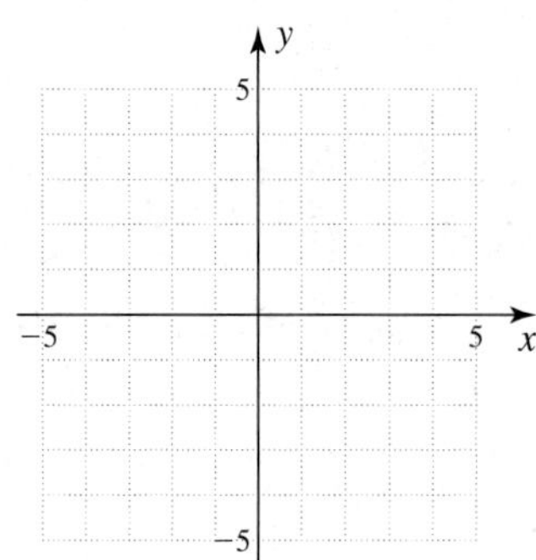

71. 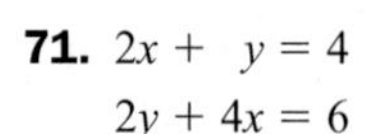
$2x + y = 4$
$2y + 4x = 6$

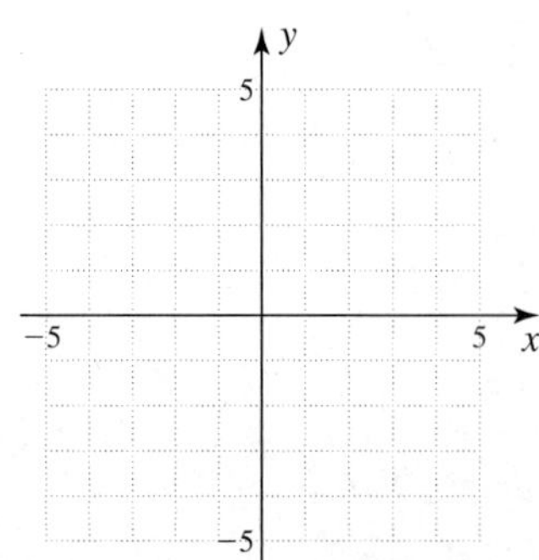

70. The monthly cost of Internet provider A is \$10 for the first 5 hours and \$3 for each hour after 5. Provider B's cost is \$15 for the first 5 hours and \$2 for each hour after 5. Make a graph of the cost C for providers A and B when h hours of airtime are used. (*Hint:* Let h run from 0 to 25 and C run from 0 to 60.)

72. $2x + y = 4$
$2y + 4x = 8$

Skill Checker

Solve.

73. $3x + 72 = 2.5x + 74$

74. $5x + 20 = 3.5x + 26$

Determine whether the given point is a solution of the equation.

75. $(3, 5)$; $2x + y = 11$

76. $(-1, 4)$; $2x - y = -6$

77. $(-1, 2)$; $2x - y = 0$

78. $(-2, 6)$; $3x - y = 0$

7.2 Solving Systems of Equations by Substitution

Objectives

Use the substitution method to:

A Solve a system of equations in two variables.

B Determine whether a system of equations is consistent, inconsistent, or dependent.

C Solve applications involving systems of equations.

To Succeed, Review How To . . .

1. Solve linear equations (pp. 137–141).
2. Determine whether an ordered pair satisfies an equation (pp. 226–228).

Getting Started

Supply, Demand, and Substitution

Does this graph look familiar? As we noted in the preceding section's *Getting Started,* the supply and the demand were about the same in the year 1980. However, this solution is only an *approximate* one because the x-scale representing the years is numbered at 5-year intervals, and it's hard to pinpoint the exact point at which the lines intersect. Suppose we have the equations for the supply and the demand between 1975 and 1985. These equations are

Supply: $y = 2.5x + 72$ $\quad$ x = the number of years elapsed after 1975

Demand: $y = 3x + 70$

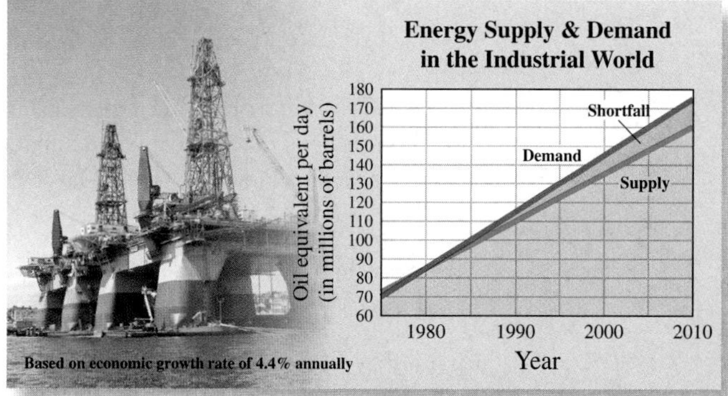

Can we now tell *exactly* where the lines meet? Not graphically! For one thing, if we let $x = 0$ in the first equation, we obtain $y = 2.5 \cdot 0 + 72$, or $y = 72$. Thus, we either need a piece of graph paper with 72 units, or we have to make each division on the graph paper 10 units, thereby *losing* accuracy. But there's a way out. We can use an *algebraic* method rather than a *graphical* one. Since we are looking for the point at which the supply (y)

is the same as the demand (y), we may *substitute* the expression for y in the demand equation—that is, $3x + 70$—into the supply equation. Thus, we have

$$\underbrace{\text{Demand }(y)}_{3x + 70} = \underbrace{\text{Supply }(y)}_{2.5x + 72}$$

$$3x = 2.5x + 2 \quad \text{Subtract 70.}$$

$$0.5x = 2 \quad \text{Subtract } 2.5x.$$

$$\frac{0.5x}{0.5} = \frac{2}{0.5} \quad \text{Divide by 0.5.}$$

$$x = 4 \quad \text{Simplify.}$$

Thus, 4 years after 1975 (or in 1979), the supply equaled the demand. At this time the demand was

$$y = 3(4) + 70 = 12 + 70 = 82 \text{ (million barrels)}$$

In this section we learn to solve equations by the ***substitution method.*** This method is recommended for solving systems in which *one* equation is solved, or can be easily solved, for one of the variables.

A > Using the Substitution Method to Solve a System of Equations

Here is a summary of the substitution method, which we just used in the *Getting Started.*

PROCEDURE

Solving a System of Equations by the Substitution Method

1. Solve one of the equations for x or y.
2. Substitute the resulting expression into the **other** equation. (Now you have an equation in one variable.)
3. Solve the new equation for the variable.
4. Substitute the value of that variable into one of the original equations and solve this equation to get the value for the second variable.
5. Check the solution by substituting the numerical values of the variables in both equations.

The idea is to solve one of the equations for a variable and substitute the result in the **other** equation. It does *not* matter which equation or which variable you pick, the final answer will be the same. However, when possible solve for the variable with a coefficient of one.

EXAMPLE 1 **Solving a system by substitution**

Solve the system:

$$x + y = 8$$
$$2x - 3y = -9$$

SOLUTION 1 We use the five-step procedure.

1. Solve one of the equations for x or y (we solve the first equation for y, since y has a coefficient of 1). $\qquad y = 8 - x$

PROBLEM 1

Solve the system:

$$x + y = 5$$
$$2x - 4y = -8$$

Answers to PROBLEMS

1. The solution is (2, 3).

2. Substitute $8 - x$ for y in $2x - 3y = -9$. $\quad 2x - 3(8 - x) = -9$

Note that you must substitute $y = 8 - x$ into the **second** equation. Substituting in the first equation gives

$$x + (8 - x) = 8$$
$$8 = 8$$

which is, of course, true.

3. Solve the new equation for the variable.

$$2x - 3(8 - x) = -9$$
$$2x - 24 + 3x = -9 \quad \text{Simplify.}$$
$$5x - 24 = -9 \quad \text{Combine like terms.}$$
$$5x = 15 \quad \text{Add 24 to both sides.}$$
$$x = 3 \quad \text{Divide by 5.}$$

4. Substitute the value of the variable $x = 3$ into one of the original equations. (We substitute in the equation $x + y = 8$.) Then solve for the second variable. Our solution is the ordered pair (3, 5).

$$3 + y = 8$$
$$y = 5$$

5. **CHECK** When $x = 3$ and $y = 5$,

$$x + y = 8$$

becomes

$$3 + 5 = 8$$
$$8 = 8$$

which is true. Then the second equation

$$2x - 3y = -9$$

becomes

$$2(3) - 3(5) = -9$$
$$6 - 15 = -9$$
$$-9 = -9$$

which is also true. Thus, our solution (3, 5) is correct.

EXAMPLE 2 Solving an inconsistent system by substitution

Solve the system:

$$x + 2y = 4$$
$$2x = -4y + 6$$

SOLUTION 2

We use the five-step procedure.

1. Solve one of the equations for one of the variables (we solve the first equation for x since x has a coefficient of 1). $\quad x = 4 - 2y$

2. Substitute $x = 4 - 2y$ into $2x = -4y + 6$.

$$2(4 - 2y) = -4y + 6$$
$$8 - 4y = -4y + 6 \quad \text{Simplify.}$$
$$8 - 4y + 4y = -4y + 4y + 6 \quad \text{Add } 4y.$$
$$8 = 6$$

(continued)

PROBLEM 2

Solve the system:

$$x - 3y = 6$$
$$2x - 6y = 8$$

Answers to PROBLEMS

2. No solution; the system is inconsistent.

3. There is no equation to solve. The result, $8 = 6$, is never true. It is a contradiction. Since our procedure is correct, we conclude that the given system has *no solution;* it is *inconsistent.*
4. We do not need step 4.
5. **CHECK** Note that if you divide the second equation by 2, you get $x = -2y + 3$ or $x + 2y = 3$ which *contradicts* the first equation, $x + 2y = 4$.

EXAMPLE 3 Solving a dependent system by substitution

Solve the system:

$$x + 2y = 4$$
$$4y + 2x = 8$$

SOLUTION 3 As before, we use the five-step procedure.

1. Solve the first equation for x. $\quad x = 4 - 2y$
2. Substitute $x = 4 - 2y$ into $4y + 2x = 8$. $\quad 4y + 2(4 - 2y) = 8$
3. There is no equation to solve. Note that in this case we have obtained the true statement $8 = 8$, regardless of the value we assign to either x or to y. $\quad 4y + 8 - 4y = 8$ Simplify. $\quad 8 = 8$
4. We do not need step 4 because the equations are *dependent;* that is, there are infinitely many solutions.
5. **CHECK** If we let $x = 0$ in the equation $x + 2y = 4$, we obtain $2y = 4$, or $y = 2$. Similarly, if we let $x = 0$ in the equation $4y + 2x = 8$, we obtain $4y = 8$, or $y = 2$, so $(0, 2)$ is a solution of both equations. It can also be shown that $x = 2$, $y = 1$ satisfies both equations. Therefore, $(2, 1)$ is another solution, and so on. Note that if you divide the second equation by 2 and rearrange, you get $x + 2y = 4$, which is identical to the first equation. Thus, any solution of the first equation is also a solution of the second equation; that is, the solution consists of all points satisfying $x + 2y = 4$. You can write this fact by writing the solution set as $\{(x, y) \mid x + 2y = 4\}$.

PROBLEM 3

Solve the system:

$$x - 3y = 6$$
$$6y - 2x = -12$$

EXAMPLE 4 Simplifying and solving a system by substitution

Solve the system:

$$-2x = -y + 2$$
$$6 - 3x + y = -4x + 5$$

SOLUTION 4 The second equation has x's and constants on both sides, so we first simplify it by adding $4x$ and subtracting 6 from both sides to obtain

$$6 - 3x + y + 4x - 6 = -4x + 5 + 4x - 6$$
$$x + y = -1$$

We now have the equivalent system

$$-2x = -y + 2$$
$$x + y = -1$$

Solving the second equation for x, we get $x = -y - 1$. Substituting $-y - 1$ for x in the first equation, we have

$$-2(-y - 1) = -y + 2$$
$$2y + 2 = -y + 2$$
$$3y = 0 \quad \text{Add } y\text{, subtract 2.}$$
$$y = 0 \quad \text{Divide by 3.}$$

PROBLEM 4

Solve the system:

$$-3x = -y + 6$$
$$6 - 3x + y = -5x + 2$$

Answers to PROBLEMS

3. Infinitely many solutions. The equations are dependent. Some solutions are $(0, -2)$, $(6, 0)$, and $(3, -1)$ 4. $(-2, 0)$; consistent system

Since $-2x = -y + 2$ and $y = 0$, we have

$$-2x = 0 + 2$$
$$x = -1$$

Thus, the system is *consistent* and its solution is $(-1, 0)$. You can verify this by substituting -1 for x and 0 for y in the two original equations.

If a system has equations that contain fractions, we clear the fractions by multiplying each side by the LCD (Remember? LCD is the lowest common denominator), and then we solve the resulting system, as shown in Example 5.

EXAMPLE 5 **Solving a system involving fractions**

Solve the system:

$$2x + \frac{y}{4} = -1$$
$$\frac{x}{4} + \frac{3y}{8} = \frac{5}{4}$$

SOLUTION 5 Multiply both sides of the first equation by 4, and both sides of the second equation by 8 (the LCM of 4 and 8) to obtain

$$4\left(2x + \frac{y}{4}\right) = (4)(-1) \quad \text{or equivalently} \quad 8x + y = -4$$
$$8\left(\frac{x}{4} + \frac{3y}{8}\right) = 8\left(\frac{5}{4}\right) \quad \text{or equivalently} \quad 2x + 3y = 10$$

Solving the first equation for y, we have $y = -8x - 4$. Now we substitute $-8x - 4$ for y in $2x + 3y = 10$:

$$2x + 3(-8x - 4) = 10$$
$$2x - 24x - 12 = 10 \quad \text{Simplify.}$$
$$-22x = 22 \quad \text{Simplify and add 12.}$$
$$x = -1 \quad \text{Divide by } -22.$$

Substituting -1 for x in $2x + \frac{y}{4} = -1$, we get $2(-1) + \frac{y}{4} = -1$ or $y = 4$. Thus, the system is *consistent* and its solution is $(-1, 4)$. Verify this!

PROBLEM 5

Solve the system:

$$2x + \frac{y}{3} = -1$$
$$\frac{x}{4} + \frac{y}{6} = \frac{1}{4}$$

B > Consistent, Inconsistent, and Dependent Systems

When we use the substitution method, one of three things can occur:

1. The equations are *consistent;* there is only *one* solution (x, y).
2. The equations are *inconsistent;* we get a contradictory (false) statement, and there will be *no* solution.
3. The equations are *dependent;* we get a statement that is true for all values of the remaining variable, and there will be *infinitely many* solutions.

Keep this in mind when you do the exercise set, and be very careful with your arithmetic!

Answers to PROBLEMS

5. $(-1, 3)$; consistent system

C > Applications Involving Systems of Equations

In Example 5 of Section 7.1 the graphs intersect at a point (x, y), where the cost is the same for both printers. We find the coordinates of the point in Example 6.

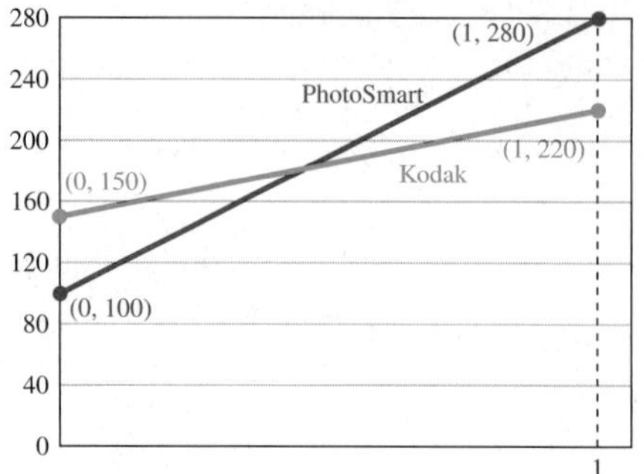

EXAMPLE 6 Finding when printing costs are the same

Find the point (x, y) at which the cost $y = 100 + 180x$ for the PhotoSmart printer and the cost $y = 150 + 70x$ for the Kodak printer are the same.

SOLUTION 6

To find when the cost y is the same for both printers, we substitute $100 + 180x$ for y in

$$y = 150 + 70x \text{ obtaining}$$

$$100 + 180x = 150 + 70x$$

$$180x = 50 + 70x \quad \text{Subtract 100.}$$

$$110x = 50 \quad \text{Subtract } 70x.$$

$$x = \frac{50}{110} \quad \text{Divide by 110.}$$

$$x = \frac{5}{11} \quad \text{Simplify.}$$

Thus, when $x = \frac{5}{11}$, the cost $y = 100 + 180\left(\frac{5}{11}\right) = \frac{2000}{11} \approx \$181.82$ is the same for both printers. The point at which this occurs is at $x = \frac{5}{11}$ and $y = 181.82$, or $\left(\frac{5}{11}, 181.82\right)$.

PROBLEM 6

Find the point (x, y) at which the annual cost y for a Kodak and a PhotoSmart of Problem 5, Section 7.1 are the same.

K: $y = 170 + 100x$

P: $y = 160 + 180x$

> Exercises 7.2

> Practice Problems > Self-Tests > Media-rich eBooks > e-Professors > Videos

< A > Using the Substitution Method to Solve a System of Equations
< B > Consistent, Inconsistent, and Dependent Systems

In Problems 1–32, use the substitution method to find the solution, if possible. Label each system as consistent (one solution), inconsistent (no solution), or dependent (infinitely many solutions). If the system is consistent, give the solution.

1. $y = 2x - 4$
$-2x = y - 4$

2. $y = 2x + 2$
$-x = y + 1$

3. $x + y = 5$
$3x + y = 9$

4. $x + y = 5$
$3x + y = 3$

5. $y - 4 = 2x$
$y = 2x + 2$

6. $y + 5 = 4x$
$y = 4x + 7$

Answers to PROBLEMS

6. When $x = \frac{1}{8}$, the cost $y = \$182.50$. The point at which this occurs is $\left(\frac{1}{8}, 182.50\right)$.

7. $x = 8 - 2y$
$x + 2y = 4$

8. $x = 4 - 2y$
$x - 2y = 0$

9. $x + 2y = 4$
$x = -2y + 4$

10. $x + 3y = 6$
$x = -3y + 6$

11. $x = 2y + 1$
$y = 2x + 1$

12. $y = 3x + 2$
$x = 3y + 2$

13. $2x - y = -4$
$4x = 4 + 2y$

14. $5x + y = 5$
$5x = 15 - 3y$

15. $x = 5 - y$
$0 = x - 4y$

16. $x = 3 - y$
$0 = 2x - y$

17. $x + 1 = y + 3$
$x - 3 = 3y - 7$

18. $x - 1 = 2y + 12$
$x + 6 = 3 - 6y$

19. $2y = -x + 4$
$8 + x - 4y = -2y + 4$

20. $y - 1 = 2x + 1$
$3x + y + 2 = 5x + 6$

21. $3x + y - 5 = 7x + 2$
$y + 3 = 4x - 2$

22. $4x + 2y + 1 = 4 + 3x + 5$
$x - 3 = 5 - 2y$

23. $4x - 2y - 1 = 3x - 1$
$x + 2 = 6 - 2y$

24. $8 + y - 4x = -2x + 4$
$2x + 3 = -y + 7$

25. $\frac{x}{6} + \frac{y}{2} = 1$
$5x - 2y = 13$

26. $\frac{x}{8} - \frac{5y}{8} = 1$
$-7x + 8y = 25$

27. $3x - y = 12$
$-\frac{x}{2} + \frac{y}{6} = -2$

28. $x - 3y = -4$
$-\frac{x}{6} + \frac{y}{2} = \frac{2}{3}$

29. $\frac{y}{4} + x = \frac{3}{8}$
$y = 8 - 4x$

30. $y = 1 - 5x$
$\frac{y}{5} + x = \frac{3}{10}$

31. $3x + \frac{y}{3} = 5$
$\frac{x}{2} - \frac{2y}{3} = 3$

32. $3y + \frac{x}{3} = 5$
$\frac{y}{2} - \frac{2x}{3} = 3$

‹ C › Applications Involving Systems of Equations

33. *Internet service costs* The Information Network charges a \$20 fee for 15 hours of Internet service plus \$3 for each additional hour while InterServe Communications charges \$20 for 15 hours plus \$2 for each additional hour.

a. Write an equation for the price p when you use h hours of Internet service with The Information Network.

b. Write an equation for the price p when you use h hours of Internet service with InterServe Communications.

c. When is the price p for both services the same?

34. *Internet service costs* TST On Ramp charges \$10 for 10 hours of Internet service plus \$2 for each additional hour.

a. Write an equation for the price p when you use h hours of Internet service with TST On Ramp.

b. When is the price of TST the same as that for InterServe Communications (see Problem 33)? (*Hint:* The algebra won't tell you—try a graph!)

35. *Cell phone costs* Phone Company A has a plan costing \$20 per month plus 60¢ for each minute m of airtime, while Company B charges \$50 per month plus 40¢ for each minute m of airtime. When is the cost for both companies the same?

36. *Cell phone charges* Sometimes phone companies charge an activation fee to "turn on" your cell phone. One company charges \$50 for the activation fee, \$40 for your cell phone, and 60¢ per minute m of airtime. Another company charges \$100 for your cell phone and 40¢ a minute of airtime. When is the cost for both companies the same?

37. *Wages and tips* Le Bon Ton restaurant pays its servers \$50 a week plus tips, which average \$10 per table. Le Magnifique pays \$100 per week but tips average only \$5 per table. How many tables t have to be served so that the weekly income of a server is the same at both restaurants?

38. *Fitness center costs* The Premier Fitness Center has a \$200 initiation fee plus \$25 per month. Bodies by Jacques has an initial charge of \$500 but charges only \$20 per month. At the end of which month is the cost the same?

Web IT go to mhhe.com/bello for more lessons

Web IT go to mhhe.com/bello for more lessons

39. *Cable company costs* One cable company charges \$35 for the initial installation plus \$20 per month. Another company charges \$20 for the initial installation plus \$35 per month. At the end of which month is the cost the same for both companies?

40. *Plumber rates* A plumber charges \$20 an hour plus \$60 for the house call. Another plumber charges \$25 an hour, but the house call is only \$50. What is the least number of hours for which the costs for both plumbers are the same?

41. *Temperature conversions* The formula for converting degrees Celsius C to degrees Fahrenheit F is

$$F = \frac{9}{5}C + 32$$

When is the temperature in degrees Fahrenheit the same as that in degrees Celsius?

42. *Temperature conversions* The formula for converting degrees Fahrenheit F to degrees Celsius C is

$$C = \frac{5}{9}(F - 32)$$

When is the temperature in degrees Celsius the same as that in degrees Fahrenheit?

43. *Supply and demand* The supply y of a certain item is given by the equation $y = 2x + 8$, where x is the number of days elapsed. If the demand is given by the equation $y = 4x$, how many days will the supply equal the demand?

44. *Supply and demand* The supply of a certain item is given by the equation $y = 3x + 8$, where x is the number of days elapsed. If the demand is given by the equation $y = 4x$, in how many days will the supply equal the demand?

45. *Supply and demand* A company has 10 units of a certain item and can manufacture 5 items each day; thus, the supply is given by the equation $y = 5x + 10$. If the demand for the item is $y = 7x$, where x is the number of days elapsed, in how many days will the demand equal the supply?

46. *Supply and demand* Clonker Manufacturing has 12 clonkers in stock. The company manufactures 3 more clonkers each day. If the clonker demand is 7 each day, in how many days will the supply equal the demand?

47. *College expenses* According to the College Board, parents will pay x dollars for tuition and room and board at public colleges this year. They are also likely to spend y dollars on textbooks, supplies, transportation, and "other."

a. The total cost for tuition, room and board (x) plus textbooks, supplies, transportation, and "other" (y) comes to \$15,127. Write an equation for this fact.

b. The tuition, room and board x is \$9127 more than the textbooks, supplies and transportation, and "other" y. Write an equation for this fact.

c. Use the substitution method to solve the systems of two equations obtained in parts **a** and **b** to find x (cost of tuition and room and board) and y (cost of textbooks, supplies, transportation, and "other").

Source: www.msnbc.msn.com.

48. *Cost of computers, books, and supplies* The "Average Cost of Attendance" page at the University of Florida estimates that the expenses for computers C plus the expenses for books and supplies B will amount to \$1820.

a. Write an equation for this fact.

b. The books and supplies B are only \$20 more than the computers C. Write an equation for this fact.

c. Use the substitution method to solve the system of equations obtained in parts **a** and **b** to find the estimated cost of computers C and the estimated cost of books and supplies B.

Source: www.sfa.ufl.edu.

49. *Transportation and personal/miscellaneous expenses* At California State University at Long Beach the expenses for transportation T plus personal/miscellaneous expenses PM amount to \$3222.

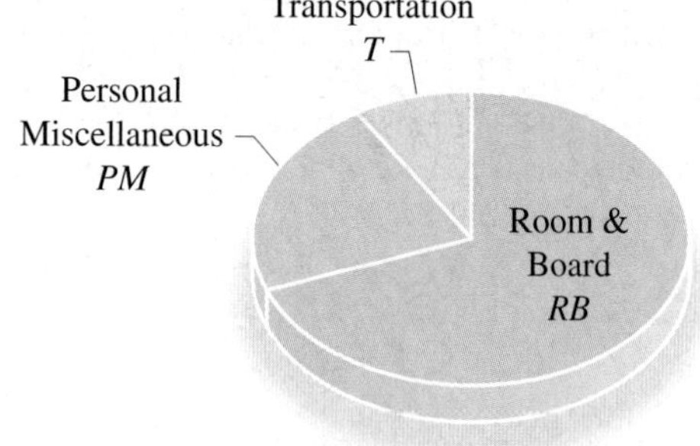

a. Write an equation for this fact.

b. The difference between PM and T is \$1350. Write an equation for this fact.

c. Use the substitution method to solve the system of equations obtained in parts **a** and **b** to find the transportation expenses T and the personal/miscellaneous expenses PM.

Source: www.csulb.edu.

50. *Textbook costs* A report released by the Student Public Interest Group indicates that the average cost F of the 22 most frequently assigned textbooks is \$65.72 more than the cost L of less expensive alternatives.

a. Write an equation for the average cost F.

b. The cost L of those less expensive counterparts is about $\frac{1}{2}$ the cost F of the most frequently assigned textbooks. Write an equation for the cost L of the less expensive counterparts.

c. Use the substitution method to solve the systems of two equations obtained in parts **a** and **b** to find the average cost F of the 22 most frequently assigned textbooks and the cost L of the less expensive counterparts.

Source: www.alternet.org.

Using Your Knowledge

Applications: Green Math

Use your knowledge and the table for finding the point (x, y) at which the printing costs are equal for the two printers specified in Problems 51–53.

	Cost	Cartridge	Yearly Cost
Canon MP560	\$100	\$ 80	$100 + 80x$ (dollars)
Epson Artisan	\$150	\$ 70	$150 + 70x$ (dollars)
Brother MFC 490	\$130	\$ 70	$130 + 70x$ (dollars)
Lexmark X4650	\$ 60	\$150	$60 + 150x$ (dollars)

51. Lexmark and Canon

52. Canon and Brother

53. Lexmark and Epson

Write On

54. If you are solving the system

$$\begin{aligned} 2x + y &= -7 \\ 3x - 2y &= 7 \end{aligned}$$

which variable would you solve for in the first step of the five-step procedure given in the text?

55. When solving a system of equations using the substitution method, how can you tell whether the system is

a. consistent?

b. inconsistent?

c. dependent?

56. In Example 5, we multiplied both sides of the first equation by 4 and both sides of the second equation by 8. Would you get the same answer if you multiplied both sides of the first equation by 8 and both sides of the second equation by 4? Why is this approach not a good idea?

Concept Checker

Fill in the blank(s) with the correct word(s), phrase, or mathematical statement.

57. When a system of equations is **consistent,** there is ______________ **solution** to the system.

58. The **solution** of a **consistent** system of equations is an ______________ pair of numbers.

59. An **inconsistent** system of equations has ______________ solutions.

60. A **dependent** system of equations has ______________ solutions.

many	**ordered**
one	**finite**
no	**infinitely many**

Mastery Test

Solve and label the system as consistent, inconsistent, or dependent. If the system is consistent, write the solution.

61. $x - 3y = 6$
$2x - 6y = 8$

62. $x - 3y = 6$
$6y - 2x = -12$

63. $x + y = 5$
$2x - 3y = -5$

64. $2y + x = 3$
$x - 3y = 0$

65. $3x - 3y + 1 = 5 + 2x$
$x - 3y = 4$

66. $5x + y = 5$
$5x + y - 10 = 5 - 2y$

67. $\frac{x}{2} - \frac{y}{4} = -1$
$2x = 2 + y$

68. $x + \frac{y}{5} = 1$
$\frac{x}{3} + \frac{y}{5} = 1$

69. A store is selling a Sony® DSS system for \$300. The basic monthly charge is \$50. An RCA® system is selling for \$500 with a \$30 monthly charge. What is the least number of months for which the prices of both systems are the same?

Skill Checker

Solve:

70. $-1.50b = -6$

71. $-0.5x = -4$

72. $-0.2y = -6$

73. $9x = -9$

74. $11y = -11$

7.3 Solving Systems of Equations by Elimination

Objectives

A Solve a system of equations in two variables by elimination.

B Determine whether a system is consistent, inconsistent, or dependent.

C Solve applications involving systems of equations.

To Succeed, Review How To . . .

1. Use the RSTUV method for solving word problems (pp. 148–149).
2. Solve linear equations (pp. 137–141).

Getting Started

Using Elimination When Buying Coffee

We've studied two methods for solving systems of equations. The *graphical method* gives us a visual model of the system and allows us to find approximate solutions. The *substitution method* gives exact solutions but is best used when either of the given equations has at least one coefficient of 1 or −1. If graphing or substitution is not desired or feasible, there is another method we can use: the *elimination method,* sometimes called the **addition** or **subtraction method.**

The man in the photo is selling coffee, ground to order. A customer wants 10 pounds of a mixture of coffee A costing \$6 per pound and coffee B costing \$4.50 per pound. If the price for the purchase is \$54, how many pounds of each of the coffees will be in the mixture?

To solve this problem, we use an idea that we've already learned: solving a system of equations. In this problem we want a precise answer, so we use the elimination method, which we will learn next.

A › Solving Systems of Equations by Elimination

To solve the coffee problem in the *Getting Started,* we first need to organize the information. The information can be summarized in a table like this:

	Price	Pounds	Total Price
Coffee A	6.00	a	$6a$
Coffee B	4.50	b	$4.50b$
Totals		$a + b$	$6a + 4.50b$

Since the customer bought a total of 10 pounds of coffee, we know that

$$a + b = 10$$

Also, since the purchase came to $54, we know that

$$6a + 4.50b = 54$$

So, the system of equations we need to solve is

$$\begin{aligned} a + b &= 10 \\ 6a + 4.50b &= 54 \end{aligned}$$

To solve this system, we shall use the **elimination method,** which consists of replacing the given system by equivalent systems until we get a system with an obvious solution. To do this, we first write the equations in the form $Ax + By = C$. Recall that an *equivalent* system is one that has the *same* solution as the given one. For example, the equation

$$A = B$$

is equivalent to the equation

$$kA = kB \quad (k \neq 0)$$

This means that you can multiply both sides of the equation $A = B$ by the same nonzero expression k and obtain the equivalent equation $kA = kB$.

Also, the system

$$\begin{aligned} A &= B \\ C &= D \end{aligned}$$

is equivalent to the system

$$\begin{aligned} A &= B \\ k_1A + k_2C &= k_1B + k_2D \quad (k_1 \text{ and } k_2 \text{ not both } 0) \end{aligned}$$

We can check this by multiplying both sides of $A = B$ by k_1 and both sides of $C = D$ by k_2 and adding. That is the theory, but how do you do it?

Let's return to the coffee problem. We multiply the first equation in the given system by -6; we then get the equivalent system

$$\begin{aligned} -6a - 6b &= -60 \\ (+)\ 6a + 4.50b &= 54 \end{aligned}$$

Multiply by -6 because we want the coefficients of a to be opposites (like -6 and 6).

Add the equations.

$$\begin{aligned} 0 - 1.50b &= -6 \\ -1.50b &= -6 \\ b &= 4 \end{aligned}$$

Divide by -1.50.

$$a + 4 = 10$$

Substitute 4 for b in $a + b = 10$.

$$a = 6$$

Solve for a.

Thus, the mixture contains 6 pounds of coffee A and 4 pounds of coffee B. This answer can be verified. If the customer bought 6 pounds of coffee A and 4 of coffee B, she did indeed buy 10 pounds. Her price for coffee A was $6 \cdot 6 = \$36$ and for coffee B, $4 \cdot 4.50 = \$18$. Thus, the entire cost was $\$36 + \$18 = \$54$, as stated.

What have we done here? Well, this technique depends on the fact that one (or both) of the equations in a system can be multiplied by a nonzero number to obtain two equivalent equations with opposite coefficients of x (or y). Here is the idea.

ELIMINATION METHOD

One or both of the equations in a system of simultaneous equations can be multiplied (or divided) by any nonzero number to obtain an *equivalent* system in which the coefficients of the *x*'s (or of the *y*'s) are opposites, thus, *eliminating x* or *y* when the equations are added.

EXAMPLE 1 Solving a consistent system by elimination

Solve the system:

$$2x + y = 1$$
$$3x - 2y = -9$$

SOLUTION 1 Remember the idea: we multiply one or both of the equations by a number or numbers that will cause either the coefficients of x or the coefficients of y to be opposites. We can do this by multiplying the first equation by 2:

$2x + y = 1$ — Multiply by 2. → $4x + 2y = 2$

$3x - 2y = -9$ — Leave as is. → $3x - 2y = -9$

Add the equations. $7x + 0 = -7$

$7x = -7$

Divide by 7. $x = -1$

Substitute -1 for x in $2x + y = 1$. $2(-1) + y = 1$

$-2 + y = 1$

Add 2. $y = 3$

Thus, the solution of the system is $(-1, 3)$.

CHECK When $x = -1$ and $y = 3$, $2x + y = 1$ becomes

$$2(-1) + 3 = 1$$
$$-2 + 3 = 1$$
$$1 = 1$$

a true statement, and $3x - 2y = -9$ becomes

$$3(-1) - 2(3) = -9$$
$$-3 - 6 = -9$$
$$-9 = -9$$

which is also true.

PROBLEM 1

Solve the system:

$$3x + y = 1$$
$$3x - 4y = 11$$

Answers to PROBLEMS

1. $(1, -2)$

B > Determining Whether a System Is Consistent, Inconsistent, or Dependent

There are *three* possibilities when solving simultaneous linear equations.

1. *Consistent* and *independent* equations have *one* solution.
2. *Inconsistent* equations have *no* solution. You can recognize them when you get a contradiction (a false statement) in your work, (See Example 2. In Example 2, we get $0 = -12$, a contradiction.)
3. *Dependent* equations have *infinitely many* solutions. You can recognize them when you get a true statement such as $0 = 0$ (See Example 3). Remember that any solution of one of these equations is a solution of the other.

Pay close attention to the position of the variables in the equations. All the equations **except** those solved by substitution should be written in the form

$$\left.\begin{aligned} ax + by &= c \\ dx + ey &= f \end{aligned}\right\} \text{This is the standard form.}$$

x column, y column, Constant terms

If the equations are not in this form, and you are not using the substitution method, rewrite them using this form. It helps to keep things straight!

Note that as it is mentioned in (2) above, not all systems have solutions. How do we find out if they don't? Let's look at the next example, which shows a **contradiction** for a system that has **no** solution.

EXAMPLE 2 Solving an inconsistent system by elimination

Solve the system:

$$2x + 3y = 3$$
$$4x + 6y = -6$$

SOLUTION 2 In this case, we try to eliminate the variable x by multiplying the first equation by -2.

$$\begin{aligned} 2x + 3y &= 3 \quad \xrightarrow{\text{Multiply by } -2.} & -4x - 6y &= -6 \\ 4x + 6y &= -6 \quad \xrightarrow{\text{Leave as is.}} & 4x + 6y &= -6 \\ & & 0 + 0 &= -12 \quad \text{Add.} \\ & & 0 &= -12 \end{aligned}$$

Of course, this is a contradiction, so there is no solution; the system is *inconsistent*.

PROBLEM 2

Solve the system:

$$3x + 2y = 1$$
$$6x + 4y = 12$$

EXAMPLE 3 Solving a dependent system by elimination

Solve the system:

$$2x - 4y = 6$$
$$-x + 2y = -3$$

PROBLEM 3

Solve the system:

$$3x - 6y = 9$$
$$-x + 2y = -3$$

(continued)

Answers to PROBLEMS

2. No solution; inconsistent

3. Infinitely many solutions; dependent

SOLUTION 3 Here we try to eliminate the variable x by multiplying the second equation by 2. We obtain

$$
\begin{array}{lcr}
2x - 4y = 6 & \xrightarrow{\text{Leave as is.}} & 2x - 4y = 6 \\
-x + 2y = -3 & \xrightarrow{\text{Multiply by 2.}} & -2x + 4y = -6 \\
 & & \overline{0 + 0 = 0} \quad \text{Add.} \\
 & & 0 = 0
\end{array}
$$

Lo and behold, we've eliminated both variables! However, notice that if we had multiplied the second equation in the original system by -2, we would have obtained

$$
\begin{array}{lcr}
2x - 4y = 6 & \xrightarrow{\text{Leave as is.}} & 2x - 4y = 6 \\
-x + 2y = -3 & \xrightarrow{\text{Multiply by } -2.} & 2x - 4y = 6
\end{array}
$$

This means that the first equation is a constant multiple of the second one; that is, they are equivalent equations. When a system of equations consists of two equivalent equations, the system is said to be *dependent,* and any solution of one equation is a solution of the other. Because of this, the system has *infinitely many* solutions. For example, if we let x be 0 in the first equation, then $y = -\frac{3}{2}$, and $(0, -\frac{3}{2})$ is a solution of the system. Similarly, if we let y be 0 in the first equation, then $x = 3$, and we obtain the solution $(3, 0)$. Many other solutions are possible; try to find some of them. As a matter of fact, any ordered pair (x, y) so that $2x - 4y = 6$ will be a solution. This fact can be written by stating that the solution set is $\{(x, y) \mid 2x - 4y = 6\}$.

Finally, in some cases, we cannot multiply just one of the equations by an integer that will cause the coefficients of one of the variables to be opposites. For example, to solve the system

$$
\begin{aligned}
2x + 3y &= 3 \\
5x + 2y &= 13
\end{aligned}
$$

we must multiply **both** equations by integers chosen so that the coefficients of one of the variables will be opposites. We can do this using either of the following methods.

METHOD 1

Solving by elimination: x or y?

To eliminate x, multiply the first equation by 5 and the second one by -2 to obtain an equivalent system.

$$
\begin{array}{lcr}
2x + 3y = 3 & \xrightarrow{\text{Multiply by 5.}} & 10x + 15y = 15 \\
5x + 2y = 13 & \xrightarrow{\text{Multiply by } -2.} & -10x - 4y = -26 \\
\text{Add.} & & \overline{0 + 11y = -11} \\
 & & 11y = -11 \\
\text{Divide by 11.} & & y = -1 \\
\text{Substitute } -1 \text{ for } y \text{ in } 2x + 3y = 3. & & 2x + 3(-1) = 3 \\
\text{Simplify.} & & 2x - 3 = 3 \\
\text{Add 3.} & & 2x = 6 \\
\text{Divide by 2.} & & x = 3
\end{array}
$$

Thus, the solution of the system is $(3, -1)$. This time we eliminated the x and solved for y. Alternatively, we can eliminate the y first, as shown next.

METHOD 2

This time, we eliminate the y:

$2x + 3y = 3$	Multiply by -2.	$-4x - 6y = -6$
$5x + 2y = 13$	Multiply by 3.	$15x + 6y = 39$
Add.		$11x + 0 = 33$
		$11x = 33$
Divide by 11.		$x = 3$
Substitute 3 for x in $2x + 3y = 3$.		$2(3) + 3y = 3$
Simplify.		$6 + 3y = 3$
Subtract 6.		$3y = -3$
Divide by 3.		$y = -1$

Thus, the solution is $(3, -1)$, as before.

CHECK: When $x = 3$ and $y = -1$, $2x + 3y = 3$ becomes

$$\begin{aligned} 2(3) + 3(-1) &= 3 \\ 6 - 3 &= 3 \\ 3 &= 3 \end{aligned}$$

which is true. Substituting $x = 3$ and $y = -1$ in $5x + 2y = 13$ yields

$$\begin{aligned} 5(3) + 2(-1) &= 13 \\ 15 - 2 &= 13 \end{aligned}$$

which is also true. Thus, $(3, -1)$ is the solution!

EXAMPLE 4 Writing in standard form and solving by elimination

Solve the system:

$$\begin{aligned} 5y + 2x &= 9 \\ 2y &= 8 - 3x \end{aligned}$$

SOLUTION 4 We first write the system in standard form—that is, the x's first, then the y's, and then the constants on the other side of the equation. The result is the equivalent system

$$\begin{aligned} 2x + 5y &= 9 \\ 3x + 2y &= 8 \end{aligned}$$

Now we multiply the first equation by 3 and the second one by -2 so that, upon addition, the x's will be eliminated.

$2x + 5y = 9$	Multiply by 3.	$6x + 15y = 27$
$3x + 2y = 8$	Multiply by -2.	$-6x - 4y = -16$
Add.		$0 + 11y = 11$
		$11y = 11$
Divide by 11.		$y = 1$
Substitute 1 for y in $2x + 5y = 9$.		$2x + 5(1) = 9$
Simplify.		$2x + 5 = 9$
Subtract 5.		$2x = 4$
Divide by 2.		$x = 2$

Thus, the solution is $(2, 1)$ and the system is *consistent*. You should verify this by substituting $x = 2$ and $y = 1$ in the original equation to make sure they satisfy both equations.

PROBLEM 4

Solve the system:

$$\begin{aligned} 5x + 4y &= 6 \\ 3y &= 4x - 11 \end{aligned}$$

Answers to PROBLEMS

4. $(2, -1)$

C > Applications Involving Systems of Equations

EXAMPLE 5 **Cell phone plans**

Do you have a cell phone? The time the phone is used, called airtime, is usually charged by the minute at two different rates: peak and off-peak. Suppose your plan charges \$0.60 for each peak-time minute and \$0.45 for each off-peak minute. If your airtime cost \$54 and you've used 100 minutes of airtime, how many minutes p of peak time and how many minutes n of off-peak time did you use?

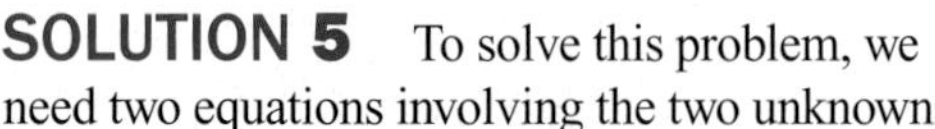

SOLUTION 5 To solve this problem, we need two equations involving the two unknowns, p and n. We know that the charges amount to \$54 and that 100 minutes of airtime were used. How can we accumulate \$54 of charges?

Since peak time costs \$0.60 per minute, peak times cost $0.60p$.

Since off-peak time costs \$0.45 per minute, off-peak times cost $0.45n$.

The total cost is \$54, so we add peak and off-peak costs:

$$0.60p + 0.45n = 54$$

Also, the total number of minutes is 100. Thus, $p + n = 100$. To try to *eliminate n,* we multiply both sides of the second equation by -0.45 and then add:

$0.60p + 0.45n = 54$	Leave as is. →	$0.60p + 0.45n = 54$
$p + n = 100$	Multiply by -0.45. →	$-0.45p - 0.45n = -45$
	Add.	$0.15p = 9$
	Divide both sides by 0.15.	$p = \dfrac{9}{0.15} = 60$

Thus, $p = 60$ minutes of peak time were used and the rest of the 100 minutes used—that is, $100 - 60 = 40$—were off-peak minutes. This means that $n = 40$. You can check that $p = 60$ and $n = 40$ by substituting in the original equations.

PROBLEM 5

Big Cell Phone has a plan that charges \$0.50 for each peak-time minute and \$0.40 for each off-peak minute. If airtime cost \$44 and 100 minutes were used, how many minutes p of peak time and how many minutes n of off-peak time were used?

Finally, let us use systems of equations to discuss polar bears! Do you think that polar bears are "threatened" or "endangered," that is, in danger of extinction? To answer that question, we should be able to estimate the size of their population. The best estimates claim that there are 20,000 to 25,000 bears worldwide, but they are spread out around the Arctic in 19 separate subpopulations, one of them in the Beaufort Sea, in Southern Alaska. A USGS (United States Geological Service) report estimated that near the Beaufort Sea, the actual polar bear population (A) plus any possible overcounts (O) would amount to 1800 bears. The overcount (O) was small because the difference between the actual bear population A and the overcount (O) was only 1252. What is the actual polar bear population A? Solve Example 6 and find out!

Answers to PROBLEMS

5. 40 minutes of peak and 60 minutes of off-peak

EXAMPLE 6 Bear population near the Beaufort Sea

We translate two statements into equations:
The polar bear population A plus overcounts (O) amount to 1800:

$$A + (O) = 1800$$

The difference between the actual bear population A and (O) is 1252:

$$A - (O) = 1252$$

Solve the system of two equations and find A and (O).

SOLUTION 6

We use the elimination method: $A + (O) = 1800$
$A - (O) = 1252$

Add. $2A = 3052$

Divide both sides by 2. $A = 1526$

Substitute $A = 1526$ in $A + (O) = 1800$

Obtaining $1526 + (O) = 1800$

Subtract 1526 from both sides. $(O) = 1800 - 1526 = 274$

Thus, the actual polar bear population A near the Beaufort Sea is **1526** and the overcount O is **274.** The USGS concluded that because the estimates were done by two different statistical methods, the two numbers, 1800 and 1526, could be used as reliable estimates of the population, but we know what the exact numbers should be!

PROBLEM 6

How many threatened and how many endangered species are there in the United States? The sum of endangered species E and threatened species T is 1320. The difference between endangered species E and threatened species T is 700. How many endangered (E) and how many threatened (T) species are there in the United States?

If you want to see a list go to http://tinyurl.com/yblgyec.

Environmentalists have requested that polar bears be listed as an endangered species, an action that would most likely place the Arctic region off limits for mineral exploration and very likely lead to strict federal regulation of greenhouse gas emissions.

› Exercises 7.3

‹A› **Solving Systems of Equations by Elimination**
‹B› **Determining Whether a System is Consistent, Inconsistent, or Dependent**

In Problems 1–30, use the elimination method to solve each system. If the system is not consistent, state whether it is inconsistent or dependent.

1. $x + y = 3$
$x - y = -1$

2. $x + y = 5$
$x - y = 1$

3. $x + 3y = 6$
$x - 3y = -6$

4. $x + 2y = 4$
$x - 2y = 8$

5. $2x + y = 4$
$4x + 2y = 0$

6. $3x + 5y = 2$
$6x + 10y = 5$

7. $2x + 3y = 6$
$4x + 6y = 2$

8. $3x - 5y = 4$
$-6x + 10y = 0$

9. $x - 5y = 15$
$x + 5y = 5$

10. $-3x + 2y = 1$
$2x + y = 4$

11. $x + 2y = 2$
$2x + 3y = -10$

12. $3x - 2y = -1$
$x + 7y = -8$

Answers to PROBLEMS

6. $E = 1010, T = 310$

Web IT go to mhhe.com/bello for more lessons

13. $3x - 4y = 10$
$5x + 2y = 34$

14. $5x - 4y = 6$
$3x + 2y = 8$

15. $11x - 3y = 25$
$5x + 8y = 2$

16. $12x + 8y = 8$
$7x - 5y = 24$

17. $2x + 3y = 21$
$3x = y + 4$

18. $2x - 3y = 16$
$x = y + 7$

19. $x = 1 + 2y$
$-y = x + 5$

20. $3y = 1 - 2x$
$3x = -4y - 1$

21. $\frac{x}{4} + \frac{y}{3} = 4$
$\frac{x}{2} - \frac{y}{6} = 3$

(*Hint:* Multiply by the LCD first.)

22. $\frac{x}{5} + \frac{y}{6} = 5$
$\frac{2x}{5} + \frac{y}{3} = -2$

(*Hint:* Multiply by the LCD first.)

23. $\frac{1}{4}x - \frac{1}{3}y = -\frac{5}{12}$
$\frac{1}{5}x + \frac{2}{5}y = 1$

(*Hint:* Multiply by the LCD first.)

24. $\frac{x}{2} + \frac{y}{2} = \frac{5}{2}$
$\frac{x}{2} - \frac{y}{3} = \frac{5}{2}$

(*Hint:* Multiply by the LCD first.)

25. $\frac{x}{8} + \frac{y}{8} = 1$
$\frac{x}{2} - \frac{y}{2} = -1$

26. $\frac{x}{5} + \frac{y}{5} = 1$
$\frac{x}{4} - \frac{y}{4} = \frac{1}{4}$

27. $\frac{x}{2} - \frac{y}{3} = 1$
$\frac{x}{2} + \frac{y}{2} = \frac{7}{2}$

28. $\frac{x}{3} + \frac{y}{2} = \frac{7}{3}$
$\frac{x}{3} - \frac{y}{2} = 0$

29. $\frac{2x}{9} - \frac{y}{2} = -1$
$x - \frac{9y}{4} = -\frac{9}{2}$

30. $-\frac{8x}{49} + \frac{5y}{49} = -1$
$\frac{2x}{3} - \frac{5y}{12} = \frac{49}{12}$

Applications: Green Math

Endangered and Threatened Species

Source: http://ecos.fws.gov/tess_public/TESSBoxscore.

31. *Endangered and threatened animal species* The total number of endangered ***E*** and threatened ***T*** animal species is 573. The difference is 245. How many endangered and how many threatened animal species are there?

32. *Endangered and threatened plant species* The total number of endangered ***E*** and threatened ***T*** plant species is 747. The difference is 455. How many endangered and how many threatened plant species are there?

C Applications Involving Systems of Equations

33. *Coffee blends* The Holiday House blends Costa Rican coffee that sells for \$8 a pound and Indian Mysore coffee that sells for \$9 a pound to make 1-pound bags of its Gourmet Blend coffee, which sells for \$8.20 a pound. How much Costa Rican and how much Indian coffee should go into each pound of the Gourmet Blend?

34. *Coffee blends* The Holiday House also blends Colombian Swiss Decaffeinated coffee that sells for \$11 a pound and High Mountain coffee that sells for \$9 a pound to make its Lower Caffeine coffee, 1-pound bags of which sell for \$10 a pound. How much Colombian and how much High Mountain should go into each pound of the Lower Caffeine mixture?

Use the following information for Problems 35–36. If you have high blood pressure, heart disease, or if you just want to maintain your health you should monitor your intake of **sodium** ($<$ 2400 milligrams per day), **total fat** ($<$ 65 grams per day), and **calories** ($<$ 2000 per day). How can we do that? Let us concentrate on daily intakes.

Source: www.pamf.org.

35. *Daily sodium intake* Suppose you go to lunch at McDonald's and dinner at Burger King. You eat **one** Burger King Fire-Grilled Chicken Salad containing c grams of sodium and **two** McDonald's Double Quarter Pounders with cheese with m milligrams of sodium each. You just ate 3100 milligrams of sodium which is over the daily recommended limit! Your sodium consumption for the day is $c + 2m = 3100$.

The next day you change your routine. Now you eat **two** Grilled Chicken Salads from Burger King at c milligrams of sodium each and just **one** Quarter Pounder with m milligrams of sodium. You just consumed 4205 milligrams of sodium, which is still over the daily recommended limit. Your sodium consumption for that day is $2c + m = 4205$.

Solve the system:

$$\begin{aligned} 2c + m &= 4205 \\ c + 2m &= 3100 \end{aligned}$$

and find c (milligrams of sodium in the Grilled Chicken Salad) and m (milligrams of sodium in the burger).

Chicken Salad

Double Quarter Pounder with Cheese

36. *Daily sodium intake* Suppose you decide to eat **two** Shrimp Garden Salads with vinaigrette dressing with s milligrams of sodium in each and **one** McDonald's hamburger with h milligrams of sodium. You are still over the daily recommended amount of sodium intake at 4070 milligrams! Your sodium consumption is $2s + h = 4070$.

Next you limit yourself to **one** Shrimp Garden Salad with s milligrams of sodium and **one** hamburger with h milligrams. Congratulations, you are finally under: you had 2300 milligrams of sodium!
Your sodium consumption is $s + h = 2300$

Solve the system:

$$\begin{aligned} s + h &= 2300 \\ 2s + h &= 4070 \end{aligned}$$

and find s (milligrams of sodium in the salad) and h (milligrams of sodium in the burger).

37. *Daily fat intake* The recommended maximum **total** fat intake is **65** grams per day. To be under that, eat chicken and salads: three Wendy's Ultimate Chicken Grilled Sandwiches at c grams per sandwich and **one** Wendy's Taco Supreme Salad containing t total grams of fat, yielding $3c + t$ or 52 grams of total fat—under the goal! The next day, try **one** Ultimate Chicken Grilled Sandwich and **two** Taco Salads with $c + 2t$ or 69 grams of total fat. We are a little over the desired 65 grams.

Solve the system:

$$\begin{aligned} 3c + t &= 52 \\ c + 2t &= 69 \end{aligned}$$

and find c (grams of total fat on Wendy's Ultimate Chicken Grill) and t (grams of total fat on the Wendy's Taco Salad).

Source: www.cspinet.org.

38. *Daily fat intake* Let's stick to chicken: two orders of McDonald's McNuggets (6 pieces per order) have m grams of total fat each and **one** McDonald's California Cobb Salad with Grilled Chicken contains c grams of total fat. So far, we consumed 50 grams or $2m + c$ total fat. **One** order of the McNuggets and **one** salad contain $m + c$ or 35 total fat calories.

Solve the system:

$$\begin{aligned} 2m + c &= 50 \\ m + c &= 35 \end{aligned}$$

and find m (total fat grams in the McNuggets) and c (total fat grams in the salad).

Source: www.cspinet.org.

39. *Caloric intake* The recommended caloric intake is less than 2000 calories per day. The difference in calories between **one** Burger King Dutch Apple pie (a calories) and **one** Burger King Chili (c calories) is $a - c = 150$ calories. The Chili (c calories) and the Apple Pie (a calories) yield a mere $c + a$ or 530 calories.

Solve the system:

$$\begin{aligned} a - c &= 150 \\ c + a &= 530 \end{aligned}$$

to find the calories c in the Chili and a in the Dutch Apple pie.

Source: www.cspinet.org.

40. *Caloric intake* Suppose we order the Wendy's Ultimate Chicken Grill Sandwich, Salad with Low Fat Honey Mustard, and Iced Tea at u calories instead of the Classic Triple with Cheese, Great Biggie Fries, and Biggie Cola at a whopping c calories. The difference is $c - u$ or 1240 calories. Even if you eat **four** Wendy's Ultimate Chicken Grill Sandwiches the caloric difference is $4u - c$ or 290.

Solve the system:

$$\begin{aligned} c - u &= 1240 \\ 4u - c &= 290 \end{aligned}$$

to find the calories u in the Wendy's Ultimate Chicken Grill Sandwich, Salad with Low Fat Honey Mustard, and Iced Tea and calories c in the Classic Triple with Cheese, Great Biggie Fries, and Biggie Cola.

Source: www.cspinet.org.

Web IT go to mhhe.com/bello for more lessons

Using Your Knowledge

Tweedledee and Tweedledum Have you ever read *Alice in Wonderland?* Do you know who the author is? It's Lewis Carroll, of course. Although better known as the author of *Alice in Wonderland,* Lewis Carroll was also an accomplished mathematician and logician. Certain parts of his second book, *Through the Looking Glass,* reflect his interest in mathematics. In this book, one of the characters, Tweedledee, is talking to Tweedledum. Here is the conversation.

Tweedledee: The sum of your weight and twice mine is 361 pounds.

Tweedledum: Contrariwise, the sum of your weight and twice mine is 360 pounds.

41. If Tweedledee weighs x pounds and Tweedledum weighs y pounds, find their weights using the ideas of this section.

Write On

42. When solving a system of equations by elimination, how would you recognize if the pair of equations is:

a. consistent? **b.** inconsistent? **c.** dependent?

43. Explain why the system

$$2x + 5y = 9$$
$$3x + 2y = 8$$

is easier to solve by elimination rather than substitution.

44. Write the procedure you use to solve a system of equations by the elimination method.

Concept Checker

Fill in the blank(s) with the correct word(s), phrase, or mathematical statement.

45. When solving a system of equations using the **elimination** method, the idea is to **multiply** the equations by a number that will yield a(n) ____________ **system** in which the **coefficients** of the ***x*'s** or the ***y*'s** are **opposites.**

46. When solving a system of equations using the **elimination** method **the system** can be one of ____________ types.

one **two**
opposite **three**
equivalent

Mastery Test

Solve the systems; if the system is not consistent, state whether it is inconsistent or dependent.

47. $2x + 5y = 9$
$4x - 3y = 11$

48. $5x - 4y = 7$
$4x + 2y = 16$

49. $2x - 5y = 5$
$2x - y = 4 + x$

50. $3x - 2y = 6$
$-6x = -4y - 12$

51. $\frac{x}{6} - \frac{y}{2} = 1$
$-\frac{x}{4} + \frac{3y}{4} = -\frac{3}{4}$

52. A 10-pound bag of coffee sells for \$114 and contains a mixture of coffee A, which costs \$12 a pound, and coffee B, which costs \$10 a pound. How many pounds of each of the coffees does the bag contain?

Skill Checker

Write an expression corresponding to the given sentence.

53. The sum of the numbers of nickels (n) and dimes (d) equals 300.

54. The difference of h and w is 922.

55. The product of 4 and $(x - y)$ is 48.

56. The quotient of x and y is 80.

57. The number m is 3 less than the number n.

58. The number m is 5 more than the number n.

7.4 Coin, General, Motion, and Investment Problems

Objectives

Solve word problems:

A Involving coins.

B Of a general nature.

C Using the distance formula $D = RT$.

D Involving the interest formula $I = PR$.

To Succeed, Review How To . . .

1. Use the RSTUV method to solve word problems (pp. 148–149).
2. Solve a system of two equations with two unknowns (pp. 570–573, 588–591, 597–598).

Getting Started

Money Problems

Patty's upset; she needs *help!* Why? Because *she* hasn't learned about systems of equations, but we have, so we can help her! In the preceding sections we studied systems of equations. We now use that knowledge to solve word problems involving two variables.

Before we tackle Patty's problem, let's get down to nickels, dimes, and quarters! Suppose you are down to your last nickel: You have 5¢.

	Follow the pattern:
	$5 \cdot 1$
If you have 2 nickels, you have $5 \cdot 2 = 10$ cents.	$5 \cdot 2$
If you have 3 nickels, you have $5 \cdot 3 = 15$ cents.	$5 \cdot 3$
If you have n nickels, you have $5 \cdot n = 5n$ cents.	$5 \cdot n$

The same thing can be done with dimes.

	Follow the pattern:
If you have 1 dime, you have $10 \cdot 1 = 10$ cents.	$10 \cdot 1$
If you have 2 dimes, you have $10 \cdot 2 = 20$ cents.	$10 \cdot 2$
If you have n dimes, you have $10 \cdot n = 10n$ cents.	$10 \cdot n$

We can construct a table that will help us summarize the information:

	Value (cents)	×	How Many	=	Total Value
Nickels	5		n		$5n$
Dimes	10		d		$10d$
Quarters	25		q		$25q$
Half-dollars	50		h		$50h$

In this section we shall use information like this and systems of equations to solve word problems.

A > Solving Coin and Money Problems

Now we are ready to help poor Patty! As usual, we use the RSTUV method. If you've forgotten how that goes, this is a good time to review it (see p. 149).

EXAMPLE 1 Patty's coin problem

Read the cartoon in the *Getting Started* again for the details of Patty's problem.

SOLUTION 1

1. Read the problem. Patty is asked how many dimes and quarters the man has.

2. Select the unknowns. Let d be the number of dimes the man has and q the number of quarters.

3. Think of a plan. We translate each of the sentences in the cartoon:

a. "A man has 20 coins consisting of dimes and quarters."

$$20 = d + q$$

b. The next sentence seems hard to translate. So, let's look at the easy part first—how much money he has now. Since he has d dimes, the table in the *Getting Started* tells us that he has $10d$ (cents). He also has q quarters, which are worth $25q$ (cents). Thus, he has

$$(10d + 25q) \text{ cents}$$

What would happen if the dimes were quarters and the quarters were dimes? We simply would change the amount the coins are worth, and he would have

$$(25d + 10q) \text{ cents}$$

Now let's translate the sentence:

$$\underbrace{25d + 10q}_{\text{If the dimes were quarters and the quarters were dimes,}} \underbrace{=}_{\text{he would have}} \underbrace{(10d + 25q) + 90}_{90¢ \text{ more than he has now.}}$$

If we put the information from parts **a** and **b** together, we have the following system of equations:

$$\begin{aligned} d + q &= 20 \\ 25d + 10q &= 10d + 25q + 90 \end{aligned}$$

Now we need to write this system in standard form—that is, with all the variables and constants in the proper columns. We do this by subtracting $10d$ and $25q$ from both sides of the second equation to obtain

$$\begin{aligned} d + q &= 20 \\ 15d - 15q &= 90 \end{aligned}$$

We then divide each term in the second equation by 15 to get

$$\begin{aligned} d + q &= 20 \\ d - q &= 6 \end{aligned}$$

PROBLEM 1

Pedro has 20 coins consisting of pennies and nickels. If the pennies were nickels and the nickels were pennies, he would have 16 cents less. How many pennies and how many nickels does Pedro have?

Answers to PROBLEMS

1. 8 pennies, 12 nickels

4. Use the elimination method to solve the problem. To eliminate q, we simply add the two equations:

$$\begin{aligned} d + q &= 20 \\ d - q &= 6 \\ \hline 2d \quad &= 26 && \text{Add.} \\ \text{Number of dimes:} \quad d &= 13 && \text{Divide by 2.} \\ 13 + q &= 20 && \text{Substitute 13 for } d \text{ in } d + q = 20. \\ \text{Number of quarters:} \quad q &= 7 \end{aligned}$$

Thus, the man has 13 dimes ($1.30) and 7 quarters ($1.75), a total of $3.05.

5. Verify the solution. If the dimes were quarters and the quarters were dimes, the man would have 13 quarters ($3.25) and 7 dimes ($0.70), a total of $3.95, which is indeed $0.90 more than the $3.05 he now has. Patty, you got your help!

Let's solve another coin problem.

EXAMPLE 2 Jack's coin problem

Jack has $3 in nickels and dimes. He has twice as many nickels as he has dimes. How many nickels and how many dimes does he have?

SOLUTION 2

As usual, we use the RSTUV method.

1. Read the problem. We are asked to find the numbers of nickels and dimes.

2. Select the unknowns. Let n be the number of nickels and d the number of dimes.

3. Think of a plan. If we translate the problem and use the table in the *Getting Started,* Jack has $3 (300 cents) in nickels and dimes:

$$300 = 5n + 10d$$

He has twice as many nickels as he has dimes:

$$n = 2d$$

We then have the system

$$\begin{aligned} 5n + 10d &= 300 \\ n &= 2d \end{aligned}$$

4. Use the substitution method to solve the problem. This time it's easy to use the substitution method.

$$5n + 10d = 300 \xrightarrow{\text{Letting } n = 2d.} 5(2d) + 10d = 300$$

$$\begin{aligned} &\text{Simplify.} & 10d + 10d &= 300 \\ &\text{Combine like terms.} & 20d &= 300 \\ &\text{Divide by 20.} & d &= 15 \\ &\text{Substitute 15 for } d \text{ in } n = 2d. & n = 2(15) &= 30 \end{aligned}$$

Thus, Jack has 15 dimes ($1.50) and 30 nickels ($1.50).

5. Verify the solution. Since Jack has $3 ($1.50 + $1.50) and he does have twice as many nickels as dimes, the answer is correct.

PROBLEM 2

Jill has $1.50 in nickels and dimes. She has twice as many dimes as she has nickels. How many nickels and how many dimes does she have?

Answers to PROBLEMS

2. 6 nickels, 12 dimes

B > Solving General Problems

We can use systems of equations to solve many problems. Here is an interesting one.

EXAMPLE 3 A heavy marriage

The greatest weight difference recorded for a married couple is 922 pounds (Mills Darden of North Carolina and his wife Mary). Their combined weight is 1118 pounds. What is the weight of each of the Dardens? (He is the heavy one.)

SOLUTION 3

1. Read the problem. We are asked to find the weight of each of the Dardens.

2. Select the unknowns. Let h be the weight of Mills and w be the weight of Mary.

3. Think of a plan. We translate the problem. The weight difference is 922 pounds:

$$h - w - 922$$

Their combined weight is 1118 pounds:

$$h + w = 1118$$

We then have the system

$$h - w = 922$$
$$h + w = 1118$$

4. Use the elimination method to solve the problem. Using the elimination method, we have

$$h - w = 922$$
$$\underline{h + w = 1118}$$
$$2h = 2040 \quad \text{Add.}$$
$$h = 1020 \quad \text{Divide by 2.}$$
$$1020 + w = 1118 \quad \text{Substitute 1020 for } h \text{ in } h + w = 1118.$$
$$w = 98 \quad \text{Subtract 1020.}$$

Thus, Mary weighs 98 pounds and Mills weighs 1020 pounds.

5. Verify the solution. $h - w$ becomes $1020 - 98 = 922$, which is true. Moreover, $1020 + 98 = 1118$ which is also true. Verify this in the *Guinness Book of World Records*.

PROBLEM 3

When a couple of astronauts stood on a scale together before a mission, their combined weight was 320 pounds. The difference in their weights was 60 pounds, and the woman was lighter than the man. What was the weight of each?

C > Solving Motion Problems

Remember the motion problems we solved in Section 2.5? They can also be done using two variables. The procedure is about the same. We write the given information in a chart labeled $R \times T = D$ and then use our RSTUV method, as demonstrated in Example 4.

EXAMPLE 4 Currents and boating

The world's strongest current is the Saltstraumen in Norway. The current is so strong that a boat that travels 48 miles downstream (with the current) in 1 hour takes 4 hours to go the same 48 miles upstream (against the current). How fast is the current flowing?

SOLUTION 4

1. Read the problem. We are asked to find the speed of the current. Note that the speed of the boat downstream has two components: the boat speed and the current speed.

PROBLEM 4

A plane travels 800 miles against a storm in 4 hours. Giving up, the pilot turns around and flies back 800 miles to the airport in only 2 hours with the aid of the tailwind. Find the speed of the wind and the speed of the plane in still air.

Answers to PROBLEMS

3. Man: 190 lb; woman: 130 lb

4. Plane speed: 300 mi/hr; wind speed: 100 mi/hr

2. Select the unknowns. Let x be the speed of the boat in still water and y be the speed of the current. Then $(x + y)$ is the speed of the boat going downstream; $(x - y)$ is the speed of the boat going upstream.

3. Think of a plan. We enter this information in a chart:

	R	$\times$	T	$=$	D	
Downstream:	$x + y$		1		48	→ $x + y = 48$
Upstream:	$x - y$		4		48	→ $4(x - y) = 48$

4. Use the elimination method to solve the problem. Our system of equations can be simplified as follows:

$$x + y = 48 \xrightarrow{\text{Leave as is.}} x + y = 48$$
$$4(x - y) = 48 \xrightarrow{\text{Divide by 4.}} x - y = 12$$

Add. $2x = 60$

Divide by 2. $x = 30$

Substitute 30 for x in $x + y = 48$. $30 + y = 48$

Subtract 30. $y = 18$

Thus, the speed of the boat in still water is $x = 30$ miles per hour, and the speed of the current is 18 miles per hour.

5. Verify the solution. We leave the verification to you.

D › Solving Investment Problems

The investment problems we solved in Section 2.5 can also be worked using two variables. These problems use the formula $I = PR$ to find the *annual* interest I on a principal P at a rate R. The procedure is similar to that used to solve distance problems and uses the same strategy: use a table to enter the information, obtain a system of two equations and two unknowns, and solve the system. We show this strategy next.

EXAMPLE 5 Clayton's credit card problem

Clayton owes a total of \$10,900 on two credit cards with annual interest rates of 12% and 18%, respectively. If he pays a total of \$1590 in interest for the year, how much does he owe on each card?

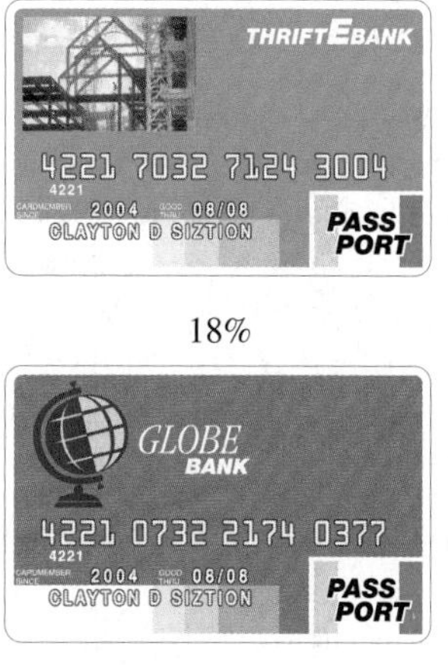

SOLUTION 5

1. Read the problem. We are asked to find the amount owed on *each* card.

2. Select the unknowns. Let x be the amount Clayton owes on the first card and y be the amount he owes on the second card.

3. Think of a plan. We make a table similar to the one in Example 4 but using the heading $P \times R = I$.

	P	$\times$	R	$=$	I
Card A	x		0.12		$0.12x$
Card B	y		0.18		$0.18y$

Since the total amount owed is \$10,900: $x + y = 10,900$

Since the total interest paid is \$1590: $0.12x + 0.18y = 1590$

(continued)

PROBLEM 5

Dorothy owes \$11,000 on two credit cards with annual interest rates of 9% and 12%. If she pays \$1140 in interest for the year, how much does she owe on each card?

Answers to PROBLEMS

5. \$6000 and \$5000, respectively

Thus, we have to solve the system

$$x + y = 10{,}900$$
$$0.12x + 0.18y = 1590$$

4. Use the substitution method to solve the problem. We solve for x in the first equation to obtain $x = 10{,}900 - y$. Now we substitute $x = 10{,}900 - y$ in

$$0.12x + 0.18y = 1590$$
$$0.12(10{,}900 - y) + 0.18y = 1590$$
$$1308 - 0.12y + 0.18y = 1590 \quad \text{Simplify.}$$
$$0.06y = 282 \quad \text{Subtract 1308 and combine } y\text{'s.}$$
$$y = 4700 \quad \text{Divide by 0.06.}$$

Since

$$x = 10{,}900 - y$$
$$x = 10{,}900 - 4700$$
$$x = 6200$$

Thus, Clayton owes \$6200 on card A and \$4700 on card B.

5. Verify the solution. Since $0.12 \cdot \$6200 + 0.18 \cdot \$4700 = \$744 + \$846 = \$1590$ (the amount Clayton paid in interest), the amounts of \$6200 and \$4700 are correct.

Which causes more CO_2 emissions: driving a car that gets 22 mpg for 20,000 miles a year or using \$200 worth of electricity each month? It depends on your location, but we use the Florida data from the American Solar Energy Society Calculator in Example 6.

EXAMPLE 6 Driving and electricity CO_2 emissions

The sum of the CO_2 emissions from driving (D) and electricity use (E) amounts to 50,000 pounds of CO_2. The difference in emissions between driving (D) and electricity (E) is 14,000 pounds. How many pounds of CO_2 are produced by driving D and electricity use E?

SOLUTION 6

1. **Read the problem.** We have to find the emissions from driving and from electricity.
2. **Select the unknowns.** Let D be the emissions from driving and E the emissions from the electricity.
3. **Think of a plan.** We translate the problem.

   The sum of the emissions $D + E$ is 50,000: $D + E = 50{,}000$

   The difference between D and E is 14,000: $D - E = 14{,}000$

   We then have the system: $D + E = 50{,}000$

   $$D - E = 14{,}000$$

PROBLEM 6

In Boston, the sum of the CO_2 emissions from driving a car that gets 22 mpg and is driven 20,000 miles a year and electricity consumption of \$200 a month is also 50,000 pounds but the difference in emissions between driving and electricity is only 4000 pounds. How many pounds of CO_2 are produced by driving and electricity use? This time you only need to plant 55 trees to offset the driving and energy consumption.

Source: http://tinyurl.com/y8bnn46.

Answers to PROBLEMS

6. 27,000 pounds are produced by driving and 23,000 by electricity use.

4. Use the elimination method to solve the problem.

$$\begin{aligned} D + E &= 50{,}000 \\ D - E &= 14{,}000 \\ \hline 2D &= 64{,}000 && \text{Add.} \\ D &= 32{,}000 && \text{Divide by 2.} \\ 32{,}000 + E &= 50{,}000 && \text{Substitute } D = 32{,}000 \text{ in } D + E = 50{,}000. \\ E &= 18{,}000 && \text{Subtract 32,000 from both sides.} \end{aligned}$$

Thus, 32,000 pounds of CO_2 are produced by driving and 18,000 by electricity use.

5. Verify the solution. $D + E = 32{,}000 + 18{,}000 = 50{,}000$ is true and $D - E = 32{,}000 - 18{,}000 = 14{,}000$ is also true.

By the way, the calculator says that to offset these emissions, you need to plant 80 trees a year!

Before you attempt to solve the Exercises, practice translating!

TRANSLATE THIS

The third step in the RSTUV procedure is to **TRANSLATE** the information into an equation. In Problems 1–10 **TRANSLATE** the sentence and match the correct translation with one of the equations A–O.

1. If you have C cents consisting of n nickels and d dimes, what is the equation relating C, n, and d?
2. If you have C cents consisting of d dimes and q quarters, what is the equation relating C, d, and q?
3. A person is h inches tall and another person is i inches tall. If the difference in their heights is 10 inches and the sum of their heights is 110 inches, what system of equations represents these facts?
4. If s is the speed of a boat in still water and c is the speed of the current, find an equation representing the distance D traveled downstream in T hours.
5. If s is the speed of a boat in still water and c is the speed of the current, find an equation representing the distance D traveled upstream in T hours.

A. $h - i = 10$; $h + i = 110$
B. $D = (s - c)T$
C. $C = 10n + 5d$
D. $C = 10d + 25q$
E. $C = 25d + 10q$
F. $I = 0.10x + 0.08(10{,}000 - x)$
G. $m = RT$
H. $I = 0.10x + 0.08y$
I. $I = 0.10x + 0.08(x - 10{,}000)$
J. $m = (R + W)T$
K. $C = 5n + 10d$
L. $10 - h = i$; $h + i = 110$
M. $m = (W - R)T$
N. $D = (s + c)T$
O. $m = (R - W)T$

6. A person invests x dollars at 10% and y dollars at 8%. If the interest from both investments is I, find an equation for I.
7. A person invests \$10,000, x dollars at 10% and the rest at 8%. Write an equation for the interest I earned on the \$10,000.
8. A plane travels m miles in T hours flying at a rate of R miles per hour. Write an equation relating m, T, and R.
9. A plane flies for T hours at R miles per hour with a tail wind of W miles per hour. If the plane travels m miles, write an equation relating T, R, W, and m.
10. A plane flies for T hours at R miles per hour with a head wind of W miles per hour. If the plane travels m miles, write an equation relating T, R, W, and m.

> Exercises 7.4

> Practice Problems > Self-Tests
> Media-rich eBooks > e-Professors > Videos

> Web IT go to mhhe.com/bello for more lessons

⟨A⟩ Solving Coin and Money Problems
⟨B⟩ Solving General Problems

In Problems 1–6, solve the money problems.

1. Mida has \$2.25 in nickels and dimes. She has four times as many dimes as nickels. How many dimes and how many nickels does she have?

2. Dora has \$5.50 in nickels and quarters. She has twice as many quarters as she has nickels. How many of each coin does she have?

3. Mongo has 20 coins consisting of nickels and dimes. If the nickels were dimes and the dimes were nickels, he would have 50¢ more than he now has. How many nickels and how many dimes does he have?

4. Desi has 10 coins consisting of pennies and nickels. Strangely enough, if the nickels were pennies and the pennies were nickels, she would have the same amount of money as she now has. How many pennies and nickels does she have?

5. Don had \$26 in his pocket. If he had only \$1 bills and \$5 bills, and he had a total of 10 bills, how many of each of the bills did he have?

6. A person went to the bank to deposit \$300. The money was in \$10 and \$20 bills, 25 bills in all. How many of each did the person have?

In Problems 7–14, find the solution.

7. The sum of two numbers is 102. Their difference is 16. What are the numbers?

8. The difference between two numbers is 28. Their sum is 82. What are the numbers?

9. The sum of two integers is 126. If one of the integers is 5 times the other, what are the integers?

10. The difference between two integers is 245. If one of the integers is 8 times the other, find the integers.

11. The difference between two numbers is 16. One of the numbers exceeds the other by 4. What are the numbers?

12. The sum of two numbers is 116. One of the numbers is 50 less than the other. What are the numbers?

13. Longs Peak is 145 feet higher than Pikes Peak. If you were to put these two peaks on top of each other, you would still be 637 feet short of reaching the elevation of Mount Everest, 29,002 feet. Find the elevations of Longs Peak and Pikes Peak.

14. Two brothers had a total of \$7500 in separate bank accounts. One of the brothers complained, and the other brother took \$250 and put it in the complaining brother's account. They now had the same amount of money! How much did each of the brothers have in the bank before the transfer?

⟨C⟩ Solving Motion Problems

In Problems 15–20, solve the motion problems.

15. A plane flying from city A to city B at 300 miles per hour arrives $\frac{1}{2}$ hour later than scheduled. If the plane had flown at 350 miles per hour, it would have made the scheduled time. How far apart are cities A and B?

16. A plane flies 540 miles with a tailwind in $2\frac{1}{4}$ hours. The plane makes the return trip against the same wind and takes 3 hours. Find the speed of the plane in still air and the speed of the wind.

17. A motorboat runs 45 miles downstream in $2\frac{1}{2}$ hours and 39 miles upstream in $3\frac{1}{4}$ hours. Find the speed of the boat in still water and the speed of the current.

18. A small plane travels 520 miles with the wind in 3 hours, 20 minutes ($3\frac{1}{3}$ hours), the same time that it takes to travel 460 miles against the wind. What is the plane's speed in still air?

19. If Bill drives from his home to his office at 40 miles per hour, he arrives 5 minutes early. If he drives at 30 miles per hour, he arrives 5 minutes late. How far is it from his home to his office?

20. An unidentified plane approaching the U.S. coast is sighted on radar and determined to be 380 miles away and heading straight toward the coast at 600 miles per hour. Five minutes ($\frac{1}{12}$ hour) later, a U.S. jet, flying at 720 miles per hour, scrambles from the coastline to meet the plane. How far from the coast does the interceptor meet the plane?

⟨D⟩ Solving Investment Problems

In Problems 21–23, solve the investment problems.

21. Fred invested \$20,000, part at 6% and the rest at 8%. Find the amount invested at each rate if the annual income from the two investments is \$1500.

22. Maria invested \$25,000, part at 7.5% and the rest at 6%. If the annual interest from the two investments amounted to \$1620, how much money was invested at each rate?

23. Dominic has a savings account that pays 5% annual interest and some certificates of deposit that pay 7% annually. His total interest from the two investments is $1100 and the total amount invested is $18,000. How much money does he have in the savings account?

Applications

24. *Hurricane damages* Two of the costliest hurricanes in U.S. history were Andrew (1992) and Hugo (1989), in that order; together they caused $33.5 billion in damages. If the difference in damages caused by Andrew and Hugo was $19.5 billion, how much damage did each of them cause?

Source: University of Colorado Natural Hazards Center.

25. *Higher education enrollments* In a recent year, the total enrollment in public and private institutions of higher education was 15 million students. If there were 8.4 million more students enrolled in public institutions than in private, how many students were enrolled in public and how many were enrolled in private institutions?

Source: U.S. Department of Education.

26. *Education expenditures* In a recent year, the education expenditures for public and private institutions amounted to $187 billion. If $49 billion more was spent in public institutions than in private, what were the education expenditures for public and for private institutions?

Source: U.S. Department of Education.

27. *Premium cable services* The total number of subscribers for Home Box Office and Showtime in a recent year was 28,700,000. If Home Box Office had 7300 more subscribers than Showtime, how many subscribers did each of the services have?

Source: National Cable Television Association.

28. *Automobile and home accidents* In a recent year, the cost of motor vehicle and home accidents reached $241.7 billion. If motor vehicle accidents caused losses that were $85.1 billion more than those caused by home accidents, what were the losses in each category?

Source: National Safety Council Accident Facts.

Applications: Green Math

Hurricane Intensities and Damages Refer to the table for Problems 29–31. The table shows the four most intense hurricanes to hit the U.S. mainland.

Rank	Hurricane	Location	Year	Category	Damage (in billions)
1.	Katrina	La./Miss.	2005	3	K
2.	Andrew	Fla./La.	1992	5	A
3.	Charley	Fla.	2004	4	C
4.	Wilma	Fla.	2005	3	W

Hurricane Katrina

29. *Hurricane damage* The combined damage from Katrina K and Andrew A reached $122.5 billion. The difference in the damage they caused was $69.5 billion. Find the damage (in billions) caused by Katrina and Andrew.

30. *Hurricane damage* The most intense hurricane recorded in the Atlantic basin was Wilma in October 2005. The combined damage of Wilma and Charley amounted to $29.4 billion, but the dollar difference in damages was a mere $0.6 billion. Find the damage (in billions) caused by Wilma W and Charley C.

31. *Hurricane wind speeds* As you can see from the table, Katrina was a category 3 hurricane (winds 111–130 miles per hour) and Andrew a category 5 (winds of more than 155 miles per hour). At landfall, Andrew's winds were 25 miles per hour stronger than Katrina's and 10 miles per hour more than the 155 miles per hour needed to make it a category 5. Find the wind speeds in mph for Katrina and Andrew.

32. *Hurricane barometric pressure* Hurricane Katrina was stronger than Andrew at landfall. How can that be? Here is the explanation: Barometric pressure is the most accurate representation of a storm's power. The lower the barometric pressure, the more intense the storm is and Katrina's barometric pressure K was 0.12 inches lower than Andrew's A. As a matter of fact, Andrew's barometric pressure was only 0.06 inches above the 27.17 needed to make it a category 5 hurricane. Find the barometric pressure K for Katrina and A for Andrew.

Most Intense U.S. Hurricanes at Landfall

Hurricane	Year	Barometric Pressure (inches)	Wind Speed (mph)
"Labor Day" (Fla. Keys)	1935	26.35	160
Camille	1969	26.84	190
Katrina	2005	K	140
Andrew	1992	A	165

Source: National Weather Service.

Write On

33. Make up a problem involving coins, and write a solution for it using the RSTUV procedure.

34. Make up a problem whose solution involves the distance formula $D = RT$, and write a solution for it using the RSTUV procedure.

35. Make up a problem whose solution involves the interest formula $I = PR$, and write a solution for it using the RSTUV procedure.

36. The problems in Examples 1–4 have precisely the information you need to solve them. In real life, however, irrelevant information is often present. This type of information is called a **red herring.** Find some problems with red herrings and point them out.

Concept Checker

Fill in the blank(s) with the correct word(s), phrase, or mathematical statement.

37. If you have ***d*** **dimes** and ***q*** **quarters,** you have ______________ cents.

38. If you have ***n*** **nickels** and ***d*** **dimes,** you have ______________ cents.

39. The **distance *D*** traveled by an object moving at a **rate *R*** for a **time *T*** equals ______________.

40. If ***P*** is the **principal** and ***R*** the **rate,** the **interest *I*** is ______________.

RD | $5n + 10d$
RT | PR
$10n + 5d$ | IR
$10d + 25q$

Mastery Test

41. Jill has $2 in nickels and dimes. She has twice as many nickels as she has dimes. How many nickels and how many dimes does she have?

42. At birth, the Stimson twins weighed a total of 35 ounces. If their weight difference was 3 ounces, what was the weight of each of the twins?

43. A plane travels 1200 miles with a tailwind in 3 hours. It takes 4 hours to travel the same distance against the wind. Find the speed of the wind and the speed of the plane in still air.

44. Harper makes two investments totaling $10,000. The first investment pays 8% annually, and the second investment pays 5%. If the annual return from both investments is $600, how much has Harper invested at each rate?

45. *Cruise ship pollution* Have you been on a cruise? In 1 week a typical cruise ship generates 247,000 gallons of sewage and oily bilge water (mostly sewage). If the sewage exceeds the bilge water by 173,000 gallons, how many gallons of sewage and how many gallons of bilge water are generated by the typical cruise ship each week? There are at least 230 cruise ships in operation with 44 more coming soon!

Skill Checker

Graph the inequalities.

46. $x + 2y < 4$

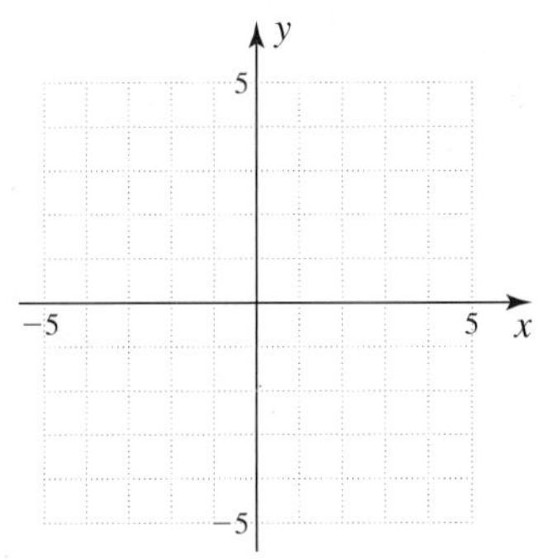

47. $x - 2y > 6$

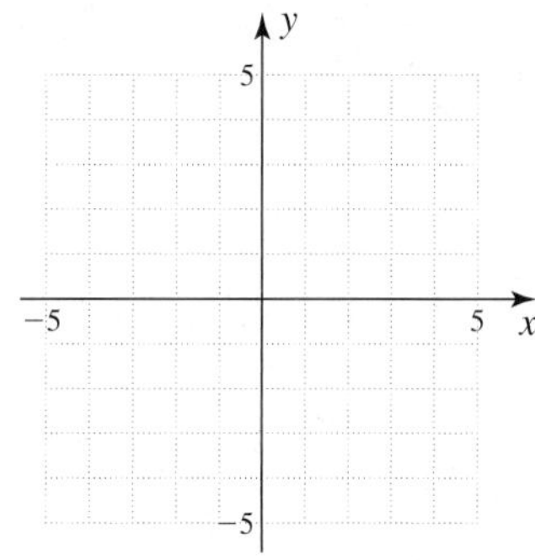

48. $x > y$

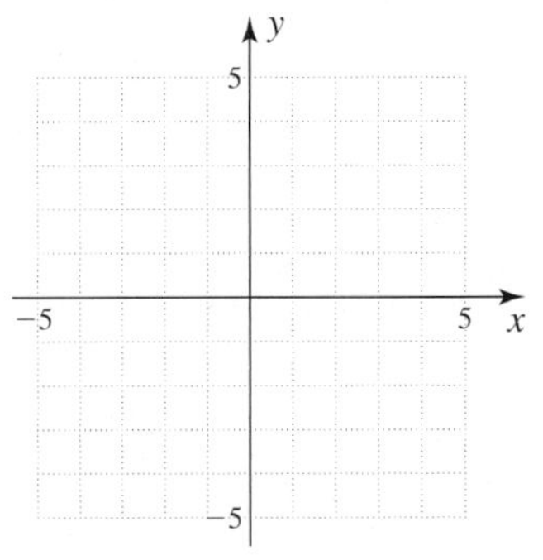

49. $x \leq 2y$

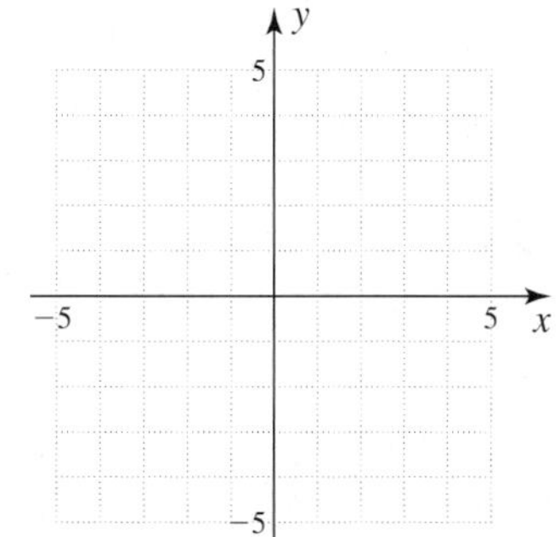

7.5 Systems of Linear Inequalities

Objective

A Solve a system of linear inequalities by graphing.

To Succeed, Review How To . . .

Graph a linear inequality (pp. 284–287).

Getting Started

Inequalities and Hospital Stays

According to the American Hospital Association, between 1980 and 2010 (inclusive), the average length of a hospital stay was $y = 7.74 - 0.09x$ (days), where x is the number of years after 1980, as shown in the graph to the right.

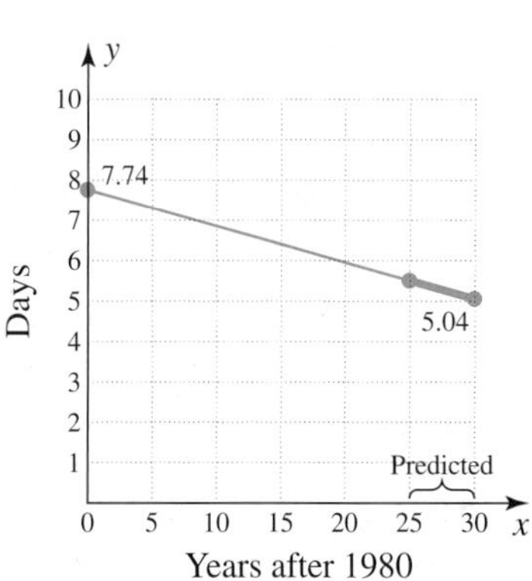

The graph of $y > 7.74 - 0.09x$ represents those longer-than-average stays. If we graph those longer-than-average stays *between* 1980 ($x = 0$) and 2010 ($x = 30$), we get the shaded area *above* the line $y = 7.74 - 0.09x$ and *between* $x = 0$ and $x = 30$. Since the line $y = 7.74 - 0.09x$ is *not* part of the graph, the line itself is shown dashed. The region satisfying the inequalities

$$y > 7.74 - 0.09x, \quad x > 0, \quad \text{and} \quad x < 30$$

is shown shaded in the graph to the right.

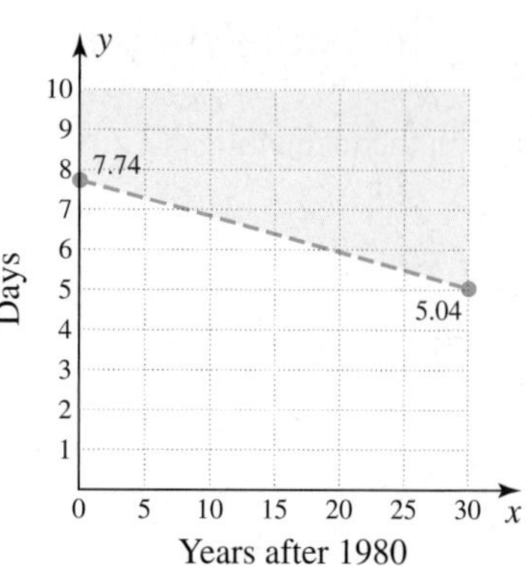

In this section we learn how to solve linear inequalities graphically by finding the set of points that satisfy *all* the inequalities in the system.

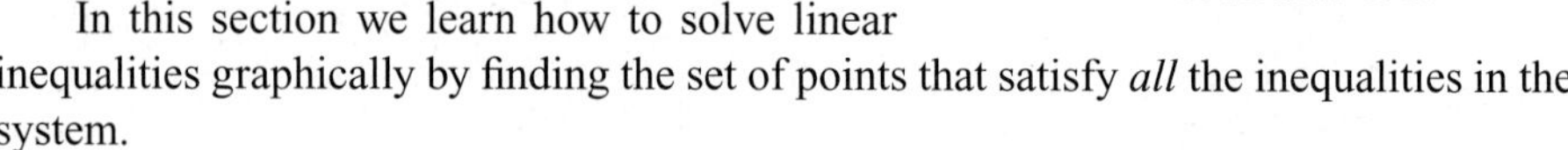

A > Solving a System of Linear Inequalities by Graphing

It turns out that we can use the procedure we studied in Section 3.7 to solve a system of linear inequalities.

PROCEDURE

Solving a System of Inequalities

Graph each inequality on the *same* set of axes using the following steps:

1. Graph the line that is the boundary of the region. If the inequality involves $\le$ or $\ge$, draw a **solid** line; if it involves $<$ or $>$, draw a **dashed** line.
2. Use any point (a, b) *not* on the line as a test point. Substitute the values of a and b for x and y in the inequality. If a *true* statement results, shade the side of the line containing the test point. If a *false* statement results, shade the other side.

The **solution set** is the set of points that satisfies *all* the inequalities in the system.

EXAMPLE 1 **Solving systems of inequalities involving horizontal and vertical lines by graphing**

Graph the solution of the system: $x \le 0$

$y \ge 2$

SOLUTION 1 Since $x = 0$ is a vertical line corresponding to the y-axis, $x \le 0$ consists of the graph of the line $x = 0$ and all points to the *left,* as shown in Figure 7.12. The condition $y \ge 2$ defines all points on the line $y = 2$ and *above,* as shown in Figure 7.13.

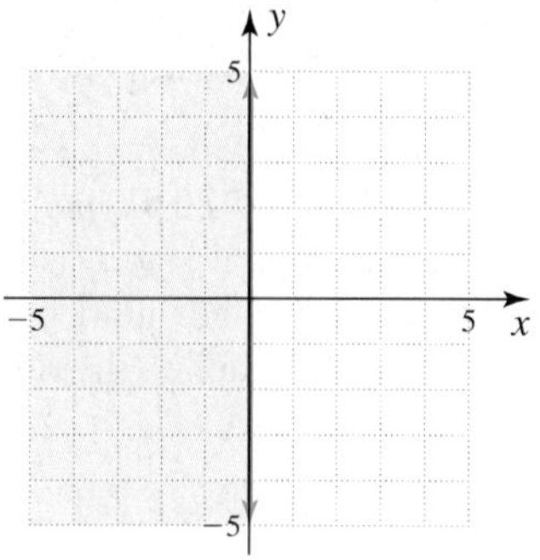

>Figure 7.12

PROBLEM 1

Graph the solution of the system:

$x \le 2$

$y \ge 0$

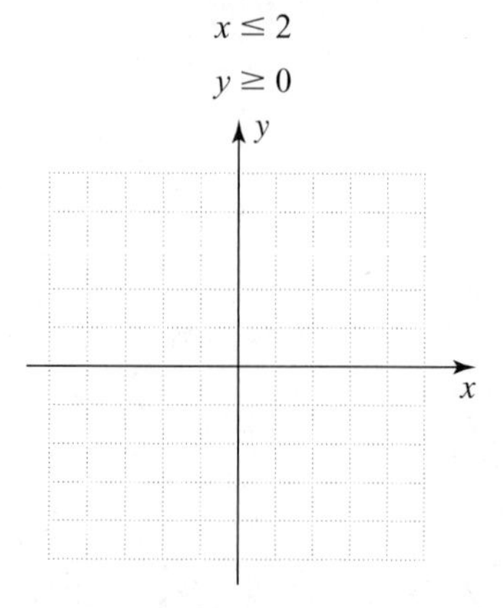

Answer on page 619

The solution set is the set satisfying *both* conditions, that is, where $x \le 0$ *and* $y \ge 2$. This set, the solution set, is the darker area in Figure 7.14.

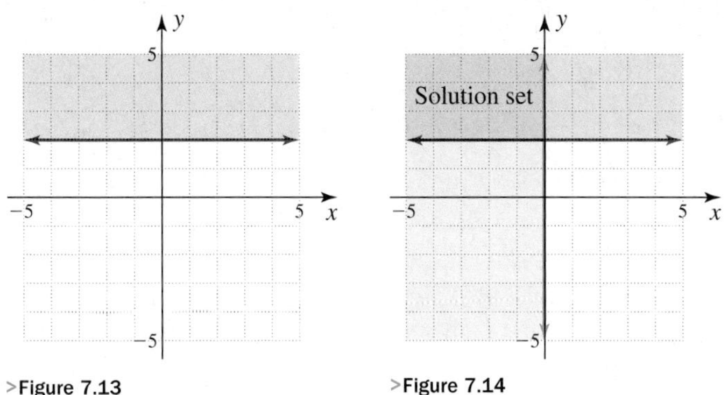

>Figure 7.13 >Figure 7.14

EXAMPLE 2 Solving systems of inequalities using a test point

Graph the solution of the system: $x + 2y \le 5$

$$x - y < 2$$

SOLUTION 2 First we graph the lines $x + 2y = 5$ and $x - y = 2$. Next we use a point (a test point), and check to see if the coordinates of the test point satisfy the inequalities. Using $(0, 0)$ as a test point, $x + 2y \le 5$ becomes $0 + 2 \cdot 0 \le 5$, a true statement. So we shade the region containing $(0, 0)$: the points *on or below* the line $x + 2y = 5$. (See Figure 7.15.) The inequality $x - y < 2$ is also satisfied by the test point $(0, 0)$, so we shade the points *above* the line $x - y = 2$. This line is drawn dashed to indicate that the points on it do *not* satisfy the inequality $x - y < 2$. (See Figure 7.16.) The solution set of the system is shown in Figure 7.17 by the darker region and the portion of the solid line forming one boundary of the region.

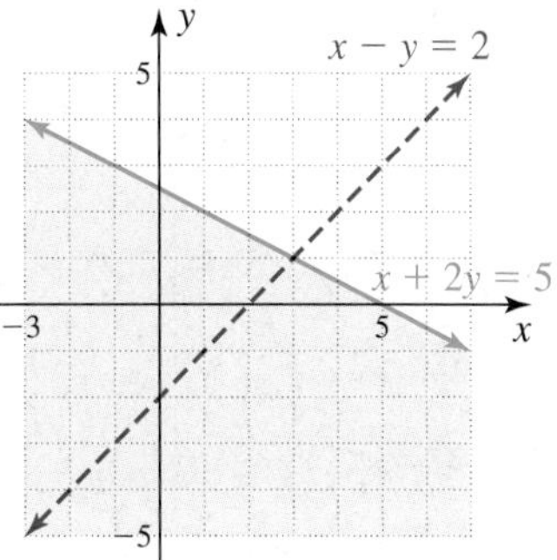

>Figure 7.15

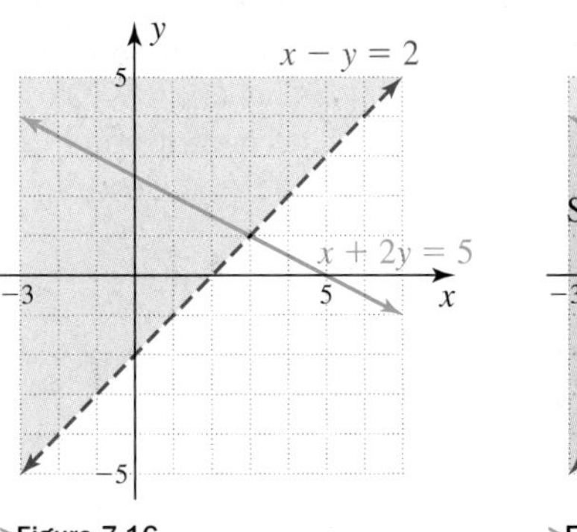

>Figure 7.16

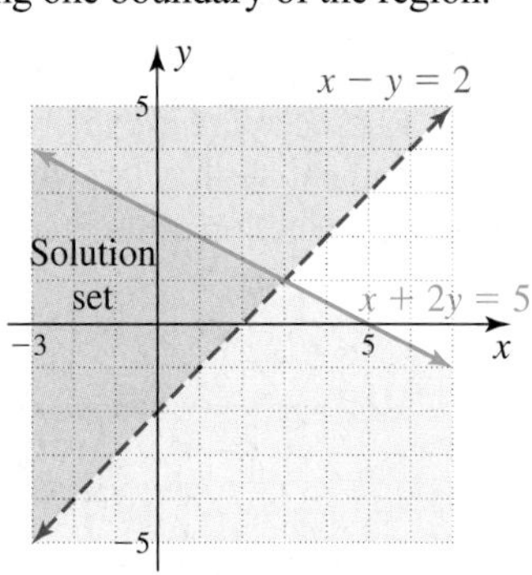

>Figure 7.17

PROBLEM 2

Graph the solution of the system:

$$2x + y \le 5$$
$$x - y < 3$$

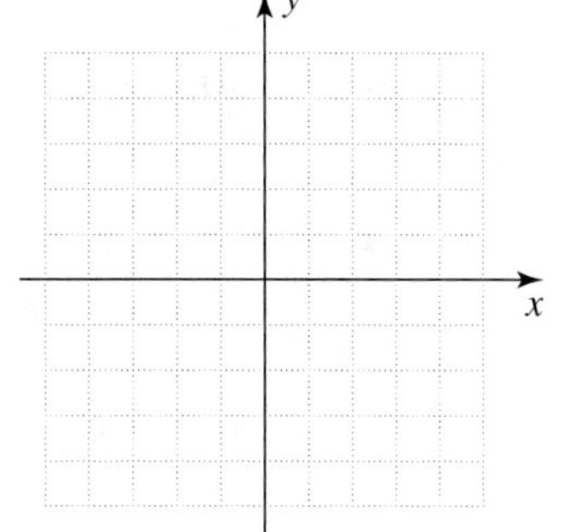

Answers to PROBLEMS

1.

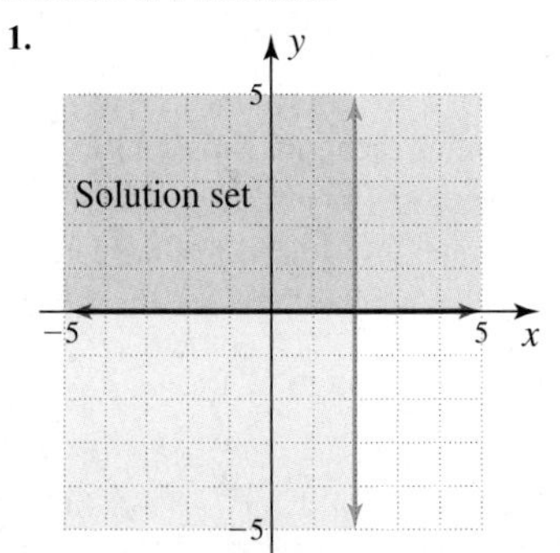

2.

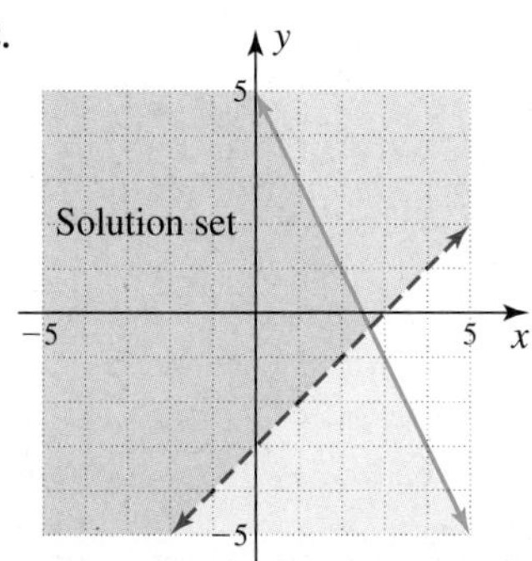

EXAMPLE 3 Solving systems of inequalities using a test point that does not satisfy both inequalities

Graph the solution of the system: $y + x \geq 2$
$y - x \leq 2$

SOLUTION 3 Graph the lines $y + x = 2$ and $y - x = 2$ and use (0, 0) as a test point. Since the test point (0, 0) does ***not*** satisfy the inequality $y + x \geq 2$, we shade the region that does *not* contain (0, 0): the points *on or above* the line $y + x = 2$. (See Figure 7.18.) The test point (0, 0) *does* satisfy the inequality $y - x \leq 2$, so we shade the points *on or below* the line $y - x = 2$. (See Figure 7.19.) The solution set of the system is the darker region in Figure 7.20 and includes parts of both lines.

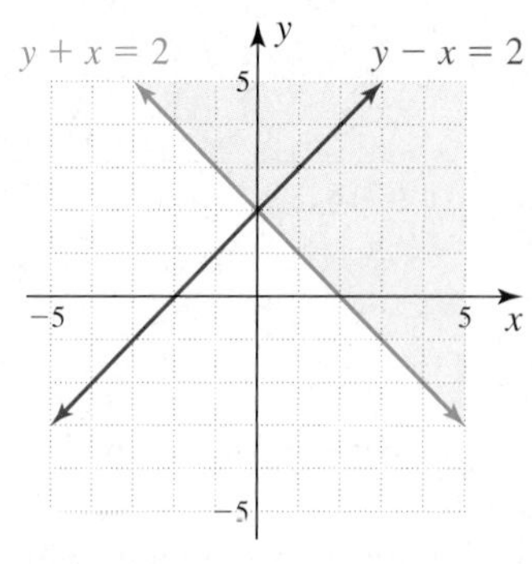

>Figure 7.18

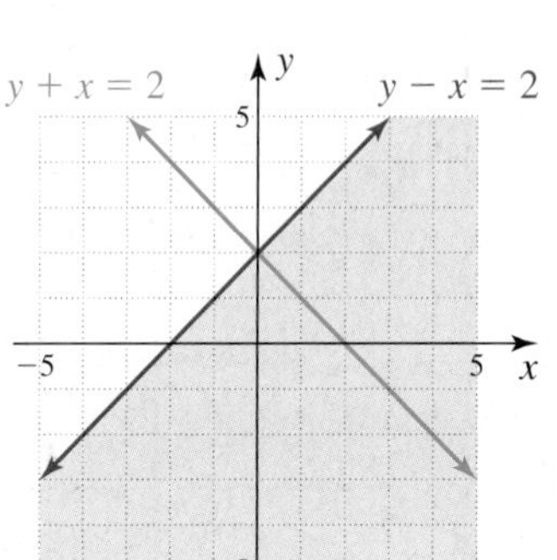

>Figure 7.19

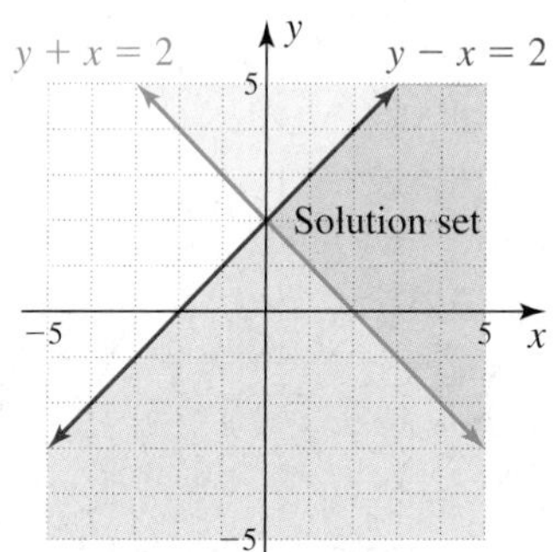

>Figure 7.20

PROBLEM 3

Graph the solution of the system:

$$y + x \leq 3$$
$$y - x \geq 3$$

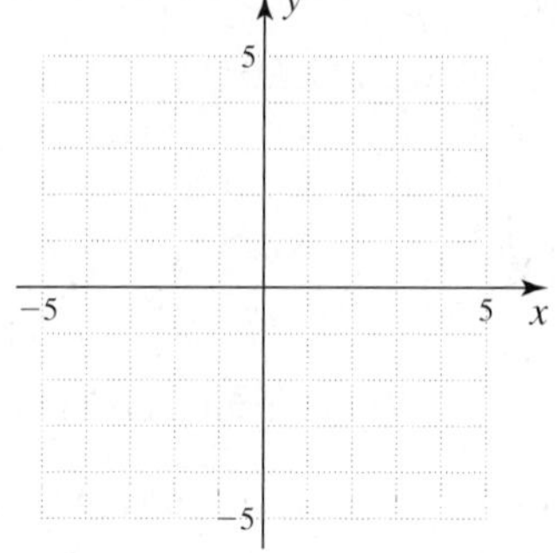

Answers to PROBLEMS

3.

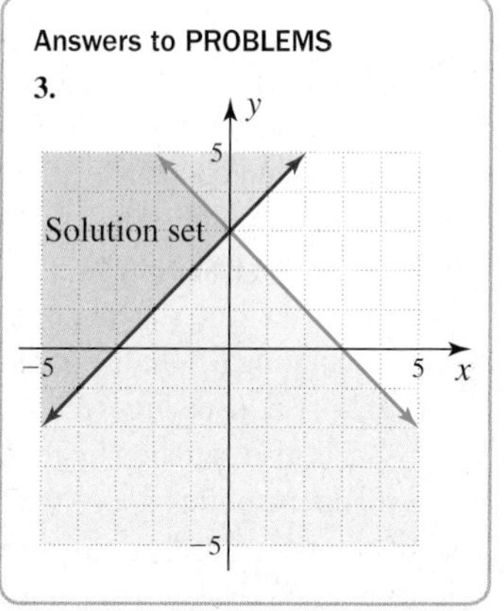

The United States is not the only polluting country in the world! One of the main CO_2 polluters is Russia. What are their future projections under the Kyoto Protocol, an international agreement setting targets for industrialized countries and the European communities in the reduction of greenhouse gases (GHG)? We shall see in Example 4.

GREEN MATH

EXAMPLE 4 Russian goals under the Kyoto Protocol

Under the worst-case scenario Russian emissions can be as high as $y = 60x + 1850$ and as low as $y = 30x + 1750$, where y is the number of megatons (millions of tons) of CO_2 emitted and x is the number of years after 2005. Graph the solution of the system

$$y \leq 60x + 1850$$
$$y \geq 30x + 1750$$
$$0 \leq x \leq 15$$

PROBLEM 4

a. In another possible scenario, Russian CO_2 emissions can be as high as 2400 megatons. Graph the line $\mathbf{y = 2400}$ between 2005 and 2020.

b. In this scenario, emissions can be as low as $\mathbf{y = 15x + 1700.}$ Graph $\mathbf{y \geq 15x + 1700,}$ where x is the number of years after 2005.

Answer on page 621

SOLUTION 4

One of the challenges when graphing real-world problems is to decide the grid for the graph! We know that x must be between 0 and 15 inclusive, but what about y? If we use $x = 15$ in $y \le 60x + 1850$ we obtain $y \le 60(15) + 1850 = 2750$. For $x = 15$ in $y \ge 30x + 1750$ we get $y \ge 30(15) + 1750 = 2200$, so we let y be between 1000 and 3000 (which includes both 2200 and 2750) at 200-unit intervals.

To graph $y = 60x + 1850$, let $x = 0$, obtaining $y = 1850$. Graph (0, 1850). Now, let $x = 15$ to obtain 2750. Graph the point (15, 2750). Join the points (0, 1850) and (15, 2750) with a red line, the graph of $y = 60x + 1850$. Now, use the point (0, 1000) as a test point. Because $0 \le 60(1000) + 1850$ is true, we shade the points *below and on the line* $y = 60x + 1850$. (We indicate this by using the red arrows in the graph.) We use a similar procedure to graph $y \ge 30x + 1750$.

Letting $x = 0$ and $y = 1750$, we graph (0, 1750).

Letting $x = 15$ and $y = 2200$, we graph (15, 2200). Join the points (0, 1750) and (15, 2200) with a blue line, the graph of $y = 30x + 1750$. Use the point (0, 1000) as a test point. Because $0 \ge 30(1000) + 1750$ is false, we shade *above and on the line* $y = 30x + 1750$. We also indicate this by using the blue arrows in the graph. The solution set of the system is the area between and including the boundaries of the two lines.

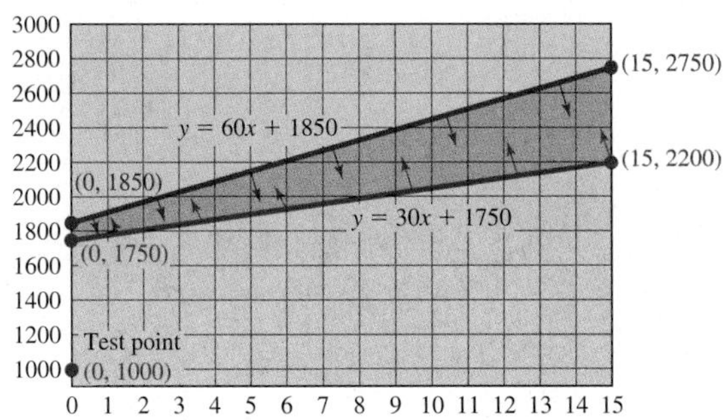

c. Use the inequalities in parts **a** and **b** to define in words the region that describes this scenario for Russia.

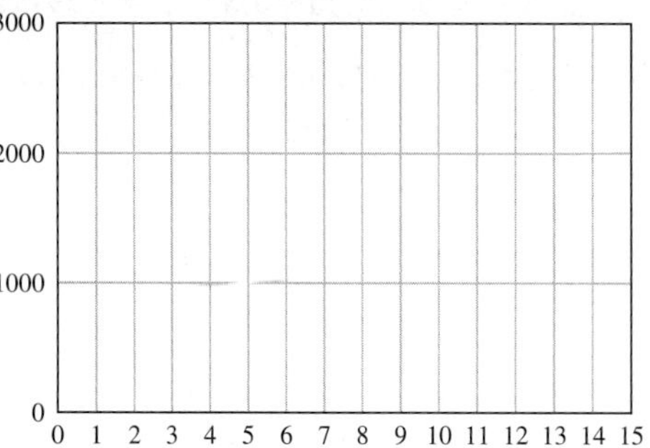

Answers to PROBLEMS

4. a. and **b.**

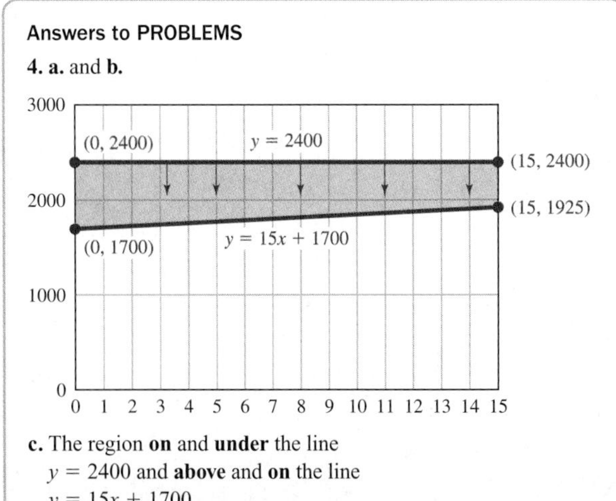

c. The region **on** and **under** the line $y = 2400$ and **above** and **on** the line $y = 15x + 1700$

> Practice Problems > Self-Tests
> Media-rich eBooks > e-Professors > Videos

› Exercises 7.5

› Web IT go to mhhe.com/bello for more lessons

‹A› Solving a System of Linear Inequalities by Graphing In Problems 1–13, graph the solution set of the system of inequalities.

1. $x \geq 0$ and $y \leq 2$

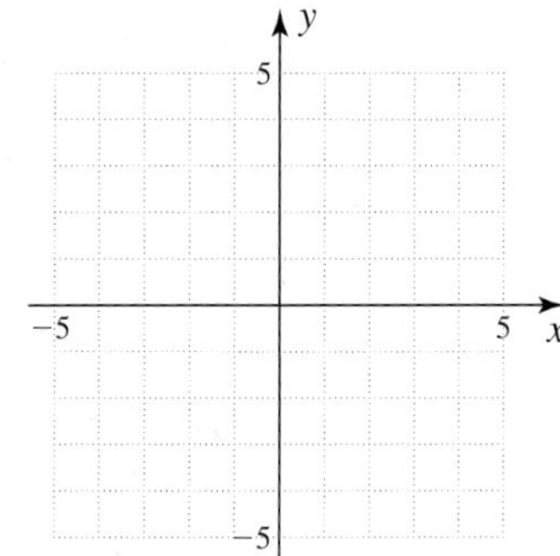

2. $x > 1$ and $y < 3$

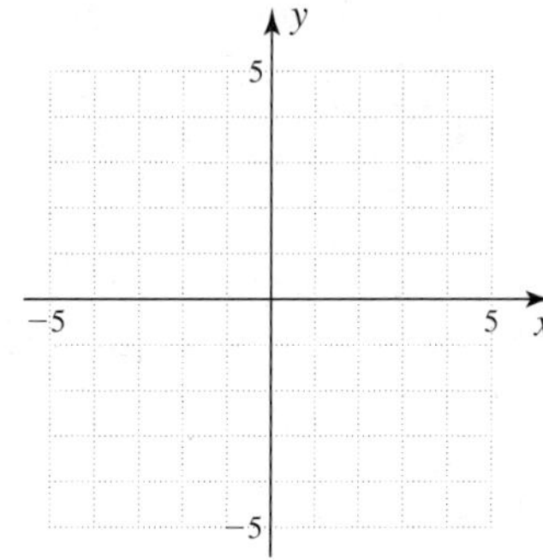

3. $x < -1$ and $y > -2$

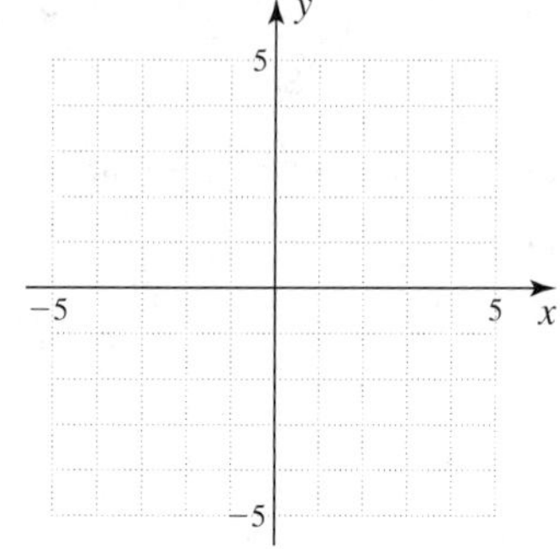

4. $x - y \geq 2$
$x + y \leq 6$

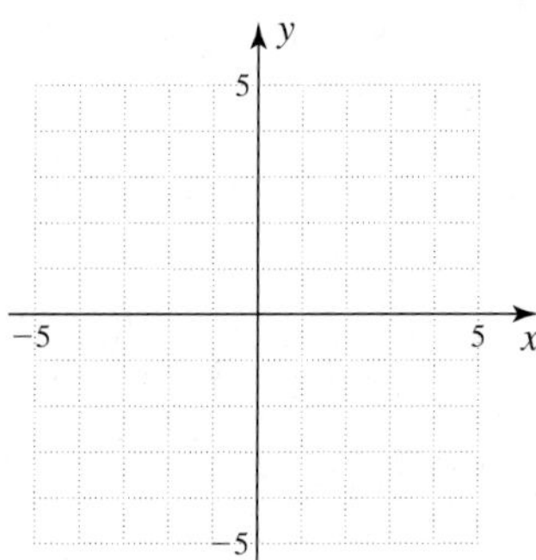

5. $x + 2y \leq 3$
$x < y$

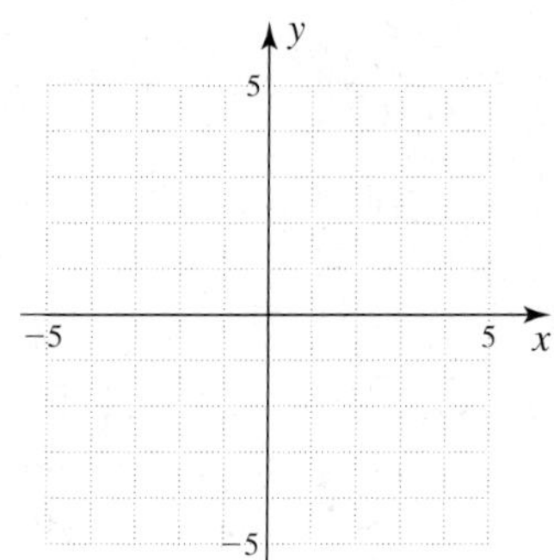

6. $5x - y > -1$
$-x + 2y \leq 6$

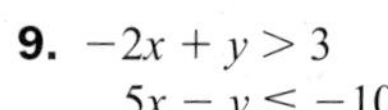

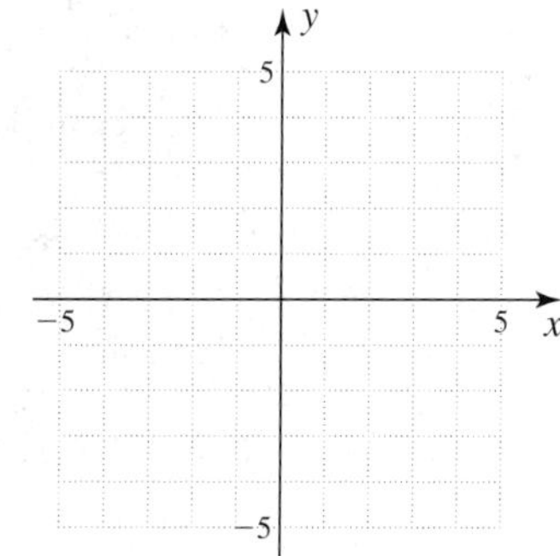

7. $4x - y > -1$
$-2x - y \leq -3$

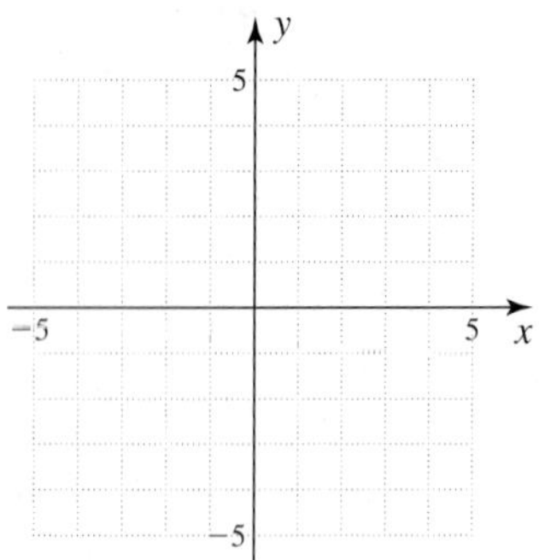

8. $2x - 3y < 6$
$4x - 3y > 12$

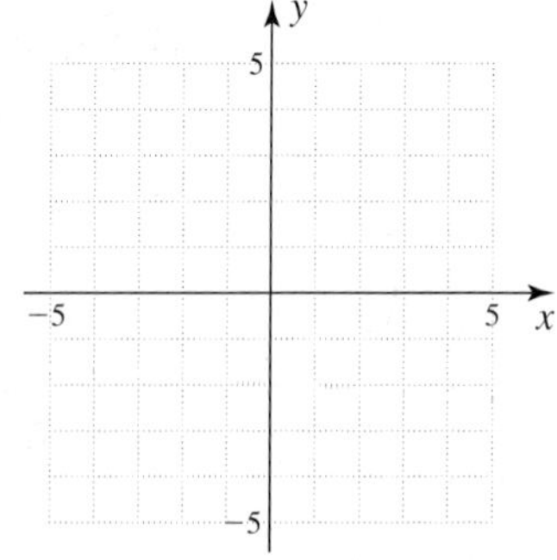

9. $-2x + y > 3$
$5x - y \leq -10$

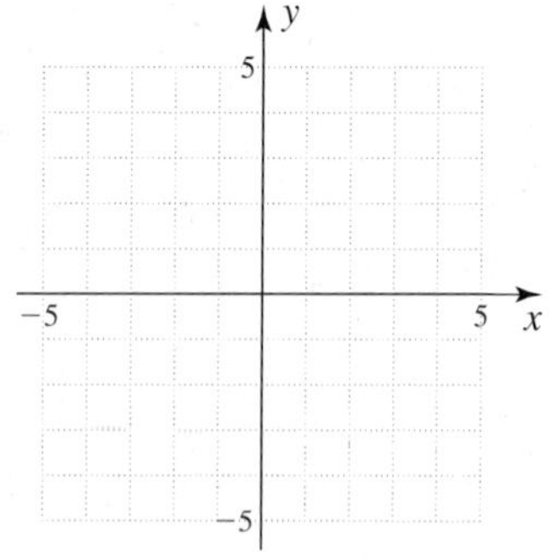

10. $2x - 5y \leq 10$
$3x + 2y < 6$

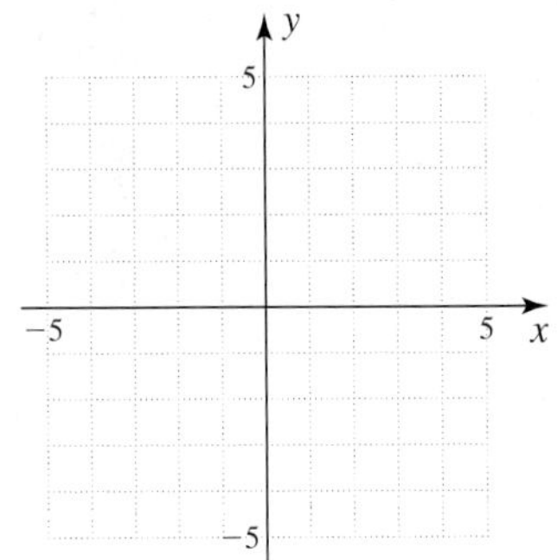

11. $2x - 3y < 5$
$x \geq y$

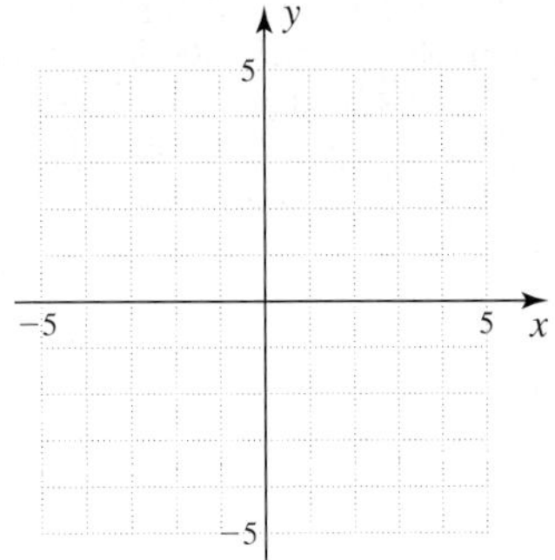

12. $x \leq 2y$
$x + y < 4$

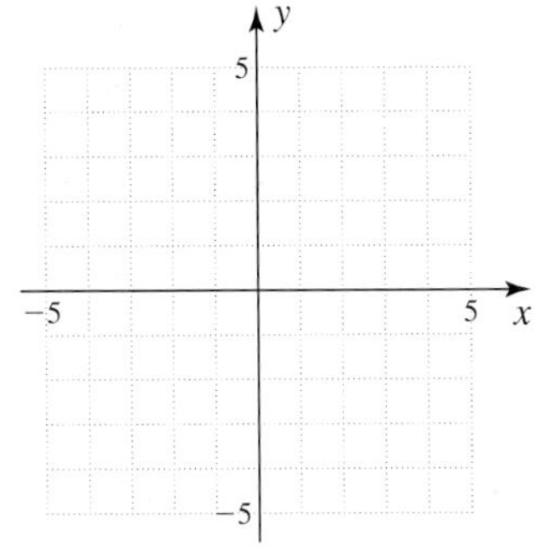

13. $x + 3y \leq 6$
$x > y$

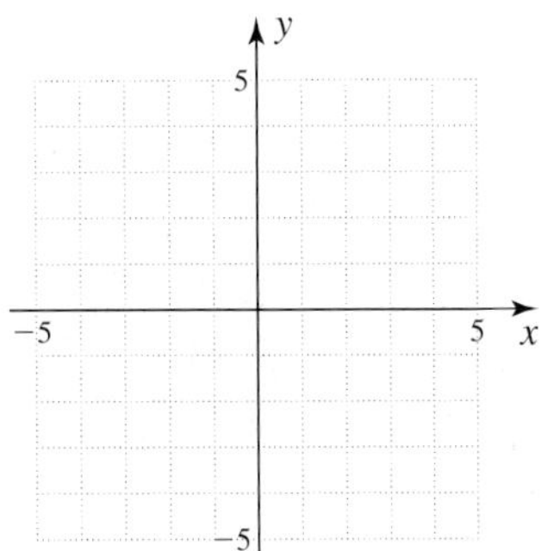

Web IT go to mhhe.com/bello for more lessons

Applications: Green Math

In Example 4 you graphed three inequalities representing a region modeling a possible scenario for the Russian goals regarding CO_2 emissions for the Kyoto Protocol. In Problems 14 and 15 we give you the regions and you are asked to find the inequalities corresponding to them.

14. *Kyoto Protocol scenario 2* Find the inequalities that define the yellow region, where x is the number of years after 2005 and y is the goal in megatons (millions of tons).

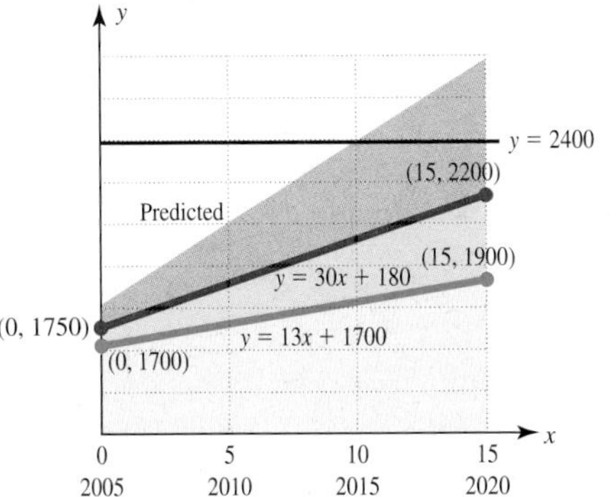

15. *Kyoto Protocol scenario 1* Find the inequalities that define the blue region, where x is the number of years after 2005 and y is the goal in megatons (millions of tons).

Applications

16. *Sugar and protein* A McDonald's Filet-O-Fish (FOF) has 8 grams of sugar, while a Burger King Tender Grilled Chicken (TGC) sandwich has 9 grams of sugar. You should limit your sugar intake to less than 40 grams per day. The RDA (Recommended Dietary Allowance) for protein is at least 50 grams per day: the FOF has 14 grams of protein and the TGC has 37 grams.

To satisfy your RDA's by eating x FOF and y TGC you have to satisfy the system:

$$8x + 9y < 40$$
$$14x + 37y \geq 50$$

Graph the system and give at least two ordered pairs that satisfy the inequalities.

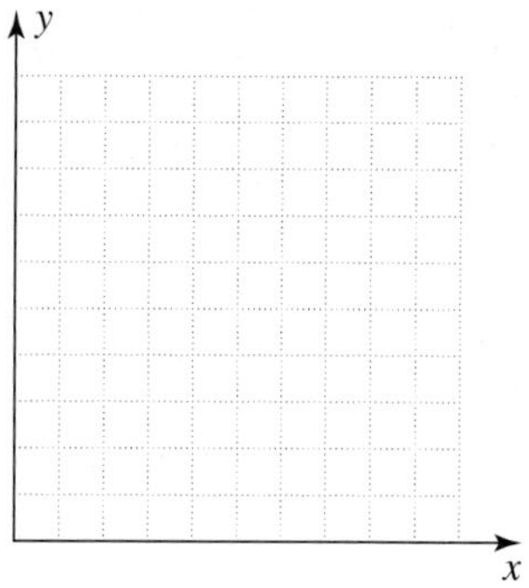

17. *Carbohydrates and protein* A McDonald's Filet-O-Fish (FOF) sandwich has 42 grams of carbohydrates and a Burger King Tender Grilled Chicken (TGC) sandwich has 53 grams of carbohydrates. Their protein contents are 14 grams and 37 grams, respectively. To satisfy the RDA (Recommended Dietary Allowance) of carbohydrates (< 300) and protein (≥ 50) when eating x FOF and y TGC you have to satisfy the system of inequalities:

$$42x + 53y < 300$$
$$14x + 37y \geq 50$$

Graph the system and give at least two ordered pairs that satisfy the inequalities.

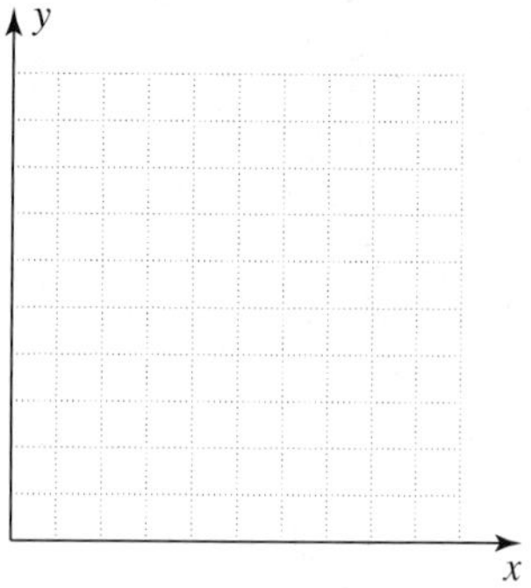

18. *Carbohydrates and sugar* A runner wants to maintain the RDA (Recommended Dietary Allowance) of carbs (< 300) by eating x McDonald's Filet-O-Fish (FOF) and y Burger King Tender Grilled Chicken (TGC). At the same time sugar levels must be kept at more than the minimum RDA of 40 grams. A system of equations satisfying these two conditions is:

$$42x + 53y < 300$$
$$8x + 9y \geq 40$$

Graph the system and give at least two ordered pairs that satisfy the inequalities.

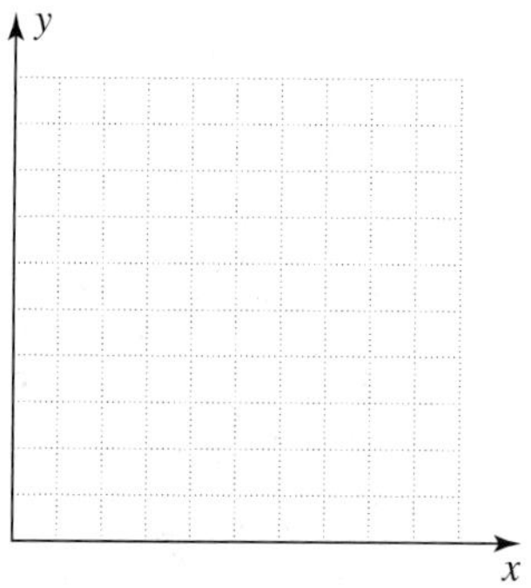

Using Your Knowledge

Inequalities in Exercise The target zone used to gauge your effort when performing aerobic exercises is determined by your pulse rate p and your age a. In this *Using Your Knowledge* the target zone is the solution set of the inequalities in Problems 19–21. What is your target zone? You have to do Problems 19–21 to find out! Graph the given inequalities on the set of axes provided.

19. $p \geq -\frac{2a}{3} + 150$

20. $p \leq -a + 190$

21. $10 \leq a \leq 70$

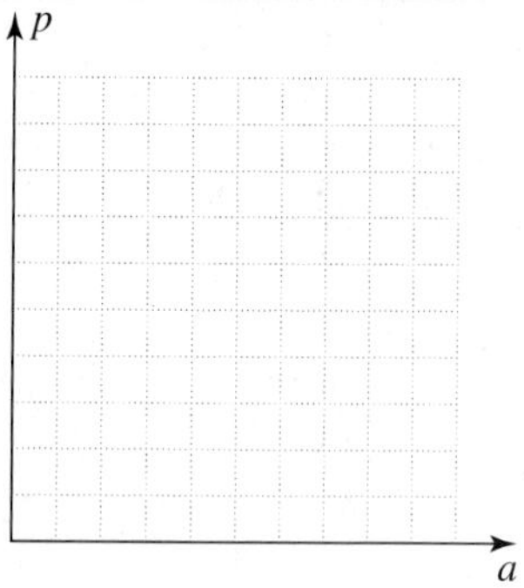

Write On

22. Write the steps you would take to graph the inequality

$$ax + by > c \quad (a \neq 0)$$

23. Describe the solution set of the inequality $x \geq k$.

24. Describe the solution set of the inequality $y < k$.

Concept Checker

Fill in the blank(s) with the correct word(s), phrase, or mathematical statement.

25. When graphing a **linear inequality** involving $\le$ or $\ge$ the **graph** is a __________ line.

26. When graphing a **linear inequality** involving $<$ or $>$ the **graph** is a __________ line.

direct **solid** **dashed**

Mastery Test

Graph the solution set for each system.

27. $x > 2$
$y < 3$

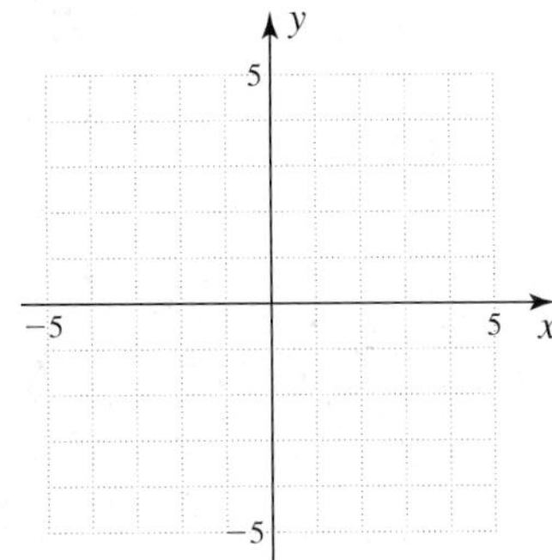

28. $x + y > 4$
$x - y \le 2$

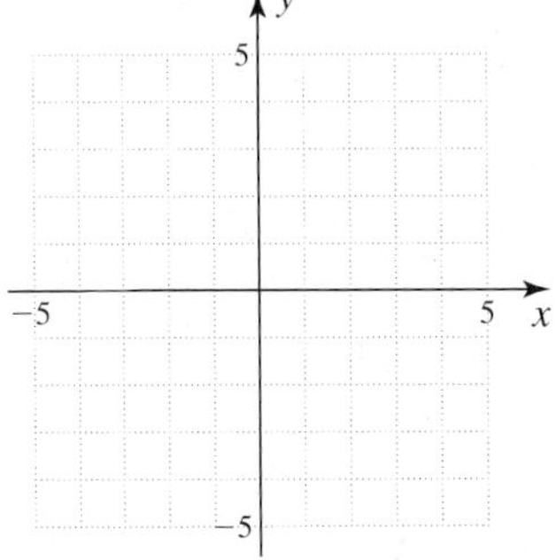

29. $3x - y < -1$
$x + 2y \le 2$

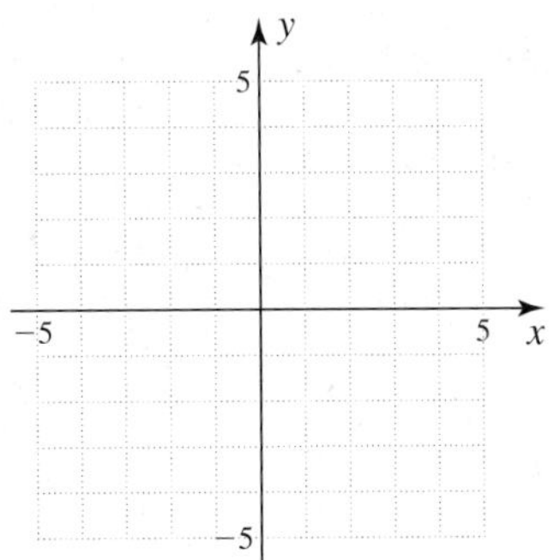

30. $2x - 3y \ge 6$
$-x + 2y < -4$

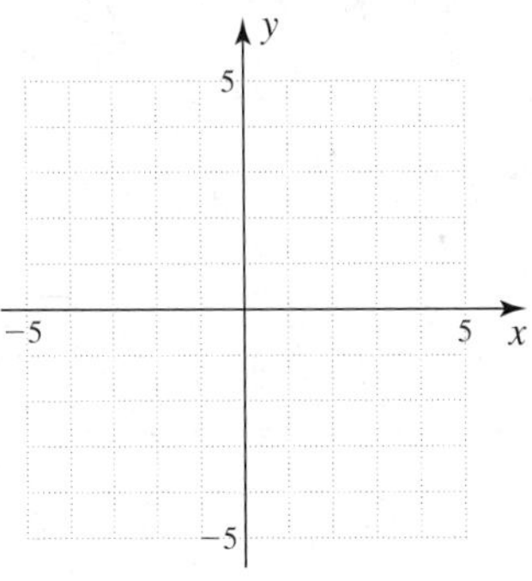

Skill Checker

Find:

31. $\sqrt{256}$

32. $\sqrt{9}$

33. $\sqrt{64}$

34. $\sqrt{144}$

Collaborative Learning

Women and Men in the Workforce

The figure on the next page shows the percentages of women (W) and men (M) in the workforce since 1955. Can you tell from the graph in which year the percentages will be the same? It will happen after 2005! If t is the number of years after 1955, W the percentage of women in the workforce, and M the percentage of men, the graphs will *intersect* at a point (a, b). The coordinates of the point of intersection (if there is one) will be the common solution of both equations. If the percentages W and M are approximated by

$$W = 0.55t + 34.4 \text{ (percent)}$$
$$M = -0.20t + 83.9 \text{ (percent)}$$

<CONTINUED>

you can find the year after 1955 when the same percentages of women and men are in the workforce by graphing M and W and locating the point of intersection.

Form three teams of students: **Team 1,** *The Estimators;* **Team 2,** *The Graphicals;* and **Team 3,** *The Substituters.*

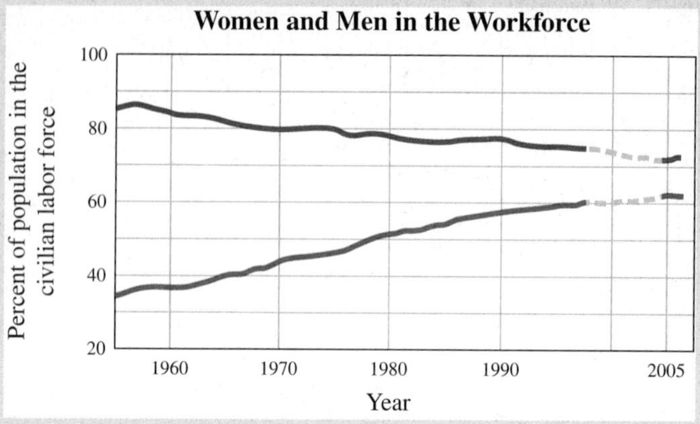

Source: U.S. Bureau of Labor Statistics.

Here are the assignments:

Team 1 Extend the lines for the percentage of men (M), the percentage of women (W), and the horizontal axis.

1. In what year do the lines seem to intersect?
2. What would be the percentage of men and women in the labor force in that year?

Team 2 Get some graph paper with a 20-by-20 grid, each grid representing 1 unit. Graph the lines

$$M = -0.20t + 83.9 \text{ (percent)}$$
$$W = 0.55t + 34.4 \text{ (percent)}$$

where t is the number of years after 1955.

1. Where do the lines intersect?
2. What does the point of intersection represent?

Team 3 The point at which the lines intersect will satisfy both equations; consequently at that point $M = W$. Substitute $-0.20t + 83.9$ for M and $0.55t + 34.4$ for W and solve for t.

1. What value did you get?
2. What is the point of intersection for the two lines, and what does it represent?

Compare the results of teams 1, 2, and 3. Do they agree? Why or why not? Which is the most accurate method for obtaining an answer for this problem?

Research Questions

1. Write a report on the content of the *Nine Chapters of the Mathematical Arts.*
2. Write a paragraph detailing how the copies of the *Nine Chapters* were destroyed.
3. Write a short biography on the Chinese mathematician Liu Hui.
4. Write a report explaining the relationship between systems of linear equations and inequalities and the simplex method.
5. Write a report on the simplex algorithm and its developer.

Summary Chapter 7

Section	Item	Meaning	Example
7.1A	System of simultaneous equations	A set of equations that may have a common solution	$x + y = 2$ $x - y = 4$ is a system of equations.
	Inconsistent system	A system with no solution	$x + y = 2$ $x + y = 3$ is an inconsistent system.
	Dependent system	A system in which both equations are equivalent	$x + y = 2$ $2x + 2y = 4$ is a dependent system.
7.2A	Substitution method	A method used to solve systems of equations by solving one equation for one variable and substituting this result in the other equation	To solve the system $x + y = 2$ $2x + 3y = 6$ solve the first equation for x: $x = 2 - y$ and substitute in the other equation: $2(2 - y) + 3y = 6$ $4 - 2y + 3y = 6$ $y = 2$ $x = 2 - 2 = 0$
7.3A	Elimination method	A method used to solve systems of equations by multiplying by numbers that will cause the coefficients of one of the variables to be opposites	To solve the system $x + 2y = 5$ $x - y = -1$ multiply the second equation by 2 and add to the first equation.
7.4	RSTUV method	A method for solving word problems consisting of **R**eading, **S**electing the variables, **T**hinking of a plan to solve the problem, **U**sing algebra to solve, and **V**erifying the answer	
7.5A	Solution set of a system of inequalities	The set of points that satisfy all inequalities in the system	The solution set of the system $x + y \geq 2$ $-x + y \leq -1$ are the shaded points in the figure. 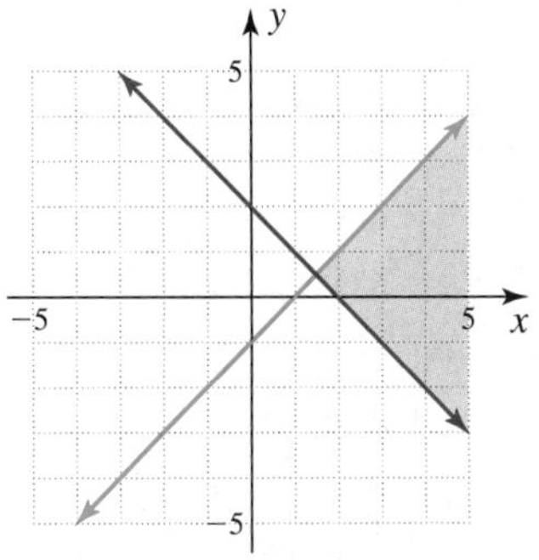 y 5 −5 5 x −5

› Review Exercises Chapter 7

(If you need help with these exercises, look in the section indicated in brackets.)

1. ‹**7.1A, B**› *Use the graphical method to solve the system (if possible).*

a. $2x + y = 4$
$y - 2x = 0$

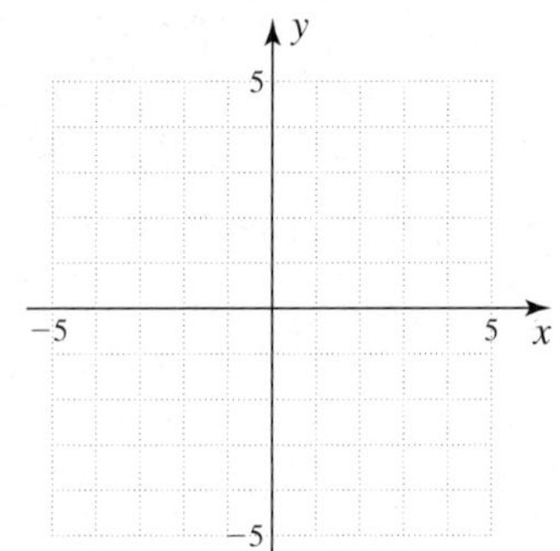

b. $x + y = 4$
$y - x = 0$

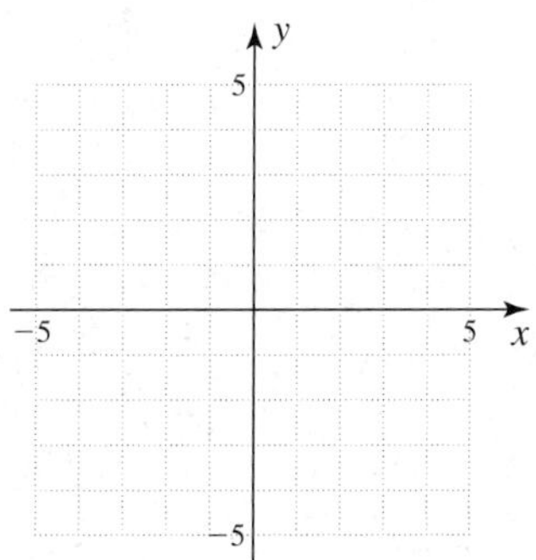

c. $x + y = 4$
$y - 3x = 0$

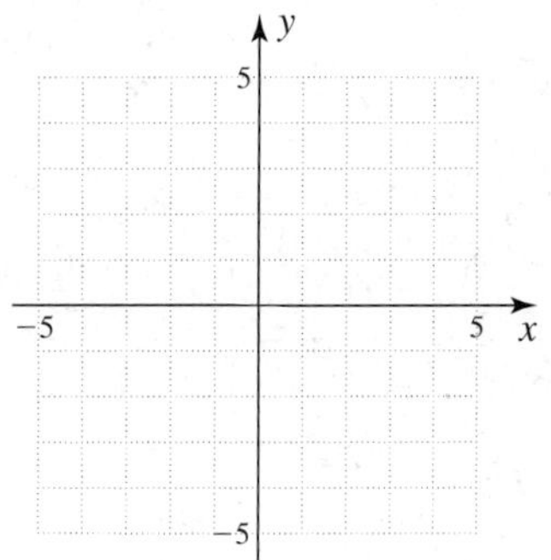

2. ‹**7.1A, B**› *Use the graphical method to solve the system (if possible).*

a. $y - 3x = 3$
$2y - 6x = 12$

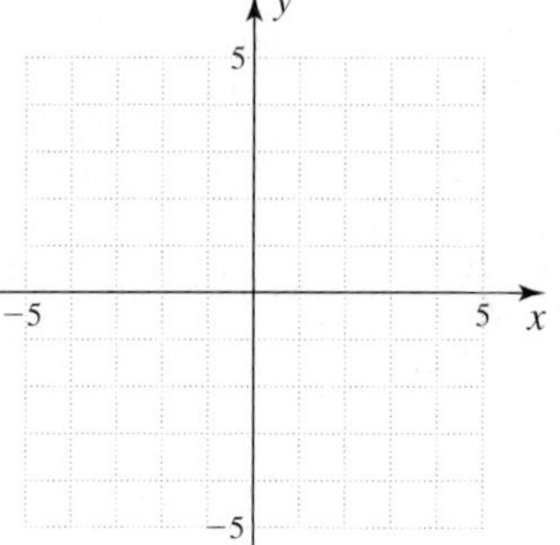

b. $y - 2x = 2$
$2y - 4x = 8$

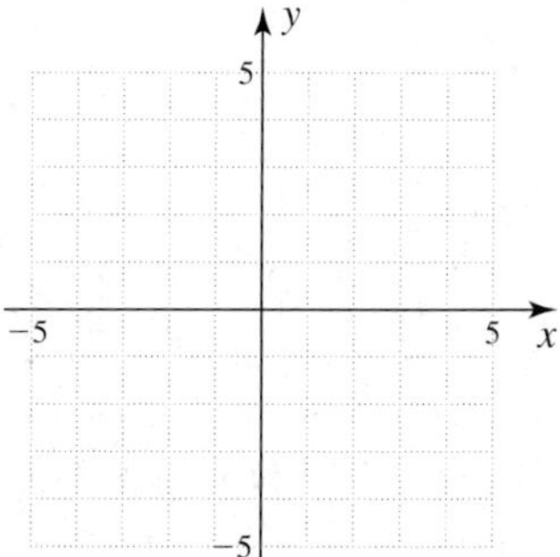

c. $y - 3x = 6$
$2y - 6x = 6$

3. ‹**7.2A, B**› *Use the substitution method to solve the system (if possible).*

a. $x + 4y = 5$
$2x + 8y = 15$

b. $x + 3y = 6$
$3x + 9y = 12$

c. $x + 4y = 5$
$2x + 13y = 15$

4. ‹**7.2A, B**› *Use the substitution method to solve the system (if possible).*

a. $x + 4y = 5$
$2x + 8y = 10$

b. $x + 3y = 6$
$3x + 9y = 18$

c. $x + 4y = 5$
$-2x - 8y = -10$

5. **⟨7.3A⟩** *Use the elimination method to solve the system (if possible).*

a. $3x + 2y = 1$
$2x + y = 0$

b. $3x + 2y = 4$
$2x + y = 3$

c. $3x + 2y = -7$
$2x + y = -4$

6. **⟨7.3B⟩** *Use the elimination method to solve the system (if possible).*

a. $2x - 3y = 6$
$-4x + 6y = -2$

b. $3x - 2y = 8$
$-9x + 6y = -4$

c. $3x - 5y = 6$
$-3x + 5y = -12$

7. **⟨7.3B⟩** *Use the elimination method to solve the system (if possible).*

a. $3y + 2x = 1$
$6y + 4x = 2$

b. $2y + 3x = 1$
$6x + 4y = 2$

c. $3y + 4x = -11$
$8x + 6y = -22$

8. **⟨7.4A⟩** *Desi has \$3 in nickels and dimes. How many nickels and how many dimes does she have if:*

a. she has the same number of nickels and dimes?

b. she has 4 times as many nickels as she has dimes?

c. she has 10 times as many nickels as she has dimes?

9. **⟨7.4B⟩** *The sum of two numbers is 180. What are the numbers if:*

a. their difference is 40?

b. their difference is 60?

c. their difference is 80?

10. **⟨7.4C⟩** *A plane flew 2400 miles with a tailwind in 3 hours. What was the plane's speed in still air if the return trip took*

a. 8 hours?

b. 10 hours?

c. 12 hours?

11. **⟨7.4D⟩** *An investor bought some municipal bonds yielding 5% annually and some certificates of deposit yielding 10% annually. If the total investment amounts to \$20,000, how much money is invested in bonds and how much in certificates of deposit (CDs) if the annual interest is*

a. \$1750?

b. \$1150?

c. \$1500?

12. **⟨7.5⟩** *Graph the solution set of the system.*

a. $x > 4$
$y < -1$

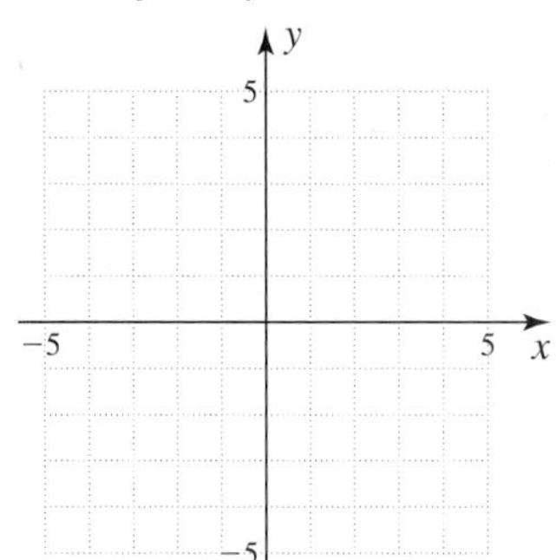

b. $x + y > 3$
$x - y < 4$

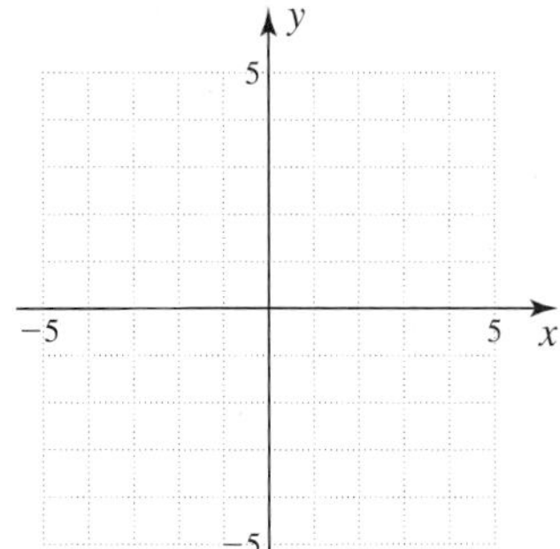

c. $2x + y \leq 4$
$x - 2y > 2$

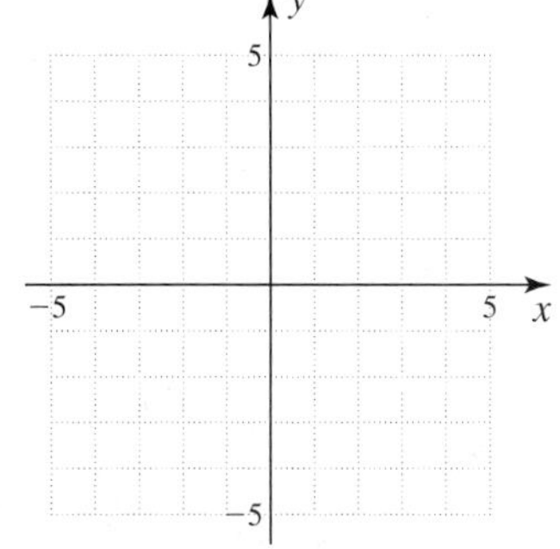

Practice Test Chapter 7

(Answers on page 631)

Visit www.mhhe.com/bello to view helpful videos that provide step-by-step solutions to several of the problems below.

1. Use the graphical method to solve the system.

$$x + 2y = 4$$
$$2y - x = 0$$

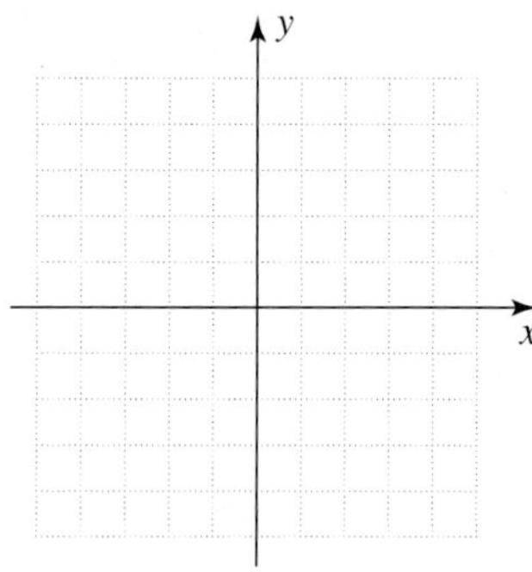

2. Use the graphical method to solve the system.

$$y - 2x = 2$$
$$2y - 4x = 8$$

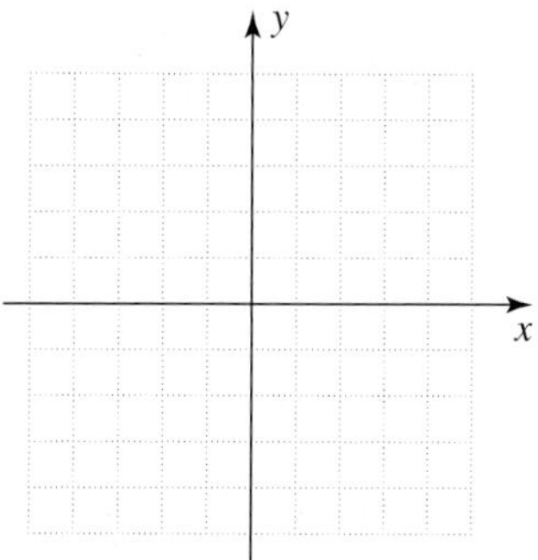

3. Use the substitution method to solve the system (if possible).

$$x + 3y = 6$$
$$2x + 6y = 8$$

4. Use the substitution method to solve the system (if possible).

$$x + 3y = 6$$
$$3x + 9y = 18$$

5. Use the elimination method to solve the system (if possible).

$$2x + 3y = -8$$
$$3x + y = -5$$

6. Use the elimination method to solve the system (if possible).

$$3x - 2y = 6$$
$$-6x + 4y = -2$$

7. Use the elimination method to solve the system (if possible).

$$2y + 3x = -12$$
$$6x + 4y = -24$$

8. Eva has \$2 in nickels and dimes. She has twice as many dimes as nickels. How many nickels and how many dimes does she have?

9. The sum of two numbers is 140. Their difference is 90. What are the numbers?

10. A plane flies 600 miles with a tailwind in 2 hours. It takes the same plane 3 hours to fly the 600 miles when flying against the wind. What is the plane's speed in still air?

11. Herbert invests \$10,000, part at 5% and part at 6%. How much money is invested at each rate if his annual interest is \$568?

12. Graph the solution set of the system.

$$x - 2y > 4$$
$$2x - y \le 6$$

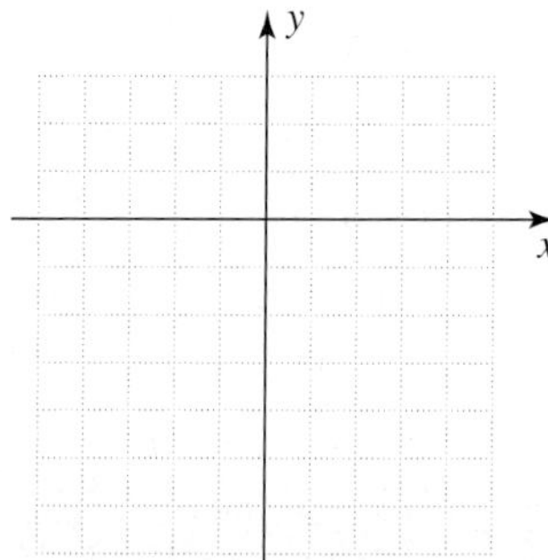

› Answers to Practice Test Chapter 7

Answer	If You Missed	Review		
	Question	Section	Examples	Page
1. The solution is (2, 1). 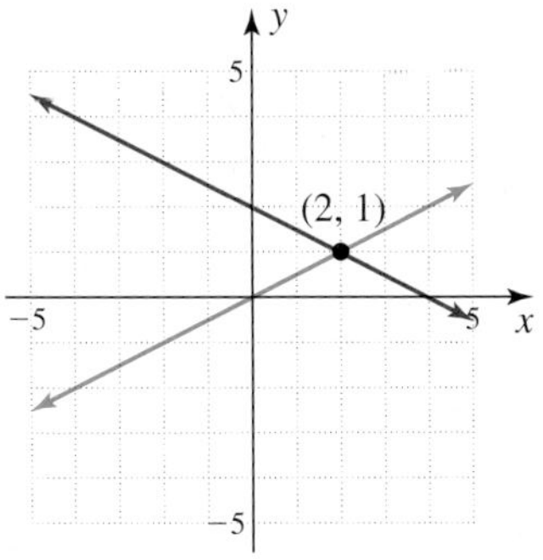	1	7.1	1	571–572
2. The lines are parallel; there is no solution (inconsistent). 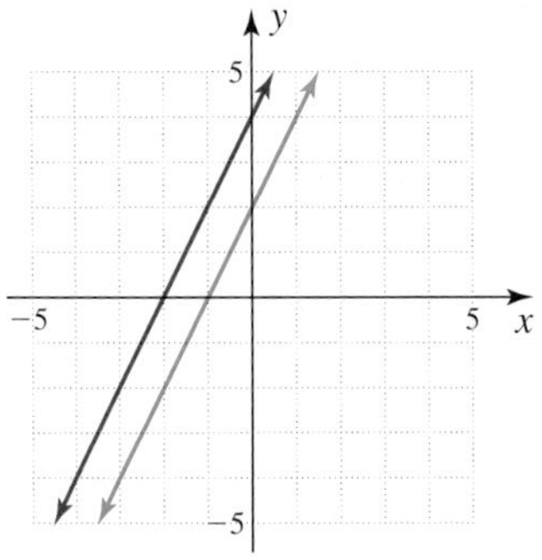	2	7.1	2	572
3. No solution (inconsistent)	3	7.2	2	589–590
4. Dependent (infinitely many solutions)	4	7.2	3	590
5. $(-1, -2)$	5	7.3	1	598
6. No solution (inconsistent)	6	7.3	2	599
7. Dependent (infinitely many solutions)	7	7.3	3	599–600
8. 8 nickels; 16 dimes	8	7.4	1, 2	608–609
9. 115 and 25	9	7.4	3	610
10. 250 mi/hr	10	7.4	4	610–611
11. \$3200 at 5%; \$6800 at 6%	11	7.4	5	611–612
12. 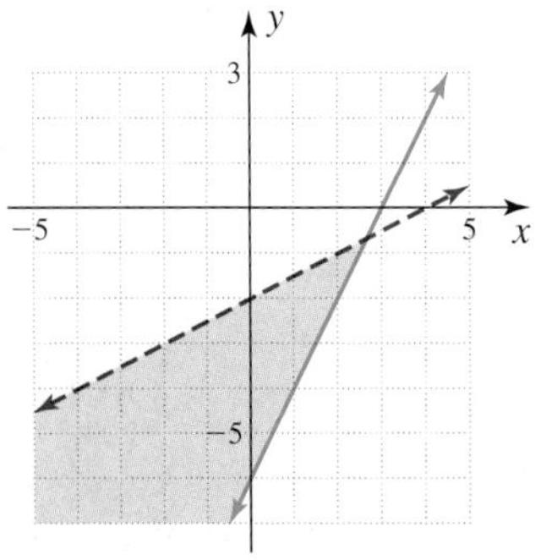	12	7.5	1, 2, 3	618–620

Cumulative Review Chapters 1–7

1. Add: $-\frac{3}{7} + \left(-\frac{1}{6}\right)$

2. Find: $(-4)^4$

3. Divide: $-\frac{1}{4} \div \left(-\frac{1}{8}\right)$

4. Evaluate $y \div 2 \cdot x - z$ for $x = 6$, $y = 24$, $z = 3$.

5. Simplify: $2x - (x + 4) - 2(x + 3)$

6. Write in symbols: The quotient of $(x - 5y)$ and z

7. Solve for x: $3 = 3(x - 2) + 3 - 2x$

8. Solve for x: $\frac{x}{4} - \frac{x}{9} = 5$

9. Graph: $-\frac{x}{3} + \frac{x}{8} \geq \frac{x - 8}{8}$

10. Graph the point $C(1, -2)$.

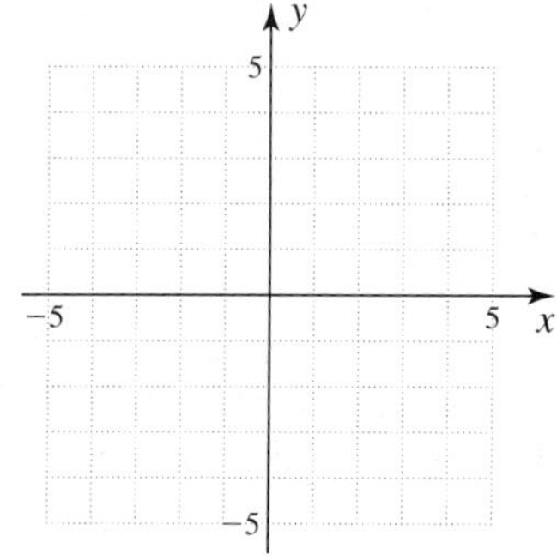

11. Determine whether the ordered pair $(-1, -1)$ is a solution of $4x + 3y = -1$.

12. Find x in the ordered pair $(x, 2)$ so that the ordered pair satisfies the equation $3x - 2y = -13$.

13. Graph: $x + y = 1$

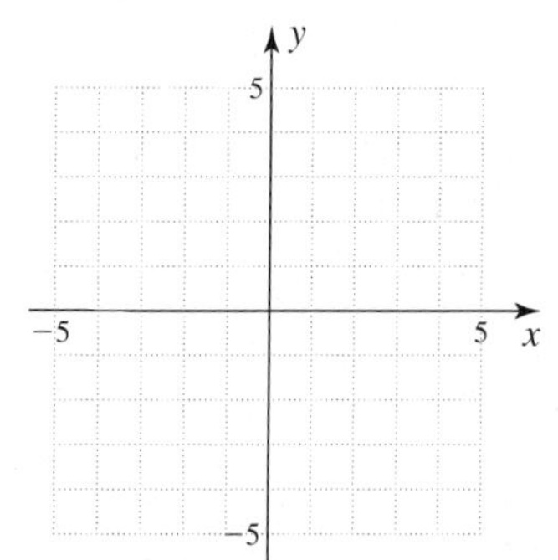

14. Graph: $2y + 8 = 0$

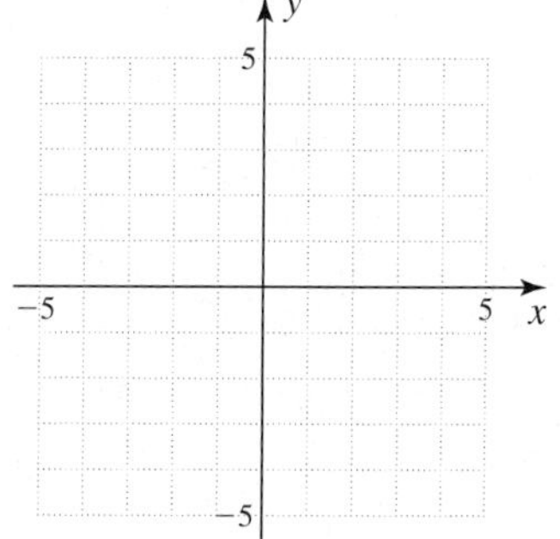

15. Find the slope of the line passing through the points $(-2, -3)$ and $(5, -4)$.

16. What is the slope of the line $12x - 4y = 14$?

17. Find the pair of parallel lines.

(1) $2y = -x + 6$

(2) $-4x - 8y = 6$

(3) $8y - 4x = 6$

18. Simplify: $(2x^4y^{-3})^{-3}$

19. Write in scientific notation: 0.000035

20. Divide and express the answer in scientific notation: $(5.46 \times 10^{-3}) \div (2.6 \times 10^4)$

21. Find (expand): $\left(6x^2 - \frac{1}{4}\right)^2$

22. Divide $(2x^3 - x^2 - 2x + 4)$ by $(x - 1)$.

23. Factor completely: $x^2 - 11x + 30$

24. Factor completely: $15x^2 - 29xy + 12y^2$

25. Factor completely: $81x^2 - 64y^2$

26. Factor completely: $-3x^4 + 12x^2$

27. Factor completely: $4x^3 - 4x^2 - 8x$

28. Factor completely: $4x^2 + 3x + 4x + 3$

29. Factor completely: $25kx^2 - 30kx + 9k$

30. Solve for x: $3x^2 + 4x = 15$

31. Write $\frac{4x}{3y}$ with a denominator of $15y^2$.

32. Reduce to lowest terms: $\frac{-9(x^2 - y^2)}{3(x - y)}$

33. Reduce to lowest terms: $\frac{x^2 - 4x - 12}{6 - x}$

34. Multiply: $(x - 5) \cdot \frac{x - 2}{x^2 - 25}$

35. Divide: $\frac{x + 4}{x - 4} \div \frac{x^2 - 16}{4 - x}$

36. Add: $\frac{9}{2(x - 9)} + \frac{3}{2(x - 9)}$

37. Subtract: $\frac{x + 2}{x^2 + x - 6} - \frac{x + 3}{x^2 - 4}$

38. Simplify: $\dfrac{\frac{4}{x} - \frac{3}{4x}}{\frac{2}{3x} - \frac{1}{2x}}$

39. Solve for x: $\frac{5x}{x - 5} - 2 = \frac{x}{x - 5}$

40. Solve for x: $\frac{x}{x^2 - 4} + \frac{2}{x - 2} = \frac{1}{x + 2}$

41. Solve for x: $\frac{x}{x + 4} - \frac{1}{5} = \frac{-4}{x + 4}$

42. Solve for x: $2 + \frac{8}{x - 2} = \frac{32}{x^2 - 4}$

43. A van travels 60 miles on 4 gallons of gas. How many gallons will it need to travel 195 miles?

44. Solve for x: $\frac{x + 1}{4} = \frac{9}{5}$

45. Sandra can paint a kitchen in 5 hours, and Roger can paint the same kitchen in 4 hours. How long would it take for both working together to paint the kitchen?

46. Find an equation of the line that passes through the point $(-6, -2)$ and has slope $m = -4$.

47. Find an equation of the line having slope 5 and y-intercept 2.

48. Graph: $4x - 3y > -12$

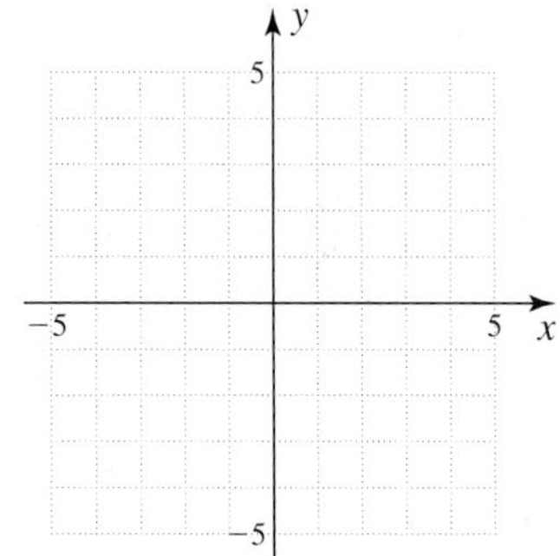

49. Graph: $-y \leq -3x + 6$

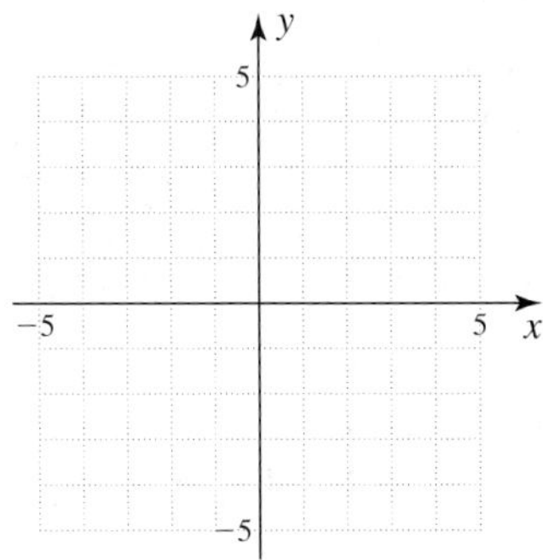

50. An enclosed gas exerts a pressure P on the walls of the container. This pressure is directly proportional to the temperature T of the gas. If the pressure is 7 pounds per square inch when the temperature is 350°F, find k.

51. If the temperature of a gas is held constant, the pressure P varies inversely as the volume V. A pressure of 1560 pounds per square inch is exerted by 4 cubic feet of air in a cylinder fitted with a piston. Find k.

52. Graph the system and find the solution (if possible):

$$x + 2y = 6$$
$$2y - x = -2$$

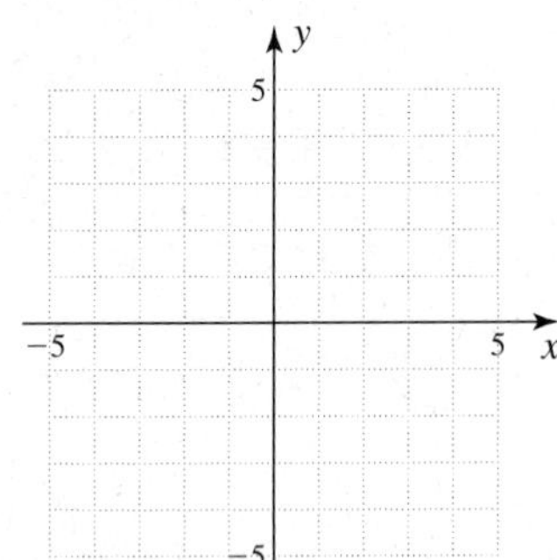

53. Graph the system and find the solution (if possible):

$$y - x = -1$$
$$2y - 2x = -4$$

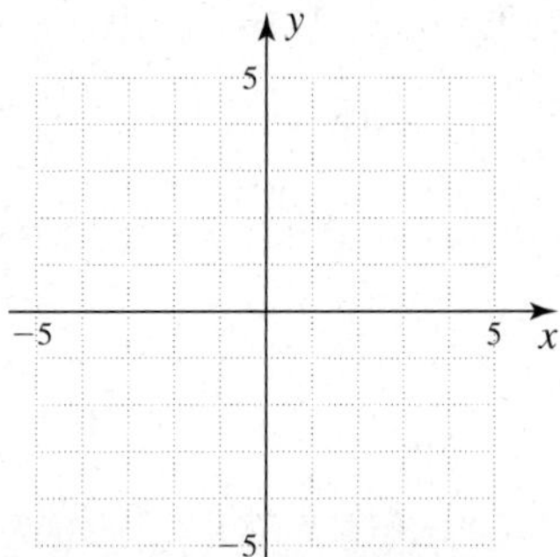

54. Solve by substitution (if possible):

$$x - 4y = 5$$
$$-3x + 12y = -17$$

55. Solve by substitution (if possible):

$$x - 4y = -13$$
$$3x - 12y = -39$$

56. Solve the system (if possible):

$$3x + 4y = 7$$
$$2x + 3y = 5$$

57. Solve the system (if possible):

$$4x - 3y = 1$$
$$4x - 3y = -4$$

58. Solve the system (if possible):

$$2y - x = 0$$
$$-2x + 4y = 0$$

59. Kaye has \$4.20 in nickels and dimes. She has three times as many dimes as nickels. How many nickels and how many dimes does she have?

60. The sum of two numbers is 165. Their difference is 95. What are the numbers?

Section

Chapter 8 eight

Roots and Radicals

The Human Side of Algebra

Beginning in the tenth century and extending into the fourteenth century, Chinese mathematics shifted to arithmetical algebra. A Chinese mathematician discovered the relation between extracting roots and the array of binomial coefficients in Pascal's triangle. This discovery and iterated multiplications were used to extend root extraction and to solve equations of higher degree than cubic, an effort that reached its apex with the work of four prominent thirteenth-century Chinese algebraists.

The square root sign $\sqrt{}$ can be traced back to Christoff Rudolff (1499–1545), who wrote it as $\surd$ with only two strokes. Rudolff thought that $\surd$ resembled the small letter r, the first letter in the word *radix,* which means root. One way of computing square roots was developed by the English mathematician Isaac Newton (1642–1727), and the process is aptly named Newton's method. With the advent of calculators and computers, Newton's method, as well as the square root tables popular a few years ago, are seldom used anymore.

Sir Isaac Newton

8.1 Finding Roots

Objectives

A Find the square root of a number.

B Square a radical expression.

C Classify the square root of a number and approximate it with a calculator.

D Find higher roots of numbers.

E Solve applications involving square roots.

To Succeed, Review How To . . .

1. Find the square of a number (p. 63).
2. Raise a number to a power (pp. 62–63).

Getting Started

Square Roots and Round Wheels

The first bicycles were built in 1839 by Kirkpatrick Macmillan of Scotland. These bicycles were heavy and unstable, so in 1870, James Starley of Great Britain set out to reduce their weight. The result, shown in the photo, was a lighter bicycle but one that was likely to tip over when going around corners. As a matter of fact, even now, the greatest speed s (in miles per hour) at which a cyclist can safely take a corner of radius r (in feet) is given by the equation

$$s = 4\sqrt{r}$$ Read "s equals 4 times the square root of r."

The concept of a *square root* is related to the concept of squaring a number. The number 36, for example, is the square of 6 because $6^2 = 36$. Six, on the other hand, is the square root of 36—that is, $\sqrt{36} = 6$. Similarly,

$$\sqrt{16} = 4 \quad \text{because} \quad 4^2 = 16$$

$$\sqrt{\frac{4}{9}} = \frac{2}{3} \quad \text{because} \quad \left(\frac{2}{3}\right)^2 = \left(\frac{4}{9}\right)$$

In this section, we learn how to find square roots.

A › Finding Square Roots

To find $\sqrt{a}$, we have to find the number b so that $b^2 = a$; that is,

$$\sqrt{a} = b \quad \text{means that} \quad b^2 = a$$

Note that since $b^2 = a$, a must be **nonnegative.**

For example,

$$\sqrt{4} = 2 \quad \text{because} \quad 2^2 = 4$$
$$\sqrt{9} = 3 \quad \text{because} \quad 3^2 = 9$$
$$\sqrt{\frac{1}{4}} = \frac{1}{2} \quad \text{because} \quad \left(\frac{1}{2}\right)^2 = \frac{1}{4}$$

The symbol $\sqrt{\ }$ is called a **radical sign,** and it indicates the **positive** square root of a number (except $\sqrt{0} = 0$). The expression under the radical sign is called the **radicand.** An algebraic expression containing a radical is called a **radical expression.** We can summarize our discussion as follows.

SQUARE ROOT

If *a* is a positive real number,

$\sqrt{a} = b$ is the **positive** square root of a so that $b^2 = a$.

$-\sqrt{a} = b$ is the **negative** square root of a so that $b^2 = a$.

If $a = 0$, $\sqrt{a} = 0$ and $\sqrt{0} = 0$ since $0^2 = 0$.

Note that the symbol $-\sqrt{\ }$ represents the *negative* square root. Thus, the square root of 4 is $\sqrt{4} = 2$, but the *negative* square root of 4 is $-\sqrt{4} = -2$. When we write $\sqrt{a}$ we mean the **positive** or **principal** square root of a.

EXAMPLE 1 Finding square roots

Find:

a. $\sqrt{121}$ **b.** $-\sqrt{100}$ **c.** $\sqrt{\frac{25}{36}}$

SOLUTION 1

a. $\sqrt{121} = 11$ Since $11^2 = 121$

b. $-\sqrt{100} = -10$ Since $10^2 = 100$

c. $\sqrt{\frac{25}{36}} = \frac{5}{6}$ Since $\left(\frac{5}{6}\right)^2 = \frac{25}{36}$

PROBLEM 1

Find:

a. $\sqrt{169}$ **b.** $-\sqrt{64}$

c. $\sqrt{\frac{81}{25}}$

B > Squaring Radical Expressions

Since $b = \sqrt{a}$ means that $b^2 = a$,

$$\sqrt{a} \cdot \sqrt{a} = (\sqrt{a})^2 = a$$

Similarly, since $b = -\sqrt{a}$ means that $b^2 = a$,

$$(-\sqrt{a}) \cdot (-\sqrt{a}) = (-\sqrt{a})^2 = a$$

Thus, we have the following rule.

RULE

Squaring a Square Root

When the square root of a nonnegative real number a is squared, the result is that **positive** real number; that is, $(\sqrt{a})^2 = a$ and $(-\sqrt{a})^2 = a$. Also, $(\sqrt{0})^2 = 0$.

Answers to PROBLEMS

1. a. 13 **b.** -8 **c.** $\frac{9}{5}$

EXAMPLE 2 Squaring expressions involving radicals

Find the square of each radical expression:

a. $\sqrt{7}$ **b.** $-\sqrt{49}$ **c.** $\sqrt{x^2 + 3}$

SOLUTION 2

a. $(\sqrt{7})^2 = 7$ **b.** $(-\sqrt{49})^2 = 49$ **c.** $(\sqrt{x^2 + 3})^2 = x^2 + 3$

PROBLEM 2

Find the square of each radical expression:

a. $\sqrt{11}$ **b.** $-\sqrt{36}$

c. $\sqrt{x^2 + 7}$

C > Classifying and Approximating Square Roots

If a number has a rational number for its square root, it's called a **perfect square.** Thus, 1, 4, 9, 16, $\frac{4}{25}$, $\frac{9}{4}$, and $\frac{81}{16}$ are examples of perfect squares. If a number is *not* a perfect square, its square root is *irrational.* Thus, $\sqrt{2}$, $\sqrt{3}$, $\sqrt{5}$, $\sqrt{\frac{3}{4}}$, and $\sqrt{\frac{5}{6}}$ are irrational. As you recall, an irrational number *cannot* be written as the ratio of two integers. Its decimal representation is nonrepeating and nonterminating. A decimal approximation for such a number can be obtained by using the 2nd x^2 keys on your calculator. Using the symbol $\approx$ (which means "is approximately equal to"), we have

$$\sqrt{2} \approx 1.4142136$$
$$\sqrt{3} \approx 1.7320508$$
$$\sqrt{5} \approx 2.236068$$

Of course, not all real numbers have a *real-number* square root. As you recall, $\sqrt{a} = b$ is *equivalent* to $b^2 = a$. Since b^2 is nonnegative, a has to be nonnegative.

SQUARE ROOT OF A NEGATIVE NUMBER

If a is negative, $\sqrt{a}$ is not a real number.

For example, $\sqrt{-7}$, $\sqrt{-4}$, and $\sqrt{-\frac{4}{5}}$ are *not* real numbers. Try some possibilities. Suppose you say $\sqrt{-4} = -2$. By the definition of square root $(-2)^2$ must be -4. But $(-2)^2 = 4$ (not -4), so $\sqrt{-4}$ is *not* a real number b because you can't find a number b such that $b^2 = -4$.

EXAMPLE 3 Classifying and approximating real numbers

Classify the given numbers as rational, irrational, or not a real number. Approximate the irrational numbers using a calculator.

a. $\sqrt{\frac{81}{25}}$ **b.** $-\sqrt{64}$ **c.** $\sqrt{13}$ **d.** $\sqrt{-64}$

SOLUTION 3

a. $\frac{81}{25} = \left(\frac{9}{5}\right)^2$ is a perfect square, so $\sqrt{\frac{81}{25}} = \frac{9}{5}$ is a rational number.

b. $64 = 8^2$ is a perfect square, so $-\sqrt{64} = -8$ is a rational number.

c. 13 is *not* a perfect square, so $\sqrt{13} \approx 3.6055513$ is irrational.

d. $\sqrt{-64}$ is *not* a real number, since there is no number that we can square and get -64 for an answer. (Try it!)

PROBLEM 3

Classify as rational, irrational, or not a real number. Approximate the irrational numbers using a calculator.

a. $\sqrt{\frac{36}{25}}$ **b.** $-\sqrt{81}$

c. $\sqrt{19}$ **d.** $\sqrt{-9}$

Answers to PROBLEMS

2. a. 11 **b.** 36 **c.** $x^2 + 7$

3. a. $\frac{6}{5}$; Rational **b.** -9; Rational

c. Irrational. $\sqrt{19} \approx 4.3588989$

d. Not real

D > Finding Higher Roots of Numbers

We've already seen that, by definition,

$$\sqrt{a} = b \quad \text{means that} \quad b^2 = a$$

Similarly,

$$\sqrt[3]{a} = b \quad \text{means that} \quad b^3 = a$$

$$\sqrt[4]{a} = b \quad \text{means that} \quad b^4 = a$$

Because of these relationships, finding the square root of a number can be regarded as the **inverse** of finding the square of the number. Similarly, the inverse of finding the **cube** or **fourth power** of a number a is to find $\sqrt[3]{a}$, the **cube root** of a, or $\sqrt[4]{a}$, the **fourth root** of a. In general,

nTH ROOT OF a

$\sqrt[n]{a}$ is called the *n*th root of *a* and the number n is the index or order of the radical.

Note that we could write $\sqrt[2]{a}$ for the square root, but the simpler notation $\sqrt{a}$ is customary because the square root is the most widely performed operation.

EXAMPLE 4 Finding higher roots of real numbers

Find each root if possible:

a. $\sqrt[4]{16}$ **b.** $\sqrt[4]{-81}$ **c.** $\sqrt[3]{8}$

d. $\sqrt[3]{-27}$ **e.** $-\sqrt[4]{16}$

SOLUTION 4

a. We need to find a number whose fourth power is 16, that is, $(?)^4 = 16$. Since $2^4 = 16$, $\sqrt[4]{16} = 2$.

b. We need to find a number whose fourth power is -81. There is no such real number, so $\sqrt[4]{-81}$ is *not* a real number.

c. We need to find a number whose cube is 8. Since $2^3 = 8$, $\sqrt[3]{8} = 2$.

d. We need to find a number whose cube is -27. Since $(-3)^3 = -27$, $\sqrt[3]{-27} = -3$.

e. We need to find a number whose fourth power is 16. Since $2^4 = 16$, $-\sqrt[4]{16} = -2$.

PROBLEM 4

Find each root if possible:

a. $\sqrt[4]{81}$

b. $\sqrt[4]{-16}$

c. $\sqrt[3]{-64}$

d. $\sqrt[3]{-216}$

e. $-\sqrt[4]{625}$

E > Applications Involving Square Roots

EXAMPLE 5 Simplifying square roots

The time t (in seconds) it takes an object dropped from a distance d (in feet) to reach the ground is given by the equation

$$t = \sqrt{\frac{d}{16}}$$

The highest regularly performed dive used to be at La Quebrada in Acapulco. How long does it take the divers to reach the water 100 feet below?

SOLUTION 5

Using

$$t = \sqrt{\frac{d}{16}} \quad \text{and} \quad d = 100$$

we have

$$t = \sqrt{\frac{100}{16}} = \frac{10}{4} = 2.5 \text{ sec}$$

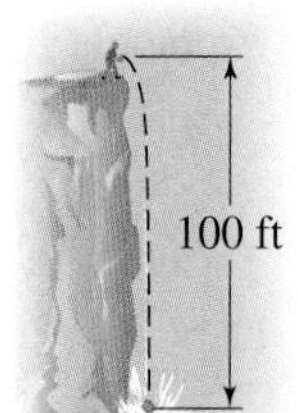

By the way, the water is only 12 feet deep, but the height of the dive has been given as high as 118 feet and as low as 87.5 feet!

PROBLEM 5

How long does it take the diver to travel 81 feet?

Answers to PROBLEMS

4. a. 3 **b.** Not real **c.** -4 **d.** -6 **e.** -5 **5.** 2.25 sec

We have already mentioned that an average tree absorbs 50 pounds of CO_2 per year, so if you want to absorb 2000 pounds of CO_2, then you need 40 trees. Suppose the 40 trees are to be planted on one acre of land (An acre is 43,560 square feet.). If each tree is given an individual square plot of land of side length S what are the dimensions of each plot? The answer is $S = \sqrt{43{,}560/40} \approx 33$ ft on each side. This means that there will be 40 square plots measuring 33 feet on each side, each containing one tree.

Note that $33 \times 33 \times 40 = 43{,}560$, so the 40 plots cover the whole acre! You can visualize this as 40 individual square plots each measuring 33 ft by 33 ft. This can be verified at http://warnell.forestry.uga.edu/service/library/for96-054/for96-054.html#tab2, but we will practice with the formula $S = \sqrt{43{,}560/n}$ in Example 6.

GREEN MATH

EXAMPLE 6 Dimensions of plots for individual trees

A car that gets 25 miles per gallon (mpg) is driven 20,000 miles and produces 1600 pounds of CO_2 a year. Assuming a tree absorbs 50 pounds of CO_2 per year, 32 trees are needed to absorb the CO_2. If we decide to plant the 32 trees in one acre of land, what are the dimensions of each of the individual square plots used for each tree? Answer to the nearest whole number.

SOLUTION 6

The dimensions of each side are

$$\begin{aligned} S &= \sqrt{43{,}560/n} \text{ with } n = 32 \\ &= \sqrt{43{,}560/32} \\ &= \sqrt{1361.25} \approx 37 \text{ ft on each side} \end{aligned}$$

Thus, the plots for each of the 32 trees should be 37 ft on each side.

PROBLEM 6

If 100 trees are to be planted in an acre of land in square individual plots, what should the dimensions of each plot be? Answer to the nearest whole number.

Note: According to Classic Landscapes, silver maple and American elms should have plots that are 11 to 16 yards on each side, while Scots pine and spruce should have plots 4 to 12 yards on each side.

Exercises 8.1

⟨A⟩ Finding Square Roots In Problems 1–8, find the square root.

1. $\sqrt{25}$ **2.** $\sqrt{36}$ **3.** $-\sqrt{9}$ **4.** $-\sqrt{81}$

5. $\sqrt{\frac{16}{9}}$ **6.** $\sqrt{\frac{25}{4}}$ **7.** $-\sqrt{\frac{4}{81}}$ **8.** $-\sqrt{\frac{9}{100}}$

In Problems 9–12, find all the square roots (positive and negative) of each number.

9. $\frac{25}{81}$ **10.** $\frac{36}{49}$ **11.** $\frac{49}{100}$ **12.** $\frac{1}{9}$

⟨B⟩ Squaring Radical Expressions In Problems 13–20, find the square of each radical expression.

13. $\sqrt{5}$ **14.** $\sqrt{8}$ **15.** $-\sqrt{11}$ **16.** $-\sqrt{13}$

17. $\sqrt{x^2 + 1}$ **18.** $-\sqrt{a^2 + 2}$ **19.** $-\sqrt{3y^2 + 7}$ **20.** $\sqrt{8z^2 + 1}$

Answers to PROBLEMS

6. 21 feet on each side

C Classifying and Approximating Square Roots In Problems 21–32, find and classify the given numbers as rational, irrational, or not a real number. Use a calculator to approximate the irrational numbers. *Note:* The number of decimals in your calculator display may be different.

21. $\sqrt{36}$ **22.** $-\sqrt{4}$ **23.** $\sqrt{-100}$ **24.** $\sqrt{-9}$

25. $-\sqrt{64}$ **26.** $\sqrt{\frac{9}{49}}$ **27.** $\sqrt{\frac{16}{9}}$ **28.** $\sqrt{-\frac{4}{81}}$

29. $-\sqrt{6}$ **30.** $\sqrt{7}$ **31.** $-\sqrt{2}$ **32.** $-\sqrt{12}$

D Finding Higher Roots of Numbers In Problems 33–40, find each root if possible.

33. $\sqrt[4]{81}$ **34.** $\sqrt[4]{-256}$ **35.** $-\sqrt[4]{81}$ **36.** $\sqrt[3]{27}$

37. $-\sqrt[3]{64}$ **38.** $-\sqrt[3]{-64}$ **39.** $-\sqrt[3]{-125}$ **40.** $-\sqrt[3]{-8}$

E Applications Involving Square Roots

41. *Ribbon Falls* Ribbon Falls in California is the tallest continuous waterfall in the United States, with a drop of about 1600 feet. How long does it take the water to travel from the top of the waterfall to the bottom? (*Hint:* See Example 5.) (Actually, the waterfall is dry from late July to early April.)

42. *Jump from airship* Have you heard about the *Hindenburg* airship? On June 22, 1936, Colonel Harry A. Froboess of Switzerland jumped almost 400 feet from the *Hindenburg* into the Bodensee. How long did his jump take? (*Hint:* See Example 5.)

43. *Highest dive into airbag* The time t (in seconds) it takes an object dropped from a distance d (in meters) to reach the ground is given by the equation

$$t = \sqrt{\frac{d}{5}}$$

The highest reported dive into an airbag is 100 meters from the top of Vegas World Hotel and Casino by stuntman Dan Koko. About how long did it take Koko to reach the airbag below?

44. *Time of highest dive* Colonel Froboess of Problem 42 jumped about 125 meters from the *Hindenburg* to the Bodensee. Using the formula in Problem 43, how long did his jump take? Is your answer consistent with that of Problem 42?

45. *Pythagorean theorem* The Greek mathematician Pythagoras discovered a theorem that states that the lengths a, b, and c of the sides of a right triangle (a triangle with a 90° angle) are related by the formula $a^2 + b^2 = c^2$. (See the diagram.) This means that $c = \sqrt{a^2 + b^2}$. If the shorter sides of a right triangle are 4 inches and 3 inches, find the length of the longest side c (called the hypotenuse).

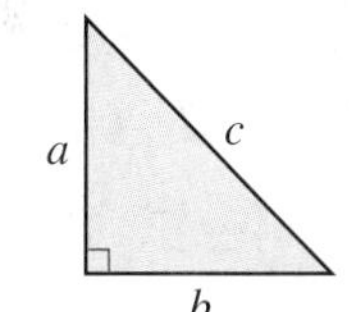

46. *Pythagorean theorem* Use the Pythagorean theorem stated in Problem 45 to find how high on the side of a building a 13-foot ladder will reach if the base of the ladder is 5 feet from the wall on which the ladder is resting.

47. *View from in-flight airplane* How far can you see from an airplane in flight? It depends on how high the airplane is. As a matter of fact, your view V_m (in miles) is modeled by the equation $V_m = 1.22\sqrt{a}$, where a is the altitude of the plane in feet. What is your view when your plane is cruising at an altitude of 40,000 feet?

48. *Dimensions of a square* The area of a square is 144 square inches. How long is each side of the square?

Applications: Green Math

49. *Plot dimensions* If 200 trees are to be planted in an acre of land in square individual plots, use the formula of Example 6 to find the dimensions of each plot. Answer to the nearest whole number.

50. *Plot dimensions* If 300 trees are to be planted in an acre of land in square individual plots, use the formula of Example 6 to find the dimensions of each plot. Answer to the nearest whole number.

Web IT go to mhhe.com/bello for more lessons

Web IT go to mhhe.com/bello for more lessons

Curving without skidding Use the following information for Problems 51 and 52. The maximum speed v (in miles per hour) a car can travel on a concrete highway curve without skidding is modeled by the equation $v = \sqrt{9r}$, where r is the radius of the curve in feet.

51. If the radius on the first curve in the photo is 100 feet, what is the maximum speed v a car can travel on this curve without skidding?

52. The radius on the second curve in the photo is only 64 feet. What is the maximum speed v a car can travel on this curve without skidding?

Velocity Use the following information for Problems 53 and 54. If an object is dropped from a height of h feet its velocity v (in feet per second) when it hits the ground is given by the equation $v = \sqrt{2gh}$, where g is the acceleration due to gravity, 32 feet per second squared (32 ft/sec^2).

53. Gumercindo was leaning over the railing of a balcony 81 feet above the street when his Blackberry slipped out of his hand. Find the Blackberry's velocity when it hit the ground.

54. Kanisha was riding the Kinda Ka roller coaster and after it went over the top at 441 feet high, her iPod slipped out of her pocket and fell to the ground. What was the velocity of the iPod when it hit the ground?

Using Your Knowledge

Interpolation Suppose you want to approximate $\sqrt{18}$ *without* using your calculator. Since $\sqrt{16} = 4$ and $\sqrt{25} = 5$, you have $\sqrt{16} = 4 < \sqrt{18} < \sqrt{25} = 5$. This means that $\sqrt{18}$ is between 4 and 5. To find a better approximation of $\sqrt{18}$, you can use a method that mathematicians call **interpolation.** Don't be scared by this name! The process is *really* simple. If you want to find an approximation for $\sqrt{18}$, follow the steps in the diagram; $\sqrt{20}$ and $\sqrt{22}$ are also approximated.

$18 - 16 = 2$
$$\sqrt{16} = 4$$
$$\sqrt{18} \approx 4 + \frac{2}{9}$$
$$\sqrt{25} = 5$$
$25 - 16 = 9$

$20 - 16 = 4$
$$\sqrt{16} = 4$$
$$\sqrt{20} \approx 4 + \frac{4}{9}$$
$$\sqrt{25} = 5$$
$25 - 16 = 9$

$22 - 16 = 6$
$$\sqrt{16} = 4$$
$$\sqrt{22} \approx 4 + \frac{6}{9}$$
$$\sqrt{25} = 5$$
$25 - 16 = 9$

As you can see,

$$\sqrt{18} \approx 4\tfrac{2}{9} \qquad \sqrt{20} \approx 4\tfrac{4}{9} \qquad \text{and} \qquad \sqrt{22} \approx 4\tfrac{6}{9} = 4\tfrac{2}{3}$$

Do you see a pattern? What is $\sqrt{24}$? Using a calculator, we obtain

$$\sqrt{18} \approx 4.2, \quad \sqrt{20} \approx 4.5, \quad \text{and} \quad \sqrt{22} \approx 4.7$$

With interpolation $\sqrt{18} \approx 4\frac{2}{9} \approx 4.2, \quad \sqrt{20} \approx 4\frac{4}{9} \approx 4.4, \quad \sqrt{22} \approx 4\frac{6}{9} \approx 4.7$

As you can see, we can get very close approximations!

Use this knowledge to approximate the following roots. Give the answer as a mixed number.

55. **a.** $\sqrt{26}$ **b.** $\sqrt{28}$ **c.** $\sqrt{30}$

56. **a.** $\sqrt{67}$ **b.** $\sqrt{70}$ **c.** $\sqrt{73}$

Write On

57. If a is a *nonnegative* number, how many square roots does a have? Explain.

58. If a is *negative*, how many square roots does a have? Explain.

59. If $a = 0$, how many square roots does a have?

60. How many real-number cube roots does a positive number have?

61. How many real-number cube roots does a negative number have?

62. How many cube roots does zero have?

63. Suppose you want to find $\sqrt{a}$. What assumption must you make about the variable a?

64. If you are given a whole number less than 200, how would you determine whether the square root of the number is rational or irrational without a calculator? Explain.

Concept Checker

Fill in the blank(s) with the correct word(s), phrase, or mathematical statement.

65. $\sqrt{a} = b$ is the **positive** square root of a so that ____________.

66. $-\sqrt{a} = b$ is the **negative** square root of a so that ____________.

67. $\sqrt{0} =$ ________.

68. For $a \geq 0$, $(\sqrt{a})^2 =$ ________.

69. For $a \geq 0$, $(-\sqrt{a})^2 =$ ________.

70. If a is **negative,** $\sqrt{a}$ is ________ **a real number.**

not defined	**a**
$a^2 = b$	**not**
$b^2 = a$	**always**
$-a$	**0**

Mastery Test

Find:

71. $\sqrt{\frac{16}{49}}$

72. $-\sqrt{\frac{36}{25}}$

73. $\sqrt{144}$

Find the square of each radical expression:

74. $-\sqrt{28}$

75. $\sqrt{17}$

76. $-\sqrt{x^2 + 9}$

Find and classify as rational, irrational, or not a real number. Approximate the irrational numbers using a calculator:

77. $-\sqrt{\frac{49}{121}}$

78. $\sqrt{-100}$

79. $\sqrt{15}$

Find if possible:

80. $\sqrt[4]{1296}$

81. $\sqrt[3]{-125}$

82. $\sqrt[4]{-1}$

83. $-\sqrt[3]{-27}$

84. The time t in seconds it takes an object dropped from a distance d (in feet) to reach the ground is given by the equation $t = \sqrt{\frac{d}{16}}$. How long does it take an object dropped from a distance of 81 feet to reach the ground?

Skill Checker

Find $\sqrt{b^2 - 4ac}$ given the specified value for each variable.

85. $a = 1, b = -4, c = 3$ **86.** $a = 1, b = -1, c = -2$ **87.** $a = 2, b = 5, c = -3$ **88.** $a = 6, b = -7, c = -3$

8.2 Multiplication and Division of Radicals

Objectives

A Multiply and simplify radicals using the product rule.

B Divide and simplify radicals using the quotient rule.

C Simplify radicals involving variables.

D Simplify higher roots.

E Solve applications involving the quotient rule.

To Succeed, Review How To . . .

1. Write a number as a product of primes (p. 6).
2. Write a number using specified powers as factors (pp. 15–16).

Getting Started

Skidding Is No Accident!

Have you seen your local police measuring skid marks at the scene of an accident? The speed s (in miles per hour) a car was traveling if it skidded d feet on a dry concrete road is given by $s = \sqrt{24d}$. If a car left a 50-foot skid mark at the scene of an accident and the speed limit was 30 miles per hour, was the driver speeding? To find the answer, we need to use the formula $s = \sqrt{24d}$. Letting $d = 50$, $s = \sqrt{24 \cdot 50} = \sqrt{1200}$. How can we simplify $\sqrt{1200}$? If we use factors that are perfect squares, we get

$$\begin{aligned}\sqrt{1200} &= \sqrt{100 \cdot 4 \cdot 3}\\ &= \sqrt{100} \cdot \sqrt{4} \cdot \sqrt{3}\\ &= 10 \cdot 2 \cdot \sqrt{3}\\ &= 20\sqrt{3}\end{aligned}$$

which is about 35 miles per hour. So the driver *was* exceeding the speed limit! In solving this problem, we have made the assumption that $\sqrt{a \cdot b} = \sqrt{a} \cdot \sqrt{b}$. Is this assumption always true? In this section you will learn that this is indeed the case!

A Using the Product Rule for Radicals

Is it true that $\sqrt{a \cdot b} = \sqrt{a} \cdot \sqrt{b}$? Let's check some examples.

$$\sqrt{4 \cdot 9} = \sqrt{36} = 6, \quad \sqrt{4} \cdot \sqrt{9} = 2 \cdot 3 = 6, \quad \text{thus,} \quad \sqrt{4 \cdot 9} = \sqrt{4} \cdot \sqrt{9}$$

Similarly,

$$\sqrt{9 \cdot 16} = \sqrt{144} = 12, \quad \sqrt{9} \cdot \sqrt{16} = 3 \cdot 4 = 12, \quad \text{thus,} \quad \sqrt{9 \cdot 16} = \sqrt{9} \cdot \sqrt{16}$$

In general, we have the following rule

PRODUCT RULE FOR RADICALS

If a and b are nonnegative numbers,

$$\sqrt{a \cdot b} = \sqrt{a} \cdot \sqrt{b}$$

The product rule can be used to *simplify* radicals—that is, to make sure that no perfect squares remain under the radical sign. Thus, $\sqrt{15}$ is simplified, but $\sqrt{18}$ is not because $\sqrt{18} = \sqrt{9 \cdot 2}$, and $\sqrt{9 \cdot 2} = \sqrt{9} \cdot \sqrt{2} = 3\sqrt{2}$ using the product rule.

EXAMPLE 1 Simplifying expressions using the product rule

Simplify:

a. $\sqrt{75}$ **b.** $\sqrt{48}$

SOLUTION 1 To use the product rule, you have to remember the square roots of a few perfect square numbers such as 4, 9, 16, 25, 36, 49, 64, and 81 and try to write the number under the radical using one of these numbers as a factor.

a. Since $\sqrt{75} = \sqrt{25 \cdot 3}$, we write $\sqrt{75} = \sqrt{25 \cdot 3} = \sqrt{25} \cdot \sqrt{3} = 5\sqrt{3}$.

b. Similarly, $\sqrt{48} = \sqrt{16 \cdot 3}$. Thus, $\sqrt{48} = 4\sqrt{3}$.

PROBLEM 1

Simplify:

a. $\sqrt{72}$ **b.** $\sqrt{98}$

We can also use the product rule to actually multiply radicals. Thus,

$$\sqrt{3} \cdot \sqrt{5} = \sqrt{3 \cdot 5} = \sqrt{15} \quad \text{and} \quad \sqrt{8} \cdot \sqrt{2} = \sqrt{16} = 4$$

EXAMPLE 2 Multiplying expressions using the product rule

Multiply:

a. $\sqrt{6} \cdot \sqrt{5}$ **b.** $\sqrt{20} \cdot \sqrt{5}$ **c.** $\sqrt{5} \cdot \sqrt{x}, x > 0$

SOLUTION 2

a. $\sqrt{6} \cdot \sqrt{5} = \sqrt{30}$ **b.** $\sqrt{20} \cdot \sqrt{5} = \sqrt{100} = 10$

c. $\sqrt{5} \cdot \sqrt{x} = \sqrt{5x}$

PROBLEM 2

Multiply:

a. $\sqrt{5} \cdot \sqrt{7}$ **b.** $\sqrt{40} \cdot \sqrt{10}$

c. $\sqrt{6} \cdot \sqrt{x}, x > 0$

B › Using the Quotient Rule for Radicals

We already know that

$$\sqrt{\frac{81}{16}} = \frac{9}{4}$$

It's also true that

$$\sqrt{\frac{81}{16}} = \frac{\sqrt{81}}{\sqrt{16}} = \frac{9}{4}$$

This rule can be stated as follows.

QUOTIENT RULE FOR RADICALS

If a and b are positive numbers,

$$\sqrt{\frac{a}{b}} = \frac{\sqrt{a}}{\sqrt{b}}$$

Answers to PROBLEMS

1. a. $6\sqrt{2}$ **b.** $7\sqrt{2}$

2. a. $\sqrt{35}$ **b.** 20 **c.** $\sqrt{6x}$

EXAMPLE 3 Simplifying expressions using the quotient rule

Simplify:

a. $\sqrt{\frac{5}{9}}$ b. $\frac{\sqrt{18}}{\sqrt{6}}$ c. $\frac{14\sqrt{20}}{7\sqrt{10}}$

SOLUTION 3

a. $\sqrt{\frac{5}{9}} = \frac{\sqrt{5}}{\sqrt{9}} = \frac{\sqrt{5}}{3}$

b. $\frac{\sqrt{18}}{\sqrt{6}} = \sqrt{\frac{18}{6}} = \sqrt{3}$

c. $\frac{14\sqrt{20}}{7\sqrt{10}} = \frac{2\sqrt{20}}{\sqrt{10}} = 2\sqrt{\frac{20}{10}} = 2\sqrt{2}$

PROBLEM 3

Simplify:

a. $\sqrt{\frac{7}{16}}$ b. $\frac{\sqrt{30}}{\sqrt{6}}$

c. $\frac{16\sqrt{30}}{8\sqrt{10}}$

C › Simplifying Radicals Involving Variables

We can simplify radicals involving variables as long as the expressions under the radical sign are defined. This means that these expressions must *not* be negative. What about the answers? Note that $\sqrt{2^2} = 2$ and $\sqrt{(-2)^2} = \sqrt{4} = 2$. Thus, the square root of a squared nonzero number is always positive. We use absolute values to indicate this.

ABSOLUTE VALUE OF A RADICAL

For any real number *a*,

$$\sqrt{a^2} = |a|$$

Of course, in examples and problems where the variables are assumed to be positive, the absolute-value bars are not necessary.

EXAMPLE 4 Simplifying radical expressions involving variables

Simplify (assume all variables represent positive real numbers):

a. $\sqrt{49x^2}$ b. $\sqrt{81n^4}$ c. $\sqrt{50y^8}$ d. $\sqrt{x^{11}}$

SOLUTION 4

a. $\sqrt{49x^2} = \sqrt{49} \cdot \sqrt{x^2}$ Use the product rule.

$= 7x$

b. $\sqrt{81n^4} = \sqrt{81} \cdot \sqrt{n^4}$ Use the product rule.

$= 9n^2$ Since $(n^2)^2 = n^4$, $\sqrt{n^4} = n^2$.

c. $\sqrt{50y^8} = \sqrt{50} \cdot \sqrt{y^8}$ Use the product rule.

$= \sqrt{25 \cdot 2} \cdot \sqrt{y^8}$ Factor 50 using a perfect square.

$= 5\sqrt{2} \cdot y^4$ Since $\sqrt{25} = 5$ and $\sqrt{y^8} = y^4$

$= 5y^4\sqrt{2}$

d. $\sqrt{x^{11}} = \sqrt{x^{10} \cdot x}$ Since $x^{10} \cdot x = x^{11}$

$= x^5\sqrt{x}$ Since $\sqrt{x^{10}} = x^5$

PROBLEM 4

Simplify (all variables are positive):

a. $\sqrt{81y^2}$ b. $\sqrt{9n^4}$

c. $\sqrt{72x^{10}}$ d. $\sqrt{y^{13}}$

Answers to PROBLEMS

3. a. $\frac{\sqrt{7}}{4}$ **b.** $\sqrt{5}$ **c.** $2\sqrt{3}$ **4. a.** $9y$ **b.** $3n^2$ **c.** $6x^5\sqrt{2}$ **d.** $y^6\sqrt{y}$

D > Simplifying Higher Roots

The product and quotient rules can be generalized for higher roots as shown in the box.

PROPERTIES OF RADICALS

For all real numbers where the indicated roots exist,

$$\sqrt[n]{a \cdot b} = \sqrt[n]{a} \cdot \sqrt[n]{b} \quad \text{and} \quad \sqrt[n]{\frac{a}{b}} = \frac{\sqrt[n]{a}}{\sqrt[n]{b}}$$

The idea when simplifying *cube* roots is to find factors that are *perfect cubes.* Similarly, to simplify *fourth roots,* we look for factors that are *perfect fourth powers.* This means that to simplify $\sqrt[4]{48}$, we need to find a perfect fourth-power factor of 48. Since $48 = 16 \cdot 3 = 2^4 \cdot 3$,

$$\sqrt[4]{48} = \sqrt[4]{2^4 \cdot 3} = 2 \cdot \sqrt[4]{3} = 2\sqrt[4]{3}$$

EXAMPLE 5 Simplifying radical expressions involving higher roots

Simplify:

a. $\sqrt[3]{54}$ b. $\sqrt[4]{162}$ c. $\sqrt[3]{\frac{27}{8}}$

SOLUTION 5

a. We are looking for a *cube* root, so we need to find a *perfect cube* factor of 54. Since $2^3 = 8$ is *not* a factor of 54, we try $3^3 = 27$:

$$\sqrt[3]{54} = \sqrt[3]{27 \cdot 2} = \sqrt[3]{3^3 \cdot 2} = 3\sqrt[3]{2}$$

b. This time we need a *perfect fourth* factor of 162. Since $2^4 = 16$ is *not* a factor of 162, we try $3^4 = 81$:

$$\sqrt[4]{162} = \sqrt[4]{81 \cdot 2} = \sqrt[4]{3^4 \cdot 2} = 3\sqrt[4]{2}$$

c. This time we are looking for *perfect cube* factors of 27 and 8. Now $3^3 = 27$ and $2^3 = 8$, so

$$\sqrt[3]{\frac{27}{8}} = \frac{\sqrt[3]{3^3}}{\sqrt[3]{2^3}} = \frac{3}{2}$$

PROBLEM 5

Simplify:

a. $\sqrt[3]{16}$ b. $\sqrt[4]{112}$ c. $\sqrt[3]{\frac{64}{27}}$

NOTE

You can find $\sqrt[3]{\frac{27}{8}}$ *without* using the quotient rule. Since $3^3 = 27$ and $2^3 = 8$, you can write

$$\sqrt[3]{\frac{27}{8}} = \sqrt[3]{\frac{3^3}{2^3}} = \sqrt[3]{\left(\frac{3}{2}\right)^3} = \frac{3}{2}$$

E > Applications Using the Quotient Rule

If an acre of land (43,560 square feet) is to be subdivided into n square plots, with one tree planted in each plot, then the side length S of each plot is given by the formula $S = \sqrt{\frac{43{,}560}{n}}$ (See Example 6, Section 8.1).When n is a perfect square, this formula can be evaluated using the quotient rule. We shall see how in Example 6.

Answers to PROBLEMS

5. a. $2\sqrt[3]{2}$ b. $2\sqrt[4]{7}$ c. $\frac{4}{3}$

GREEN MATH

EXAMPLE 6 Quotient rule and planting trees

If 81 trees are to be planted on an acre of land in square individual plots, use the formula $S = \sqrt{\frac{43{,}560}{n}}$ to find the dimensions of each plot. Answer to the nearest whole number.

SOLUTION 6

We let $n = 81$ in $S = \sqrt{\frac{43{,}560}{n}}$ obtaining $S = \sqrt{\frac{43{,}560}{81}}$.

Using the quotient rule, $\sqrt{\frac{43{,}560}{81}} = \frac{\sqrt{43{,}560}}{\sqrt{81}} = \frac{\sqrt{43{,}560}}{9} \approx 23.$

Thus, the dimension of each square plot is about 23 feet on each side.

Note that you get the same answer if you first divide 43,560 by 81 and then take the square root of the result but the computation is easier if you use the quotient rule.

PROBLEM 6

If 121 trees are to be planted on an acre of land in square individual plots, what should the dimensions of each plot be? Answer to the nearest whole number.

Calculator Corner

Finding Roots of Numbers

You can find the root of a number using your calculator. Start by graphing $y = \sqrt{x}$. To find $\sqrt{25}$, use TRACE to move the cursor slowly through the x-values—$x = 1, x = 2, x = 3$, and so on until you get to $x = 25$. The y-value for $x = 25$ is $y = \sqrt{25} = 5$. (See Window 1.) Did you notice as you traced to the right on the curve $y = \sqrt{x}$ that the y-values for $\sqrt{1}$, $\sqrt{2}$, $\sqrt{3}$, and so on were displayed? In fact, what you have here is a built-in square root table! How would you obtain $\sqrt{8}$ as displayed in Window 2?

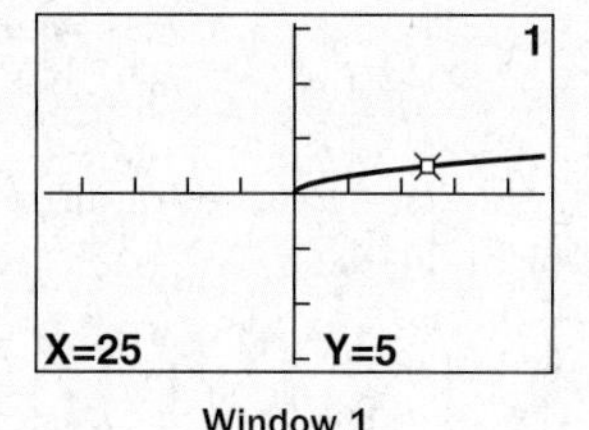

Window 1

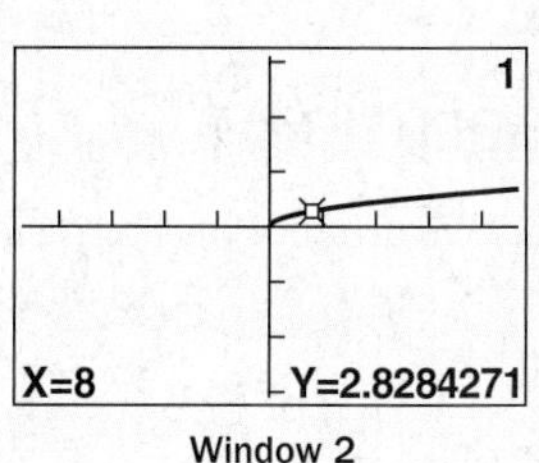

Window 2

Finally, there are some other things to learn from the graph in Window 1. Can you obtain $\sqrt{-2}$? Why not? Can you find fourth roots by graphing $y = \sqrt[4]{x}$? Note that the calculator doesn't have a fourth-root button. If you know that $\sqrt[4]{x} = x^{1/4}$, then you can graph $y = x^{1/4}$ and use a similar procedure to find the fourth root of a number.

Now look at Window 3, where $\sqrt[4]{81}$ is shown. What is the value for $\sqrt[4]{81}$? Can you get $\sqrt[4]{-1}$? Why not? Can you now discover how to obtain $\sqrt[3]{8}$ and $\sqrt[3]{-8}$ with your calculator? Do you get the same answer as you would doing it with paper and pencil?

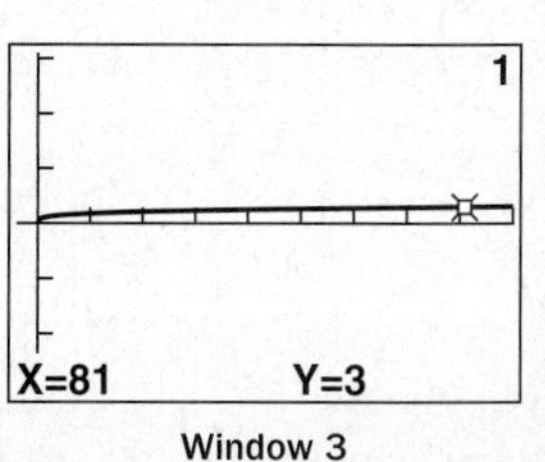

Window 3

› Exercises 8.2

‹ A › Using the Product Rule for Radicals In Problems 1–20, simplify.

1. $\sqrt{45}$	**2.** $\sqrt{128}$	**3.** $\sqrt{125}$	**4.** $\sqrt{175}$	**5.** $\sqrt{180}$
6. $\sqrt{162}$	**7.** $\sqrt{200}$	**8.** $\sqrt{245}$	**9.** $\sqrt{384}$	**10.** $\sqrt{486}$
11. $\sqrt{320}$	**12.** $\sqrt{80}$	**13.** $\sqrt{600}$	**14.** $\sqrt{324}$	**15.** $\sqrt{361}$
16. $\sqrt{648}$	**17.** $\sqrt{700}$	**18.** $\sqrt{726}$	**19.** $\sqrt{432}$	**20.** $\sqrt{507}$

Answers to PROBLEMS

6. 19 feet on each side

In Problems 21–30, find the product. Assume all variables represent positive numbers.

21. $\sqrt{3} \cdot \sqrt{5}$ **22.** $\sqrt{5} \cdot \sqrt{11}$ **23.** $\sqrt{27} \cdot \sqrt{3}$ **24.** $\sqrt{2} \cdot \sqrt{32}$ **25.** $\sqrt{7} \cdot \sqrt{7}$

26. $\sqrt{15} \cdot \sqrt{15}$ **27.** $\sqrt{3} \cdot \sqrt{x}$ **28.** $\sqrt{5} \cdot \sqrt{y}$ **29.** $\sqrt{2a} \cdot \sqrt{18a}$ **30.** $\sqrt{3b} \cdot \sqrt{12b}$

⟨B⟩ Using the Quotient Rule for Radicals In Problems 31–40, find the quotient. Assume all variables represent positive numbers.

31. $\sqrt{\frac{2}{25}}$ **32.** $\sqrt{\frac{5}{81}}$ **33.** $\frac{\sqrt{20}}{\sqrt{5}}$ **34.** $\frac{\sqrt{27}}{\sqrt{9}}$ **35.** $\frac{\sqrt{27}}{\sqrt{3}}$

36. $\frac{\sqrt{72}}{\sqrt{2}}$ **37.** $\frac{15\sqrt{30}}{3\sqrt{10}}$ **38.** $\frac{52\sqrt{28}}{26\sqrt{4}}$ **39.** $\frac{18\sqrt{40}}{3\sqrt{10}}$ **40.** $\frac{22\sqrt{40}}{11\sqrt{10}}$

⟨C⟩ Simplifying Radicals Involving Variables In Problems 41–50, simplify. Assume all variables represent positive numbers.

41. $\sqrt{100a^2}$ **42.** $\sqrt{16b^2}$ **43.** $\sqrt{49a^4}$ **44.** $\sqrt{64x^8}$

45. $-\sqrt{32a^6}$ **46.** $-\sqrt{18b^{10}}$ **47.** $\sqrt{m^{13}}$ **48.** $\sqrt{n^{17}}$

49. $-\sqrt{27m^{11}}$ **50.** $-\sqrt{50n^7}$

⟨D⟩ Simplifying Higher Roots In Problems 51–60, simplify.

51. $\sqrt[3]{40}$ **52.** $\sqrt[3]{108}$ **53.** $\sqrt[3]{-16}$ **54.** $\sqrt[3]{-48}$

55. $\sqrt[3]{\frac{8}{27}}$ **56.** $\sqrt[3]{\frac{27}{64}}$ **57.** $\sqrt[4]{48}$ **58.** $\sqrt[4]{243}$

59. $\sqrt[3]{\frac{64}{27}}$ **60.** $\frac{\sqrt[4]{16}}{\sqrt[4]{81}}$

Applications

⟨E⟩ Applications Using the Quotient Rule

Horizon Use the following information for Problems 61–62. Because of the Earth's curvature, the distance you can see in each direction is bounded by a circle called the **horizon.** The size of this circle depends on how high you are: the greater your height h (in feet), the larger the circle. The distance d (in miles) you can see when your height is h is given by the equation:

$$d = \frac{\sqrt{3h}}{\sqrt{2}}$$

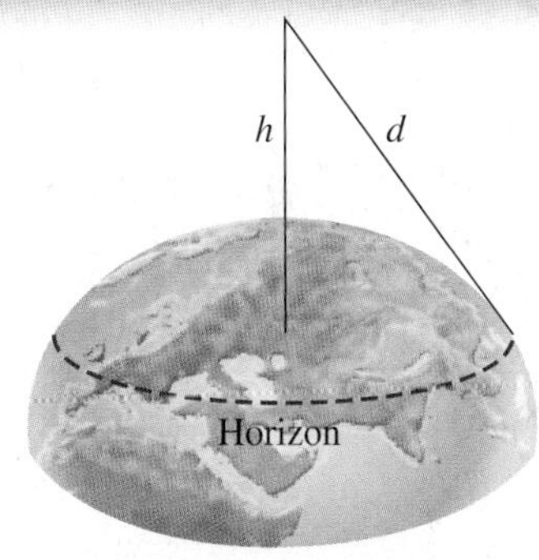

61. Suppose you are on the second floor of a building that is 54 feet high. How far in each direction can you see from the top of this building? (*Hint:* Use the quotient rule of radicals to simplify.)

62. Suppose you are at the top of a tower that is 486 feet tall. How far in each direction can you see from the top of this tower?

Web IT go to mhhe.com/bello for more lessons

Applications: Green Math

63. *Plot dimensions* If 144 trees are to be planted in an acre of land in square individual plots, use $\sqrt{\frac{43,560}{n}}$ and the quotient formula to find the dimensions of each plot. Answer to the nearest whole number.

Write On

64. Explain why $\sqrt{(-2)^2}$ is not -2.

65. Explain why $\sqrt[3]{(-2)^3} = -2$.

66. Which of the following is in simplified form? Explain.

a. $-\sqrt{18}$ **b.** $\sqrt{41}$ **c.** $\sqrt{144}$

67. Assume that p is a prime number.

a. Is $\sqrt{p}$ rational or irrational?

b. Is $\sqrt{p}$ in simplified form? Explain.

Concept Checker

Fill in the blank(s) with the correct word(s), phrase, or mathematical statement.

68. If a and b are **nonnegative** numbers, $\sqrt{a \cdot b} =$ ________________.

69. If a and b are **positive** numbers, $\sqrt{\frac{a}{b}} =$ ____________.

70. For **any** real number a, $\sqrt{a^2} =$ ____________.

71. For **all real** numbers where the roots exist, $\sqrt[n]{a \cdot b} =$ ________________.

a

$|a|$

$\sqrt{a} \cdot \sqrt{b}$

$\frac{\sqrt{a}}{\sqrt{b}}$

$\sqrt[n]{a} \cdot \sqrt[n]{b}$

Mastery Test

Simplify (assume all variables represent positive numbers):

72. $\sqrt[4]{80}$

73. $\sqrt[3]{135}$

74. $\sqrt[4]{16x^4}$

75. $\sqrt[3]{\frac{27}{64}}$

76. $\frac{\sqrt[4]{81}}{\sqrt[4]{16}}$

77. $\sqrt{100x^6}$

78. $\sqrt{32x^7}$

79. $\frac{\sqrt{28}}{\sqrt{14}}$

80. $\frac{\sqrt{72}}{\sqrt{2}}$

81. $\sqrt{\frac{4}{9}}$

82. $\frac{\sqrt{17}}{\sqrt{36}}$

83. $\sqrt{72} \cdot \sqrt{2}$

84. $\sqrt{18} \cdot \sqrt{3x^2}$

Skill Checker

Combine like terms:

85. $5x + 7x$

86. $8x^2 - 3x^2$

87. $9x^3 + 7x^3 - 2x^3$

8.3 Addition and Subtraction of Radicals

Objectives

A › Add and subtract like radicals.

B › Use the distributive property to simplify radicals.

C › Rationalize the denominator in an expression.

To Succeed, Review How To . . .

1. Combine like terms (pp. 89–92).
2. Simplify radicals (pp. 644–648).
3. Use the distributive property (pp. 81–82, 91–92).

Getting Started

A Broken Pattern

In the preceding section we learned the product rule for radicals: $\sqrt{a \cdot b} = \sqrt{a} \cdot \sqrt{b}$, which we know means that the square root of a product is the product of the square roots. Similarly, the square root of a quotient is the quotient of the square roots; that is,

$$\sqrt{\frac{a}{b}} = \frac{\sqrt{a}}{\sqrt{b}}$$

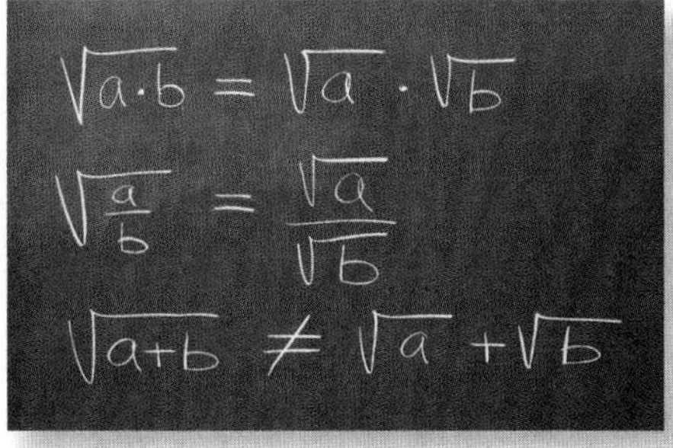

Is the square root of a sum or difference the sum or difference of the square roots? Let's look at an example. Is $\sqrt{9 + 16}$ the same as $\sqrt{9} + \sqrt{16}$? First, $\sqrt{9 + 16} = \sqrt{25} = 5$. But $\sqrt{9} + \sqrt{16} = 3 + 4 = 7$. Thus,

$$\sqrt{9 + 16} \neq \sqrt{9} + \sqrt{16} \quad \text{because} \quad 5 \neq 3 + 4$$

In general,

$$\sqrt{a + b} \neq \sqrt{a} + \sqrt{b}$$

So, what can be done with sums and differences involving radicals? We will learn that next.

A › Adding and Subtracting Radicals

Expressions involving radicals can be handled using simple arithmetic rules. For example, like terms can be combined,

$$3x + 7x = 10x$$

and

$$3\sqrt{6} + 7\sqrt{6} = 10\sqrt{6}$$

Similarly,

$$9x - 2x = 7x$$

and

$$9\sqrt{3} - 2\sqrt{3} = 7\sqrt{3}$$

Test your understanding of this idea in Example 1.

EXAMPLE 1 Adding and subtracting expressions involving radicals

Simplify:

a. $6\sqrt{7} + 9\sqrt{7}$ **b.** $8\sqrt{5} - 2\sqrt{5}$

SOLUTION 1

a. $6\sqrt{7} + 9\sqrt{7} = 15\sqrt{7}$ (just like $6x + 9x = 15x$)

b. $8\sqrt{5} - 2\sqrt{5} = 6\sqrt{5}$ (just like $8x - 2x = 6x$)

PROBLEM 1

Simplify:

a. $8\sqrt{5} + 2\sqrt{5}$ **b.** $7\sqrt{6} - 2\sqrt{6}$

Of course, you may have to simplify before combining **like radical terms**—that is, terms in which the *radical factors* are *exactly* the same. (For example, $4\sqrt{3}$ and $5\sqrt{3}$ are like radical terms.) Here is a problem that may seem difficult:

$$\sqrt{48} + \sqrt{75}$$

In this case, $\sqrt{48}$ and $\sqrt{75}$ are *not* like terms, and only like terms can be combined. However,

$$\sqrt{48} = \sqrt{16 \cdot 3} = \sqrt{16} \cdot \sqrt{3} = 4\sqrt{3}$$
$$\sqrt{75} = \sqrt{25 \cdot 3} = \sqrt{25} \cdot \sqrt{3} = 5\sqrt{3}$$

Now,

$$\sqrt{48} + \sqrt{75} = 4\sqrt{3} + 5\sqrt{3}$$
$$= 9\sqrt{3}$$

EXAMPLE 2 Adding and subtracting expressions involving radicals

Simplify:

a. $\sqrt{80} + \sqrt{20}$ **b.** $\sqrt{75} + \sqrt{12} - \sqrt{147}$

SOLUTION 2

a.
$$\sqrt{80} = \sqrt{16 \cdot 5} = \sqrt{16} \cdot \sqrt{5} = 4\sqrt{5}$$
$$\sqrt{20} = \sqrt{4 \cdot 5} = \sqrt{4} \cdot \sqrt{5} = 2\sqrt{5}$$
$$\sqrt{80} + \sqrt{20} = 4\sqrt{5} + 2\sqrt{5} = 6\sqrt{5}$$

b.
$$\sqrt{75} = \sqrt{25 \cdot 3} = \sqrt{25} \cdot \sqrt{3} = 5\sqrt{3}$$
$$\sqrt{12} = \sqrt{4 \cdot 3} = \sqrt{4} \cdot \sqrt{3} = 2\sqrt{3}$$
$$\sqrt{147} = \sqrt{49 \cdot 3} = \sqrt{49} \cdot \sqrt{3} = 7\sqrt{3}$$
$$\sqrt{75} + \sqrt{12} - \sqrt{147} = 5\sqrt{3} + 2\sqrt{3} - 7\sqrt{3} = 0$$

PROBLEM 2

Simplify:

a. $\sqrt{150} + \sqrt{24}$

b. $\sqrt{20} + \sqrt{80} - \sqrt{45}$

B > Using the Distributive Property to Simplify Expressions

Now let's see how we can use the distributive property to simplify an expression.

EXAMPLE 3 Multiplying expressions involving radicals

Simplify:

a. $\sqrt{5}(\sqrt{40} - \sqrt{2})$ **b.** $\sqrt{2}(\sqrt{2} - \sqrt{3})$

PROBLEM 3

Simplify:

a. $\sqrt{3}(\sqrt{45} - \sqrt{2})$

b. $\sqrt{5}(\sqrt{5} - \sqrt{3})$

Answers to PROBLEMS

1. a. $10\sqrt{5}$ **b.** $5\sqrt{6}$ **2. a.** $7\sqrt{6}$ **b.** $3\sqrt{5}$ **3. a.** $3\sqrt{15} - \sqrt{6}$ **b.** $5 - \sqrt{15}$

SOLUTION 3

a. Using the distributive property,

$\sqrt{5}(\sqrt{40} - \sqrt{2}) = \sqrt{5}\sqrt{40} - \sqrt{5}\sqrt{2}$

$= \sqrt{200} - \sqrt{10}$ Since $\sqrt{5}\sqrt{40} = \sqrt{200}$ and $\sqrt{5}\sqrt{2} = \sqrt{10}$

$= \sqrt{100 \cdot 2} - \sqrt{10}$ Since $\sqrt{100 \cdot 2} = \sqrt{100} \cdot \sqrt{2} = 10\sqrt{2}$

$= 10\sqrt{2} - \sqrt{10}$

b. $\sqrt{2}(\sqrt{2} - \sqrt{3}) = \sqrt{2}\sqrt{2} - \sqrt{2}\sqrt{3}$ Use the distributive property.

$= 2 - \sqrt{6}$ Since $\sqrt{2}\sqrt{2} = 2$

C > Rationalizing Denominators

In Chapter 9 the solution of some quadratic equations will be of the form

$$\sqrt{\frac{9}{5}}$$

If we use the quotient rule for radicals, we obtain

$$\sqrt{\frac{9}{5}} = \frac{\sqrt{9}}{\sqrt{5}} = \frac{3}{\sqrt{5}}$$

The expression $\frac{3}{\sqrt{5}}$ contains the square root of a nonperfect square, which is an irrational number. To simplify $\frac{3}{\sqrt{5}}$, we **rationalize the denominator.** This means we remove all radicals from the denominator.

PROCEDURE

Rationalizing Denominators

Method 1. Multiply *both* the *numerator* and *denominator* of the fraction by the **square root** in the denominator; or

Method 2. Multiply numerator and denominator by the square root of a number that makes the denominator the square root of a perfect square.

Thus, to rationalize the denominator in $\frac{3}{\sqrt{5}}$, we multiply the numerator and denominator by $\sqrt{5}$ to obtain

$$\frac{3}{\sqrt{5}} = \frac{3 \cdot \sqrt{5}}{\sqrt{5} \cdot \sqrt{5}}$$

$$= \frac{3\sqrt{5}}{5}$$ Since $\sqrt{5} \cdot \sqrt{5} = 5$

Note that the idea in rationalizing

$$\sqrt{\frac{a}{b}} = \frac{\sqrt{a}}{\sqrt{b}}$$

is to make the denominator $\sqrt{b}$ a square root of a perfect square. Multiplying numerator and denominator by $\sqrt{b}$ *always* works, but you can save time if you find a factor smaller than $\sqrt{b}$ that will make the denominator a square root of a perfect square. Thus, when rationalizing $\frac{\sqrt{3}}{\sqrt{8}}$, you could *first* multiply numerator and denominator by $\sqrt{8}$. However, it's *better* to multiply numerator and denominator by $\sqrt{2}$, as shown in Example 4.

EXAMPLE 4 Rationalizing denominators: Two methods

Write with a rationalized denominator: $\sqrt{\frac{3}{8}}$

SOLUTION 4

Method 1. Use the quotient rule and then multiply numerator and denominator by $\sqrt{8}$.

$$\sqrt{\frac{3}{8}} = \frac{\sqrt{3}}{\sqrt{8}} \quad \text{Use the quotient rule.}$$

$$= \frac{\sqrt{3}\cdot\sqrt{8}}{\sqrt{8}\cdot\sqrt{8}} \quad \text{Multiply numerator and denominator by } \sqrt{8}.$$

$$= \frac{\sqrt{24}}{8} \quad \text{Since } \sqrt{3}\cdot\sqrt{8} = \sqrt{24} \text{ and } \sqrt{8}\cdot\sqrt{8} = 8$$

$$= \frac{\sqrt{4\cdot 6}}{8} \quad \text{Since } 24 = 4\cdot 6 \text{ and 4 is a perfect square}$$

$$= \frac{2\cdot\sqrt{6}}{8} \quad \text{Since } \sqrt{4} = 2$$

$$= \frac{\sqrt{6}}{4} \quad \text{Divide numerator and denominator by 2.}$$

Method 2. If you noticed that multiplying numerator and denominator by $\sqrt{2}$ yields $\sqrt{16} = 4$ in the denominator, you would have obtained

$$\sqrt{\frac{3}{8}} = \frac{\sqrt{3}}{\sqrt{8}} \quad \text{Use the quotient rule.}$$

$$= \frac{\sqrt{3}\cdot\sqrt{2}}{\sqrt{8}\cdot\sqrt{2}} \quad \text{Multiply by } \sqrt{2} \text{ so the denominator is } \sqrt{16} = 4.$$

$$= \frac{\sqrt{6}}{\sqrt{16}} \quad \text{Since } \sqrt{3}\cdot\sqrt{2} = \sqrt{6}$$

$$= \frac{\sqrt{6}}{4} \quad \text{Since } \sqrt{16} = 4$$

You get the same answer as in Method 1 by multiplying by $\sqrt{2}$. As you can see fewer steps were involved when simplifying! *You can save time by looking for factors that make the denominator the square root of a perfect square.*

PROBLEM 4

Write with a rationalized denominator: $\sqrt{\frac{5}{12}}$

EXAMPLE 5 Rationalizing denominators by making them perfect squares

Write with a rationalized denominator: $\sqrt{\frac{x^2}{32}}$ $(x > 0)$

SOLUTION 5

Since

$$\sqrt{\frac{x^2}{32}} = \frac{\sqrt{x^2}}{\sqrt{32}}$$

we could multiply numerator and denominator by $\sqrt{32}$. However, multiplying by $\sqrt{2}$ gives a denominator of $\sqrt{64} = 8$. Thus,

$$\sqrt{\frac{x^2}{32}} = \frac{\sqrt{x^2}\cdot\sqrt{2}}{\sqrt{32}\cdot\sqrt{2}}$$

$$= \frac{x\sqrt{2}}{\sqrt{64}}$$

$$= \frac{x\sqrt{2}}{8}$$

PROBLEM 5

Write with a rationalized denominator:

$$\sqrt{\frac{x^2}{50}} \quad (x > 0)$$

Answers to PROBLEMS

4. $\frac{\sqrt{15}}{6}$ 5. $\frac{x\sqrt{2}}{10}$

Some scientists claim that the severity and duration of storms is influenced by climate change. Do you live in an area where they have frequent summer storms? Can you predict how long they will last? The formula that approximates the time t (in hours) a storm will last based on the diameter d (in miles) of the storm is given by $t = \sqrt{\left(\frac{d}{6}\right)^3}$. We will use this formula to find the predicted time a storm lasts.

GREEN MATH

EXAMPLE 6 Storm duration predictions

How long will a storm with a diameter of 8 miles last? Give the answer with a rationalized denominator and as an approximation.

SOLUTION 6

We substitute $d = 8$ in $\sqrt{\left(\frac{d}{6}\right)^3}$ obtaining $\sqrt{\left(\frac{8}{6}\right)^3} = \sqrt{\left(\frac{4}{3}\right)^3}$. Here are the steps to write the answer in rationalized form:

$$\sqrt{\left(\frac{4}{3}\right)^3} = \sqrt{\left(\frac{4^3}{3^3}\right)} \quad \text{Using the power rule for quotients}$$

$$= \sqrt{\left(\frac{4^3 \cdot 3}{3^3 \cdot 3}\right)} \quad \text{To make the denominator a perfect square}$$

$$= \sqrt{\left(\frac{64 \cdot 3}{3^4}\right)} \quad \text{Since } 4^3 = 64 \text{ and } 3^3 \cdot 3 = 3^4$$

$$= \frac{\sqrt{64} \cdot \sqrt{3}}{\sqrt{3^4}}$$

$$= \frac{8\sqrt{3}}{3^2} \quad \sqrt{64} = 8, \sqrt{\frac{1}{3^4}} = \frac{1}{3^2}$$

$$= \frac{8\sqrt{3}}{9}$$

$$\approx 1.54 \quad \text{Using a calculator}$$

Thus, the storm will last a little more than $1\frac{1}{2}$ hours.

PROBLEM 6

Use the formula to predict the duration t of a storm that has a diameter of 9 miles. Give the answer with a rationalized denominator and as an approximation.

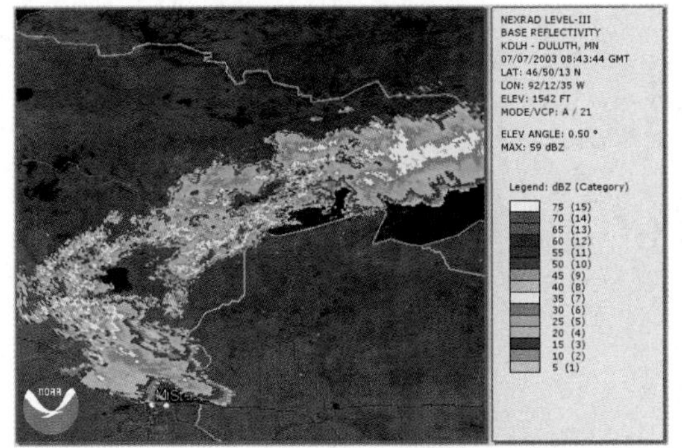

The water from thunderstorms carries a lot of sediments and pollutants with it. One such storm carried over 12,000 pounds of habitat smothering, gill fouling mud past a sensor in Tischer Creek (Duluth, Minnesota) in just a few hours.

Exercises 8.3

> Practice Problems > Self-Tests > Media-rich eBooks > e-Professors > Videos

⟨A⟩ Adding and Subtracting Radicals In Problems 1–16, perform the indicated operations and simplify.

1. $6\sqrt{7} + 4\sqrt{7}$
2. $4\sqrt{11} + 9\sqrt{11}$
3. $9\sqrt{13} - 4\sqrt{13}$
4. $6\sqrt{10} - 2\sqrt{10}$
5. $\sqrt{32} + \sqrt{50} - \sqrt{72}$
6. $\sqrt{12} + \sqrt{27} - \sqrt{75}$
7. $\sqrt{162} + \sqrt{50} - \sqrt{200}$
8. $\sqrt{48} + \sqrt{75} - \sqrt{363}$
9. $9\sqrt{48} - 5\sqrt{27} + 3\sqrt{12}$
10. $3\sqrt{32} - 5\sqrt{8} + 4\sqrt{50}$
11. $5\sqrt{7} - 3\sqrt{28} - 2\sqrt{63}$
12. $3\sqrt{28} - 6\sqrt{7} - 2\sqrt{175}$
13. $-5\sqrt{3} + 8\sqrt{75} - 2\sqrt{27}$
14. $-6\sqrt{99} + 6\sqrt{44} - \sqrt{176}$
15. $-3\sqrt{45} + \sqrt{20} - \sqrt{5}$
16. $-5\sqrt{27} + \sqrt{12} - 5\sqrt{48}$

Answers to PROBLEMS

6. $t = \frac{3\sqrt{6}}{4} \approx 1.84$ or a little less than 2 hours.

‹B› Using the Distributive Property to Simplify Expressions In Problems 17–30, simplify.

17. $\sqrt{10}(\sqrt{20} - \sqrt{3})$

18. $\sqrt{10}(\sqrt{30} - \sqrt{2})$

19. $\sqrt{6}(\sqrt{14} + \sqrt{5})$

20. $\sqrt{14}(\sqrt{18} + \sqrt{3})$

21. $\sqrt{3}(\sqrt{3} - \sqrt{2})$

22. $\sqrt{6}(\sqrt{6} - \sqrt{5})$

23. $\sqrt{5}(\sqrt{2} + \sqrt{5})$

24. $\sqrt{3}(\sqrt{2} + \sqrt{3})$

25. $\sqrt{6}(\sqrt{2} - \sqrt{3})$

26. $\sqrt{5}(\sqrt{15} - \sqrt{27})$

27. $2(\sqrt{2} - 5)$

28. $\sqrt{5}(3 - \sqrt{5})$

29. $\sqrt{2}(\sqrt{6} - 3)$

30. $\sqrt{3}(\sqrt{6} - 4)$

‹C› Rationalizing Denominators In Problems 31–50, rationalize the denominator. Assume all variables are positive real numbers.

31. $\frac{3}{\sqrt{6}}$

32. $\frac{6}{\sqrt{7}}$

33. $\frac{-10}{\sqrt{5}}$

34. $\frac{-9}{\sqrt{3}}$

35. $\frac{\sqrt{8}}{\sqrt{2}}$

36. $\frac{\sqrt{48}}{\sqrt{3}}$

37. $\frac{-\sqrt{2}}{\sqrt{5}}$

38. $\frac{-\sqrt{3}}{\sqrt{7}}$

39. $\frac{\sqrt{2}}{\sqrt{8}}$

40. $\frac{\sqrt{3}}{\sqrt{12}}$

41. $\frac{\sqrt{x^2}}{\sqrt{18}}$

42. $\frac{\sqrt{a^4}}{\sqrt{32}}$

43. $\frac{\sqrt{a^2}}{\sqrt{b}}$

44. $\frac{\sqrt{x^4}}{\sqrt{y}}$

45. $\sqrt{\frac{3}{10}}$

46. $\sqrt{\frac{2}{27}}$

47. $\sqrt{\frac{x^2}{32}}$

48. $\sqrt{\frac{x}{18}}$

49. $\sqrt{\frac{x^4}{20}}$

50. $\sqrt{\frac{x^6}{72}}$

Web IT go to mhhe.com/bello for more lessons

››› Applications: Green Math

51. *Meteorology* Use the formula $t = \sqrt{\left(\frac{d}{6}\right)^3}$ (t is time in hours, d is diameter in miles) and follow the procedure of Example 6 to find:

a. How long will a storm 6 miles in diameter last?

b. How long will a storm 10 miles in diameter last? Give the answer with a rationalized denominator and as an approximation.

52. *Meteorology* A storm 3 miles in diameter is threatening a baseball game, which must be resumed within an hour or the game will have to be postponed.

a. How long will the storm last? (Give the answer with a rationalized denominator).

b. Will the game be resumed or postponed?

53. *Sphere* The radius r of a sphere is given by the equation $r = \sqrt{\frac{S}{4\pi}}$, where S is the surface area of the sphere. Write an expression for r with a rationalized denominator.

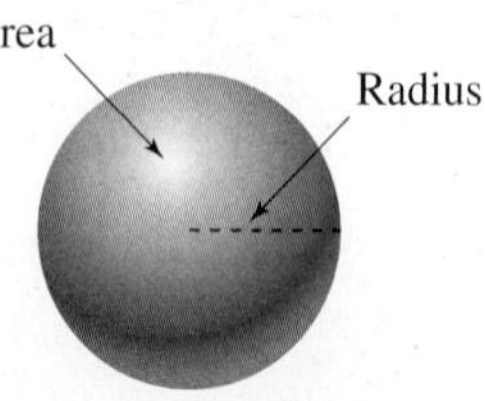

54. *Cone* The radius r of a cone is given by the equation $r = \sqrt{\frac{3V}{\pi h}}$, where V is the volume of the cone and h is its height. Write an expression for r with a rationalized denominator.

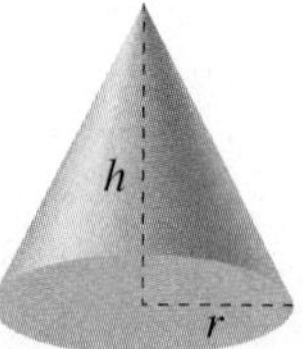

Using Your Knowledge

"Radical" Shortcuts Suppose you want to rationalize the denominator in the expression $\sqrt{\frac{3}{32}}$. Using the quotient rule, we can write

$$\sqrt{\frac{3}{32}} = \frac{\sqrt{3}}{\sqrt{32}} = \frac{\sqrt{3}\cdot\sqrt{32}}{\sqrt{32}\cdot\sqrt{32}} = \frac{\sqrt{96}}{32}$$

$$= \frac{\sqrt{16\cdot 6}}{32} = \frac{4\cdot\sqrt{6}}{32} = \frac{\sqrt{6}}{8}$$

A shorter way is as follows:

$$\sqrt{\frac{3}{32}} = \sqrt{\frac{3\cdot 2}{32\cdot 2}} = \sqrt{\frac{6}{64}} = \frac{\sqrt{6}}{8}$$

55. Use this shorter procedure to do Problems 45–50.

Write On

56. In Problems 41–50, we specified that *all* variables should be positive real numbers. Specify which variables *have to be* positive real numbers and in which problems this must occur.

57. In Example 4, we rationalized the denominator in the expression

$$\sqrt{\frac{3}{8}} = \frac{\sqrt{3}}{\sqrt{8}}$$

by *first* using the quotient rule and writing

$$\sqrt{\frac{3}{8}} = \frac{\sqrt{3}}{\sqrt{8}}$$

and then multiplying numerator and denominator by $\sqrt{2}$. How can you do the example without first using the quotient rule?

58. Write an explanation of what is meant by like radicals.

Concept Checker

Fill in the blank(s) with the correct word(s), phrase, or mathematical statement.

59. Like radical terms are terms in which the radical **factors** are ____________ the same.

60. To **rationalize** the denominator means to **remove** all ____________ from the **denominator.**

almost **radicals**
numbers **exactly**

Mastery Test

Simplify:

61. $8\sqrt{3} + 5\sqrt{3}$ **62.** $7\sqrt{5} - 2\sqrt{5}$ **63.** $\sqrt{18} + 3\sqrt{2}$

64. $\sqrt{32} - 3\sqrt{2}$ **65.** $7\sqrt{18} + 5\sqrt{2} - 7\sqrt{8}$ **66.** $3(\sqrt{5} - 2)$

67. $\sqrt{3}(5 - \sqrt{3})$ **68.** $\sqrt{18}(\sqrt{2} - 2)$

Rationalize the denominator, assuming all variables are positive real numbers:

69. $\frac{3}{\sqrt{7}}$ **70.** $\sqrt{\frac{x^6}{10}}$ **71.** $\frac{\sqrt{x}}{\sqrt{2}}$ **72.** $\frac{\sqrt{x^2}}{\sqrt{20}}$

Skill Checker

Multiply:

73. $(x + 3)(x - 3)$ **74.** $(a + b)(a - b)$ **75.** $\frac{6x + 12}{3}$ **76.** $\frac{4x - 8}{2}$

8.4 Simplifying Radicals

Objectives

A Simplify a radical expression involving products, quotients, sums, or differences.

B Use the conjugate of a number to rationalize the denominator of an expression.

C Reduce a fraction involving a radical by factoring.

D Use radicals to solve application problems.

To Succeed, Review How To . . .

1. Find the root of an expression (pp. 639–640).
2. Add, subtract, multiply, and divide radicals (pp. 644–645, 651–652).
3. Rationalize the denominator in an expression (pp. 653–654).
4. Expand binomials of the form $(x \pm y)^2$ (pp. 378–382).

Getting Started

Breaking the Sound Barrier

How fast can this plane travel? The answer is classified information, but it exceeds twice the speed of sound (747 miles per hour). It is said that the plane's speed is more than Mach 2. The formula for calculating the Mach number is

$$M = \sqrt{\frac{2}{\gamma}}\sqrt{\frac{P_2 - P_1}{P_1}}$$

where P_1 and P_2 are air pressures and γ is a constant. This expression can be simplified by multiplying the radical expressions and then rationalizing the denominator (we discuss how to do this in the *Using Your Knowledge*). In this section you will learn how to simplify more complicated radical expressions such as the one on page 659 by using the techniques we've discussed in the last three sections. What did we do in those sections? We found the roots of algebraic expressions, multiplied and divided radicals, added and subtracted radicals, and rationalized denominators. You will see that the procedure used to *completely* simplify a radical involves steps in which you will perform these tasks in precisely the order in which the topics were studied.

A › Simplifying Radical Expressions

In the preceding sections you were asked to "simplify" expressions involving radicals. To make this idea more precise and to help you simplify radical expressions, we use the following rules.

RULES

Simplifying Radical Expressions

1. Whenever possible, write the *rational-number* representation of a radical expression. For example, write

$$\sqrt{81} \text{ as } 9, \quad \sqrt{\frac{4}{9}} \text{ as } \frac{2}{3} \quad \text{and} \quad \sqrt[3]{\frac{1}{8}} \text{ as } \frac{1}{2}$$

2. Use the product rule $\sqrt{x} \cdot \sqrt{y} = \sqrt{xy}$ to write indicated products as a single radical. For example, write

$$\sqrt{6} \text{ instead of } \sqrt{2} \cdot \sqrt{3} \quad \text{and} \quad \sqrt{2ab} \text{ instead of } \sqrt{2a} \cdot \sqrt{b}$$

3. Use the quotient rule

$$\frac{\sqrt{x}}{\sqrt{y}} = \sqrt{\frac{x}{y}}$$

to write indicated quotients as a single radical. For example, write

$$\frac{\sqrt{6}}{\sqrt{2}} \text{ as } \sqrt{3} \quad \text{and} \quad \frac{\sqrt[3]{10}}{\sqrt[3]{5}} \text{ as } \sqrt[3]{2}$$

4. If a radicand has a perfect square as a factor, write the radical expression as the product of the square root of the perfect square and the radical of the other factor. A similar statement applies to cubes and higher roots. For example, write

$$\sqrt{18} = \sqrt{9 \cdot 2} \text{ as } 3\sqrt{2} \quad \text{and} \quad \sqrt[3]{54} = \sqrt[3]{27 \cdot 2} \text{ as } 3\sqrt[3]{2}$$

5. Combine like radicals whenever possible. For example,

$$2\sqrt{5} + 8\sqrt{5} = (2 + 8)\sqrt{5} = 10\sqrt{5}$$
$$9\sqrt{11} - 2\sqrt{11} = (9 - 2)\sqrt{11} = 7\sqrt{11}$$

6. Rationalize the denominator of algebraic expressions. For example,

$$\frac{3}{\sqrt{2}} = \frac{3 \cdot \sqrt{2}}{\sqrt{2} \cdot \sqrt{2}} = \frac{3\sqrt{2}}{2}$$

Now let's use these rules.

EXAMPLE 1 Simplifying radicals: Sums and differences

Simplify:

a. $\sqrt{9 + 16} - \sqrt{4}$ b. $9\sqrt{6} - \sqrt{2} \cdot \sqrt{3}$ c. $\dfrac{\sqrt{6x^3}}{\sqrt{2x^2}}$, $x > 0$

SOLUTION 1

In each case, a rule applies.

a. Since $\sqrt{9 + 16} = \sqrt{25} = 5$ and $\sqrt{4} = 2$, Use Rules 1 and 5.

$\sqrt{9 + 16} - \sqrt{4} - 5 - 2 = 3$

b. $9\sqrt{6} - \sqrt{2} \cdot \sqrt{3} = 9\sqrt{6} - \sqrt{6} = (9 - 1)\sqrt{6} = 8\sqrt{6}$ Use Rules 2 and 5.

c. $\dfrac{\sqrt{6x^3}}{\sqrt{2x^2}} = \sqrt{\dfrac{6x^3}{2x^2}} = \sqrt{3x}$ Use Rule 3.

PROBLEM 1

Simplify:

a. $\sqrt{60 + 4} - \sqrt{9}$

b. $8\sqrt{10} - \sqrt{2}\sqrt{5}$

c. $\dfrac{\sqrt{10x^3}}{\sqrt{5x^2}}$, $x > 0$

EXAMPLE 2 Simplifying quotients

Simplify:

a. $\dfrac{\sqrt[3]{256}}{\sqrt[3]{2}}$ b. $\dfrac{9}{\sqrt[3]{4}}$

SOLUTION 2

A rule applies in each case.

a. $\dfrac{\sqrt[3]{256}}{\sqrt[3]{2}} = \sqrt[3]{128}$ Use Rule 3.

$= \sqrt[3]{64 \cdot 2}$ Since $64 \cdot 2 = 128$ and 64 is a perfect cube

$= 4\sqrt[3]{2}$ Use Rule 4.

PROBLEM 2

Simplify:

a. $\dfrac{\sqrt[3]{108}}{\sqrt[3]{2}}$ b. $\dfrac{7}{\sqrt[3]{9}}$

(continued)

Answers to PROBLEMS

1. a. 5 b. $7\sqrt{10}$ c. $\sqrt{2x}$ 2. a. $3\sqrt[3]{2}$ b. $\dfrac{7\sqrt[3]{3}}{3}$

b. We have to make the denominator the cube root of a perfect cube. To do this, we multiply numerator and denominator by $\sqrt[3]{2}$. Note that the denominator will be $\sqrt[3]{4} \cdot \sqrt[3]{2} = \sqrt[3]{8} = 2$. Thus, we have

$$\frac{9}{\sqrt[3]{4}} = \frac{9 \cdot \sqrt[3]{2}}{\sqrt[3]{4} \cdot \sqrt[3]{2}}$$
$$= \frac{9 \cdot \sqrt[3]{2}}{\sqrt[3]{8}}$$
$$= \frac{9\sqrt[3]{2}}{2}$$

Do you recall the FOIL method? It and the special products can also be used to simplify expressions involving radicals. We do this next.

EXAMPLE 3 **Using FOIL to simplify products**

Simplify:

a. $(\sqrt{2} + 5\sqrt{3})(\sqrt{2} - 4\sqrt{3})$ **b.** $(\sqrt{3} + 2\sqrt{5})(\sqrt{3} - 2\sqrt{5})$

SOLUTION 3 Using the FOIL method, we have

a. $(\sqrt{2} + 5\sqrt{3})(\sqrt{2} - 4\sqrt{3})$

$$= \underbrace{\sqrt{2} \cdot \sqrt{2}}_{F} + \underbrace{\sqrt{2}(-4\sqrt{3})}_{O} + \underbrace{5\sqrt{3}(\sqrt{2})}_{I} + \underbrace{(5\sqrt{3})(-4\sqrt{3})}_{L}$$

$= 2 - 4\sqrt{6} + 5\sqrt{6} - 20 \cdot 3$ Use the product rule.

$= 2 + \sqrt{6} - 60$ Combine radicals.

$= -58 + \sqrt{6}$ Since $2 - 60 = -58$

b. Using the special products formula SP4 (p. 380), we have

$$(X + A)(X - A) = X^2 - A^2$$
$$(\sqrt{3} + 2\sqrt{5})(\sqrt{3} - 2\sqrt{5}) = (\sqrt{3})^2 - (2\sqrt{5})^2$$
$$= 3 - [(2)^2(\sqrt{5})^2]$$
$$= 3 - (4)(5)$$
$$= 3 - 20$$
$$= -17$$

PROBLEM 3

Simplify:

a. $(\sqrt{3} + 5\sqrt{2})(\sqrt{3} - 4\sqrt{2})$

b. $(2 + 3\sqrt{5})(2 - 3\sqrt{5})$

B › Using Conjugates to Rationalize Denominators

In Example 3(b), the product of the sum $\sqrt{3} + 2\sqrt{5}$ and the difference $\sqrt{3} - 2\sqrt{5}$ is the rational number -17. This is no coincidence! The expressions $\sqrt{3} + 2\sqrt{5}$ and $\sqrt{3} - 2\sqrt{5}$ are **conjugates** of each other. In general, the expressions $a\sqrt{b} + c\sqrt{d}$ and $a\sqrt{b} - c\sqrt{d}$ are conjugates of each other. Their product is obtained by using the special products formula SP4 (p. 380) and always results in a rational number. Here is one way of using conjugates to simplify radical expressions.

Answers to PROBLEMS

3. a. $-37 + \sqrt{6}$ **b.** -41

PROCEDURE

Using Conjugates to Simplify Radical Expressions

To simplify an algebraic expression with two terms in the denominator, at least one of which is a square root, multiply both numerator and denominator by the conjugate of the denominator.

EXAMPLE 4 Using conjugates to rationalize denominators

Simplify:

a. $\dfrac{7}{\sqrt{5}+1}$ **b.** $\dfrac{3}{\sqrt{5}-\sqrt{3}}$

PROBLEM 4

Simplify:

a. $\dfrac{5}{\sqrt{7}+1}$ **b.** $\dfrac{5}{\sqrt{6}-\sqrt{2}}$

SOLUTION 4

a. The denominator $\sqrt{5}+1$ has two terms, one of which is a radical. To simplify

$$\frac{7}{\sqrt{5}+1}$$

multiply numerator and denominator by the conjugate of the denominator, which is $\sqrt{5}-1$. [*Note:* $(\sqrt{5}+1)(\sqrt{5}-1)=(\sqrt{5})^2-(1)^2$.] Thus,

$$\frac{7}{\sqrt{5}+1}=\frac{7\cdot(\sqrt{5}-1)}{(\sqrt{5}+1)(\sqrt{5}-1)}$$ Multiply the numerator and denominator by $\sqrt{5}-1$.

$$=\frac{7\sqrt{5}-7}{(\sqrt{5})^2-(1)^2}$$ Use the distributive property and SP4.

$$=\frac{7\sqrt{5}-7}{5-1}$$ Since $(\sqrt{5})^2=5$

$$=\frac{7\sqrt{5}-7}{4}$$

b. This time we multiply the numerator and denominator of the fraction by the conjugate of $\sqrt{5}-\sqrt{3}$, which is $\sqrt{5}+\sqrt{3}$.

$$\frac{3}{\sqrt{5}-\sqrt{3}}=\frac{3(\sqrt{5}+\sqrt{3})}{(\sqrt{5}-\sqrt{3})(\sqrt{5}+\sqrt{3})}$$ Multiply the numerator and denominator by $\sqrt{5}+\sqrt{3}$.

$$=\frac{3\sqrt{5}+3\sqrt{3}}{(\sqrt{5})^2-(\sqrt{3})^2}$$ Use the distributive property and SP4.

$$=\frac{3\sqrt{5}+3\sqrt{3}}{5-3}$$ Since $(\sqrt{5})^2=5$ and $(\sqrt{3})^2=3$

$$=\frac{3\sqrt{5}+3\sqrt{3}}{2}$$

C > Reducing Fractions Involving Radicals by Factoring

In the next chapter we will encounter solutions of quadratic equations written as

$$\frac{8+\sqrt{20}}{4}$$

Answers to PROBLEMS

4. a. $\dfrac{5\sqrt{7}-5}{6}$ **b.** $\dfrac{5\sqrt{6}+5\sqrt{2}}{4}$

To simplify these expressions, first note that $\sqrt{20} = \sqrt{4 \cdot 5} = 2\sqrt{5}$. Thus,

$$\frac{8 + \sqrt{20}}{4} = \frac{8 + 2\sqrt{5}}{4} \quad \text{Since } \sqrt{20} = 2\sqrt{5}$$

$$= \frac{2 \cdot (4 + \sqrt{5})}{2 \cdot 2} \quad \text{Factor the numerator and denominator.}$$

$$= \frac{4 + \sqrt{5}}{2} \quad \text{Divide by 2.}$$

EXAMPLE 5 **Reducing fractions by factoring first**

Simplify:

a. $\frac{-4 + \sqrt{8}}{2}$ **b.** $\frac{-8 + \sqrt{28}}{4}$

SOLUTION 5

a. Since $\sqrt{8} = \sqrt{4 \cdot 2} = 2\sqrt{2}$, we have

$$\frac{-4 + \sqrt{8}}{2} = \frac{-4 + 2\sqrt{2}}{2}$$

$$= \frac{2 \cdot (-2 + \sqrt{2})}{2} \quad \text{Factor the numerator and denominator.}$$

$$= -2 + \sqrt{2} \quad \text{Divide numerator and denominator by 2.}$$

b. Since $\sqrt{28} = \sqrt{4 \cdot 7} = 2\sqrt{7}$, we have

$$\frac{-8 + \sqrt{28}}{4} = \frac{-8 + 2\sqrt{7}}{4}$$

$$= \frac{2(-4 + \sqrt{7})}{2 \cdot 2} \quad \text{Factor the numerator and denominator.}$$

$$= \frac{-4 + \sqrt{7}}{2} \quad \text{Divide numerator and denominator by 2.}$$

PROBLEM 5

Simplify:

a. $\frac{-9 + \sqrt{18}}{3}$

b. $\frac{-12 + \sqrt{24}}{6}$

D › Applications Involving Radicals

Earthquakes, volcanic eruptions, giant landslides, and tsunamis (a very large ocean wave caused by an underwater earthquake or volcanic eruption) may become more frequent as global warming changes the earth's crust, scientists said in London. How fast do these waves travel? We shall see next.

Source: http://tinyurl.com/y8gq756.

EXAMPLE 6 **Speed of tsunamis and radicals**

The speed S (in meters per second, m/sec) of a tsunami is given by the equation $S = \sqrt{g} \cdot \sqrt{d}$, where $g \approx 10$ m/sec^2 is the acceleration due to gravity and d is the average depth of the water in meters.

a. Find the speed S of a tsunami when the average depth d of the water is 40 meters.

b. On December 26, 2004, a giant tsunami hit Indonesia. The average depth of the ocean at that location is about 4000 m. How fast was that tsunami moving?

PROBLEM 6

Use the formula of Example 6 to find the speed of a tsunami that hit Samoa in 2009 if the depth of the water around Samoa is assumed to be

a. 1000 m **b.** 160 m

Answers to PROBLEMS

5. a. $-3 + \sqrt{2}$ **b.** $\frac{-6 + \sqrt{6}}{3}$ **6. a.** 100 m/sec or about 224 mph **b.** 40 m/sec which is almost 90 mph

SOLUTION 6

a. We use Rule 2 on page 659 and simplify $S = \sqrt{g} \cdot \sqrt{d}$ to $\sqrt{gd}$. Letting $g = 10$ and $d = 40$, $S = \sqrt{gd} = \sqrt{10 \cdot 40} = \sqrt{400} = 20$ m/sec.

b. Using the formula $S = \sqrt{gd}$ with $g = 10$ and $d = 4000$, we get $S = \sqrt{gd} = \sqrt{10 \cdot 4000} = \sqrt{40{,}000} = 200$ m/sec.

Note that 200 m/sec ≈ 450 mph. You can check this at http://www.unitarium.com/speed or you can do the conversion yourself!

› Exercises 8.4

> Practice Problems > Self-Tests > Media-rich eBooks > e-Professors > Videos

‹A› Simplifying Radical Expressions In Problems 1–36, simplify. Assume all variables represent positive real numbers.

1. $\sqrt{36} + \sqrt{100}$ **2.** $\sqrt{9} + \sqrt{25}$ **3.** $\sqrt{144 + 25}$

4. $\sqrt{24^2 + 10^2}$ **5.** $\sqrt{4} - \sqrt{36}$ **6.** $\sqrt{64} - \sqrt{121}$

7. $\sqrt{13^2 - 12^2}$ **8.** $\sqrt{17^2 - 8^2}$ **9.** $15\sqrt{10} + \sqrt{90}$

10. $8\sqrt{7} + \sqrt{28}$ **11.** $14\sqrt{11} - \sqrt{44}$ **12.** $3\sqrt{13} - \sqrt{52}$

13. $\sqrt[3]{54} - \sqrt[3]{8}$ **14.** $\sqrt[3]{81} - \sqrt[3]{16}$ **15.** $5\sqrt[3]{16} - 3\sqrt[3]{54}$

16. $\sqrt[3]{250} - \sqrt[3]{128}$ **17.** $\sqrt{\frac{9x^2}{x}}$ **18.** $\sqrt{\frac{16x^5}{x^2}}$

19. $\sqrt{\frac{81y^7}{16y^5}}$ **20.** $\sqrt{\frac{4x^3y^4}{3z}}$ **21.** $\sqrt{\frac{64a^4b^6}{3ab^4}}$

22. $\sqrt{\frac{25a^5b^6}{7a^2b^4c^2}}$ **23.** $\sqrt[3]{\frac{8b^6c^{10}}{27bc}}$ **24.** $\sqrt[3]{\frac{64ab^4}{125a^4b}}$

25. $\frac{\sqrt[3]{500}}{\sqrt[3]{2}}$ **26.** $\frac{\sqrt[3]{243}}{\sqrt[3]{3}}$ **27.** $\frac{6}{\sqrt[3]{9}}$

28. $\frac{7}{\sqrt[3]{2}}$ **29.** $(\sqrt{3} + 6\sqrt{5})(\sqrt{3} - 4\sqrt{5})$

30. $(\sqrt{5} + 6\sqrt{2})(\sqrt{5} - 3\sqrt{2})$ **31.** $(\sqrt{2} + 3\sqrt{3})(\sqrt{2} + 3\sqrt{3})$

32. $(3\sqrt{2} + \sqrt{3})(3\sqrt{2} + \sqrt{3})$ **33.** $(5\sqrt{2} - 3\sqrt{3})(5\sqrt{2} - \sqrt{3})$

34. $(7\sqrt{5} - 4\sqrt{2})(7\sqrt{5} - 4\sqrt{2})$ **35.** $(\sqrt{13} + 2\sqrt{2})(\sqrt{13} - 2\sqrt{2})$

36. $(\sqrt{17} + 3\sqrt{5})(\sqrt{17} - 3\sqrt{5})$

‹B› Using Conjugates to Rationalize Denominators In Problems 37–50, rationalize the denominator.

37. $\frac{3}{\sqrt{2} + 1}$ **38.** $\frac{5}{\sqrt{5} + 1}$ **39.** $\frac{4}{\sqrt{7} - 1}$

40. $\frac{6}{\sqrt{7} - 2}$ **41.** $\frac{\sqrt{2}}{2 + \sqrt{3}}$ **42.** $\frac{\sqrt{3}}{3 + \sqrt{2}}$

43. $\frac{\sqrt{5}}{2 - \sqrt{3}}$ **44.** $\frac{\sqrt{6}}{3 - \sqrt{5}}$ **45.** $\frac{\sqrt{5}}{\sqrt{2} + \sqrt{3}}$

› Web IT go to mhhe.com/bello for more lessons

46. $\frac{\sqrt{2}}{\sqrt{5}+\sqrt{3}}$

47. $\frac{\sqrt{3}}{\sqrt{5}-\sqrt{2}}$

48. $\frac{6}{\sqrt{6}-\sqrt{2}}$

49. $\frac{\sqrt{3}+\sqrt{2}}{\sqrt{3}-\sqrt{2}}$

50. $\frac{\sqrt{5}-\sqrt{2}}{\sqrt{5}+\sqrt{2}}$

‹ C › Reducing Fractions Involving Radicals by Factoring In Problems 51–62, reduce the fraction.

51. $\frac{-8+\sqrt{16}}{2}$

52. $\frac{-4+\sqrt{36}}{4}$

53. $\frac{-6-\sqrt{4}}{6}$

54. $\frac{-8-\sqrt{16}}{8}$

55. $\frac{2+2\sqrt{3}}{6}$

56. $\frac{6+2\sqrt{7}}{8}$

57. $\frac{-2+2\sqrt{23}}{4}$

58. $\frac{-6-2\sqrt{6}}{4}$

59. $\frac{-6+3\sqrt{10}}{9}$

60. $\frac{-20+5\sqrt{10}}{15}$

61. $\frac{-8+\sqrt{28}}{6}$

62. $\frac{15-\sqrt{189}}{6}$

‹ D › Applications Involving Radicals

› › › Applications: Green Math

63. *Tsunami speed* Tsunamis are caused by an underwater earthquake or a volcanic eruption. An earthquake of magnitude 8.8 occurred February 27, 2010, near Concepcion, Chile. If the average depth of the water in the Pacific is 4280 m, how fast was the resulting tsunami traveling? Use Rule 4 on page 659 and the formula from Example 6 to write your answer.

See the quake report at http://tinyurl.com/ybt4ae8.

64. *Tsunami speed* An earthquake of magnitude 4.2 occurred February 14, 2010, near San Diego, California. If the depth of the water is 30 m, how fast was the resulting tsunami traveling? Use Rule 4 on page 659 and the formula from Example 6 to write your answer.

See the quake report at http://tinyurl.com/yjfbnu2.

› › › Applications

The **golden mean** or **golden ratio** is the number $\varphi = \frac{\sqrt{5}+1}{2}$. Here is an amazing property of the golden mean: **the golden mean φ equals its reciprocal $\left(\frac{1}{\varphi}\right)$ plus 1, that is,** $\varphi = \frac{1}{\varphi} + 1$. We will prove this, step by step, in Problems 65–69.

65. Find the reciprocal $\frac{1}{\varphi}$ of $\frac{\sqrt{5}+1}{2}$.

66. From Problem 65, $\frac{1}{\varphi} = \frac{2}{\sqrt{5}+1}$. Rationalize the denominator in $\frac{2}{\sqrt{5}+1}$.

67. From Problem 66, $\frac{1}{\varphi} = \frac{\sqrt{5}-1}{2}$. Find $\frac{\sqrt{5}-1}{2} + 1$.

68. From Problem 67, $\frac{\sqrt{5}-1}{2} + 1 = \frac{\sqrt{5}+1}{2}$. Rewrite $\frac{\sqrt{5}-1}{2} + 1 = \frac{\sqrt{5}+1}{2}$ substituting $\varphi = \frac{\sqrt{5}+1}{2}$ and $\frac{1}{\varphi} = \frac{\sqrt{5}-1}{2}$.

69. Write in words: $\varphi = \frac{1}{\varphi} + 1$. Use "The golden mean" for φ. Is that the statement we wanted to show?

In Problems 70–74, we will prove that "If we have a number x so that its reciprocal plus 1 is the number x," then that number must be

$$\varphi = \frac{\sqrt{5}+1}{2}$$

70. Let x be a real number. Translate: $\frac{1}{x} + 1 = x$.

71. Add $\frac{1}{x} + 1$.

72. From Problem 71, $\frac{1}{x} + 1 = \frac{x+1}{x}$, substitute this result in $\frac{1}{x} + 1 = x$.

73. **a.** Use "cross products" to rewrite $\frac{x+1}{x} = x$ as an equivalent equation.

b. Rewrite the equation obtained in **a** with all the terms on the right and 0 on the left.

74. Verify that a solution of $x^2 - x - 1 = 0$ is $x = \frac{\sqrt{5}+1}{2}$. Suggestion: Find x^2, then find $-x$, then find $x^2 - x - 1$. What do you get?

Using Your Knowledge

Simplifying Mach Numbers The Mach number M mentioned in the *Getting Started* is given by the expression

$$\sqrt{\frac{2}{\gamma}}\sqrt{\frac{P_2 - P_1}{P_1}}$$

75. Write this expression as a single radical.

76. Rationalize the denominator of the expression obtained in Problem 75.

Write On

77. Write in your own words the procedure you use to simplify any expression containing radicals.

78. Explain the difference between rationalizing the denominator in an algebraic expression whose denominator has only one term involving a radical and one whose denominator has two terms, at least one of which involves a radical.

79. Suppose you wish to rationalize the denominator in the expression

$$\frac{1}{\sqrt{2} + 1}$$

and you decide to multiply numerator and denominator by $\sqrt{2} + 1$. Would you obtain a rational denominator? What should you multiply by?

Concept Checker

Fill in the blank(s) with the correct word(s), phrase, or mathematical statement.

80. The **conjugate** of $a\sqrt{b} + c\sqrt{d}$ is ______________.

81. $(a\sqrt{b} + c\sqrt{d})(a\sqrt{b} - c\sqrt{d})$ = ______________.

$a\sqrt{b} + c\sqrt{d}$ $a\sqrt{b} - c\sqrt{d}$

$a^2b - c^2d$ $a^2b + c^2d$

Mastery Test

Simplify:

82. $\frac{4 + \sqrt{36}}{8}$

83. $\frac{-4 + \sqrt{28}}{2}$

84. $\frac{-6 - \sqrt{72}}{2}$

85. $\frac{3}{\sqrt{2} + 2}$

86. $\frac{\sqrt{2} + \sqrt{3}}{\sqrt{5} - \sqrt{2}}$

87. $\frac{3}{\sqrt[3]{2}}$

88. $\frac{5}{\sqrt[3]{9}}$

89. $\frac{\sqrt[3]{256}}{\sqrt[3]{2}}$

90. $\frac{\sqrt{12x^4}}{\sqrt{3x^2}}$ $(x > 0)$

91. $\frac{\sqrt[3]{16a^5}}{\sqrt[3]{2a^3}}$ $(a > 0)$

92. $(\sqrt{7} + \sqrt{3})(\sqrt{7} - \sqrt{3})$

93. $(\sqrt{5} + 2\sqrt{3})(\sqrt{5} - 3\sqrt{3})$

94. $(\sqrt{2} + \sqrt{3})(3\sqrt{2} - 5\sqrt{3})$

Skill Checker

Find the square of each radical expression:

95. $\sqrt{x - 1}$

96. $\sqrt{x + 7}$

97. $2\sqrt{x}$

98. $-3\sqrt{y}$

99. $\sqrt{x^2 + 2x + 1}$

100. $\sqrt{x^2 - 2x + 7}$

Factor completely:

101. $x^2 - 3x$

102. $x^2 + 4x$

103. $x^2 - 3x + 2$

104. $x^2 + 4x + 3$

8.5 Applications: Solving Radical Equations

Objectives

A Solve equations with one square root term containing the variable.

B Solve equations with two square root terms containing the variable.

C Solving applications involving radical equations.

To Succeed, Review How To . . .

1. Square a radical expression (pp. 637–638).
2. Square a binomial (pp. 378–382).
3. Solve quadratic equations by factoring (pp. 456–461).

Getting Started

Your Weight and Your Life

Has your doctor said that you are a little bit overweight? What does that mean? Can it be quantified? The "threshold weight" T (in pounds) for a man between 40 and 49 years of age is defined as "the crucial weight above which the mortality risk rises astronomically." In plain language, this means that if you get too fat, you are almost certainly going to die sooner as a result! The formula linking T and the height h in inches is

$$12.3\sqrt[3]{T} = h$$

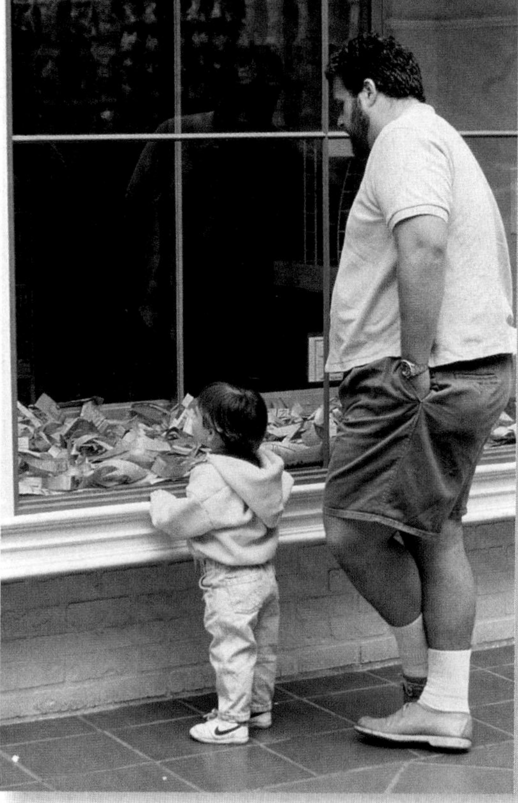

Can you solve for T in this equation? To start, we divide both sides of the equation by 12.3 so that the radical term $\sqrt[3]{T}$ is *isolated* (by itself) on one side of the equation. We obtain

$$\sqrt[3]{T} = \frac{h}{12.3}$$

Now we can cube each side of the equation to get

$$(\sqrt[3]{T})^3 = \left(\frac{h}{12.3}\right)^3$$

or

$$T = \left(\frac{h}{12.3}\right)^3$$

Thus, if a 40-year-old man is 73.8 inches tall, his threshold weight is

$$T = \left(\frac{73.8}{12.3}\right)^3 = 6^3 = 216 \text{ lb}$$

In this section we shall use a new technique to solve equations involving radicals—raising both sides of an equation to a power. We did just that when we *cubed* both sides of the equation

$$\sqrt[3]{T} = \frac{h}{12.3}$$

so we could solve for T. We will use this method of solving radical equations in this section.

A > Solving Equations with One Square Root Term Containing the Variable

The equations $12.3\sqrt[3]{T} = h$, $\sqrt{x + 1} = 2$ and $\sqrt{x + 1} - x = -1$ are examples of **radical equations.** A **radical equation** has variables in one or more radicals.

The properties of equality studied in Chapter 2 are not enough to solve equations such as $\sqrt{x + 1} = 2$. A new property that we need to solve these types of equations is stated as follows.

RAISING BOTH SIDES OF AN EQUATION TO A POWER

If both sides of the equation $A = B$ are *squared,* all solutions of $A = B$ are *among* the solutions of the new equation $A^2 = B^2$.

Note that this property can yield a new equation that has *more* solutions than the original equation. For example, the equation $x = 3$ has one solution, 3. If we square both sides of the equation $x = 3$, we have

$$x^2 = 3^2$$
$$x^2 = 9$$

which has *two* solutions, $x = 3$ and $x = -3$. The -3 *does not* satisfy the original equation $x = 3$, and it is called an **extraneous** solution. Because of this, we must check our answers carefully when we solve equations with radicals by substituting the answers in the *original* equation and discarding any extraneous solutions.

EXAMPLE 1 Solving equations in which the radical expression is isolated

Solve:

a. $\sqrt{x + 3} = 4$ **b.** $\sqrt{x + 3} = x + 3$

SOLUTION 1

a. We shall proceed in steps.

1. $\sqrt{x + 3} = 4$ The square root term is isolated.
2. $(\sqrt{x + 3})^2 = 4^2$ Square each side.
3. $x + 3 = 16$ Simplify.
4. $x = 13$ Subtract 3.
5. Now we check this answer in the original equation:

$\sqrt{13 + 3} \stackrel{?}{=} 4$ Substitute $x = 13$ in the original equation.

$\sqrt{16} = 4$ A true statement

Thus, $x = 13$ is the only solution of $\sqrt{x + 3} = 4$.

PROBLEM 1

Solve:

a. $\sqrt{x + 1} = 5$

b. $\sqrt{x + 1} = x + 1$

(continued)

Answers to PROBLEMS

1. a. 24 **b.** -1; 0

b. We again proceed in steps.

1.	$\sqrt{x+3} = x + 3$	The square root term is isolated.
2.	$(\sqrt{x+3})^2 = (x+3)^2$	Square each side.
3.	$x + 3 = x^2 + 6x + 9$	Simplify.
4.	$0 = x^2 + 5x + 6$	Subtract $x + 3$.
	$0 = (x+3)(x+2)$	Factor.
	$x + 3 = 0$ or $x + 2 = 0$	Set each factor equal to zero.
	$x = -3$ $\quad$ $x = -2$	Solve each equation.

Thus, the proposed solutions are -3 and -2.

5. We check these proposed solutions in the original equation.

If $x = -3$,

$$\sqrt{-3+3} \overset{?}{=} -3 + 3$$

$$\sqrt{0} = 0 \qquad \text{True}$$

If $x = -2$,

$$\sqrt{-2+3} \overset{?}{=} -2 + 3$$

$$\sqrt{1} = 1 \qquad \text{True}$$

Thus, -3 and -2 are the solutions of $\sqrt{x+3} = x + 3$.

EXAMPLE 2 **Solving equations by first isolating the radical expression**

Solve: $\sqrt{x+1} - x = -1$

SOLUTION 2 First we must isolate the radical term $\sqrt{x+1}$ by adding x to both sides of the equation.

	$\sqrt{x+1} - x = -1$	Given
1.	$\sqrt{x+1} = x - 1$	Add x.
2.	$(\sqrt{x+1})^2 = (x-1)^2$	Square each side.
3.	$x + 1 = x^2 - 2x + 1$	Simplify.
4.	$0 = x^2 - 3x$	Subtract $x + 1$.
	$0 = x(x-3)$	Factor.
	$x = 0$ or $x - 3 = 0$	Set each factor equal to zero.
	$x = 0$ $\quad$ $x = 3$	Solve each equation.

Thus, the proposed solutions are 0 and 3.

5. The check is as follows:

If $x = 0$,

$$\sqrt{0+1} - 0 \overset{?}{=} -1$$

$$\sqrt{1} - 0 \overset{?}{=} -1 \qquad \text{False}$$

If $x = 3$,

$$\sqrt{3+1} - 3 \overset{?}{=} -1$$

$$\sqrt{4} - 3 = -1 \qquad \text{True}$$

Thus, the equation $\sqrt{x+1} - x = -1$ has *one* solution, 3.

PROBLEM 2

Solve: $\sqrt{x+3} - x = -3$

Answers to PROBLEMS

2. 6

We can now generalize the steps we've been using to solve equations that have one square root term containing the variable.

PROCEDURE

Solving Radical Equations

1. Isolate the square root terms containing the variable.
2. Square both sides of the equation.
3. Simplify and repeat steps 1 and 2 if there is a square root term containing the variable.
4. Solve the resulting linear or quadratic equation.
5. Check all proposed solutions in the original equation.

B › Solving Equations with Two Square Root Terms Containing the Variable

The equations $\sqrt{y+4} = \sqrt{2y+3}$ and $\sqrt{y+7} - 3\sqrt{2y-3} = 0$ are different from the equations we have just solved because they have *two* square root terms containing the variable. However, we can still solve them using the five-step procedure.

EXAMPLE 3 Solving equations using the five-step procedure

Solve:

a. $\sqrt{y+4} = \sqrt{2y+3}$

b. $\sqrt{y+7} - 3\sqrt{2y-3} = 0$

SOLUTION 3

a. Since the square root terms containing the variable are isolated, we first square each side of the equation and then solve for y.

1. $\sqrt{y+4} = \sqrt{2y+3}$ The radicals are isolated.
2. $(\sqrt{y+4})^2 = (\sqrt{2y+3})^2$ Square both sides.
3. $y + 4 = 2y + 3$ Simplify.
4. $4 = y + 3$ Subtract y.
   $1 = y$ Subtract 3.

Thus, the proposed solution is 1.

5. Let's check this: If $y = 1$,

$$\sqrt{1+4} \stackrel{?}{=} \sqrt{2 \cdot 1 + 3}$$

$$\sqrt{5} = \sqrt{5}$$ True

Thus, the solution of $\sqrt{y+4} = \sqrt{2y+3}$ is 1.

b. We start by isolating the square root terms containing the variable by adding $3\sqrt{2y-3}$ to both sides.

$\sqrt{y+7} - 3\sqrt{2y-3} = 0$ Given

1. $\sqrt{y+7} = 3\sqrt{2y-3}$ Add $3\sqrt{2y-3}$.
2. $(\sqrt{y+7})^2 = (3\sqrt{2y-3})^2$ Square both sides.
3. $y + 7 = 3^2(2y-3)$ Simplify.

PROBLEM 3

Solve:

a. $\sqrt{x+4} = \sqrt{2x+1}$

b. $\sqrt{x+3} - 3\sqrt{2x-11} = 0$

(continued)

Answers to PROBLEMS

3. a. 3 **b.** 6

4.
$$\begin{aligned} y + 7 &= 9(2y - 3) && \text{Since } 3^2 = 9 \\ y + 7 &= 18y - 27 && \text{Simplify.} \\ 7 &= 17y - 27 && \text{Subtract } y. \\ 34 &= 17y && \text{Add 27.} \\ 2 &= y && \text{Divide by 17.} \end{aligned}$$

5. And our check: If $y = 2$,

$$\sqrt{2 + 7} - 3\sqrt{2 \cdot 2 - 3} \stackrel{?}{=} 0$$
$$\sqrt{9} - 3\sqrt{1} \stackrel{?}{=} 0$$
$$3 - 3 = 0 \quad \text{True}$$

Thus, the solution of $\sqrt{y + 7} - 3\sqrt{2y - 3} = 0$ is 2.

C › Applications Involving Radical Equations

Let's see how square roots can be used in a real-world application.

Have you called the Peanut Hotline lately? No, not the peanuts that you eat at baseball games but the peanuts used as a packing material! "The Plastic Loose Fill Council is a program that delivers savings to the environment by directing consumers to local packing businesses willing to accept used fill for their own packing needs." How successful is this program collecting and recycling these materials? We shall see next.

Source: http://tinyurl.com/yk53ufr.

EXAMPLE 4 Peanut recycling

The amount A of peanut packing material collected (in millions of pounds) can be approximated by $A = \sqrt{126t + 625}$, where t is the number of years after 2004.

a. How many million pounds were collected in 2004?

b. In what year will 50 million pounds of materials be collected?

SOLUTION 4

a. In 2004, $t = 0$. Substituting $t = 0$ in A, we have

$$A = \sqrt{126(0) + 625} = \sqrt{625} = 25 \text{ million pounds}$$

b. To find the year in which 50 million pounds of materials will be collected, we have to solve the equation $\sqrt{126t + 625} = 50$.

Here are the steps:

$$\begin{aligned} \sqrt{126t + 625} &= 50 && \text{Given} \\ 126t + 625 &= 2500 && \text{Square both sides.} \\ 126t &= 1875 && \text{Subtract 625.} \\ t &= \frac{1875}{126} \approx 15 && \text{Divide by 126.} \end{aligned}$$

Thus, 15 years after 2004, that is, in 2019 they will be collecting 50 million pounds of peanuts! You can check this result by substituting 15 for t in $\sqrt{126t + 625}$ obtaining $\sqrt{126(15) + 625} = \sqrt{2515} \approx 50$.

PROBLEM 4

Use the formula in Example 4 to find:

a. how many million pounds were collected in 2008.

b. the year in which 60 million pounds of materials will be collected.

Answers to PROBLEMS

4. a. $\sqrt{126(4) + 625} \approx 33.6$ million pounds

b. $23.6 \approx 24$ years after 2004, in 2028

› Exercises 8.5

‹A› Solving Equations with One Square Root Term Containing the Variable

In Problems 1–20, solve the given equations.

1. $\sqrt{x} = 4$ **2.** $\sqrt{x} = -3$ **3.** $\sqrt{x-1} = -2$

4. $\sqrt{x+1} = 2$ **5.** $\sqrt{y} - 2 = 0$ **6.** $\sqrt{y} + 3 = 0$

7. $\sqrt{y+1} - 3 = 0$ **8.** $\sqrt{y-1} + 2 = 0$ **9.** $\sqrt{x+1} = x - 5$

10. $\sqrt{x+14} = x + 2$ **11.** $\sqrt{x+4} = x + 2$ **12.** $\sqrt{x+9} = x - 3$

13. $\sqrt{x-1} - x = -3$ **14.** $\sqrt{x-2} - x = -4$ **15.** $y - 10 - \sqrt{5y} = 0$

16. $y - 3 - \sqrt{4y} = 0$ **17.** $\sqrt{y} + 20 = y$ **18.** $\sqrt{y} + 12 = y$

19. $4\sqrt{y} = y + 3$ **20.** $6\sqrt{y} = y + 5$

‹B› Solving Equations with Two Square Root Terms Containing the Variable

In Problems 21–30, solve the given equation.

21. $\sqrt{y+3} = \sqrt{2y-3}$ **22.** $\sqrt{y+7} = \sqrt{3y+3}$ **23.** $\sqrt{3x+1} = \sqrt{2x+6}$

24. $\sqrt{4x-3} = \sqrt{3x-2}$ **25.** $2\sqrt{x+5} = \sqrt{8x+4}$ **26.** $3\sqrt{x+2} = 2\sqrt{x+7}$

27. $\sqrt{4x-1} - \sqrt{x+10} = 0$ **28.** $\sqrt{3x+6} - \sqrt{5x+4} = 0$ **29.** $\sqrt{3y-2} - \sqrt{2y+3} = 0$

30. $\sqrt{5y+7} - \sqrt{3y+11} = 0$

‹C› Applications Involving Radical Equations

31. *Radius of a sphere* The radius r of a sphere is given by the equation

$$r = \sqrt{\frac{S}{4\pi}}$$

where S is the surface area. If the radius of a sphere is 2 feet, what is its surface area? Use $\pi \approx 3.14$.

32. *Radius of a cone* The radius r of a cone is given by the equation

$$r = \sqrt{\frac{3V}{\pi h}}$$

where V is the volume of the cone and h is its height. If a 10-centimeter-high ice cream cone has a radius of 2 centimeters, what is the volume of the ice cream in the cone? Use $\pi \approx 3.14$ and round to one decimal place.

33. *Time to fall* The time t (in seconds) it takes a body to fall d feet is given by the equation

$$t = \sqrt{\frac{d}{16}}$$

How far would a body fall in 3 seconds?

34. *Velocity of a falling body* After traveling d feet, the velocity v (in feet per second) of a falling body starting from rest is given by the equation $v = \sqrt{64d}$. If a body that started from rest is traveling at 44 feet per second, how far has it fallen?

35. *Length of pendulum cycle* A pendulum of length L feet takes

$$t = 2\pi\sqrt{\frac{L}{32}} \text{ (seconds)}$$

to go through a complete cycle. If a pendulum takes 2 seconds to go through a complete cycle, how long is the pendulum? Use $\pi \approx \frac{22}{7}$ and round to two decimal places.

››› Applications: Green Math

36. *Vehicle emissions* According to Environmental Protection Agency (EPA) figures, the estimated amount A of particulate matter (in short tons) emitted by transportation sources (cars, buses, and so on) is modeled by the equation $A = \sqrt{0.2t + 5.20}$, where t is the number of years after 2000. In what year would the amount of particulate matter emitted by transportation sources reach 3 short tons?

37. *Sulfur oxide emissions* According to EPA figures, the amount A of sulfur oxides (in short tons) emitted by transportation sources is modeled by the equation $A = \sqrt{1 + 0.04y}$, where y is the number of years after 1989. In what year would the amount of sulfur oxides emitted by transportation sources reach 1.2 short tons?

38. *Distance seen from tall buildings* The maximum distance d in kilometers you can see from a tall building is modeled by the equation $d = 110\sqrt{h}$, where h is the height of the building in kilometers. If the maximum distance you can see from the tallest building in the world is 77 kilometers, how high is the building? Write your answer in decimal form.

39. *Speed* The speed v (in miles per hour) a car can travel in a concrete highway curve without skidding is modeled by the equation $v = \sqrt{9r}$, where r is the radius of the curve in feet. What is the radius r if the speed limit is 30 miles per hour?

40. *Radius of a curve* Using the equation in Problem 39, find the radius r of the curve when the speed limit is 45 miles per hour.

41. *Terminal velocity* If air resistance is neglected, the terminal velocity v of a body (in meters per second) depends on the height h of the body above the ground and is given by the equation $v = \sqrt{20h + v_0}$, where $v_0 = 5\text{m/sec}$ is the initial velocity of the object. If the terminal velocity v of an object is 25 meters per second, how far has the object fallen?

42. *Velocity* If the velocity v of an object (in feet per second) is modeled by the equation $v = \sqrt{64h + v_0}$, where v_0 is the initial velocity and h is the height (in feet), how far has an object with an initial velocity of 16 feet per second fallen when the velocity of the object is 20 feet per second?

Web IT go to mhhe.com/bello for more lessons

Using Your Knowledge

Working with Higher Roots Step 2 in the five-step procedure for solving radical equations directs us to "square each side of the equation." Thus, to solve $\sqrt{x} = 2$, we square each side of the equation to obtain 4 as our solution. If we have an equation of the form $\sqrt[3]{x} = 2$, we can **cube** each side of the equation to obtain $2^3 = 8$. You can check that 8 is the correct solution by substituting 8 for x in $\sqrt[3]{x} = 2$ to obtain $\sqrt[3]{8} = 2$, a true statement.

Use this knowledge to solve the following equations.

43. $\sqrt[3]{x} = 3$ **44.** $\sqrt[3]{x} = -4$ **45.** $\sqrt[3]{x+1} = 2$ **46.** $\sqrt[3]{x-1} = -2$

Generalize the idea used in Problems 43–46 to solve the following problems.

47. $\sqrt[4]{x} = 2$ **48.** $\sqrt[4]{x+1} = 1$ **49.** $\sqrt[4]{x-1} = -2$ **50.** $\sqrt[4]{x-1} = 2$

Write On

51. Consider the equation $\sqrt{x} + 3 = 0$. What should the first step be in solving this equation? If you follow the rest of the steps in the procedure to solve radical equations, you should conclude that this equation has no real-number solution. Can you write an explanation of why this is so *after the first step* in the procedure?

52. Consider the equation $\sqrt{x+1} = -\sqrt{x+2}$. Write an explanation of why this equation has no real-number solutions; then follow the procedure given in the text to prove that this is the case.

53. Write your explanation of a "proposed" solution for solving equations involving radicals. Why do you think they are called "proposed" solutions?

54. Why is it necessary to check proposed solutions in the original equation when solving equations involving radicals?

Concept Checker

Fill in the blank(s) with the correct word(s), phrase, or mathematical statement.

55. If **both** sides of the equation $A = B$ are **squared,** all **solutions of** $A = B$ are ______________ the **solutions** of the new equation $A^2 = B^2$.

56. A(n) ______________ **solution** is a **possible (proposed)** solution that **does not** satisfy the **original** equation.

equal to **extraneous**

among **extra**

Mastery Test

57. The total monthly cost C (in millions of dollars) of running daily flights between two cities is modeled by the equation $C = \sqrt{0.3p + 1}$, where p is the number of passengers in thousands. If the monthly cost C for a certain month was \$2 million, what was the number of passengers for the month?

Solve if possible:

58. $\sqrt{y + 6} = \sqrt{2y + 3}$

59. $\sqrt{3y + 10} = \sqrt{y + 14}$

60. $\sqrt{4x - 3} - \sqrt{x + 3} = 0$

61. $\sqrt{5x + 1} - \sqrt{x + 9} = 0$

62. $\sqrt{x + 1} - x = -5$

63. $\sqrt{x + 2} - x = -4$

64. $\sqrt{x + 2} = 3$

65. $\sqrt{x - 1} = -2$

66. $\sqrt{x - 3} = x - 3$

67. $\sqrt{x - 2} = x - 2$

Skill Checker

Find:

68. $\sqrt{169}$

69. $\sqrt{49}$

70. $\sqrt{\frac{81}{16}}$

71. $\sqrt{\frac{5}{16}}$

72. $\sqrt{\frac{7}{4}}$

Collaborative Learning

The Golden Ratio

Form two groups of students. Each group should find a different picture of the Mona Lisa and compute the ratio of length to width of the frame. Are the answers about the same for all groups?

Each group should compare their answer to the golden ratio, $\frac{\sqrt{5} + 1}{2} \approx 1.618$. Are the answers close?

Let us now construct a golden rectangle:

Group 1 Start with a 1-by-1 square. What is the ratio of length to width?

Group 2 Add another 1-by-1 square to the right of the original square. What is the ratio of length to width?

Group 1 Add a 2-by-2 square under the previous rectangle. What is the ratio of length to width?

Group 2 Add a 3-by-3 square to the right of the previous rectangle. What is the ratio of length to width?

Group 1 Add a 5-by-5 square under the previous rectangle. What is the ratio of length to width?

Are the ratios approximating the golden ratio?

Now, let us take some measurements. Group 1 will take the smaller measurements (x) and Group 2 the larger measurements (y). Record all measurements to the nearest tenth of a centimeter.

Group 1 (x) Smaller Measurement	Group 2 (y) Larger Measurement
Height of belly button from the floor	Total height
Belly button to top of head	Belly button height from floor
Chin to top of head	Belly button to chin

Look at the ratio of each individual measurement y to x in each row. Are the ratios close to the golden ratio? Compute the average of the individual measurements in each row (for example, the average of the heights of belly buttons in row 1 and the average of total heights in row 1). Look at the ratios of the averages of the y (larger) measurements to the averages of the x (smaller) measurements (for example, average of total heights to average of heights of belly buttons). Are they now closer to the golden ratio?

Research Questions

1. Write a paragraph about Christoff Rudolff, the inventor of the square root sign, and indicate where the square root sign was first used.
2. A Chinese mathematician discovered the relationship between extracting roots and the array of binomial coefficients in Pascal's triangle. Write a paragraph about this mathematician.
3. Name the four Chinese algebraists who discovered the relationship between root extraction and the coefficients of Pascal's triangle and who then extended the idea to solve higher-than-cubic equations.
4. Write a paragraph about Newton's method for finding square roots and illustrate its use by finding the square root of 11, for example.

Summary Chapter 8

Section	Item	Meaning	Example
8.1A	$\sqrt{a}$ $-\sqrt{a}$	$\sqrt{a} = b$ is equivalent to $b^2 = a$. $-\sqrt{a} = b$ is equivalent to $b^2 = a$.	$\sqrt{4} = 2$ because $2^2 = 4$. $-\sqrt{4} = -2$ because $(-2)^2 = 4$.
8.1C	$\sqrt{a}$ if a is negative	If a is negative, $\sqrt{a}$ is not a real number.	$\sqrt{-16}$ and $\sqrt{-7}$ are not real numbers.
8.1D	$\sqrt[n]{a}$	The nth root of a	$\sqrt[3]{8} = 2$, $\sqrt[4]{81} = 3$
8.2A	Product rule for radicals	$\sqrt{a \cdot b} = \sqrt{a} \cdot \sqrt{b}$	$\sqrt{16 \cdot 9} = \sqrt{16} \cdot \sqrt{9} = 4 \cdot 3 = 12$ $\sqrt{18} = \sqrt{9 \cdot 2} = \sqrt{9} \cdot \sqrt{2} = 3\sqrt{2}$
8.2B	Quotient rule for radicals	$\sqrt{\frac{a}{b}} = \frac{\sqrt{a}}{\sqrt{b}}$	$\sqrt{\frac{9}{4}} = \frac{\sqrt{9}}{\sqrt{4}} = \frac{3}{2}$
8.2C	$\sqrt{a^2} = \lvert a \rvert$	The square root of a real number a is the absolute value of a.	$\sqrt{5^2} = \lvert 5 \rvert$, $\sqrt{(-3)^2} = \lvert -3 \rvert = 3$
8.2D	Properties of radicals	For real numbers where the roots exist, $\sqrt[n]{a \cdot b} = \sqrt[n]{a}\,\sqrt[n]{b}$ $\sqrt[n]{\frac{a}{b}} = \frac{\sqrt[n]{a}}{\sqrt[n]{b}}$	$\sqrt[4]{80} = \sqrt[4]{2^4 \cdot 5} = 2\sqrt[4]{5}$ $\sqrt[3]{\frac{8}{27}} = \frac{\sqrt[3]{8}}{\sqrt[3]{27}} = \frac{\sqrt[3]{2^3}}{\sqrt[3]{3^3}} = \frac{2}{3}$
8.3C	Rationalizing the denominator	Multiply the numerator and denominator of the fraction by the square root in the denominator.	$\frac{1}{\sqrt{3}} = \frac{1 \cdot \sqrt{3}}{\sqrt{3} \cdot \sqrt{3}} = \frac{\sqrt{3}}{3}$
8.4B	Conjugate	$a + b$ and $a - b$ are conjugates.	To rationalize the denominator of $\frac{1}{\sqrt{5} - \sqrt{3}}$, multiply the numerator and denominator of $\frac{1}{\sqrt{5} - \sqrt{3}}$ by the conjugate of $\sqrt{5} - \sqrt{3}$, which is $\sqrt{5} + \sqrt{3}$.
8.5A, B	Raising both sides of an equation to a power	If both sides of the equation $A = B$ are squared, all solutions of $A = B$ are among the solutions of the new equation $A^2 = B^2$.	If both sides of the equation $\sqrt{x} = 3$ are squared, all solutions of $\sqrt{x} = 3$ are among the solutions of $(\sqrt{x})^2 = 3^2$, that is, $x = 9$.

Review Exercises Chapter 8

(If you need help with these exercises, look in the sections indicated in brackets.)

1. ⟨ **8.1A, C** ⟩ *Find the root if possible.*
   a. $\sqrt{81}$ **b.** $\sqrt{-64}$
   c. $\sqrt{\frac{36}{25}}$

2. ⟨ **8.1A, C** ⟩ *Find the root if possible.*
   a. $-\sqrt{36}$ **b.** $-\sqrt{\frac{64}{25}}$
   c. $\sqrt{-\frac{9}{4}}$

3. ⟨ **8.1B** ⟩ *Find the square of each radical expression.*
   a. $\sqrt{8}$ **b.** $\sqrt{25}$ **c.** $\sqrt{17}$

4. ⟨ **8.1B** ⟩ *Find the square of each radical expression.*
   a. $-\sqrt{36}$ **b.** $-\sqrt{17}$ **c.** $-\sqrt{64}$

5. ⟨ **8.1B** ⟩ *Find the square of each radical expression.*
   a. $\sqrt{x^2 + 1}$ **b.** $\sqrt{x^2 + 4}$
   c. $-\sqrt{x^2 + 5}$

6. ⟨ **8.1C** ⟩ *Find and classify each number as rational, irrational, or not a real number. Approximate the irrational numbers using a calculator.*
   a. $\sqrt{11}$ **b.** $-\sqrt{25}$
   c. $\sqrt{-9}$

7. ⟨ **8.1C** ⟩ *Classify each number as rational, irrational, or not a real number.*
   a. $\sqrt{\frac{9}{4}}$ **b.** $-\sqrt{\frac{9}{4}}$
   c. $\sqrt{-\frac{9}{4}}$

8. ⟨ **8.1D** ⟩ *Find each root if possible.*
   a. $\sqrt[3]{64}$ **b.** $\sqrt[3]{-8}$ **c.** $-\sqrt[4]{81}$

9. ⟨ **8.1D** ⟩ *Find each root if possible.*
   a. $\sqrt[4]{16}$ **b.** $-\sqrt[4]{16}$ **c.** $\sqrt[4]{-16}$

10. ⟨ **8.1E** ⟩ *If an object is dropped from a distance d (in feet), it takes*
    $$t = \sqrt{\frac{d}{16}}$$
    seconds to reach the ground. How long does it take an object to reach the ground if it is dropped from:
    a. 121 feet **b.** 144 feet **c.** 169 feet

11. ⟨ **8.2A** ⟩ *Simplify.*
    a. $\sqrt{32}$ **b.** $\sqrt{48}$ **c.** $\sqrt{196}$

12. ⟨ **8.2A** ⟩ *Multiply.*
    a. $\sqrt{3} \cdot \sqrt{7}$ **b.** $\sqrt{12} \cdot \sqrt{3}$ **c.** $\sqrt{5}\,\sqrt{y}, y > 0$

13. ⟨ **8.2B** ⟩ *Simplify.*
    a. $\sqrt{\frac{3}{16}}$ **b.** $\sqrt{\frac{5}{36}}$ **c.** $\sqrt{\frac{9}{4}}$

14. ⟨ **8.2B** ⟩ *Simplify.*
    a. $\frac{\sqrt{8}}{\sqrt{2}}$ **b.** $\frac{\sqrt{21}}{\sqrt{3}}$ **c.** $\frac{6\sqrt{50}}{2\sqrt{10}}$

15. ⟨ **8.2C** ⟩ *Simplify. Assume all variables represent positive real numbers.*
    a. $\sqrt{36x^2}$ **b.** $\sqrt{100y^4}$ **c.** $\sqrt{81n^8}$

16. ⟨ **8.2C** ⟩ *Simplify. Assume all variables represent positive real numbers.*
    a. $\sqrt{72y^{10}}$ **b.** $\sqrt{147z^8}$ **c.** $\sqrt{48x^{12}}$

17. ⟨ **8.2C** ⟩ *Simplify. Assume all variables represent positive real numbers.*
    a. $\sqrt{y^{15}}$ **b.** $\sqrt{y^{13}}$ **c.** $\sqrt{50n^7}$

18. ⟨ **8.2D** ⟩ *Simplify.*
    a. $\sqrt[3]{24}$ **b.** $\sqrt[3]{\frac{8}{27}}$ **c.** $\sqrt[3]{-\frac{125}{64}}$

19. ⟨ **8.2D** ⟩ *Simplify.*
    a. $\sqrt[4]{81}$ **b.** $\sqrt[4]{48}$ **c.** $\sqrt[4]{80}$

20. ⟨ **8.3A** ⟩ *Add and simplify.*
    a. $7\sqrt{3} + 8\sqrt{3}$ **b.** $\sqrt{32} + 5\sqrt{2}$ **c.** $\sqrt{12} + \sqrt{48}$

21. ⟨ **8.3A** ⟩ *Subtract and simplify.*

a. $9\sqrt{11} - 6\sqrt{11}$ **b.** $\sqrt{50} - 4\sqrt{2}$

c. $\sqrt{108} - \sqrt{75}$

22. ⟨ **8.3B** ⟩ *Simplify.*

a. $\sqrt{3}(\sqrt{20} - \sqrt{2})$ **b.** $\sqrt{5}(\sqrt{5} - \sqrt{3})$

c. $\sqrt{7}(\sqrt{7} - \sqrt{98})$

23. ⟨ **8.3C** ⟩ *Write with a rationalized denominator.*

a. $\sqrt{\frac{5}{8}}$ **b.** $\sqrt{\frac{x^2}{50}}$, $x > 0$ **c.** $\sqrt{\frac{y^2}{27}}$, $y > 0$

24. ⟨ **8.4A** ⟩ *Simplify.*

a. $\sqrt{32} + 4 - \sqrt{9}$ **b.** $\sqrt{18} + 7 - \sqrt{4}$

c. $\sqrt{60} + 4 - \sqrt{16}$

25. ⟨ **8.4A** ⟩ *Simplify.*

a. $8\sqrt{15} - \sqrt{3} \cdot \sqrt{5}$ **b.** $7\sqrt{6} - \sqrt{2} \cdot \sqrt{3}$

c. $9\sqrt{14} - 2\sqrt{7} \cdot \sqrt{2}$

26. ⟨ **8.4A** ⟩ *Simplify.*

a. $\frac{\sqrt[3]{162}}{\sqrt[3]{2}}$ **b.** $\frac{\sqrt[3]{135}}{\sqrt[3]{5}}$ **c.** $\frac{\sqrt[3]{192}}{\sqrt[3]{24}}$

27. ⟨ **8.4A** ⟩ *Simplify by rationalizing the denominator.*

a. $\frac{7}{\sqrt[3]{4}}$ **b.** $\frac{5}{\sqrt[3]{9}}$ **c.** $\frac{9}{\sqrt[3]{25}}$

28. ⟨ **8.4A** ⟩ *Multiply and simplify.*

a. $(\sqrt{3} + 3\sqrt{2})(\sqrt{3} - 5\sqrt{2})$

b. $(\sqrt{7} + 3\sqrt{5})(\sqrt{7} - 2\sqrt{5})$

29. ⟨ **8.4A** ⟩ *Multiply and simplify.*

a. $(\sqrt{7} + 2\sqrt{3})(\sqrt{7} - 2\sqrt{3})$

b. $(\sqrt{11} + 3\sqrt{5})(\sqrt{11} - 3\sqrt{5})$

30. ⟨ **8.4B** ⟩ *Simplify by rationalizing the denominator.*

a. $\frac{3}{\sqrt{3} + 1}$ **b.** $\frac{5}{\sqrt{2} - 1}$

31. ⟨ **8.4B** ⟩ *Simplify by rationalizing the denominator.*

a. $\frac{7}{\sqrt{3} - \sqrt{2}}$ **b.** $\frac{2}{\sqrt{5} - \sqrt{2}}$

32. ⟨ **8.4C** ⟩ *Simplify.*

a. $\frac{-8 + \sqrt{8}}{2}$ **b.** $\frac{-16 + \sqrt{12}}{4}$

33. ⟨ **8.5A** ⟩ *Solve.*

a. $\sqrt{x + 2} = 3$ **b.** $\sqrt{x - 2} = -2$

34. ⟨ **8.5A** ⟩ *Solve.*

a. $\sqrt{x + 5} = x - 1$ **b.** $\sqrt{x + 10} = x - 2$

35. ⟨ **8.5A** ⟩ *Solve.*

a. $\sqrt{x + 4} - x = -2$ **b.** $\sqrt{x + 2} - x = -4$

36. ⟨ **8.5B** ⟩ *Solve.*

a. $\sqrt{y + 5} = \sqrt{3y - 3}$

b. $\sqrt{y + 5} = \sqrt{2y + 5}$

37. ⟨ **8.5B** ⟩ *Solve.*

a. $\sqrt{y + 8} - 3\sqrt{2y - 1} = 0$

b. $\sqrt{y + 9} - 3\sqrt{2y + 1} = 0$

38. ⟨ **8.5C** ⟩ *The total daily cost C (thousand dollars) of producing a certain product is given by the equation $C = \sqrt{0.2x + 1}$, where x is the number of items in hundreds. How many items were produced on a day in which the cost was:*

a. \$3(thousand) **b.** \$7(thousand)

Practice Test Chapter 8

(Answers on page 678)

Visit www.mhhe.com/bello to view helpful videos that provide step-by-step solutions to several of the problems below.

1. Find.

a. $\sqrt{169}$ b. $-\sqrt{\frac{49}{81}}$

2. Find the square of each radical expression.

a. $-\sqrt{121}$ b. $\sqrt{x^2 + 7}$

3. Classify each number as rational, irrational, or not a real number, and simplify if possible.

a. $\sqrt{17}$ b. $-\sqrt{36}$

c. $\sqrt{-100}$ d. $\sqrt{\frac{100}{49}}$

4. Find each root if possible.

a. $\sqrt[4]{81}$ b. $-\sqrt[4]{625}$

c. $\sqrt[3]{-8}$ d. $\sqrt[4]{-16}$

5. A diver jumps from a cliff 20 meters high. If the time t (in seconds) it takes an object dropped from a distance d (in meters) to reach the ground is given by the equation

$$t = \sqrt{\frac{d}{5}}$$

how long does it take the diver to reach the water?

6. Simplify.

a. $\sqrt{125}$ b. $\sqrt{54}$

7. Multiply.

a. $\sqrt{3} \cdot \sqrt{11}$ b. $\sqrt{11} \cdot \sqrt{y}, y > 0$

8. Simplify.

a. $\sqrt{\frac{7}{16}}$ b. $\frac{21\sqrt{50}}{7\sqrt{5}}$

9. Simplify.

a. $\sqrt{144n^2}, n > 0$ b. $\sqrt{32y^7}, y > 0$

10. Simplify.

a. $\sqrt[4]{96}$ b. $\sqrt[3]{\frac{-125}{8}}$

11. Simplify.

a. $9\sqrt{13} + 7\sqrt{13}$ b. $14\sqrt{6} - 3\sqrt{6}$

12. Simplify.

a. $\sqrt{28} + \sqrt{63}$ b. $\sqrt{40} + \sqrt{90} - \sqrt{160}$

13. Multiply and simplify.

a. $\sqrt{3}(\sqrt{18} - \sqrt{5})$ b. $\sqrt{5}(\sqrt{5} - \sqrt{7})$

14. Write $\sqrt{\frac{3}{20}}$ with a rationalized denominator.

15. Write $\sqrt{\frac{y^2}{50}}, y > 0$ with a rationalized denominator.

16. Simplify.

a. $8\sqrt{14} - \sqrt{7} \cdot \sqrt{2}$ b. $\frac{\sqrt{12x^3}}{\sqrt{4x^2}}, x > 0$

17. Simplify by rationalizing the denominator.

a. $\frac{\sqrt[3]{500}}{\sqrt[3]{2}}$ b. $\frac{3}{\sqrt[3]{25}}$

18. Multiply and simplify.

a. $(\sqrt{3} + 6\sqrt{2})(\sqrt{3} - 2\sqrt{2})$

b. $(\sqrt{10} - 2\sqrt{20})(\sqrt{10} + 2\sqrt{20})$

19. Simplify by rationalizing the denominator.

a. $\frac{11}{\sqrt{3} + 1}$ b. $\frac{2}{\sqrt{5} - \sqrt{2}}$

20. Simplify.

a. $\frac{-6 + \sqrt{18}}{3}$ b. $\frac{-8 + \sqrt{8}}{4}$

21. Solve.

a. $\sqrt{x + 1} = 2$ b. $\sqrt{x + 6} = x + 6$

22. Solve $\sqrt{x + 4} - x = 2$.

23. Solve $\sqrt{y + 3} = \sqrt{2y + 1}$.

24. Solve $\sqrt{y + 6} - 3\sqrt{2y - 5} = 0$.

25. The average length L of a long-distance call (in minutes) has been approximated by the equation $L = \sqrt{t + 4}$, where t is the number of years after 1995. In how many years would you expect the average length of a call to be 3 minutes?

Answers to Practice Test Chapter 8

Answer		If You Missed	Review		
		Question	Section	Examples	Page
1. **a.** 13	**b.** $-\frac{7}{9}$	1	8.1	1	637
2. **a.** 121	**b.** $x^2 + 7$	2	8.1	2	638
3. **a.** Irrational	**b.** Rational; -6	3	8.1	3	638
c. Not a real number	**d.** Rational; $\frac{10}{7}$				
4. **a.** 3	**b.** -5	4	8.1	4	639
c. -2	**d.** Not a real number				
5. 2 sec		5	8.1	5	639
6. **a.** $5\sqrt{5}$	**b.** $3\sqrt{6}$	6	8.2	1	645
7. **a.** $\sqrt{33}$	**b.** $\sqrt{11y}$	7	8.2	2	645
8. **a.** $\frac{\sqrt{7}}{4}$	**b.** $3\sqrt{10}$	8	8.2	3	646
9. **a.** $12n$	**b.** $4y^3\sqrt{2y}$	9	8.2	4	646
10. **a.** $2\sqrt[4]{6}$	**b.** $\frac{-5}{2}$	10	8.2	5	647
11. **a.** $16\sqrt{13}$	**b.** $11\sqrt{6}$	11	8.3	1	652
12. **a.** $5\sqrt{7}$	**b.** $\sqrt{10}$	12	8.3	2	652
13. **a.** $3\sqrt{6} - \sqrt{15}$	**b.** $5 - \sqrt{35}$	13	8.3	3	652–653
14. $\frac{\sqrt{15}}{10}$		14	8.3	4	654
15. $\frac{y\sqrt{2}}{10}$		15	8.3	5	654
16. **a.** $7\sqrt{14}$	**b.** $\sqrt{3x}$	16	8.4	1	659
17. **a.** $5\sqrt[3]{2}$	**b.** $\frac{3\sqrt[3]{5}}{5}$	17	8.4	2	659–660
18. **a.** $-21 + 4\sqrt{6}$	**b.** -70	18	8.4	3	660
19. **a.** $\frac{11\sqrt{3} - 11}{2}$	**b.** $\frac{2\sqrt{5} + 2\sqrt{2}}{3}$	19	8.4	4	661
20. **a.** $-2 + \sqrt{2}$	**b.** $\frac{-4 + \sqrt{2}}{2}$	20	8.4	5	662
21. **a.** $x = 3$	**b.** $x = -5$ or $x = -6$	21	8.5	1	667–668
22. $x = 0$		22	8.5	2	668
23. $y = 2$		23	8.5	3	669–670
24. $y = 3$		24	8.5	3	669–670
25. $t = 5$ yr		25	8.5	4	670

Cumulative Review Chapters 1–8

1. Add: $-\frac{2}{9} + \left(-\frac{1}{8}\right)$

2. Find: $(-3)^4$

3. Divide: $-\frac{1}{6} \div \left(-\frac{7}{12}\right)$

4. Evaluate $y \div 5 \cdot x - z$ for $x = 5$, $y = 50$, $z = 3$.

5. Simplify: $2x - (x + 4) - 2(x + 3)$

6. Write in symbols: The quotient of $(m + 3n)$ and p

7. Solve for x: $2 = 5(x - 3) + 1 - 4x$

8. Solve for x: $\frac{x}{2} - \frac{x}{3} = 1$

9. Graph: $-\frac{x}{6} + \frac{x}{2} \leq \frac{x - 2}{2}$

10. Graph the point $C(-1, -3)$.

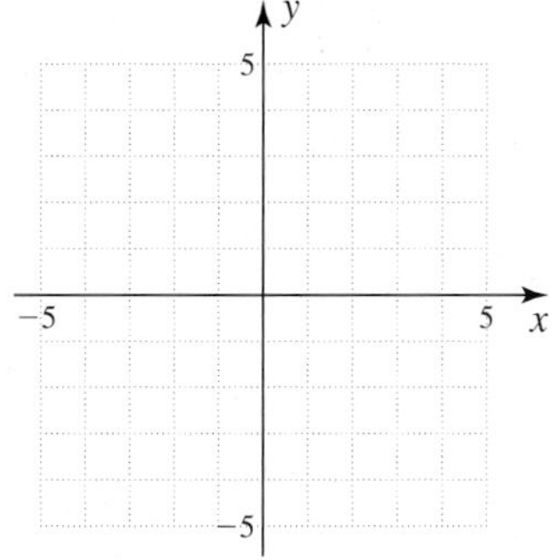

11. Determine whether the ordered pair $(-1, 3)$ is a solution of $4x - y = -1$.

12. Find x in the ordered pair $(x, 3)$ so that the ordered pair satisfies the equation $2x - 4y = -10$.

13. Graph: $5x + y = 5$

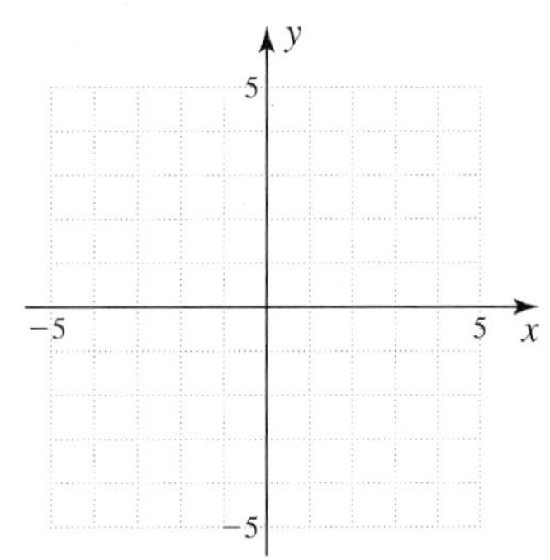

14. Graph: $4y - 8 = 0$

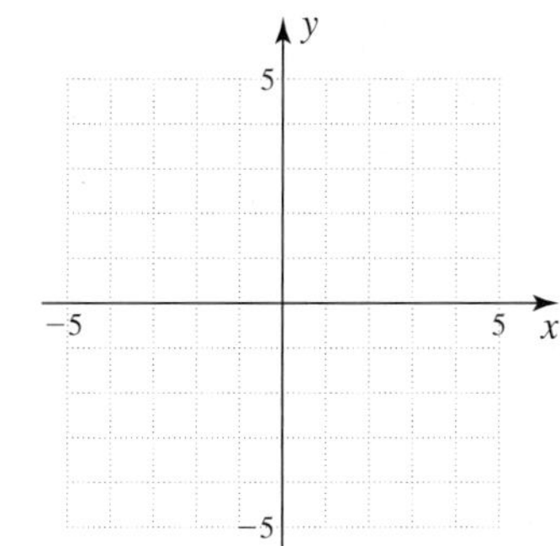

15. Find the slope of the line passing through the points (0, 4) and (6, 1).

16. What is the slope of the line $3x - 3y = -8$?

17. Find the pair of parallel lines.

(1) $15x - 12y = 4$

(2) $12y + 15x = 4$

(3) $-4y = -5x + 4$

18. Simplify: $(2x^4y^{-3})^{-4}$

19. Write in scientific notation: 0.00000025

20. Divide and express the answer in scientific notation: $(2.72 \times 10^{-4}) \div (1.6 \times 10^4)$

21. Find (expand): $\left(4x^2 - \frac{1}{2}\right)^2$

22. Divide $(2x^3 - 7x^2 + x + 9)$ by $(x - 3)$.

23. Factor completely: $x^2 - 4x + 3$

24. Factor completely: $9x^2 - 27xy + 20y^2$

25. Factor completely: $4x^2 - 25y^2$

26. Factor completely: $-5x^4 + 5x^2$

27. Factor completely: $4x^3 - 8x^2 - 12x$

28. Factor completely: $3x^2 + 4x + 9x + 12$

29. Factor completely: $16kx^2 + 8kx + k$

30. Solve for x: $4x^2 + 17x = 15$

31. Write $\frac{2x}{5y}$ with a denominator of $15y^2$.

32. Reduce to lowest terms: $\frac{-16(x^2 - y^2)}{4(x - y)}$

33. Reduce to lowest terms: $\frac{x^2 + 4x - 21}{3 - x}$

34. Multiply: $(x - 8) \cdot \frac{x + 4}{x^2 - 64}$

35. Divide: $\frac{x + 5}{x - 5} \div \frac{x^2 - 25}{5 - x}$

36. Add: $\frac{7}{3(x + 5)} + \frac{5}{3(x + 5)}$

37. Subtract: $\frac{x + 4}{x^2 + x - 20} - \frac{x + 5}{x^2 - 16}$

38. Simplify: $\dfrac{\frac{2}{3x} + \frac{1}{2x}}{\frac{1}{x} + \frac{1}{4x}}$

39. Solve for x: $\frac{4x}{x - 4} + 1 = \frac{3x}{x - 4}$

40. Solve for x: $\frac{x}{x^2 - 4} + \frac{2}{x - 2} = \frac{1}{x + 2}$

41. Solve for x: $\frac{x}{x + 6} - \frac{1}{7} = \frac{-6}{x + 6}$

42. Solve for x: $1 + \frac{2}{x - 5} = \frac{20}{x^2 - 25}$

43. A van travels 100 miles on 4 gallons of gas. How many gallons will it need to travel 625 miles?

44. Solve for x: $\frac{x - 5}{5} = \frac{5}{4}$

45. Janet can paint a kitchen in 3 hours and James can paint the same kitchen in 4 hours. How long would it take to paint the kitchen if they worked together?

46. Find an equation of the line that passes through the points (6, 4) and has slope $m = 5$.

47. Find an equation of the line having slope 4 and y-intercept 3.

48. Graph: $x - 5y < -5$

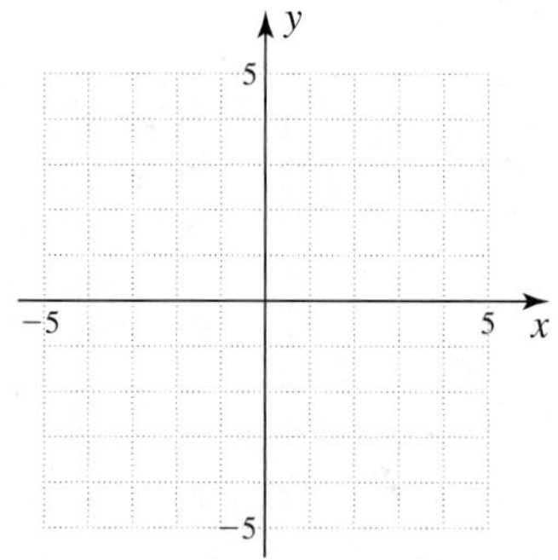

49. Graph:
$-y \geq -5x - 5$

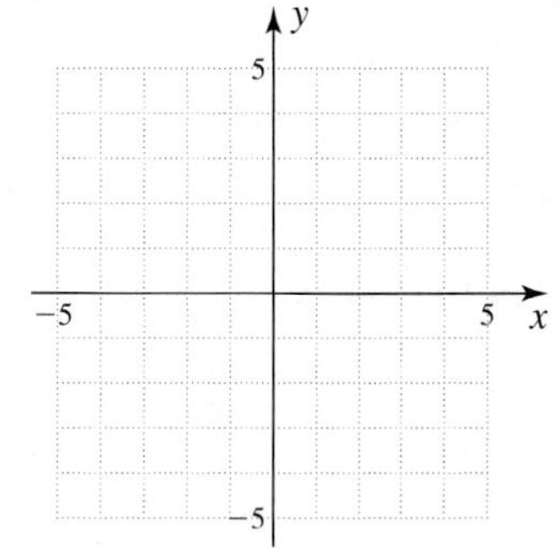

50. An enclosed gas exerts a pressure P on the walls of the container. This pressure is directly proportional to the temperature T of the gas. If the pressure is 3 lb/in.^{2} when the temperature is 240°F, find k.

51. If the temperature of a gas is held constant, the pressure P varies inversely as the volume V. A pressure of 1800 lb/in.^{2} is exerted by 6 ft^3 of air in a cylinder fitted with a piston. Find k.

52. Graph the system and find the solution (if possible):

$$x + 4y = 16$$
$$4y - x = 12$$

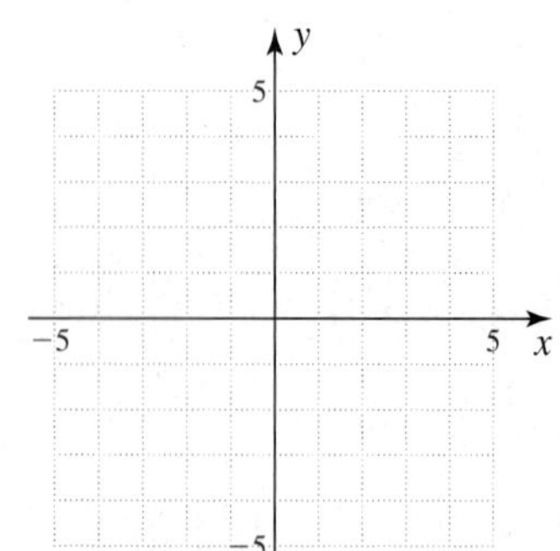

53. Graph the system and find the solution if possible:

$$y + 3x = -3$$
$$2y + 6x = -12$$

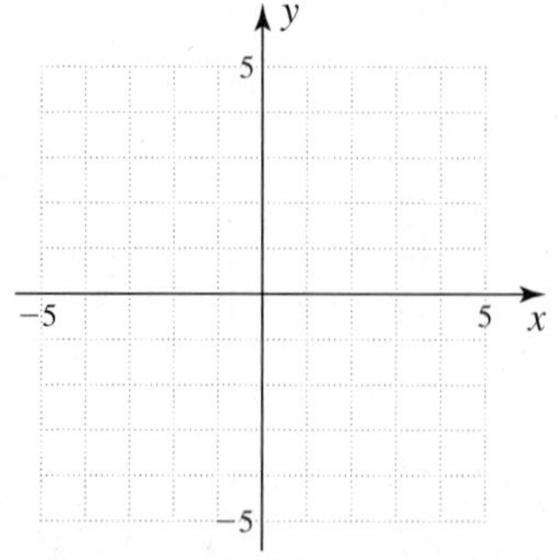

54. Solve by substitution (if possible):

$$\begin{aligned} x + 4y &= -18 \\ -2x - 8y &= 32 \end{aligned}$$

55. Solve by substitution (if possible):

$$\begin{aligned} x + 3y &= 10 \\ -2x - 6y &= -20 \end{aligned}$$

56. Solve the system (if possible):

$$\begin{aligned} x - 2y &= 3 \\ 2x - y &= 0 \end{aligned}$$

57. Solve the system (if possible):

$$\begin{aligned} 5x + 4y &= -18 \\ -10x - 8y &= 3 \end{aligned}$$

58. Solve the system (if possible):

$$\begin{aligned} 4y + 3x &= 11 \\ 6x + 8y &= 22 \end{aligned}$$

59. Sara has $3.50 in nickels and dimes. She has three times as many dimes as nickels. How many nickels and how many dimes does she have?

60. The sum of two numbers is 215. Their difference is 85. What are the numbers?

61. Evaluate: $\sqrt[3]{-64}$

62. Simplify: $\sqrt[7]{(-3)^7}$

63. Simplify: $\sqrt{\dfrac{7}{243}}$

64. Simplify: $\sqrt[3]{128a^8b^9}$

65. Add and simplify: $\sqrt{48} + \sqrt{12}$

66. Perform the indicated operations: $\sqrt[3]{2x}\left(\sqrt[3]{4x^2} - \sqrt[3]{81x}\right)$

67. Rationalize the denominator: $\dfrac{\sqrt{3}}{\sqrt{2p}}$

68. Find the product: $\left(\sqrt{125} + \sqrt{343}\right)\left(\sqrt{245} + \sqrt{175}\right)$

69. Rationalize the denominator: $\dfrac{\sqrt{x}}{\sqrt{x} - \sqrt{5}}$

70. Reduce: $\dfrac{6 + \sqrt{18}}{3}$

71. Solve: $\sqrt{x + 3} = -2$

72. Solve: $\sqrt{x - 6} - x = -6$

Section

Chapter

9 nine

Quadratic Equations

The Human Side of Algebra

The study of quadratic equations dates back to antiquity. Scores of clay tablets indicate that the Babylonians of 2000 B.C. were already familiar with the quadratic formula that you will study in this chapter. Their solutions using the formula are actually verbal instructions that amounted to using the formula

$$x = \sqrt{\left(\frac{a}{2}\right)^2 + b} - \frac{a}{2}$$

to solve the equation $x^2 + ax = b$.

By Euclid's time (circa 300 B.C.), Greek geometry had reached a stage of development where geometric algebra could be used to solve quadratic equations. This was done by reducing them to the geometric equivalent of one of the following forms:

$$x(x + a) = b^2$$
$$x(x - a) = b^2$$
$$x(a - x) = b^2$$

These equations were then solved by applying different theorems dealing with specific areas.

Later, the Arabian mathematician Muhammed ibn Musa al-Khwarizmi (read more about him in the Exercises for Section 9.2) (circa A.D. 820) divided quadratic equations into three types:

$$x^2 + ax = b$$
$$x^2 + b = ax$$
$$x^2 = ax + b$$

with only positive coefficients admitted. All of these developments form the basis for our study of quadratic equations.

9.1 Solving Quadratic Equations by the Square Root Property

Objectives

Solve quadratic equations of the form

A › $X^2 = A$

B › $(AX \pm B)^2 = C$

To Succeed, Review How To . . .

1. Find the square roots of a number (pp. 46, 636–637).
2. Simplify expressions involving radicals (pp. 646–647, 658–662).

Getting Started

Square Chips

The Pentium chip is used in many computers. Because of space limitations, the chip is very small and covers an area of only 324 square millimeters. If the chip is square, how long is each side? As you recall, the area of a square is obtained by multiplying the length X of its side by itself. If we assume that the length of the side of the chip is X millimeters, then its area is X^2. The area is also 324 square millimeters; we thus have the equation $X^2 = 324$.

Suppose the side of a square is 5 units. Then the area is 5^2 or 25 square units. Now, suppose a side of the square is X units. What is the area? X^2. Now, let's go backwards. If we know that the area of a square is 100 square units, what is the length of a side? The corresponding equation is $X^2 = 100$ and the solution is $X = 10$.

5 units	X units
Area 25 square units	Area X^2 square units

In this section, we will learn how to solve quadratic equations that can be written in the form $X^2 = A$ or $(AX \pm B)^2 = C$ by introducing a new property called the *square root property of equations.*

A › Solving Quadratic Equations of the Form $X^2 = A$

The equation $X^2 = 16$ is a *quadratic equation.* In general, a **quadratic equation** is an equation that can be written in the form $ax^2 + bx + c = 0$. How can we solve $X^2 = 16$? First, the equation tells us that a certain number X multiplied by itself gives 16 as a result. Obviously, one possible answer is $X = 4$ because $4^2 = 16$. But wait, what about $X = -4$? It's also true that $(-4)^2 = (-4)(-4) = 16$. Thus, the solutions of the equation $X^2 = 16$ are 4 and -4.

In mathematics, the number 4 is called the **positive square root** of 16, and -4 is called the **negative square root** of 16. These roots are usually denoted by

$$\sqrt{16} = 4 \qquad \text{Read "the positive square root of 16 is 4."}$$

and

$$-\sqrt{16} = -4 \qquad \text{Read "the negative square root of 16 is } -4\text{."}$$

Many of the equations we are about to study have irrational numbers for their solutions. For example, the equation $x^2 = 3$ has two *irrational* solutions. How do we obtain these solutions? By taking the square root of both sides of the equation:

$$x^2 = 3$$

Then,

$$x = \pm\sqrt{3} \qquad \text{Note that } (\sqrt{3})^2 = 3 \text{ and } (-\sqrt{3})^2 = 3.$$

The notation $\pm\sqrt{3}$ is a shortcut to indicate that x can be $\sqrt{3}$ or $-\sqrt{3}$.

NOTE

With a calculator, the answers to the equation $x^2 = 3$ can be approximated as $x \approx \pm 1.7320508$.

On the other hand, the equation $x^2 = 100$ has *rational* roots. To solve this equation, we proceed as follows:

$$x^2 = 100$$

Then,

$$x = \pm\sqrt{100}$$
$$x = \pm 10$$

Thus, the solutions are 10 and -10. Here is the property we just used.

SQUARE ROOT PROPERTY OF EQUATIONS

If A is a positive number and $X^2 = A$, then $X = \pm\sqrt{A}$; that is,

$$X = \sqrt{A} \quad \text{or} \quad X = -\sqrt{A}$$

EXAMPLE 1 Solving quadratic equations using the square root property

Solve:

a. $x^2 = 36$ **b.** $x^2 - 49 = 0$ **c.** $x^2 = 10$

SOLUTION 1

a. Given: $x^2 = 36$. Then,

$$x = \pm\sqrt{36} \qquad \text{Use the square root property.}$$
$$x = \pm 6$$

Thus, the solutions of the equation $x^2 = 36$ are 6 and -6, since $6^2 = 36$ and $(-6)^2 = 36$. (Both solutions are rational numbers.)

b. Given: $x^2 - 49 = 0$. Unfortunately, this equation is not of the form $X^2 = A$. However, by adding 49 to both sides of the equation, we can remedy this situation.

$$x^2 - 49 = 0 \qquad \text{Given}$$
$$x^2 = 49 \qquad \text{Add 49.}$$
$$x = \pm\sqrt{49} \qquad \text{Use the square root property.}$$
$$x = \pm 7$$

(continued)

PROBLEM 1

Solve:

a. $x^2 = 81$

b. $x^2 - 1 = 0$

c. $x^2 = 13$

Answers to PROBLEMS

1. a. $x = \pm 9$ **b.** $x = \pm 1$

c. $x = \pm\sqrt{13}$

Thus, the solutions of $x^2 - 49 = 0$ are 7 and -7. (Both solutions are rational numbers.)

Note that the equation $x^2 - 49 = 0$ can also be solved by factoring to write $x^2 - 49 = 0$ as $(x + 7)(x - 7) = 0$. Thus, $x + 7 = 0$ or $x - 7 = 0$; that is, $x = -7$ or $x = 7$, as before.

c. Given: $x^2 = 10$. Then, using the square root property,

$$x = \pm\sqrt{10}$$

Since 10 does not have a rational square root, the solutions of the equation $x^2 = 10$ are written as $\sqrt{10}$ and $-\sqrt{10}$. (Here, both solutions are irrational.)

In solving Example 1(b), we added 49 to both sides of the equation to obtain an equivalent equation of the form $X^2 = A$. Does this method work for more complicated examples? The answer is yes. In fact, when solving any quadratic equation in which the *only* exponent of the variable is 2, we can always transform the equation into an equivalent one of the form $X^2 = A$. Here is the idea.

PROCEDURE

Solving Quadratic Equations of the Form $AX^2 - B = 0$

To solve any equation of the form

$$AX^2 - B = 0$$

write

$$AX^2 = B \quad \text{Add } B.$$

and

$$X^2 = \frac{B}{A} \quad \text{Divide by } A.$$

so that, using the square root property,

$$X = \pm\sqrt{\frac{B}{A}}, \quad \frac{B}{A} \geq 0$$

Thus, to solve the equation

$$16x^2 - 81 = 0$$

we write it in the form $AX^2 = B$ or equivalently $X^2 = \frac{B}{A}$:

$$16x^2 - 81 = 0 \quad \text{Given}$$

$$16x^2 = 81 \quad \text{Add 81.}$$

$$x^2 = \frac{81}{16} \quad \text{Divide by 16.}$$

$$x = \pm\sqrt{\frac{81}{16}} \quad \text{Use the square root property.}$$

$$x = \pm\frac{9}{4}$$

The solutions are $\frac{9}{4}$ and $-\frac{9}{4}$. Since $16 \cdot \left(\frac{9}{4}\right)^2 - 81 = 16 \cdot \frac{81}{16} - 81 = 0$ and $16\left(-\frac{9}{4}\right)^2 - 81 = 16 \cdot \frac{81}{16} - 81 = 0$, our result is correct. Just remember to first rewrite the equation in the form $X^2 = \frac{B}{A}$.

EXAMPLE 2 **Solving a quadratic equation of the form $AX^2 - B = 0$**

Solve: $36x^2 - 25 = 0$

SOLUTION 2

$$36x^2 - 25 = 0 \quad \text{Given}$$
$$36x^2 = 25 \quad \text{Add 25.}$$
$$x^2 = \frac{25}{36} \quad \text{Divide by 36.}$$
$$x = \pm\sqrt{\frac{25}{36}} \quad \text{Use the square root property.}$$
$$x = \pm\frac{5}{6}$$

The solutions are $\frac{5}{6}$ and $-\frac{5}{6}$. (Here, both solutions are rational numbers.)

PROBLEM 2

Solve: $9x^2 - 16 = 0$

Of course, not all equations of the form $X^2 = A$ have solutions that are real numbers. For example, to solve the equation $x^2 + 64 = 0$, we write

$$x^2 + 64 = 0 \quad \text{Given}$$
$$x^2 = -64 \quad \text{Subtract 64.}$$

But there is no *real* number whose square is -64. If you square a nonzero real number, the answer is always positive. Thus, x^2 is positive and can never equal -64. The equation $x^2 + 64 = 0$ has *no real-number* solution.

As we have seen, not all equations have *rational-number* solutions—that is, solutions of the form $\frac{a}{b}$, where a and b are integers, $b \neq 0$. For example, the equation

$$16x^2 - 5 = 0$$

is solved as follows:

$$16x^2 - 5 = 0 \quad \text{Given}$$
$$16x^2 = 5 \quad \text{Add 5.}$$
$$x^2 = \frac{5}{16} \quad \text{Divide by 16.}$$
$$x = \pm\sqrt{\frac{5}{16}} \quad \text{Use the square root property.}$$

However, $\sqrt{\frac{5}{16}}$ is *not* a rational number even though the denominator, 16, is the square of 4. As you recall, using the quotient rule for radicals, we may write

$$\pm\sqrt{\frac{5}{16}} = \pm\frac{\sqrt{5}}{\sqrt{16}} = \pm\frac{\sqrt{5}}{4}$$

Thus, the solutions of the equation $16x^2 - 5 = 0$ are

$$\frac{\sqrt{5}}{4} \quad \text{and} \quad -\frac{\sqrt{5}}{4}$$

Both solutions are irrational numbers.

Answers to PROBLEMS

2. $\frac{4}{3}$ and $-\frac{4}{3}$

EXAMPLE 3 **Solving a quadratic equation using the quotient rule**

Solve:

a. $4x^2 - 7 = 0$ **b.** $8x^2 + 49 = 0$

SOLUTION 3

a.

$$\begin{aligned} 4x^2 - 7 &= 0 && \text{Given} \\ 4x^2 &= 7 && \text{Add 7.} \\ x^2 &= \frac{7}{4} && \text{Divide by 4.} \\ x &= \pm\sqrt{\frac{7}{4}} && \text{Use the square root property.} \\ x &= \pm\frac{\sqrt{7}}{\sqrt{4}} && \text{Use the quotient rule for radicals.} \\ x &= \pm\frac{\sqrt{7}}{2} \end{aligned}$$

Thus, the solutions of $4x^2 - 7 = 0$ are $\frac{\sqrt{7}}{2}$ and $-\frac{\sqrt{7}}{2}$. They are both irrational numbers.

b.

$$\begin{aligned} 8x^2 + 49 &= 0 && \text{Given} \\ 8x^2 &= -49 && \text{Subtract 49.} \\ x^2 &= -\frac{49}{8} && \text{Divide by 8.} \end{aligned}$$

Since the square of a real number x cannot be negative and $-\frac{49}{8}$ is negative, this equation has *no* real-number solution.

PROBLEM 3

Solve:

a. $49x^2 - 3 = 0$

b. $10x^2 + 9 = 0$

B > Solving Quadratic Equations of the Form $(AX \pm B)^2 = C$

We've already mentioned that to solve an equation in which the only exponent of the variable is 2, we must transform the equation into an equivalent one of the form $X^2 = A$. Now consider the equation

$$(x - 2)^2 = 9$$

If we think of $(x - 2)$ as X, we have an equation of the form $X^2 = 9$, which we just learned how to solve! Thus, we have the following.

$$\begin{aligned} (x - 2)^2 &= 9 && \text{Given} \\ X^2 &= 9 && \text{Write } x - 2 \text{ as } X. \\ X &= \pm\sqrt{9} && \text{Use the square root property.} \\ X &= \pm 3 \\ x - 2 &= \pm 3 && \text{Write } X \text{ as } x - 2. \\ x &= 2 \pm 3 && \text{Add 2.} \end{aligned}$$

Hence,

$$x = 2 + 3 \quad \text{or} \quad x = 2 - 3$$

The solutions are $x = 5$ and $x = -1$.

Clearly, by thinking of $(x - 2)$ as X, we can solve a more complicated equation. In the same manner, we can solve $9(x - 2)^2 - 5 = 0$:

Answers to PROBLEMS

3. a. $\frac{\sqrt{3}}{7}$ and $-\frac{\sqrt{3}}{7}$

b. No real-number solution

$$9(x-2)^2 - 5 = 0 \quad \text{Given}$$
$$9(x-2)^2 = 5 \quad \text{Add 5.}$$
$$(x-2)^2 = \frac{5}{9} \quad \text{Divide by 9.}$$
$$x - 2 = \pm\sqrt{\frac{5}{9}} \quad \text{Think of } x-2 \text{ as } X \text{ and use the square root property.}$$
$$x - 2 = \pm\frac{\sqrt{5}}{\sqrt{9}} \quad \text{Use the quotient rule for radicals.}$$
$$x - 2 = \pm\frac{\sqrt{5}}{3}$$
$$x = 2 \pm \frac{\sqrt{5}}{3} \quad \text{Add 2.}$$

The solutions are

$$2 + \frac{\sqrt{5}}{3} = \frac{6+\sqrt{5}}{3} \quad \text{and} \quad 2 - \frac{\sqrt{5}}{3} = \frac{6-\sqrt{5}}{3}$$

EXAMPLE 4 Solving quadratic equations of the form $(AX + B)^2 = C$

Solve:

a. $(x+3)^2 = 9$ **b.** $(x+1)^2 - 4 = 0$ **c.** $25(x+2)^2 - 3 = 0$

SOLUTION 4

a. $(x+3)^2 = 9$ Given

$x + 3 = \pm\sqrt{9}$ *Think* of $(x+3)$ as X.

$x + 3 = \pm 3$

$x = -3 \pm 3$ Subtract 3.

$x = -3 + 3$ or $x = -3 - 3$

$x = 0$ or $x = -6$

The solutions are 0 and -6.

b. $(x+1)^2 - 4 = 0$ Given

$(x+1)^2 = 4$ Add 4 (to have an equation of the form $X^2 = A$).

$x + 1 = \pm\sqrt{4}$ *Think* of $(x+1)$ as X.

$x + 1 = \pm 2$

$x = -1 \pm 2$ Subtract 1.

$x = -1 + 2$ or $x = -1 - 2$

$x = 1$ or $x = -3$

The solutions are 1 and -3.

c. $25(x+2)^2 - 3 = 0$ Given

$25(x+2)^2 = 3$ Add 3.

$(x+2)^2 = \frac{3}{25}$ Divide by 25.

$x + 2 = \pm\sqrt{\frac{3}{25}}$ *Think* of $(x+2)$ as X.

$x + 2 = \pm\frac{\sqrt{3}}{\sqrt{25}} = \pm\frac{\sqrt{3}}{5}$ Since $\sqrt{\frac{3}{25}} = \frac{\sqrt{3}}{\sqrt{25}}$

$x = -2 \pm \frac{\sqrt{3}}{5}$ Subtract 2.

The solutions are

$$-2 + \frac{\sqrt{3}}{5} = \frac{-10+\sqrt{3}}{5} \quad \text{and} \quad -2 - \frac{\sqrt{3}}{5} = \frac{-10-\sqrt{3}}{5}.$$

PROBLEM 4

Solve:

a. $(x+6)^2 = 36$

b. $(x+2)^2 - 9 = 0$

c. $16(x+1)^2 - 5 = 0$

Answers to PROBLEMS

4. a. The solutions are 0 and -12.

b. The solutions are 1 and -5.

c. The solutions are

$$-1 + \frac{\sqrt{5}}{4} = \frac{-4+\sqrt{5}}{4}$$

and

$$-1 - \frac{\sqrt{5}}{4} = \frac{-4-\sqrt{5}}{4}.$$

NOTE

It is easy to spot quadratic equations that have no real-number solution. Here is how: if the equation can be written in the form $(\text{Expression})^2 = \text{Negative number}$ [that is, $(\quad)^2 = \text{Negative number}$], the equation has no solution.

As we mentioned before, if we square any real number, the result is not negative. For this reason, an equation such as

$$(x - 4)^2 = -5$$ If $(x - 4)$ represents a real number, $(x - 4)^2$ *cannot* be negative. But -5 is negative, so $(x - 4)^2$ and -5 can *never* be equal.

has *no* real-number solution. Similarly,

$$(x - 3)^2 + 8 = 0$$

has no real-number solution, since

$$(x - 3)^2 + 8 = 0 \quad \text{is equivalent to} \quad (x - 3)^2 = -8$$

by subtracting 8. We use this idea in Example 5.

EXAMPLE 5 **Solving a quadratic equation when there is no real-number solution**

Solve: $9(x - 5)^2 + 1 = 0$

SOLUTION 5

$9(x - 5)^2 + 1 = 0$ Given

$9(x - 5)^2 = -1$ Subtract 1.

$(x - 5)^2 = -\frac{1}{9}$ Divide by 9.

Since $(x - 5)$ is to be a real number, $(x - 5)^2$ *can never* be negative. But $-\frac{1}{9}$ is negative. Thus, the equation $(x - 5)^2 = -\frac{1}{9}$ [which is equivalent to $9(x - 5)^2 + 1 = 0$] has *no* real-number solution.

PROBLEM 5

Solve: $16(x - 3)^2 + 7 = 0$

EXAMPLE 6 **Solving a quadratic equation of the form $(AX - B)^2 = C$**

Solve: $3(2x - 3)^2 = 54$

SOLUTION 6 We want to write the equation in the form $X^2 = A$.

$3(2x - 3)^2 = 54$ Given

$(2x - 3)^2 = \frac{54}{3} = 18$ Divide by 3.

$X^2 = 18$ Write $2x - 3$ as X.

$X = \pm\sqrt{18}$ Use the square root property.

$X = \pm 3\sqrt{2}$ Simplify the radical ($\sqrt{18} = \sqrt{9 \cdot 2} = 3\sqrt{2}$).

$2x - 3 = \pm 3\sqrt{2}$ Write X as $2x - 3$.

$2x = 3 \pm 3\sqrt{2}$ Add 3.

$x = \frac{3 \pm 3\sqrt{2}}{2}$ Divide by 2.

The solutions are

$$\frac{3 + 3\sqrt{2}}{2} \quad \text{and} \quad \frac{3 - 3\sqrt{2}}{2}$$

PROBLEM 6

Solve: $2(3x - 2)^2 = 36$

Answers to PROBLEMS

5. No real-number solution

6. The solutions are $\frac{2 \pm 3\sqrt{2}}{3}$

EXAMPLE 7 **Application: BMI and weight**

Your body mass index (BMI) is a measurement of your healthy weight relative to your height. According to the National Heart and Lung Institute, the formula for your BMI is given by the equation $\text{BMI} = \frac{705W}{H^2}$, where W is your weight in pounds and H is your height in inches. If a person has a BMI of 20, which is in the "normal" range, and weighs 140 pounds, how tall is the person?

SOLUTION 7 Substituting 20 for BMI and 140 for W, we have

$$20 = \frac{705 \cdot 140}{H^2}$$

Multiplying both sides by H^2 $\quad 20H^2 = 705 \cdot 140$

Dividing both sides by 20 $\quad H^2 = 705 \cdot 7$

Or $\quad H^2 = 4935$

Thus, $\quad H = \sqrt{4935} \approx 70$ inches

PROBLEM 7

What is the height of a person weighing 180 pounds with a BMI of 30, which is in the "overweight" range?

On February 27, 2010, an earthquake of magnitude 8.8 struck off the coast of Chile creating fears that a huge tsunami (a very large ocean wave caused by an underwater earthquake or volcanic eruption) would result. How long would it take for such a wave to reach Hawaii? We can find out if we know that $S^2 = 32d$, where S is the speed of the tsunami in feet per second (ft/sec) and the average ocean depth is $d = 13{,}100$ ft and then solve for S. We do that next.

EXAMPLE 8 **Speed of a tsunami**

The speed of a tsunami is given by $S^2 = 32d$. Find the speed S of the Chilean tsunami if the average ocean depth is $d = 13{,}100$ ft.

SOLUTION 8 Using the square root property of equations,

If $\quad S^2 = 32d \quad$ then $\quad S = \sqrt{32 \cdot d} = \sqrt{32 \cdot 13{,}100} = \sqrt{419{,}200} \approx$ **647 ft/sec**

which is about 441 miles per hour. You can verify this by going to http://www.unitarium.com/speed, or you can figure it out yourself! By the way, it took about 15 hours for the tsunami to reach Hawaii, but fortunately, the waves generated were not higher than 6 ft or so.

PROBLEM 8

The speed of a tsunami in meters per second (m/sec) is also given by $S^2 = 10d$. Find the speed S of the Chilean tsunami if the average depth of the ocean is $d = 4000$ meters.

› Exercises 9.1

‹A› Solving Quadratic Equations of the Form $X^2 = A$ In Problems 1–20, solve the given equation.

1. $x^2 = 100$ **2.** $x^2 = 1$ **3.** $x^2 = 0$

4. $x^2 = 121$ **5.** $y^2 = -4$ **6.** $y^2 = -16$

7. $x^2 = 7$ **8.** $x^2 = 3$ **9.** $x^2 - 9 = 0$

10. $x^2 - 64 = 0$ **11.** $x^2 - 3 = 0$ **12.** $x^2 - 5 = 0$

Answers to PROBLEMS

7. $\sqrt{4230} \approx 65$ inches 8. 200 m/sec. This is about 447 mi/hr, which is very close to the 441 miles obtained in Example 8.

13. $25x^2 - 1 = 0$
14. $36x^2 - 49 = 0$
15. $100x^2 - 49 = 0$
16. $81x^2 - 36 = 0$
17. $25y^2 - 17 = 0$
18. $9y^2 - 11 = 0$
19. $25x^2 + 3 = 0$
20. $49x^2 + 1 = 0$

B Solving Quadratic Equations of the Form $(AX \pm B)^2 = C$ In Problems 21–60, solve the given equation.

21. $(x + 1)^2 = 81$
22. $(x + 3)^2 = 25$
23. $(x - 2)^2 = 36$
24. $(x - 3)^2 = 16$
25. $(z - 4)^2 = -25$
26. $(z + 2)^2 = -16$
27. $(x - 9)^2 = 81$
28. $(x - 6)^2 = 36$
29. $(x + 4)^2 = 16$
30. $(x + 7)^2 = 49$
31. $25(x + 1)^2 - 1 = 0$
32. $16(x + 2)^2 - 1 = 0$
33. $36(x - 3)^2 - 49 = 0$
34. $9(x - 1)^2 - 25 = 0$
35. $4(x + 1)^2 - 25 = 0$
36. $49(x + 2)^2 - 16 = 0$
37. $9(x - 1)^2 - 5 = 0$
38. $4(x - 2)^2 - 3 = 0$
39. $16(x + 1)^2 + 1 = 0$
40. $25(x + 2)^2 + 16 = 0$
41. $x^2 = \frac{1}{81}$
42. $x^2 = \frac{1}{9}$
43. $x^2 - \frac{1}{16} = 0$
44. $x^2 - \frac{1}{36} = 0$
45. $6x^2 - 24 = 0$
46. $3x^2 - 75 = 0$
47. $2(v + 1)^2 - 18 = 0$
48. $3(v - 2)^2 - 48 = 0$
49. $8(x - 1)^2 - 18 = 0$
50. $50(x + 3)^2 - 72 = 0$
51. $4(2y - 3)^2 = 32$
52. $2(3y - 1)^2 = 24$
53. $8(2x - 3)^2 - 64 = 0$
54. $5(3x - 1)^2 - 60 = 0$
55. $3\left(\frac{1}{2}x + 1\right)^2 = 54$
56. $4\left(\frac{1}{3}x - 1\right)^2 = 80$
57. $2\left(\frac{1}{3}x - 1\right)^2 - 40 = 0$
58. $3\left(\frac{1}{3}x - 2\right)^2 - 54 = 0$
59. $5\left(\frac{1}{2}y - 1\right)^2 + 60 = 0$
60. $6\left(\frac{1}{3}y - 2\right)^2 + 72 = 0$

Applications

61. *Curves on roads* Naomi was traveling on a curved concrete highway of radius 64 feet when her car began to skid. The velocity v (in miles per hour) at which skidding starts to happen on a curve of radius r is given by the equation $v^2 = 9r$. How fast was Naomi traveling when she started to skid? If the speed limit was 25 miles per hour, was she speeding?

62. *Skid marks* Have you seen your local law enforcement person measuring skid marks at the scene of an accident? The speed s (in miles per hour) a car was traveling if it skidded d feet after the brakes were applied on a dry concrete road is given by the equation $s^2 = 24d$, where d is the length of the skid mark. Find the speed s of a car leaving a 50-foot skid mark after the brakes were applied. Give the final answer to the nearest mph.

Applications: Green Math

63. Follow the procedure of Example 8 and use the formula $S^2 = 10d$ to find the speed S in meters per second (m/sec) of the Haitian Tsunami of January 12, 2010, if the average ocean depth d in the Caribbean Sea around Haiti is 2400 m. Fortunately, no tsunami developed and the warnings were canceled. Write your answer in simplified form and give an approximation.

64. *Investing in quadratics* Tran invested \$100 at r percent compounded for 2 years. At the end of the 2 years, he received \$121. Use the formula $(r + 1)^2 = \frac{A}{P}$, where A is the amount received at the end of the two years, and P is the principal (original amount), to find r.

Lenders Earn 8% to 29%

Using Your Knowledge

Going in Circles The area A of a circle of radius r is given by $A = \pi r^2$. Find the radius of a circle with the specified area.

65. 25π square inches

66. 12π square feet

The surface area A of a sphere of radius r is given by $A = 4\pi r^2$. Find the radius of a sphere with the specified surface area.

67. 49π square feet

68. 81π square inches

Write On

69. Explain why the equation $x^2 + 6 = 0$ has no real-number solution.

70. Explain why the equation $(x + 1)^2 + 3 = 0$ has no real-number solution.

Consider the equation $X^2 = A$.

71. What can you say about A if the equation has no real-number solution?

72. What can you say about A if the equation has exactly one solution?

73. What can you say about A if the equation has two solutions?

74. What types of solution does the equation have if A is a prime number?

75. What types of solution does the equation have if A is a positive perfect square?

Concept Checker

Fill in the blank(s) with the correct word(s), phrase, or mathematical statement.

76. A **quadratic equation** is an equation that **can be written** in the **form** ______________.

77. The **positive square root** of **81** is _____.

78. The **negative square root** of **81** is ______.

79. If A is a **positive** number and $X^2 = A$, then $X =$ ______.

−9	**−A**
A	$ax^2 + bx + c = 0$
9	**±A**
$ax^2 + bx + c$	

Mastery Test

Solve if possible:

80. $16(x - 3)^2 - 7 = 0$

81. $3(x - 1)^2 - 24 = 0$

82. $(x + 5)^2 = 25$

83. $(x + 2)^2 - 9 = 0$

84. $16(x + 1)^2 - 5 = 0$

85. $49x^2 - 3 = 0$

86. $10x^2 + 9 = 0$

87. $9x^2 - 16 = 0$

88. $x^2 = 121$

89. $x^2 - 1 = 0$

90. $x^2 = 13$

Skill Checker

Expand:

91. $(x + 7)^2$

92. $(x + 5)^2$

93. $(x - 3)^2$

94. $(x - 5)^2$

9.2 Solving Quadratic Equations by Completing the Square

Objective

A Solve a quadratic equation by completing the square.

To Succeed, Review How To . . .

1. Recognize a quadratic equation (p. 684).
2. Expand $(x \pm a)^2$ (pp. 378–381).

Getting Started

Completing the Square for Round Baseballs

The man has just batted the ball straight up at 96 feet per second. At the end of t seconds, the height h of the ball will be

$$h = -16t^2 + 96t$$

How long will it be before the ball reaches 44 feet? To solve this problem, we let $h = 44$ to obtain

$$-16t^2 + 96t = 44$$

$$t^2 - 6t = -\frac{44}{16} = -\frac{11}{4} \quad \text{Divide by } -16 \text{ and simplify.}$$

This equation is a quadratic equation (an equation that can be written in the form $ax^2 + bx + c = 0, a \neq 0$). Can we use the techniques we studied in Section 9.1 to solve it? The answer is yes, if we can write the equation in the form

$$(t - N)^2 = A$$

N and A are the numbers we need to find to solve the problem.

To do this, however, we should know a little bit more about a technique used in algebra called **completing the square.** We will present several examples and then generalize the results. Why? Because when we generalize the process and learn how to solve the quadratic equation $ax^2 + bx + c = 0$ by completing the square, we can use the formula obtained, called the *quadratic formula,* to solve *any* equation of the form $ax^2 + bx + c = 0$, by simply substituting the values of a, b, and c in the quadratic formula. We will do this in the next section, but right now we need to review how to expand binomials, because completing the square involves that process.

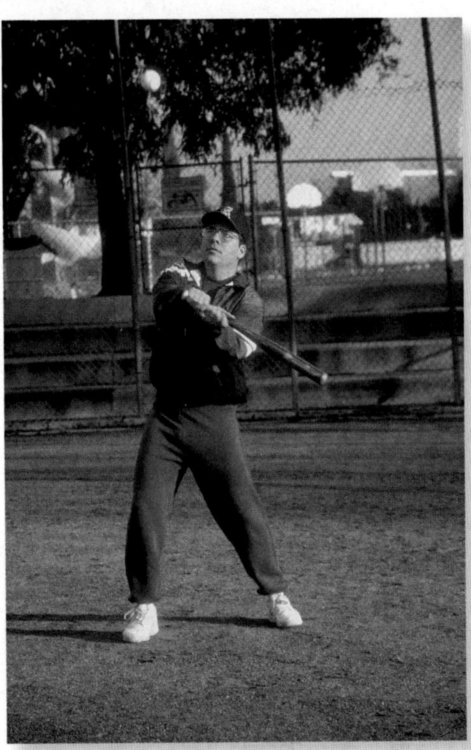

A > Solving Quadratic Equations by Completing the Square

Recall that

$$\underbrace{(X}_{\text{First term}} + \underbrace{A)^2}_{\text{Second term}} = \underbrace{X^2}_{\text{First term squared}} + \underbrace{2AX}_{\text{Coefficient of } X} + \underbrace{A^2}_{\text{Second term squared}}$$

Thus,

$$(x + 7)^2 = x^2 + 14x + 7^2$$
$$(x + 2)^2 = x^2 + 4x + 2^2$$
$$(x + 5)^2 = x^2 + 10x + 5^2$$

Do you see any relationship between the coefficients of x (14, 4, and 10, respectively) and the last terms? Perhaps you will see it better if we write it in a table.

Coefficient of *X*	Last Term Squared
14	7^2
4	2^2
10	5^2

It seems that half the coefficient of x gives the number to be squared for the last term. Thus,

$$\frac{14}{2} = 7$$
$$\frac{4}{2} = 2$$
$$\frac{10}{2} = 5$$

Now, what numbers would you add to complete the given squares?

$$(x + 3)^2 = x^2 + 6x + \square$$
$$(x + 4)^2 = x^2 + 8x + \square$$
$$(x + 6)^2 = x^2 + 12x + \square$$

The correct answers are $\left(\frac{6}{2}\right)^2 = 3^2 = 9$, $\left(\frac{8}{2}\right)^2 = 4^2 = 16$, and $\left(\frac{12}{2}\right)^2 = 6^2 = 36$. We then have

$$(x + 3)^2 = x^2 + 6x + 3^2$$
$6 \div 2 = 3$

The last term is always the square of half of the coefficient of the middle term.

$$(x + 4)^2 = x^2 + 8x + 4^2$$
$8 \div 2 = 4$

$$(x + 6)^2 = x^2 + 12x + 6^2$$
$12 \div 2 = 6$

You can also find the missing number to complete the square by using a diagram. For example, to complete the square in $(x + 3)^2$ complete the diagram:

	x	$+ 3$
x	x^2	$3x$
$+ 3$	$3x$	?

You need 3^2 or 9 in the lower right corner, so $(x + 3)^2 = x^2 + 6x + 9$, as before. You can see this technique used in Problems 43–46.

Here is the procedure we have just used to complete the square.

PROCEDURE

Completing the Square $x^2 + bx + \square$

1. Find the coefficient of the x term. (b)
2. Divide the coefficient by 2. $\left(\frac{b}{2}\right)$
3. Square this number to obtain the last term. $\left(\frac{b}{2}\right)^2 = \frac{b^2}{4}$

The idea is to add a term $\square$ to the binomial $x^2 + bx$ to change it into a perfect square trinomial.

Thus, to complete the square in

$$x^2 + 16x + \square$$

we proceed as follows:

1. Find the coefficient of the x term $\longrightarrow$ 16.
2. Divide the coefficient by 2 $\longrightarrow$ 8.
3. Square this number to obtain the last term $\longrightarrow$ 8^2.

Hence,

$$x^2 + 16x + \boxed{8^2} = (x + 8)^2$$

Now consider

$$x^2 - 18x + \square$$

Our steps to fill in the blank are as before:

1. Find the coefficient of the x term $\longrightarrow -18$.
2. Divide the coefficient by 2 $\longrightarrow -9$.
3. Square this number to obtain the last term $\longrightarrow (-9)^2$.

Hence,

$$x^2 - 18x + (-9)^2$$
$$= x^2 - 18x + 9^2$$

Recall that $(-9)^2 = (9)^2$; they are both 81.

$$= x^2 - 18x + 81 = (x - 9)^2$$

EXAMPLE 1 Completing the square

Find the missing term to complete the square:

a. $x^2 + 20x + \square$ **b.** $x^2 - x + \square$

SOLUTION 1

a. We use the three-step procedure:

1. The coefficient of x is 20.
2. $\frac{20}{2} = 10$
3. The missing term is $\boxed{10^2} = \boxed{100}$. Hence, $x^2 + 20x + 100 = (x + 10)^2$.

b. Again we use the three-step procedure:

1. The coefficient of x is -1.
2. $\frac{-1}{2} = -\frac{1}{2}$
3. The missing term is $\boxed{\left(-\frac{1}{2}\right)^2} = \boxed{\frac{1}{4}}$. Hence, $x^2 - x + \frac{1}{4} = \left(x - \frac{1}{2}\right)^2$.

PROBLEM 1

Find the missing terms:

a. $(x + 11)^2 = x^2 + 22x + \square$

b. $\left(x - \frac{1}{4}\right)^2 = x^2 - \frac{1}{2}x + \square$

Can we use the patterns we've just studied to look for further patterns? Of course! For example, how would you fill in the blanks in

$$x^2 + 16x + \square = (\quad)^2$$

Here the coefficient of x is 16, so $\left(\frac{16}{2}\right)^2 = 8^2$ goes in the box. Since

Same

$$X^2 + 2AX + A^2 = (X + A)^2$$

Same

$$x^2 + 16x + 8^2 = (x + 8)^2$$

Similarly,

$$x^2 - 6x + \square = (\quad)^2$$

is completed by reasoning that the coefficient of x is -6, so

$$\left(-\frac{6}{2}\right)^2 = (-3)^2 = (3)^2$$

Answers to PROBLEMS

1. a. $\left(\frac{22}{2}\right)^2 = 11^2 = 121$

b. $\left(\frac{-\frac{1}{2}}{2}\right)^2 = \left(-\frac{1}{4}\right)^2 = \frac{1}{16}$

goes in the box. Now

$$X^2 - 2AX + A^2 = (X - A)^2$$

Same / Same

Since the middle term on the left has a negative sign, the sign inside the parentheses must be negative.

Hence,

$$x^2 - 6x + \square = (\quad)^2$$

becomes

$$x^2 - 6x + 3^2 = (x - 3)^2$$

EXAMPLE 2 Completing the square

Find the missing terms:

a. $x^2 - 10x + \square = (\quad)^2$ **b.** $x^2 + 3x + \square = (\quad)^2$

SOLUTION 2

a. The coefficient of x is -10; thus, the number in the box should be $\left(-\frac{10}{2}\right)^2 = (-5)^2 = 5^2$, and we should have

Same

$$x^2 - 10x + 5^2 = (x - 5)^2$$

b. The coefficient of x is 3; thus, we have to add $\left(\frac{3}{2}\right)^2$. Then

$$x^2 + 3x + \left(\frac{3}{2}\right)^2 = \left(x + \frac{3}{2}\right)^2$$

PROBLEM 2

Find the missing terms:

a. $x^2 - 12x + \square = (\quad)^2$

b. $x^2 + 5x + \square = (\quad)^2$

We are finally ready to solve the equation from the *Getting Started* that involves the time it takes the ball to reach 44 feet, that is,

$$t^2 - 6t = -\frac{11}{4}$$

As you recall, t represents the time it takes the ball to reach 44 feet (p. 694).

Since the coefficient of t is -6, we must add $\left(-\frac{6}{2}\right)^2 = (-3)^2 = 3^2$ to both sides of the equation. We then have

$$t^2 - 6t + 3^2 = -\frac{11}{4} + 3^2$$

$$(t - 3)^2 = -\frac{11}{4} + \frac{36}{4}$$ Note that $3^2 = 9 = \frac{36}{4}$.

$$(t - 3)^2 = \frac{25}{4}$$

Then

$$(t - 3) = \pm\sqrt{\frac{25}{4}} = \pm\frac{5}{2}$$

$$t = 3 \pm \frac{5}{2}$$

$$t = 3 + \frac{5}{2} = \frac{11}{2} \quad \text{or} \quad t = 3 - \frac{5}{2} = \frac{1}{2}$$

This means that the ball reaches 44 feet after $\frac{1}{2}$ second (on the way up) and after $\frac{11}{2} = 5\frac{1}{2}$ seconds (on the way down).

Answers to PROBLEMS

2. a. $x^2 - 12x + 6^2 = (x - 6)^2$

b. $x^2 + 5x + \left(\frac{5}{2}\right)^2 = \left(x + \frac{5}{2}\right)^2$

Here is a summary of the steps needed to solve a quadratic equation by completing the square. The solution of our original equation follows so you can see how the steps are carried out.

PROCEDURE

Solving a Quadratic Equation by Completing the Square

1. Write the equation with the variables in descending order on the left and the constants on the right.
2. If the coefficient of the squared term is not 1, divide each term by this coefficient.
3. Add the square of one-half of the coefficient of the first-degree term to both sides.
4. Rewrite the left-hand side as the square of a binomial.
5. Use the square root property to solve the resulting equation.

For the equation from the *Getting Started,* we have:

1. $-16t^2 + 96t = 44$

2. $\frac{-16t^2}{-16} + \frac{96t}{-16} = \frac{44}{-16}$

$t^2 - 6t = -\frac{11}{4}$

3. $t^2 - 6t + \left(-\frac{6}{2}\right)^2 = -\frac{11}{4} + \left(-\frac{6}{2}\right)^2$

$t^2 - 6t + 3^2 = -\frac{11}{4} + 3^2$

4. $(t - 3)^2 = \frac{25}{4}$

5. $(t - 3) = \pm\sqrt{\frac{25}{4}}$

$t - 3 = \pm\frac{5}{2}$

$t = 3 \pm \frac{5}{2}$

Thus, $t = 3 + \frac{5}{2} = 5\frac{1}{2}$ or $t = 3 - \frac{5}{2} = \frac{1}{2}$.

We now use this procedure in another example.

EXAMPLE 3 **Solving a quadratic equation by completing the square**

Solve: $4x^2 - 16x + 7 = 0$

SOLUTION 3 We use the five-step procedure just given.

$4x^2 - 16x + 7 = 0$ Given

1. Subtract 7. $4x^2 - 16x = -7$
2. Divide by 4 so the coefficient of x^2 is 1. $x^2 - 4x = -\frac{7}{4}$
3. Add $\left(-\frac{4}{2}\right)^2 = (-2)^2 = 2^2$. $x^2 - 4x + 2^2 = -\frac{7}{4} + 2^2$
4. Rewrite. $(x - 2)^2 = -\frac{7}{4} + \frac{16}{4} = \frac{9}{4}$

$(x - 2)^2 = \frac{9}{4}$

(continued)

PROBLEM 3

Solve by completing the square:

$4x^2 - 24x + 27 = 0$

Answers to PROBLEMS

3. The solutions are $3 \pm \frac{3}{2}$, that is, $\frac{9}{2}$ and $\frac{3}{2}$.

5. Use the square root property to solve the resulting equation.

$$(x - 2) = \pm\sqrt{\frac{9}{4}}$$

$$x - 2 = \pm\frac{3}{2}$$

$$x = 2 \pm \frac{3}{2}$$

That is,

$$x = \frac{4}{2} + \frac{3}{2} = \frac{7}{2} \quad \text{or} \quad x = \frac{4}{2} - \frac{3}{2} = \frac{1}{2}$$

Thus, the solutions of $4x^2 - 16x + 7 = 0$ are $\frac{7}{2}$ and $\frac{1}{2}$.

Finally, we must point out that in many cases the answers you will obtain when you solve quadratic equations by completing the square are *not* rational numbers. (Remember? A rational number can be written as $\frac{a}{b}$, a and b integers, $b \neq 0$.) As a matter of fact, quadratic equations with rational solutions can be solved by factoring, often a simpler method.

EXAMPLE 4 **Solving a quadratic equation by completing the square**

Solve: $36x + 9x^2 + 31 = 0$

SOLUTION 4 We proceed using the five-step procedure.

$$36x + 9x^2 + 31 = 0 \quad \text{Given}$$

1. Subtract 31 and write in descending order. $9x^2 + 36x = -31$
2. Divide by 9. $x^2 + 4x = -\frac{31}{9}$
3. Add $\left(\frac{4}{2}\right)^2 = 2^2$. $x^2 + 4x + 2^2 = -\frac{31}{9} + 2^2 = -\frac{31}{9} + \frac{36}{9}$
4. Rewrite as a perfect square. $(x + 2)^2 = \frac{5}{9}$
5. Use the square root property. $(x + 2) = \pm\sqrt{\frac{5}{9}}$

$$x + 2 = \pm\frac{\sqrt{5}}{3}$$

$$x = -2 \pm \frac{\sqrt{5}}{3} = \frac{-6 \pm \sqrt{5}}{3}$$

Thus, the solutions of $36x + 9x^2 + 31 = 0$ are $\frac{-6 + \sqrt{5}}{3}$ and $\frac{-6 - \sqrt{5}}{3}$.

PROBLEM 4

Solve:

$$4x^2 + 24x + 31 = 0$$

An article from CBC (Canadian Broadcasting Corporation) News reported that the number of polar bears in the Davis Straits (a strait between Canada and Greenland), has increased from 1400 in 1997 to 1650 in 2004. Using these figures the number of polar bears N can be approximated by $\mathbf{N = 10x^2 - 40x + 1400,}$ where x is the number of years after 1997. When will the bear population double to 2800 bears? Just solve $\mathbf{10x^2 - 40x + 1400 = 2800}$ for x! We do this in Example 5.

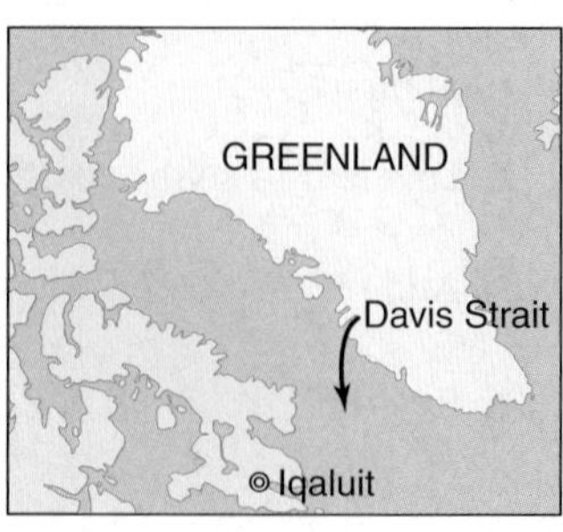

Source: http://tinyurl.com/yll5qz7.

Answers to PROBLEMS

4. The solutions are $-3 \pm \frac{\sqrt{5}}{2} = \frac{-6 \pm \sqrt{5}}{2}$

EXAMPLE 5 Bear population estimates

Solve the equation $10x^2 - 40x + 1400 = 2800$ for x, the number of years after 1997, by completing the square. This will tell you in how many years the bear population will double.

SOLUTION 5 We use the 5 steps in the procedure on page 699.

	$10x^2 - 40x + 1400 = 2800$	Given
1.	$10x^2 - 40x = 1400$	Subtract 1400 from both sides.
2.	$x^2 - 4x = 140$	Divide each term by 10.
3.	$x^2 - 4x + 4 = 140 + 4$	Add $\left(\frac{1}{2} \cdot 4\right)^2 = 2^2 = 4$.
4.	$(x - 2)^2 = 144$	Factor: $x^2 - 4x + 4 = (x - 2)^2$.
5.	$x - 2 = \pm\sqrt{144} = \pm 12$	Use the square root property.
	$x = \pm 12 + 2$	Add 2 to both sides.
	$x = +12 + 2$ or $-12 + 2$	
	$x = 14$ (Discard $-12 + 2$, which is negative. Why?)	

This means that $x = 14$ years after 1997 (in 2011) the number of bears in the Davis Strait will double to 2800.

PROBLEM 5

A more accurate polar bear count uses $N = 11x^2 - 44x + 1430$ as the number N of polar bears x years after 1997. Under this model in how many years will the polar bear population reach 3245 bears?

Answers to PROBLEMS

5. In 15 years (2012)

Exercises 9.2

> Practice Problems > Self-Tests > Media-rich eBooks > e-Professors > Videos

A **Solving Quadratic Equations by Completing the Square** In Problems 1–20, find the missing term(s) to make the expression a perfect square.

1. $x^2 + 18x + \square$ **2.** $x^2 + 2x + \square$ **3.** $x^2 - 16x + \square$ **4.** $x^2 - 4x + \square$

5. $x^2 + 7x + \square$ **6.** $x^2 + 9x + \square$ **7.** $x^2 - 3x + \square$ **8.** $x^2 - 7x + \square$

9. $x^2 + x + \square$ **10.** $x^2 - x + \square$ **11.** $x^2 + 4x + \square = (\quad)^2$ **12.** $x^2 + 6x + \square = (\quad)^2$

13. $x^2 + 3x + \square = (\quad)^2$ **14.** $x^2 + 9x + \square = (\quad)^2$ **15.** $x^2 - 6x + \square = (\quad)^2$ **16.** $x^2 - 24x + \square = (\quad)^2$

17. $x^2 - 5x + \square = (\quad)^2$ **18.** $x^2 - 11x + \square = (\quad)^2$ **19.** $x^2 - \frac{3}{2}x + \square = (\quad)^2$ **20.** $x^2 - \frac{5}{2}x + \square = (\quad)^2$

In Problems 21–40, solve the given equation if possible.

21. $x^2 + 4x + 1 = 0$ **22.** $x^2 + 2x + 7 = 0$ **23.** $x^2 + x - 1 = 0$

24. $x^2 + 2x - 1 = 0$ **25.** $x^2 + 3x - 1 = 0$ **26.** $x^2 - 3x - 4 = 0$

Web IT go to mhhe.com/bello for more lessons

27. $x^2 - 3x - 3 = 0$

28. $x^2 - 3x - 1 = 0$

29. $4x^2 + 4x - 3 = 0$

30. $2x^2 + 10x - 1 = 0$

31. $4x^2 - 16x = 15$

32. $25x^2 - 25x = -6$

33. $4x^2 - 7 = 4x$

34. $2x^2 - 18 = -9x$

35. $2x^2 + 1 = 4x$

36. $2x^2 + 3 = 6x$

37. $(x + 3)(x - 2) = -4$

38. $(x + 4)(x - 1) = -6$

39. $2x(x + 5) - 1 = 0$

40. $2x(x - 4) = 2(9 - 8x) - x$

Applications: Green Math

41. *Future Polar Bear Population* Use the approximation $N = 10x^2 - 40x + 1400$ of Example 5 to find the time it takes to more than triple the polar bear population of 2004 to 4970 polar bears.

42. *More Polar Bears* Use the more accurate approximation $N = 11x^2 - 44x + 1430$ to find the time it takes to triple the polar bear population of 2004 to 4950 polar bears.

A visual method to complete the square In the ninth century, an Arab mathematician named al-Khwarizmi (from which the word "algorithm" was derived) developed a method to complete the square in his book (*Hisab al-jabr w'al-muqabala, The Compendious Book on Calculation by Completion and Balancing*). To solve the equation

$$x^2 + 10x = 39$$

he used a square like the one on the left and then literally completed it by adding 25 to the lower right corner.

A page from al-Khwarizmi's book

x^2	$5x$
$5x$	

	x	5
x	x^2	$5x$
5	$5x$	25

The shaded area of the squares on the left is $x^2 + 10x$, so adding the 25 to the $x^2 + 10x = 39$ yields

$$x^2 + 10x + 25 = 39 + 25$$

or $(x + 5)^2 = 64$

Taking roots $x + 5 = \pm 8$

Subtracting 5 $x = \pm 8 - 5$

Which means $x = 3$ and $x = -13$.

In Problems 43–44, add a term to the binomial to change it into a trinomial square. Use the diagram to find the term that needs to be added to complete the square.

43. $x^2 + 6x$

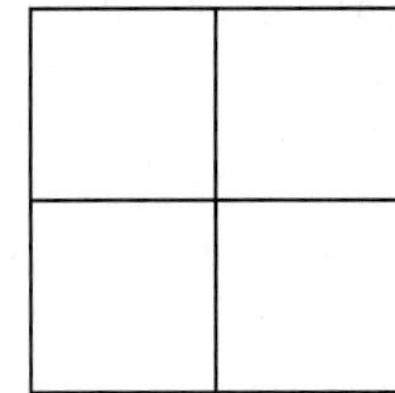

44. $x^2 + 8x$

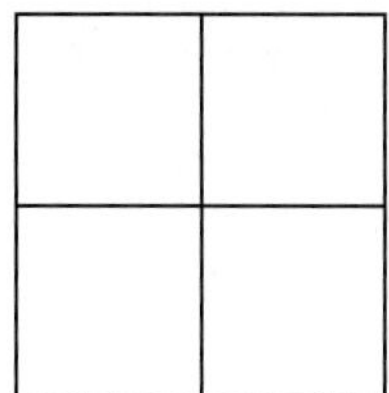

In Problems 45–46, use the results of Problems 43–44 to solve the equations.

45. $x^2 + 6x = 16$

46. $x^2 + 8x = 9$

Using Your Knowledge

Finding a Maximum or a Minimum Many applications of mathematics require finding the maximum or the minimum of certain algebraic expressions. Thus, a certain business may wish to find the price at which a product will bring *maximum* profits, whereas engineers may be interested in *minimizing* the amount of carbon monoxide produced by automobiles.

Now suppose you are the manufacturer of a certain product whose average manufacturing cost $\overline{C}$ (in dollars), based on producing x (thousand) units, is given by the expression

$$\overline{C} = x^2 - 8x + 18$$

How many units should be produced to *minimize* the cost per unit? If we consider the right-hand side of the equation, we can complete the square and leave the equation unchanged by adding and subtracting the appropriate number. Thus,

$$\overline{C} = x^2 - 8x + 18$$

$$\overline{C} = (x^2 - 8x + \quad) + 18$$

$$\overline{C} = (x^2 - 8x + 4^2) + 18 - 4^2$$

Note that we have added and subtracted the square of one-half of the coefficient of x, $\left(\frac{-8}{2}\right)^2 = 4^2$.

Then

$$\overline{C} = (x - 4)^2 + 2$$

Since the smallest possible value that $(x - 4)^2$ can have is 0 for $x = 4$, let $x = 4$ to give the minimum cost, $\overline{C} = 2$.

Use your knowledge about completing the square to solve the following problems.

47. A manufacturer's average cost $\overline{C}$ (in dollars), based on manufacturing x (thousand) items, is given by the equation

$$\overline{C} = x^2 - 4x + 6$$

a. How many units should be produced to minimize the cost per unit?

b. What is the minimum average cost per unit?

48. The demand D for a certain product depends on the number x (in thousands) of units produced and is given by the equation

$$D = x^2 - 2x + 3$$

What is the number of units that have to be produced so that the demand is at its lowest?

49. Have you seen people adding chlorine to their pools? This is done to reduce the number of bacteria present in the water. Suppose that after t days, the number of bacteria per cubic centimeter is given by the expression

$$B = 20t^2 - 120t + 200$$

In how many days will the number of bacteria be at its lowest?

Write On

50. What is the first step you would perform to solve the equation $6x^2 + 9x = 3$ by completing the square?

51. Can you solve the equation $x^3 + 2x^2 = 5$ by completing the square? Explain.

52. Describe the procedure you would use to find the number that has to be added to $x^2 + 5x$ to make the expression a perfect square trinomial.

53. Make a perfect square trinomial that has a term of $-6x$ and explain how you constructed your trinomial.

Concept Checker

Fill in the blank(s) with the correct word(s), phrase, or mathematical statement.

multiply **left** **right** **divide**

54. The **first** step to **complete the square** when solving an equation is to write the equation with the variables in **descending order** on the __________ side of the equation and the constants on the __________ side of the equation.

55. When **completing the square,** if the **coefficient** of the **square** term is **not 1,** __________ each term by this coefficient.

Mastery Test

Find the missing term:

56. $(x + 11)^2 = x^2 + 22x + \square$

57. $\left(x - \frac{1}{4}\right)^2 = x^2 - \frac{1}{2}x + \square$

58. $x^2 - 12x + \square = (\quad)^2$

59. $x^2 + 5x + \square = (\quad)^2$

Solve:

60. $4x^2 - 24x + 27 = 0$

61. $4x^2 + 24x + 33 = 0$

Skill Checker

Write each equation in the standard form $ax^2 + bx + c = 0$, where a, b, and c are integers. (*Hint:* For equations involving fractions, first multiply each term by the LCD.)

62. $10x^2 + 5x = 12$

63. $x^2 = 2x + 2$

64. $9x = x^2$

65. $\frac{x^2}{2} + \frac{3}{5} = \frac{x}{4}$

66. $\frac{x}{4} = \frac{3}{5} - \frac{x^2}{2}$

67. $\frac{x}{3} = \frac{3}{4}x^2 - \frac{1}{6}$

9.3 Solving Quadratic Equations by the Quadratic Formula

Objectives

A Write a quadratic equation in the form $ax^2 + bx + c = 0$ and identify a, b, and c.

B Solve a quadratic equation using the quadratic formula.

To Succeed, Review How To . . .

1. Solve a quadratic equation by completing the square (pp. 695–701).
2. Write a quadratic equation in standard form (p. 458).

Getting Started

Name That Formula!

As you were going through the preceding sections of this chapter, you probably wondered whether there is a surefire method for solving any quadratic equation. Fortunately, there is! As a matter of fact, this new technique incorporates the method of completing the square. As you recall, in each of the problems we followed the same procedure. Why not use the method of completing the square and solve the equation $ax^2 + bx + c = 0$ once and for all? We shall do that now. (You can refer to the procedure we gave to solve equations by completing the square on page 699.) Here is how the quadratic formula is derived:

"My final recommendation is that we reject all these proposals and continue to call it the quadratic formula."

Courtesy of Thelma Castellano

$$ax^2 + bx + c = 0, \quad a \neq 0 \qquad \text{Given}$$

Rewrite with the constant c on the right.
$$ax^2 + bx = -c$$

Divide each term by a.
$$x^2 + \frac{b}{a}x = -\frac{c}{a}$$

Add the square of one-half the coefficient of x.
$$x^2 + \frac{b}{a}x + \left(\frac{b}{2a}\right)^2 = \left(\frac{b}{2a}\right)^2 - \frac{c}{a}$$

Rewrite the left side as a perfect square.
$$\left(x + \frac{b}{2a}\right)^2 = \frac{b^2}{4a^2} - \frac{c}{a}$$

$$\left(x + \frac{b}{2a}\right)^2 = \frac{b^2}{4a^2} - \frac{4ac}{4a^2}$$

Write the right side as a single fraction.
$$\left(x + \frac{b}{2a}\right)^2 = \frac{b^2 - 4ac}{4a^2}$$

Now, take the square root of both sides.
$$x + \frac{b}{2a} = \frac{\pm\sqrt{b^2 - 4ac}}{2a}$$

Subtract $\frac{b}{2a}$ from both sides.
$$x = -\frac{b}{2a} \pm \frac{\sqrt{b^2 - 4ac}}{2a}$$

Combine fractions.
$$x = \frac{-b \pm \sqrt{b^2 - 4ac}}{2a}$$

There you have it, *the* surefire method for solving quadratic equations! We call this the *quadratic formula,* and we shall use it in this section to solve quadratic equations.

Do you realize what we've just done? We've given ourselves a powerful tool! *Any* time we have a quadratic equation in the standard form $ax^2 + bx + c = 0$, we can solve the equation by simply substituting a, b, and c in the formula.

THE QUADRATIC FORMULA

The solutions of $ax^2 + bx + c = 0 \quad (a \neq 0)$ are

$$x = \frac{-b \pm \sqrt{b^2 - 4ac}}{2a}$$

that is,

$$x = \frac{-b + \sqrt{b^2 - 4ac}}{2a} \quad \text{and} \quad x = \frac{-b - \sqrt{b^2 - 4ac}}{2a}$$

This formula is so important that you should memorize it right now. Before using it, however, you must remember to do two things:

1. First, write the given equation in the **standard form** $ax^2 + bx + c = 0$; then
2. Determine the values of a, b, and c.

A › Writing Quadratic Equations in Standard Form

In order to solve the equation $x^2 = 5x - 2$, we must first write the equation in standard form and then determine the values of a, b, and c as follows.

1. We must *first* write it in the standard form $ax^2 + bx + c = 0$ by subtracting $5x$ and adding 2; we then have

$$x^2 - 5x + 2 = 0 \quad \text{In standard form}$$

2. When the equation is in standard form, it's very easy to find the values of a, b, and c:

$$\underbrace{x^2}_{a=1} \quad - \quad \underbrace{5x}_{b=-5} \quad + \quad \underbrace{2}_{c=2} = 0$$

Recall that $x^2 = 1x^2$; thus, the coefficient of x^2 is 1.

If the equation contains fractions, multiply each term by the least common denominator (LCD) to clear the fractions. For example, consider the equation

$$\frac{x}{4} = \frac{3}{5} - \frac{x^2}{2}$$

Since the LCD of 4, 5, and 2 is 20, we multiply each term by 20:

$$20 \cdot \frac{x}{4} = \frac{3}{5} \cdot 20 - \frac{x^2}{2} \cdot 20$$

$$5x = 12 - 10x^2$$

$$10x^2 + 5x = 12 \quad \text{Add } 10x^2.$$

$$10x^2 + 5x - 12 = 0 \quad \text{Subtract 12.}$$

Now the equation is in standard form:

$$\underbrace{10x^2}_{a=10} \quad + \quad \underbrace{5x}_{b=5} \quad - \quad \underbrace{12}_{c=-12} = 0$$

You can get a lot of practice by completing this table:

Given	Standard Form	*a*	*b*	*c*
1. $2x^2 + 7x - 4 = 0$				
2. $x^2 = 2x + 2$				
3. $9x = x^2$				
4. $\frac{x^2}{4} + \frac{2}{3}x = -\frac{1}{3}$				

STOP! Do not proceed until you complete the table!

We will now use these four equations as examples.

B › Solving Quadratic Equations with the Quadratic Formula

EXAMPLE 1 Solving a quadratic equation using the quadratic formula

Solve using the quadratic formula: $2x^2 + 7x - 4 = 0$

SOLUTION 1

1. The equation is already written in standard form:

$$\underbrace{2x^2}_{a=2} + \underbrace{7x}_{b=7} \underbrace{- 4}_{c=-4} = 0$$

2. As Step 1 shows, it is clear that $a = 2$, $b = 7$, and $c = -4$.
3. Substituting the values of a, b, and c in the quadratic formula, we obtain

$$x = \frac{-7 \pm \sqrt{(7)^2 - 4(2)(-4)}}{2(2)}$$

$$x = \frac{-7 \pm \sqrt{49 + 32}}{4}$$

$$x = \frac{-7 \pm \sqrt{81}}{4}$$

$$x = \frac{-7 \pm 9}{4}$$

Thus,

$$x = \frac{-7 + 9}{4} = \frac{2}{4} = \frac{1}{2} \quad \text{or} \quad x = \frac{-7 - 9}{4} = \frac{-16}{4} = -4$$

The solutions are $\frac{1}{2}$ and -4.

PROBLEM 1

Solve using the quadratic formula:

$$3x^2 + 2x - 5 = 0$$

In Example 1, you could also solve $2x^2 + 7x - 4 = 0$ by factoring $2x^2 + 7x - 4$ and writing $(2x - 1)(x + 4) = 0$. You will get the same answers!

NOTE

Try factoring first unless otherwise directed.

EXAMPLE 2 Writing in standard form before using the quadratic formula

Solve using the quadratic formula: $x^2 = 2x + 2$

SOLUTION 2

We proceed by steps as in Example 1.

1. To write the equation in standard form, subtract $2x$ and then subtract 2 to obtain

PROBLEM 2

Solve using the quadratic formula:

$$x^2 = 4x + 4$$

(continued)

Answers to PROBLEMS

1. The solutions are 1 and $-\frac{5}{3}$. **2.** The solutions are $2 \pm 2\sqrt{2}$.

$$\underbrace{x^2}_{a=1} \quad - \quad \underbrace{2x}_{b=-2} \quad - \quad \underbrace{2 = 0}_{c=-2}$$

2. As Step 1 shows, $a = 1$, $b = -2$, and $c = -2$.
3. Substituting these values in the quadratic formula, we have

$$x = \frac{-(-2) \pm \sqrt{(-2)^2 - 4(1)(-2)}}{2(1)}$$

$$x = \frac{2 \pm \sqrt{4 + 8}}{2}$$

$$x = \frac{2 \pm \sqrt{12}}{2}$$

$$x = \frac{2 \pm \sqrt{4 \cdot 3}}{2}$$

$$x = \frac{2 \pm 2\sqrt{3}}{2}$$

Thus,

$$x = \frac{2 + 2\sqrt{3}}{2} = \frac{2(1 + \sqrt{3})}{2} = 1 + \sqrt{3}$$

or

$$x = \frac{2 - 2\sqrt{3}}{2} = \frac{2(1 - \sqrt{3})}{2} = 1 - \sqrt{3}$$

The solutions are $1 + \sqrt{3}$ and $1 - \sqrt{3}$.

Note that $x^2 - 2x - 2$ is *not* factorable. You *must* know the quadratic formula or you must complete the square to solve this equation!

EXAMPLE 3 Rewriting a quadratic equation before using the quadratic formula

Solve using the quadratic formula: $9x = x^2$

SOLUTION 3 We proceed in steps.

1. Subtracting $9x$, we have

$$0 = x^2 - 9x$$

or

$$\underbrace{x^2}_{a=1} \quad - \quad \underbrace{9x}_{b=-9} \quad + \quad \underbrace{0 = 0}_{c=0}$$

2. As Step 1 shows, $a = 1$, $b = -9$, and $c = 0$ (because the c term is missing).
3. Substituting these values in the quadratic formula, we obtain

$$x = \frac{-(-9) \pm \sqrt{(-9)^2 - 4(1)(0)}}{2(1)}$$

$$x = \frac{9 \pm \sqrt{81 - 0}}{2}$$

$$x = \frac{9 \pm \sqrt{81}}{2}$$

$$x = \frac{9 \pm 9}{2}$$

Thus,

$$x = \frac{9 + 9}{2} = \frac{18}{2} = 9 \quad \text{or} \quad x = \frac{9 - 9}{2} = \frac{0}{2} = 0$$

The solutions are 9 and 0.

PROBLEM 3

Solve using the quadratic formula:

$$6x = x^2$$

Answers to PROBLEMS

3. The solutions are 0 and 6.

NOTE

$x^2 - 9x = x(x - 9) = 0$ could have been solved by factoring. Always try factoring first unless otherwise directed!

How do we know that the original equation could have been solved by factoring? We know because the answers are rational numbers (fractions or whole numbers). If your answer is a rational number, the equation could have been solved by factoring. Can we determine this in advance? Yes, and we will explain how to do so before the end of the section.

EXAMPLE 4 Solving a quadratic equation involving fractions

Solve using the quadratic formula:

$$\frac{x^2}{4} + \frac{2}{3}x = -\frac{1}{3}$$

SOLUTION 4

1. We have to write the equation in standard form, but first we clear fractions by multiplying by the LCM of 4 and 3, that is, by 12:

$$12 \cdot \frac{x^2}{4} + 12 \cdot \frac{2}{3}x = -\frac{1}{3} \cdot 12$$

$$3x^2 + 8x = -4$$

We then add 4 to obtain

$$\underbrace{3x^2}_{a=3} + \underbrace{8x}_{b=8} + \underbrace{4}_{c=4} = 0$$

2. As Step 1 shows, $a = 3$, $b = 8$, and $c = 4$.
3. Substituting in the quadratic formula gives

$$x = \frac{-8 \pm \sqrt{(8)^2 - 4(3)(4)}}{2(3)}$$

$$x = \frac{-8 \pm \sqrt{64 - 48}}{6}$$

$$x = \frac{-8 \pm \sqrt{16}}{6}$$

$$x = \frac{-8 \pm 4}{6}$$

Thus,

$$x = \frac{-8 + 4}{6} = \frac{-4}{6} = -\frac{2}{3} \quad \text{or} \quad x = \frac{-8 - 4}{6} = \frac{-12}{6} = -2$$

The solutions are $-\frac{2}{3}$ and -2. (Can you tell whether the original equation was factorable by looking at the answers?)

PROBLEM 4

Solve using the quadratic formula:

$$\frac{x^2}{4} - \frac{3}{8}x = \frac{1}{4}$$

Now, a final word of warning. As you recall, some quadratic equations do not have real-number solutions. This is still true, even when you use the quadratic formula. The next example shows how this happens.

Answers to PROBLEMS

4. The solutions are 2 and $-\frac{1}{2}$.

EXAMPLE 5 Recognizing a quadratic equation with no real-number solution

Solve using the quadratic formula: $2x^2 + 3x = -3$

SOLUTION 5

1. We add 3 to write the equation in standard form. We then have

$$\underbrace{2x^2}_{a\,=\,2} + \underbrace{3x}_{b\,=\,3} + \underbrace{3}_{c\,=\,3} = 0$$

2. As Step 1 shows, $a = 2$, $b = 3$, and $c = 3$. Now,

$$x = \frac{-3 \pm \sqrt{(3)^2 - 4(2)(3)}}{2(2)}$$

$$x = \frac{-3 \pm \sqrt{9 - 24}}{4}$$

$$x = \frac{-3 \pm \sqrt{-15}}{4}$$

Thus,

$$x = \frac{-3 + \sqrt{-15}}{4} \quad \text{or} \quad x = \frac{-3 - \sqrt{-15}}{4}$$

But wait! If we check the definition of square root, we can see that $\sqrt{a}$ is defined only for $a \geq 0$. Hence, $\sqrt{-15}$ is **not** a real number, and the equation $2x^2 + 3x = -3$ has no *real-number solution.*

PROBLEM 5

Solve using the quadratic formula:

$$3x^2 + 2x = -1$$

As you can see from Example 5, the number -15 under the radical determines the type of roots the equation has. The expression $b^2 - 4ac$ under the radical in the quadratic formula is called the **discriminant.**

THE DISCRIMINANT

If $b^2 - 4ac \geq 0$, the equation has real-number solutions.

If $b^2 - 4ac < 0$, the equation has no real-number solutions.

Note that in Example 5, $a = 2$, $b = 3$, and $c = 3$. Thus,

$$b^2 - 4ac = 3^2 - 4(2)(3) = 9 - 24 = -15 < 0$$

so the equation has **no** real-number solutions.

Is the discriminant good for anything else? Of course. The discriminant can also suggest the method you use to solve quadratic equations. Here are the suggestions:

Start by writing the equation in standard form. Then, if $b^2 - 4ac \geq 0$ is a perfect square, try factoring. (*Note:* Some perfect squares are 0, 1, 4, 9, 16, and so on.)

In Example 1, $2x^2 + 7x - 4 = 0$ and $b^2 - 4ac = 81$, so we could try to solve the equation by factoring.

In Example 2, $x^2 - 2x - 2 = 0$ and $b^2 - 4ac = 12$. The equation cannot be factored! Use the quadratic formula.

In Example 3, $x^2 - 9x = 0$ and $b^2 - 4ac = 81$. Try factoring!

In Example 4, $3x^2 + 8x + 4 = 0$ and $b^2 - 4ac = 16$. Try to solve by factoring.

In Example 5, $2x^2 + 3x + 3 = 0$ and $b^2 - 4ac = -15$, so you know there are **no** real-number solutions.

Answers to PROBLEMS

5. No real-number solution

We list the best method to solve each type of quadratic equation in the following chart.

Type of Equation	Method to Solve
$X^2 = A$	Square root property
$(AX \pm B)^2 = C$	Square root property
$ax^2 + bx + c = 0$	Factor if $b^2 - 4ac$ is a perfect square.
$ax^2 + bx + c = 0$	Quadratic formula if $b^2 - 4ac$ is not a perfect square

Note: If $b^2 - 4ac < 0$, there is no real-number solution. You do not even have to try a method!

Model Year	Passenger Car
2010	27.5
2011	31.2
2012	32.8

The Corporate Average Fuel Economic (CAFE) standards were first enacted by Congress in 1975 with the purpose of reducing energy consumption by increasing the fuel efficiency (mpg) of cars and light trucks. The standards for passenger cars in model years 2010–2012 are shown in the table, but you have to drive more efficiently to get good mileage! The approximation $M = -0.01x^2 + x + 7$ miles per gallon, where x is the speed of the car in miles per hour, gives a formula for the fuel efficiency of an average car.

To get this fuel efficiency:

1. Observe the speed limit.
2. Remove excess weight.
3. Use cruise control.

Source: http://www.fueleconomy.gov/Feg/driveHabits.shtml.

EXAMPLE 6 Fuel efficiency and speed

Use the approximation $M = -0.01x^2 + x + 7$ to calculate the speed x a car should travel (on average) to get the $M = 31.2$ mpg efficiency for passenger cars in 2011.

SOLUTION 6 We have to solve the quadratic equation $-0.01x^2 + x + 7 = 31.2$. Here is the work

$$-0.01x^2 + x + 7 = 31.2 \quad \text{Given}$$
$$-0.01x^2 + x - 24.2 = 0 \quad \text{Subtract 31.2 from both sides.}$$
$$\mathbf{x^2 - 100x + 2420 = 0} \quad \text{Multiply each term by } -100.$$

We have $a = 1$, $b = -100$, and $c = 2420$.

Which method is best to solve this equation? Follow the recommendations in the chart by first finding the discriminant $b^2 - 4ac$.

$$b^2 - 4ac = (-100)^2 - 4(1)(2420) = 10{,}000 - 9680 = 320$$

Since 320 is not a perfect square, the chart suggests that we use the quadratic formula:

$$x = \frac{-(-100) \pm \sqrt{(-100)^2 - 4(1)(2420)}}{2} = \frac{100 \pm \sqrt{320}}{2} = \frac{100 \pm \sqrt{64 \cdot 5}}{2}$$
$$= \frac{100 \pm 8\sqrt{5}}{2} = 50 \pm 4\sqrt{5} = 50 \pm 8.9$$

This means that the speed $x = 50 + 8.9 = 58.9$ or $x = 50 - 8.9 = 41.1$. These are the average speeds at which you should travel to obtain the 31.2 mpg shown in the table.

PROBLEM 6

Use $M = -0.01x^2 + x + 7$ to find the speed x a car should travel (on average) to obtain the 27.5 mpg efficiency for passenger cars in 2010.

Answers to PROBLEMS

6. $x = 50 \pm 15\sqrt{2} \approx 50 \pm 21.2$, that is, $x = 71.2$ mph or $x = 28.8$ mph

Exercises 9.3

Web IT go to mhhe.com/bello for more lessons

A Writing Quadratic Equations in Standard Form
B Solving Quadratic Equations with the Quadratic Formula

In Problems 1–30, write the given equation in standard form, then solve the equation using the quadratic formula.

1. $x^2 + 3x + 2 = 0$
2. $x^2 + 4x + 3 = 0$
3. $x^2 + x - 2 = 0$
4. $x^2 + x - 6 = 0$

5. $2x^2 + x - 2 = 0$
6. $2x^2 + 7x + 3 = 0$
7. $3x^2 + x = 2$
8. $3x^2 + 2x = 5$

9. $2x^2 + 7x = -6$
10. $2x^2 - 7x = -6$
11. $7x^2 = 12x - 5$
12. $-5x^2 = 16x + 8$

13. $5x^2 = 11x - 4$
14. $7x^2 = 12x - 3$
15. $\frac{x^2}{5} - \frac{x}{2} = -\frac{3}{10}$
16. $\frac{x^2}{7} + \frac{x}{2} = -\frac{3}{4}$

17. $\frac{x^2}{2} = \frac{3x}{4} - \frac{1}{8}$
18. $\frac{x^2}{10} = \frac{x}{5} + \frac{3}{2}$
19. $\frac{x^2}{8} = -\frac{x}{4} - \frac{1}{8}$
20. $\frac{x^2}{3} = -\frac{x}{3} - \frac{1}{12}$

21. $6x = 4x^2 + 1$
22. $6x = 9x^2 - 4$
23. $3x = 1 - 3x^2$
24. $3x = 2x^2 - 5$

25. $x(x + 2) = 2x(x + 1) - 4$
26. $x(4x - 7) - 10 = 6x^2 - 7x$
27. $6x(x + 5) = (x + 15)^2$

28. $6x(x + 1) = (x + 3)^2$
29. $(x - 2)^2 = 4x(x - 1)$
30. $(x - 4)^2 = 4x(x - 2)$

Applications

31. *Customer service cost* The customer service department at a store has determined that the cost C of serving x customers is modeled by the equation $C = (0.1x^2 + x + 50)$ dollars. How many customers can be served if they are willing to spend \$250?

32. *Customer service cost* The cost C of serving x customers at a boat dealership is given by the equation $C = (x^2 + 10x + 100)$ dollars. If the dealership spent \$1300 serving customers, how many customers were served?

Applications: Green Math

33. *Reaching the CAFE standards for 2012* Find the discriminant of $-0.01x^2 + x + 7 = 32.8$ to calculate the speed x a car should travel to get the $M = 32.8$ mpg CAFE standard for the year 2012, then follow the suggestions on page 711 to solve the equation.

34. *Lowering the 2012 CAFE standards* Solve the equation $-0.01x^2 + x + 7 = 32$ to calculate the speed x a car should travel to get $M = 32$ mpg by first multiplying both sides by -100, then writing the result in standard form.

Using Your Knowledge

Deriving the Quadratic Formula and Deciding When to Use It In this section, we derived the quadratic formula by completing the square. The procedure depends on making the x coefficient a equal to 1. But there's another way to derive the quadratic formula. See whether you can identify the change being made to $ax^2 + bx + c = 0$ in each step.

$$ax^2 + bx + c = 0 \quad \text{Given}$$

35. $4a^2x^2 + 4abx + 4ac = 0$

36. $4a^2x^2 + 4abx = -4ac$

37. $4a^2x^2 + 4abx + b^2 = b^2 - 4ac$

38. $(2ax + b)^2 = b^2 - 4ac$

39. $2ax + b = \pm\sqrt{b^2 - 4ac}$

40. $2ax = -b \pm \sqrt{b^2 - 4ac}$

41. $x = \dfrac{-b \pm \sqrt{b^2 - 4ac}}{2a}$

The quadratic formula is not the only way to solve quadratic equations. We've actually studied four methods for solving them. The following table lists these methods and suggests the best use for each.

Method	When Used
1. The square root property	When you can write the equation in the form $X^2 = A$
2. Factoring	When the equation is factorable. Use the *ac* test (from Section 5.3) to find out!
3. Completing the square	You can always use this method, but it requires more steps than the other methods.
4. The quadratic formula	This formula can always be used, but you should try factoring first.

In Problems 42–56, solve the equation by the method of your choice.

42. $x^2 = 144$

43. $x^2 - 17 = 0$

44. $x^2 - 2x = -1$

45. $x^2 + 4x = -4$

46. $x^2 - x - 1 = 0$

47. $y^2 - y = 0$

48. $y^2 + 5y - 6 = 0$

49. $x^2 + 5x + 6 = 0$

50. $5y^2 = 6y - 1$

51. $(z + 2)(z + 4) = 8$

52. $(y - 3)(y + 4) = 18$

53. $y^2 = 1 - \frac{8}{3}y$

54. $x^2 = \frac{3}{2}x + 1$

55. $3z^2 + 1 = z$

56. $\frac{1}{2}x^2 - \frac{1}{2}x = \frac{1}{4}$

Write On

In the quadratic formula, the expression $D = b^2 - 4ac$ appears under the radical, and it is called the **discriminant** of the equation $ax^2 + bx + c = 0$.

57. How many solutions does the equation have if $D = 0$? Is (are) the solution(s) rational or irrational?

58. How many solutions does the equation have if $D > 0$?

59. How many solutions does the equation have if $D < 0$?

60. Suppose that one solution of $ax^2 + bx + c = 0$ is

$$x = \frac{-b + \sqrt{D}}{2a}$$

What is the other solution?

Concept Checker

Fill in the blank(s) with the correct word(s), phrase, or mathematical statement.

61. The **solutions** of the equation $ax^2 + bx + c = 0$ are ____________________.

62. When **solving** the equation $ax^2 + bx + c = 0$, if $b^2 - 4ac \geq 0$ the equation has ________ number solutions.

no

$\frac{b \pm \sqrt{b^2 - 4ac}}{2a}$

$\frac{-b \pm \sqrt{b^2 - 4ac}}{2a}$

real

Mastery Test

Solve if possible:

63. $3x^2 + 2x = -1$

64. $6x = x^2$

65. $x^2 = 4x + 8$

66. $3x^2 + 2x - 5 = 0$

67. $\frac{x^2}{4} - \frac{3}{8}x = \frac{1}{4}$

68. $\frac{2}{3}x^2 + x = -1$

Skill Checker

Find:

69. $(-2)^2$

70. -2^2

71. $(-1)^2$

72. -1^2

9.4 Graphing Quadratic Equations

Objectives

A Graph quadratic equations.

B Find the intercepts, the vertex, and graph parabolas involving factorable quadratic expressions.

To Succeed, Review How To . . .

1. Evaluate an expression (pp. 62–63, 69–72).
2. Graph points in the plane (p. 211).
3. Factor a quadratic (pp. 420–422, 432–434).

Getting Started

Parabolic Water Streams

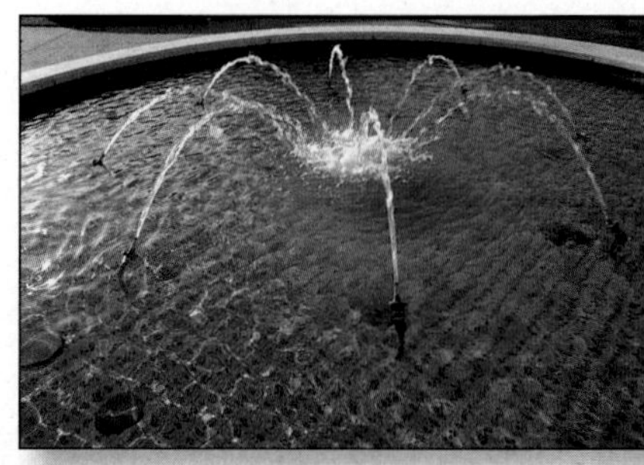

Look at the streams of water from the fountain. What shape do they have? This shape is called a **parabola.** The *graph* of a quadratic equation of the form $y = ax^2 + bx + c$ is a parabola. The simplest of these equations is $y = x^2$. This equation can be graphed in the same way as lines were graphed—that is, by selecting values for x and then finding the corresponding y-values as shown in the table on the left. The usual shortened version is shown in the next table.

x-value	y-value
$x = -2$	$y = x^2 = (-2)^2 = 4$
$x = -1$	$y = x^2 = (-1)^2 = 1$
$x = 0$	$y = x^2 = (0)^2 = 0$
$x = 1$	$y = x^2 = (1)^2 = 1$
$x = 2$	$y = x^2 = (2)^2 = 4$

x	y
−2	4
−1	1
0	0
1	1
2	4

We graph these points on a coordinate system and draw a smooth curve through them. The result is the graph of the parabola $y = x^2$. Note that the arrows at the end indicate that the curve goes on indefinitely.

In this section, we shall learn how to graph parabolas.

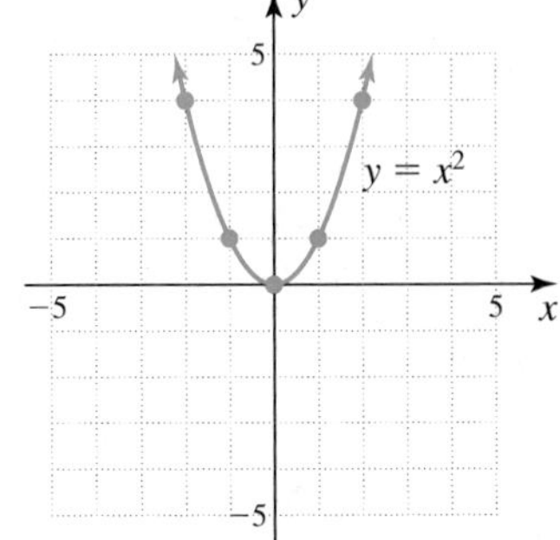

A > Graphing Quadratic Equations

The graph of a quadratic equation of the form $y = ax^2 + bx + c$ is a **parabola.** Now that we know how to graph $y = x^2$, what would happen if we graph $y = -x^2$? To start, we could alter the table in the *Getting Started* by using the negative of the y-value on $y = x^2$, as shown next.

EXAMPLE 1 Graphing $y = ax^2$ when a is negative

Graph: $y = -x^2$

SOLUTION 1 We could always make a table of x- and y-values as before. However, note that for any x-value, the y-value will be the *negative* of the y-value on the parabola $y = x^2$. (If you don't believe this, go ahead and make the table and check it.) Thus, the parabola $y = -x^2$ has the same shape as $y = x^2$, but it's turned in the *opposite* direction (opens *downward*). The graph of $y = -x^2$ is shown in Figure 9.1.

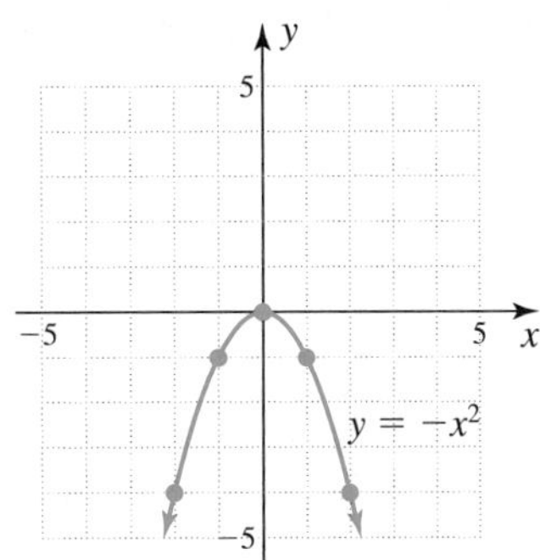

>Figure 9.1

PROBLEM 1

Graph: $y = -2x^2$

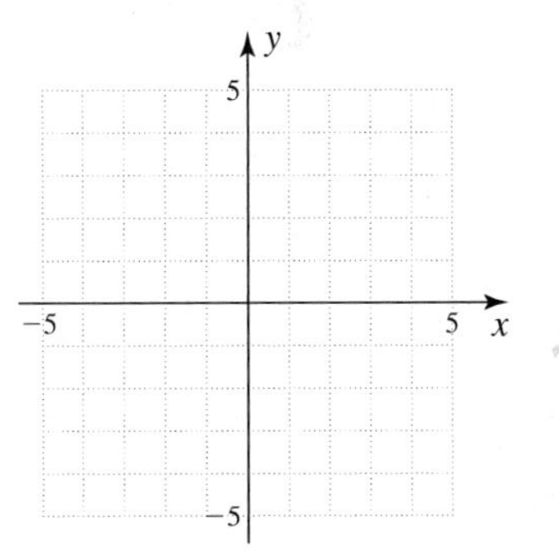

As you can see from the two preceding graphs, when the coefficient of x^2 is positive (as in $y = x^2 = 1x^2$), the parabola opens *upward,* but when the coefficient of x^2 is negative (as in $y = -x^2 = -1x^2$), the parabola opens *downward.* In general, we have the following definition.

GRAPH OF A QUADRATIC EQUATION

The graph of a quadratic equation of the form $y = ax^2 + bx + c$ is a parabola that

1. Opens upward if $a > 0$

2. Opens downward if $a < 0$

Answers to PROBLEMS

1.

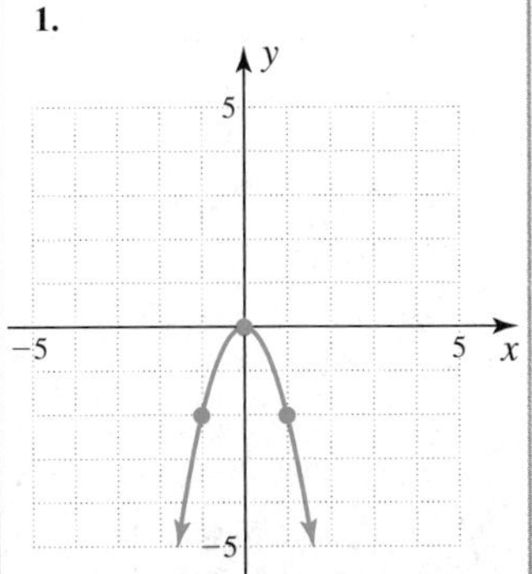

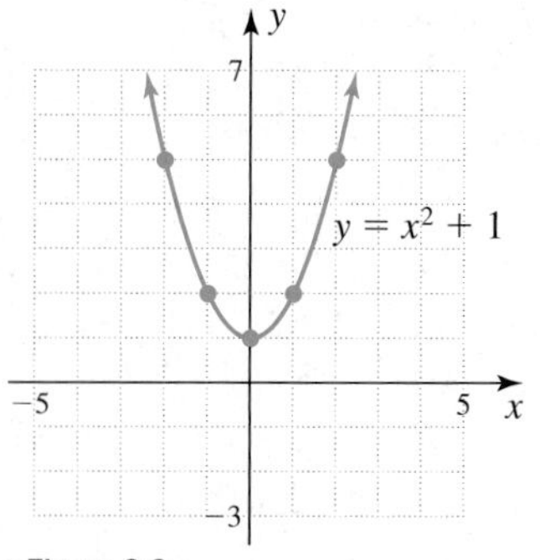

>Figure 9.2

Now, what do you think will happen if we graph the parabola $y = x^2 + 1$? Two things: First, the parabola opens upward, since the coefficient of x^2 is understood to be 1. Second, all of the points will be 1 unit higher than those for the same value of x on the parabola $y = x^2$. Thus, we can make the graph of $y = x^2 + 1$ by following the pattern of $y = x^2$. Similarly, the graph of $y = x^2 + 2$ is 2 units higher than for the parabola $y = x^2$. The graphs of $y = x^2 + 1$ and $y = x^2 + 2$ are shown in Figures 9.2 and 9.3, respectively.

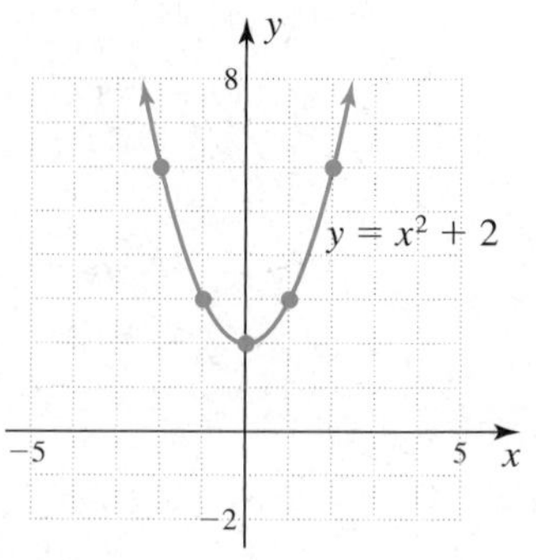

>Figure 9.3

You can verify that these graphs are correct by graphing several of the points for $y = x^2 + 1$ and $y = x^2 + 2$ that are shown in the following tables:

$y = x^2 + 1$

x	y
0	1
1	2
−1	2

$y = x^2 + 2$

x	y
0	2
1	3
−1	3

EXAMPLE 2 Graphing a parabola that opens downward

Graph: $y = -x^2 - 2$

SOLUTION 2 Since the coefficient of x^2 (which is understood to be −1) is *negative,* the parabola opens downward. It is also 2 units *lower* than the graph of $y = -x^2$. Thus, the graph of $y = -x^2 - 2$ is as shown in Figure 9.4.

You can verify the graph by checking the points in the table for $y = -x^2 - 2$:

$y = -x^2 - 2$

x	y
0	−2
1	−3
−1	−3

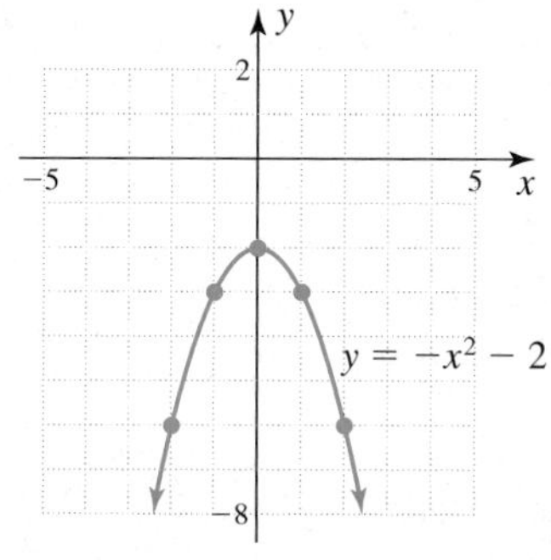

>Figure 9.4

PROBLEM 2

Graph: $y = -x^2 - 1$

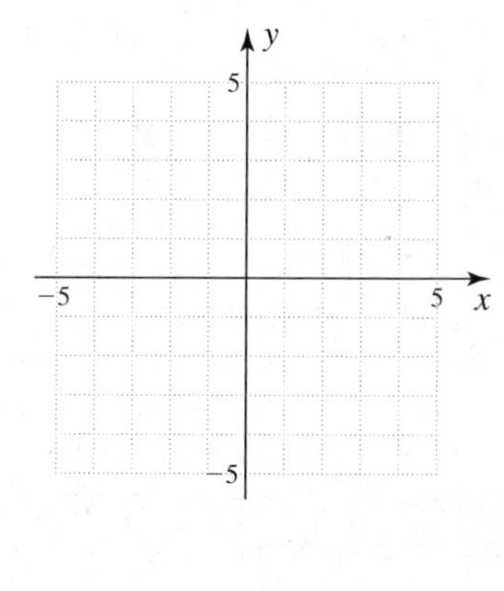

So far, we've graphed parabolas only of the form $y = ax^2 + b$. How do you think the graph of $y = (x - 1)^2$ looks? As before, we make a table of values. For example,

For $x = -1$, $y = (-1 - 1)^2 = (-2)^2 = 4$
For $x = 0$, $y = (0 - 1)^2 = (-1)^2 = 1$
For $x = 1$, $y = (1 - 1)^2 = (0)^2 = 0$
For $x = 2$, $y = (2 - 1)^2 = 1^2 = 1$

$y = (x - 1)^2$

x	y
−1	4
0	1
1	0
2	1

Answers to PROBLEMS

2.

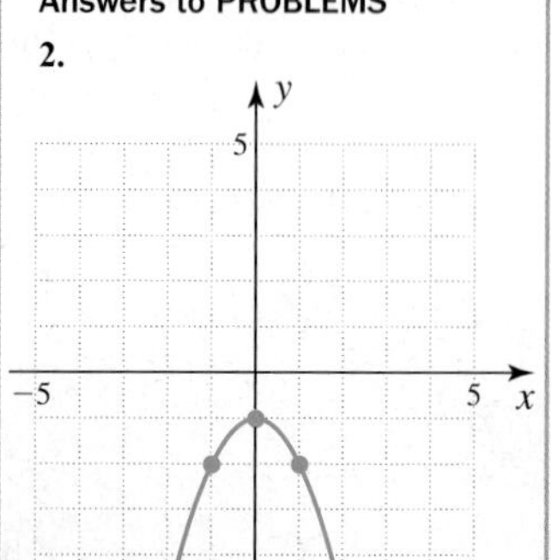

The completed graph is shown in Figure 9.5. Note that the shape of the graph is identical to that of $y = x^2$, but it is shifted 1 unit to the *right.*

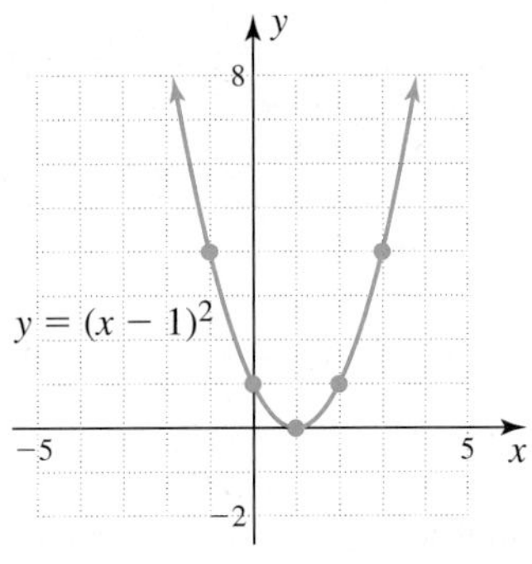

>Figure 9.5

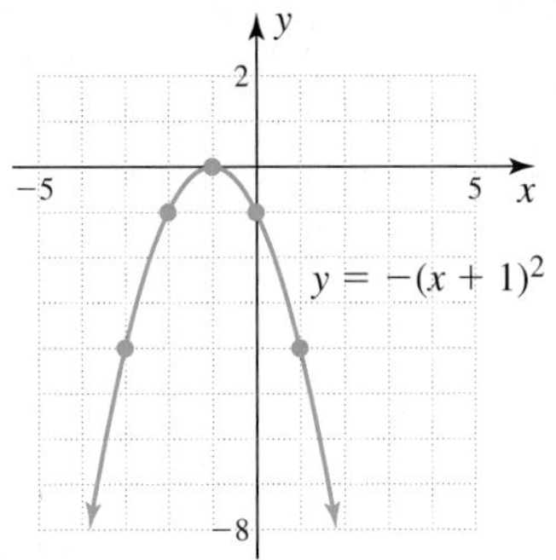

>Figure 9.6

Similarly, the graph of $y = -(x + 1)^2$ is identical to that of $y = -x^2$ but shifted 1 unit to the *left,* as shown in Figure 9.6.

You can verify this using the table for $y = -(x + 1)^2$:

$y = -(x + 1)^2$

x	y
0	−1
−1	0
−2	−1

EXAMPLE 3 Graphing by vertical and horizontal shifts

Graph: $y = -(x - 1)^2 + 2$

SOLUTION 3 The graph of this equation is identical to the graph of $y = -x^2$ except for its position. The parabola $y = -(x - 1)^2 + 2$ opens downward (because of the first negative sign), and it's shifted 1 unit to the right (because of the −1) and 2 units up (because of the +2):

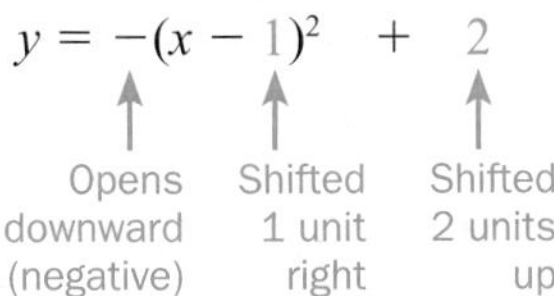

>Figure 9.7

Figure 9.7 shows the finished graph of $y = -(x - 1)^2 + 2$.

As usual, you can verify this using the table for $y = -(x - 1)^2 + 2$:

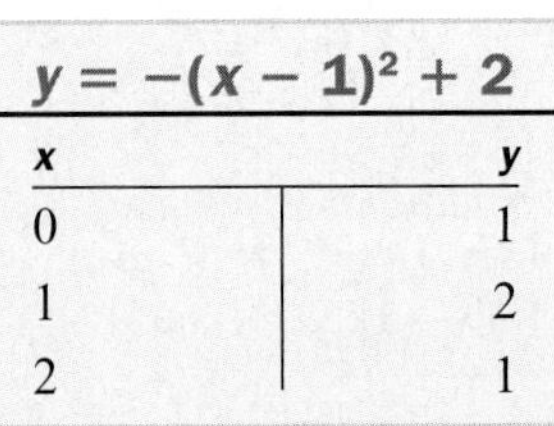

$y = -(x - 1)^2 + 2$

x	y
0	1
1	2
2	1

PROBLEM 3

Graph: $y = -(x - 2)^2 - 1$

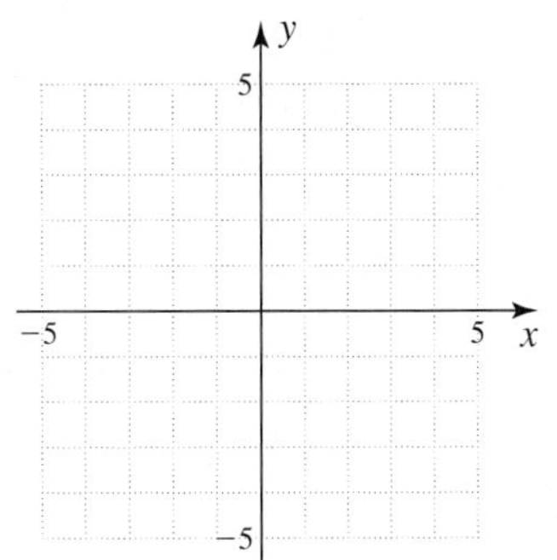

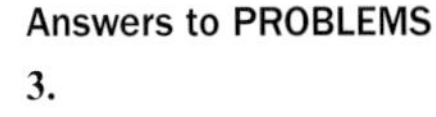

3.

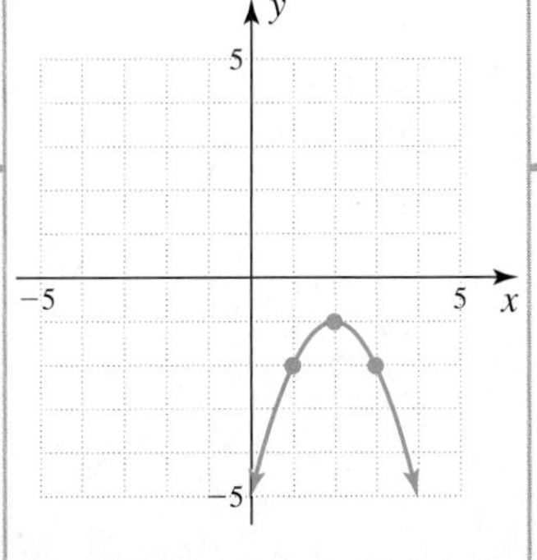

In conclusion, we follow these directions for changing the graph of $y = ax^2$:

$$y = a(x \pm b)^2 \pm c \quad b \text{ and } c \text{ positive}$$

- a: Opens upward for $a > 0$, downward for $a < 0$
- b: Shifts the graph right ($-b$) or left ($+b$)
- c: Moves the graph up ($+c$) or down ($-c$)

Note that as a increases, the parabola "stretches." (We shall examine this fact in the *Calculator Corner*.)

B › Graphing Parabolas Using Intercepts and the Vertex

You've probably noticed that the graph of a parabola is **symmetric;** that is, if you draw a vertical line through the **vertex** (the high or low point on the parabola) and fold the graph along this line, the two halves of the parabola coincide. If a parabola crosses the x-axis, we can use the x-intercepts to find the vertex. For example, to graph $y = x^2 + 2x - 8$, we start by finding the x- and y-intercepts:

For $x = 0$, $\quad y = 0^2 + 2 \cdot 0 - 8 = -8$ $\quad$ The y-intercept

For $y = 0$, $\quad 0 = x^2 + 2x - 8$

$$0 = (x + 4)(x - 2)$$

Thus,

$$x = -4 \quad \text{or} \quad x = 2 \qquad \text{The } x\text{-intercepts}$$

We enter the points $(0, -8)$, $(-4, 0)$, and $(2, 0)$ in a table like this:

x	y	
0	−8	y-intercept
−4	0	x-intercepts
2	0	x-intercepts
?	?	Vertex

How can we find the vertex? Since the parabola is symmetric, the x-coordinate of the vertex is *exactly* halfway between the x-intercepts -4 and 2, so

$$x = \frac{-4 + 2}{2} = \frac{-2}{2} = -1$$

We then find the y-coordinate by letting $x = -1$ in $y = x^2 + 2x - 8$ to obtain

$$y = (-1)^2 + 2 \cdot (-1) - 8 = -9$$

Thus, the vertex is at $(-1, -9)$. The table now looks like this:

x	y	
0	−8	y-intercept
−4	0	x-intercepts
2	0	x-intercepts
−1	−9	Vertex

We can draw the graph using these four points or plot one or two more points. We plotted $(1, -5)$ and $(-3, -5)$. The graph is shown in Figure 9.8.

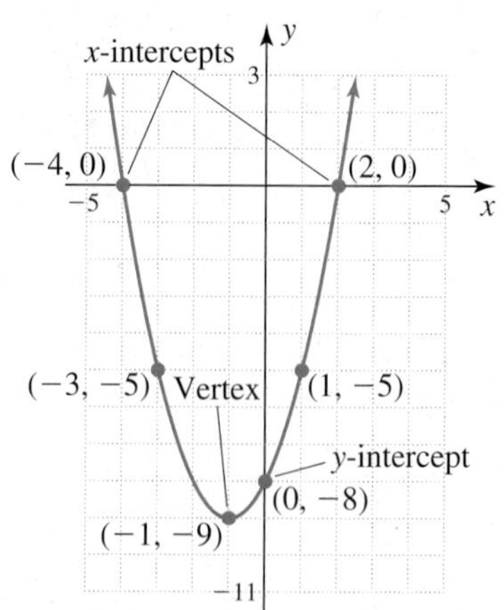

>Figure 9.8

Here is the complete procedure.

PROCEDURE

Graphing a Factorable Quadratic Equation

1. Find the y-intercept by letting $x = 0$ and then finding y.
2. Find the x-intercepts by letting $y = 0$, factoring the equation, and solving for x.
3. Find the vertex by averaging the solutions of the equation found in Step 2 (the average is the x-coordinate of the vertex) and substituting in the equation to find the y-coordinate of the vertex.
4. Plot the points found in Steps 1–3 and one or two more points, if desired. The curve drawn through the points found in Steps 1–4 is the graph.

Now let's use this procedure to graph a parabola.

EXAMPLE 4 Graphing parabolas using the intercepts and vertex

Graph: $y = -x^2 + 2x + 8$

SOLUTION 4

We use the four steps discussed.

1. Since $-x^2 = -1 \cdot x^2$ and -1 is negative, the parabola opens downward. We then let $x = 0$ to obtain $y = -(0)^2 + 2 \cdot 0 + 8 = 8$. Thus, $(0, 8)$ is the y-intercept.
2. We find the x-intercepts by letting $y = 0$ and solving for x. We have

$$0 = -x^2 + 2x + 8$$

$$0 = x^2 - 2x - 8 \quad \text{Multiply both sides by } -1 \text{ to make the factorization easier.}$$

$$0 = (x - 4)(x + 2)$$

$$x = 4 \quad \text{or} \quad x = -2$$

We now have the x-intercepts: $(4, 0)$ and $(-2, 0)$.

3. The x-coordinate of the vertex is found by averaging 4 and -2 to obtain

$$x = \frac{4 + (-2)}{2} = \frac{2}{2} = 1$$

Letting $x = 1$ in $y = -x^2 + 2x + 8$, we find $y = -(1)^2 + 2 \cdot (1) + 8 = 9$, so the vertex is at $(1, 9)$.

4. We plot these points and the two additional points $(2, 8)$ and $(3, 5)$. The completed graph is shown in Figure 9.9.

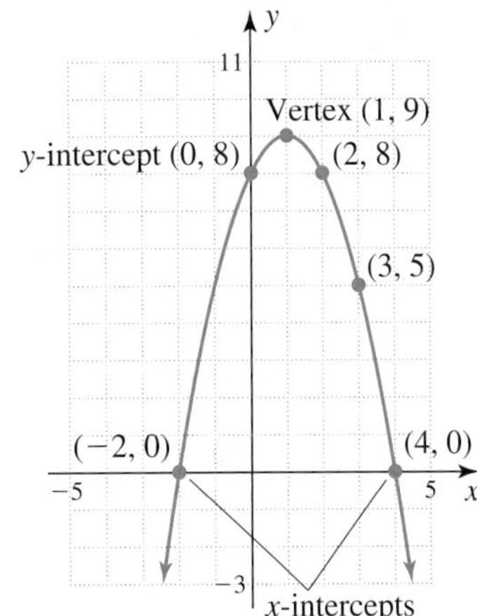

>Figure 9.9

PROBLEM 4

Graph: $y = -x^2 - 2x + 8$

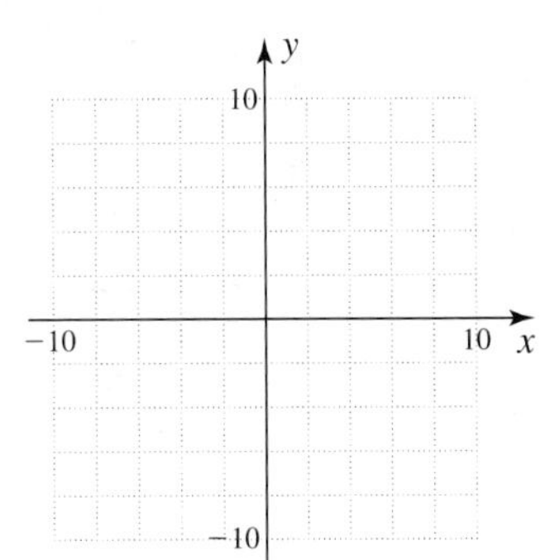

Answers to PROBLEMS

4.

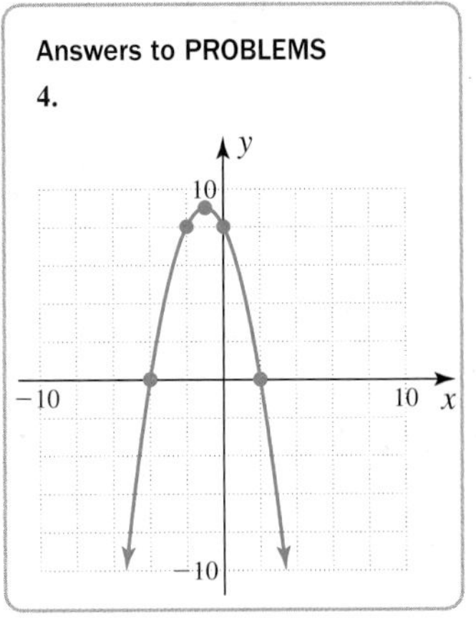

Actually, there is another way of finding the vertex of a parabola (without having to take averages). The technique involves completing the square! Suppose we wish to graph the equation

$$y = ax^2 + bx + c$$

Factor a from the first two terms $\quad y = a\left(x^2 + \frac{b}{a}x\right) + c$

Complete the square on $x^2 + \frac{b}{a}x$

by subtracting $\left(\frac{b}{2a}\right)^2$ and adding $\left(\frac{b}{2a}\right)^2$ $\quad y = a\left[x^2 + \frac{b}{a}x + \left(\frac{b}{2a}\right)^2\right] + c - a\left(\frac{b}{2a}\right)^2$

Rewrite as a binomial square $\quad y = a\left(x + \frac{b}{2a}\right)^2 + c - a\left(\frac{b}{2a}\right)^2$

If $a > 0$ the parabola is ∪ shaped and y is a **minimum** when $x + \frac{b}{2a} = 0$, that is, when $x = -\frac{b}{2a}$. This means that the vertex is at $x = -\frac{b}{2a}$.

If $a < 0$ the parabola is ∩ shaped and y is a **maximum** when $x + \frac{b}{2a} = 0$, that is, when $x = -\frac{b}{2a}$. In either case, we have the following:

VERTEX OF A PARABOLA

The vertex of the parabola $y = ax^2 + bx + c$ has coordinate $x = -\frac{b}{2a}$. You *do not* have to memorize the y-coordinate of the vertex. After you find the x-coordinate, substitute $x = -\frac{b}{2a}$ in $y = ax^2 + bx + c$ and find y.

For example, to find the vertex of the parabola $y = -x^2 + 2x + 8$ of Example 4 simply let $x = -\frac{b}{2a} = -\frac{2}{2(-1)} = 1$. Substituting $x = 1$, we get $y = -(1)^2 + 2(1) + 8 = 9$. So the vertex is (1, 9), as before.

The graph on the right shows the gas mileage y of a car based on its speed x in mph. How was that graph constructed? A fairly accurate approximation for fuel economy is $y = -0.01x^2 + x + 5$, where x is the speed in miles per hour with the graph starting at $x = 5$. You should recognize the general shape of the graph as that of a *parabola*. Let us create our own graph and see how close we come to the original!

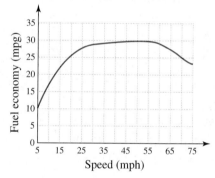

Source: http://www.fueleconomy.gov/FEG/driveHabits.shtml.

GREEN MATH

EXAMPLE 5 Graphing fuel economy based on speed

Graph $\mathbf{y = -0.01x^2 + x + 5}$, where x is the speed (mph) and y the fuel economy in miles per gallon (mpg).

SOLUTION 5 Let us look at some facts about $y = -0.01x^2 + x + 5$.

1. $y = -0.01x^2 + x + 5$ is of the form $y = ax^2 + bx + c$, with $a = -0.01$, $b = 1$, and $c = 5$.

$$y = -0.01\,x^2 + 1\,x + 5$$

$a = -0.01$ $\quad b = 1$ $\quad c = 5$

so its graph is a *parabola*.

PROBLEM 5

The polar bear population in Davis Strait can be approximated by $N = 10x^2 - 40x + 1400$, where x is the number of years after 1997.

a. What is the shape of the graph?

b. Does it open upward or downward?

c. What was the polar bear population in 1997?

d. What was the polar bear population in 2007?

Answer on page 721

2. Since the coefficient of x^2 (-0.01) is negative, the *parabola* opens downward.

3. The x coordinate of the vertex is $x = \frac{-b}{2a} = \frac{-1}{2(-0.01)} = \frac{1}{0.02} = \frac{1}{2/100} = 50$ and $y = -0.01(50)^2 + 50 + 5 = -0.01(2500) + 50 + 5 = -25 + 50 + 5 = 30.$

This means that the vertex is at (50, 30).

4. When $x = 5$, $y = -0.01(5)^2 + 5 + 5$
$= -0.01(25) + 5 + 5$
$= -0.25 + 10 = 9.75.$

5. When $x = 70$, $y = -0.01(70)^2 + 70 + 5$
$= -0.01(4900) + 70 + 5$
$= -49 + 70 + 5 = 26.$

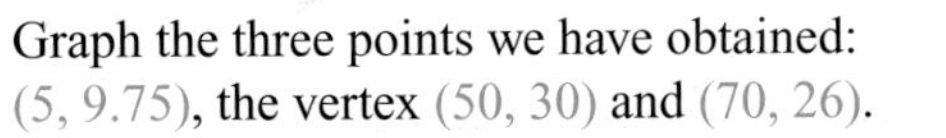

Graph the three points we have obtained: (5, 9.75), the vertex (50, 30) and (70, 26).
Join them with a smooth parabola opening downward as shown. As you can see the result is very similar to the original graph.

e. What are the coordinates of the vertex of the parabola?

f. Use the information in parts **a** to **e** to graph $N = 10x^2 - 40x + 1400$, where x is the number of years after 1997.

Calculator Corner

Graphing Parabolas

If you have a graphing calculator, you can graph parabolas very easily. Let's explore the effect of different values of a, b, and c in the equation

$$y = a(x \pm b) \pm c$$

Using a standard window, graph $y = x^2$, $y = x^2 + 3$, and $y = x^2 - 3$. Clearly, adding the positive number k "moves" the graph of the parabola k units *up*. Similarly, subtracting the positive number k "moves" the graph of the parabola k units *down*, as shown in Window 1. Graph $y = -x^2$, $y = -x^2 + 3$, and $y = -x^2 - 3$ to check the effect of having a negative sign in front of x^2.

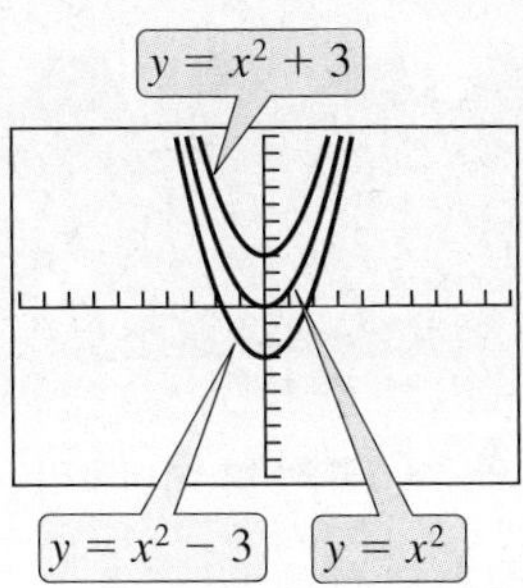

Window 1

Now, let's try some stretching exercises! Look at the graphs of $y = x^2$, $y = 2x^2$, and $y = 5x^2$ using a decimal window. As you can see in Window 2, as the coefficient of x^2 increases, the parabola "stretches"!

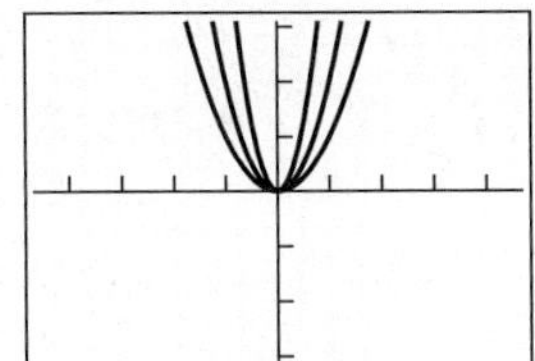

Window 2

Can you find the vertex and intercepts with your calculator? Yes, they're quite easy to find using a calculator. For example, let's use a decimal window to graph

$$y = x^2 - x - 2$$

which is shown in Window 3. You can use the TRACE key to find the vertex $(0.5, -2.25)$. Better yet, some calculators find the minimum of a function by pressing 2nd TRACE 3 and following the prompts. The **minimum,** of course, occurs at the vertex, as shown in Window 4. On the other hand, the **maximum** of $y = -x^2 + x + 2$ occurs at the vertex. Can you find this maximum?

(continued)

Answers to PROBLEMS

5. a. Parabola **b.** Upward **c.** 1400 **d.** 2000 **e.** (2, 1360) **f.** The graph is shown.

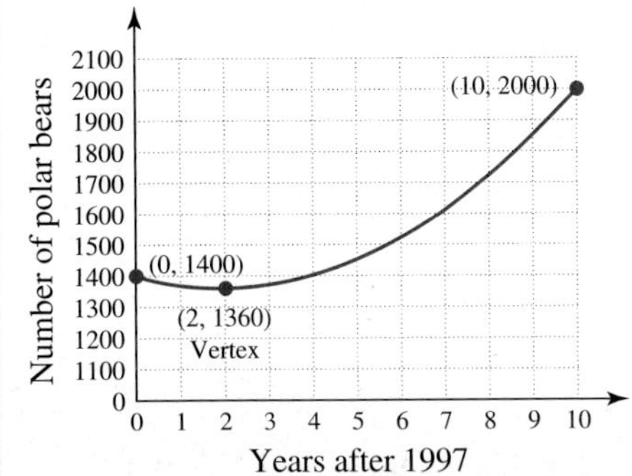

Let's return to the parabola $y = x^2 - x - 2$. This parabola has two x-intercepts, as shown in Window 5. Using your TRACE key and a decimal window, you can find both of them! Window 5 shows that one of the intercepts occurs at $x = -1$. The other x-intercept (not shown) occurs when $x = 2$. Now solve the equation $0 = x^2 - x - 2$. What are the solutions of this equation? Do you see a relationship between the graph of $y = ax^2 + bx + c$ and the solutions of $0 = ax^2 + bx + c$? Can you devise a procedure so that you can solve quadratic equations using a calculator?

If you're lucky, your calculator also has a "root" or "zero" feature. With this feature, you can graph $y = x^2 - x - 2$ and ask the calculator to find the root. The prompts ask "Left bound?" Use the ◀ key to move the cursor to a point left of the intercept (where this graph is above the x-axis). Find a point below the x-axis for the "Right bound" and press ENTER when asked for a "Guess." The root -1 is shown in Window 6.

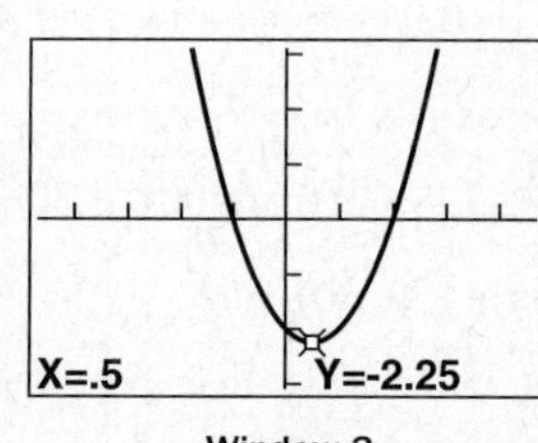

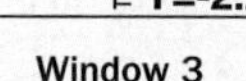
Window 3

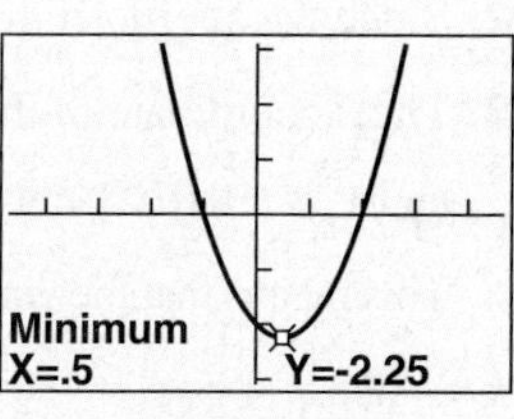

Window 4

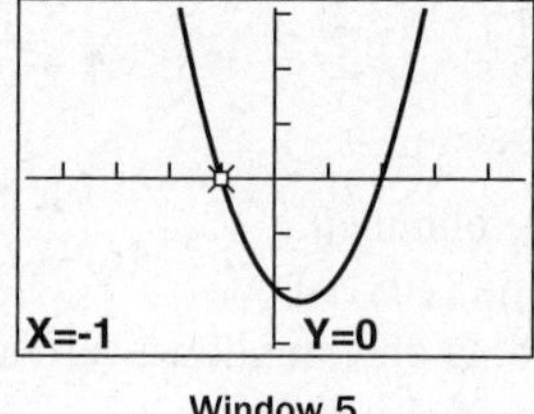

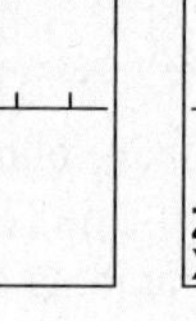
Window 5

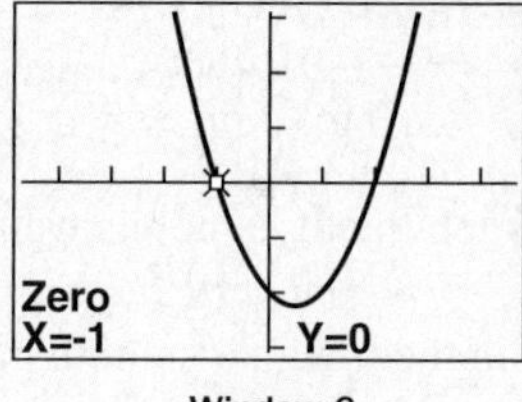

Window 6

› Exercises 9.4

› Web IT go to mhhe.com/bello for more lessons

‹ A › Graphing Quadratic Equations In Problems 1–16, graph the given equation.

1. $y = 2x^2$

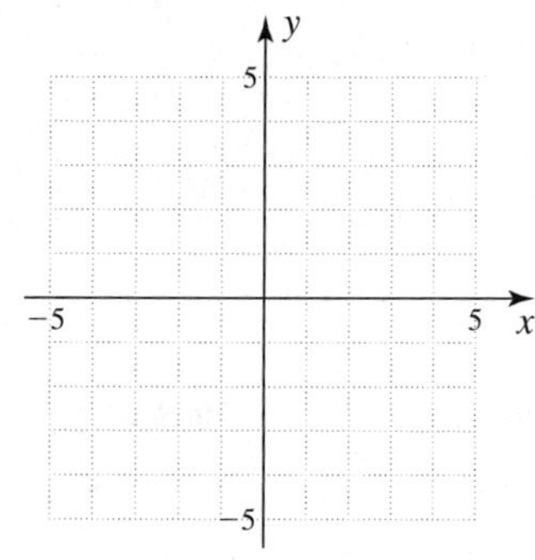

2. $y = 2x^2 + 1$

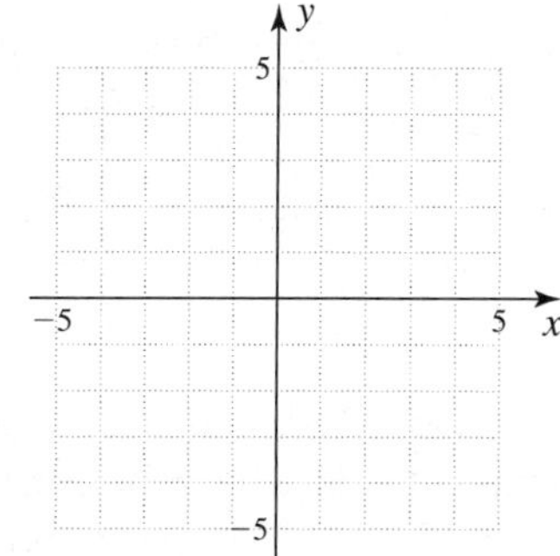

3. $y = 2x^2 - 1$

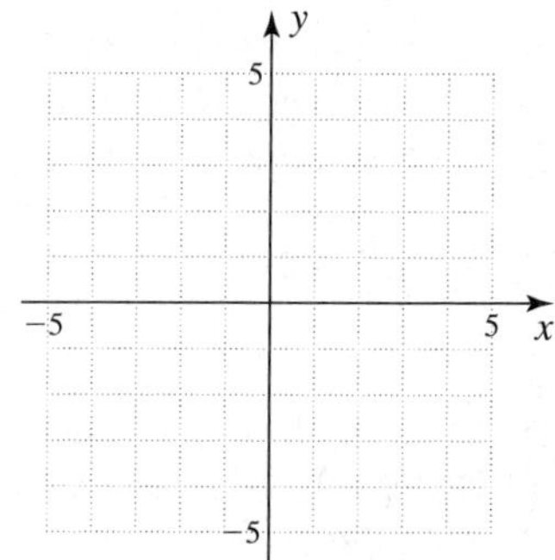

4. $y = 2x^2 - 2$

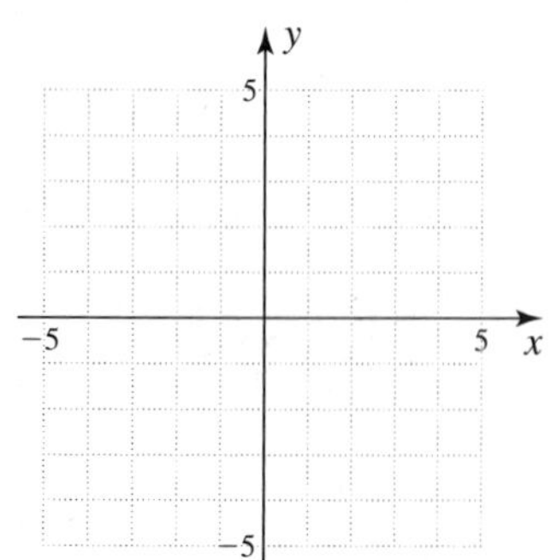

5. $y = -2x^2 + 2$

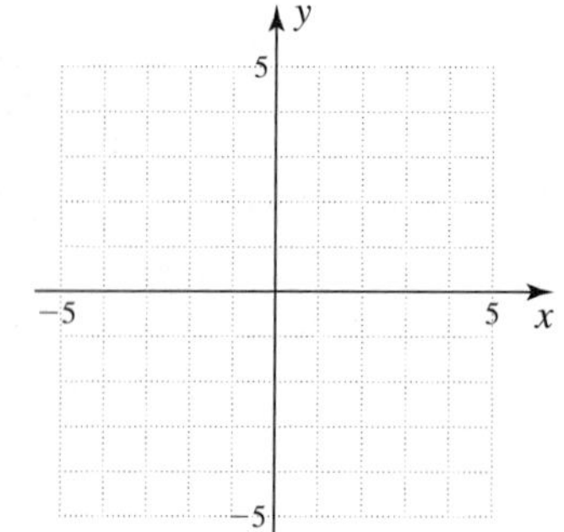

6. $y = -2x^2 - 1$

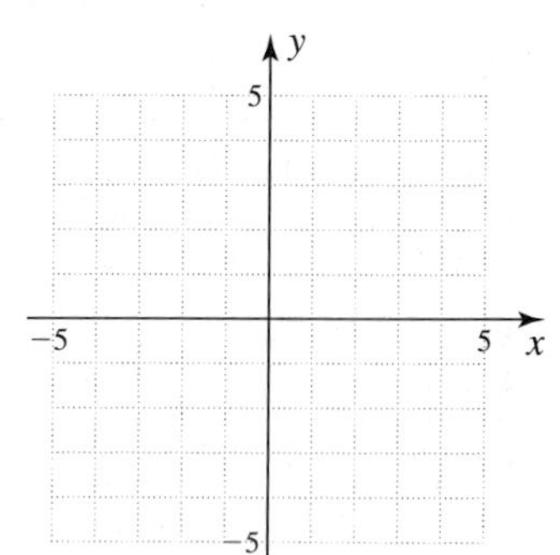

7. $y = -2x^2 - 2$

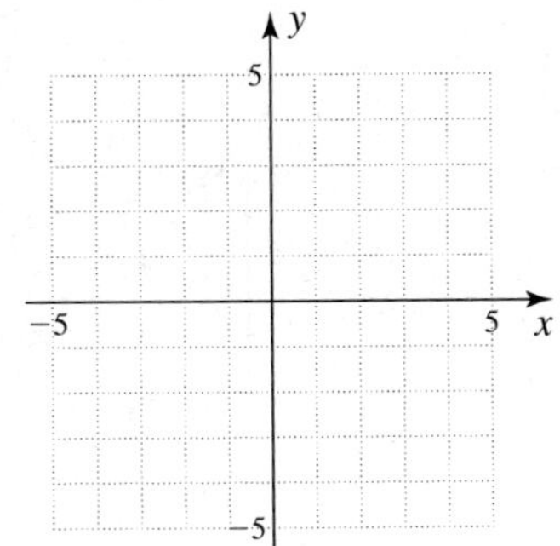

8. $y = -2x^2 + 1$

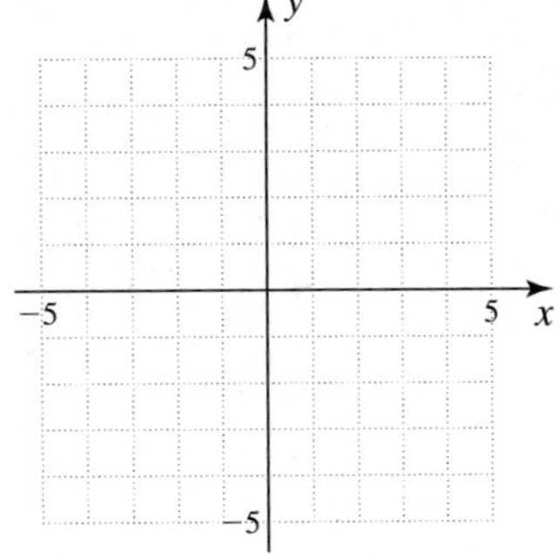

9. $y = (x - 2)^2$

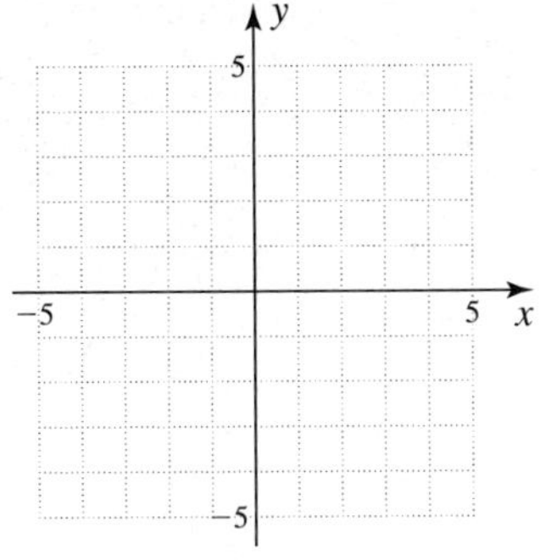

10. $y = (x - 2)^2 + 2$

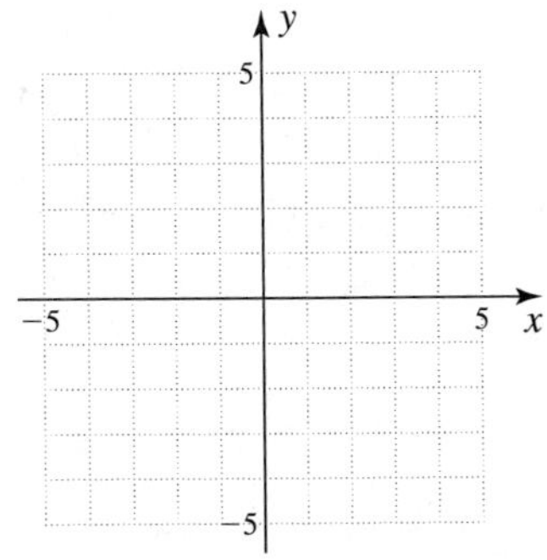

11. $y = (x - 2)^2 - 2$

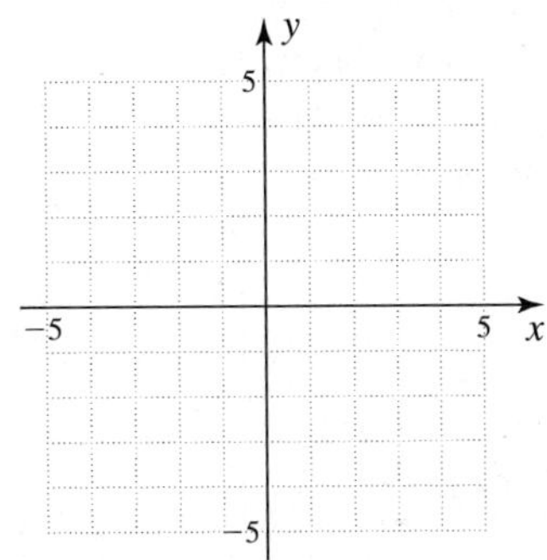

12. $y = (x - 2)^2 - 1$

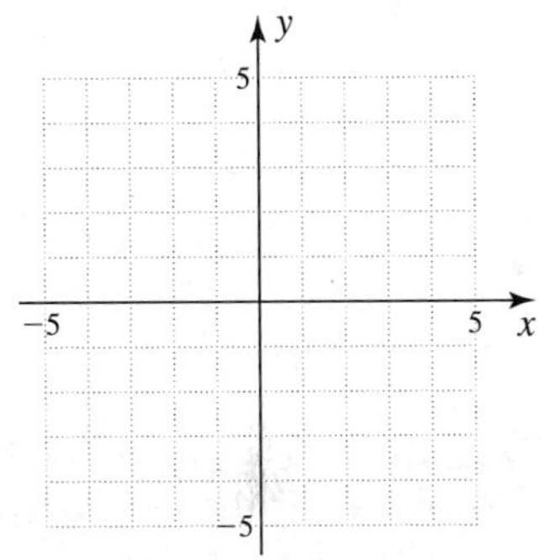

13. $y = -(x - 2)^2$

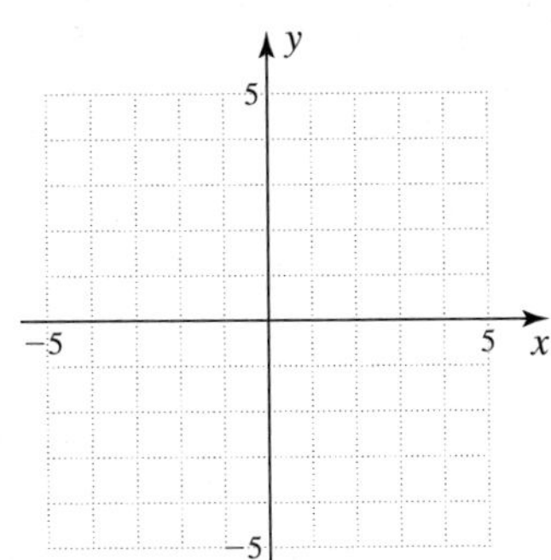

14. $y = -(x - 2)^2 + 2$

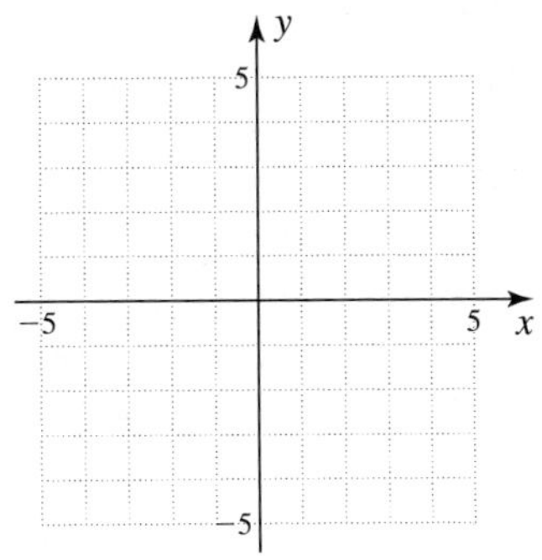

15. $y = -(x - 2)^2 - 2$

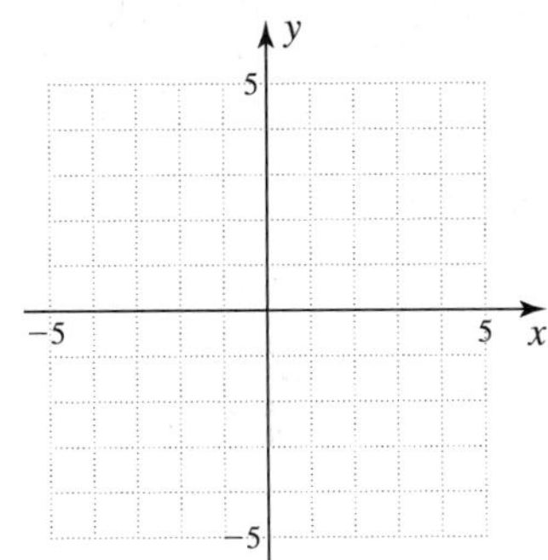

16. $y = -(x - 2)^2 - 3$

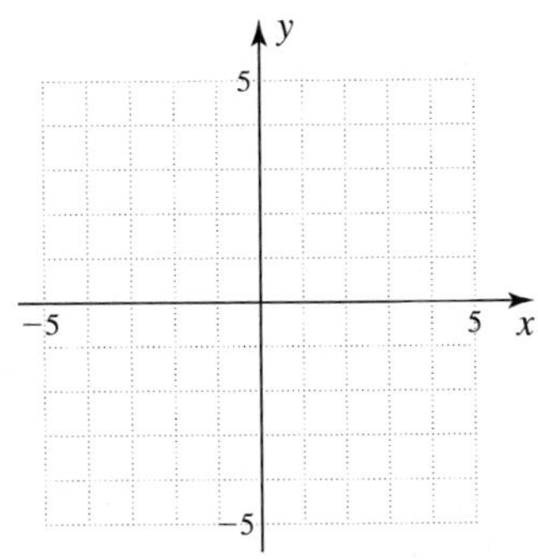

Web IT go to **mhhe.com/bello** for more lessons

Web IT go to mhhe.com/bello for more lessons

⟨B⟩ Graphing Parabolas Using Intercepts and the Vertex In Problems 17–22, find the x-intercepts, the y-intercept, the vertex, and then draw the graph.

17. $y = x^2 + 4x + 3$

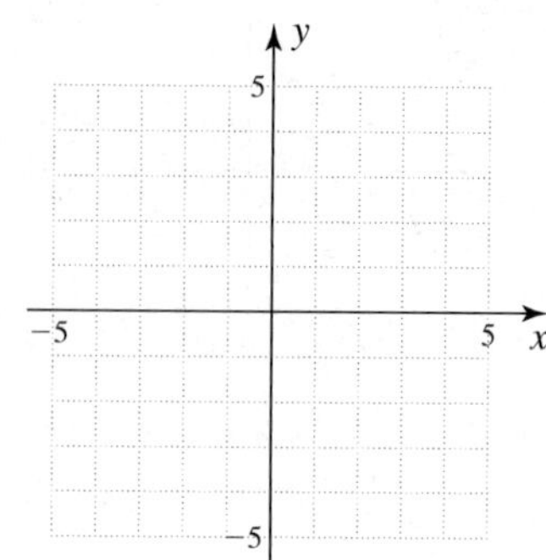

18. $y = x^2 - 4x + 4$

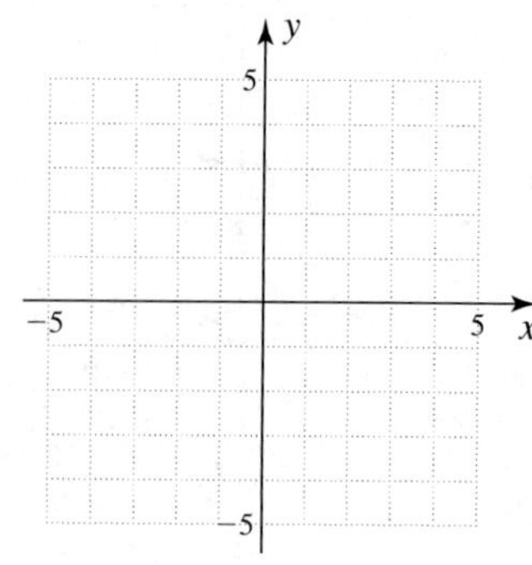

19. $y = x^2 + 2x - 3$

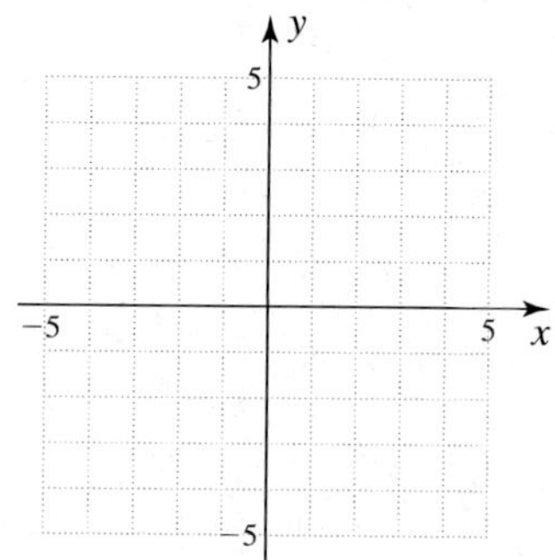

20. $y = -x^2 - 4x - 3$

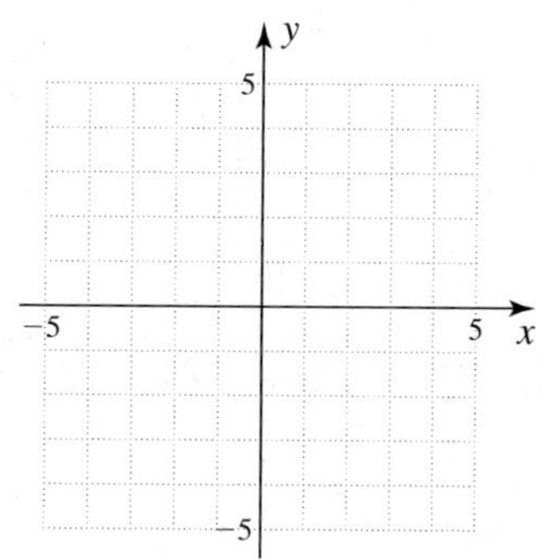

21. $y = -x^2 + 4x - 3$

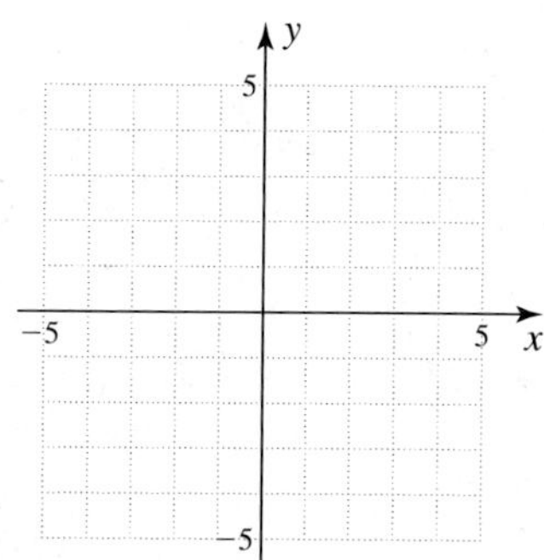

22. $y = -x^2 - 2x + 3$

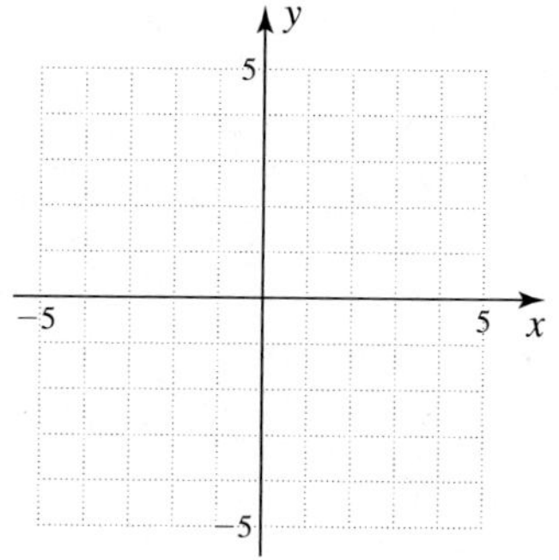

Applications

Travel from Earth to Moon The following graph looks somewhat like one-half of a parabola. It indicates the time it takes to travel from the earth to the moon at different speeds. Use it to solve Problems 23–26.

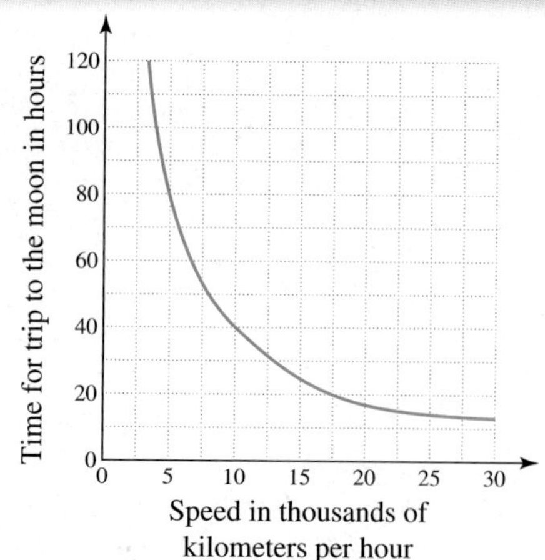

23. How long does the trip take if you travel at 10,000 kilometers per hour?

24. How long does the trip take if you travel at 25,000 kilometers per hour?

25. *Apollo 11* averaged about 6400 kilometers per hour when returning from the moon. How long was the trip?

26. What should the speed of the shuttle be if you wish to make the trip in 50 hours?

27. *Maximizing profit* The profit P (in dollars) for a company is modeled by the equation $P = -5000 + 8x - 0.001x^2$, where x is the number of items produced each month. How many items does the company have to produce to obtain maximum profit? What is this profit? (*Hint:* The maximum point in the graph occurs at the vertex.)

28. *Maximizing revenue* The revenue R for Shady Glasses is given by the equation $R = 1500p - 75p^2$, where p is the price of each pair of sunglasses (R and p in dollars). What should the price be to maximize revenue? (*Hint:* The maximum point in the graph occurs at the vertex.)

Applications: Green Math

29. *Polar bears in Davis Strait* An alternate approximation for the number of polar bears in Davis Strait is $N = 11x^2 - 44x + 1400$, where x is the number of years after 1997.

a. What is the shape of the graph of $11x^2 - 44x + 1400$?

b. Does it open upward or downward?

c. What was the polar bear population in 1997?

d. What was the polar bear population in 2004?

e. What was the polar bear population in 2007?

f. What are the coordinates of the vertex of the graph?

g. Use the information in parts a-f to graph $N = 11x^2 - 44x + 1400$ where x is the number of years after 1997.

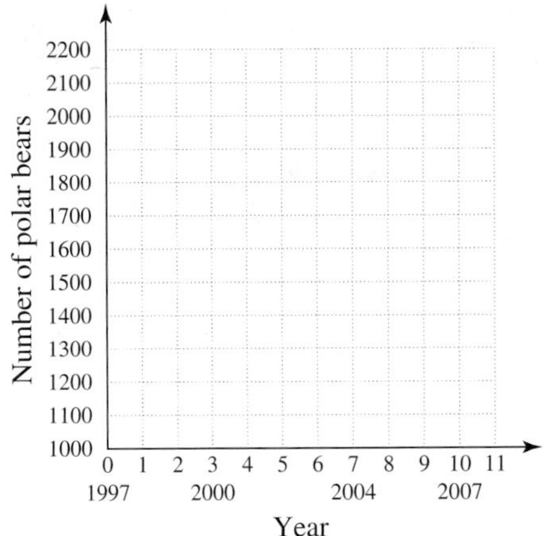

30. *Polar bears in western Hudson Bay* The table gives the estimated polar bear population in Western Hudson Bay, which can be approximated by $N = 4x^2 + 140x + 5000$, where x is the number of years after 1950.

Polar Bear Population Estimates

1950s	5000
1965–1970	8000–10,000
1984	25,000
2005	20,000–25,000

Sources: New York Times; Covebear.com; International Bear Association; International Wildlife; IUCN, Polar Bear Study Group.

Source: http://tinyurl.com/p5qsjg.

Graph $N = 4x^2 + 140x + 5000$.

a. What is the shape of the graph?

b. Does it open upward or downward?

c. What was the polar bear population in 1950?

d. What was the polar bear population in 2000?

e. The table predicts polar bear populations of 5000 in the 1950s, 8000–10,000 in 1965–1970, and 20,000–25,000 in 2005. How close is the graph to these predictions?

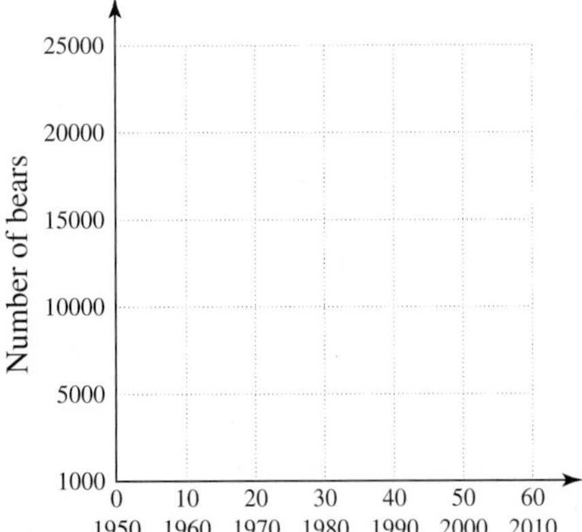

Using Your Knowledge

Maximizing Profits The ideas studied in this section are used in business to find ways to maximize profits. For example, if a manufacturer can produce a certain item for \$10 each, and then sell the item for x dollars, the profit per item will be $x - 10$ dollars. If it is then estimated that consumers will buy $60 - x$ items per month, the total profit will be

$$\text{Total profit} = \begin{pmatrix}\text{Number of}\\ \text{items sold}\end{pmatrix}\begin{pmatrix}\text{Profit per}\\ \text{item}\end{pmatrix}$$
$$= (60 - x)(x - 10)$$

The graph for the total profit is this parabola:

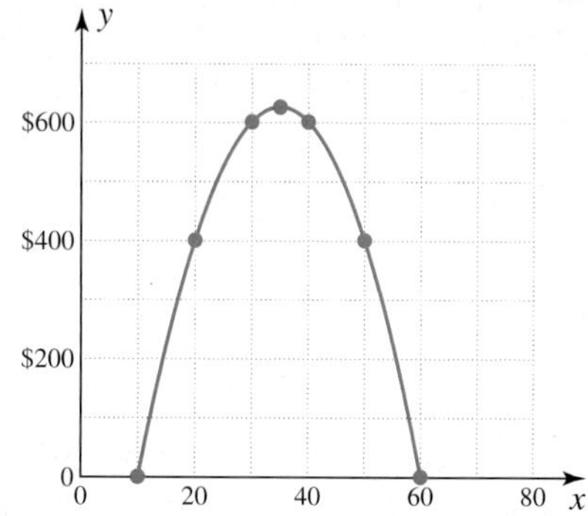

When will the profits be at a maximum? When the manufacturer produces 35 items. Note that 35 is exactly halfway between

$$10 \quad \text{and} \quad 60 \qquad \left(\frac{10 + 60}{2} = 35\right)$$

At this price, the total profits will be

$$P_T = (60 - 35)(35 - 10)$$
$$= (25)(25)$$
$$= \$625$$

Use your knowledge to answer the following questions.

31. What price would maximize the profits of a certain item costing \$20 each if $60 - x$ items are sold each month (where x is the selling price of each item)?

32. Sketch the graph of the resulting parabola.

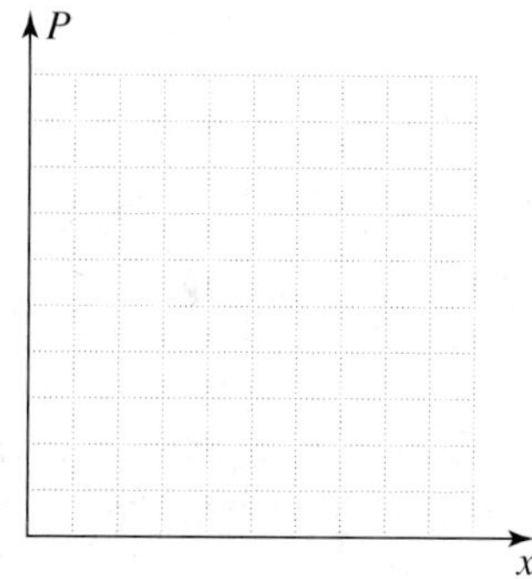

33. What will the maximum profits be?

Write On

34. What is the vertex of a parabola?

For Problems 35–40, consider the graph of the parabola $y = ax^2$.

35. What can you say if $a > 0$?

36. What can you say if $a < 0$?

37. If a is a positive integer, what happens to the graph of $y = ax^2$ as a increases?

38. What causes the graph of $y = ax^2$ to be wider or narrower than that of $y = x^2$?

39. What happens to the graph of $y = ax^2$ if you add the positive constant k? What happens if k is negative?

40. If a parabola has two x-intercepts and the vertex is at (1, 1), does the parabola open upward or downward? Explain.

Concept Checker

Fill in the blank(s) with the correct word(s), phrase, or mathematical statement.

41. The graph of $y = ax^2 + bx + c$ is a **parabola** and opens **upward** (∪) when ______________.

42. The graph of $y = ax^2 + bx + c$ is a **parabola** and opens **downward** (∩) when ______________.

$a < 0$
$a = 0$
$a > 0$

Mastery Test

Graph:

43. $y = -(x - 2)^2 - 1$

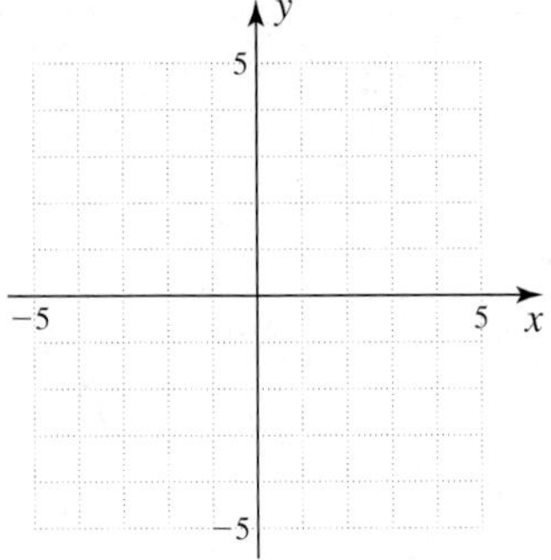

44. $y = -x^2 - 1$

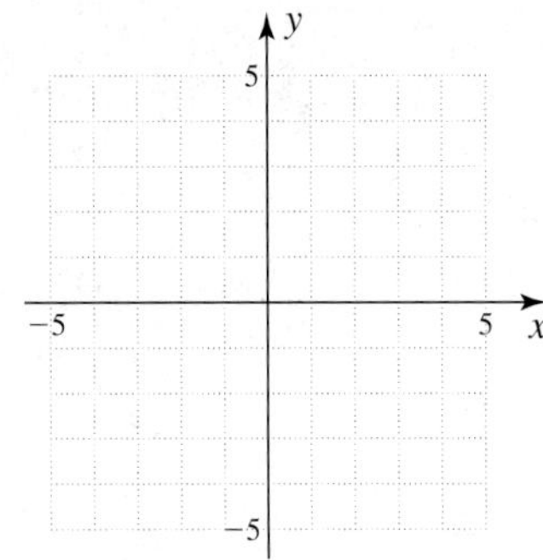

45. $y = -2x^2$

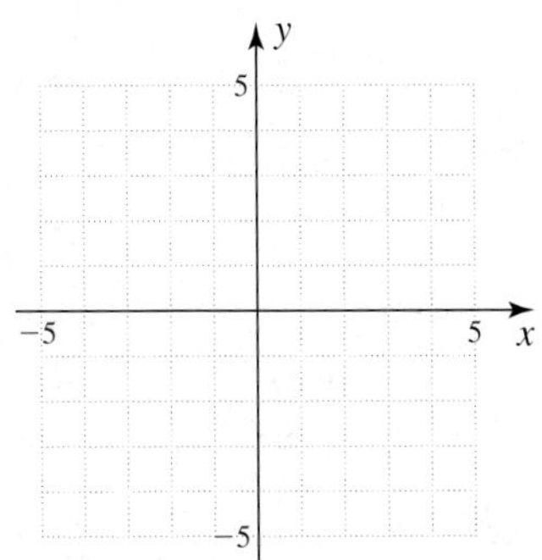

46. $y = -2x^2 + 3$

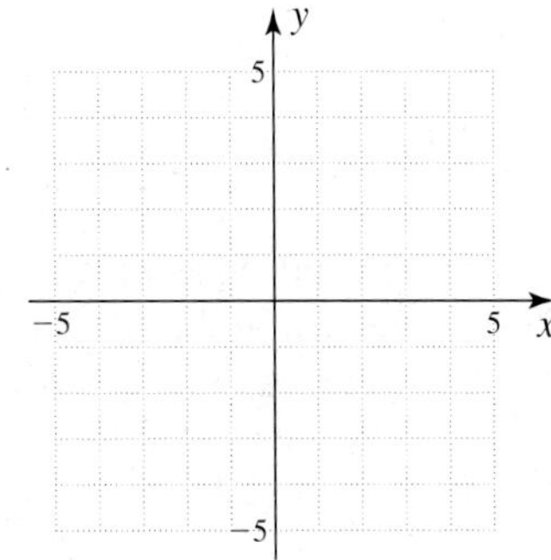

Graph and label the vertex and intercepts:

47. $y = -x^2 - 2x + 3$

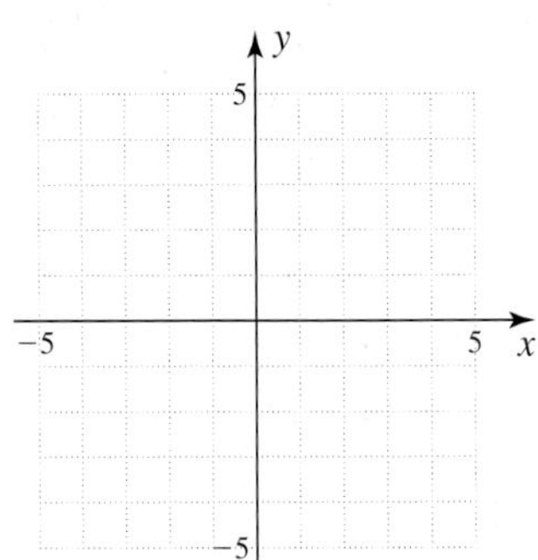

48. $y = x^2 + 3x + 2$

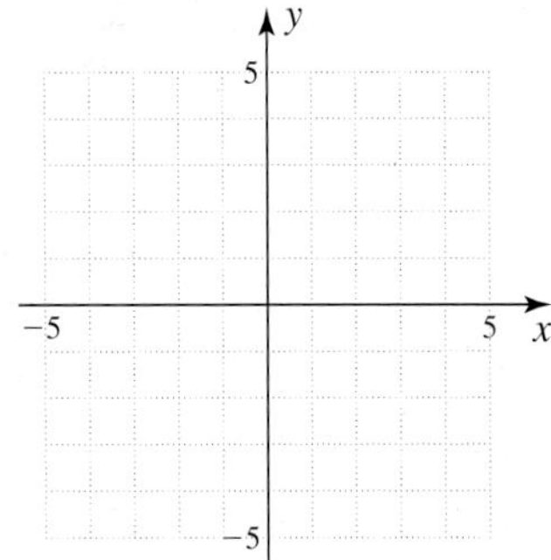

Skill Checker

Find:

49. $3^2 + 4^2$

50. $5^2 + 12^2$

51. $\frac{120}{0.003}$

52. $\frac{150}{0.005}$

9.5 The Pythagorean Theorem and Other Applications

Objectives

A › Use the Pythagorean theorem to solve right triangles.

B › Solve word problems involving quadratic equations.

To Succeed, Review How To . . .

1. Find the square of a number (p. 62).
2. Divide by a decimal (pp. 23–24).
3. Solve a quadratic by factoring (pp. 456–460).

Getting Started

A Theorem Not Dumb Enough!

By Permission of John L. Hart FLP, and Creators Syndicate, Inc.

In this section we discuss several applications involving quadratic equations. One of them pertains to the theorem mentioned in the cartoon. This theorem, first proved by Pythagoras, will be stated next.

A › Using the Pythagorean Theorem

Before using the Pythagorean theorem, you have to know what a right triangle is! A right triangle is a triangle containing one right angle (90°), and the hypotenuse is the side opposite the 90° angle. If that is the case, here is what the theorem says:

PYTHAGOREAN THEOREM

The square of the hypotenuse of a right triangle equals the sum of the squares of the other two sides; that is,

$$c^2 = a^2 + b^2$$

where c is the hypotenuse of the triangle, and a and b are the two remaining sides (called the legs).

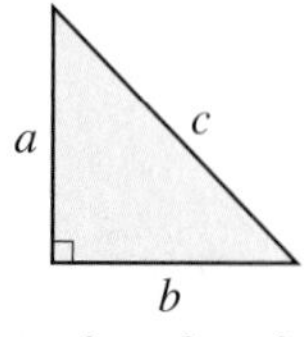

$a^2 + b^2 = c^2$

>Figure 9.10

Figure 9.10 shows a right triangle with sides a, b, and c. So if $a = 3$ units and $b = 4$ units, then the hypotenuse c is such that

$$c^2 = 3^2 + 4^2$$
$$c^2 = 9 + 16$$
$$c^2 = 25$$
$$c = \pm\sqrt{25} = \pm 5$$

Since c represents length, c cannot be negative, so the length of the hypotenuse is 5 units.

EXAMPLE 1 Using the Pythagorean theorem

The distance from the bottom of the wall to the base of the ladder in Figure 9.11 is 5 feet, and the distance from the floor to the top of the ladder is 12 feet. Can you find the length of the ladder?

>Figure 9.11

SOLUTION 1 We use the RSTUV method given earlier.

1. Read the problem. We are asked to find the length of the ladder.

2. Select the unknown. Let L represent the length of the ladder.

3. Think of a plan. First we draw a diagram (see Figure 9.12) that gives the appropriate dimensions. Then we translate the problem using the Pythagorean theorem:

$$\underbrace{\begin{pmatrix}\text{Distance from}\\ \text{wall to base of}\\ \text{ladder}\end{pmatrix}^2}_{5^2} + \underbrace{\begin{pmatrix}\text{Distance from}\\ \text{floor to top of}\\ \text{ladder}\end{pmatrix}^2}_{12^2} = \underbrace{\begin{pmatrix}\text{Length of}\\ \text{ladder}\end{pmatrix}^2}_{L^2}$$

4. Use algebra to solve the problem. We use algebra to solve this equation:

$$\begin{aligned} 5^2 + 12^2 &= L^2 \\ 25 + 144 &= L^2 \\ L^2 &= 169 \\ L &= \pm 13 \end{aligned}$$

Since L is the length of the ladder, L is positive, so $L = 13$.

5. Verify the solution. This solution is correct because $5^2 + 12^2 = 13^2$ since $25 + 144 = 169$.

PROBLEM 1

Find the length of the ladder if the distance from the wall to the base of the ladder is 5 feet but the height of the top of the ladder is 14 feet.

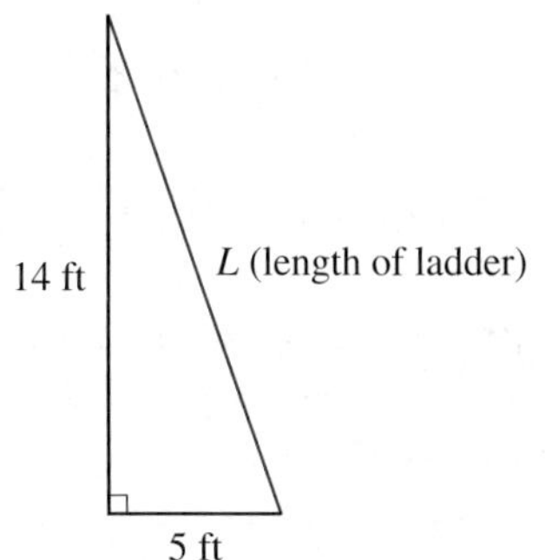

>Figure 9.12

B > Solving Word Problems Involving Quadratic Equations

Many applications of quadratic equations come from the field of engineering. For example, it is known that the pressure p (in pounds per square foot) exerted by a wind blowing at v miles per hour is given by the equation

$$p = 0.003v^2$$ Since v is raised to the second power, the equation is a quadratic.

If the pressure p is known, the equation $p = 0.003v^2$ is an example of a quadratic equation in v. We show how this information is used in the next example.

Answers to PROBLEMS

1. $\sqrt{221}$ ft

EXAMPLE 2 Determining wind speed

A wind pressure gauge at Commonwealth Bay registered 120 pounds per square foot during a gale. If the pressure $p = 0.003\,v^2$, where v is the wind speed, what was the wind speed at that time?

SOLUTION 2 We will again use the RSTUV method.

1. Read the problem. We are asked to find the wind speed.

2. Select the unknown. Since we know that $p = 120$ and that $p = 0.003v^2$, we let v be the wind speed.

3. Think of a plan. We substitute for p to obtain

$$0.003v^2 = 120$$

4. Use algebra to solve the problem. The easiest way to solve this equation is to divide by 0.003 first. We then have

$$v^2 = \frac{120}{0.003}$$

$$v^2 = \frac{120 \cdot 1000}{0.003 \cdot 1000} = \frac{120{,}000}{3} = 40{,}000$$

You can also divide 120 by 0.003 to obtain

$$0.003\overline{)120{,}000} = 40{,}000 \quad (12)$$

$$v = \pm\sqrt{40{,}000} = \pm 200$$

Thus, the wind speed v was 200 miles per hour. (We discard the negative answer as not suitable.)

5. Verify the solution. We leave the verification to you.

PROBLEM 2

Find the wind speed at the Golden Gate Bridge when a pressure gauge is registering 4.8 lb/ft².

It can be shown in physics that when an object is dropped or thrown downward with an initial velocity v_0, the distance d (in meters) traveled by the object in t seconds is given by the formula

$$d = 5t^2 + v_0t$$

This is an approximate formula. A more exact formula is $d = 4.9t^2 + v_0t$.

Suppose an object is dropped from a height of 125 meters. How long would it be before the object hits the ground? The solution to this problem is given in Example 3.

EXAMPLE 3 Falling the distance

Use the formula $d = 5t^2 + v_0t$ to find the time it takes an object dropped from a height of 125 meters to hit the ground.

PROBLEM 3

Use the formula $d = 5t^2 + v_0t$ to find the time it takes an object dropped from a height of 180 meters to hit the ground.

Answers to PROBLEMS

2. 40 mi/hr

3. 6 sec

SOLUTION 3 Since the object is dropped, the initial velocity is $v_0 = 0$, and we are given $d = 125$; we then have

$$125 = 5t^2$$
$$5t^2 = 125$$
$$t^2 = 25 \quad \text{Divide by 5.}$$
$$t = \pm\sqrt{25} = \pm 5$$

Since the time must be positive, the correct answer is $t = 5$; that is, it takes 5 seconds for the object to hit the ground.

Finally, quadratic equations are also used in business. Perhaps you already know what it means to "break even." In business, the break-even point is the point at which the revenue R equals the cost C of the manufactured goods. In symbols,

$$R = C$$

Now suppose a company produces x (thousand) items. If each item sells for \$3, the revenue R (in thousands of dollars) can be expressed by the equation

$$R = 3x$$

If the manufacturing cost C (in thousands of dollars) is given by the equation

$$C = x^2 - 3x + 5$$

and it is known that the company always produces more than 1000 items, the break-even point occurs when

$$R = C$$
$$3x = x^2 - 3x + 5$$
$$x^2 - 6x + 5 = 0$$
$$(x - 5)(x - 1) = 0 \quad \text{Factor.}$$
$$x - 5 = 0 \quad \text{or} \quad x - 1 = 0 \quad \text{Use the principle of zero product.}$$
$$x = 5 \qquad\qquad x = 1$$

But since it's known that the company produces more than 1 (thousand) item(s), the break-even point occurs when the company manufactures 5 (thousand) items.

EXAMPLE 4 **Finding the break-even point**

Find the number x of items (in thousands) that have to be produced to break even when the revenue R (in thousands of dollars) of a company is given by the equation $R = 3x$ and the cost (also in thousands of dollars) is given by the equation $C = 2x^2 - 3x + 4$.

SOLUTION 4 In order to break even,

$$R = C$$
$$3x = 2x^2 - 3x + 4$$
$$2x^2 - 6x + 4 = 0$$
$$x^2 - 3x + 2 = 0 \quad \text{Divide by 2.}$$
$$(x - 1)(x - 2) = 0 \quad \text{Factor.}$$
$$x - 1 = 0 \quad \text{or} \quad x - 2 = 0 \quad \text{Use the property zero product.}$$
$$x = 1 \qquad\qquad x = 2$$

This means that the company breaks even when it produces either 1 (thousand) or 2 (thousand) items.

PROBLEM 4

Find the number x of items (in thousands) that have to be produced to break even when the revenue R of a company (in thousands of dollars) is given by the equation $R = 4x$ and the cost C (also in thousands of dollars) is given by the equation $C = x^2 - 3x + 6$.

Answers to PROBLEMS

4. 1 (thousand) or 6 (thousand) items

Do you recycle paper, bottles, and metals? How many million tons of materials have been recovered for recycling? The amount can be approximated by $R = 0.02x^2 + 2x + 70$, where x is the number of years after 2000. *Paper and paperboard* are the most commonly recycled items, but the amount of paper and paperboard generated is on the decline, as you can see by the approximation $P = -0.06x^2 - 0.3x + 90$. Will the amount P of paper and paperboard generated ever equal the total amount R of materials recovered for recycling? To find out we could try to solve the equation $-0.06x^2 - 0.3x + 90 = 0.02x^2 + 2x + 70$, but the resulting coefficients would make using the quadratic formula tedious. Instead, we solve the equation graphically in Example 5.

Source: http://www.epa.gov/epawaste/nonhaz/municipal/pubs/msw07-rpt.pdf.

GREEN MATH

EXAMPLE 5 Paper and paperboard recovered for recycling

a. Graph $P = -0.06x^2 - 0.3x + 90$ and $R = 0.02x^2 + 2x + 70$ on the same coordinate axis and find when P and R will be equal.

b. How can we interpret this result?

SOLUTION 5

a. We let x correspond to the number of years after 2000, starting with $x = 0$ and y be from 70 to 95 million tons. When $x = \mathbf{0}$, $P = -0.06(\mathbf{0})^2 - 0.3(\mathbf{0}) + 90 = 90$ and $R = 0.02(\mathbf{0})^2 + 2(\mathbf{0}) + 70 = 70$, so we graph the points $(0, 90)$ and $(0, 70)$. As you can see $P = R$ when the two parabolas intersect and this occurs when $x = 7$ and $y \approx 85$, that is, at $(7, 85)$. You can verify that this approximation is correct by evaluating $R = 0.02(7)^2 + 2(7) + 70 = 84.98 \approx 85$ and $P = -0.06(7)^2 - 0.3(7) + 90 \approx 85$.

b. The result means that 7 years after 2000 (in 2007) the material recovered for recycling and the amount of paper and paperboard generated was about the same, 85 million tons.

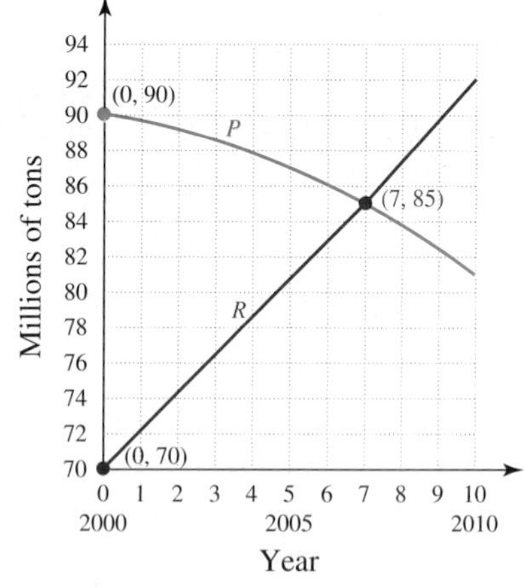

PROBLEM 5

The amount of millions of tons of ferrous metals F (iron and steel in appliances, furniture and containers) can be approximated by $F = 0.01x^2 + 0.05x + 9$ and the amount G of millions of tons of glass containers (beer, soft drinks, wine, jars) can be approximated by $G = -0.003x^2 + 0.1x + 10$, where x is the number of years after 2000 and F and G are in millions of tons.

a. Graph F and G on the same coordinate axes and find when F and G will be equal.

b. How can we interpret the result?

Answers to PROBLEMS

5. a. Let x be from 0 to 15 and y be from 0 to 12. The parabolas intersect at $x = 11$ and $y \approx 10.7$.

b. This means that after 11 years (2011) the amount of ferrous metals and the glass generated will be about the same, 10.7 million tons.

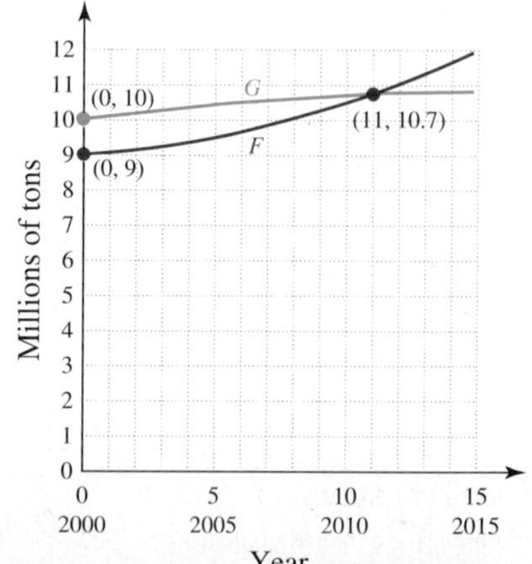

› Exercises 9.5

> Practice Problems > Self-Tests > Media-rich eBooks > e-Professors > Videos

‹ A › Using the Pythagorean Theorem In Problems 1–10, let a and b represent the lengths of the sides of a right triangle and c the length of the hypotenuse. Find the missing side.

1. $a = 12, c = 13$

2. $a = 4, c = 5$

3. $a = 5, c = 15$

4. $a = b, c = 4$

5. $b = 3, a = \sqrt{6}$

6. $b = 7, a = 9$

7. $a = \sqrt{5}, b = 2$

8. $a = 3, b = \sqrt{7}$

9. $c = \sqrt{13}, a = 3$

10. $c = \sqrt{52}, a = 4$

‹ B › Solving Word Problems Involving Quadratic Equations In Problems 11–24, solve the resulting quadratic equation.

11. How long is a wire extending from the top of a 40-foot telephone pole to a point on the ground 30 feet from the base of the pole?

12. Repeat Problem 11 where the point on the ground is 16 feet away from the pole.

13. The pressure p (in pounds per square foot) exerted on a surface by a wind blowing at v miles per hour is given by the equation

$$p = 0.003v^2$$

Find the wind speed when a wind pressure gauge recorded a pressure of 30 pounds per square foot.

14. Repeat Problem 13 where the pressure is 2.7 pounds per square foot.

15. An object is dropped from a height of 320 meters. How many seconds does it take for this object to hit the ground? (See Example 3.)

16. Repeat Problem 15 where the object is dropped from a height of 45 meters.

17. An object is thrown downward with an initial velocity of 3 meters per second. How long does it take for the object to travel 8 meters? (See Example 3.)

18. The revenue of a company is given by the equation $R = 2x$, where x is the number of units produced (in thousands). If the cost is given by the equation $C = 4x^2 - 2x + 1$, how many units have to be produced before the company breaks even?

19. Repeat Problem 18 where the revenue is given by the equation $R = 5x$ and the cost is given by the equation $C = x^2 - x + 9$.

20. If P dollars are invested at r percent compounded annually, then at the end of 2 years the amount will have grown to $A = P(1 + r)^2$. At what rate of interest r will \$1000 grow to \$1210 in 2 years? (*Hint:* $A = 1210$ and $P = 1000$.)

21. A rectangle is 2 feet wide and 3 feet long. Each side is increased by the same amount to give a rectangle with twice the area of the original one. Find the dimensions of the new rectangle. (*Hint:* Let x feet be the amount by which each side is increased.)

22. The hypotenuse of a right triangle is 4 centimeters longer than the shortest side and 2 centimeters longer than the remaining side. Find the dimensions of the triangle.

23. Repeat Problem 22 where the hypotenuse is 16 centimeters longer than the shortest side and 2 centimeters longer than the remaining side.

24. The square of a certain positive number is 5 more than 4 times the number itself. Find this number.

› Web IT go to **mhhe.com/bello** *for more lessons*

››› *Applications*

Economists and managers use quadratic equations to determine **market equilibrium,** the point at which the demand D for a quantity q of a product equals the supply S.

25. *Supply and demand* The supply S of custom vans produced is given by the equation $S = q^2 + 6q + 20$ and the demand is given by the equation $D = -4q + 220$.

a. Find q the number of vans produced at which the supply S equals the demand D **(market equilibrium).**

b. How many vans can be supplied at this point?

26. *Supply and demand* Windy manufacturing produces q sailboats a month. The supply S for the boats is given by the equation $S = -3q^2 - 6q + 86$ and the demand is given by the equation $D = \frac{1}{4}q^2 + 10$.

a. How many boats are produced at market equilibrium?

b. What is the demand for the boats at market equilibrium?

>>> Applications: Green Math

27. The *Annual Energy Outlook* prepared by the Energy Information Administration (EIA) presents long-term projections of carbon dioxide (CO_2) emissions **in millions of metric tons** through 2030. (One metric ton is 2204 pounds.) Which do you think produces more CO_2: cooking $\boldsymbol{C = 0.002x^2 + 0.3x + 30}$, or the use of personal computers and related equipment $\boldsymbol{P = -0.008x^2 + 0.6x + 30}$, where x is the number of years after 2006?

a. What is the shape of the graph of C?

b. What is the shape of the graph of P?

c. How many millions of metric tons of CO_2 were produced each by C and by P in 2006 ($x = 0$)?

d. How many millions of metric tons of CO_2 will be produced each by C and by P in 2021 ($x = 15$)?

e. How many millions of metric tons of CO_2 will be produced each by C and by P in 2036 ($x = 30$)?

f. In what year will the CO_2 emissions for P and C be the same?

g. Use the answers for parts **a** to **e** to graph C and P.

h. Which of the two activities (C or P) produced the most CO_2 emissions from 2006 to 2036?

Source: http://www.eia.doe.gov/oiaf/aeo/pdf/0383(2009).pdf.

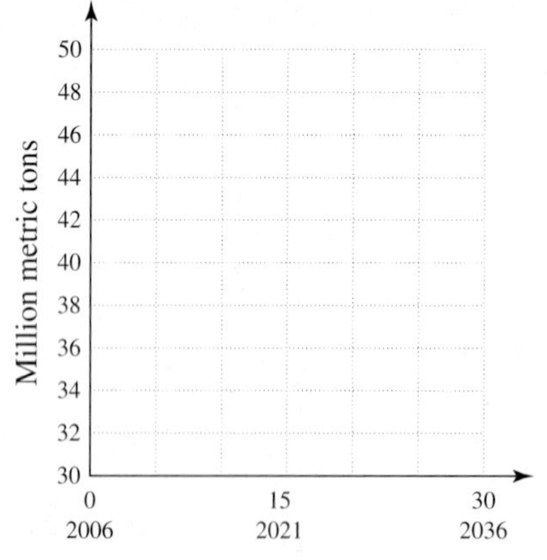

28. *PC or not PC?* Which type of computer do you use, a PC or another type? Which one do you think produces the most CO_2 emissions over a 10-year period? Here are the approximations for the emissions for PCs and other types $\boldsymbol{O}$ of computers in millions of tons of CO_2 for x years after 2006. $\mathbf{PC} = \boldsymbol{0.01x^2 + 0.2x + 50}$, $\boldsymbol{O = -0.05x^2 + 3x + 40}$.

a. What is the shape of the graph of **PC**?

b. What is the shape of the graph of O?

c. How many millions of tons of CO_2 were produced each by **PC** and by O in 2006 ($x = 0$)?

d. How many millions of tons of CO_2 will be produced each by **PC** and by O in 2016 ($x = 10$)?

e. Use the answers for parts **a** to **e** to graph **PC** and O.

f. In what year were the CO_2 emissions about the same?

g. Which of the two activities (**PC** or O) produced the most CO_2 emissions from 2006 to 2016?

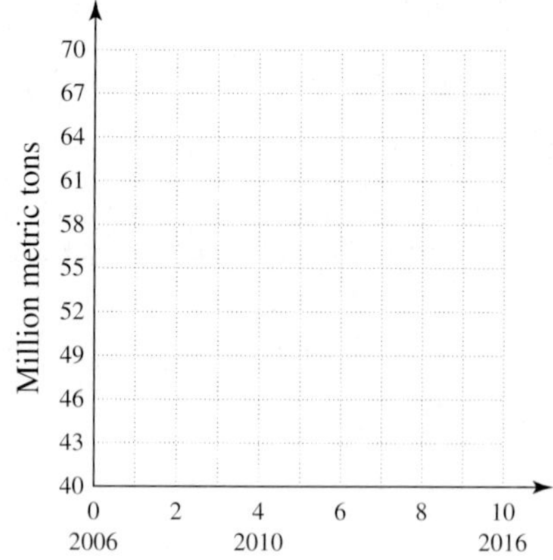

Using Your Knowledge

Solving for Variables Many formulas require the use of some of the techniques we've studied when solving for certain unknowns in these formulas. For example, consider the formula

$$v^2 = 2gh$$

How can we solve for v (the speed) in this equation? In the usual way, of course! Thus, we have

$$v^2 = 2gh$$
$$v = \pm\sqrt{2gh}$$

Since the speed v is positive, we simply write

$$v = \sqrt{2gh}$$

Use this idea to solve for the indicated variables in the given formulas.

29. Solve for c in the formula $E = mc^2$.

30. Solve for r in the formula $A = \pi r^2$.

31. Solve for r in the formula $F = \frac{GMm}{r^2}$.

32. Solve for v in the formula $\text{KE} = \frac{1}{2}mv^2$.

33. Solve for P in the formula $I = kP^2$.

Write On

34. When solving for the hypotenuse c in a right triangle whose sides were 3 and 4 units, respectively, a student solved the equation $3^2 + 4^2 = c^2$ and obtained the answer 5. Later, the same student was given the equation $3^2 + 4^2 = x^2$ and gave the same answer. The instructor said that the answer was not complete. Explain.

35. In Example 2, one of the answers was discarded. Explain why.

36. Referring to Example 4, explain why you "break even" when $R = C$.

Concept Checker

Fill in the blank(s) with the correct word(s), phrase, or mathematical statement.

37. A **right triangle** is a **triangle** containing one _______________ **angle.**

38. According to the **Pythagorean Theorem,** in a **right** triangle with legs of length a and b and hypotenuse c, _______________.

obtuse	$a^2 + c^2 = b^2$
right	$a^2 + b^2 = c^2$

Mastery Test

39. Find the number x of hundreds of items that have to be produced to break even when the revenue R of a company (in thousands of dollars) is given by the equation $R = 4x$ and the cost C (also in thousands of dollars) is given by the equation $C = x^2 - 3x + 6$.

40. Find the time t (in seconds) it takes an object dropped from a height of 245 meters to hit the ground by using the formula $d = 5t^2 + v_0t$ (meters).

41. The pressure p (in pounds per square foot) exerted by a wind blowing at v miles per hour is given by the equation

$$p = 0.003v^2$$

Find the wind speed at the Golden Gate Bridge when a pressure gauge is registering 7.5 pounds per square foot.

42. Find the length L of the ladder in the diagram if the distance from the wall to the base of the ladder is 5 feet and the height to the top of the ladder is 13 feet.

Skill Checker

43. Evaluate $2x + 3$ when $x = -2$.

44. Evaluate $x^3 - 2x^2 + 3x - 4$ when $x = -3$.

9.6 Functions

Objectives

A Find the domain and range of a relation.

B Determine whether a given relation is a function.

C Use function notation.

D Solve applications involving functions.

To Succeed, Review How To . . .

1. Evaluate an expression (pp. 62–63, 69–72).
2. Use set notation (p. 45).

Getting Started

Functions for Fashions

Did you know that women's clothing sizes are getting smaller? According to a recent J. C. Penney catalog, "Simply put, you will wear one size smaller than before." Is there a relationship between the new sizes and waist size? The table gives a $1\frac{1}{2}$-inch leeway for waist sizes. Let's use this table to consider the first number in the waist sizes corresponding to different dress sizes—that is, 30, 32, 34, 36. For Petite sizes 14–22, the waist size is 16 inches more than the dress size. If we wish to formalize this relationship, we can write

$$w(s) = s + 16 \quad \text{Read "}w\text{ of }s\text{ equals }s\text{ plus 16."}$$

What would be the waist size of a woman who wears size 14? It would be

$$w(14) = 14 + 16 = 30$$

For size 16,

$$w(16) = 16 + 16 = 32$$

and so on. It works! Can you do the same for hip sizes? If you get

$$h(s) = s + 26.5$$

you are on the right track. As you can see, there is a *relationship* between hip size and dress size.

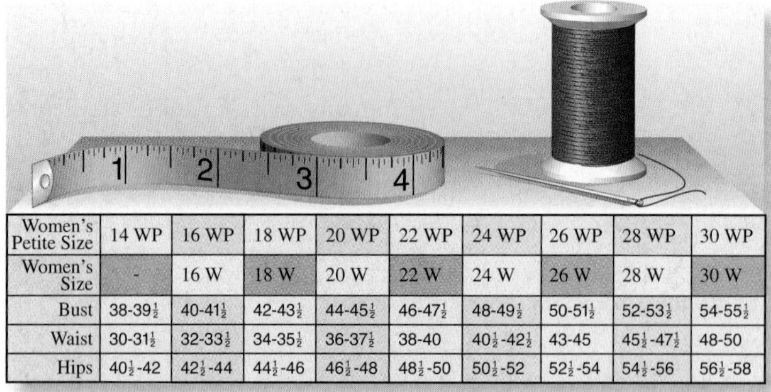

Women's Petite Size	14 WP	16 WP	18 WP	20 WP	22 WP	24 WP	26 WP	28 WP	30 WP
Women's Size	-	16 W	18 W	20 W	22 W	24 W	26 W	28 W	30 W
Bust	38-39½	40-41½	42-43½	44-45½	46-47½	48-49½	50-51½	52-53½	54-55½
Waist	30-31½	32-33½	34-35½	36-37½	38-40	40½-42½	43-45	45½-47½	48-50
Hips	40½-42	42½-44	44½-46	46½-48	48½-50	50½-52	52½-54	54½-56	56½-58

The word *relation* might remind you of members of your family—parents, brothers, sisters, and so on—but in mathematics, relations are expressed using ordered pairs. Thus, the relationship between Petite sizes 14–22 in the table and the waist size can be expressed by the set of ordered pairs:

$$w = \{(14, 30), (16, 32), (18, 34), (20, 36), (22, 38)\}$$

Note that we've specified that the waist size corresponding to the dress size is the first number in the row labeled "Waist." Thus, for every dress size, there is only *one* waist size. Relations that have this property are called *functions*. In this section, we shall learn about *relations* and *functions,* and how to evaluate them using notation such as $w(s) = s + 16$ and $h(s) = s + 26.5$.

A > Finding Domains and Ranges

In previous chapters, we studied ordered pairs that satisfy linear equations, graphed these ordered pairs, and then found the graph of the corresponding line. Thus, to find the graph of $y = 3x - 1$, we assigned the value 0 to x to obtain the corresponding y-value -1. Thus, a y-value of -1 is paired to an x-value of 0, which results in the ordered pair $(0, -1)$ as assigned by the rule $y = 3x - 1$.

RELATION, DOMAIN, AND RANGE

In mathematics, a relation is a set of ordered pairs. The domain of the relation is the set of all possible *x*-values, and the range of the relation is the set of all possible *y*-values.

EXAMPLE 1 Finding the domain and range of a relation given as a set of ordered pairs

Find the domain and range of the relation:

$$S = \{(4, -3), (2, -5), (-3, 4)\}$$

SOLUTION 1 The domain D is the set of all possible x-values. Thus, $D = \{4, 2, -3\}$.

The range R is the set of all possible y-values. Thus, $R = \{-3, -5, 4\}$.

PROBLEM 1

Find the domain and range of the relation:

$$T = \{(6, -4), (4, -6), (-1, 3)\}$$

In many cases, relations are defined by a **rule** for finding the y-value for a given x-value. Thus, the relation

$$\underbrace{S = \{}_{\text{"The set}} \quad \underbrace{(x, y)}_{\text{of all } (x, y)} \quad \underbrace{|}_{\text{such that}} \quad \underbrace{y = 3x - 1, x \text{ a natural number less than } 5\}}_{y = 3x - 1 \text{ and } x \text{ is a natural number less than 5."}}$$

defines—that is, gives a *rule* about—a set of ordered pairs obtained by assigning a natural number less than 5 to x and obtaining the corresponding y-value. Thus,

For $x = 1$, $y = 3(1) - 1 = 2$ and $(1, 2)$ is part of the relation.
For $x = 2$, $y = 3(2) - 1 = 5$ and $(2, 5)$ is part of the relation.
For $x = 3$, $y = 3(3) - 1 = 8$ and $(3, 8)$ is part of the relation.
For $x = 4$, $y = 3(4) - 1 = 11$ and $(4, 11)$ is part of the relation.

Can you see the pattern? For $S = \{(1, 2), (2, 5), (3, 8), (4, 11)\}$, the domain is $\{1, 2, 3, 4\}$, and the range is $\{2, 5, 8, 11\}$.

Answers to PROBLEMS

1. $D = \{6, 4, -1\}$; $R = \{-4, -6, 3\}$

NOTE

Unless otherwise specified, the *domain* of a relation is the largest set of real numbers that can be substituted for x that result in a real number for y. The *range* is then determined by the rule of the relation.

For example, for the relation

$$S = \left\{(x, y) \,\middle|\, y = \frac{1}{x}\right\}$$

the domain consists of all real numbers *except zero* since you can substitute any real number for x in $y = \frac{1}{x}$ *except zero* because division by zero is not defined. The range is also the set of real numbers *except zero* because the rule $y = \frac{1}{x}$ will never yield a value of zero for any given value of x.

EXAMPLE 2 **Finding the domain and range of a relation given as an equation**

Find the domain and range of the relation:

$$\left\{(x, y) \,\middle|\, y = \frac{1}{x - 1}\right\}$$

SOLUTION 2 The domain is the set of real numbers *except* 1, since

$$\frac{1}{1 - 1} = \frac{1}{0} \quad \text{which is undefined.}$$

The range is the set of real numbers *except* 0, since

$$y = \frac{1}{x - 1}$$

is never 0.

PROBLEM 2

Find the domain and range of the relation:

$$\left\{(x, y) \,\middle|\, y = \frac{1}{x + 1}\right\}$$

B > Determining Whether a Relation Is a Function

In mathematics, an important type of relation is one where for each element in the domain, there corresponds one and only one element in the range. Such relations are called *functions*.

FUNCTION

A function is a set of ordered pairs in which each domain value has exactly one range value; that is, no two different ordered pairs have the same first coordinate.

Note that the graph of a function will never have two points stacked vertically, that is, two points that can be connected with a vertical line. (See the *Using Your Knowledge.*)

Answers to PROBLEMS

2. The domain is the set of all real numbers except -1. The range is the set of all real numbers except 0.

EXAMPLE 3 **Determining whether a relation is a function**

Determine whether the given relations are functions:

a. $\{(3, 4), (4, 3), (4, 4), (5, 4)\}$ **b.** $\{(x, y) \mid y = 5x + 1\}$

SOLUTION 3

a. The relation is not a function because the two ordered pairs (4, 3) and (4, 4) have the same first coordinate.

b. The relation $\{(x, y) \mid y = 5x + 1\}$ is a function because if x is any real number, the expression $y = 5x + 1$ yields one and only one value, so there will never be two ordered pairs with the same first coordinate.

PROBLEM 3

Determine whether the relations are functions:

a. $\{(5, 6), (6, 5), (6, 6), (7, 6)\}$

b. $\{(x, y) \mid y = 3x + 2\}$

C > Using Function Notation

We often use letters such as f, F, g, G, h, and H to designate functions. Thus, for the relation in Example 3(b), we use set notation to write

$$f = \{(x, y) \mid y = 5x + 1\}$$

because we know this relation to be a function. Another very commonly used notation for denoting the range value that corresponds to a given domain value x is $f(x)$. (This is usually read, "f of x.")

The $f(x)$ notation, called **function notation,** is quite convenient because it denotes the value of the function for the given value of x. For example, if

$$f(x) = 2x + 3$$

then

$$f(1) = 2(1) + 3 = 5$$
$$f(0) = 2(0) + 3 = 3$$
$$f(-6) = 2(-6) + 3 = -9$$
$$f(4) = 2(4) + 3 = 11$$
$$f(a) = 2(a) + 3 = 2a + 3$$
$$f(w + 2) = 2(w + 2) + 3 = 2w + 7$$

and so on. Whatever appears between the parentheses in $f(\ \ )$ is to be substituted for x in the rule that defines $f(x)$.

Instead of describing a function in set notation, we frequently say, "the function defined by $f(x) = \ldots$" where the three dots are replaced by the expression for the value of the function. For instance, "the function defined by $f(x) = 5x + 1$" has the same meaning as "the function $f = \{(x, y) \mid y = 5x + 1\}$."

EXAMPLE 4 **Evaluating functions**

Let $f(x) = 3x + 5$. Find:

a. $f(4)$ **b.** $f(2)$ **c.** $f(2) + f(4)$ **d.** $f(x + 1)$

SOLUTION 4 Since $f(x) = 3x + 5$,

a. $f(4) = 3 \cdot 4 + 5 = 12 + 5 = 17$

b. $f(2) = 3 \cdot 2 + 5 = 6 + 5 = 11$

c. Since $f(2) = 11$ and $f(4) = 17$,

$$f(2) + f(4) = 11 + 17 = 28$$

d. $f(x + 1) = 3(x + 1) + 5 = 3x + 8$

PROBLEM 4

Let $f(x) = 4x - 5$. Find:

a. $f(3)$ **b.** $f(4)$

c. $f(3) + f(4)$ **d.** $f(x - 1)$

Answers to PROBLEMS

3. a. Not a function because (6, 5) and (6, 6) have the same first coordinate. **b.** A function. If x is a real number, $y = 3x + 2$ yields one and only one value.

4. a. 7 **b.** 11 **c.** 18 **d.** $4x - 9$

EXAMPLE 5 Evaluating functions

A function g is defined by $g(x) = x^3 - 2x^2 + 3x - 4$. Find:

a. $g(2)$ **b.** $g(-3)$ **c.** $g(2) - g(-3)$

SOLUTION 5 Since $g(x) = x^3 - 2x^2 + 3x - 4$,

a. $g(2) = 2^3 - 2(2)^2 + 3(2) - 4$
$= 8 - 8 + 6 - 4 = 2$

b. $g(-3) = (-3)^3 - 2(-3)^2 + 3(-3) - 4$
$= -27 - 18 - 9 - 4 = -58$

c. $g(2) - g(-3) = 2 - (-58) = 60$

PROBLEM 5

If $g(x) = x^3 - 3x^2 + 2x - 1$, find:

a. $g(3)$ **b.** $g(-2)$

c. $g(3) - g(-2)$

In Examples 4 and 5, we evaluated a specified function. Sometimes, as was the case in *Getting Started,* we must find the function, as shown in Example 6.

EXAMPLE 6 Finding the relationship between ordered pairs

Consider the ordered pairs (2, 6), (3, 9), (1.2, 3.6), and $\left(\frac{2}{5}, \frac{6}{5}\right)$. There is a functional relationship $y = f(x)$ between the numbers in each pair. Find $f(x)$ and use it to fill in the missing numbers in the pairs (__, 12), (__, 3.3), and (5, __).

SOLUTION 6 The given pairs are of the form (x, y). A close examination reveals that each of the y's in the pairs is 3 times the corresponding x; that is, $y = 3x$ or $f(x) = 3x$. Now in each of the ordered pairs (__, 12), (__, 3.3), and (5, __), the y-value must be three times the x-value. Thus,

(__, 12) = (4, 12)
(__, 3.3) = (1.1, 3.3)
(5, __) = (5, 15)

PROBLEM 6

Consider the ordered pairs (3, 6), (4, 8), (1.1, 2.2), and $\left(\frac{3}{5}, \frac{6}{5}\right)$. There is a functional relationship $y = f(x)$ between the numbers in each pair. Find $f(x)$ and use it to fill in the missing numbers in the pairs (__, 10), (__, 4.4), and (6, __).

D > Applications Involving Functions

In recent years, aerobic exercises such as jogging, swimming, bicycling, and roller blading have been taken up by millions of Americans. To see whether you are exercising too hard (or not hard enough), you should stop from time to time and take your pulse to determine your heart rate. The idea is to keep your rate within a range known as the **target zone,** which is determined by your age. Example 7 explains how to find the **lower limit** of your target zone.

EXAMPLE 7 Exercises and functions

The lower limit L (heartbeats per minute) of your target zone is a function of your age a (in years) and is given by the equation

$$L(a) = -\frac{2}{3}a + 150$$

a. Find the value of L for a person who is 30 years old.

b. Find the value of L for a person who is 45 years old.

SOLUTION 7

a. We need to find $L(30)$, and because

$$L(a) = -\frac{2}{3}a + 150,$$

$$L(30) = -\frac{2}{3}(30) + 150$$
$$= -20 + 150 = 130$$

This result means that a 30-year-old person should try to attain at least 130 heartbeats per minute while exercising.

PROBLEM 7

a. Find L for a 21-year-old person.

b. Find L for a 33-year-old person.

Answers to PROBLEMS

5. a. 5 **b.** −25 **c.** 30

6. $f(x) = 2x$; (5, 10), (2.2, 4.4), (6, 12)

7. a. $L = 136$ **b.** $L = 128$

b. Here, we want to find $L(45)$. Proceeding as before, we obtain

$$\begin{aligned} L(45) &= -\frac{2}{3}(45) + 150 \\ &= -30 + 150 = 120 \end{aligned}$$

(Find the value of L for your own age.)

CURRENT LEAK ESTIMATES	
USGS	504,000 GAL/DAY
OUTSIDE ESTIMATES	1,050,000 GAL/DAY
BP (WORST CASE)	2,520,000 GAL/DAY
EXPERTS' WORST CASE	4,200,000 GAL/DAY

Gulf Leak Meter from PBS Newshour

The worst ecological disaster in U.S. history began on April 20, 2010, and was caused by an explosion in the oil rig *Deepwater Horizon,* which resulted in the death of 11 people and triggered a massive oil spill into the Gulf of Mexico. Estimates of the exact amount of the oil spilled varies based on the source (see chart) but it is clearly a **function** of the number of days n it lasted and the leak rate of the oil, estimated by the United States Geological Service (USGS) to be 504,000 gallons each day! The function u that estimates the amount of oil spilled in n days can be defined as:

$$u(n) = 504{,}000n \quad \text{gallons/day}$$

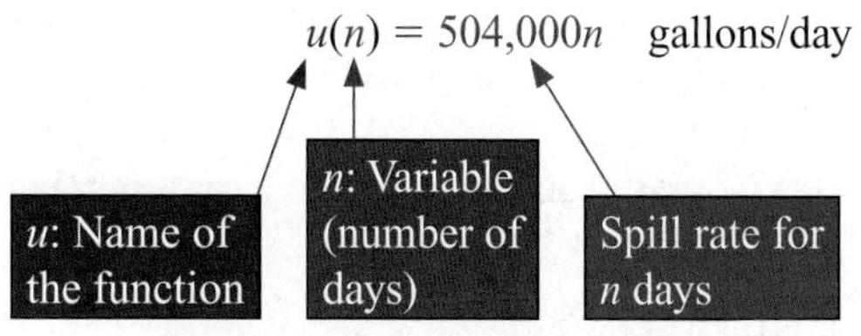

To find out how much oil spilled the first month (30 days), let $n = 30$:

$$u(30) = 504{,}000(30) = 15{,}120{,}000 \text{ gallons}$$

EXAMPLE 8 Gulf oil spill functions

a. Using the function $u(n) = 504{,}000n$ estimate how many gallons were spilled during the first 100 days.

b. Outside estimates indicate that the leak may be 1.05 million gallons each day! (See the chart.) Define a function o that will estimate the amount of oil spilled in n days using the 1.05 million gallons a day estimate.

c. Use the function in part **b** to estimate how many gallons of oil were spilled in 60 days.

SOLUTION 8

a. To estimate how many gallons were spilled in 100 days, let $n = 100$ in $u(n) = 504{,}000n$, obtaining $u(100) = 504{,}000(100) = 50{,}400{,}000$. This means that more than 50 million gallons of oil were spilled the first 100 days.

b. Using the 1.05 million gallon per day estimate, o is defined as

$$o(n) = 1.05n \text{ million gallons/day}$$

c. Letting $n = 60$ in $o(n) = 1.05n$, we obtain

$$o(60) = 1.05(60) = 63 \text{ million gallons}$$

PROBLEM 8

a. A British Petroleum (BP) worst-case scenario uses the function $b(n) = 2.520n$ million gallons a day to estimate the leak. Using this estimate, how many gallons were spilled the first 100 days?

b. Define a function e that will estimate the amount of oil spilled in n days using the expert's worst case of 4.2 million gallons a day estimate.

c. Using the function in part **b** estimate how many gallons of oil were spilled in 60 days.

Answers to PROBLEMS

8. a. 252,000,000 gallons

b. $e(n) = 4.2n$ million gallons

c. 252 million gallons

Calculator Corner

Entering Functions

The idea of a function is so important in mathematics that even your calculator will only accept functions when you use the graphing feature. Thus, if you are given an equation with variables x and y and you can solve for y, you have a function. What does this mean?

Let's take $2x + 4y = 4$. Solving for y, we obtain $y = -\frac{1}{2}x + 1$, which can be entered and graphed. Similarly, $y = x^2 + 1$ can be entered and graphed (see Window 1). We cannot solve for y in the equation $x^2 + y^2 = 4$. The equation $x^2 + y^2 = 4$, whose graph happens to be a circle of radius 2, does not describe a function.

Your calculator can also evaluate functions. For example, to find the value $L(30)$ in Example 7, let's start by using x's instead of a's. Store the 30 in the calculator's memory by pressing 30 STO▸ X,T,θ,n ENTER, then enter the function $-\frac{2}{3}x + 150$ ENTER; the answer, 130, is then displayed as shown in Window 2.

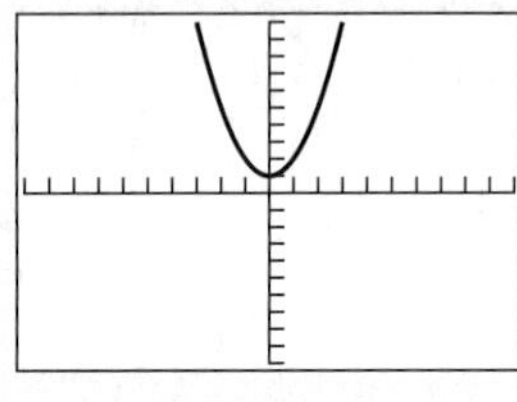

Window 1

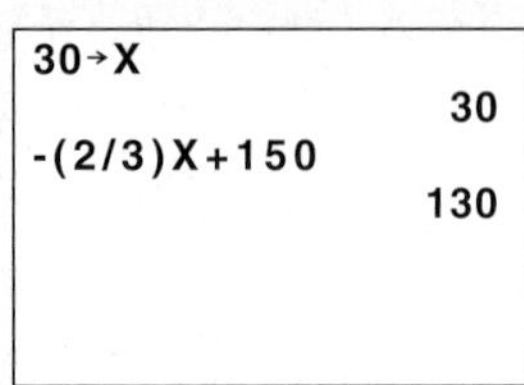

Window 2

Exercises 9.6

A Finding Domains and Ranges In Problems 1–14, find the domain and the range of the given relation. [*Hint:* Remember that you cannot divide by zero.]

1. $\{(1, 2), (2, 3), (3, 4)\}$

2. $\{(3, 1), (2, 1), (1, 1)\}$

3. $\{(1, 1), (2, 2), (3, 3)\}$

4. $\{(4, 1), (5, 2), (6, 1)\}$

5. $\{(x, y) \mid y = 3x\}$

6. $\{(x, y) \mid y = 2x + 1\}$

7. $\{(x, y) \mid y = x + 1\}$

8. $\{(x, y) \mid y = 1 - 2x\}$

9. $\{(x, y) \mid y = x^2\}$

10. $\{(x, y) \mid y = 2 + x^2\}$

11. $\{(x, y) \mid y^2 = x\}$ (*Hint:* y^2 is never negative.)

12. $\{(x, y) \mid x = 1 + y^2\}$

13. $\left\{(x, y) \,\middle|\, y = \frac{1}{x - 3}\right\}$

14. $\left\{(x, y) \,\middle|\, y = \frac{1}{x - 2}\right\}$

In Problems 15–22, find the domain and the range of the relation. List the ordered pairs in each relation.

15. $\{(x, y) \mid y = 2x$, x an integer between -1 and 2, inclusive$\}$

16. $\{(x, y) \mid y = 2x - 1$, x a counting number not greater than 5$\}$

17. $\{(x, y) \mid y = 2x - 3$, x an integer between 0 and 4, inclusive$\}$

18. $\{(x, y) \mid y = \frac{1}{x}$, x an integer between 1 and 5, inclusive$\}$

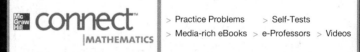

19. $\{(x, y) \mid y = \sqrt{x}, x = 0, 1, 4, 9, 16, \text{ or } 25\}$

20. $\{(x, y) \mid y \le x + 1, x \text{ and } y \text{ positive integers less than } 4\}$

21. $\{(x, y) \mid y > x, x \text{ and } y \text{ positive integers less than } 5\}$

22. $\{(x, y) \mid 0 < x + y < 5, x \text{ and } y \text{ positive integers less than } 4\}$

‹B› Determining Whether a Relation Is a Function In Problems 23–30, decide whether the given relation is a function. State the reason for your answer in each case.

23. $\{(0, 1), (1, 2), (2, 3)\}$

24. $\{(1, 2), (2, 1), (1, 3), (3, 1)\}$

25. $\{(-1, 1), (-2, 2), (-3, 3)\}$

26. $\{(-1, 1), (-1, 2), (-1, 3)\}$

27. $\{(x, y) \mid y = 5x + 6\}$

28. $\{(x, y) \mid y = 3 - 2x\}$

29. $\{(x, y) \mid x = y^2\}$

30. $\{(x, y) \mid y = x^2\}$

‹C› Using Function Notation In Problems 31–35, find the indicated value of the function.

31. A function f is defined by $f(x) = 3x + 1$. Find:
a. $f(0)$ **b.** $f(2)$ **c.** $f(-2)$

32. A function g is defined by $g(x) = -2x + 1$. Find:
a. $g(0)$ **b.** $g(1)$ **c.** $g(-1)$

33. A function F is defined by $F(x) = \sqrt{x - 1}$. Find:
a. $F(1)$ **b.** $F(5)$ **c.** $F(26)$

34. A function G is defined by $G(x) = x^2 + 2x - 1$. Find:
a. $G(0)$ **b.** $G(2)$ **c.** $G(-2)$

35. A function f is defined by $f(x) = 3x + 1$. Find:
a. $f(x + h)$
b. $f(x + h) - f(x)$
c. $\dfrac{f(x + h) - f(x)}{h}, \quad h \ne 0$

36. Given the ordered pairs: (2, 1), (6, 3), (9, 4.5), and (1.6, 0.8), there is a simple functional relationship, $y = f(x)$, between the numbers in each pair. What is $f(x)$? Use this to fill in the missing number in the pairs (____, 7.5), (____, 2.4), and (____, $\frac{1}{7}$).

37. Given the ordered pairs: $(\frac{1}{2}, \frac{1}{4})$, (1.2, 1.44), (5, 25), and (7, 49), there is a simple functional relationship, $y = g(x)$, between the numbers in each pair. What is $g(x)$? Use this to fill in the missing number in the pairs $(\frac{1}{4}, \_\_\_\_)$, (2.1, ____), and (____, 64).

38. Given that $f(x) = x^3 - x^2 + 2x$, find:
a. $f(-1)$ **b.** $f(-3)$ **c.** $f(2)$

39. If $g(x) = 2x^3 + x^2 - 3x + 1$, find:
a. $g(0)$ **b.** $g(-2)$ **c.** $g(2)$

‹D› Applications Involving Functions

40. *Fahrenheit and Celsius temperatures* The Fahrenheit temperature reading F is a function of the Celsius temperature reading C. This function is given by

$$F(C) = \frac{9}{5}C + 32$$

a. If the temperature is 15°C, what is the Fahrenheit temperature?
b. Water boils at 100°C. What is the corresponding Fahrenheit temperature?
c. The freezing point of water is 0°C or 32°F. How many Fahrenheit degrees below freezing is a temperature of −10°C?
d. The lowest temperature attainable is −273°C; this is the zero point on the absolute temperature scale. What is the corresponding Fahrenheit temperature?

41. *Upper limit of target zone* Refer to Example 7. The **upper limit** U of your target zone when exercising is also a function of your age a (in years), and is given by

$$U(a) = -a + 190$$

Find the highest safe heart rate for a person who is
a. 50 years old
b. 60 years old

Web IT go to mhhe.com/bello for more lessons

42. *Target zone* Refer to Example 7 and Problem 41. The target zone for a person a years old consists of all the heart rates between $L(a)$ and $U(a)$, inclusive. Thus, if a person's heart rate is R, that person's target zone is described by $L(a) \le R \le U(a)$. Find the target zone for a person who is

a. 30 years old **b.** 45 years old

43. *Ideal weight for men* The ideal weight w (in pounds) of a man is a function of his height h (in inches). This function is defined by

$$w(h) = 5h - 190$$

a. If a man is 70 inches tall, what should his weight be?

b. If a man weighs 200 pounds, what should his height be?

44. *Car rental costs* The cost C in dollars of renting a car for 1 day is a function of the number m of miles traveled. For a car renting for \$20 per day and 20¢ per mile, this function is given by

$$C(m) = 0.20m + 20$$

a. Find the cost of renting a car for 1 day and driving 290 miles.

b. If an executive paid \$60.60 after renting a car for 1 day, how many miles did she drive?

45. *Pressure below ocean surface* The pressure P (in pounds per square foot) at a depth of d feet below the surface of the ocean is a function of the depth. This function is given by

$$P(d) = 63.9d$$

Find the pressure on a submarine at a depth of:

a. 10 feet **b.** 100 feet

46. *Distance traveled by dropped ball* If a ball is dropped from a point above the surface of the earth, the distance s (in meters) that the ball falls in t seconds is a function of t. This function is given by

$$s(t) = 4.9t^2$$

Find the distance that the ball falls in:

a. 2 seconds **b.** 5 seconds

47. *Distance traveled by falling object* The function $S(t) = \frac{1}{2}gt^2$ gives the distance that an object falls from rest in t seconds. If S is measured in feet, then the gravitational constant, g, is approximately 32 feet/second per second. Find the distance that the object will fall in:

a. 3 seconds

b. 5 seconds

48. *Estimation of gravitational constant* An experiment, carefully carried out, showed that a ball dropped from rest fell 64.4 feet in 2 seconds. What is a more accurate value of g than that given in Problem 47?

Applications: Green Math

49. *Oil recovery* A containment cap installed by BP collects about 441,000 gallons of oil and gas flowing from the well and transports them to the *Discoverer Enterprise* drillship on the surface.

a. Define a function r that will estimate the amount of oil that can be recovered in n days using the 441,000 gallons a day estimate.

b. Use the function in part **a** to estimate how many gallons of oil can be recovered in 60 days.

50. *More oil recovery* According to BP's worst-case scenario the amount of oil spilled in n days can be estimated by the function $b(n) = 2.520n$ million gallons, while the function $r(n) = 441{,}000n$ gallons represents the amount of oil that can be recovered in n days.

a. The function $a(n) = b(n) - r(n)$ represents the **actual** amount of oil spilling into the gulf daily. Find $a(n)$.

b. How many gallons will actually spill into the gulf over a 60-day period?

Using Your Knowledge

The Vertical Line Test The graph of an equation can be used to determine whether a relation is a function by applying the **vertical line test.** Since a function is a set of ordered pairs in which no two ordered pairs have the same first coordinate, the graph of a function will never have two points "stacked" vertically; that is, the graph of a function *never* contains two points that can be connected with a vertical line.

Thus, the first two graphs shown here represent functions (no two points on the graph can be connected with a vertical line), but the last two do not.

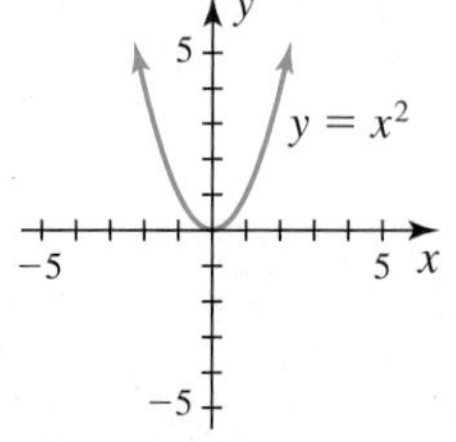

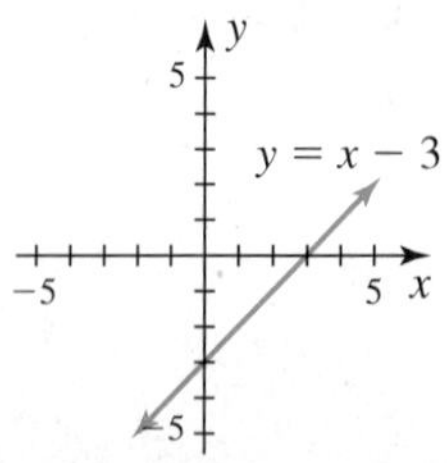

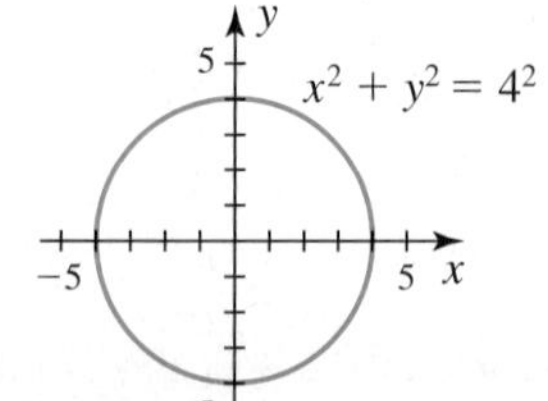

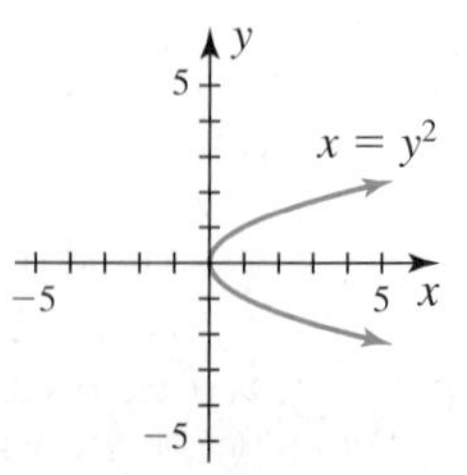

In Problems 51–54, use the vertical line test to determine whether the graph represents a function.

51.

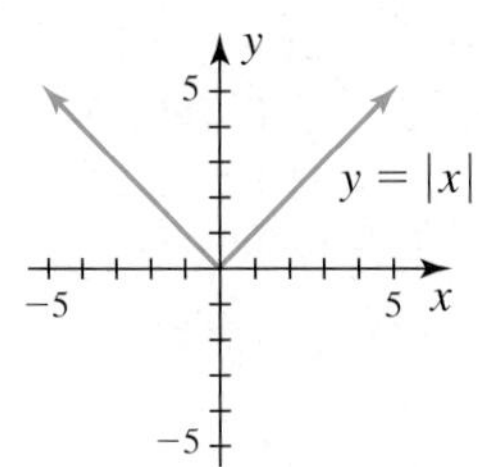

52.

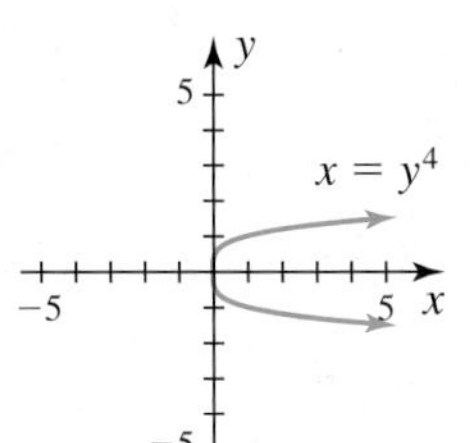

53.

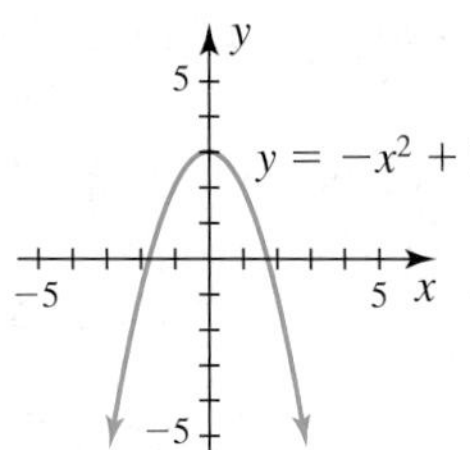

54.

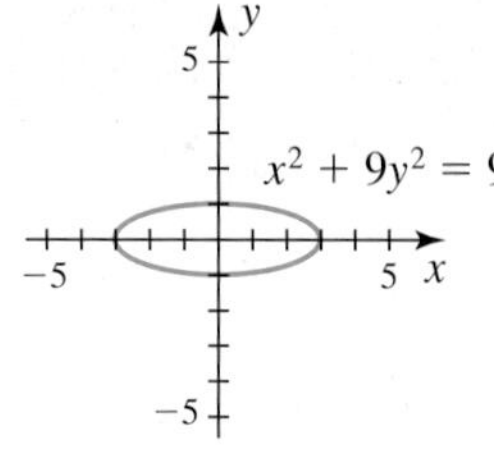

Write On

55. Is every relation a function? Explain.

56. Is every function a relation? Explain.

57. Explain how you determine the domain of a function defined by a set of ordered pairs.

58. Explain how you determine the domain of a function defined by a formula.

Concept Checker

Fill in the blank(s) with the correct word(s), phrase, or mathematical statement.

range	**domain**
relation	**function**
ordered	

59. A **relation** is a set of ________________ **pairs.**

60. The ________________ of a **relation** is the set of all possible ***x*-values.**

61. The ________________ of a **relation** is the set of all possible ***y*-values.**

62. A ________________ is a **set of ordered** pairs in which each **domain value** has **exactly one range value.**

Mastery Test

In Problems 63–66, find the domain and range of the relation.

63. $\{(-5, 5), (-6, 6), (-7, 7)\}$

64. $\{(5, -5), (-5, 5), (3, -3), (-3, 3)\}$

65. $\left\{(x, y) \,\middle|\, y = \frac{1}{x-3}\right\}$

66. $\{(x, y) \mid y = -x + 2\}$

In Problems 67–71, determine whether the relation is a function, and explain why or why not.

67. $\{(-4, 5), (-5, 5), (-6, 5)\}$

68. $\{(5, -3), (5, -4), (5, -5)\}$

69. $\{(x, y) \mid y = 2x + 3\}$

70. $\{(x, y) \mid y = x^2\}$

71. $\{(x, y) \mid x = 3y^2\}$

72. If $h(x) = x^2 - 3$, find $h(-2)$.

73. If $g(x) = x^3 - 1$, find $g(-1)$.

74. The monthly cost for searches (questions) in an online service is \$9.95 with 100 free searches. After that, the cost is \$0.10 per search. The monthly cost for the service (in dollars) can be represented by the following function, where q is the number of searches:

For $q \le 100$: $f(q) = 9.95$

For $q > 100$: $f(q) = 9.95 + 0.10(q - 100)$

a. What would your monthly cost be if you made 99 searches during the month.

b. What would your monthly cost be if you made 130 searches during the month.

Collaborative Learning

This exercise concerns an emergency ejection from a fighter jet and how it could relate to completing the square.

The height H (in feet relative to the ground) of a pilot after t seconds is given by the equation

$$H = -16t^2 + 608t + 4482$$

At about 10,000 feet a malfunction occurs. Eject! Eject!

We want to answer three questions:

1. What is the maximum height (relative to the ground) reached by the pilot after being ejected?
2. How many feet above the jet was the pilot ejected?
3. When did the pilot reach the ground with the aid of his or her parachute?

WINDOW
Xmin=0
Xmax=50
Xscl=10
Ymin=-2000
Ymax=12000
Yscl=2000
Xres=1

Form three teams to investigate the incident.

Team 1 is in charge of graphing the path of the pilot and the point of ejection using a graphing calculator. A possible viewing window to graph the path is shown. What expression Y_1 does the team have to graph? What is the equation of the line Y_2 shown in the screen and representing the point at which the pilot ejected?

Team 2 is in charge of answering the first two questions.

Start with: $H = -16(\quad)^2 + C$

Here is a hint: $H = -16(t^2 - 38t) + 4482$

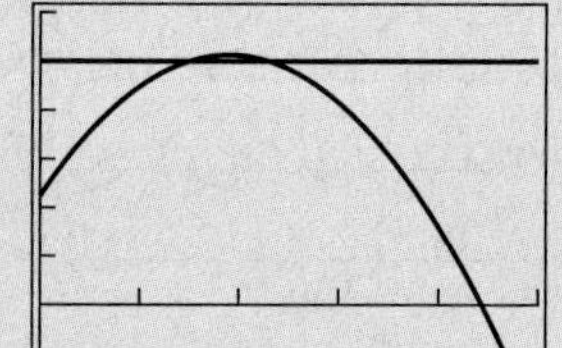

Now, complete the square inside the parentheses to put the equation in the standard form for a parabola. What is the maximum for this parabola? This answers question 1. Since we know that the pilot ejected at 10,000 feet, how many feet above the jet was the pilot ejected? This answers question 2.

Team 3 has to answer the third question. When the pilot reaches the ground, what is H? The point (t, H) is a solution (zero) of the equation whose graph is shown and it occurs when $H = 0$. To find this point press 2nd TRACE 2. The value of t is the time it took the pilot to land back on earth!

Research Questions

1. Write a short paper detailing the work and techniques used by the Babylonians to solve quadratic equations.
2. Show that the solution of the equation $x^2 + ax = b$ obtained by the Babylonians and the solution obtained by using the quadratic formula developed in this chapter are identical.
3. Write a short paper detailing the methods used by Euler to solve quadratic equations.
4. Write a short paper detailing the methods used by the Arabian mathematician al-Khwarizmi to solve quadratic equations.

Summary Chapter 9

Section	Item	Meaning	Example
9.1A	Quadratic equation	An equation that can be written in the form $ax^2 + bx + c = 0$	$x^2 = 16$ is a quadratic equation since it can be written as $x^2 - 16 = 0$.
	$\sqrt{a}$, the principal square root of a, $a \geq 0$	$\sqrt{a}$ is a nonnegative number b such that $b^2 = a$	$\sqrt{16} = 4$ since $4^2 = 16$
	Irrational number	A number that cannot be written in the form $\frac{a}{b}$, where a and b are integers and b is not 0	$\sqrt{2}$, $\sqrt{5}$, and $3\sqrt{7}$ are irrational.
	Real numbers	The rational and the irrational numbers	-7, $-\frac{3}{2}$, 0, $\sqrt{5}$, and 17 are real numbers.
	Square root property of equations	If A is a positive number and $X^2 = A$, then $X = \pm\sqrt{A}$.	If $X^2 = 7$, then $X = \pm\sqrt{7}$.
9.2A	Completing the square	A method used to solve quadratic equations	To solve $2x^2 + 4x - 1 = 0$ **1.** Add 1. $2x^2 + 4x = 1$ **2.** Divide by 2. $x^2 + 2x = \frac{1}{2}$ **3.** Add 1^2. $x^2 + 2x + 1 = \frac{3}{2}$ **4.** Rewrite. $(x + 1)^2 = \frac{3}{2}$ **5.** Solve. $x + 1 = \pm\sqrt{\frac{3}{2}}$ $x + 1 = \pm\frac{\sqrt{6}}{2}$ $x = -1 \pm \frac{\sqrt{6}}{2} = \frac{-2 \pm \sqrt{6}}{2}$
9.3	Quadratic formula	The solutions of $ax^2 + bx + c = 0$ are $x = \frac{-b \pm \sqrt{b^2 - 4ac}}{2a}$, $a \neq 0$	The solutions of $2x^2 + 3x + 1 = 0$ are $\frac{-3 \pm \sqrt{3^2 - 4 \cdot 2 \cdot 1}}{2 \cdot 2}$ that is, -1 and $-\frac{1}{2}$.
9.3A	Standard form of a quadratic equation	$ax^2 + bx + c = 0$ is the standard form of a quadratic equation, $a \neq 0$	The standard form of $x^2 + 2x = -11$ is $x^2 + 2x + 11 = 0$.
9.4	Graph of a quadratic equation	The graph of $y = ax^2 + bx + c$ ($a \neq 0$) is a parabola.	
9.5	Pythagorean Theorem	The square of the hypotenuse of a right triangle equals the sum of the squares of the other two sides.	If a and b are the lengths of the sides and h is the length of the hypotenuse, $c^2 = a^2 + b^2$.

(continued)

Section	Item	Meaning	Example
9.6A	Relation	A set of ordered pairs	$\{(2, 5), (2, 6), (3, 7)\}$ is a relation.
	Domain	The domain of a relation is the set of all possible x-values.	The domain of the relation $\{(2, 5), (2, 6), (3, 7)\}$ is $\{2, 3\}$.
	Range	The range of a relation is the set of all possible y-values.	The range of the relation $\{(2, 5), (2, 6), (3, 7)\}$ is $\{5, 6, 7\}$.
9.6B	Function	A set of ordered pairs in which no two ordered pairs have the same x-value	The relation $\{(1, 2), (3, 4)\}$ is a function.

› Review Exercises Chapter 9

(If you need help with these exercises, look at the section indicated in brackets.)

1. ‹ **9.1A** › *Solve.*

a. $x^2 = 1$

b. $x^2 = 100$

c. $x^2 = 81$

2. ‹ **9.1A** › *Solve.*

a. $16x^2 - 25 = 0$

b. $25x^2 - 9 = 0$

c. $64x^2 - 25 = 0$

3. ‹ **9.1A** › *Solve.*

a. $7x^2 + 36 = 0$

b. $8x^2 + 49 = 0$

c. $3x^2 + 81 = 0$

4. ‹ **9.1B** › *Solve.*

a. $49(x + 1)^2 - 3 = 0$

b. $25(x + 2)^2 - 2 = 0$

c. $16(x + 1)^2 - 5 = 0$

5. ‹ **9.2A** › *Find the missing term in the given expression.*

a. $(x + 3)^2 = x^2 + 6x + \square$

b. $(x + 7)^2 = x^2 + 14x + \square$

c. $(x + 6)^2 = x^2 + 12x + \square$

6. ‹ **9.2A** › *Find the missing terms in the given expression.*

a. $x^2 - 6x + \square = (\quad)^2$

b. $x^2 - 10x + \square = (\quad)^2$

c. $x^2 - 12x + \square = (\quad)^2$

7. ‹ **9.2A** › *Find the number that must divide each term in the given equation so that the equation can be solved by the method of completing the square.*

a. $7x^2 - 14x = -4$

b. $6x^2 - 18x = -2$

c. $5x^2 - 15x = -3$

8. ‹ **9.2A** › *Find the term that must be added to both sides of the given equation so that the equation can be solved by the method of completing the square.*

a. $x^2 - 4x = -4$

b. $x^2 - 6x = -9$

c. $x^2 - 12x = -3$

9. ⟨ **9.3B** ⟩ *Solve.*

a. $dx^2 + ex + f = 0, d \neq 0$

b. $gx^2 + hx + i = 0, g \neq 0$

c. $jx^2 + kx + m = 0, j \neq 0$

10. ⟨ **9.3B** ⟩ *Solve.*

a. $2x^2 - x - 1 = 0$

b. $2x^2 - 2x - 5 = 0$

c. $2x^2 - 3x - 3 = 0$

11. ⟨ **9.3B** ⟩ *Solve.*

a. $3x^2 - x = 1$

b. $3x^2 - 2x = 2$

c. $3x^2 - 3x = 2$

12. ⟨ **9.3B** ⟩ *Solve.*

a. $9x = x^2$

b. $4x = x^2$

c. $25x = x^2$

13. ⟨ **9.3B** ⟩ *Solve.*

a. $\frac{x^2}{9} - x = -\frac{4}{9}$

b. $\frac{x^2}{5} - \frac{x}{2} = \frac{3}{10}$

c. $\frac{x^2}{3} + \frac{x}{6} = -\frac{1}{2}$

14. ⟨ **9.4A** ⟩ *Graph.*

a. $y = x^2 + 1$

b. $y = x^2 + 2$

c. $y = x^2 + 3$

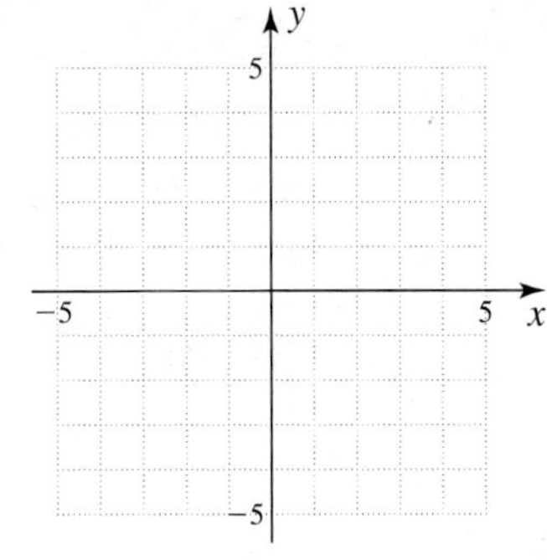

15. ⟨ **9.4A** ⟩ *Graph.*

a. $y = -x^2 - 1$

b. $y = -x^2 - 2$

c. $y = -x^2 - 3$

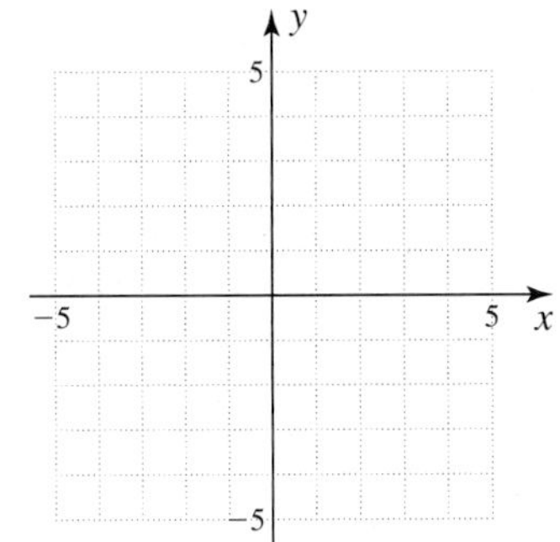

16. ⟨ **9.4A** ⟩ *Graph.*

a. $y = -(x - 2)^2$

b. $y = -(x - 3)^2$

c. $y = -(x - 4)^2$

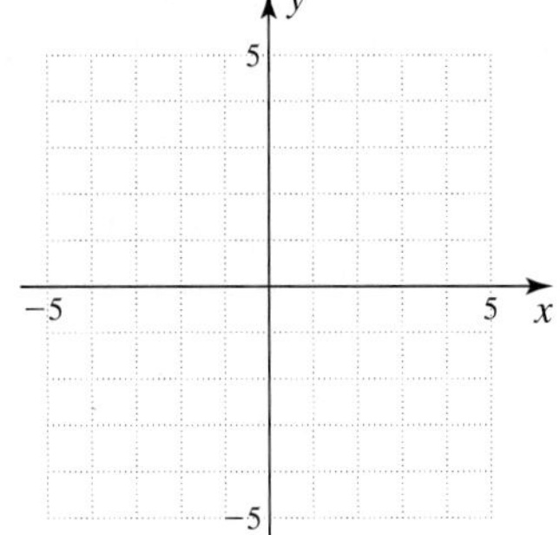

17. ⟨ **9.4A** ⟩ *Graph.*

a. $y = (x - 2)^2 + 1$

b. $y = (x - 2)^2 + 2$

c. $y = (x - 2)^2 + 3$

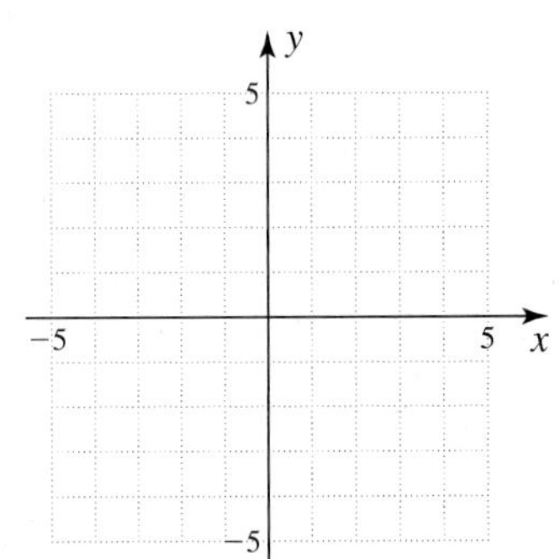

18. ⟨ **9.4B** ⟩ *Graph and label the vertex and intercepts.*

a. $y = -x^2 + 6x - 8$

b. $y = -x^2 + 6x - 5$

c. $y = -x^2 + 6x$

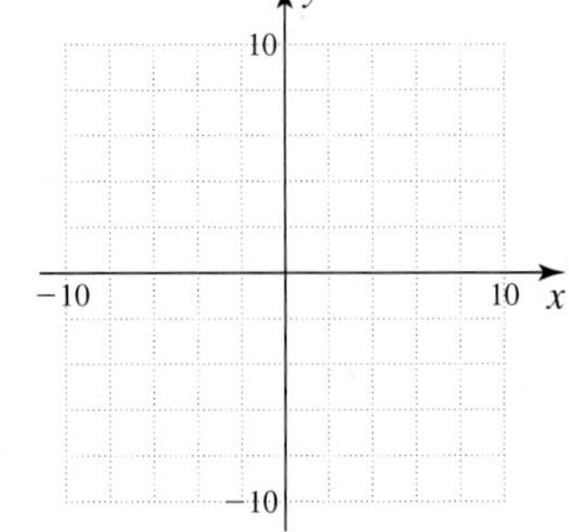

19. ⟨ **9.5A** ⟩ *Find the length of the hypotenuse of a right triangle if the lengths of the two sides are:*

a. 5 inches and 12 inches

b. 2 inches and 3 inches

c. 4 inches and 5 inches

20. ⟨ **9.5B** ⟩ *After t seconds, the distance d (in meters) traveled by an object thrown downward with an initial velocity v_0 is given by the equation*

$$d = 5t^2 + v_0 t$$

Find the number of seconds it takes an object to hit the ground if the object is dropped ($v_0 = 0$) from a height of:

a. 125 meters **b.** 245 meters

c. 320 meters

21. ⟨ **9.6A** ⟩ *Find the domain and range.*

a. $\{(-3, 1), (-4, 1), (-5, 2)\}$

b. $\{(2, -4), (-1, 3), (-1, 4)\}$

c. $\{(-1, 2), (-1, 3), (-1, 4)\}$

22. ⟨ **9.6A** ⟩ *Find the domain and range.*

a. $\{(x, y) \mid y = 2x - 3\}$

b. $\{(x, y) \mid y = -3x - 2\}$

c. $\{(x, y) \mid y = x^2\}$

23. ⟨ **9.6B** ⟩ *Which of the following are functions?*

a. $\{(-3, 1), (-4, 1), (-5, 2)\}$

b. $\{(2, -4), (-1, 3), (-1, 4)\}$

c. $\{(-1, 2), (-1, 3), (-1, 4)\}$

24. ⟨ **9.6C** ⟩ *If $f(x) = x^3 - 2x^2 + x - 1$, find:*

a. $f(2)$

b. $f(-2)$

c. $f(1)$

25. ⟨ **9.6D** ⟩ *The average price P(n) of books depends on the number n of millions of books sold and is given by the function*

$$P(n) = 25 - 0.3n \text{ (dollars)}$$

a. Find the average price of a book when 10 million copies are sold.

b. Find the average price of a book when 20 million copies are sold.

c. Find the average price of a book when 30 million copies are sold.

Practice Test Chapter 9

(Answers on pages 753–754)

Visit www.mhhe.com/bello to view helpful videos that provide step-by-step solutions to several of the problems below.

1. Solve $x^2 = 64$.

2. Solve $49x^2 - 25 = 0$.

3. Solve $6x^2 + 49 = 0$.

4. Solve $36(x + 1)^2 - 7 = 0$.

5. Solve $9(x - 3)^2 + 1 = 0$.

6. Find the missing term in the expression $(x + 4)^2 = x^2 + 8x + \square$.

7. The missing terms in the expressions $x^2 - 8x + \square = (\quad)^2$ are ______ and ______, respectively.

8. To solve the equation $8x^2 - 32x = -5$ by completing the square, the first step will be to divide each term by ______.

9. To solve the equation $x^2 - 8x = -15$ by completing the square, one has to add ______ to both sides of the equation.

10. The solutions of $ax^2 + bx + c = 0$ are ______.

11. Solve $2x^2 - 3x - 2 = 0$.

12. Solve $x^2 = 3x - 2$.

13. Solve $16x = x^2$.

14. Solve $\frac{x^2}{2} + \frac{5}{4}x = -\frac{1}{2}$.

15. Graph $y = 3x^2$.

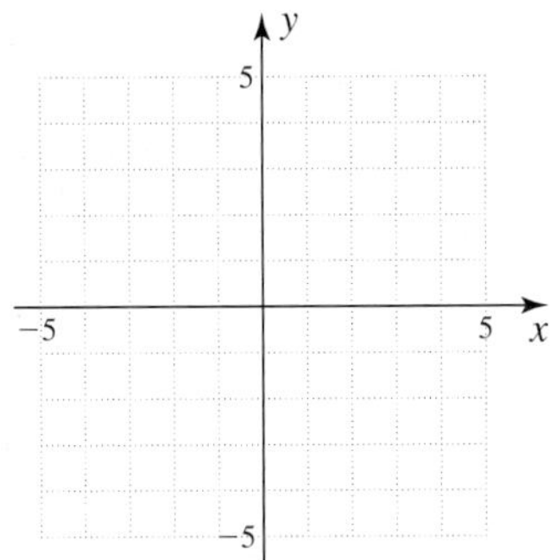

16. Graph $y = (x - 1)^2 + 1$.

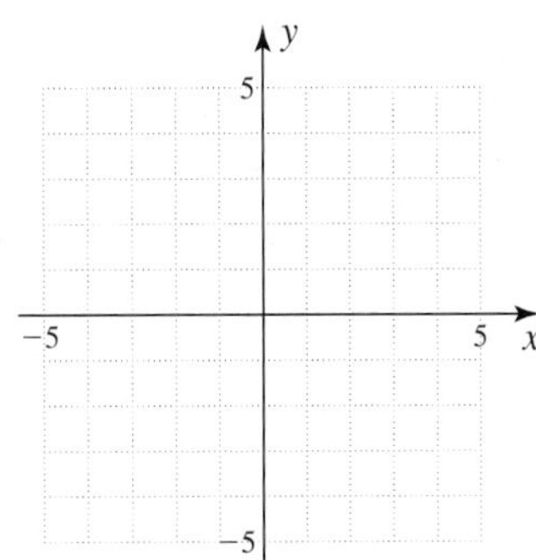

17. Graph $y = -(x - 1)^2 + 1$.

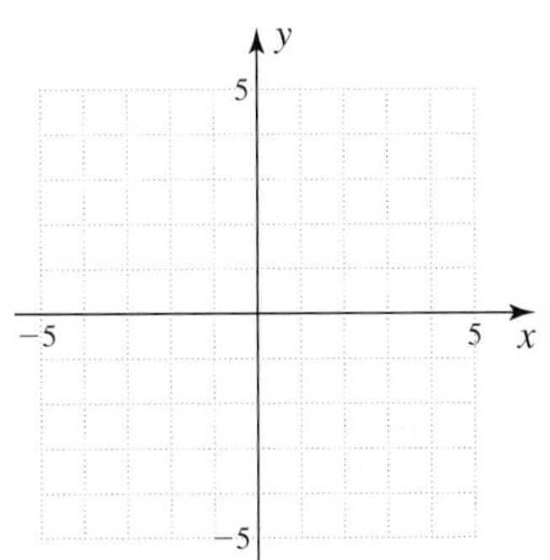

18. Graph $y = -x^2 - 2x + 8$. Label the vertex and intercepts.

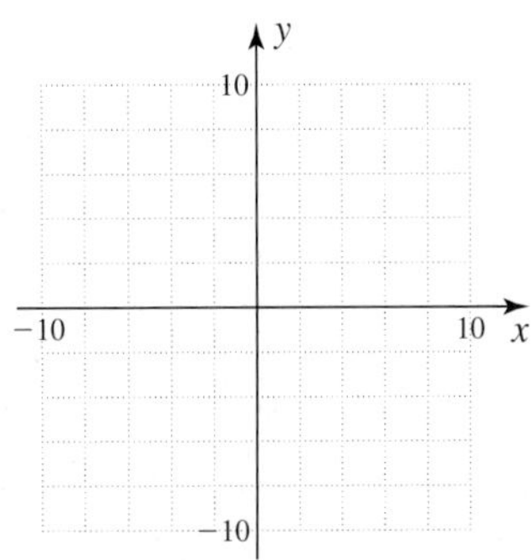

19. Find the length of the hypotenuse of a right triangle if the lengths of the two sides are 2 inches and 5 inches.

20. The formula $d = 5t^2 + v_0t$ gives the distance d (in meters) an object thrown downward with an initial velocity v_0 will have gone after t seconds. How long would it take an object dropped ($v_0 = 0$) from a distance of 180 meters to hit the ground?

21. Find the domain and range of the relation

$$\{(-1, 1), (-2, 1), (-3, 2)\}$$

22. Find the domain and range of the relation

$$\left\{(x, y) \,\middle|\, y = \frac{1}{x - 8}\right\}$$

23. State whether each of the following relations is a function.

a. $\{(2, -4), (-1, 3), (-1, 4)\}$
b. $\{(2, -4), (-1, 3), (1, -3)\}$

24. If $f(x) = x^3 + 2x^2 - x - 1$, find $f(-2)$.

25. The average price $P(n)$ of books depends on the number n of millions of books sold and is given by the function

$$P(n) = 25 - 0.4n \text{ (dollars)}$$

Find the average price of a book when 20 million copies are sold.

Answers to Practice Test Chapter 9

Answer	If You Missed	Review		
	Question	Section	Examples	Page
1. ± 8	1	9.1	1	685–686
2. $\pm\frac{5}{7}$	2	9.1	1, 2	685–687
3. No real-number solution	3	9.1	3	688
4. $-1 \pm \frac{\sqrt{7}}{6} = \frac{-6 \pm \sqrt{7}}{6}$	4	9.1	4	689
5. No real-number solution	5	9.1	5	690
6. 16	6	9.2	1	697
7. 16; $x - 4$	7	9.2	2	698
8. 8	8	9.2	3, 4	699–700
9. 16	9	9.2	3, 4	699–700
10. $x = \frac{-b \pm \sqrt{b^2 - 4ac}}{2a}$	10	9.3	1	707
11. 2; $-\frac{1}{2}$	11	9.3	1	707
12. 1; 2	12	9.3	2	707–708
13. 0; 16	13	9.3	3	708
14. $-\frac{1}{2}$; -2	14	9.3	4	709
15. 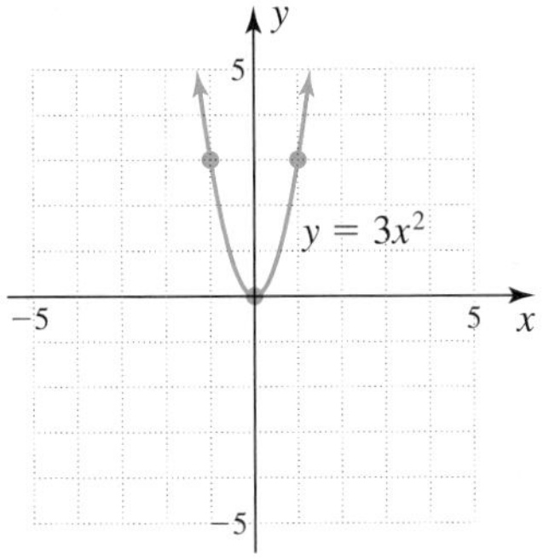	15	9.4	1	715
16. 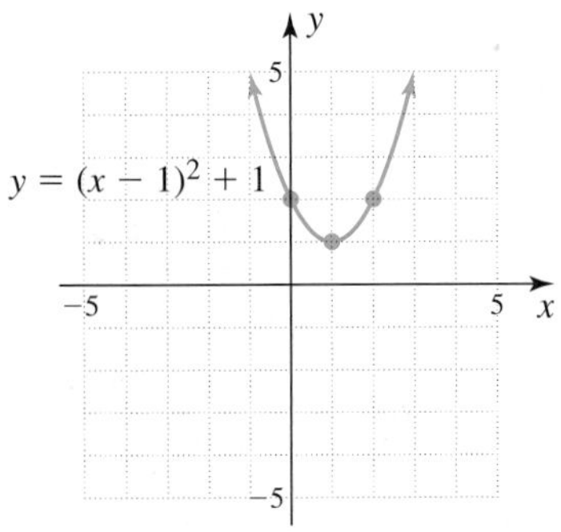	16	9.4	3	717

Answer	If You Missed	Review		
	Question	Section	Examples	Page
17. 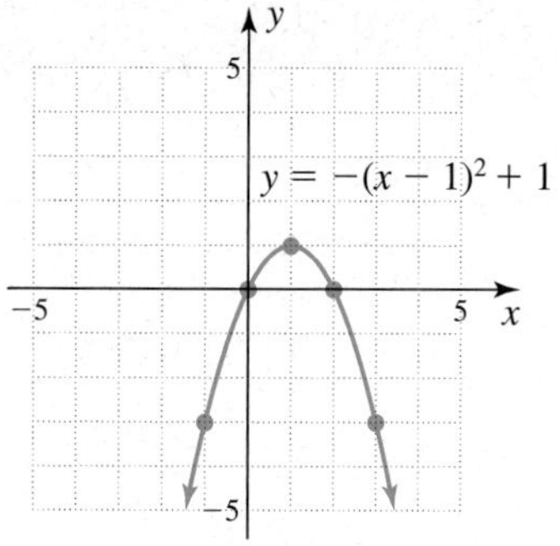	17	9.4	3	717
18. 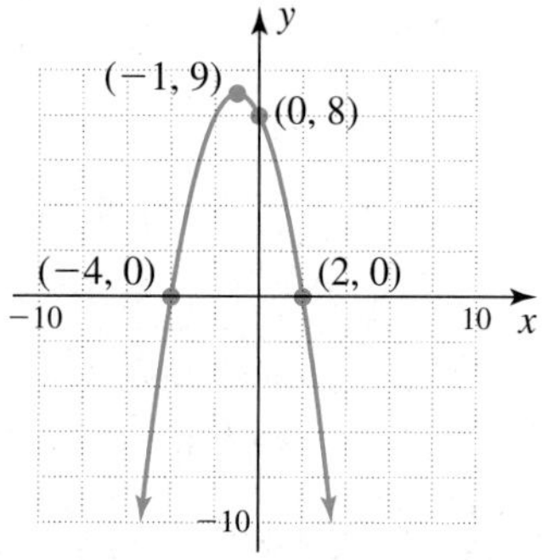	18	9.4	4	719
19. $\sqrt{29}$ in.	19	9.5	1	729
20. 6 sec	20	9.5	2, 3	730, 731
21. $D = \{-1, -2, -3\}$; $R = \{1, 2\}$	21	9.6	1	737
22. Domain: All real numbers except 8; Range: All real numbers except zero.	22	9.6	2	738
23. **a.** Not a function **b.** A function	23	9.6	3	739
24. $f(-2) = 1$	24	9.6	4, 5	739, 740
25. $17	25	9.6	7	740, 741

Cumulative Review Chapters 1–9

1. Add: $-\frac{3}{8} + \left(-\frac{1}{7}\right)$

2. Find: $(-4)^4$

3. Divide: $-\frac{3}{8} \div \left(-\frac{5}{24}\right)$

4. Evaluate $y \div 4 \cdot x - z$ for $x = 4, y = 16, z = 2$.

5. Simplify: $2x - (x + 4) - 2(x + 3)$

6. Write in symbols: The quotient of $(x - 5y)$ and z

7. Solve for x: $5 = 3(x - 1) + 5 - 2x$

8. Solve for x: $\frac{x}{8} - \frac{x}{9} = 1$

9. Graph: $-\frac{x}{4} + \frac{x}{8} \geq \frac{x - 8}{8}$

10. Graph the point $C(-1, -2)$.

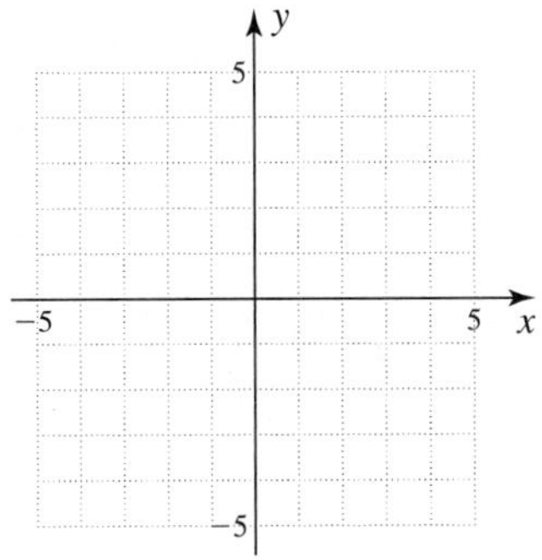

11. Determine whether the ordered pair $(-2, -3)$ is a solution of $2x - y = -1$.

12. Find x in the ordered pair $(x, 2)$ so that the ordered pair satisfies the equation $2x - 3y = -10$.

13. Graph: $2x + y = 4$

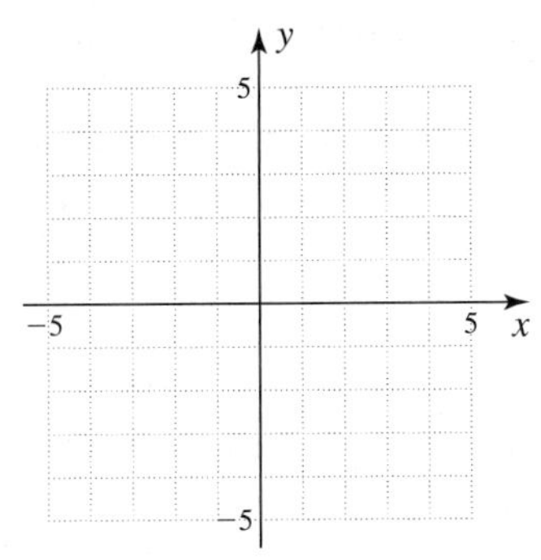

14. Graph: $2y - 8 = 0$

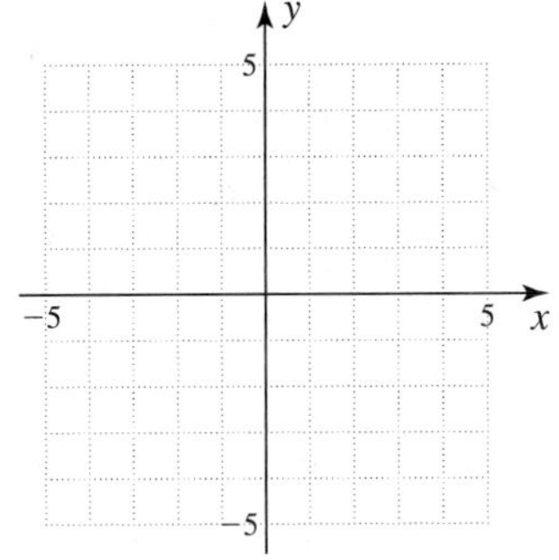

15. Find the slope of the line passing through the points $(-5, -8)$ and $(6, 6)$.

16. What is the slope of the line $8x - 4y = -13$?

17. Find the pair of parallel lines.

(1) $8y + 12x = 7$

(2) $12x - 8y = 7$

(3) $-2y = -3x + 7$

18. Simplify: $(3x^3y^{-5})^4$

19. Write in scientific notation: 0.0050

20. Divide and express the answer in scientific notation: $(1.65 \times 10^{-4}) \div (1.1 \times 10^3)$

21. Expand: $\left(2x^2 - \frac{1}{5}\right)^2$

22. Divide $(2x^3 + 5x^2 - 4x - 2)$ by $(x + 3)$.

23. Factor completely: $x^2 - 8x + 15$

24. Factor completely: $12x^2 - 25xy + 12y^2$

25. Factor completely: $25x^2 - 36y^2$

26. Factor completely: $-5x^4 + 80x^2$

27. Factor completely: $3x^3 - 3x^2 - 18x$

28. Factor completely: $4x^2 + 12x + 5x + 15$

29. Factor completely: $25kx^2 + 20kx + 4k$

30. Solve for x: $3x^2 + 13x = 10$

31. Write $\frac{6x}{5y}$ with a denominator of $10y^3$.

32. Reduce to lowest terms: $\dfrac{-4(x^2 - y^2)}{4(x - y)}$

33. Reduce to lowest terms: $\dfrac{x^2 - 4x - 21}{7 - x}$

34. Multiply: $(x - 4) \cdot \dfrac{x - 3}{x^2 - 16}$

35. Divide: $\dfrac{x + 2}{x - 2} \div \dfrac{x^2 - 4}{2 - x}$

36. Add: $\dfrac{5}{4(x + 1)} + \dfrac{7}{4(x + 1)}$

37. Subtract: $\dfrac{x + 7}{x^2 + x - 56} - \dfrac{x + 8}{x^2 - 49}$

38. Simplify: $\dfrac{\frac{1}{x} + \frac{3}{2x}}{\frac{2}{3x} - \frac{1}{4x}}$

39. Solve for x: $\dfrac{x}{x - 3} + 6 = \dfrac{5x}{x - 3}$

40. Solve for x: $\dfrac{x}{x^2 - 16} + \dfrac{4}{x - 4} = \dfrac{1}{x + 4}$

41. Solve for x: $\dfrac{x}{x + 4} - \dfrac{1}{5} = -\dfrac{4}{x + 4}$

42. Solve for x: $2 + \dfrac{8}{x - 2} = \dfrac{32}{x^2 - 4}$

43. A bus travels 250 miles on 10 gallons of gas. How many gallons will it need to travel 725 miles?

44. Solve for x: $\dfrac{x - 8}{4} = \dfrac{6}{5}$

45. Janet can paint a kitchen in 4 hours and James can paint the same kitchen in 5 hours. How long would it take for both working together to paint the kitchen?

46. Find an equation of the line that passes through the point $(-2, -4)$ and has slope $m = 6$.

47. Find an equation of the line having slope -2 and y-intercept -5.

48. Graph: $3x - 2y < -6$

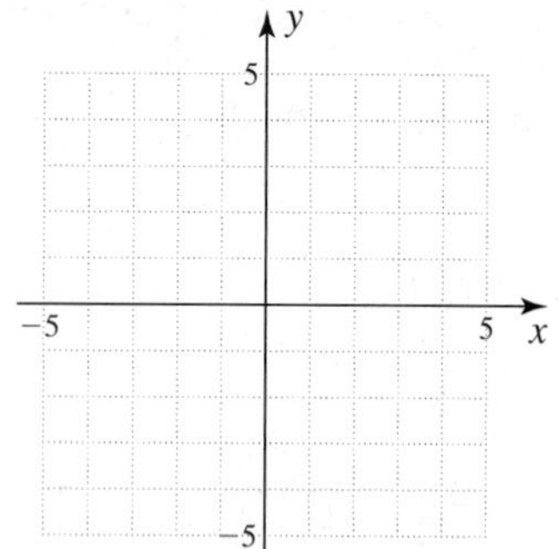

49. Graph: $-y \geq -5x + 5$

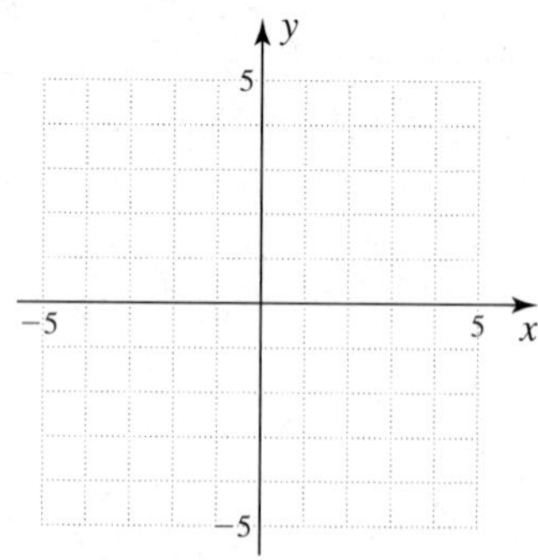

50. An enclosed gas exerts a pressure P on the walls of the container. This pressure is directly proportional to the temperature T of the gas. If the pressure is 7 pounds per square inch when the temperature is 490°F, find k.

51. If the temperature of a gas is held constant, the pressure P varies inversely as the volume V. A pressure of 1960 pounds per square inch is exerted by 7 cubic feet of air in a cylinder fitted with a piston. Find k.

52. Graph the system and find the solution if possible:

$$x + 3y = -12$$
$$2y - x = -2$$

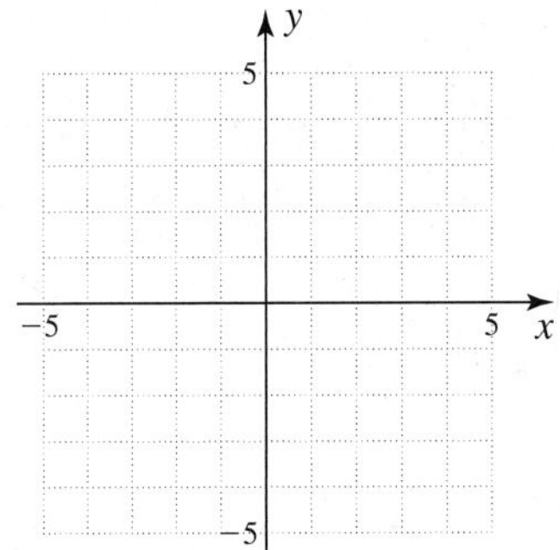

53. Graph the system and find the solution if possible:

$$y + 3x = 3$$
$$3y + 9x = 18$$

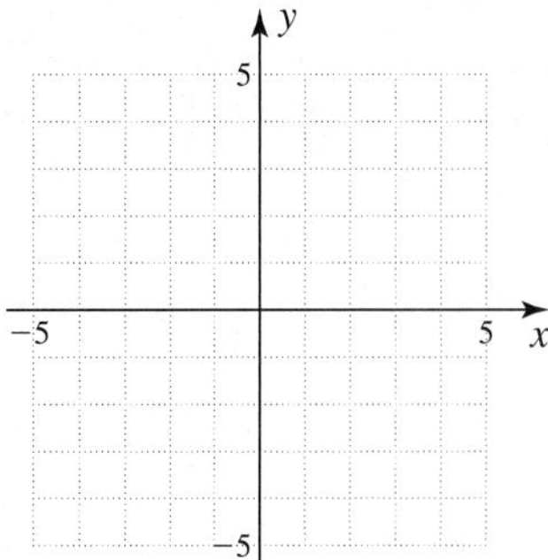

54. Solve (if possible) by substitution:

$$x + 3y = 10$$
$$4x + 12y = 42$$

55. Solve (if possible) by substitution:

$$x - 3y = 3$$
$$-3x + 9y = -9$$

56. Solve the system (if possible):

$$3x - 4y = 3$$
$$2x - 3y = 3$$

57. Solve the system (if possible):

$$4x + 3y = 18$$
$$8x + 6y = -3$$

58. Solve the system (if possible):

$$3y + x = -13$$
$$-4x - 12y = 52$$

59. Kaye has \$2.50 in nickels and dimes. She has twice as many dimes as nickels. How many nickels and how many dimes does she have?

60. The sum of two numbers is 170. Their difference is 110. What are the numbers?

61. Evaluate: $\sqrt[5]{243}$

62. Simplify: $\sqrt[4]{(-5)^4}$

63. Simplify: $\sqrt{\frac{5}{243}}$

64. Rationalize the denominator: $\frac{\sqrt{2}}{\sqrt{7j}}$

65. Add: $\sqrt{32} + \sqrt{18}$

66. Perform the indicated operations: $\sqrt[3]{3x}\left(\sqrt[3]{9x^2} - \sqrt[3]{16x}\right)$

67. Find the product: $\left(\sqrt{125} + \sqrt{63}\right)\left(\sqrt{245} + \sqrt{175}\right)$

68. Reduce: $\frac{20 - \sqrt{32}}{4}$

69. Solve: $\sqrt{x + 4} = -3$

70. Solve: $\sqrt{x - 4} - x = -4$

71. Solve for x: $9x^2 - 4 = 0$

72. Solve for x: $64x^2 + 9 = 0$

73. Solve for x: $36(x - 4)^2 - 7 = 0$

74. Find the missing term: $(x + 2)^2 = x^2 + 4x + $ ____

75. Find the number that must divide each term in the equation so that the equation can be solved by completing the square: $4x^2 + 8x = 14$

76. Find the number that must be added to both sides of the equation so that the equation can be solved by the method of completing the square: $x^2 + 4x = 5$

77. Solve for x: $jx^2 + kx + m = 0$

78. Solve for x: $2x^2 + 3x - 9 = 0$

79. Solve for x: $16x = x^2$

80. Graph: $y = -(x + 3)^2 + 2$

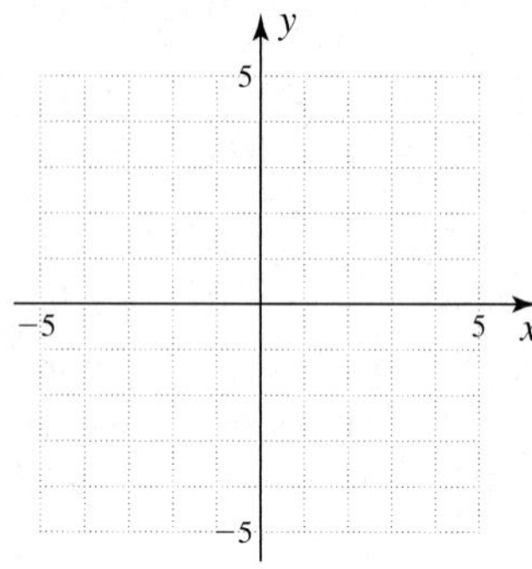

81. The distance d (in meters, m) traveled by an object thrown downward with an initial velocity of v_0 after t seconds is given by the formula $d = 5t^2 + v_0 t$. Find the number of seconds it takes an object to hit the ground if the object is dropped ($v_0 = 0$) from a height of 20 meters.

Selected Answers

The brackets preceding answers for the Chapter Review Exercises indicate the Chapter, Section, and Objective for you to review for further study. For example, [3.4C] appearing before answers means those exercises correspond to Chapter 3, Section 4, Objective C.

Chapter R

Exercises R.1

1. $\frac{28}{1}$ **3.** $\frac{-42}{1}$ **5.** $\frac{0}{1}$ **7.** $\frac{-1}{1}$ **9.** 3 **11.** 42 **13.** 25 **15.** 21 **17.** 32 **19.** 6 **21.** 14 **23.** 30 **25.** 3 **27.** 1 **29.** 2 **31.** $\frac{5}{4}$ **33.** $\frac{1}{4}$ **35.** $\frac{7}{3}$ **37.** $\frac{2}{3}$ **39.** $\frac{4}{1} = 4$ **41. (a)** $\frac{1000}{45}$; **(b)** $\frac{200}{9}$; **(c)** $22\frac{2}{9}$; **(d)** $\frac{1000}{24} = \frac{125}{3} = 41\frac{2}{3}$ **43.** $\frac{41}{100}$ **45. (a)** $\frac{2}{6}$; **(b)** 2 pieces; **(c)** $\frac{2}{6} = \frac{1}{3}$; **(d)** They ate the same amount. **47.** $\frac{1}{8}$ **49.** $\frac{5}{8}$ **51. (a)** 2880 seconds; **(b)** 120 shots; **(c)** 24 **53.** $\frac{2}{15}$ **55.** Nat'l & Int'l News, $\frac{1}{2}$ **57.** $\frac{7}{30}$ **59.** Most: News & Beyond the Bay, $\frac{13}{30}$; Least: Traffic, $\frac{1}{10}$

Exercises R.2

1. $\frac{14}{9}$ **3.** $\frac{7}{5}$ **5.** $\frac{2}{1}$ or 2 **7.** $\frac{8}{1}$ or 8 **9.** $\frac{39}{7}$ **11.** $\frac{35}{3}$ **13.** $\frac{2}{3}$ **15.** $\frac{3}{2}$ **17.** $\frac{16}{5}$ **19.** $\frac{2}{5}$ **21.** $\frac{3}{5}$ **23.** $\frac{1}{1}$ or 1 **25.** $\frac{13}{8}$ **27.** $\frac{17}{15}$ **29.** $\frac{23}{6}$ **31.** $\frac{31}{10}$ **33.** $\frac{193}{28}$ **35.** $\frac{3}{8}$ **37.** $\frac{1}{6}$ **39.** $\frac{11}{20}$ **41.** $\frac{34}{75}$ **43.** $\frac{9}{20}$ **45.** $\frac{5}{4}$ **47.** $\frac{11}{18}$ **49.** $\frac{37}{15}$ **51.** 75 lb **53.** 5600 **55.** $16\frac{5}{8}$ **57.** $\frac{3}{10}$ **59.** $7\frac{7}{10}$ **61.** 4 pkg of hot dogs, 5 pkg of buns **67.** 8 **69.** $\frac{1}{5}$

Exercises R.3

1. $4 + \frac{7}{10}$ **3.** $5 + \frac{6}{10} + \frac{2}{100}$ **5.** $10 + 6 + \frac{1}{10} + \frac{2}{100} + \frac{3}{1000}$ **7.** $40 + 9 + \frac{1}{100} + \frac{2}{1000}$ **9.** $50 + 7 + \frac{1}{10} + \frac{4}{1000}$ **11.** $\frac{9}{10}$ **13.** $\frac{3}{50}$ **15.** $\frac{3}{25}$ **17.** $\frac{27}{500}$ **19.** $\frac{213}{100}$ **21.** \$989.07 **23.** 919.154 **25.** 182.103 **27.** 4.077 **29.** 26.85 **31.** \$2.38 **33.** 3.024 **35.** 6.844 **37.** 9.0946 **39.** 12.0735 **41.** 5.6396 **43.** 95.7 **45.** 0.024605 **47.** 12.90516 **49.** 0.002542 **51.** 0.6 **53.** 6.4 **55.** 1700 **57.** 80 **59.** 0.046 **61.** 0.2 **63.** 0.875 **65.** 0.1875 **67.** $0.\overline{2}$ **69.** $0.\overline{54}$ **71.** $0.\overline{27}$ **73.** $0.1\overline{6}$ **75.** $1.\overline{1}$ **77.** 0.33 **79.** 0.05 **81.** 3 **83.** 0.118 **85.** 0.005 **87.** 5% **89.** 39% **91.** 41.6% **93.** 0.3% **95.** 100% **97.** $\frac{3}{10}$ **99.** $\frac{3}{50}$ **101.** $\frac{7}{100}$ **103.** $\frac{9}{200}$ **105.** $\frac{1}{75}$ **107.** 60% **109.** 50% **111.** $83\frac{1}{3}$% **113.** 50% **115.** $133\frac{1}{3}$% **117.** 27.6 **119.** 26.7467 **121.** 35.250 **123.** 52.38 **125.** 74.84601 **127.** 0.955 **129.** $\frac{1}{2}$; 0.5; 50% **131.** 70% **133. (a)** 40%; **(b)** $\frac{2}{5}$ **135. (a)** $\frac{49}{100}$; **(b)** 0.49 **137.** 5% **139.** 1% **141.** 30

Chapter 1

Exercises 1.1

1. $a + c$ **3.** $3x + y$ **5.** $9x + 17y$ **7.** $3a - 2b$ **9.** $-2x - 5$ **11.** $7a$ **13.** $\frac{1}{7}a$ **15.** bd **17.** xyz **19.** $-b(c + d)$ **21.** $(a - b)x$ **23.** $(x - 3y)(x + 7y)$ **25.** $(c - 4d)(x + y)$ **27.** $\frac{y}{3x}$ **29.** $\frac{2b}{a}$ **31.** $\frac{a}{x + y}$ **33.** $\frac{a - b}{c}$ **35.** $\frac{y}{x}$ **37.** $\frac{p - q}{p + q}$ **39.** $\frac{x + 2y}{x - 2y}$ **41.** 16 **43.** 61 **45.** 9 **47.** 3 **49.** 56 **51.** 6 **53.** 50 **55.** 1 **57.** $\frac{M}{m}$ **59.** $\frac{A}{50}$ **61.** $V = IR$ **63.** $P = P_A + P_B + P_C$ **65.** $D = RT$ **67.** $E = mc^2$ **69.** $c^2 = a^2 + b^2$ **71.** Multiplication **75.** $a + b$ **77.** ab or $a \cdot b$ **79.** evaluating **81.** $2x - y$ **83.** $7x + 4y$ **85.** 5 **87.** $\frac{1}{3}$ **89.** 0 **91.** 0

Exercises 1.2

1. −4 **3.** 49 **5.** $-\frac{7}{3}$ **7.** 6.4 **9.** $-3\frac{1}{7}$ **11.** −0.34 **13.** $0.\overline{5}$ **15.** $-\sqrt{7}$ **17.** $-\pi$ **19.** 2 **21.** 48 **23.** 3 **25.** $\frac{4}{5}$ **27.** 3.4 **29.** $1\frac{1}{2}$ **31.** $-\frac{3}{4}$ **33.** $-0.\overline{5}$ **35.** $-\sqrt{3}$ **37.** $-\pi$ **39.** Natural, whole, integer, rational, real **41.** Rational, real **43.** Whole, integer, rational, real **45.** Rational, real **47.** Rational, real **49.** Irrational, real **51.** Rational, real **53.** Rational, real **55.** 8 **57.** 8 **59.** 0, 8 **61.** $-5, \frac{1}{5}, 0, 8, 0.\overline{1}, 3.666\ldots$ **63.** True **65.** False; $|0| = 0$ **67.** True **69.** False; $\frac{3}{5}$ is not an integer. **71.** False; 0.12345 … is not rational. **73.** True **75.** +20 yards **77.** −1312 feet from sea level **79.** −4°F to +45°F **81.** +55% **83.** −3.1% **85.** +1000 **87.** +750 **89.** about −43°F **91.** about −40°F **95.** All numbers with either terminating or repeating decimal representations are rational numbers. **99.** 0 **101.** irrational **103.** $\sqrt{19}$ **105.** $8\frac{1}{4}$ **107.** $0.\overline{4}$ **109.** $-\frac{1}{2}$ **111.** Integer, rational, real **113.** Rational, real **115.** Whole, integer, rational, real **117.** 0 **119.** 6.4 **121.** $\frac{31}{12}$

Exercises 1.3

1. 6 **3.** −4 **5.** 1 **7.** −7 **9.** 0 **11.** 3 **13.** 13 **15.** 2 **17.** −9 **19.** −12 **21.** 3.1 **23.** −4.7 **25.** −5.4 **27.** −8.6 **29.** $\frac{3}{7}$ **31.** $-\frac{1}{2}$ **33.** $-\frac{1}{12}$ **35.** $\frac{7}{12}$ **37.** $-\frac{13}{21}$ **39.** $-\frac{31}{18}$ **41.** −16 **43.** −20 **45.** −6 **47.** 16 **49.** −4 **51.** −2.6 **53.** 5.2 **55.** −3.7 **57.** $\frac{4}{7}$ **59.** $-\frac{29}{12}$ **61.** −7 **63.** 9 **65.** −4 **67.** 3500°C **69.** 57.7°F **71.** 14°C **73.** −380 **75.** −355 **77.** 799 yr **79.** 284 yr **81.** 2262 yr **87.** larger **89.** inverse **91.** −3.1 **93.** −7.7 **95.** −20 **97.** 1.2 **99.** −0.3 **101.** −5.7 **103.** $\frac{14}{5}$ **105.** $\frac{2}{5}$ **107.** $\frac{703}{20}$ or $35\frac{3}{20}$ **109.** $\frac{36}{121}$

Exercises 1.4

1. 36 **3.** −40 **5.** −81 **7.** −36 **9.** −54 **11.** −7.26 **13.** 2.86 **15.** $-\frac{25}{42}$ **17.** $\frac{1}{4}$ **19.** $-\frac{15}{4}$ **21.** −81 **23.** 25 **25.** −125 **27.** 1296 **29.** $-\frac{1}{32}$ **31.** 7 **33.** −5 **35.** −3 **37.** 0 **39.** Undefined **41.** 0 **43.** 5 **45.** 5 **47.** −2 **49.** −6 **51.** $-\frac{21}{20}$ **53.** $\frac{4}{7}$ **55.** $-\frac{5}{7}$ **57.** −0.5 or $-\frac{1}{2}$ **59.** $0.1\overline{6}$ or $\frac{1}{6}$ **61.** 15 calories; gain **63.** 5 calories; loss **65.** 6 min **67.** 15.4 mi/hr each second **69.** \$4.45 **71.** 8710 pounds **73.** 1675 pounds **75.** −1.35 **77.** −3.045 **79.** 3.875 **85.** negative **87.** negative **89.** 0.25 or $\frac{1}{4}$ **91.** −33 **93.** 55 **95.** 64 **97.** −7.04 **99.** $\frac{12}{35}$ **101.** $-\frac{7}{20}$ **103.** $-\frac{1}{2}$ **105.** −13 **107.** 14

Exercises 1.5

1. 26 **3.** 13 **5.** 53 **7.** 5 **9.** 3 **11.** 10 **13.** 7 **15.** 3 **17.** 8 **19.** 10 **21.** 10 **23.** 20 **25.** 27 **27.** −31 **29.** 36 **31.** −5 **33.** −5 **35.** 25 **37.** −15 **39.** −24 **41.** 87 **43. (a)** 144; **(b)** 126

45. (a) 115 lb; (b) 130 lb **47.** (a) \$99.99; (b) \$77.99 **49.** 160 **51.** 5 milligrams **53.** $1\frac{1}{3}$ tablets every 12 hours **57.** PEMDAS **59.** exponents **61.** division **63.** subtraction **65.** 0 **67.** 18 **69.** 38 **71.** 10 **73.** 119 **75.** 1 **77.** 0 **79.** 1

Exercises 1.6

1. Commutative property of addition
3. Commutative property of multiplication
5. Associative property of addition
7. Commutative property of multiplication
9. Associative property of addition
11. Associative property of addition; 5
13. Commutative property of multiplication; 7
15. Commutative property of addition; 6.5
17. Associative property of multiplication; 2 **19.** $13 + 2x$
21. Multiplicative inverse property **23.** Identity element for addition **25.** 0 **27.** 1 **29.** 3 **31.** 5 **33.** a **35.** -2 **37.** -2
39. $24 + 6x$ **41.** $8x + 8y + 8z$ **43.** $6x + 42$ **45.** $ab + 5b$ **47.** $30 + 6b$ **49.** $-4x - 4y$ **51.** $-9a - 9b$ **53.** $-12x - 6$ **55.** $-\frac{3a}{2} + \frac{6}{7}$ **57.** $-2x + 6y$ **59.** $-2.1 - 3y$ **61.** $-4a - 20$ **63.** $-6x - xy$ **65.** $-8x + 8y$ **67.** $-6a + 21b$ **69.** $0.5x + 0.5y - 1.0$ **71.** $\frac{6}{5}a - \frac{6}{5}b + 6$ **73.** $-2x + 2y - 8$ **75.** $-0.3x - 0.3y + 1.8$ **77.** $-\frac{5}{2}a + 5b - \frac{5}{2}c + \frac{5}{2}$ **79.** (a) $4.2x + 465$ (b) 465 million pounds (c) 507 million pounds **81.** $12.6x + 1395 - 4.2x - 465 = 8.4x + 930$ **83.** (a) $A = a(b + c)$; (b) $A_1 = ab$; (c) $A_2 = ac$; (d) The distributive property **85.** (a) $5h - 195$; 145 lb; (b) $3h - 46$; 170 lb **87.** 266 **89.** 276 **93.** No **95.** Yes **97.** $a \cdot (b \cdot c) = (a \cdot b) \cdot c$ **99.** $a \cdot b = b \cdot a$ **101.** $a \cdot 1 = 1 \cdot a = a$ **103.** $\frac{1}{a}$ **105.** $a(b + c) = ab + ac$ **107.** $-4a + 20$ **109.** $4a + 24$ **111.** 7 **113.** $16 + 4x$ **115.** 0 **117.** 1 **119.** 2 **121.** Identity element for multiplication
123. Identity element for addition
125. Associative property for multiplication
127. Associative property for addition
129. Associative property for multiplication; 2
131. Commutative property for multiplication; 1
133. Associative property for addition; 3 **135.** -7 **137.** $2x + 2$

Exercises 1.7

Translate This

1. H **3.** M **5.** N **7.** L **9.** G

1. $11a$ **3.** $-5c$ **5.** $12n^2$ **7.** $-7ab^2$ **9.** $3abc$ **11.** $1.3ab$ **13.** $0.2x^2y - 0.9xy^2$ **15.** $3ab^2c$ **17.** $-6ab + 7xy$ **19.** $\frac{3}{5}a + \frac{4}{7}a^2b$ **21.** $11x$ **23.** $8ab$ **25.** $-7a^2b$ **27.** 0 **29.** 0 **31.** $10xy - 4$ **33.** $-2R + 6$ **35.** $-L - 2W$ **37.** $-3x - 1$ **39.** $\frac{x}{9} + 2$ **41.** $6a + 2b$ **43.** $3x - 4y$ **45.** $x - 5y + 36$ **47.** $-2x^3 + 9x^2 - 3x + 12$ **49.** $x^2 - \frac{2}{5}x + \frac{1}{2}$ **51.** $10a - 18$ **53.** $-23a + 18b + 9$ **55.** $-4.8x + 3.4y + 5$ **57.** (a) 131.6; (b) 122.1 **59.** $F = \frac{n}{4} + 40$; 2 terms **61.** $I = Prt$; 1 term **63.** $P = a + b + c$; 3 terms **65.** $A = \frac{bh}{2}$; 1 term **67.** $K = Cmv^2$; 1 term **69.** $d = 16t^2$; 1 term **71.** $C = 12W - 500$; 2 terms **73.** (a) $C = 55h$; (b) $C = 85h$; (c) $C = 160h$ **75.** $4S$; $P = 4S$ **77.** 18 ft, 6 in. **79.** $2h + 2w + L$ **81.** $(7x + 2)$ in. **87.** $a + b$ **89.** = **91.** $A = LW$; 1 term **93.** $14y - 11$ **95.** $-3ab^3$ **97.** $-3x^2 + 12x + 6$ **99.** 11 **101.** $-2x - 10$

Review Exercises

1. [1.1A] (a) $a + b$; (b) $a - b$; (c) $7a + 2b - 8$ **2.** [1.1A] (a) $3m$; (b) $-mnr$; (c) $\frac{1}{7}m$; (d) $8m$ **3.** [1.1A] (a) $\frac{m}{9}$; (b) $\frac{9}{n}$ **4.** [1.1A] (a) $\frac{m+n}{r}$; (b) $\frac{m+n}{m-n}$ **5.** [1.1B] (a) 12; (b) 6; (c) 36 **6.** [1.1B] (a) 3; (b) 9; (c) 7 **7.** [1.2A] (a) 5; (b) $-\frac{2}{3}$; (c) -0.37 **8.** [1.2B] (a) 8; (b) $3\frac{1}{2}$; (c) -0.76 **9.** [1.2C] (a) Integer, rational, real; (b) Whole, integer, rational, real; (c) Rational, real; (d) Irrational, real; (e) Irrational, real; (f) Rational, real **10.** [1.3A] (a) 2; (b) -0.8; (c) $-\frac{11}{20}$; (d) -2.2; (e) $\frac{1}{4}$ **11.** [1.3B] (a) -20; (b) -2.4; (c) $-\frac{17}{12}$ **12.** [1.3C] (a) 30; (b) -15 **13.** [1.4A] (a) -35; (b) -18.4; (c) -19.2; (d) $\frac{12}{35}$ **14.** [1.4B] (a) 16; (b) -9; (c) $-\frac{1}{27}$; (d) $\frac{1}{27}$ **15.** [1.4C] (a) -4; (b) 2; (c) $-\frac{2}{3}$; (d) $\frac{3}{2}$ **16.** [1.5A] (a) 10; (b) 0 **17.** [1.5B] (a) 8; (b) -41 **18.** [1.5C] 152 **19.** [1.6A] (a) Commutative property of addition; (b) Associative property of multiplication; (c) Associative property of addition **20.** [1.6B] (a) $4x - 4$; (b) $-8x + 2$ **21.** [1.6C, E] (a) 1; (b) $-\frac{3}{4}$; (c) -3.7; (d) $\frac{2}{3}$; (e) 0; (f) -5 **22.** [1.6E] (a) $-3a - 24$; (b) $-4x + 20$; (c) $-x + 4$ **23.** [1.7A] (a) $-5x$; (b) $-10x$; (c) $-8x$; (d) $3a^2b$ **24.** [1.7B] (a) $12a - 5$; (b) $5x - 9$ **25.** [1.7C] (a) $m = \frac{p}{n}(2.75)$; one term; (b) $W = \frac{11h}{2} - 220$; two terms

Chapter 2

Exercises 2.1

1. Yes **3.** Yes **5.** Yes **7.** No **9.** No **11.** $x = 14$ **13.** $m = 19$ **15.** $y = \frac{10}{3}$ **17.** $k = 21$ **19.** $z = \frac{11}{12}$ **21.** $x = \frac{7}{2}$ **23.** $c = 0$ **25.** $x = -3$ **27.** $y = 0$ **29.** $c = -2.5$ **31.** $p = -9$ **33.** $x = -3$ **35.** $m = 6$ **37.** $y = 18$ **39.** $a = -7$ **41.** $c = -8$ **43.** $x = -2$ **45.** $g = -3$ **47.** No solution **49.** All real numbers **51.** $b = 6$ **53.** $p = \frac{20}{3}$ **55.** $r = \frac{9}{8}$ **57.** \$9.41 **59.** 184.7 **61.** 88.1 million tons; 145.9 million tons **63.** (a) \$1534; (b) \$2644 **65.** \$25,190 **67.** (a) No; (b) 110 lb **69.** (a) No; (b) 2 kg **71.** (a) No; (b) 1 kg **73.** (a) Yes; (b) $W = 56.2 + 1.41(70 - 60) = 56.2 + 14.1 = 70.3$ **75.** (a) No; (b) 100 over **77.** No **79.** No **85.** $a + c = b + c$ **87.** No solution **89.** $x = 9$ **91.** $x = 5.7$ **93.** $x = 4$ **95.** $x = \frac{7}{9}$ **97.** $z = 7$ **99.** No solution **101.** No **103.** -20 **105.** $-\frac{1}{2}$ **107.** $\frac{2}{3}$ **109.** 48 **111.** 40

Exercises 2.2

1. $x = 35$ **3.** $x = -8$ **5.** $b = -15$ **7.** $f = 6$ **9.** $v = \frac{4}{3}$ **11.** $x = -\frac{15}{4}$ **13.** $z = 11$ **15.** $x = -7$ **17.** $c = -7$ **19.** $x = 7$ **21.** $y = -\frac{11}{3}$ **23.** $a = -0.6$ **25.** $t = \frac{3}{2}$ **27.** $x = -2.25$ **29.** $C = -8$ **31.** $a = 12$ **33.** $y = -0.5$ **35.** $p = 0$ **37.** $t = -30$ **39.** $x = -0.02$ **41.** $y = 12$ **43.** $x = 21$ **45.** $x = 20$ **47.** $t = 24$ **49.** $x = 1$ **51.** $c = 15$ **53.** $W = 10$ **55.** $x = 4$ **57.** $x = 10$ **59.** $x = 3$ **61.** 12 **63.** 28 **65.** 50% **67.** 150 **69.** 20 **71.** 7.6% **73.** 9.5% **75.** 3.8% **77.** 5.3% **79.** 3.8% **81.** (a) $85h = 3500$; (b) $h = 41\frac{3}{17}$ **83.** (a) $260h = 3500$; (b) $h = 13\frac{6}{13}$ **85.** (a) $450h = 3500$;

(b) $h = 7\frac{7}{9}$ **87.** 30% **91.** Multiply by the reciprocal of $-\frac{3}{4}$; answers may vary. **93.** $\frac{a}{c} = \frac{b}{c}$ **95.** LCM **97.** 20% **99.** $x = 6$ **101.** $y = 30$ **103.** $y = 10$ **105.** $y = 14$ **107.** $x = -14$ **109.** $4x - 24$ **111.** $40 - 5y$ **113.** $54 - 27y$ **115.** $-15x + 20$ **117.** 4 **119.** 3

Exercises 2.3

1. $x = 4$ **3.** $y = 1$ **5.** $z = 2$ **7.** $y = \frac{14}{5}$ **9.** $x = 3$ **11.** $x = -4$ **13.** $v = -8$ **15.** $m = -4$ **17.** $z = -\frac{2}{3}$ **19.** $x = 0$ **21.** $a = \frac{1}{13}$ **23.** $c = 35.6$ **25.** $x = -\frac{11}{80}$ **27.** $x = \frac{9}{2}$ **29.** $x = -20$ **31.** $x = -5$ **33.** $h = 0$ **35.** $w = -1$ **37.** $x = -3$ **39.** $x = \frac{5}{2}$ **41.** All real numbers **43.** $x = 2$ **45.** $x = 10$ **47.** $x = \frac{27}{20}$ **49. (a)** No solution; **(b)** $x = \frac{69}{7}$ **51.** $r = \frac{C}{2\pi}$ **53.** $y = \frac{6 - 3x}{2}$ **55.** $s = \frac{A - \pi r^2}{\pi r}$ **57.** $V_2 = \frac{P_1V_1}{P_2}$ **59.** $H = \frac{f + Sh}{S}$ **61. (a)** $L = \frac{S + 22}{3}$; **(b)** $L = \frac{S + 21}{3}$ **63.** $H = \frac{W + 200}{5}$ **65.** 11 inches **67.** size 8 **69.** 15 hr **71. (a)** 0.723; **(b)** 0.68 **73. (a)** 1.569; **(b)** 1.61 **75.** 20 years **77.** $66\frac{2}{3}$ miles **79. (a)** $S = 1.2C$; **(b)** \$9.60 **83.** $x = 0$ and division by 0 is undefined. **85.** literal **87.** $m = 150$ **89.** $x = -1$ **91.** $x = 1$ **93.** $x = 1$ **95.** $\frac{a + b}{c}$ **97.** $a(b + c)$ **99.** $a - bc$

Exercises 2.4

Translate This

1. D **3.** I **5.** B **7.** C **9.** J

1. 44, 46, 48 **3.** −10, −8, −6 **5.** −13, −12 **7.** 7, 8, 9 **9.** Any consecutive integers **11.** \$20, \$44 **13.** 39 to 94 **15.** \$104, \$43 **17.** 37, 111, 106 **19.** 21, 126 **21.** 38% from home, 13% from work **23.** 5% **25.** 222 feet **27.** 3 sec **29.** 304 and 676 **31.** 10 minutes **33. (a)** 22 mi; **(b)** The limo is cheaper. **35.** 20° **37.** 45° **39.** 42° **41.** 157 million tons **43.** 140 million tons **45.** 5.6 million tons **47.** 236 million tons **49.** 12.1% **51.** 84 **55.** RSTUV **57.** complementary **59.** 81, 83, 85 **61.** Pizza: 230 calories, Shake: 300 calories **63.** $T = \frac{20}{11}$ **65.** $T = 8$ **67.** $x = 8$ **69.** $P = 3500$ **71.** $x = 1000$

Exercises 2.5

Translate This

1. G **3.** A **5.** M **7.** D **9.** E

1. 50 mi/hr **3.** 2 meters/min **5.** $\frac{1}{2}$ hr **7.** 4 hr **9.** 10 mi **11.** 6 hr **13.** 1920 mi **15.** 24 in./sec first 180 sec, 1.2 in./sec last 60 sec **17.** 6 L **19.** 52 lb copper, 28 lb zinc **21.** 20 lb **23.** 32 oz of each **25.** 12 quarts **27.** Impossible; it would be 120% of the concentrate. **29.** \$5000 at 6%, \$15,000 at 8% **31.** \$8000 **33.** \$6000 at 5%, \$4000 at 6% **35.** 20 miles per day **41.** $I = Pr$ **43.** 10 gal **45.** 5 hr **47.** 158 **49.** 100 **51.** $7\frac{2}{3}$

Exercises 2.6

Translate This

1. C **3.** I **5.** F **7.** K **9.** D

1. (a) 120; **(b)** 275 mi; **(c)** $R = \frac{D}{T}$; **(d)** 60 mi/hr **3. (a)** 110 pounds; **(b)** $H = \frac{W + 190}{5}$; **(c)** 78 inches tall (6 ft, 6 in.) **5. (a)** $C = \frac{5}{9}(F - 32)$; **(b)** 43 **7. (a)** $\text{EER} = \frac{\text{Btu}}{w}$; **(b)** 9; **(c)** Btu = (EER)(w); **(d)** 20,000 **9. (a)** $S = C + M$; **(b)** \$67; **(c)** $M = S - C$; **(d)** \$8.25 **11. (a)** 60 cm; **(b)** 90 cm **13. (a)** 62.8 in.; **(b)** $r = \frac{C}{2\pi}$; **(c)** 10 in. **15. (a)** 13.02 m²; **(b)** $W = \frac{A}{L}$; **(c)** 6 m **17.** 35° each **19.** 140° each **21.** 175° each **23.** 70°, 20° **25.** 23°, 67° **27.** 142°, 38° **29.** 69°, 111° **31.** 75 m **33.** 4.5 in. **35. (a)** $A = C - 10$; **(b)** 40 **37. (a)** 13.74 million; **(b)** $t = \frac{N - 9.74}{0.40}$ or $t = \frac{5(N - 9.74)}{2}$; **(c)** 1995 **39. (a)** \$700; **(b)** $x = \frac{A - 420}{28}$; **(c)** 21 yr (2021) **41. (a)** 67 in.; **(b)** $r = \frac{H - 34}{3.3}$; **(c)** 11 in. **43. (a)** 149 cm; **(b)** $t = \frac{H - 62}{2.9}$; **(c)** 31 cm **45.** 90 ft **51.** 50% **53.** $P = 2L + 2W$ **55.** $A = \pi r^2$ **57.** equal **59.** 44°, 46° **61.** 128°, 52° **63.** 46°, 134° **65. (a)** 126.48 cm; **(b)** $h = \frac{H - 71.48}{2.75}$; **(c)** 25 cm **67. (a)** 75 in.²; **(b)** $b = \frac{2A}{h}$; **(c)** 4 in. **69. (a)** 5°C; **(b)** $F = \frac{9}{5}C + 32$; **(c)** 68°F **71.** $x = 4$ **73.** $x = 3$ **75.** $x = -6$

Exercises 2.7

1. < **3.** > **5.** < **7.** > **9.** <

11. $x \le 1$
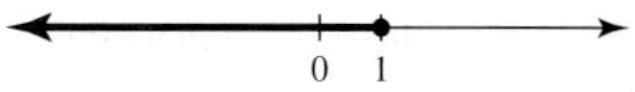

13. $y \le 2$
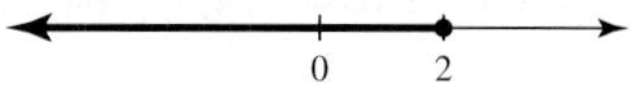

15. $x > 3$

17. $a \le 3$

19. $z \le -4$
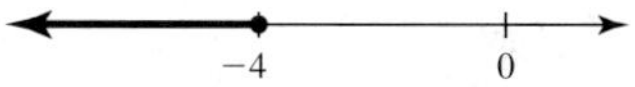

21. $x \le -1$
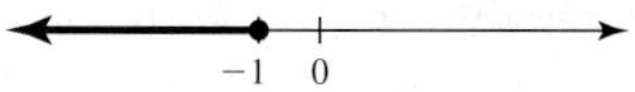

23. $x < 0$
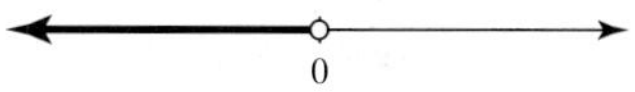

25. $x \le -\frac{11}{80}$

27. $x \ge -20$
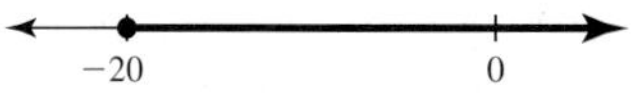

29. $x \ge -2$
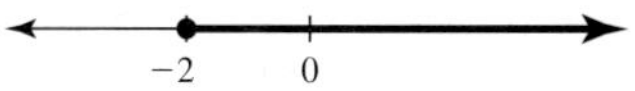

31. $2 < x < 3$ or $(2, 3)$

33. $1 < x < 3$ or $(1, 3)$
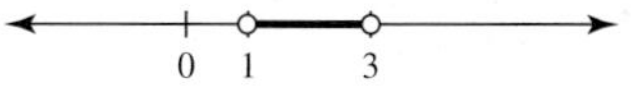

35. $-2 < x < 5$ or $(-2, 5)$

37. $-5 < x < 1$ or $(-5, 1)$

39. $3 \le x < 5$ or $[3, 5)$

41. 20°F < t < 40°F **43.** \$12,000 < s < \$13,000 **45.** $2 \le e \le 7$ **47.** \$3.50 < c < \$4.00 **49.** $a < 41$ ft **51.** $E = 26$ **53.** $N = 5\frac{5}{17} \approx 5$; in 2015 **55.** $J = 5$ ft = 60 in. **57.** $F = S - 3$ **59.** $S = 6$ ft 5 in. = 77 in. **61.** $B > 74$ in. or $B > 6$ ft 2 in. **63.** When multiplying or dividing by a negative number **67.** $x - a < y - a$ **69.** $ax > ay$ **71.** $\frac{x}{a} > \frac{y}{a}$

73. $-3 \le x \le -1$ or $[-3, -1]$
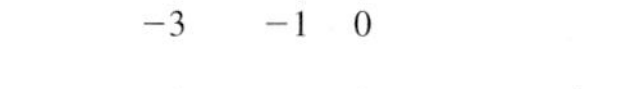

75. $x > 3$
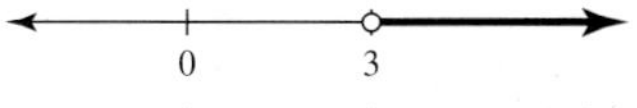

77. $x > 3$
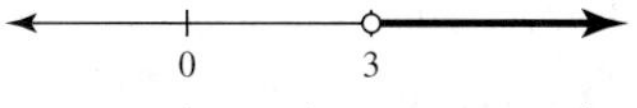

79. $x \ge 2$
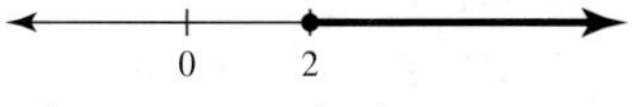

81. $x < 1$
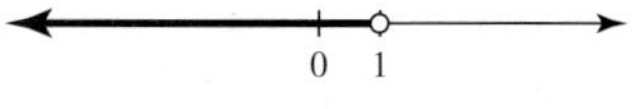

83. < **85.** >

87.

89.

91.

93.

95.

Review Exercises

1. [2.1A] **(a)** No; **(b)** Yes; **(c)** Yes **2.** [2.1B] **(a)** $x = \frac{2}{3}$; **(b)** $x = 1$; **(c)** $x = \frac{2}{3}$ **3.** [2.1B] **(a)** $x = \frac{2}{9}$; **(b)** $x = \frac{4}{7}$; **(c)** $x = \frac{1}{6}$ **4.** [2.1C] **(a)** $x = 5$; **(b)** $x = 0$; **(c)** $x = 3$ **5.** [2.1C] **(a)** No solution; **(b)** $x = -\frac{1}{8}$; **(c)** No solution **6.** [2.1C] **(a)** All real numbers; **(b)** All real numbers; **(c)** All real numbers **7.** [2.2A] **(a)** $x = -15$; **(b)** $x = -14$; **(c)** $x = -2$ **8.** [2.2B] **(a)** $x = 12$; **(b)** $x = 15$; **(c)** $x = 9$ **9.** [2.2C] **(a)** $x = 6$; **(b)** $x = \frac{24}{7}$; **(c)** $x = 20$ **10.** [2.2C] **(a)** $x = 12$; **(b)** $x = 28$; **(c)** $x = 40$ **11.** [2.2C] **(a)** $x = 17$; **(b)** $x = 7$; **(c)** $x = 9$ **12.** [2.2D] **(a)** 20%; **(b)** 10%; **(c)** 20% **13.** [2.2D] **(a)** 50; **(b)** $33\frac{1}{3}$; **(c)** $33\frac{1}{3}$ **14.** [2.3A] **(a)** $x = -3$; **(b)** $x = -4$; **(c)** $x = -5$ **15.** [2.3B] **(a)** $h = \frac{2A}{b}$; **(b)** $r = \frac{C}{2\pi}$; **(c)** $b = \frac{3V}{h}$ **16.** [2.4A] **(a)** 32, 52; **(b)** 14, 33; **(c)** 29, 52 **17.** [2.4B] **(a)** Chicken: 278, pie: 300; **(b)** Chicken: 291, pie: 329; **(c)** Chicken: 304, pie: 346 **18.** [2.4C] **(a)** 35°; **(b)** 30°; **(c)** 25° **19.** [2.5A] **(a)** 4 hours; **(b)** $1\frac{1}{2}$ hours; **(c)** 2 hours **20.** [2.5B] **(a)** 10 pounds; **(b)** 15 pounds; **(c)** 25 pounds **21.** [2.5C] **(a)** \$10,000 at 6%, \$20,000 at 5%; **(b)** \$20,000 at 7%, \$10,000 at 9%; **(c)** \$25,000 at 6%, \$5000 at 10% **22.** [2.6A] **(a)** $m = \frac{C-3}{3.05}$; 8 minutes; **(b)** $m = \frac{C-3}{3.15}$; 10 minutes; **(c)** $m = \frac{C-2}{3.25}$; 6 minutes **23.** [2.6C] **(a)** 100°, 80°; **(b)** Both 60°; **(c)** 65°, 25° **24.** [2.7A] **(a)** $<$; **(b)** $>$; **(c)** $<$

25. [2.7B] **(a)** $x < 3$;

(b) $x < 2$;

(c) $x < 1$

26. [2.7B] **(a)** $x \geq 4$;

(b) $x \geq 2$;

(c) $x \geq 5$

27. [2.7B] **(a)** $x \geq \frac{1}{2}$;

(b) $x \geq \frac{4}{7}$;

(c) $x \geq \frac{5}{3}$

28. [2.7C] **(a)** $-3 < x \leq 2$;

(b) $-3 < x \leq 2$;

(c) $-2 < x \leq 1$

Cumulative Review Chapters 1–2

1. 7 **2.** $9\frac{9}{10}$ **3.** $-\frac{32}{63}$ **4.** 8.2 **5.** -8.64 **6.** -16 **7.** $\frac{21}{5}$ **8.** 69 **9.** Commutative property of multiplication **10.** $30x + 42$ **11.** cd **12.** $-3x - 11$ **13.** $\frac{a-4b}{c}$ **14.** Yes **15.** $x = 13$ **16.** $x = 9$ **17.** $x = 15$ **18.** $x = 8$ **19.** $b = \frac{S}{6a^2}$ **20.** 60 and 95 **21.** \$350 from stocks and \$245 from bonds **22.** 24 hr **23.** \$2000 in bonds and \$3000 in certificates of deposit **24.** $x \geq 6$

Chapter 3

Exercises 3.1

1.

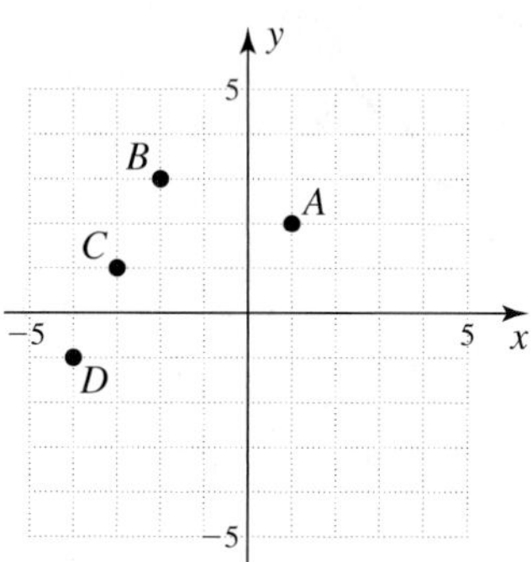

3.

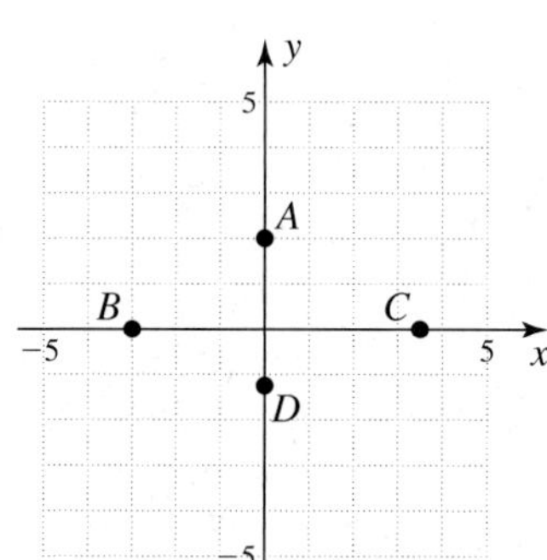

5.

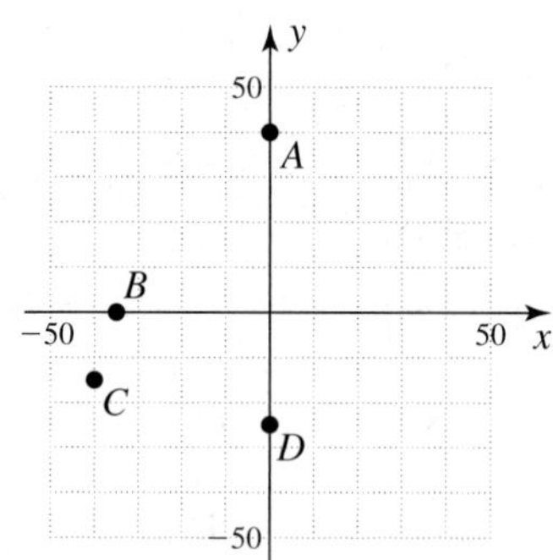

7. $A(0, 2\frac{1}{2})$; $B(-3, 1)$; $C(-2, -2)$; $D(0, -3\frac{1}{2})$; $E(1\frac{1}{2}, -2)$; Quadrant II: B; Quadrant III: C; Quadrant IV; E; y-axis: A, D **9.** $A(10, 10)$; $B(-30, 20)$; $C(-25, -20)$; $D(0, -40)$; $E(15, -15)$; Quadrant I: A; Quadrant II: B; Quadrant III: C; Quadrant IV: E; y-axis: D **11.** (20, 140) **13.** (45, 147) **15.** 2.68 **17.** 1.94 **19.** about 4.61 **21.** 254.1 **23.** 15 **25.** \$950 **27.** \$800 **29.** 70 **31.** 3 A.M. Answers may vary. **33.** 55 **35.** Before 3 A.M.: 195;

after 3 A.M.: 288 **37.** Before 3 A.M.: 162; after 3 A.M.: 13
39. (a) Auto batteries; **(b)** Tires **41.** 26.5%; 1.43% **43.** \$500 **45.** \$25
47. and **49.**

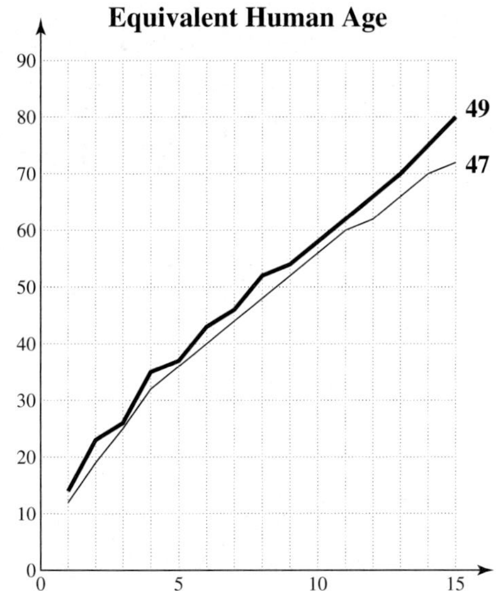

51. 75–100 lb **53.** Less/younger, or more/older **55.** Ages 6, 7, 9, 12, 13, and 15 **57. (a)** About \$120,000; **(b)** About \$50,000; **(c)** About 23 years
59. (a)

Altitude (kilometers)	Temperature Change	Ordered Pair
1	−7°C	(1, −7)
2	−14°C	(2, −14)
3	−21°C	(3, −21)
4	−28°C	(4, −28)
5	−35°C	(5, −35)

(b)

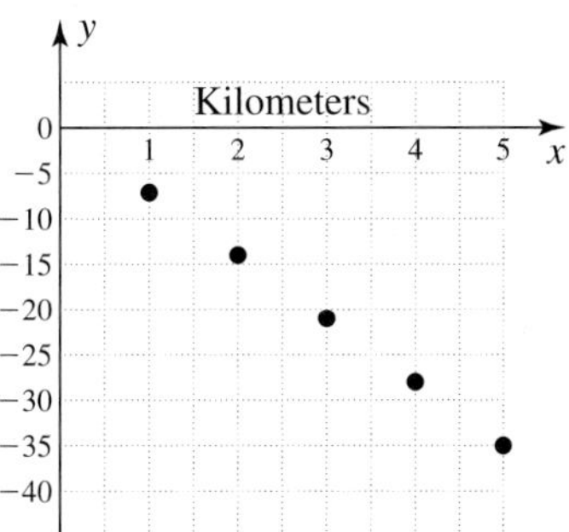

61. 86°F **63.** Less than 10% **65.** 12°F **67.** a is negative, b is positive; no **69.** $a = c$ and $b = d$ **71.** abscissa
73. and **75.**

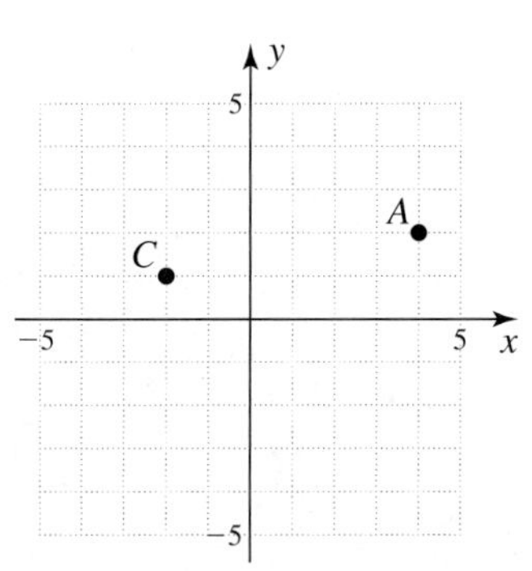

77.

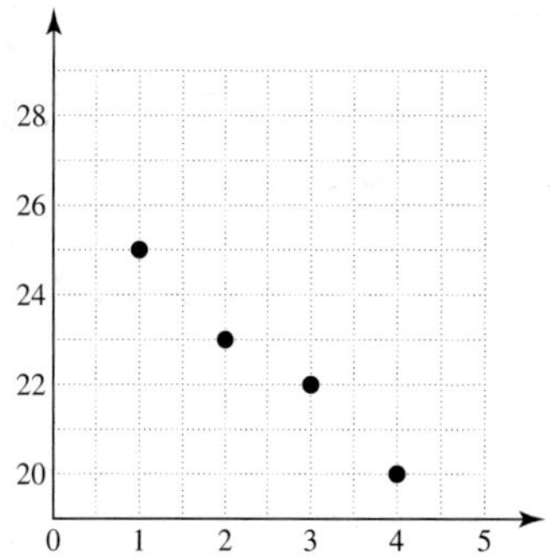

79. 11 human years **81. (a)** Quadrant II; **(b)** Quadrant IV; **(c)** Quadrant I; **(d)** Quadrant III **83.** $x = 3$ **85.** $y = 9$ **87.** $y = 3$

Exercises 3.2

1. Yes **3.** Yes **5.** No **7.** 0 **9.** −2 **11.** −3 **13.** 3 **15.** 2
17. $2x + y = 4$

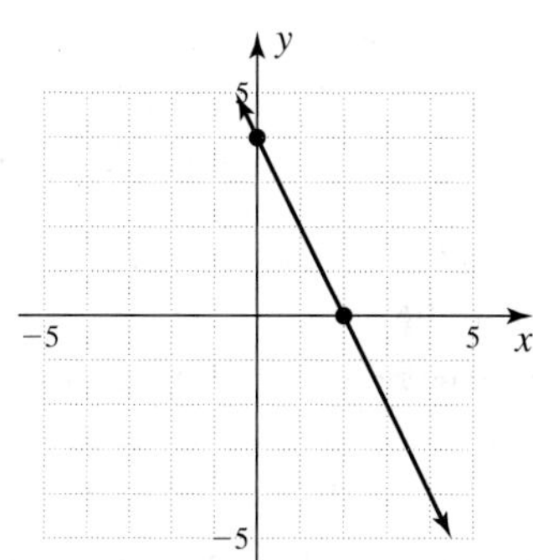

19. $-2x - 5y = -10$

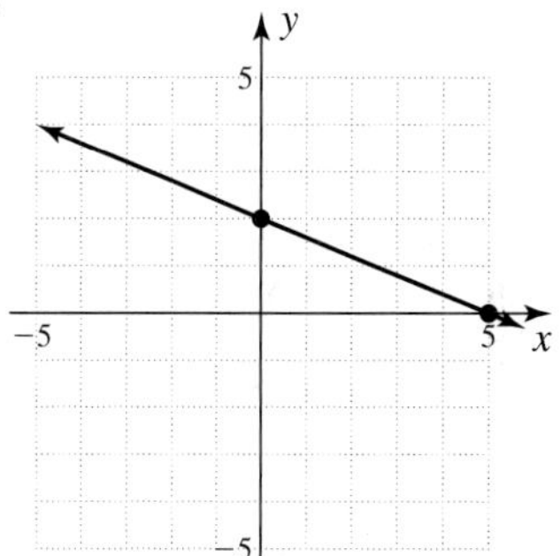

21. $y + 3 = 3x$

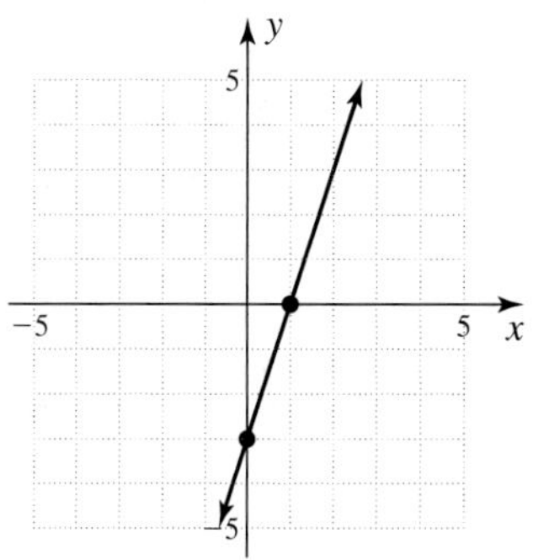

23. $6 = 3x - 6y$

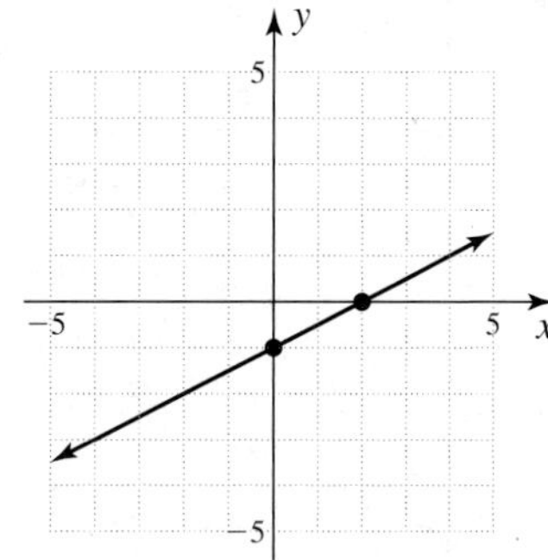

25. $-3y = 4x + 12$

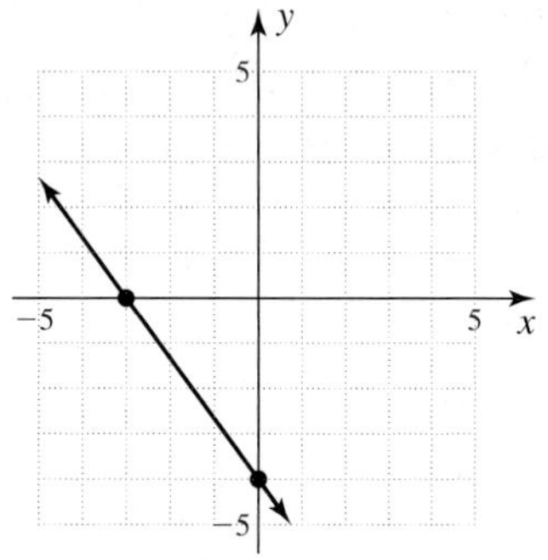

27. $-2y = -x + 4$

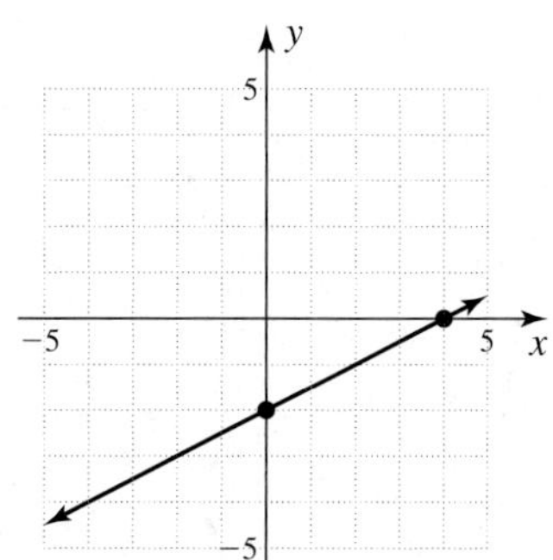

29. $-3y = -6x + 3$

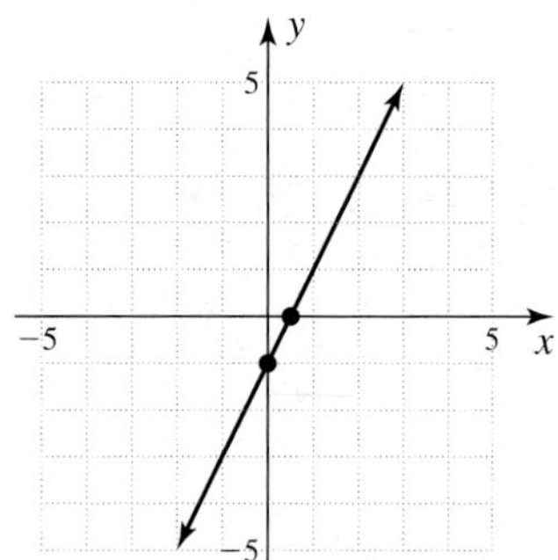

31. $y = 2x + 4$

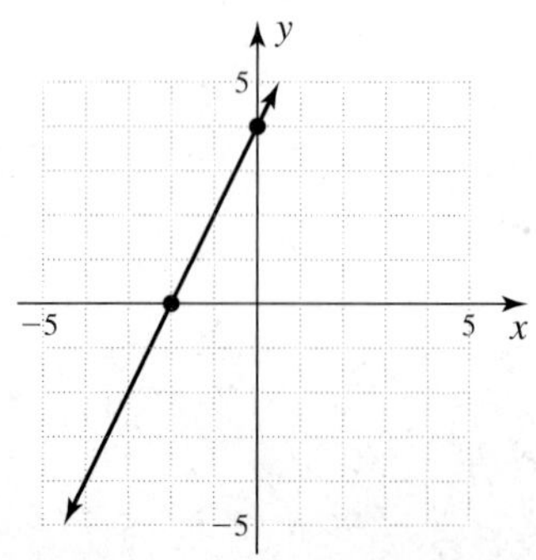

33. $y = -2x + 4$

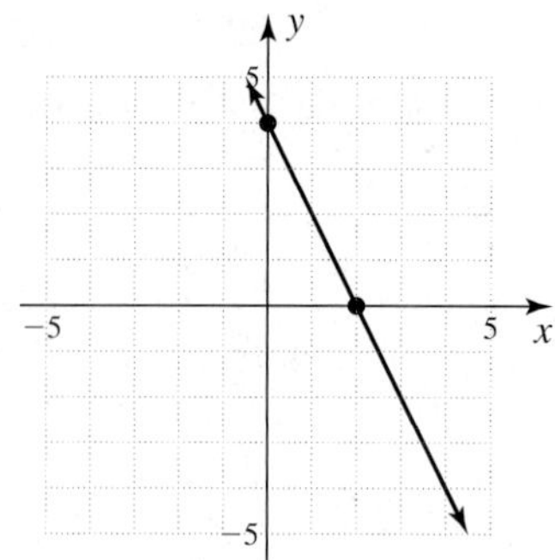

35. $y = -3x - 6$

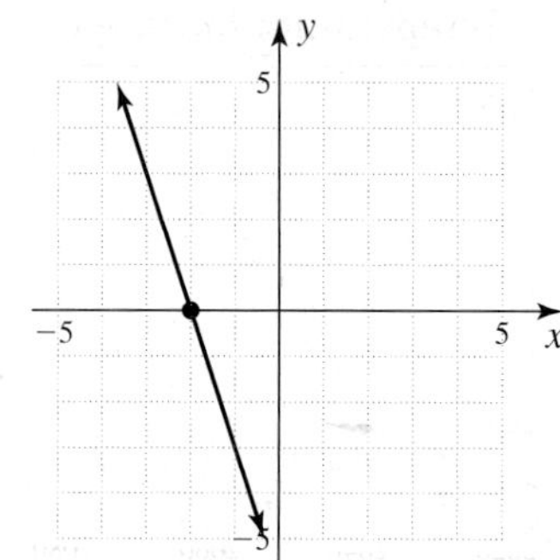

37. $y = \frac{1}{2}x - 2$

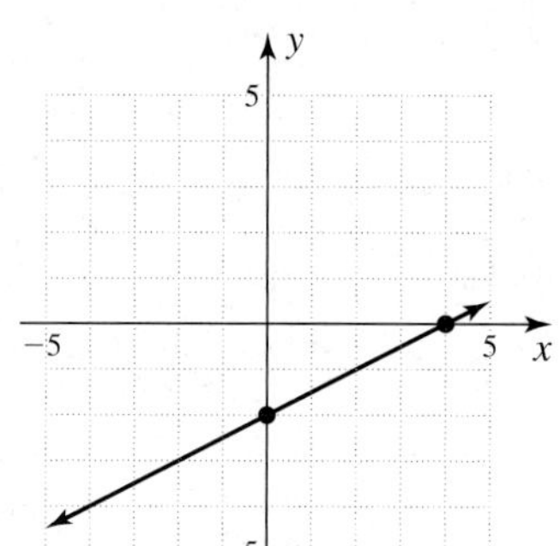

39. $y = -\frac{1}{2}x - 2$

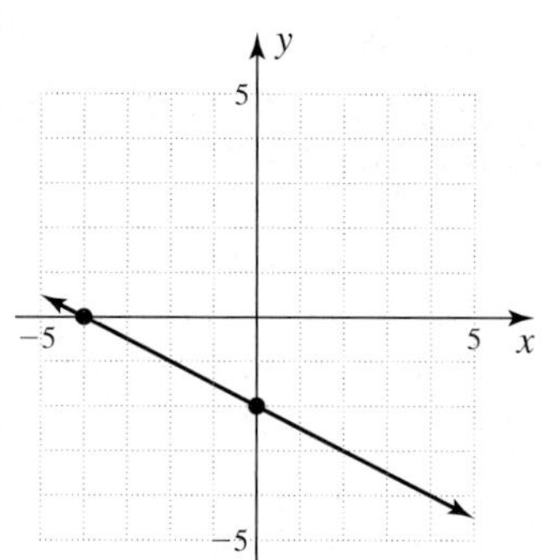

41. **(a)** -9; **(b)** 2;

(c)

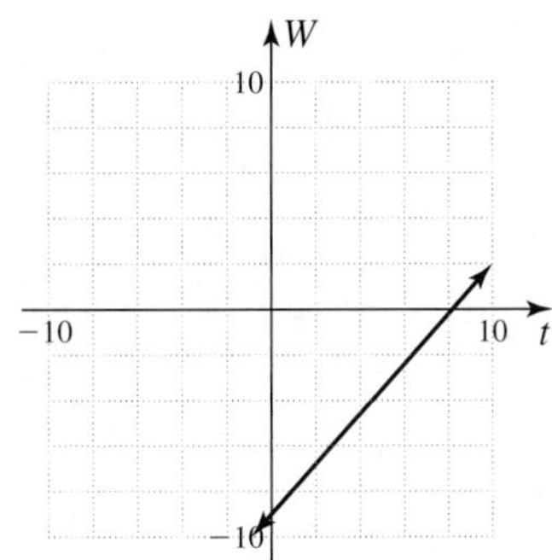

43. **(a)** (2040, 1); **(b)** (2100, 2.5);

(c)

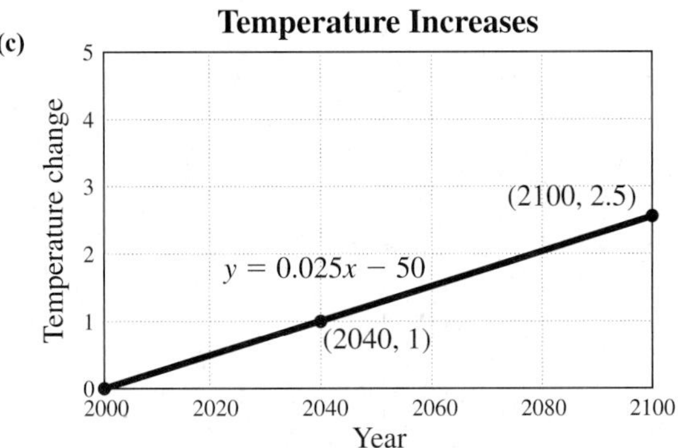

45. **(a)** 56°F; **(b)** 20°F; **(c)** 60°F;

(d)

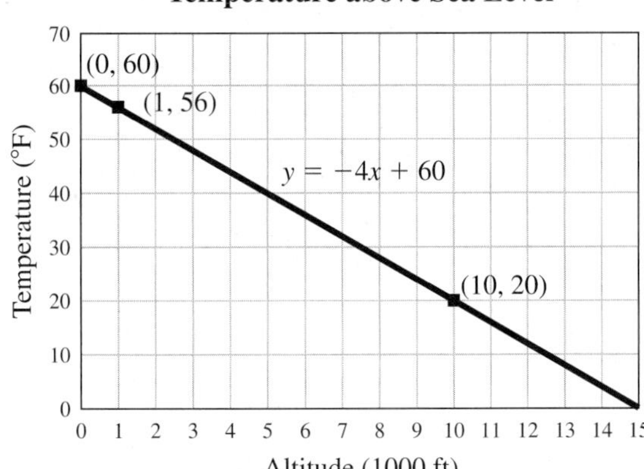

(e) The temperature is negative.

47. **(a)** \$1400; **(b)** \$2200; **(c)** \$200;

(d)

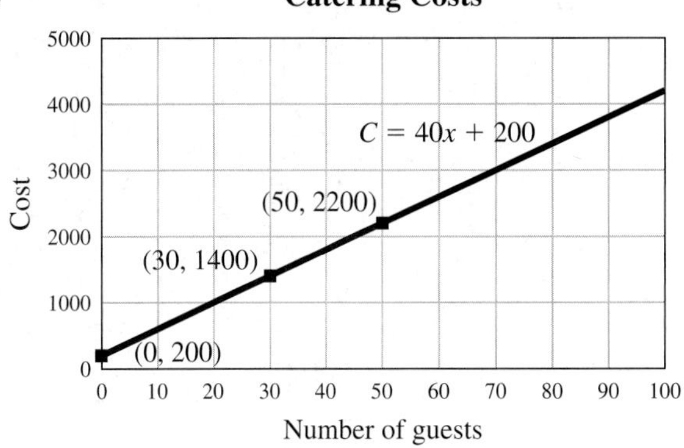

49. Yes **51.** No; 16 in. **53.** Yes **59.** straight **61.** **(a)** No; **(b)** No

63. $y = -3x + 6$

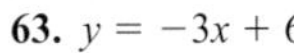

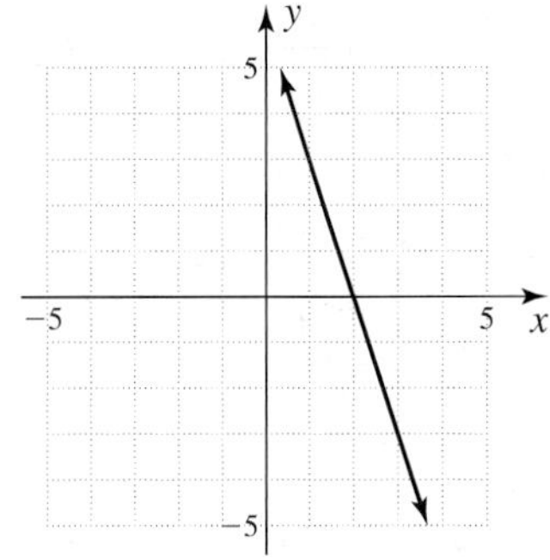

65. **(a)** -18; **(b)** -5;

(c)

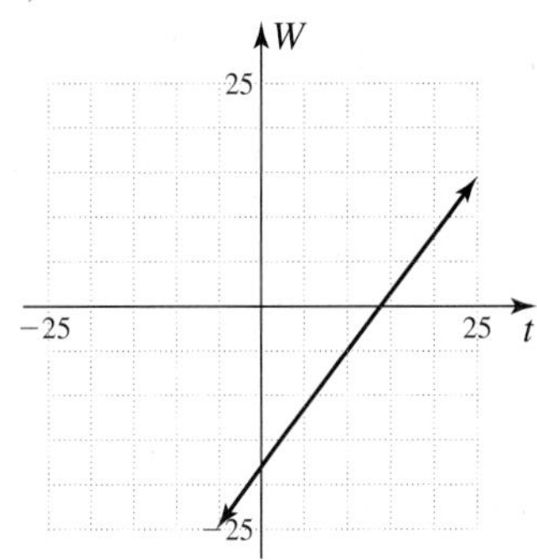

67. $x = -2$ **69.** $t = 0$

Exercises 3.3

1. $x + 2y = 4$

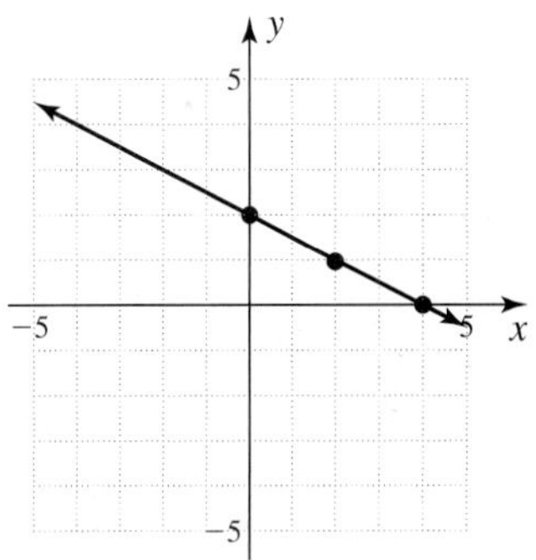

3. $-5x - 2y = -10$

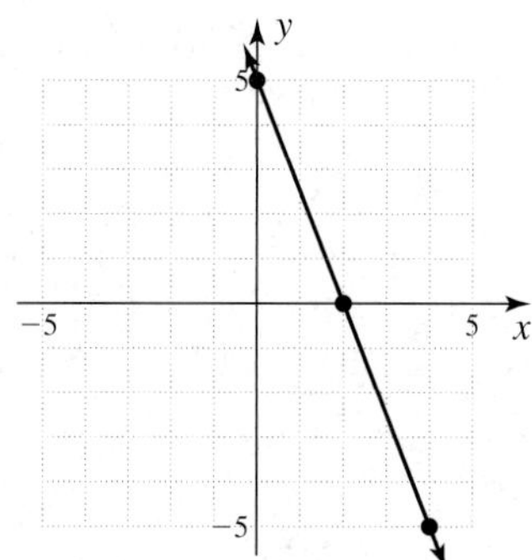

5. $y - 3x - 3 = 0$

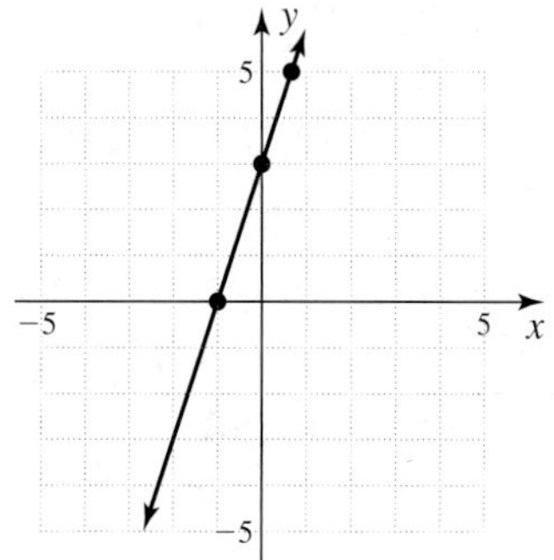

7. $6 = 6x - 3y$

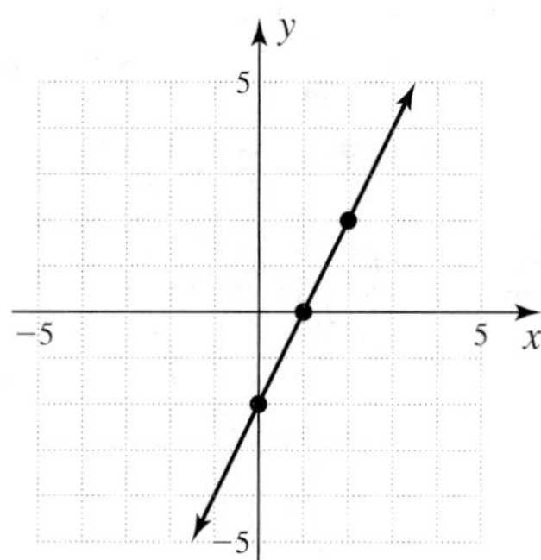

9. $3x + 4y + 12 = 0$

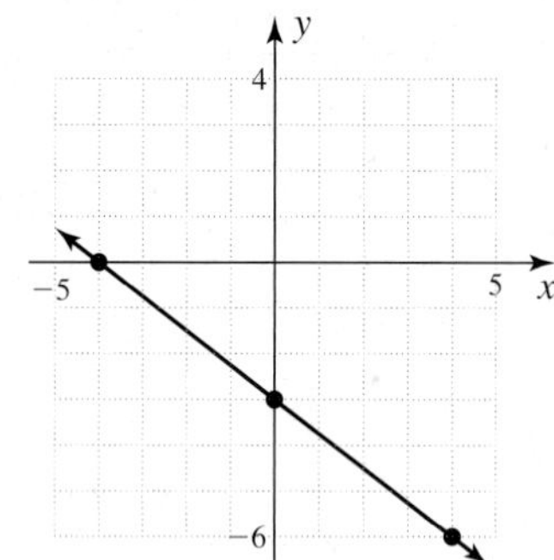

11. $3x + y = 0$

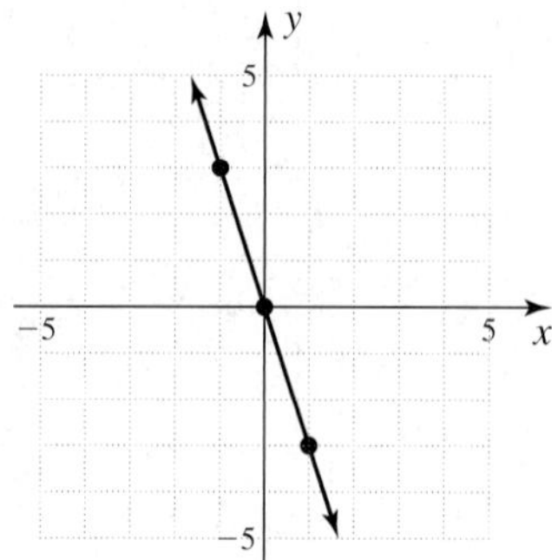

13. $2x + 3y = 0$

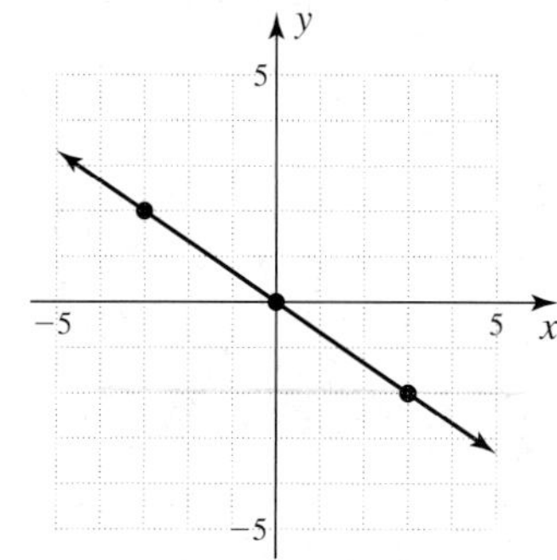

15. $-2x + y = 0$

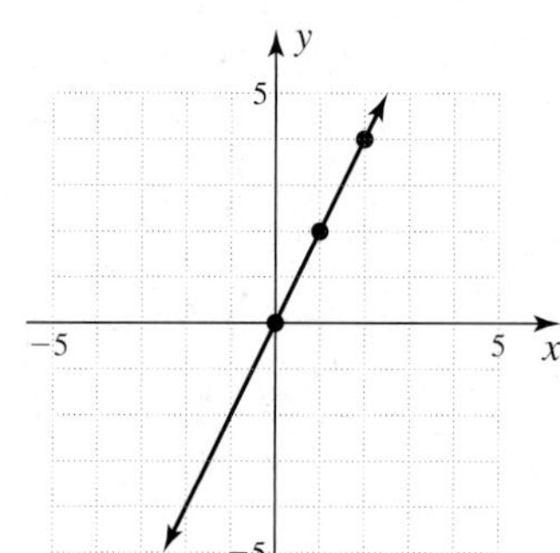

17. $2x - 3y = 0$

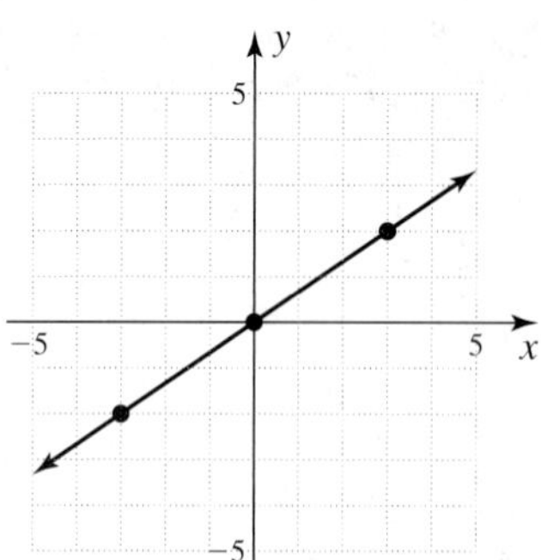

19. $-3x = -2y$

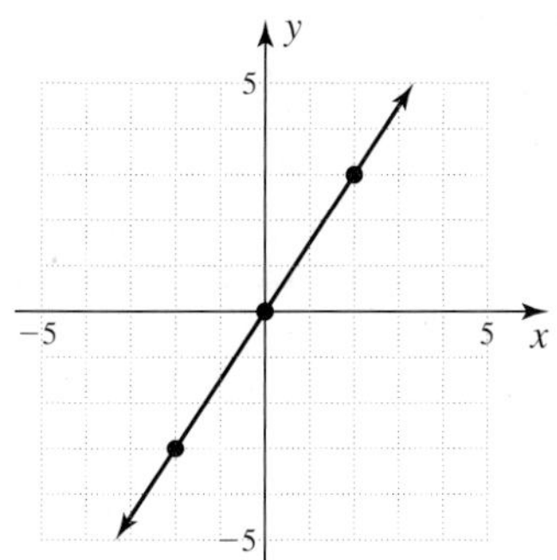

21. $y = -4$

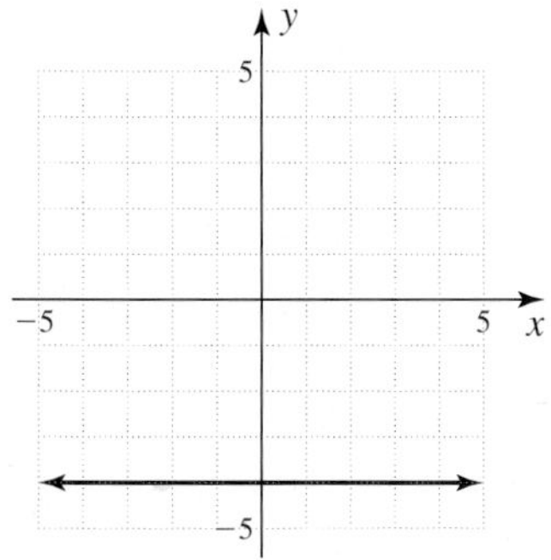

23. $2y + 6 = 0$

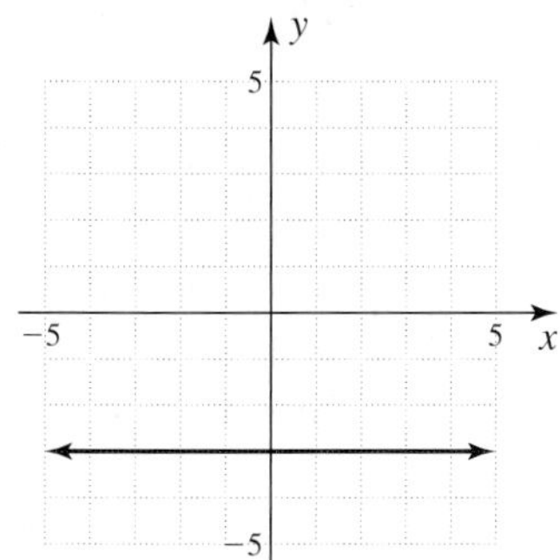

25. $x = -\frac{5}{2}$

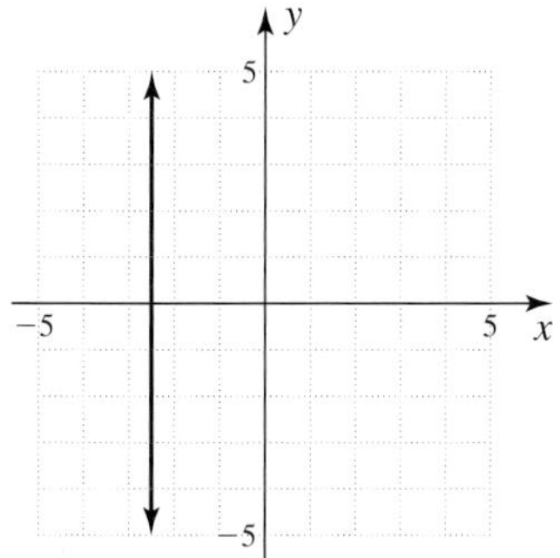

27. $2x + 4 = 0$

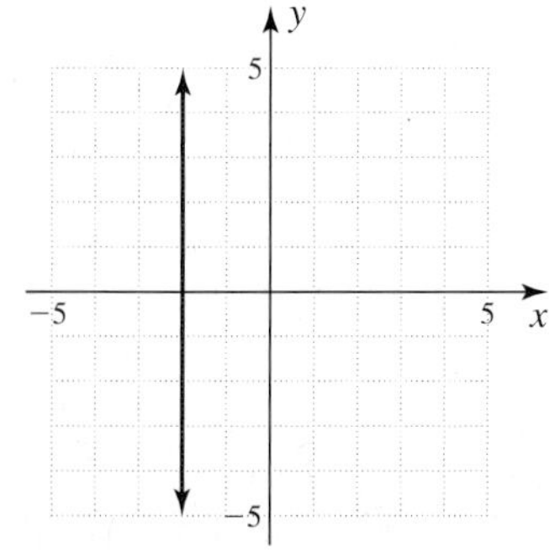

29. $2x - 9 = 0$

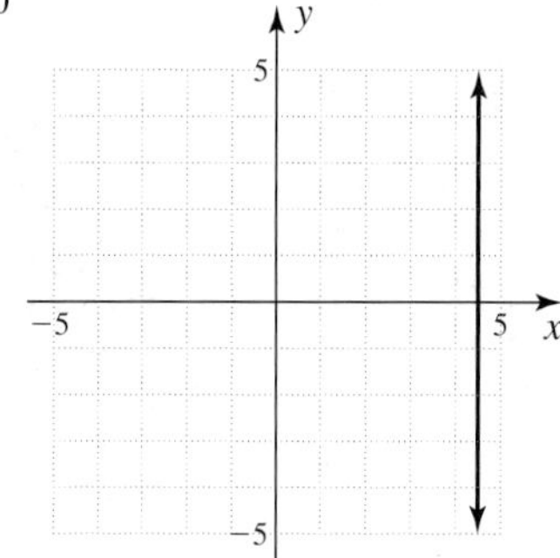

31. (a) (0, 1250); **(b)** (2, 1500); **(c)** (6, 2000);

(d)

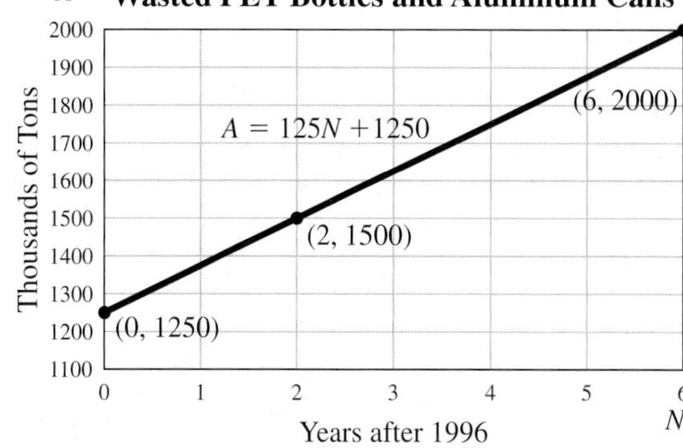

(e) 2000

33. (a) 190 g; **(b)** 200 g; **(c)** 210 g;

(d)

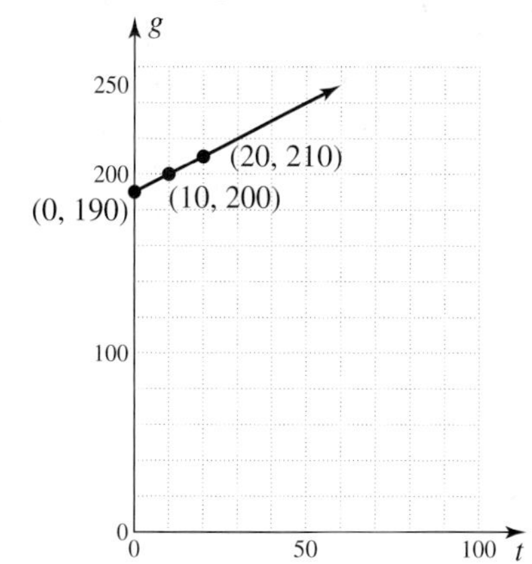

35. (a) 18; **(b)** 36;

(c)

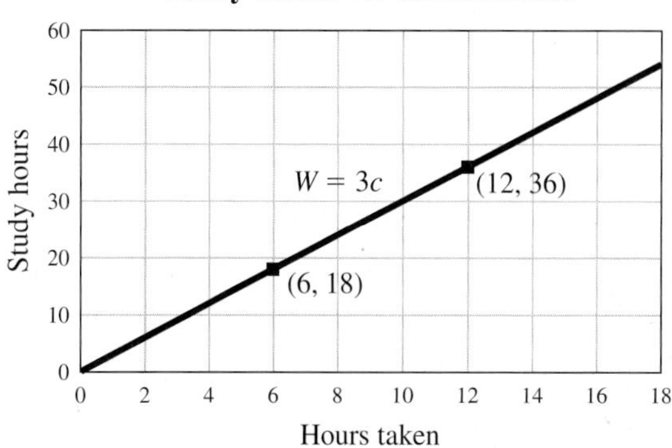

37. **(a)** 0; **(b)** 624 lb/ft²; **(c)** 2496 lb/ft²;
(d)

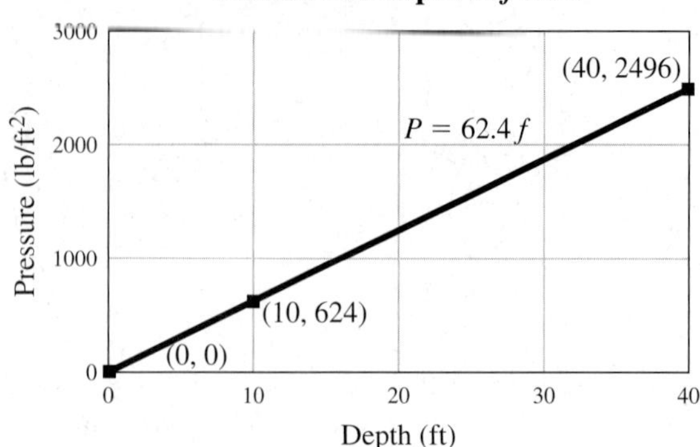

39. 215
41.

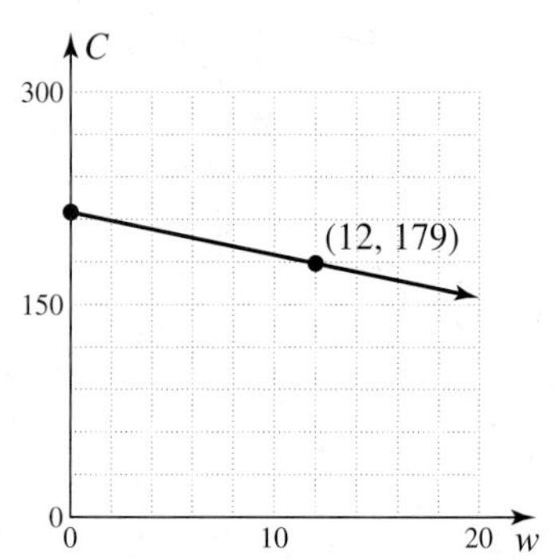

43. A horizontal line **45.** A line through the origin **47.** Two
49. $(0, y)$ **51.** origin **53.** vertical
55. $-2x + y = 4$

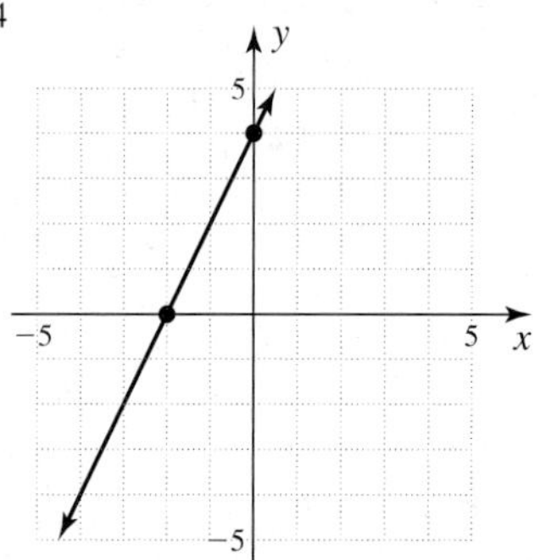

57. $4 + 2y = 0$

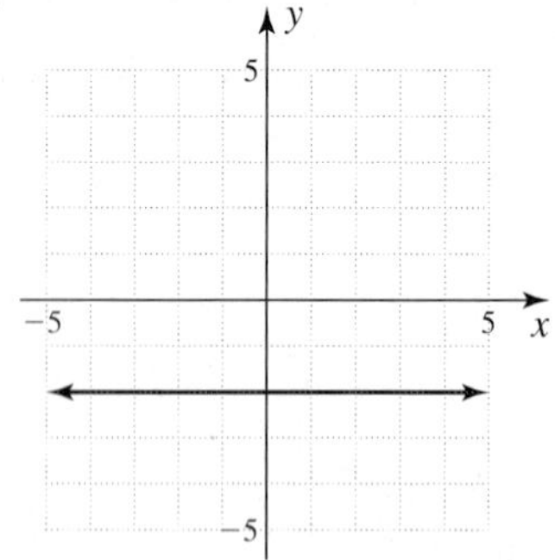

59. $-x + 4y = 0$

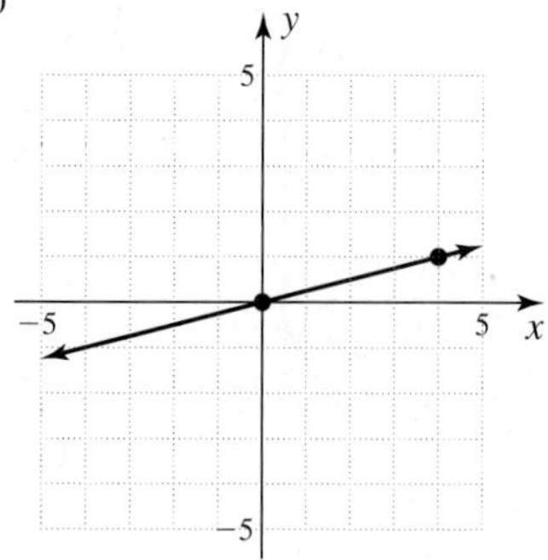

61. **(a)** C: 30, A: -30,

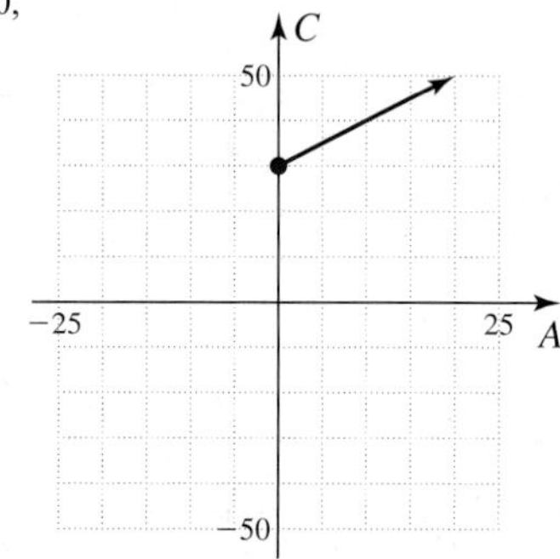

(b) Quadrant I **63.** 9 **65.** -9

Exercises 3.4

1. $m = 1$ **3.** $m = -1$ **5.** $m = -\frac{1}{8}$ **7.** $m = \frac{1}{4}$ **9.** $m = 0$
11. $m = 0$ **13.** Undefined **15.** $m = 3$ **17.** $m = -\frac{2}{3}$
19. $m = -\frac{1}{3}$ **21.** $m = \frac{2}{5}$ **23.** $m = 0$ **25.** Undefined
27. Parallel **29.** Perpendicular **31.** Parallel **33.** Parallel
35. Neither; the lines coincide **37.** Perpendicular
39. **(a)** 0.36; **(b)** The annual population increase (persons per square mile) in non-coastal areas; **(c)** 48; **(d)** Yes **41.** **(a)** 0.15; **(b)** Increasing; **(c)** The slope 0.15 represents the annual increase in the life span of American women. **43.** **(a)** -32; **(b)** Decreasing; **(c)** The slope -32 represents the decrease in the velocity of the ball. **45.** **(a)** 0.4; **(b)** Increasing; **(c)** The slope 0.4 represents the annual increase in the consumption of fat. **47.** **(a)** From year 0 to year 1; **(b)** \$160; **(c)** $m = -160$; **(d)** The decrease from year 0 to year 1
49. **(a)** From year 3 to year 4; **(b)** $m = 52$; **(c)** $m = 43$; **(d)** The slope for year 3 to 4

51.

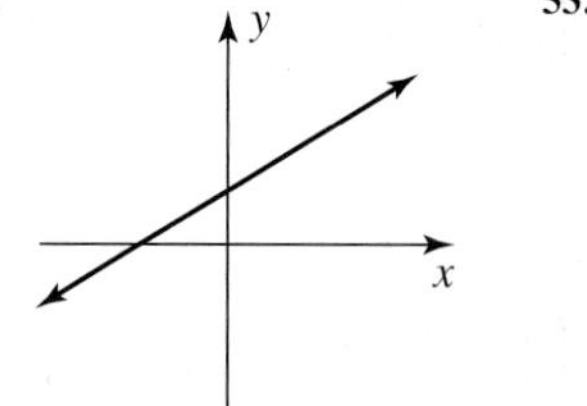

53.

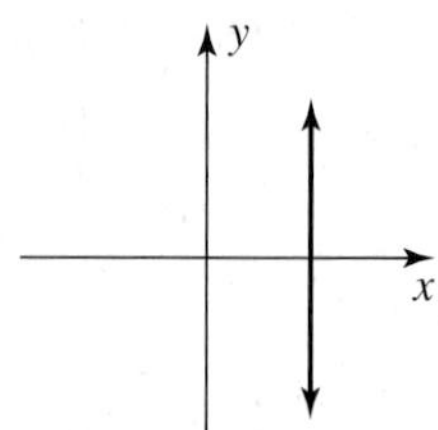

57. $\frac{y_2 - y_1}{x_2 - x_1}$ **59.** parallel **61.** $m = -1$ **63.** Undefined **65.** $m = -\frac{3}{2}$
67. Parallel **69.** Perpendicular **71.** $2x + 8$ **73.** $-2x - 2$

Exercises 3.5

1. $y = \frac{1}{2}x + \frac{3}{2}$

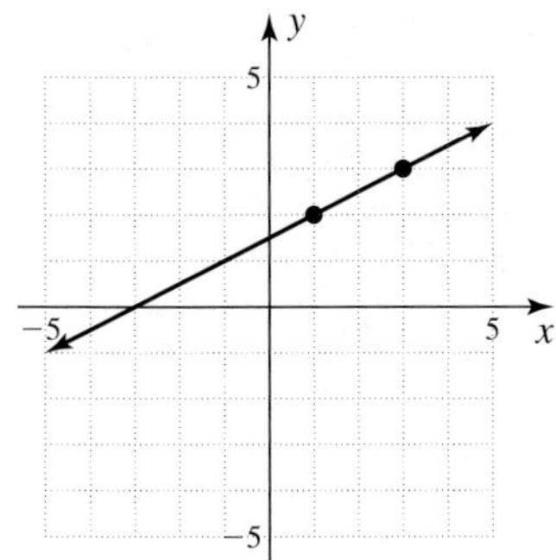

3. $y = -x + 6$

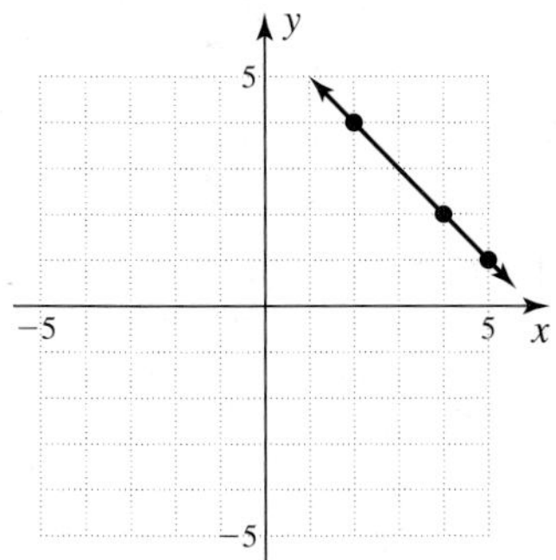

5. $y = 5$

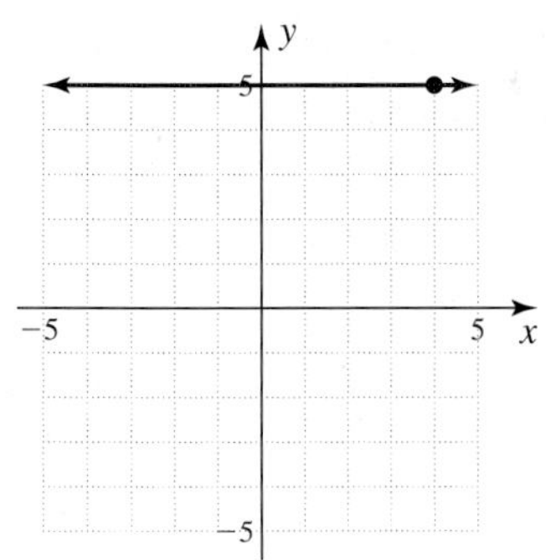

7. $y = 2x - 3$ **9.** $y = -4x + 6$ **11.** $y = \frac{3}{4}x + \frac{7}{8}$ **13.** $y = 2.5x - 4.7$
15. $y = -3.5x + 5.9$
17. $y = \frac{1}{4}x + 3$

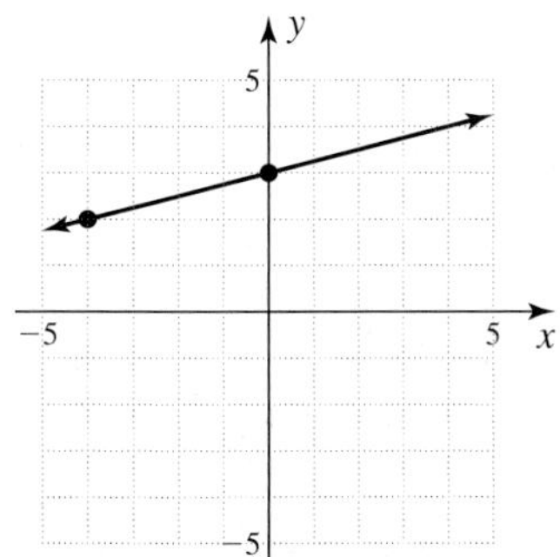

19. $y = -\frac{3}{4}x - 2$

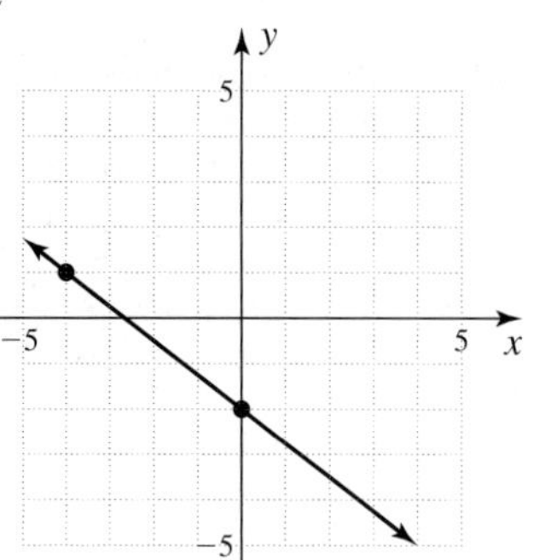

21. $x - y = -1$

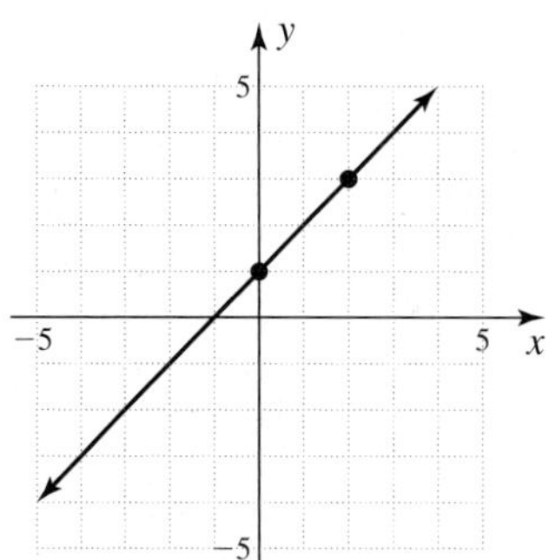

23. $3x - y = 4$

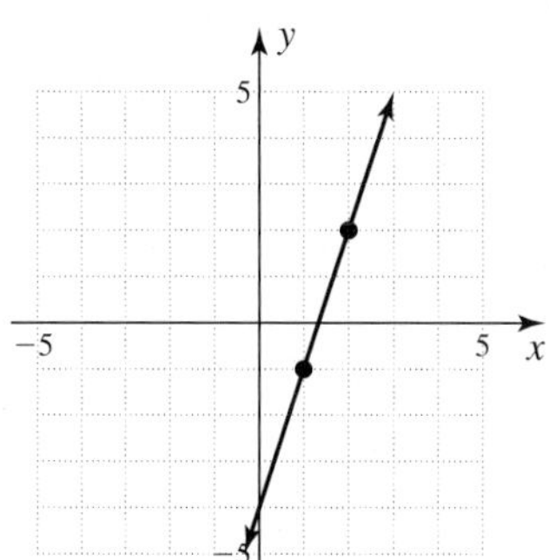

25. $4x + 3y = 12$

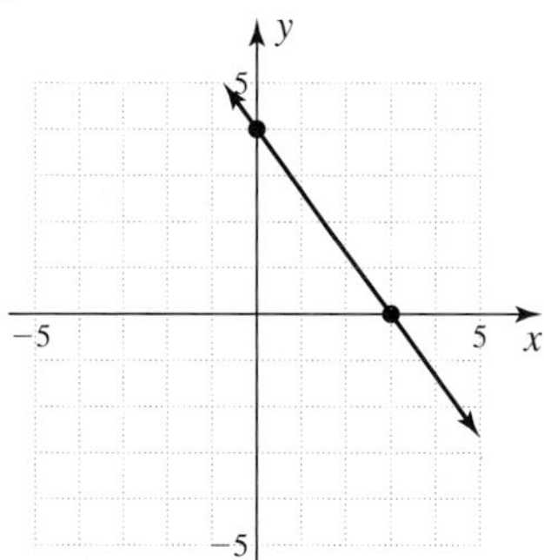

27. $x = 3$

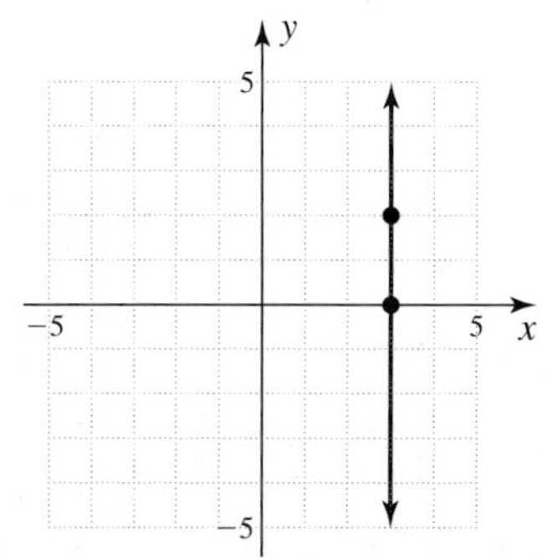

29. $y = -3$

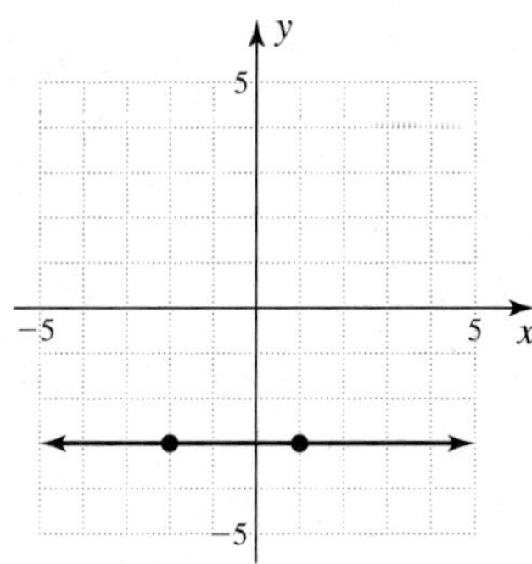

31. $y = x + 3$ **33.** $y = 2x$ **35.** $y = -2x - 3$ **37. (a)** $y = 0.02x + 2.50$; **(b)** \$0.03, \$7.00 **39. (a)** \$2.65; **(b)** \$2.40; **(c)** In 25 days **41.** The y-axis **43.** $Ax + 0y = C$ **47.** $y = mx + b$ **49.** $Ax + By = C$

51. $y = -\frac{3}{4}x + 2$

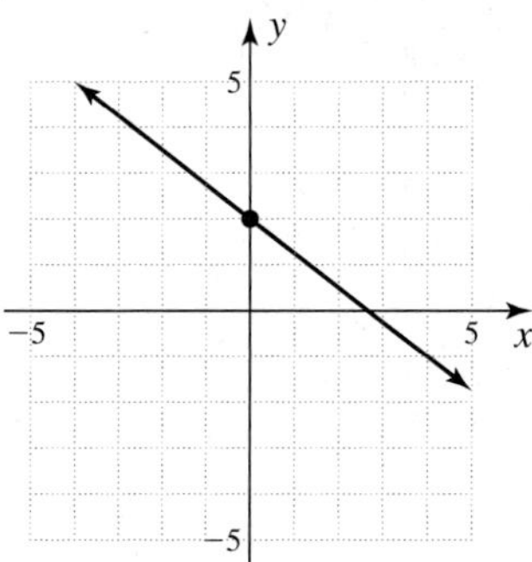

53. $2x - 3y = 9$

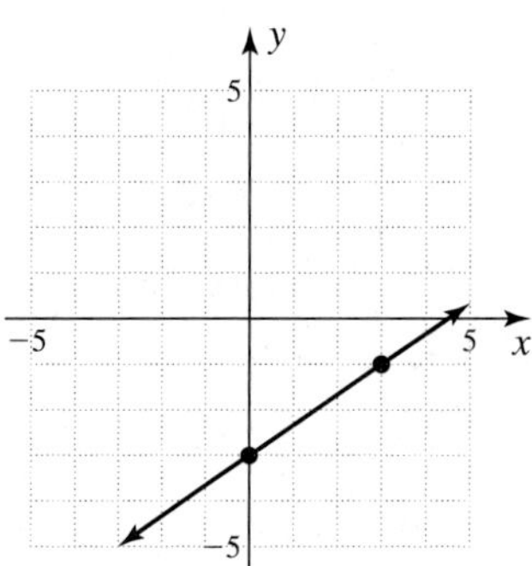

55. $5x + y = -11$

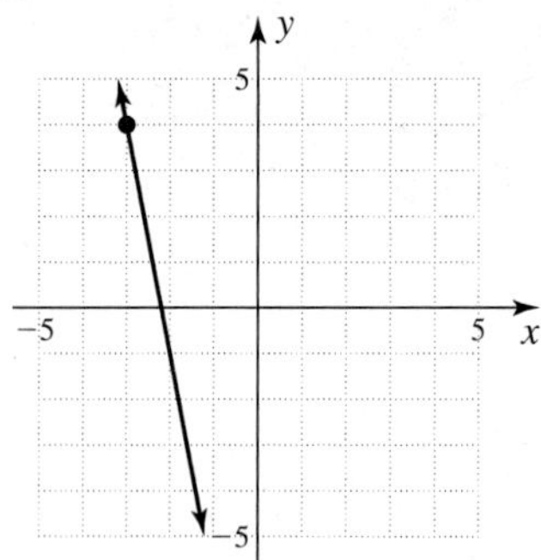

57. $y = 0.35x + 50$ **59.** $4x - y = 12$

Exercises 3.6

1. $C = 1.7m + 0.3$; \$51.30 **3.** 6 mi **5. (a)** \$2; **(b)** $m - 1$; **(c)** \$1.70; **(d)** $1.70(m - 1)$; **(e)** $C = 2 + 1.70(m - 1)$ or $C = 1.7m + 0.3$; yes **7. (a)** $C = 175$, if $m \le 60$; **(b)** $C = 175$, if $m \le 60$; **(c)** $C = \frac{175}{m}$, if $m \le 60$ **9. (a)** $m = \frac{6}{5}$; **(b)** $m = \frac{6}{5}$; **(c)** Yes; **(d)** $y = \frac{6}{5}x - 11$; **(e)** $y = \frac{6}{5}x - 11$; yes; **(f)** -5

11. (a) $C = 37.5h + 100$; **(b)** 3 hr **13. (a)** $C = 2.3m + 40$; **(b)** 70 min; **(c)** $C = 7.2m$; **(d)** About 8 min **15. (a)** $C = 0.35m + 50$; **(b)** 110 min **17.** 100 min; after 100 min **19. (a)** $y = 1.95x + 29$; **(b)** $m = 1.95$; **(c)** $b = 29$ **21.** $y = -0.1x + 8$; 5 days; 4 days **23. (a)** 3 human yr corresponds to 30 dog yr; 9 human yr corresponds to 60 dog yr; **(b)** $m = 5$; **(c)** $d = 5h + 15$; **(d)** 35 dog yr; **(e)** 10 yr; **(f)** 1.2 yr **25. (a)** $m = 0.0255$; **(b)** $c = 0.0255b - 0.0245$; **(c)** 0.0775; 0.1285; **(d)** 4

27.

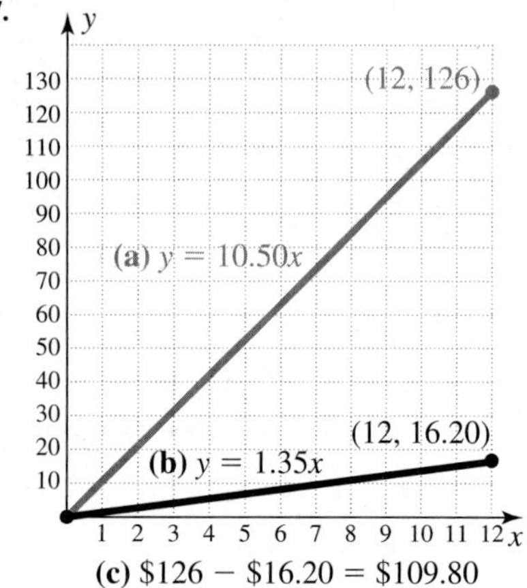

29. (a) $H = 6A + 77$; **(b)** $m = 6$; **(c)** $b = 77$; **(d)** $H = 89$; **(e)** $H = 149$; **(f)**

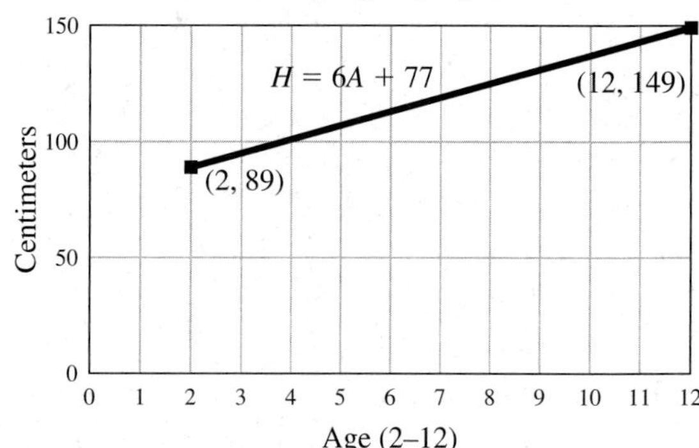

31. (a) $H = 62.92 + 2.39f$; **(b)** $m = 2.39$; **(c)** $b = 62.92$

37.

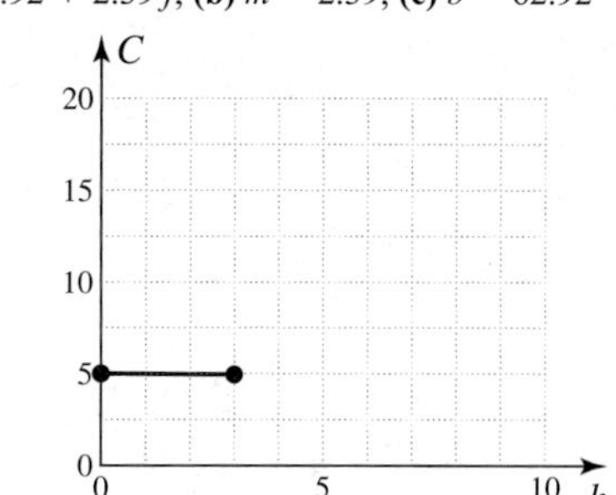

39.

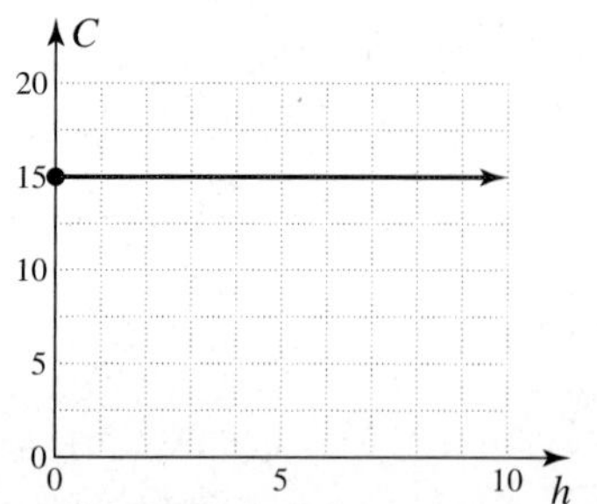

41. $C = 0.15m + 40$; \$107.50 **43. (a)** 30 hr cost \$24; **(b)** 33 hr cost \$26.70; **(c)** $C = 0.9x - 3$ **45.** $>$ **47.** $>$ **49.** $<$

Exercises 3.7

1. $2x + y > 4$

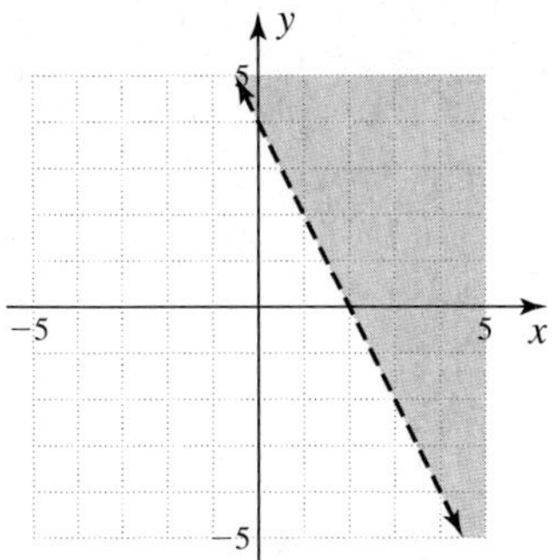

3. $-2x - 5y \le 10$

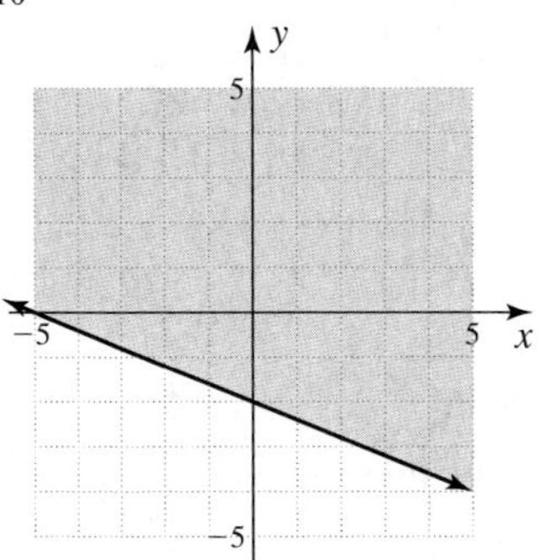

5. $y \ge 3x - 3$

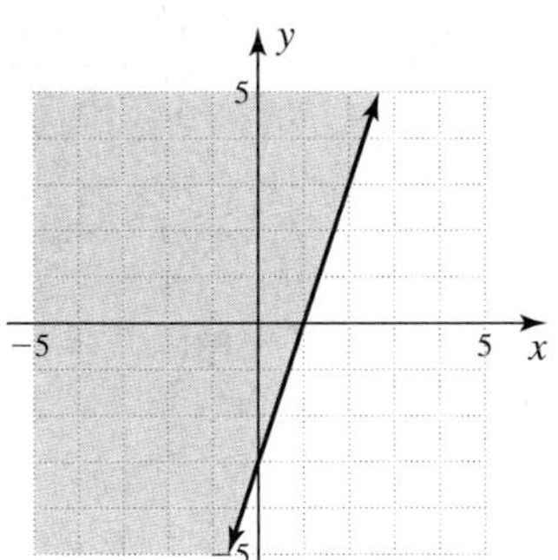

7. $6 < 3x - 6y$

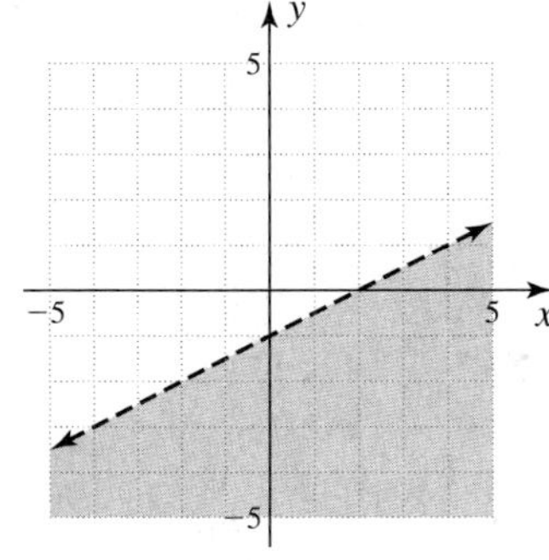

9. $3x + 4y \ge 12$

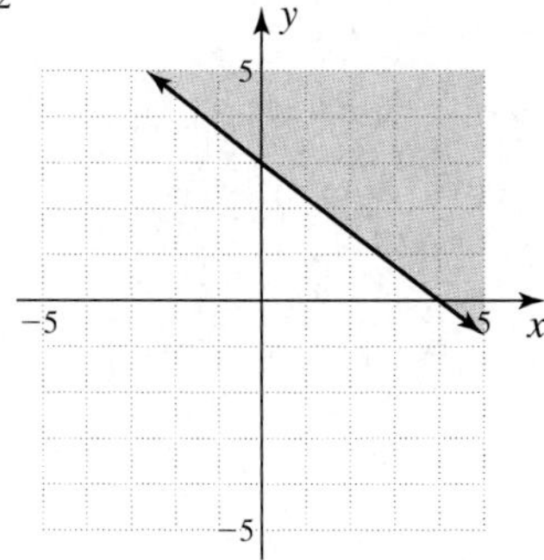

11. $10 < -2x + 5y$

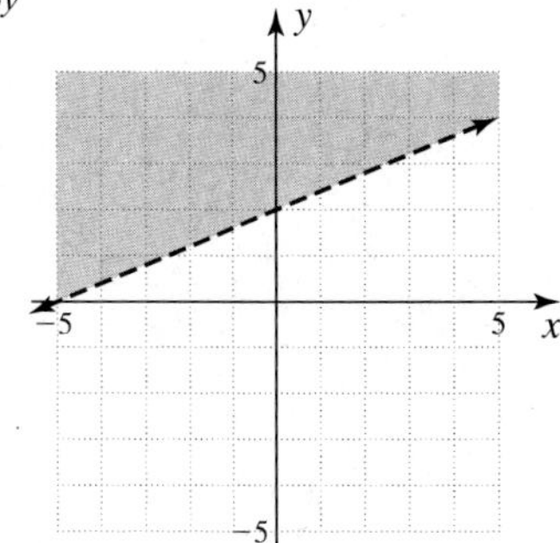

13. $x \ge 2y - 4$

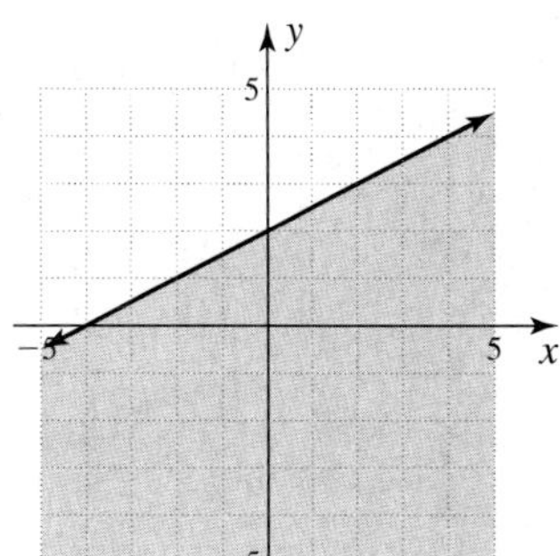

15. $y < -x + 5$

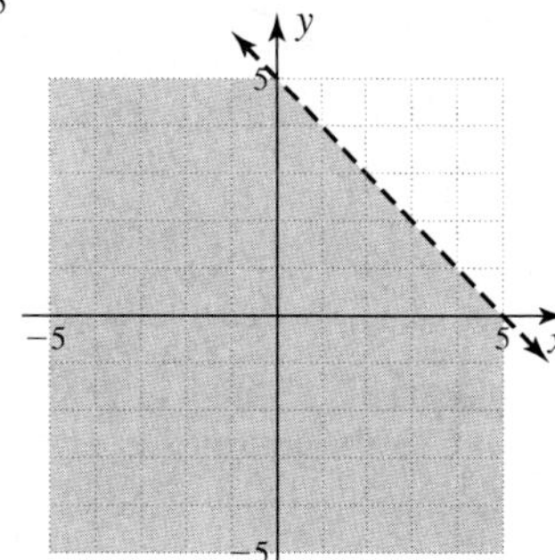

17. $2y < 4x + 5$

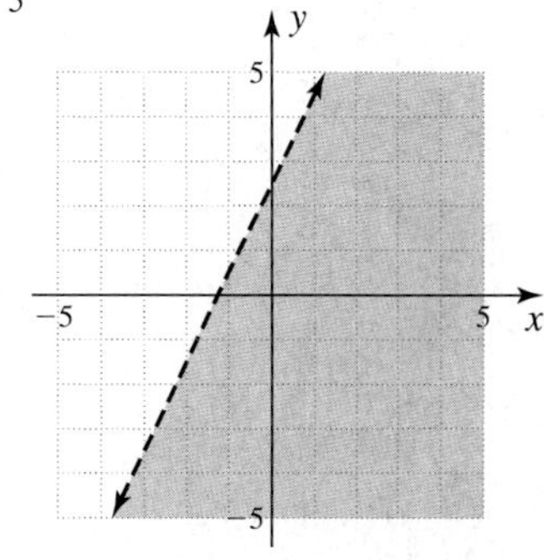

19. $x > 1$

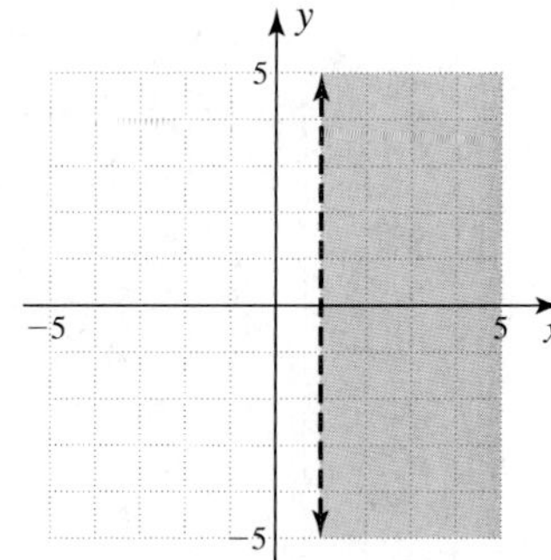

21. $x \le \frac{5}{2}$

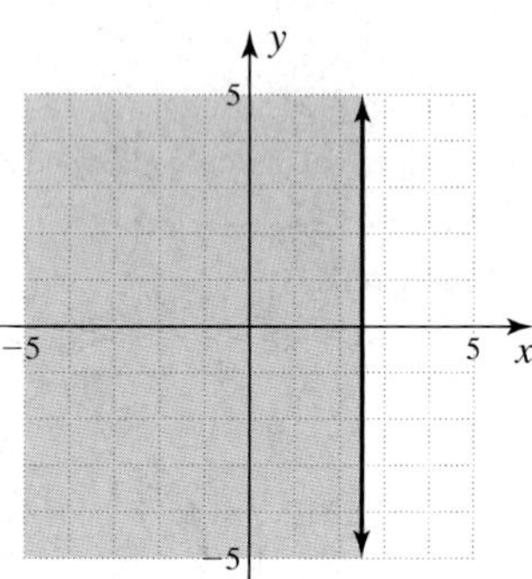

23. $y \le -\frac{3}{2}$

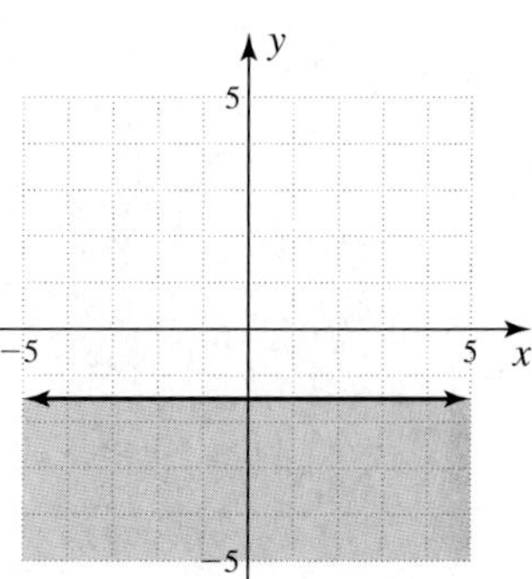

25. $x - \frac{2}{3} > 0$

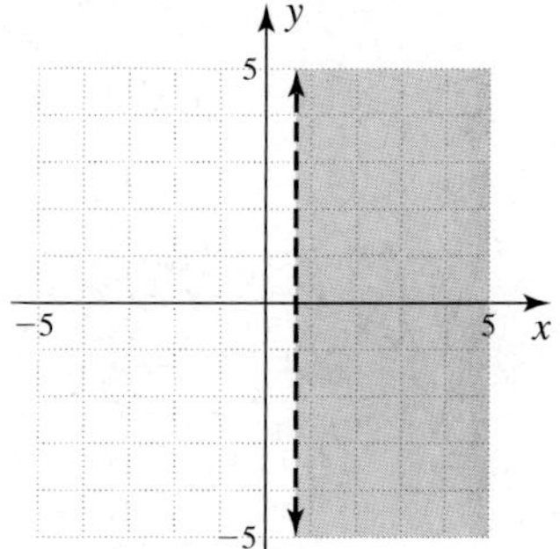

27. $y + \frac{1}{3} \ge \frac{2}{3}$

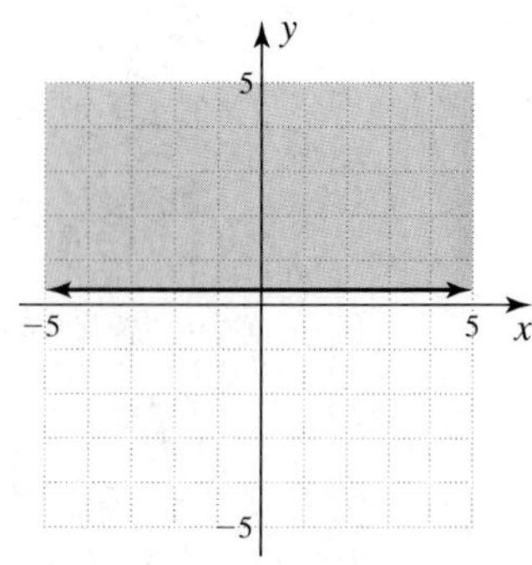

29. $2x + y < 0$

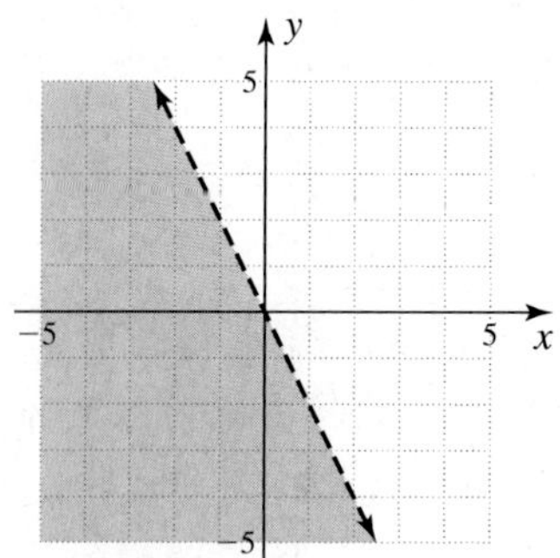

31. $y - 3x > 0$

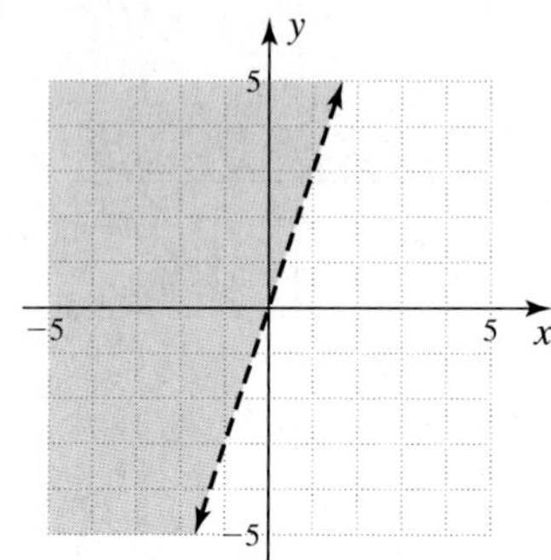

33. rectangle

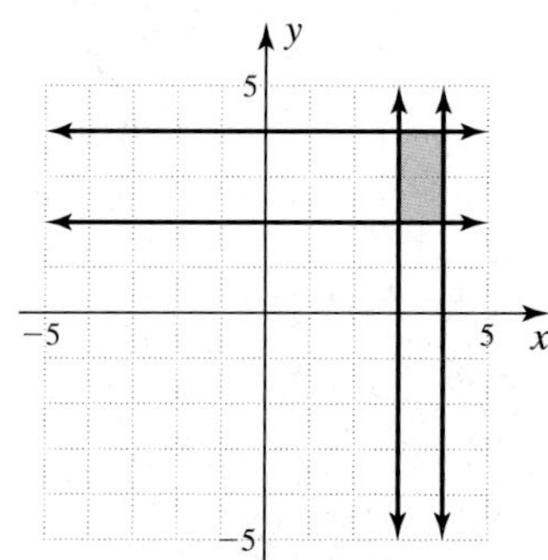

35. (a) $1990 \le x < 2010$; **(b)** $2010 < x \le 2100$; **(c)** In 2010 **37.** The years between 1990 and 2010, including 1990. Answers may vary.

39.

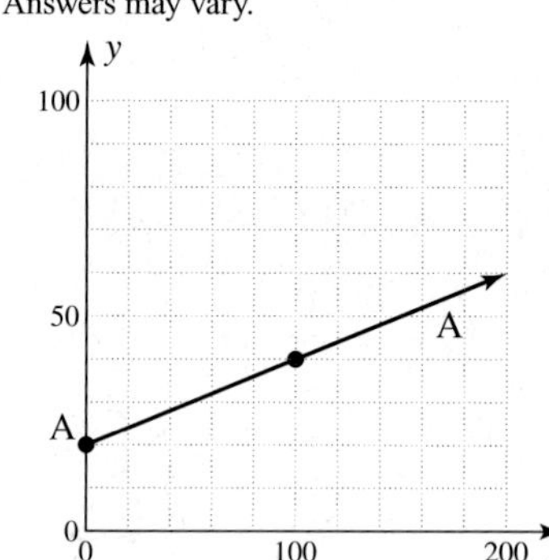

41. When you travel 100 mi **43.** When traveling less than 100 mi

49. solid **51.** dashed

53. $y \geq -4$

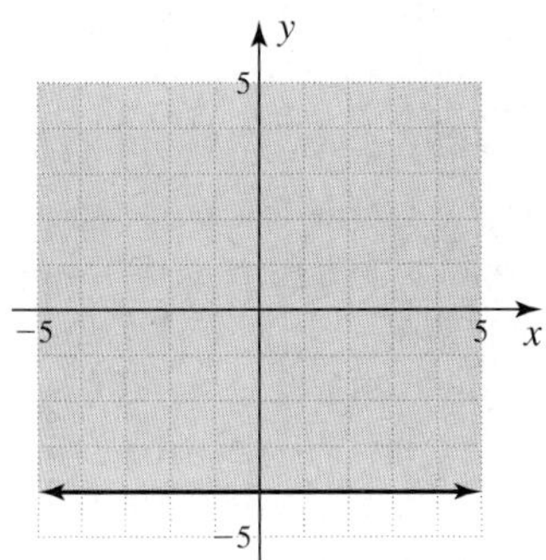

55. $y - 3 > 0$

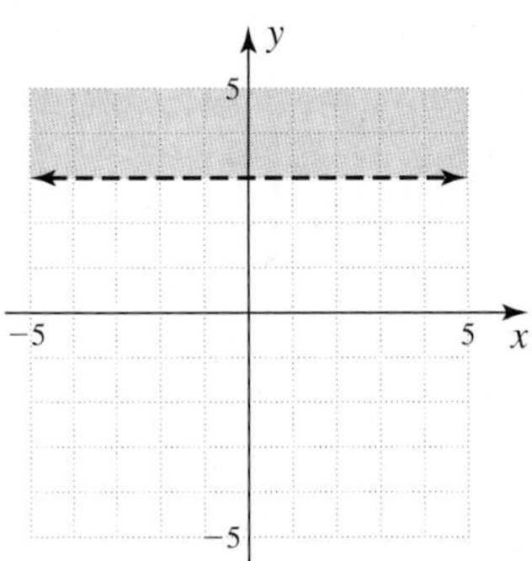

57. $y \leq -4x + 8$

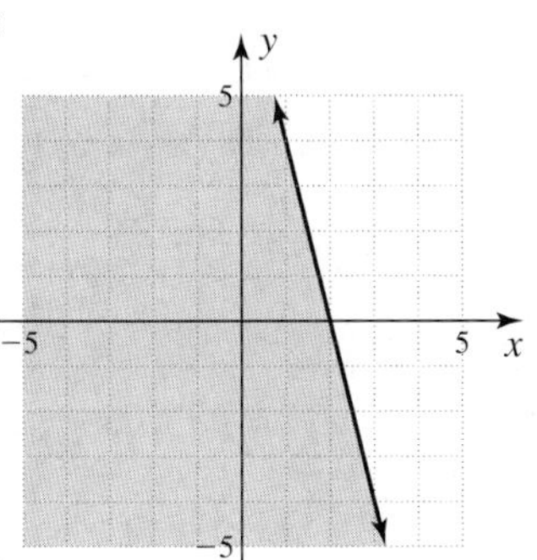

59. $y < 3x$

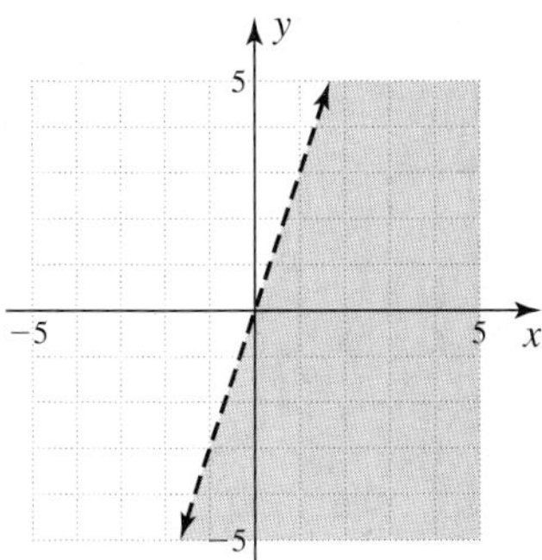

61. $3x + y \leq 0$

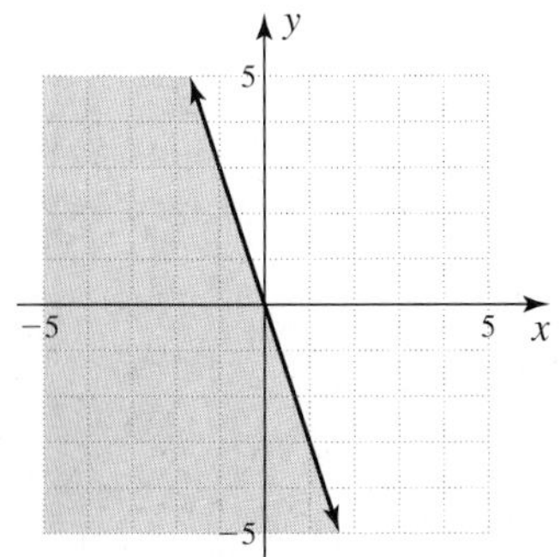

63. -24 **65.** -21 **67.** -4

Review Exercises

1. [3.1A]

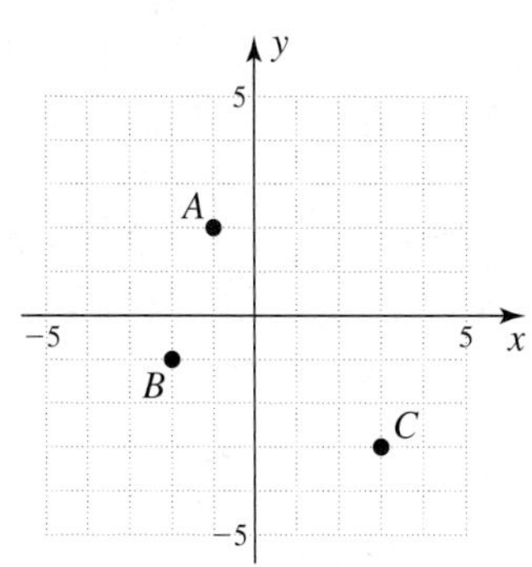

2. [3.1B] A: $(1, 1)$; B: $(-1, -2)$; C: $(3, -3)$

3. [3.1C] **(a)** For a wind speed of 10 mi/hr, the wind chill temperature is −10°F; **(b)** −22°F; **(c)** −45°F

4. [3.1D] **(a)** \$13; **(b)** \$15; **(c)** \$25

5. [3.1E] **(a)** Quadrant II; **(b)** Quadrant III; **(c)** Quadrant IV

6. [3.1F] **(a)**

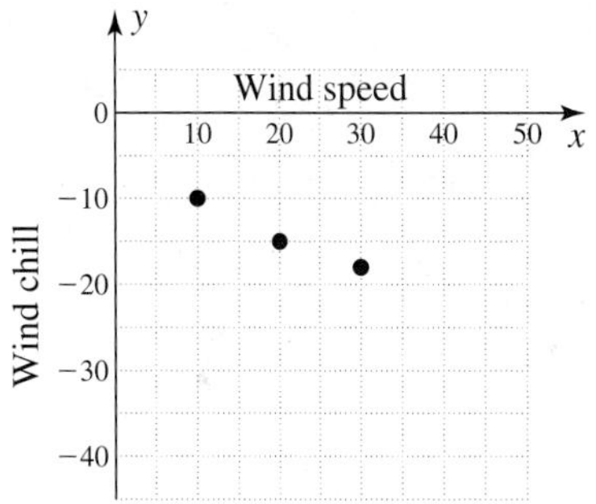

(b) For a wind speed of 10 mi/hr, the wind chill temperature is −10°F; **(c)** $s = 20$ **7.** [3.2A] **(a)** No; **(b)** No; **(c)** Yes

8. [3.2B] **(a)** $x = 3$; **(b)** $x = 4$; **(c)** $x = 2$

9. [3.2C] **a.** $x + y = 4$
b. $x + y = 2$
c. $x + 2y = 2$

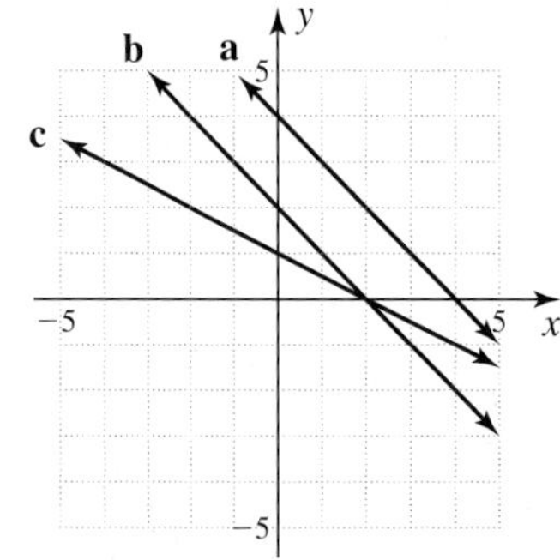

10. [3.2C]

a. $y = \frac{3}{2}x + 3$
b. $y = -\frac{3}{2}x + 3$
c. $y = \frac{3}{4}x + 4$

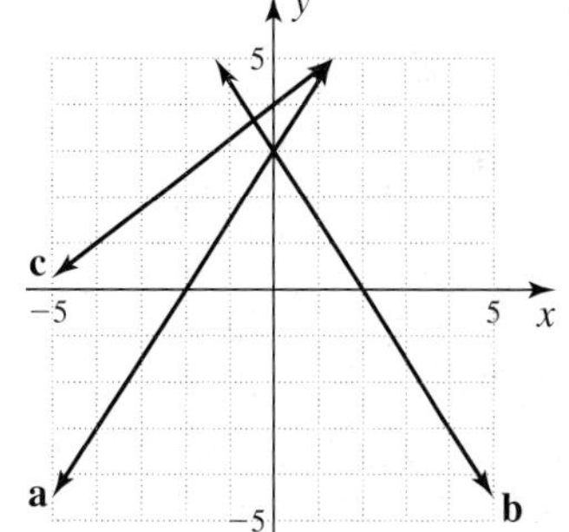

11. [3.2D]

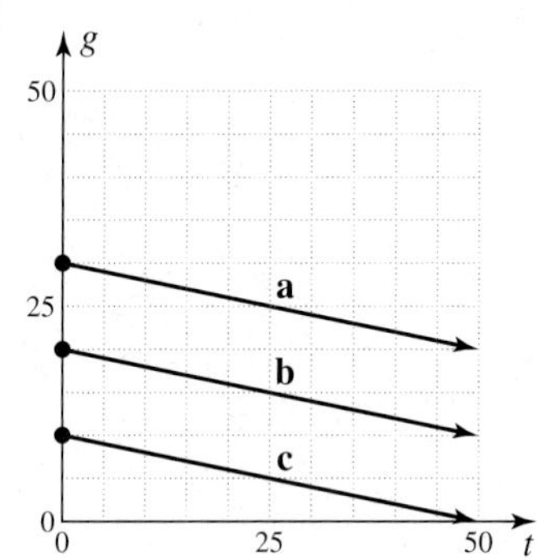

a. $g = 30 - 0.2t$
b. $g = 20 - 0.2t$
c. $g = 10 - 0.2t$

12. [3.3A]

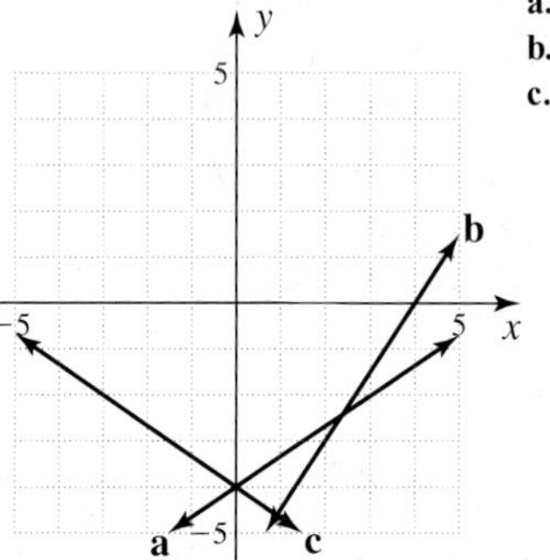

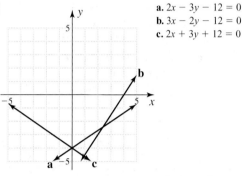

a. $2x - 3y - 12 = 0$
b. $3x - 2y - 12 = 0$
c. $2x + 3y + 12 = 0$

13. [3.3B]

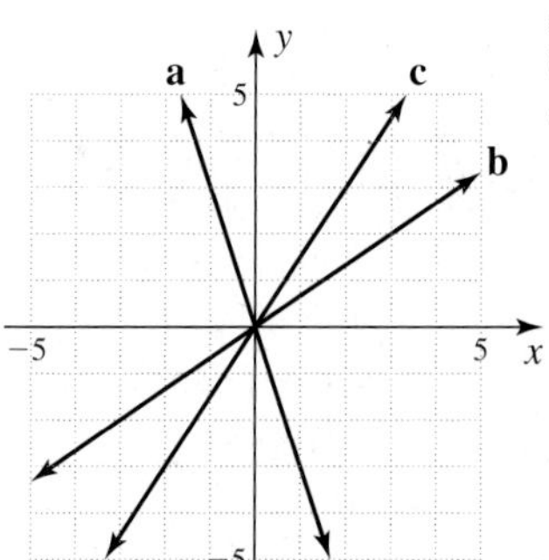

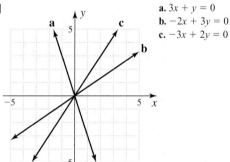

a. $3x + y = 0$
b. $-2x + 3y = 0$
c. $-3x + 2y = 0$

14. [3.3C]

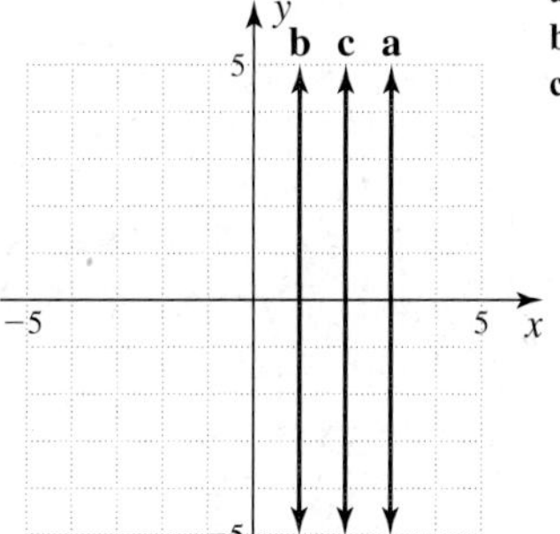

a. $2x - 6 = 0$
b. $2x - 2 = 0$
c. $2x - 4 = 0$

15. [3.3C]

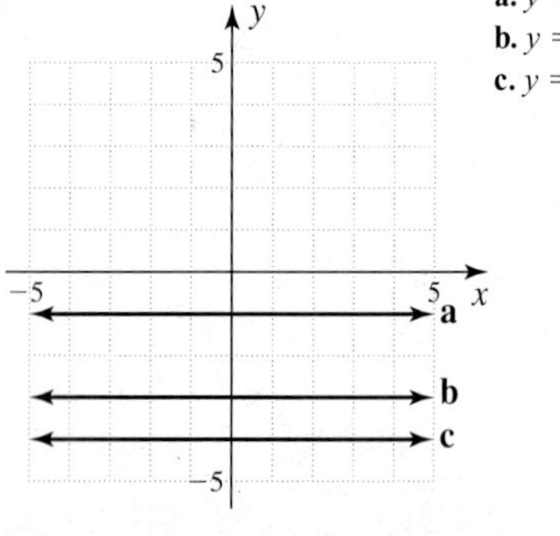

a. $y = -1$
b. $y = -3$
c. $y = -4$

16. [3.4A] **(a)** Undefined; **(b)** 0; **(c)** Undefined

17. [3.4A] **(a)** $m = 1$; **(b)** $m = \frac{7}{3}$; **(c)** $m = -4$

18. [3.4B] **(a)** $m = -\frac{3}{2}$; **(b)** $m = -\frac{1}{4}$; **(c)** $m = \frac{2}{3}$

19. [3.4C] **(a)** Neither; **(b)** Perpendicular; **(c)** Parallel

20. [3.4D] **(a)** $m = 0.6$; **(b)** The change (increase) in the number of theaters per year; **(c)** 600

21. [3.5A] **(a)** $2x + y = 1$; **(b)** $3x + y = 4$; **(c)** $4x + y = 7$

22. [3.5B] **(a)** $y = 5x - 2$; **(b)** $y = 4x + 7$; **(c)** $y = 6x - 4$

23. [3.5C] **(a)** $x - y = -3$; **(b)** $x - 2y = -5$; **(c)** $2x + 3y = 8$

24. [3.6A] **(a)** $C = 0.2m + 3$; \$6; **(b)** $C = 0.2m + 5$; \$8; **(c)** $C = 0.3m + 5$; \$9.50 **25.** [3.6B] **(a)** $C = 0.4m + 30$; \$150; **(b)** $C = 0.3m + 40$; \$70; **(c)** $C = 0.2m + 50$; \$130

26. [3.6C] **(a)** $C = 0.40m + 1$; **(b)** $C = 0.40m + 2$; **(c)** $C = 0.40m + 3$

27. [3.7A] **(a)** $2x - 4y < -8$

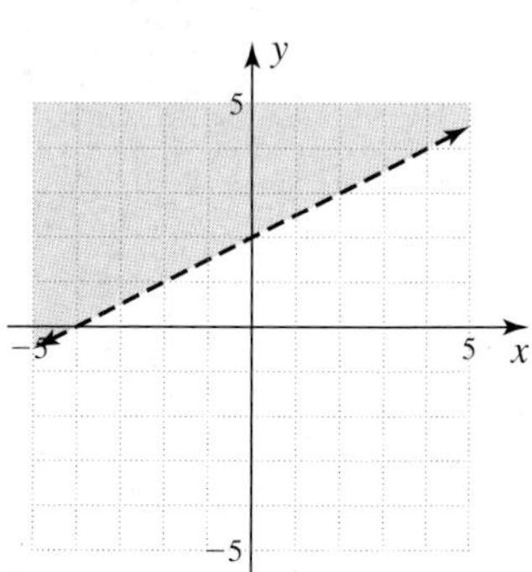

(b) $3x - 6y < -12$

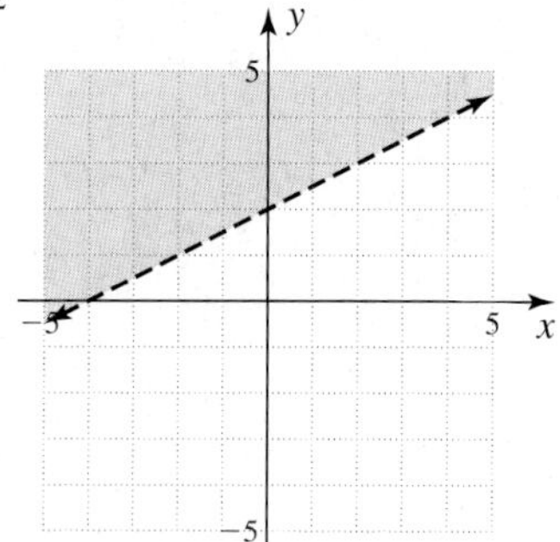

(c) $4x - 2y < -8$

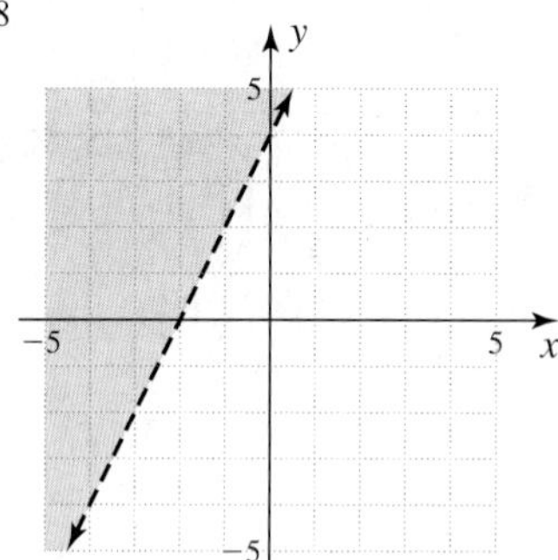

28. [3.7A] **(a)** $-y \leq -2x + 2$

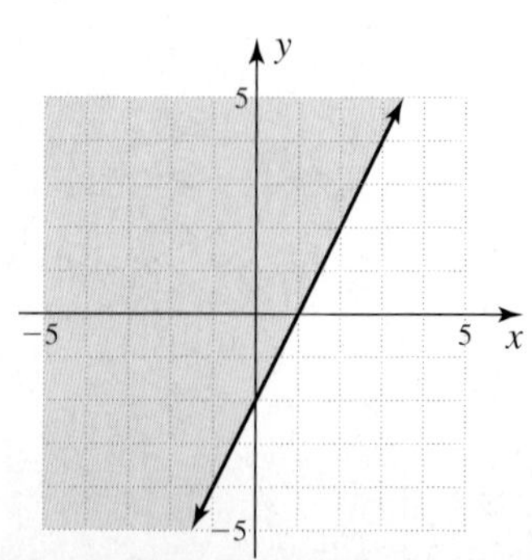

(b) $-y \leq -2x + 4$

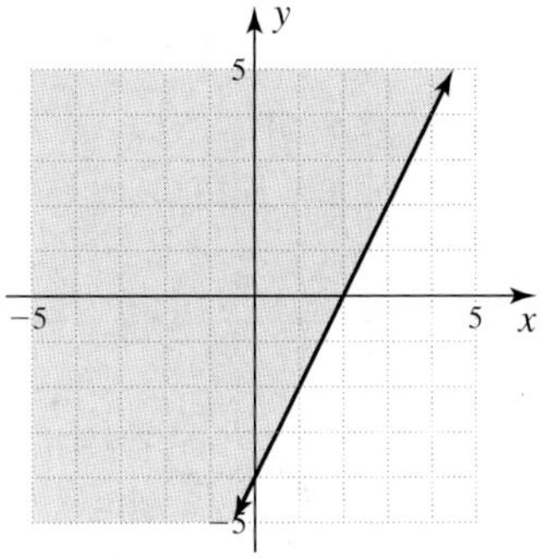

(c) $-y \leq -x + 3$

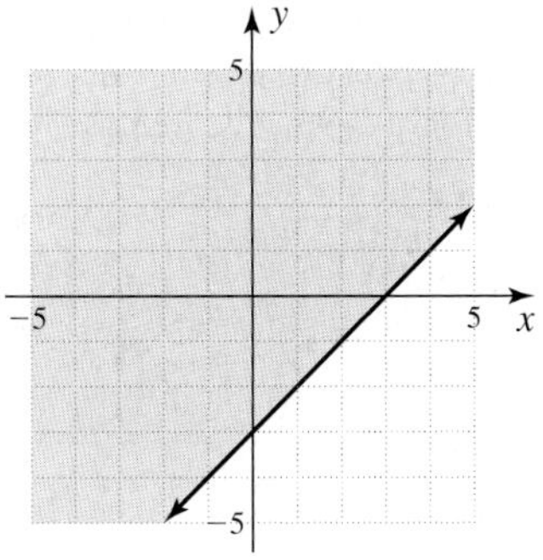

29. [3.7A] **(a)** $2x + y > 0$

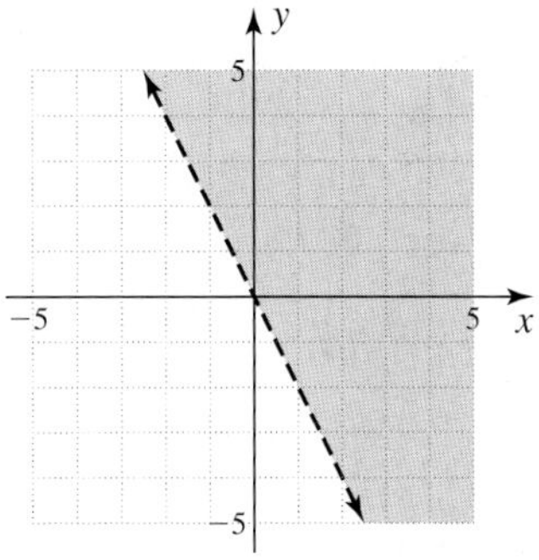

(b) $3x + y > 0$

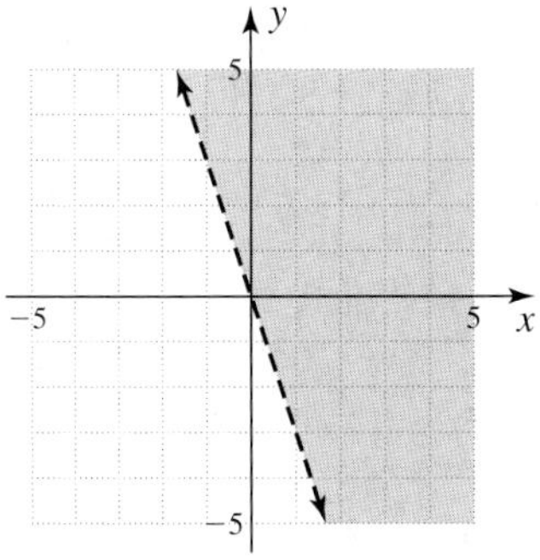

(c) $3x - y < 0$

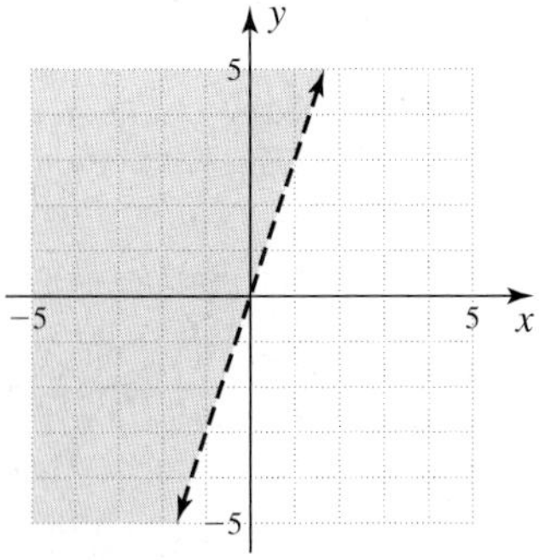

30. [3.7A] **(a)** $x \geq -4$

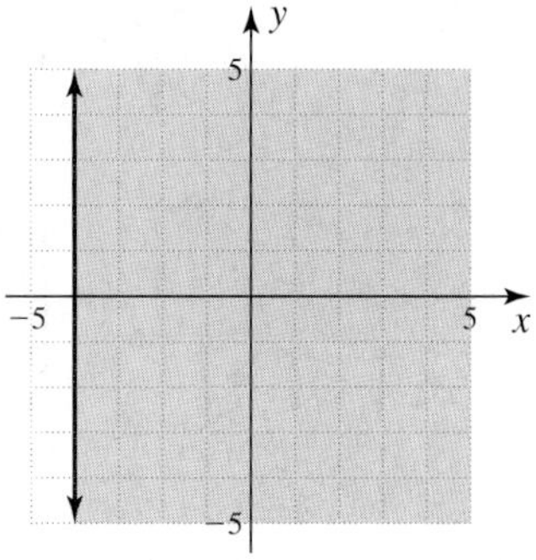

(b) $y - 4 < 0$

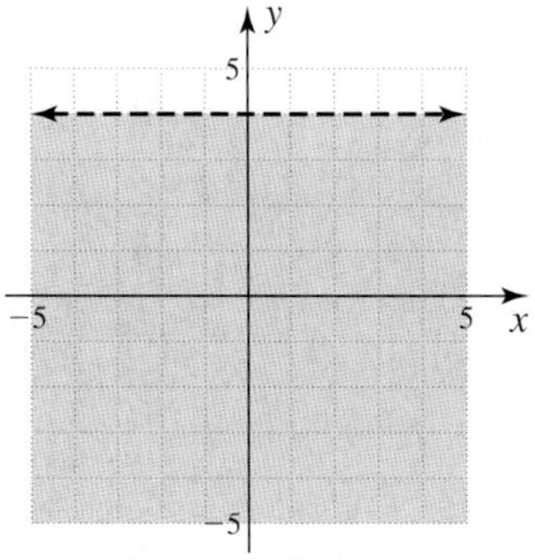

(c) $2y - 4 \geq 0$

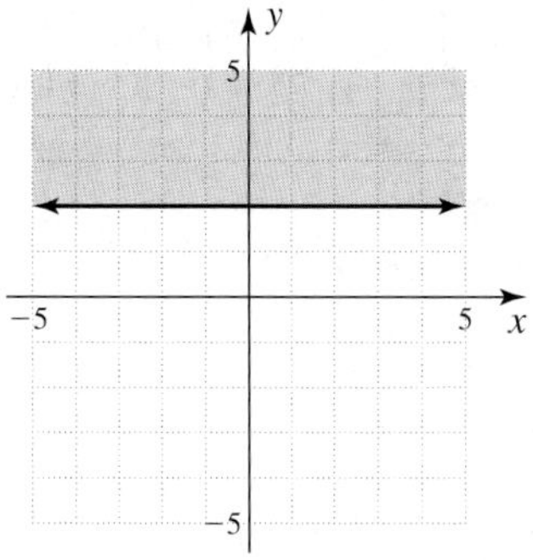

Cumulative Review Chapters 1–3

1. 1 **2.** $3\frac{1}{7}$ **3.** $-\frac{13}{24}$ **4.** 13.0 **5.** −19.76 **6.** 1296 **7.** 12 **8.** 12 **9.** Commutative property of addition **10.** $18x - 21$ **11.** cd^2 **12.** $-x - 14$ **13.** $\frac{m+n}{p}$ **14.** No **15.** $x = 6$ **16.** $x = 9$ **17.** $x = 54$ **18.** $x = 10$ **19.** $d = \frac{S}{7c^2}$ **20.** 65 and 105

21. \$435 from stocks and \$190 from bonds **22.** 2 hr

23. \$12,000 in bonds, \$13,000 in certificates of deposit

24.

25.

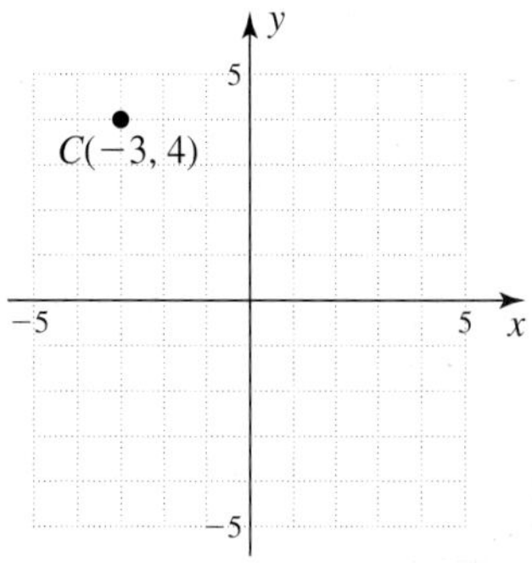

26. (2, −4) **27.** No **28.** $x = -2$

29. $2x + y = 8$

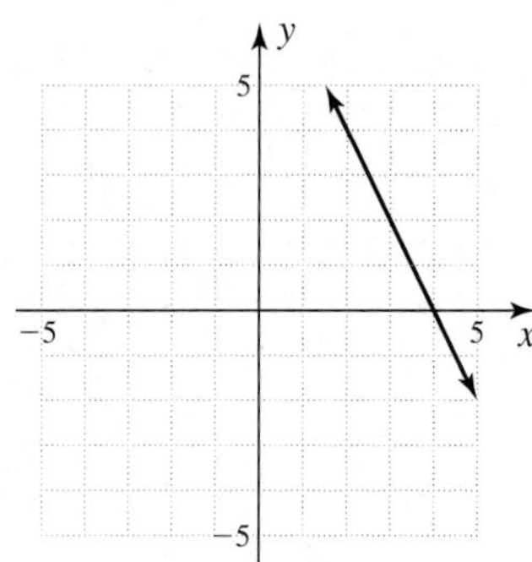

30. $4x - 8 = 0$

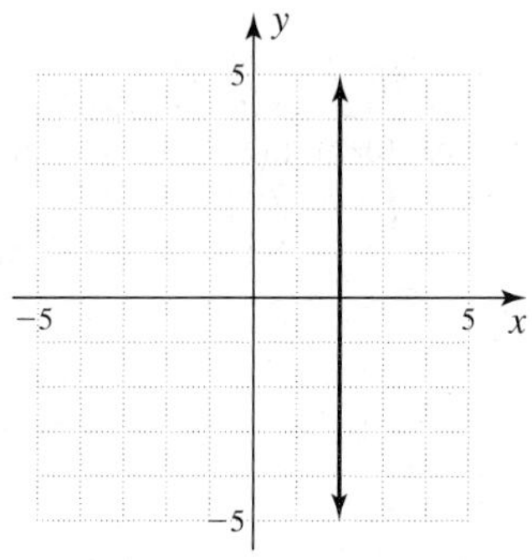

31. 9 **32.** 2 **33.** (1) and (3) **34. (a)** Undefined; **(b)** 0 **35.** $-\frac{25}{72}$ **36.** 8.2 **37.** 16 **38.** 2 **39.** 5 **40.** $6x - 11$ **41.** $\frac{d + 4e}{f}$ **42.** $x = 3$ **43.** $x = 63$ **44.** 35 and 75 **45.** \$4000 in bonds; \$5000 in certificates of deposit

46.

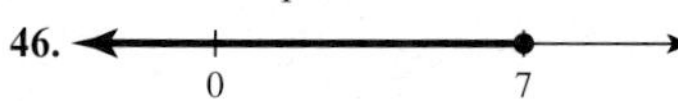

47.

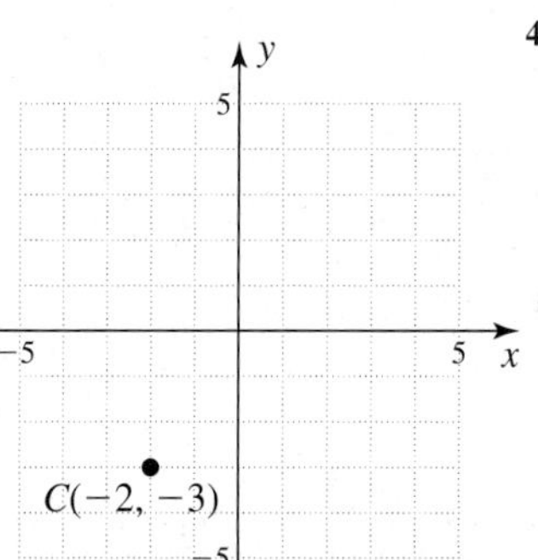

48. No **49.** $x = -2$

50. $3x + y = 3$

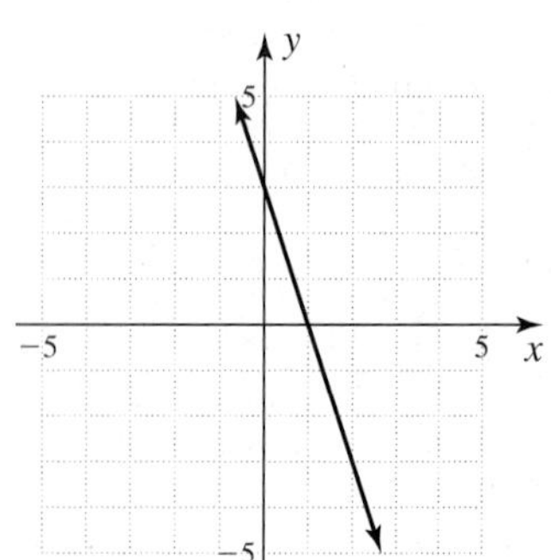

51. $3x + 9 = 0$

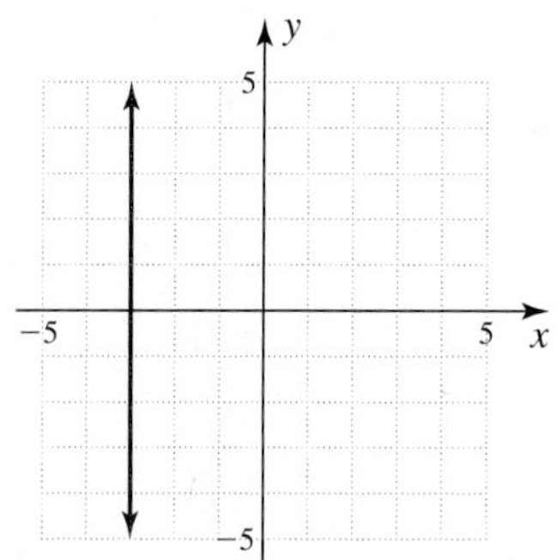

52. $-\frac{7}{3}$ **53.** 3 **54.** (1) and (3) **55. (a)** 0; **(b)** Undefined **56.** $y = -3x + 6$ **57.** $y = 3x - 1$

58. $6x - y < -6$

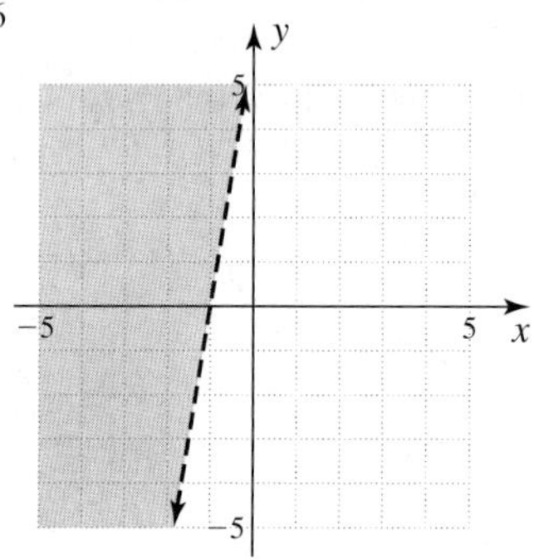

59. $-y \geq -6x - 6$

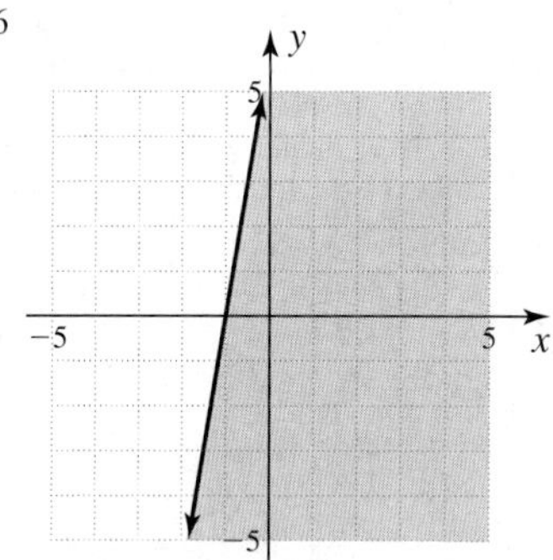

Chapter 4

Exercises 4.1

1. $24x^3$ **3.** $30a^4b^3$ **5.** $3x^3y^3$ **7.** $-\frac{b^5c}{5}$ **9.** $\frac{3x^3y^3z^6}{2}$ **11.** $-30x^5y^3z^8$ **13.** $15a^4b^3c^4$ **15.** $24a^3b^5c^5$ **17.** x^4 **19.** $-\frac{a^2}{2}$ **21.** $2x^3y^2$ **23.** $-\frac{x^3y^2}{2}$ **25.** $\frac{2a^3y^4}{3}$ **27.** $\frac{3ab^3c}{4}$ **29.** $\frac{3a^4}{2}$ **31.** $-x^6y$ **33.** $-x^2y^2$ **35.** $2^6 = 64$ **37.** $3^2 = 9$ **39.** x^9 **41.** y^6 **43.** $-a^6$ **45.** $8x^9y^6$ **47.** $4x^4y^6$ **49.** $-27x^9y^6$ **51.** $9x^{12}y^6$ **53.** $-8x^{12}y^{12}$ **55.** $\frac{16}{81}$ **57.** $\frac{27x^6}{8y^9}$ **59.** $\frac{16x^8}{81y^{12}}$ **61.** $12x^6y^3$ **63.** $-576a^2b^3$ **65.** $-144a^4b^6$ **67.** 96 **69.** x^5y^2 **71.** 36.8×10^8 kg = 3,680,000,000 kg **73.** 36.96×10^8 kg = 3,696,000,000 kg **75.** $V = \frac{2}{3}x^3$ **77.** 144 sandwiches **79.** \$0.0075 or 0.75¢ **81.** $V = x^3$ **83.** 16 in.^{3} **85.** ≈ \$23.63 **87.** $V = \frac{9}{16}x^3$ in.^{3} **89.** 37.5 servings **91.** $V = \pi r^2h$ **93.** $V = \frac{1}{3}\pi r^2h$ **95. (a)** 123,454,321; **(b)** 12,345,654,321 **97. (a)** $5^2 = 25$; **(b)** $7^2 = 49$ **99.** The bases are different. **103.** x^{m+n} **105.** negative **107.** negative **109.** x^{mn} **111.** $\frac{x^m}{y^m}$ **113.** $30a^{11}$ **115.** $-12x^4yz^5$ **117.** $-5x^2y^6$ **119.** $\frac{ab^7c}{4}$ **121.** y^{20} **123.** $\frac{27y^6}{x^{15}}$ **125.** 4 **127.** 64 **129.** $\frac{1}{8}$ **131.** $\frac{81}{16}$

Exercises 4.2

1. $\frac{1}{4^2} = \frac{1}{16}$ **3.** $\frac{1}{5^3} = \frac{1}{125}$ **5.** $8^2 = 64$ **7.** x^7 **9.** $\frac{3^3}{4^2} = \frac{27}{16}$ **11.** $\frac{3^4}{5^2} = \frac{81}{25}$ **13.** $\frac{b^6}{a^5}$ **15.** $\frac{y^9}{x^9}$ **17.** 2^{-3} **19.** y^{-5} **21.** q^{-5} **23.** 3 **25.** 4 **27.** $\frac{1}{16}$ **29.** $\frac{1}{216}$ **31.** $\frac{1}{64}$ **33.** x^2 **35.** y^2 **37.** $\frac{1}{a^5}$ **39.** $\frac{1}{x^2}$ **41.** $\frac{1}{x^2}$ **43.** $\frac{1}{a^5}$ **45.** 1 **47.** $3^5 = 243$ **49.** $\frac{1}{4^3} = \frac{1}{64}$ **51.** $\frac{1}{y^2}$ **53.** x^3 **55.** $\frac{1}{x^2}$ **57.** $\frac{1}{x^7}$ **59.** x^3 **61.** $\frac{a^2}{b^6}$ **63.** $-\frac{8b^6}{27a^3}$ **65.** a^8b^4 **67.** $\frac{1}{x^{15}y^6}$ **69.** $x^{27}y^6$ **71.** $\frac{1}{2^4}, \frac{1}{16}$ **73. (a)** $\frac{1}{1000}$; **(b)** $\frac{1}{10^3}$; **(c)** 10^{-3}; **(d)** 10^{-3} **75.** 6.242 billion **77.** 4.922 billion **79. (a)** \$954.81; **(b)** \$1106.89 **81.** \$61.68 **83. (a)** \$11,616; **(b)** \$22,011.18 **85. (a)** \$13,312.80; **(b)** \$28,299.40 **87. (a)** 50% per day; **(b)** $A = 40(1 + 0.50)^n$; **(c)** 203; **(d)** 11,677; **(e)** Answers may vary. **91.** No **93.** 1 **95.** x^n **97.** x^{m+n} **99.** $\frac{6^2}{7^2} = \frac{36}{49}$ **101.** $9^2 = 81$ **103.** 7^{-6} **105.** $5^2 = 25$ **107.** z^2 **109.** $\frac{81x^8}{16y^{24}}$ **111. (a)** \$3645.00; **(b)** \$6172.84 **113.** 839 **115.** 0.0816

Exercises 4.3

1. 5.5×10^7 **3.** 3×10^8 **5.** 1.9×10^9 **7.** 2.4×10^{-4} **9.** 2×10^{-9} **11.** 153 **13.** 171,000,000 **15.** 6,850,000,000 **17.** 0.23 **19.** 0.00025 **21.** 1.5×10^{10} **23.** 3.06×10^4 **25.** 1.24×10^{-4} **27.** 2×10^3 **29.** 2.5×10^{-3} **31. (a)** 8.0592×10^9; **(b)** 8,059,200,000 **33.** 4.55 **35.** 2×10^7 **37.** Now: 2,000,000; later: 12,000,000 **39.** About 294,355 **41. (a)** Mercury; **(b)** Mars; **(c)** Mercury, Venus, Earth, Mars **43. (a)** Jupiter; **(b)** Uranus; **(c)** Uranus, Neptune, Saturn, Jupiter **45.** 2.99792458×10^8 **47.** 3.09×10^{13} **49.** 3.27 **53.** 10, integer **55.** powers **57.** Distance: 2.39×10^5; mass: 1.2456×10^{-1} **59.** $6 \times 10^0 = 6$ **61.** $1.32 \times 10^8 = 132{,}000{,}000$ **63.** 102 **65.** 8 **67.** 4

Exercises 4.4

1. Binomial; 1 **3.** Monomial; 1 **5.** Trinomial; 2 **7.** Monomial; 0 **9.** Binomial; 3 **11.** $8x^3 - 3x$; 3 **13.** $8x^2 + 4x - 7$; 2 **15.** $x^2 + 5x$; 2 **17.** $x^3 - x^2 + 3$; 3 **19.** $4x^5 - 3x^3 + 2x^2$; 5 **21. (a)** 4; **(b)** -8 **23. (a)** 7; **(b)** 7 **25. (a)** 9; **(b)** 13 **27. (a)** 9; **(b)** -3 **29. (a)** 3; **(b)** -1 **31. (a)** $(-16t^2 + 150)$ ft; **(b)** 134 ft; **(c)** 86 ft **33. (a)** $(-4.9t^2 + 200)$ m; **(b)** 195.1 m; **(c)** 180.4 m **35. (a)** 251; **(b)** About 610; about 1318 **37. (a)** About 11; **(b)** 11.75; The car gets 11.75 mpg at 5 mph; **(c)** About 30.1 and 30 respectively; **(d)** $P(50) = 32$, $P(55) = 31.75$; The car gets 32 mpg at 50 mph and 31.75 mpg at 55 mph; **(e)** Answers will vary; 11.75 and 11 are close, 32 and 31.75 are also close to the values in the graph. **39. (a)** 28°; **(b)** $-122°$; 122° below 0 is unreasonable. **41. (a)** 34 million; **(b)** $N(5) = 34.6$; the result is close. **43. (a)** \$21,236; **(b)** $C(5) = \$21{,}237$; close! **(c)** \$26,402 **45. (a)** \$2191; **(b)** $C(5) = \$2192$; close! **(c)** \$2742 **47. (a)** \$12,127; **(b)** $C(5) = \$12{,}129$; close! **(c)** \$15,819 **49.** $(-32t - 10)$ ft/sec **51. (a)** -11.8 m/sec; **(b)** -21.6 m/sec **55.** No; negative exponent on x. **57.** 0; $7^4 = 2401x^0$ **59.** polynomial **61.** binomial **63.** 12 **65.** 18 **67.** 8 **69.** No degree **71.** $5x^2 - 3x - 8$ **73.** Monomial **75.** Polynomial **77. (a)** 0.115; **(b)** 0.105; **(c)** 0.1149; very close; **(d)** 0.1057; very close **79.** $-7ab$ **81.** $3x^2y$ **83.** $8xy^2$

Exercises 4.5

1. $12x^2 + 5x + 6$ **3.** $-2x^2 - x - 8$ **5.** $-3x^2 + 7x - 5$ **7.** $-x^2 - 5$ **9.** $x^3 - 2x^2 - x - 2$ **11.** $2x^4 - 8x^3 + 2x^2 + x + 5$ **13.** $-\frac{3}{5}x^2 + x + 1$ **15.** $0.4x^2 + 0.3x - 0.4$ **17.** $-2x^2 - 2x + 3$ **19.** $3x^4 + x^3 - 5x - 1$ **21.** $-5x^4 + 5x^3 - 2x^2 - 2$ **23.** $5x^3 - 4x^2 + 5x - 1$ **25.** $-\frac{1}{7}x^3 - \frac{2}{3}x^2 + 5x$ **27.** $-\frac{3}{7}x^3 + \frac{1}{9}x^2 + x - 6$ **29.** $-4x^4 - 4x^3 - 3x^2 - x + 2$ **31.** $4x^2 + 7$ **33.** $-x^2 - 4x - 6$ **35.** $4x^2 - 7x + 6$ **37.** $-3x^3 + 6x^2 - 2x$ **39.** $7x^3 - x^2 + 2x - 6$ **41.** $3x^2 - 7x + 7$ **43.** 0 **45.** $4x^3 - 3x^2 - 7x + 6$ **47.** $3x^3 - 2x^2 + x - 8$ **49.** $-5x^3 - 5x^2 + 4x - 9$ **51.** $2x^2 + 6x$ **53.** $3x^2 + 4x$ **55.** $27x^2 + 10x$ **57. (a)** 31 mpg; **(b)** 25 mpg; **(c)** $E(t)$; **(d)** $-0.01t^2 + 0.16t + 6$; **(e)** $41 - 35.8 = 5.2$ mpg; Same answer: 5.2 mpg **59. (a)** $T(t) + B(t) + R(t) = 77t^2 + 337.5t + 8766$; **(b)** \$8766; **(c)** \$31,153.50 **61.** $T(t) + B(t) + R(t) - S(t) = -485.5t^2 + 25t + 6141$ **63. (a)** $-14t^2 + 73t + 1078$; **(b)** \$1078; **(c)** \$408 **65. (a)** $0.3t^3 - 3t^2 + 6t + 40$; **(b)** \$40,000; \$100,000 **67.** $R = x - 50$ **69.** $R = 2x$ **71.** 30 **77.** $a - b$ **79.** $5x^2 + 3x + 1$ **81.** $4x^3 + 10x^2 + 1$ **83.** $4x^2 + 8x$ **85.** $-6x^5$ **87.** $-6x^9$ **89.** $6y - 24$

Exercises 4.6

1. $45x^5$ **3.** $-10x^3$ **5.** $6y^3$ **7.** $3x + 3y$ **9.** $10x - 5y$ **11.** $-8x^2 + 12x$ **13.** $x^5 + 4x^4$ **15.** $4x^2 - 4x^3$ **17.** $3x^2 + 3xy$ **19.** $-8xy^2 + 12y^3$ **21.** $x^2 + 3x + 2$ **23.** $y^2 - 5y - 36$ **25.** $x^2 - 5x - 14$ **27.** $x^2 - 12x + 27$ **29.** $y^2 - 6y + 9$ **31.** $6x^2 + 7x + 2$ **33.** $6y^2 + y - 15$ **35.** $10z^2 + 43z - 9$ **37.** $6x^2 - 34x + 44$ **39.** $16z^2 + 8z + 1$ **41.** $6x^2 + 11xy + 3y^2$ **43.** $2x^2 + xy - 3y^2$ **45.** $10z^2 + 13yz - 3y^2$ **47.** $12x^2 - 11xz + 2z^2$ **49.** $4x^2 - 12xy + 9y^2$ **51.** $6 + 17x + 12x^2$ **53.** $6 - 7x - 3x^2$ **55.** $8 - 16x - 10x^2$ **57.** $(x^2 + 7x + 10)$ square units **59.** $(96t - 16t^2)$ ft **61.** $V_2CP + V_2PR - V_1CP - V_1PR$ **63.** 1200 ft² **65.** $800 + 160 + 40 + 200 = 1200$ ft² **67. (a)** $S_1 = 20x$ ft²; **(b)** $S_2 = xy$ ft²; **(c)** $S_3 = 40y$ ft² **69.** $800 + 20x + xy + 40y$; yes **71.** $Y = 1000x - x^2$ **73. (a)** $R = 1000p - 30p^2$; **(b)** \$8000 **75. (a)** $124 + 1.2t$; **(b)** 124 million pounds; **(c)** 136 million pounds **77. (a)** $0.03t^2 + 4.3t + 124$; **(b)** 124 million pounds; **(c)** 170 million pounds **79.** Yes **81.** No; consider the results in problem 82. **83.** first **85.** inner **87.** $-35x^6$ **89.** $x^2 + 4x - 21$ **91.** $9x^2 + 9x - 4$ **93.** $10x^2 - 19xy + 6y^2$ **95.** $6x - 18y$ **97.** $64x^2$ **99.** $9x^2$ **101.** $9A^2$

Exercises 4.7

1. $x^2 + 2x + 1$ **3.** $4x^2 + 4x + 1$ **5.** $9x^2 + 12xy + 4y^2$ **7.** $x^2 - 2x + 1$ **9.** $4x^2 - 4x + 1$ **11.** $9x^2 - 6xy + y^2$ **13.** $36x^2 - 60xy + 25y^2$ **15.** $4x^2 - 28xy + 49y^2$ **17.** $x^2 - 4$ **19.** $x^2 - 16$ **21.** $9x^2 - 4y^2$ **23.** $x^2 - 36$ **25.** $x^2 - 144$ **27.** $9x^2 - y^2$ **29.** $4x^2 - 49y^2$ **31.** $x^4 + 7x^2 + 10$ **33.** $x^4 + 2x^2y + y^2$ **35.** $9x^4 - 12x^2y^2 + 4y^4$ **37.** $x^4 - 4y^4$ **39.** $4x^2 - 16y^4$ **41.** $x^3 + 4x^2 + 8x + 15$ **43.** $x^3 + 3x^2 - x + 12$ **45.** $x^3 + 2x^2 - 5x - 6$ **47.** $x^3 - 8$ **49.** $-x^3 + 2x^2 - 3x + 2$ **51.** $-x^3 + 8x^2 - 15x - 4$ **53.** $2x^3 + 6x^2 + 4x$ **55.** $3x^3 + 3x^2 - 6x$ **57.** $4x^3 - 12x^2 + 8x$ **59.** $5x^3 - 20x^2 - 25x$ **61.** $x^3 + 15x^2 + 75x + 125$ **63.** $8x^3 + 36x^2 + 54x + 27$ **65.** $8x^3 + 36x^2y + 54xy^2 + 27y^3$ **67.** $16t^4 + 24t^2 + 9$ **69.** $16t^4 + 24t^2u + 9u^2$ **71.** $9t^4 - 2t^2 + \frac{1}{9}$ **73.** $9t^4 - 2t^2u + \frac{1}{9}u^2$ **75.** $9x^4 - 25$ **77.** $9x^4 - 25y^4$ **79.** $16x^6 - 25y^6$ **81.** $\frac{(x^3 + 8x^2 + 20x + 16)\pi}{3} = \frac{1}{3}\pi x^3 + \frac{8}{3}\pi x^2 + \frac{20}{3}\pi x + \frac{16}{3}\pi$ **83.** $\frac{4(x^3 + 3x^2 + 3x + 1)\pi}{3} = \frac{4}{3}\pi x^3 + 4\pi x^2 + 4\pi x + \frac{4}{3}\pi$ **85.** $(x^3 + 4x^2 + 5x + 2)\pi = \pi x^3 + 4\pi x^2 + 5\pi x + 2\pi$ **87. (a)** $t^2 - 20t + 100$; **(b)** $0.003t^2 - 0.06t + 1.3$; **(c)** \$3.70; **(d)** \$1.30; **(e)** Increasing **89.** $T_1^4 - T_2^4$ **91.** $Kt_n^2 - 2Kt_nt_a + Kt_a^2$ **93.** $v_i^2 - v_0^2$ **95. (a)** 9; **(b)** 5; **(c)** No **97. (a)** x^2; **(b)** xy; **(c)** y^2; **(d)** xy **99.** They are equal. **103.** When the binomials are the sum and difference of the same two terms. **105.** $X^2 + XB + AX + AB$ **107.** $X^2 - 2AX + A^2$ **109.** $x^2 + 14x + 49$ **111.** $4x^2 + 20xy + 25y^2$ **113.** $x^4 + 4x^2y + 4y^2$ **115.** $9x^2 - 4y^2$ **117.** $15x^2 + 2x - 24$ **119.** $21x^2 - 34xy + 8y^2$ **121.** $x^3 + 3x^2 + 3x + 2$ **123.** $x^4 + x^3 - x^2 - 2x - 2$ **125.** $x^3 + 6x^2y + 12xy^2 + 8y^3$ **127.** $3x^5 - 3x$ **129.** $2x$ **131.** $-\frac{1}{2}$

Exercises 4.8

1. $x + 3y$ **3.** $2x - y$ **5.** $-2y + 8 - \frac{4}{y}$ **7.** $10x + 8$ **9.** $3x - 2$ **11.** $x + 2$ **13.** $(y - 2)$ R -1 **15.** $x + 6$ **17.** $3x - 4$ **19.** $(2y - 5)$ R -1 **21.** $x^2 - x - 1$ **23.** $y^2 - y - 1$ **25.** $(4x^2 + 3x + 7)$ R 12 **27.** $x^2 + 2x + 4$ **29.** $4y^2 + 8y + 16$

31. $(x^2 + x + 1)$ R 3 **33.** $x^3 + x^2 - x + 1$ **35.** $(m^2 - 8)$ R 10 **37.** $x^2 + xy + y^2$ **39.** $(x^2 + 2x + 4)$ R 16 **41.** **(a)** $2 + \frac{20}{x}$; **(b)** \$22; **(c)** \$4 **43.** **(a)** $(x + 2)$ lb; **(b)** 7 lb **45.** **(a)** $\frac{20(1.01)^3 - 20}{0.01} = 60.602$ pounds; **(b)** $\frac{612.08}{60.602} \approx 10$ trees **47.** $\overline{C(x)} = 3x + 5$ **49.** $\overline{P(x)} = 50 + x - \frac{7000}{x}$ **55.** divisor, quotient, remainder **57.** $2x^2 + 2x + 3$ **59.** $4x^2 - 3x$ **61.** $2y - 4 - \frac{1}{y^2}$ **63.** 180 **65.** 120 **67.** 180

Review Exercises

1. [4.1A] **(a)** **(i)** $-15a^3b^4$; **(ii)** $-24a^3b^5$; **(iii)** $-35a^3b^4$; **(b)** **(i)** $6x^3y^3z^5$; **(ii)** $12x^3y^4z^3$; **(iii)** $20x^2y^3z^4$ **2.** [4.1B] **(a)** **(i)** $-2x^5y^4$; **(ii)** $-6x^6y^3$; **(iii)** $-2x^7y^3$; **(b)** **(i)** $\frac{x^5y^6}{2}$; **(ii)** $\frac{xy^7}{2}$; **(iii)** $\frac{xy^6}{3}$ **3.** [4.1C] **(a)** $2^6 = 64$; **(b)** $2^4 = 16$; **(c)** $3^4 = 81$ **4.** [4.1C] **(a)** y^6; **(b)** x^6; **(c)** a^{20} **5.** [4.1C] **(a)** $16x^2y^6$; **(b)** $8x^6y^3$; **(c)** $27x^6y^6$ **6.** [4.1C] **(a)** $-8x^3y^9$; **(b)** $9x^4y^6$; **(c)** $-8x^6y^6$ **7.** [4.1C] **(a)** $\frac{4y^4}{x^8}$; **(b)** $\frac{27x^3}{y^9}$; **(c)** $\frac{16x^8}{y^{16}}$ **8.** [4.1C] **(a)** $32x^{12}y^4$; **(b)** $-72x^4y^9$; **(c)** $256x^6y^{16}$ **9.** [4.2A] **(a)** $\frac{1}{2^3} = \frac{1}{8}$; **(b)** $\frac{1}{3^4} = \frac{1}{81}$; **(c)** $\frac{1}{5^2} = \frac{1}{25}$ **10.** [4.2A] **(a)** x^4; **(b)** y^3; **(c)** z^5 **11.** [4.2B] **(a)** x^{-5}; **(b)** y^{-7}; **(c)** z^{-8} **12.** [4.2C] **(a)** **(i)** $2^3 = 8$; **(ii)** $2^2 = 4$; **(iii)** $3^3 = 27$; **(b)** **(i)** $\frac{1}{y^8}$; **(ii)** $\frac{1}{y^5}$; **(iii)** $\frac{1}{y^6}$ **13.** [4.2C] **(a)** **(i)** $\frac{1}{x^4}$; **(ii)** $\frac{1}{x^6}$; **(iii)** $\frac{1}{x^8}$; **(b)** **(i–iii)** 1; **(c)** **(i)** x; **(ii)** x^3; **(iii)** x^2 **14.** [4.2C] **(a)** $\frac{9x^2}{4y^{14}}$; **(b)** $\frac{8x^6y^{21}}{27}$; **(c)** $\frac{4}{9x^2y^8}$ **15.** [4.3A] **(a)** **(i)** 4.4×10^7; **(ii)** 4.5×10^6; **(iii)** 4.6×10^5; **(b)** **(i)** 1.4×10^{-3}; **(ii)** 1.5×10^{-4}; **(iii)** 1.6×10^{-5} **16.** [4.3B] **(a)** **(i)** 2.2×10^5; **(ii)** 9.3×10^6; **(iii)** 1.24×10^8; **(b)** **(i)** 5; **(ii)** 6; **(iii)** 7 **17.** [4.4A] **(a)** Trinomial; **(b)** Monomial; **(c)** Binomial **18.** [4.4B] **(a)** 4; **(b)** 2; **(c)** 2 **19.** [4.4C] **(a)** $9x^4 + 4x^2 - 8x$; **(b)** $4x^2 - 3x - 3$; **(c)** $-4x^2 + 3x + 8$ **20.** [4.4D] **(a)** 284; **(b)** 156; **(c)** -100 **21.** [4.5A] **(a)** $5x^2 - x - 10$; **(b)** $-5x^2 + 15x + 2$; **(c)** $9x^2 - 7x + 1$ **22.** [4.5B] **(a)** $-x^2 - 7x + 4$; **(b)** $7x^2 - 7x + 3$; **(c)** $-5x^2 + 4x - 11$ **23.** [4.6A] **(a)** $-18x^7$; **(b)** $-40x^9$; **(c)** $-27x^{11}$ **24.** [4.6B] **(a)** $-2x^3 - 4x^2y$; **(b)** $-6x^4 - 9x^3y$; **(c)** $-20x^4 - 28x^3y$ **25.** [4.6C] **(a)** $x^2 + 15x + 54$; **(b)** $x^2 + 5x + 6$; **(c)** $x^2 + 16x + 63$ **26.** [4.6C] **(a)** $x^2 + 4x - 21$; **(b)** $x^2 + 4x - 12$; **(c)** $x^2 + 4x - 5$ **27.** [4.6C] **(a)** $x^2 - 4x - 21$; **(b)** $x^2 - 4x - 12$; **(c)** $x^2 - 4x - 5$ **28.** [4.6C] **(a)** $6x^2 - 13xy + 6y^2$; **(b)** $20x^2 - 27xy + 9y^2$; **(c)** $8x^2 - 26xy + 15y^2$ **29.** [4.7A] **(a)** $4x^2 + 12xy + 9y^2$; **(b)** $9x^2 + 24xy + 16y^2$; **(c)** $16x^2 + 40xy + 25y^2$ **30.** [4.7B] **(a)** $4x^2 - 12xy + 9y^2$; **(b)** $9x^2 - 12xy + 4y^2$; **(c)** $25x^2 - 20xy + 4y^2$ **31.** [4.7C] **(a)** $9x^2 - 25y^2$; **(b)** $9x^2 - 4y^2$; **(c)** $9x^2 - 16y^2$ **32.** [4.7D] **(a)** $x^3 + 4x^2 + 5x + 2$; **(b)** $x^3 + 5x^2 + 8x + 4$; **(c)** $x^3 + 6x^2 + 11x + 6$ **33.** [4.7E] **(a)** $3x^3 + 9x^2 + 6x$; **(b)** $4x^3 + 12x^2 + 8x$; **(c)** $5x^3 + 15x^2 + 10x$ **34.** [4.7E] **(a)** $x^3 + 6x^2 + 12x + 8$; **(b)** $x^3 + 9x^2 + 27x + 27$; **(c)** $x^3 + 12x^2 + 48x + 64$ **35.** [4.7E] **(a)** $25x^4 - 5x^2 + \frac{1}{4}$; **(b)** $49x^4 - 7x^2 + \frac{1}{4}$; **(c)** $81x^4 - 9x^2 + \frac{1}{4}$ **36.** [4.7E] **(a)** $9x^4 - 4$; **(b)** $9x^4 - 16$; **(c)** $4x^4 - 25$ **37.** [4.8A] **(a)** $2x^2 - x$; **(b)** $4x^2 - 2x$; **(c)** $4x^2 - 2x$ **38.** [4.8B] **(a)** $x + 6$; **(b)** $x + 7$; **(c)** $x + 8$ **39.** [4.8B] **(a)** $4x^2 - 4x - 4$; **(b)** $6x^2 - 6x - 6$; **(c)** $2x^2 - 2x - 2$ **40.** [4.8B] **(a)** $(2x^2 + 6x - 2)$ R 2; **(b)** $(2x^2 + 6x - 3)$ R 3; **(c)** $(3x^2 + 3x - 1)$ R 4 **41.** [4.8B] **(a)** $(x^2 + x)$ R $(4x + 1)$; **(b)** $(x^2 + x)$ R $(5x + 1)$; **(c)** $(x^2 + x)$ R $(6x + 1)$

Cumulative Review Chapters 1–4

1. 7 **2.** $9\frac{9}{10}$ **3.** $-\frac{5}{18}$ **4.** 13.6 **5.** -6.24 **6.** -16 **7.** $\frac{6}{7}$ **8.** 47 **9.** Associative property of multiplication **10.** $6x - 24$ **11.** $-7xy^3$ **12.** $4x + 10$ **13.** $\frac{m + 3n}{p}$ **14.** No **15.** $x = 15$ **16.** $x = 49$ **17.** $x = 36$ **18.** $x = 6$ **19.** $b = \frac{S}{6a^2}$ **20.** 40 and 60 **21.** \$430 from stocks and \$185 from bonds **22.** 18 hr **23.** \$8000 in bonds; \$9000 in certificates of deposit

24.

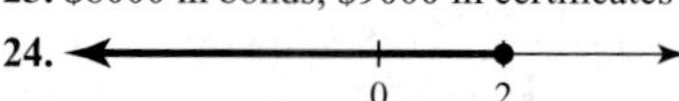

25.

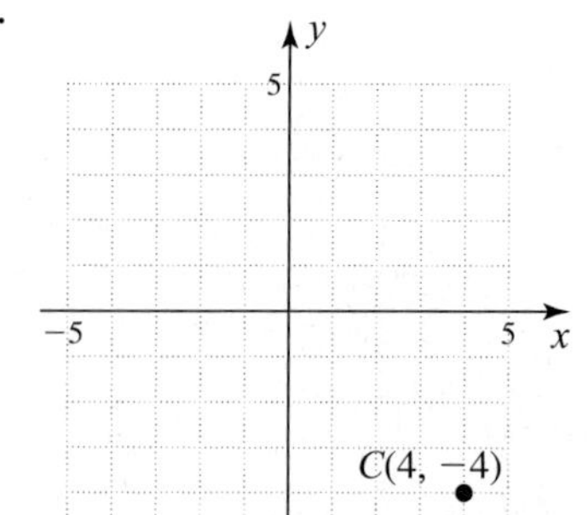

26. $(3, -2)$ **27.** Yes **28.** $x = 1$

29. $5x + y = 5$

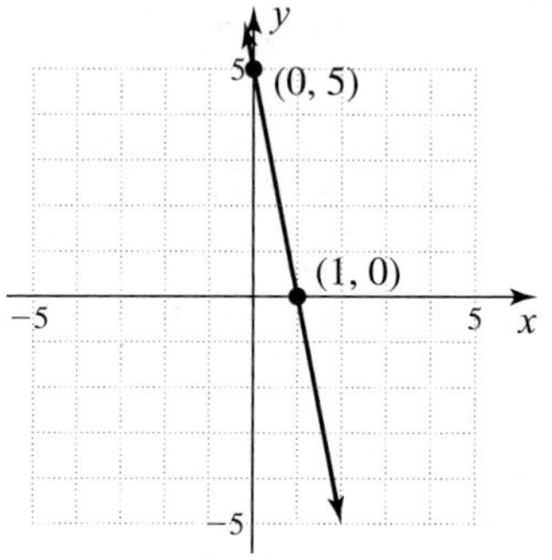

30. $3y - 15 = 0$

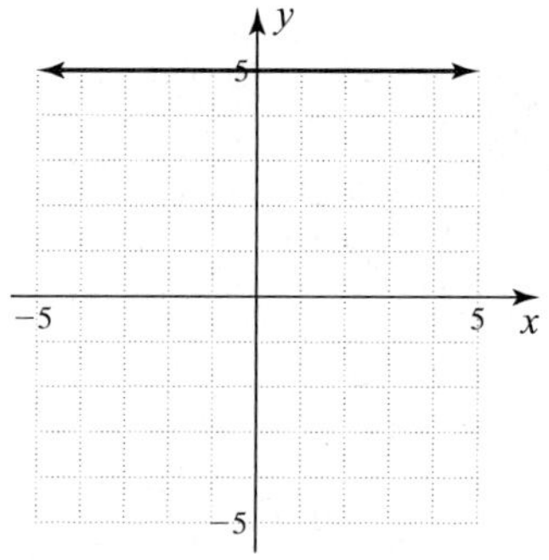

31. $\frac{7}{5}$ **32.** 2 **33.** (2) and (3) **34.** $18x^7y^3$ **35.** $-\frac{5x}{y^2}$ **36.** x^3 **37.** x^5 **38.** $\frac{64x^{12}}{y^{12}}$ **39.** 8.0×10^6 **40.** 4.2×10^{-8} **41.** Monomial **42.** 2 **43.** $-3x^3 - 3x^2 - x + 4$ **44.** 32 **45.** $4x^3 + 9x - 3$ **46.** $-9x^3 - 12xy$ **47.** $9x^2 + 12xy + 4y^2$ **48.** $16x^2 - 9y^2$ **49.** $25x^4 - 5x^2 + \frac{1}{4}$ **50.** $16x^4 - 81$ **51.** $(3x^2 - 5x + 4)$ R 3 **52.** $y - 6 = 4(x + 2)$ or $y = 4x + 14$ **53.** $y = -2x + 5$ **54.** $2x - 3y < -6$

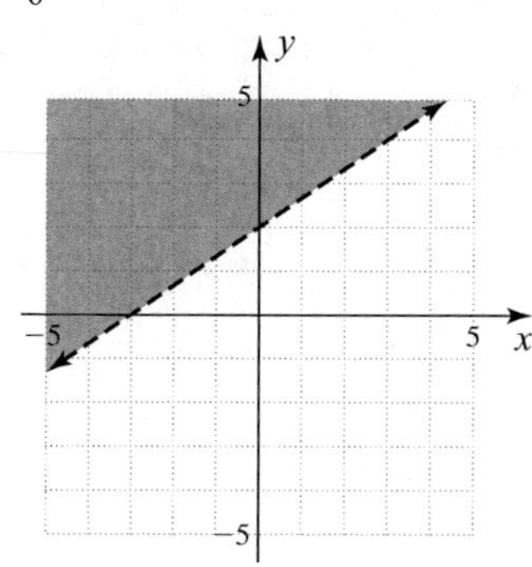

55. $3x - y \geq 0$

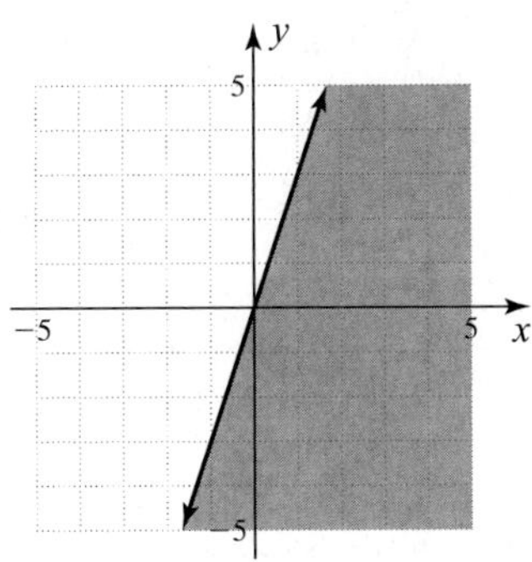

Chapter 5

Exercises 5.1

1. 4 **3.** 8 **5.** 1 **7.** a^3 **9.** x^3 **11.** $5y^6$ **13.** $2x^3$ **15.** $3bc$ **17.** 3 **19.** $9ab^3z$ **21.** $3(x + 5)$ **23.** $9(y - 2)$ **25.** $-5(y - 4)$ **27.** $-3(x + 9)$ **29.** $4x(x + 8)$ **31.** $6x(1 - 7x)$ **33.** $-5x^2(1 + 5x^2)$ **35.** $3x(x^2 + 2x + 3)$ **37.** $9y(y^2 - 2y + 3)$ **39.** $6x^2(x^4 + 2x^3 - 3x^2 + 5)$ **41.** $8y^3(y^5 + 2y^2 - 3y + 1)$ **43.** $\frac{1}{7}(4x^3 + 3x^2 - 9x + 3)$ **45.** $\frac{1}{8}y^2(7y^7 + 3y^4 - 5y^2 + 5)$ **47.** $(x + 4)(3 - y)$ **49.** $(y - 2)(x - 1)$ **51.** $(t + s)(c - 1)$ **53.** $4x^3(1 + x - 3x^2)$ **55.** $6y^7(1 - 2y^2 - y^4)$ **57.** $(x + 2)(x^2 + 1)$ **59.** $(y - 3)(y^2 + 1)$ **61.** $(2x + 3)(2x^2 + 1)$ **63.** $(3x - 1)(2x^2 + 1)$ **65.** $(y + 2)(4y^2 + 1)$ **67.** $(2a + 3)(a^2 + 1)$ **69.** $(x^2 + 4)(3x^2 + 1)$ **71.** $(2y^2 + 3)(3y^2 + 1)$ **73.** $(y^2 + 3)(4y^2 + 1)$ **75.** $(a^2 - 2)(3a^2 - 2)$ **77.** $(6a - 5b)(1 + 2d)$ **79.** $(x^2 - y)(1 - 3z)$ **81.** **(a)** $\alpha L(t_2 - t_1)$; **(b)** $t_2 - t_1$ **83.** **(a)** 5000; Same; **(b)** 8000; Same; **(c)** 24,800; It is in the range 20,000–25,000; **(d)** Yes **85.** $2\pi r(h + r)$ **87.** $P(Q_2 - Q_1)$ **89.** $V(k - PV)$ **91.** $f(u + v_s)$ **93.** $-w(\ell - z)$ **95.** $a(a + 2s)$ **97.** $-16(t^2 - 5t - 15)$ **101.** reverse **103.** smallest **105.** $(3x^2 + 1)(2x^2 - 3)$ **107.** $(2x - 3)(3x^2 - 1)$ **109.** $3x^2(x^4 - 2x^3 + 4x^2 + 9)$ **111.** $(2x + y)(6x - 5y)$ **113.** $-3(y - 7)$ **115.** $(x + b)(5x + 6y)$ **117.** 2 **119.** $x^2 + 8x + 15$ **121.** $x^2 - 5x + 4$ **123.** $2x^2 - x - 3$

Exercises 5.2

1. $(y + 2)(y + 4)$ **3.** $(x + 2)(x + 5)$ **5.** $(y + 5)(y - 2)$ **7.** $(x + 7)(x - 2)$ **9.** $(x + 1)(x - 7)$ **11.** $(y + 2)(y - 7)$ **13.** $(y - 2)(y - 1)$ **15.** $(x - 4)(x - 1)$ **17.** Not factorable (prime) **19.** Not factorable (prime) **21.** $(x + a)(x + 2a)$ **23.** $(z + 3b)(z + 3b) = (z + 3b)^2$ **25.** $(r + 4a)(r - 3a)$ **27.** $(x + 10a)(x - a)$ **29.** $-(b - y)(b + 3y)$ **31.** $(m - 2a)(m + a)$ **33.** $2t(t + 1)(t + 4)$ **35.** $ax^2(ax + 3a + 2)$ **37.** $b^3x^5(x - 3)(x + 4)$ **39.** $2c^5z^4(z - 3)(z + 5)$ **41.** **(a)** $-5t^2 + 5t + 10$; **(b)** $-5(t - 2)(t + 1)$; **(c)** $t = 2$ and $t = -1$; **(d)** 2 seconds **43.** **(a)** $-16(t - 5)(t + 3)$; **(b)** $t = 5$ and $t = -3$; **(c)** 5 seconds **45.** **(a)** $-16(t + 1)(t - 2)$; **(b)** 32 ft; **(c)** 0 ft **47.** **(a)** $(x + 4)(x - 5)$; **(b)** 20 **49.** $(D + 6)(D + 2)$ **51.** $-(x - 2L)(x - 2L) = -(x - 2L)^2$ **53.** **(a)** $-16(t + 3)(t - 8)$; **(b)** 2005: $A(0) = 384$; 2006: $A(1) = 448$; 2007: $A(2) = 480$; **(c)** Increasing; **(d)** 384 **63.** positive **65.** positive, negative **67.** $(y - 2)(3y^2 - 1)$ **69.** $(y - 3)(2y^2 - 1)$ **71.** $(x + 4y)(x + 3y)$ **73.** $(x - 2y)(x - 5y)$ **75.** Not factorable (prime) **77.** $2y^3(1 + 3y^2 + 5y^4)$ **79.** **(a)** -12; **(b)** -4 **81.** **(a)** 4; **(b)** -5

Exercises 5.3

1. $(2x + 3)(x + 1)$ **3.** $(2x + 3)(3x + 1)$ **5.** $(2x + 1)(3x + 4)$ **7.** $(x + 2)(2x - 1)$ **9.** $(x + 6)(3x - 2)$ **11.** $(4y - 3)(y - 2)$ **13.** $2(2y^2 - 4y + 3)$ **15.** $2(3y + 1)(y - 2)$ **17.** $(3y + 2)(4y - 3)$ **19.** $3(3y + 1)(2y - 3)$ **21.** $(3x + 1)(x + 2)$ **23.** $(5x + 1)(x + 2)$ **25.** Not factorable (prime) **27.** $(3x + 1)(x - 2)$ **29.** $(5x + 2)(3x - 1)$ **31.** $4(2x + y)(x + 2y)$ **33.** $(2x + 3y)(3x - y)$ **35.** $(7x - 3y)(x - y)$ **37.** $(3x + y)(5x - 2y)$ **39.** Not factorable (prime) **41.** $(3r + 5)(4r - 1)$ **43.** $(11t + 2)(2t - 3)$ **45.** $3(3x - 2)(2x - 1)$ **47.** $a(3b + 1)(2b + 1)$ **49.** $x^3y(6x + y)(x + 4y)$ **51.** $-(3x + 2)(2x + 1)$ **53.** $-(3x + 2)(3x - 1)$ **55.** $-(4m + n)(2m - 3n)$ **57.** $-(8x - y)(x - y)$ **59.** $-x(x + 3)(x + 2)$ **61.** $(2g + 9)(g - 4)$ **63.** $(2R - 1)(R - 1)$ **65.** **(a)** $-(3t + 250)(t - 20)$; 250 gallons; **(b)** $-(2t + 11)(t - 20)$; 11 gallons **67.** $5(m + 49)(m - 5)$ **69.** $-5(t - 1)(t + 2)$ **71.** $(3x - 5)(x - 4)$ **73.** $(2L - 3)(L - 3)$ **75.** $(2L - x)(L - x)$ **77.** **(a)** $(2x + 3)(3x + 1)$; **(b)** $(3x + 1)(2x + 3)$; **(c)** Both **79.** ac, b **81.** $(3x + 2)(x - 2)$ **83.** $(2x + 3y)(x - 2y)$ **85.** $2(4x - 1)(2x + 1)$ **87.** $x^2(3x^2 + 5x - 3)$ **89.** $x^2 + 16x + 64$ **91.** $9x^2 - 12x + 4$ **93.** $4x^2 + 12xy + 9y^2$ **95.** $9x^2 - 25y^2$ **97.** $x^4 - 16$

Exercises 5.4

1. Yes **3.** No **5.** Yes **7.** No **9.** Yes **11.** $(x + 1)^2$ **13.** $3(x + 5)^2$ **15.** $(3x + 1)^2$ **17.** $(3x + 2)^2$ **19.** $(4x + 5y)^2$ **21.** $(5x + 2y)^2$ **23.** $(y - 1)^2$ **25.** $3(y - 4)^2$ **27.** $(3x - 1)^2$ **29.** $(4x - 7)^2$ **31.** $(3x - 2y)^2$ **33.** $(5x - y)^2$ **35.** $(x + 7)(x - 7)$ **37.** $(3x + 7)(3x - 7)$ **39.** $(5x + 9y)(5x - 9y)$ **41.** $(x^2 + 1)(x + 1)(x - 1)$ **43.** $(4x^2 + 1)(2x + 1)(2x - 1)$ **45.** $\left(\frac{1}{3}x + \frac{1}{4}\right)\left(\frac{1}{3}x - \frac{1}{4}\right)$ **47.** $\left(\frac{1}{2}z + 1\right)\left(\frac{1}{2}z - 1\right)$ **49.** $\left(1 + \frac{1}{2}s\right)\left(1 - \frac{1}{2}s\right)$ **51.** $\left(\frac{1}{2} + \frac{1}{3}y\right)\left(\frac{1}{2} - \frac{1}{3}y\right)$ **53.** Not factorable (prime) **55.** $3x(x + 2)(x - 2)$ **57.** $5t(t + 2)(t - 2)$ **59.** $5t(1 + 2t)(1 - 2t)$ **61.** $(7x + 2)^2$ **63.** $(x + 10)(x - 10)$ **65.** $(x + 10)^2$ **67.** $(3 + 4m)(3 - 4m)$ **69.** $(3x - 5y)^2$ **71.** $(z^2 + 4)(z + 2)(z - 2)$ **73.** $3x(x + 5)(x - 5)$ **75.** $(9 + x)(9 - x)$ **77.** $(C + kp)(C - kp)$ **79.** **(a)** $\frac{31(1.02)^2 - 31}{0.02} = 62.62$ pounds; **(b)** $I(x + 1)(x - 1)$; **(c)** $\frac{I(x + 1)(x - 1)}{x - 1} = I(x + 1)$ **(d)** $I(x + 1) = 31(2.02) = 62.62$; Same answer **81.** $D(x) = (x - 7)^2$ **83.** $C(x) = (x + 6)^2$ **87.** $(x + 2)$ **89.** $(X + A)^2$ **91.** $(X + A)(X - A)$ **93.** $(x + 1)(x - 1)$ **95.** $(3x + 5y)(3x - 5y)$ **97.** $(3x - 4y)^2$ **99.** $(4x + 3y)^2$ **101.** $(3x + 5)^2$ **103.** Not factorable (prime) **105.** $\left(\frac{1}{9} + \frac{1}{2}x\right)\left(\frac{1}{9} - \frac{1}{2}x\right)$ **107.** $2x(3x + 5y)(3x - 5y)$ **109.** Yes; $(x + 3)^2$ **111.** No **113.** Yes; $(2x - 5y)^2$ **115.** $R^2 - r^2$ **117.** $6(x + 1)(x - 4)$ **119.** $2(x + 3)(x - 3)$

Exercises 5.5

1. $(x + 2)(x^2 - 2x + 4)$ **3.** $(2m - 3)(4m^2 + 6m + 9)$ **5.** $(3m - 2n)(9m^2 + 6mn + 4n^2)$ **7.** $s^3(4 - s)(16 + 4s + s^2)$ **9.** $x^4(3 + 2x)(9 - 6x + 4x^2)$ **11.** $3(x - 3)(x + 2)$ **13.** $(5x + 1)(x + 2)$ **15.** $3x(x^2 + 2x + 7)$ **17.** $2x^2(x^2 - 2x - 5)$ **19.** $2x^2(2x^2 + 6x + 9)$ **21.** $(x + 2)(3x^2 + 1)$ **23.** $(x + 1)(3x^2 + 2)$ **25.** $(x + 1)(2x^2 - 1)$ **27.** $3(x + 4)^2$ **29.** $k(x + 2)^2$ **31.** $4(x - 3)^2$ **33.** $k(x - 6)^2$ **35.** $3x(x + 2)^2$ **37.** $2x(3x + 1)^2$ **39.** $3x^2(2x - 3)^2$ **41.** $(x^2 + 1)(x + 1)(x - 1)$ **43.** $(x^2 + y^2)(x + y)(x - y)$ **45.** $(x^2 + 4y^2)(x + 2y)(x - 2y)$ **47.** $-(x + 3)^2$ **49.** $-(x + 2)^2$ **51.** $-(2x + y)^2$ **53.** $-(3x + 2y)^2$ **55.** $-(2z - 3y)^2$ **57.** $-2x(3x + 2y)^2$ **59.** $-2x(3x + 5y)^2$ **61.** $-x(x + 1)(x - 1)$ **63.** $-x^2(x + 2)(x - 2)$ **65.** $-x^2(2x + 3)(2x - 3)$ **67.** $-2x(x - 2)(x^2 + 2x + 4)$ **69.** $-2x^2(2x + 1)(4x^2 - 2x + 1)$ **71.** $-16(t + 3)(t - 8)$ **73.** **(a)** $P(1 - r)^2$; **(b)** \$18,062.50 **75.** $\frac{2\pi A}{360}(R_1 + Kt)$ **77.** $\frac{3S}{2bd^3}(d + 2z)(d - 2z)$ **81.** $x^2 - 9$ can be factored as $(x + 3)(x - 3)$ **83.** $(X - A)(X^2 + AX + A^2)$ **85.** look **87.** $5x^2(x^2 - 2x + 4)$ **89.** $(3x + 7)(2x - 5)$ **91.** $(3t - 4)(9t^2 + 12t + 16)$ **93.** $(x^2 + 9)(x + 3)(x - 3)$ **95.** $-(3x + 5y)^2$ **97.** $(4y + 3x)(16y^2 - 12xy + 9x^2)$ **99.** $-x^2(x + y)(x^2 - xy + y^2)$ **101.** $(2x + 3)(5x - 1)$ **103.** $(2x + 1)(x - 3)$

Exercises 5.6

1. $x = -3$ or $x = -\frac{1}{2}$ **3.** $x = 1$ or $x = -\frac{3}{2}$ **5.** $y = 3$ or $y = \frac{2}{3}$ **7.** $y = 1$ or $y = -\frac{1}{3}$ **9.** $x = -\frac{4}{3}$ or $x = -\frac{1}{2}$ **11.** $x = 2$ or $x = -\frac{1}{3}$ **13.** $x = \frac{4}{5}$ or $x = -2$ **15.** $x = 1$ or $x = \frac{8}{5}$ **17.** $y = 5$ or $y = \frac{2}{3}$

19. $y = -2$ or $y = -\frac{1}{2}$ **21.** $x = -\frac{1}{3}$ **23.** $y = 4$ **25.** $x = -\frac{2}{3}$
27. $y = \frac{5}{2}$ **29.** $x = -5$ **31.** $x = 4$ or $x = 1$ **33.** $x = -2$ or $x = -\frac{5}{2}$
35. $x = 1$ **37.** $x = \frac{1}{2}$ or $x = -\frac{1}{2}$ **39.** $y = \frac{5}{2}$ or $y = -\frac{5}{2}$
41. $z = 3$ or $z = -3$ **43.** $x = \frac{7}{5}$ or $x = -\frac{7}{5}$ **45.** $m = 0$ or $m = 5$
47. $n = 0$ or $n = 5$ **49.** $y = -3$ or $y = -8$ **51.** $y = 9$ or $y = 7$
53. $v = 2$ or $v = -1$ or $v = -2$ **55.** $m = 2$ or $m = 1$ or $m = 4$
57. $n = 4$ or $n = -1$ or $n = -2$ **59.** $x = 1$ or $x = -3$
61. (a) $m = -49$ or $m = 5$; **(b)** No; $m = 5$ **63. (a)** 1 sec;
(b) $t = 1$ or $t = -2$; No; $t = 1$ **65.** (a) When $t = 10$ (2013);
(b) When $t = 20$ (2023) **69.** One **71.** $A = 0, B = 0$ **73.** $x = 3$
or $x = -\frac{2}{3}$ **75.** $x = -2$ or $x = \frac{1}{5}$ **77.** $x = 0$ or $x = 1$ **79.** $x = 4$
or $x = -5$ or $x = 1$ **81.** $m = -\frac{2}{3}$ or $m = -1$ **83.** $2H^2 + 6H + 9$
85. $H^2 - 6H - 27$

Exercises 5.7

Translate This

1. K **3.** B **5.** A **7.** G **9.** J

1. 8, 9 or 1, 2 **3.** 3, 5 or −3, −1 **5.** 20,000 million liters
7. $b = 10$ in., $h = 8$ in. **9.** $L = 15$ in., $h = 10$ in.
11. $x = 4$; $r = 7$ units **13.** $L = 50$ ft, $W = 5$ ft **15.** About 247 ft
17. 6 in., 8 in., 10 in. **19.** 9 in., 12 in., 15 in. **21.** 30 mi/hr
23. (a) 35 mi/hr; **(b)** Yes **25.** 40 mi/hr **27.** 5 teams **29.** 8 delegates
31. 30 students **33.** 1 sec **35.** 1 sec **37.** The first formula
41. $a^2 + b^2 = c^2$ **43.** 3, 5 or −7, −5 **45.** 5 in., 12 in., 13 in.
47. 50 mi/hr **49.** $\frac{15}{18}$ **51.** $18(x - 2y)$ **53.** $(x + 6)(x - 1)$

Review Exercises

1. [5.1A] **(a)** 30; **(b)** 6; **(c)** 1 **2.** [5.1B] **(a)** $6x^5$; **(b)** $2x^8$; **(c)** x^3
3. [5.1C] **(a)** $5x^3(4 - 11x^2)$; **(b)** $7x^4(2 - 5x^2)$; **(c)** $8x^7(2 - 5x^2)$
4. [5.1C] **(a)** $\frac{1}{7}x^2(3x^4 - 5x^3 + 2x^2 - 1)$; **(b)** $\frac{1}{9}x^3(4x^4 - 2x^3 + 2x^2 - 1)$;
(c) $\frac{1}{8}x^5(3x^4 - 7x^3 + 3x^2 - 1)$ **5.** [5.1D] **(a)** $(x - 7)(3x^2 - 1)$;
(b) $(x + 6)(3x^2 + 1)$; **(c)** $(x - 2y)(4x^2 + 1)$
6. [5.2A] **(a)** $(x + 7)(x + 1)$; **(b)** $(x - 9)(x + 1)$; **(c)** $(x + 5)(x + 1)$
7. [5.2A] **(a)** $(x - 5)(x - 2)$; **(b)** $(x - 7)(x - 2)$; **(c)** $(x + 4)(x - 2)$
8. [5.3B] **(a)** $(2x + 3)(3x - 2)$; **(b)** $(2x + 1)(3x - 1)$;
(c) $(2x + 5)(3x - 1)$ **9.** [5.3B] **(a)** $(3x - y)(2x - 5y)$;
(b) $(3x - 2y)(2x - y)$; **(c)** $(3x - 4y)(2x - y)$
10. [5.4B] **(a)** $(x + 2)^2$; **(b)** $(x + 5)^2$; **(c)** $(x + 4)^2$
11. [5.4B] **(a)** $(3x + 2y)^2$; **(b)** $(3x + 5y)^2$; **(c)** $(3x + 4y)^2$
12. [5.4B] **(a)** $(x - 2)^2$; **(b)** $(x - 3)^2$; **(c)** $(x - 6)^2$
13. [5.4B] **(a)** $(2x - 3y)^2$; **(b)** $(2x - 5y)^2$; **(c)** $(2x - 7y)^2$
14. [5.4C] **(a)** $(x + 6)(x - 6)$; **(b)** $(x + 7)(x - 7)$; **(c)** $(x + 9)(x - 9)$
15. [5.4C] **(a)** $(4x + 9y)(4x - 9y)$; **(b)** $(5x + 8y)(5x - 8y)$;
(c) $(3x + 10y)(3x - 10y)$ **16.** [5.5A] **(a)** $(m + 5)(m^2 - 5m + 25)$;
(b) $(n + 4)(n^2 - 4n + 16)$; **(c)** $(y + 2)(y^2 - 2y + 4)$
17. [5.5A] **(a)** $(2y - 3x)(4y^2 + 6xy + 9x^2)$;
(b) $(4y - 5x)(16y^2 + 20xy + 25x^2)$;
(c) $(2m - 5n)(4m^2 + 10mn + 25n^2)$
18. [5.5B] **(a)** $3x(x^2 - 2x + 9)$; **(b)** $3x(x^2 - 2x + 10)$;
(c) $4x(x^2 - 2x + 8)$ **19.** [5.5B] **(a)** $2x(x - 2)(x + 1)$;
(b) $3x(x - 3)(x + 1)$; **(c)** $4x(x - 4)(x + 1)$
20. [5.5B] **(a)** $(x + 4)(2x^2 + 1)$; **(b)** $(x + 5)(2x^2 + 1)$;
(c) $(x + 6)(2x^2 + 1)$ **21.** [5.5B] **(a)** $k(3x + 2)^2$; **(b)** $k(3x + 5)^2$;
(c) $k(2x + 5)^2$ **22.** [5.5D] **(a)** $-3x^2(x + 3)(x - 3)$;
(b) $-4x^2(x + 4)(x - 4)$; **(c)** $-5x^2(x + 2)(x - 2)$
23. [5.5D] **(a)** $-(x + y)(x^2 - xy + y^2)$;
(b) $-(2m + 3n)(4m^2 - 6mn + 9n^2)$;
(c) $-(4n + m)(16n^2 - 4mn + m^2)$
24. [5.5D] **(a)** $-(y - x)(y^2 + xy + x^2)$;
(b) $-(2m - 3n)(4m^2 + 6mn + 9n^2)$;
(c) $-(4t - 5s)(16t^2 + 20st + 25s^2)$

25. [5.5D] **(a)** $-(4x^2 + 12xy - 9y^2)$; **(b)** $-(25x^2 + 30xy - 9y^2)$;
(c) $-(16x^2 + 24xy - 9y^2)$ **26.** [5.6A] **(a)** $x = 5$ or $x = -1$;
(b) $x = 6$ or $x = -1$; **(c)** $x = 7$ or $x = -1$
27. [5.6A] **(a)** $x = -\frac{5}{2}$ or $x = 2$; **(b)** $x = -\frac{5}{2}$ or $x = 1$;
(c) $x = -\frac{3}{2}$ or $x = 1$ **28.** [5.6A] **(a)** $x = 1$ or $x = \frac{4}{3}$;
(b) $x = 3$ or $x = \frac{7}{2}$; **(c)** $x = 3$ or $x = -1$ **29.** [5.7A] **(a)** 4, 6;
(b) 2, 4 or −2, 0; **(c)** 10, 12 **30.** [5.7C] 18 in., 24 in., 30 in.

Cumulative Review Chapters 1–5

1. $-\frac{13}{24}$ **2.** 16.4 **3.** −31.35 **4.** −25 **5.** 3 **6.** 69
7. Associative property of multiplication **8.** xy^4 **9.** $-x - 11$
10. $\frac{d + 5e}{f}$ **11.** $x = 3$ **12.** $x = 63$ **13.** $x = 6$ **14.** 40 and 65
15. 10 hr **16.** \$4000 in bonds; \$2000 in certificates of deposit
17.

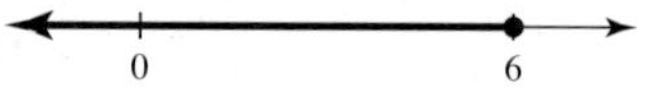

18.

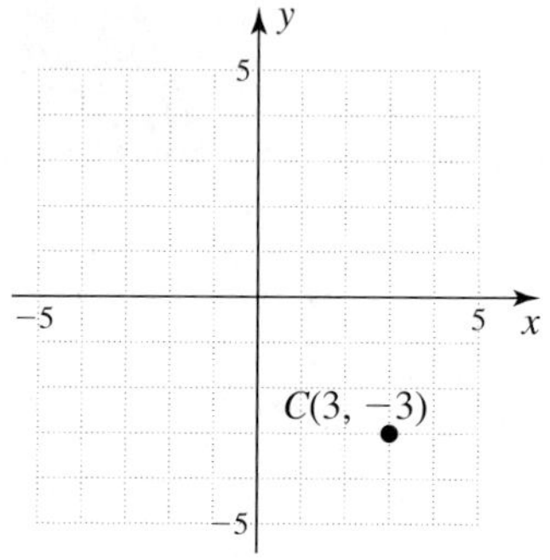

19. No **20.** $x = 2$
21. $x + y = 4$

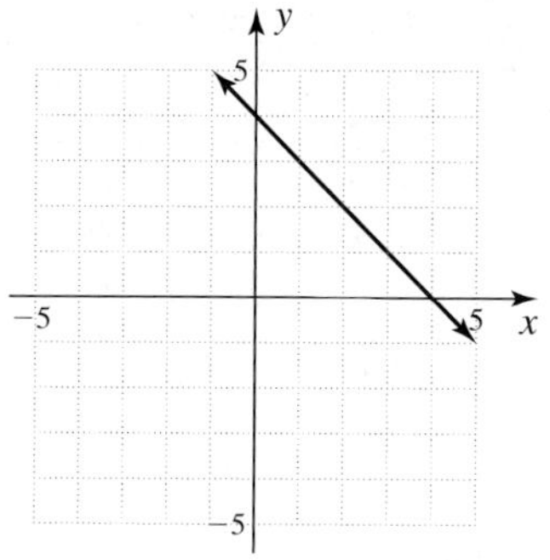

22. $4y - 20 = 0$

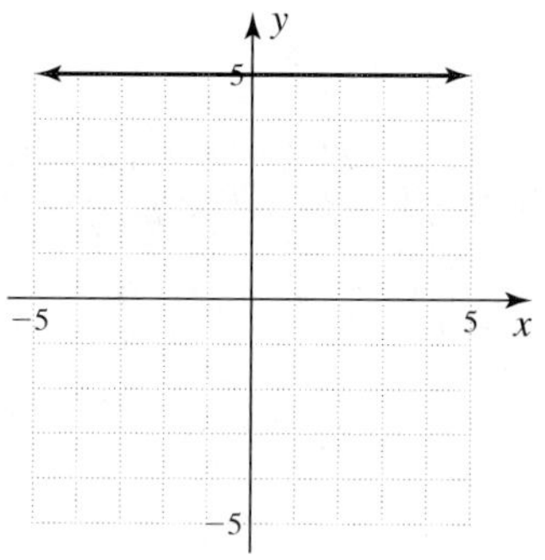

23. $\frac{2}{11}$ **24.** 3 **25.** (1) and (2) **26.** $y - 3 = 6(x + 5)$ or $y = 6x + 33$
27. $y = -3x + 4$

28.

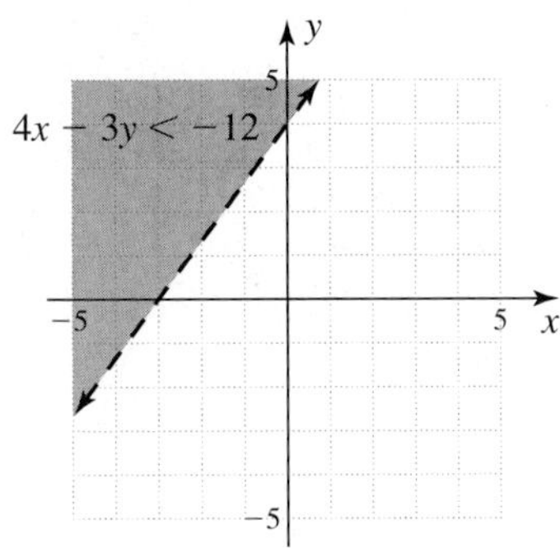

29.

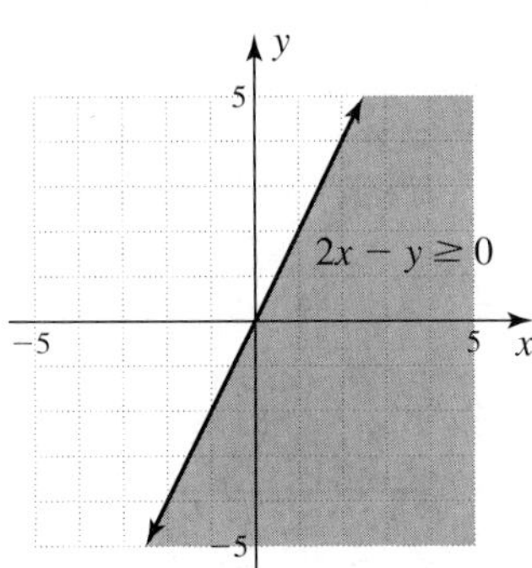

30. $-\frac{4x^2}{y^4}$ **31.** x^2 **32.** $\frac{1}{x^4}$ **33.** $\frac{y^9}{27x^9}$ **34.** 4.8×10^{-5} **35.** 4.60×10^{-5}
36. 2 **37.** 1 **38.** $3x^3 - 7x^2 - 13$ **39.** $-32x^6 - 16x^4y$
40. $9x^2 + 12xy + 4y^2$ **41.** $25x^2 - 49y^2$ **42.** $4x^4 - \frac{4}{5}x^2 + \frac{1}{25}$
43. $25x^4 - 81$ **44.** $(2x^2 + 5x + 5)$ R 1 **45.** $2x^6(6 - 7x^3)$
46. $\frac{1}{5}x^3(4x^4 - 3x^3 + 4x^2 - 2)$ **47.** $(x - 3)(x - 9)$
48. $(5x - 2y)(4x - 3y)$ **49.** $(5x + 7y)(5x - 7y)$
50. $-5x^2(x + 4)(x - 4)$ **51.** $3x(x + 1)(x - 3)$ **52.** $(2x + 5)(x + 3)$
53. $k(3x + 1)^2$ **54.** $x = -5$ or $x = \frac{3}{4}$

Chapter 6

Exercises 6.1

1. $x = 7$ **3.** $y = -4$ **5.** $x = 3$ or $x = -3$ **7.** None **9.** $y = 2$ or $y = 4$
11. $x = -2$ **13.** $x = 3$ **15.** $x = -3$ **17.** $y = -2$ or $y = 8$
19. $x = 0$ or $x = -1$ **21.** $\frac{9}{21}$ **23.** $\frac{-16}{22}$ **25.** $\frac{20xy}{24y^3}$ **27.** $\frac{-9xy^3}{21y^4}$
29. $\frac{4x(x-2)}{x^2 - x - 2}$ or $\frac{4x^2 - 8x}{x^2 - x - 2}$ **31.** $\frac{-5x(x-2)}{x^2 + x - 6}$ or $\frac{-5x^2 + 10x}{x^2 + x - 6}$
33. $\frac{x^2}{2y^3}$ **35.** $\frac{-3y^4}{x}$ **37.** $\frac{1}{2xy^3}$ **39.** $\frac{5x}{y^2}$ **41.** $\frac{x - y}{3}$
43. $-3(x - y)$ or $3y - 3x$ **45.** $\frac{-1}{4(x - y)}$ or $\frac{1}{4y - 4x}$
47. $\frac{1}{2(x + 2)}$ or $\frac{1}{2x + 4}$ **49.** $\frac{1}{x + y}$ **51.** $\frac{1}{2}$ **53.** $\frac{1}{3}$ **55.** $\frac{-1}{1 + 2y}$
57. $\frac{-1}{1 + 2y}$ **59.** $\frac{-1}{x + 2}$ **61.** -1 **63.** $-(x + 5)$ or $-x - 5$
65. $-(x - 2)$ or $2 - x$ **67.** $\frac{-1}{x + 6}$ **69.** $\frac{1}{x - 2}$
71. **(a)** \$50 million; \$130 million; \$80 million; **(b)** \$30 million; \$78 million; \$63 million; **(c)** \$20 million; \$52 million; \$17 million; **(d)** The fraction of the total advertising that is spent on national advertising **73.** **(a)** 10%; **(b)** 21.25%; **(c)** The fraction of the total advertising that is spent on TV advertising **75.** **(a)** \$42.75 million; **(b)** \$110.25 million; **(c)** $x = 100$; You can't remove 100% of the pollutants. **77.** **(a)** \$400 million; **(b)** \$900 million **79.** **(a)** $\frac{1.1t + 33}{2.8t + 281}$; **(b)** In 2000: $\frac{33}{281} \approx 12\%$; In 2050: $\frac{88}{421} \approx 21\%$; **(c)** Increasing **81.** $\frac{2}{3}$
83. **(a)** 4 to 1; **(b)** 2500 rpm **85.** **(a)** Yes; **(b)** No **87.** $P(x) = -Q(x)$
89. $\frac{A \cdot C}{B \cdot C}$ **91.** $\frac{-a}{b}$ **93.** -1 **95.** $x - 3$ **97.** $\frac{5}{2}$ **99.** $\frac{-1}{x - 4}$ or $\frac{1}{4 - x}$
101. $2(x + y)$ or $2x + 2y$ **103.** $\frac{9xy}{24y^3}$ **105.** $x = 2$ or $x = -2$
107. $\frac{2}{3}$ **109.** $(x + 3)(x - 1)$ **111.** $(x - 5)(x - 2)$

Exercises 6.2

1. $\frac{8x}{3y}$ **3.** $\frac{-4xy}{3}$ **5.** $\frac{3y}{2}$ **7.** $-8xy$ **9.** $\frac{x + 1}{x + 7}$ **11.** $\frac{2 - 2x}{x - 2}$
13. $\frac{3x + 15}{x + 1}$ **15.** $-(x + 2)$ or $-x - 2$ **17.** $\frac{x + 3}{x - 3}$ **19.** $\frac{1}{4}$
21. $\frac{2x^2 - 2x}{x + 4}$ **23.** $\frac{y^2 + 5y + 6}{y + 6}$ **25.** $\frac{y^2}{2 - 3y}$ **27.** $\frac{5x - 2}{x - 2}$ **29.** $2y - 5$
31. $\frac{x - 1}{x + 2}$ **33.** $\frac{x - 5}{5x - 15}$ **35.** $\frac{x + 4}{x - 3}$ **37.** $\frac{-5x - 20}{2x + 6}$ **39.** $\frac{x - 2}{x - 1}$
41. $\frac{x + 3}{1 - x^2}$ **43.** $\frac{2x - 4}{x - 3}$ **45.** $\frac{x^2 + 11x + 30}{4}$ **47.** $\frac{x^2 - 4x + 4}{x^2 + 2x - 3}$
49. $\frac{x + 6}{5}$ **51.** $x - 1$ **53.** $\frac{x^2 - 6x + 5}{x^2 - 6x + 8}$ **55.** 1 **57.** $\frac{x + y}{x}$
59. $\frac{xy + 2x + 6y + 3y^2}{xy - 2x - 12y + 6y^2}$ **61.** $\frac{x^2 - x - 12}{2x^2 - x - 1}$ **63.** $\frac{a}{a + 3}$ **65.** **(a)** 28,000;
(b) Hybrid gas cost: $\frac{28{,}000}{35} \times \$3 = \$2400$
Regular car cost: $\frac{20{,}000}{25} \times \$3 = \$2400$; There are *no* savings.
67. $\frac{1000}{x}$ **69.** $\frac{5t(t + 2)}{t^2 + 5}$ or $\frac{5t^2 + 10t}{t^2 + 5}$ **71.** $\frac{RR_T}{R - R_T}$ **73.** $C_R = \frac{60{,}000 + 9000x}{x}$
79. $\frac{AD}{BC}$ **81.** $\frac{x^2 + 4x + 3}{x^2 - 4}$ **83.** $\frac{1}{6 - x}$ **85.** $x^2 - 6x + 9$ **87.** $2 - x$
89. $12x$ **91.** $\frac{x^2 - x - 12}{2x^2 - x - 1}$ **93.** $\frac{51}{40}$ **95.** $\frac{19}{40}$ **97.** $\frac{7}{3}$

Exercises 6.3

1. **(a)** $\frac{5}{7}$; **(b)** $\frac{11}{x}$ **3.** **(a)** $\frac{2}{3}$; **(b)** $\frac{4}{x}$ **5.** **(a)** 2; **(b)** $\frac{5}{x}$
7. **(a)** 2; **(b)** $\frac{2}{3(x + 1)}$ **9.** **(a)** $\frac{4}{3}$; **(b)** $\frac{5x}{2(x + 1)}$
11. **(a)** $\frac{5}{12}$; **(b)** $\frac{56 - 3x}{8x}$ **13.** **(a)** $\frac{12}{35}$; **(b)** $\frac{x^2 + 36}{9x}$
15. **(a)** $\frac{2}{15}$; **(b)** $\frac{5}{14(x - 1)}$ **17.** **(a)** $\frac{53}{56}$; **(b)** $\frac{8x - 1}{(x + 1)(x - 2)}$
19. **(a)** $\frac{13}{24}$; **(b)** $\frac{3x + 12}{(x - 2)(x + 1)}$ **21.** $\frac{2x^2 - 2x - 6}{(x + 4)(x - 4)(x - 1)}$
23. $\frac{5x^2 + 19x}{(x + 5)(x - 2)(x + 3)}$ **25.** $\frac{6x - 4y}{(x + y)^2(x - y)}$ **27.** $\frac{10 - x}{(x + 5)(x - 5)}$
29. $\frac{-6x - 10}{(x + 2)(x + 1)(x + 3)}$ **31.** $\frac{y^2}{(y + 1)(y - 1)}$ **33.** $\frac{5 - 4y - 2y^2}{(y + 4)(y - 4)}$ or $\frac{-2y^2 - 4y + 5}{(y + 4)(y - 4)}$
35. $\frac{2x^2 - x + 3}{(x - 2)(x + 1)^2}$ **37.** 0 **39.** $\frac{a^2 + 3a + 3}{(a + 2)(a^2 - 2a + 4)}$ **41.** **(a)** $\frac{f + u}{f + u}$; **(b)** 1
43. $\frac{g_m + 8}{g_m^2}$ **45.** **(a)** $\frac{0.55t + 20}{1.85t + 251}$; **(b)** 7.97%; **(c)** 9.46%
47. $\frac{-w_0x^3 + 3w_0L^2x - 2w_0L^3}{6L}$ **49.** $\frac{p^2 - 2gm^2rM}{2mr^2}$ **51.** 1.5 **53.** $1.41\overline{6}$
55. $0.0024\overline{6}$ or 0.0025 **59.** LCD **61.** $\frac{R(t) - P(t)}{G(t)} = \frac{0.02t^2 - 0.34t + 1.42}{0.04t^2 + 2.34t + 90}$
63. $\frac{2x + 18}{(x - 1)(x + 3)}$ **65.** $\frac{2}{x - 2}$ **67.** $\frac{7x + 9}{(x + 2)(x - 2)(x + 1)}$ **69.** $\frac{9}{20}$ **71.** 42
73. $x^2 - 1$

Exercises 6.4

1. $\frac{1}{5}$ **3.** $\frac{1}{3}$ **5.** $\frac{ab - a}{b + a}$ **7.** $\frac{b + a}{b - a}$ **9.** $\frac{6b + 4a}{48b - 9a}$ **11.** $\frac{26}{5}$
13. $\frac{x}{2x - 1}$ **15.** 2 **17.** $\frac{1}{x - y}$ **19.** $\frac{x + 2}{x + 1}$ **21.** $\frac{x - 5}{4}$
23. $\frac{1}{2(x - 4)}$ **25.** **(a)** $\frac{5t + 400}{-5t + 4630} = \frac{t + 80}{-5(t - 926)}$ **(b)** $\frac{400}{4630} \approx 8.6\%$ and $\frac{425}{4605} \approx 9.2\%$
27. $\frac{6}{25}$ yr **29.** $11\frac{43}{50}$ yr **31.** $84\frac{11}{1000}$ yr **33.** $4\frac{3}{5}$ yr **35.** $248\frac{27}{50}$ yr
37. A fraction that has one or more fractions in its numerator, denominator or both. **43.** LCD **45.** $\frac{11}{6}$ **47.** $\frac{w + 4}{w + 1}$ **49.** $\frac{m + 5}{m - 4}$
51. $w = 124$ **53.** $x = -3$ **55.** $x = 3$

Exercises 6.5

1. $x = 6$ **3.** $x = 4$ **5.** $x = -6$ **7.** $x = 12$ **9.** $x = 4$
11. $x = 5$ **13.** $x = 2$ **15.** $x = 2$ **17.** $x = -11$ **19.** $x = -2$
21. No solution **23.** $x = \frac{-3}{2}$ **25.** $x = -6$ **27.** $x = -8$
29. No solution **31.** No solution **33.** $x = -3$ **35.** $x = -\frac{5}{2}$
37. $x = \frac{1}{7}$ **39.** $x = \frac{-3}{5}$ **41.** $z = \frac{45}{2}$ **43.** $y = \frac{5}{7}$ **45.** $v = \frac{2}{5}$
47. $z = 3$ **49.** $x = 4$ **51.** $V_2 = \frac{P_1V_1}{P_2}$ **53.** $V_2 = 4$ L **55.** 75%
57. $h = \frac{2A}{b_1 + b_2}$ **59.** $Q_1 = \frac{PQ_2}{P + 1}$ **61.** $f = \frac{ab}{a + b}$ **65.** term **67.** $x = -3$
69. $x = -4$ **71.** $x = -2$ **73.** $F = \frac{9}{5}C + 32$ **75.** $x = \frac{33}{4}$ **77.** $h = \frac{12}{5}$
79. $R = 15$

Exercises 6.6

1. 9250 **3.** 100 **5.** 6 **7.** 3.6 **9.** **(a)** 2250; **(b)** 36
11. **(a)** 3 to 1; **(b)** 35 to 1; **(c)** Minneapolis; climate **13.** 37 million
15. $5\frac{1}{7}$ hr **17.** $1\frac{7}{8}$ min **19.** 6 hr **21.** 10 mi/hr **23.** 100 mi/hr
25. 200 **27.** 169 **29.** $x = 5\frac{1}{3}; y = 6\frac{2}{3}$ **31.** $DE = 13\frac{5}{7}$ in.
33. $x = 16; y = 8$ **35.** $x = 12; y = 12$ **37.** $a = 24; b = 30$
39. $x = 8; y = 8\frac{4}{5}$ **41.** $x = 10; y = 25$ **43.** 1.5 in. × 2 in.
45. **(a)** Approximately 32 hr; **(b)** Approximately $6\frac{1}{2}$ days **47.** $18\frac{2}{3}$ lb
49. 6 ft **51.** 2650 **57.** $ad = bc$ **59.** similar **61.** $3\frac{1}{13}$ hr **63.** 200
65. $k = 5$ **67.** $k = 240$

Exercises 6.7

1. **(a)** $S = \frac{W}{16}$; **(b)** $k = \frac{1}{16}$; **(c)** 10 lb **3.** **(a)** $R = \frac{W}{10}$; **(b)** $k = \frac{1}{10}$;
(c) 16 **5.** **(a)** $R = kt$; **(b)** $k = 45$; **(c)** 2.4 min **7.** **(a)** $T = kh^3$;
(b) $k \approx 0.00057$; **(c)** 240 lb **9.** **(a)** $f = \frac{k}{d}$; **(b)** $k = 4$; **(c)** 16
11. 10.8 in.^{3} **13.** **(a)** $w = \frac{k}{s}$; **(b)** $k = 7200$; **(c)** 720 words
15. **(a)** $b = \frac{k}{a}$; **(b)** $k = 2970$; **(c)** 90 (per 1000 women) **17.** **(a)** $d = ks$;
(b) $k \approx 17.63$; **(c)** The time it took to drive d mi at s mph
19. **(a)** $C = 4(F - 37)$ or $C = 4F - 148$; **(b)** 212 chirps per minute
21. **(a)** $g = kA$; **(b)** $k = \frac{20}{31}$; **(c)** 200 million gallons
23. **(a)** BAC $= k(N - 1)$; **(b)** $k = 0.026$; **(c)** 0.104; **(d)** 4
25. **(a)** BAC $= \frac{7.8}{W}$; **(b)** 0.03; **(c)** 97.5 lb; **(d)** Less than 0.08
27. **(a)** $T = 15S$; **(b)** 4 **33.** directly **35.** **(a)** $C = kn$; **(b)** $k = 100$;
(c) 2 strips **37.** **(a)** $v = \frac{k}{t}$; **(b)** $k = 120$; **(c)** The distance traveled
39. $x + 2y = 4$

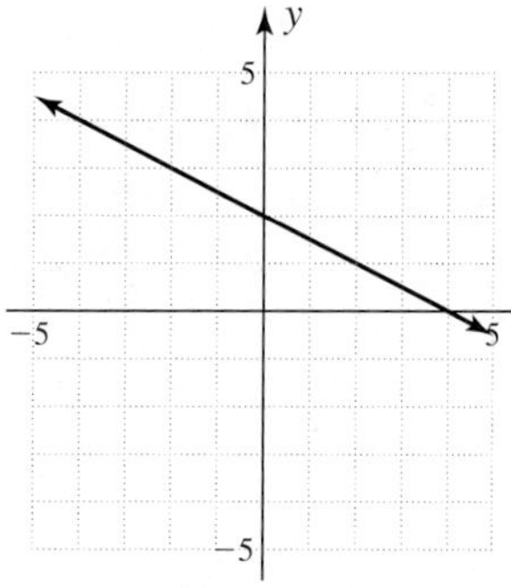

41. $y - 2x = 4$

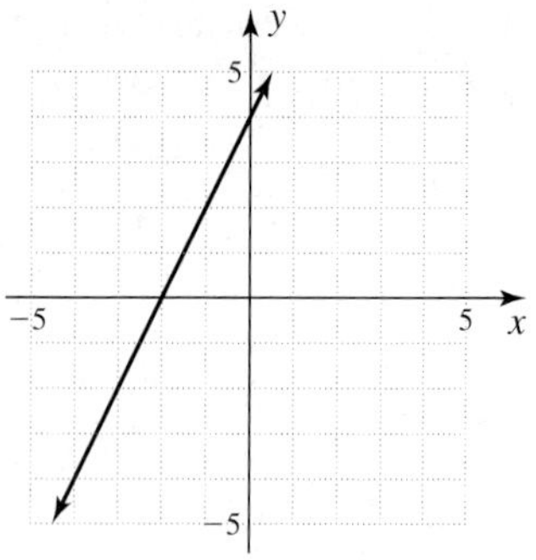

43. Not parallel

Review Exercises

1. [6.1B] **(a)** $\frac{10xy}{16y^2}$; **(b)** $\frac{12xy}{16y^3}$; **(c)** $\frac{10x^2y}{15x^5}$
2. [6.1C] **(a)** $-3(x - y)$ or $3y - 3x$;
(b) $-2(x - y)$ or $2y - 2x$; **(c)** $4(x - y)$ or $4x - 4y$
3. [6.1A, C] **(a)** $\frac{-1}{x + 1}$; 0, −1; **(b)** $\frac{-1}{x - 1}$ or $\frac{1}{1 - x}$; 0, 1;
(c) $\frac{-1}{1 - x}$ or $\frac{1}{x - 1}$; 0, 1 **4.** [6.1C] **(a)** $-(x + 3)$ or $-x - 3$;
(b) $-(x + 2)$ or $-x - 2$; **(c)** $-(x + 3)$ or $-x - 3$
5. [6.2A] **(a)** $\frac{2xy}{3}$; **(b)** $\frac{3xy}{2}$; **(c)** $8xy^2$
6. [6.2A] **(a)** $\frac{x + 2}{x + 3}$; **(b)** $\frac{x + 1}{x + 5}$; **(c)** $\frac{x + 5}{x + 4}$
7. [6.2B] **(a)** $\frac{x - 3}{x + 2}$; **(b)** $\frac{x - 4}{x + 1}$; **(c)** $\frac{x - 5}{x + 4}$
8. [6.2B] **(a)** $\frac{-1}{x - 5}$ or $\frac{1}{5 - x}$; **(b)** $\frac{-1}{x - 1}$ or $\frac{1}{1 - x}$; **(c)** $\frac{-1}{x - 2}$ or $\frac{1}{2 - x}$
9. [6.3A] **(a)** $\frac{2}{x - 1}$; **(b)** $\frac{2}{x - 2}$; **(c)** $\frac{1}{x + 1}$
10. [6.3A] **(a)** $\frac{2}{x + 1}$; **(b)** $\frac{2}{x + 2}$; **(c)** $\frac{2}{x + 3}$
11. [6.3B] **(a)** $\frac{3x - 2}{(x + 2)(x - 2)}$; **(b)** $\frac{4x - 2}{(x + 1)(x - 1)}$; **(c)** $\frac{5x - 9}{(x + 3)(x - 3)}$
12. [6.3B] **(a)** $\frac{-6x - 10}{(x + 1)(x + 2)(x + 3)}$; **(b)** $\frac{7x + 1}{(x - 2)(x + 1)^2}$;
(c) $\frac{-4x}{(x + 2)(x + 1)(x - 1)}$ **13.** [6.4A] **(a)** $\frac{6}{17}$; **(b)** $\frac{14}{27}$; **(c)** $\frac{6}{25}$
14. [6.5A] **(a)** $x = 3$; **(b)** $x = 7$; **(c)** $x = 6$
15. [6.5A] **(a)** $x = -3$; **(b)** $x = -5$; **(c)** $x = -6$
16. [6.5A] **(a)** No solution; **(b)** No solution; **(c)** No solution
17. [6.5A] **(a)** $x = -6$ or $x = \frac{13}{3}$; **(b)** $x = -7$ or $x = \frac{11}{2}$;
(c) $x = -8$ or $x = \frac{33}{5}$ **18.** [6.5B] **(a)** $a_1 = \frac{A(1 - b^n)}{1 - b}$; **(b)** $b_1 = \frac{B(1 - c)}{1 - c^n}$;
(c) $c_1 = \frac{C(d + 1)}{(1 - d)^n}$ **19.** [6.6A] **(a)** $10\frac{1}{2}$ gal; **(b)** $13\frac{1}{2}$ gal; **(c)** 18 gal
20. [6.6A] **(a)** $x = 18$; **(b)** $x = 10\frac{2}{5}$; **(c)** $x = \frac{-13}{5}$
21. [6.6B] **(a)** $3\frac{3}{7}$ hr; **(b)** $4\frac{4}{9}$ hr; **(c)** $3\frac{3}{5}$ hr
22. [6.6B] **(a)** 4 mi/hr; **(b)** 8 mi/hr; **(c)** 12 mi/hr
23. [6.6B] **(a)** 32; **(b)** 35; **(c)** 40
24. [6.6B] **(a)** 1125; **(b)** 1250; **(c)** 1375
25. [6.6B] **(a)** $10\frac{1}{2}$; **(b)** $10\frac{2}{3}$; **(c)** $2\frac{2}{3}$ **26.** [6.7A] **(a)** $C = km$; 150 calories;
(b) $C = km$; 420 calories; **(c)** $C = km$; 45 calories
27. [6.7B] **(a)** $F = \frac{30}{L}$; $F = 5$; **(b)** $F = 3$; **(c)** $F = 2$

Cumulative Review Chapters 1–6

1. $\frac{-17}{30}$ **2.** 2.7 **3.** 25 **4.** 2 **5.** 69 **6.** $6x - 14$
7. $\frac{a - b}{c}$ **8.** $x = 4$ **9.** $x = 63$ **10.** 30 and 65
11. \$9000 in bonds; \$3000 in certificates of deposit
12.

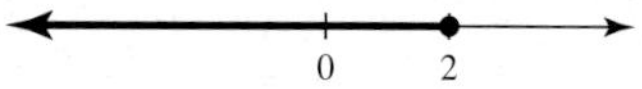

13.

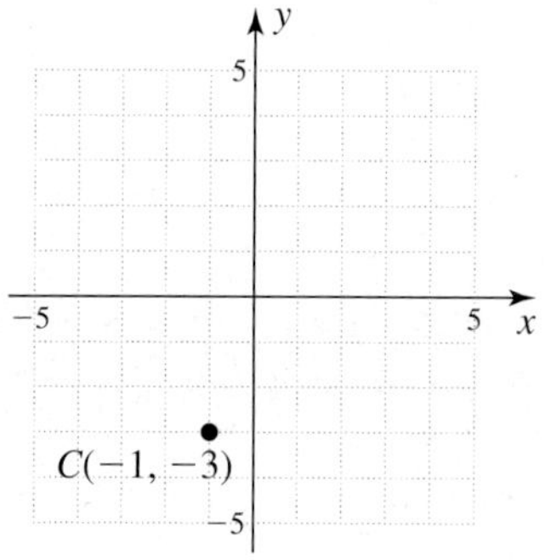

14. Yes **15.** $x = 3$
16.

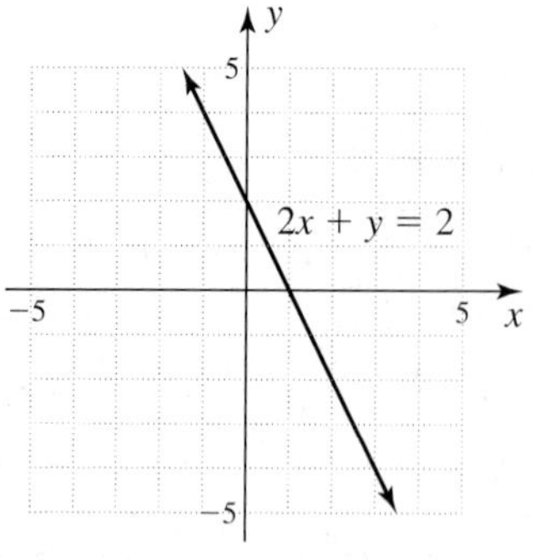

17.

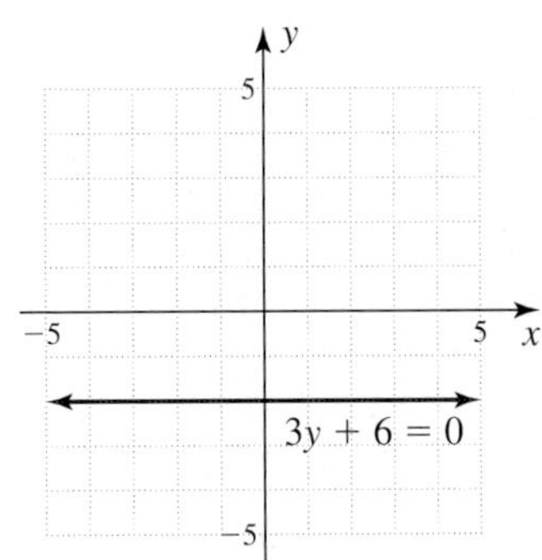

18. $\frac{7}{16}$ **19.** 2 **20.** (2) and (3) **21.** $y - 3 = 6(x + 4)$ or $y = 6x + 27$
22. $y = -2x + 5$
23. $3x - 4y < -12$

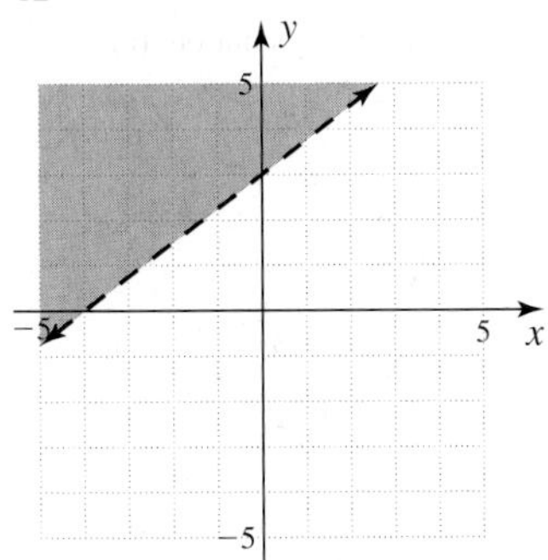

24. $3x - y \geq 0$

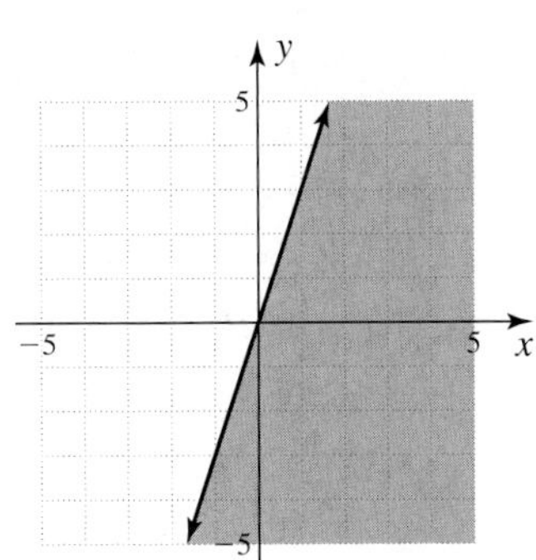

25. x **26.** $\frac{1}{x^2}$
27. $\frac{y^8}{81x^{12}}$ **28.** 3.6×10^{-4} **29.** 1.80×10^{-5}
30. -4 **31.** $-12x^6 + 9x^4 - 4$ **32.** $25x^4 - 2x^2 + \frac{1}{25}$
33. $81x^4 - 4$ **34.** $(3x^2 - x + 4)$ R 2 **35.** $3x^2(2 - 3x^2)$
36. $\frac{1}{5}x^3(2x^4 - 4x^3 + 4x^2 - 1)$ **37.** $(x - 7)(x - 8)$
38. $(5x - 4y)(3x - 5y)$ **39.** $(4x + 3y)(4x - 3y)$
40. $-4x^2(x + 1)(x - 1)$ **41.** $3x(x + 1)(x - 3)$ **42.** $(x + 1)(4x + 3)$
43. $k(4x - 1)^2$ **44.** $x = -3, x = \frac{5}{2}$ **45.** $\frac{14xy^2}{12y^3}$
46. $-2(x + y)$ or $-2x - 2y$ **47.** $-(x + 5)$ or $-x - 5$ **48.** $\frac{x + 3}{x + 7}$
49. $\frac{-1}{x - 3}$ or $\frac{1}{3 - x}$ **50.** $\frac{6}{x + 8}$ **51.** $\frac{-2x - 11}{(x + 6)(x + 5)(x - 5)}$ **52.** 42
53. $x = 3$ **54.** $x = -8$ **55.** No solution **56.** $x = -5$ **57.** 23 gal
58. $x = \frac{-70}{11}$ **59.** $3\frac{3}{13}$ hr **60.** $k = \frac{1}{50}$ **61.** $k = 13{,}720$

Chapter 7

Exercises 7.1

1. Consistent; (1, 3)
① $x + y = 4$; ② $x - y = -2$

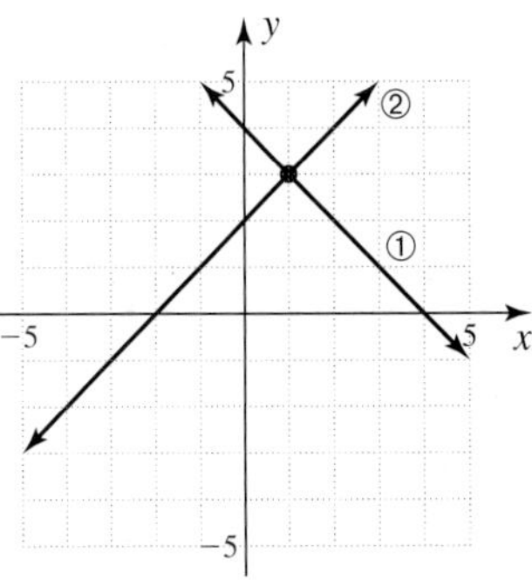

3. Consistent; (−2, 1)
① $x + 2y = 0$; ② $x - y = -3$

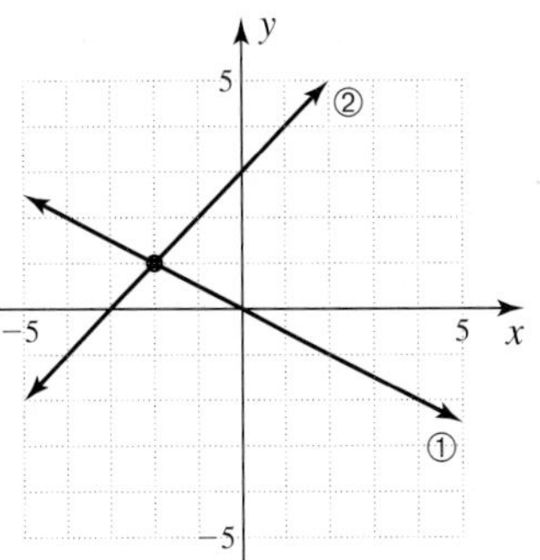

5. Dependent
$3x - 2y = 6$; $6x - 4y = 12$

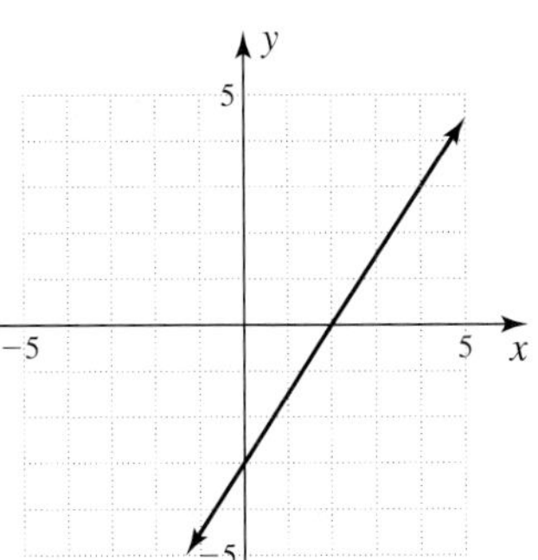

7. Dependent
$3x - y = -3$; $y - 3x = 3$

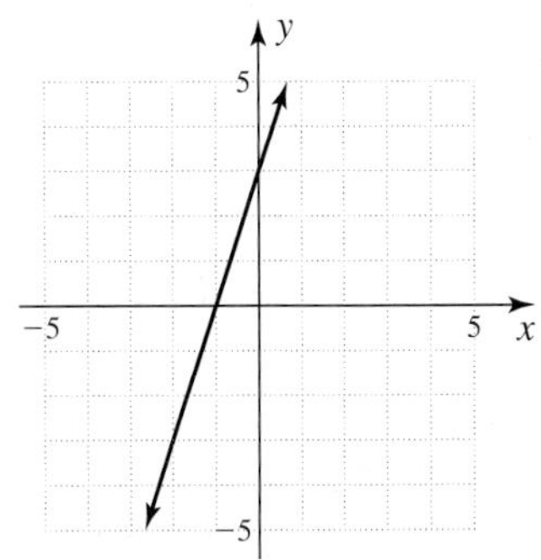

9. Inconsistent
① $2x - y = -2$; ② $y = 2x + 4$

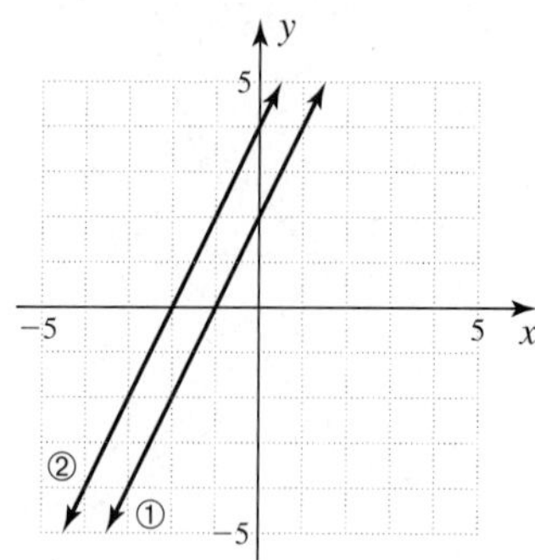

11. Consistent; $(-2, -2)$
① $y = -2$; ② $2y = x - 2$

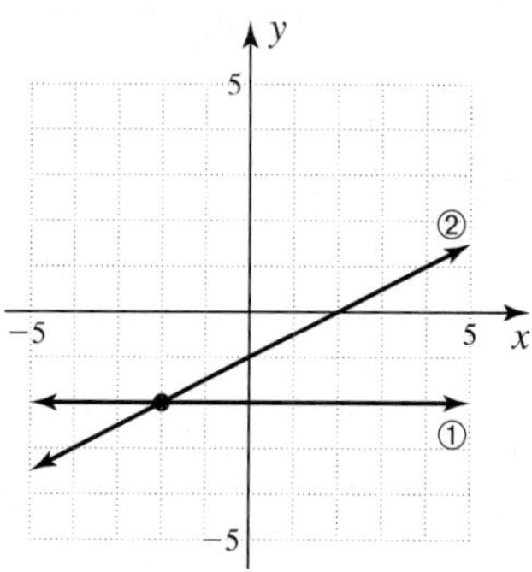

13. Consistent; $(3, 2)$
① $x = 3$; ② $y = 2x - 4$

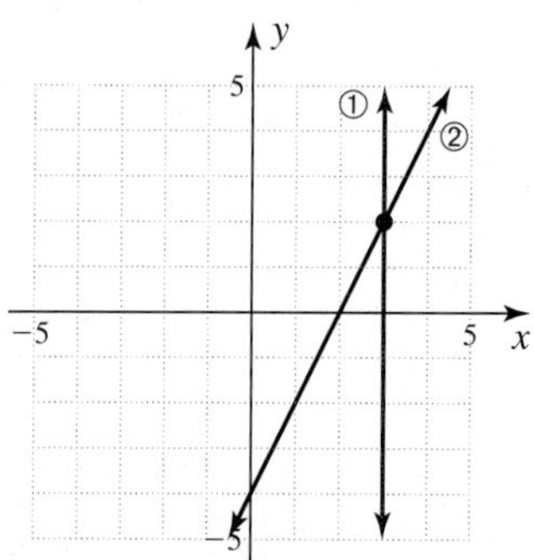

15. Consistent; $(1, 2)$
① $x + y = 3$; ② $2x - y = 0$

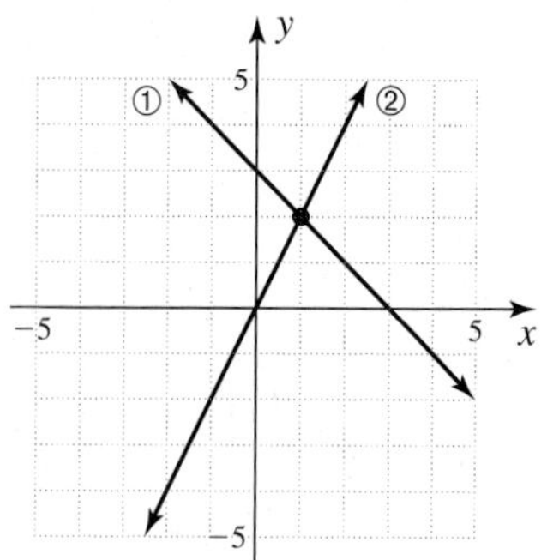

17. Consistent; $(0, 5)$
① $5x + y = 5$; ② $5x = 15 - 3y$

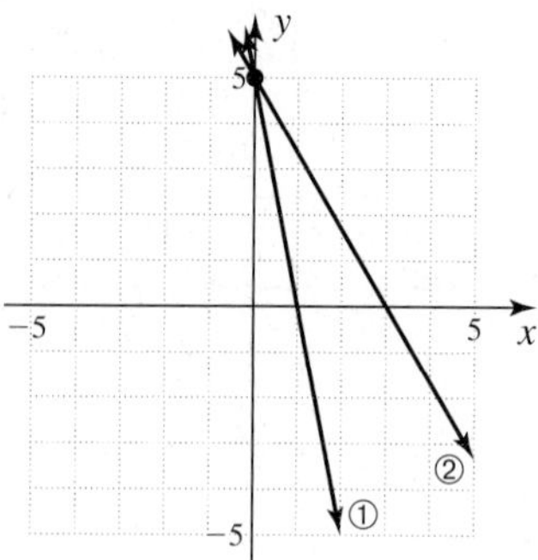

19. Dependent
$3x + 4y = 12$; $8y = 24 - 6x$

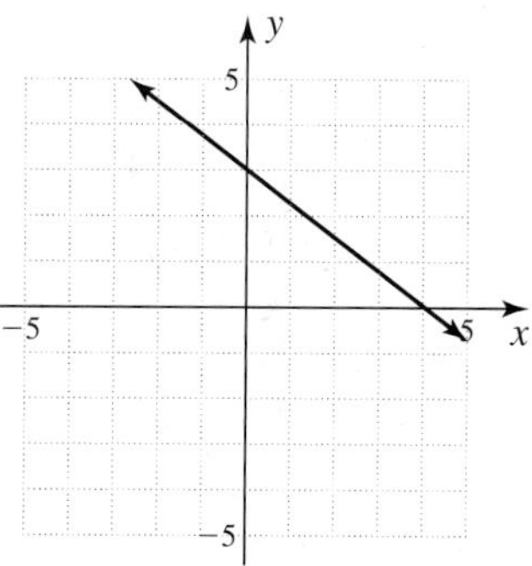

21. Consistent; $(0, 3)$
① $y = x + 3$; ② $y = -x + 3$

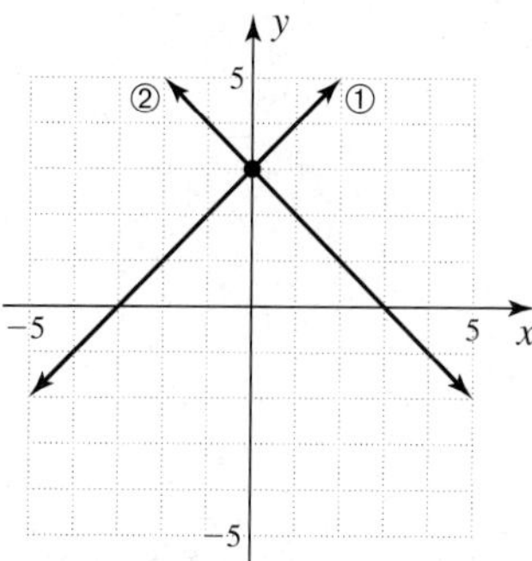

23. Consistent; $(1, 0)$
① $y = 2x - 2$; ② $y = -3x + 3$

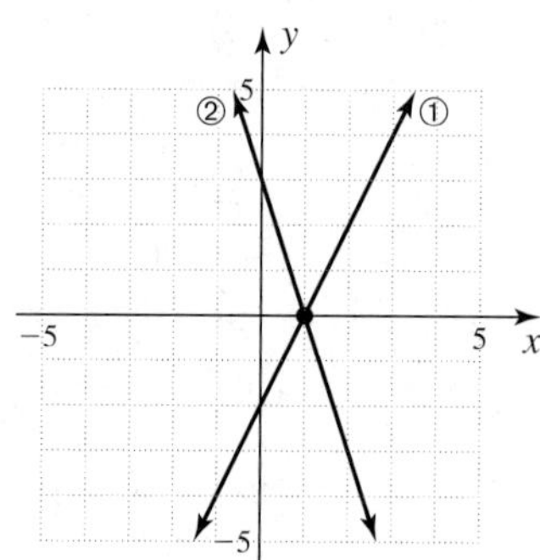

25. Consistent; $(-2, -3)$
① $-2x = 4$; ② $y = -3$

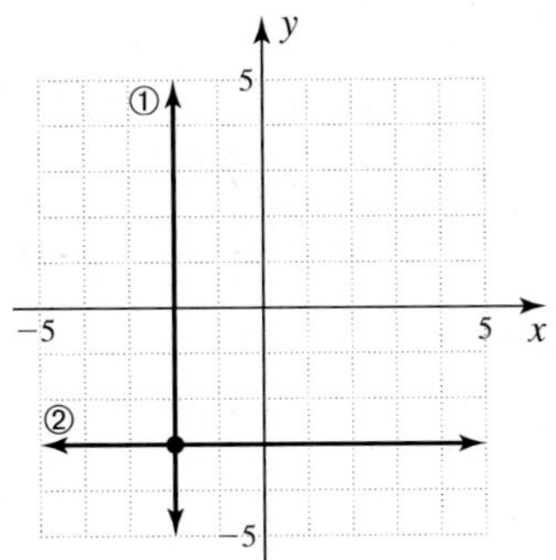

27. Consistent; $(3, -3)$
① $y = -3$; ② $y = -3x + 6$

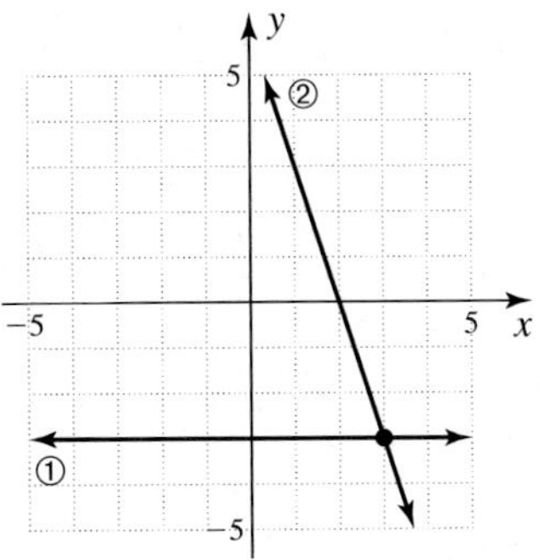

29. Inconsistent
① $x + 4y = 4$; ② $y = -\frac{1}{4}x + 2$

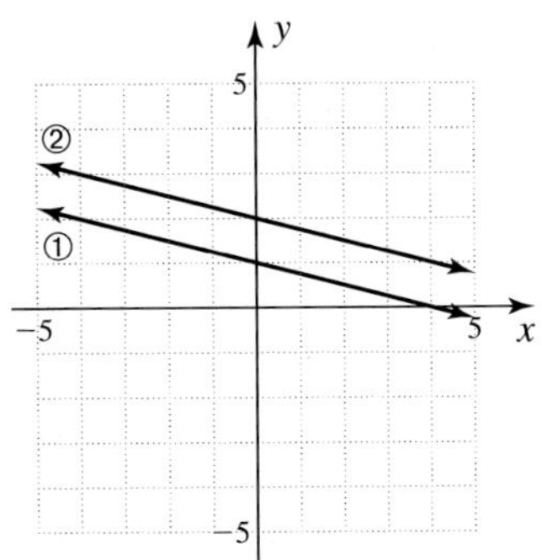

31. (a) $C = 20 + 35m$; **(b)**

m	C
6	230
12	440
18	650

(c)

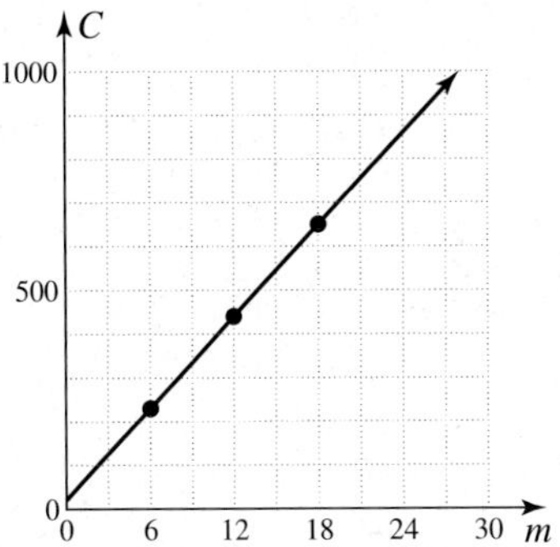

33.

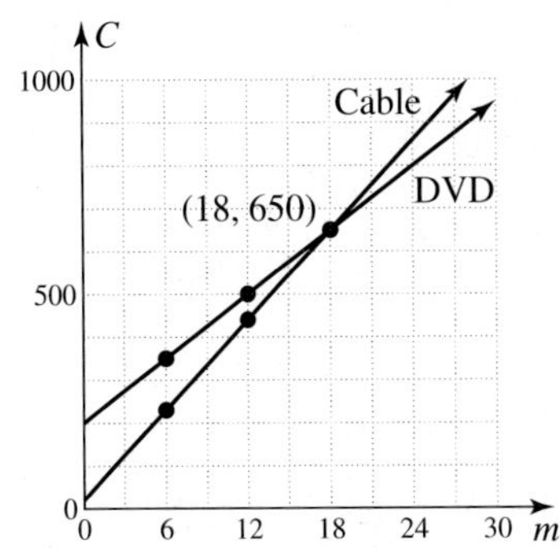

35. DVD player and rental option is cheaper if used more than 18 months. (The options are equal if used for 18 months.)

37. (a) $W = 100 + 3t$; **(b)**

t	W
5	115
10	130
15	145
20	160

(c)

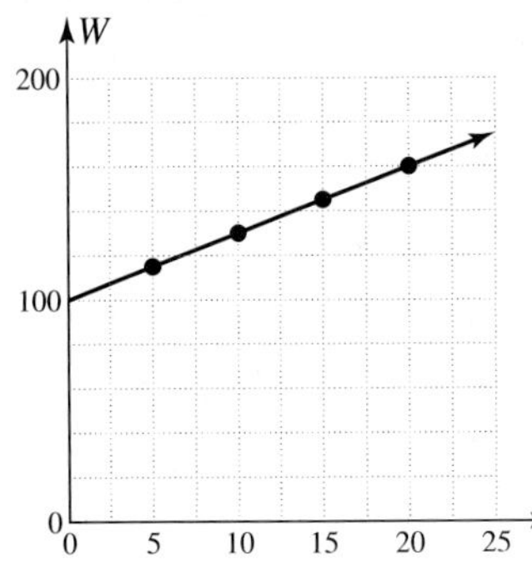

39. (a) $C = 0.60m$; **(b)** $C = 0.45m + 45$;

(c)

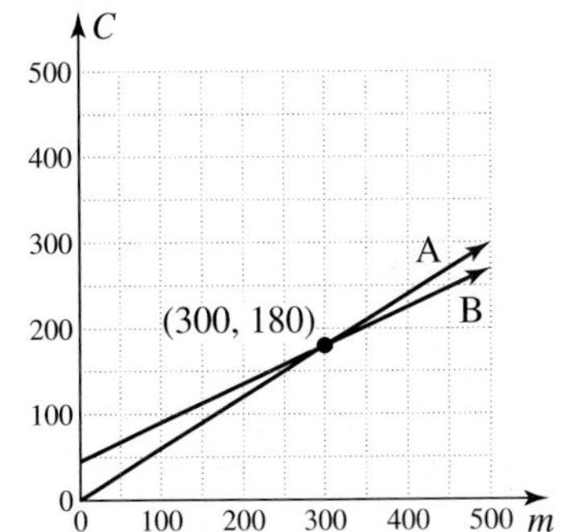

41. (a) $C_B = 5x$; **(b)** $C_H = 6y$; **(c)** $x + y$; $x + y = 50$; **(d)** $5x + 6y = 270$; **(e)** 30 baskets, 20 buffets

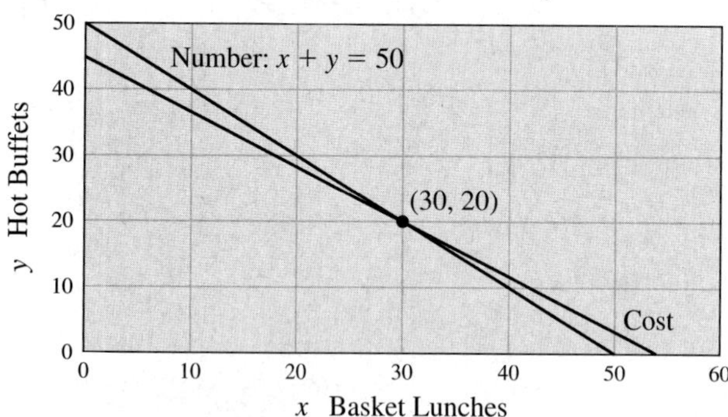

43. (a) $C_B = 6x$; **(b)** $C_C = 7y$; **(c)** $x + y$; $x + y = 20$; **(d)** $6x + 7y = 126$; **(e)** 14 croissants, 6 omelets

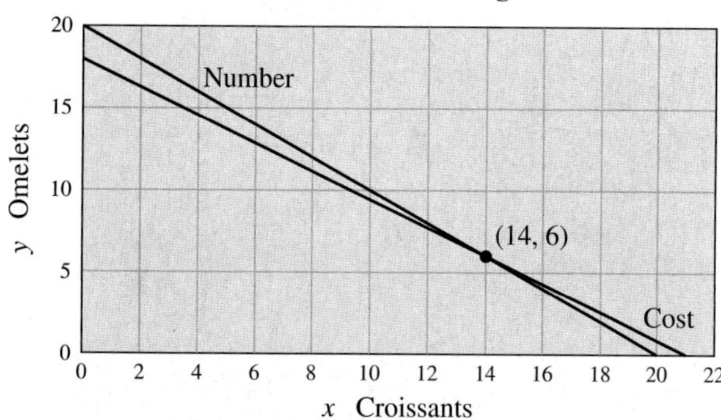

45. (a) $x + y = 16$; **(b)** $P_Q = 2x$; **(c)** $P_W = y$; **(d)** $2x + y = 24$; **(e)** Both 8 grams

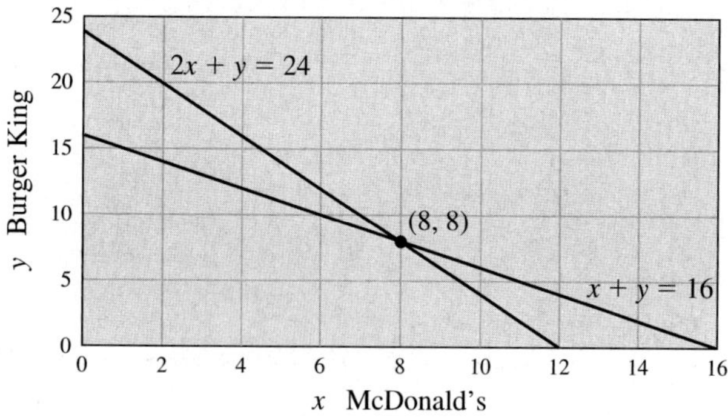

47.

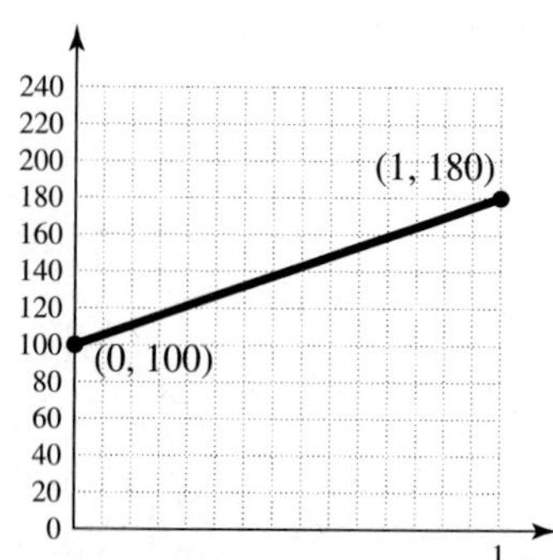

49. (a) Canon; **(b)** C: \$180; E: \$220; **(c)** \$40 if using Canon

51.

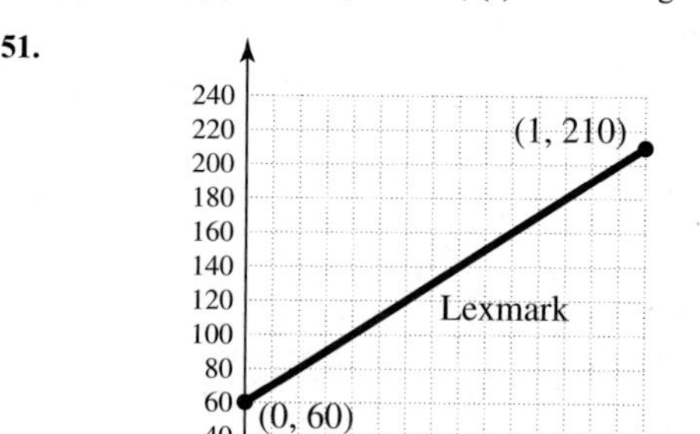

53. (a)

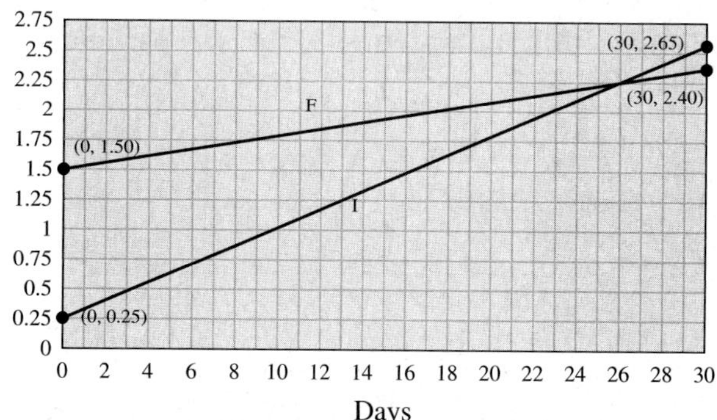

(b) Incandescent; **(c)** 25 days

57. No solution **59.** One solution **61.** one **63.** no **65.** infinitely

67. No solution; ① $y - 3x = 3$; ② $2y = 6x + 12$

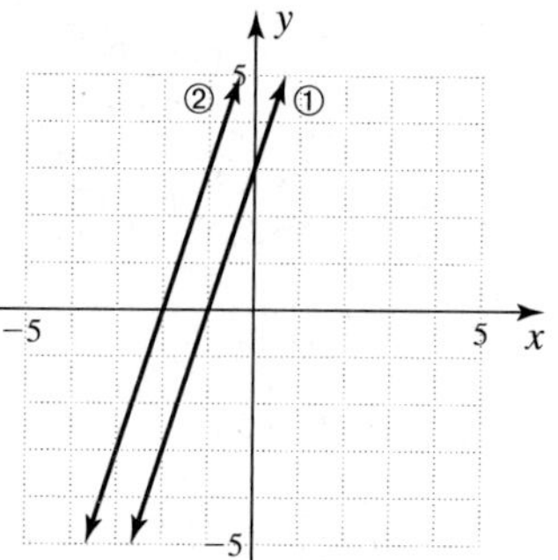

69. Consistent; (2, 2)
① $x + y = 4$; ② $2x - y = 2$

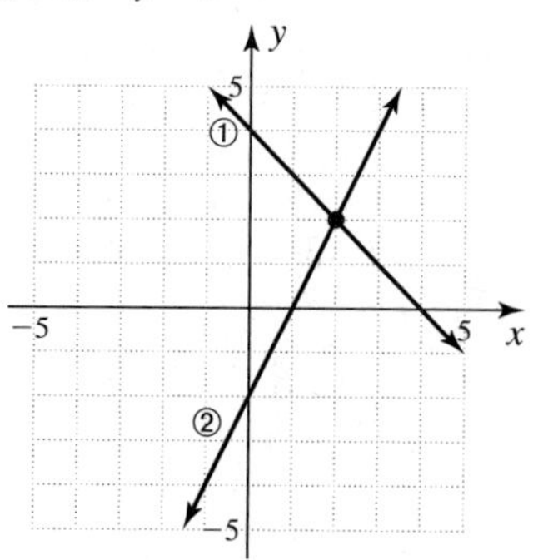

71. Inconsistent; no solution
① $2x + y = 4$; ② $2y + 4x = 6$

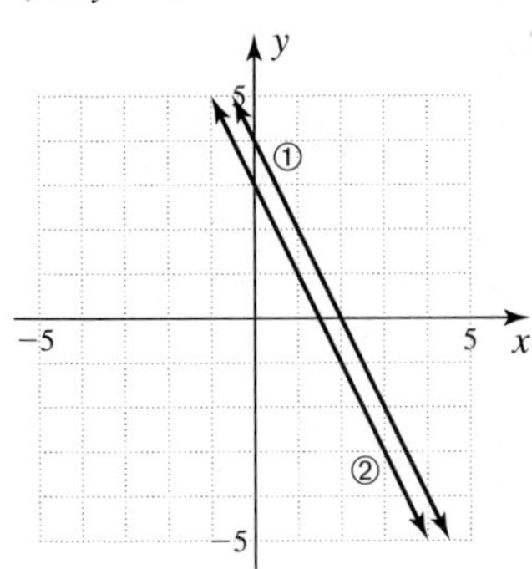

73. $x = 4$ **75.** Yes **77.** No

Exercises 7.2

1. Consistent; (2, 0) **3.** Consistent; (2, 3) **5.** Inconsistent; no solution **7.** Inconsistent; no solution **9.** Dependent; infinitely many solutions **11.** Consistent; $(-1, -1)$ **13.** Inconsistent; no solution **15.** Consistent; (4, 1) **17.** Consistent; (5, 3) **19.** Consistent; (0, 2) **21.** Inconsistent; no solution **23.** Consistent; (2, 1) **25.** Consistent; (3, 1) **27.** Dependent; infinitely many solutions **29.** Inconsistent; no solution **31.** Consistent; $(2, -3)$ **33. (a)** $p = 20 + 3(h - 15)$ when $h > 15$; **(b)** $p = 20 + 2(h - 15)$ when $h > 15$; **(c)** When $h \leq 15$ hours, $p = \$20$. **35.** When 150 minutes are used **37.** 10 tables **39.** 1st **41.** At $-40°$, $F = C$. **43.** 4 days **45.** 5 days **47. (a)** $x + y = \$15{,}127$; **(b)** $x = y + 9127$; **(c)** $x = \$12{,}127$, $y = \$3000$ **49. (a)** $T + PM = \$3222$; **(b)** $PM - T = \$1350$; **(c)** $PM = \$2286$, $T = \$936$ **51.** $(\frac{4}{7}, 145.71)$ **53.** $(\frac{9}{8}, 228.75)$ **57.** one **59.** no **61.** Inconsistent **63.** Consistent; (2, 3) **65.** Dependent **67.** Inconsistent **69.** 10 months **71.** $x = 8$ **73.** $x = -1$

Exercises 7.3

1. (1, 2) **3.** (0, 2) **5.** Inconsistent **7.** Inconsistent **9.** $(10, -1)$ **11.** $(-26, 14)$ **13.** (6, 2) **15.** $(2, -1)$ **17.** (3, 5) **19.** $(-3, -2)$ **21.** (8, 6) **23.** (1, 2) **25.** (5, 3) **27.** (4, 3) **29.** Dependent **31.** $E = 409$, $T = 164$ **33.** 0.8 lb Costa Rican; 0.2 lb Indian Mysore **35.** $c = 1770$, $m = 665$ **37.** $c = 7$, $t = 31$ **39.** $c = 190$, $a = 340$ **41.** Tweedledee $120\frac{2}{3}$ lb; Tweedledum $119\frac{2}{3}$ lb **45.** equivalent **47.** $\left(\frac{41}{13}, \frac{7}{13}\right)$ **49.** (5, 1) **51.** Inconsistent **53.** $n + d = 300$ **55.** $4(x - y) = 48$ **57.** $m = n - 3$

Exercises 7.4

Translate This

1. K **3.** A **5.** B **7.** F **9.** J

1. 5 nickels; 20 dimes **3.** 15 nickels; 5 dimes **5.** 4 fives; 6 ones **7.** 59; 43 **9.** 105; 21 **11.** Impossible **13.** Pikes Peak = 14,110 ft; Longs Peak = 14,255 ft **15.** 1050 mi **17.** Boat speed = 15 mi/hr; current speed = 3 mi/hr **19.** 20 mi **21.** \$5000 at 6%; \$15,000 at 8% **23.** \$8000 **25.** Public: 11.7 million; private: 3.3 million **27.** HBO: 14,353,650; Showtime: 14,346,350 **29.** $K = 96$, $A = 26.5$ **31.** $K = 140$, $A = 165$ **37.** $10d + 25q$ **39.** RT **41.** 20 nickels; 10 dimes **43.** Wind speed = 50 mi/hr; plane speed = 350 mi/hr **45.** 210,000 gallons sewage, 37,000 gallons bilge water.

47. $x - 2y > 6$

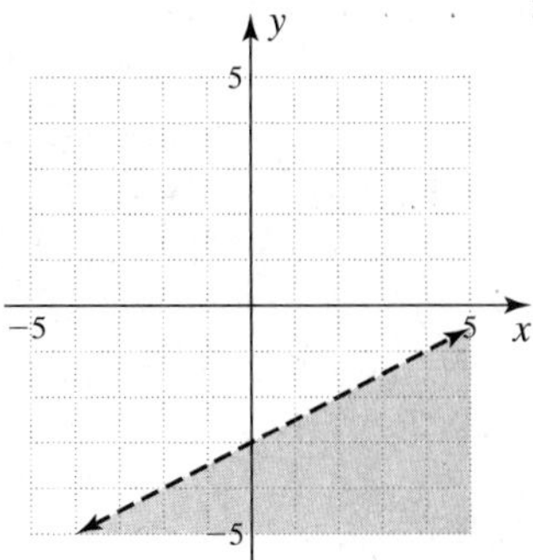

49. $x \leq 2y$

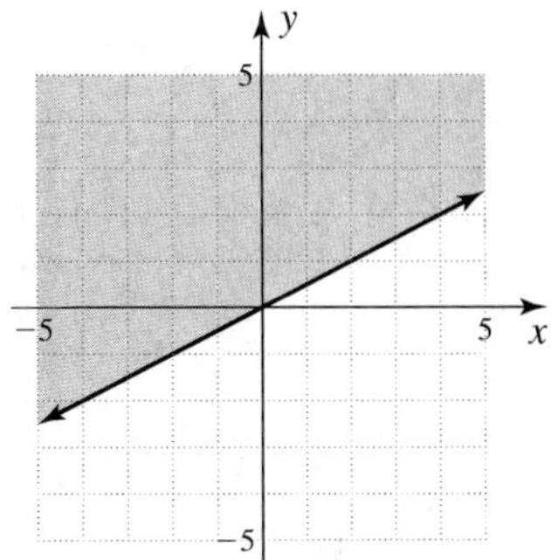

Exercises 7.5

1. $x \geq 0$ and $y \leq 2$

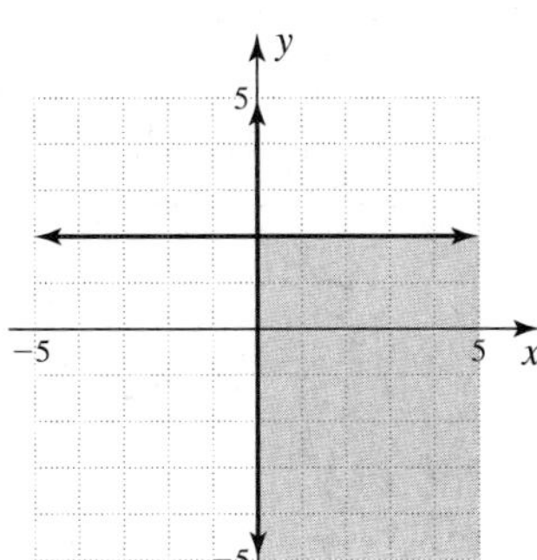

3. $x < -1$ and $y > -2$

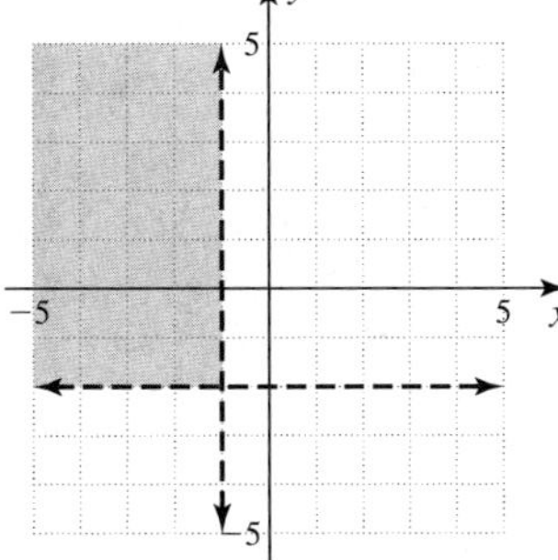

5. $x + 2y \leq 3$; $x < y$

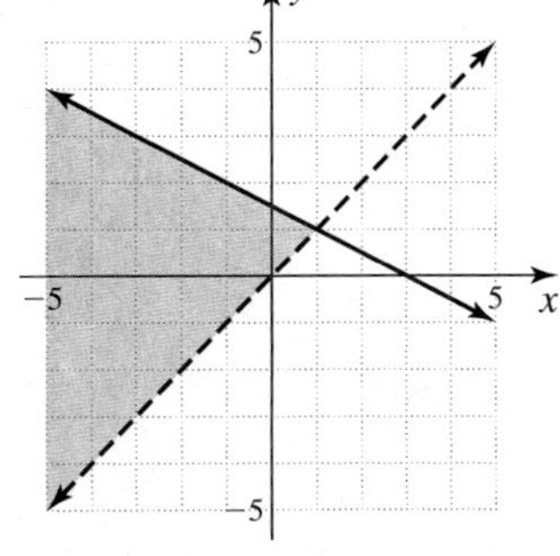

7. $4x - y > -1$; $-2x - y \le -3$

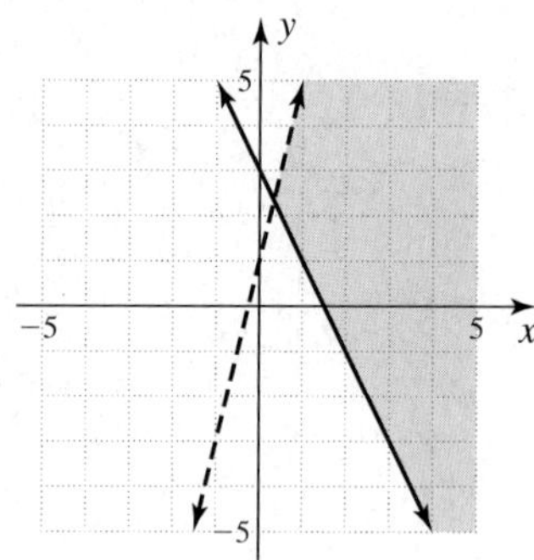

9. $-2x + y > 3$; $5x - y \le -10$

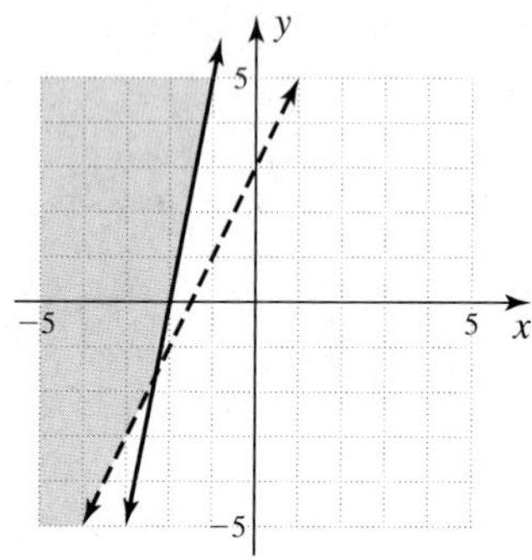

11. $2x - 3y < 5$; $x \ge y$

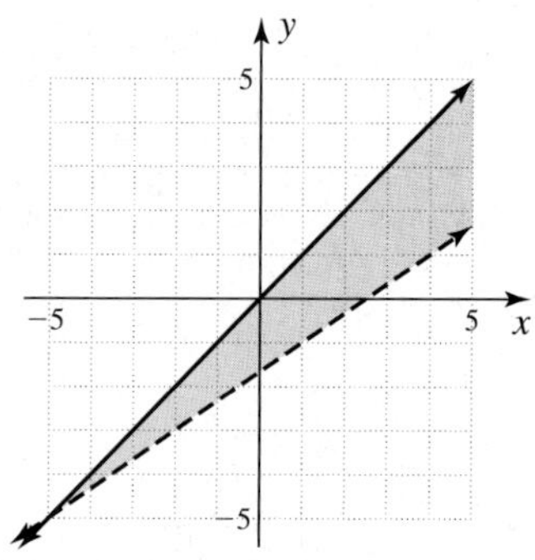

13. $x + 3y \le 6$; $x > y$

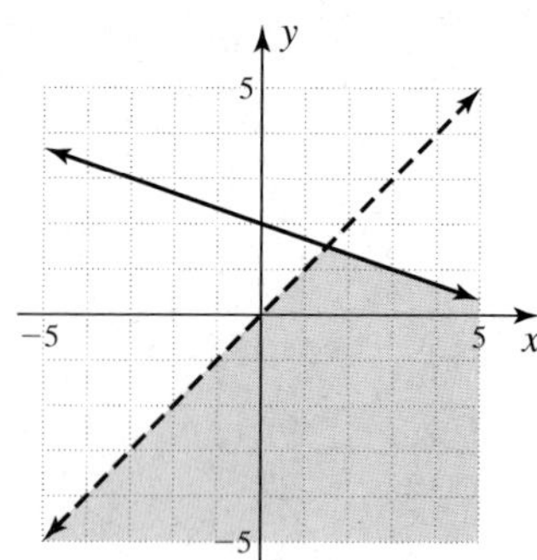

15. $0 \le x \le 15$
$y \ge 13x + 1700$
$y \le 30x + 180$

17. For example, (1, 3), (2, 3)

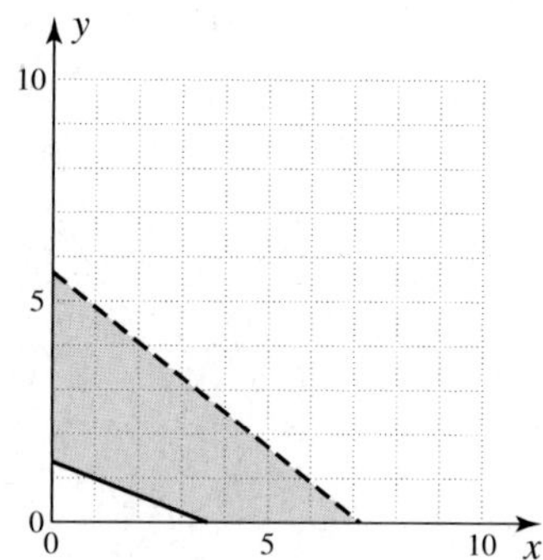

19. & **21.**

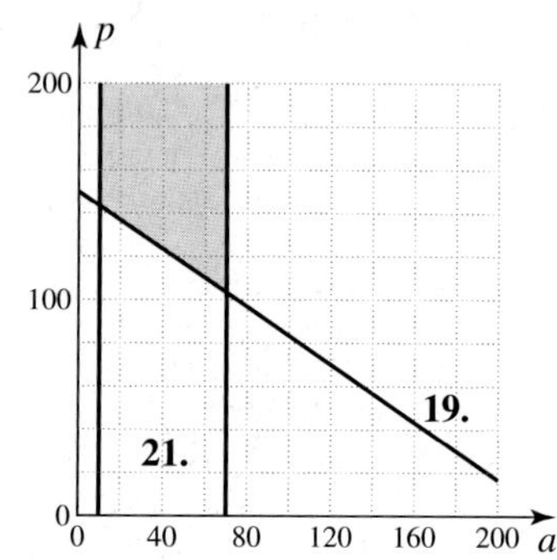

25. solid

27. $x > 2$ and $y < 3$

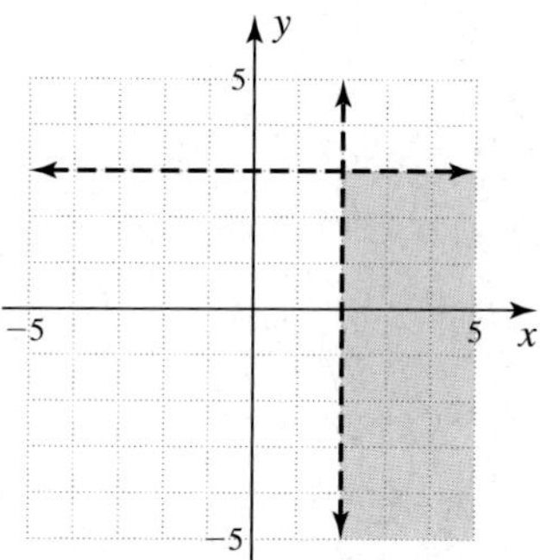

29. $3x - y < -1$; $x + 2y \le 2$

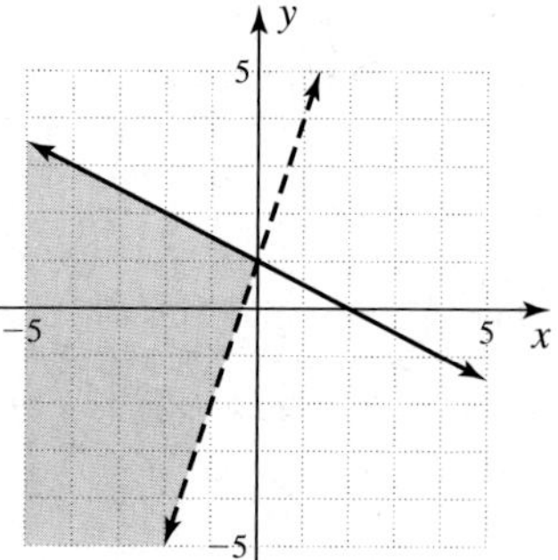

31. 16 **33.** 8

Review Exercises

1. [7.1A, B]
(a) Solution: (1, 2)
$2x + y = 4; y - 2x = 0$

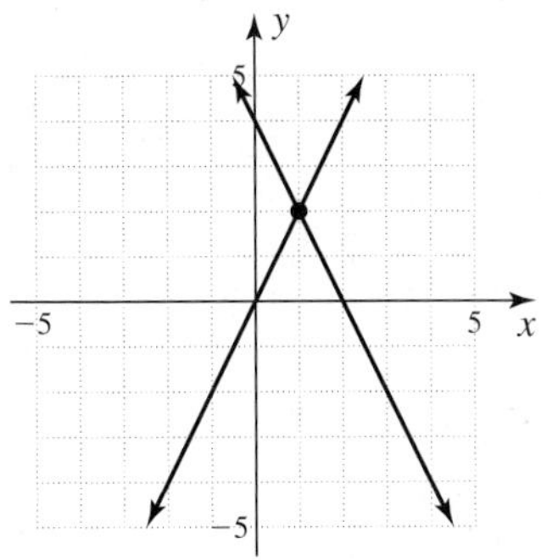

b. Solution: (2, 2)
$x + y = 4; y - x = 0$

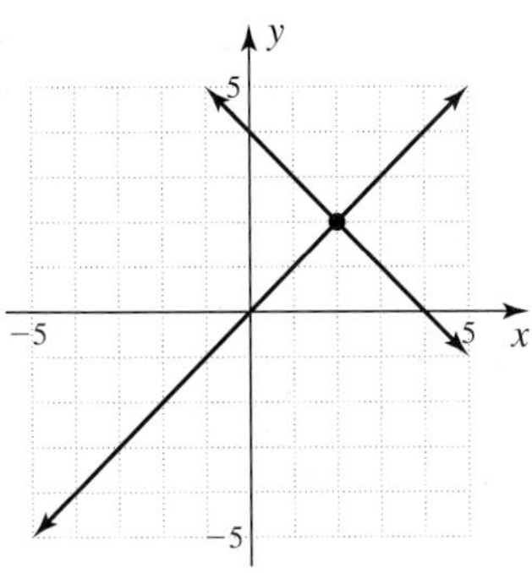

c. Solution: (1, 3)
$x + y = 4; y - 3x = 0$

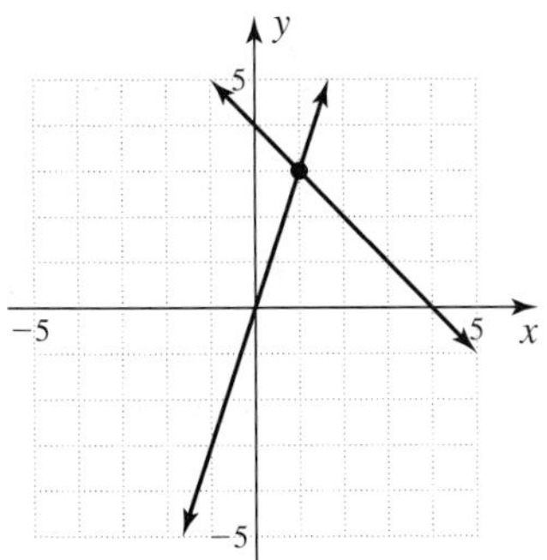

2. [7.1A, B]
(a) Inconsistent; no solution
$y - 3x = 3; 2y - 6x = 12$

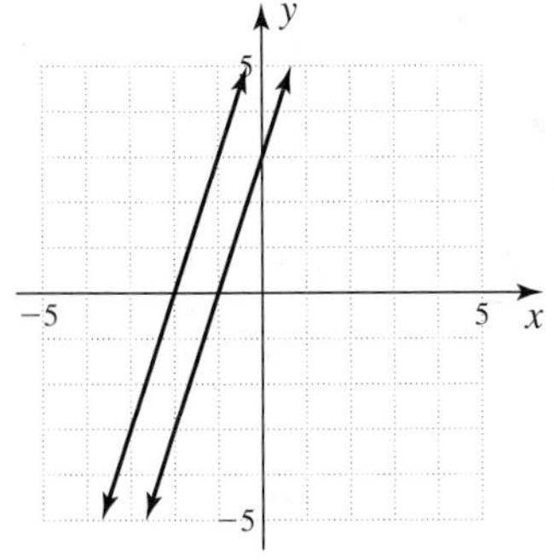

(b) Inconsistent; no solution
$y - 2x = 2; 2y - 4x = 8$

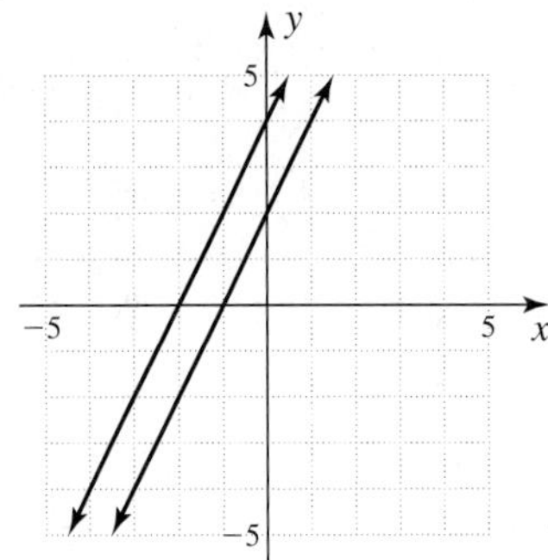

(c) Inconsistent; no solution
$y - 3x = 6; 2y - 6x = 6$

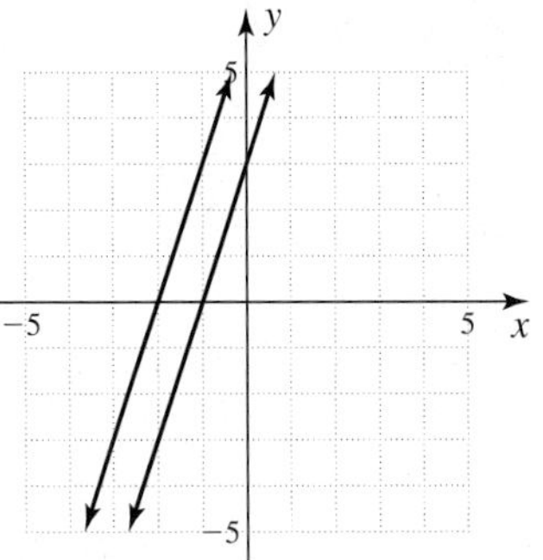

3. [7.2A, B] **(a)** Inconsistent; no solution; **(b)** Inconsistent; no solution; **(c)** (1, 1) **4.** [7.2A, B] **(a)** Dependent; infinitely many solutions; **(b)** Dependent; infinitely many solutions; **(c)** Dependent; infinitely many solutions **5.** [7.3A] **(a)** $(-1, 2)$; **(b)** $(2, -1)$; **(c)** $(-1, -2)$ **6.** [7.3, B] **(a)** Inconsistent; no solution; **(b)** Inconsistent; no solution; **(c)** Inconsistent; no solution
7. [7.3B] **(a)** Dependent; infinitely many solutions; **(b)** Dependent; infinitely many solutions; **(c)** Dependent; infinitely many solutions
8. [7.4A] **(a)** 20 nickels; 20 dimes; **(b)** 40 nickels; 10 dimes; **(c)** 50 nickels; 5 dimes **9.** [7.4B] **(a)** 70; 110; **(b)** 60; 120; **(c)** 50; 130
10. [7.4C] **(a)** 550 mi/hr; **(b)** 520 mi/hr; **(c)** 500 mi/hr
11. [7.4D] **(a)** Bonds: \$5000; CDs: \$15,000; **(b)** Bonds: \$17,000; CDs: \$3000; **(c)** Bonds: \$10,000; CDs: \$10,000
12. [7.5] **(a)** $x > 4; y < -1$

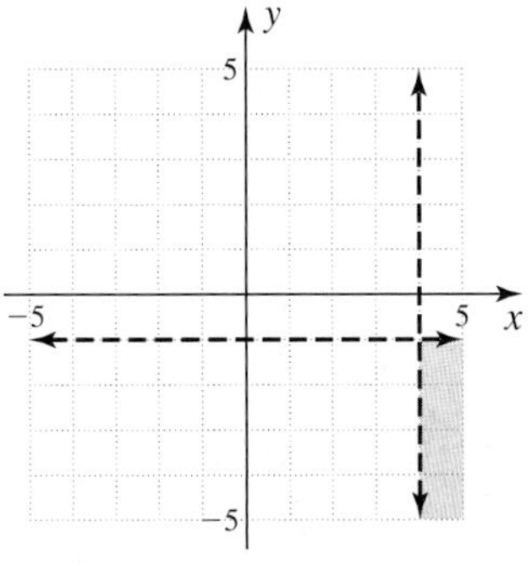

(b) $x + y > 3; x - y < 4$

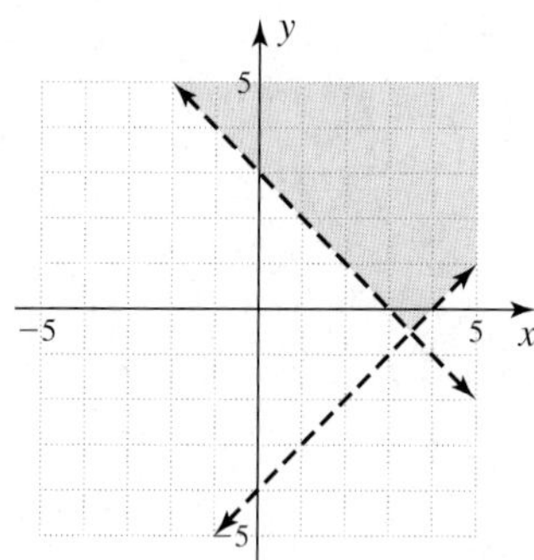

(c) $2x + y \le 4; x - 2y > 2$

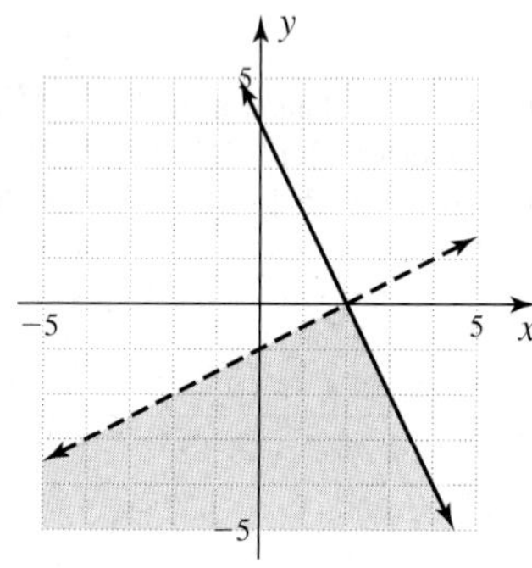

Cumulative Review Chapters 1–7

1. $-\frac{25}{42}$ **2.** 256 **3.** 2 **4.** 69 **5.** $-x - 10$ **6.** $\frac{x - 5y}{z}$ **7.** $x = 6$

8. $x = 36$ **9.**

10.

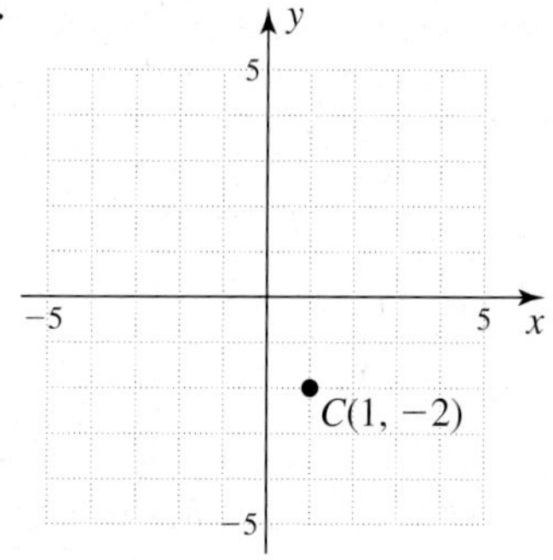

11. No **12.** $x = -3$

13. $x + y = 1$

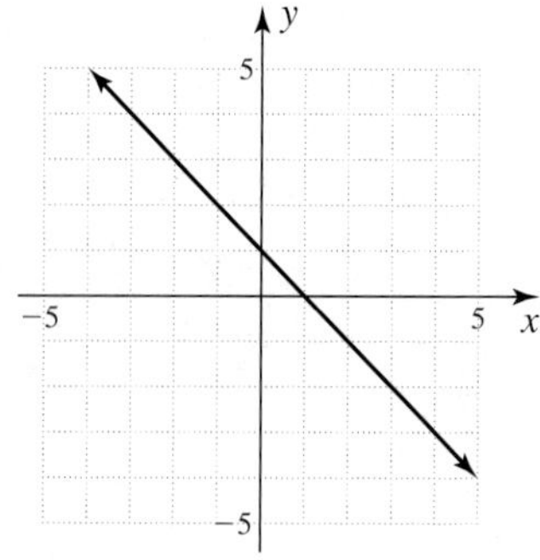

14. $2y + 8 = 0$

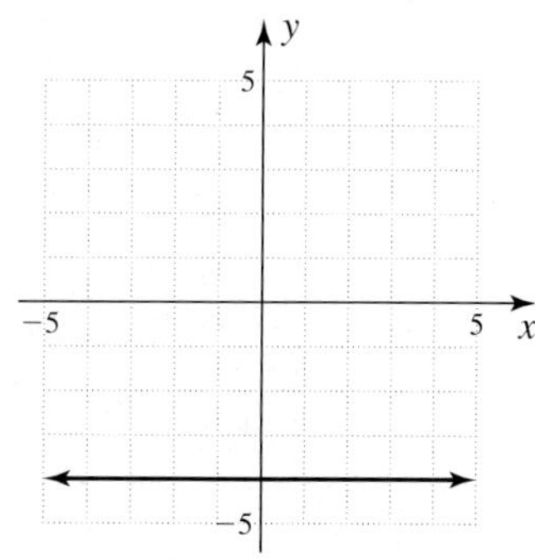

15. $-\frac{1}{7}$ **16.** 3 **17.** (1) and (2) **18.** $\frac{y^9}{8x^{12}}$ **19.** 3.5×10^{-5} **20.** 2.10×10^{-7} **21.** $36x^4 - 3x^2 + \frac{1}{16}$ **22.** $(2x^2 + x - 1)$ R 3 **23.** $(x - 6)(x - 5)$ **24.** $(3x - 4y)(5x - 3y)$ **25.** $(9x + 8y)(9x - 8y)$ **26.** $-3x^2(x + 2)(x - 2)$ **27.** $4x(x + 1)(x - 2)$ **28.** $(4x + 3)(x + 1)$ **29.** $k(5x - 3)^2$ **30.** $x = -3; x = \frac{5}{3}$ **31.** $\frac{20xy}{15y^2}$ **32.** $-3(x + y)$ **33.** $-(x + 2)$ **34.** $\frac{x - 2}{x + 5}$ **35.** $\frac{1}{4 - x}$ **36.** $\frac{6}{x - 9}$ **37.** $\frac{-2x - 5}{(x + 3)(x + 2)(x - 2)}$ **38.** $\frac{39}{2}$ **39.** $x = -5$ **40.** $x = -3$ **41.** No solution **42.** $x = -6$ **43.** 13 gal **44.** $x = \frac{31}{5}$ **45.** $2\frac{2}{9}$ hr **46.** $y + 2 = -4(x + 6)$ or $y = -4x - 26$ **47.** $y = 5x + 2$

48. $4x - 3y > -12$

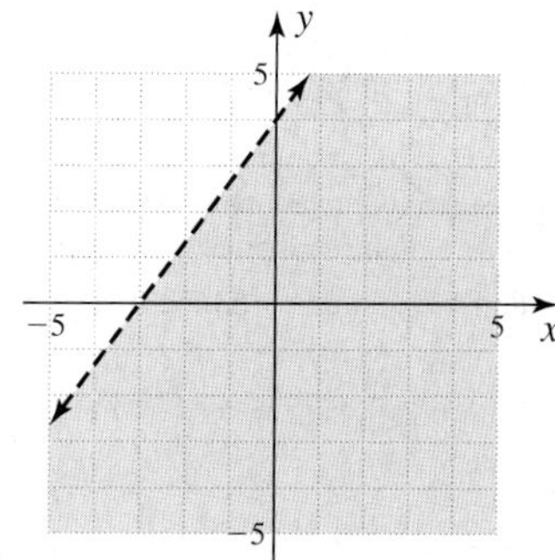

49. $-y \le -3x + 6$

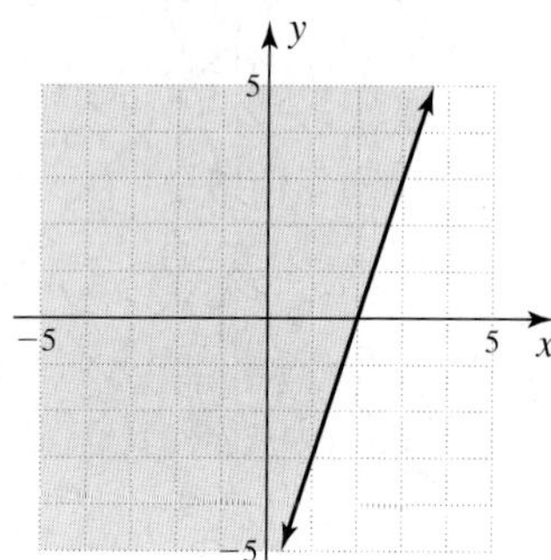

50. $\frac{1}{50}$ **51.** 6240

52. $x + 2y = 6; 2y - x = -2$

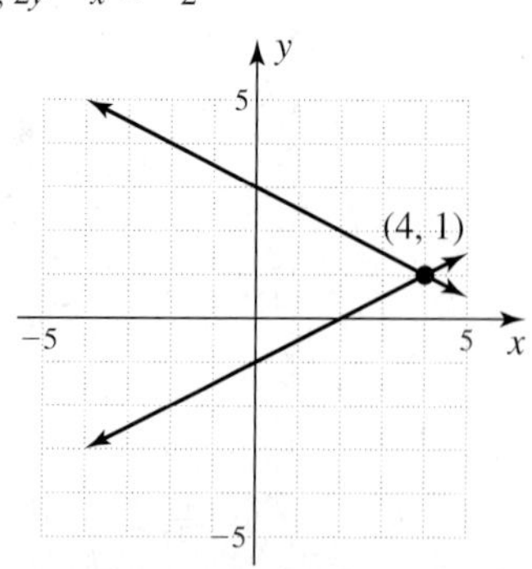

53. $y - x = -1$; $2y - 2x = -4$. No solution (inconsistent)

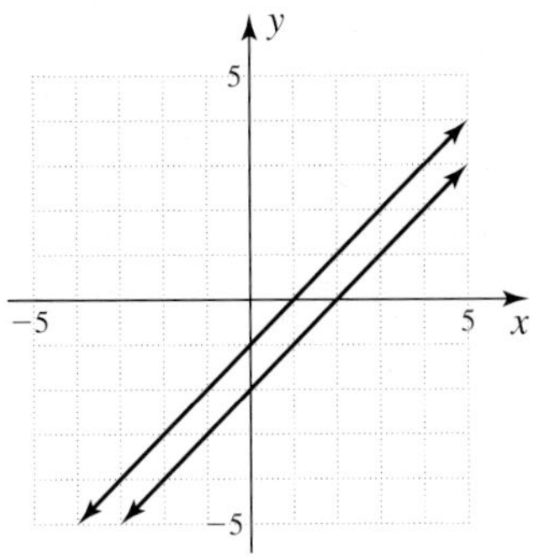

54. Inconsistent; no solution
55. Dependent; infinitely many solutions **56.** (1, 1)
57. Inconsistent; no solution
58. Dependent; infinitely many solutions
59. 12 nickels and 36 dimes
60. 35 and 130

Chapter 8

Exercises 8.1

1. 5 **3.** −3 **5.** $\frac{4}{3}$ **7.** $-\frac{2}{9}$ **9.** $\frac{5}{9}$ and $-\frac{5}{9}$ **11.** $\frac{7}{10}$ and $-\frac{7}{10}$ **13.** 5 **15.** 11 **17.** $x^2 + 1$ **19.** $3y^2 + 7$ **21.** 6; rational **23.** Not a real number **25.** −8; rational **27.** $\frac{4}{3}$; rational **29.** −2.449489743; irrational **31.** −1.414213562; irrational **33.** 3 **35.** −3 **37.** −4 **39.** 5 **41.** 10 sec **43.** $\sqrt{20}$ sec ≈ 4.5 sec **45.** 5 in. **47.** 244 mi **49.** 15 ft on each side **51.** 30 mi/hr **53.** 72 ft/sec **55. (a)** $5\frac{1}{11}$; **(b)** $5\frac{3}{11}$; **(c)** $5\frac{5}{11}$ **57.** One **59.** One **61.** One **63.** $a \geq 0$ **65.** $b^2 = a$ **67.** 0 **69.** a **71.** $\frac{4}{7}$ **73.** 12 **75.** 17 **77.** $-\frac{7}{11}$; rational **79.** 3.872983346; irrational **81.** −5 **83.** 3 **85.** 2 **87.** 7

Exercises 8.2

1. $3\sqrt{5}$ **3.** $5\sqrt{5}$ **5.** $6\sqrt{5}$ **7.** $10\sqrt{2}$ **9.** $8\sqrt{6}$ **11.** $8\sqrt{5}$ **13.** $10\sqrt{6}$ **15.** 19 **17.** $10\sqrt{7}$ **19.** $12\sqrt{3}$ **21.** $\sqrt{15}$ **23.** 9 **25.** 7 **27.** $\sqrt{3x}$ **29.** $6a$ **31.** $\frac{\sqrt{2}}{5}$ **33.** 2 **35.** 3 **37.** $5\sqrt{3}$ **39.** 12 **41.** $10a$ **43.** $7a^2$ **45.** $-4a^3\sqrt{2}$ **47.** $m^6\sqrt{m}$ **49.** $-3m^5\sqrt{3m}$ **51.** $2\sqrt[3]{5}$ **53.** $-2\sqrt[3]{2}$ **55.** $\frac{2}{3}$ **57.** $2\sqrt[4]{3}$ **59.** $\frac{4}{3}$ **61.** 9 mi **63.** 17 ft on each side **67. (a)** Irrational; **(b)** Yes **69.** $\frac{\sqrt{a}}{\sqrt{b}}$ **71.** $\sqrt[n]{a} \cdot \sqrt[n]{b}$ **73.** $3\sqrt[3]{5}$ **75.** $\frac{3}{4}$ **77.** $10x^3$ **79.** $\sqrt{2}$ **81.** $\frac{2}{3}$ **83.** 12 **85.** $12x$ **87.** $14x^3$

Exercises 8.3

1. $10\sqrt{7}$ **3.** $5\sqrt{13}$ **5.** $3\sqrt{2}$ **7.** $4\sqrt{2}$ **9.** $27\sqrt{3}$ **11.** $-7\sqrt{7}$ **13.** $29\sqrt{3}$ **15.** $-8\sqrt{5}$ **17.** $10\sqrt{2} - \sqrt{30}$ **19.** $2\sqrt{21} + \sqrt{30}$ **21.** $3 - \sqrt{6}$ **23.** $\sqrt{10} + 5$ **25.** $2\sqrt{3} - 3\sqrt{2}$ **27.** $2\sqrt{2} - 10$ **29.** $2\sqrt{3} - 3\sqrt{2}$ **31.** $\frac{\sqrt{6}}{2}$ **33.** $-2\sqrt{5}$ **35.** 2 **37.** $\frac{-\sqrt{10}}{5}$ **39.** $\frac{1}{2}$ **41.** $\frac{x\sqrt{2}}{6}$ **43.** $\frac{a\sqrt{b}}{b}$ **45.** $\frac{\sqrt{30}}{10}$ **47.** $\frac{x\sqrt{2}}{8}$ **49.** $\frac{x^2\sqrt{5}}{10}$ **51. (a)** 1 hr; **(b)** $\frac{5\sqrt{15}}{9} \approx 2.15$ hr **53.** $r = \frac{\sqrt{\pi S}}{2\pi}$ **59.** exactly **61.** $13\sqrt{3}$ **63.** $6\sqrt{2}$ **65.** $12\sqrt{2}$ **67.** $5\sqrt{3} - 3$ **69.** $\frac{3\sqrt{7}}{7}$ **71.** $\frac{\sqrt{2x}}{2}$ **73.** $x^2 - 9$ **75.** $2x + 4$

Exercises 8.4

1. 16 **3.** 13 **5.** −4 **7.** 5 **9.** $18\sqrt{10}$ **11.** $12\sqrt{11}$ **13.** $3\sqrt[3]{2} - 2$ **15.** $\sqrt[3]{2}$ **17.** $3\sqrt{x}$ **19.** $\frac{9y}{4}$ **21.** $\frac{8ab\sqrt{3a}}{3}$ **23.** $\frac{2bc^3\sqrt[3]{b^2}}{3}$ **25.** $5\sqrt[3]{2}$ **27.** $2\sqrt[3]{3}$ **29.** $2\sqrt{15} - 117$ **31.** $6\sqrt{6} + 29$ **33.** $59 - 20\sqrt{6}$ **35.** 5 **37.** $3\sqrt{2} - 3$ **39.** $\frac{2\sqrt{7} + 2}{3}$ **41.** $2\sqrt{2} - \sqrt{6}$ **43.** $2\sqrt{5} + \sqrt{15}$ **45.** $-\sqrt{10} + \sqrt{15}$ **47.** $\frac{\sqrt{15} + \sqrt{6}}{3}$ **49.** $5 + 2\sqrt{6}$ **51.** −2 **53.** $\frac{-4}{3}$ **55.** $\frac{1 + \sqrt{3}}{3}$ **57.** $\frac{-1 + \sqrt{23}}{2}$ **59.** $\frac{-2 + \sqrt{10}}{3}$ **61.** $\frac{-4 + \sqrt{7}}{3}$ **63.** $\sqrt{42{,}800} = 20\sqrt{107}$ m/sec **65.** $\frac{2}{\sqrt{5} + 1}$ **67.** $\frac{\sqrt{5} + 1}{2}$ **69.** The golden mean equals its reciprocal plus 1; yes. **71.** $\frac{x + 1}{x}$ **73. (a)** $x + 1 = x^2$; **(b)** $x^2 - x - 1 = 0$ **75.** $\sqrt{\frac{2(P_2 - P_1)}{\gamma P_1}}$ **79.** No; $\sqrt{2} - 1$ **81.** $a^2b - c^2d$ **83.** $-2 + \sqrt{7}$ **85.** $\frac{6 - 3\sqrt{2}}{2}$ **87.** $\frac{3\sqrt[3]{4}}{2}$ **89.** $4\sqrt[3]{2}$ **91.** $2\sqrt[3]{a^2}$ **93.** $-13 - \sqrt{15}$ **95.** $x - 1$ **97.** $4x$ **99.** $x^2 + 2x + 1$ **101.** $x(x - 3)$ **103.** $(x - 2)(x - 1)$

Exercises 8.5

1. $x = 16$ **3.** No solution **5.** $y = 4$ **7.** $y = 8$ **9.** $x = 8$ **11.** $x = 0$ **13.** $x = 5$ **15.** $y = 20$ **17.** $y = 25$ **19.** $y = 9$ or $y = 1$ **21.** $y = 6$ **23.** $x = 5$ **25.** $x = 4$ **27.** $x = \frac{11}{3}$ **29.** $y = 5$ **31.** $S = 50.24$ ft^2 **33.** 144 ft **35.** 3.24 ft **37.** 2000 **39.** $r = 100$ ft **41.** 31 m **43.** $x = 27$ **45.** $x = 7$ **47.** $x = 16$ **49.** No solution **55.** among **57.** 10 thousand **59.** $y = 2$ **61.** $x = 2$ **63.** $x = 7$ **65.** No solution **67.** $x = 2$ or $x = 3$ **69.** 7 **71.** $\frac{\sqrt{5}}{4}$

Review Exercises

1. [8.1A, C] **(a)** 9; **(b)** Not a real number; **(c)** $\frac{6}{5}$ **2.** [8.1A, C] **(a)** −6; **(b)** $-\frac{8}{5}$; **(c)** Not a real number **3.** [8.1B] **(a)** 8; **(b)** 25; **(c)** 17 **4.** [8.1B] **(a)** 36; **(b)** 17; **(c)** 64 **5.** [8.1B] **(a)** $x^2 + 1$; **(b)** $x^2 + 4$; **(c)** $x^2 + 5$ **6.** [8.1C] **(a)** Irrational; 3.3166; **(b)** Rational; −5; **(c)** Not a real number **7.** [8.1C] **(a)** Rational; **(b)** Rational; **(c)** Not a real number **8.** [8.1D] **(a)** 4; **(b)** −2; **(c)** −3 **9.** [8.1D] **(a)** 2; **(b)** −2; **(c)** Not a real number **10.** [8.1E] **(a)** $2\frac{3}{4}$ sec; **(b)** 3 sec; **(c)** $3\frac{1}{4}$ sec **11.** [8.2A] **(a)** $4\sqrt{2}$; **(b)** $4\sqrt{3}$; **(c)** 14 **12.** [8.2A] **(a)** $\sqrt{21}$; **(b)** 6; **(c)** $\sqrt{5y}$ **13.** [8.2B] **(a)** $\frac{\sqrt{3}}{4}$; **(b)** $\frac{\sqrt{5}}{6}$; **(c)** $\frac{3}{2}$ **14.** [8.2B] **(a)** 2; **(b)** $\sqrt{7}$; **(c)** $3\sqrt{5}$ **15.** [8.2C] **(a)** $6x$; **(b)** $10y^2$; **(c)** $9n^4$ **16.** [8.2C] **(a)** $6y^5\sqrt{2}$; **(b)** $7z^4\sqrt{3}$; **(c)** $4x^6\sqrt{3}$ **17.** [8.2C] **(a)** $y^7\sqrt{y}$; **(b)** $y^6\sqrt{y}$; **(c)** $5n^3\sqrt{2n}$ **18.** [8.2D] **(a)** $2\sqrt[3]{3}$; **(b)** $\frac{2}{3}$; **(c)** $-\frac{5}{4}$ **19.** [8.2D] **(a)** 3; **(b)** $2\sqrt[4]{3}$; **(c)** $2\sqrt[4]{5}$ **20.** [8.3A] **(a)** $15\sqrt{3}$; **(b)** $9\sqrt{2}$; **(c)** $6\sqrt{3}$ **21.** [8.3A] **(a)** $3\sqrt{11}$; **(b)** $\sqrt{2}$; **(c)** $\sqrt{3}$ **22.** [8.3B] **(a)** $2\sqrt{15} - \sqrt{6}$; **(b)** $5 - \sqrt{15}$; **(c)** $7 - 7\sqrt{14}$ **23.** [8.3C] **(a)** $\frac{\sqrt{10}}{4}$; **(b)** $\frac{x\sqrt{2}}{10}$; **(c)** $\frac{y\sqrt{3}}{9}$ **24.** [8.4A] **(a)** 3; **(b)** 3; **(c)** 4 **25.** [8.4A] **(a)** $7\sqrt{15}$; **(b)** $6\sqrt{6}$; **(c)** $7\sqrt{14}$ **26.** [8.4A] **(a)** $3\sqrt[3]{3}$; **(b)** 3; **(c)** 2 **27.** [8.4A] **(a)** $\frac{7\sqrt[3]{2}}{2}$; **(b)** $\frac{5\sqrt[3]{3}}{3}$; **(c)** $\frac{9\sqrt[3]{5}}{5}$ **28.** [8.4A] **(a)** $-27 - 2\sqrt{6}$; **(b)** $-23 + \sqrt{35}$ **29.** [8.4A] **(a)** −5; **(b)** −34 **30.** [8.4B] **(a)** $\frac{3\sqrt{3} - 3}{2}$; **(b)** $5\sqrt{2} + 5$ **31.** [8.4B] **(a)** $7\sqrt{3} + 7\sqrt{2}$; **(b)** $\frac{2\sqrt{5} + 2\sqrt{2}}{3}$ **32.** [8.4C] **(a)** $-4 + \sqrt{2}$; **(b)** $\frac{-8 + \sqrt{3}}{2}$ **33.** [8.5A] **(a)** $x = 7$; **(b)** No solution **34.** [8.5A] **(a)** $x = 4$; **(b)** $x = 6$ **35.** [8.5A] **(a)** $x = 5$; **(b)** $x = 7$ **36.** [8.5B] **(a)** $y = 4$; **(b)** $y = 0$ **37.** [8.5B] **(a)** $y = 1$; **(b)** $y = 0$ **38.** [8.5C] **(a)** 40 thousand or 40,000; **(b)** 240 thousand or 240,000

Cumulative Review Chapters 1–8

1. $-\frac{25}{72}$ **2.** 81 **3.** $\frac{2}{7}$ **4.** 47 **5.** $-x - 10$ **6.** $\frac{m + 3n}{p}$ **7.** $x = 16$ **8.** $x = 6$ **9.**

10.

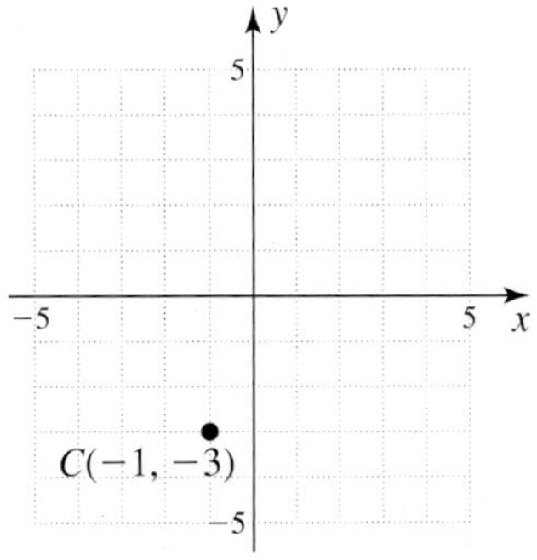

11. No **12.** $x = 1$
13. $5x + y = 5$

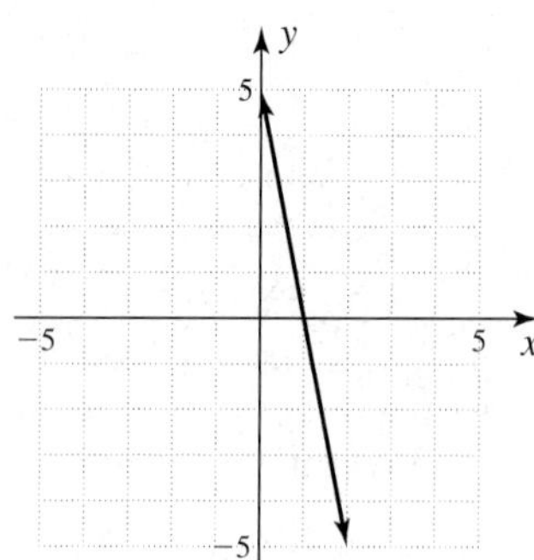

14. $4y - 8 = 0$

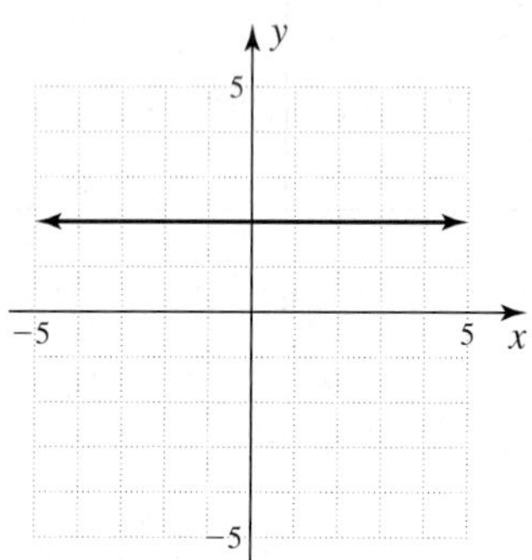

15. $-\frac{1}{2}$ **16.** 1 **17.** (1) and (3) **18.** $\frac{y^{12}}{16x^{16}}$ **19.** 2.5×10^{-7}
20. 1.7×10^{-8} **21.** $16x^4 - 4x^2 + \frac{1}{4}$ **22.** $(2x^2 - x - 2)$ R 3
23. $(x - 1)(x - 3)$ **24.** $(3x - 5y)(3x - 4y)$ **25.** $(2x + 5y)(2x - 5y)$
26. $-5x^2(x + 1)(x - 1)$ **27.** $4x(x + 1)(x - 3)$ **28.** $(3x + 4)(x + 3)$
29. $k(4x + 1)^2$ **30.** $x = -5; x = \frac{3}{4}$ **31.** $\frac{6xy}{15y^2}$ **32.** $-4(x + y)$
33. $-(x + 7)$ **34.** $\frac{x + 4}{x + 8}$ **35.** $\frac{1}{5 - x}$ **36.** $\frac{4}{x + 5}$ **37.** $\frac{-2x - 9}{(x + 5)(x + 4)(x - 4)}$
38. $\frac{14}{15}$ **39.** $x = 2$ **40.** $x = -3$ **41.** No solution **42.** $x = -7$
43. 25 gal **44.** $x = \frac{45}{4}$ **45.** $1\frac{5}{7}$ hr **46.** $y - 4 = 5(x - 6)$ or $y = 5x - 26$ **47.** $y = 4x + 3$
48. $x - 5y < -5$

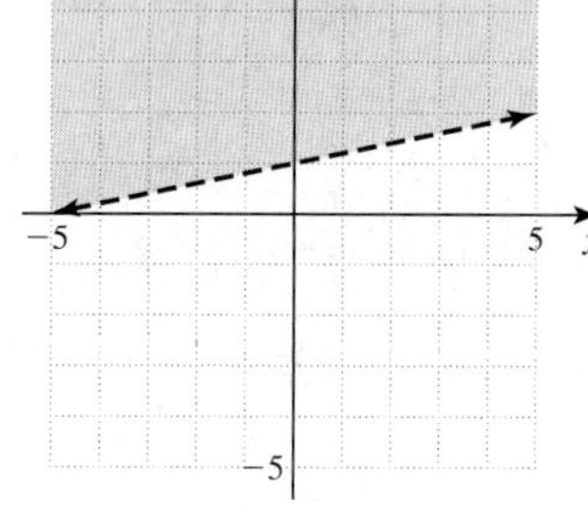

49. $-y \geq -5x - 5$

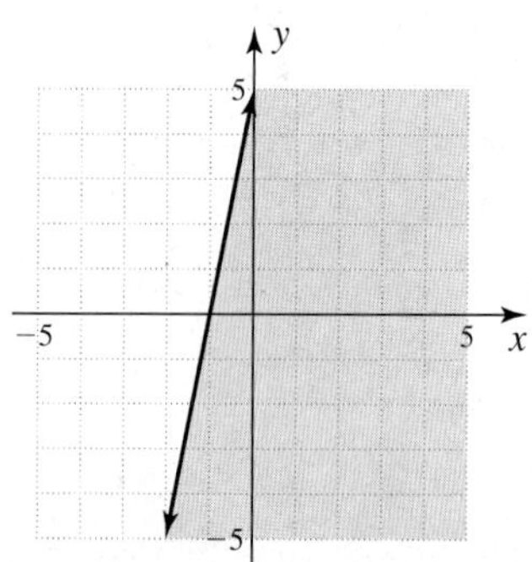

50. $\frac{1}{80}$ **51.** 10,800
52. $x + 4y = 16; 4y - x = 12$

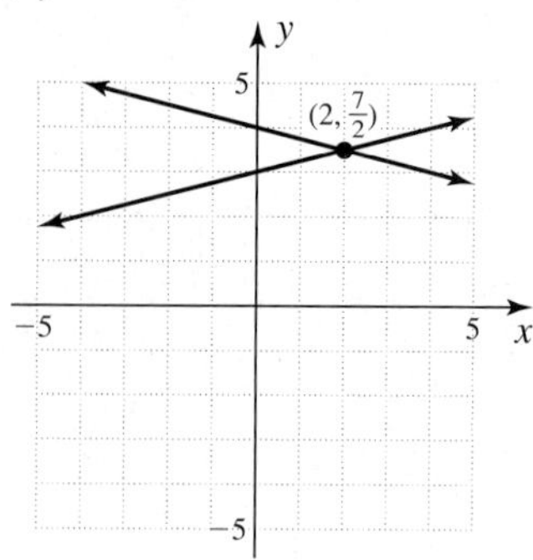

53. $y + 3x = -3; 2y + 6x = -12$; No solution

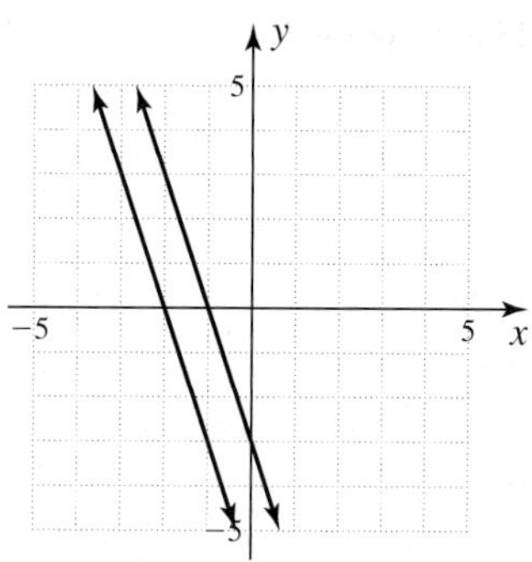

54. Inconsistent; no solution **55.** Dependent; infinitely many solutions
56. $(-1, -2)$ **57.** Inconsistent; no solution
58. Dependent; infinitely many solutions **59.** 10 nickels and 30 dimes
60. 65 and 150 **61.** -4 **62.** -3 **63.** $\frac{\sqrt{21}}{27}$ **64.** $4a^2b^3\sqrt[3]{2a^2}$
65. $6\sqrt{3}$ **66.** $2x - 3\sqrt[3]{6x^2}$ **67.** $\frac{\sqrt{6p}}{2p}$ **68.** $420 + 74\sqrt{35}$
69. $\frac{x + \sqrt{5x}}{x - 5}$ **70.** $2 + \sqrt{2}$ **71.** No real-number solution
72. $x = 6; x = 7$

Chapter 9

Exercises 9.1

1. $x = \pm 10$ **3.** $x = 0$ **5.** No real-number solution **7.** $x = \pm\sqrt{7}$
9. $x = \pm 3$ **11.** $x = \pm\sqrt{3}$ **13.** $x = \pm\frac{1}{5}$ **15.** $x = \pm\frac{7}{10}$
17. $y = \pm\frac{\sqrt{17}}{5}$ **19.** No real-number solution **21.** $x = 8$ or $x = -10$
23. $x = 8$ or $x = -4$ **25.** No real-number solution **27.** $x = 18$ or $x = 0$ **29.** $x = 0$ or $x = -8$ **31.** $x = -\frac{4}{5}$ or $x = -\frac{6}{5}$ **33.** $x = \frac{25}{6}$ or $x = \frac{11}{6}$ **35.** $x = \frac{3}{2}$ or $x = -\frac{7}{2}$ **37.** $x = 1 \pm \frac{\sqrt{5}}{3} = \frac{3 \pm \sqrt{5}}{3}$
39. No real-number solution **41.** $x = \pm\frac{1}{9}$ **43.** $x = \pm\frac{1}{4}$
45. $x = \pm 2$ **47.** $v = 2$ or $v = -4$ **49.** $x = \frac{5}{2}$ or $x = -\frac{1}{2}$
51. $y = \frac{3}{2} \pm \sqrt{2} = \frac{3 \pm 2\sqrt{2}}{2}$ **53.** $x = \frac{3}{2} \pm \sqrt{2} = \frac{3 \pm 2\sqrt{2}}{2}$
55. $x = -2 \pm 6\sqrt{2}$ **57.** $x = 3 \pm 6\sqrt{5}$ **59.** No real-number solution
61. 24 mph; no **63.** $40\sqrt{15} \approx 155$ m/sec **65.** 5 in. **67.** $3\frac{1}{2}$ ft
71. $A < 0$ **73.** $A > 0$ **75.** Rational **77.** 9 **79.** $\pm A$
81. $x = 1 \pm 2\sqrt{2}$ **83.** $x = 1$ or $x = -5$ **85.** $x = \pm\frac{\sqrt{3}}{7}$
87. $x = \pm\frac{4}{3}$ **89.** $x = \pm 1$ **91.** $x^2 + 14x + 49$ **93.** $x^2 - 6x + 9$

Exercises 9.2

1. 81 **3.** 64 **5.** $\frac{49}{4}$ **7.** $\frac{9}{4}$ **9.** $\frac{1}{4}$ **11.** $4; x + 2$ **13.** $\frac{9}{4}; x + \frac{3}{2}$
15. $9; x - 3$ **17.** $\frac{25}{4}; x - \frac{5}{2}$ **19.** $\frac{9}{16}; x - \frac{3}{4}$ **21.** $x = -2 \pm \sqrt{3}$

23. $x = -\frac{1}{2} \pm \frac{\sqrt{5}}{2} = \frac{-1 \pm \sqrt{5}}{2}$ **25.** $x = -\frac{3}{2} \pm \frac{\sqrt{13}}{2} = \frac{-3 \pm \sqrt{13}}{2}$
27. $x = \frac{3}{2} \pm \frac{\sqrt{21}}{2} = \frac{3 \pm \sqrt{21}}{2}$ **29.** $x = \frac{1}{2}$ or $x = -\frac{3}{2}$
31. $x = 2 \pm \frac{\sqrt{31}}{2} = \frac{4 \pm \sqrt{31}}{2}$ **33.** $x = \frac{1}{2} \pm \sqrt{2} = \frac{1 \pm 2\sqrt{2}}{2}$
35. $x = 1 \pm \frac{\sqrt{2}}{2} = \frac{2 \pm \sqrt{2}}{2}$ **37.** $x = 1$ or $x = -2$
39. $x = -\frac{5}{2} \pm \frac{3\sqrt{3}}{2} = \frac{-5 \pm 3\sqrt{3}}{2}$ **41.** 21 years
43. $x^2 + 6x + 9 = (x + 3)^2$ **45.** $x = 2$ and $x = -8$
47. (a) 2 (thousand) = 2000; **(b)** \$2 **49.** 3 days **51.** No; x is cubed in the equation. **53.** $x^2 - 6x + 9$ **55.** divide **57.** $\frac{1}{16}$
59. $\frac{25}{4}$; $x + \frac{5}{2}$ **61.** $x = -3 \pm \frac{\sqrt{3}}{2} = \frac{-6 \pm \sqrt{3}}{2}$ **63.** $x^2 - 2x - 2 = 0$
65. $10x^2 - 5x + 12 = 0$ **67.** $9x^2 - 4x - 2 = 0$

Exercises 9.3

1. $x = -2$ or $x = -1$ **3.** $x = -2$ or $x = 1$ **5.** $x = \frac{-1 \pm \sqrt{17}}{4}$
7. $x = \frac{2}{3}$ or $x = -1$ **9.** $x = -\frac{3}{2}$ or $x = -2$ **11.** $x = \frac{5}{7}$ or $x = 1$
13. $x = \frac{11 \pm \sqrt{41}}{10}$ **15.** $x = \frac{3}{2}$ or $x = 1$ **17.** $x = \frac{3 \pm \sqrt{5}}{4}$ **19.** $x = -1$
21. $x = \frac{3 \pm \sqrt{5}}{4}$ **23.** $x = \frac{-3 \pm \sqrt{21}}{6}$ **25.** $x = 2$ or $x = -2$
27. $x = \pm 3\sqrt{5}$ **29.** $x = \pm\frac{2\sqrt{3}}{3}$ **31.** 40 **33.** The discriminant is $\sqrt{-0.032}$; there is no solution. **35.** Multiply each term by $4a$.
37. Add b^2 on both sides. **39.** Take the square root of each side.
41. Divide each side by $2a$. **43.** $x = \pm\sqrt{17}$ **45.** $x = -2$
47. $y = 0$ or $y = 1$ **49.** $x = -3$ or $x = -2$ **51.** $z = 0$ or $z = -6$
53. $y = \frac{1}{3}$ or $y = -3$ **55.** No real-number solution **57.** One; rational
59. None **61.** $\frac{-b \pm \sqrt{b^2 - 4ac}}{2a}$ **63.** No real-number solution
65. $x = 2 \pm 2\sqrt{3}$ **67.** $x = -\frac{1}{2}$ or $x = 2$ **69.** 4 **71.** 1

Exercises 9.4

1. $y = 2x^2$

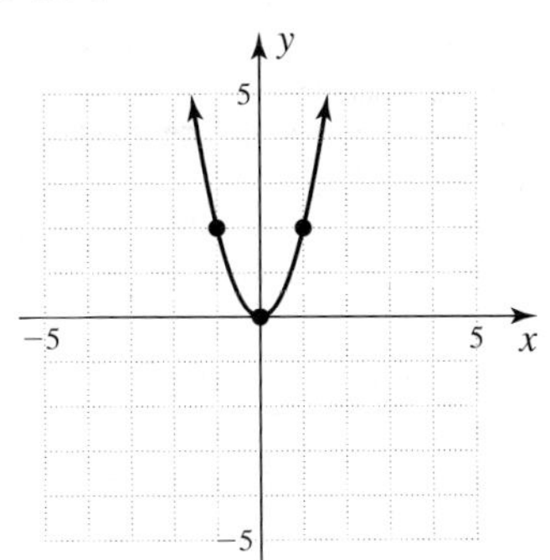

3. $y = 2x^2 - 1$

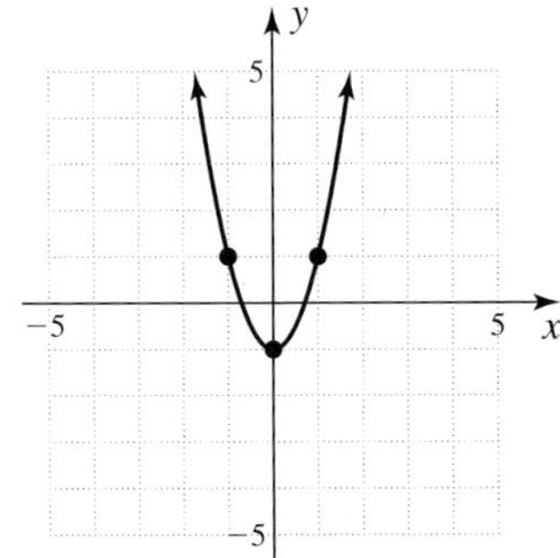

5. $y = -2x^2 + 2$

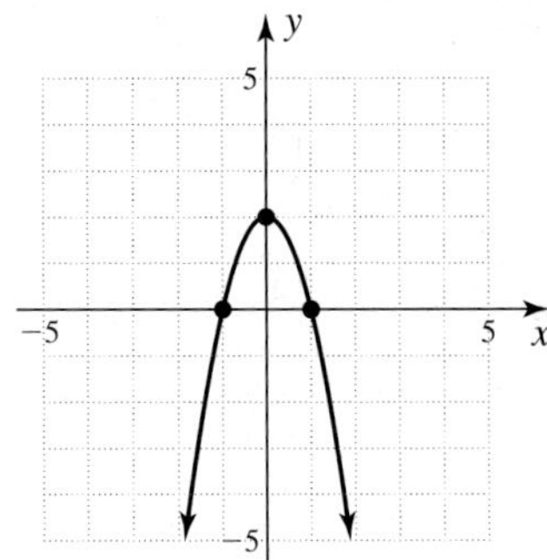

7. $y = -2x^2 - 2$

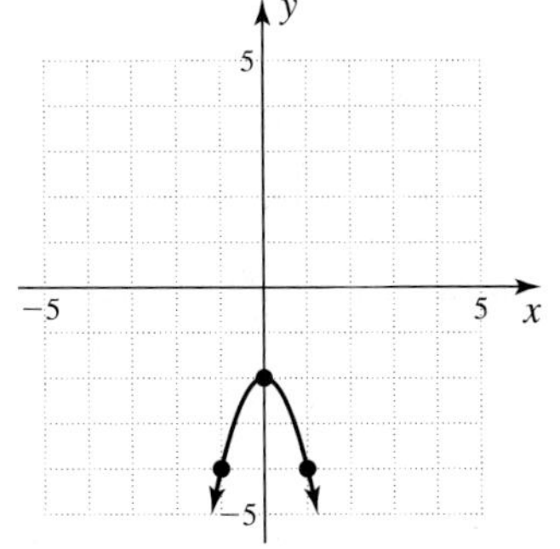

9. $y = (x - 2)^2$

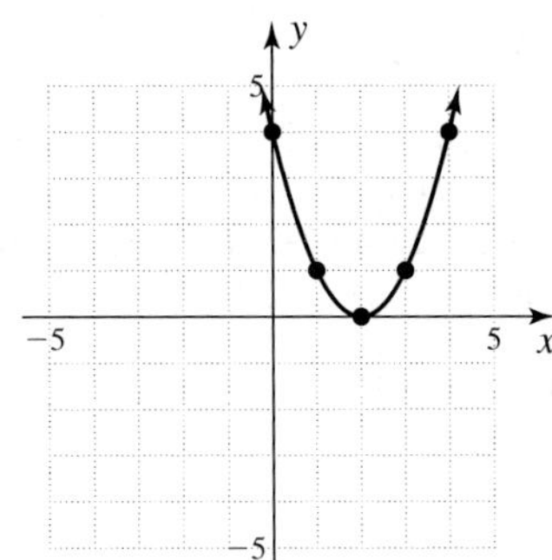

11. $y = (x - 2)^2 - 2$

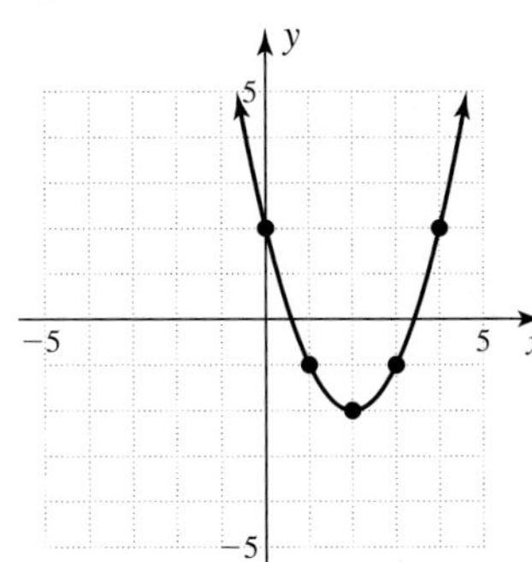

13. $y = -(x - 2)^2$

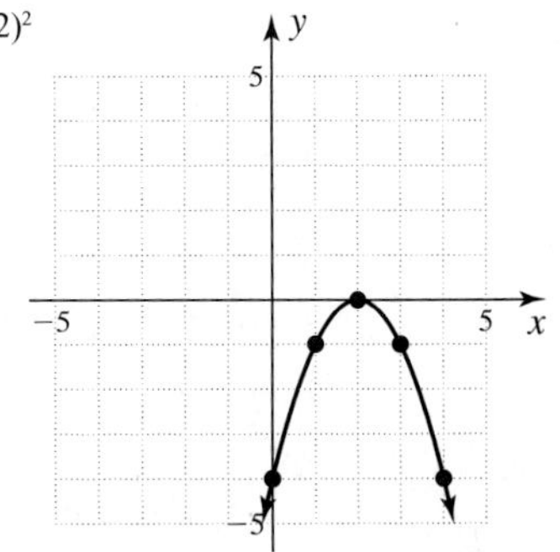

15. $y = -(x - 2)^2 - 2$

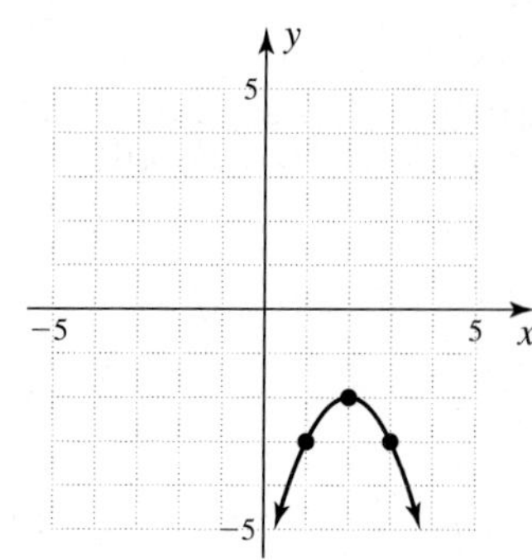

17. $y = x^2 + 4x + 3$

x-int: $(-3, 0), (-1, 0)$; y-int: $(0, 3)$; $V(-2, -1)$

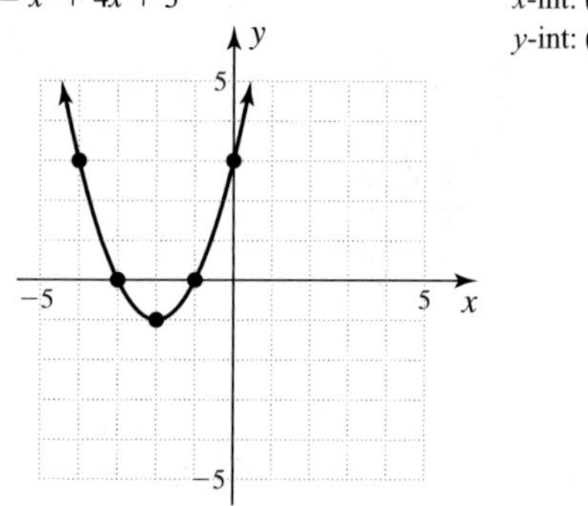

19. $y = x^2 + 2x - 3$

x-int: $(-3, 0), (1, 0)$; y-int: $(0, -3)$; $V(-1, -4)$

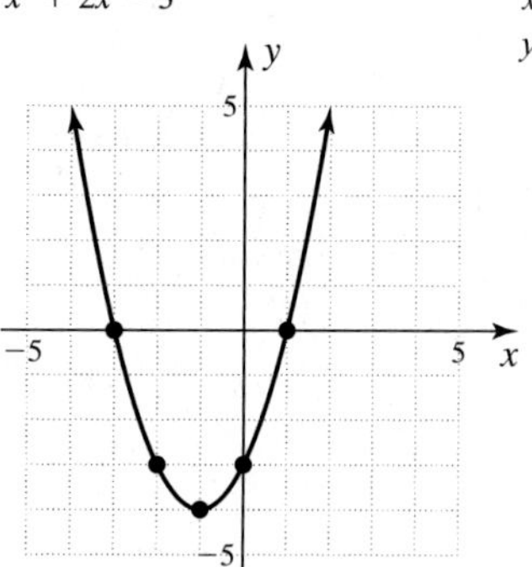

21. $y = -x^2 + 4x - 3$

x-int: $(3, 0), (1, 0)$; y-int: $(0, -3)$; $V(2, 1)$

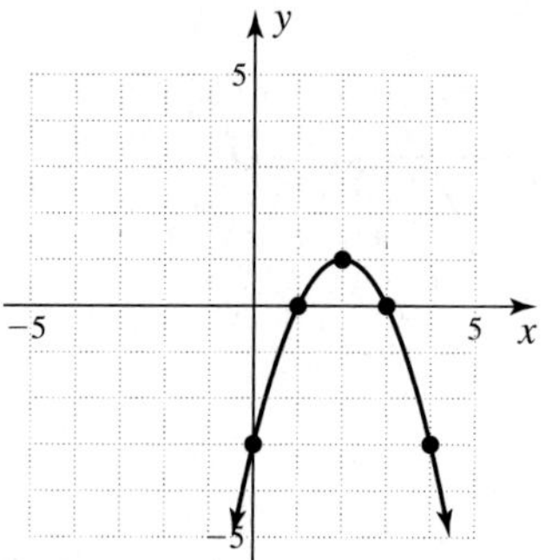

23. About 40 hr **25.** About 60 hr **27.** $x = 4000$; $P = \$11,000$
29. (a) Parabola; **(b)** Upward; **(c)** 1400 polar bears; **(d)** 1631 polar bears; **(e)** 2060 polar bears; **(f)** (2, 1356);

(g)

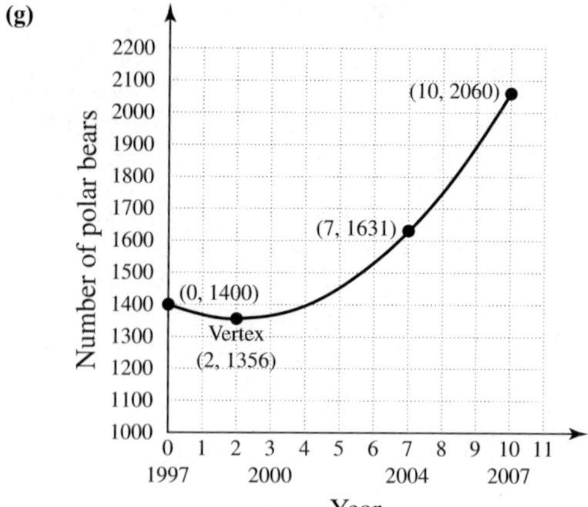

31. \$40 **33.** \$400
35. The parabola opens upward. **37.** The graph "stretches" upward.
39. The graph is shifted upward; the graph is shifted downward.
41. $a > 0$
43. $y = -(x - 2)^2 - 1$

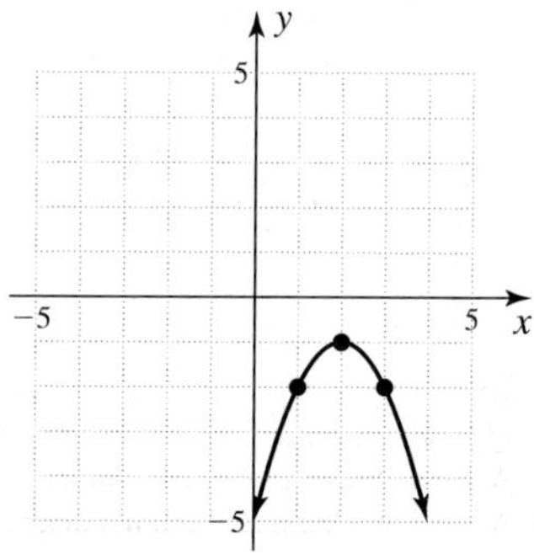

45. $y = -2x^2$

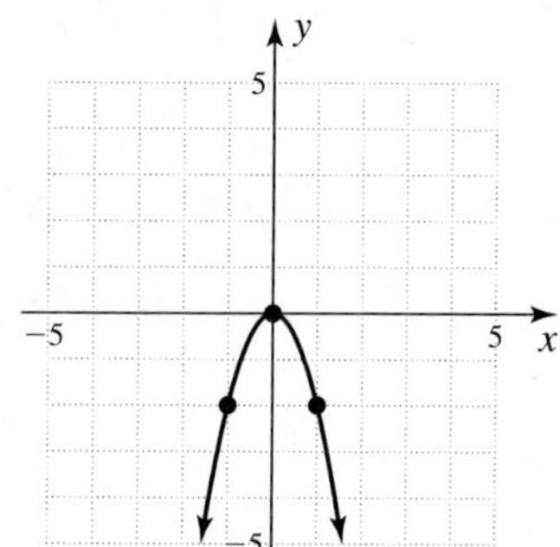

47. $y = -x^2 - 2x + 3$

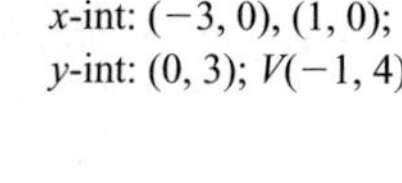
x-int: $(-3, 0), (1, 0)$; y-int: $(0, 3)$; $V(-1, 4)$

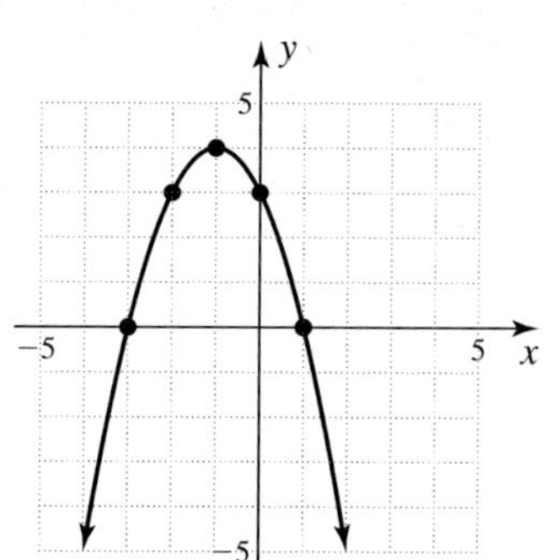

49. 25 **51.** 40,000

Exercises 9.5

1. $b = 5$ **3.** $b = 10\sqrt{2}$ **5.** $c = \sqrt{15}$ **7.** $c = 3$ **9.** $b = 2$ **11.** 50 ft **13.** 100 mi/hr **15.** 8 sec **17.** 1 sec **19.** 3000 units **21.** 3 ft by 4 ft **23.** 10 cm; 24 cm; 26 cm **25.** **(a)** $q = 10$; **(b)** 180 **27.** **(a)** A parabola opening upward; **(b)** A parabola opening downward; **(c)** 30; **(d)** About 35 and 37, respectively; **(e)** 40.8; **(f)** In 2036; **(g)** the graph is shown; **(h)** Personal computers

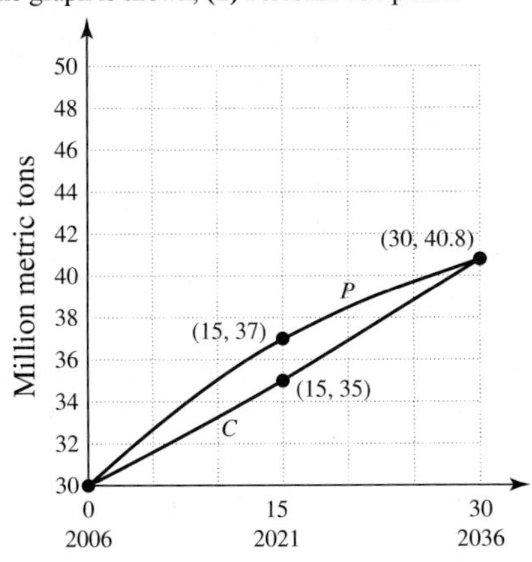

29. $c = \sqrt{\frac{E}{m}} = \frac{\sqrt{Em}}{m}$ **31.** $r = \sqrt{\frac{GMm}{F}} = \frac{\sqrt{FGMm}}{F}$ **33.** $p = \sqrt{\frac{I}{k}} = \frac{\sqrt{Ik}}{k}$ **37.** right **39.** 1 or 6 (hundred) **41.** 50 mi/hr **43.** -1

Exercises 9.6

1. $D = \{1, 2, 3\}$; $R = \{2, 3, 4\}$ **3.** $D = \{1, 2, 3\}$; $R = \{1, 2, 3\}$
5. $D =$ All real numbers; $R =$ All real numbers
7. $D =$ All real numbers; $R =$ All real numbers
9. $D =$ All real numbers; $R =$ All nonnegative real numbers
11. $D =$ All nonnegative real numbers; $R =$ All real numbers
13. $D =$ All real numbers except 3; $R =$ All real numbers except 0
15. $D = \{-1, 0, 1, 2\}$; $R = \{-2, 0, 2, 4\}$; $(-1, -2), (0, 0), (1, 2), (2, 4)$
17. $D = \{0, 1, 2, 3, 4\}$; $R = \{-3, -1, 1, 3, 5\}$; $(0, -3), (1, -1), (2, 1), (3, 3), (4, 5)$ **19.** $D = \{0, 1, 4, 9, 16, 25\}$; $R = \{0, 1, 2, 3, 4, 5\}$; $(0, 0), (1, 1), (4, 2), (9, 3), (16, 4), (25, 5)$ **21.** $D = \{1, 2, 3\}$; $R = \{2, 3, 4\}$; $(1, 2), (1, 3), (1, 4), (2, 3), (2, 4), (3, 4)$
23. A function; one y-value for each x-value **25.** A function; one y-value for each x-value **27.** A function; one y-value for each x-value **29.** Not a function; two y-values for each positive x-value **31.** **(a)** $f(0) = 1$; **(b)** $f(2) = 7$; **(c)** $f(-2) = -5$
33. **(a)** $F(1) = 0$; **(b)** $F(5) = 2$; **(c)** $F(26) = 5$
35. **(a)** $f(x + h) = 3x + 3h + 1$; **(b)** $f(x + h) - f(x) = 3h$; **(c)** $\frac{f(x+h) - f(x)}{h} = 3, h \neq 0$ **37.** $y = g(x) = x^2$; $\left(\frac{1}{4}, \frac{1}{16}\right)$, $(2.1, 4.41)$, $(\pm 8, 64)$ **39.** **(a)** $g(0) = 1$; **(b)** $g(-2) = -5$; **(c)** $g(2) = 15$
41. **(a)** 140; **(b)** 130 **43.** **(a)** 160 lb; **(b)** 78 in. **45.** **(a)** 639 lb/ft^2; **(b)** 6390 lb/ft^2 **47.** **(a)** 144 ft; **(b)** 400 ft **49.** **(a)** $r(n) = 441{,}000n$; **(b)** 26,460,000 gallons **51.** A function **53.** A function **55.** No **59.** ordered **61.** range **63.** $D = \{-5, -6, -7\}$; $R = \{5, 6, 7\}$
65. $D =$ All real numbers except 3; $R =$ All real numbers except 0
67. A function; one y-value for each x-value **69.** A function; one y-value for each x-value **71.** Not a function; two y-values for each positive x-value **73.** $g(-1) = -2$

Review Exercises

1. [9.1A] **(a)** $x = \pm 1$; **(b)** $x = \pm 10$; **(c)** $x = \pm 9$
2. [9.1A] **(a)** $x = \pm\frac{5}{4}$; **(b)** $x = \pm\frac{3}{5}$; **(c)** $x = \pm\frac{5}{8}$
3. [9.1A] **(a)** No real-number solution; **(b)** No real-number solution; **(c)** No real-number solution
4. [9.1B] **(a)** $x = -1 \pm \frac{\sqrt{3}}{7} = \frac{-7 \pm \sqrt{3}}{7}$; **(b)** $x = -2 \pm \frac{\sqrt{2}}{5} = \frac{-10 \pm \sqrt{2}}{5}$; **(c)** $x = -1 \pm \frac{\sqrt{5}}{4} = \frac{-4 \pm \sqrt{5}}{4}$
5. [9.2A] **(a)** 9; **(b)** 49; **(c)** 36 **6.** [9.2A] **(a)** 9; $x - 3$; **(b)** 25; $x - 5$; **(c)** 36; $x - 6$ **7.** [9.2A] **(a)** 7; **(b)** 6; **(c)** 5 **8.** [9.2A] **(a)** 4; **(b)** 9; **(c)** 36 **9.** [9.3B] **(a)** $x = \frac{-e \pm \sqrt{e^2 - 4df}}{2d}$; **(b)** $x = \frac{-h \pm \sqrt{h^2 - 4gi}}{2g}$; **(c)** $x = \frac{-k \pm \sqrt{k^2 - 4jm}}{2j}$
10. [9.3B] **(a)** $x = 1$ or $x = -\frac{1}{2}$; **(b)** $x = \frac{1 \pm \sqrt{11}}{2}$; **(c)** $x = \frac{3 \pm \sqrt{33}}{4}$
11. [9.3B] **(a)** $x = \frac{1 \pm \sqrt{13}}{6}$; **(b)** $x = \frac{1 \pm \sqrt{7}}{3}$; **(c)** $x = \frac{3 \pm \sqrt{33}}{6}$
12. [9.3B] **(a)** $x = 0$ or $x = 9$; **(b)** $x = 0$ or $x = 4$; **(c)** $x = 0$ or $x = 25$
13. [9.3B] **(a)** $x = \frac{9 \pm \sqrt{65}}{2}$; **(b)** $x = -\frac{1}{2}$ or $x = 3$; **(c)** No real-number solution

14. [9.4A]

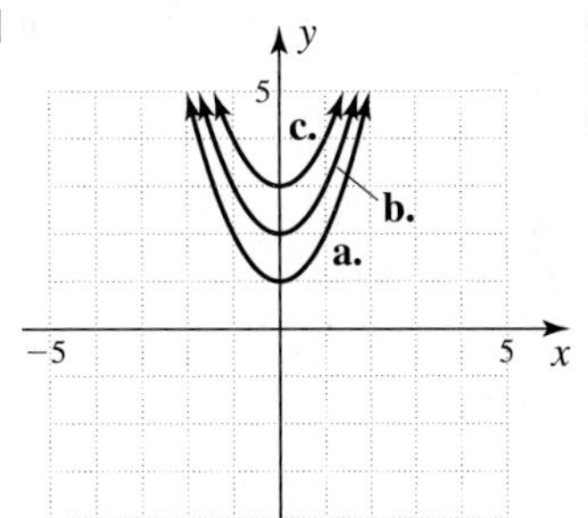

(a) $y = x^2 + 1$; **(b)** $y = x^2 + 2$; **(c)** $y = x^2 + 3$

15. [9.4A]

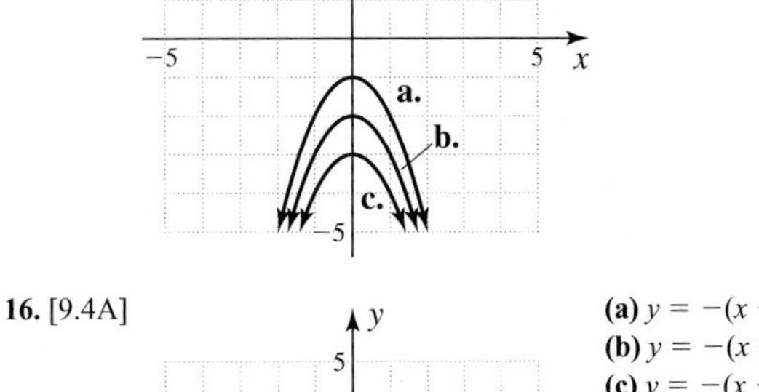

(a) $y = -x^2 - 1$; **(b)** $y = -x^2 - 2$; **(c)** $y = -x^2 - 3$

16. [9.4A]

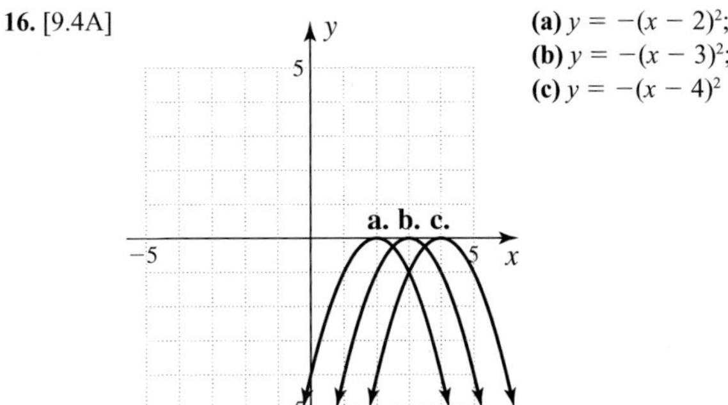

(a) $y = -(x - 2)^2$; **(b)** $y = -(x - 3)^2$; **(c)** $y = -(x - 4)^2$

17. [9.4A]

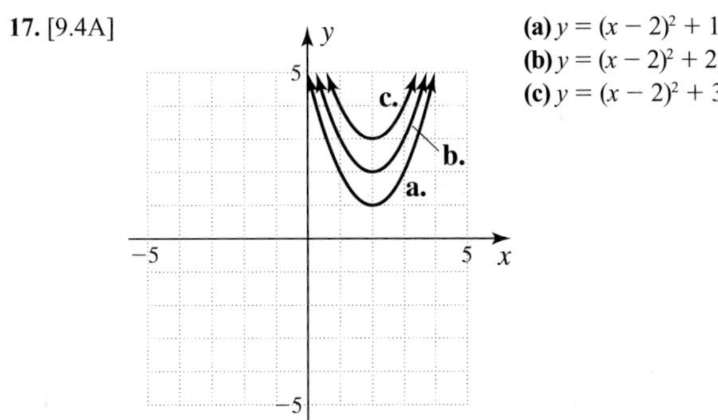

(a) $y = (x - 2)^2 + 1$; **(b)** $y = (x - 2)^2 + 2$; **(c)** $y = (x - 2)^2 + 3$

18. [9.4B]

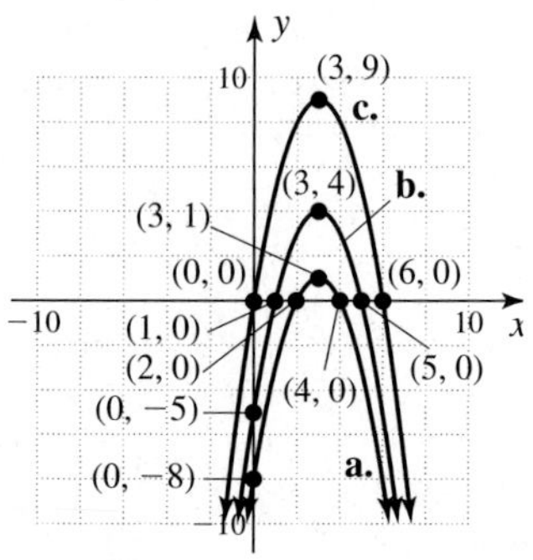

(a) $y = -x^2 + 6x - 8$ **(b)** $y = -x^2 + 6x - 5$ **(c)** $y = -x^2 + 6x$

19. [9.5A] **(a)** 13 in.; **(b)** $\sqrt{13}$ in.; **(c)** $\sqrt{41}$ in. **20.** [9.5B] **(a)** 5 sec; **(b)** 7 sec; **(c)** 8 sec **21.** [9.6A] **(a)** $D = \{-3, -4, -5\}$; $R = \{1, 2\}$; **(b)** $D = \{2, -1\}$; $R = \{-4, 3, 4\}$; **(c)** $D = \{-1\}$; $R = \{2, 3, 4\}$
22. [9.6A] **(a)** $D =$ All real numbers; $R =$ All real numbers; **(b)** $D =$ All real numbers; $R =$ All real numbers; **(c)** $D =$ All real numbers; $R =$ All nonnegative real numbers
23. [9.6B] **(a)** A function; **(b)** Not a function; **(c)** Not a function
24. [9.6C] **(a)** $f(2) = 1$; **(b)** $f(-2) = -19$; **(c)** $f(1) = -1$
25. [9.6D] **(a)** \$22; **(b)** \$19; **(c)** \$16

Cumulative Review Chapters 1–9

1. $-\frac{29}{56}$ **2.** 256 **3.** $\frac{9}{5}$ **4.** 14 **5.** $-x - 10$ **6.** $\frac{x - 5y}{z}$ **7.** $x = 3$
8. $x = 72$ **9.**

10.

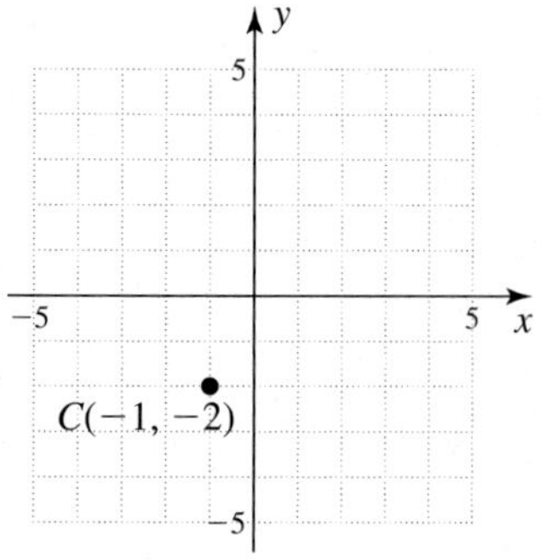

11. Yes **12.** $x = -2$

13. $2x + y = 4$

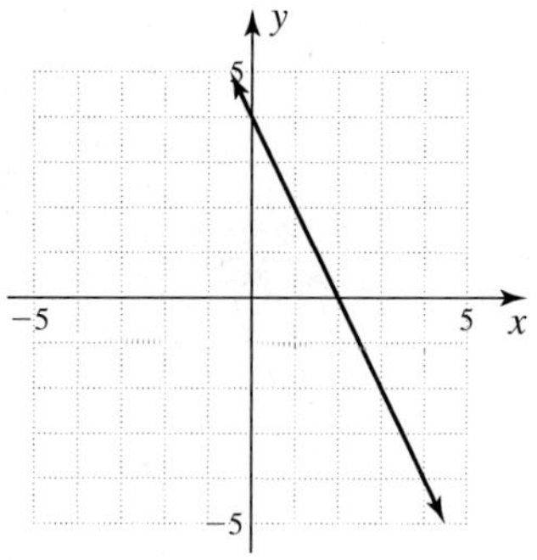

14. $2y - 8 = 0$

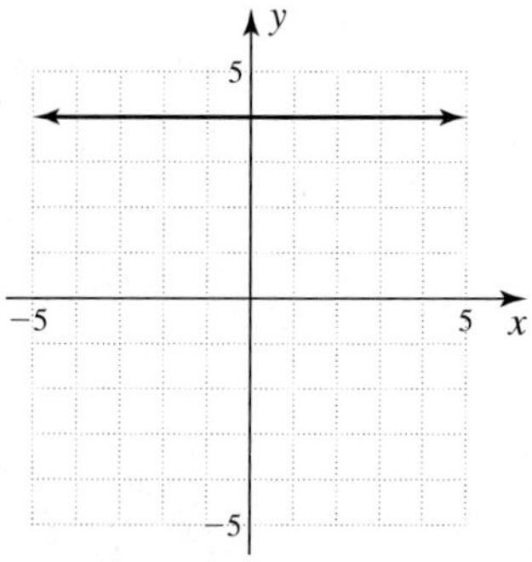

15. $\frac{14}{11}$ **16.** 2 **17.** (2) and (3) **18.** $\frac{81x^{12}}{y^{20}}$ **19.** 5.0×10^{-3}
20. 1.5×10^{-7} **21.** $4x^4 - \frac{4}{5}x^2 + \frac{1}{25}$ **22.** $(2x^2 - x - 1)$ R 1
23. $(x - 5)(x - 3)$ **24.** $(4x - 3y)(3x - 4y)$ **25.** $(5x + 6y)(5x - 6y)$
26. $-5x^2(x + 4)(x - 4)$ **27.** $3x(x + 2)(x - 3)$ **28.** $(x + 3)(4x + 5)$
29. $k(5x + 2)^2$ **30.** $x = -5$; $x = \frac{2}{3}$ **31.** $\frac{12xy^2}{10y^3}$ **32.** $-(x + y)$
33. $-(x + 3)$ **34.** $\frac{x - 3}{x + 4}$ **35.** $\frac{1}{2 - x}$ **36.** $\frac{3}{x + 1}$
37. $\frac{-2x - 15}{(x + 8)(x + 7)(x - 7)}$ **38.** 6 **39.** $x = 9$ **40.** $x = -5$
41. No solution **42.** $x = -6$ **43.** 29 gal **44.** $x = \frac{64}{5}$ **45.** $2\frac{2}{9}$ hr
46. $y + 4 = 6(x + 2)$ or $y = 6x + 8$ **47.** $y = -2x - 5$
48. $3x - 2y < -6$

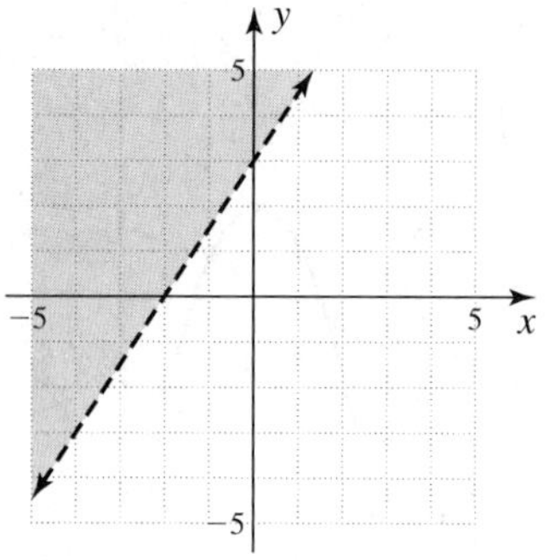

49. $-y \geq -5x + 5$

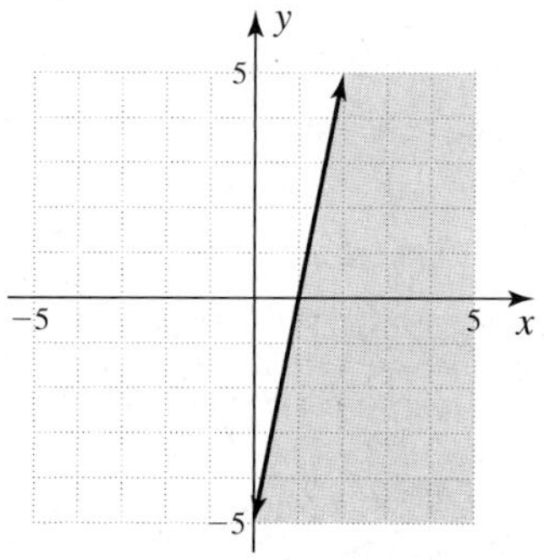

50. $\frac{1}{70}$ **51.** 13,720
52. $x + 3y = -12$; $2y - x = -2$

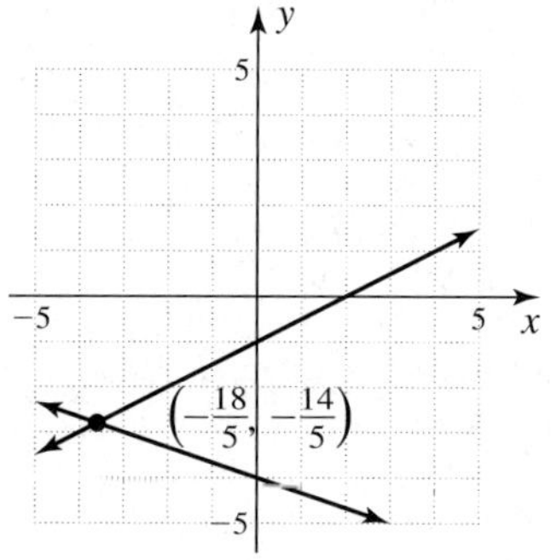

53. $y + 3x = 3$; $3y + 9x = 18$ No solution

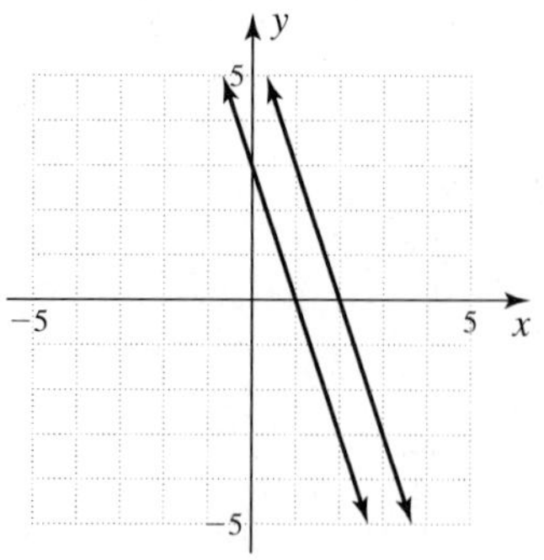

54. Inconsistent; no solution **55.** Dependent; infinitely many solutions **56.** $(-3, -3)$ **57.** Inconsistent; no solution **58.** Dependent; infinitely many solutions **59.** 10 nickels and 20 dimes **60.** 30 and 140 **61.** 3 **62.** 5 **63.** $\frac{\sqrt{15}}{27}$ **64.** $\frac{\sqrt{14j}}{7j}$ **65.** $7\sqrt{2}$ **66.** $3x - 2\sqrt[3]{6x^2}$ **67.** $280 + 46\sqrt{35}$ **68.** $5 - \sqrt{2}$ **69.** No real-number solution **70.** $x = 4; x = 5$ **71.** $x = \frac{2}{3}; x = -\frac{2}{3}$ **72.** No real-number solution **73.** $x = \frac{24 \pm \sqrt{7}}{6}$ or $x = 4 + \frac{\sqrt{7}}{6}; x = 4 - \frac{\sqrt{7}}{6}$ **74.** 4 **75.** 4 **76.** 4 **77.** $x = \frac{-k + \sqrt{k^2 - 4jm}}{2j}; x = \frac{-k - \sqrt{k^2 - 4jm}}{2j}$ **78.** $x = \frac{3}{2}; x = -3$ **79.** $x = 0; x = 16$ **80.** $y = -(x + 3)^2 + 2$

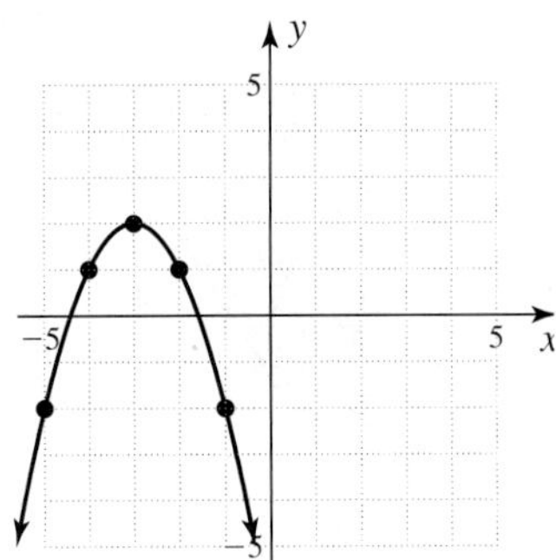

81. 2 sec

Photo Credits

Chapter R

Page p. 2(left): © Thinkstock/Getty RF; **p. 2(right):** Corbis RF; **p. 8(left):** © Brand X Pictures/Getty RF; **p. 8(right):** © Chris Speedie/Getty Images; **p. 9, 12:** Courtesy Ignacio Bello; **p. 20:** © The McGraw-Hill Companies, Inc./Dr. Thomas D. Porter, photographer.

Chapter 1

Page 60: © Corbis RF.

Chapter 2

Opener: © The British Museum; **p. 110:** © Corbis RF; **p. 137:** © Jose Carrilo/Photo Edit; **p. 148:** © Mary Kate Denny/Photo Edit; **p. 166(left):** U.S. Coast Guard Photo; **p. 179:** © Fred Bellett/Tampa Tribune.

Chapter 3

Opener: © Stapleton Collection/Corbis; **p. 275:** © Eunice Harris/Index Stock/Photolibrary.

Chapter 4

Opener: © Bibliotheque national de France; **p. 325:** © Vol. OS23/PhotoDisc/Getty RF; **p. 327(all):** Courtesy Ignacio Bello; **p. 329:** © Doug Struthers/Stone/Getty Images; **p. 336:** © Steve Gschmeissner/SPL/Photo Researchers, Inc.; **p. 339:** Courtesy Ignacio Bello; **p. 343:** NASA/JPL-Caltech/R. Hurt (SSC/Caltech); **p. 345**(Neptune): NASA, L. Sromovsky, and P. Fry (University of Wisconsin-Madison); **p. 345**(Saturn): NASA, ESA, J. Clarke (Boston University), and Z. Levay (STScI); **p. 345**(Uranus): NASA, ESA, L. Sromovsky and P. Fry (University of Wisconsin), H. Hammel (Space Science Institute), and K. Rages (SETI Institute); **p. 345**(Jupiter): STScl/NASA; **p. 345**(Mercury. Earth): NASA; **p. 345**(Venus): USGS MAGELLAN IMAGING RADAR; **p. 345**(Mars): David Crisp and the WFPC2 Science Team (Jet Propulsion Laboratory/California Institute of Technology), and NASA; **p. 347:** © Dallas and John Heaton/Jupiterimages/Index Stock/Stock Connection; **p. 368:** Courtesy Ignacio Bello; **p. 390:** © Corbis RF.

Chapter 5

Opener: © Archivo Iconografico, S.A./Corbis; **p. 424:** Photo by Dawn Noren, collected under Marine Mammal Protection Act General Authorization Number 781-1725-01 issued to the Northwest Fisheries Science Center. NOAA Fisheries; **p. 427:** Courtesy Ignacio Bello; **p. 439:** © Morton Beebe/Corbis; **p. 447:** © Martin Rotker/Photo Researchers, Inc.; **p. 452(left):** Courtesy of Brendan Borrell; **p. 452(right):** © Ryan Pyle/Corbis; **p. 456:** Courtesy Ignacio Bello; **p. 472:** © The State of Queensland Australia, 2010.

Chapter 6

Opener: © The British Museum/Topham-HIP/The Image Works; **p. 491:** Courtesy Ignacio Bello; **p. 501:** © Jean-Pierre Mueller/AFP/Getty Images; **p. 514:** © Tony Freeman/Photo Edit; **p. 525:** Courtesy of Adler Planetarium and Astronomy Museum, Chicago, IL; **p. 530:** © 2006 Calvin J. Hamilton; **p. 531:** Courtesy Ignacio Bello; **p. 538:** © Corbis RF; **p. 539:** NASA/JSC; **p. 548(both):** © Nick Hanna; **p. 551:** Courtesy Ignacio Bello.

Chapter 7

Opener: Courtesy National Library; **p. 582:** © Vol. **30**/Getty RF; **p. 583(left):** © Brand X/Corbis RF; **p. 583(right):** © Foodcollection/Getty RF; **p. 596:** Courtesy Ignacio Bello; **p. 602:** © Amanda Byrd/Alaska Stock; **p. 605(left, right):** © Foodcollection/Getty RF; **p. 615:** NASA.

Chapter 8

Opener: © Stock Montage/Hulton Archive/Getty Images; **p. 636:** © AP Photo/Steve Miller; **p. 641:** © Hal Whipple; **p. 642(top):** © Corbis RF; **p. 642(right):** © AP Photo/Mike Derer; **p. 644:** Courtesy Ignacio Bello; **p. 649:** © Vol. **245**/Getty RF; **p. 655:** NOAA; **p. 656(left):** © Vol. **44**/Getty RF; **p. 656(right):** © Brian Bahr/Getty Images; **p. 658:** Courtesy Ignacio Bello; **p. 666:** © Tony Freeman/Photo Edit; **p. 672:** © The McGraw-Hill Companies, Inc./Photo by Connie Mueller.

Chapter 9

Opener: © Bettmann/Corbis; **p. 684:** © Tony Freeman/Photo Edit; **p. 692:** © Corbis RF; **p. 693:** © David R. Frazier/Photo Researchers, Inc.; **p. 696:** © Tony Freeman/Photo Edit; **p. 712(top):** © PhotoDisc/Getty RF; **p. 712(bottom):** © Digital Vision/Getty RF; **p. 714:** Courtesy Ignacio Bello; **p. 732:** © The McGraw-Hill Companies, Inc./Mark Dierker, photographer; **p. 734:** © Photolink/PhotoDisc/GettyRF; **p. 746:** Courtesy NASA Dryden Flight Research Center, Tony Landis, photographer.

Index

Index

K

L

M

N

O

P

Q

Index